HUMAN PHYSIOLOGY AND MECHANISMS OF DISEASE

Third Edition

ARTHUR C. GUYTON, M.D.

Chairman and Professor of the
Department of Physiology and Biophysics,
University of Mississippi, School of Medicine

1982

W. B. SAUNDERS COMPANY
Philadelphia   London   Toronto   Mexico City   Sydney   Tokyo

W. B. Saunders Company: West Washington Square
Philadelphia, PA   19105

1 St. Anne's Road
Eastbourne, East Sussex BN21 3UN, England

1 Goldthorne Avenue
Toronto, Ontario M8Z 5T9, Canada

Cedro 512
Mexico 4, D.F. Mexico

9 Waltham Street
Artarmon, N.S.W. 2064, Australia

Ichibancho, Central Bldg., 22-1
Chiyoda-ku, Tokyo 102, Japan

Library of Congress Cataloging in Publication Data

Guyton, Arthur C.

Human physiology and mechanisms of disease

Includes bibliographies and index.

1. Human physiology.   2. Physiology, Pathological.
  I. Title.

QP34.5.G87 1982   612   81–8595

ISBN 0-7216-4384-1   AACR2

Listed here are the latest translated editions of this book together with
the language of the translation and the publisher.

Portuguese (*2nd Edition*) — Editora Interamericana Ltda.
                    Rio de Janeiro, Brazil

Spanish (*2nd Edition*) — Nueva Editorial Interamericana, S.A.

Human Physiology and Mechanisms of Disease   ISBN   0-7216-4384-1

Last digit is the print number: 9   8   7   6   5   4   3   2   1

PREFACE

This textbook is written for those college and professional students, both medical and paramedical, who require more than an elementary introduction to the physiology of the human body, yet who cannot afford the time to study one of the more formidable textbooks. A special attempt has been made to give wide coverage to all aspects of human physiology and also to present the material at a level that is acceptable to the previous training of most college and professional students. Yet, throughout the text I have still attempted to discuss bodily mechanisms in the light of well known physical and chemical laws, not merely to describe physiological functions as if they were unrelated to other scientific disciplines.

Though a major share of the information used to develop the present book has come from basic experiments in animals, another very large body of knowledge has come also from human experiments, both planned experiments in normal human beings and unplanned experiments caused by disease. For instance, a major share of our knowledge of the regulation of blood glucose, of the mechanisms of carbohydrate metabolism, and of the intricacies of fat metabolism has come from study of diabetes mellitus, a disease that alters these physiological functions profoundly and that is widespread among the human population. By the same token, literally thousands of "human" experiments proceed each day in the fields of high blood pressure, congestive heart failure, gastrointestinal disturbances, respiratory diseases, and so forth. I have discussed the physiology of these abnormalities partly because study of the diseases themselves can be particularly enlightening but even more because they illustrate many important basic physiological concepts.

I hope that this text can convey the understanding that the human body is one of the most complex and yet one of the most beautiful of all functional mechanisms. I hope that students will begin to realize, as a special example, that the human brain is itself a computer having capabilities and functions that all the electronic computers in the world cannot at present achieve. I hope that they will understand that each individual living cell, the basic functional component of the body, carries within its nucleus all the genetic information required to create an entire human being, and that at the same time the genetic pool is in reality a myriad of control systems that regulate a symphony of literally thousands of chemical reactions within each cell. I could go on detailing the miracles of the human body. That, indeed, is the purpose of this entire text, and its success will be measured by the degree of excitement that it leaves with the student for further study in the field of physiology or for a lifetime of physiological thinking.

A vast amount of labor goes into the development and publication of a text. For the figures in the text I am especially indebted to Ms. Tomiko Mita and Mrs. Carolyn Hull, and for the great quality and quantity of secreterial services

I owe my gratitude to Mrs. Billie Howard, Mrs. Jane Strickland, Ms. Gwendolyn Robbins, and Mrs. Laveda Morgan. Finally, I extend my appreciation to the staff of the W. B. Saunders Company for its continued excellence in all publication matters.

ARTHUR C. GUYTON, M.D.
Jackson, Mississippi

CONTENTS

PART VI   THE BODY FLUIDS AND KIDNEYS

PART VII   RESPIRATION

PART VIII   AVIATION, SPACE, AND DEEP SEA
DIVING PHYSIOLOGY

PART IX   THE NERVOUS SYSTEM

PART X   THE SPECIAL SENSES

PART XI   THE GASTROINTESTINAL TRACT

PART XII   METABOLISM AND TEMPERATURE REGULATION

PART XIII   ENDOCRINOLOGY AND REPRODUCTION

Part I

INTRODUCTION TO PHYSIOLOGY: THE CELL AND GENERAL PHYSIOLOGY

1

Functional Organization of the Human Body and Control of the "Internal Environment"

In human physiology we attempt to explain the specific characteristics and mechanisms of the human body that make it a living being. The very fact that we are alive is almost beyond our own control, for hunger makes us seek food and fear makes us seek refuge. Sensations of cold make us provide warmth, and other forces cause us to seek fellowship and to reproduce. Thus, the human being is actually an automaton, and the fact that we are sensing, feeling, and knowledgeable beings is part of this automatic sequence of life; these special attributes allow us to exist under widely varying conditions that otherwise would make life impossible.

CELLS AS THE LIVING UNITS OF THE BODY

The basic living unit of the body is the cell, and each organ is actually an aggregate of many different cells held together by intercellular supporting structures. Each type of cell is specially adapted to perform one particular function. For instance, the red blood cells, 25 trillion in all, transport oxygen from the lungs to the tissues. Though this type of cell is perhaps the most abundant, there are approximately another 50 trillion cells. The entire body, then, contains about 75 trillion cells.

However much the many cells of the body differ from each other, all of them have certain basic characteristics that are alike. For instance, each cell requires nutrition for maintenance of life, and all cells utilize almost identically the same types of nutrients. All cells use oxygen as one of the major substances from which energy is derived; the oxygen combines with carbohydrate, fat, or protein to release the energy required for cell function. The general mechanisms for changing nutrients into energy are basically the same in all cells, and all cells also deliver end-products of their chemical reactions into the surrounding fluids.

Almost all cells also have the ability to reproduce, and whenever cells of a particular type are destroyed from one cause or another, the remaining cells of this type usually divide again and again until the appropriate number is replenished.

THE EXTRACELLULAR FLUID — THE INTERNAL ENVIRONMENT

About 56 per cent of the adult human body is fluid. Though most of this fluid is inside the cells and is called *intracellular fluid*, about one-third of it is in the spaces outside the cells and is called *extracellular fluid*. This extracellular fluid is in constant motion throughout the body. It is rapidly mixed by the blood circulation and by diffusion between the blood and the tissue fluids, and in the extracellular fluid are the ions and nutrients needed by the cells for maintenance of cellular function. Therefore, all cells live in essentially the same environment, the extracellular fluid, for which reason the extracellular fluid is often called the *internal environment* of the body, or the *milieu intérieur*, a term introduced a hundred years ago by the great 19th century French physiologist Claude Bernard.

The body's cells are capable of living, growing,

2

and providing their special functions so long as the proper concentrations of oxygen, glucose, the different ions, amino acids, and fatty substances are available in this internal environment.

Differences Between Extracellular and Intracellular Fluids. The extracellular fluid contains large amounts of sodium, chloride, and bicarbonate ions, plus nutrients for the cells, such as oxygen, glucose, fatty acids, and amino acids. It also contains carbon dioxide, which is being transported from the cells to the lungs to be excreted, and other cellular products, which are being transported to the kidneys for excretion.

The intracellular fluid differs significantly from the extracellular fluid; particularly, it contains large amounts of potassium, magnesium, and phosphate ions instead of the sodium and chloride ions found in the extracellular fluid. Special mechanisms for transporting ions through the cell membranes maintain these differences. These mechanisms will be discussed in detail in Chapter 4.

"HOMEOSTATIC" MECHANISMS OF THE MAJOR FUNCTIONAL SYSTEMS

HOMEOSTASIS

The term *homeostasis* is used by physiologists to mean *maintenance of static, or constant, conditions in the internal environment.* Essentially all the organs and tissues of the body perform functions that help to maintain these constant conditions. For instance, the lungs provide oxygen to the extracellular fluid to replenish continually the oxygen that is being used by the cells, the kidneys maintain constant ion concentrations, and the gut provides nutrients. A large segment of this text is concerned with the manner in which each organ or tissue contributes to homeostasis. To begin this discussion, the different functional systems of the body and their homeostatic mechanisms will be outlined briefly; then the basic theory of the control systems that cause the functional systems to operate in harmony with each other will be discussed.

THE EXTRACELLULAR FLUID TRANSPORT SYSTEM

Extracellular fluid is transported to all parts of the body in two different stages. The first stage entails movement of blood around and around the circulatory system, and the second, movement of fluid between the blood capillaries and the cells. Figure 1–1 illustrates the overall circula-

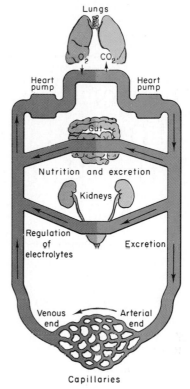

Figure 1–1. General organization of the circulatory system.

tion of blood, showing that the heart is actually two separate pumps, one of which propels blood through the lungs and the other through the systemic circulation. All the blood in the circulation traverses the entire circuit of the circulation an average of once each minute at rest and as many as six times each minute when a person becomes extremely active.

As blood passes through the capillaries, continual fluid exchange occurs between the plasma portion of the blood and the interstitial fluid in the intercellular spaces surrounding the capillaries. This process is illustrated in Figure 1–2. Note that the capillaries are porous so that large amounts of fluid and its dissolved constituents can *diffuse* back and forth between the blood and the tissue spaces, as illustrated by the arrows. This process of diffusion is caused by kinetic motion of the molecules in both the plasma and the interstitial fluid. That is, fluid and dissolved molecules are continually moving and bouncing in all directions, through the pores, through the tissue spaces, and so forth. Almost no cell is located more than 25 to 50 microns from a capillary, which insures diffusion of almost any substance from the capillary to the cell within a few seconds. Thus, the extracellular fluid throughout the body, both that of the plasma and that in the interstitial spaces, is continually mixed and thereby maintains almost complete homogeneity.

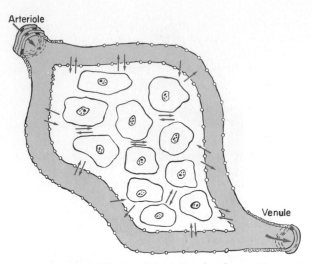

Figure 1–2. Diffusion of fluids through the capillary walls and through the interstitial spaces.

ORIGIN OF NUTRIENTS IN THE EXTRACELLULAR FLUID

The Respiratory System. Figure 1–1 shows that each time the blood passes through the body it also flows through the lungs. The blood picks up oxygen in the alveoli, thus acquiring the oxygen needed by the cells. The membrane between the alveoli and the lumen of the pulmonary capillaries is only 0.4 to 2.0 microns in thickness, and oxygen diffuses through this membrane into the blood in exactly the same manner that water, nutrients, and excreta diffuse through the tissue capillaries.

The Gastrointestinal Tract. Figure 1–1 also shows that a large portion of the blood pumped by the heart passes through the walls of the gastrointestinal organs. Here, different dissolved nutrients, including carbohydrates, fatty acids, amino acids, and others, are absorbed into the extracellular fluid.

The Liver and Other Organs That Perform Primarily Metabolic Functions. Not all substances absorbed from the gastrointestinal tract can be used in their absorbed form by the cells. The liver changes the chemical compositions of many of these to more usable forms, and other tissues of the body — the fat cells, the gastrointestinal mucosa, the kidneys, and the endocrine glands — help to modify the absorbed substances or store them until they are needed at a later time.

Musculoskeletal System. Sometimes the question is asked: How does the musculoskeletal system fit into the homeostatic functions of the body? The answer to this is obvious and simple: Were it not for this system, the body could not move to the appropriate place at the appropriate time to obtain the foods required for nutrition.

The musculoskeletal system also provides mobility for protection against adverse surroundings, without which the entire body, and along with it all the homeostatic mechanisms, could be destroyed instantaneously.

REMOVAL OF METABOLIC END-PRODUCTS

Removal of Carbon Dioxide by the Lungs. At the same time that blood picks up oxygen in the lungs, carbon dioxide is released from the blood into the alveoli, and the respiratory movement of air into and out of the alveoli carries the carbon dioxide to the atmosphere. Carbon dioxide is the most abundant of all the end-products of metabolism.

The Kidneys. Passage of the blood through the kidneys removes most substances from the plasma that are not needed by the cells. These substances especially include different end-products of cellular metabolism and excesses of ions and water that might have accumulated in the extracellular fluid. The kidneys perform their function by first filtering large quantities of plasma through the glomeruli into the tubules and then reabsorbing into the blood those substances needed by the body, such as glucose, amino acids, appropriate amounts of water, and many of the ions. However, substances not needed by the body, especially the metabolic end-products such as urea, generally are not reabsorbed but instead pass on through the renal tubules into the urine.

REGULATION OF BODY FUNCTIONS

The Nervous System. The nervous system is composed of three major parts: the *sensory portion,* the *central nervous system* (or *integrative portion*), and the *motor portion.* Sensory receptors detect the state of the body or the state of the surroundings. For instance, receptors present everywhere in the skin apprise one every time an object touches him at any point. The eyes are sensory organs that give one a visual image of the surrounding area. The ears also are sensory organs. The central nervous system is composed of the brain and spinal cord. The brain can store information, generate thoughts, create ambition, and determine reactions that the body should perform in response to the sensations. Appropriate signals are then transmitted through the motor portion of the nervous system to carry out the person's desires.

A large segment of the nervous system is the *autonomic system.* It operates at a subconscious level and controls many functions of the internal organs, including the action of the heart, the movements of the gastrointestinal tract, and the secretion by different glands.

The Hormonal System of Regulation. Located in the body are eight major endocrine glands that secrete chemical substances, the *hormones.* Hormones are transported in the extracellular fluid to all parts of the body to help regulate function. For instance, thyroid hormone increases the rates of most chemical reactions in all cells. In this way thyroid hormone helps to set the tempo of bodily activity. Likewise, insulin controls glucose metabolism; adrenocortical hormones control ion and protein metabolism; and parathormone controls bone metabolism. Thus, the hormones are a system of regulation that complements the nervous system. The nervous system, in general, regulates muscular and secretory activities of the body, while the hormonal system regulates mainly the metabolic functions.

REPRODUCTION

Reproduction sometimes is not considered to be a homeostatic function. But it does help to maintain static conditions by generating new beings to take the place of ones that are dying. This perhaps sounds like a far-fetched usage of the term homeostasis, but it does illustrate that, in the final analysis, essentially all structures of the body are so organized that they help to maintain continuity of life.

THE CONTROL SYSTEMS OF THE BODY

The human body has literally thousands of control systems in it. The most intricate of all these are the genetic control systems that operate within all cells to control intracellular function, a subject that will be discussed in detail in Chapter 3. But many other control systems operate within the organs to control functions of the individual parts of the organs, while others operate throughout the entire body to control the interrelationships between the different organs. For instance, the respiratory system, operating in association with the nervous system, regulates the concentration of carbon dioxide in the extracellular fluid; the liver and the pancreas regulate the concentration of glucose in the extracellular fluid; and the kidneys regulate the concentrations of hydrogen, sodium, potassium, phosphate, and other ions in the extracellular fluid.

An Example of a Control Mechanism: Regulation of Arterial Pressure. Several different systems contribute to the regulation of arterial pressure. One of these, the *baroreceptor system,* is very simple and is an excellent example of a control mechanism. In the walls of most of the great arteries of the upper body, especially the bifurcation region of the carotids and the arch of the aorta, are many nerve receptors, called *baroreceptors,* which are stimulated by stretch of the arterial wall. When the arterial pressure becomes great, these baroreceptors are stimulated excessively, and impulses are transmitted to the medulla of the brain. Here the impulses inhibit the *vasomotor center,* which in turn decreases the number of impulses transmitted through the sympathetic nervous system to the heart and blood vessels. Lack of these impulses causes diminished pumping activity by the heart and increased ease of blood flow through the peripheral vessels, both of which lower the arterial pressure back toward normal. Conversely, a fall in arterial pressure relaxes the stretch receptors, allowing the vasomotor center to become more active than usual and thereby causing the arterial pressure to rise back toward normal.

Negative Feedback Nature of Control Systems. Most of the control systems of the body act by a process of *negative feedback,* which can be explained best by analyzing the baroreceptor pressure-regulating mechanism that was just discussed. In this mechanism, it is clear that a high pressure causes a series of reactions that promote a lowered pressure, or a low pressure causes a series of reactions that promote an elevated pressure. In both instances these effects are opposite to, or *negative* to, the initiating stimulus, hence the term "negative feedback."

Essentially all other control mechanisms of the body also operate by the process of negative feedback. For instance, if the oxygen concentration in the body fluids falls too low, the mechanisms for controlling oxygen automatically return the oxygen back to a higher level. Thus, the effect is *negative* to the initiating stimulus. Likewise, elevated carbon dioxide concentration in the body fluids causes increased respiration, which then removes the excess carbon dioxide. Again, the response is negative to the stimulus. Essentially all of the endocrine control systems also operate in this manner. For instance, when the concentration of potassium falls too low in the extracellular fluid, the adrenal glands decrease their secretion of the hormone aldosterone, and lack of this hormone decreases the rate of potassium excretion by the kidneys into the urine. Therefore, potassium accumulates in the extracellular fluid until its concentration returns to normal. This is still another example of negative feedback.

Thus, in general, if some factor becomes excessive or too little, a control system initiates *negative feedback*, which consists of a series of changes that returns the factor toward a certain mean value, thus maintaining homeostasis.

Amplification, or Gain, of a Control System. The degree of effectiveness with which a control system maintains constant conditions is called the *amplification,* or *gain,* of the system.

For instance, let us assume that a large volume of blood is suddenly transfused into a person and that this immediately raises the arterial pressure from a normal value of 100 mm. Hg up to 160 mm. Hg. However, within 15 to 30 seconds the baroreceptor control mechanism becomes fully operative, and the arterial pressure is reduced back to 120 mm. Hg. Thus, the pressure is corrected 40 mm. Hg, while the final abnormality is only 20 mm. Hg instead of the 60 mm. Hg that would have occurred without the control system. The gain of the mechanism is calculated by the following equation:

$$\text{Gain} = \frac{\text{Amount of correction of abnormality}}{\text{Amount of abnormality still remaining}}$$

In the above example the correction is −40 mm. Hg, and the amount of abnormality still remaining is 20 mm. Hg; therefore, the gain of the baroreceptor system for control of arterial pressure is approximately −2.

The gains of different control systems of the body vary markedly, with gains as low as −1 to −2 for control of arterial pressure by a hormone called renin, which is released from the kidney, and as high as −30 for control of body temperature in the face of changing atmospheric temperature. In other words, the pressure-controlling ability of the renin mechanism is only moderate, while the temperature-controlling ability of the temperature feedback system is very great.

AUTOMATICITY OF THE BODY

The purpose of this chapter has been to point out, first, the overall organization of the body and, second, the means by which the different parts of the body operate in harmony. To summarize, the body is actually a *social order of about 75 trillion cells* organized into different functional structures, some of which are called *organs.* Each functional structure provides its share in the maintenance of homeostatic conditions in the extracellular fluid, which is often called the *internal environment.* As long as normal conditions are maintained in this internal environment, the cells of the body will continue to live and function properly. Thus, each cell benefits from homeostasis, and each in turn contributes its share toward the maintenance of homeostasis. This reciprocal interplay provides continuous automaticity of the body until one or more functional systems lose their ability to contribute their share of function. When this happens, all the cells of the body suffer. Extreme dysfunction leads to death, while moderate dysfunction leads to sickness.

REFERENCES

Adolph, E. F.: Origins of Physiological Regulations. New York, Academic Press, 1968.

Adolph, E. F.: Physiological adaptations: Hypertrophies and superfunctions. *Am. Sci., 60*:608, 1972.

Bernard, C.: Lectures on the Phenomena of Life Common to Animals and Plants. Springfield, Ill., Charles C Thomas, 1974.

Cannon, W. B.: The Wisdom of the Body. New York, W. W. Norton & Co., 1932.

Frisancho, A. R.: Human Adaptation. St. Louis, C. V. Mosby Co., 1979.

Guyton, A. C., *et al.*: Circulatory Physiology II: Dynamics and Control of the Body Fluids. Philadelphia, W. B. Saunders Co., 1975.

Huffaker, C. B. (ed.): Biological Control. New York, Plenum Press, 1974.

Iberall, A. S., and Guyton, A. C. (eds.): *Proc. Int. Symp. on Dynamics and Controls in Physiological Systems.* Regulation and Control in Physiological Systems. ISA, Pittsburgh, 1973.

Jones, R. W.: Principles of Biological Regulation: An Introduction to Feedback Systems. New York, Academic Press, 1973.

McIntosh, J. E. A., and McIntosh, R. P.: Mathematical Modelling and Computers in Endocrinology. New York, Springer-Verlag, 1980.

Miller, S. L., and Orgel, L. E.: The Origins of Life on the Earth. Englewood Cliffs, N.J., Prentice-Hall, 1974.

Randall, J. E.: Microcomputers and Physiological Simulation. Reading, Mass., Addison-Wesley Publishing Co., 1980.

Reeve, E. B., and Guyton, A. C.: Physical Bases of Circulatory Transport: Regulation and Exchange. Philadelphia, W. B. Saunders Co., 1967.

Rusak, B., and Zucker, I.: Neural regulation of circadian rhythms. *Physiol. Rev.,* 59:449, 1979.

Söderberg, U.: Neurophysiological aspects of homeostasis. *Annu. Rev. Physiol.,* 26:271, 1964.

Sweetser, W.: Human Life (Aging and Old Age). New York, Arno Press, 1979.

Toates, F. M.: Control Theory in Biology and Experimental Psychology. London, Hutchinson Education Ltd., 1975.

Weston, L.: Body Rhythm: The Circadian Rhythms Within You. New York, Harcourt Brace Jovanovich, 1979.

2

The Cell and Its Function

Each of the 75 trillion cells in the human being is a living structure that can survive indefinitely and, in most instances, can even reproduce itself, provided its surrounding fluids contain appropriate nutrients. A typical cell, as seen by the light microscope, is illustrated in Figure 2–1. Its two major parts are the *nucleus* and the *cytoplasm*. The nucleus is separated from the cytoplasm by a *nuclear membrane*, and the cytoplasm is separated from the surrounding fluids by a *cell membrane*.

The different substances that make up the cell are collectively called *protoplasm*. Protoplasm is composed mainly of five basic substances: water, electrolytes, proteins, lipids, and carbohydrates.

PHYSICAL STRUCTURE OF THE CELL

The cell is not merely a bag of fluid, enzymes, and chemicals; it also contains highly organized physical structures called *organelles*, which are just as important to the function of the cell as are the cell's chemical constituents. For instance, without one of the organelles, the *mitochondria*, more than 95 per cent of the energy supply of the cell would cease immediately. Some principal organelles of the cell are illustrated in Figure 2–2, including the *cell membrane, nuclear membrane, endoplasmic reticulum, mitochondria,* and *lysosomes*. Others not shown in the figure are the *Golgi complex, centrioles, cilia,* and *microtubules*.

THE MEMBRANOUS STRUCTURES OF THE CELL

Essentially all physical structures of the cell are lined by membranes composed primarily of lipids and proteins. The lipids provide a barrier that prevents free movement of water and water-soluble substances from one cell compartment to the other. The protein molecules, on the other hand, interrupt the continuity of the lipid barrier and therefore provide pathways for passage of various substances through the membrane. The different membranes include the *cell membrane,* the *nuclear membrane,* the *membrane of the endoplasmic reticulum,* and the *membranes of the mitochondria, lysosomes, Golgi complex,* and so forth.

The Cell Membrane. The cell membrane, which completely envelops the cell, is a very thin, elastic structure only 7.5 to 10 nanometers thick. It is composed almost entirely of proteins and lipids; the approximate composition is proteins, 55 per cent; phospholipids, 25 per cent; cholesterol, 13 per cent; other lipids, 4 per cent; and carbohydrates, 3 per cent.

The Lipid Barrier of the Cell Membrane. Figure 2–3 illustrates that the basic structure of the cell membrane is a *lipid bilayer,* which is a thin film of lipids only two molecules thick that is continuous over the entire cell surface. Interspersed in this lipid film are large globular protein molecules.

The lipid bilayer is composed almost entirely of phospholipids and cholesterol. It is almost entirely impermeable to water and to the usual water-soluble substances such as ions, glucose,

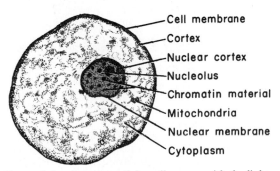

Figure 2–1. Structure of the cell as seen with the light microscope.

Cell membrane
Cortex
Nuclear cortex
Nucleolus
Chromatin material
Mitochondria
Nuclear membrane
Cytoplasm

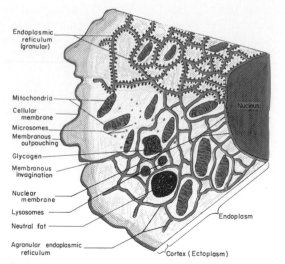

Figure 2–2. Organization of the cytoplasmic compartment of the cell.

urea, and others. On the other hand, fat-soluble substances such as oxygen, carbon dioxide, and alcohols can penetrate this portion of the membrane.

A special feature of the lipid bilayer is that it is a lipid *fluid* and not a solid. Therefore, portions of the membrane can literally flow from one point to another in the membrane. Proteins or other substances dissolved in or floating in the lipid bilayer tend to diffuse to all areas of the cell membrane.

The Cell Membrane Proteins. Figure 2–3 illustrates globular masses floating in the lipid bilayer.

These are the cell proteins, most of which are *glycoproteins.* Two types of proteins occur: the *integral proteins* that protrude all the way through the cell and the *peripheral proteins* that are attached only to the surface of the membrane and do not penetrate. The integral proteins provide structural pathways through which water and water-soluble substances, especially the ions, can diffuse between the extracellular and intracellular fluid. However, these proteins have selective properties that cause preferential diffusion of some substances more than others. Some of them can also act as enzymes.

The peripheral proteins occur either entirely or almost entirely on the inside of the membrane, and they are normally attached to one of the integral proteins. These peripheral proteins function almost entirely as enzymes.

The Membrane Carbohydrates. The membrane carbohydrates occur almost invariably on the outside of the membrane; they are the "glyco" portion of protruding glycoprotein molecules. These carbohydrate moieties are the portion of the cell membrane that enters into immune reactions, as we shall discuss in Chapter 6, and they often act as receptor substances for binding hormones, such as insulin, that stimulate specific types of activity in the cells.

The Nuclear Membrane. The nuclear membrane, illustrated in Figure 2–7, is actually two membranes, one surrounding the other with a wide space in between. Each membrane is almost identical to the cell membrane, having a basic lipid bilayer structure with globular proteins

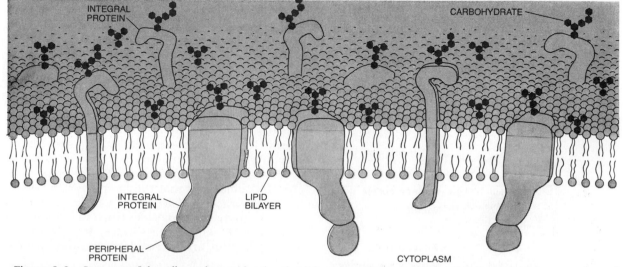

Figure 2–3. Structure of the cell membrane, showing that it is composed mainly of a lipid bilayer but with large numbers of protein molecules protruding through the layer. Also, carbohydrate moieties are attached to the protein molecules on the outside of the membrane and additional protein molecules on the inside. (From Lodish and Rothman: *Sci. Am., 240*:48, 1979. ©1979 by Scientific American, Inc.)

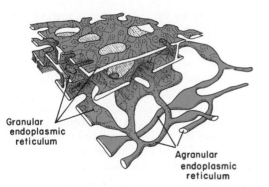

Figure 2–4. Structure of the endoplasmic reticulum. (Modified from De Robertis, Saez, and De Robertis: Cell Biology. 6th ed. Philadelphia, W. B. Saunders Company, 1975.)

floating in the lipid fluid. At many points the two membranes fuse with each other, and at these points the nuclear membrane is so permeable that almost all dissolved or suspended substances, including even very large, newly formed ribosomes, can move with ease between the fluids of the nucleus and the cytoplasm.

The Endoplasmic Reticulum. Figure 2–2 illustrates in the cytoplasm a continuous network of tubular and flat vesicular structures, constructed of lipid bilayer–protein membranes, called the *endoplasmic reticulum.* The total surface area of this structure in some cells — the liver cells, for instance — can be as much as 30 to 40 times as great as the cell membrane area. The detailed structure of this organelle is illustrated in Figure 2–4. The space inside the tubules and vesicles is filled with *endoplasmic matrix,* a fluid medium that is different from the fluid outside the endoplasmic reticulum. Electron micrographs show that the space inside the endoplasmic reticulum is connected with the space between the two membranes of the double nuclear membrane.

Substances formed in different parts of the cell enter the spaces of the endoplasmic reticulum and are then conducted to other parts of the cell. Also, the vast surface area of the reticulum, as well as its many enzyme systems, provides the machinery for a major share of the metabolic functions of the cell.

Ribosomes and the Granular Endoplasmic Reticulum. Attached to the outer surfaces of many parts of the endoplasmic reticulum are large numbers of small granular particles called *ribosomes.* Where these are present, the reticulum is frequently called the *granular endoplasmic reticulum.* The ribosomes are composed mainly of ribonucleic acid, which functions in the synthesis of protein in the cells, as is discussed later in this and the following chapter.

The Agranular Endoplasmic Reticulum. Part of the endoplasmic reticulum has no attached ribosomes. This part is called the *agranular,* or *smooth, endoplasmic reticulum.* The agranular reticulum functions in the synthesis of lipid substances and in many other enzymatic processes of the cell.

Golgi Complex. The Golgi complex, illustrated in Figure 2–5, is closely related to the endoplasmic reticulum. It has membranes similar to those of the agranular endoplasmic reticulum. It is usually composed of four or more stacked layers of thin, flat vesicles lying near the nucleus. This complex is very prominent in secretory cells; in these it is located on the side of the cell from which the secretory substances will be extruded.

The Golgi complex functions mainly in association with the endoplasmic reticulum. As illustrated in Figure 2–5, small "transport vesicles" continually pinch off from the endoplasmic reticulum and shortly thereafter fuse with the Golgi complex. In this way substances are transported from the endoplasmic reticulum to the Golgi complex. The transported substances are then processed in the Golgi complex to form secretory vesicles, lysosomes, or other cytoplasmic components that will be discussed later in the chapter.

THE CYTOPLASM AND ITS ORGANELLES

The cytoplasm is filled with both minute and large dispersed particles and organelles ranging in size from a few nanometers to 3 microns in size. The clear fluid portion of the cytoplasm in which the particles are dispersed is called *hyaloplasm;* this contains mainly dissolved proteins, electrolytes, glucose, and small quantities of phospholipids, cholesterol, and esterified fatty acids.

Among the large dispersed particles in the cytoplasm are neutral fat globules, glycogen

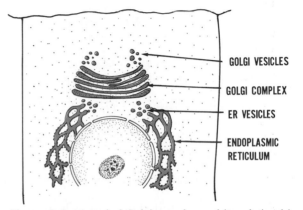

GOLGI VESICLES

GOLGI COMPLEX

ER VESICLES

ENDOPLASMIC RETICULUM

Figure 2–5. A typical Golgi complex and its relationship to the endoplasmic reticulum and the nucleus.

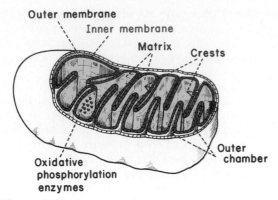

Figure 2–6. Structure of a mitochondrion. (Modified from De Robertis, Saez, and De Robertis: Cell Biology. 6th ed. Philadelphia, W. B. Saunders Company, 1975.)

granules, ribosomes, secretory granules, and two especially important organelles — the *mitochondria* and *lysosomes* — which are discussed below.

The Mitochondria

The mitochondria are called the "powerhouses" of the cell because they extract energy from the nutrients and oxygen and in turn provide it in a usable form to energize essentially all cellular functions. The number of mitochondria per cell varies from less than a hundred to many thousand, depending upon the amount of energy required by each cell. Furthermore, the mitochondria are concentrated in those portions of the cell that are responsible for the major share of its energy metabolism.

The basic structure of the mitochondrion is illustrated in Figure 2–6, which shows it to be composed mainly of two lipid bilayer-protein membranes: an *outer membrane* and an *inner membrane*. Many infoldings of the inner membrane form *shelves*, onto which the oxidative enzymes of the cell are attached. In addition, the inner cavity of the mitochondrion is filled with a gel *matrix* containing large quantities of dissolved enzymes that are necessary for extracting energy from nutrients. These enzymes operate in association with the oxidative enzymes on the shelves to cause oxidation of the nutrients, thereby forming carbon dioxide and water. The liberated energy is used to synthesize a high-energy substance called *adenosine triphosphate (ATP)*. ATP is then transported out of the mitochondrion, and it diffuses throughout the cell to release its energy wherever it is needed for performing cellular functions. The function of ATP is so important to the cell that it is discussed in detail later in the chapter.

Mitochondria are self-replicative, which means that one mitochondrion can form a second one, a third one, and so on, whenever there is need in the cell for increased amounts of ATP.

The Lysosomes

The lysosomes provide an intracellular digestive system that allows the cell to digest and thereby remove unwanted substances and structures, especially damaged or foreign structures, such as bacteria. The lysosome, illustrated in Figure 2–2, is 250 to 750 nanometers in diameter and is surrounded by a typical lipid bilayer membrane. It is filled with large numbers of small granules, which are protein aggregates of hydrolytic (digestive) enzymes. A hydrolytic enzyme is capable of splitting an organic compound into two or more parts by combining hydrogen from a water molecule with part of the compound and by combining the hydroxyl portion of the water molecule with the other part of the compound. For instance, protein is hydrolyzed to form amino acids, and glycogen is hydrolyzed to form glucose. More than 40 different *acid hydrolases* have been found in lysosomes, and the principal substances that they digest are proteins, nucleic acids, mucopolysaccharides, lipids, and glycogen.

Ordinarily, the membrane surrounding the lysosome prevents the enclosed hydrolytic enzymes from coming in contact with other substances in the cell. However, many different conditions of the cell will break the membranes of some of the lysosomes, allowing release of the enzymes. These enzymes then split the organic substances with which they come in contact into small, highly diffusible substances, such as amino acids and glucose. Some of the more specific functions of lysosomes are discussed later in the chapter.

Other Cytoplasmic Structures and Organelles

The cytoplasm of each cell contains two pairs of *centrioles*, which are small cylindrical structures that play a major role in cell division, as will be discussed in Chapter 3. Also, most cells contain small *lipid droplets* and *glycogen granules* that play important roles in energy metabolism of the cell. Certain cells contain highly specialized structures such as the cilia of ciliated cells, which are actually outgrowths from the cytoplasm, and the *myofibrils* of muscle cells. All of these are discussed in detail at different points in this text.

THE NUCLEUS

The nucleus is the control center of the cell. It controls both the chemical reactions that occur in the cell and reproduction of the cell. Briefly, the nucleus contains large quantities of *deoxyribonucle-*

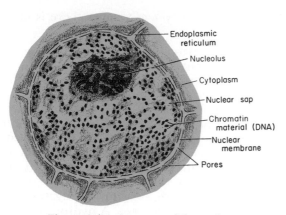

Figure 2–7. Structure of the nucleus.

ic acid, which we have called *genes* for many years. The genes determine the characteristics of the protein enzymes of the cytoplasm and in this way control cytoplasmic activities. To control reproduction, the genes first reproduce themselves, and after this is accomplished the cell splits by a special process called *mitosis* to form two daughter cells, each of which receives one of the two sets of genes. These activities of the nucleus are considered in detail in the following chapter.

The appearance of the nucleus under the microscope does not give much of a clue to the mechanisms by which it performs its control activities. Figure 2–7 illustrates the light microscopic appearance of the interphase nucleus (period between mitoses), showing darkly staining *chromatin material* throughout the *nuclear sap*. During mitosis, the chromatin material becomes readily identifiable as part of the highly structured *chromosomes,* which can be seen easily with the light microscope. Even during the interphase of cellular activity the chromatin material is still organized into fibrillar chromosomal structures, but this is impossible to see except in a few types of cells.

Nucleoli. The nuclei of many cells contain one or more lightly staining structures called nucleoli. The nucleolus is simply a protein structure that contains a large amount of *ribonucleic acid* of the type found in ribosomes. The nucleolus becomes considerably enlarged when a cell is actively synthesizing proteins. The genes of a particular chromosome pair synthesize the ribonucleic acid and then store it in the nucleolus, beginning with a loose fibrillar RNA that later condenses to form granular ribosomes. These in turn migrate through the nuclear membrane pores into the cytoplasm, where most of them become attached to the endoplasmic reticulum and there play an essential role for the formation of proteins, as we shall discuss in the following chapter.

FUNCTIONAL SYSTEMS OF THE CELL

In the remainder of this chapter we will discuss most of the functional systems of the cell that make it a living organism. However, two cellular functions are so important that they will be the subjects of the following two chapters: (1) control of protein synthesis and of other cellular functions by the genes in the nucleus, and (2) transport of substances through the cell membrane. But in the present chapter, let us begin by discussing the means by which the cell ingests substances, then consider the energy systems of the cell and the synthesis of new substances in the cell.

INGESTION AND DIGESTION BY THE CELL

If a cell is to live and grow, it must obtain nutrients and other substances from the surrounding fluids. Substances can pass through a cell membrane in three separate ways: (1) by *diffusion* through the pores in the membrane or through the membrane matrix itself; (2) by *active transport* through the membrane, a mechanism in which enzyme systems and special carrier substances "carry" the substances through the membrane; and (3) by *endocytosis*, a mechanism by which the membrane actually engulfs particulate matter or extracellular fluid and its contents. The important subject of transport of substances by diffusion and active transport, the means by which most nutrients and other substances enter the cell, will be considered in Chapter 4. Endocytosis is a specialized cellular function that merits mention here.

Endocytosis — Phagocytosis and Pinocytosis. Phagocytosis means the ingestion of large particulate matter by a cell, such as the ingestion of (1) a bacterium, (2) some other cell, or (3) particles of degenerating tissue. Pinocytosis, on the other hand, means ingestion of minute quantities of extracellular fluid and dissolved substances in the form of minute vesicles. The pinocytic vesicles are so small that they were not discovered until the advent of the electron microscope. However, phagocytosis has been known since the time of the earliest studies using the light microscope.

Thus, phagocytosis and pinocytosis are both types of *endocytosis*, and their mechanisms are essentially identical except for the sizes and natures of the ingested vesicles.

Phagocytosis occurs when certain objects contact the cell membrane. In general, those objects that have an electronegative charge are rejected,

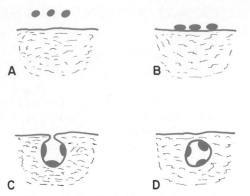

Figure 2–8. Mechanism of pinocytosis.

while those that have an electropositive charge are especially susceptible to phagocytosis. The difference presumably results from the fact that the phagocytic cells themselves normally are electronegatively charged and therefore repel other electronegative objects. Most *normal* particulate objects in the extracellular fluid are also negatively charged; on the other hand, damaged tissues and also foreign invaders that have been especially prepared for phagocytosis by attachment to antibodies (a process called *opsonization,* which will be discussed in Chapters 5 and 6) usually acquire positive charges and are therefore phagocytized.

Pinocytosis also occurs in response to certain types of substances that contact the cell membrane. The two most important are proteins and strong electrolyte solutions. It is especially significant that proteins cause pinocytosis, because pinocytosis is the only means by which proteins can pass through the cell membrane.

Figure 2–8 illustrates the successive steps of pinocytosis, showing first three molecules of protein attaching to the membrane by the simple process of adsorption. The presence of these proteins then causes the surface properties of the membrane to change in such a way that it invaginates and then rapidly closes over the proteins. Immediately thereafter, the invaginated portion of the membrane breaks away from the surface of the cell, forming a *pinocytic vesicle.* *Phagocytic vesicles* are formed in a similar manner.

What causes the cell membrane to go through the necessary contortions for forming the pinocytic and phagocytic vesicles remains a mystery. However, it is known that this process requires energy from within the cell; this is supplied by adenosine triphosphate, the high-energy substance that will be discussed elsewhere in this chapter. Also, endocytosis requires the presence of calcium ions in the extracellular fluid and probably a contractile function by *microfilaments* immediately beneath the cell membrane.

The Digestive Organelle of the Cell – The Lysosomes

Almost immediately after a pinocytic or phagocytic vesicle appears inside a cell, one or more lysosomes become attached to the vesicle and empty their hydrolases into the vesicle, as illustrated in Figure 2–9. Thus, *a digestive vesicle* is formed, in which the hydrolases begin hydrolyzing the proteins, glycogen, lipids, nucleic acids, mucopolysaccharides, and other substances in the vesicle. The products of digestion are small molecules of amino acids, glucose, fatty acids, phosphates, and so forth that can diffuse through the membrane of the vesicle into the cytoplasm. What is left of the digestive vesicle, called the *residual body,* represents the undigestible substances. In most instances this is finally excreted through the cell membrane by a process called *exocytosis,* which is essentially the opposite of endocytosis.

Thus, the lysosomes may be called the *digestive organelles* of the cells.

Regression of Tissues and Autolysis of Cells. Often, tissues of the body regress to a much smaller size than previously. For instance, this occurs in the uterus following pregnancy, in muscles during long periods of inactivity, and in mammary glands at the end of the period of lactation. Lysosomes are responsible for most of this regression. However, the mechanism by which lack of activity in a tissue causes the lysosomes to increase their activity is still unknown.

Another very special role of the lysosomes is the removal of damaged cells or damaged portions of cells from tissues — cells damaged by heat, cold, trauma, chemicals, or any other factor. Damage to the cell causes lysosomes to rupture, and the released hydrolases begin immediately to digest the surrounding organic substances. If the damage is slight, only a portion of the cell will be removed, followed by repair of the cell. However, if the damage is severe, the entire cell will be digested, a process called *autolysis.*

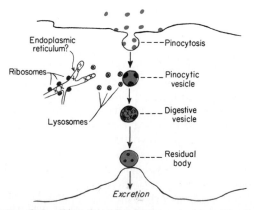

Figure 2–9. Digestion of substances in pinocytic vesicles by enzymes derived from lysosomes.

The lysosomes also contain bactericidal agents that can kill phagocytized bacteria before they can cause cellular damage. And in lysosomes are stored enzymes that, upon release into the cytoplasm, can dissolve lipid droplets and glycogen granules, making the lipid and glycogen available for use elsewhere in the cell or elsewhere in the body. In the absence of these enzymes, which results from occasional genetic disorders, extreme quantities of lipids or of glycogen often accumulate in the cells of many organs, especially the liver, and lead to early death.

EXTRACTION OF ENERGY FROM NUTRIENTS — FUNCTION OF THE MITOCHONDRIA

The principal nutrients from which cells extract energy are oxygen and one or more of the foodstuffs. Figure 2–10 shows oxygen and the foodstuffs — glucose, fatty acids, and amino acids — all entering the cell. Inside the cell, the foodstuffs react chemically with the oxygen under the influence of various enzymes that

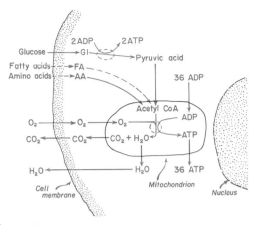

Figure 2–10. Formation of adenosine triphosphate in the cell, showing that most of the ATP is formed in the mitochondria.

control their rates of reactions and channel the energy that is released in the proper direction.

Formation of Adenosine Triphosphate (ATP). The energy released from the nutrients is used to form adenosine triphosphate, generally called ATP, the formula for which is:

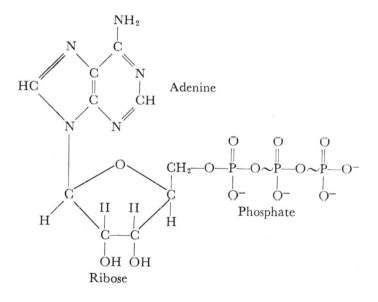

ATP is a *nucleotide* composed of the nitrogenous base *adenine*, the pentose sugar *ribose*, and three *phosphate radicals*. The last two phosphate radicals are connected with the remainder of the molecule by so-called *high-energy phosphate bonds*, which are represented by the symbol ~. Each of these bonds contains about 8000 calories of energy per mole of ATP under the physical conditions of the body (7000 calories under standard conditions), which is much greater than the energy stored in the average chemical bond of most other organic compounds, thus giving rise to the term "high-energy" bond. Furthermore, the high-energy phosphate bond is very labile so that it can be split instantly on demand whenever energy is required to promote other cellular reactions.

When ATP releases its energy, a phosphoric acid radical is split away, and *adenosine diphosphate (ADP)* is formed. Then, energy derived from the cellular nutrients causes the ADP and phosphoric acid to recombine to form new ATP, the entire process continuing over and over again. For these

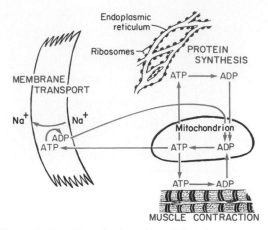

Figure 2–11. Use of adenosine triphosphate to provide energy for three major cellular functions: membrane transport, protein synthesis, and muscle contraction.

reasons ATP has been called the *energy currency* of the cell, for it can be spent and remade again and again.

Chemical Processes in the Formation of ATP — Role of the Mitochondria. Most of the ATP formed in the cell is synthesized in the mitochondria, a process which is discussed in detail in Chapter 45. However, the mechanism is basically the following: The different foods are first digested in the person's digestive tract to form glucose, fatty acids, and amino acids. Mainly in these forms, they are then delivered to the cells. In the cells they are eventually converted into the compound *acetyl co-A*, which in turn is split into hydrogen atoms and carbon dioxide. The carbon dioxide diffuses out of the mitochondrion and eventually out of the cell. The hydrogen atoms combine with carrier substances and are carried to the surfaces of the shelves that protrude into the mitochondrion, as shown in Figure 2–6. Attached to these shelves are the so-called *oxidative enzymes* and also protruding globules of *ATPase,* the enzyme that catalyzes the conversion of ADP to ATP. The oxidative enzymes, by a series of sequential reactions, cause the hydrogen atoms to combine with oxygen. During the course of these reactions, the energy released from the combination of hydrogen with oxygen is used to activate the ATPase and drive the reaction to manufacture tremendous quantities of ATP from ADP. The ATP is then transported out of the mitochondrion into all parts of the cytoplasm and nucleoplasm, where its energy is used to energize the functions of the cell.

The formation of ATP is so important to the function of the cell that many more details of this subject will be presented in Chapters 45 through 47.

Uses of ATP for Cellular Function. ATP is used to promote three categories of cellular func-

tions: (1) *membrane transport,* (2) *synthesis of chemical compounds* throughout the cell, and (3) *mechanical work*. These three different uses of ATP are illustrated in Figure 2–11: (a) to supply energy for the transport of sodium through the cell membrane, (b) to promote protein synthesis by the ribosomes, and (c) to supply the energy needed during muscle contraction.

In addition to membrane transport of sodium, energy from ATP is required for transport of potassium ions and, in certain cells, calcium ions, phosphate ions, chloride ions, urate ions, hydrogen ions, and still many other special substances. Membrane transport is so important to cellular function that some cells, the renal tubular cells for instance, utilize as much as 80 per cent of the ATP formed in the cells for this purpose alone.

In addition to synthesizing proteins, cells also synthesize phospholipids, cholesterol, purines, pyrimidines, and a great host of other substances. Synthesis of almost any chemical compound requires energy. For instance, a single protein molecule might be composed of as many as several thousand amino acids attached to each other by peptide linkages; the formation of each of these linkages requires the breakdown of three high-energy bonds; thus many thousand ATP molecules must release their energy as each protein molecule is formed. Indeed, some cells utilize as much as 75 per cent of all the ATP formed in the cell simply to synthesize new chemical compounds; this is particularly true during the growth phase of cells.

The final major use of ATP is to supply energy for special cells to perform mechanical work. We shall see in Chapter 9 that each contraction of a muscle fibril requires expenditure of tremendous quantities of ATP. Other cells perform mechanical work in two additional ways, by *ciliary* or *ameboid motion*, both of which will be described later in this chapter. The source of energy for all these types of mechanical work is ATP.

In summary, therefore, ATP is always available to release its energy rapidly and almost explosively wherever in the cell it is needed. To replace the ATP used by the cell, other much slower chemical reactions break down carbohydrates, fats, and proteins and use the energy derived from these to form new ATP.

SYNTHESIS AND FORMATION OF CELLULAR STRUCTURES BY THE ENDOPLASMIC RETICULUM AND THE GOLGI COMPLEX

The synthesis of most intracellular substances begins in the endoplasmic reticulum, but the products formed in the endoplasmic reticulum are then passed on to the Golgi complex, where they are further processed prior to release into the cell. But first, let us note the specific products that are synthesized in the specific portions of the endoplasmic reticulum and the Golgi complex.

Formation of Proteins by the Granular Endoplasmic Reticulum. The granular endoplasmic reticulum is characterized by the presence of large numbers of ribosomes attached to the outer surfaces of the reticulum membrane. As we shall discuss in the following chapter, protein molecules are synthesized within the structure of the ribosomes. Furthermore, the ribosomes extrude many of the synthesized protein molecules not into the hyaloplasm but instead through the endoplasmic reticular wall into the endoplasmic matrix.

Within the endoplasmic matrix, the protein molecules are further processed during the next few minutes. In the presence of the enzymes in the endoplasmic reticular wall, the simple protein molecules are often folded and are also modified in other ways. In addition, most of them are rapidly conjugated with carbohydrate moieties to form glycoproteins.

Synthesis of Lipids, and Other Functions of the Smooth Endoplasmic Reticulum. The smooth endoplasmic reticulum synthesizes mainly lipids, including phospholipids and cholesterol, rather than proteins. The phospholipids and cholesterol are rapidly incorporated into the lipid bilayer of the endoplasmic reticulum itself, thus causing the smooth portion of the endoplasmic reticulum to grow continually. However, small vesicles continually break away from the smooth endoplasmic reticulum; we shall see later that these vesicles mainly migrate rapidly to the Golgi apparatus.

Other significant functions of the smooth endoplasmic reticulum are these:

(1) It contains the enzymes that control glycogen breakdown when glycogen is to be used for energy.

(2) It contains a vast number of enzymes that are capable of detoxifying substances that are damaging to the cell, such as drugs. It achieves this by coagulation, oxidation, hydrolysis, and conjugation with glycuronic acid, and in other ways.

(3) It can synthesize a few carbohydrate moieties that are usually conjugated with protein molecules to form glycoproteins.

Synthetic Functions of the Golgi Complex. Though the major function of the Golgi complex is to process substances already formed in the endoplasmic reticulum, it also has the capability of synthesizing certain carbohydrates that cannot be formed in the endoplasmic retic-

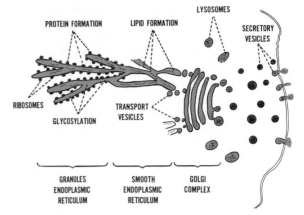

Figure 2–12. Formation of proteins, lipids, and cellular vesicles by the endoplasmic reticulum and Golgi complex.

ulum. This is especially true of siliac acid, fructose, and galactose. In addition, it can cause the formation of saccharide polymers, the most important of which are hyaluronic acid and chondroitin sulfate. A few of the many functions of hyaluronic acid and chrondroitin sulfate in the body are these: (1) They are the major components of proteoglycans secreted in mucus and other glandular secretions. (2) They are the major components of the ground substance in the interstitial spaces, acting as a filler between collagen fibers and cells. (3) They are principal components of the organic matrix in both cartilage and bone.

Processing of Endoplasmic Secretions by the Golgi Complex — Formation of Intracellular Vesicles. Figure 2–12 summarizes the major functions of the endoplasmic reticulum and Golgi complex and also shows the formation of secretory vesicles by the Golgi complex. As substances are formed in the endoplasmic reticulum, especially the proteins, they are transported through the tubules toward the portions of the smooth endoplasmic reticulum that lie nearest the Golgi complex. At this point small "transport" vesicles of smooth endoplasmic reticulum continually break away and diffuse to the *proximal layers* of the Golgi complex, carrying inside the vesicles the synthesized proteins and other products. These vesicles instantly fuse with the Golgi complex, and their contained substances enter the vesicular spaces of the Golgi complex. Here, a few additional carbohydrate moieties are usually added to the secretions, but usually the main function of the Golgi complex is to compact the endoplasmic reticular secretions into highly concentrated packets. As the secretions pass toward the distal layers of the Golgi complex the compaction and processing proceed, and finally at the distal layer both small and large vesicles continually break away from the Golgi complex, carrying with them the compacted secretory substances, and they then diffuse throughout the cell.

To give one an idea of the timing of these processes, when a glandular cell is bathed in radioactive amino acids, newly formed radioactive protein molecules can be detected in the granular endoplasmic reticulum within 3 to 5 minutes. Within 20 minutes the newly formed proteins are present in the Golgi complex, and within 1 to 2 hours radioactive proteins are secreted from the surface of the cell.

In a highly secretory cell, the vesicles that are formed by the Golgi complex are mainly *secretory vesicles*, containing especially the protein substances that are to be secreted through the surface of the cell. These vesicles diffuse to the surface, fuse with the cell membrane, and empty their substances to the exterior by a mechanism called *exocytosis*, which is essentially the opposite of endocytosis.

On the other hand, some of the vesicles are destined for intracellular use. For instance, specialized portions of the Golgi complex form the *lysosomes* that have already been discussed.

CELL MOVEMENT

By far the most important type of cell movement that occurs in the body is that of the specialized muscle cells in skeletal, cardiac, and smooth muscle, which constitute almost 50 per cent of the entire body mass. The specialized functions of these cells will be discussed in Chapters 9 through 11. However, two other types of movement occur in other cells, *ameboid movement* and *ciliary movement*.

Ameboid Motion. Ameboid motion means movement of an entire cell in relation to its surroundings, such as the movement of white blood cells through tissues. Typically, ameboid motion begins with protrusion of a *pseudopodium* from one end of the cell. The pseudopodium projects far out away from the cell body, and then the remainder of the cell moves toward the pseudopodium. Formerly, it was believed that the protruding pseudopodium attached itself far away from the cell and then pulled the remainder of the cell toward it. However, recent studies have changed this idea to a "streaming" concept, as illustrated in Figure 2–13. It is believed that ameboid movement is caused in the following way: The outer portion of the cytoplasm is in a *gel* state and is called the *ectoplasm*, while the central portion of the cytoplasm is in a *sol* state and is called *endoplasm*. In the gel are numerous microfilaments composed of *actomyosin*, which is a

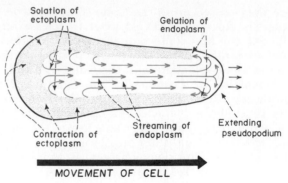

Figure 2–13. Ameboid motion by a cell.

highly contractile protein essentially the same as that found in muscle. Therefore, normally there is a continual tendency for the ectoplasm to contract. However, in response to a chemical or physical stimulus the ectoplasm at one end of the cell becomes thin, causing a pseudopodium to bulge outward in the direction of the chemotactic source. But the ectoplasm at the opposite end of the cell contracts, causing the endoplasm to "stream" into the pseudopodium. On reaching the pseudopodial end of the cell, the endoplasm turns toward the sides of the cell to form new ectoplasm. Therefore, at one end of the cell, ectoplasm is continually being solated while new ectoplasm is being formed at the other end. The continuous repetition of this process makes the cell move in the direction in which the pseudopodium projects. One can readily see that this streaming movement inside the cell is analogous to the revolving track of a Caterpillar tractor.

Types of Cells That Exhibit Ameboid Motion. The most common cells to exhibit ameboid motion in the human body are the *white blood cells*, which move out of the blood into the tissues in the form of tissue *macrophages* or *microphages.* However, many other types of cells can move by ameboid motion under certain circumstances. For instances, fibroblasts will move into any damaged area to help repair the damage, and even some of the germinal cells of the skin, though ordinarily completely sessile cells, will move by ameboid motion toward a cut area to repair the rent.

Control of Ameboid Motion — "Chemotaxis." The most important factor that usually initiates ameboid motion is the appearance of certain chemical substances in the tissues. This phenomenon is called *chemotaxis*, and the chemical substance causing it to occur is called a *chemotactic substance.*

Movement of Cilia. A second type of cellular motion, *ciliary movement*, is the bending of cilia along the surface of cells in the respiratory tract

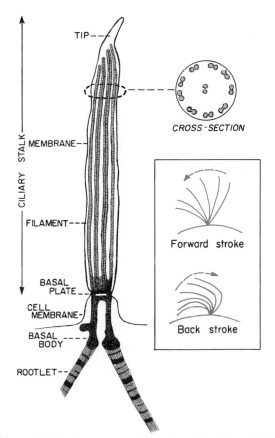

Figure 2–14. Structure and function of the cilium. (Modified from Satir: *Sci. Am., 204*:108, 1961. ©1961 by Scientific American, Inc. All rights reserved.)

and in the fallopian tubes of the reproductive tract. As illustrated in Figure 2–14, a cilium looks like a minute, sharp-pointed hair that projects 3 to 4 microns from the surface of the cell. Many cilia can project from a single cell.

The cilium is covered by an outcropping of the cell membrane, and it is supported by 11 microtubules, 9 double tubules located around the periphery of the cilium and 2 single tubules down the center, as shown in the cross-section illustrated in Figure 2–14.

In the inset of Figure 2–14, movement of the cilium is illustrated. The cilium moves forward with a sudden rapid stroke, 10 to 60 times per second, bending sharply where it projects from the surface of the cell. Then it moves backward very slowly in a whiplike manner. The rapid forward movement pushes the fluid lying adjacent to the cell in the direction that the cilium moves, then slow whiplike movement in the other direction has almost no effect on the fluid. As a result, fluid is continually propelled in the direction of the forward stroke. Since most ciliated cells have large numbers of cilia on their surfaces,

and since all the cilia are oriented in the same direction, this is a very effective means for moving fluids from one part of the surface to another; for instance, for moving mucus out of the lungs or for moving the ovum along the fallopian tube.

Mechanism of Ciliary Movement. Though not all aspects of ciliary movement are as yet clear, the postulated mechanism is the following: First, the nine double tubules and the two single tubules are all linked to each other by a complex of protein cross-linkages. Second, it is believed that release of energy from ATP in contact with the ATPase arms causes the arms to "crawl" along the surfaces of the adjacent tubules. If the front tubules crawl outward while the back tubules remain stationary, this obviously will cause bending.

Since many cilia on a cell surface contract simultaneously in a wavelike manner, it is presumed that some synchronizing signal — perhaps an electrochemical signal over the cell surface — is transmitted from cilium to cilium.

REFERENCES

Allen, R. D., and Allen, N. S.: Cytoplasmic streaming in amoeboid movement. *Annu. Rev. Biophys. Bioeng.,* 7:469, 1978.

Andresen, C. C.: Endocytosis in freshwater amebas. *Physiol. Rev., 57*:371, 1977.

Bulger, R. E., and Strum, J. M.: The Functioning Cytoplasm. New York, Plenum Press, 1974.

Capaldi, R. A.: A dynamic model of cell membranes. *Sci. Am., 230*(3):26, 1974.

Chance, B., *et al.*: Hydroperoxide metabolism in mammalian organs. *Physiol. Rev., 59*:527, 1979.

Cherkin, A., *et al.* (eds.): Physiology and Cell Biology of Aging. New York, Raven Press, 1979.

De Robertis, E. D. P., *et al.*: Cell Biology, 6th Ed. Philadelphia, W. B. Saunders Co., 1975.

Dingle, J. T.: Lysosomes in Biology and Pathology. New York, American Elsevier Publishing Co., 1973.

Fawcett, D. W.: The Cell. Philadelphia, W. B. Saunders Co., 1966.

Flickinger, C. J., *et al.*: Medical Cell Biology. Philadelphia, W. B. Saunders Co., 1979.

Fowler, S., and Wolinsky, H.: Lysosomes in vascular smooth muscle cells. *In* Bohr, D. F., *et al.* (eds.): *Handbook of Physiology.* Sec. 2, Vol. II. Baltimore, Williams & Wilkins Co., 1980, p. 133.

Giese, A. C.: Cell Physiology, 5th Ed. Philadelphia, W. B. Saunders Co., 1979.

Goldman, R. D., *et al.*: Cytoplasmic fibers in mammalian cells: Cytoskeletal and contractile elements. *Annu. Rev. Physiol., 41*:703, 1979.

Hammersen, F.: Histology: A Color Atlas of Cytology, Histology, and Microscopic Anatomy. Baltimore, Urban & Schwarzenberg, 1980.

Harris, H.: Nucleus and Cytoplasm, 3rd Ed. New York, Oxford University Press, 1974.

Hayflick, L.: The cell biology of human aging. *Sci. Am., 242*(1): 58, 1980.

Hinkle, P. C., and McCarty, R. E.: How cells make ATP. *Sci. Am. 238*(3):104, 1978.

Jakoby, W. B., and Pastan, I. H. (eds.): Cell Culture. New York, Academic Press, 1979.

Koshland, D. E., Jr.: Bacterial chemotaxis in relation to neurobiology. *Annu. Rev. Neurosci., 3*:43, 1980.

Lodish, H. F., and Rothman, J. E.: The assembly of cell membranes. *Sci. Am., 240*(1):48, 1979.

Marchesi, V. T., *et al.* (eds.): Cell Surface Carbohydrates and Biological Recognition. New York, A. R. Liss, 1978.

Masters, C., and Holmes, R.: Peroxisomes: New aspects of cell physiology and biochemistry. *Physiol. Rev., 57*:816, 1977.

Metcalfe, J. C. (ed.): Biochemistry of Cell Walls and Membranes II. Baltimore, University Park Press, 1978.

Nicholls, P. (ed.): Membrane Proteins. New York, Pergamon Press, 1978.

Reid, E. (ed.): Plant Organelles. New York, Halsted Press, 1979.

Satir, P.: How cilia move. *Sci. Am., 231*(4):44, 1974.

Singer, S. J.: The molecular organization of membranes. *Annu. Rev. Physiol., 43*:805, 1974.

Sloane, B. F.: Isolated membranes and organelles from vascular smooth muscle. *In* Bohr, D. F., *et al.* (eds.): Handbook of Physiology. Sec. 2, Vol. II, Baltimore, Williams & Wilkins Co., 1980, p. 121.

Staehelin, L. A., and Hull, B. E.: Junctions between living cells. *Sci. Am., 238*(5):140, 1978.

Stephens, R. E., and Edds, K. T.: Microtubules: Structure, chemistry, and function. *Physiol. Rev., 56*:709, 1976.

Tseng, H.: Atlas of Ultrastructure. New York, Appleton-Century-Crofts, 1980.

Wallach, D. F. H.: Plasma Membranes and Disease. New York, Academic Press, 1979.

Wilkinson, P. C.: Chemotaxis and Inflammation. New York, Churchill Livingstone, 1973.

Williamson, J. R.: Mitochondrial function in the heart. *Annu. Rev. Physiol., 41*:485, 1979.

3

Genetic Control of Cell Function — Protein Synthesis and Cell Reproduction

Almost everyone knows that the genes control heredity from parents to children, but most persons do not realize that the same genes control the reproduction and the day-by-day function of all cells. The genes control function of the cell by determining what substances will be synthesized within the cell — what structures, what enzymes, what chemicals.

Figure 3–1 illustrates the general schema by which the genes control cellular function. Each gene, which is a nucleic acid called *deoxyribonucleic acid (DNA)*, automatically controls the formation of another nucleic acid, *ribonucleic acid (RNA)*, which spreads throughout the cell and controls the formation of a specific protein. Some proteins are *structural proteins*, which, in association with various lipids, form the structures of the various organelles that were discussed in the preceding chapter. But by far the majority of the proteins are *enzymes* that catalyze the different chemical reactions in the cells. For instance, enzymes promote all the oxidative reactions that supply energy to the cell, and they promote the synthesis of various chemicals such as lipids, glycogen, adenosine triphosphate, and others.

THE GENES

The genes, of which there are about 100,000 different types in human cells, are contained in long, double-stranded, helical molecules of *deoxyribonucleic acid (DNA)* having molecular weights usually measured in the millions. A very short segment of such a molecule is illustrated in Figure 3–2. This molecule is composed of several simple chemical compounds arranged in a regular pattern explained in the following few paragraphs.

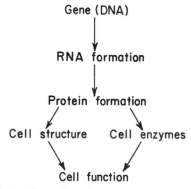

Gene (DNA)

↓

RNA formation

↓

Protein formation

Cell structure   Cell enzymes

Cell function

Figure 3–1. General schema by which the genes control cell function.

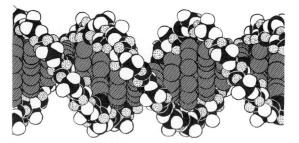

Figure 3–2. The helical, double-stranded structure of the gene. The outside strands are composed of phosphoric acid and the sugar deoxyribose. The internal molecules connecting the two strands of the helix are purine and pyrimidine bases; these determine the "code" of the gene.

PHOSPHORIC ACID:

DEOXYRIBOSE:

BASES:

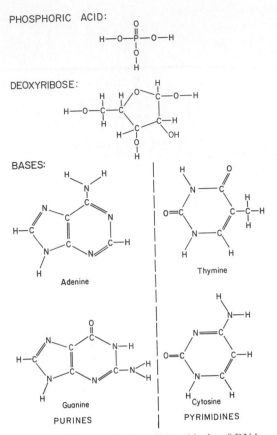

Adenine

Thymine

Guanine

Cytosine

PURINES

PYRIMIDINES

Figure 3–3. The basic building blocks of DNA.

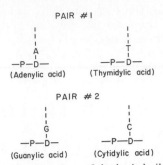

PAIR #1

A
—P—D—
(Adenylic acid)

T
—P—D—
(Thymidylic acid)

PAIR #2

G
—P—D—
(Guanylic acid)

C
—P—D—
(Cytidylic acid)

Figure 3–5. Combinations of the basic building blocks of DNA to form nucleotides. (P, phosphoric acid; D, deoxyribose. The four nucleotide bases are A, adenine; T, thymine; G, guanine; and C, cytosine.) Note that four different types of nucleotides make up DNA.

The Basic Building Blocks of DNA. Figure 3–3 illustrates the basic chemical compounds involved in the formation of DNA. These include (1) *phosphoric acid*, (2) a sugar called *deoxyribose*, and (3) four nitrogenous bases: *adenine, guanine, thymine,* and *cytosine*. The phosphoric acid and deoxyribose form the two helical strands of DNA, and the bases lie between the strands and connect them together.

The Nucleotides. The first stage in the formation of DNA is the combination of one molecule of phosphoric acid, one molecule of deoxyribose, and one of the four bases to form a nucleotide. Four separate nucleotides are thus formed, one for each of the four bases: *adenylic, thymidylic, guanylic,* and *cytidylic acids*. Figure 3–4 illustrates the chemical structure of adenylic acid, and Figure 3–5 illustrates simple symbols for all the four basic nucleotides that form DNA.

Note also in Figure 3–5 that the nucleotides are separated into two *complementary pairs:* (1) Adenylic acid and thymidylic acid form one pair, and (2) guanylic acid and cytidylic acid form the other pair. The *bases* of each pair can attach loosely (by hydrogen bonding) to each other, thus providing the means by which the two strands of the DNA helix are bound together.

Organization of the Nucleotides to Form DNA. Figure 3–6 illustrates the manner in which multiple numbers of nucleotides are bound together to form the DNA strands. Note that these are combined in such a way that phosphoric acid and deoxyribose alternate with each other in the two separate strands, and these strands are held together by the respective complementary pairs of bases. Thus, in Figure 3–6 the sequence of complementary pairs of bases is CG, CG, GC, TA, CG, TA, GC, AT, and AT. However, the bases are bound together by very loose hydrogen bonding, represented in the figure by dashed lines. Because of the looseness of

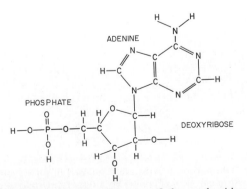

ADENINE

PHOSPHATE

DEOXYRIBOSE

Figure 3–4. Adenylic acid, one of the nucleotides that make up DNA.

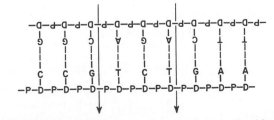

Figure 3–6. Combination of deoxyribose nucleotides to form DNA.

these bonds, the two strands can pull apart with ease, and they do so many times during the course of their function in the cell.

Now, to put the DNA of Figure 3–6 into its proper physical perspective, one needs merely to pick up the two ends and twist them into a helix. Ten pairs of nucleotides are present in each full turn of the helix in the DNA molecule, as illustrated in Figure 3–2.

THE GENETIC CODE

The importance of DNA lies in its ability to control the formation of other substances in the cell. It does this by means of the so-called genetic code. When the two strands of a DNA molecule are split apart, this exposes the bases projecting to the side of each strand. It is these projecting bases that form the code.

Research studies in the past few years have demonstrated that the so-called *code words* consist of "triplets" of bases — that is, each three successive bases are a code word. And the successive code words control the sequence of amino acids in a protein molecule during its synthesis in the cell. Note in Figure 3–6 that each of the two strands of the DNA molecule carries its own genetic code. For instance, the top strand, reading from left to right, has the genetic code GGC, AGA, CTT, the code words being separated from each other by the arrows. As we follow this genetic code through Figures 3–7 and 3–8, we shall see that these three code words are responsible for placement of the three amino acids *proline*, *serine*, and *glutamic acid* in a molecule of protein. Furthermore, these three amino acids will be lined up in the protein molecule in exactly the same way that the genetic code is lined up in this strand of DNA.

RIBONUCLEIC ACID (RNA) — THE PROCESS OF TRANSCRIPTION

Since almost all DNA is located in the nucleus of the cell and yet most of the functions of the cell are carried out in the cytoplasm, some means must be available for the genes of the nucleus to control the chemical reactions of the cytoplasm. This is achieved through the intermediary of another type of nucleic acid, ribonucleic acid (RNA), the formation of which is controlled by the DNA of the nucleus, the process being called *transcription*. The RNA is then transported into the cytoplasmic cavity, where it controls protein synthesis.

Three separate types of RNA are important to protein synthesis: *messenger RNA, transfer RNA,* and *ribosomal RNA*. Before we describe the function of these different RNAs in the synthesis of proteins, let us see how DNA controls the formation of RNA.

Synthesis of RNA. One strand of the DNA molecule, which contains the genes, acts as a template for synthesis of RNA molecules. (The other strand of the DNA has no genetic function but does function for replication of the gene itself, which will be discussed later in the chapter.) The code words in DNA cause the formation of *complementary* code words called *codons* in RNA. The stages of RNA synthesis are as follows:

The Basic Building Blocks of RNA. The basic building blocks of RNA are almost the same as those of DNA except for two differences. First, the sugar deoxyribose is not used in the formation of RNA. In its place is another sugar of very slightly different composition, *ribose*. Second, thymine is replaced by another base, *uracil*.

Formation of RNA Nucleotides. The basic building blocks of RNA first form nucleotides exactly as described above for the synthesis of DNA. Here again, four separate nucleotides are used in the formation of RNA. These nucleotides contain the bases *adenine, guanine, cytosine,* and *uracil,* respectively, the uracil replacing the thymine found in the four nucleotides that make up DNA.

Activation of the Nucleotides. The next step in the synthesis of RNA is activation of the nucleotides. This occurs by the addition to each nucleotide of two phosphate radicals to form triphosphate. These last two phosphates are combined with the nucleotide by *high-energy phosphate bonds* derived from the ATP of the cell.

The result of this activation process is that large quantities of energy are made available to each of the nucleotides, and this energy is used in promoting the subsequent chemical reactions that eventuate in the formation of the RNA chain.

Assembly of the RNA Molecule from Activated Nucleotides Using the DNA Strand as a Template — the Process of Transcription. The next stage in the formation of RNA is separation of the two strands of the DNA molecule. Then, one of these strands is used as a template on which the RNA molecule is assembled. It is this strand that contains the genes while the other strand remains genetically inactive. Assembly of the RNA molecule is accomplished in the manner illustrated in Figure 3–7 under the influence of the enzyme *RNA polymerase*. The steps of this procedure are (1) temporary bonding of an RNA base with each DNA base, (2) bonding of the successive RNA bases with each other, and (3) splitting of the RNA strand away from the DNA strand.

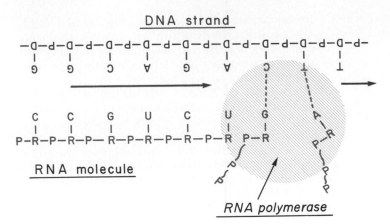

Figure 3–7. Combination of ribose nucleotides with a strand of DNA to form a molecule of ribonucleic acid (RNA) that carries the DNA code from the gene to the cytoplasm.

It should be remembered that there are four different types of DNA bases and also four different types of RNA nucleotide bases. Furthermore, these always combine with each other in specific combinations. Therefore, the code that is present in the DNA strand is transmitted in *complementary* form to the RNA molecule. The ribose nucleotide bases always combine with the deoxyribose bases in the following combinations:

DNA base	RNA base
guanine	cytosine
cytosine	guanine
adenine	uracil
thymine	adenine

Once the RNA molecules are formed, they diffuse out of the nucleus and into all parts of the cytoplasm, where they perform further functions. The type of RNA called *messenger RNA* carries the genetic code to the cytoplasm for formation of proteins. The *ribosomal RNA* is used for the formation of ribosomes, which are the physical and chemical structures on which protein molecules are actually assembled. *Transfer RNA* is utilized to carry activated amino acids to the ribosomes, where the protein molecules are assembled.

MESSENGER RNA

Messenger RNA molecules are long straight strands that are suspended in the cytoplasm. These molecules are usually composed of several hundred to several thousand nucleotides in unpaired strands, and they contain *codons* that are exactly complementary to the code words of the genes. Figure 3–8 illustrates a small segment of a molecule of messenger RNA. Its codons are CCG, UCU, and GAA. These are the codons for proline, serine, and glutamic acid. The transcription of these codons from the DNA molecule was demonstrated in Figure 3–7.

RNA Codons. Table 3–1 gives the RNA codons for the 20 common amino acids found in protein molecules. Note that several of the amino acids are represented by more than one codon; some codons represent such signals as "start manufacturing a protein molecule" or "stop manufacturing a protein molecule." In Table 3–1, these two codons are designated CI for "chain-initiating" and CT for "chain-terminating."

TRANSFER RNA

Another type of RNA that plays a prominent role in protein synthesis is called *transfer RNA* because it transfers amino acid molecules to protein molecules as the protein is synthesized. Each type of transfer RNA combines specifically with only one of the 20 amino acids that are incorporated into proteins. The transfer RNA then acts as a *carrier* to transport its specific type of amino acid to the ribosomes where protein molecules are formed.

Transfer RNA, containing only about 80 nucleotides, is a relatively small molecule in comparison with messenger RNA. It is a folded chain of nucleotides with a cloverleaf appearance similar to that illustrated in Figure 3–9. At one end of the molecule is always an adenylic acid that attaches to the amino acid.

The specific prosthetic group in the transfer

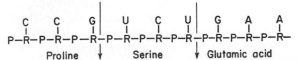

Figure 3–8. Portion of a ribonucleic acid molecule, showing three "code" words, CCG, UCU, and GAA, which represent the three amino acids *proline, serine,* and *glutamic acid.*

TABLE 3-1 RNA CODONS FOR THE DIFFERENT AMINO ACIDS AND FOR START AND STOP

Amino Acid	RNA Codons					
Alanine	GCU	GCC	GCA	GCG		
Arginine	CGU	CGC	CGA	CGG	AGA	AGG
Asparagine	AAU	AAC				
Aspartic acid	GAU	GAC				
Cysteine	UGU	UGC				
Glutamic acid	GAA	GAG				
Glutamine	CAA	CAG				
Glycine	GGU	GGC	GGA	GGG		
Histidine	CAU	CAC				
Isoleucine	AUU	AUC	AUA			
Leucine	CUU	CUC	CUA	CUG	UUA	UUG
Lysine	AAA	AAG				
Methionine	AUG					
Phenylalanine	UUU	UUC				
Proline	CCU	CCC	CCA	CCG		
Serine	UCU	UCC	UCA	UCG		
Threonine	ACU	ACC	ACA	ACG		
Tryptophan	UGG					
Tyrosine	UAU	UAC				
Valine	GUU	GUC	GUA	GUG		
Start (CI)	AUG	GUG				
Stop (CT)	UAA	UAG	UGA			

RNA that allows it to recognize a specific codon in a messenger RNA strand is called an *anticodon*, and this is located approximately in the middle of the transfer RNA molecule (at the bottom of the cloverleaf configuration illustrated in Figure 3–9). During formation of a protein molecule, the anticodon bases combine loosely by hydrogen bonding with the codon bases of the messenger RNA. In this way the respective amino acids are lined up one after another along the messenger RNA chain, thus establishing the appropriate sequence of amino acids in the protein molecule.

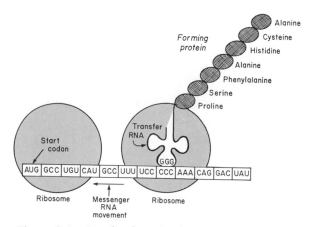

Figure 3–9. Postulated mechanism by which a protein molecule is formed in ribosomes in association with messenger RNA and soluble RNA.

RIBOSOMAL RNA

The third type of RNA in the cell is ribosomal RNA; it constitutes about 60 per cent of the ribosome. The remainder of the ribosome is protein, containing as many as 50 different types of proteins, both structural proteins and enzymes needed in the manufacture of protein molecules.

The ribosome is the physical and chemical structure in the cytoplasm on which protein molecules are actually synthesized. However, it always functions in association with both the other types of RNA as well: transfer RNA transports amino acids to the ribosomes for incorporation into the developing protein molecules, while messenger RNA provides the information necessary for sequencing the amino acids in proper order for each specific type of protein to be manufactured.

Formation of Ribosomes in the Nucleolus. The DNA molecules for formation of ribosomal RNA are all located in a single chromosomal pair of the nucleus. However, this chromosomal pair contains many duplicates of these ribosomal genes because of the large amount of ribosomal RNA required for cellular function.

As the ribosomal RNA forms, it collects in the *nucleolus*, a specialized structure lying adjacent to the chromosome. The ribosomal RNA is specially processed in the nucleolus and combined with "ribosomal proteins" to form granular condensation products that are primordial forms of the ribosomes. These are then released from the nucleolus, and they migrate through the large "pores" of the nuclear membrane to almost all parts of the cytoplasm, most of them eventually attaching to the surfaces of the endoplasmic reticulum.

FORMATION OF PROTEINS IN THE RIBOSOMES — THE PROCESS OF TRANSLATION

When a molecule of messenger RNA comes in contact with a ribosome, it travels through it, beginning at a predetermined end specified by an appropriate sequence of RNA bases. However, the protein molecule does not begin to form until a "start" (or "chain-initiating") codon enters the ribosome. Then, as illustrated in Figure 3–9, while the messenger RNA travels through the ribosome, a protein molecule is formed — a process called *translation*. Thus, the ribosome reads the code of the messenger RNA in much the same way that a tape is "read" as it passes through the playback head of a tape recorder. Then,

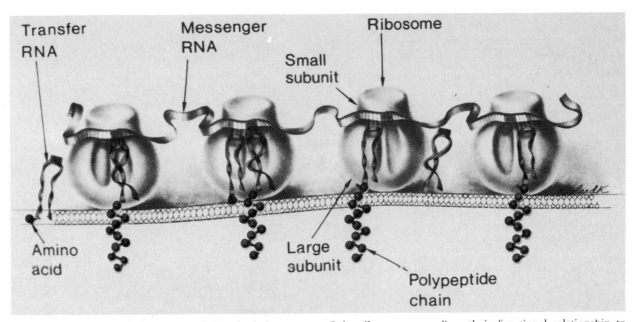

Figure 3–10. An artist's concept of the physical structure of the ribosomes as well as their functional relationship to messenger RNA, transfer RNA, and the endoplasmic reticulum during the formation of protein molecules. (From Bloom and Fawcett: A Textbook of Histology. 10th ed. Philadelphia, W. B. Saunders Company, 1975.)

when a "stop" (or "chain-terminating") codon slips past the ribosome, the end of a protein molecule is signaled, and the entire protein molecule is freed into the cytoplasm.

It is especially important that a messenger RNA can cause the formation of a protein molecule in any ribosome, and that there is no specificity of ribosomes for given types of protein. The ribosome seems to be simply the physical structure in which or on which the chemical reactions take place.

Figure 3–10 shows the functional relationship of messenger RNA to the ribosomes and also the manner in which the ribosomes attach to the membrane of the endoplasmic reticulum. Note the process of translation occurring in several ribosomes at the same time in response to the same strand of messenger RNA. And note also the newly forming polypeptide chains passing through the endoplasmic reticulum membrane into the endoplasmic matrix, thus generating the protein molecules.

Peptide Linkage. The successive amino acids in the newly forming protein chain combine with each other according to the following typical reaction:

In this chemical reaction, a hydroxyl radical is removed from the COOH portion of one amino acid while a hydrogen of the NH_2 portion of the other amino acid is removed. These combine to form water, and the two reactive sites left on the two successive amino acids combine, resulting in a single molecule. This process is called *peptide linkage*.

SYNTHESIS OF OTHER SUBSTANCES IN THE CELL

Many thousand protein enzymes formed in the manner just described control essentially all the other chemical reactions that take place in cells. These enzymes promote synthesis of lipids, glycogen, purines, pyrimidines, and hundreds of other substances. We will discuss many of these synthetic processes in relation to carbohydrate, lipid, and protein metabolism in Chapters 45 and 46. It is by means of all these different substances that the many functions of the cells are performed.

CONTROL OF GENETIC FUNCTION AND BIOCHEMICAL ACTIVITY IN CELLS

There are basically two different methods by which the biochemical activities in the cell are controlled. One of these is called *genetic regulation*, in which the activities of the genes themselves are

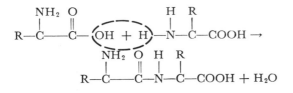

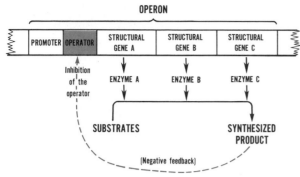

Figure 3–11. Function of the operon in controlling biosynthesis. Note that the synthesized product exerts a negative feedback to inhibit function of the operon, in this way automatically controlling the concentration of the product itself.

controlled, and the other *enzyme regulation*, in which the activity rates of the enzymes within the cell are controlled.

GENETIC REGULATION

Gene function is controlled in several different ways. Some genes are normally dormant but can be activated by *inducer substances*. Other genes are naturally active but can be inhibited by *repressor substances*. As an illustration, let us describe one of the mechanisms for genetic control.

The Operon and Its Control of Biochemical Synthesis. The synthesis of a cellular biochemical product usually requires a series of reactions, and each of these reactions is catalyzed by a specific enzyme. Formation of all the enzymes needed for the synthetic process is in turn usually controlled by a sequence of genes all located in series one after the other on the same chromosomal DNA strand. This area of the DNA strand is called an *operon*, and the genes responsible for forming the respective enzymes are called the *structural genes*. In Figure 3–11, three respective structural genes are illustrated in an operon, and it is shown that they control the formation of three respective enzymes utilized in a particular biochemical synthetic process.

The rate at which the operon functions to transcribe RNA, and therefore to set into motion the enzymatic system for the biochemical process, is determined by the presence of two other small segments on the DNA strand called, respectively, the *promoter* and the *operator;* these are also shown in Figure 3–11. Each of these is a specific sequence of DNA nucleotides, but they do not themselves serve as templates to cause the formation of RNA. Instead, they merely function as control units of the operon.

The promoter first binds with *RNA polymerase,*

which is an enzyme that moves along the operon to cause transcription of the appropriate messenger RNAs. However, lying between the promoter and the structural genes is the operator; this is a control gate that can be opened or closed. If the gate is open, the RNA polymerase will travel along the operon and cause the transcription process; but if the gate is closed, the RNA polymerase becomes blocked at the promoter level and the operon remains dormant.

In Figure 3–11, it is shown that the presence of a critical amount of the synthesized product in the cellular cytoplasm will cause negative feedback to the operator to inhibit it — that is, to close the gate. Therefore, whenever there is already enough of the required product, the operon becomes dormant. On the other hand, as the synthesized product becomes degraded in the cell and its concentration falls, the operator gate opens and the operon once again becomes active. In this way, the concentration of the synthesized product is automatically controlled.

Other Mechanisms for Control of Transcription by the Operon. Variations in the basic mechanism for control of the operon have been discovered with rapidity in the last few years. Without going into detail for all these, let us merely list some of the mechanisms of control:

(1) An *inducer* substance from outside the cell sometimes activates the operator.

(2) A *regulatory gene* elsewhere in the cell nucleus sometimes regulates the operator.

(3) Inhibitors or inducers sometimes control many different operators at the same time.

(4) Some synthetic processes are controlled not at the DNA level but instead at the RNA level to control the translation process for formation of proteins by messenger RNA.

CONTROL OF ENZYME ACTIVITY

In the same way that inhibitors and activators can affect the genetic regulatory system, so also can the enzymes themselves be directly controlled by other inhibitors or activators. This, then, represents a second category of mechanisms by which cellular biochemical functions can be controlled.

Enzyme Inhibition. A great many of the chemical substances formed in the cell have a direct feedback effect of inhibiting the respective enzyme systems that synthesize them. Almost always the synthesized product acts on the first enzyme in a sequence, rather than on the subsequent enzymes. One can readily recognize the importance of inhibiting this first enzyme to prevent buildup of intermediary products that will not be utilized.

This process of enzyme inhibition is another example of negative feedback control; it is responsible for controlling the intracellular concentrations of some of the amino acids that are not controlled by the genetic mechanism as well as the concentrations of the purines, the pyrimidines, vitamins, and other substances.

Enzyme Activation. Enzymes that are either normally inactive or that have been inactivated by some inhibitor substance can often be activated. An example of this is the action of cyclic adenosine monophosphate (AMP) in causing glycogen to split so that energy derived from the released glucose molecules can be used to form high-energy ATP, as discussed in the previous chapter. When most of the adenosine triphosphate has been depleted in a cell, a considerable amount of cyclic AMP begins to be formed as a breakdown product of the ATP; the presence of this cyclic AMP indicates that the cellular reserves of ATP have approached a low ebb. However, the cyclic AMP immediately activates the glycogen-splitting enzyme phosphorylase, liberating glucose molecules that are rapidly used for replenishment of the ATP stores. Thus, in this case the cyclic AMP acts as an enzyme activator and thereby helps to control intracellular ATP concentration.

In summary, there are two different methods by which the cells control proper proportions and proper quantities of different cellular constituents: (1) the mechanism of genetic regulation, and (2) the mechanism of enzyme regulation. The genes can be either activated or inhibited; likewise, the enzymes can be either activated or inhibited. Most often, these regulatory mechanisms function as feedback control systems that continually monitor the cell's biochemical composition and make corrections as needed. But, on occasion, substances from without the cell (especially some of the hormones that will be discussed later in this text) also control the intracellular biochemical reactions by activating or inhibiting one or more of the intracellular control systems.

CELL REPRODUCTION

Cell reproduction is another example of the pervading, ubiquitous role that the DNA-genetic system plays in all life processes. It is the genes and their internal regulatory mechanisms that determine the growth characteristics of the cells and also when or whether these cells will divide to form new cells. In this way, this all-important genetic system controls each stage in the development of the human being from the single-cell fertilized ovum to the whole functioning body. Thus, if there is any central theme to life it is the DNA-genetic system.

As is true of almost all other events in the cell, reproduction also begins in the nucleus itself. The first step is *replication (duplication) of all DNA in the chromosomes*. The next step is *mitosis*, which consists, first, of division of the two sets of DNA between two separate nuclei and, second, splitting of the cell itself to form two new daughter cells.

The complete life cycle of a cell that is not inhibited in some way is about 10 to 30 hours from reproduction to reproduction, and the period of mitosis lasts for approximately half an hour. The period between mitoses is called *interphase*. However, in the body there are almost always inhibitory controls that slow or stop the uninhibited life cycle of the cell and give cells life cycle periods that vary from as little as 10 hours for stimulated bone marrow cells to an entire lifetime of the human body for nerve and striated muscle cells.

REPLICATION OF THE DNA

The DNA begins to be reproduced about 5 hours before mitosis takes place, forming two exact *replicates* of all DNA, which respectively become the DNA in each of the two new daughter cells that will be formed in mitosis.

Chemical and Physical Events. The DNA is duplicated in almost exactly the same way that RNA is formed from DNA. First, the two strands of the DNA helix of the gene pull apart. Second, each of these strands combines with deoxyribose nucleotides of the four types described early in this chapter, and complementary DNA strands are formed. The only difference between this formation of the new strands of DNA and the formation of an RNA strand is that the new strands of DNA remain attached to the old strands that have formed them, thus forming two new double-stranded DNA helixes.

THE CHROMOSOMES AND THEIR REPLICATION

The chromosomes consist of two major parts: the DNA and protein. The protein, in turn, is composed mainly of small molecules. Many of these are *histones*, which probably serve to fold or otherwise compact the DNA strands into manageable sizes. On the other hand, the *nonhistone chromosomal proteins* are major components of the genetic regulatory system, acting as activators, inhibitors, and enzymes.

Recent experiments indicate that all the DNA of a particular chromosome is arranged in one long double helix and that the genes are attached end-on-end with each other. Such a molecule in the human being has a molecular weight of about 60 billion and if spread out linearly would be approximately 7.5 cm. long, or several thousand times as long as the diameter of the nucleus itself; but the experiments also indicate that this long double helix is folded or coiled like a spring and is held in this position by its linkages to the histone molecules.

Replication of the chromosomes follows as a natural result of replication of the DNA strand. When the new double helix separates from the original double helix, it presumably carries some of the old protein with it or combines with new protein, the DNA acting as the backbone of the newly replicated chromosomes.

Number of Chromosomes in the Human Cell. Each human cell contains 46 chromosomes arranged in 23 pairs. In general, the genes in the two chromosomes of each pair are identical or almost identical with each other, so that it is usually stated that the different genes exist in pairs, though occasionally this is not the case.

MITOSIS

The actual process by which the cell splits into two new cells is called mitosis. Once the genes have been duplicated and each chromosome has split to form two new chromosomes, each of which is now called a *chromatid,* mitosis follows automatically, almost without fail, within about an hour.

The Mitotic Apparatus. One of the first events of mitosis takes place in the cytoplasm in or around the small structures called *centrioles.* As illustrated in Figure 3–12, two pairs of centrioles lie close to each other near one pole of the nucleus. Each centriole is a small cylindrical body about 0.4 micron long and about 0.15 micron in diameter, consisting mainly of nine parallel, tubule-like structures arranged around the inner wall of the cylinder.

At the beginning of mitosis the two pairs of centrioles begin to move apart from each other. This is caused by protein microtubules growing between the respective pairs and actually pushing them apart. At the same time, microtubules grow radially away from each of the pairs, forming a spiny star, called the *aster,* in each end of the cell. Some of the spines penetrate the nucleus and will play a role in separating the two sets of DNA helixes, the chromatids, during mitosis. The set of microtubules connecting the two centriole pairs is called the *spindle,* and the entire set of microtubules plus the two pairs of centrioles is called the *mitotic apparatus.*

Prophase. The first stage of mitosis, called *prophase,* is shown in Figure 3–12A, B, and C.

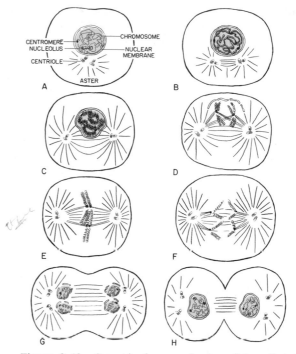

Figure 3–12. Stages in the reproduction of the cell. A, B, and C, prophase; D, prometaphase; E, metaphase; F, anaphase; G and H, telophase. (Redrawn from Mazia: *Sci. Am.,* 205:102, 1961. © by Scientific American, Inc. All rights reserved.)

While the spindle is forming, the *chromatin material* of the nucleus (the DNA), which in interphase consists of long, loosely coiled strands, becomes shortened into well-defined chromosomes.

Prometaphase. During this stage (Figure 3–12D) the nuclear envelope disintegrates, and microtubules from the forming mitotic apparatus become attached to the chromosomes.

Metaphase. During metaphase (Fig. 3–12E) the centriole pairs are pushed far apart by the growing spindle, and the chromosomes are thereby pulled tightly by the attached microtubules to the very center of the cell, lining up in the equatorial plane of the mitotic spindle.

Anaphase. With still further growth of the spindle, the chromatids in each pair of chromosomes are now broken apart, a stage of mitosis called anaphase (Fig. 3–12F).

Telophase. In telophase (Fig. 3–12G and H) the mitotic spindle grows still longer, pulling the two sets of daughter chromosomes completely apart. Then the mitotic apparatus dissolves and a new nuclear membrane develops around each set of chromosomes, this membrane being formed from portions of the endoplasmic reticulum that are already present in the cytoplasm. Shortly thereafter, the cell pinches in two midway between the two nuclei. This is caused by a contractile ring of *microfilaments* composed of *actin* *osin*, the two contractile proteins ... develops at the juncture of the ... ng cells and pinches them off from

CONTROL OF CELL GROWTH AND REPRODUCTION

Cell growth and reproduction usually go together; growth normally leads to replication of the DNA of the nucleus, followed a few hours later by mitosis.

In the normal human body, regulation of cell growth and reproduction is mainly a mystery. We know that certain cells grow and reproduce all the time, such as the blood-forming cells of the bone marrow, the germinal layers of the skin, and the epithelium of the gut. However, many other cells, such as smooth muscle cells, do not reproduce for many years. And a few cells, such as the neurons and striated muscle cells, do not reproduce during the entire life of the person.

If there is an insufficiency of some types of cells in the body, these will grow and reproduce very rapidly until appropriate numbers of them are again available. For instance, seven-eighths of the liver can be removed surgically, and the cells of the remaining one-eighth will grow and divide until the liver mass returns almost to normal. The same effect occurs for almost all glandular cells, for cells of the bone marrow, the subcutaneous tissue, the intestinal epithelium, and almost any other tissue except highly differentiated cells, such as nerve and muscle cells.

We know very little about the mechanisms that maintain proper numbers of the different types of cells in the body. However, experimental studies have shown that control substanc s called *chalones* are secreted by the different cells and cause feedback effects to stop or slow their growth and mitosis when too many of them have been formed. We know that cells of any type removed from the body and grown in tissue culture can grow and reproduce rapidly and indefinitely if the medium in which they grow is continually replenished. Yet they will stop growing when even small amounts of their own secretions are allowed to collect in the medium, which supports the idea that control substances limit cellular growth.

CANCER

Cancer is a disease that attacks the basic life process of the cell, in almost all instances altering the cell's *genome* (the total genetic complement of the cell) and leading to wild and spreading growth of the cancerous cells. The cause of the altered genome is a *mutation* (alteration) of one or more genes; or mutation of a large segment of a DNA strand containing many genes; or, in some instances, addition or loss of large segments of chromosomes.

But what is it that causes the mutations? When one realizes that many trillions of new cells are formed each year in the human being, this question should probably better be asked in the following form: Why is it that we do not develop literally millions or billions of mutant cancerous cells? The answer is the incredible precision with which DNA chromosomal strands are replicated in each cell before mitosis takes place. Indeed, even after each new strand is formed, the veracity of the replication process is "proofread" several different times. If any mistakes have been made, the new strand is cut and repaired before the mitotic process is allowed to proceed. Yet, despite all these precautions, probably one newly formed cell in every few hundred thousand to every few million still has significant mutant characteristics. We know this because it has been ascertained that each gene in a human offspring has the probability of 1 in 100,000 of being a mutant when compared with the genes of the parents.

Thus, chance alone is all that is required for mutations to take place. However, other factors that increase the probability of mutation include (1) *ionizing radiation*, (2) certain types of chemical substances called *carcinogens*, (3) some *viruses*, (4) *physical irritation*, and (5) *hereditary predisposition*.

Invasive Characteristic of the Cancer Cell. The two major differences between the cancer cell and the normal cell are these: (1) The cancer cell does not respect usual cellular growth limits; the reason is that it presumably does not secrete the appropriate *chalones* that are responsible for stopping excess growth of normal cells. (2) Cancer cells are far less adhesive to each other than are normal cells. Therefore, they have a tendency to wander through the tissues, to enter the blood stream, and to be transported all through the body where they form nidi for numerous new cancerous growths.

Why Do Cancer Cells Kill? The answer to this is very simple: Cancer tissue competes with normal tissues for nutrients. Because cancer cells continue to proliferate indefinitely, their number multiplying day by day, one can readily understand that the cancer cells will soon demand essentially all the nutrition available to the body. As a result the normal tissues gradually suffer nutritive death.

REFERENCES

Baum, H., and Gergely, J. (eds.): Molecular Aspects of Medicine. New York, Pergamon Press, 1978.

Butler, J. G., and Klug, A.: The assembly of a virus. *Sci. Am.,* 239(5):62, 1978.

Clark, B. F. C., *et al.* (eds.): Gene Expression: Protein Synthesis and Control, RNA Synthesis and Control, Chromatin Structure and Function. New York, Pergamon Press, 1978.

Cohen, S. N.: The manipulation of genes. *Sci. Am.,* 233(1):24, 1975.

Cummings, D. J. *et al.* (eds.): Extrachromosomal DNA. New York, Academic Press, 1979.

Dickerson, R. E.: Chemical evolution and the origin of life. *Sci. Am.,* 239(3):70, 1978.

Fiddes, J. C.: The nucleotide sequence of a viral DNA. *Sci. Am.,* 237(6):54, 1977.

Frankel, E.: DNA, The Ladder of Life. New York, McGraw-Hill, 1978.

Friedman, D. L.: Role of cyclic nucleotides in cell growth and differentiation. *Physiol. Rev.,* 56:652, 1976.

Hiatt, H. H., *et al.* (eds.): Origins of Human Cancer. Cold Spring Harbor, N.Y., Cold Spring Harbor Laboratory, 1977.

Horton, J. D. (ed.): Development and Differentiation of Vertebrates. New York, Elsevier/North-Holland, 1980.

Kaplan, J. G.: Membrane cation transport and the control of proliferation of mammalian cells. *Annu. Rev. Physiol.,* 40:19, 1978.

Kastrup, K. W., and Neilsen, J. H. (eds.): Growth Factors: Cellular Growth Processes, Growth Factors, Hormonal Control of Growth. New York, Pergamon Press, 1978.

Kornberg, A.: DNA Replication. San Francisco, W. H. Freeman, 1980.

Maniatis, T., and Ptashne, M.: A DNA operator-repressor system. *Sci. Am.,* 234(1):64, 1976.

Molineaux, I., and Kohiyama, M. (eds.): DNA Synthesis: Present and Future. New York, Plenum Press, 1978.

Nicolson, G. L.: Cancer Metastasis. *Sci. Am.,* 240(3):66, 1979.

Rich, A., and Kim, S. H.: The three-dimensional structure of transfer RNA. *Sci. Am.,* 238(1):52, 1978.

Russell, T. R., *et al.* (eds.): From Gene to Protein: Information Transfer in Normal and Abnormal Cells. New York, Academic Press, 1979.

Sandberg, A. A.: The Chromosomes in Human Cancer and Leukemia. New York, Elsevier/North-Holland, 1979.

Schopf, J. W.: The evolution of the earliest cells. *Sci. Am.,* 239(3):110, 1978.

Sobell, H. M.: Symmetry in nucleic acid structure and its role in protein-nucleic acid interactions. *Annu. Rev. Biophys. Bioeng.,* 5:307, 1976.

Söll, D., *et al.* (eds.): Transfer RNA: Biological Aspects. Cold Spring Harbor, N.Y., Cold Spring Harbor Laboratory, 1979.

Stein, G. S., *et al.*: Chromosomal proteins and gene regulation. *Sci. Am.,* 232(2):46, 1975.

Walker, R. T., *et al.* (eds.): Nucleoside Analogues: Chemistry, Biology, and Medical Applications. New York, Plenum Press, 1979.

Weissman, S. M.: Gene structure and function. *In* Bondy, R. K., and Rosenberg, L. E. (eds.): Metabolic Control and Disease. 8th Ed. Philadelphia, W. B. Saunders Co., 1980, p. 1.

Wolpert, L.: Pattern formation in biological development. *Sci. Am.,* 239(4):154, 1978.

Wu, R. (ed.): Recombinant DNA. New York, Academic Press, 1979.

Transport Through the Cell Membrane

The fluid inside the cells of the body, called *intracellular fluid,* is very different from that outside the cells, called *extracellular fluid.* The extracellular fluid is the fluid of the blood, in which the red cells are dispersed (the *plasma*), and also the fluid in the spaces between the tissue cells (the *interstitial fluid*). It is the extracellular fluid that supplies the cells with nutrients and other substances needed for cellular function.

Figure 4–1 gives the compositions of both the extracellular and intracellular fluids. Note that the extracellular fluid contains large quantities of *sodium* but only small quantities of *potassium.* Exactly the opposite is true of the intracellular fluid. Also, the extracellular fluid contains large quantities of chloride, while the intracellular fluid contains phosphates and proteins. These differences between the components of the intracellular and extracellular fluids are extremely important to the life of the cell. It is the purpose of this chapter to explain how these differences are brought about by the transport mechanisms in the cell membrane.

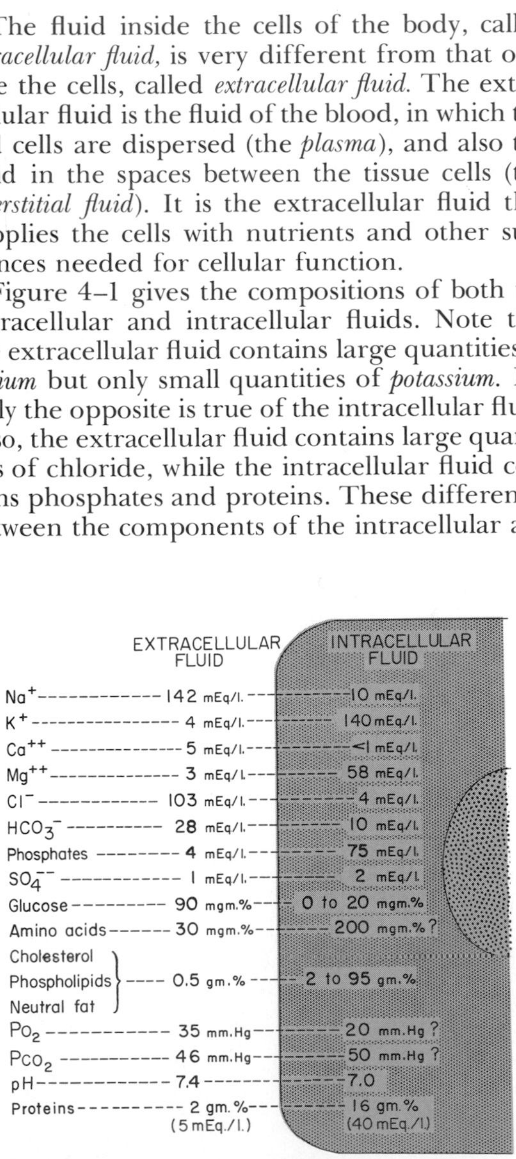

Substances are transported through the cell membrane by two major processes, *diffusion* and *active transport.* Basically, diffusion means free movement of substances in a random fashion caused by the normal kinetic motion of matter, whereas active transport means movement of substances against an energy gradient, such as that between a low concentration state and a high concentration state, a process that requires chemical energy to cause the movement.

DIFFUSION

All molecules and ions in the body fluids, including both water molecules and dissolved substances, are in constant motion, each particle moving its own separate way. Motion of these particles is what physicists call heat — the greater the motion, the higher is the temperature — and motion never ceases under any conditions except absolute zero temperature. When a moving molecule, A, approaches a stationary molecule, B, the electrostatic and nuclear forces of molecule A repel molecule B, adding some of the energy of motion to molecule B. Consequently, molecule B gains kinetic energy of motion while molecule A slows down, losing some of its kinetic energy. Thus, as shown in Figure 4–2, a single molecule in solution bounces among the other molecules first in one direction, then another, then another, and so forth, bouncing randomly hundreds to millions of times each second. At times it travels a far distance before striking the next molecule, but at other times only a short distance.

This continual movement of molecules among

	EXTRACELLULAR FLUID	INTRACELLULAR FLUID
Na$^+$	142 mEq/l.	10 mEq/l.
K$^+$	4 mEq/l.	140 mEq/l.
Ca^{++}	5 mEq/l.	<1 mEq/l.
Mg^{++}	3 mEq/l.	58 mEq/l.
Cl$^-$	103 mEq/l.	4 mEq/l.
HCO$_3^-$	28 mEq/l.	10 mEq/l.
Phosphates	4 mEq/l.	75 mEq/l.
SO$_4^{--}$	1 mEq/l.	2 mEq/l.
Glucose	90 mgm.%	0 to 20 mgm.%
Amino acids	30 mgm.%	200 mgm.% ?
Cholesterol Phospholipids Neutral fat	0.5 gm.%	2 to 95 gm.%
Po$_2$	35 mm.Hg	20 mm.Hg ?
Pco$_2$	46 mm.Hg	50 mm.Hg ?
pH	7.4	7.0
Proteins	2 gm.% (5 mEq./l.)	16 gm.% (40 mEq./l.)

Figure 4–1. Chemical compositions of extracellular and intracellular fluids.

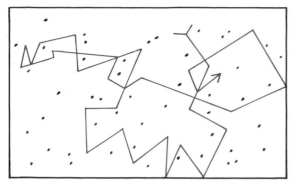

Figure 4–2. Diffusion of a fluid molecule during a fraction of a second.

each other in liquids or in gases is called *diffusion*. Ions diffuse in exactly the same manner as whole molecules, and even suspended colloid particles diffuse in a similar manner, except that because of their very large sizes they diffuse far less rapidly than molecular substances.

KINETICS OF DIFFUSION — THE CONCENTRATION DIFFERENCE

When a large amount of dissolved substance is placed in a solvent at one end of a chamber, it immediately begins to diffuse toward the opposite end. If the same amount of substance is placed in the opposite end of the chamber, it begins to diffuse toward the first end, the same amount diffusing in each direction. As a result, the *net rate of diffusion* from one end to the other is zero. If however, the concentration of the substance is greater at one end of the chamber than at the other end, the net rate of diffusion from the area of high concentration to lower concentration is directly proportional to the larger concentration minus the lower concentration. The total concentration change along the axis of the chamber is called a *concentration difference,* and the concentration difference divided by the distance is called the *concentration* or *diffusion gradient.*

If we consider all the different factors that affect the rate of diffusion of a substance from one area to another, they are the following:

(1) The greater the concentration difference, the greater is the rate of diffusion.

(2) The less the square root of the molecular weight, the greater is the rate of diffusion.

(3) The shorter the distance, the greater is the rate.

(4) The greater the cross-section of the chamber in which diffusion is taking place, the greater is the rate of diffusion.

(5) The greater the temperature, the greater is the molecular motion and also the greater is the rate of diffusion. All these can be expressed in an approximate formula, found at the bottom of this page, for diffusion in solutions.

DIFFUSION THROUGH THE CELL MEMBRANE

The cell membrane is essentially a sheet of lipid material, called the *lipid matrix,* with interspersed islands of globular protein molecules in the matrix. Some of these protein molecules penetrate all the way through the membrane and thus form membrane "pores," as was discussed in Chapter 2. Therefore, two different methods by which substances can diffuse through the membrane are (1) becoming dissolved in the lipid matrix and diffusing through it in the same way that diffusion occurs in water, or (2) diffusing through the minute pores that pass directly through the membrane.

Diffusion in the Dissolved State Through the Lipid Portion of the Membrane

A few substances are soluble in the lipid of the cell membrane. These include especially oxygen, carbon dioxide, alcohol, and fatty acids. When one of these comes in contact with the membrane it immediately becomes dissolved in the lipid, and the molecule continues its random motion within the substance of the membrane in exactly the same way that it undergoes random motion in the surrounding fluids.

The primary factor that determines how rapidly a substance can diffuse through the lipid matrix of the cell membrane is its solubility in lipids. If it is very soluble, it becomes dissolved in the membrane very easily and therefore passes on through. On the other hand, almost no substances that dissolve very poorly in lipids, such as water, glucose, and the electrolytes, pass through the lipid matrix.

Facilitated Diffusion Through the Lipid Matrix

Some substances are very insoluble in lipids and yet can still pass through the lipid matrix by a

$$\text{Diffusion rate} \propto \frac{\text{Concentration difference} \times \text{Cross-sectional area} \times \text{Temperature}}{\sqrt{\text{Molecular weight}} \times \text{Distance}}$$

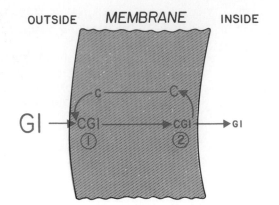

Figure 4–3. Facilitated diffusion of glucose through the cell membrane.

process called *carrier-mediated* or *facilitated diffusion*. This is the means by which some sugars and amino acids, in particular, cross the membrane. The most important of the sugars is glucose, the membrane transport of which is illustrated in Figure 4–3. This shows that glucose (G1) combines with a *carrier* substance (C) at point 1 to form the compound CG1. This combination is soluble in the lipid, so that it can diffuse to the other side of the membrane, where the glucose breaks away from the carrier (point 2) and passes to the inside of the cell, while the carrier moves back to the outside surface to pick up still more glucose. Thus, the effect of the carrier is to make the glucose soluble in the membrane; without it, glucose cannot pass through the membrane.

The rate at which a substance passes through a membrane by facilitated diffusion depends on the difference in concentration of the substance on the two sides of the membrane, the amount of carrier available, and the rapidity with which the chemical (or physical) reactions can take place.

Diffusion Through the Membrane Pores

Some substances, such as water and many of the dissolved ions, go through holes in the cell membrane called *membrane pores*, believed to be caused by large protein molecules penetrating all the way through the cell membrane, as discussed earlier. A large portion of these pores behave as if they were minute round holes approximately 0.8 nanometer (8 Angstroms) in diameter and as if the total area of the pores equaled approximately 1/5000 of the total surface area of the cell. Despite this very minute total area of the pores, molecules and ions diffuse so rapidly that the entire volume of fluid in some types of cells — the red blood cell for instance — can easily pass through the pores within one hundredth of a second.

Figure 4–4 illustrates a schematized structure of one type of pore, indicating that its surface

may be lined with electrical charges. This figure shows several small particles passing through the pore and also shows that the maximum diameter of the particle that can pass through is approximately equal to the diameter of the pore itself, about 0.8 nanometer.

Effect of Pore Size on Diffusion Through the Pore — Permeability. Table 4–1 gives the effective diameters of various substances in comparison with the diameter of a pore, and it also gives the relative permeability of the pores for the different substances. *The permeability can be defined as the rate of transport through the membrane for a given concentration difference.* Note that some substances, such as the water molecule, urea molecule, and chloride ion, are considerably smaller than the pore. All these pass through most pores, particularly those of red cells, with great ease. The rates of diffusion of urea and chloride ions through the membrane are somewhat less than that of water, which is in keeping with the fact that their effective diameters are slightly greater than that of water.

Table 4–1 also shows that most of the sugars, including glucose, have effective diameters that are slightly greater than that of the pores. For this reason essentially none of the sugars can pass through the pores; instead, those that do enter the cell pass through the lipid matrix by the process of facilitated diffusion.

Effect of the Electrical Charge of Ions on Their Ability to Diffuse Through Membrane Pores. The electrical charges of ions often markedly affect their ability to diffuse through pores, sometimes impeding and sometimes enhancing their diffusion. For instance, one set of pores, called *sodium channels*, seems especially to allow easy diffusion of sodium ions. It is believed that these pores are lined with strongly *negative*

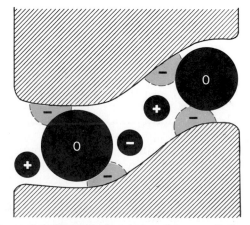

Figure 4–4. Postulated structure of the pore in the mammalian red cell membrane, showing the sphere of influence exerted by charges along the surface of the pore. (Modified from Solomon: *Sci. Am. 203*:146, 1960. ©1960 by Scientific American, Inc.)

TABLE 4–1 RELATIONSHIP OF EFFECTIVE DIAMETERS OF DIFFERENT SUBSTANCES TO PORE DIAMETER AND RELATIVE PERMEABILITIES*

Substance	Diameter Å	Ratio to Pore Diameter	Approximate Relative Permeability
Water molecule	3	0.38	50,000,000
Urea molecule	3.6	0.45	1,500,000
Hydrated chloride ion			
(red cell)	3.86	0.48	500,000
(nerve membrane)	—	—	0.04
Hydrated potassium ion			
(red cell)	3.96	0.49	1.1
(nerve membrane)	—	—	0.02
Hydrated sodium ion			
(red cell)	5.12	0.64	1
(nerve membrane)	—	—	0.0003
Lactate ion	5.2	0.65	?
Glycerol molecule	6.2	0.77	?
Ribose molecule	7.4	0.93	?
Pore size	8 (Ave.)	1.00	—
Galactose	8.4	1.03	?
Glucose	8.6	1.04	0.4
Mannitol	8.6	1.04	?
Sucrase	10.4	1.30	?
Lactose	10.8	1.35	?

*These data have been gathered from different sources but relate primarily to the red cell membrane. Other cell membranes have different characteristics.

charges that attract sodium ions into the pores and then shuttle the sodium ions from one negative charge to the next, thus allowing rapid movement through the pores. On the other hand, larger-sized positive ions, such as potassium ions, have considerable difficulty passing through these channels, mainly because of their diameters.

A second factor that affects ion diffusion through pores is the hydration energy of ions, which retards movement through pores. Most ions are loosely bound with water, and most or all of the water molecules bound with the ions must be removed before the particle size is small enough to allow passage through the pores. On the other hand, the smaller the ion, the greater is the hydration energy. For instance, the hydration energy for sodium is much greater than for potassium and in most cell membranes there seems to be a large population of either uncharged pores or poorly charged pores that are relatively impermeable to sodium ions because of their high hydration energy but are relatively permeable to potassium ions because of their lower hydration energy. These pores are called *potassium channels*.

Net Diffusion Through the Cell Membrane and Effect of a Concentration Difference

From the preceding discussion, it is evident that many different substances can diffuse either through the lipid matrix of the cell membrane or through the pores. It should be noted, however, that substances that diffuse in one direction can also diffuse in the opposite direction. Usually it is not the total quantity of substances diffusing in both directions through the cell membrane that is important to the cell but instead the *net quantity* diffusing either into or out of the cell.

In addition to the permeability of the membrane, which has already been discussed, the most important factor that determines the rate of net diffusion of a substance is the concentration difference of the substance across the membrane. Figure 4–5 illustrates a membrane with a substance in high concentration on the outside and low concentration on the inside. The rate at which the substance diffuses *inward* is proportional to the concentration of molecules on the outside, for this concentration determines how many of the molecules strike the outside of the pore each second. On the other hand, the rate at which the molecules diffuse *outward* is proportional to their concentration inside the membrane. Obviously, therefore, the rate of net diffusion into the cell is proportional to the concentration on the outside *minus* the concentration on the inside, or

$$\text{Net diffusion} \propto P(C_o - C_i)$$

in which C_o is the concentration on the outside, C_i is the concentration on the inside, and P is the permeability of the membrane for the substance.

Net Movement of Water Across Cell Membranes — Osmosis Across Semipermeable Membranes

By far the most abundant substance to diffuse through the cell membrane is water. It should be recalled again that enough water ordinarily diffuses in each direction through the red cell membrane per second to equal about *100 times the volume of the cell itself*. Yet, *normally*, the amount that diffuses in the two directions is so precisely balanced that not even the slightest *net* movement of water occurs. Therefore, the volume of the cell remains constant. However, under certain conditions, a *concentration difference for water* can devel-

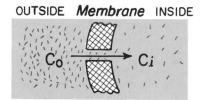

Figure 4–5. Effect of concentration difference on diffusion of molecules and ions through a cell membrane.

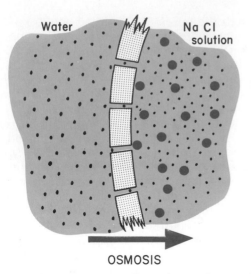

Figure 4–6. Osmosis at a cell membrane when a sodium chloride solution is placed on one side of the membrane and water on the other side.

op across a membrane, just as concentration differences for other substances can also occur. When this happens, net movement of water does occur across the cell membrane, causing the cell either to swell or to shrink, depending on the direction of the net movement. This process of net movement of water caused by the concentration difference is called *osmosis*.

To give an example of osmosis, let us assume that we have the conditions shown in Figure 4–6, with pure water on one side of the cell membrane and a solution of sodium chloride on the other side. Referring back to Table 4–1, we see that water molecules pass through the cell membrane with extreme ease, while sodium and chloride ions pass through only with difficulty. As a result, in the example of Figure 4–6, more water molecules strike the pores on the left side where there is pure water than on the right side where the water concentration has been reduced. Thus, net movement of water occurs from left to right — that is, osmosis occurs from left to right.

Osmotic Pressure

If in Figure 4–6 pressure were applied to the sodium chloride solution, osmosis of water into this solution could be slowed or even stopped because the pressure itself can force molecules and ions through the membrane in the opposite direction. The amount of pressure required to stop the osmosis completely is called the osmotic pressure of the sodium chloride solution.

The principle of a pressure difference opposing osmosis is illustrated in Figure 4–7, which shows a semipermeable membrane separating two columns of fluid, one containing water and the other containing a solution of water and some solute that will not penetrate the membrane. Osmosis of water from chamber B into chamber A causes the levels of the fluid columns to become farther and farther apart, until eventually a pressure difference is developed that is great enough to oppose the osmotic effect. The pressure difference across the membrane at this time is equal to the osmotic pressure of the solution containing the nondiffusible solute.

Lack of Effect of Molecular and Ionic Mass on Osmotic Pressure — Importance of Numbers of Osmotic Particles (or of Molar Concentration). The osmotic pressure exerted by nondiffusible particles in a solution, whether they are molecules or ions, is determined by the *numbers* of particles per unit volume of fluid and not the mass of the particles. The reason for this is that each particle in a solution, regardless of its mass, exerts, on the average, the same amount of pressure against the membrane. That is, all particles are bouncing among each other with, on the average, equal energy. If some particles have greater kinetic energy of movement than others, their impact with the low-energy particles will impart part of their energy to these, thus bringing all particles to the same average energy level (and therefore to the same pressure level). Consequently, the factor that determines the osmotic pressure of a solution is the concentration of the solution in terms of numbers of particles and not in terms of mass of the solute.

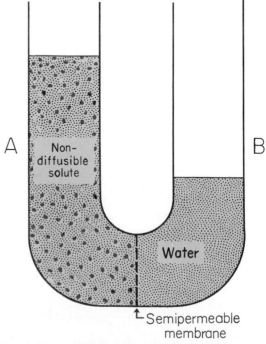

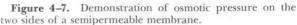

Figure 4–7. Demonstration of osmotic pressure on the two sides of a semipermeable membrane.

Osmolality. Since the amount of osmotic pressure exerted by a solute is proportional to the concentration of the solute in numbers of molecules or ions, expressing the solute concentration in terms of mass is of no value in determining osmotic pressure. To express the concentration in terms of numbers of particles, the unit called the *osmol* is used in place of grams.

One osmol is the number of particles (molecules) in 1 gram molecular weight of undissociated solute. Thus, 180 grams of glucose is equal to 1 osmol of glucose, because glucose does not dissociate. On the other hand, if the solute dissociates into 2 ions, 1 gram molecular weight of the solute equals 2 osmols, because the number of osmotically active particles is now twice as great as in the case in the undissociated solute. Therefore, 1 gram molecular weight of sodium chloride, 58.5 gm., is equal to 2 osmols.

A solution that has 1 osmol of solute dissolved in each kilogram of water is said to have an *osmolality* of 1 osmol per kilogram, and a solution that has 1/1000 osmol dissolved per kilogram has an osmolality of 1 milliosmol per kilogram. The normal osmolality of the extracellular and intracellular fluids is about 300 milliosmols per kilogram.

Relationship of Osmolality to Osmotic Pressure. At normal body temperature, 38° C, a concentration of 1 osmol per kilogram will cause *19,3000 mm. Hg* osmotic pressure in the solution. Likewise, 1 milliosmol per kilogram concentration is equivalent to *19.3 mm. Hg* osmotic pressure.

ACTIVE TRANSPORT

Often only a minute concentration of a substance is present in the extracellular fluid, and yet a large concentration of the substance is required in the intracellular fluid. For instance, this is true of potassium ions. Conversely, other substances frequently enter cells and must be removed even though their concentrations inside are far less than outside. This is true of sodium ions.

From the discussion thus far it is evident that *no substances can diffuse against a concentration gradient,* or, as is often said, "uphill." To cause movement of substances uphill, energy must be imparted to the substance. This is analogous to the compression of air by a pump. Compression causes the concentration of the air molecules to increase, but to create this greater concentration, energy must be imparted to the air molecules by the piston of the pump as they are compressed. Likewise, as molecules are transported through a cell membrane from a dilute solution to a concentrated solution, energy must be imparted to the molecules. When a cell membrane moves molecules

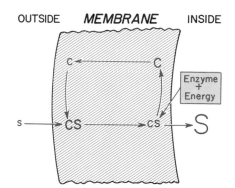

Figure 4–8. Basic mechanism of active transport.

uphill against a concentration gradient (or uphill against an electrical or pressure gradient) the process is called *active transport.*

Among the different substances that are actively transported through cell membranes are sodium ions, potassium ions, calcium ions, iron ions, hydrogen ions, chloride ions, iodide ions, urate ions, several different sugars, and some amino acids.

BASIC MECHANISM OF ACTIVE TRANSPORT

The mechanism of active transport is believed to be similar for most substances and to depend on transport by *carriers.* Figure 4–8 illustrates the basic mechanism, showing a substance S entering the outside surface of the membrane, where it combines with carrier C. At the inside surface of the membrane, S separates from the carrier and is released to the inside of the cell. C then moves back to the outside to pick up more S.

One will immediately recognize the similarity between this mechanism of active transport and that of facilitated diffusion discussed earlier in the chapter and illustrated in Figure 4–3. The difference, however, is that *energy is imparted to the system* in the course of active transport, so that transport can occur *against a concentration gradient* (or against an electrical or pressure gradient).

Though the mechanism by which energy is utilized to cause active transport is not entirely known, we do know some features of this process:

First, the energy is delivered to the inside surface of the membrane from high-energy substances, principally ATP, inside the cytoplasm of the cell.

Second, a specific "carrier" molecule (or combination of molecules) is required to transport each type of substance or each class of similar substances.

Third, a specific enzyme (or enzymes) is required to promote the chemical reactions between the carrier and each transported substance.

A special characteristic of active transport is that the mechanism *saturates* when the concentration of the substance to be transported is very high. This saturation results from limitation either of quantity of carrier available to transport the substance or of enzymes to promote the chemical reactions. This principle of saturation also applies to carrier-mediated facilitated diffusion, which was discussed earlier in the chapter.

Chemical Nature of Carrier Substances. Carrier substances are all believed to be proteins, conjugated proteins, or loose physical combinations of more than one protein molecule. Several different carrier systems exist in cell membranes, each of which transports only certain specific substances. One carrier system, for instance, transports sodium to the outside of the membrane and probably transports potassium to the inside at the same time. Another system actively transports sugars through the membranes of intestinal and renal tubular epithelial cells, while still other carrier systems transport different ones of the amino acids.

Energetics of Active Transport. In terms of calories, the amount of energy required to concentrate 1 osmol of substance ten-fold is about 1400 calories. One can see that the energy expenditure for concentrating substances in cells or for removing substances from cells against a concentration gradient can be tremendous. Some cells, such as those lining the renal tubules, expend almost 100 per cent of their energy for this purpose alone.

ACTIVE TRANSPORT OF SODIUM AND POTASSIUM

Referring back to Figure 4–1, one sees that the sodium concentration outside the cell is very high in comparison with its concentration inside, and the converse is true of potassium. Also, Table 4–1 shows that minute quantities of sodium and potassium can diffuse through the pores of the cell. If such diffusion should take place over a long period of time, the concentrations of the two ions would eventually become equal inside and outside the cell unless there were some means to remove the sodium from the inside and to transport potassium back in.

Fortunately, a mechanism for active transport of sodium and potassium ions is present in all cell membranes of the body. It is called the *sodium-potassium pump*. The basic principles of this pump are illustrated in Figure 4–9. The carrier for this mechanism transports sodium from inside the

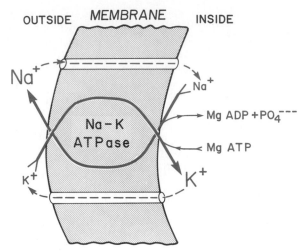

Figure 4–9. Postulated mechanism for active transport of sodium and potassium through the cell membrane, showing coupling of the two transport mechanisms and delivery of energy to the system at the inner surface of the membrane.

cell to the outside and potassium from the outside to the inside. And, because this carrier also has the capability of splitting ATP molecules and utilizing the energy from this source to promote the sodium and potassium transport, the carrier also acts as an enzyme and is called *sodium-potassium ATPase*. This ATPase is composed of two protein molecules, one a globulin with a molecular weight of 95,000 and the other a glycoprotein with a molecular weight of 55,000. The larger molecule actually binds with both the sodium and potassium ions and also with the ATP, but the smaller molecule is also necessary to provide some facilitating function not yet understood.

Note in Figure 4–9 that energy released from ATP at the inner surface of the cell membrane causes potassium ions to split away from the sodium-potassium ATPase carrier molecule and simultaneously causes sodium ions to bind. Then, at the outside surface of the membrane, the sodium ions split away from the carrier while potassium ions bind.

One of the peculiarities of this sodium-potassium transport system is that it normally transports three sodium ions to the inside of the membrane for each two potassium ions transported to the outside. Another important feature of the pump is that it is strongly activated by an increase in sodium ion concentration inside the cell, the activity increasing in proportion to (sodium concentration)^3. This effect is extremely important because it allows even a slight excess buildup of sodium ions inside the cell to activate the pump very strongly and thereby return the intracellular sodium concentration back to its normal low level.

The sodium transport mechanism is so important to many different functioning systems of the body — such as to nerve and muscle fibers for transmission of impulses, various glands for the secretion of different substances, and all cells of the body to prevent cellular swelling — that it is frequently called the *sodium pump*. We will discuss the sodium pump at many places in this text.

ACTIVE TRANSPORT OF OTHER IONS

Calcium and magnesium are probably transported by all cell membranes in much the same manner that sodium and potassium are transported, and certain cells of the body have the ability to transport still other ions. For instance, the glandular cell membranes of the thyroid gland can transport large quantities of iodide ion; the epithelial cells of the intestine can transport sodium, chloride, calcium, iron, hydrogen, and probably many other ions; and the epithelial cells of the renal tubules can transport calcium, magnesium, chloride, sodium, potassium, and a number of other ions.

TRANSPORT OF SUGARS — FACILITATED DIFFUSION AND THE SODIUM "CO-TRANSPORT" MECHANISM

Facilitated diffusion of glucose and certain other sugars occurs in essentially all cells of the body, but transport of sugars against concentration gradients occurs in only a few places in the body. For instance, in the intestine and renal tubules, glucose and several other monosaccharides are continually transported into the blood even though their concentrations may be minute in the lumens. Thus, essentially none of these sugars is lost in either the intestinal excreta or the urine.

The mechanism of glucose (and other sugar) transport through the epithelial cells of the intestinal mucosa and renal tubules is neither pure diffusion nor active transport but a curious mixture of the two. This mechanism, called *secondary active transport* or *sodium co-transport*, functions in the following way:

First, let us remember that the epithelial cell has two functionally distinct sides, a brush border that lines the lumen of the gut or renal tubule and a base that lies adjacent to the absorptive capillaries. The basal part of the cell transports sodium out of the cell cytoplasm and into the sublying capillaries. This occurs by the usual process of active transport for sodium. The result is marked depletion of sodium ions in the interior of the epithelial cell. This in turn creates a diffusion gradient for sodium ions through the brush border from the lumen of the intestine toward the interior of the epithelial cell. Consequently, sodium ions attempt to diffuse down this gradient into the cell. However, the brush border is reasonably impermeable to sodium except when the sodium combines with a carrier molecule, one type of which is the so-called *sodium-glucose carrier*. This carrier is peculiar in that it will not transport the sodium by itself but must also transport a glucose molecule at the same time. That is, when bound with both sodium and glucose, the carrier will then transport both to the interior of the cell. One can readily see that it is the sodium gradient across the brush border membrane that provides the energy to promote this transport of both the sodium and the glucose. Thus, even when glucose is present in the lumen of the intestine (or of the renal tubule) in very low concentration, it can still be transported to the interior of the epithelial cell.

Once the glucose has entered the interior of the epithelial cell, it crosses the basal side of the cell by the usual process of facilitated diffusion, in the same manner that glucose crosses essentially all other membranes of the body.

Though not all sugars are transported, almost all monosaccharides that are important to the body *are* transported, including *glucose, galactose, fructose, mannose, xylose, arabinose,* and *sorbose.* On the other hand, the disaccharides such as sucrose, lactose, and maltose are not transported at all.

Effect of Insulin on Glucose Transport. Perhaps the most important function of insulin is its ability to increase the rate of facilitated diffusion of glucose into most cells of the body, especially into muscle cells and fat cells. It is in this way that insulin controls the body's metabolism of carbohydrates, sometimes increasing the metabolism as much as 5- to 10-fold.

TRANSPORT OF AMINO ACIDS — FACILITATED DIFFUSION AND "SODIUM CO-TRANSPORT"

Most, if not all, amino acids, like glucose, are transported to the inside of essentially all cells of the body by facilitated diffusion mechanisms.

Sodium co-transport of amino acids also occurs through a few membranes of the body: the epithelia of the intestines, renal tubules, and some exocrine glands. This involves at least four different carrier systems for transporting, respectively, the following different groups of amino acids: (1) neutral amino acids, (2) dibasic amino acids, (3) imino acids, and (4) dicarboxylic acids. These transport systems will be discussed further in relation to intestinal absorption in Chapter 44. It should be noted again that in the sodium co-transport mechanism it is the concentration gradient of sodium between the intestinal lumen and the interior of the cell that provides the energy for transport of the amino acid molecules, as explained earlier in the chapter.

Hormonal Regulation of Amino Acid Transport. At least four different hormones are important in controlling amino acid transport: (1) *Growth hormone,* secreted by the adenohypophysis, increases amino acid transport into essentially all cells. (2) *Insulin* and (3) *glucocorticoids* increase amino acid transport at least into liver cells and possibly into other cells as well, though much less is known about these. (4) *Estradiol,* the most important of the female sex hormones, causes transport of amino acids into the musculature of the uterus, thereby promoting development of this organ.

Thus, several of the hormones exert much, if not most, of their effects in the body by controlling active transport of amino acids into all or certain cells.

REFERENCES

Andreoli, T. E., *et al.* (eds.): Membrane Physiology. New York, Plenum Press, 1980.

Avery, J. (ed.): Membrane Structure and Mechanisms of Biological Energy Transduction. New York, Plenum Press, 1974.

Ellory, C., and Lew, V. L. (eds.): Membrane Transport in Red Cells. New York, Academic Press, 1977.

Fettiplace, R., and Haydon, D. A.: Water permeability of lipid membranes. *Physiol. Rev., 60*:510, 1980.

Finn, A. L.: Changing concepts of transepithelial sodium transport. *Physiol. Rev., 56*:453, 1976.

Gilles, R. (ed.): Mechanisms of Osmoregulation: Maintenance of Cell Volume. New York, John Wiley & Sons, 1979.

Glynn, I. M., and Karlish, S. J. D.: The sodium pump. *Annu. Rev. Physiol., 37*:13, 1975.

Gregor, H. P., and Gregor, C. D.: Synthetic-membrane technology. *Sci. Am., 239*(1):112, 1978.

Gupta, B. L., *et al.* (eds.): Transport of Ions and Water in Animals. New York, Academic Press, 1977.

Guyton, A. C., *et al.:* Circulatory Physiology II: Dynamics and Control of the Body Fluids. Philadelphia, W. B. Saunders Co., 1975.

Hemmings, W. A. (ed.): Protein Transmission Through Living Membranes. New York, Elsevier/North-Holland, 1979.

Jacob, H. S. (ed.): Blood Cell Membranes. New York, Grune & Stratton, 1979.

Keynes, R. D.: Ion channels in the nerve-cell membrane. *Sci. Am., 240*(3):126, 1979.

Korn, E. D.: Transport. New York, Plenum Publishing Corp., 1975.

Lodish, H. F., and Rothman, J. E.: The assembly of cell membranes. *Sci. Am., 240*(1):48, 1979.

Macknight, A. D. C., and Leaf, A.: Regulation of cellular volume. *Physiol. Rev., 57*:510, 1977.

Metcalfe, J. C. (ed.): Biochemistry of Cell Walls and Membranes II. Baltimore, University Park Press, 1978.

Parsons, D. S. (ed.): Biological Membranes. New York, Oxford University Press, 1975.

Schultz, S. G.: Principles of electrophysiology and their application to epithelial tissues. *Int. Rev. Physiol., 4*:69, 1974.

Singer, S. J.: The molecular organization of membranes. *Annu. Rev. Physiol., 43*:805, 1974.

Ullrich, K. L.: Sugar, amino acid, and Na^+ cotransport in the proximal tubule. *Annu. Rev. Physiol., 41*:181, 1979.

Wallick, E. T., *et al.*: Biochemical mechanism of the sodium pump. *Annu. Rev. Physiol., 41*:397, 1979.

Wright, E. M., and Diamond, J. M.: Anion selectivity in biological systems. *Physiol. Rev., 57*:109, 1977.

Part II

BLOOD CELLS, IMMUNITY, AND BLOOD CLOTTING

5

Red Blood Cells, White Blood Cells, and Resistance of the Body to Infection

With this chapter we begin a discussion of the blood cells and of other cells closely related to those of the blood: the cells of the reticuloendothelial system and of the lymphatic system.

THE RED BLOOD CELLS

The major function of red blood cells is to transport hemoglobin, which in turn carries oxygen from the lungs to the tissues. Normal red blood cells are biconcave disks having a mean diameter of approximately 8 microns and a thickness at the thickest point of 2 microns and in the center of 1 micron or less. The shapes of red blood cells can change remarkably as the cells pass through capillaries. Actually, the red blood cell is a "bag" that can be deformed into almost any shape. Furthermore, because the normal cell has a great excess of cell membrane for the quantity of material inside, deformation does not stretch the membrane and consequently does not rupture the cell as would be the case with many other cells.

In normal men the average number of red blood cells per cubic millimeter is 5,200,000 and in normal women 4,700,000. Also, the altitude at which the person lives affects the number of red blood cells; this is discussed later.

Quantity of Hemoglobin in the Cells and Transport of Oxygen. When the hematocrit (defined as the percentage of the blood that is red cells — normally 40 to 45 per cent) and the quantity of hemoglobin in each respective cell are normal, the blood contains an average of 15 grams of hemoglobin. As will be discussed in connection with the transport of oxygen in Chapter 28, each gram of pure hemoglobin is capable of combining with approximately 1.39 ml. of oxygen. Therefore, in a normal person, over 20 ml. of oxygen can be carried in combination with hemoglobin in each 100 ml. of blood.

GENESIS OF THE RED BLOOD CELL

The red blood cells are derived from a cell known as the *hemocytoblast,* illustrated in Figure 5–1. New hemocytoblasts are continually being formed from primordial stem cells located throughout the bone marrow.

As illustrated in Figure 5–1, the hemocytoblast first forms the *basophil erythroblast,* which begins the synthesis of hemoglobin. The erythroblast then becomes a *polychromatophil erythroblast,* so called because of a mixture of basophilic material and the red hemoglobin. Following this, the nucleus of the cell shrinks while still greater quantitiss of hemoglobin are formed, and the cell becomes a *normoblast.* Finally, after the cytoplasm of the normoblast has become filled with hemoglobin the nucleus becomes extremely small and is extruded. At the same time, the endoplasmic reticulum is being reabsorbed. The cell at this stage of development is called a *reticulocyte* because it still contains a small amount of basophilic endoplasmic reticulum interspersed among the hemoglobin in the cytoplasm. While the cells are in this reticulocyte stage, they pass into the blood capillaries by *diapedesis* (squeezing through the pores of the membrane).

The remaining endoplasmic reticulum in the reticulocyte continues to produce a small amount of hemoglobin for one to two days, but by the end

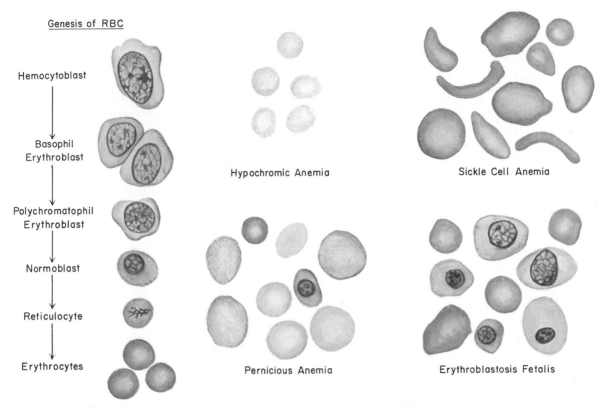

Figure 5–1. Genesis of red blood cells, and red blood cells in different types of anemia.

of that time the reticulum completely disappears. In normal blood, the total proportion of reticulocytes among all the cells is slightly less than 1 per cent. Once the reticulum has been completely reabsorbed, the cell is then the mature erythrocyte.

Tissue Oxygenation as the Basic Regulator of Red Blood Cell Production

Any condition that decreases the quantity of oxygen transported to the tissue ordinarily increases the rate of red blood cell production. Thus, when a person becomes extremely *anemic* as a result of hemorrhage or any other condition, the bone marrow immediately begins to produce large quantities of red blood cells. Also, destruction of major portions of the bone marrow by any means, especially x-ray therapy, causes hyperplasia of the remaining bone marrow in an attempt to supply the demand for red blood cells in the body.

At very *high altitudes,* where the quantity of oxygen in the air is greatly decreased, insufficient oxygen is transported to the tissues, and red cells are produced so rapidly that their number in the blood is considerably increased.

Therefore, it is obvious that it is not the concentration of red blood cells in the blood that controls the rate of red cell production, but instead it is the functional ability of the cells to transport oxygen to the tissues in relation to the tissue demand for oxygen.

Erythropoietin, Its Formation in Response to Hypoxia, and Its Function in Regulating Red Blood Cell Production. Erythropoietin is a glycoprotein hormone that appears in the blood in hypoxia and in turn acts on the bone marrow to increase the rate of red blood cell production.

There is no direct response of the bone marrow to hypoxia. Instead, hypoxia stimulates red blood cell production only through this mechanism of erythropoietin.

The kidneys play an essential role in the formation of erythropoietin, as follows: When the kidneys become hypoxic, they release an enzyme called *renal erythropoietin factor.* This is secreted into the blood, where it acts within a few minutes on one of the plasma globulins to split away the glycoprotein erythropoietin molecule. Erythropoietin, in turn, circulates in the blood for about one day and during this time acts on the bone marrow to cause erythropoiesis.

In the complete absence of the kidneys, only minute amounts of erythropoietin are formed elsewhere in the body. Therefore, in the absence of the kidneys a person usually becomes very anemic because of extremely low levels of circulating erythropoietin.

Effect of Erythropoietin on Erythrogenesis. Though erythropoietin begins to be formed almost immediately when an animal or person is placed in an atmosphere of low oxygen, almost no new red blood cells appear in the circulating blood within the first two days; and it is only after five or more days that the maximum rate of new red cell production is reached. Thereafter, cells continue to be produced as long as the person remains in the low oxygen state or until he has produced enough red blood cells to carry adequate amounts of oxygen to his tissues despite the low oxygen.

Upon removal of a person from a state of low oxygen, his rate of oxygen transport to the tissues rises above normal, which causes his rate of erythropoietin formation to decrease to zero almost instantaneously and his rate of red blood cell production to fall essentially to zero within several days. Erythropoietin has an especially strong effect of increasing the rate of division of hemocytoblasts, which is the initial stage of erythrogenesis. It is possible that it increases the rate of conversion of stem cells into hemocytoblasts as well, though this is not yet certain.

In the complete absence of erythropoietin, few red blood cells are formed by the bone marrow. At the other extreme, when extreme quantities of erythropoietin are formed, the rate of red blood cell production can rise to as high as 6 to 8 times normal. Therefore, the erythropoietin control mechanism for red blood cell production is an extremely powerful one.

Vitamins Needed for Formation of Red Blood Cells

The Maturation Factor — Vitamin B_{12} (Cyanocobalamin). Vitamin B_{12} is an essential nutrient for all cells of the body, and growth of tissues in general is greatly depressed when this vitamin is lacking. This results from the fact that vitamin B_{12} is required for synthesis of DNA. Lack of this vitamin causes failure of nuclear maturation and division and therefore greatly inhibits the rate of red blood cell production.

Maturation Failure Caused by Poor Absorption of Vitamin B_{12} — Pernicious Anemia. The most common cause of maturation failure is not lack of vitamin B_{12} in the diet but instead failure to absorb vitamin B_{12} from the gastrointestinal tract. This often occurs in the disease called *pernicious anemia,* in which the basic abnormality is an *atrophic gastric mucosa* that fails to secrete normal gastric secretions. The parietal cells of the gastric glands secrete a substance (molecular weight

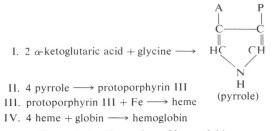

I. 2 α-ketoglutaric acid + glycine $\longrightarrow$

II. 4 pyrrole $\longrightarrow$ protoporphyrin III

III. protoporphyrin III + Fe $\longrightarrow$ heme

IV. 4 heme + globin $\longrightarrow$ hemoglobin

Figure 5–2. Formation of hemoglobin.

about 50,000) called *intrinsic factor,* which combines with vitamin B_{12}. In this bound state the B_{12} is protected from digestion by the gastrointestinal enzymes until it is absorbed.

Once vitamin B_{12} has been absorbed from the gastrointestinal tract it is stored in large quantities in the liver and then released slowly as needed to the bone marrow and other tissues of the body. The normal liver can store enough vitamin B_{12} to last many months because only 1 microgram is required each day.

Effect of Folic Acid (Pteroylglutamic Acid) on Red Cell Maturation. Occasionally maturation failure anemia results from deficiency of folic acid (another member of the vitamin B complex) instead of deficiency of vitamin B_{12}. Folic acid, like B_{12}, is required for formation of DNA but in a different way. It promotes formation of one of the nucleotides required for DNA synthesis. Folic acid is also required for RNA synthesis.

FORMATION OF HEMOGLOBIN

Synthesis of hemoglobin begins in the erythroblasts and continues through the normoblast and reticulocyte stages. Figure 5–2 shows the basic chemical steps in the formation of hemoglobin. From tracer studies with isotopes it is known that the heme portion of hemoglobin is synthesized mainly from acetic acid and glycine and that most of this synthesis occurs in the mitochondria. The initial stage of the synthesis is formation of a pyrrole compound. In turn, four pyrrole compounds combine to form a protoporphyrin compound, which then combines with iron to form the heme molecule. Finally, four heme molecules combine with one molecule of globin, a globulin that is synthesized in the ribosomes of the endoplasmic reticulum, to form hemoglobin, the formula for which is shown in Figure 5–3. Hemoglobin has a molecular weight of 64,458.

Combination of Hemoglobin with Oxygen. The most important feature of the hemoglobin molecule is its ability to combine loosely and reversibly with oxygen. This ability will be

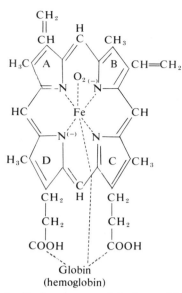

Figure 5–3. Basic structure of the hemoglobin molecule, showing one of the four heme complexes bound with the central globin core of the hemoglobin molecule.

discussed in detail in Chapter 28 in relation to respiration, for the primary function of hemoglobin in the body depends upon its ability to combine with oxygen in the lungs and then to release this oxygen readily in the tissue capillaries, where the gaseous tension of oxygen is much lower than in the lungs.

Oxygen *does not* combine with the two positive valences of the ferrous iron in the hemoglobin molecule. Instead, it binds loosely with one of the six "coordination" valences of the iron atom. This is an extremely loose bond, so the combination is easily reversible.

IRON METABOLISM

Because iron is important for formation of hemoglobin, myoglobin in the muscle, and other substances, it is essential to understand the means by which iron is utilized in the body.

The total quantity of iron in the body averages about 4 grams, approximately 65 per cent of which is present in the form of hemoglobin. About 4 per cent is present in the form of myoglobin, 1 per cent in the form of the various heme compounds that control intracellular oxidation, 0.1 per cent combined with the protein transferrin in the blood plasma, and 15 to 30 per cent stored in the liver, mainly in the form of ferritin.

Transport and Storage of Iron. Transport, storage, and metabolism of iron in the body are illustrated in Figure 5–4, which may be explained as follows: When iron is absorbed from the small intestine, it immediately combines with a globulin, *transferrin,* and is transported in this bound form in the blood plasma. The iron is very loosely combined with the globulin molecule and, consequently, can be released to any of the tissue cells at any point in the body. Excess iron in the blood is deposited *especially in the liver cells,* where about 60 per cent of the excess is stored. There it combines mainly with a protein, *apoferritin,* to form *ferritin.* Apoferritin has a molecular weight of approximately 460,000, and varying quantities of iron can combine in clusters of iron radicals with this large molecule; therefore, ferritin may contain only a small amount of iron or a relatively large amount.

When the quantity of iron in the plasma falls very low, iron is removed quite easily from the ferritin. The iron is then transported to the portions of the body where it is needed.

When red blood cells have lived their life span and are destroyed, the hemoglobin released from the cells is ingested by the reticuloendothelial cells. There free iron is liberated, and it can then either be stored in the ferritin pool or be reused for formation of hemoglobin.

Figure 5–4. Iron transport and metabolism.

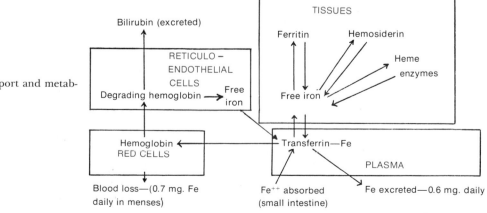

Daily Loss of Iron.   About 0.6 mg. of iron is excreted each day by the male, mainly into the feces. Additional quantities of iron are lost whenever bleeding occurs. In the female, the menstrual loss of blood brings the average iron loss to a value of approximately 1.3 mg. per day.

Obviously, the average quantity of iron derived from the diet each day must at least equal that lost from the body.

Absorption of Iron from the Gastrointestinal Tract.   Iron is absorbed almost entirely in the upper part of the small intestine, mainly in the duodenum. It is absorbed by an active absorptive process, though the precise mechanism of this active absorption is unknown.

Regulation of Total Body Iron by Alteration of Rate of Absorption.   When essentially all the apoferritin in the body has become saturated with iron, it becomes difficult for the transferrin of the blood to release iron to the tissues. As a consequence, the transferrin, which is normally only one-third saturated with iron, now becomes almost fully bound with iron and will accept almost no new iron from the intestinal mucosal cells. Then, as a final stage of this process, the buildup of excess iron in the mucosal cells themselves depresses active absorption of iron from the intestinal lumen and at the same time slightly enhances the rate of excretion of iron from the mucosa.

DESTRUCTION OF RED BLOOD CELLS

When red blood cells are delivered from the bone marrow into the circulatory system they normally circulate an average of 120 days before being destroyed. Even though mature red cells do not have a nucleus and also have neither mitochondria nor endoplasmic reticulum, nevertheless they do still have cytoplasmic enzymes that are capable to metabolizing glucose by the glycolytic process, thus forming small amounts of ATP. The ATP in turn provides the necessary energy to keep the red cell viable and the cell membrane pliable. However, these metabolic systems of the red cell become progressively less active with time, and the cells become progressively more fragile.

Once the red cell membrane becomes very fragile, the cell may rupture during passage through some tight spot of the circulation. Many of the red cells fragment in the spleen where the cells squeeze through the red pulp of the spleen. When the spleen is removed, the number of abnormal cells and old cells circulating in the blood increases considerably.

Destruction of Hemoglobin.   The hemoglobin released from the cells when they burst is phagocytized almost immediately by reticuloendothelial cells. During the next few days, they release the iron from the hemoglobin back into the blood to be reused as described earlier. The heme portion of the hemoglobin molecule is converted by the reticuloendothelial cell, through a series of stages, into the bile pigment *bilirubin,* which is released into the blood and later secreted by the liver into the bile; this will be discussed in relation to liver function in Chapter 43.

ANEMIA

Anemia means a deficiency of red blood cells, which can be caused either by too rapid loss or by too slow production of red blood cells. Some of the common causes of anemia are:

1. *Blood loss.*

2. *Bone marrow aplasia,* in which the bone marrow is destroyed. Common causes of this are drug poisoning or gamma ray irradiation — for instance, exposure to radiation from a nuclear bomb blast.

3. *Maturation failure* because of lack of vitamin B_{12} or folic acid, as was previously explained.

4. *Hemolysis of red cells* resulting from many possible causes, such as (a) drug poisoning, (b) hereditary diseases such as sickle cell disease, spherocytosis, or others that make the red cell membranes friable, and (c) erythroblastosis fetalis, a disease of the newborn in which antibodies from the mother destroy red cells in the baby (this will be discussed in Chapter 6).

Effects of Anemia on the Circulatory System.   The viscosity of the blood, which will be discussed in detail in Chapter 14, is dependent almost entirely on the concentration of red blood cells. In severe anemia the blood viscosity may decrease to less than one-half the normal value. This in turn decreases the resistance to blood flow in the peripheral vessels so that blood returns to the heart in far greater than normal quantities. As a consequence, the cardiac output increases to as much as two or more times normal.

Moreover, hypoxia due to diminished transport of oxygen by the blood causes the tissue vessels to dilate, allowing further increased return of blood to the heart, increasing the cardiac output to a higher level than ever. Thus, one of the major effects of anemia is greatly *increased work load on the heart.*

The increased cardiac output in anemia partially offsets many of the effects of anemia, for even though each unit quantity of blood carries only small quantities of oxygen, the rate of blood flow may be increased to such an extent that almost normal quantities of oxygen are delivered

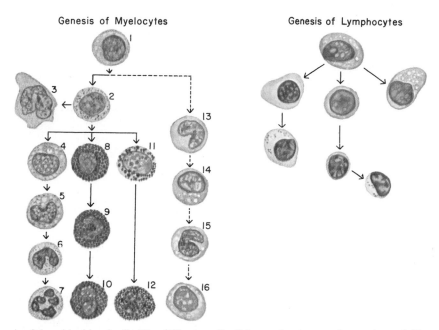

Figure 5–5. Genesis of the white blood cells. The different cells of the myelogenous series are 1, myeloblast; 2, promyelocyte; 3, megakaryocyte; 4, neutrophil myelocyte; 5, young neutrophil metamyelocyte; 6, "band" neutrophil metamyelocyte; 7, polymorphonuclear neutrophil; 8, eosinophil myelocyte; 9, eosinophil metamyelocyte; 10, polymorphonuclear eosinophil; 11, basophil myelocyte; 12, polymorphonuclear basophil; 13–16, stages of monocyte formation.

to the tissues. However, when a person begins to exercise, the heart is not capable of pumping much greater quantities of blood than it is already pumping. Consequently, during exercise, which greatly increases the tissue demand for oxygen, extreme tissue hypoxia results, and acute cardiac failure often ensues.

WHITE BLOOD CELLS AND RESISTANCE OF THE BODY TO INFECTION

Our bodies are constantly exposed to bacteria, viruses, fungi, and parasites, which occur especially in the skin, the mouth, the respiratory passageways, the colon, the mucous membranes of the eyes, and even the urinary tract. Many of these agents are capable of causing serious disease if they invade the deeper tissues. In addition, we are exposed intermittently to other highly infectious bacteria and viruses besides those that are normally present in our bodies, and these cause lethal diseases such as pneumonia, streptococcal infection, and typhoid fever.

Fortunately, our bodies have a special system for combating the different infectious and toxic agents. This is composed of the *white blood cells* (also called *leukocytes*), the *tissue macrophage system* (frequently but incorrectly called the reticuloendothelial system), and the *lymphoid tissue*. These tissues function in two different ways to prevent disease: (1) by actually destroying invading agents by the process of phagocytosis and (2) by forming antibodies and sensitized lymphocytes, one or both of which may destroy the invader.

The Types of White Blood Cells. Six different types of white blood cells are normally found in the blood. These are *polymorphonuclear neutrophils, polymorphonuclear eosinophils, polymorphonuclear basophils, monocytes, lymphocytes,* and *plasma cells*. In addition, there are large numbers of *platelets,* which are fragments of a seventh type of white cell found in the bone marrow, the *megakaryocyte*. The three types of polymorphonuclear cells have a granular appearance, as illustrated in Figure 5–5, for which reason they are called *granulocytes,* or in clinical terminology they are often called simply "polys."

The granulocytes and the monocytes protect the body against invading organisms by ingesting them — that is, by the process of *phagocytosis*. One of the functions of lymphocytes also is to attach to specific invading organisms and to destroy them; this is part of the immunity system and will be discussed in the following chapter. Finally, the function of platelets is to activate the blood clotting mechanism, which will be discussed in Chapter 7.

Concentrations of the Different White Blood Cells in the Blood. The adult human being has approximately 7000 white blood cells per cubic millimeter of blood. The normal percentages of the different types of white blood cells are approximately the following:

Polymorphonuclear neutrophils	62.0%
Polymorphonuclear eosinophils	2.3%
Polymorphonuclear basophils	0.4%
Monocytes	5.3%
Lymphocytes	30.0%

The number of platelets, which are only cell fragments, in each cubic millimeter of blood is normally about 300,000.

GENESIS OF THE LEUKOCYTES

Figure 5–5 illustrates the stages in the development of white blood cells. The polymorphonuclear cells and monocytes are normally formed only in the bone marrow. On the other hand, lymphocytes and plasma cells are produced in the various lymphogenous organs, including the lymph glands, the spleen, the thymus, the tonsils, and various lymphoid rests in the gut and elsewhere.

Some of the white blood cells formed in the bone marrow, especially the granulocytes, are stored within the marrow until they are needed in the circulatory system. Then when the need arises, various factors that are discussed later cause them to be released.

As illustrated in Figure 5–5, megakaryocytes are also formed in the bone marrow and are part of the myelogenous group of bone marrow cells. These megakaryocytes fragment in the bone marrow, the small fragments, known as *platelets* or *thrombocytes,* passing then into the blood.

PROPERTIES OF NEUTROPHILS, MONOCYTES, AND MACROPHAGES

It is mainly the neutrophils and the monocytes that attack and destroy invading bacteria, viruses, and other injurious agents. The neutrophils are mature cells that can attack and destroy bacteria and viruses even in the circulating blood. Immature monocytes have very little capability to fight infectious agents, but once they enter the tissues they begin to swell, often increasing their diameters as much as five-fold, to as great as 80 microns, a size that can be seen with the naked eye. Also, extremely large numbers of lysosomes and mitochondria develop in the cytoplasm, giving the cytoplasm the appearance of a bag filled with granules. These cells are now called *macrophages*, and they represent the mature form of the monocytes.

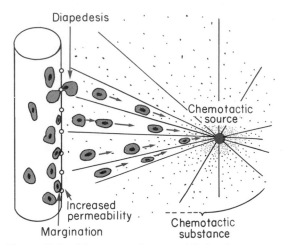

Figure 5–6. Movement of neutrophils by the process of *chemotaxis* toward an area of tissue damage.

Diapedesis. Neutrophils and monocytes can squeeze through the pores of the blood vessels by the process of diapedesis. That is, even though a pore is much smaller than the size of the cell, a small portion of the cell slides through the pore at a time, the portion sliding through being momentarily constricted to the size of the pore, as illustrated in Figure 5–6.

Ameboid Motion. Both neutrophils and macrophages move through the tissues by ameboid motion, which was described in Chapter 2. Some of the cells can move through the tissues at rates as great as 40 microns per minute — that is, the neutrophils can often move three times their own length each minute.

Chemotaxis. Chemical substances in the tissues often cause both neutrophils and macrophages to move toward the source of the chemical. This phenomenon is known as *chemotaxis.* When a tissue becomes inflamed, a number of different products can cause chemotaxis, including (a) some of the bacterial toxins, (b) degenerative products of the inflamed tissues themselves, and (c) still other substances.

Phagocytosis. The most important function of the neutrophils and macrophages is phagocytosis.

Obviously, the phagocytes must be selective of the material that is phagocytized, or otherwise some of the normal cells and structures of the body would be ingested. Whether or not phagocytosis will occur depends especially upon three selective procedures. First, if the surface of a particle is rough, the likelihood of phagocytosis is increased. Second, most natural substances of the body have electronegative surface charges that repel the phagocytes, which also carry electronegative surface charges. On the other hand, dead tissues and foreign particles are frequently electropositive and are therefore subject to phagocytosis. Third, the body has a specific means for recognizing certain foreign materials. This is the function of the immune system that will be described in the following chapter. The immune system develops antibodies against infectious agents like bacteria. These antibodies then adhere to the bacterial membranes and thereby make the bacteria especially suspectible to phagocytosis. In this case, the antibody is called an *opsonin*.

Enzymatic Digestion of the Phagocytized Particles. Once a foreign particle has been phagocytized, lysosomes immediately come in contact with the phagocytic vesicle, and their membranes fuse with those of the vesicle, thereby dumping many digestive enzymes of the lysosomes into the vesicle. Thus, the phagocytic vesicle now becomes a *digestive vesicle,* and digestion of the phagocytized particle begins immediately.

THE TISSUE MACROPHAGE SYSTEM (THE RETICULOENDOTHELIAL SYSTEM)

In the above paragraphs we have described the macrophages mainly as mobile cells that are capable of wandering through the tissues. However, a vast majority of the monocytes, on entering the tissues and after becoming macrophages, become attached to the tissues and remain attached for months or even years unless they are called upon to perform specific protective functions. They have the same capabilities as the mobile macrophages to phagocytize large quantities of bacteria, viruses, necrotic tissue, or other foreign particles in the tissue. And, when appropriately stimulated, they can break away from their attachments and become mobile macrophages that respond to chemotaxis and all the other stimuli related to the inflammation process.

The combination of mobile macrophages and fixed tissue macrophages is collectively called the *reticuloendothelial system.* The reason for this name is that it was formerly believed that a major share of the blood vessel endothelial cells could perform phagocytic functions similar to those performed by the macrophage system. However, recent studies have disproved this. Therefore, the reticuloendothelial system is actually a misnomer. Yet, because the term is so widely used, it should be remembered that it is almost synonymous with the tissue macrophage system.

The tissue macrophages in various tissues differ in appearances because of environmental differences, and they are known by different names: *Kupffer cells* in the liver; *reticulum cells* in

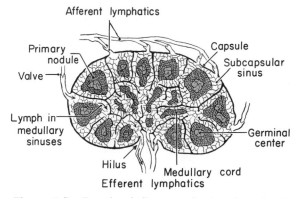

Figure 5–7. Functional diagram of a lymph node. (Redrawn from Ham: Histology, Philadelphia, J. B. Lippincott Company, 1971.)

lymph nodes, spleen, and bone marrow; *alveolar macrophages* in the alveoli of the lungs; *tissue histiocytes, clasmatocytes,* or *fixed macrophages,* in the subcutaneous tissues; and *microglia* in the brain. Let us describe briefly the function of tissue macrophages in two areas of the body that are especially exposed to infectious agents.

Macrophages of the Lymph Nodes (Reticular Cells). Essentially no particulate matter that enters the tissues can be absorbed directly through the capillary membranes into the blood. Instead, if the particles are not destroyed locally in the tissues, they enter the lymph and flow through the lymphatic vessels to the lymph nodes located intermittently along the course of the lymphatics. The foreign particles are trapped there in a meshwork of sinuses lined by tissue macrophages called *reticular cells.*

Figure 5–7 illustrates the general organization of the lymph node, showing lymph entering by way of the *afferent lymphatics,* flowing through the *medullary sinuses,* and finally passing out of the *hilus* into the *efferent lymphatics.* Large numbers of reticular cells line the sinuses, and, if any particles enter the sinuses, these cells phagocytize them and prevent general dissemination throughout the body.

Macrophages (Kupffer Cells) in the Liver Sinuses. Still another favorite route by which bacteria invade the body is through the gastrointestinal tract. Large numbers of bacteria constantly pass through the gastrointestinal mucosa into the portal blood. However, before this blood enters the general circulation, it must pass through the sinuses of the liver; these sinuses are lined with tissue macrophages called *Kupffer cells,* illustrated in Figure 5–8. These cells form such an effective particulate filtration system that almost none of the bacteria from the gastrointestinal tract succeed in passing from the portal blood into the general systemic circulation. Indeed, motion pic-

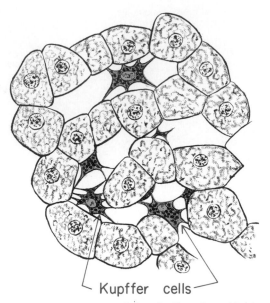

Figure 5–8. Kupffer cells lining the liver sinusoids, showing phagocytosis of India ink particles. (Redrawn from Copenhaver et al.: Bailey's Textbook of Histology. Baltimore, Williams and Wilkins Company, 1969.)

tures of phagocytosis by Kupffer cells have demonstrated phagocytosis of single bacteria in less than 1/100 second.

INFLAMMATION AND FUNCTION OF NEUTROPHILS AND MACROPHAGES

Inflammation is a complex of sequential changes in the tissue in response to injury. When tissue injury occurs, whether it is caused by bacteria, trauma, chemicals, heat, or any other phenomenon, large quantities of *histamine, bradykinin, serotonin,* and other substances are liberated by the damaged tissue into the surrounding fluids. These, especially the histamine, increase the local blood flow and also increase the permeability of the venous capillaries and venules, allowing large quantities of fluid and protein, including fibrinogen, to leak into the tissues. Local extracellular edema results, and the extracellular fluid and lymphatic fluid both clot because of the coagulating effect of tissue exudates on the leaking fibrinogen. Thus, *brawny edema* develops in the spaces surrounding the injured cells.

The "Walling Off" Effect of Inflammation. It is clear that the above described effects of the inflammation process "wall off" the area of injury from the remaining tissues. The tissue spaces and the lymphatics in the inflamed area are blocked by fibrinogen clots so that fluid barely flows through the spaces. Therefore, walling off the area of injury delays the spread of bacteria or toxic products.

The Macrophage and Neutrophil Response to Inflammation

Soon after inflammation begins, the inflamed area becomes invaded by both neutrophils and macrophages, and these set about performing their scavenger functions to rid the tissue of infectious or toxic agents. However, the macrophage and neutrophil responses occur in several different stages.

The Tissue Macrophages at the First Line of Defense. The macrophages that are already present in the tissues, whether they are the histiocytes in the subcutaneous tissues, the alveolar macrophages in the lungs, the microglia in the brain, or so forth, immediately begin their phagocytic actions. Therefore, they are the first line of defense against infection during the first hour or so. However, their numbers often are not very great.

Neutrophilia and Neutrophil Invasion of the Inflamed Area — the Second Line of Defense. The term *neutrophilia* means an increase in the number of neutrophils in the blood. The term "leukocytosis" is also often used to mean the same as neutrophilia, though this term actually means excess number of all white cells, whatever their types.

Within a few hours after the onset of acute inflammation, the number of neutrophils in the blood sometimes increases as much as four- to five-fold — to as high as 15,000 to 25,000 per cubic millimeter. This results from a combination of chemical substances that are released from the inflamed tissues, collectively called *leukocytosis-inducing factor.* This factor diffuses from the inflamed tissue into the blood and is carried to the bone marrow. There it is believed to dilate the venous sinusoids of the marrow, thus causing release of many leukocytes, especially neutrophils, that are stored in these venous sinusoids. In this way large numbers of neutrophils are almost immediately transferred from the bone marrow storage pool into the circulating blood.

Movement of Neutrophils to the Area of Inflammation. Products from the inflamed tissues also cause neutrophils to move from the circulation into the inflamed area. They do this in three ways, which were illustrated in Figure 5–6:

First, they damage the capillary walls and thereby cause neutrophils to stick, which is the process called *margination.*

Second, they greatly increase the permeability of the capillaries and small venules, and this allows the neutrophils to pass by *diapedesis* into the tissue spaces.

Third, the phenomenon of *chemotaxis* causes the neutrophils to migrate toward the injured tissues, as was described earlier.

Thus, within several hours after tissue damage

begins, the area becomes well supplied with neutrophils. Since neutrophils are already mature cells, they are ready to begin their scavenger functions immediately for removal of foreign matter from the inflamed tissues.

Macrophage Proliferation and the Monocyte Response — the Third Line of Defense.   Still a third line of defense is a slow but long-continuing increase in the number of macrophages as well. This results partly from reproduction of the already present tissue macrophages but also from migration of large numbers of monocytes into the inflamed area. Though the monocytes are still immature cells and are not capable of phagocytosis when they first enter the tissues, over a period of 8 to 12 hours they swell markedly, form greatly increased quantities of cytoplasmic lysosomes, exhibit increased ameboid motion, and move chemotactically toward the damaged tissues.

Next, the rate of production of monocytes by the bone marrow also increases. This (as well as increased production of neutrophils) results from stimulation by yet undefined stimulating factors. In long-term chronic infection there is progressively increasing production of monocytes, which increases the ratio of macrophages to neutrophils in the tissues. Therefore, the long-term chronic defense against infection is mainly a macrophage response rather than a neutrophil response.

The macrophages can phagocytize far more bacteria than can the neutrophils, and they can also ingest large quantities of necrotic tissue.

Formation of Pus

When the neutrophils and macrophages engulf large numbers of bacteria and necrotic tissue, essentially all of the neutrophils and many if not most of the macrophages themselves eventually die. After several days, a cavity is often excavated in the inflamed tissues, containing varying portions of necrotic tissue, dead neutrophils, and dead macrophages. Such a mixture is commonly known as *pus*.

Ordinarily, pus formation continues until all infection is suppressed. Sometimes the pus cavity eats its way to the surface of the body and in this way empties itself. At other times the pus cavity remains closed even after tissue destruction has ceased. When this happens the dead cells and necrotic tissue in the pus gradually autolyze over a period of days, and the end-products of autolysis are usually absorbed into the surrounding tissues until most of the evidence of tissue damage is gone.

THE EOSINOPHILS

The eosinophils normally constitute 1 to 3 per cent of all the blood leukocytes. They are weak phagocytes, and they exhibit chemotaxis. They also have a special propensity to collect at sites of antigen-antibody reactions in the tissues and a special capability to phagocytize and digest the combined antigen-antibody complex after the immune process has performed its function. Also, the total number of eosinophils increases greatly in the circulating blood during allergic reactions, following foreign protein injections, and during infections with parasites. It is possible that eosinophils help to remove foreign proteins, whatever their source.

THE BASOPHILS

The basophils in the circulating blood are very similar to, though maybe not identical with, the large *mast* cells located immediately outside many of the capillaries in the body. These cells liberate *heparin* into the blood, a substance that can prevent blood coagulation. It is probable that the basophils in the circulating blood perform similar functions within the blood stream, or it is even possible that the blood simply transports basophils to tissues where they then become mast cells and perform the function of heparin liberation.

The mast cells and basophils also release histamine as well as smaller quantities of bradykinin and serotonin. Indeed, it is mainly the mast cells in inflamed tissues that release these substances during inflammation.

AGRANULOCYTOSIS

A clinical condition known as "agranulocytosis" occasionally occurs, in which the bone marrow stops producing white blood cells, leaving the body unprotected against bacteria and other agents that might invade the tissues. Within two days after the bone marrow stops producing white blood cells, ulcers may appear in the mouth and the colon, or the person develops some form of severe respiratory infection. Bacteria from the ulcers then rapidly invade the surrounding tissues and the blood. Without treatment, death often ensues three to six days after acute agranulocytosis begins.

Irradiation of the body by gamma rays caused by a nuclear explosion, or exposure to drugs or chemicals containing benzene or anthracene nuclei, is quite likely to cause aplasia of the bone marrow. Indeed, some of the common drugs, such as the sulfonamides, chloramphenicol, thiouracil (used to treat thyrotoxicosis), and even the various barbiturate hypnotics, on occasion cause agranulocytosis.

After irradiation injury to the bone marrow, a large number of stem cells usually remain undestroyed. Therefore, the patient properly treated with antibiotics and other drugs to ward off infection will usually develop enough new bone marrow within several weeks to several months that his blood cell concentrations can return to normal.

THE LEUKEMIAS

Ordinarily, leukemias are divided into two general types: the *lymphogenous leukemias* and the *myelogenous leukemias*. The lymphogenous leukemias are caused by cancerous production of lymphoid cells, beginning first in a lymph node or other lymphogenous tissue and then spreading to other areas of the body. The second type of leukemia, myelogenous leukemia, begins by cancerous production of young myelogenous cells (early forms of neutrophils, monocytes, or others) in the bone marrow and then spreads throughout the body so that white blood cells are produced in many extramedullary organs.

Effects of Leukemia on the Body

The leukemic cells of the bone marrow may reproduce so greatly that they invade the surrounding bone, causing pain and eventually a tendency to easy fracture. Almost all leukemias spread to the spleen, the lymph nodes, the liver, and other especially vascular regions, regardless of whether the origin of the leukemia is in the bone marrow or in the lymph nodes.

Very common effects in leukemia are the development of infections, severe anemia, and a bleeding tendency caused by thrombocytopenia (lack of platelets). These effects result mainly from displacement of the normal bone marrow by the leukemic cells.

Finally, the leukemic tissues reproduce new cells so rapidly that tremendous demands are made on the body for foodstuffs, especially the amino acids and vitamins. Thus, while the leukemic tissues grow the other tissues are debilitated. Obviously, after metabolic starvation has continued long enough, that alone is sufficient to cause death.

REFERENCES

Red Blood Cells

Baum, S. J., and Ledney, G. D. (eds.): Experimental Hematology Today, 1978. New York, Springer-Verlag, 1978.
Bessis, M., *et al.* (eds.): Red Cell Rheology. New York, Springer-Verlag, 1978.
Beutler, E.: Hemolytic Anemia in Disorders of Red Cell Metabolism. New York, Plenum Press, 1978.
Bottomley, S. S., and Whitcomb, W. H.: Erythropoiesis. *In* Frohlich, E. D. (ed.): Pathophysiology, 2nd Ed. Philadelphia, J. B. Lippincott Co., 1976, p. 567.
Cokelet, G. R., *et al.* (eds.): Erythrocyte Mechanics and Blood Flow. New York, A. R. Liss, 1979.
Dunn, C. D. R.: The Differentiation of Haemopoietic Stem Cells. Baltimore, Williams & Wilkins, 1971.
Ellory, C., and Lew, V. L. (eds.): Membrane Transport in Red Cells. New York, Academic Press, 1977.

Erslev, A. J., and Gabuzda, T. G.: Pathophysiology of Blood. Philadelphia, W. B. Saunders Co., 1979.
Heimpel, H. *et al.* (eds.): International Symposium on Aplastic Anemia. New York, Springer-Verlag, 1979.
Kass, L.: Bone Marrow Interpretation. Philadelphia, J. B. Lippincott Co., 1979.
Kelemen, E., *et al.*: Atlas of Human Hemopoietic Development. New York, Springer-Verlag, 1979.
Konigsberg, W.: Protein structure and molecular dysfunction: Hemoglobin. *In* Bondy, P. K., and Rosenberg, L. E. (eds.): Metabolic Control and Disease, 8th Ed. Philadelphia, W. B. Saunders Co., 1980, p. 27.
Lux, S. E., *et al.* (eds.): Normal and Abnormal Red Cell Membranes. New York, A. R. Liss, 1979.
Maclean, N.: Haemoglobin. London, Edward Arnold, 1978.
Munro, H. N., and Linder, M. C.: Ferritin: Structure, biosynthesis, and role in iron metabolism. *Physiol. Rev., 58*:317, 1978.
Peschle, C.: Erythropoiesis. *Annu. Rev. Med., 31*:303, 1980.
Petz, L. D., and Garratty, G.: Acquired Immune Hemolytic Anemias. New York, Churchill Livingstone, 1979.
Platt, W. R.: Color Atlas and Textbook of Hematology. Philadelphia, J. B. Lippincott Co., 1978.
Zagalak, B., and Friedrich, W. (eds.): Vitamin B12. New York, Walter De Gruyter, 1979.

White Blood Cells

Allison, A. C., *et al.*: Inflammation. New York, Springer-Verlag, 1978.
Crowther, D. G. (ed.): Leukemia and Non-Hodgkin Lymphoma. New York, Pergamon Press, 1979.
Escobar, M. R., and Friedman, H. (eds.): Macrophages and Lymphocytes; Nature, Functions and Interaction. New York, Plenum Press, 1979.
Friedman, H., *et al.* (eds.): The Reticuloendothelial System. New York, Plenum Press, 1979.
Gadebusch, H. J. (ed.): Phagocytosis and Cellular Immunity. West Palm Beach, Fla., CRC Press, 1979.
Gowans, J. L. (in honour of): Blood Cells and Vessel Walls: Functional Interactions. Princeton, N.J., Excerpta Medica, 1980.
Güttler, F., *et al.* (eds.): Inborn Errors of Immunity and Phagocytosis. Baltimore, University Park Press, 1979.
Houck, J. C. (ed.): Chemical Messengers of the Inflammatory Process. New York, Elsevier/North-Holland, 1980.
Janoff, A.: Neutrophil chemotaxis and mediation of tissue damage. *In* Kaley, G., and Altura, B. M. (eds.): Microcirculation. Vol. III. Baltimore, University Park Press, 1977.
Kaley, G.: Mechanisms of inflammation. *In* Kaley, G., and Altura, B. M. (eds.): Microcirculation. Vol. III. Baltimore, University Park Press, 1977.
Klebanoff, S. J., and Clark, R. A.: The Neutrophil: Function and Clinical Disorders. New York, Elsevier/North-Holland, 1978.
Kokubun, Y., and Kobayashi, N. (eds.): Phagocytosis, Its Physiology and Pathology. Baltimore, University Park Press, 1979.
Lichtman, M. A. (ed.): Hematology and Oncology. New York, Grune & Stratton, 1980.
Lisiewicz, J.: Human Neutrophils. Bowie, Md., Charles Press Publishers, 1979.
Movat, H. Z. (ed.): Inflammatory Reaction. New York, Springer-Verlag, 1979.
Quastel, M. R. (ed.): Cell Biology and Immunology of Leukocyte Function. New York, Academic Press, 1979.
Simone, J. V. (ed.): Acute Leukemia. Philadelphia, W. B. Saunders Co., 1978.
Spivak, J. L. (ed.): Fundamentals of Clinical Hematology. Hagerstown, Md., Harper & Row, 1980.
Van Furth, R. (ed.): Mononuclear Phagocytes: Functional Aspects of Mononuclear Phagocytes. Boston, M. Nijhoff, 1979.

6

Immunity, Allergy, Blood Groups, Transfusion, and Transplantation

INNATE IMMUNITY

The human body has the ability to resist almost all types of organisms or toxins that tend to damage the tissues and organs. This capacity is called *immunity*. Much of the immunity is caused by a special immunity system that forms antibodies and sensitized lymphocytes that attack and destroy the specific organisms or toxins. This type of immunity is *acquired immunity*. However, an additional portion of the immunity, called *innate immunity*, results from such processes as these:

1. Phagocytosis of bacteria and other invaders by white blood cells and cells of the tissue macrophage system, as described in the previous chapter.

2. Destruction of organisms swallowed into the stomach by the digestive enzymes.

3. Resistance of the skin to invasion by organisms.

4. Presence in the blood of certain chemical compounds that attach to foreign organisms or toxins and destroy them.

This innate immunity makes the human body resistant to such diseases as dysentery, some paralytic virus diseases of animals, hog cholera, cattle plague, and distemper, a viral disease that kills a large percentage of dogs that become afflicted with it. On the other hand, animals are resistant to many human diseases, such as poliomyelitis, mumps, human cholera, measles, and syphilis, which are very destructive or even lethal to the human being.

ACQUIRED IMMUNITY (OR ADAPTIVE IMMUNITY)

In addition to its innate immunity, the human body also has the ability to develop extremely powerful specific immunity against individual invading agents such as lethal bacteria, viruses, toxins, and even foreign tissues from other animals. This is called *acquired immunity* or *adaptive immunity*.

Acquired immunity is important for protection against invading organisms to which the body does not have innate or natural immunity. The body does not block the invading organisms upon first exposure. However, within a few days to a few weeks after exposure the special immune system does then develop extremely powerful resistance to the invader. Furthermore, the resistance is highly specific for that particular invader and not for others.

Acquired immunity can often bestow extreme protection. For instance, certain toxins, such as the paralytic toxin of botulinum or the tetanizing toxin of tetanus, can be protected against in doses as high as 100,000 times the amount that would be lethal without immunity. For this reason, the process known as "vaccination" is extremely important in protecting human beings against diseases and against toxins, as will be explained in the course of this chapter.

TWO BASIC TYPES OF ACQUIRED IMMUNITY

Two basic, but closely allied, types of acquired immunity occur in the body. In one of these the body develops circulating *antibodies*, which are globulin molecules that are capable of attacking the invading agent. This type of immunity is called *humoral immunity*.

The second type of acquired immunity is achieved through the formation of large numbers of highly specialized lymphocytes that are specifically sensitized against the foreign agent. These *sensitized lymphocytes* have the special capability of attaching to the foreign agent and destroying it. This type of immunity is called *cellular immunity* or, sometimes, *lymphocytic immunity*.

We shall see shortly that both the antibodies and the sensitized lymphocytes are formed in the lymphoid tissue of the body. First, let us discuss the initiation of the immune process by *antigens*.

ANTIGENS

Each toxin or each type of invading organism contains one or more specific chemical compounds in its makeup that are different from all other compounds. In general, these are proteins, large polysaccharides, or large lipoprotein complexes, and it is they that cause the acquired immunity. These substances are called *antigens*.

Likewise, tissues such as a transplanted heart from other human beings or animals contain numerous antigens that can elicit the immune process that in turn causes subsequent destruction of the transplant.

For a substance to be antigenic it usually must have a high molecular weight, 8000 or greater. Furthermore, the process of antigenicity depends upon regularly recurring prosthetic radicals on the surface of the large molecule, which explains why proteins and polysaccharides are almost always antigenic, for they both have this type of stereochemical characteristic.

ROLE OF LYMPHOID TISSUE IN ACQUIRED IMMUNITY

Acquired immunity is the product of the body's lymphoid tissue. In persons who have a genetic lack of lymphoid tissue or whose lymphoid tissue has been destroyed by radiation or by chemicals, no acquired immunity whatsoever can develop.

The lymphoid tissue is located most extensively in the lymph nodes, but it is also found in special lymphoid tissues such as that of the spleen, submucosal areas of the gastrointestinal tract, and, to a slight extent, in the bone marrow. As pointed out in the previous chapter, the lymph nodes and other lymphoid tissue are distributed very advantageously in the body to intercept the invading organisms or toxins before they can spread too widely.

Two Types of Lymphocytes That Promote, Respectively, Cellular Immunity and Humoral Immunity — the T and the B Lymphocytes. Though most of the lymphocytes in normal lymphoid tissue look alike when studied under the microscope, these cells are distinctly divided into two major populations. One of the populations, called the *T lymphocytes,* is responsible for forming the sensitized lymphocytes that provide cellular immunity, and the other, called the *B lymphocytes*, for forming the antibodies that provide humoral immunity.

The T Lymphocytes — Their Formation in the Thymus. The T lymphocytes are called by that name because they originate initially from stem cells in the thymus gland, as illustrated in Figure 6–1, not in the lymph tissue itself. Most of the formation of T lymphocytes occurs shortly before the birth of the baby and during the few months after birth. Therefore, removal of the fetal thymus gland after this period of time usually will not seriously impair the T lymphocytic immunity system (the system necessary for cellular immunity). However, removal of the fetal thymus several months before birth can completely prevent development of all cellular immunity.

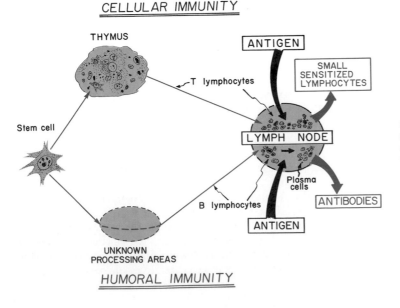

Figure 6–1. Formation of antibodies and sensitized lymphocytes by a lymph node in response to antigens. This figure also shows the origin of *thymic* (T) and *bursal* (B) lymphocytes, which are responsible for the cellular and humoral immune processes of the lymph nodes.

The B Lymphocytes —Their Formation in the Bursa of Fabricius. The B lymphocytes are named for a special organ found in birds called the bursa of Fabricius, because these lymphocytes originate from the stem cells in that organ. These lymphocytes also are formed mainly during the last few months of fetal life and the first few months after birth.

In mammals, no specific organ similar to the bursa of Fabricius has been found. However, lymphoid tissue, mainly in the fetal liver, is believed to perform the same function to generate the early B lymphocytes.

Spread of the T and B Lymphocytes to the Lymphoid Tissue. After formation of the early T and B lymphocytes in the thymus and the fetal liver, they first circulate freely in the blood for a few hours but then become entrapped by the reticulum meshwork in the lymphoid tissue. Thereafter, they continue to reproduce and to grow in the lymphoid tissue throughout the body.

Mechanisms for Determining Specificity of Sensitized Lymphocytes and Antibodies — Lymphocyte Clones

Earlier in this chapter it was pointed out that the lymphoid tissue can form sensitized lymphocytes and antibodies, each of which can react highly specifically against particular types of invading agents. This effect is believed to occur in the following way:

Specificity of the Sensitized Lymphocytes or Antibodies Formed by a Single Type of Lymphocyte — Lymphocyte Clones. Because it is known that the lymphocytes of the lymphoid tissue can form 10,000 to 100,000 different types of sensitized lymphocytes or antibodies, each specific for a different antigen, it is also almost certain that just as many different types of precursor lymphocytes pre-exist in the lymph nodes for formation of those many specific types of sensitized lymphocytes or antibodies.

All the lymphocytes of one specific type in the lymphoid tissue — those that form one specific type of sensitized lymphocyte or one specific type of antibody — are called a *clone of lymphocytes*. That is, all of the lymphocytes in each clone are alike and are probably derived originally from one or a few early lymphocytes of the specific type.

Origin of the Many Clones of Lymphocytes. The way in which the many different clones of lymphocytes are originally formed is not known, but there are two main theories for the origin. The first theory suggests that each clone is genetically determined — that is, there is a separate gene for each clone.

The second theory assumes that the lymphocytes in the thymus or in the B cell processing area simply differentiate wildly into a whole host of random clones of lymphocytes.

Excitation of a Clone of Lymphocytes. Each clone of lymphocytes is responsive to only a single type of antigen (or to a group of antigens that have almost exactly the same stereochemical characteristics). When excited by the clone's specific antigen, all the cells of the clone proliferate madly, forming tremendous numbers of progeny, and these in turn lead to the formation of large quantities of antibodies if the clone is B lymphocytes, or to the formation of sensitized lymphocytes if the clone is T lymphocytes. During this process, the total number of lymphocytes in the lymphoid tissue increases markedly.

Tolerance of the Acquired Immunity System to One's Own Tissues — Role of the Thymus and the B-Cell Processor

Obviously, if a person should become immune to his or her own tissues, the process of acquired immunity would destroy the individual's own body. Fortunately, the immune mechanism normally "recognizes" a person's own tissues as being completely distinctive from those of invaders, and his immunity system forms very few antibodies or sensitized lymphocytes against his own antigens. This phenomenon is known as *tolerance* to the body's own tissues.

Mechanism of Tolerance. It is believed that tolerance develops during the processing of the lymphocytes in the thymus and in the B lymphocyte processing area. The reason for this belief is that injecting a strong antigen into a fetus while the lymphocytes are being processed in these two areas will prevent the development of clones of lymphocytes in the lymphoid tissue that are specific for the injected antigen. Also, experiments have shown that specific immature lymphocytes in the thymus, when exposed to a strong antigen, become lymphoblastic, proliferate considerably, and then combine with the stimulating antigen — an effect that is believed to cause the cells themselves to be destroyed before they can migrate to and colonize the lymphoid tissue.

Therefore, it is believed that during the processing of lymphocytes in the thymus and in the B lymphocyte processing area, all those clones of lymphocytes that are specific for the body's own tissues are self-destroyed because of their continual exposure to the body's antigens.

Failure of the Tolerance Mechanism — Autoimmune Diseases. Unfortunately, people frequently lose some of their immune tolerance to their own tissues. This occurs to a greater extent the older a person becomes. It usually results from destruction of some of the body's tissues, which releases considerable quantities of antigens that circulate in the body and cause acquired immunity in the form of either sensitized lymphocytes or antibodies.

Some common diseases that result from autoimmunity include *rheumatic fever*, in which the body becomes immunized against tissues in the heart and joints following exposure to a specific type of streptococcal toxin; one type of *glomerulonephritis*, a renal disease in which the person becomes immunized against the basement membranes of his glomeruli; *myasthenia gravis*, in which immunity develops against the muscle membrane portion of the neuromuscular junction, causing paralysis; and *lupus erythematosus*, in which the person becomes immunized against many different body tissues at the same time. The disease causes extensive damage, often causing rapid death.

Specific Attributes of the B Lymphocyte System — Humoral Immunity and the Antibodies

Formation of Antibodies by the Plasma Cells. Prior to exposure to a specific antigen, the clones of B lymphocytes remain dormant in the lymphoid tissue. However, upon entry of a foreign antigen, the lymphoid tissue macrophages phagocytize the antigen and then present it to the adjacent lymphocytes. Those lymphocytes specific for that antigen immediately enlarge and take on the appearance of *lymphoblasts*. Some of those then further differentiate to form a total of about 500 *plasma cells* for each lymphoblast during the next four days. The mature plasma cell then produces gamma globulin antibodies at an extremely rapid rate — about 2000 molecules per second for each plasma cell. The antibodies are secreted into the lymph and are carried to the circulating blood. This process continues for several weeks until death of the plasma cells.

Formation of "Memory" Cells — Difference Between the Primary Response and the Secondary Response. Some of the lymphoblasts formed by activation of a clone of B lymphocytes do not go on to form plasma cells but, instead, form moderate numbers of new B lymphocytes similar to those of the original clone. In other words, the population of the specifically activated clone becomes greatly enhanced. And the new B lymphocytes are added to the original lymphocytes of the clone. These then remain dormant in the lymphoid tissue until activated once again by a new quantity of the same antigen. They are called *memory cells*. Obviously, subsequent exposure to the same antigen will then cause a much

more rapid and much more potent antibody response.

The Nature of the Antibodies. The antibodies are gamma globulins called *immunoglobulins*, and they have molecular weights between approximately 150,000 and 900,000.

All of the immunoglobulins are composed of combinations of *light* and *heavy polypeptide chains*, most of which are a combination of two light and two heavy chains, as illustrated in Figure 6–2. Some of the immunoglobulins, though, have combinations of greater than two heavy and two light chains, which give rise to the immunoglobulins of a much larger molecular weight. Yet, in all of these, each heavy chain is paralleled by a light chain at one of its ends, thus forming a heavy-light pair, and there are always at least two such pairs in each immunoglobulin molecule.

Figure 6–2 shows a designated end of each of the light and each of the heavy chains, called the "variable portion," and the remainder of each chain is called the "constant portion." The variable portion is different for each specificity of antibody, and it is this portion that allows the antibody to attach specifically to a particular type of antigen.

Specificity of Antibodies. Each antibody that is specific for a particular antigen has a different organization of amino acid residues in the variable portions of both the light and heavy chains. These have a specific steric shape for each antigen specificity, so that when an antigen comes in contact with it, the prosthetic radicals of the antigen fit as a mirror image with those of the antibody, thus allowing a rapid and tight chemical or physical bond between the antibody and the antigen.

Note, especially, in Figure 6–2 that there are two variable sites on the antibody for attachment of antigens. Thus, most antibodies are *bivalent*. However, a small proportion of the antibodies, which have high molecular weight combinations of light and heavy chains, have more than two reactive sites.

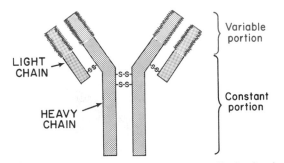

Figure 6–2. Structure of the typical IgG antibody, showing it to be composed of two heavy polypeptide chains and two light polypeptide chains. The antigen binds at two different sites on the variable portions of the chains.

Mechanisms of Action of Antibodies

Antibodies can act in three different ways to protect the body against the invading agents: (1) direct attack on the invader, (2) activation of the complement system that then destroys the invader, or (3) activation of the anaphylactic system that changes the local environment around the invading antigen and in this way prevents its virulence.

Direct Action of Antibodies on Invading Agents. Because of the bivalent nature of the antibodies and the multiple antigen sites on most invading agents, the antibodies can inactive the invading agent in one of several ways, as follows:

1. *Agglutination*, in which multiple antigenic agents are bound together into a clump.

2. *Precipitation*, in which the complex of soluble antigen (such as tetanus toxin) and antibody becomes insoluble and precipitates.

3. *Neutralization,* in which the antibodies cover the toxic sites of the antigenic agent.

4. *Lysis*, in which some very potent antibodies are capable of directly attacking membranes of cellular agents and thereby causing rupture of the cell.

However, the direct actions of antibodies attacking the antigenic invaders probably are not strong enough, under normal conditions, to play a major role in protecting the body against the invader. Most of the protection comes through the *amplifying* effects of the complement and anaphylactic effector systems described below.

The Complement System for Antibody Action. Complement is a system of nine different enzyme precursors (designated C-1 through C-9) plus several other associated substances that are found normally in the plasma and other body fluids, but the enzymes are normally inactive. However, when an antibody combines with an antigen, a reactive site on the "constant" portion of the antibody becomes uncovered, or activated, and this in turn sets into motion a "cascade" of sequential reactions in the complement system, illustrated in Figure 6–3. The activated enzymes then attack the invading agent in several different ways as well as initiate local tissue reactions that also provide protection against damage by the invader. Among the more important effects that occur are the following:

1. *Lysis.* The proteolytic enzymes of the complement system digest portions of the cell membrane, thus causing rupture of cellular agents such as bacteria or other types of invading cells.

2. *Opsonization and phagocytosis.* The complement enzymes attack the surfaces of bacteria and other antigens, making these highly susceptible to phagocytosis by neutrophils and tissue macrophages. This process is called *opsonization*. It often enhances the number of bacteria that can be destroyed many hundredfold.

3. *Chemotaxis.* One or more of the complement products cause chemotaxis of neutrophils and macrophages, thus greatly enhancing the number of these phagocytes in the local region of the antigenic agent.

4. *Agglutination.* The complement enzymes also change the surfaces of some of the antigenic agents so that they adhere to each other, thus causing agglutination.

5. *Neutralization of viruses.* The complement enzymes frequently attack the molecular structures of viruses and thereby render them nonvirulent.

6. *Inflammatory effects.* The complement products elicit a local inflammatory reaction, leading to hyperemia, coagulation of proteins in the tissues, and other aspects of the inflammation process, thus preventing movement of the invading agent through the tissues.

Activation of the Anaphylactic System by Antibodies. Some of the antibodies attach to the membranes of cells in the tissues and blood. Among the most important cells are the *mast cells* in tissues surrounding the blood vessels and the

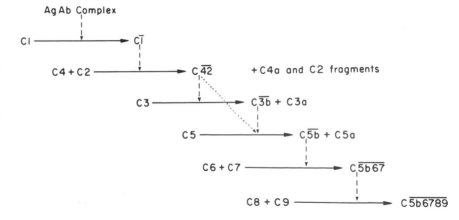

Figure 6–3. Cascade of reactions during activation of the classic pathway of complement. (From Alexander and Good: Fundamentals of Clinical Immunology. Philadelphia, W. B. Saunders Company, 1977.)

basophils circulating in the blood. When an antigen reacts with one of the antibody molecules attached to the cell, there is immediate swelling and then rupture of the cell, with the release of a large number of factors that affect the local environment. Such factors include especially *histamine* but also other substances that cause local inflammatory reactions. These effects in turn are believed to help immobilize the antigenic invader.

Special Attributes of the T Lymphocyte System — Cellular Immunity and Sensitized Lymphocytes

Release of Sensitized Lymphocytes from Lymphoid Tissue and Formation of Memory Cells. Upon exposure to the proper antigens, sensitized lymphocytes are released from lymphoid tissue in ways that parallel antibody release. The only real difference is that instead of releasing antibodies, whole sensitized lymphocytes are formed and released into the lymph. These then pass into the circulation, where they remain a few minutes to a few hours at most, filtering out of the circulation into all the tissues of the body.

Also, lymphocyte *memory cells* are formed in the same way that memory cells are formed in the humoral antibody system. Therefore, upon subsequent exposure to the same antigen, the release of sensitized lymphocytes occurs much more rapidly and much more powerfully than in the first response.

Mechanism of Sensitization of T Lymphocytes. It is believed that T lymphocytes become sensitized against specific antigens by forming on their surfaces a type of "antibody." This antibody is composed of a *variable unit* similar to the variable portion of the humoral antibody, but it has no constant portion. Instead, multiple variable units are attached directly to the cell membrane of the T lymphocyte.

Persistence of Cellular Immunity. An important difference between cellular immunity and humoral immunity is its persistence. Humoral antibodies rarely persist more than a few months, or at most, a few years. On the other hand, sensitized lymphocytes probably have an indefinite life span and seem to persist until they eventually come in contact with their specific antigen. There is reason to believe that such sensitized lymphocytes might persist as long as ten years in some instances.

Types of Organisms Resisted by Sensitized Lymphocytes. Although the humoral antibody mechanism for immunity is especially efficacious against more acute bacterial diseases, the cellular immunity system is activated much more potently by the more slowly developing bacterial diseases such as tuberculosis, brucellosis, and so forth. Also, this system is active against cancer cells, cells of transplanted organs, and fungus organisms, all of which are far larger than bacteria.

Mechanism of Action of Sensitized Lymphocytes

The sensitized lymphocyte, on coming in contact with its specific antigen, combines with the antigen. This combination in turn leads to a sequence of reactions whereby the sensitized lymphocyte destroys the invader. As is also true of the humoral immunity system, the sensitized lymphocyte destroys the invader either directly or indirectly.

Direct Destruction of the Invader. Figure 6–4 illustrates sensitized lymphocytes that have bound with antigens in the membrane of an invading cell such as a cancer cell, a heart transplant cell, or a parasitic cell of another type. The immediate effect of this attachment is swelling of the sensitized lymphocyte and release of cytotoxic substances from the lymphocyte to attack the invading cell. The cytotoxic substances are probably mainly lysosomal enzymes manufactured in the lymphocytes. However, these direct effects of the sensitized lymphocyte in destroying the invading cell are relatively weak in comparison with the indirect effects, as follows:

The Indirect "Amplifying" Mechanisms of Cellular Immunity. When the sensitized lymphocytes combine with their specific antigens, they release a number of different substances into the surrounding tissues that lead to a se-

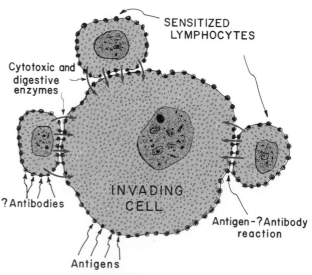

Figure 6–4. Direct destruction of an invading cell by sensitized lymphocytes.

quence of reactions. Some of these reactions are the following:

Release of Transfer Factor. The sensitized lymphocytes release a polypeptide substance with a molecular weight of 2000 to 8000, called *transfer factor*. This then reacts with other small lymphocytes in the tissues that are of the nonsensitized variety. On entering these lymphocytes, the transfer factor causes them to take on characteristics similar to those of the original sensitized T cells and to attack the invader along with the original cells.

Attraction and Activation of Macrophages. A second product of the activated sensitized lymphocyte is a *macrophage chemotaxic factor* that causes as many as 1000 macrophages to enter the vicinity of the activated sensitized lymphocyte. These macrophages then play a major role in removing the foreign antigenic invader.

Thus, it is by a combination of a weak direct effect of the sensitized lymphocytes on the antigen invader and much more powerful indirect reactions that the cellular immunity system destroys the invader.

VACCINATION

The process of vaccination has been used for many years to cause acquired immunity against specific diseases. A person can be vaccinated by injecting dead organisms that are no longer capable of causing disease but that still have their chemical antigens. This type of vaccination is used to protect against typhoid fever, whooping cough, diphtheria, and many other types of bacterial diseases. Also, immunity can be achieved against toxins that have been treated with chemicals so that their toxic nature has been destroyed even though their antigens for causing immunity are still intact. This procedure is used in vaccinating against tetanus, botulism, and other similar toxic diseases. And, finally, a person can be vaccinated by infecting him with live organisms that have been "attenuated." That is, these organisms either have been grown in special culture mediums or have been passed through a series of animals until they have mutated enough that they will not cause disease but do still carry the specific antigens. This procedure is used to protect against poliomyelitis, yellow fever, measles, smallpox, and many other viral diseases.

PASSIVE IMMUNITY

Thus far, all the acquired immunity that we have discussed has been *active immunity*. That is, the person's body develops either antibodies or sensitized lymphocytes in response to invasion by a foreign antigen. However, temporary immunity can be achieved in a person without injecting any antigen whatsoever. This is done by infusing antibodies, sensitized lymphocytes, or both from someone else or from some other animal that has been actively immunized against the antigen. The antibodies will last for two to three weeks, and during that time the person is protected against the invading disease. Sensitized lymphocytes will last for a few weeks if transfused from another person, and for a few hours to a few days if transfused from an animal. The transfusion of antibodies or lymphocytes to confer immunity is called *passive immunity*.

ALLERGY

One of the important side effects of immunity is the development, under some conditions, of allergy. There are at least three different types of allergy, two of which can occur in any person, and a third that occurs only in persons who have a specific allergic tendency.

ALLERGIES THAT OCCUR IN NORMAL PEOPLE

Delayed-Reaction Allergy. This type of allergy frequently causes skin eruptions in response to certain drugs or chemicals, particularly some cosmetics and household chemicals, to which one's skin is often exposed. Another example of such an allergy is the skin eruption caused by exposure to poison ivy.

Delayed-reaction allergy is caused by sensitized lymphocytes and not by antibodies. In the case of poison ivy, the toxin of poison ivy in itself does not cause much harm to the tissues. However, upon repeated exposure it does cause the formation of sensitized lymphocytes. Then the sensitized lymphocytes diffuse into the skin in sufficient numbers to combine with the poison ivy toxin and elicit a cellular immunity type of reaction. Remembering that cellular immunity can cause release of many toxic substances from the sensitized lymphocytes, as well as extensive invasion of the tissues by macrophages and their subsequent effects, one can well understand that the eventual result of some delayed-reaction allergies can be serious tissue damage.

Allergies Caused by Reaction Between Antibodies and Antigens. When a person becomes strongly immunized against an antigen and has developed a very high titer of antibodies, subsequent sudden exposure of that person to a high concentration of the same antigen can cause a serious tissue reaction. The antigen-antibody complex that is formed precipitates, and some of it deposits as granules in the walls of the small blood vessels. These granules also activate the complement system, setting off extensive release of proteolytic enzymes. The result of these two effects is severe inflammation and destruction of the small blood vessels.

One manifestation of this type of reaction is *serum sickness.* Serum injected into a person can cause subsequent formation of antibodies. When these begin to appear, they react with the protein of the injected serum and elicit a widespread antigen-antibody reaction throughout the body. Fortunately, this reaction occurs slowly over a period of days as the antibodies are formed, and usually it is not lethal. However, it can be lethal on occasion, and on other occasions it can cause widespread inflammation and edema throughout the body with development of a circulatory shocklike syndrome.

ALLERGIES IN THE "ALLERGIC" PERSON

Some persons have an "allergic" tendency. This phenomenon is genetically passed on from parent to child, and it is characterized by the presence of large quantities of *IgE antibodies.* These antibodies are called *reagins* or *sensitizing antibodies* to distinguish them from the more common IgG antibodies. When an *allergen* (defined as an antigen that reacts specifically with a specific type of IgE reagin antibody) enters the body, an allergen-reagin reaction takes place, and a subsequent allergic reaction takes place.

The IgE antibodies (the reagins) attach to cells throughout the body, especially to mast cells and basophils; therefore, the allergen-reagin reaction damages the cells. The result is *anaphylactoid types of immune reactions.* These result primarily from the rupture of the mast cells and basophils with consequent release of *histamine, slow-reacting substance of anaphylaxis, eosinophil chemotactic substance, lysosomal enzymes,* and other less important substances.

Among the different types of allergic reaction of this type are:

Anaphylaxis. When a specific allergen is injected directly into the circulation it can react in widespread areas of the body with the basophils of the blood and the mast cells located immediately outside the small blood vessels. Therefore, the anaphylactic type of reaction occurs everywhere. The histamine released into the circulation causes widespread peripheral vasodilatation as well as increased permeability of the capillaries and marked loss of plasma from the circulation.

Often, persons experiencing this reaction die of circulatory shock within a few minutes unless treated with norepinephrine to oppose the effects of the histamine.

Urticaria. Urticaria results from antigen entering specific skin areas and causing localized anaphylactoid reactions. *Histamine* released locally causes (a) vasodilatation that induces an immediate *red flare* and (b) increased permeability of the capillaries that leads to swelling of the skin in another few minutes. The swellings are commonly called "hives." Administration of antihistamine drugs to a person prior to exposure will prevent the hives.

Hay Fever. In hay fever, the allergen-reagin reaction occurs in the nose. *Histamine* released in response to this causes local vascular dilatation with resultant increased capillary pressure, and it also causes increased capillary permeability. Both of these effects cause rapid fluid leakage into the tissues of the nose, and the nasal linings become swollen and secretory. Here again, use of antihistaminic drugs can prevent this swelling reaction.

Asthma. In asthma, the allergen-reagin reaction occurs in the bronchioles of the lungs. Here, the most injurious product released from the mast cells seems to be *slow-reacting substance of anaphylaxis,* which causes spasm of the bronchiolar smooth muscle. Consequently, the person has difficulty breathing. Unfortunately, administration of antihistaminics has little effect on the course of asthma, because histamine does not appear to be the major factor eliciting the asthmatic reaction.

BLOOD GROUPS

ANTIGENICITY AND IMMUNE REACTIONS OF BLOOD CELLS

When blood transfusions from one person to another were first attempted the transfusions were successful in some instances, but, in many more, immediate or delayed agglutination and hemolysis of the red blood cells occurred. Soon it was discovered that the bloods of different persons usually have different antigenic and immune properties so that antibodies in the plasma of one blood react with antigens on the red cells of another. And these reactions sometimes are severe enough to cause death.

Though several hundred different antigens have been found in the blood cells of different persons, two particular groups of antigens are more likely than the others to cause blood transfusion reactions. These are the so-called *O-A-B* system of antigens and the *Rh* system. Bloods are divided into different groups and types in accordance with the types of antigens present in the cells.

O-A-B BLOOD GROUPS

The A and B Antigens — Called "Agglutinogens"

Two different but related antigens — type A and type B — occur on the surfaces of the red blood cells in different persons. Because of the way these antigens are inherited, people may have neither of them in their cells, or may have one or both simultaneously.

As will be discussed below, some bloods also contain strong antibodies that react specifically with either the type A or the type B antigens in the cells, causing agglutination and hemolysis. Because the type A and type B antigens in the cells make the cells susceptible to agglutination, these antigens are called *agglutinogens*.

The Four Major O-A-B Blood Groups. In transfusing blood from one person to another, the bloods of donors and recipients are normally classified into four major O-A-B groups, as illustrated in Table 6–1, depending on the presence or absence of the two agglutinogens. When neither A nor B agglutinogen is present, the blood group is *group O*. When only type A agglutinogen is present, the blood is *group A*. When only type B agglutinogen is present, the blood is *group B*. And when both A and B agglutinogens are present, the blood is *group AB*.

Relative Frequencies of the Different Blood Types. The prevalence of the different blood types among Caucasoids is approximately as follows:

Type	Per cent
O	47
A	41
B	9
AB	3

It is obvious from these percentages that the O and A genes occur frequently but the B gene is infrequent.

TABLE 6–1 THE BLOOD GROUPS, WITH THEIR GENOTYPES AND THEIR CONSTITUENT AGGLUTINOGENS AND AGGLUTININS

Genotypes	Blood Groups	Agglutinogens	Agglutinins
OO	O	—	Anti-A and Anti-B
OA or AA	A	A	Anti-B
OB or BB	B	B	Anti-A
AB	AB	A and B	—

The Agglutinins

When type A agglutinogen *is not present* in a person's red blood cells, antibodies known as "anti-A" agglutinins develop in his plasma. Also, when type B agglutinogen *is not present* in the red blood cells, antibodies known as "anti-B" agglutinins develop in the plasma.

Thus referring once again to Table 6–1, it will be observed that group O blood, though containing no agglutinogens, does contain both *anti-A* and *anti-B agglutinins*, while group A blood contains type A agglutinogens and *anti-B agglutinins*, and group B blood contains type B agglutinogens and *anti-A agglutinins*. Finally, group AB blood contains both A and B agglutinogens but no agglutinins at all.

Origin of the Agglutinins in the Plasma. The agglutinins are gamma globulins, as are other antibodies, and are produced by the same cells that produce antibodies to any other antigens.

It is difficult to understand how agglutinins are produced in individuals who do not have the respective antigenic substances in their red blood cells. However, small amounts of group A and B antigens enter the body in the food, in bacteria, and in other ways, and these substances initiate the development of the anti-A or anti-B agglutinins. The newborn baby has few if any agglutinins, showing that agglutinin formation occurs almost entirely after birth.

The Agglutination Process in Transfusion Reactions

When bloods are mismatched so that anti-A or anti-B agglutinins are mixed with red blood cells containing A or B agglutinogens respectively, the red cells agglutinate by the following process: The agglutinins attach themselves to the red blood cells. Because the agglutinins are bivalent or polyvalent, a single agglutinin can attach to two different red blood cells at the same time, thereby causing the cells to adhere to each other. This causes the cells to clump. Then these clumps plug small blood vessels throughout the circulatory system. During the ensuing few hours to few days, the phagocytic white blood cells and the reticuloendothelial system destroy the agglutinated cells, releasing hemoglobin into the plasma.

Blood Typing

Prior to giving a transfusion, it is necessary to determine the blood group of the recipient and that of the donor so that the bloods will be appropriately matched. This "typing" of blood is performed as follows:

A usual method of blood typing is the slide technique. In using this technique a drop or more

TABLE 6–2 BLOOD TYPING—SHOWING AGGLUTINATION OF CELLS OF THE DIFFERENT BLOOD GROUPS WITH ANTI-A AND ANTI-B AGGLUTININS

Red Blood Cells	Sera	
	Anti-A	Anti-B
O	−	−
A	+	−
B	−	+
AB	+	+

of blood is removed from the person to be typed. This is then diluted approximately 50 times with saline so that clotting will not occur. Two separate drops of this suspension are placed on a microscope slide, and a drop of anti-A agglutinin serum is mixed with one of the drops of cell suspension while a drop of anti-B agglutinin serum is mixed with the second drop. After several minutes have been allowed for the agglutination process to take place, the slide is observed under a microscope to determine whether or not the cells have clumped. If they have clumped, one knows that an immune reaction has resulted between the serum and the cells.

Table 6–2 illustrates the reactions that occur with each of the four different types of blood. Group O red blood cells have no agglutinogens and therefore do not react with either the anti-A or the anti-B serum. Group A blood has A agglutinogens and therefore agglutinates with anti-A agglutinins. Group B blood has B agglutinogens and agglutinates with the anti-B serum. Group AB blood has both A and B agglutinogens and agglutinates with both types of serum.

THE Rh BLOOD TYPES

In addition to the O-A-B blood group system, several other systems are sometimes important in the transfusion of blood. The most important of these is the Rh system. The one major difference between the O-A-B system and the Rh system is the following: In the O-A-B system, the agglutinins responsible for causing transfusion reactions develop spontaneously, while in the Rh system spontaneous agglutinins almost never occur. Instead, the person must first be massively exposed to an Rh antigen, usually by transfusion of blood, before enough agglutinins to cause a significant transfusion reaction are developed.

The Rh Antigens — "Rh Positive" and "Rh Negative" Persons. There are six common types of Rh antigens, each of which is called an Rh factor, but only three of these — known as C, D, and E, Rh antigens — are usually antigenic

enough to cause significant development of anti-Rh antibodies that are capable of causing transfusion reactions. Therefore, anyone who has any one of these three antigens, or any combination of them, is said to be *Rh positive*. A person who has no C, D, or E antigens is said to be Rh negative.

Approximately 85 per cent of all Caucasoids are Rh positive and 15 per cent, Rh negative. In American blacks, the percentage of Rh positives is about 95.

Typing Bloods for Rh Factors. Typing for Rh factors is performed in a manner similar to that used in typing the A-B-O agglutinogens. This is usually achieved using from four to six different anti-Rh sera. However, the anti-Rh antibodies are far less potent in their capability of causing red cell agglutination than are the anti-A and the anti-B antibodies. Therefore, to cause agglutination in the presence of the Rh antibodies, a small amount of protein must be added to the reactant mixture; this protein provides cross-linkages between the antibodies after they attach to the red blood cells.

The Rh Immune Response

Formation of Anti-Rh Agglutinins. When red blood cells containing one or more Rh positive factors (C, D, or E), or even protein breakdown products of such cells, are injected into an Rh negative person, anti-Rh agglutinins develop very slowly, the maximum concentration of agglutinins occurring approximately two to four months later. On repeated exposure to the Rh factor, the Rh negative person eventually becomes strongly "sensitized" to the Rh factor — that is, he or she develops a very high titer of anti-Rh agglutinins. On subsequent transfusion of Rh positive blood into the same person, who is now immunized against the Rh factor, a transfusion reaction can occur and can be as severe as the reactions that occur with types A and B bloods.

Erythroblastosis Fetalis. Erythroblastosis fetalis is a disease of the fetus and newborn infant characterized by progressive agglutination and subsequent phagocytosis of the red blood cells. In most instances of erythroblastosis fetalis the mother is Rh negative, and the father is Rh positive; the baby has inherited the Rh positive characteristic from the father, and the mother has developed anti-Rh agglutinins.

Effect of the Mother's Antibodies on the Fetus. After anti-Rh antibodies have formed in the mother, they diffuse very slowly through the placental membrane and cause slow agglutination of the fetus's blood. The agglutinated red blood cells gradually hemolyze, releasing hemoglobin into the blood. The reticuloendothelial cells then convert the hemoglobin into bilirubin, which causes yellowness (jaundice) of the skin. The antibodies probably also attack and damage many of the other cells of the body.

Clinical Picture of Erythroblastosis. The newborn, jaundiced, erythroblastotic baby is usually anemic at birth, and the anti-Rh agglutinins from the mother usually circulate in the baby's blood for one to two months after birth, destroying more and more red blood cells. Therefore, the hemoglobin level of the untreated erythroblastotic baby often falls for approximately the first 45 days after birth, and, if the level falls below 6 to 8 grams per cent, the baby usually dies.

The hemopoietic tissues of the baby attempt to replace the hemolyzing red blood cells. Even the liver and the spleen become greatly enlarged and produce red blood cells in the same manner that they normally do only during the middle of gestation. Because of the very rapid production of cells, many early forms, including many nucleated blastic forms, are emptied into the circulatory system, and it is because of the presence of these in the blood that the disease has been called "erythroblastosis fetalis."

Though the severe anemia of erythroblastosis fetalis is usually the cause of death, many children who barely survive the anemia exhibit permanent mental impairment or damage to motor areas of the brain because of precipitation of bilirubin in the neuronal cells, causing their destruction, a condition called *kernicterus.*

Treatment of the Erythroblastotic Baby. The usual treatment for erythroblastosis fetalis is to replace the newborn infant's blood one or more times with Rh negative blood. Approximately 400 ml. of Rh negative blood is infused over a period of several hours while the baby's own Rh positive blood is being removed. By the time the Rh negative cells are replaced with the baby's own Rh positive cells, a process that requires six or more weeks, the anti-Rh agglutinins that had come from the mother will have been destroyed.

OTHER BLOOD FACTORS

Many other antigenic proteins besides the O, A, B, and Rh factors are present in red blood cells of different persons, but these only rarely cause transfusion reactions and therefore are mainly of academic and legal importance. Some of these different blood factors are the M, N, O, S, s, P, Kell, Lewis, Duffy, Kidd, Diego, and Lutheran factors.

Blood Typing in Legal Medicine. In the past three decades the use of blood typing has become an important legal procedure in cases of disputed parentage. Including all the blood types, as many as 50 common blood group genes can be deter-

mined for each person by blood-typing procedures. After the mother's and child's genes have been determined, many of the father's genes, and his corresponding blood factors, are then known immediately, because any gene present in a child but not present in the mother must be present in the father. If a suspected man is missing any one of the necessary blood factors, he is not the father. A falsely charged man can be cleared in about 75 per cent of the cases using the usually available antisera for blood typing, and in essentially all cases using all possible types of antisera.

TRANSFUSION

Indications for Transfusion. The most important reason for transfusion is decreased blood volume in persons with circulatory shock. Transfusions are often used as well for treating anemia or to supply the recipient with some other constituent of whole blood besides red blood cells, such as to supply a thrombocytopenic patient with new platelets. Also, hemophilic patients can be rendered temporarily nonhemophilic by plasma transfusion.

Transfusion Reactions Resulting From Mismatched Blood Groups

If blood of one blood group is transfused to a recipient of another blood group, a transfusion reaction is likely to occur in which the red blood cells *of the donor blood* are agglutinated. It is very rare that the transfused blood ever causes agglutination *of the recipient's cells,* for the following reason: The plasma portion of the donor blood immediately becomes diluted by all the plasma of the recipient, thereby decreasing the titer of the infused agglutinins to a level too low to cause agglutination. On the other hand, the infused blood does not dilute the agglutinins in the recipient's plasma to a major extent. Therefore, the recipient's agglutinins can still agglutinate the donor cells.

Hemolysis of Red Cells Following Transfusion Reactions. All transfusion reactions resulting from mismatched blood groups eventually cause hemolysis of the red blood cells. Occasionally, the antibodies are potent enough and are composed of the appropriate class of immunoglobulins to cause immediate hemolysis, but more frequently the cells agglutinate first and then are mainly entrapped in the peripheral vessels. Over a period of hours to days the entrapped cells are phagocytized, and hemoglobin is thereby liberated into the circulatory system. A small quantity of hemoglobin can become attached to one of the plasma proteins, *haptoglobin*, and continue to circulate in the blood without causing any harm. However, above the threshold value of 100 mg. of hemoglobin per 100 ml. of plasma, the excess hemoglobin remains in the free form and diffuses out of the circulation into the tissue spaces or through the renal glomeruli into the kidney tubules, as is discussed below. The hemoglobin remaining in the circulation or passing into the tissue space is gradually ingested by phagocytic cells and converted into bilirubin, which will be discussed in Chapter 43. The concentration of bilirubin in the body fluids sometimes rises high enough to cause *jaundice* — that is, the person's tissues become tinged with yellow pigment. But, if liver function is normal, jaundice usually does not appear unless more than 300 to 500 ml. of blood is hemolyzed in less than a day.

Acute Kidney Shutdown Following Transfusion Reactions. One of the most lethal effects of transfusion reactions is acute kidney shutdown, which can begin within a few minutes to a few hours and sometimes continue until the person dies of renal failure.

The kidney shutdown seems to result from three different causes. First, the antigen-antibody reaction of the transfusion reaction releases toxic substances from the hemolyzing blood that cause powerful renal vasoconstriction. Second, the loss of circulating red cells along with production of toxic substances from the cells and from the immune reaction often causes circulatory shock; the arterial blood pressure falls very low, and the renal blood flow and urinary output decrease. Third, if the total amount of free hemoglobin in the circulating blood is greater than the quantity which can bind with haptoglobin, much of the excess leaks through the glomerular membranes into the kidney tubules. Then, as water is reabsorbed from the tubules, the tubular hemoglobin concentration rises so high that it precipitates and blocks many of the tubules; this is especially true if the urine is acidic. Thus, renal vasoconstriction, circulatory shock, and tubular blockage all add together to cause acute renal shutdown. If the shutdown is complete, the patient dies within a week to 12 days unless treated with the artificial kidney. If he is treated, the kidneys will usually return to full function in a few weeks with no permanent consequences.

TRANSPLANTATION OF TISSUES AND ORGANS

Relation of Genotypes to Transplantation. In this modern age of surgery, many attempts are being made to transplant tissues and organs from one person to another, or, occasionally, from lower animals to the human being. Many of the different antigenic proteins of red blood cells that cause transfusion reactions, plus still many more, are present in the other cells of the body as well. Consequently, any foreign cell transplanted into a recipient can cause immune responses and immune reactions. In other words, most recipients are just as able to resist invasion by foreign cells as to resist invasion by foreign bacteria.

Isografts, Allografts, and Xenografts. A transplant of a tissue or whole organ from one identical twin to another is called an *isograft*. A transplant from one human being to another or from any animal to another animal of the same species is called an *allograft*. Finally, a transplant from a lower animal to a human being or from an animal of one species to one of another species is called a *xenograft*.

In the case of isografts, cells in the transplant will almost always live indefinitely if an adequate blood supply is provided, but in the case of allografts and xenografts, immune reactions almost always occur, causing death of all the cells in the graft three to ten weeks after transplantation unless some specific therapy is used to prevent the immune reaction. However, when the tissues are properly "typed" and are very similar prior to transplant, completely successful allografts occasionally result. The greater the difference in antigenic structure, the more rapid and the more severe are the immune reactions to the graft.

Some of the different tissues and organs that have been transplanted either experimentally or for temporary benefit from one person to another are skin, cornea, kidney, heart, liver, glandular tissue, bone marrow, and lung. Many kidney allografts have been successful for as long as five to ten years, a rare liver or heart transplant for one to five years, and lung transplants for one month.

Attempts to Overcome the Antigen-Antibody Reactions in Transplanted Tissue

Because of the extreme potential importance of transplanting certain tissues and organs, such as skin, kidneys, and lungs, serious attempts have been made to prevent the antigen-antibody reactions associated with transplants. The following specific procedures have met with certain degrees of clinical or experimental success.

Tissue Typing. In the same way that red blood cells can be typed to prevent reactions between recipient and donor, so also is it possible to "type" tissues to help prevent graft rejection, though thus far this procedure has met with far less success than has been achieved in red blood cell typing. The most important antigens that cause graft rejection are a group of antigens called the HLA antigens. These are a group of 50 or more different antigens in the tissue cell membranes. The best success has been tissue-type matches between members of the same family. Of course, the match in identical twins is exact.

Glucocorticoid Therapy (Cortisone, Hydrocortisone, and ACTH). The glucocorticoid hormones from the adrenal gland greatly suppress the formation of both antibodies and immunologically competent lymphocytes. Therefore, administration of large quantities of these, or of ACTH which causes the adrenal gland to produce glucocorticoids, helps tremendously in preventing transplant rejection and is a mainstay of many treatment programs.

Suppression of Antibody Formation. Occasionally, a person has naturally suppressed antibody formation resulting from (a) congenital agammaglobulinemia, in which case gamma globulins are not produced, and (b) destructive diseases of the lymphocytic system. Transplants of allografts into such individuals are occasionally successful, or at least their destruction is delayed. Also, irradiative destruction of most of the lymphoid tissue by either x-rays or gamma rays renders a person much more receptive than usual to an allograft. And treatment with certain drugs, such as azathioprine (Imuran), which suppresses antibody formation, also increases the likelihood of success; indeed, this, along with glucocorticoids, is the basis for most immunosuppressive therapy. Unfortunately, all of these procedures also leave the person relatively unprotected from disease.

Use of Antilymphocyte Serum. It was pointed out earlier in the chapter that grafted tissues are usually destroyed by lymphocytes that become sensitized against the graft. These lymphocytes invade the graft and then cause the cells of the graft to swell, their membranes to become very permeable, and finally their cell membranes to rupture. Simultaneously, macrophages move in to clean up the debris. Within a few days to a few weeks after this process begins, the tissue often is completely destroyed even though the graft had been completely viable and functioning normally only a short time earlier.

Therefore, an effective procedure for preventing rejection of grafted tissues has been to inoculate the recipient with antilymphocyte serum. This serum is made in animals by injecting human lymphocytes into them; the antibodies that develop in these animals will then attack human lymphocytes. When this serum is injected into the transplanted recipient, the number of circulating small lymphocytes can be decreased to as little as 5 to 10 per cent of normal, and there is a resulting decrease in the intensity of the graft rejection reaction. Unfortunately, the procedure does not continue to work well after the first few injections of the antiserum because the recipient soon begins to build up antibodies against the animal antiserum itself.

Summary. To summarize, transplantation of living tissues in human beings up to the present has been mainly an experiment except in the case of kidney transplants, which are now successful in more than 50 per cent of cases, but only when massive immunosuppressive treatment is used simultaneously. However, when someone succeeds in blocking the immune response of the recipient to a donor organ without at the same time destroying the recipient's specific immunity for disease, this story will change overnight.

REFERENCES

Immunity and Allergy

Amos, D. B., *et al.* (eds.): Immune Mechanisms and Disease. New York, Academic Press, 1979.

Baram, P., *et al.* (eds.): Immunologic Tolerance and Macrophage Function. New York, Elsevier/North-Holland, 1979.

Bellanti, J. A.: Immunology II. Philadelphia, W. B. Saunders Co., 1978.

Benacerraf, B., and Unanue, E. R.: Textbook of Immunology. Baltimore, Williams & Wilkins, 1979.

Burke, D. C.: The status of interferon. *Sci. Am.*, *236*(4):42, 1977.

Capra, J. D., and Edmundson, A. B.: The antibody combining site. *Sci. Am.*, *236*(1):50, 1977.

Cohen, A. S. (ed.): Rheumatology and Immunology. New York, Grune & Stratton, 1979.

Dick, G. (ed.): Immunological Aspects of Infectious Diseases. Baltimore, University Park Press, 1978.

Fudenberg, H. H., and Smith, C. L. (eds.): The Lymphocyte in Health and Disease. New York, Grune & Stratton, 1979.

Ham, A. W., *et al.*: Blood Cell Formation and the Cellular Basis of Immune Responses. Philadelphia, J. B. Lippincott Co., 1979.

Johnson, F. (ed.): Allergy, Including IgE in Diagnosis and Treatment. Chicago, Year Book Medical Publishers, 1979.

Kaplan, J. G. (ed.): The Molecular Basis of Immune Cell Function. New York, Elsevier/North-Holland, 1979.

Koffler, D.: The Immunology of Rheumatoid Diseases. Summit, N.J., CIBA Pharmaceutical Company, 1979.

Marchesi, V. T., *et al.* (eds.): Cell Surface Carbohydrates and Biological Recognition. New York, A. R. Liss, 1978.

Metzger, H.: Early molecular events in antigen-antibody cell activation. *Annu. Rev. Pharmacol. Toxicol.*, *19*:427, 1979.

Mygind, N.: Nasal Allergy. Oxford, Blackwell Scientific Publications, 1978.

Nahmias, A. J., and O'Reilly, R. (eds.): Immunology of Human Infection. New York, Plenum Press, 1979.

Pernis, B., and Vogel, H. J. (eds.): Cells of Immunoglobulin Synthesis. New York, Academic Press, 1979.

Raff, M. C.: Cell-surface immunology. *Sci. Am.*, *234*(5):30, 1976.

Samter, M. (ed.): Immunological Diseases. Boston, Little, Brown, 1978.

Schiff, G. M.: Active immunization for adults. *Annu. Rev. Med.*, *31*:411, 1980.

Schwartz, L. M.: Compendium of Immunology. New York, Van Nostrand Reinhold, 1979.

Sercarz, E. E., and Cunningham, A. J. (eds.): Strategies of Immune Regulation. New York, Academic Press, 1979.

Stuart, F. P., and Fitch, F. W. (eds.): Immunological Tolerance and Enhancement. Baltimore, University Park Press, 1979.

Terry, W. D., and Yamamura, Y. (eds.): Immunobiology and Immunotherapy of Cancer. New York, Elsevier/North-Holland, 1979.

Voller, A., and Friedman, H. (eds.): New Trends and Developments in Vaccines. Baltimore, University Park Press, 1978.

Blood Groups, Transfusions, and Transplantation

Ballantyne, D. L., and Converse, J. M.: Experimental Skin Grafts and Transplantation Immunity: A Recapitulation. New York, Springer-Verlag, 1979.

Carpenter, C. B., and Miller, W. V. (eds.): Clinical Histocompatibility Testing. New York, Grune & Stratton, 1977.

Chatterjee, S. N. (ed.): Symposium on Organ Transplantation. Philadelphia, W. B. Saunders Co., 1978.

Chatterjee, S. N. (ed.): Renal Transplantation: A Multidisciplinary Approach. New York, Raven Press, 1980.

Cunningham, B. A.: The structure and function of histocompatibility antigens. *Sci. Am.*, *234*(4):96, 1977.

Dick, H. F., and Kissmeyer-Nielsen, F. (eds.): Histocompatibility Techniques. New York, Elsevier/North-Holland, 1979.

Festenstein, H., and Démant, P.: HLA and H-2: Basic Immunogenetics, Biology, and Clinical Relevance. London, Edward Arnold, 1978.

Hubbell, R. C. (ed.): Advances in Blood Transfusion. Arlington, Va., American Blood Commission, 1979.

Mohn, J. F., *et al.* (eds.): Human Blood Groups. New York, S. Karger, 1977.

Mollison, P. L.: Blood Transfusion in Clinical Medicine, 5th Ed. Philadelphia, J. B. Lippincott Co., 1972.

Nickander, R., *et al.*: Nonsteroidal anti-inflammatory agents. *Annu. Rev. Pharmacol. Toxicol.*, *19*:469, 1979.

Rosenfield, R. E., *et al.*: Genetic model for the Rh blood-group system. *Proc. Natl. Acad. Sci. USA*, *70*:1303, 1973.

Selwood, N., and Hedges, A.: Transplantation Antigens: A Study in Serological Data Analysis. New York, John Wiley & Sons, 1978.

Stiller, C. R., *et al.* (eds.): Immunologic Monitoring of the Transplant Patient. New York, Grune & Stratton, 1978.

Touraine, J. L., *et al.* (eds.): Transplantation and Clinical Immunology. New York, Elsevier/North-Holland, 1980.

Unanue, E. R.: Cellular events following binding of antigen to lymphocytes. *Am. J. Pathol.*, *77*:2, 1974.

7

Hemostasis and Blood Coagulation

EVENTS IN HEMOSTASIS

The term hemostasis means prevention of blood loss. Whenever a vessel is severed or ruptured, hemostasis is achieved by several different mechanisms, including (1) vascular spasm, (2) formation of a platelet plug, (3) blood coagulation, and (4) growth of fibrous tissue into the blood clot to close the hole in the vessel permanently.

VASCULAR SPASM

Immediately after a blood vessel is cut or ruptured, the wall of the vessel contracts; this instantaneously reduces the flow of blood from the vessel rupture. The contraction results from both nervous reflexes and local myogenic spasm. The nervous reflexes presumably are initiated by impulses originating in the traumatized vessel or in nearby tissues. However, most of the spasm probably results from local myogenic contraction of the blood vessels initiated by direct damage to the vascular wall, which presumably causes transmission of action potentials along the vessel wall for several centimeters and results in constriction of the vessel. The more of the vessel that is traumatized, the greater is the degree of spasm; this means that a sharply cut blood vessel usually bleeds much more than does a vessel ruptured by crushing. This local vascular spasm lasts for as long as 20 to 30 minutes, during which time the ensuing processes of platelet plugging and blood coagulation can take place.

FORMATION OF THE PLATELET PLUG

The second event in hemostasis is an attempt by the platelets to plug the rent in the vessel. To understand this, it is important that we first understand the nature of platelets themselves.

Platelets are minute round or oval discs about 2 microns in diameter. They are fragments of *megakaryocytes,* which are extremely large cells of the hemopoietic series formed in the bone marrow. The megakaryocytes disintegrate into platelets while they are still in the bone marrow and release the platelets into the blood. The normal concentration of platelets in the blood is between 200,000 and 400,000 per cubic millimeter.

Mechanism of the Platelet Plug. Platelet repair of vascular openings is based on several important functions of the platelet itself: When platelets come in contact with a damaged vascular surface, such as the collagen fibers in the vascular wall or even damaged endothelial cells, they immediately change their characteristics drastically. They begin to swell, they assume irregular forms with numerous irradiating processes protruding from their surfaces, they become sticky so that they stick to the collagen fibers, and they secrete large quantities of *ADP* and enzymes that cause formation of *thromboxane A* in the plasma. The ADP and thromboxane A, in turn, act on nearby platelets to activate them as well, and the stickiness of these additional platelets causes them to adhere to the originally activated platelets. Therefore, a circular process of activation of successively increasing numbers of platelets occurs; these accumulate to form a *platelet plug.*

If the rent in a vessel is small, the platelet plug by itself can stop blood loss completely, but if there is a large hole, a blood clot in addition to the platelet plug is required to stop the bleeding.

The platelet plugging mechanism is extremely important to close the minute ruptures in very small blood vessels that occur hundreds of times daily, including those through the endothelial cells themselves. A person who has very few platelets develops literally hundreds of small hemorrhagic areas under his skin and throughout his internal tissues, but this does not occur in the normal person.

BLOOD COAGULATION IN THE RUPTURED VESSEL

The third mechanism for hemostasis is formation of a blood clot. The clot begins to develop in 15 to 20 seconds if the trauma of the vascular wall has been severe and in one to two minutes if the trauma has been minor. Activator substances both from the traumatized vascular wall and from platelets and blood proteins adhering to the traumatized vascular wall initiate the clotting process. The physical events of this process are illustrated in Figure 7–1, and the chemical events will be discussed in detail later in the chapter.

Within 3 to 6 minutes after rupture of a vessel, the entire cut or broken end of the vessel is filled with clot. After 30 minutes to an hour, the clot retracts; this closes the vessel still further. Platelets play an important role in this clot retraction, as will also be discussed later in the chapter.

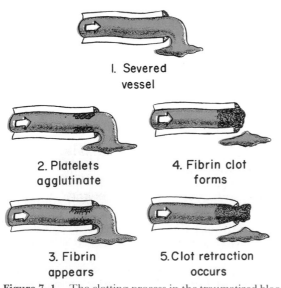

I. Severed vessel

2. Platelets agglutinate

3. Fibrin appears

4. Fibrin clot forms

5. Clot retraction occurs

Figure 7–1. The clotting process in the traumatized blood vessel.

FIBROUS ORGANIZATION OR DISSOLUTION OF THE BLOOD CLOT

Once a blood clot has formed, it can follow two separate courses: It can become invaded by fibroblasts, which subsequently form connective tissue all through the clot; or it can dissolve. The usual course for a clot that forms in a small hole of a vessel wall is invasion by fibroblasts, beginning within a few hours after the clot is formed and continuing to complete organization of the clot into fibrous tissue within approximately seven to ten days. On the other hand, when a large amount of blood coagulates to form one large blood clot, such as blood that has leaked into tissues, special substances within the clot itself become activated, and these then function as enzymes to dissolve the clot itself, as will be discussed later in the chapter.

MECHANISM OF BLOOD COAGULATION

General Mechanism. Almost all research workers in the field of blood coagulation agree that clotting takes place in three essential steps:

First, a substance or complex of substances called *prothrombin activator* is formed in response to rupture of the vessel or damage to the blood itself.

Second, the prothrombin activator catalyzes the conversion of prothrombin into *thrombin*.

Third, the thrombin acts as an enzyme to convert fibrinogen into *fibrin threads* that enmesh platelets, blood cells, and plasma to form the clot itself.

Let us first discuss the conversion of prothrombin to thrombin; then later we will come back to the initiating stages in the clotting process by which prothrombin activator is formed.

CONVERSION OF PROTHROMBIN TO THROMBIN

After prothrombin activator has been formed as a result of rupture of the blood vessel or as a result of damage to special activator substances in the blood itself, the prothrombin activator then causes conversion of prothrombin to thrombin, which in turn causes polymerization of fibrinogen molecules into fibrin threads within another 10 to 15 seconds. Thus the rate-limiting factor in

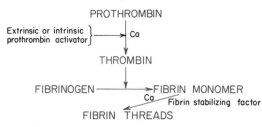

Figure 7–2. Schema for conversion of prothrombin to thrombin, and polymerization of fibrinogen to form fibrin threads.

causing blood coagulation is usually the formation of prothrombin activator and not the subsequent reactions beyond that point.

Prothrombin and Thrombin. Prothrombin is a plasma protein having a molecular weight of 68,700. It is present in normal plasma in a concentration of about 15 mg. per 100 ml. It is an unstable protein that is split, in the presence of prothrombin activator and calcium ions, into smaller compounds, one of which is *thrombin,* which has a molecular weight of 33,700, almost exactly half that of prothrombin. This reaction is shown in Figure 7–2.

Prothrombin is formed continually by the liver, and it is continually being used throughout the body for blood clotting. If the liver fails to produce prothrombin, its concentration in the plasma falls too low within 24 hours to provide normal blood coagulation. Vitamin K is required by the liver for normal formation of prothrombin; therefore, either lack of vitamin K or the presence of liver disease that prevents normal prothrombin formation can often decrease the prothrombin level so low that a bleeding tendency results.

CONVERSION OF FIBRINOGEN TO FIBRIN — FORMATION OF THE CLOT

Fibrinogen. Fibrinogen is a high molecular weight protein (340,000) occurring in the plasma in quantities of 100 to 700 mg. per 100 ml. It is formed in the liver, and liver disease occasionally decreases the concentration of circulating fibrinogen, as it does the concentration of prothrombin, which was pointed out previously.

Because of its large molecular size, very little fibrinogen normally leaks into the interstitial fluids, and, since it is one of the essential factors in the coagulation process, interstitial fluids ordinarily coagulate poorly if at all. Yet, when the permeability of the capillaries becomes pathologically increased, fibrinogen does then appear in the tissue fluids in sufficient quantities to allow clotting in much the same way that plasma and whole blood clot.

Action of Thrombin on Fibrinogen to Form Fibrin. Thrombin is a protein *enzyme* with proteolytic capabilities. It acts on fibrinogen to remove two low molecular weight peptides from each molecule of fibrinogen, forming a molecule of *fibrin monomer,* which has the automatic capability of polymerizing with other fibrin monomer molecules. Therefore, many fibrin monomer molecules polymerize within seconds into *long fibrin threads* that form the *reticulum* of the clot.

In the early stages of this polymerization, the fibrin threads are not cross-linked with each other, and the resultant clot is weak and can be broken apart with ease. However, still another process occurs during the following few minutes that greatly strengthens the fibrin reticulum. This involves a substance called *fibrin-stabilizing factor,* which is normally present in small amounts in the plasma globulins but which is also released from platelets entrapped in the clot. This factor operates as an enzyme to cause covalent cross-linking bonds between the adjacent fibrin threads, thus adding tremendously to the three-dimensional strength of the fibrin meshwork.

The Blood Clot. The clot is composed of a meshwork of fibrin threads running in all directions and entrapping blood cells, platelets, and plasma. The fibrin threads adhere to damaged surfaces of blood vessels; therefore, the blood clot becomes adherent to any vascular opening and thereby prevents blood loss.

Clot Retraction — Serum. Within a few minutes after a clot is formed, it begins to contract and usually expresses most of the fluid from the clot within 30 to 60 minutes. The fluid expressed is called *serum,* because all of its fibrinogen and most of the other clotting factors have been removed; in this way, serum differs from plasma. Serum obviously cannot clot because of lack of these factors.

Platelets are necessary for clot retraction to occur. Therefore, failure of clot retraction is an indication that the number of platelets in the circulating blood is low. Electron micrographs of platelets in blood clots show that they become attached to the fibrin threads in such a way that they actually bond different threads together. Furthermore, platelets entrapped in the clot continue to release procoagulant substances, one of which is fibrin stabilizing factor that causes more and more cross-linking bondage between the adjacent fibrin threads.

As the clot retracts, the edges of the broken blood vessel are pulled together, thus contributing to the ultimate state of hemostasis.

THE CIRCULAR PROCESS OF CLOT FORMATION

Once a blood clot has started to develop, it normally extends within minutes into the surrounding blood. That is, the clot itself initiates a circular process to promote more clotting. One of the most important causes of this is the fact that the proteolytic action of thrombin allows it to act on many of the other blood clotting factors in addition to fibrinogen. For instance, thrombin has a direct proteolytic effect on prothrombin itself, tending to split it into still more thrombin. Therefore, once a critical amount of thrombin has been formed, a circular process develops that causes still more blood clotting and more thrombin to be formed; thus, the blood clot continues to grow until something stops its growth.

Fortunately, the circular process of continued clot formation stops wherever blood is flowing, because flowing blood carries the thrombin and the other procoagulants released during the clotting process away so rapidly that their concentrations cannot rise high enough to promote further clotting.

INITIATION OF COAGULATION: FORMATION OF PROTHROMBIN ACTIVATOR

Now that we have discussed the clotting process initiated by the formation of thrombin from prothrombin, we must turn to the more complex mechanisms that cause formation of prothrombin activator.

There are two basic ways in which prothrombin activator can be formed: (1) by the *extrinsic pathway* that begins with trauma to the vascular wall or to the tissues outside the blood vessels, and (2) by the *intrinsic pathway* that begins in the blood itself.

In both the extrinsic and intrinsic pathways a series of different plasma proteins, especially beta-globulins, play major roles. These are called *blood clotting factors,* and for the most part they are inactive forms of proteolytic enzymes. When converted to the active forms, their enzymatic actions cause the successive reactions of the clotting process.

Most of the clotting factors are designated by Roman numerals, as listed in Table 7–1. In the sections dealing with intrinsic and extrinsic pathways we will specifically discuss blood clotting factor V and factors VII through XII.

The Extrinsic Mechanism for Initiating Clotting

The extrinsic mechanism for initiating the formation of prothrombin activator begins with blood coming in contact with traumatized vascular wall or extravascular tissues and occurs according to the following three basic steps, as illustrated in Figure 7–3.

1. Release of Tissue Factor and Tissue Phospholipids. The traumatized tissue releases two factors that set the clotting process into motion. These are (a) *tissue factor,* which is a proteolytic enzyme, and (b) *tissue phospholipids,* which are mainly phospholipids of the tissue cell membranes.

TABLE 7–1 CLOTTING FACTORS IN THE BLOOD AND THEIR SYNONYMS

Clotting Factor	Synonym
Fibrinogen	Factor I
Prothrombin	Factor II
Tissue thromboplastin	Factor III
Calcium	Factor IV
Factor V	Proaccelerin; labile factor; Ac-globulin; Ac-G
Factor VII	Serum prothrombin conversion accelerator; SPCA; convertin; stable factor
Factor VIII	Antihemophilic factor; AHF; antihemophilic globulin; AHG; antihemophilic factor A
Factor IX	Plasma thromboplastin component; PTC; Christmas factor; antihemophilic factor B
Factor X	Stuart factor; Stuart-Prower factor; antihemophilic factor C
Factor XI	Plasma thromboplastin antecedent; PTA; antihemophilic factor C
Factor XII	Hageman factor; antihemophilic factor D
Factor XIII	Fibrin stabilizing factor
Prothrombin activator	Thrombokinase; complete thromboplastin

EXTRINSIC PATHWAY

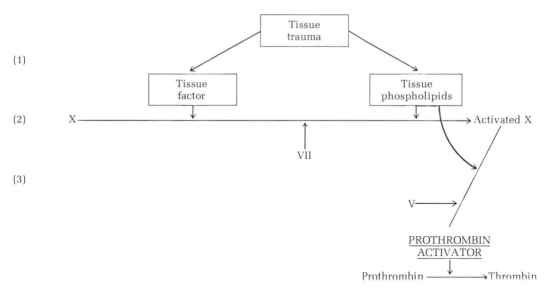

Figure 7–3. The extrinsic pathway for initiating blood clotting.

2. Activation of Factor X to Form Activated Factor X — Role of Factor VII and Tissue Factor. The tissue factor complexes with blood coagulation factor VII, and this complex, in the presence also of tissue phospholipids, acts enzymatically on factor X to form *activated factor X*.

3. Effect of Activated Factor X to Form Prothrombin Activator — Role of Factor V. The activated factor X complexes immediately with the tissue phospholipids released from the traumatized tissue and also with factor V to form the complex called *prothrombin activator*. Within a few seconds this splits prothrombin to form thrombin, and the clotting process proceeds as has already been explained.

The Intrinsic Mechanism for Initiating Clotting

The second mechanism for initiating the formation of prothrombin activator, and therefore for initiating clotting, begins with trauma to the blood itself and continues through the following series of cascading reactions, as illustrated in Figure 7–4.

1. Activation of Factor XII and Release of Platelet Phospholipids by Blood Trauma. Trauma to the blood alters two important clotting factors in the blood — factor XII and the platelets. When factor XII is disturbed, such as by coming into contact with collagen or with a wettable surface such as glass, it takes on a new configuration that converts it into a proteolytic enzyme called "activated factor XII."

Simultaneously, the blood trauma also damages the platelets, either because of adherence to collagen or to a wettable surface or by damage in other ways, and this releases platelet phospholipid, frequently called *platelet factor III*, which also plays a role in subsequent clotting reactions.

2. Activation of Factor XI. The activated factor XII acts enzymatically on factor XI to activate it as well, which is the second step in the intrinsic pathway.

3. Activation of Factor IX by Activated Factor XI. The activated factor XI then acts enzymatically on factor IX to activate this factor, also.

4. Activation of Factor X — Role of Factor VIII. The activated factor IX, *acting in concert with factor VIII* and with the platelet phospholipids from the traumatized platelets, activates factor X. It is clear that when either factor VIII or platelets are in short supply, this step is deficient. Factor VIII is the factor that is missing in the person who has classical *hemophilia*, for which reason it is called *antihemophilic factor*. Platelets are the clotting factor that is lacking in the bleeding disease called *thrombocytopenia*.

5. Action of Activated Factor X to Form Prothrombin Activator — Role of Factor V. This step in the intrinsic pathway is essentially the same as the last step in the extrinsic pathway. That is, activated factor X combines with factor V and platelet phospholipids to form the complex called *prothrombin activator*. The only difference is

INTRINSIC PATHWAY

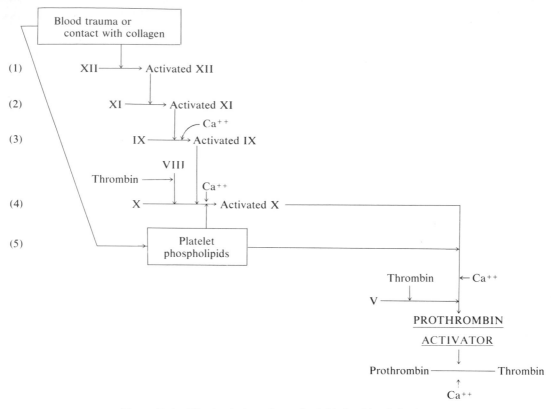

Figure 7–4. The intrinsic pathway for initiating blood clotting.

that the phospholipids in this instance come from the traumatized platelets rather than from traumatized tissues. The prothrombin activator in turn initiates within seconds the cleavage of prothrombin to form thrombin, thereby setting into motion the final clotting process, as described earlier.

Role of Calcium Ions in the Intrinsic and Extrinsic Pathways

Except for the first two steps in the intrinsic pathway, calcium ions are required for promotion of all of the reactions. Therefore, in the absence of calcium ions, blood clotting will not occur. Fortunately, in the living body the calcium ion concentration never falls low enough to affect significantly the kinetics of blood clotting.

On the other hand, when blood is removed from a person, it can be prevented from clotting by reducing the calcium ion concentration below the threshold level for clotting, either by deionizing the calcium by reacting it with substances such as *citrate ion* or by precipitating the calcium with substances such as the *oxalate ion*.

Summary of Blood Clotting Initiation

It is clear from the above schemas of the intrinsic and extrinsic systems for initiating blood clotting that clotting is initiated after rupture of blood vessels by both of the pathways. The tissue factor and tissue phospholipids initiate the extrinsic pathway, while contact of factor XII and the platelets with collagen in the vascular wall initiates the intrinsic pathway.

In contrast, when blood is removed from the body and held in a test tube, it is the intrinsic pathway alone that must elicit the clotting. This usually results from contact of factor XII and platelets with the wall of the tube, which activates both of them and initiates the intrinsic mechanism. If the surface of the container is very "nonwettable," as it would be if siliconized, blood clotting can sometimes be prevented for an hour or more.

Intravascular clotting sometimes results also from other factors that activate the intrinsic pathway. For instance, antigen-antibody reactions sometimes initiate the clotting process.

An especially important difference between the extrinsic and intrinsic pathways is that the extrinsic pathway is explosive; once initiated, its speed of occurrence is limited only by the amount of tissue factor and tissue phospholipids released from the traumatized tissues, and by the quantities of factors X, VII, and V in the blood. With severe tissue trauma, clotting can occur in as little as 15 seconds. On the other hand, the intrinsic pathway is much slower to proceed, usually requiring 2 to 6 minutes to cause clotting.

PREVENTION OF BLOOD CLOTTING IN THE NORMAL VASCULAR SYSTEM — THE INTRAVASCULAR ANTICOAGULANTS

Endothelial Surface Factors. Probably the two most important factors for preventing clotting in the normal vascular system are, first, the smoothness of the endothelium, which prevents contact activation of the intrinsic clotting system, and, second, a monomolecular layer of negatively charged protein, adsorbed to the inner surface of the endothelium, that repels the clotting factors and platelets, thereby preventing activation of clotting. When the endothelial wall is damaged, its smoothness and its negative electrical charge are both lost, which is believed to set off the intrinsic pathway of clotting.

The Antithrombin Action of Fibrin and of Antithrombin-Heparin Cofactor. Among the most important anticoagulants in the blood itself are those that remove thrombin from the blood. The two most powerful of these are the *fibrin threads* that are formed during the process of clotting and an alpha globulin called *antithrombin III* or also *antithrombin-heparin cofactor.*

While a clot is forming, approximately 85 to 90 per cent of the thrombin formed from the prothrombin becomes adsorbed to the fibrin threads as they develop. This obviously helps to prevent the spread of thrombin into the remaining blood.

The thrombin that does not adsorb to the fibrin threads soon combines with antithrombin-heparin cofactor, which blocks the effect of the thrombin on the fibrinogen and then inactivates the bound thrombin during the next 12 to 20 minutes.

Heparin. Small amounts of heparin, a powerful anticoagulant, are normally present in the blood. Heparin is a conjugated polysaccharide found in the cytoplasm of many types of cells, including even the cytoplasm of unicellular animals. Especially large quantities of heparin are formed by the basophilic *mast cells* located in the pericapillary connective tissue throughout the body, and the heparin then diffuses into the circulatory system. The *basophil cells* of the blood, which seem to be functionally almost identical with the mast cells, also release some heparin into the plasma.

Mast cells are extremely abundant in the tissue surrounding the capillaries of the lungs and to a lesser extent the capillaries of the liver. It is easy to understand why large quantities of heparin might be needed in these areas, for the capillaries of the lungs and liver receive many embolic clots formed in the slowly flowing venous blood; sufficient formation of heparin can prevent further growth of the clots.

Mechanism of Heparin Action. Heparin prevents blood coagulation almost entirely by combining with antithrombin-heparin cofactor, which makes this factor combine with thrombin 1000 times as rapidly as normally. Therefore, in the presence of an excess of heparin, the removal of thrombin from the circulating blood is almost instantaneous. This complex of heparin and antithrombin-heparin cofactor also reacts in a similar way with several other activated coagulation factors of both the intrinsic and extrinsic pathways, thus inactivating their proteolytic (and blood clotting) functions as well.

LYSIS OF BLOOD CLOTS — PLASMIN

The plasma proteins contain a euglobulin called *plasminogen* or *profibrinolysin*, which when activated becomes a substance called *plasmin* or *fibrinolysin*. Plasmin is a proteolytic enzyme that digests the fibrin threads and other substances in the surrounding blood, such as fibrinogen, factor V, factor VIII, prothrombin, and factor XII. Therefore, whenever plasmin is formed in a blood clot, it can cause lysis of the clot.

When a clot is formed, a large amount of plasminogen is incorporated in the clot along with other plasma proteins. However, this will not become plasmin and will not cause lysis of the clot until it is activated. Fortunately, the tissues and blood contain substances that can activate plasminogen to plasmin, including (1) thrombin, (2) activated factor XII, (3) lysosomal enzymes from damaged tissues, and (4) factors from the vascular endothelium. Within a day or two after blood has leaked into a tissue and clotted, these activators cause the formation of enough plasmin that it in turn dissolves the clot.

Significance of the Fibrinolysin System. The lysis of blood clots allows slow clearing (over a period of several days) of extraneous blood from the tissues and sometimes even allows reopening of clotted vessels. Unfortunately, reopening of large vessels occurs only rarely. But an important function of the fibrinolysin system is to remove very minute clots from the millions of tiny peripheral vessels that eventually would all become occluded were there no way to cleanse them.

CONDITIONS THAT CAUSE EXCESSIVE BLEEDING IN HUMAN BEINGS

Excessive bleeding can result from deficiency of any one of the many different blood clotting factors. Three particular types of bleeding tendencies that have been studied to the greatest extent will be discussed: (1) bleeding caused by vitamin K deficiency, (2) hemophilia, and (3) thrombocytopenia (platelet deficiency).

Decreased Prothrombin, Factor VII, Factor IX, and Factor X Caused by Vitamin K Deficiency. Hepatitis, cirrhosis, acute yellow atrophy, and other diseases of the liver can all depress the formation of prothrombin and factors VII, IX, and X so greatly that the patient develops a severe tendency to bleed.

Another cause of depressed levels of these substances is vitamin K deficiency. Vitamin K is necessary for some of the intermediate stages in the formation of all of them.

Hemophilia. The term *hemophilia* is loosely applied to several different hereditary deficiencies of coagulation, all of which cause bleeding tendencies hardly distinguishable from one another. By far the most common cause of hemophilia is deficiency of factor VIII.

Many persons with hemophilia die in early life, though many others with less severe bleeding have a normal life span. Very commonly, the person's joints become severely damaged because of repeated joint hemorrhage following exercise or trauma.

Transfusion of normal fresh plasma or of the appropriate purified protein clotting factor — factor VIII, for instance — into the hemophilic person usually relieves his bleeding tendency for a few days.

Thrombocytopenia. Thrombocytopenia means the presence of a very low quantity of platelets in the circulatory system. Persons with thrombocytopenia, like hemophiliacs, have a tendency to bleed, except that the bleeding is usually from many small capillaries rather than from larger vessels, as in hemophilia. As a result, small punctate hemorrhages occur throughout all the body tissues. The skin of such a person displays many small, purplish blotches, giving the disease the name *thrombocytopenic purpura*.

Most persons with thrombocytopenia have the disease known as *idiopathic thrombocytopenia*, which means simply "thrombocytopenia of unknown cause." However, in the past few years it has been discovered that in most of these persons specific antibodies are destroying the platelets. Usually these antibodies result from development of autoimmunity to the person's own platelets.

Relief from bleeding for one to four days can often be effected in the thrombocytopenic patient by giving *fresh whole blood transfusions* containing viable platelets.

THROMBOEMBOLIC CONDITIONS IN THE HUMAN BEING

Thrombi and Emboli. An abnormal clot that develops in a blood vessel is called a *thrombus*. Once a clot has developed, continued flow of blood past the clot is likely to break it away from its attachment, and such freely flowing clots are known as *emboli*. Emboli are especially likely to plug blood vessels of the lungs and the brain.

The causes of thromboembolic conditions in the human being are usually 2-fold: First, any *roughened endothelial surface of a vessel* — as may be caused by arteriosclerosis, infection, or trauma — is likely to initiate the clotting process. Second, blood often clots when it *flows very slowly* through blood vessels, for small quantities of thrombin and other procoagulants are always being formed. These are generally removed from the blood by the reticuloendothelial cells, mainly the Kupffer cells of the liver. If the blood is flowing too slowly, the concentrations of the procoagulants in local areas often rise high enough to initiate clotting, but when the blood flows rapidly they are rapidly mixed with large quantities of blood and are removed during passage through the liver.

Femoral Thrombosis and Massive Pulmonary Embolism. Because clotting almost always occurs when blood·flow is blocked for many hours in any vessel of the body, the immobility of bed patients plus the practice of propping the knees up with underlying pillows often causes intravascular clotting because of blood stasis in one or more of the leg veins for hours at a time. Then the clot grows, especially in the direction of the slowly moving blood, sometimes growing the entire length of the leg veins and occasionally even up into the common iliac vein and inferior vena cava. Then, about one time in ten, a large part of the clot disengages and flows freely with the venous blood into the right side of the heart and thence into the pulmonary arteries to cause *massive pulmonary embolism*. If the clot is large enough to occlude both the pulmonary arteries, immediate death ensues. If only one pulmonary artery or a smaller branch is blocked, death may not occur, or the embolism may lead to death a few hours to several days later because of further growth of the clot within the pulmonary vessels.

Disseminated Intravascular Clotting. Occasionally, the clotting mechanism becomes activated in widespread areas of the circulation, giving rise to the condition called *disseminated intravascular clotting*. Frequently, the clots are small but numerous, and they plug a large share of the small peripheral blood vessels. This effect occurs especially in septicemic shock, in which either circulating bacteria or bacterial toxins — especially *endotoxins* — activate the clotting mechanisms. The plugging of the small peripheral vessels greatly diminishes the delivery of oxygen and other nutrients to the tissues — a situation which exacerbates the shock. It is partly for this reason that full-blown septicemic shock is lethal in 85 per cent or more of the patients.

A peculiar effect of disseminated intravascular clotting is that the patient frequently begins to bleed. The reason for this is that so many of the clotting factors are removed by the widespread clotting that too few procoagulants remain to allow normal hemostasis of the remaining blood.

Anticoagulants for Clinical Use. In some thromboembolic conditions it is desirable to delay the coagulation process to a certain degree. Therefore, various anticoagulants have been developed for treatment of these conditions. The one most useful clinically is heparin.

PREVENTION OF BLOOD COAGULATION OUTSIDE THE BODY

Blood removed from the body can be prevented from clotting by:

1. Collecting the blood in *siliconized containers,* which prevent contact activation of factors XI and XII that initiates the intrinsic clotting mechanism.

2. Mixing *heparin* with the blood.

3. *Decreasing the calcium ions* in the blood. For instance, soluble *oxalate* compounds mixed in very small quantity with a sample of blood cause precipitation of calcium oxalate from the plasma and thereby block blood coagulation. Calcium deionizing agents also are used for preventing coagulation, including *sodium, ammonium, or potassium citrate,* or edetate (EDTA). These substances combine with calcium in the blood to cause unionized calcium compounds, and the lack of ionic calcium prevents coagulation. Citrate anticoagulants have a very important advantage oxalate anticoagulants, for oxalate is tox body, whereas small quantities of citrate injected intravenously. After injection, the citrate ion is removed from the blood within a few minutes by the liver, is polymerized into glucose, and is then metabolized in the usual manner. Consequently, 500 ml. of blood that has been rendered incoagulable by sodium citrate can ordinarily be injected into a recipient within a few minutes without any dire consequences. Therefore, citrate is the anticoagulant used in transfusion blood.

BLOOD COAGULATION TESTS

Bleeding time. When a sharp knife is used to pierce the tip of the finger or lobe of the ear, bleeding ordinarily lasts three to six minutes.

Clotting Time. A method widely used for determining clotting time is to collect blood in a chemically clean glass test tube and then to tip the tube back and forth approximately every 30 seconds until the blood has clotted. By this method, the normal clotting time ranges between five and eight minutes.

Prothrombin Time. The prothrombin time gives an indication of the total quantity of prothrombin in the blood. The method for determining prothrombin time is the following:

Blood removed from the patient is immediately oxalated so that none of the prothrombin can change into thrombin. At any time later, a large excess of calcium ions and tissue extract is suddenly mixed with the oxalated blood. The calcium nullifies the effect of the oxalate, and the tissue extract activates the prothrombin-to-thrombin reaction. The time required for coagulation to take place, known as the "prothrombin time," depends on the prothrombin concentration. The normal prothrombin time is approximately 12 seconds, and clotting is often severely impaired when this time increases to more than 24 seconds.

REFERENCES

Biggs, R. (ed.): The Treatment of Haemophilia A and B and Von Willebrand's Disease. Philadelphia, J. B. Lippincott Co., 1978.

Collen, D., *et al.* (eds.): The Physiological Inhibitors of Blood Coagulation and Fibrinolysis. New York, Elsevier/North-Holland, 1979.

Cooper, H. A., *et al.*: The platelet: Membrane and surface reactions. *Annu. Rev. Physiol.* 38:501, 1976.

Engelbert, H.: Heparin. New York, S. Karger, 1978.

Gowans, J. L. (in honour of): Blood Cells and Vessel Walls: Functional Interactions. Princeton, N.J., Excerpta Medica, 1980.

Hirsh, J., *et al.*: Concepts in Hemostasis and Thrombosis. New York, Churchill Livingstone, 1979.

Jacques, L. B.: Heparin. *Sem. Thromb. Hemostasis, 4*:275, 1978.

Joist, J. H., and Sherman, L. A. (eds.): Venous and Arterial Thrombosis: Pathogenesis, Diagnosis, Prevention and Therapy. New York, Grune & Stratton, 1979.

Kline, D. L., and Reddy, K. N. N. (eds.): Fibrinolysis. Boca Raton, Fla., CRC Press, 1980.

Lewis, J. H., *et al.*: Bleeding Disorders. Garden City, N.Y., Medical Examination Publishing Co., 1979.

Malinovsky, N. N., and Kozlov, V. A.: Anticoagulant and Thrombolytic Therapy in Surgery. St. Louis, C. V. Mosby, 1979.

Markwardt, F. (ed.): Fibinolytics and Antifibrinolytics. New York, Springer-Verlag, 1978.

McDuffie, N. M. (ed.): Heparin; Structure, Cellular Functions, and Clinical Application. New York, Academic Press, 1979.

Mitchell, J. R. A., and Domenet, J. G. (eds.): Thromboembo-lism: A New Approach to Therapy. New York, Academic Press, 1977.

Murano, G., and Bick, R. L. (eds.): Basic Concepts of Hemostasis and Thrombosis: Clinical Laboratory Evaluation of Thrombohemorrhagic Phenomena. Boca Raton, Fla., CRC Press, 1980.

Mustard, J. F., *et al.*: Prostaglandins and platelets. *Annu. Rev. Med., 31*:89, 1980.

Quick, A. J.: The Hemorrhagic Diseases and the Pathology of Hemostasis. Springfield, Ill., Charles C Thomas, 1974.

Suttie, J. W. (ed.): Vitamin K Metabolism and Vitamin K–Dependent Proteins. Baltimore, University Park Press, 1979.

Suttie, J. W., and Jackson, C. M.: Prothrombin structure, activation, and biosynthesis. *Physiol. Rev., 57*:1, 1977.

Thomson, J. M. (ed.): Blood Coagulation and Haemostasis: A Practical Guide. London, Churchill Livingstone, 1979.

Wall, R. T., and Harker, L. A.: The endothelium and thrombosis. *Annu. Rev. Med., 31*:361, 1980.

Part III
NERVE AND MUSCLE

8

Membrane Potentials, Action Potentials, Excitation, and Rhythmicity

Electrical potentials exist across the membranes of essentially all cells of the body, and some cells, such as nerve and muscle cells, are "excitable" — that is, capable of self-generation of electrochemical impulses at their membranes and, in some instances, utilization of these impulses to transmit signals along the membranes. In still other types of cells, such as glandular cells, macrophages, and ciliated cells, changes in membrane potentials probably play significant roles in controlling many of the cell's functions. However, the present discussion is concerned with membrane potentials generated both at rest and during action by nerve and muscle cells.

BASIC PHYSICS OF MEMBRANE POTENTIALS

Before beginning this discussion, let us first recall that the fluids both inside and outside the cells are electrolytic solutions containing approximately 155 mEq. per liter of positive ions and the same concentration of negative ions. Generally, a very minute excess of negative ions (anions) accumulates immediately inside the cell membrane along its inner surface, and an equal number of positive ions (cations) accumulates immediately outside the membrane. The effect of this is the establishment of a *membrane potential* between the inside and outside of the cell.

The two basic means by which membrane potentials can develop are (1) active transport of ions through the membrane, thus creating an imbalance of negative and positive charges on the two sides of the membrane, and (2) diffusion of ions through the membrane as a result of ion concentration differences between the two sides of the membrane, thus also creating an imbalance of charges.

MEMBRANE POTENTIALS CAUSED BY ACTIVE TRANSPORT — THE "ELECTROGENIC PUMP"

Figure 8–1A illustrates how the process of active transport can create a membrane potential. In this figure equal concentrations of anions, which are *negatively charged,* are present both inside and outside the nerve fiber. However, the sodium "pump," which was discussed in Chapter 4, has transported some of the *positively charged* sodium ions to the exterior of the fiber. Thus, more negatively charged anions than positively charged sodium ions remain inside the nerve fiber, causing negativity on the inside. A pump such as this, which causes the development of a membrane potential, is called an *electrogenic pump.*

In Chapter 4, it was pointed out that the sodium pump is also a potassium pump. That is, the same ATPase that acts as a carrier to transport sodium out of the cell also transports potassium inward at the same time. However, this pump normally transports three sodium ions outward for every two potassium ions inward. Thus, there is always more transfer of positively charged ions outward than inward. Therefore, as illustrated in Figure 8–1B, operation of this pump can still create electronegativity inside the nerve fiber membrane.

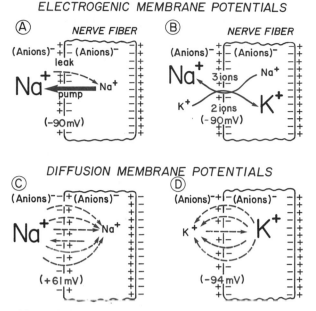

Figure 8–1. A, Establishment of a membrane potential as a result of active transport of sodium ions out of the nerve fiber. B, Establishment of a membrane potential as a result of sodium and potassium pumping through the nerve membrane by the sodium-potassium electrogenic pump, three sodium ions passing out of the membrane for each two potassium ions entering the membrane. C, Establishment of a diffusion membrane potential caused by permeability of the membrane only to sodium ions and impermeability to all other ions. D, Establishment of a diffusion membrane potential because of permeability of the membrane only to potassium ions and impermeability to all other ions. Note that the internal membrane potential is positive in the case of sodium ion permeability and negative in the case of potassium ion permeability because of opposite concentration gradients.

MEMBRANE POTENTIALS CAUSED BY DIFFUSION

Figure 8–1C and 1D illustrates the nerve fiber when there is no active transport of either sodium or potassium. In Figure 8–1C the sodium concentration is very great outside the membrane, while that inside is very low. Furthermore, the membrane is very permeable to the sodium ions but not to the anions. Because of the large sodium concentration gradient from the outside toward the inside, there is a strong tendency for sodium ions to diffuse inward. As they do so, they carry positive charges to the inside, thus creating a state of electropositivity inside the membrane, and electronegativity on the outside because of the negative anions that remain behind.

Figure 8–1D illustrates the same effect as that in Figure 8–1C but with a high concentration of potassium ions inside the membrane and a low potassium concentration outside. These ions are also positively charged. Also, the membrane is

highly permeable to the potassium ions but impermeable to the anions. Diffusion of the potassium ions to the outside creates a membrane potential now of opposite polarity, with negativity inside and positivity outside.

Thus, in both Figure 8–1C and 1D, we see that a concentration difference of ions across a semipermeable membrane can, under appropriate conditions, cause the creation of a membrane potential. In later sections of this chapter, we shall also see that many of the membrane potential changes observed during the course of nerve and muscle impulse transmission result from the occurrence of rapidly changing diffusion membrane potentials.

Relationship of the Diffusion Potential to the Concentration Difference — the Nernst Equation. When a concentration difference of a single type of positive ions across a membrane causes diffusion of ions through the membrane, thus creating a membrane potential, the magnitude of the potential is determined by the following formula (at body temperature, 38° C.):

$$\text{EMF (millivolts)} = -61 \log \frac{\text{Conc. inside}}{\text{Conc. outside}}$$

Thus, when the concentration of positive ions on the inside of a membrane is 10 times that on the outside, the log of 10 is 1, and the potential difference calculates to be −61 millivolts. This equation is called the *Nernst equation*.

However, two conditions are necessary for this *Nernst potential* to develop as a result of diffusion: (1) The membrane must be semipermeable, allowing one species of ion to diffuse through the pores while all other ions do not diffuse. (2) The concentration of the diffusible ions must be greater on one side of the membrane than on the other side.

ORIGIN OF THE NERVE CELL MEMBRANE POTENTIAL

Distinction Between Original Development of a Membrane Potential and the Instantaneous Re-establishment of the Membrane Potential after Nerve Impulse Transmission. Let us assume that in the beginning, the concentrations of all ions are the same inside and outside the nerve fiber. Under these conditions there will be no membrane potential. However, the normal cell will automatically develop a membrane potential. How does this come about? The answer to this lies in the function of the sodium-potassium pump and in the electrogenicity of this pump. That is, the continual pumping of more positive charges to the outside of the membrane than to the inside (three sodium ions pumped outward for every two potassium ions inward) eventually leads to the negative membrane potential inside the cell.

But now let us write another scenario: First, assume that the sodium-potassium pump is non-functional but that there is already a high potassium concentration inside the cell and low potassium outside. Second, assume that the cell membrane is highly permeable to potassium while poorly permeable to sodium. Therefore, large numbers of potassium ions will diffuse to the outside, while only a few sodium ions will diffuse to the inside. This results in far more transfer of positive ions to the outside than in the other direction. Consequently, once again a negative membrane potential develops inside the nerve membrane. It is in this manner that the negative membrane potential is re-established within less than a millisecond immediately following transmission of a nerve impulse along the nerve membrane.

Therefore, from the outset it is very important for the student to recognize the difference between the original development of the membrane potential, which is an electrogenic phenomenon, and the instantaneous re-establishment of the membrane potential following each nerve impulse (that is, following each action potential), which is a diffusion phenomenon.

Maximum Potential That Can Be Achieved by the Electrogenic Sodium-Potassium Pump. Even though the normal resting membrane is only slightly permeable to sodium ions, nevertheless, as progressively more sodium ions are pumped out of the nerve fiber they begin to diffuse back into the nerve fiber for two reasons: (1) A sodium concentration gradient develops from the outside of the fiber toward the inside, and (2) a negative membrane potential develops inside the fiber and attracts the positively charged sodium ions inward. Eventually, there comes a point at which the inward diffusion equals the outward pumping by the sodium pump. When this occurs, the pump has reached its maximum capability of causing net transfer of sodium ions to the outside. As illustrated in Figure 8–2, this occurs when the sodium ion concentration inside the nerve fiber falls to about 14 mEq. per liter (in contrast to 142 mEq. per liter in the extracellular fluid outside the fiber), and when the membrane potential inside the fiber falls to about −90 millivolts. Therefore, this −90 millivolts becomes the *resting potential* of the nerve membrane.

The Potassium Ion Concentration Gradient Across the Membrane. At the same time that the sodium-potassium pump pumps sodium ions

to the exterior, it pumps about two thirds as many potassium ions to the interior. However, the resting membrane is 50 to 100 times as permeable to potassium ions as to sodium ions, which means that each time potassium ions are pumped in they tend to diffuse back out of the fiber almost immediately. Therefore, the potassium pump by itself can build up only a slight excess of potassium ions inside the nerve fiber.

Yet, it is known that potassium ions do build up to a high concentration inside the nerve fiber. How could this be, if it is not caused by the potassium pump? The answer is the high degree of negativity created inside the fiber by the pumping of sodium ions to the outside. The −90 millivolts inside the fiber attract the positive potassium ions from the exterior to the inside, and this attraction is essential for most of the buildup of potassium ions inside the fiber. Figure 8–2 illustrates that the potassium concentration inside the nerve membrane averages 140 mEq. per liter in contrast to its concentration outside in the extracellular fluid of only 4 mEq. per liter — a 35-fold difference.

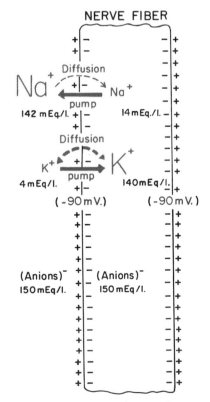

Figure 8–2. Establishment of a membrane potential of −90 millivolts in the normal resting nerve fiber and development of concentration differences of sodium and potassium ions between the two sides of the membrane. The dashed arrows represent diffusion and the solid arrows represent active transport (pumps).

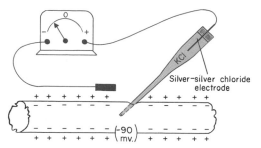

Figure 8–3. Measurement of the membrane potential of the nerve fiber using a microelectrode.

Membrane Potentials Measured in Nerve and Muscle Fibers. Figure 8–3 illustrates a method for measuring the resting membrane potential. A micropipet is made from a minute capillary glass tube so that the tip of the pipet has a diameter of only 0.25 to 2 microns. Inside the pipet is a solution of very concentrated potassium chloride, which acts as an electrical conductor. The fiber whose membrane is to be measured is pierced with the pipet, and electrical connections are made from the pipet to an appropriate meter, as illustrated in the figure. The resting membrane potential measured in many different nerve and skeletal muscle fibers of mammals has usually ranged between −70 and −100 millivolts, with −90 millivolts as a reasonable average of the different measurements.

THE ACTION POTENTIAL

In the above section we described the original development of the *resting membrane potential* and pointed out that it is normally about 90 millivolts. Now we will describe how membrane potentials are employed by nerve fibers to transmit nerve impulses, which is the means by which information signals are transmitted from one part of the nervous system to another. This is achieved by means of *action potentials,* which are abrupt pulselike changes in the membrane potential, lasting a few ten thousandths to a few thousandths of a second as will be illustrated in Figures 8–5, 8–6, 8–8, and 8–9. And, for reasons that will be discussed, the action potential moves along the length of the nerve, thus giving rise to the nerve signals.

An action potential can be elicited in a nerve fiber by almost any factor that suddenly increases the permeability of the membrane to sodium ions — such factors as electrical stimulation, mechanical compression of the fiber, application of chemicals to the membrane, or almost any other event that disturbs the normal resting state of the membrane.

Role of Sodium and Potassium Membrane Diffusion Potentials in the Generation of the Action Potential. In the above discussion of the original development of the resting membrane potential, it became clear that the sodium-potassium pump is basically responsible for generating this resting potential. Furthermore, the sodium-potassium pump creates large sodium and potassium differences across the membranes, with very high sodium concentration outside the membrane and very high potassium concentration inside. But it is important to emphasize that once the sodium and potassium concentration differences have been created across the cell membrane, the occurrence of action potentials does not depend upon operation of the sodium-potassium pump. Indeed, many thousands of impulses can be transmitted even after the pump has been poisoned before the sodium and potassium concentration differences "run down." The action potential results from rapid changes in membrane permeability to sodium and potassium ions — the sodium permeability increasing about 5000-fold at the onset of the action potential, followed instantaneously by return of the sodium permeability to normal and then by a great increase in the potassium permeability. As a result, the membrane potential changes rapidly from its normal negative value to an instantaneous positive value, then equally rapidly back to its negative value. These changes will be discussed in the following sections.

Membrane Depolarization and Membrane Repolarization — the Two Stages of the Action Potential. The action potential occurs in two separate stages called *membrane depolarization* and *membrane repolarization,* each of which may be explained by referring to Figure 8–4. Figure 8–4A illustrates the resting state of the membrane, with negativity inside and positivity outside. When the permeability of the membrane to sodium ions suddenly increases, many of the sodium ions that are present in very high concentration outside the fiber rush to the inside, carrying enough positive charges to the inside to cause complete disappearance of the normal negative resting potential, and usually enough charges actually to develop a positive state inside the fiber. This sudden loss of normal negative potential inside the fiber is called *depolarization.* The positive potential that develops momentarily inside the fiber is called the *reversal potential.*

Almost immediately after depolarization takes place, the pores of the membrane again become almost impermeable to sodium ions but at the same time considerably more permeable than

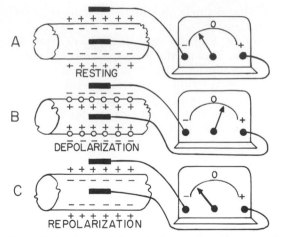

Figure 8–4. Sequential events during the action potential, showing: A, the normal resting potential, B, development of a reversal potential during depolarization, and C, re-establishment of the normal resting potential during repolarization.

sodium diffusion in the resting state, as has been explained. Yet, when the gates are widely opened, the permeability of the sodium channels can increase as much as 5000-fold, and the permeability of the potassium channels, which are already partially opened, can increase an additional 50-fold. It is this opening and closing of the gates to the sodium and potassium channels that is responsible for the depolarization and repolarization processes.

A Positive Feedback Regenerative Process that Opens the Sodium Gates to Cause Depolarization. The initial event causing the action potential is a slight increase in the permeability of the membrane to sodium ions — that is, a slight opening of the sodium gates. When these gates open even a slight amount, a positive feedback effect immediately begins; that is, in some way not completely understood, the initial movement of sodium ions into the fiber initiates a regenerative cycle that opens the gates still wider and

normal to potassium ions. Therefore, sodium ions stop moving to the inside of the fiber, and, instead, potassium ions move to the outside because of the high potassium concentration on the inside. Thus, because the potassium ions are positively charged, the excess positive charges inside the fiber are transferred back out of the fiber, and the normal negative resting membrane potential returns. This effect, called *repolarization*, is illustrated in Figure 8–4C. The curve of Figure 8–5 shows the sequence of the recorded action potential inside the fiber membrane. Now let us explain in more detail *how* this sequence of events occurs.

Sodium "Channels" and Potassium "Channels." The sodium and potassium ions diffuse mainly through separate types of pores. The pores through which the sodium ions diffuse are called *sodium channels,* and those through which the potassium ions diffuse are called *potassium channels.* The actual anatomical structures of these channels are not known, but they have physiological properties *as if* the sodium channels were oval-shaped pores with dimensions of about 0.3 by 0.5 nanometers, and the potassium channels were round pores with dimensions of about 0.3 by 0.3 nanometers.

Each channel is believed to be guarded by an electrically charged "gate" that can open and close the channel. Under resting conditions, the gates of the sodium channels are almost completely closed, while those of the potassium channels are only partially closed; therefore, the potassium channels are some 50 to 100 times as permeable as the sodium channels. This difference allows far more potassium diffusion than

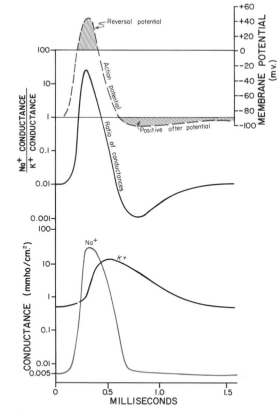

Figure 8–5. Changes in sodium and potassium conductances during the course of the action potential. Note that sodium conductance increases several thousand-fold during the early stages of the action potential, while potassium conductance increases only about 30-fold during the latter stages of the action potential and for a short period thereafter. (Curves constructed from data in Hodgkin and Huxley papers but transposed from squid axon to apply to the membrane potentials of large mammalian nerve fibers.)

wider until the sodium channel permeability increases within a few ten thousandths of a second about 5000-fold. This effect is illustrated by the abrupt rise in the sodium conductance curve (conductance = permeability × ion concentration) in Figure 8–5. At this instant the membrane is approximately 20 to 30 times as permeable to sodium as to potassium, because only the sodium channels open widely during the depolarization process. The result is that far more sodium ions now diffuse to the interior of the fiber than potassium ions to the exterior, carrying positive changes to the inside and leaving negative anions on the outside. Therefore, the potential inside the fiber suddenly changes from negative to positive, thus causing the *reversal potential*. This is illustrated by the action potential at the top of Figure 8–5.

Mechanism of Repolarization — Closure of the Sodium Channels and Opening of the Potassium Channels. Once the sodium channels become widely opened and the positive reversal potential appears inside the fiber, the sodium and potassium channel permeabilities make an about-face. The exact cause of this is not known, but it is known that the positive interior potential now causes electrical repulsion of the incoming sodium ions until their incoming movement slows or stops. At this point the sodium channels begin to close, but just as rapidly the potassium channels begin to open, presumably because the positive potential inside the membrane is now pushing potassium ions to the exterior at an extremely rapid rate. Loss of potassium ions to the exterior carries positive charges to the exterior, and the potential on the interior of the fiber now returns toward the normal resting negative state. As the potential becomes more negative, a positive feedback regenerative cycle develops for further opening of the potassium channels, as illustrated by the increasing potassium conductance in Figure 8–5, in exactly the same way that a similar cycle had occurred during depolarization to open the sodium channels. More and more potassium ions pass to the exterior, and the electrical potential inside the fiber, as illustrated by the action potential in the figure, returns to the normal resting level of approximately −90 mv. Then the potassium channels also close back to their normal levels of permeability during the next few milliseconds.

Once the membrane potential returns back to its negative resting level, it then remains at this level until the membrane is disturbed once again.

Quantity of Ions Lost from the Nerve Fiber During the Action Potential. The actual quantity of ions that must pass through the nerve membrane to cause the action potential — that is,

to cause a 135 millivolt increase in the membrane potential and then to return this potential back to its normal resting level — is extremely slight. For large myelinated nerve fibers only about 1/100,000 to 1/500,000 of the ions normally inside the fiber are exchanged during this process. Nevertheless, this does cause a very slight increase of sodium ions inside the fiber and a corresponding very slight decrease in potassium ions. As we shall see later in the chapter, the active transport processes restore these ions within a few milliseconds or seconds.

Relationship of the Action Potential to the Potassium and Sodium Nernst Potentials. Now that we have explained the events that transpire during the course of the action potential, let us review, and illustrate in Figure 8–6, the relationship of this action potential to the potassium and sodium Nernst potentials. It was pointed out earlier in the chapter that under resting conditions the nerve membrane is highly permeable to potassium but only slightly permeable to sodium. Using the equation presented earlier for calculating the Nernst potential across a membrane, we find the following:

$$K^+ \text{ Nernst Potential} = -61 \times$$

$$\log \frac{140 \text{ mEq./l.}}{4 \text{ mEq./l.}} = -94 \text{ millivolts}$$

This value is very near to the normal resting potential of −90 millivolts. The reason is that the resting membrane is very permeable to potassium ions but almost impermeable to sodium.

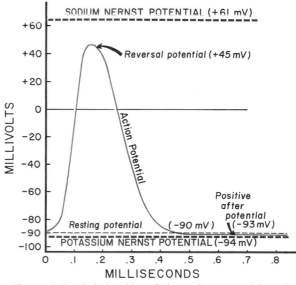

Figure 8–6. Relationship of the action potential to the potassium and sodium Nernst potentials.

Now let us calculate the Nernst potential for sodium:

$$Na^+ \text{ Nernst Potential} = -61 \times$$

$$\log \frac{14 \text{ mEq./l.}}{140 \text{ mEq./l.}} = +61 \text{ millivolts}$$

Note the change to a positive value. Note also that this value is near to that of the reversal potential, +45 millivolts, the reason being that the membrane is about 20 times as permeable to sodium as to potassium at the instant immediately after depolarization.

PROPAGATION OF THE ACTION POTENTIAL

In the preceding paragraphs we have discussed the action potential as it occurs at one spot on the membrane. However, an action potential elicited at any one point on an excitable membrane usually excites adjacent portions of the membrane, resulting in propagation of the action potential. The mechanism of this is illustrated in Figure 8–7. Figure 8–7A shows a normal resting nerve fiber, and Figure 8–7B shows a nerve fiber that has been excited in its mid-portion — that is, the mid-portion has suddenly developed increased permeability to sodium. The arrows illustrate a local circuit of current flow between the depolarized and the resting membrane areas; positive current flows inward through the depolarized membrane and outward through the resting membrane, thus completing a circuit. In some way not understood, *the outward current flow through this resting membrane now increases its permeability to sodium as well,* which immediately allows sodium ions to diffuse inward through the membrane, thus setting up the vicious circle of increas-

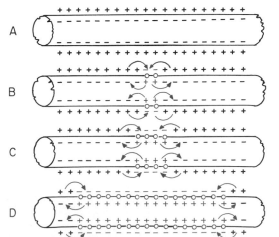

Figure 8–7. Propagation of action potentials in both directions along a conductive fiber.

ing sodium permeability discussed earlier in the chapter. As a result, depolarization occurs at this area of the membrane too. Therefore, as illustrated in Figure 8–7 C and D, successive portions of the membrane become depolarized. And these newly depolarized areas cause local circuits of current flow still farther along the membrane, causing progressively more and more depolarization. Thus, the depolarization process travels in both directions along the entire extent of the fiber. The transmission of the depolarization process along a nerve of muscle fiber is called a *nerve* or *muscle impulse.*

The All-or-Nothing Principle. It is obvious that once an action potential has been elicited at any point on the membrane of a normal fiber, the depolarization process will travel over the entire membrane. This is called the all-or-nothing principle, and it applies to all normal excitable tissues. Occasionally, though, when the fiber is in an abnormal state the impulse will reach a point on the membrane at which the action potential does not generate sufficient voltage to stimulate the adjacent area of the membrane. When this occurs the spread of depolarization will stop. Therefore, for normal propagation of an impulse to occur, the ratio of action potential to threshold for excitation, called the *safety factor,* must at all times be greater than unity.

Propagation of Repolarization. The action potential normally lasts almost the same length of time at each point along a fiber. Therefore, repolarization normally occurs first at the point of original stimulus and then spreads progressively along the membrane, moving in the same direction that depolarization had previously spread but a few ten thousandths of a second later.

"RECHARGING" THE FIBER MEMBRANE — IMPORTANCE OF ENERGY METABOLISM

Transmission of each impulse along the nerve fiber reduces the concentration differences of sodium and potassium between the inside and outside of the membrane. For a single action potential, this effect is so minute that it cannot even be measured. Indeed, 100,000 to 500,000 impulses can be transmitted by a large nerve fiber before the concentration differences have run down to the point that action potential conduction ceases. Yet, even so, with time it becomes necessary to re-establish these sodium and potassium membrane concentration differences. This is achieved by the action of the sodium and potassium pump in exactly the same way as that described in the first part of the chapter for establishment of the original resting potential. That is, the sodium ions that, during the action potentials, have diffused to the interior of the cell and the potassium ions that have diffused to the exterior are returned to their original state by the sodium and potassium pump. Since this pump requires energy for operation, this process of "recharging" the nerve fiber is an active metabolic one, utilizing energy derived from the adenosine triphosphate energy "currency" system of the cell.

A special feature of the sodium–potassium ATPase membrane pumping system is that its degree of activity is very strongly stimulated by excess sodium ions inside the cell membrane. In fact, when the internal sodium concentration rises from 10 to 20 mEq. per liter, the activity of the pump does not merely double but instead increases approximately 8-fold. Therefore, it can easily be understood how the recharging process of the nerve fiber can rapidly be set into motion whenever the concentration differences of sodium and potassium across the membrane begin to "run down."

The Positive After-Potential. Once the membrane potential has returned to its resting value, it then becomes a millivolt or so more negative than its normal resting value; this excess negativity is called the *positive after-potential*. It lasts from 50 milliseconds to as long as many seconds.

The first part of this positive after-potential is caused by the excess permeability of the nerve membrane to potassium ions at the end of the action potential. However, the prolonged continuance of this potential is caused principally by the electrogenic pumping of excess sodium outward through the nerve fiber membrane, which is the recharging process that was discussed above.

(The student might wonder why greater negativity in the membrane potential is called a positive rather than a negative after-potential. The reason is that these potentials were first measured *outside* the nerve fibers rather than inside, and all potential changes on the outside are of exactly opposite polarity, whereas modern terminology expresses membrane potentials in terms of the inside potential.)

PLATEAU IN THE ACTION POTENTIAL

In some instances the excitable membrane does not repolarize immediately after depolarization, but instead the potential remains on a plateau near the peak of the spike sometimes for many milliseconds before repolarization begins. Such a plateau is illustrated in Figure 8–8, from which one can readily see that the plateau greatly pro-

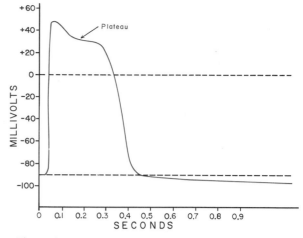

Figure 8–8.   An action potential from a Purkinje fiber of the heart, showing a "plateau."

longs the period of depolarization. It is this type of action potential that occurs in the heart, where the plateau lasts for as long as two- to three-tenths second and causes contraction of the heart muscle during this entire period of time.

The cause of the action potential plateau is probably a combination of several different factors. First, there is delay in closure of the sodium channels, which allows extra sodium ions to continue flowing into the fiber. Second, there is a small amount of calcium current flowing into the fiber at the same time, and these two currents together maintain the positive state inside the membrane that causes the plateau. However, third, probably equally important is the fact that the permeability of the potassium channels *decreases* about 5-fold at the onset of the action potential in excitable membranes that exhibit plateaus, and this prevents rapid outflow of potassium ions to the outside of the fiber and therefore delays the repolarization process.

Yet, when closure of the sodium channels does begin, the process proceeds unabated; simultaneously, the potassium permeability of the membrane increases 100-fold or more. Therefore, sodium and calcium ions stop diffusing to the interior of the fiber, while potassium ions diffuse outward extremely rapidly. Consequently, the membrane potential returns quickly to its normal negative level, as illustrated in Figure 8–8, by the rapid decline of the potential at the end of the plateau.

RHYTHMICITY OF CERTAIN EXCITABLE TISSUES — REPETITIVE DISCHARGE

All excitable tissues can discharge repetitively if the threshold for stimulation is reduced low enough. For instance, even nerve fibers and skeletal muscle fibers, which normally are highly stable, discharge repetitively when they are placed in a solution containing the drug veratrine or when the calcium ion concentration falls below a critical value. Repetitive discharges, or rhythmicity, occur normally in the heart, in most smooth muscle, and also in many of the neurons of the central nervous system. It is these rhythmical discharges that cause the heart beat, that cause peristalsis, and that cause such nervous events as the rhythmical control of breathing.

The Re-Excitation Process Necessary for Rhythmicity. For rhythmicity to occur, the membrane, even in its natural state, must be already permeable enough to sodium ions to allow automatic membrane depolarization. Thus, Figure 8–9 shows the "resting" membrane potential is only −60 to −70 millivolts. This is not enough negative voltage to keep the sodium channels closed. That is, (a) sodium ions flow inward, (b) this further increases the membrane permeability, (c) still more sodium ions flow inward, (d) the permeability increases more, and so forth, thus eliciting the regenerative process of sodium channel opening until an action potential is generated. Then a few milliseconds to a few seconds later a new action potential begins.

The delay between succeeding action potentials can be explained as follows: Referring back to Figure 8–5, note that toward the end of all action potentials, and continuing for a short period thereafter, the membrane becomes excessively permeable to potassium. The excessive outflow of potassium ions carries tremendous numbers of positive charges to the outside of the membrane, creating inside the fiber considerably more negativity than would otherwise occur for a short period after the preceding action potential is over. That is, the excessive potassium permeability at this time overbalances the naturally high sodium permeability. This excess negativity is called *hyperpolarization*; it is illustrated in Figure

8–9. As long as this state exists, re-excitation will not occur; but gradually the excess potassium conductance (and the state of hyperpolarization) disappears, thereby allowing the onset of a new action potential.

Figure 8–9 illustrates this relationship between repetitive action potentials and potassium conductance. The state of hyperpolarization is established immediately after each preceding action potential, but it gradually recedes, and the membrane potential correspondingly increases until it reaches the *threshold* for excitation; then suddenly a new action potential results, the process occurring again and again.

SPECIAL ASPECTS OF IMPULSE TRANSMISSION IN NERVES

Myelinated and Unmyelinated Nerve Fibers. In a typical small nerve trunk, a few very large nerve fibers compose most of the cross-sectional area but many more small fibers lie between the large ones. The large fibers are *myelinated* and the small ones are *unmyelinated*. The average nerve trunk contains about twice as many unmyelinated fibers as myelinated fibers.

Figure 8–10 illustrates a typical myelinated fiber. The central core of the fiber is the *axon,* and the membrane of the axon is the actual *conductive membrane*. The axon is filled in its center with *axoplasm,* which is a viscid intracellular fluid. Surrounding the axon is a *myelin sheath* caused by Schwann cells wrapping their cell membranes around the axon. This sheath is approximately as thick as the axon itself, and about once every millimeter along the extent of the axon the myelin sheath is interrupted by a *node of Ranvier*.

The myelin sheath is an excellent insulator that increases the resistance to ion flow through the membrane approximately 5000-fold. However, at the nodes of Ranvier, ions can flow with ease between the extracellular fluid and the axon.

Saltatory Conduction in Myelinated Fibers from Node to Node. Action potentials are conducted from node to node by the myelinated nerve rather than continuously along the entire fiber as occurs in the unmyelinated fiber. This process is called *saltatory conduction.* That is, electrical current flows through the surrounding extracellular fluids and also through the axoplasm from node to node, exciting successive nodes one after another. Thus, the impulse jumps down the fiber, which is the origin of the term "saltatory."

Saltatory conduction is of value for two reasons: First, by causing the depolarization process to jump long intervals along the axis of the nerve fiber, this mechanism greatly increases the velocity of nerve transmission in myelinated fibers.

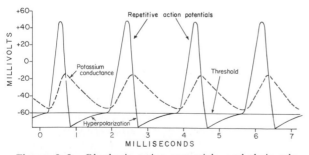

Figure 8–9. Rhythmic action potentials, and their relationship to potassium conductance and to the state of hyperpolarization.

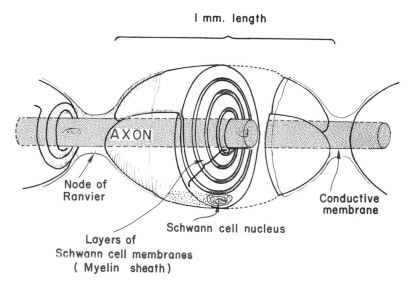

Figure 8–10. The myelin sheath and its formation by Schwann cells. (Modified from Elias and Pauley: Human Microanatomy. Philadelphia, F. A. Davis Company, 1966.)

Second, saltatory conduction conserves energy for the axon, for only the nodes depolarize, allowing several hundred times less loss of ions than would otherwise be necessary.

VELOCITY OF CONDUCTION IN NERVE FIBERS

The velocity of conduction in nerve fibers varies from as little as 0.5 meter per second in very small unmyelinated fibers up to as high as 130 meters per second (the length of a football field) in very large myelinated fibers. The velocity increases approximately with the fiber diameter in myelinated nerve fibers and approximately with the square root of fiber diameter in unmyelinated fibers.

EXCITATION — THE PROCESS OF ELICITING THE ACTION POTENTIAL

Chemical Stimulation—Acetylcholine. Certain chemicals can stimulate a nerve fiber by increasing the membrane permeability. Such chemicals include acids, bases, almost any salt solution of very strong concentration, and, most importantly, the substance *acetylcholine.* Many nerve fibers, when stimulated, secrete acetylcholine at their endings where they synapse with other neurons or where they end on muscle fibers. The acetylcholine in turn stimulates the successive neuron or muscle fiber by opening pores in the membrane with diameters of 0.6 to 0.7 nanometer, large enough for sodium (as well as most other ions) to go through with ease. This

is discussed in much greater detail in Chapter 10, and it is one of the most important means by which nerve and muscle fibers are stimulated. Likewise, *norepinephrine* secreted by sympathetic nerve endings can stimulate cardiac muscle fibers and some smooth muscle fibers, and still other hormonal substances can stimulate successive neurons in the central nervous system.

Mechanical Stimulation. Crushing, pinching, or pricking a nerve fiber can cause a sudden surge of sodium influx and, for obvious reasons, can elicit an action potential. Even slight pressure on some specialized nerve endings can stimulate these events; they will be discussed in Chapter 32 in relation to sensory perception.

Electrical Stimulation. Electrical stimulation also can initiate an action potential. An electrical charge artificially induced across the membrane causes excess flow of ions through the membrane; this in turn can initiate an action potential. However, a *negative* electrode placed on the outside of a nerve fiber is much more likely to excite the fiber than is a positive electrode. The reason is that the negative electrode attracts positive charges in the outward direction through the membrane, an effect that makes the membrane more permeable to sodium ions. In fact, a *positive* electrode can at times actually decrease the membrane excitability because it repels the positive charges.

The Refractory Period. A second action potential cannot occur in an excitable fiber as long as the membrane is still depolarized from the preceding action potential. Therefore, even an electrical stimulus of maximum strength applied before the first spike potential is almost over will not elicit a second one. This interval of inexcitability is called the *absolute refractory period.* The absolute refractory period of large myelinated nerve fibers is about 1/2500 second. Therefore, one can readily calculate that such a fiber can carry a maximum of about 2500 impulses per second. Following the absolute refractory period is a *relative refractory period* lasting about one quarter as long. During this period, stronger than normal stimuli are required to excite the fiber.

FACTORS THAT ALTER MEMBRANE EXCITABILITY

Low Calcium Tetany.   An extremely important potentiator of excitability is low concentration of calcium ions in the extracellular fluids. Calcium ions normally decrease the permeability of the membrane to sodium. If sufficient calcium ions are not available, however, the permeability becomes increased, and as a result, the membrane excitability greatly increases — sometimes so greatly that many spontaneous impulses result and cause muscular spasm. This condition is known as low calcium tetany. It often occurs in patients who have lost their parathyroid glands and who therefore cannot maintain normal calcium ion concentrations. This condition will be discussed in Chapter 53.

Local Anesthetics.   Many substances are used clinically as local anesthetics, including *cocaine, procaine, tetracaine,* and many other drugs. These act directly on the membrane, decreasing its permeability to sodium and, therefore, also reducing membrane excitability. When the excitability has been reduced so low that the ratio of *action potential strength to excitability threshold* (called the "safety factor") is reduced below unity, a nerve impulse fails to pass through the anesthetized area.

RECORDING MEMBRANE POTENTIALS AND ACTION POTENTIALS

The Cathode Ray Oscilloscope.   For practical purposes the only type of meter that is capable of responding accurately to the very rapid membrane potential changes of most excitable fibers is the cathode ray oscilloscope.

Figure 8–11 illustrates the basic components of a cathode ray oscilloscope. The cathode ray tube itself is composed basically of an *electron gun* and a *fluorescent surface* against which electrons are fired. Where the electrons hit the surface, the fluorescent material glows. If the electron beam is moved across the surface, the spot of glowing light also moves and draws a fluorescent line on the screen.

In addition to the electron gun and fluorescent surface, the cathode ray tube is provided with two sets of plates: one set, called the *horizontal deflection plates,* positioned on either side of the electron beam, and the other set, called the *vertical deflection plates,* positioned above and below the beam. If a negative charge is applied to the left-hand plate and a positive charge to the right-hand plate, the electron beam will be repelled away from the left plate and attracted toward the right plate, thus pulling the beam toward the right, and this will cause the spot of light on the fluorescent surface of the cathode ray screen to move to the right. Likewise, positive and negative charges can be applied to the vertical deflection plates to move the beam up or down.

Since electrons travel at extremely rapid velocity and since the plates of the cathode ray tube can be alternately charged positively or negatively within less than a millionth of a second, it is

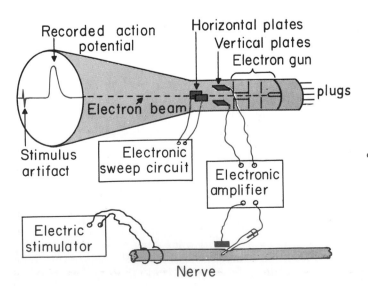

Figure 8–11.   The cathode ray oscilloscope for recording transient action potentials.

obvious that the spot of light on the face of the tube can also be moved to almost any position in less than a millionth of a second. For this reason, the cathode ray tube oscilloscope can be considered to be an inertialess meter capable of recording with extreme fidelity almost any change in membrane potential.

To use the cathode ray tube for recording action potentials, two electrical circuits must be employed. These are (1) an *electronic sweep circuit* that controls the voltages on the horizontal deflection plates and (2) an *electronic amplifier* that controls the voltages on the vertical deflection plates. The sweep circuit automatically causes the spot of light to begin at the left-hand side and move slowly toward the right. When the spot reaches the right side it jumps back immediately to the left-hand side and starts a new trace.

The electronic amplifier amplifies signals that come from the nerve. If a change in membrane potential occurs while the spot of light is moving across the screen, this change in potential will be amplified and will cause the spot to rise above or fall below the mean level of the trace, as illustrated in the figure. In other words, the sweep circuit provides the lateral movement of the electron beam while the amplifier provides the vertical movement in direct proportion to the changes in membrane potentials picked up by appropriate electrodes.

Figure 8–11 also shows an electric stimulator used to stimulate the nerve. When the nerve is stimulated, a small *stimulus artifact* usually appears on the oscilloscope screen prior to the action potential.

Recording the Monophasic Action Potential. Throughout this chapter "monophasic" action potentials have been shown in the different diagrams. To record these, an electrode such as that illustrated earlier in Figure 8–3 must be inserted into the interior of the fiber. Then, as the action potential spreads down the fiber, the changes in the potential inside the fiber are recorded as illustrated earlier in Figures 8–5, 8–6, 8–8, and 8–9.

Recording Biphasic Action Potentials. When one wishes to record impulses from a whole nerve trunk, it is not feasible to place electrodes inside the nerve fibers. Therefore, the usual method of recording is to place two electrodes on the outside of fibers. However, the record that is obtained is then biphasic (but sometimes even triphasic) for the following reasons: When an action potential moving down the nerve fiber reaches the first electrode, it becomes charged negatively, while the second electrode is still unaffected. This causes the oscilloscope to record in the negative direction. Then as the action potential is receding down the nerve, there comes a point when the

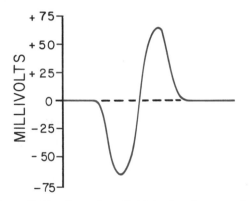

Figure 8–12. Recording of a biphasic action potential.

membrane beneath the first electrode becomes repolarized while the second electrode is still negative, and the oscilloscope records in the opposite direction. When these changes are recorded by the oscilloscope, a graphic record such as that illustrated in Figure 8–12 is recorded, showing a potential change first in one direction and then in the opposite direction.

REFERENCES

Adrian, R. H.: Charge movement in the membrane of striated muscle. *Annu. Rev. Biophys. Bioeng.,* 7:85, 1978.

Aguayo, A. J., and Karpati, G. (eds.). Current Topics in Nerve and Muscle Research. New York, Elsevier/North-Holland, 1979.

Almers, W.: Gating currents and charge movements in excitable membranes. *In* Adrian, R. H., *et al.* (eds.): Reviews of Physiology, Biochemistry, and Pharmacology. New York, Springer-Verlag, 1978, p. 96.

Baker, P. F., and Reuter, H.: Calcium Movement in Excitable Cells. New York, Pergamon Press, 1975.

Brinley, F. J., Jr.: Calcium buffering in squid axons. *Annu. Rev. Biophys. Bioeng.,* 7:363, 1978.

Carmeliet, E., and Vereecke, J.: Electrogenesis of the action potential and automaticity. *In* Berne, R. M., *et al.* (eds.): Handbook of Physiology. Sec. 2, Vol. 1. Baltimore, Williams & Wilkins, 1979, p. 269.

Ceccarelli, B., and Clementi, F. (eds.): Neurotoxins, Tools in Neurobiology. New York, Raven Press, 1979.

Cohen, L. B., and De Weer, P.: Structural and metabolic processes directly related to action potential propagation. *In* Brookhart, J. M., and Mountcastle, V. B. (eds.): Handbook of Physiology. Sec. 1, Vol. 1. Baltimore, Williams & Wilkins, 1977, p. 137.

Finkelstein, A., and Mauro, A.: Physical principles and formalisms of electrical excitability. *In* Brookhart, J. M., and Mountcastle, V. B. (eds.): Handbook of Physiology. Sec. 1, Vol. 1. Baltimore, Williams & Wilkins, 1977, p. 161.

Fozzard, H. A.: Conduction of the action potential. *In* Berne, R. M., *et al.* (eds.): Handbook of Physiology. Sec. 2, Vol. 1., Baltimore, Williams & Wilkins, 1979, p. 335.

Greene, L. A., and Shooter, E. M.: The nerve growth factor: Biochemistry, synthesis, and mechanism of action. *Annu. Rev. Neurosci., 3*:353, 1980.

Guroff, G.: Molecular Neurobiology. New York, Marcel Dekker, 1979.

Hille, B.: Gating in sodium channels of nerve. *Annu. Rev. Physiol., 38*:139, 1976.

Hille, B.: Ionic basis of resting and action potentials. *In* Brookhart, J. M., and Mountcastle, V. B. (eds.): Handbook of Physiology. Sec. 1, Vol. 1. Baltimore, Williams & Wilkins, 1977, p. 99.

Hodgkin, A. L.: The Conduction of the Nervous Impulse. Springfield, Ill., Charles C Thomas, 1963.

Hodgkin, A. L., and Huxley, A. F.: Quantitative description of membrane current and its application to conduction and excitation in nerve. *J. Physiol. (Lond.), 117*:500, 1952.

Jewett, D. L., and McCarroll, H. R. (eds.): Nerve Repair and Regeneration: Its Clinical and Experimental Basis. St. Louis, C. V. Mosby, 1979.

Keynes, R. D.: Ion channels in the nerve-cell membrane. *Sci. Am. 240*(3):126, 1979.

Kristensson, K.: Retrograde transport of macromolecules in axons. *Annu. Rev. Pharmacol. Toxicol., 18*:97, 1978.

Montal, M.: Experimental membranes and mechanisms of bioenergy transductions. *Annu. Rev. Biophys. Bioeng., 5*:119, 1976.

Neher, E., and Stevens, C. F.: Conductance fluctuations and ionic pores in membranes. *Annu. Rev. Biophys. Bioeng., 6*:345, 1977.

Nelson, P. G.: Nerve and muscle cells in culture. *Physiol. Rev., 55*:1, 1975.

Patrick, J., *et al.*: Biology of cultured nerve and muscle. *Annu. Rev. Neurosci., 1*:417, 1978.

Pfenninger, K. H.: Organization of neuronal membrane. *Annu. Rev. Neurosci., 1*:445, 1978.

Rall, W.: Core conductor theory and cable properties of neurons. *In* Brookhart, J. M., and Mountcastle, V. B. (eds.): Handbook of Physiology. Sec. 1, Vol. 1. Baltimore, Williams & Wilkins, 1977, p. 39.

Ritchie, J. M.: A pharmacological approach to the structure of sodium channels in myelinated axons. *Annu. Rev. Neurosci., 2*:341, 1979.

Schultz, S. G.: Principles of electrophysiology and their application to epithelial tissues. *In* MTP International Review of Science: Physiology. Baltimore, University Park Press, 1974, Vol. 4, p. 69.

Schwartz, J. H.: Axonal transport: components, mechanisms, and specificity. *Annu. Rev. Neurosci., 2*:476, 1979.

Stevens, C. F.: The neuron. *Sci. Am., 241*(3):54, 1979.

Trautwein, W.: Membrane currents in cardiac muscle fibers. *Physiol. Rev., 53*:793, 1973.

Ulbricht, W.: Ionic channels and gating currents in excitable membranes. *Annu. Rev. Biophys. Bioeng., 6*:7, 1977.

Walker, J. L., and Brown, H. M.: Intracellular ionic activity measurements in nerve and muscle. *Physiol. Rev., 57*:729, 1977.

Wallick, E. T., *et al.*: Biochemical mechanisms of the sodium pump. *Annu. Rev. Physiol., 41*:397, 1979.

Waxman, S. G. (ed.): Physiology and Pathobiology of Axons. New York, Raven Press, 1978.

9

Contraction of Skeletal Muscle

Approximately 40 per cent of the body is skeletal muscle and another 5 to 10 per cent is smooth and cardiac muscle. Many of the same principles of contraction apply to all these different types of muscle, but in the present chapter the function of skeletal muscle is mainly considered, while the specialized functions of smooth muscle will be discussed in the following chapter, and cardiac muscle in Chapter 11.

PHYSIOLOGIC ANATOMY OF SKELETAL MUSCLE

THE SKELETAL MUSCLE FIBER

Figure 9–1 illustrates the organization of skeletal muscle, showing that all skeletal muscles are made of numerous fibers ranging between 10 and 80 microns in diameter. Each of these fibers in turn is made up of successively smaller subunits, also illustrated in Figure 9–1, that will be described in subsequent paragraphs.

The Sarcolemma. The sarcolemma is the cell membrane of the muscle fiber. However, the sarcolemma consists of a true cell membrane, called the *plasma membrane,* and a thin layer of polysaccharide material similar to that of the basement membrane surrounding blood capillaries; thin collagen fibrillae are also present in the outer layer of the sarcolemma. At the ends of the muscle fibers, these surface layers of the sarcolemma fuse with tendon fibers, which in turn collect into bundles to form the muscle tendons, and thence insert into the bones.

Myofibrils; Actin and Myosin Filaments. Each muscle fiber contains several hundred to several thousand myofibrils, which are illustrated by the many small open dots in the cross-sectional view of Figure 9–1C. Each myofibril (Figure 9–1D) in turn has, lying side-by-side, about 1500 *myosin filaments* and 3000 *actin filaments,* which are large polymerized protein molecules that are responsible for muscle contraction. These can be seen in longitudinal view in the electron micrograph of Figure 9–2, and are represented diagrammatically in Figure 9–1E. The thick filaments are *myosin* and the thin filaments are *actin.* Note that the myosin and actin filaments partially interdigitate and thus cause the myofibrils to have alternate light and dark bands. The light bands, which contain only actin filaments, are called *I bands.* The dark bands, which contain the myosin filaments as well as the ends of the actin filaments where they overlap the myosin, are called *A bands.* Note also the small projections from the sides of the myosin filaments. These are called *cross-bridges.* They protrude from the surfaces of the myosin filaments along the entire extent of the filament, except in the very center. It is interaction between these cross-bridges and the actin filaments that causes contraction.

Figure 9–1E also shows that the actin filaments are attached to the so-called Z membrane or Z *disc,* and the filaments extend on either side of the Z membrane to interdigitate with the myosin filaments. The Z membrane also passes from myofibril to myofibril, attaching the myofibrils to each other all the way across the muscle fiber.

The portion of a myofibril (or of the whole muscle fiber) that lies between two successive Z membranes is called a *sarcomere.* When the muscle fiber is at its normal fully stretched resting length, the length of the sarcomere is about 2.0 microns. At this length, the actin filaments completely

SKELETAL MUSCLE

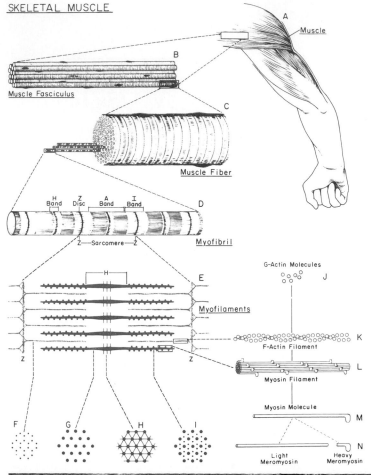

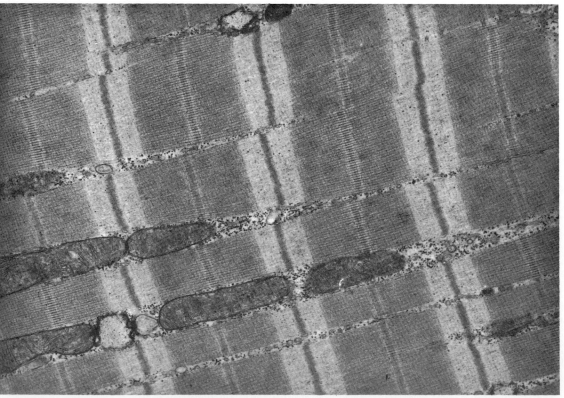

Figure 9–1. Organization of skeletal muscle, from the gross to the molecular level. F, G, H, and I are cross-sections at the levels indicated. (From Bloom and Fawcett: A Textbook of Histology. Philadelphia, W. B. Saunders Company, 1975. Drawn by Sylvia Colard Keene.)

Figure 9–2. Electron micrograph of muscle myofibrils, showing the detailed organization of actin and myosin filaments. Note the mitochondria lying between the myofibrils. (From Fawcett: The Cell, 2nd Ed. Philadelphia, W. B. Saunders Company, 1981.)

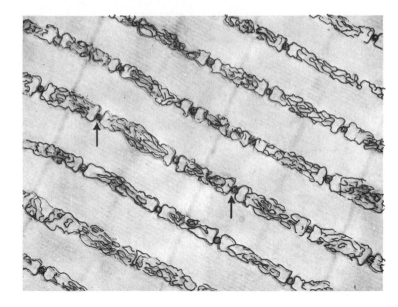

Figure 9–3. Sarcoplasmic reticulum surrounding the myofibril, showing the longitudinal system paralleling the myofibrils. Also shown in cross-section are the T tubules that lead to the exterior of the fiber membrane and that contain extracellular fluid (arrows). (From Fawcett: The Cell, 2nd Ed. Philadelphia, W. B. Saunders Company, 1981.)

overlap the myosin filaments and are just beginning to overlap each other. We shall see later that it is at this length that the sarcomere also is capable of generating its greatest force of contraction.

The Sarcoplasm. The myofibrils are suspended inside the muscle fiber in a matrix called *sarcoplasm*, which is composed of the usual intracellular constituents. The fluid of the sarcoplasm contains large quantities of potassium, magnesium, phosphate, and protein enzymes. Also present are tremendous numbers of *mitochondria* that lie between and parallel to the myofibrils, a condition that indicates the great need of the contracting myofibrils for large amounts of ATP formed by the mitochondria.

The Sarcoplasmic Reticulum. Also in the sarcoplasm is an extensive endoplasmic reticulum, which in the muscle fiber is called the *sarcoplasmic reticulum.* This reticulum has a special organization that is extremely important in the control of muscle contraction, which will be discussed later in the chapter. The electron micrograph of Figure 9–3 illustrates the arrangement of this sarcoplasmic reticulum and shows how extensive it can be. The more rapidly contracting types of muscle have an especially extensive sarcoplasmic reticulum, indicating that this structure is important in causing rapid muscle contraction, as will also be discussed later.

MOLECULAR MECHANISM OF MUSCLE CONTRACTION

Sliding Mechanism of Contraction. Figure 9–4 illustrates the basic mechanism of muscle contraction. It shows the relaxed state of a sarcomere (above) and the contracted state (below). In the relaxed state, the ends of the actin filaments derived from two successive Z membranes barely overlap each other while at the same time completely overlapping the myosin filaments. On the other hand, in the contracted state these actin filaments have been pulled inward among the myosin filaments so that they now overlap each other to a major extent. Also, the Z membranes have been pulled by the actin filaments up to the ends of the myosin filaments. Indeed, the actin filaments can be pulled together so tightly that the ends of the myosin filaments actually buckle during very intense contraction. Thus, muscle contraction occurs by a *sliding filament mechanism.*

But what causes the actin filaments to slide inward among the myosin filaments? Unfortunately, we do not completely know the answer to this question. However, in the next few sections we will describe what is known about the details of the molecular processes of contraction.

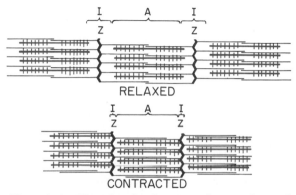

Figure 9–4. The relaxed and contracted states of a myofibril, showing sliding of the actin filaments into the channels between the myosin filaments.

MOLECULAR CHARACTERISTICS OF
THE CONTRACTILE FILAMENTS

The Myosin Filament. The myosin filament is composed of approximately 200 myosin molecules, each having a molecular weight of 490,000. Figure 9–5, section A, illustrates an individual molecule; section B illustrates the organization of the molecules to form a myosin filament as well as its interaction with the ends of two actin filaments.

The myosin molecule is composed of two parts: *light meromyosin* and *heavy meromyosin.* The light meromyosin consists of two peptide strands wound around each other in a helix. The heavy meromyosin in turn consists of two parts: first, a double helix similar to that of the light meromyosin; second, a *head* attached to the end of the double helix. The head itself is a composite of two globular protein masses.

It is believed that the myosin molecule is especially flexible at two points — at the juncture between the light meromyosin and the heavy meromysin and between the body of the heavy meromyosin and the head. These two areas are called *hinges.*

In section B of Figure 9–5 the central portion of a myosin filament is illustrated. The body of this filament is composed of parallel strands of light meromyosin. The heavy meromyosin portions of the myosin molecules protrude from all sides of the myosin filament, as illustrated in the figure. These protrusions constitute the *cross-bridges.* The heads of the cross-bridges lie in apposition to the actin filaments, whereas the rod portions of the cross-bridges act as arms that allow the heads either to extend far outward from the body of the myosin filament or to lie close to the body.

Note also that the arms of the cross-bridges extend in both directions away from the center-most part of the filament. Therefore, in the very center of the myosin filament, for a length of about 0.2 micron, there are no cross-bridge heads.

The total length of the myosin filament is 1.6 microns, and the 200 myosin molecules allow the formation of 100 pairs of cross-bridges — 50 pairs on each end of the myosin filament.

Now, to complete the picture, the myosin filament is twisted so that it makes one complete revolution for each three pairs of cross-bridges. Each twist has a length of 42.9 nanometers, with each pair of cross-bridges axially displaced from the previous pair by 120 degrees.

The Actin Filament. The actin filament is also complex. It is composed of three different components: *actin, tropomyosin,* and *troponin.*

The backbone of the actin filament is a double-stranded F-actin protein molecule, illustrated in Figure 9–6. The two strands are wound in a helix in the same manner as the myosin molecule, but with a complete revolution every 70 nanometers.

Each strand of the double F-actin helix is composed of polymerized G-actin molecules, each having a molecular weight of 47,000. There are approximately 13 of these molecules in each revolution of each strand of helix. Attached to each one of the G-actin molecules is one molecule of ADP. It is believed that these ADP molecules are the active sites on the actin filaments with which the cross-bridges of the myosin filaments interact to cause muscle contraction. The active sites on the two F-actin strands of the double helix are staggered, giving one active site on the overall actin filament approximately every 2.7 nanometers.

The Tropomyosin Strands. The actin filament also contains two additional protein strands that are polymers of *tropomyosin* molecules, each molecule having a molecular weight of 70,000 and

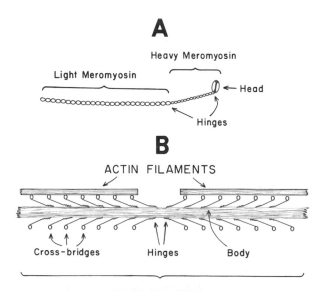

A

Heavy Meromyosin

Light Meromyosin

← Head

Hinges

B

ACTIN FILAMENTS

Cross-bridges   Hinges   Body

MYOSIN FILAMENT

Figure 9–5. A, The myosin molecule. B, Combination of many myosin molecules to form a myosin filament. Also shown are the cross-bridges and the interaction between the heads of the cross-bridges and adjacent actin filaments.

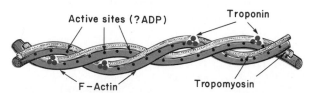

Active sites (?ADP)   Troponin

F–Actin   Tropomyosin

Figure 9–6. The actin filament, composed of two helical strands of F-actin and two tropomyosin strands that lie in the grooves between the actin strands. Attaching the tropomyosin to the actin are several troponin complexes.

extending a length of 40 nanometers. It is believed that each tropomyosin strand is loosely attached to an F-actin strand and that in the resting state it physically covers the active sites of the actin strands so that interaction cannot occur between the actin and myosin to cause contraction.

Troponin and Its Role in Muscle Contraction. Attached approximately two-thirds the distance along each tropomyosin molecule is a complex of three globular protein molecules called *troponin*. One of the globular proteins has a strong affinity for actin, another for tropomyosin, and a third for calcium ions. This complex is believed to attach the tropomyosin to the actin. The strong affinity of the troponin for calcium ions is believed to initiate the contraction process, as will be explained in the following section.

Interaction of Myosin and Actin Filaments to Cause Contraction

Inhibition of the Actin Filament by the Troponin-Tropomyosin Complex; Activation by Calcium Ions. A pure actin filament without the presence of the troponin-tropomyosin complex normally binds strongly with myosin molecules. But if the troponin-tropomyosin complex is added to the actin filament, this binding does not take place. Therefore, it is believed that the active sites on the actin filament of the relaxed muscle are inhibited (or perhaps physically covered) by the troponin-tropomyosin complex. Consequently, they cannot interact with the myosin filaments to cause contraction. Before contraction can take place the inhibitory effect of the troponin-tropomyosin complex must itself be inhibited.

In the presence of large amounts of calcium ions the inhibitory effect of the troponin-tropomyosin on the actin filaments is itself inhibited. When calcium ions combine with the calcium binding subunit of troponin, the troponin complex supposedly undergoes a conformational change that in some way tugs on the tropomyosin protein strand and supposedly moves the tropomyosin strand deeper into the groove between the two actin strands. This is believed to "uncover" the active sites of the actin, thus allowing contraction to proceed. Though this is a hypothetical mechanism, nevertheless it does emphasize that the normal relationship between the tropomyosin-troponin complex and actin is altered by calcium ions — a condition that leads to contraction.

Interaction Between the "Activated" Actin Filament and the Myosin Molecule — The Ratchet Theory of Contraction. It is believed that as soon as the actin filament becomes acti-

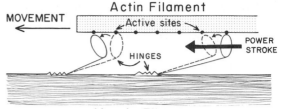

Figure 9–7. The ratchet mechanism for contraction of the muscle.

vated by the calcium ions, the heads of the cross-bridges from the myosin filaments immediately become attracted to the active sites of the actin filament and this in some way causes contraction to occur. Though the precise manner by which this interaction between the cross-bridges and the actin causes contraction is still unknown, a suggested hypothesis is the so-called *ratchet theory of contraction*.

Figure 9–7 illustrates postulated ratchet mechanism, showing the heads of two cross-bridges attaching to and disengaging from the active sites of an actin filament. It is believed that when the head attaches to an active site this attachment simultaneously causes profound changes in the intramolecular forces in the head and arm of the cross-bridge. The new alignment of forces causes the head to tilt toward the arm and to drag the actin filament along with it. This tilt of the head of the cross-bridge is called the *power stroke*. Then, immediately after tilting, the head automatically splits away from the active site and returns to its normal perpendicular direction. In this position it combines with an active site further down along the actin filament; then, a similar tilt takes place again to cause a new power stroke, and the actin filament moves another step. Thus, the heads of the cross-bridges bend back and forth, and step by step pull the actin filament toward the center of the myosin filament. Thus, the movements of the cross-bridges use the active sites of the actin filaments as cogs of a *ratchet*.

Each one of the cross-bridges is believed to operate independently of all others, each attaching and pulling in a continuous, alternating ratchet cycle. Therefore, the greater the number of cross-bridges in contact with the actin filament at any given time, the greater, theoretically, is the force of contraction.

ATP as the Source of Energy for Contraction. When a muscle contracts against a load, work is performed, and energy is required. It is found that large amounts of ATP are cleaved to form ADP during the contraction process. Although we still do not know exactly how ATP is used to provide the energy for contraction, it is believed that once the head of the cross-bridge has completed its power stroke, the tilted position of the head now exposes a site in the head where ATP can bind. Therefore, one molecule of ATP binds with the head, and this binding in turn causes detachment of the head from the active site. In addition, the ATP is itself immediately cleaved by a very potent ATPase activity of the heavy meromyosin. The energy released supposedly tilts the head back to its normal perpendicular condition and theoretically "cocks" the head in this position. Then, when the "cocked" head, with its stored energy derived from the cleaved ATP, binds with a new active site on the actin filament, it becomes uncocked and once again provides another power stroke.

Thus, the process proceeds again and again until the actin filament pulls the Z membrane up against the ends of the myosin filaments or until the load on the muscle becomes too great for further pulling to occur.

RELATIONSHIP BETWEEN ACTIN AND MYOSIN FILAMENT OVERLAP AND TENSION DEVELOPED BY THE CONTRACTING MUSCLE

Figure 9–8 illustrates the relationship between the length of sarcomere and the tension developed by a single, contracting, *isolated* muscle fiber. To the right are illustrated different degrees of overlap of the myosin and actin filaments at different sarcomere lengths. At point D on the diagram, the actin filament has pulled all the way out to the end of the myosin filament with no overlap at all. At this point, the tension developed by the activated muscle is zero. Then, as the

sarcomere shortens and the actin filament overlaps the myosin filament more and more, the tension increases progressively until the sarcomere length decreases to about 2.2 microns. At this point the actin filament has already overlapped all the cross-bridges of the myosin filament but has not yet reached the center of the myosin filament. Upon further shortening, the sarcomere maintains full tension until point B at a sarcomere length of approximately 2.0 microns. It is at this point that the ends of the two actin filaments begin to overlap. As the sarcomere length falls from 2 microns down to about 1.65 microns, at point A, the strength of contraction decreases. It is at this point that the two Z membranes of the sarcomere abut the ends of the myosin filaments. Then, as contraction proceeds to still shorter sarcomere lengths, the ends of the myosin filaments are actually crumpled, and, as illustrated in Figure 9–8, the strength of contraction also decreases precipitously. This diagram illustrates that the greater the number of cross-bridges pulling the actin filaments, the greater is the strength of contraction.

INITIATION OF MUSCLE CONTRACTION: EXCITATION-CONTRACTION COUPLING

THE MUSCLE ACTION POTENTIAL

Initiation of contraction in skeletal muscle begins with action potentials in the muscle fibers. These cause release of calcium ions from the sarcoplasmic reticulum. It is the calcium ions that in turn initiate the chemical events of the contractile process.

Almost everything discussed in Chapter 8 regarding initiation and conduction of action potentials in nerve fibers applies equally well to skeletal muscle fibers, except for quantitative differences. Some of the quantitative aspects of muscle potentials are the following:

1. Resting membrane potential: Approximately −90 millivolts in skeletal fibers — the same as in large myelinated nerve fibers.

2. Duration of action potential: 1 to 5 milliseconds in skeletal muscle — about five times as long as in large myelinated nerves.

3. Velocity of conduction: 3 to 5 meters per second — about $1/18$ the velocity of conduction in the large myelinated nerve fibers that excite skeletal muscle.

Excitation of Skeletal Muscle Fibers by Nerves. In normal function of the body, skeletal muscle fibers are excited by large myelinated nerve fibers. These attach to the skeletal muscle

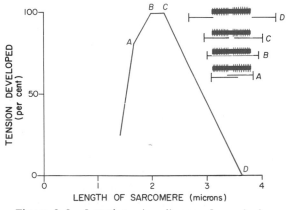

Figure 9–8. Length-tension diagram for a single sarcomere, illustrating maximum strength of contraction when the sarcomere is 2.0 to 2.2 microns in length. At the upper right are shown the relative positions of the actin and myosin filaments at different sarcomere lengths from point A to point D. (Modified from Gordon, Huxley, and Julian: *J. Physiol.*, 171:28P, 1964.)

fibers at the neuromuscular junction, which will be discussed in detail in the following chapter. Except for 2 per cent of the muscle fibers, there is only one neuromuscular junction to each muscle fiber; this junction is located near the middle of the fiber. Therefore, the action potential spreads from the middle of the fiber towards its two ends. This spreading is important because it allows nearly coincident contraction of all sarcomeres of the muscles so that they can all contract together rather than separately.

SPREAD OF THE ACTION POTENTIAL TO THE INTERIOR OF THE MUSCLE FIBER BY WAY OF THE TRANSVERSE TUBULE SYSTEM

The skeletal muscle fiber is so large that action potentials spreading along the membrane cause almost no current flow deep within the fiber. Yet, to cause contraction, these electrical currents must penetrate to the vicinity of all the separate myofibrils. This is achieved by transmission of

the action potentials along *transverse tubules* (T tubules) that penetrate all the way through the muscle fiber from one side to the other. The T tubule action potentials in turn cause the sarcoplasmic reticulum to release calcium ions in the immediate vicinity of all the myofibrils, and it is these calcium ions that in turn cause contraction. Now, let us describe this system in greater detail.

The Transverse Tubule–Sarcoplasmic Reticulum System. Figure 9–9 illustrates a group of myofibrils surrounded by the transverse tubule-sarcoplasmic reticulum system. Note that the transverse tubules penetrate all the way from one side of the muscle fiber to the opposite side — also, that where the T tubules originate from the cell membrane they are open to the exterior. Therefore, they communicate with the fluid surrounding the muscle fiber and contain extracellular fluid in their lumens. In other words, the T tubules are internal extensions of the cell membrane. Therefore, when an action potential spreads over a muscle fiber membrane, it spreads

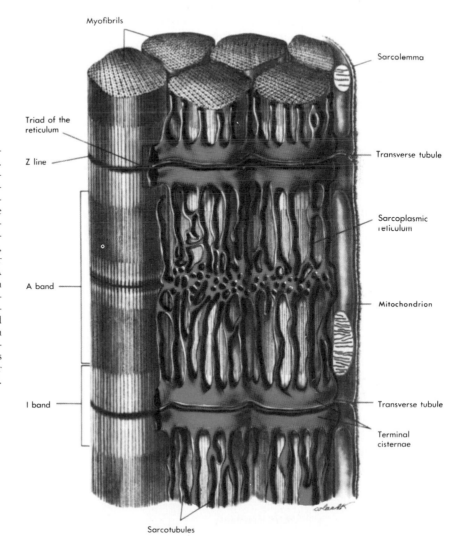

Figure 9–9. The transverse tubule–sarcoplasmic reticulum system. Note the *longitudinal tubules* that terminate in large *cisternae*. The cisternae in turn abut the transverse tubules. Note also that the transverse tubules communicate with the outside of the cell membrane. This illustration was drawn from frog muscle, which has one transverse tubule per sarcomere, located at the Z line. A similar arrangement is found in mammalian heart muscle, but mammalian skeletal muscle has two transverse tubules per sarcomere, located at the A-I junctions. (From Bloom and Fawcett: A Textbook of Histology. Philadelphia, W. B. Saunders Company, 1975. Modified after Peachey: *J. Cell Biol. 25*:209, 1965. Drawn by Sylvia Colard Keene.)

Myofibrils

Sarcolemma

Triad of the reticulum

Z line

Transverse tubule

A band

Sarcoplasmic reticulum

Mitochondrion

I band

Transverse tubule

Terminal cisternae

Sarcotubules

along the T tubules to the deep interior of the muscle fiber as well.

Figure 9–9 shows the extensiveness of the *sarcoplasmic reticulum* as well. This is composed of two major parts: (1) long *longitudinal tubules* that terminate in (2) large chambers called *terminal cisternae*. This reticulum is also seen in the electron micrograph of Figure 9–3.

In cardiac muscle there is a single T tubule network for each sarcomere, located at the level of the Z membrane as illustrated in Figure 9–9. However, in skeletal muscle there are two T tubule networks for each sarcomere located near the two ends of the myosin filaments, which are the points where the actual mechanical forces of muscle contraction are created.

RELEASE OF CALCIUM IONS BY THE CISTERNAE OF THE SARCOPLASMIC RETICULUM

One of the special features of the sarcoplasmic reticulum is that it contains calcium ions in very high concentration, and many of these ions are released when the adjacent T tubule is excited.

Figure 9–10 shows that the action potential of the T tubule causes current flow through the cisternae where they abut the T tubule. This current flow in turn causes rapid release of calcium ions from the cisternae into the surrounding sarcoplasm. Presumably this release results from the opening of calcium pores similar to the opening of sodium pores at the onset of the action potential, though the actual mechanism is still unknown.

The calcium ions that are thus released from the cisternae diffuse to the adjacent myofibrils, where they bind strongly with troponin, as discussed in an earlier section, and this in turn elicits the muscle contraction.

The Calcium Pump for Removing Calcium Ions at the End of Contraction. Once the calcium ions have been released from the cisternae and have diffused to the myofibrils, muscle contraction will then continue as long as the calcium ions are still present in high concentration in the sarcoplasmic fluid. However, a continually active calcium pump located in the walls of the sarcoplasmic reticulum pumps calcium ions out of the sarcoplasmic fluid back into the vesicular cavities of the reticulum. This pump can concentrate the calcium ions about 2000-fold inside the reticulum, a condition that allows massive build-up of calcium in the sarcoplasmic reticulum and also causes almost total depletion of calcium ions in the fluid of the myofibrils. Therefore, except immediately after an action potential, the calcium ion concentration in the myofibrils is kept at an extremely low level (about 10^{-7} molar). This level is too little to elicit contraction, but full excitation of the T tubule–sarcoplasmic reticulum system causes enough release of calcium ions to increase the concentration in the myofibrillar fluid to as high as 2×10^{-4} molar concentration, which causes maximum muscle contraction. Immediately thereafter, the calcium pump depletes the calcium ions again, and muscle contraction ceases.

The total duration of this calcium "pulse" in the usual skeletal muscle fiber lasts about $1/30$ of a second, though in heart muscle it lasts for as long as 0.3 second. It is during this calcium pulse that muscle contraction occurs. If the contraction is to continue without interruption for longer intervals, a series of such pulses must be initiated by a continuous series of repetitive action potentials, as will be discussed in more detail later in the chapter.

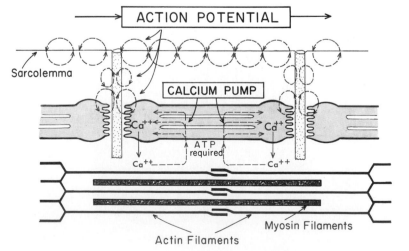

Figure 9–10. Excitation-contraction coupling in the muscle, showing an action potential that causes release of calcium ions from the sarcoplasmic reticulum and then re-uptake of the calcium ions by a calcium pump.

CHARACTERISTICS OF CONTRACTION BY THE WHOLE MUSCLE

Isometric Versus Isotonic Contraction. Muscle contraction is said to be *isometric* when the muscle does not shorten during contraction and *isotonic* when it shortens but the tension on the muscle remains constant.

There are several basic differences between isometric and isotonic contractions. First, isometric contraction does not require much sliding of myofibrils among each other, but *force* is developed. Second, in isotonic contraction sliding does occur, and a load is moved, allowing the performance of external work, an effect that causes, for reasons yet unknown, much greater expenditure of chemical energy by the muscle.

Muscles can contract both isometrically and istonically in the body, but most contractions are actually a mixture of the two. When standing, a person tenses the quadriceps muscles to tighten the knee joints and to keep the legs stiff. This is isometric contraction. On the other hand, when a person lifts a weight using the biceps, this is mainly an isotonic contraction. Finally, contractions of leg muscles during running are a mixture of isometric and isotonic contractions — isometric mainly to keep the limbs stiff when the legs hit the ground and isotonic mainly to move the limbs.

Characteristics of Isometric Muscle Contractions Recorded from Different Muscles. The body has many different sizes of skeletal muscles — from the very small stapedius muscle, only a few millimeters in length and a millimeter or so in diameter, up to the very large quadriceps muscle, with fibers almost a half meter in length. Furthermore, the fibers may be as small as 10 microns in diameter or as large as 80 microns. And, finally, the energetics of muscle contraction vary considerably from one muscle to another. These different physical and chemical characteristics often manifest themselves in the form of different characteristics of contraction, some muscles contracting rapidly and others slowly.

Figure 9–11 illustrates isometric contractions of three different types of skeletal muscles caused by single action potentials: an ocular muscle, which has a duration of contraction of less than $1/100$ second; the gastrocnemius muscle, which has a duration of contraction of about $1/30$ second; and the soleus muscle, which has a duration of contraction of about $1/10$ second. It is interesting that these durations of contractions are adapted to the function of each of the respective muscles, for ocular movements must be extremely rapid to maintain fixation of the eyes upon specific ob-

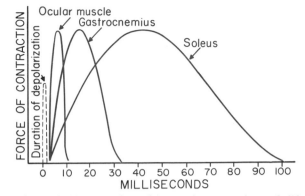

Figure 9–11. Duration of isometric contractions of different types of mammalian muscles, showing also a latent period between the action potential and muscle contraction.

jects, the gastrocnemius muscle must contract moderately rapidly to provide sufficient velocity of limb movement for running and jumping, while the soleus muscle is concerned principally with slow reactions for continual support of the body against gravity.

MECHANICS OF SKELETAL MUSCLE CONTRACTION

THE MOTOR UNIT

Each motor nerve fiber that leaves the spinal cord usually innervates many different muscle fibers, the number depending on the type of muscle. All the muscle fibers innervated by a single motor nerve fiber are called a *motor unit*. In general, small muscles that react rapidly and whose control is exact have few muscle fibers (as few as 2 to 3 in some of the laryngeal muscles) in each motor unit and have a relatively large number of nerve fibers going to each muscle. On the other hand, the large muscles that do not require very fine degree of control, such as the gastrocnemius muscle, may have several hundred muscle fibers in a motor unit. An average figure for all the muscles of the body can be considered to be about 150 muscle fibers to the motor unit.

SUMMATION OF MUSCLE CONTRACTION

Summation means the adding together of individual muscle twitches to make strong and concerted muscle movements. In general, summation occurs in two different ways: (1) by increasing the number of motor units contracting simultaneously and (2) by increasing the rapidity of contraction of individual motor units. These are called, respectively, *multiple motor unit summation* and *wave summation* (or spatial summation and temporal summation).

Multiple Motor Unit Summation. Even within a single muscle, the numbers of muscle fibers and their sizes in the different motor units vary tremendously, so that one motor unit may be as much as 50 times as strong as another. The smaller motor units are far more easily excited than are the larger ones because they are innervated by smaller nerve fibers whose cell bodies in the spinal cord have a naturally high level of excitability. This effect causes the gradations of muscle strength during weak muscle contraction to occur in very small steps. Yet, the steps become progressively greater as the intensity of contraction increases because the larger motor units then begin to contract.

Wave Summation. Figure 9–12 illustrates the principles of wave summation, showing in the lower left-hand corner several muscle contractions caused by single action potentials followed by successive muscle contractions at various frequencies. When the frequency of contractions rises above 10 per second, the first muscle contraction is not completely over by the time the second one begins. Furthermore, at increasingly rapid rates of contraction, the degree of summation of successive contractions becomes greater and greater, because the successive contractions appear at earlier times following the preceding contraction.

Tetanization. When a muscle is stimulated at progressively greater frequencies, a frequency is finally reached at which the successive contractions fuse together and cannot be distinguished one from the other. This state is called *tetanization.*

Tetanization results partly from the viscous properties of the muscle and partly from the fact that the successive pulsatile states of activation of the muscle fiber fuse into a long continual state of activation. Once the critical frequency for tetanization is reached, further increase in rate of stimulation increases the force of contraction

only a few more per cent, as shown in Figure 9–12.

Asynchronous Summation of Motor Units. Actually it is rare for either multiple motor unit summation or wave summation to occur separately from the other in normal muscle function. Instead, *the different motor units fire asynchronously;* that is, while one is contracting another is relaxing; then another fires, followed by still another, and so forth. Consequently, even when motor units fire as infrequently as 5 times per second, the muscle contraction, though weak, is nevertheless very smooth.

Maximum Strength of Contraction. The maximum strength of tetanic contraction of a muscle is about 3.5 kilograms per square centimeter cross-section of muscle, or 50 pounds per square inch. Since a quadriceps muscle can at times have as much as 16 square inches of muscle belly, as much as 800 pounds of tension may at times be applied to the patellar tendon.

MUSCLE FATIGUE

Prolonged and strong contraction of a muscle leads to the well-known state of muscle fatigue. This results simply from inability of the contractile and metabolic processes of the muscle fibers to continue supplying the same work output. The nerve continues to function properly, the nerve impulses pass normally through the neuromuscular junction into the muscle fiber, and even normal action potentials spread over the muscle fibers, but the contraction becomes weaker and weaker because of reduction of ATP formation in the muscle fibers themselves.

Interruption of blood flow through a contracting muscle leads to almost complete muscle fatigue in less than a minute because of the obvious loss of nutrient supply.

MUSCLE HYPERTROPHY

Forceful muscular activity causes the muscle size to increase, a phenomenon called hypertrophy. The diameters of the individual muscle fibers increase, the sarcoplasm increases, and the fibers gain in various nutrient and intermediary metabolic substances, such as adenosine triphosphate, creatine phosphate, glycogen, intracellular lipids, and even many additional mitochondria. It is likely that the myofibrils also increase in size and perhaps in numbers as well, but this has not been proved. Briefly, muscular hypertrophy increases both the motive power of the muscle and the nutrient mechanisms for maintaining increased motive power.

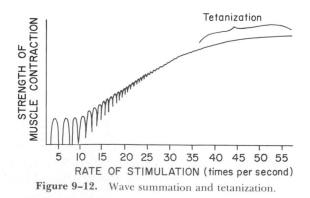

Figure 9–12. Wave summation and tetanization.

Weak muscular activity, even when sustained over long periods of time, does not result in significant hypertrophy. Instead, hypertrophy results mainly from *very* forceful muscle activity, though the activity might occur for only a few minutes each day. For this reason, strength can be developed in muscles much more rapidly when strong "isometric" exercise is used rather than simply prolonged mild exercise. Indeed, essentially no enlargement of the muscle fibers occurs unless the muscle contracts to at least 75 per cent of its maximum tension.

On the other hand, prolonged muscle activity does increase muscle endurance, causing increases in the oxidative enzymes, myoglobin, and even blood capillaries — all of which are essential to increased muscle metabolism.

MUSCLE ATROPHY

Muscle atrophy is the reverse of muscle hypertrophy; it results any time a muscle is not used or even when a muscle is used only for very weak contractions. Atrophy is particularly likely to occur when limbs are placed in casts, thereby preventing muscular contraction. As little as one month of disuse can sometimes decrease the muscle size to one-half normal.

Atrophy Caused by Muscle Denervation. When a muscle is denervated it immediately begins to atrophy, and the muscle continues to decrease in size for several years. If the muscle becomes reinnervated during the first three to four months, full function of the muscle usually returns, but after four months of denervation some of the muscle fibers usually will have degenerated. Reinnervation after two years rarely results in return of any function at all. Pathological studies show that the muscle fibers have by that time been replaced by fat and fibrous tissue.

THE ELECTROMYOGRAM

Each time an action potential passes along a muscle fiber a small portion of the electrical current spreads away from the muscle as far as the skin. If many muscle fibers contract simultaneously, the summated electrical potentials at the skin may be very great. If two electrodes are placed on the skin or needle electrodes are inserted into the muscle, an electrical recording called the electromyogram can be made when the muscle is stimulated. Figure 9–13 illustrates a typical electromyographic recording of the gastrocnemius muscle during a moderate contraction. Electromyograms are frequently used clinically to discern abnormalities of muscle excitation. Two such abnormalities are muscle *fasciculation* and *fibrillation*.

REFERENCES

Basmajian, J. V.: Muscles Alive; Their Functions Revealed by Electromyography. Baltimore, Williams & Wilkins, 1978.

Bourne, G. H. (ed.): The Structure and Function of Muscle, 2nd Ed. New York, Academic Press, 1973.

Buchthal, F., and Schmalbruch, H.: Motor unit of mammalian muscle. *Physiol. Rev.*, 60:90, 1980.

Caputo, C.: Excitation and contraction processes in muscle. *Annu. Rev. Biophys. Bioeng.*, 7:63, 1978.

Cohen, C.: The protein switch of muscle contraction. *Sci. Am.*, 233(5):36, 1975.

Costantin, L. L.: Activation in striated muscle. *In* Brookhart, J. M., and Mountcastle, V. B. (eds.): Handbook of Physiology. Sec. 1, Vol. 1. Baltimore, Williams & Wilkins, 1977, p. 215.

Curtin, N. A., and Woledge, R. C.: Energy changes in muscular contraction. *Physiol. Rev.*, 58:690, 1978.

Ebashi, S.: Excitation-contraction coupling. *Annu. Rev. Physiol.*, 38:293, 1976.

Endo, M.: Calcium release from the sarcoplasmic reticulum. *Physiol. Rev.*, 57:71, 1977.

Fabiato, A., and Fabiato, F.: Calcium and cardiac excitation contraction coupling. *Annu. Rev. Physiol.*, 41:473, 1979.

Fuchs, F.: Striated muscle. *Annu. Rev. Physiol.*, 36:461, 1974.

Holloszy, J. O., and Booth, F. W.: Biochemical adaptations to endurance exercise in muscle. *Annu. Rev. Physiol.*, 38:273, 1976.

Homsher, E., and Kean, C. J.: Skeletal muscle energetics and metabolism. *Annu. Rev. Physiol.*, 40:93, 1978.

Huxley, A. F., and Gordon, A. M.: Striation patterns in active and passive shortening of muscle. *Nature (Lond.)*, 193:280, 1962.

Huxley, H. E.: Muscular contraction and cell motility. *Nature*, 243:445, 1973.

Lyman, R. W.: Kinetic analysis of myosin and actinomysin ATPase. *Annu. Rev. Biophys. Bioeng.*, 8:145, 1979.

Mauro, A., and Bischoff, R. (eds.): Muscle Regeneration. New York, Raven Press, 1979.

Nelson, P. G.: Nerve and muscle cells in culture. *Physiol. Rev.*, 55:1, 1975.

Northrip, J. W., et al.: Introduction to Biomechanic Analysis of Sport. Dubuque, Iowa, W. C. Brown Co., 1979.

Rasch, P. J., and Burke, R. K.: Kinesiology and Applied Anatomy: The Science of Human Movement. Philadelphia, Lea & Febiger, 1978.

Sugi, H., and Pollack, G. H. (eds.): Cross-Bridge Mechanism in Muscle Contraction. Baltimore, University Park Press, 1979.

Tada, M., et al.: Molecular mechanism of active calcium transport by sarcoplasmic reticulum. *Physiol. Rev.*, 58:1, 1978.

Toida, N., et al.: Obliquely striated muscle. *Physiol. Rev.*, 55:700, 1975.

Tregear, R. T., and Marston, S. B.: The crossbridge theory. *Annu. Rev. Physiol.*, 41:723, 1979.

Walton, J. N. (ed.): Disorders of Voluntary Muscle. New York, Churchill Livingstone, 1974.

Winter, D. A.: Biomechanics of Human Movement. New York, John Wiley & Sons, 1979.

Figure 9–13. Electromyogram recorded during contraction of the gastrocnemius muscle.

10

Neuromuscular Transmission; and Function of Smooth Muscle

TRANSMISSION OF IMPULSES FROM NERVES TO SKELETAL MUSCLE FIBERS: THE NEUROMUSCULAR JUNCTION

The skeletal muscles are innervated by large myelinated nerve fibers that originate in the large motoneurons of the anterior horns of the spinal cord. It was pointed out in the previous chapter that each nerve fiber normally branches many times and stimulates from three to several hundred skeletal muscle fibers. The nerve ending makes a junction, called the *neuromuscular junction,* with the muscle fiber approximately at the fiber's midpoint so that the action potential in the fiber travels in both directions. Normally there is only one such junction per muscle fiber.

Physiologic Anatomy of the Neuromuscular Junction. Figure 10–1A and B illustrates the neuromuscular junction between a large myelinated nerve fiber and a skeletal muscle fiber. The nerve fiber branches at its end to form a complex of branching nerve *terminals* called the *end-plate,* which invaginates into the muscle fiber but lies entirely outside the muscle fiber plasma membrane.

Figure 10–1C shows an electron micrographic sketch of the juncture between a single-branch axon terminal and the muscle fiber membrane. The invagination of the membrane is called the *synaptic gutter* or *synaptic trough,* and the space between the terminal and the fiber membrane, about 20 to 30 nanometers wide, is called the *synaptic cleft.* This cleft is filled with a gelatinous

"ground" substance through which diffuses extracellular fluid. At the bottom of the gutter are numerous *folds* of the muscle membrane, which form *subneural clefts* that greatly increase the surface area at which the synaptic transmitter can act. In the axon terminal are many mitochondria that supply energy mainly for synthesis of the excitatory transmitter *acetylcholine* that, in turn, excites the muscle fiber. The acetylcholine is synthesized in the cytoplasm of the terminal but is rapidly absorbed into many small synaptic vesicles, approximately 300,000 of which are normally in all the terminals of a single end-plate. On the surfaces of the subneural clefts are aggregates of the enzyme *cholinesterase,* which is capable of destroying acetylcholine, as is explained in further detail below.

Secretion of Acetylcholine by the Axon Terminals. When a nerve impulse reaches the neuromuscular junction, about 300 vesicles of acetylcholine are released by the terminals into the synaptic clefts and subneural clefts. This release of vesicles is caused by movement of calcium ions from the extracellular fluid into the terminals when the action potential depolarizes their membranes. In the absence of calcium or in the presence of excess magnesium in the extracellular fluid, the release of acetylcholine is greatly depressed.

Destruction of the Released Acetylcholine by Cholinesterase. Within approximately 2 to 3 milliseconds after acetylcholine is released into the cleft, a small portion of it diffuses out of the gutter and no longer acts on the muscle fiber membrane, but the greater bulk of it is destroyed

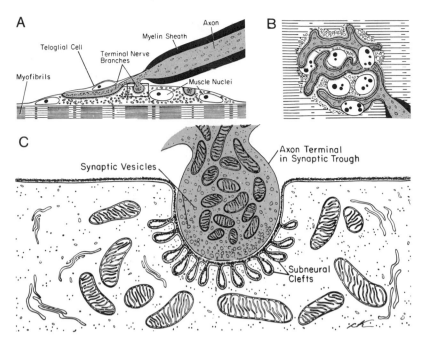

Figure 10–1. Different views of the motor end-plate. A, Longitudinal section through the end-plate. B, Surface view of the end-plate. C, Electronmicrographic appearance of the contact point between one of the axon terminals and the muscle fiber membrane, representing the rectangular area shown in A. (From Bloom and Fawcett: A Textbook of Histology. Philadelphia, W. B. Saunders Company, 1975. Modified from R. Couteaux.)

by the cholinesterase on the surfaces of the subneural clefts. The very short period of time that the acetylcholine remains in contact with the muscle fiber membrane — 2 to 3 milliseconds — is almost always sufficient to excite the muscle fiber, and yet the rapid removal of the acetylcholine prevents re-excitation after the muscle fiber has recovered from the first action potential.

The "End-Plate Potential" and Excitation of the Skeletal Muscle Fiber. Even though the acetylcholine released into the space between the end-plate and the muscle membrane lasts for only a minute fraction of a second, during this period of time it can affect the muscle membrane sufficiently to open large pores with diameters of 0.6 to 0.7 nanometers (6 to 7 Angstroms) in the membrane. This makes the membrane very permeable to sodium ions (as well as most other ions), allowing rapid influx of sodium into the muscle fiber. As a result, the membrane potential rises *in the local area of the end-plate* as much as 50 to 75 millivolts, creating a *local potential* called the *end-plate potential.*

The mechanism by which acetylcholine increases the permeability of the muscle membrane is probably the following: It is believed that the muscle membrane contains a special protein molecule called an *acetylcholine receptor substance* to which the acetylcholine binds. Under the influence of the acetylcholine this receptor supposedly undergoes a conformational change that increases the permeability of the membrane to ions. Rapid influx of sodium ions ensues, and this elicits the end-plate potential that is responsible for initiation of the action potential at the muscle fiber membrane.

Ordinarily, each impulse that arrives at the neuromuscular junction creates an end-plate current flow about three to four times that required to stimulate the muscle fiber. Therefore, the normal neuromuscular junction is said to have a very high *safety factor.* However, artificial stimulation of the nerve fiber at rates greater than 100 times per second for several minutes often diminishes the number of vesicles of acetylcholine released with each impulse so much that impulses then often fail to pass into the muscle fiber. This is called *fatigue* of the neuromuscular junction. However, under normal functioning conditions, such fatigue rarely occurs, because almost never do the spinal nerves stimulate even the most active neuromuscular junctions more than 100 times per second.

Drugs that Affect Transmission at the Neuromuscular Junction. Many different compounds, including *methacholine, carbachol,* and *nicotine,* have the same effect on the muscle fiber as does acetylcholine. The difference between these drugs and acetylcholine is that they are not destroyed by cholinesterase or are destroyed very slowly, so that when once applied to the muscle fiber the action persists for many minutes to several hours, often causing a state of spasm.

One group of drugs, known as the *curariform drugs,* can prevent passage of impulses from the end-plate into the muscle. Thus, D-tubocurarine affects the membrane, probably by competing with acetylcholine for the receptor sites of the membrane, so that the acetylcholine cannot increase the permeability of the membrane sufficiently to initiate a depolarization wave.

Three particularly well-known drugs, *neostigmine, physostigmine*, and *diisopropyl fluorophosphate*, inactivate cholinesterase so that the cholinesterase normally in the muscle fibers will not hydrolyze acetylcholine released at the end-plate. As a result, acetylcholine increases in quantity with successive nerve impulses so that extreme amounts of acetylcholine can accumulate and repetitively stimulate the muscle fiber. This causes *muscular spasm* when even a few nerve impulses reach the muscle; death due to laryngeal spasm, which smothers the person, can ensue. Neostigmine and physostigmine combine with cholinesterase to inactivate it for several hours, after which they are displaced from the cholinesterase so that it once again becomes active. On the other hand, diisopropyl fluorophosphate, which has military potential as a very powerful "nerve" gas, actually inactivates cholinesterase for several weeks, making it a particularly lethal drug.

MUSCLE PARALYSIS CAUSED BY MYASTHENIA GRAVIS

The disease *myasthenia gravis*, which occurs rarely in human beings, causes the person to become paralyzed because of inability of the neuromuscular junctions to transmit signals from the nerve fibers to the muscle fibers. Pathologically, the number of subneural clefts in the synaptic gutter is reduced and the synaptic cleft itself is widened as much as 50 per cent. Antibodies that attack the muscle fibers have been demonstrated in the bloods of many of these patients. Therefore, it is believed that myasthenia gravis is an autoimmune disease in which patients have developed antibodies against their own muscles, and one of the effects is partial destruction of the receptor membrane of the neuromuscular junction.

Regardless of the cause, the end-plate potentials developed in the muscle fibers are too weak to stimulate the muscle fibers adequately. If the disease is intense enough, the patient dies of paralysis — in particular, of paralysis of the respiratory muscles. However, the disease can usually be ameliorated with several different drugs, such as *neostigmine*, that are capable of inactivating or destroying cholinesterase. This prevents the acetylcholine secreted by the end-plate from being destroyed immediately. Therefore, the quantity of acetylcholine present at the membrane increases progressively until finally the end-plate potential rises above threshold value for stimulating the muscle fiber.

CONTRACTION OF SMOOTH MUSCLE

In the previous chapter and thus far in the present chapter, the discussion has been concerned with skeletal muscle. We now turn to smooth muscle, which is composed of far smaller fibers — usually 2 to 5 microns in diameter and only 50 to 200 microns in length — in contrast to the skeletal muscle fibers that are as much as 20 times as large (in diameter) and thousands of times as long. Nevertheless, many of the same principles of contraction apply to smooth muscle the same as to skeletal muscle. Most important, the same chemical substances cause contraction in smooth muscle as in skeletal muscle, but the physical arrangement of smooth muscle fibers is entirely different, as we shall see.

TYPES OF SMOOTH MUSCLE

The smooth muscle of each organ is distinctive from that of most other organs in several different ways: physical dimensions, organization into bundles or sheets, response to different types of stimuli, characteristics of its innervation, and function. Yet, for the sake of simplicity, smooth muscle can generally be divided into two major types, which are illustrated in Figure 10–2: *multiunit smooth muscle* and *visceral smooth muscle*.

Multiunit Smooth Muscle. This type of smooth muscle is composed of discrete smooth muscle fibers. Each fiber operates entirely independently of the others and is often innervated by a single nerve ending, as occurs for skeletal muscle fibers. This is in contrast to visceral smooth muscle, which is controlled to a greater extent by non-nervous stimuli. An additional characteristic is that they rarely exhibit spontaneous contractions.

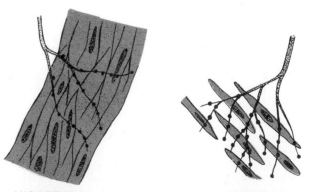

VISCERAL   MULTIUNIT

Figure 10–2. Visceral and multiunit smooth muscle fibers.

Some examples of multiunit smooth muscle found in the body are the smooth muscle fibers of the ciliary muscle of the eye, the iris of the eye, the nictitating membrane that covers the eyes in some lower animals, the piloerector muscles that cause erection of the hairs when stimulated by the sympathetic nervous system, and the smooth muscle of many of the larger blood vessels.

Visceral Smooth Muscle. Visceral smooth muscle fibers are usually arranged in sheets or bundles, and their cell membranes contact each other at multiple points to form many *gap junctions,* or *nexi*, through which ions can flow with ease from the inside of one smooth muscle fiber to the next one. Therefore, when one portion of visceral muscle is stimulated, the action potential is conducted to the surrounding fibers as well. Thus, the fibers form a *functional syncytium* that usually contracts in large areas at once. Visceral smooth muscle is found in most of the organs of the body, especially the walls of the gut, the bile ducts, the ureters, the uterus, and so forth.

THE CONTRACTILE PROCESS IN SMOOTH MUSCLE

The Chemical Basis for Contraction. Chemical studies have shown that *actin* and *myosin* filaments derived from smooth muscle interact with each other in the same way as do actin and myosin derived from skeletal muscle. Furthermore, the contractile process is activated by calcium ions, and ATP is degraded to ADP to provide the energy for contraction.

On the other hand, there are major differences in the physical organization of smooth muscle and skeletal muscle, as well as differences in other aspects of smooth muscle function such as excitation-contraction coupling, control of the contractile process by calcium ions, duration of contraction, and amount of energy required for the contractile process.

The Physical Basis for Smooth Muscle Contraction. The physical organization of the smooth muscle cell is illustrated in Figure 10–3, which shows large numbers of actin filaments attached to so-called *dense bodies*. Some of these bodies in turn are attached to the cell membrane, while others are dispersed throughout the sarcoplasm. There also appear to be enough cross-attachments from one dense body to another to hold them in relatively fixed positions within the cell. Interspersed among the actin filaments are a few thick filaments about 2.5 times the diameter of the thin actin filaments. These are assumed to be myosin filaments. However, there are only one-twelfth to one-fifteenth as many of these "myosin filaments" as actin filaments.

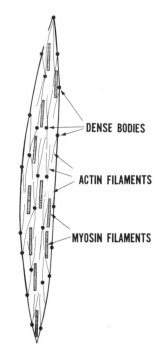

Figure 10–3. Arrangement of actin and myosin filaments in the smooth muscle cell. Note the attachment of the actin filaments to "dense bodies," some of which are themselves attached to the cell membrane.

Despite the relative paucity of myosin filaments, it is assumed that they have sufficient cross-bridges to attract the many actin filaments and cause contraction by the sliding filament mechanism in essentially the same way as in skeletal muscle. And it is especially interesting that the maximum strength of contraction of smooth muscle is approximately equal to that of skeletal muscle, about 2 to 3 kilograms per square centimeter of cross-sectional area of the muscle.

Slowness of Contraction and Relaxation of Smooth Muscle. A typical smooth muscle tissue will begin to contract 50 to 100 milliseconds after it is excited and will reach full contraction about half a second later. Then the contraction declines in another 1 to 2 seconds, giving a total contraction time of 1 to 3 seconds, which is about 30 times as long as the single contraction of skeletal muscle.

Energy Required to Sustain Smooth Muscle Contraction. As little as one five-hundredth as much energy is required to sustain the same tension of contraction in smooth muscle as in skeletal muscle. This presumably results from the very slow activity of smooth muscle myosin ATPase and also from the fact that there are far fewer myosin filaments in smooth muscle than in skeletal muscle.

This economy of energy utilization by smooth muscle is exceedingly important to overall function of the body, because organs such as the intestines, the urinary bladder, the gallbladder, and other viscera must maintain moderate degrees of muscle contractile tone day in and day out.

MEMBRANE POTENTIALS AND ACTION POTENTIALS IN SMOOTH MUSCLE

Smooth muscle exhibits membrane potentials and action potentials similar to those that occur in skeletal muscle fibers. However, in the normal resting state, the membrane potential is usually about −50 to −60 millivolts, or about 30 millivolts less negative than in skeletal muscle.

Action Potentials in Visceral Smooth Muscle. Typical *spike* action potentials, such as those seen in skeletal muscle, occur in most types of visceral smooth muscle. The duration of this type of action potential is 10 to 50 milliseconds, as illustrated in Figure 10–4A. Such action potentials can be elicited in many ways, such as electrical stimulation, the action of hormones on the smooth muscle, the action of transmitter substances from nerve fibers, or spontaneous generation in the muscle fiber itself, as discussed below.

Action Potentials with Plateaus. Action potentials with plateaus also occur in some smooth muscle, such as the action potential shown in Figure 10–5. The onset of this action potential is similar to that of the typical spike potential. However, instead of rapid repolarization of the muscle fiber membrane, the repolarization is de-

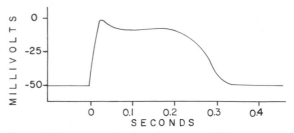

Figure 10–5. Monophasic action potential from a smooth muscle fiber of the rat uterus.

layed for several hundred to several thousand milliseconds. Plateaus as long as 30 seconds have been recorded. The importance of the plateau is that it can account for the prolonged periods of contraction that occur in some types of smooth muscle.

Slow Wave Potentials in Visceral Smooth Muscle and Spontaneous Generation of Action Potentials. Some smooth muscle is self-excitatory. That is, action potentials arise within the smooth muscle itself without an extrinsic stimulus. This is usually associated with a basic *slow wave rhythm* of the membrane potential. A typical slow wave of this type is illustrated in Figure 10–4B. The slow wave itself is not an *action* potential. It is not a self-regenerative process that spreads progressively over the membranes of the muscle fibers. Instead, it is believed to result from waxing and waning of the pumping of sodium outward through the muscle fiber membrane; the membrane potential becomes more negative when sodium is pumped rapidly and less negative when the sodium pump becomes less active.

The importance of the slow waves lies in the fact that they can initiate action potentials. The slow waves themselves cannot cause muscle contraction, but when the potential of the slow wave rises above the level of approximately −35 millivolts (the approximate threshold for eliciting action potentials in most visceral smooth muscle), an action potential develops and spreads over the visceral smooth muscle mass, and then contraction does occur. Figure 10–4B illustrates this effect, showing that at each peak of the slow wave, one or more action potentials occur. This effect can obviously promote rhythmical contractions of the smooth muscle mass synchronized with the slow waves. Therefore, the slow waves are frequently called *pacemaker waves*. This type of activity is especially prominent in tubular types of smooth muscle masses, such as in the gut, the ureter, and so forth. In Chapter 42 we shall see that this type of activity controls the rhythmical contractions of the gut.

Excitation of Visceral Smooth Muscle by Stretch. When visceral smooth muscle is stretched sufficiently, spontaneous action poten-

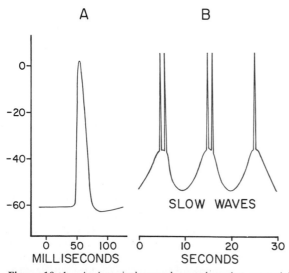

Figure 10–4. A, A typical smooth muscle action potential (spike potential) elicited by an external stimulus. B, A series of spike action potentials elicited by the rhythmical slow electrical waves occurring spontaneously in the smooth muscle wall of the intestine.

tials are usually generated. These result from a combination of the normal slow wave potentials plus a decrease in the membrane potential caused by the stretch itself. This response to stretch is an especially important function of visceral smooth muscle because it allows a hollow organ that is excessively stretched to contract automatically and therefore to resist the stretch. For instance, when the gut is overstretched by intestinal contents, a local automatic contraction sets up a peristaltic wave that moves the contents away from the over-stretched intestine.

Depolarization of Multiunit Smooth Muscle Without Action Potentials

The smooth muscle fibers of multiunit smooth muscle normally contract only in response to nerve stimuli. The nerve endings secrete acetylcholine in the case of some multiunit smooth muscles and norepinephrine in the case of others. In both instances, these transmitter substances cause depolarization of the smooth muscle membrane, and this response in turn elicits the contraction. However, true action potentials with self-regeneration most often do not develop. The presumed reason is that the fibers are too small to generate an action potential. Yet, even without an action potential in the multiunit smooth muscle fibers, the local depolarization caused by the nerve transmitter substance itself spreads "electrotonically" over the entire fiber and is all that is needed to cause the muscle contraction.

EXCITATION-CONTRACTION COUPLING — ROLE OF CALCIUM IONS

In the previous chapter it was pointed out that the actual contractile process in skeletal muscle is activated by calcium ions. This is also true in smooth muscle. However, the source of the calcium ions differs in smooth muscle because the sarcoplasmic reticulum of smooth muscle is poorly developed, in contrast to the sarcoplasmic reticulum of skeletal muscle, which is very extensive and is the source of very nearly 100 per cent of the contraction-inducing calcium ions.

In some types of smooth muscle, most of the calcium ions that cause contraction enter the muscle fiber from the extracellular fluid at the time of the action potential. There is a reasonably high concentration of calcium ions in the extracellular fluid, and the action potential itself is caused at least partly by influx of calcium ions along with sodium ions into the muscle fiber.

Yet, in some smooth muscle there is a moderately developed sarcoplasmic reticulum. However, there are no T tubules. Instead, the cisternae

of the reticulum abut the cell membrane. Therefore, it is believed that the membrane action potentials in these smooth muscle fibers cause release of calcium ions from these cisternae, thereby providing a greater degree of contraction than would occur on the basis of calcium ions entering through the cell membrane alone.

The Calcium Pump. To cause relaxation of the smooth muscle contractile elements, it is necessary to remove the calcium ions. This removal is achieved by a calcium pump that pumps the calcium ions out of the smooth muscle fiber and back into the extracellular fluid, or pumps the calcium ions into the sarcoplasmic reticulum. However, this pump is very slow-acting in comparison with the fast-acting sarcoplasmic reticulum pump in skeletal muscle. Therefore, the duration of smooth muscle contraction is often in the order of seconds rather than in tens of milliseconds, as occurs for skeletal muscle.

NEUROMUSCULAR JUNCTIONS OF SMOOTH MUSCLE

Physiologic Anatomy of Smooth Muscle Neuromuscular Junctions. Neuromuscular junctions of the type found on skeletal muscle fibers do not occur in smooth muscle. Instead, the nerve fibers generally branch diffusely on top of a sheet of muscle fibers, as was illustrated in Figure 10–2. In a few instances the endings make direct *contact junctions* with the smooth muscle cells, but usually the fibers do not make direct contact with the smooth muscle fibers at all but instead form so-called *diffuse junctions* that secrete their transmitter substance into the interstitial fluid a few microns away from the muscle cells; the transmitter substance then diffuses to the cells.

The axons innervating smooth muscle fibers also do not have typical end-feet as observed in the end-plate on skeletal muscle fibers. Instead, the fine terminal axons have multiple varicosities spread along their axes. At these points the Schwann cells are interrupted so that transmitter substance can be secreted through the walls of the varicosities. In the varicosities are vesicles similar to those present in the skeletal muscle end-plate containing transmitter substance. However, in contrast to the vesicles of skeletal muscle junctions that contain only acetylcholine, the vesicles of the autonomic nerve fiber varicosities contain acetylcholine in some fibers and norepinephrine in others.

Excitatory and Inhibitory Transmitter Substances at the Smooth Muscle Neuromuscular Junction. Two different transmitter substances known to be secreted by the autonomic nerves innervating smooth muscle are *acetylcholine* and *norepinephrine*. Acetylcholine is an excitatory substance for smooth muscle fibers in some organs but an inhibitory substance for smooth muscle in other organs. And when acetylcholine excites a muscle fiber, norepinephrine ordinarily inhibits it, or when acetylcholine inhibits a fiber norepinephrine excites it.

It is believed that *receptor substances* in the membranes of the different smooth muscle fibers determine which will excite them, acetylcholine or norepinephrine. Thus, there are *excitatory receptors* and *inhibitory receptors*. These receptor substances will be discussed in more detail in Chapter 38 in relation to function of the autonomic nervous system.

Excitation of Action Potentials in Smooth Muscle Fibers—The Junctional Potential. When an action potential reaches the terminal of an excitatory nerve fibril, there is a typical latent period of 50 milliseconds before any change in the membrane potential of the smooth muscle fiber can be detected. Then the local potential rises to a maximal level in approximately 100 milliseconds and lasts for almost a second. This complete sequence of potential changes is called the *junctional potential;* it is analogous to the end-plate potential of the skeletal muscle fibers except that its duration is 20 to 100 times as long.

If the junctional potential rises to the threshold level for discharge of the smooth muscle membrane, an action potential will occur in the smooth muscle fiber in exactly the same way that an action potential occurs in a skeletal muscle fiber. A typical smooth muscle fiber has a normal resting membrane potential of −50 to −60 millivolts, and the threshold potential at which the action potential occurs is about −30 to −35 millivolts.

Inhibition at the Smooth Muscle Neuromuscular Junction. When a transmitter substance at the nerve ending interacts with an inhibitory receptor instead of an excitatory receptor, the membrane potential of the muscle fiber becomes more negative than ever, for instance −70 millivolts; that is, it becomes *hyperpolarized* and therefore becomes much more difficult to excite than is usually the case.

SMOOTH MUSCLE CONTRACTION WITHOUT ACTION POTENTIALS — EFFECT OF LOCAL TISSUE FACTORS AND HORMONES

Though we have thus far discussed smooth muscle contraction elicited only by nervous sig-

nals and smooth muscle membrane action potentials, we must quickly disavow the fact that all smooth muscle contraction occurs in this way. In fact, probably half or more of all smooth muscle contraction is initiated not by action potentials but by stimulatory factors acting directly on the smooth muscle contractile machinery. The two types of non-nervous and non–action potential stimulating factors most often involved are (1) local tissue factors and (2) various hormones.

Some of the local tissue factors that affect smooth muscle contraction are these:

1. Lack of oxygen in the local tissues causes smooth muscle relaxation and therefore vasodilatation.

2. Excess carbon dioxide causes vasodilatation.

3. Increased hydrogen ion concentration also causes increased vasodilatation.

And such factors as lactic acid, potassium ion concentration, diminished calcium ion concentration, and decreased body temperature will also cause local vasodilatation.

Some of the more important hormones that affect smooth muscle contraction are norepinephrine, epinephrine, acetylcholine, angiotensin, vasopressin, oxytocin, serotonin, and histamine.

A hormone will cause contraction of smooth muscle when the smooth muscle cells contain an *excitatory receptor* for the respective hormone. However, the hormone will cause inhibition instead of contraction if the cells contain an *inhibitory receptor* rather than an excitatory receptor.

Mechanism of Muscle Excitation by Local Tissue Factors and Hormones. It is believed that most of the local tissue factors and hormones that cause smooth muscle contraction do so by activating the calcium mechanism for control of the contractile process. Some of these factors change the membrane potential a moderate amount but without necessarily causing an action potential, and this increases the flow of calcium ions to the interior of the cell. However, most of them can activate contraction even when the membrane potential is not altered and even when calcium ions are not available to enter the cell. In these circumstances calcium ions probably are released from the sarcoplasmic reticulum.

Tone of Smooth Muscle

Smooth muscle can maintain a state of long-term, steady contraction that has been called either *tonus* contraction of smooth muscle or simply *smooth muscle tone*. This is an important feature of smooth muscle contraction because it allows prolonged or even indefinite continuance of the smooth muscle function. For instance, the arterioles are maintained in a state of tonic contraction almost throughout the entire life of the person. Likewise, tonic contraction in the gut wall maintains steady pressure on the contents of the gut, and tonic contraction of the urinary bladder wall maintains a moderate amount of pressure on the urine in the bladder.

Tonic contractions of smooth muscle can be caused in either of two ways:

1. They are sometimes caused by *summation of individual contractile pulses;* each contractile pulse is initiated by a separate action potential in the same way that tetanic contractions are produced in skeletal muscle.

2. However, most smooth muscle tonic contractions probably result from *prolonged direct smooth muscle excitation* without action potentials, usually caused by local tissue factors or circulating hormones. For instance, prolonged tonic contractions of the blood vessels without the mediation of action potentials are regularly caused by angiotensin, vasopressin, or norepinephrine, and these play an important role in the long-term regulation of arterial pressure, as will be discussed in Chapters 17 and 18.

REFERENCES

Bolton, T. B.: Mechanisms of action of transmitters and other substances on smooth muscle. *Physiol. Rev., 59*:606, 1979.

Bulbring, E., *et al.* (eds.): Physiology of Smooth Muscles; Twenty-sixth International Congress of Physiological Sciences. New York, Raven Press, 1975.

Ceccarelli, B., and Hurlbut, W. P.: Vesicle hypothesis of the release of quanta of acetylcholine. *Physiol. Rev., 60*:396, 1980.

Chamley-Campbell, J., *et al.*: Smooth muscle cell in culture. *Physiol. Rev., 59*:1, 1979.

Daniel, E. E., and Sarna, S.: The generation and conduction of activity in smooth muscle. *Annu. Rev. Pharmacol. Toxicol., 18*:145, 1978.

Fambrough, D. M.: Control of acetylcholine receptors in skeletal muscle. *Physiol. Rev., 59*:165, 1979.

Gage, P. W.: Generation of end-plate potentials. *Physiol. Rev., 56*:177, 1976.

Guyton, A. C., and MacDonald, M. A.: Physiology of botulinus toxin. *Arch. Neurol. Psychiat., 57*:578, 1947.

Guyton, A. C., and Reeder, R. C.: The dynamics of curarization. *J. Pharmacol. Exp. Ther., 97*:322, 1949.

Hartshorne, D. J., and Gorecka, A.: Biochemistry of the contractile proteins of smooth muscle. *In* Bohr, D. F., *et al.* (eds.): Handbook of Physiology. Sec. 2, Vol. II. Baltimore, Williams & Wilkins, 1980, p. 93.

Johansson, B.: Vascular smooth muscle biophysics. *In* Kaley, G., and Altura, B. M. (eds.): Microcirculation. Vol. II. Baltimore, University Park Press, 1977.

Johansson, B., and Somlyo, A. P.: Electrophysiology and excitation-contraction coupling. *In* Bohr, D. F., *et al.* (eds.): Handbook of Physiology. Sec. 2, Vol. II. Baltimore, Williams & Wilkins, 1980, p. 301.

Lester, H. A.: The response to acetylcholine. *Sci. Am., 236*(2): 106, 1977.

Lunt, G. G., and Marchbanks, R. M.: The Biochemistry of Myasthenia Gravis and Muscular Dystrophy. New York, Academic Press, 1978.

McGeachie, J. K.: Smooth Muscle Regeneration; A Review and Experimental Study. Basel, S. Karger, 1975.

Murphy, R. A.: Filament organization and contractile function in vertebrate smooth muscle. *Annu. Rev. Physiol., 41*: 737, 1979.

Paul, R. J.: Chemical energetics of vascular smooth muscle. *In* Bohr, D. F., *et al.* (eds.): Handbook of Physiology. Sec. 2, Vol. II. Baltimore, Williams & Wilkins, 1980, p. 201.

Paul, R. J., and Rüegg, J. C.: Biochemistry of vascular smooth muscle: Energy metabolism and proteins. *In* Kaley, G., and Altura, B. M. (eds.): Microcirculation. Vol. II. Baltimore, University Park Press, 1977.

Pepeu, G., *et al.* (eds.): Receptors for Neurotransmitters and Peptide Hormones. New York, Raven Press, 1980.

Prosser, C. L.: Evolution and diversity of nonstriated muscles. *In* Bohr, D. F., *et al.* (eds.): Handbook of Physiology. Sec. 2, Vol. II. Baltimore, Williams & Wilkins, 1980, p. 635.

Somlyo, A. P.: Ultrastructure of vascular smooth muscle. *In* Bohr, D. F., *et al.* (eds.): Handbook of Physiology. Sec. 2, Vol. II. Baltimore, Williams & Wilkins, 1980, p. 33.

Part IV
THE HEART

11

Heart Muscle; the Heart as a Pump

With this chapter we begin discussion of the heart and circulatory system. In the present chapter we will explain how the heart operates as a pump: that is, explain the functions of heart muscle, of the valves, and of the various chambers of the heart.

PHYSIOLOGY OF CARDIAC MUSCLE

PHYSIOLOGIC ANATOMY OF CARDIAC MUSCLE

Figure 11–1 illustrates a typical histologic picture of cardiac muscle, showing the cardiac muscle fibers arranged in a latticework, the fibers dividing, then recombining, and then dividing again. One notes immediately from this figure that cardiac muscle is *striated,* the same as skeletal muscle. Furthermore, cardiac muscle has typical myofibrils that contain *actin* and *myosin filaments*, also almost identical to those found in skeletal muscle, and these filaments inter-digitate and slide along each other during the process of contraction in the same manner as occurs in skeletal muscle. (See Chapter 9.)

Cardiac Muscle as Syncytium. The angulated dark areas crossing the cardiac muscle fibers in Figure 11–1 are called *intercalated discs;* however, they are actually cell membranes that separate individual cardiac muscle cells from each other. That is, cardiac muscle fibers are made up of many cardiac muscle cells connected in series with each other. Yet electrical resistance through the intercalated disc is only one four-hundredth the resistance through the outside membrane of the cardiac muscle fiber, because the cell membranes fuse with each other to form "tight junctions" that allow almost completely free diffusion of ions. Therefore, from a functional point of view, ions move with ease along the axes of the cardiac muscle fibers so that action potentials travel from one cardiac muscle cell to another, past the intercalated discs, without significant hindrance. Therefore, cardiac muscle is a *syncytium,* in which the cardiac muscle cells are so tightly bound that when one of these cells becomes excited, the action potential spreads to all of them, spreading from cell to cell and spreading throughout the latticework interconnections.

The heart is composed of two separate syncytiums, the *atrial syncytium* and the *ventricular syncytium.* These are separated from each other by the fibrous tissue surrounding the valvular rings, but an action potential can be conducted from the atrial syncytium into the ventricular syncytium by way of a specialized conductive system, the *A-V bundle,* which will be discussed in detail in the following chapter.

All-or-Nothing Principle as Applied to the Heart. Because of the syncytial nature of cardiac muscle, stimulation of any single atrial muscle fiber causes the action potential to travel over the entire atrial muscle mass, and similarly, stimulation of any single ventricular fiber causes excitation of the entire ventricular muscle mass. If the A-V bundle is intact, the action potential passes also from the atria to the ventricles. This is called the all-or-nothing principle, and it is precisely the same as that discussed in Chapter 8 for nerve fibers. However, because the cardiac muscle fibers interconnect with each other, the all-or-nothing principle applies to the entire functional syncytium of the heart rather than to single muscle fibers as in the case of skeletal muscle fibers.

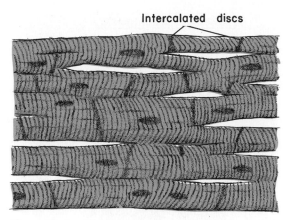

Figure 11–1. The "syncytial" nature of cardiac muscle.

ACTION POTENTIALS IN CARDIAC MUSCLE

The *resting membrane potential* of normal cardiac muscle is approximately −85 to −95 millivolt (mv.) and approximately −90 to −100 mv. in the specialized conductive fibers, the Purkinje fibers, which are discussed in the following chapter.

The *voltage of the action potential* recorded in ventricular muscle, shown at the bottom of Figure 11–2, is 105 mv., which means that the membrane potential rises from its normally very negative value to a slightly positive value of about +20 mv. After the initial *spike* the membrane remains depolarized for 0.15 for atrial muscle to 0.3 second for ventricular muscle, exhibiting a *plateau* as illustrated in the figure, followed at the end of the plateau by abrupt repolarization. The presence of this plateau in the action potential causes the action potential to last 10 to 30 times as long in cardiac muscle as in skeletal muscle and causes a correspondingly increased period of contraction.

At this point we must ask the question: Why does the action potential of cardiac muscle have a plateau, while that of skeletal muscle does not? There are at least two major differences between the membrane properties of these two types of muscle that presumably account for the plateau in cardiac muscle. First, a moderate quantity of calcium ions diffuses to the inside of the cardiac muscle fiber during the action potential, while only a very small quantity diffuses into skeletal muscle. Furthermore, the calcium ion influx does not occur only at the onset of the action potential, as is true for sodium ions, but instead continues for 0.2 to 0.3 second. The plateau occurs during this prolonged influx of calcium ions.

The second major functional difference between cardiac muscle and skeletal muscle that helps to account for the plateau is this: Immediately after the onset of the action potential the permeability of the cardiac muscle membrane for potassium *decreases* about five-fold, an effect that does not occur in skeletal muscle. It is believed that this decreased potassium permeability is caused by the excess calcium influx noted above. The decreased potassium permeability greatly decreases the outflux of potassium ions for the next 0.2 to 0.3 second, which prevents rapid repolarization of the membrane and thus also contributes to the plateau.

Velocity of Conduction in Cardiac Muscle. The velocity of conduction of the action potential in both atrial and ventricular muscle fibers is about 0.3 to 0.5 meter per second, or about 1/250 the velocity in very large nerve fibers and about one tenth the velocity in skeletal muscle fibers.

Refractory Period of Cardiac Muscle. Cardiac muscle, like all excitable tissue, is refractory to restimulation during the action potential. Therefore, the normal refractory period of the ventricle is 0.25 to 0.3 second, which is approximately the duration of the action potential. There is an additional *relative refractory period* of about 0.05 second during which the muscle is more difficult than normal to excite but nevertheless can be excited.

The refractory period of atrial muscle is much shorter than that for the ventricles (about 0.15 second). Therefore, the rhythmical rate of contraction of the atria can be much faster than that of the ventricles.

EXCITATION-CONTRACTION COUPLING—FUNCTION OF CALCIUM IONS AND OF THE T TUBULES

The term excitation-contraction coupling means the mechanism by which the action poten-

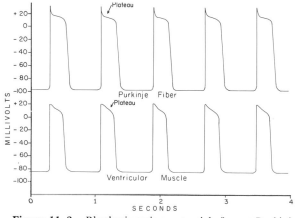

Figure 11–2. Rhythmic action potentials from a Purkinje fiber and from a ventricular muscle fiber, recorded by means of microelectrodes.

tial causes the myofibrils of muscle to contract. As is true for skeletal muscle, when an action potential passes over the cardiac muscle membrane, the action potential also spreads to the interior of the cardiac muscle fiber along the membranes of the T tubules. The T tubule action potentials in turn cause instantaneous release of calcium ions into the muscle sarcoplasm from the cisternae of the sarcoplasmic reticulum. Then the calcium ions diffuse in another few thousandths of a second into the myofibrils, where they catalyze the chemical reactions that promote sliding of the actin and myosin filaments along each other; this in turn produces the muscle contraction.

Thus far, this mechanism of excitation-contraction coupling is the same as that for skeletal muscle, but at this point a major difference begins to appear. In addition to the calcium ions that are released into the sarcoplasm from the cisternae of the sarcoplasmic reticulum, large quantities of calcium ions also diffuse during the action potential from the T tubules into the sarcoplasm, mainly because these T tubules in heart muscle are much larger and contain many times as much calcium as in skeletal muscle. This extra supply of calcium from the T tubules is at least one of the factors that prolongs the cardiac muscle action potential and maintains cardiac muscle contraction for as long as a third of a second rather than one-tenth that time, as occurs in skeletal muscle.

At the end of the plateau of the action potential the supply of calcium ions to the interior of the muscle fiber is suddenly cut off, and the calcium ions in the sarcoplasm are rapidly pumped back into the sarcoplasmic reticulum and T tubules — thus, the contraction ends until a new action potential occurs.

It is especially interesting that the strength of contraction of cardiac muscle depends partly upon the concentration of calcium ions in the extracellular fluids, which normally is not the case for skeletal muscle. The probable reason for this is that the quantity of calcium ions in the T tubules themselves are actually filled with extracellular fluid. Consequently, the availability of calcium ions to cause cardiac muscle contraction directly depends upon the extracellular fluid calcium.

THE CARDIAC CYCLE

Figure 11–3 illustrates that the heart is in reality four separate pumps: two *primer pumps,* the *atria,* and two *power pumps,* the *ventricles.*

The period from the end of one heart contraction to the end of the next is called the *cardiac*

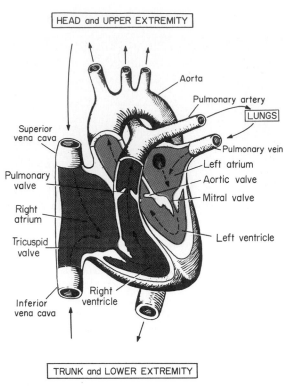

Figure 11–3. Structure of the heart, and course of blood flow through the heart chambers.

cycle. Each cycle is initiated by spontaneous generation of an action potential in the S-A node, as will be explained in detail in the following chapter. This node is about 0.3 square centimeter in area, and it is located in the posterior wall of the right atrium near the opening of the superior vena cava. The action potential then travels rapidly through both atria and thence through the A-V bundle into the ventricles. However, because of a special arrangement of the conducting system from the atria into the ventricles, there is a delay of more than 0.1 second between passage of the cardiac impulse from the atria into the ventricles. This allows the atria to contract ahead of the ventricles, thereby pumping blood into the ventricles prior to the very strong ventricular contraction. Thus, the atria act as primer pumps for the ventricles, and the ventricles then provide the major source of power for moving blood through the vascular system.

SYSTOLE AND DIASTOLE

The cardiac cycle consists of a period of relaxation called *diastole* followed by a period of contraction called *systole.*

Figure 11–4 illustrates the different events during the cardiac cycle. The top three curves show the pressure changes in the aorta, the left ventri-

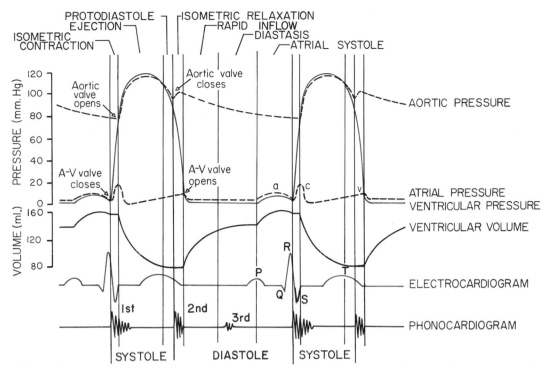

Figure 11-4. The events of the cardiac cycle, showing changes in left atrial pressure, left ventricular pressure, aortic pressure, ventricular volume, the electrocardiogram, and the phonocardiogram.

cle, and the left atrium, respectively. The fourth curve depicts the changes in ventricular volume, the fifth the electrocardiogram, and the sixth a phonocardiogram, which is a recording of the sounds produced by the heart as it pumps. It is especially important that the student study in detail the diagram of this figure and understand the causes of all the events illustrated. These are explained as follows:

RELATIONSHIP OF THE ELECTROCARDIOGRAM TO THE CARDIAC CYCLE

The electrocardiogram in Figure 11-4 shows the *P, Q, R, S,* and *T* waves, which will be discussed in Chapter 13. These are electrical voltages generated by the heart and recorded by the electrocardiograph from the surface of the body. The *P wave* is caused by the *spread of depolarization* through the atria, and this is followed by atrial contraction, which causes a slight rise in the atrial pressure curve immediately after the P wave. Approximately 0.16 second after the onset of the P wave, the *QRS waves* appear as a result of depolarization of the ventricles, which initiates contraction of the ventricles and causes the ventricular pressure to begin rising. Therefore, the QRS complex begins slightly before the onset of ventricular systole.

Finally, one observes the *ventricular T wave* in the electrocardiogram. This represents the stage of repolarization of the ventricles at which time the ventricular muscle fibers begin to relax. Therefore, the T wave occurs slightly prior to the end of ventricular contraction.

FUNCTION OF THE ATRIA AS PUMPS

Blood normally flows continually from the great veins into the atria; approximately 70 per cent flows directly through the atria into the ventricles even before the atria contract. Then, atrial contraction causes an additional 30 per cent filling of the ventricles. Therefore, the atria simply function as primer pumps that increase the ventricular pumping effectiveness by approximately 30 per cent. Yet, the heart can continue to operate quite satisfactorily under normal resting conditions even without this extra 30 per cent effectiveness because it normally has the capability of pumping 300 to 500 per cent more blood than is required by the body anyway.

Pressure Changes in the Atria — The a, c, and v Waves. In the atrial pressure curve of Figure 11-4 three pressure elevations called the *a, c,* and *v atrial pressure waves* can be noted, during each of which the atrial pressure rises 3 to 8 mm. Hg.

The *a wave* is caused by atrial contraction. The *c wave* occurs when the ventricles begin to contract, and it is caused mainly by (1) bulging of the A-V valves backward toward the atria because of increasing pressure in the ventricles, and (2) pulling on the atrial muscle by the contracting ventricles. The *v wave* occurs toward the end of ventricular contraction; it results from slow build-up of blood in the atria while the A-V valves are closed during ventricular contraction.

FUNCTION OF THE VENTRICLES AS PUMPS

Filling of the Ventricles. During ventricular systole large amounts of blood accumulate in the atria because of the closed A-V valves. Therefore, just as soon as systole is over and the ventricular pressures fall again to their low diastolic values, the high pressures in the atria immediately push the A-V valves open and allow blood to flow rapidly into the ventricles, as shown by the ventricular volume curve in Figure 11-4. This is called the *period of rapid filling of the ventricles*.

The period of rapid filling lasts approximately the first third of diastole. During the middle third of diastole only a small amount of blood normally flows into the ventricles; this is blood that continues to empty into the atria from the veins and passes on through the atria directly into the ventricles.

During the latter third of diastole, the atria contract and give an additional thrust to the inflow of blood into the ventricles; this accounts for approximately 30 per cent of the filling of the ventricles during each heart cycle.

Emptying of the Ventricles During Systole. *Period of Isometric (Isovolumic) Contraction.* When ventricular contraction begins, the ventricular pressure abruptly rises, as shown in Figure 11-4, causing the A-V valves to close. Then an additional 0.02 to 0.03 second is required for the ventricle to build up sufficient pressure to push the semilunar (aortic and pulmonary) valves open against the pressures in the aorta and pulmonary artery. Therefore, during this period of time, contraction is occurring in the ventricles, but there is no emptying. This period is called the period of isometric or isovolumic contraction, meaning by these terms that tension is increasing in the muscle but no shortening of the muscle fibers is occurring.

Period of Ejection. When the left ventricular pressure rises slightly above 80 mm. Hg and the right ventricular pressure slightly above 8 mm Hg, the ventricular pressures now push the semilunar valves open. Immediately, blood begins to pour out of the ventricles, with about 60 per cent of the emptying occurring during the first quarter of systole and usually most of the remaining

40 per cent during the next two quarters. These three-quarters of systole are called the period of ejection.

Protodiastole. During the last one-fourth of ventricular systole, little blood flows from the ventricles into the large arteries; yet the ventricular musculature remains contracted. This period is called protodiastole. The arterial pressure falls during this period, because large quantities of blood are flowing from the arteries through the peripheral vessels.

Period of Isometric (Isovolumic) Relaxation. At the end of systole, ventricular relaxation begins suddenly, allowing the intraventricular pressures to fall rapidly. The elevated pressures in the large arteries immediately push blood back toward the ventricles, which snaps the aortic and pulmonary valves closed For another 0.03 to 0.06 second, the ventricular muscle continues to relax, and the intraventricular pressures fall rapidly back to their very low diastolic levels. Then the A-V valves open to begin a new cycle of ventricular pumping.

END-DIASTOLIC AND END-SYSTOLIC VOLUMES OF THE VENTRICLES

During diastole, filling of the ventricles normally increases the volume of each ventricle to about 120 to 130 ml. This volume is known as the *end-diastolic volume*. Then, as the ventricles empty during systole the volume decreases about 70 ml., which is called the *stroke volume*. The remaining volume in each ventricle, about 50 to 60 ml., is called the *end-systolic volume*.

When the heart contracts strongly, the end-systolic volume can fall to as little as 10 to 30 ml. And when large amounts of blood flow into the ventricles during diastole, their end-diastolic volumes can become as great as 200 to 250 ml. in the normal heart. Therefore, by both increasing the end-diastolic volume and decreasing the end-systolic volume, the stroke volume output can be increased to more than double normal.

FUNCTION OF THE VALVES

The Atrioventricular Valves. The *A-V valves* (the *tricuspid* and the *mitral* valves) prevent backflow of blood from the ventricles to the atria during systole, and the *semilunar valves* (the *aortic* and *pulmonary* valves) prevent backflow from the aorta and pulmonary arteries into the ventricles during diastole. All these valves close and open *passively*. That is, they close when a backward pressure gradient pushes blood backward, and they open when a forward pressure gradient forces blood in the forward direction. For obvious anatomical reasons, the thin, filmy A-V valves require almost no backflow to cause closure, while the much heavier semilunar valves require rather strong backflow for a few milliseconds.

THE AORTIC PRESSURE CURVE

When the left ventricle contracts, the entry of blood into the aorta and other arteries causes the walls of these arteries to stretch and the pressure to rise to a normal *systolic pressure* of 120 mm. Hg. Then, at the end of systole, the pressure in the aorta falls slowly throughout diastole because the blood stored in the distended elastic arteries continues to flow through the peripheral vessels back to the veins. Before the ventricle contracts again, the aortic pressure usually falls to a *diastolic pressure* level of approximately 80 mm. Hg, which is two-thirds the maximal pressure of 120 mm. Hg (systolic pressure) occurring during ventricular contraction.

The pressure curve in the pulmonary artery is similar to that in the aorta, except that the pressures are much less, as will be discussed in Chapter 15.

RELATIONSHIP OF THE HEART SOUNDS TO HEART PUMPING

When listening to the heart with a stethoscope, one does not hear the opening of the valves, for this is a relatively slowly developing process that makes no noise. However, when the valves close, the vanes of the valves and the surrounding fluids vibrate under the influence of the sudden pressure differentials that develop, giving off sound that travels in all directions through the chest. When the ventricles first contract, one hears a sound that is caused by closure of the A-V valves. The vibration is low in pitch and relatively long continued and is known as the *first heart sound*. When the aortic and pulmonary valves close, one hears a relatively rapid snap, for these valves close extremely rapidly, and the surroundings vibrate for only a short period of time. This sound is known as the *second heart sound*. The precise causes of these sounds will be discussed in Chapter 20 in relation to auscultation.

STROKE VOLUME OUTPUT OF THE HEART

The stroke volume output of the heart is the quantity of blood pumped from each ventricle with each heartbeat. Normally, this is about 70

ml., but under conditions compatible with life it can decrease to as little as a few ml. per beat and can increase to about 140 ml. per beat in the normal heart and to over 200 ml. per beat in persons with very large hearts, such as in some athletes.

ENERGY FOR CARDIAC CONTRACTION

Heart muscle, like skeletal muscle, utilizes chemical energy to provide the work of contraction. This energy is derived mainly from metabolism of fatty acids and to a lesser extent from metabolism of other nutrients, especially lactate and glucose. The different reactions that liberate this energy will be discussed in detail in Chapters 45 and 46.

Efficiency of Cardiac Contraction. During muscular contraction most of the chemical energy is converted into heat and a small portion into work output. The ratio of work output to chemical energy expenditure is called the efficiency of cardiac contraction, or simply *efficiency of the heart*. The efficiency of the normal heart is between 20 and 25 per cent.

REGULATION OF CARDIAC FUNCTION

When a person is at rest, the heart pumps only 4 to 6 liters of blood each minute. However, during severe exercise it may be required to pump as much as 4 to 7 times this amount. The present section discusses the means by which the heart can adapt itself to such extreme increases in cardiac output.

The two basic means by which the volume pumped by the heart is regulated are (1) intrinsic autoregulation of pumping in response to changes in volume of blood flowing into the heart and (2) reflex control of heart rate and heart strength of contraction by the autonomic nervous system.

INTRINSIC AUTOREGULATION OF CARDIAC PUMPING—THE FRANK-STARLING LAW OF THE HEART

In Chapter 19 we shall see that one of the major factors determining the amount of blood pumped by the heart each minute is the rate of blood flow into the heart from the veins, which is called *venous return*. That is, each peripheral tissue of the body controls its own blood flow, and all the blood flow through all the peripheral tissues returns by way of the veins to the right atrium. The heart in turn automatically pumps this incoming blood on into the systemic arteries so that it can flow around the circuit again. Thus, the heart adapts from moment to moment or even second to second to widely varying inputs of blood.

This intrinsic ability of the heart to adapt itself to changing loads of inflowing blood is called the *Frank-Starling law of the heart,* in honor of Frank and Starling, two great physiologists of over half a century ago. Basically, the Frank-Starling law states that the greater the heart is filled during diastole, the greater will be the quantity of blood pumped into the aorta. Another way to express this law is: *Within physiological limits, the heart pumps all the blood that comes to it without allowing excessive damming of blood in the veins.* In other words, the heart can pump either a small amount of blood or a large amount, depending on the amount that flows into it from the veins; and it automatically adapts to whatever this load might be as long as the total quantity does not rise above the physiological limit that the heart can pump.

Mechanism of the Frank-Starling Law. The primary mechanism by which the heart adapts to changing inflow of blood is the following: When the cardiac muscle becomes stretched an extra amount, as it does when increased amounts of blood enter the heart chambers, the stretched muscle contracts with a greatly increased force, thereby automatically pumping the extra blood into the arteries. This ability of stretched muscle to contract with increased force is characteristic of all striated muscle and not simply of cardiac muscle. As was pointed out in Chapter 9, the increased force of contraction is caused by the fact that the actin and myosin filaments are brought to a more nearly optimal degree of interdigitation for achieving contraction.

Failure of Arterial Pressure Load to Alter Cardiac Output. One of the most important

features of the Frank-Starling law of the heart is that, within reasonable limits, changes in arterial pressure load against which the heart pumps have almost no effect on the rate at which blood is pumped by the heart each minute (the cardiac output). This effect is illustrated in Figure 11–5, which shows the effect on the cardiac output when the arterial pressure was progressively increased by constricting the arteries. The significance of this effect is this: Regardless of the arterial pressure, the most important factor determining the amount of blood pumped by the heart is still the right atrial pressure generated by the entry of blood into the heart. In fact, only when the pressure rises above approximately 170 mm. Hg does the arterial pressure load then cause the heart to begin to fail.

Ventricular Function Curves

One of the best ways to express the functional ability of the ventricles to pump blood is by ventricular function curves, as shown in Figure 11–6. This figure illustrates a type of ventricular function curve called the *minute ventricular output curve.* The two curves in the figure represent the function of the two respective ventricles of the human heart based on data extrapolated from lower animals. As each atrial pressure rises, the respective ventricular volume output per minute also increases.

Thus, ventricular function curves are another way of expressing the Frank-Starling law of the heart. That is, as the ventricles fill to higher atrial pressures, the strength of cardiac contraction increases, causing the heart to pump increased quantities of blood into the arteries. In later chapters we shall see that ventricular function curves are exceedingly important in analyzing overall function of the circulation, for it is by such means that one can express the quantitative capabilities of the heart as a pump.

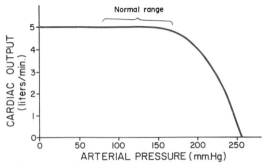

Figure 11–5. Constancy of cardiac output even in the face of wide changes in arterial pressure. Only when the arterial pressure rises above the normal operating pressure range does the pressure load cause the heart to begin to fail.

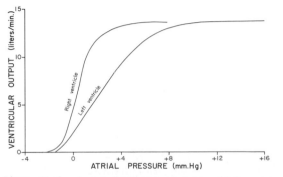

Figure 11–6. Approximate normal right and left ventricular output curves for the human heart as extrapolated from data obtained in dogs.

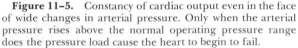

CONTROL OF THE HEART BY NERVES

The heart is well supplied with both sympathetic and parasympathetic (vagal) nerves, as illustrated in Figure 11–7. These nerves affect cardiac pumping in two ways: (1) by changing the heart rate and (2) by changing the strength of contraction of the heart. The effect of nerve stimulation on heart rate and rhythm will be discussed in the following chapter. For the present, suffice it to say that parasympathetic stimulation decreases heart rate, and sympathetic stimulation increases the rate. The range of control is from as little as 20 to 30 heart beats per minute with maximum vagal stimulation, up to as high as 250 or, rarely, 300 heart beats per minute with maximum sympathetic stimulation.

Effect of Heart Rate on Function of the Heart as a Pump.   In general, the more times the heart beats per minute, the more blood it can pump, but there are important limitations to this effect. For instance, once the heart rate rises above a critical level, the heart strength itself decreases, presumably because of overutilization of metabolic substrates in the cardiac muscle. In addition, the period of diastole between the contractions becomes so reduced that blood does not have time to flow adequately from the atria into the ventricles. For these reasons, when the heart rate is increased artificially by electrical stimulation, the heart has its peak ability to pump large quantities of blood at a heart rate between 100 and 150 beats per minute.

Nervous Regulation of Contractile Strength of the Heart.   The two atria are especially well supplied with large numbers of both sympathetic and parasympathetic nerves, but the ventricles are supplied mainly by sympathetic nerves and far fewer parasympathetic fibers. In general, sympathetic stimulation increases the strength of heart muscle contraction, whereas parasympathetic stimulation decreases the strength of contraction.

Under normal conditions the sympathetic nerve fibers to the heart continually discharge at a slow rate that maintains a strength of ventricular contraction about 20 per cent above its strength with no sympathetic stimulation at all. Therefore, one method by which the nervous system can decrease the strength of ventricular contraction is simply to slow or stop the transmission of sympathetic impulses to the heart. On the other hand, maximal sympathetic stimulation can increase the strength of ventricular contraction to approximately 100 per cent greater than normal.

Maximal parasympathetic stimulation of the heart decreases ventricular contractile strength about 30 per cent. Thus, the parasympathetic effect is relatively small in contrast with the sympathetic effect.

EFFECT OF HEART DEBILITY ON CARDIAC FUNCTION—THE HYPOEFFECTIVE HEART

Any factor that damages the heart, whether it is damage to the myocardium, to the valves, to the conducting system, or otherwise, is likely to make the heart a poorer pump, and the heart under these conditions is called a hypoeffective heart. Figure 11–8 illustrates by the very dark curve the normal cardiac function curve and by the three curves below it the effect of different degrees of hypoeffectiveness on cardiac function. Obviously,

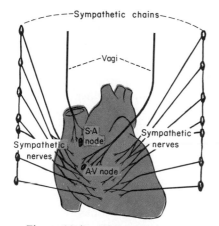

Figure 11–7.   The cardiac nerves.

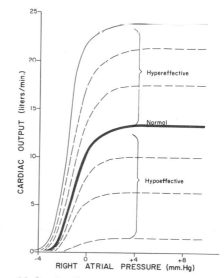

Figure 11–8.   Cardiac output curves for various degrees of hypo- and hypereffective hearts. (From Guyton, Jones, and Coleman: Circulatory Physiology: Cardiac Output and Its Regulation. Philadelphia, W. B. Saunders Company, 1973.)

the more serious the damage, the less will be the cardiac output at any given right atrial pressure.

Different factors that can cause a hypoeffective heart include

Myocardial infarction
Valvular heart disease
Vagal stimulation of the heart
Inhibition of the sympathetics to the heart
Congenital heart disease
Myocarditis
Cardiac anoxia
Diphtheritic or other types of myocardial damage

EFFECT OF EXERCISE ON THE HEART—THE HYPEREFFECTIVE HEART

Chronic heavy exercise over a period of many weeks or months leads to hypertrophy of the cardiac muscle and also to enlargement of the ventricular chambers. As a result, the overall strength of the heart becomes greatly enhanced, and the effectiveness of the heart as a pump increases. The upper three function curves of Figure 11–8 illustrate the effect of different degrees of cardiac hypereffectiveness on the cardiac function curve; maximal degrees of hypereffectiveness can increase pumping by the heart more than 100 per cent.

Other factors besides hypertrophy that can cause a hypereffective heart are:

Sympathetic stimulation (50 to 100 per cent increase)
Parasympathetic inhibition (10 to 20 per cent increase)

EFFECT OF VARIOUS IONS ON HEART FUNCTION

In the discussion of membrane potentials in Chapter 8, it was pointed out that three particular cations — potassium, sodium, and calcium — all have marked effects on action potential transmission, and in Chapter 9 it was noted that calcium ions play an especially important role in initiating the muscle contractile process. Therefore, it is to be expected that the concentrations of these three ions in the extracellular fluids will also have important effects on cardiac function.

Effect of Potassium Ions. Excess potassium in the extracellular fluids causes the heart to become extremely dilated and flaccid and slows the heart rate. Very large quantities can also block conduction of the cardiac impulse from the atria to the ventricles through the A-V bundle.

Elevation of potassium concentration to only 8 to 12 mEq./liter — two to three times the normal value — will usually cause such weakness of the heart and abnormal rhythm that it will cause death.

All these effects of potassium excess are caused by the decreased negativitiy of the resting membrane potential that results from the high potassium concentration in the extracellular fluids. As the membrane potential decreases, the intensity of the action potential also decreases, which makes the contraction of the heart progressively weaker, for the strength of the action potential determines to a great extent the strength of contraction.

Effect of Calcium Ions. An excess of calcium ions causes effects almost exactly opposite to those of excess potassium ions, causing the heart to go into spastic contraction. This is caused by the direct effect of calcium ions of exciting the cardiac contractile process, as explained earlier in the chapter. Conversely, a deficiency of calcium ions causes cardiac flaccidity, similar to the effect of potassium.

However, the calcium ion concentration rarely changes sufficiently during life to alter cardiac function greatly, for greatly diminished calcium ion concentration will usually kill a person because of tetany before it will significantly affect the heart, and elevation of the calcium ion concentration to a level that will significantly affect the heart almost never occurs because calcium ions are precipitated in bone, or occasionally elsewhere in the body's tissues, as insoluble calcium salts before such a level can be reached.

Effect of Sodium Ions. An excess of sodium ions depresses cardiac function, an effect similar to that of potassion ions but for an entirely different reason. Sodium ions compete with calcium ions at some yet unexplained point in the contractile process of muscle in such a way that the greater the sodium ion concentration in the extracellular fluids the less the effectiveness of the calcium ions in causing contraction when an action potential occurs. Yet, from a practical point of view, the sodium ion concentration in the extracellular fluids probably never becomes high enough, even in serious pathological conditions, to cause significant change in cardiac strength.

However, very low sodium concentration, as occurs in water intoxication, often causes death because of cardiac fibrillation, a phenomenon explained in the following chapter.

EFFECT OF TEMPERATURE ON THE HEART

Increased temperature causes greatly increased heart rate, and decreased temperature causes greatly decreased rate. These effects presumably result from increased permeability of the muscle membrane at higher temperatures to the different ions, resulting in acceleration of the self-excitation process.

Contractile strength of the heart is often enhanced temporarily by a moderate increase in temperature, but prolonged elevation of the temperature exhausts the heart and causes weakness.

REFERENCES

Alpert, N. R., et al.: Heart muscle mechanics. *Annu. Rev. Physiol.*, 41:521, 1979.

Baan, J., et al. (eds.): Cardiac Dynamics. Boston, M. Nijhoff, 1979.

Bishop, V. S., et al.: Factors influencing cardiac performance. *Int. Rev. Physiol.*, 9:239, 1976.

Brady, A. J.: Mechanical properties of cardiac fibers. *In* Berne, R. M., et al. (eds.): Handbook of Physiology. Sec. 2, Vol. 1. Baltimore, Williams & Wilkins, 1979, p. 461.

Braunwald, E., and Ross, J., Jr.: Control of cardiac performance. *In* Berne, R. M., et al. (eds.): Handbook of Physiology. Sec. 2, Vol. I. Baltimore, Williams & Wilkins, 1979, p. 533.

Cowley, A. W., Jr., and Guyton, A. C.: Heart rate as a determinant of cardiac output in dogs with arteriovenous fistula. *Am. J. Cardiol.*, 28:321, 1971.

Ebashi, S.: Modern concepts of myocardial contraction. *In* Hayase, S., and Murao, S. (eds.): Cardiology: Proceedings of the VIII World Congress of Cardiology, Tokyo, 1978. New York, Elsevier/North-Holland, 1979, p. 92.

Fabiato, A., and Fabiato, F.: Calcium and cardiac excitation-contraction coupling. *Annu. Rev. Physiol.*, 41:473, 1979.

Guyton, A. C.: Determination of cardiac output by equating venous return curves with cardiac response curves. *Physiol. Rev.*, 35:123, 1955.

Guyton, A. C., et al.: Circulatory Physiology: Cardiac Output and Its Regulation, 2nd Ed. Philadelphia, W. B. Saunders Co., 1973.

Langer, G. A., and Brady, A. J.: The Mammalian Myocardium. New York, John Wiley & Sons, 1974.

Langer, G. A., et al.: The myocardium. *Int. Rev. Physiol.*, 9:191, 1976.

Lüllmann, H., and Peters, T.: Action of Cardiac Glycosides on the Excitation-Contraction Coupling in Heart Muscle. New York, Fischer, 1979.

Mirsky, I.: Elastic properties of the myocardium: A quantitative approach with physiological and clinical applications. *In* Berne, R. M., et al. (eds.): Handbook of Physiology. Sec. 2, Vol. I. Baltimore, Williams & Wilkins, 1979, p. 497.

Parmley, W. W., and Talbot, W.: Heart as a pump. *In* Berne, R. M., et al. (eds.): Handbook of Physiology. Sec. 2, Vol. I. Baltimore, Williams & Wilkins, 1979, p. 429.

Sarnoff, S. J.: Myocardial contractility as described by ventricular function curves. *Physiol. Rev.*, 35:107, 1955.

Sommer, J. R., and Johnson, E. A.: Ultrastructure of cardiac muscle. *In* Berne, R. M., et al. (eds.): Handbook of Physiology. Sec. 2, Vol. I. Baltimore, Williams & Wilkins, 1979, p. 113.

Starling, E. H.: The Linacre Lecture on the Law of the Heart. London, Longmans Green & Co., 1918.

Trautwein, W.: Membrane currents in cardiac muscle fibers. *Physiol. Rev.*, 53:793, 1973.

Weisfeldt, M. L. (ed.): The Heart in Old Age: Its Function and Response to Stress. New York, Raven Press, 1980.

Winegrad, S.: Electromechanical coupling in heart muscle. *In* Berne, R. M., et al. (eds.): Handbook of Physiology. Sec. 2, Vol. I. Baltimore, Williams & Wilkins, 1979, p. 393.

12

Rhythmic Excitation of the Heart

The heart is endowed with a special system (1) for generating rhythmical impulses to cause rhythmical contraction of the heart muscle and (2) for conducting these impulses rapidly throughout the heart. Many of the ills of the heart, especially the cardiac arrhythmias, are based on abnormalities of this special excitatory and conductive system.

THE SPECIAL EXCITATORY AND CONDUCTIVE SYSTEM OF THE HEART

The adult human heart normally contracts at a rhythmic rate of about 72 beats per minute. Figure 12–1 illustrates the special excitatory and conductive system of the heart that controls these cardiac contractions. The figure shows: (A) the *S-A node,* in which the normal rhythmic self-excitatory impulse is generated, (B) the *internodal pathways* that conduct the impulse from the S-A node to the A-V node, (C) the *A-V node,* in which the impulse from the atria is delayed before passing into the ventricles, (D) the *A-V bundle,* which conducts the impulse from the atria into the ventricles, and (E) the *left* and *right bundles of Purkinje fibers,* which conduct the cardiac impulse to all parts of the ventricles.

THE SINO-ATRIAL NODE

The sino-atrial (S-A) node is a small, crescent strip of specialized muscle approximately 3 mm. wide and 1 cm. long; it is located in the posterior wall of the right atrium immediately beneath and medial to the opening of the superior vena cava. The fibers of this node are each 3 to 5 microns in diameter, in contrast to a diameter of 15 to 20 microns for the surrounding atrial muscle fibers. However, the S-A fibers are continuous with the atrial fibers so that any action potential that begins in the S-A node spreads immediately into the atria.

Automatic Rhythmicity of the Sino-Atrial Fibers. Most cardiac fibers have the capability of *self-excitation,* a process that can cause automatic rhythmical contraction. This is especially true of the fibers of the heart's specialized conducting system. The portion of this system that displays self-excitation to the greatest extent is the fibers of the S-A node. For this reason, the sino-atrial (S-A) node ordinarily controls the rate of beat of the entire heart, as will be discussed in detail later in this chapter.

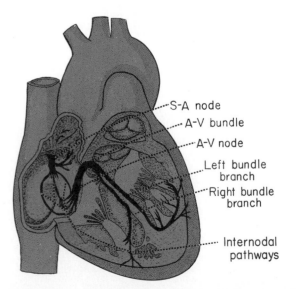

Figure 12–1. The S-A node and the Purkinje system of the heart.

120

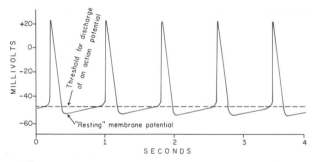

Figure 12-2.   Rhythmic discharge of an S-A nodal fiber.

The sino-atrial fibers are somewhat different from most other cardiac muscle fibers, having a resting membrane potential of only −55 to −60 millivolts in comparison with −85 to −95 millivolts in most of the other fibers. This low resting potential is caused by a natural leakiness of the membranes to sodium ions. And it is also this leakage of sodium that causes the self-excitation of the S-A fibers. Figure 12–2 illustrates the rhythmical repetitiveness of the action potentials in the S-A node.

The mechanism of this rhythmicity was explained in Chapter 8. Briefly, it is the following: Immediately after each action potential is over, the membrane is more permeable to potassium ions than normally. Therefore, excessive numbers of potassium ions, carrying positive charges, diffuse out of the cell. This makes the interior of the cell temporarily more negative than usual, a state called *hyperpolarization.* However, this does not last long because the membrane becomes progressively less permeable to potassium during the next few tenths of a second, and the natural leakiness of the membrane to sodium returns and causes the membrane potential to "drift" slowly back toward a less negative value. This is illustrated after each one of the action potentials in Figure 12–2. After a few tenths of a second, the potential drifts enough to reach the threshold level for excitation of the fiber, and a new action potential occurs.

This process continues over and over throughout the life of the person, thereby providing rhythmical excitation of the S-A nodal fibers at a normal resting rate of about 72 times per minute.

INTERNODAL PATHWAYS AND TRANSMISSION OF THE CARDIAC IMPULSE THROUGH THE ATRIA

The ends of the S-A nodal fibers fuse with the surrounding atrial muscle fibers, and action potentials originating in the S-A node travel out-

ward into these fibers. In this way, the action potential spreads through the entire atrial muscle mass and eventually also to the A-V node. The velocity of conduction in the atrial muscle is approximately 0.3 meter per second. However, conduction is somewhat more rapid in several small bundles of atrial muscle fibers, some of which pass directly from the S-A node to the A-V node and conduct the cardiac impulse at a velocity of 0.45 to 0.6 meter per second. These bundles, called *internodal pathways,* are illustrated in Figure 12–1.

THE ATRIOVENTRICULAR (A-V) NODE AND THE PURKINJE SYSTEM

Delay in Transmission at the A-V Node.   Fortunately, the conductive system is organized so that the cardiac impulse will not travel from the atria into the ventricles too rapidly; this allows time for the atria to empty their contents into the ventricles before ventricular contraction begins. It is primarily the A-V node and its associated conductive fibers that delay this transmission of the cardiac impulse from the atria into the ventricles.

Figure 12–3 shows diagrammatically the different parts of the A-V node and its connections with the atrial internodal pathway fibers and the A-V bundle. The cardiac impulse, after traveling through the internodal pathway, reaches the A-V node approximately 0.04 second after its origin in the S-A node. However, between this time and the time that the impulse emerges in the A-V bundle, another 0.11 second elapses. About one half of this time lapse occurs in the *junctional fibers,* which are very small fibers that connect the normal atrial fibers with the fibers of the node

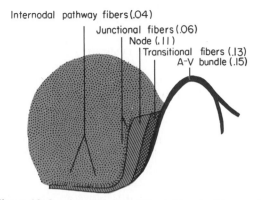

Figure 12-3.   Organization of the A-V node. The numbers represent the interval of time from the origin of the impulse in the S-A node. The values have been extrapolated to the human being. (This figure is based on studies in lower animals discussed and illustrated in Hoffman and Cranefield: Electrophysiology of the Heart. New York, McGraw-Hill Book Company, 1960.)

itself. The velocity of conduction in these fibers is about 0.02 meter per second (about one twenty-fifth that in normal cardiac muscle), which greatly delays entrance of the impulse into the A-V node. After entering the node proper, the velocity of conduction in the *nodal fibers* is still quite low, only 0.1 meter per second, about one-fourth the conduction velocity in normal cardiac muscle. Therefore, a further delay in transmission occurs as the impulse travels through the A-V node into the *transitional fibers* and finally into the *A-V bundle*.

TRANSMISSION IN THE PURKINJE SYSTEM

The *Purkinje fibers* that lead from the A-V node through the A-V bundle and into the ventricles have functional characteristics quite the opposite of those of the A-V nodal fibers; they are very large fibers, even larger than the normal ventricular muscle fibers, and they transmit impulses at a velocity of 1.5 to 4.0 meters per second, a velocity about 6 times that in the usual cardiac muscle and 150 times that in the junctional fibers. This allows almost immediate transmission of the cardiac impulse throughout the entire ventricular system.

Distribution of the Purkinje Fibers in the Ventricles. The Purkinje fibers, after originating in the A-V node, form the A-V bundle, which then threads through the fibrous tissue between the valves of the heart and thence into the ventricular system, as shown in Figure 12–1. The A-V bundle then divides almost immediately into the *left* and *right bundle branches* that lie beneath the endocardium of the respective sides of the septum. Each of these branches spreads downward toward the apex of the respective ventricle, but then divides into small branches that spread around each ventricular chamber and finally back toward the base of the heart along the lateral wall. The terminal Purkinje fibers then penetrate into the muscle mass to terminate on the muscle fibers.

From the time that the cardiac impulse first enters the A-V bundle until it reaches the terminations of the Purkinje fibers, the total time that lapses is only 0.03 second; therefore, once a cardiac impulse enters the Purkinje system, it spreads almost immediately to the entire endocardial surface of the ventricular muscle.

TRANSMISSION OF THE CARDIAC IMPULSE IN THE VENTRICULAR MUSCLE

Once the cardiac impulse has reached the ends of the Purkinje fibers, it is then transmitted through the ventricular muscle mass by the ventricular muscle fibers themselves. The velocity of transmission is now only 0.4 to 0.5 meter per second, one-sixth that in the Purkinje fibers.

The cardiac muscle is coiled around the heart in a double spiral with fibrous septa between the spiraling layers; therefore, the cardiac impulse does not necessarily travel directly outward toward the surface of the heart but instead angulates toward the surface along the directions of the spirals. Because of this, transmission from the endocardial surface to the epicardial surface of the ventricle requires as much as another 0.03 second, approximately equal to the time required for transmission through the entire Purkinje system. Thus, the total time for transmission of the cardiac impulse from the origin of the Purkinje system to the last of the ventricular muscle fibers in the normal heart is about 0.06 second.

SUMMARY OF THE SPREAD OF THE CARDIAC IMPULSE THROUGH THE HEART

Figure 12–4 illustrates in summary form the transmission of the cardiac impulse through the human heart. The numbers on the figure represent the intervals of time in fractions of a second that lapse between the origin of the cardiac impulse in the S-A node and its appearance at each respective point in the heart. Note that the impulse spreads at moderate velocity through the atria but is delayed more than 0.1 second in the A-V nodal region before appearing in the A-V bundle. Once it has entered the bundle, it spreads rapidly through the Purkinje fibers to the entire endocardial surfaces of the ventricles. Then the impulse spreads slowly through the ventricular muscle to the epicardial surfaces.

It is extremely important that the student learn in detail the course of the cardiac impulse through the heart and the times of its appearance

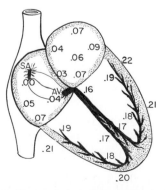

Figure 12–4. Transmission of the cardiac impulse through the heart, showing the time of appearance (in fractions of a second) of the impulse in different parts of the heart.

in each separate part of the heart, for a quantitative knowledge of this process is essential to the understanding of electrocardiography, which is discussed in the following chapter.

CONTROL OF EXCITATION AND CONDUCTION IN THE HEART

THE S-A NODE AS THE PACEMAKER OF THE HEART

In the above discussion of the genesis and transmission of the cardiac impulse through the heart, it was stated that the impulse normally arises in the S-A node. However, this need not be the case under abnormal conditions, for other parts of the heart can exhibit rhythmic contraction in the same way that the fibers of the S-A node can; this is particularly true of the A-V nodal and Purkinje fibers.

The A-V nodal fibers, when not stimulated from some outside source, discharge at an intrinsic rhythmic rate of 40 to 60 times per minute, and the Purkinje fibers discharge at a rate of somewhere between 15 to 40 times per minute. These rates are in contrast to the normal rate of the S-A node of 70 to 80 times per minute.

Therefore, the question that we must ask is: Why does the S-A node, rather than the A-V node or the Purkinje fibers, control the heart's rhythmicity? The answer to this is simply that the rate of the S-A node is considerably greater than that of either the A-V node or the Purkinje fibers. Each time the S-A node discharges, its impulse is conducted into both the A-V node and the Purkinje fibers, discharging their excitable membranes. Then these tissues, as well as the S-A node, recover from the action potential and become hyperpolarized. But the S-A node loses this hyperpolarization much more rapidly than does either of the other two and emits a new impulse before either one of them can reach its own threshold for self-excitation. The new impulse again discharges both the A-V node and Purkinje fibers. This process continues on and on, the S-A node always exciting these other potentially self-excitatory tissues before self-excitation can actually occur.

Thus, the S-A node controls the beat of the heart because its rate of rhythmic discharge is greater than that of any other part of the heart. Therefore, it is said that the S-A node is the normal *pacemaker* of the heart.

Abnormal Pacemakers — The Ectopic Pacemaker. Occasionally some other part of the heart develops a rhythmic discharge rate that is more rapid than that of the S-A node. For instance, this frequently occurs in the A-V node or in the Purkinje fibers. In either of these cases, the pacemaker of the heart shifts from the S-A node to the A-V node or to the excitable Purkinje fibers. Under rare conditions a point in the atrial or ventricular muscle develops excessive excitability and becomes the pacemaker.

A pacemaker elsewhere than the S-A node is called an *ectopic pacemaker*. Obviously, an ectopic pacemaker causes an abnormal sequence of contraction of the different parts of the heart.

ROLE OF THE PURKINJE SYSTEM IN CAUSING SYNCHRONOUS CONTRACTION OF THE VENTRICULAR MUSCLE

It is clear from the above description of the Purkinje system that the cardiac impulse arrives at almost all portions of the ventricles within a very narrow span of time, exciting the first ventricular muscle fiber only 0.06 second ahead of excitation of the last ventricular muscle fiber. Since the ventricular muscle fibers remain contracted for a total period of 0.30 second, one can see that this rapid spread of excitation throughout the entire ventricular muscle mass causes all portions of the ventricular muscle in both ventricles to contract at almost exactly the same time. Effective pumping by the two ventricular chambers requires this synchronous type of contraction. If the cardiac impulse traveled through the ventricular muscle very slowly, then much of the ventricular mass would contract prior to contraction of the remainder, in which case the overall pumping effect would be greatly depressed.

Refractory Period of the Purkinje Fibers and Its Functional Significance. The action potential and refractory period of the Purkinje fibers last about 25 per cent longer than the action potential and refractory period of the usual ventricular cardiac muscle. This has very important functional importance for maintaining the normal heart rhythm, as follows: By the time the Purkinje fibers are no longer refractory, the ventricular muscle fibers have already long been out of the refractory state. As a result, all portions of the ventricular muscle are already available to accept a new action potential and therefore are ready to contract. We shall see later in the chapter that if some portions of the ventricular muscle are still in a state of refractoriness when a new depolarization wave travels over the ventricles, there is great likelihood that the heart will develop ventricular fibrillation, a condtion in which there are incoordinate action potentials traveling in many diverse directions at the same time and never stopping — a condition that is lethal.

CONTROL OF HEART RHYTHMICITY AND CONDUCTION BY THE AUTONOMIC NERVES

The heart is supplied with both sympathetic and parasympathetic nerves, as illustrated in Figure 11–7 of the previous chapter. The parasympathetic nerves are distributed mainly to the S-A and A-V nodes, to a lesser extent to the muscle of the two atria, and even less to the ventricular muscle. The sympathetic nerves are distributed to these same areas but with a strong representation to the ventricular muscle as well as to the other parts of the heart.

Effect of Parasympathetic (Vagal) Stimulation on Cardiac Rhythm and Conduction — "Ventricular Escape." Stimulation of the parasympathetic nerves to the heart (the vagi) causes the hormone acetylcholine to be released at the vagal endings. This hormone has two major effects on the heart. First, it decreases the rate of rhythm of the S-A node, and second, it decreases the excitability of the A-V junctional fibers between the atrial musculature and the A-V node, thereby slowing transmission of the cardiac impulse into the ventricles. Very strong stimulation of the vagi can completely stop the rhythmic activity of the S-A node or completely block transmission of the cardiac impulse through the A-V junction. In either case, rhythmic impulses are no longer transmitted into the ventricles. The ventricles stop beating for 4 to 10 seconds, but then some point in the Purkinje fibers, usually in the A-V bundle, develops a rhythm of its own and causes ventricular contraction at a rate of 15 to 40 beats per minute. This phenomenon is called *ventricular escape.*

Mechanism of the Vagal Effects. The acetylcholine released at the vagal nerve endings greatly increases the permeability of the fiber membranes to potassium, which allows rapid leakage of potassium to the exterior. This causes increased negativity inside the fibers, an effect called *hyperpolarization,* which makes excitable tissue much less excitable, for reasons explained in Chapter 8.

Effect of Sympathetic Stimulation on Cardiac Rhythm and Conduction. Sympathetic stimulation causes essentially the opposite effects on the heart to those caused by vagal stimulation as follows: First, it increases the rate of S-A nodal discharge. Second, it increases the rate of conduction and the excitability in all portions of the heart. Third, it increases greatly the force of contraction of all the cardiac musculature, both atrial and ventricular, as discussed in the previous chapter.

In short, sympathetic stimulation increases the overall activity of the heart. Maximal stimulation can almost triple the rate of heartbeat and can increase the strength of heart contraction as much as two-fold.

Mechanism of the Sympathetic Effect. Stimulation of the sympathetic nerves releases the hormone norepinephrine at the sympathetic nerve endings. This hormone in turn increases the permeability of the fiber membrane to sodium and calcium. In the S-A node, an increase of sodium permeability causes increased self-excitation, which obviously increases the heart rate.

In the A-V node, increased sodium permeability makes it easier for each fiber to excite the succeeding fiber, thereby decreasing the conduction time from the atria to the ventricles.

The increase in permeability to calcium ions is at least partially responsible for the increase in contractile strength of the cardiac muscle under the influence of sympathetic stimulation, because calcium ions play a powerful role in exciting the contractile process of the myofibrils.

ABNORMAL RHYTHMS OF THE HEART

PREMATURE CONTRACTIONS — ECTOPIC FOCI

Often, a small area of the heart becomes much more excitable than normal and causes abnormal impulses to be generated in between the normal impulses. A depolarization wave spreads outward from the irritable area and initiates *premature contraction* of the heart. The focus at which the abnormal impulse is generated is called an *ectopic focus.*

The usual cause of an ectopic focus is an irritable area of cardiac muscle resulting from a local area of ischemia (too little coronary blood flow to the muscle), overuse of stimulants such as caffeine or nicotine, lack of sleep, anxiety, or some other debilitating state.

HEART BLOCK

Occasionally, transmission of the impulse through the heart is blocked at a critical point in the conductive system. One of the most common of these points is between the atria and the ventricles; this condition is called *atrioventricular block.* It can result from localized damage or depression of the *A-V junctional* fibers or of the *A-V bundle.* The causes include different types of infectious processes, excessive stimulation by the vagus nerves (which depresses conductivity in the junctional fibers), localized destruction of the

A-V bundle as a result of a coronary infarct, pressure on the A-V bundle by arteriosclerotic plaques, or depression caused by various drugs.

FLUTTER AND FIBRILLATION

Frequently, either the atria or the ventricles begin to contract extremely rapidly and often incoordinately. The low frequency, more coordinate contractions up to 200 to 300 beats per minute are generally called *flutter,* and the very high frequency, incoordinate contractions, *fibrillation.* Both of these are usually caused by a phenomenon called a *circus movement.*

The Circus Movement. A circus movement occurs when an impulse begins in one part of the heart muscle, spreads in a circuitous pathway through the heart, and then returns to the originally excited muscle to stimulate it again. Thus the excitatory signal continues again and again around the circle, never stopping. This effect is illustrated in Figure 12–5, which depicts several small cardiac muscle strips cut in the form of circles. If such a strip is stimulated at the 12 o'clock position *so that the impulse travels in only one direction,* the impulse spreads progressively around the circle until it returns to the 12 o'clock position. If the originally stimulated muscle fibers are still in a refractory state, the impulse then dies out, for refractory muscle cannot transmit a second impulse. However, three different conditions can cause this impulse to continue to travel around the circle.

First, if the *length of the pathway around the circle is long,* by the time the impulse returns to the 12 o'clock position the originally stimulated muscle will no longer be refractory, and the impulse will continue around the circle again and again.

Second, if the length of the pathway remains

constant but the *velocity of conduction becomes decreased* enough, an increased interval of time will elapse before the impulse returns to the 12 o'clock position. By this time the originally stimulated muscle might be out of the refractory state, and the impulse can continue around the circle again and again.

Third, the *refractory period of the muscle might become greatly shortened.* In this case also, the impulse can continue around and around the circle.

All three of these conditions occur in different pathological states of the human heart as follows: (1) A long pathway frequently occurs in dilated hearts. (2) Decreased rate of conduction frequently results from blockage of the Purkinje system, ischemia of the muscle, heart muscle disease, and many other factors. (3) A shortened refractory period frequently occurs in response to various drugs, such as epinephrine, or following repetitive electrical stimulation. Thus, in many different cardiac disturbances circus movements can cause abnormal cardiac rhythmicity that completely ignores the pacesetting effects of the S-A node.

Atrial Flutter Resulting from a Circus Pathway. The left-hand panel of Figure 12–6 illustrates a circus pathway around and around the atria from top to bottom, the pathway lying to the left of the superior and inferior venae cavae. Such circus pathways often develop in the human heart when the atria become greatly dilated as a result of valvular heart disease. The rate of flutter is usually 200 to 350 times per minute.

The "Chain Reaction" Mechanism of Fibrillation. Fibrillation, whether it occurs in the atria or in the ventricles, is a very different condition from flutter. One can see many separate contractile waves spreading in different directions over the cardiac muscle at the same time in either atrial or ventricular fibrillation. One of the best ways to explain the mechanism of fibrillation is to describe the initiation of fibrillation by stimulation with 60 cycle alternating electrical current.

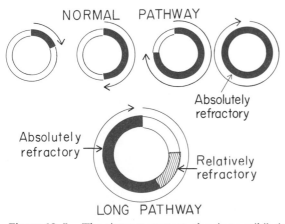

Figure 12–5. The circus movement, showing annihilation of the impulse in the short pathway and continued propagation of the impulse in the long pathway.

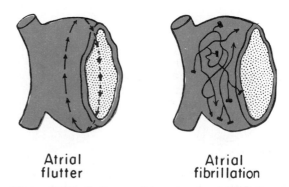

Figure 12–6. Pathways of impulses in atrial flutter and atrial fibrillation.

Fibrillation Caused by 60 Cycle Alternating Current. At a central point in the ventricles of heart A in Figure 12–7, a 60 cycle electrical stimulus is applied through a stimulating electrode. The first cycle of the electrical stimulus causes a depolarization wave to spread in all directions, leaving all the muscle beneath the electrode in a refractory state. After about 0.25 second, this muscle begins to come out of the refractory state, some portions of the muscle coming out of refractoriness prior to other portions. This state of events is depicted in heart A by many light patches, which represent excitable cardiac muscle, and dark patches, which represent still refractory muscle. New stimuli from the electrode can now cause impulses to travel in certain directions through the heart but not in all directions. Furthermore, when a depolarization wave reaches a refractory area in the heart, it travels to both sides around the area. Thus, a single impulse becomes two impulses. Then when each of these reaches another refractory area it, too, divides to form still two more impulses. In this way many different impulses are continually being formed in the heart by a progressive *chain reaction* until, finally, there are many small depolarization waves traveling in many different directions at the same time. Obviously, this irregular pattern of impulse travel causes a *circuitous route for the impulses to travel, greatly lengthening the conductive pathway, which leads to fibrillation.* One can readily see that a vicious circle has been initiated: More and more impulses are formed, these cause more and more patches of refractory muscle, and the refractory patches cause more and more division of the impulses. Thus, heart B in Figure 12–7 illustrates the final state that develops in fibrillation.

Atrial Fibrillation. Atrial fibrillation is completely different from atrial flutter because the circus movement does not travel in a regular pathway. Instead, many different excitation waves can be seen to travel over the surface of the atria at the same time. Atrial fibrillation occurs frequently when the atria become greatly overdilated — in fact, many times as frequently as flutter. When flutter does occur, it usually becomes fibrillation after a few days or weeks. To the right in Figure 12–6 are illustrated the pathways of fibrillatory impulses traveling through the atria.

Obviously, atrial fibrillation results in complete incoordination of atrial contraction so that atrial pumping ceases entirely. However, the effectiveness of the heart as a pump is reduced only 25 to 30 per cent, which is well within the "cardiac reserve" of all but severely weakened hearts. For this reason, atrial fibrillation can continue for many years without serious cardiac debility.

Ventricular Fibrillation. Ventricular fibrillation is extremely important because at least one quarter of all persons die in ventricular fibrillation. For instance, the hearts of most patients with coronary infarcts fibrillate shortly before death. In only a few instances on record have fibrillating human ventricles been known to return of their own accord to a rhythmic beat.

The likelihood of circus movements in the ventricles and, consequently, of ventricular fibrillation is greatly increased when the ventricles are dilated or when the rapidly conducting *Purkinje system* is blocked so that impulses cannot be transmitted rapidly. Also, *electric shock,* particularly with 60 cycle electric current, as discussed above, or *ectopic foci* are common initiating causes of ventricular fibrillation. The ventricular contractions become so fine and asynchronous that they pump no blood whatsoever. Therefore, death is immediate.

Electrical Defibrillation of the Ventricles. Though a weak alternating current almost invariably throws the ventricles into fibrillation, a very strong electrical current passed through the ventricles for a short interval of time can stop fibrillation by throwing all the ventricular muscle into refractoriness simultaneously. This is accomplished by passing intense current through electrodes placed on the heart or on the chest wall over the heart. The current penetrates most of the fibers of the ventricles, thus stimulating essentially all parts of the ventricles simultaneously and causing them to become refractory. All impulses stop, and the heart then remains quiescent for three to five seconds, after which it begins to beat again, with the S-A node or, often, some other part of the heart becoming the pacemaker. Occasionally, however, the same irritable focus that originally threw the ventricles into fibrillation is still present, and fibrillation begins again immediately.

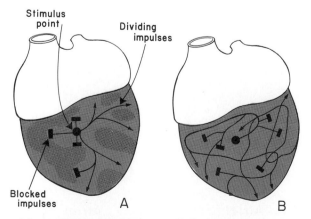

Figure 12–7. A, Initiation of fibrillation in a heart when patches of refractory musculature are present. B, Continued propagation of *fibrillatory impulses* in the fibrillating ventricle.

Unless defibrillated within one minute after fibrillation begins, the heart is usually too weak to be revived. However, it is still possible to revive the heart by preliminarily pumping it by hand and then defibrillating it later. In this way small quantities of blood are delivered into the aorta, and a renewed coronary blood supply develops. After a few minutes, electrical defibrillation often becomes possible. Indeed, fibrillating hearts have been pumped by hand as long as 90 minutes before defibrillation. In recent years, a technique of pumping the heart without opening the chest has been developed; this technique consists of intermittent, very powerful thrusts of pressure on the chest wall.

Lack of blood flow to the brain for more than five to ten minutes usually results in permanent mental impairment or even total destruction of the brain. Even though the heart should be revived, the person might die from the effects of brain damage or live with permanent mental impairment.

CARDIAC ARREST

When cardiac metabolism becomes greatly disturbed as a result of any one of many possible conditions, the rhythmic contractions of the heart occasionally stop. One of the most common causes of cardiac arrest is hypoxia of the heart, for severe hypoxia prevents the muscle fibers from maintaining normal ionic differentials across their membranes.

Occasionally, patients with severe myocardial disease develop cardiac arrest, which obviously can lead to death. In many cases, however, rhythmic electrical impulses from an implanted electronic cardiac "pacemaker" have been used successfully to keep patients alive for many years.

REFERENCES

Cranefield, P. F.: The Conduction of the Cardiac Impulse. Mount Kisco, N.Y., Futura Publishing Co., 1975.

Durrer, D., et al.: Human cardiac electrophysiology. In Dickinson, C. J., and Marks, J. (eds.): Developments in Cardiovascular Medicine. Lancaster, England, MTP Press, 1978, p. 53.

Ewy, G. A.: Cardiac Arrest and Resuscitation: Defibrillators and Defibrillation. Chicago, Year Book Medical Publishers, 1978.

Fozzard, H. A.: Heart: Excitation-contraction coupling. Annu. Rev. Physiol., 39:201, 1977.

Guyton, A. C., and Satterfield, J.: Factors concerned in electrical defibrillation of the heart, particularly through the unopened chest. Am. J. Physiol., 167:81, 1951.

Irisawa, H.: Comparative physiology of the cardiac pacemaker mechanism. Physiol. Rev., 58:461, 1978.

Jones, P.: Cardiac Pacing. New York, Appleton-Century-Crofts, 1980.

Josephson, M. E., and Seides, S. F.: Clinical Cardiac Electrophysiology Techniques and Interpretations. Philadelphia, Lea & Febiger, 1979.

Kulbertus, H. E., and Wellens, H. J. J. (eds.): Sudden Death. Hingham, Mass., Kluwer Boston, 1980.

Levy, M. N., and Martin, P. J.: Neural control of the heart. In Berne, R. M., et al. (eds.): Handbook of Physiology. Sec. 2, Vol. I. Baltimore, Williams & Wilkins, 1979, p. 581.

Narula, O. S. (ed.): Cardiac Arrhythmias: Electrophysiology, Diagnosis, and Management. Baltimore, Williams & Wilkins, 1979.

Nobel, D.: The Initiation of the Heartbeat. New York, Oxford University Press, 1979.

Pick, A., and Langendorf, R.: Interpretation of Complex Arrhythmias. Philadelphia, Lea & Febiger, 1980.

Reuter, H.: Properties of two inward membrane currents in the heart. Annu. Rev. Physiol., 41:413, 1979.

Samet, P., and El-Sherif, N. (eds.): Cardiac Pacing, 2nd ed. New York, Grune & Stratton, 1979.

Sperelakis, N.: Origin of the cardiac resting potential. In Berne, R. M., et al. (eds.): Handbook of Physiology. Sec. 2, Vol. I. Baltimore, Williams & Wilkins, 1979, p. 187.

Sperelakis, N.: Propagation mechanisms in heart. Annu. Rev. Physiol., 41:441, 1979.

Stull, J. T., and Mayer, S. E.: Biochemical mechanisms of adrenergic and cholinergic regulation of myocardial contractility. In Berne, R. M., et al. (eds.): Handbook of Physiology. Sec. 2, Vol. I. Baltimore, Williams & Wilkins, 1979, p. 741.

Varraile, P., and Naclerio, E. A.: Cardiac Pacing: A Concise Guide to Clinical Practice. Philadelphia, Lea & Febiger, 1979.

Vasselle, M.: Electrogenesis of the plateau and pacemaker potential. Annu. Rev. Physiol., 41:425, 1979.

13

The Electrocardiogram

Transmission of the impulse through the heart has been discussed in detail in Chapter 12. As the impulse passes through the heart, electrical currents spread into the tissues surrounding the heart, and a small proportion of them spreads all the way to the surface of the body. If electrodes are placed on the body surface on opposite sides of the heart, the electrical potentials generated by the heart can be recorded; the recording is the *electrocardiogram*. A normal electrocardiogram for two beats of the heart is illustrated in Figure 13–1.

CHARACTERISTICS OF THE NORMAL ELECTROCARDIOGRAM

The normal electrocardiogram is composed of a P wave, a "QRS complex," and a T wave. The QRS complex is actually three separate waves, the Q wave, the R wave, and the S wave, all of which are caused by passage of the cardiac impulse through the ventricles. In the normal electrocardiogram, the Q and S waves are often much less prominent than the R wave and sometimes are actually absent, but even so, the wave is still known as the QRS complex, or simply the QRS wave.

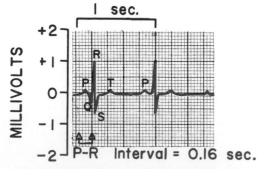

Figure 13–1. The normal electrocardiogram.

The P wave is caused by electrical currents generated as the atria depolarize prior to atrial contraction, and the QRS complex is caused by currents generated when the ventricles depolarize prior to their contraction. Therefore, both the P wave and the components of the QRS complex are *depolarization waves*. The T wave is caused by currents generated as the ventricles recover from the state of depolarization. This process occurs in ventricular muscle about 0.25 second after depolarization, and this wave is known as a *repolarization wave*. Thus, the electrocardiogram is composed of both depolarization and repolarization waves. The principles of depolarization and repolarization were discussed in Chapter 8. However, the distinction between depolarization waves and repolarization waves is so important in electrocardiography that further clarification is needed as follows:

DEPOLARIZATION WAVES VERSUS REPOLARIZATION WAVES

Figure 13–2 illustrates a muscle fiber in four different stages of depolarization and repolarization. During the process of depolarization the normal negative potential inside the fiber is lost and the membrane potential actually reverses; that is, it becomes slightly positive inside and negative outside.

In Figure 13–2A the process of depolarization, illustrated by the positivity inside and negativity outside, is traveling from left to right, and the first half of the fiber is already depolarized while the remaining half is still polarized. This causes the meter to record positively. To the right of the muscle fiber is illustrated a record of the potential between the electrodes as recorded by a high-speed recording meter at this particular stage of depolarization.

In Figure 13–2B depolarization has extended

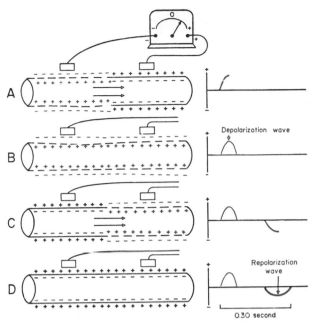

Figure 13-2. Recording the *depolarization wave* and the *repolarization wave* from a cardiac muscle fiber.

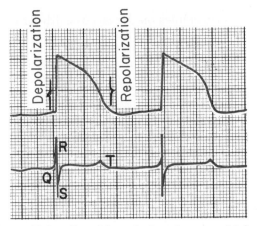

Figure 13-3. *Above:* Monophasic action potential from a ventricular muscle fiber during normal cardiac function, showing rapid depolarization and then repolarization occurring slowly during the plateau stage but very rapidly toward the end. *Below:* Electrocardiogram recorded simultaneously.

over the entire muscle fiber, and the recording to the right has now returned to the zero base line because both electrodes are in areas of equal negativity. The completed wave is a *depolarization wave* because it results from spread of depolarization along the extent of the muscle fiber.

Figure 13-2C illustrates the repolarization process, during which the recording, as illustrated to the right, becomes negative.

Finally, in Figure 13-2D the muscle fiber has completely repolarized, and the recording returns once more to the zero level. This completed negative wave is a *repolarization wave* because it results from spread of the repolarization process over the muscle fiber.

Relationship of the Monophasic Action Potential of Cardiac Muscle to the QRS and T Waves. The monophasic action potential of ventricular muscle, which was discussed in the preceding chapter, normally lasts between 0.25 and 0.30 second. The top part of Figure 13-3 illustrates a monophasic action potential recorded from a microelectrode inserted into the inside of a single ventricular muscle fiber. The upsweep of this action potential is caused by *depolarization,* and the return of the potential to the base line is caused by *repolarization.* Note below the simultaneous recording of the electrocardiogram from this same ventricle, which shows the QRS wave appearing at the beginning of the monophasic action potential and the T wave at the end. Note especially that *no potential at all is recorded in the electrocardiogram when the ventricular muscle is either completely polarized or completely depolarized.*

The Electrocardiogram During Ventricular Repolarization — The T Wave. Because the septum and the endocardium depolarize first, it seems logical that they should also repolarize first at the end of ventricular contraction, but this is not the normal case because the septum and endocardium have longer periods of depolarization than other areas of the ventricles. On the other hand, the outer portions of the ventricles have short periods of depolarization. Therefore, *the portion of ventricular muscle to repolarize first is that located exteriorly near the apex of the heart,* and the endocardial surfaces normally repolarize last. The reason for this abnormal sequence of repolarization is reputed to be that high pressure in the ventricles during contraction greatly reduces coronary blood flow to the endocardium, thereby slowing the repolarization process on the endocardial surfaces. But whatever the cause, *the predominant direction of current flow through the heart during* repolarization *of the ventricles is from base to apex, which is the same predominant direction of current flow during* depolarization. *As a result, the T wave in the normal electrocardiogram is positive, which is also the polarity of most of the normal QRS complex.*

VOLTAGE AND TIME CALIBRATION OF THE ELECTROCARDIOGRAM

All recordings of electrocardiograms are made with appropriate calibration lines on the recording paper.

As illustrated in Figure 13-1, the horizontal lines are spaced so that ten small divisions (1 cm.) in the vertical direction in the standard electrocardiogram represent one millivolt.

The vertical lines on the electrocardiogram are time calibration lines. Each inch (2.5 cm.) of the standard electrocardiogram is one second. Each inch in turn is usually broken into five segments by dark vertical lines, the distance between which represents 0.20 second. The intervals between the dark vertical lines are broken into five smaller intervals by thin lines, and the distance between each two of the smaller lines represents 0.04 second.

The P-Q Interval. The duration of time between the beginning of the P wave and the beginning of the QRS wave is the interval between the beginning of contraction of the atria and the beginning of contraction of the ventricles. This period of time is called the P-Q interval. The normal P-Q interval is approximately 0.16 second. This interval is sometimes also called the P-R interval, because the Q wave frequently is absent.

The Q-T Interval. Contraction of the ventricle lasts essentially between the Q wave and the end of the T wave. This interval of time is called the Q-T interval and ordinarily is approximately 0.30 second.

RECORDING ELECTROCARDIOGRAMS — THE PEN RECORDER

The electrical currents generated by the cardiac muscle during each beat of the heart regularly change potentials and polarity in less than 0.03 second. Therefore, it is essential that any apparatus for recording electrocardiograms be capable of responding rapidly to these changes in electrical potentials. The most usual type of recorder now used is the pen recorder. This instrument writes the electrocardiogram with a pen directly on a moving sheet of paper. The pen is often a thin tube connected at one end to an inkwell, with its recording end connected to a powerful electromagnet system that is capable of moving the pen back and forth at high speed. As the paper moves forward, the pen records the electrocardiogram. The movement of the pen in turn is controlled by appropriate amplifiers connected to electrocardiographic electrodes on the subject.

FLOW OF CURRENT AROUND THE HEART DURING THE CARDIAC CYCLE

Figure 13–4 illustrates the ventricular muscle mass lying within the chest. Even the lungs, though filled with air, conduct electricity to a surprising extent, and fluids of the other tissues surrounding the heart conduct electricity even

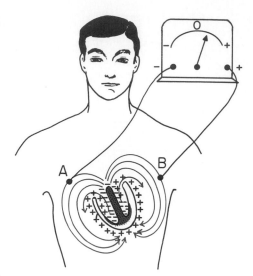

Figure 13–4. Flow of current in the chest around a partially depolarized heart.

more easily. Therefore, the heart is actually suspended in a conductive medium. When one portion of the ventricles becomes electronegative with respect to the remainder, electrical current flows from the depolarized area to the polarized area in large circuitous routes, as noted in the figure.

It will be recalled from the discussion of the Purkinje system in Chapter 12 that the cardiac impulse first arrives in the ventricles in the walls of the septum and almost immediately thereafter on the endocardial surfaces of the remainder of the ventricles, as shown by the shaded areas and the negative signs in Figure 13–4. This provides electronegativity on the insides of the ventricles and electropositivity on the outer walls of the ventricles. If one algebraically averages all the lines of current flow (the elliptical lines in Figure 13–4), he finds that the average current flow is from the base of the heart toward the apex. During most of the cycle of depolarization, the current continues to flow in this direction as the impulse spreads from the endocardial surface outward through the ventricular muscle. However, immediately before the depolarization wave has completed its course through the ventricles, the direction of current flow reverses for about 0.01 second, flowing then from the apex toward the base because the very last part of the heart to become depolarized is the outer walls of the ventricles near their base.

If a meter is connected to the surface of the body as shown in Figure 13–4, the electrode nearer the base will be negative with respect to the electrode nearer the apex, so that the meter shows a potential between the two electrodes. In making electrocardiographic recordings, various standard positions for placement of electrodes are used, and whether the polarity of the recording during each cardiac cycle is positive or negative is determined by the orientation of electrodes with respect to the current flow in the heart. Some of the conventional electrode systems, commonly called *electrocardiographic leads,* are discussed below.

ELECTROCARDIOGRAPHIC LEADS

THE THREE STANDARD LIMB LEADS

Figure 13–5 illustrates electrical connections between the limbs and the electrocardiograph for recording electrocardiograms from the so-called "standard" limb leads. The electrocardiograph in each instance is illustrated by a mechanical meter in the diagram, though the actual electrocardiograph is a high-speed recording meter.

Lead I. In recording limb lead I, the negative terminal of the electrocardiograph is connected to the right arm, and the positive terminal to the left arm. Therefore, when the point on the chest where the right arm connects to the chest is electronegative with respect to the point where

the left arm connects, the electrocardiograph records positively — that is, above the zero voltage line in the electrocardiogram. When the opposite is true, the electrocardiograph records below the line.

Lead II. In recording limb lead II, the negative terminal of the electrocardiograph is connected to the right arm and the positive terminal to the left leg. Thus, when the right arm is negative with respect to the left leg, the electrocardiograph records positively.

Lead III. In recording limb lead III, the negative terminal of the electrocardiograph is connected to the left arm and the positive terminal to the left leg. This means that the electrocardiograph records positively when the left arm is negative with respect to the left leg.

Normal Electrocardiograms Recorded by the Three Standard Leads. Figure 13–6 illustrates simultaneous recordings of the electrocardiogram in leads I, II, and III. It is obvious from this figure that the electrocardiograms in these three standard leads are very similar to each other, for they all record positive P waves and positive T waves, and the major proportion of the QRS complex is positive in each electrocardiogram.

Because all normal electrocardiograms are very similar to each other, it does not matter greatly which electrocardiographic lead is recorded when one wishes to diagnose the different cardiac arrhythmias, for diagnosis of arrhythmias depends mainly on the time relationships between the different waves of the cardiac cycle. On the other hand, when one wishes to determine the extent and type of damage in the ventricles or in

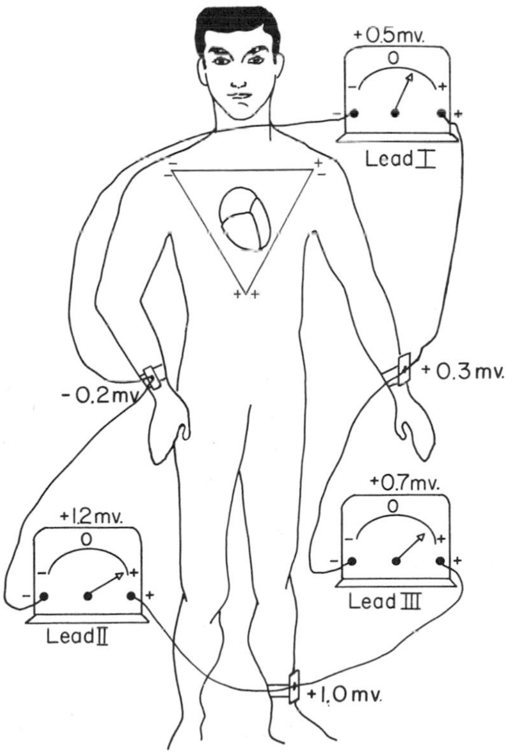

Figure 13–5. Conventional arrangement of electrodes for recording the standard electrocardiographic leads. Einthoven's triangle is superimposed on the chest.

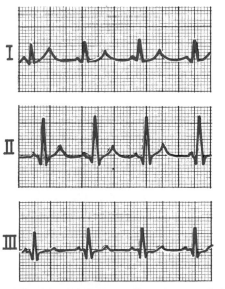

Figure 13–6. Normal electrocardiograms recorded from the three standard electrocardiographic leads.

the atria, it does matter greatly which leads are recorded, for abnormalities of the cardiac muscle change the patterns of the electrocardiograms markedly in some leads and yet may not affect other leads.

Electrocardiographic interpretation of these two types of conditions — cardiac arrhythmias and cardiac myopathies — are discussed separately in later sections of this chapter.

CHEST LEADS (PRECORDIAL LEADS)

Often electrocardiograms are recorded with one electrode placed on the anterior chest over the heart, as illustrated by the six separate points in Figure 13–7. This electrode is connected to the positive terminal of the electrocardiograph, and the negative electrode, called the *indifferent electrode,* is normally connected simultaneously through electrical resistances to the right arm, left arm, and left leg, as also shown in the figure. Usually six different standard chest leads are recorded from the anterior chest wall, the chest electrode being placed respectively at the six points illustrated in the diagram. The different leads recorded by the method illustrated in Figure 13–7 are known as leads V_1, V_2, V_3, V_4, V_5, and V_6.

Figure 13–8 illustrates the electrocardiograms of the normal heart as recorded in the six standard chest leads. Because the heart surfaces are close to the chest wall, each chest lead records mainly the electrical potential of the cardiac musculature immediately beneath the electrode. Therefore, relatively minute abnormalities in the ventricles, particularly in the anterior ventricular wall, frequently cause marked changes in the electrocardiograms recorded from chest leads.

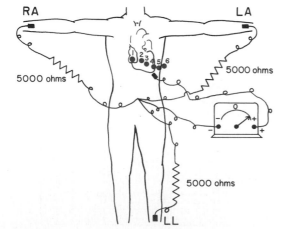

Figure 13–7. Connections of the body with the electrocardiograph for recording chest leads.

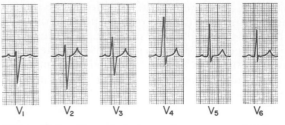

Figure 13–8. Normal electrocardiograms recorded from the six standard chest leads.

ELECTROCARDIOGRAPHIC INTERPRETATION OF CARDIAC ARRHYTHMIAS

The rhythmicity of the heart and some abnormalities of rhythmicity were discussed in Chapter 12. The major purpose of the present section is to describe the electrocardiograms recorded in a few conditions known clinically as "cardiac arrhythmias."

ATRIOVENTRICULAR BLOCK

The only means by which impulses can ordinarily pass from the atria into the ventricles is through the *A-V bundle*, which is also known as the *bundle of His.* Different conditions that can either decrease the rate of conduction of the impulse through this bundle or totally block the impulse are

1. *Ischemia of the A-V junctional fibers.*
2. *Compression of the A-V bundle* by scar tissue or by calcified portions of the heart.
3. *Inflammation of the A-V bundle or fibers of the A-V junction.*
4. *Extreme stimulation of the heart by the vagi.*

Incomplete Heart Block. When conduction through the A-V junction is slowed from a normal value of 0.16 second to as great as 0.25 to 0.50 second, the action potentials traveling through the A-V junctional fibers are sometimes strong enough to pass on into the A-V node and at other times are not strong enough. Often the impulse passes into the ventricles on one heartbeat and fails to pass on the next one or two beats, thus alternating between conduction and nonconduction. In this instance, the atria beat at a considerably faster rate than the ventricles, and it is said that there are "dropped beats" of the ventricles. This condition is called *incomplete heart block.*

Figure 13–9 illustrates P-R intervals as long as 0.30 second, and it also illustrates one dropped beat as a result of failure of conduction from the atria to the ventricles.

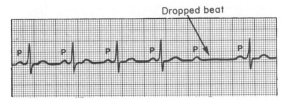

Figure 13–9. Partial atrioventricular block (lead V₃).

At times every other beat of the ventricles is dropped, so that a "2:1 rhythm" develops in the heart, with the atria beating twice for every single beat of the ventricles. Sometimes other rhythms, such as 3:2 or 3:1, also develop.

Complete Atrioventricular Block. When the condition causing poor conduction in the A-V bundle becomes extremely severe, complete block of the impulse from the atria into the ventricles occurs. In this instance the P waves become completely dissociated from the QRS-T complexes, as illustrated in Figure 13–10. Note that the rate of rhythm of the atria in this electrocardiogram is approximately 100 beats per minute, while the rate of ventricular beat is less than 40 per minute. Furthermore, there is no relationship whatsoever between the rhythm of the atria and that of the ventricles, for the ventricles have "escaped" from control by the atria, and they are beating at their own natural rate.

Stokes-Adams Syndrome — Ventricular Escape. In some patients with atrioventricular block, total block comes and goes — that is, all impulses are conducted from the atria into the ventricles for a period of time, and then suddenly no impulses at all are transmitted. Particularly does this condition occur in hearts with borderline ischemia.

Immediately after A-V conduction is first blocked, the ventricles stop contracting entirely for about 5 to 10 seconds. Then some part of the Purkinje system beyond the block, usually in the A-V bundle itself, begins discharging rhythmically at a rate of 15 to 40 times per minute and acting as the pacemaker of the ventricles. This is called *ventricular escape.* Because the brain cannot remain active for more than 3 to 5 seconds without blood supply, patients usually faint between block of conduction and "escape" of the ventricles. These periodic fainting spells are known as the Stokes-Adams syndrome.

PREMATURE BEATS

A premature beat is a contraction of the heart prior to the time when normal contraction would have been expected. This condition is also frequently called *extrasystole.*

Most premature beats result from *ectopic foci* in the heart, which emit abnormal impulses at odd times during the cardiac rhythm. The possible causes of ectopic foci are (1) local areas of ischemia, (2) small calcified plaques at different points in the heart, which press against the adjacent cardiac muscle so that some of the fibers are irritated, and (3) toxic irritation of the A-V node, Purkinje system, or myocardium caused by drugs, nicotine, caffeine, and so forth.

Atrial Premature Beats. Figure 13–11 illustrates an electrocardiogram showing a single atrial premature beat. The P wave of this beat is relatively normal and the QRS complex is also normal, but the interval between the preceding beat and the premature beat is shortened. Also, the interval between the premature beat and the next succeeding beat is slightly prolonged. The reason is that the premature beat originated in the atrium at some distance from the S-A node, and the impulse of the premature beat had to travel through a short distance of atrial muscle before it discharged the S-A node. Consequently, the S-A node discharged late in the cycle and made the succeeding heartbeat also late in appearing.

Ventricular Premature Beats. The electrocardiogram of Figure 13–12 illustrates a series of ventricular premature beats alternating with normal beats. Ventricular premature beats cause several special effects in the electrocardiogram, as follows: First, the QRS complex is usually considerably prolonged. The reason is that the impulse is conducted mainly through the muscle of the ventricles rather than through the Purkinje system.

Second, the QRS complex has a very high voltage, for the following reason: When the normal impulse passes through the heart, it passes through both ventricles approximately simultaneously; consequently, the depolarization waves of the two sides of the heart partially neutralize each other. However, when a ventricular prema-

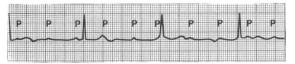

Figure 13–10. Complete atrioventricular block (lead II).

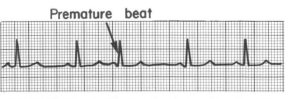

Figure 13–11. Atrial premature contraction (lead I).

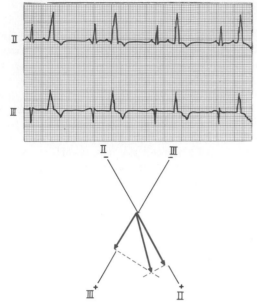

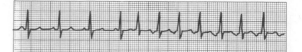

Figure 13–13. Atrial paroxysmal tachycardia — onset in middle of record (lead I).

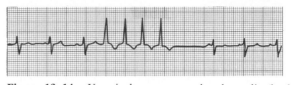

Figure 13–14. Ventricular paroxysmal tachycardia (lead III).

Figure 13–12. Premature ventricular contractions (PVCs) illustrated by the large abnormal QRS-T complexes (leads II and III). Axis of the premature contractions is plotted in accord with the principles of vectorial analysis.

ture beat occurs, the impulse travels in only one direction so that there is no such neutralization effect.

Third, following almost all ventricular premature beats, the T wave has a potential opposite to that of the QRS complex because the *slow conduction of the impulse* through the cardiac muscle causes the area first depolarized to repolarize first as well. As a result, the direction of current flow in the heart during repolarization is opposite to that during depolarization. This is not true of the normal T wave, as was explained earlier in the chapter.

Some ventricular premature beats are benign in their origin and result from simple factors such as cigarettes, coffee, lack of sleep, various mild toxic states, and even emotional irritability. On the other hand, a large share of ventricular premature beats result from an actual pathologic condition of the heart. For instance, many ventricular premature beats occur following coronary thrombosis because of stray impulses originating around the borders of the infarcted area of the heart.

PAROXYSMAL TACHYCARDIA

Abnormalities in any portion of the heart, including the atria, the Purkinje system, and the ventricles, can sometimes cause rapid rhythmic discharge of impulses which spread in all directions throughout the heart. Because of the rapid

rhythm in the irritable focus, this focus becomes the pacemaker of the heart.

Atrial Paroxysmal Tachycardia. Figure 13–13 illustrates a sudden increase in rate of heartbeat from approximately 95 beats per minute to approximately 150 beats per minute. Close analysis of the electrocardiogram shows that an inverted P wave occurs before each of the QRS-T complexes during the paroxysm of rapid heartbeat, though this P wave is partially superimposed on the normal T wave of the preceding beat. This indicates that the origin of this particular paroxysmal tachycardia is in the atrium, but because the P wave is abnormal, the origin is not near the S-A node.

Ventricular Paroxysmal Tachycardia. Figure 13–14 illustrates a typical short paroxysm of ventricular tachycardia. The electrocardiogram of ventricular paroxysmal tachycardia has the appearance of a series of ventricular premature beats occurring one after another without any normal beats interspersed.

Ventricular paroxysmal tachycardia is usually a serious condition for two reasons. First, this type of tachycardia usually does not occur unless considerable damage is present in the ventricles. Second, ventricular tachycardia predisposes to ventricular fibrillation, which is almost invariably fatal.

ABNORMAL RHYTHMS RESULTING FROM CIRCUS MOVEMENTS

The circus movement phenomenon was discussed in detail in Chapter 12, and it was pointed out that these movements can cause atrial flutter, atrial fibrillation, and ventricular fibrillation.

Atrial Flutter. Figure 13–15 illustrates lead II of the electrocardiogram in atrial flutter. The rate of atrial contraction (P waves) is approxi-

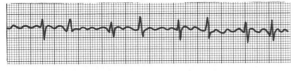

Figure 13–15. Atrial flutter — 2:1 and 3:1 rhythm (lead II).

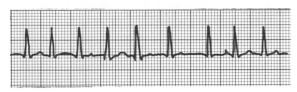

Figure 13–16. Atrial fibrillation (lead I).

mately 300 times per minute, while the rate of ventricular contraction (QRS-T waves) is only 125 times per minute. From the record it will be seen that sometimes a 2:1 rhythm occurs and sometimes a 3:1 rhythm. In other words, the atria beat two or three times for every one impulse that is conducted through the A-V bundle into the ventricles.

Atrial Fibrillation. Figure 13–16 illustrates the electrocardiogram during atrial fibrillation. As was discussed in Chapter 12, numerous impulses spread in all directions through the atria during atrial fibrillation. The intervals between impulses arriving at the A-V node are extremely variable. Therefore, an impulse may arrive at the A-V node immediately after the node itself is out of its refractory period from its previous discharge, or it may not arrive there for several tenths of a second. Consequently, the rhythm of the ventricles is very irregular, many of the ventricular beats falling quite close together and many far apart, the overall ventricular rate being 125 to 150 beats per minute in most instances. On the other hand, the QRS-T complexes are entirely normal unless there is some simultaneous pathological condition of the ventricles.

The pumping efficiency of the heart in atrial fibrillation is considerably depressed because the ventricles often do not have sufficient time to become filled between beats.

Ventricular Fibrillation. In ventricular fibrillation the electrocardiogram is extremely bizarre, as shown in Figure 13–17, and ordinarily shows no tendency toward a rhythm of any type. The irregularity of this electrocardiogram is what would be expected from multiple impulses traveling in all directions through the heart, as explained in the previous chapter. Obviously, this condition is immediately lethal.

ELECTROCARDIOGRAPHIC INTERPRETATION IN CARDIAC MYOPATHIES

From the discussion in Chapter 12 of impulse transmission through the heart, it is obvious that any change in the pattern of this transmission can cause abnormal electrical currents around the heart and, consequently, can alter the shapes of the waves in the electrocardiogram. For this reason, almost all serious abnormalities of the heart muscle can be detected by analyzing the contours of the different waves in the different electrocardiographic leads. The purpose of the present section is to present several representative electrocardiograms when the muscle of the heart, especially of the ventricles, functions abnormally.

THE MEAN ELECTRICAL AXIS OF THE VENTRICLES

It was shown in Figure 13–4 that during most of the cycle of ventricular depolarization, current normally flows from the base of the ventricle toward the apex. This preponderant direction of current flow during depolarization is called the *mean electrical axis of the ventricles.* The mean electrical axis of the normal ventricles is 59 degrees (zero degrees is toward the left side of the heart, and the axis is measured clockwise from this direction). However, in certain pathological conditions of the heart, the direction of current flow is changed markedly — sometimes even to opposite poles of the heart.

HYPERTROPHY OF ONE VENTRICLE

When one ventricle becomes greatly hypertrophied, the principal direction of current flow through the heart during depolarization — that is, the axis of the heart — shifts toward the hypertrophied ventricle for two reasons: First, far greater quantity of muscle exists on the hypertrophied side of the heart than on the other side, and this allows excess generation of electrical current on that side. Second, more time is required for the depolarization wave to travel through the hypertrophied ventricle than

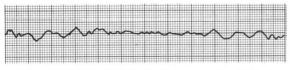

Figure 13–17. Ventricular fibrillation (lead II).

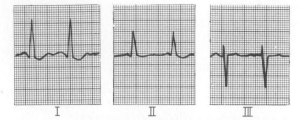

Figure 13–18. Left axis deviation in hypertensive heart disease. Note the slightly prolonged QRS complex.

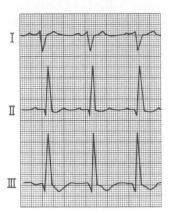

Figure 13–19. Right axis deviation due to right bundle branch block. Note the greatly prolonged QRS complex.

through the normal ventricle. Consequently, the normal ventricle becomes depolarized considerably in advance of the hypertrophied ventricle, and this causes strong current flow from the normal side of the heart toward the hypertrophied side. Thus the axis deviates toward the hypertrophied ventricle.

Left Axis Deviation Resulting From Hypertrophy of the Left Ventricle. Figure 13–18 illustrates the three standard leads of an electrocardiogram in which current flow is strongly positive in lead I and strongly negative in lead III. This means that the major current flow in the heart is mainly in the direction of lead I, which is from right arm toward left arm; and the current flow is opposite to the direction of lead III, which is from left arm toward left leg. That is, the axis of the heart points upward toward the left shoulder. This is called *left axis deviation* because it is to the left of the normal axis of the heart pointing downward and to the left in the chest.

The electrocardiogram of Figure 13–18 is typical of that resulting from increased muscular mass of the left ventricle. In this instance the axis deviation was caused by *high blood pressure,* which caused the left ventricle to hypertrophy in order to pump blood against the elevated systemic arterial pressure. However, a similar picture of left axis deviation occurs when the left ventricle hypertrophies as a result of aortic valvular stenosis, aortic valvular regurgitation, or any of a number of congenital heart conditions in which the left ventricle enlarges while the right side of the heart remains relatively normal in size.

Right axis deviation occurs when the right ventricle enlarges. That is, the potential in lead I becomes negative while that in lead III becomes strongly positive.

BUNDLE BRANCH BLOCK

Ordinarily, the two lateral walls of the ventricles depolarize at almost the same time, because both the left and right bundle branches transmit the cardiac impulse to the endocardial surfaces of the two ventricular walls at almost the same

instant. As a result, the currents flowing from the walls of the two ventricles almost exactly neutralize each other. However, if one of the major bundle branches is blocked, depolarization of the two ventricles does not occur even nearly simultaneously, and the depolarization currents do not neutralize each other. As a result, axis deviation occurs, as follows:

Right or Left Bundle Branch Block. When the right bundle branch is blocked, the left ventricle depolarizes far more rapidly than the right ventricle (because the normal left bundle still conducts a rapid signal to the left ventricle), so that the left becomes electronegative while the right remains electropositive. Very strong current flows with its negative end toward the left ventricle and its positive end toward the right ventricle. In other words, intense right axis deviation occurs because the positive end of the current flow is to the right of the normal downward and leftward flow.

Right axis deviation (denoted especially by the negative QRS in lead I) caused by right bundle branch block is illustrated in Figure 13–19, which also shows a prolonged QRS complex because of blocked conduction.

Left bundle branch block causes the opposite effect, namely, left axis deviation but also prolonged QRS complex.

CURRENT OF INJURY

Many different cardiac abnormalities, especially those that damage the heart muscle itself, cause part of the heart to remain *depolarized all the time.* When this occurs, current flows between the pathologically depolarized and the normally polarized areas. This is called a *current of injury.* The most common cause of a current of injury is *ischemia of the muscle caused by coronary occlusion.*

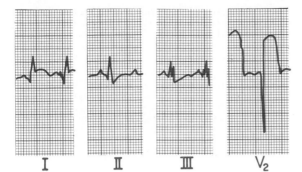

Figure 13-20. Current of injury in acute anterior wall infarction. Note the intense current of injury in lead V₂.

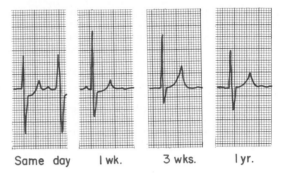

Figure 13-22. Recovery of the myocardium following moderate posterior wall infarction, illustrating disappearance of the current of injury (lead V₃).

Effect of Current of Injury on the QRS Complex — The S-T Segment Shift. It will be recalled that in the normal heart no current flows around the heart when the heart is totally polarized during the T-P interval or when the heart is totally depolarized during the S-T interval. Therefore, in the normal electrocardiogram, both the T-P and the S-T intervals record on the same level. However, when there is a current of injury, the heart cannot completely repolarize during the T-P interval. For this reason, the potential level of the T-P interval is different from that of the S-T interval. This effect is called the *S-T segment shift,* which unfortunately is a misnomer because the abnormality is actually in the T-P interval.

Acute Anterior Wall Infarction. Figure 13-20 illustrates the electrocardiogram in the three standard leads and in one chest lead recorded from a patient with acute anterior wall cardiac infarction caused by coronary thrombosis. The most important diagnostic feature of this electrocardiogram is the current of injury in the chest lead as denoted by the S-T segment shift (the broad wave at the top of the record following the QRS complex).

Posterior Wall Infarction. Figure 13-21 illustrates the three standard leads and one chest lead from a patient with posterior wall infarction. The major diagnostic feature of this electrocardiogram is the S-T segment shift in the chest lead and in leads II and III.

Recovery from Coronary Thrombosis. Figure 13-22 illustrates the chest lead from a patient with posterior wall infarction, showing the change in this chest lead from the day of the attack to one week later, then three weeks later, and finally one year later. From this electrocardiogram it can be seen that a slight current of injury (S-T segment shift) is present immediately after the acute attack, but after approximately one week the current of injury has diminished considerably and after three weeks it is completely gone. After that, the electrocardiogram changes slightly during the following year because of progressive repair of the damaged heart.

REFERENCES

Burch, G. E., and Winsor, T.: A Primer of Electrocardiography, 6th Ed. Philadelphia, Lea & Febiger, 1972.

Chou, T.: Electrocardiography in Clinical Practice. New York, Grune & Stratton, 1979.

Chung, E. K.: Ambulatory Electrocardiography. New York, Springer-Verlag, 1979.

Chung, E. K.: Electrocardiography: Practical Applications With Vectorial Principles, 2nd Ed. Hagerstown, Md., Harper & Row, 1980.

Cranefield, P. F., and Wit, A. L.: Cardiac arrhythmias. *Annu. Rev. Physiol.,* 41:459, 1979.

Fletcher, G. F.: Dynamic Electrocardiographic Recording. Mt. Kisco, N.Y., Futura Publishing Co., 1979.

Goldberger, A. L.: Myocardial Infarction: Electrocardiographic Differential Diagnosis, 2nd Ed. St. Louis, C. V. Mosby, 1979.

Guyton, A. C., and Crowell, J. W.: A stereovectorcardiograph. *J. Lab. Clin. Med., 40:*726, 1952.

Halhuber, M. J., *et al.:* ECG, An Introductory Course: A Practical Introduction to Clinical Electrocardiography. New York, Springer-Verlag, 1979.

Harris, C. C.: A Primer of Cardiac Arrhythmias: A Self-Instructional Program. St. Louis, C. V. Mosby, 1979.

Josephson, M. E., and Seides, S. F.: Clinical Cardiac Electrophysiology Techniques and Interpretations. Philadelphia, Lea & Febiger, 1979.

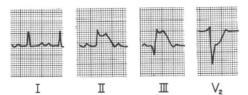

Figure 13-21. Current of injury in acute posterior wall, apical infarction.

Mangiola, S., and Ritota, M. C.: Cardiac Arrhythmias. Philadelphia, J. B. Lippincott Co., 1974.

Moe, G. K.: Mechanisms of cardiac dysrhythmias. *In* Frohlich,
E. D. (ed.): Pathophysiology, 2nd Ed. Philadelphia, J. B.
Lippincott Co., 1976, p. 83.

Narula, O. S. (ed.): Cardiac Arrhythmias: Electrophysiology,
Diagnosis, and Management. Baltimore, Williams &
Wilkins, 1979.

Phibbs, B.: The Cardiac Arrhythmias. St. Louis, C. V. Mosby,
1978.

Pick, A., and Langendorf, R.: Interpretation of Complex
Arrhythmias. Philadelphia, Lea & Febiger, 1980.

Pordy, L.: Computer Electrocardiography: Present Status and
Criteria. Mt. Kisco, N.Y., Futura Publishing Co., 1977.

Scher, A. M., and Spach, M. S.: Cardiac depolarization and repolarization and the electrocardiogram. *In* Berne, R. M.,
et al. (eds.): Handbook of Physiology. Sec. 2, Vol. I. Baltimore, Williams & Wilkins, 1979, p. 357.

Watanabe, Y., and Sloman, J. G.: Antiarrhythmic agents. *In*
Hayase, S., and Murao, S. (eds.): Cardiology: Proceedings
of the VIII World Congress of Cardiology, Tokyo, 1978.
New York, Elsevier/North-Holland, 1979, p. 920.

Part V
THE CIRCULATION

14

Physics of Blood, Blood Flow, and Pressure: Hemodynamics

Figure 14–1 illustrates the general plan of the circulation, showing the two major subdivisions, the *systemic circulation* and the *pulmonary circulation*. In the figure the arteries of each subdivision are represented by a single distensible chamber and all the veins by another even larger distensible chamber, and the arterioles and capillaries represent very small connections between the arteries and veins. Blood flows with almost no resistance in all the larger vessels of the circulation, but this is not the case in the arterioles and capillaries, where considerable resistance does occur. To cause blood to flow through these small "resistance" vessels, the heart pumps blood into the arteries under high pressure — normally at a systolic pressure of about 120 mm. Hg in the systemic system and 22 mm. Hg in the pulmonary system.

As a first step toward explaining the overall function of the circulation, this chapter will discuss the physical characteristics of blood itself and then the physical principles of blood flow through the vessels, including especially the interrelationships among pressure, flow, and resistance. The study of these interrelationships and other basic physical principles of blood circulation is called *hemodynamics*.

THE PHYSICAL CHARACTERISTICS OF BLOOD

Blood is a viscous fluid composed of *cells* and *plasma*. More than 99 per cent of the cells are red blood cells; this means that for practical purposes the white blood cells play almost no role in determining the physical characteristics of the blood.

THE HEMATOCRIT

The per cent of the blood that is cells is called the hematocrit. Thus, if a person has a hematocrit of 40, 40 per cent of the blood volume is cells and the remainder is plasma. The average hematocrit of normal man is about 42, while that of normal woman is about 38. These values vary tremendously, depending upon whether or not the person has anemia, the degree of bodily activity, and the altitude at which the person resides.

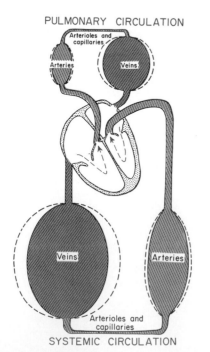

Figure 14–1. Representation of the circulation, showing both the distensible and the resistive portions of the systemic and pulmonary circulations.

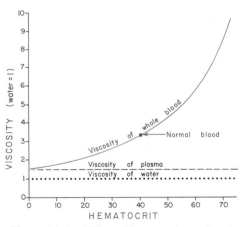

Figure 14–2. Effect of hematocrit on viscosity.

Effect of Hematocrit on Blood Viscosity. The greater the percentage of cells in the blood — that is, the greater the hematocrit — the more friction there is between successive layers of blood, and this friction determines viscosity. Therefore, the viscosity of blood increases drastically as the hematocrit increases, as illustrated in Figure 14–2. If we arbitrarily consider the viscosity of water to be 1, then the viscosity of whole blood at normal hematocrit is about 3; this means that 3 times as much pressure is required to force whole blood through a given tube than to force water through the same tube. Note that when the hematocrit rises to 60 or 70, which it often does in polycythemia, the blood viscosity can become as great as 10 times that of water, and its flow through blood vessels is greatly retarded.

Another factor that affects blood viscosity is the concentration and types of proteins in the plasma, but these effects are so much less important than the effect of hematocrit that they are not significant considerations in most hemodynamic studies. The viscosity of blood plasma is about 1.5 times that of water.

PLASMA

Plasma is part of the extracellular fluid of the body. It is almost identical with the interstitial fluid found between the tissue cells except for one major difference: Plasma contains about 7 per cent protein, while interstitial fluid contains an average of only 2 per cent protein. The reason for this difference is that plasma protein leaks only slightly through the capillary pores into the interstitial spaces. As a result, most of the plasma protein is held in the circulatory system, and that which does leak is eventually returned to the circulation by the lymph vessels.

The Types of Protein in Plasma. The plasma protein is divided into three major types, as follows:

	Grams *Per Cent*
Albumin	4.5
Globulins	2.5
Fibrinogen	0.3

The primary function of the *albumin* (and of the other types of protein to a lesser extent) is to cause osmotic pressure at the capillary membrane. This pressure, called *colloid osmotic pressure*, prevents the fluid of the plasma from leaking out of the capillaries into the interstitial spaces. This function is so important that it will be discussed in detail in Chapter 22.

The globulins are divided into three major types: alpha, beta, and gamma globulins. The *alpha* and *beta globulins* perform diverse functions in the circulation, such as transporting other substances by combining with them, acting as substrates for formation of other substances, and transporting protein itself from one part of the body to another. The *gamma globulins*, and to a lesser extent the *beta globulins*, play a special role in protecting the body against infection, for it is these globulins that are mainly the *antibodies* that resist infection and toxicity, thus providing the body with what we call *immunity*. The function of immunity was discussed in detail in Chapter 6.

The *fibrinogen* of plasma is of basic importance in blood clotting and was discussed in Chapter 7.

INTERRELATIONSHIPS AMONG PRESSURE, FLOW, AND RESISTANCE

Flow through a blood vessel is determined entirely by two factors: (1) the *pressure difference* between the two ends of the vessel, which is the force that pushes the blood through the vessel, and (2) the impediment to blood flow through the vessel, which is called vascular *resistance*. Figure 14–3 illustrates these relationships, showing a blood vessel segment located anywhere in the circulatory system.

P_1 represents the pressure at the origin of the vessel; at the other end the pressure is P_2. The flow through the vessel can be calculated as follows:

$$Q = \frac{\Delta P}{R} \qquad (1)$$

in which Q is blood flow, ΔP is the pressure difference $(P_1 - P_2)$ between the two ends of the vessel, and R is the resistance.

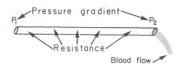

Figure 14–3. Relationships among pressure, resistance, and blood flow.

It should be noted especially that it is the *difference* in pressure between the two ends of the vessel that determines the rate of flow and not the absolute pressure in the vessel. For instance, if the pressure at both ends of the segment were 100 mm. Hg and yet no difference existed between the two ends, there would be no flow despite the presence of the 100 mm. Hg pressure.

The above formula expresses the most important of all the relationships that the student needs to understand to comprehend the hemodynamics of the circulation. Because of the extreme importance of this formula the student should also become familiar with its other two algebraic forms:

$$\Delta P = Q \times R \qquad (2)$$

$$R = \frac{\Delta P}{Q} \qquad (3)$$

BLOOD FLOW

Blood flow means simply the quantity of blood that passes a given point in the circulation in a given period of time. Ordinarily, blood flow is expressed in *milliliters* or *liters per minute,* but it can be expressed in milliliters per second or in any other unit of flow.

The overall blood flow in the circulation of an adult person at rest is about 5000 ml. per minute. This is called the *cardiac output* because it is the amount of blood pumped by the heart in a unit period of time.

Methods for Measuring Blood Flow. Many different mechanical or mechanoelectrical devices can be inserted in series with a blood vessel, or in some instances applied to the outside of the vessel, to measure flow. These are called simply *flowmeters.*

The Electromagnetic Flowmeter. In recent years several new devices have been developed that can be used to measure blood flow in a vessel without opening it. One of the most important of these is the electromagnetic flowmeter, the principles of which are illustrated in Figure 14–4. Figure 14–4A shows generation of electromotive force in a wire that is moved rapidly through a magnetic field. This is the well-known principle for production of electricity by the electric generator. Figure 14–4B shows that exactly the same principle applies for generation of electromotive force in blood when it moves through a magnetic field. In this case, a blood vessel is placed between the poles of a strong magnet, and electrodes are placed on the two sides of the vessel perpendicular to the magnetic lines of force. When blood flows through the vessel, electrical voltage proportional to the rate of flow is generated between the two electrodes, and this is recorded using an appropriate meter or electronic apparatus. Figure 14–4C illustrates an actual "probe" that is placed on a large blood vessel to record its blood flow.

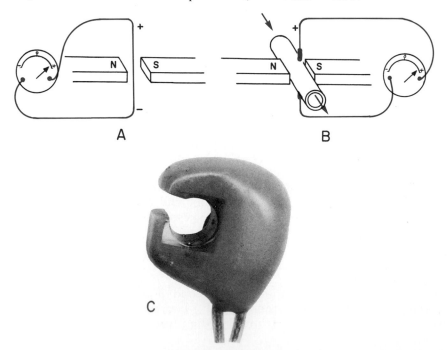

Figure 14–4. A flowmeter of the electromagnetic type, showing (A) generation of an electromotive force in a wire as it passes through an electromagnetic field, (B) generation of an electromotive force in electrodes on a blood vessel when the vessel is placed in a strong magnetic field and blood flows through the vessel, and (C) a modern electromagnetic flowmeter "probe" for continued implantation around blood vessels.

An additional advantage of the electromagnetic flowmeter is that it can record changes in flow that occur in less than 0.01 second, allowing accurate recording of pulsatile changes in flow as well as steady flow.

The Ultrasonic Doppler Flowmeter.   Another type of flowmeter that can be applied to the outside of the vessel and that has many of the same advantages as the electromagnetic flowmeter is the ultrasonic Doppler flowmeter, illustrated in Figure 14–5. A minute piezoelectric crystal is mounted in the wall of the device. This crystal, when energized with an appropriate electronic apparatus, transmits sound of a frequency of several million cycles per second downstream along the flowing blood. A portion of the sound is reflected by the flowing red blood cells, and the reflected sound waves travel backward from the blood toward the crystal. However, these reflected waves have a lower frequency than the transmitted wave because the red cells are moving away from the transmitter crystal. This is called the Doppler effect. It is the same effect that one experiences when a train approaches a listener and passes by while blowing the whistle. Once the whistle has passed by the person, the pitch of the sound from the whistle suddenly becomes much lower than when the train is approaching. An electronic apparatus determines the frequency difference between the transmitted wave and the reflected wave, thus determining the velocity of blood flow.

Laminar Flow of Blood in Vessels.   When blood flows at a steady rate through a long, smooth vessel, it flows in *streamlines,* with each layer of blood remaining the same distance from the wall. Also, the central portion of the blood stays in the center of the vessel. This type of flow is called *laminar flow* or *streamline flow,* and it is opposite to *turbulent flow,* explained in the following section.

Turbulent Flow of Blood Under Some Conditions.   When the rate of blood flow becomes too great, when it passes by an obstruction in a vessel, when it makes a sharp turn, or when it passes over a rough surface, the flow may then become *turbulent* rather than streamline. Turbulent flow means that the blood flows crosswise in the vessel as well as along the vessel, usually forming whorls

in the blood called *eddy currents*. These are similar to the whirlpools that one frequently sees in a rapidly flowing river at a point of obstruction.

When eddy currents are present, blood flows with much greater resistance than when the flow is streamline because the eddies add tremendously to the overall friction of flow in the vessel.

BLOOD PRESSURE

The Standard Units of Pressure.   Blood pressure is almost always measured in *millimeters of mercury (mm. Hg)* because the mercury manometer (shown in Figure 14–6) has been used as the standard reference for measuring blood pressure throughout the history of physiology. Actually, blood pressure means the *force exerted by the blood against any unit area of the vessel wall.* When one says that the pressure in a vessel is 50 mm. Hg, this means that the force exerted is sufficient to push a column of mercury up to a level 50 mm. high. If the pressure is 100 mm. Hg, it will push the column of mercury up to 100 mm.

Occasionally, pressure is measured in *centimeters of water.* A pressure of 10 cm. of water means a pressure sufficient to raise a column of water to a height of 10 cm. *One millimeter of mercury equals 1.36 cm. of water* because the specific gravity of mercury is 13.6 times that of water, and 1 cm. is 10 times as great as 1 mm. Dividing 13.6 by 10, we derive the factor 1.36.

Measurement of Blood Pressure Using the Mercury Manometer.   Figure 14–6 illustrates a

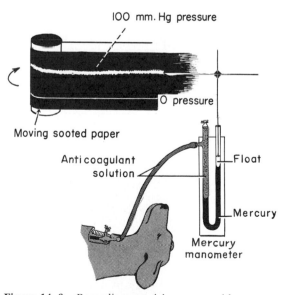

Figure 14–6.   Recording arterial pressure with a mercury manometer, a method that has been used in the manner shown above for recording pressure throughout the history of physiology.

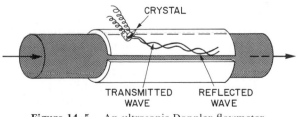

Figure 14–5.   An ultrasonic Doppler flowmeter.

standard mercury manometer for measuring blood pressure. A cannula or catheter is inserted into an artery, a vein, or even the heart, and the pressure from the cannula or catheter is transmitted to the left-hand side of the manometer, where it pushes the mercury down while raising the right-hand mercury column. The difference between the two levels of mercury is approximately equal to the pressure in the circulation in terms of millimeters of mercury. (To be more exact, it is equal to 104 per cent of the true pressure because of the weight of the water on the left-hand column of mercury.)

High-Fidelity Methods for Measuring Blood Pressure. Unfortunately, the mercury in the mercury manometer has so much *inertia* that it cannot rise and fall rapidly. For this reason the mercury manometer, though excellent for recording steady pressures, cannot respond to pressure changes that occur more rapidly than approximately one cycle every 2 to 3 seconds. Whenever it is desired to record rapidly changing pressures, some other type of pressure recorder is needed. Figure 14–7 demonstrates the basic principles of three electronic pressure *transducers* commonly used for converting pressure into electrical signals and then recording the pressure on a high-speed electrical recorder. Each of these transducers employs a very thin and highly stretched metal membrane, which forms one wall of the fluid chamber. The fluid chamber in turn

is connected through a needle or catheter with the vessel in which the pressure is to be measured. Pressure variations in the vessel cause changes of pressure in the chamber beneath the membrane. When the pressure is high the membrane bulges outward slightly, and when the pressure is low it returns toward its resting position.

In Figure 14–7A a simple metal plate is placed a few thousandths of an inch above the membrane. When the membrane bulges outward, the *capacitance* between the plate and membrane increases. In Figure 14–7B a small iron slug is displaced upward into a coil, and this changes the *inductance* of the coil. And in Figure 14–7C a very thin resistance wire is stretched and its resistance increases. Each of these changes can be recorded by means of an electronic system.

Using some of these high fidelity types of recording systems, pressure cycles up to 500 cycles per second have been recorded accurately, and in common use are recorders capable of registering pressure changes as rapidly as 10 to 50 cycles per second.

RESISTANCE TO BLOOD FLOW

Units of Resistance. Resistance is the impediment to blood flow in a vessel, but it cannot be measured by any direct means. Instead, resistance must be calculated from measurements of blood flow and pressure difference in the vessel. If the pressure difference between two points in a vessel is 1 mm. Hg and the flow is 1 ml./sec., then the resistance is said to be 1 *peripheral resistance unit,* usually abbreviated *PRU.*

Total Peripheral Resistance and Total Pulmonary Resistance. The rate of blood flow through the circulatory system when a person is at rest is close to 100 ml./sec., and the pressure difference from the systemic arteries to the systemic veins is about 100 mm. Hg. Therefore, in round figures the resistance of the entire systemic circulation, called the *total peripheral resistance,* is approximately 100/100 or 1 PRU. In some conditions in which the blood vessels throughout the body become strongly constricted, the total peripheral resistance rises to as high as 4 PRU, and when the vessels become greatly dilated it can fall to as little as 0.2 PRU.

In the pulmonary system the mean arterial pressure averages 16 mm. Hg and the mean left atrial pressure averages 4 mm. Hg, giving a net pressure difference of 12 mm. Therefore, in round figures the *total pulmonary resistance* at rest is calculated at about 0.12 PRU. This can increase in disease conditions to as high as 1 PRU and can fall in certain physiological states, such as exercise, to as low as 0.03 PRU.

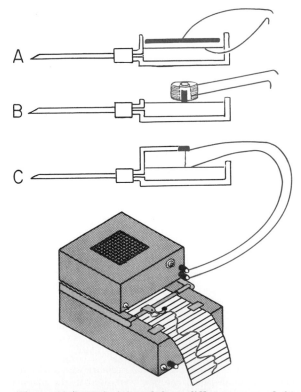

Figure 14–7. Principles of three different types of electronic transducers for recording rapidly changing blood pressures.

"Conductance" of Blood in a Vessel and Its Relationship to Resistance. Conductance is a measure of the blood flow through a vessel for a given pressure difference. This is generally expressed in terms of ml./sec./mm. Hg pressure, but it can also be expressed in terms of liters/sec./mm. Hg or in any other units of blood flow and pressure.

It is immediately evident that conductance is the reciprocal of resistance in accordance with the following equation:

$$\text{Conductance} = \frac{1}{\text{Resistance}} \quad (4)$$

Effect of Vascular Diameter on Conductance. Slight changes in the diameter of a vessel cause tremendous changes in its ability to conduct blood when the blood flow is streamline. This is illustrated forcefully by the experiment in Figure 14–8, which shows three separate vessels with relative diameters of 1, 2, and 4 but with the same pressure difference of 100 mm. Hg between the two ends of the vessels. Though the diameters of these vessels increase only four-fold, the respective flows are 1, 16, and 256 ml./mm., which is a 256-fold increase in flow. Thus, the conductance of the vessel increases in proportion to the *fourth power of the diameter*, in accordance with the following formula:

$$\text{Conductance} \propto \text{Diameter}^4 \quad (5)$$

Poiseuille's Law. The quantity of blood that will flow through a vessel in a given period of time is given by the following equation, which is known as Poiseuille's law:

$$Q = \frac{\pi \Delta P r^4}{8 \eta l} \quad (6)$$

in which Q is blood flow, ΔP is pressure difference, r is radius, η is blood viscosity, and l is length.

Note particularly in this equation that the rate of blood flow is directly proportional to the *fourth power of the radius* of the vessel, which illustrates once again that the diameter of a blood vessel plays by far the greatest role of all factors in determining the rate of blood flow through the vessel.

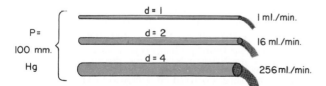

Figure 14–8. Demonstration of the effect of vessel diameter on blood flow.

Effect of Pressure on Vascular Resistance — Critical Closing Pressure

Since all blood vessels are distensible, increasing the pressure inside the vessels causes the vascular diameters also to increase. This in turn reduces the resistance of the vessel. Conversely, reduction in vascular pressure increases the resistance.

The middle curve of Figure 14–9 illustrates the normal effect on blood flow through a small tissue vascular bed caused by changing the arterial pressure. As the arterial pressure falls from 130 mm. Hg, the flow decreases rapidly at first because of two factors: (1) the decreasing pressure difference between the artery and the vein of the tissue, and (2) the decreasing diameters of the vessels. At 20 mm. Hg blood flow ceases entirely. This point at which the blood stops flowing is called the *critical closing pressure,* because at this point the small vessels, the arterioles in particular, close so completely that all flow through the tissue ceases.

Effect of Sympathetic Inhibition and Stimulation on Vascular Flow. Essentially all blood vessels of the body are normally stimulated by sympathetic impulses even under resting conditions. Different circulatory reflexes, which will be discussed in Chapter 17, can cause these impulses either to disappear, called *sympathetic inhibition,* or to increase to many times their normal rate, called *sympathetic stimulation.* Sympathetic impulses in most parts of the body cause an increase in tone of the vascular smooth muscle. Therefore, as illustrated by the dashed curves of Figure 14–9, sympathetic inhibition allows far more blood to flow through a tissue for a given pressure than is normally true, whereas sympathetic stimulation vastly decreases the rate of blood flow.

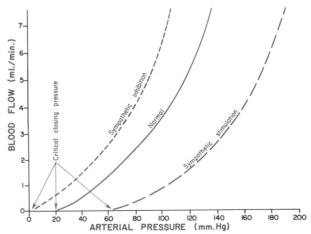

Figure 14–9. Effect of arterial pressure on blood flow through a blood vessel at different degrees of vascular tone, showing also the effect of vascular tone on critical closing pressure.

VASCULAR DISTENSIBILITY — PRESSURE-VOLUME CURVES

The diameter of blood vessels, unlike that of metal pipes and glass tubes, increases as the internal pressure increases, because blood vessels are *distensible*. However, the vascular distensibilities differ greatly in different segments of the circulation, and as we shall see, this affects significantly the operation of the circulatory system under many changing physiological conditions.

Units of Vascular Distensibility. Vascular distensibility is normally expressed as the fractional increase in volume for each millimeter mercury rise in pressure in accordance with the following formula:

$$\text{Vascular distensibility} = \frac{\text{Increase in volume}}{\text{Increase in pressure} \times \text{Original volume}} \quad (7)$$

That is, if 1 mm. Hg causes a vessel originally containing 10 ml. of blood to increase its volume by 1 ml., then the distensibility is 0.1 per mm. Hg or 10 per cent per mm. Hg.

Difference in Distensibility of the Arteries and the Veins. Anatomically, the walls of arteries are far stronger than those of veins. Consequently, the veins, on the average, are about 6 to 10 times as distensible as the arteries. That is, a given rise in pressure will cause about 6 to 10 times as much extra blood to fill a vein as an artery of comparable size.

VASCULAR COMPLIANCE (OR CAPACITANCE)

Usually in hemodynamic studies it is much more important to know the *total quantity of blood* that can be stored in a given portion of the circulation for each mm. Hg pressure rise than to know the distensibility of the individual vessels. This value is called *compliance* or *capacitance*, which are physical terms meaning the increase in volume that causes a given increase in pressure as follows:

$$\text{Vascular compliance} = \frac{\text{Increase in volume}}{\text{Increase in pressure}} \quad (8)$$

Compliance and distensibility are quite different. A highly distensible vessel that has a very slight volume may have far less compliance than a much less distensible vessel that has a very large volume, for *compliance is equal to distensibility × volume*.

The compliance of a vein is about 24 times that of its corresponding artery because it is about 8 times as distensible and it has a volume about 3 times as great ($8 \times 3 = 24$).

VOLUME-PRESSURE CURVES OF THE ARTERIAL AND VENOUS CIRCULATIONS

A convenient method for expressing the relationship of pressure to volume in a vessel or in a large portion of the circulation is the so-called *volume-pressure curve* (also frequently called the *pressure-volume curve*). The two solid curves of Figure 14–10 represent respectively the volume-pressure curves of the normal arterial and venous systems, showing that when the arterial system, including the larger arteries, small arteries, and arterioles, is filled with approximately 750 ml. of blood the mean arterial pressure is 100 mm. Hg, but when it is filled with only 500 ml. the pressure falls to zero.

The volume of blood normally in the entire venous tree is about 2500 ml., and tremendous changes in this volume are required to change the venous pressure only a few millimeters of mercury.

Difference in Compliance of the Arterial and Venous Systems. Referring once again to Figure 14–10, one can see that a change of 1 mm. Hg requires a very large change in venous volume but much less change in arterial volume. That is, the *compliance of the venous system is far greater than the compliance of the arteries — about 24 times as great.*

This difference in compliance is particularly important because it means that tremendous amounts of blood can be stored in the veins with only slight changes in pressure. Therefore, the veins are frequently called the *storage areas* of the circulation.

Effect of Sympathetic Stimulation or Sympathetic Inhibition on the Volume-Pressure Relationships of the Arterial and Venous Systems. Also shown in Figure 14–10 are the volume-pressure curves of the arterial and venous systems during moderate sympathetic stimulation and during sympathetic inhibition. It is evident that sympathetic stimulation, with its concomitant increase in smooth muscle tone in the vascular walls, increases the pressure at each volume of the arteries or veins, while on the other hand, sympathetic inhibition decreases the pressure at each respective volume. Obviously, an increase in vascular tone throughout the systemic circulation can cause large volumes of blood to shift into the heart, which is a major way in which pumping by the heart is increased.

Sympathetic control of vascular capacity is also especially important during hemorrhage. En-

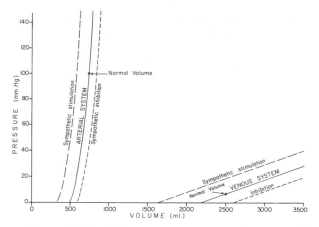

Figure 14–10. Volume-pressure curves of the systemic arterial and venous systems, showing also the effects of sympathetic stimulation and sympathetic inhibition.

hancement of the sympathetic tone of the vessels, especially of the veins, reduces the dimensions of the circulatory system, and the circulation continues to operate almost normally even when as much as 25 per cent of the total blood volume has been lost.

THE MEAN CIRCULATORY FILLING PRESSURE

The mean circulatory filling pressure is a measure of the degree of filling of the circulatory system. That is, it is the pressure that would be measured in the circulation if one could instantaneously stop all blood flow and bring all the pressures in the circulation immediately to equilibrium. The mean circulatory filling pressure has been measured reasonably accurately in dogs within two to five seconds after the heart has been stopped. To do this the heart is thrown into fibrillation by an electrical stimulus, and blood is pumped rapidly from the systemic arteries to the veins to cause equilibrium between the two major chambers of the circulation.

The mean circulatory filling pressure measured in the above manner in the anesthetized dog is almost exactly 7 mm. Hg and almost never varies more than 1 mm. from this value. It is believed to be about the same value in the human being. However, many different factors can change it, especially including change in the blood volume and increased or decreased sympathetic stimulation.

The mean circulatory filling pressure is *one of the major factors that determines the rate at which blood flows from the vascular tree into the right atrium of the heart, which in turn determines the cardiac output.* This is so important that it will be explained in detail in Chapter 19.

DELAYED COMPLIANCE (STRESS-RELAXATION) OF VESSELS

The term "delayed compliance" means that a vessel exposed to constantly increased pressure becomes progressively enlarged, much of this extra enlargement occurring in the first minutes, but some continuing to develop for days or weeks.

Delayed compliance occurs only slightly in the arteries but to a much greater extent in the veins. As a result, prolonged elevation of venous pressure can often double the blood volume in the venous tree. This is a valuable mechanism by which the circulation can accommodate much extra blood when necessary, such as following too large a transfusion. Also, delayed compliance in the reverse direction — that is, progressive decrease in vessel size when the pressure falls — is one of the ways by which the circulation automatically adjusts over a period of minutes or hours to diminished blood volume after serious hemorrhage.

REFERENCES

Bessis, M., *et al.* (eds.): Red Cell Rheology. New York, Springer-Verlag, 1978.

Cokelet, G. R., *et al.* (eds.): Erythrocyte Mechanics and Blood Flow. New York, A. R. Liss, 1979.

Dobrin, P. B.: Mechanical properties of arteries. *Physiol. Rev.,* 58:397, 1978.

Fung, Y. C.: Introduction to biophysical aspects of microcirculation. *In* Kaley, G., and Altura, B. M. (eds.): Microcirculation. Vol. I. Baltimore, University Park Press, 1977, p. 253.

Fung, Y. C.: Rheology of blood in microvessels. *In* Kaley, G., and Altura, B. M. (eds.): Microcirculation. Vol. I. Baltimore, University Park Press, 1977, p. 279.

Ghista, D. N., *et al.* (eds.): Theoretical Foundations of Cardiovascular Processes. New York, S. Karger, 1979.

Gow, B. S.: Circulatory correlates: Vascular impedance, resistance, and capacity. *In* Bohr, D. F., *et al.* (eds.): Handbook of Physiology. Sec. 2, Vol. 2. Baltimore, Williams & Wilkins, 1980, p. 353.

Green, H. D.: Circulation: Physical principles. *In* Glasser, O. (ed.): Medical Physics. Chicago, Year Book Medical Publishers, 1944.

Gross, J. F., and Popel, A. (eds.): Mathematics of Microcirculation Phenomena. New York, Raven Press, 1980.

Guyton, A. C.: Arterial Pressure and Hypertension. Philadelphia, W. B. Saunders Co., 1980.

Guyton, A. C., and Greganti, F. P.: A physiologic reference point for measuring circulatory pressures in the dog — particularly venous pressure. *Am. J. Physiol.,* 185:137, 1956.

Guyton, A. C., *et al.*: Pressure-volume curves of the entire arterial and venous systems in the living animal. *Am. J. Physiol.,* 184:253, 1956.

Guyton, A. C., *et al.*: Cardiac Output and Its Regulation. Philadelphia, W. B. Saunders Co., 1973.

Hwang, M. H. C., *et al.* (eds.): Quantitative Cardiovascular Studies: Clinical Research Applications of Engineering Principles. Baltimore, University Park Press, 1978.

James, D. G. (ed.): Circulation of the Blood. Baltimore, University Park Press, 1978.

Lee, J.: Pressure-flow relationships of single vessels and organs. *In* Kaley, G., and Altura, B. M. (eds.): Microcirculation. Vol. I. Baltimore, University Park Press, 1977, p. 335.

Manning, R. D., Jr., *et al.*: Essential role of mean circulatory filling pressure in salt-induced hypertension. *Am. J. Physiol., 236*:R40, 1979.

Murphy, R. A.: Mechanics of vascular smooth muscle. *In* Bohr, D. F., *et al.* (eds.): Handbook of Physiology. Sec. 2, Vol. 2. Baltimore, Williams & Wilkins, 1980, p. 325.

Pedley, T. J.: The Fluid Mechanics of Large Blood Vessels. New York, Cambridge University Press, 1979.

Roach, M. R.: Biophysical analyses of blood vessel walls and blood flow. *Annu. Rev. Physiol., 39*:51, 1977.

Schmid-Schönbein, H.: Microrheology of erythrocytes, blood viscosity, and the distribution of blood flow in the microcirculation. *Int. Rev. Physiol., 9*:1, 1976.

Schneck, D. J., and Vawter, D. L. (eds.): Biofluid Mechanics. New York, Plenum Press, 1980.

Stehbens, W. E. (ed.): Hemodynamics and the Blood Vessel Wall. Springfield, Ill., Charles C Thomas, 1978.

15

The Systemic and Pulmonary Circulations

The circulation is divided into the *systemic circulation* and the *pulmonary circulation*. Though the vascular system in each separate tissue of the body has its own special characteristics, some general principles of vascular function nevertheless apply in all parts of the circulation. It is the purpose of the present chapter to discuss these general principles.

The Functional Parts of the Circulation. Before attempting to discuss the details of function in the circulation, it is important to understand the overall role of each of its parts, as follows:

The function of the *arteries* is to transport blood *under high pressure* to the tissues. For this reason the arteries have strong vascular walls, and blood flows rapidly in the arteries.

The *arterioles* are the last small branches of the arterial system, and they act as *control valves* through which blood is released into the capillaries. The arteriole has a strong muscular wall that is capable of closing the arteriole completely or of allowing it to be dilated several fold, thus having the capability of vastly altering blood flow to the capillaries.

The function of the *capillaries* is to exchange fluid, oxygen, carbon dioxide, nutrients, electrolytes, hormones, and other substances between the blood and the interstitial spaces or air spaces of the lungs. For this role, the capillary walls are very thin and permeable to small molecular substances.

The *venules* collect blood from the capillaries; they gradually coalesce into progressively larger veins.

The *veins* function as conduits for transport of blood from the tissues back to the heart. Since the pressure in the venous system is very low, the venous walls are thin. Even so, they are muscular, and this allows them to contract or expand and thereby to act as a reservoir for extra blood.

Quantities of Blood in the Different Parts of the Circulation. Figure 15–1 shows that approximately 84 per cent of the entire blood volume of the body is in the systemic circulation, with 64 per cent in the veins, 15 per cent in the arteries, and 5 per cent in the capillaries. The heart contains 7 per cent of the blood, and the pulmonary vessels, 9 per cent. Most surprising is the very low blood volume in the capillaries, only a small percentage of the total, for it is here that the most important function of the circulation occurs, namely, diffusion of substances back and forth between the blood and the tissues or the pulmonary air spaces. This function is so important that it will be discussed in detail in Chapter 22.

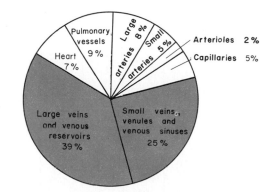

Figure 15–1. Percentage of the total blood volume in each portion of the circulatory system.

THE SYSTEMIC CIRCULATION

Pressures and Resistances in the Various Portions of the Systemic Circulation. Because the heart pumps blood continually into the aorta, the pressure in the aorta is obviously high, averaging approximately 100 mm. Hg. And, also, because the pumping by the heart is pulsatile, the arterial pressure fluctuates between a *systolic level* of 120 mm. Hg and a *diastolic level* of 80 mm. Hg, as illustrated in Figure 15–2. As the blood flows through the systemic circulation, its pressure falls progressively to approximately 0 mm. Hg by the time it reaches the right atrium.

The decrease in arterial pressure in each part of the systemic circulation is directly proportional to the vascular resistance. In the aorta the resistance is almost zero; therefore, the mean arterial pressure at the end of the aorta is still almost 100 mm. Hg. Likewise, the resistance in the large arteries is very slight, so that the mean arterial pressure in arteries as small as 3 mm. in diameter is still 95 to 97 mm. Hg. Then the resistance begins to increase rapidly in the very small arteries, causing the pressure to drop to approximately 85 mm. Hg at the beginning of the arterioles.

The resistance of the *arterioles* is greatest of that in any part of the systemic circulation, accounting for about half the resistance in the entire systemic circulation. Thus, the pressure decreases about 55 mm. Hg in the arterioles so that the pressure of the blood as it leaves the arterioles to enter the capillaries is only about 30 mm. Hg. Arteriolar resistance is so important to the regulation of blood flow in different tissues of the body that it is discussed in detail later in the chapter and also in the following few chapters, which consider the regulation of the systemic circulation.

The pressure at the arterial ends of the *systemic capillaries* is normally about 30 mm. Hg, and at the venous ends about 10 mm. Hg. Therefore, the pressure decrease in the capillaries is only 20 mm. Hg, which illustrates that the capillary resistance is about two-fifths that of the arterioles.

The pressure at the beginning of the venous system, that is, at the *venules,* is about 10 mm. Hg, and this decreases to almost exactly 0 mm. Hg at the right atrium. This large decrease in pressure in the veins indicates that the veins have far more resistance than one would expect for vessels of their large size. Much of this resistance is caused by compression of the veins from the outside, which keeps many of them collapsed a large share of the time. This effect is discussed later in the chapter.

PRESSURE PULSES IN THE ARTERIES

Since the heart is a pulsatile pump, blood enters the arteries intermittently with each heart beat, causing *pressure pulses* in the arterial system. In the normal young adult the pressure at the height of a pulse, the *systolic pressure,* is about 120 mm. Hg, and at its lowest point, the *diastolic pressure,* about 80 mm. Hg. The difference between these two pressures, about 40 mm. Hg, is called the *pulse pressure.*

Figure 15–3 illustrates a typical, idealized *pressure pulse curve* recorded in the ascending aorta of a human being, showing a very rapid rise in arterial pressure during ventricular systole, followed by a maintained high level of pressure for

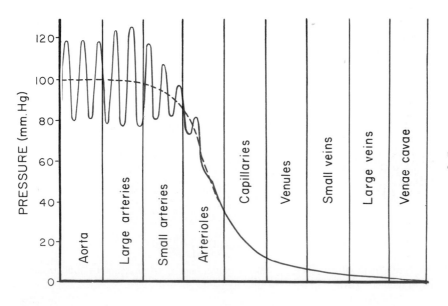

Figure 15–2. Blood pressures in the different portions of the systemic circulatory system.

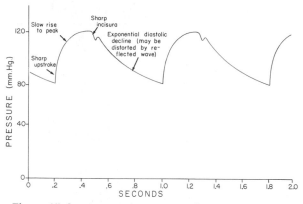

Figure 15–3. A normal pressure pulse contour recorded from the ascending aorta. (From Opdyke: *Fed. Proc., 11:*734, 1952.)

0.2 to 0.3 second. This is terminated by a sharp *incisura* or *notch* at the end of systole, followed by a slow decline of pressure back to the diastolic level. The incisura occurs immediately before the aortic valve closes and is caused as follows: When the ventricle relaxes, the intraventricular pressure begins to fall rapidly, and backflow of blood from the aorta into the ventricle allows the aortic pressure also to begin falling. However, the backflow suddenly snaps the aortic valve closed.

Factors that Affect the Pulse Pressure

There are two major factors that affect the pulse pressure: (1) the *stroke volume output* of the heart and (2) the *compliance* of the arterial tree; a third, less important factor is the character of ejection from the heart during systole.

In general, the greater the stroke volume output, the greater is the amount of blood that must be accommodated in the arterial tree with each heartbeat and, therefore, the greater is the pressure rise during systole and the pressure fall during diastole, thus causing a greater pulse pressure.

On the other hand, the greater the compliance of the arterial system, the less will be the rise in pressure for a given stroke volume of blood pumped into the arteries. In effect, then, the pulse pressure is determined approximately by the *ratio of stroke volume output to compliance of the arterial tree.* Therefore, any condition of the circulation that affects either of these two factors will also affect the pulse pressure.

Abnormal Pressure Pulse Contours

Some conditions of the circulation cause abnormal contours of the pressure pulse wave in addition to altering the pulse pressure. Especially distinctive among these are arteriosclerosis, aortic regurgitation, and patent ductus arteriosus.

Arteriosclerosis. In arteriosclerosis the arteries become fibrous and sometimes calcified, thereby resulting in greatly reduced arterial compliance and also resulting in markedly increased pulse pressure. The middle curve of Figure 15–4 illustrates a characteristic aortic pressure pulse contour in arteriosclerosis, showing a markedly elevated systolic pressure and a great increase in pulse pressure.

Aortic Regurgitation. In aortic regurgitation, which means a leaking aortic valve, much of the blood that is pumped into the aorta during systole flows back into the left ventricle during diastole. However, this backflow is compensated for by a much greater than normal stroke volume output during systole. Thus, as illustrated in the lower curve of Figure 15–4, the pulse pressure is greatly increased. Also, if the aortic valve does not close at all, there is no incisura.

Patent Ductus Arteriosus. This is an abnormal condition in which the ductus arteriosus, which carries blood from the pulmonary artery to the aorta during fetal life, fails to close after birth. Instead, the blood now flows backward from the aorta through the open ductus into the pulmonary artery, allowing very rapid runoff of blood from the arterial tree after each heartbeat and a greatly decreased diastolic pressure. However, this is compensated for by a greater than normal stroke volume output so that the systolic pressure rises much higher than normal. These effects give the pressure pulse contour shown by the lowest curve in Figure 15–4.

Damping of the Pressure Pulse in the Small Arteries and Arterioles. The pressure pulse becomes less and less intense as it passes through the small arteries and arterioles, until it becomes almost absent in the capillaries. This was illustrated in Figure 15–2 and is called *damping* of the pressure pulse.

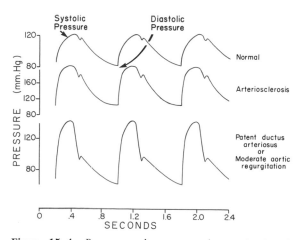

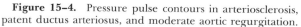

Figure 15–4. Pressure pulse contours in arteriosclerosis, patent ductus arteriosus, and moderate aortic regurgitation.

Damping of the pressure pulse is caused mainly by a combined effect of vascular distensibility and vascular resistance. That is, for a pressure wave to travel from one area of an artery to another area, a small amount of blood must flow between the two areas. The resistance in the small arteries and arterioles is great enough that this small flow of blood, and consequently the transmission of the pressure wave, is greatly impeded. At the same time, the distensibility of the small arteries is great enough that the small amount of blood that is caused to flow during each pressure pulse produces progressively less pressure rise and fall the more distal the pulse wave proceeds in the vessels.

The Radial Pulse

Clinically, it has been the habit for many years for a physician to feel the radial pulse of each patient. This is done to determine the rate of the heartbeat or, frequently, because of the psychic contact that it gives the doctor with the patient. Under certain circumstances, however, the character of the pulse can also be of value in the diagnosis of circulatory diseases. Especially important, a weak pulse at the radial artery often indicates greatly decreased central pulse pressure, such as occurs when the stroke volume output is low.

THE SYSTEMIC ARTERIOLES AND CAPILLARIES

Blood flow in each tissue is controlled almost entirely by the degree of contraction or dilatation of the arterioles, and it is in the capillaries that the important process of exchange between blood and the interstitial fluid occurs. These two segments of the systemic circulation are so important that they will receive special discussion in chapters to follow: arteriolar regulation of blood flow in Chapters 16 and 17, and capillary exchange phenomena in Chapters 22 and 23.

Upon leaving the small arteries, blood courses through the arterioles, which are only a few millimeters in length and have diameters of 8 to 50 microns. Each arteriole branches many times to supply 10 to 100 capillaries.

There are approximately 10 billion capillaries in the peripheral tissues, having more than 500 square meters of surface area. The thickness of the capillary wall is usually less than 1 micron, and there are small pores in the wall (as will be explained in Chapter 22) through which substance can diffuse.

Though the capillary walls are very thin and therefore very weak, their diameters are also very small. It is a law of physics, the *law of Laplace*, that the wall tension is directly proportional to the *pressure times the diameter*. Therefore, since the diameter is extremely minute, the tension developed in the wall is also extremely minute, which explains why the very thin-walled capillaries can withstand the pressure therein.

Exchange of Fluid and Solutes Through the Capillary Membrane

The detailed dynamics of fluid exchange through the capillary membrane will be presented in Chapter 22, but it is important to introduce this subject briefly here.

The capillary membrane is highly permeable to water as well as to all the substances dissolved in plasma and tissue fluids *except the plasma proteins.* This failure of the plasma proteins to go through the pores of the capillaries allows the proteins to cause osmotic pressure, called *colloid osmotic pressure* or *oncotic pressure,* at the membrane.

Therefore, two different types of pressure gradients can cause fluid movement through the capillary membrane: (1) the hydrostatic pressure gradient between the inside and outside of the membrane and (2) the colloid osmotic pressure gradient between the two sides. The greater the difference between the intracapillary hydrostatic pressure and the hydrostatic pressure in the tissue spaces surrounding the capillaries, the greater will be the tendency for fluid to move out of the capillaries into the interstitial spaces. On the other hand, the greater the difference between the plasma colloid osmotic pressure and the tissue fluid colloid osmotic pressure, the greater will be the osmotic tendency for fluid to move from the tissue spaces into the capillary. Under normal conditions, the hydrostatic and osmotic pressures are approximately in equilibrium, so that net exchange of fluid volume across the capillary membrane is very slight — only a minute normal outward flow, which provides barely enough fluid for formation of the lymph that returns by way of the lymphatic vessels to the circulation.

On the other hand, any significant increase in the capillary pressure from its normal value will cause loss of fluid out of the circulation into the tissue spaces, or a decrease in capillary pressure will cause osmotic movement of fluid into the circulation from the tissue spaces. Therefore, we will often refer to loss of fluid from the circulation when the capillary pressure rises too high, or to gain of fluid into the circulation when the capillary pressure falls below normal.

Another important feature of capillary function is two-way diffusion through the capillary membrane of dissolved substances between the plasma and tissue fluids. Thus, sodium ions diffuse in both directions in approximately equal amounts, so that the sodium concentration remains almost exactly the same in both the blood and the tissue fluids. Likewise, when the cells of the tissues deplete the oxygen in the tissue fluids, oxygen diffuses from the blood toward the cells. Conversely, when the cells form excess carbon dioxide, it diffuses toward the blood. In this way, the capillaries provide nutrition to the cells and remove the end-products of metabolism from the cells. All these functions will be discussed in much more detail in Chapter 22.

THE SYSTEMIC VEINS AND THEIR FUNCTIONS

For years the veins have been considered to be nothing more than passageways for flow of blood into the heart, but it is rapidly becoming apparent that they perform many functions that are necessary to the operation of the circulation. They are capable of constricting and enlarging, of storing large quantities of blood and making this blood available when it is required by the remainder of the circulation, of actually propelling blood forward by means of a so-called "venous pump," and even of helping to regulate cardiac output, a function so important that it will be described in detail in Chapter 19.

Right Atrial Pressure and Its Relation to Venous Pressure

To understand the various functions of the veins, it is first necessary to know something about the pressures in the veins and how they are regulated. Blood from all the systemic veins flows into the right atrium; therefore, the pressure in the right atrium is frequently called the *central venous pressure*. The pressures in the peripheral veins depend to a great extent on the level of this pressure; that is, anything that affects right atrial pressure usually affects venous pressure everywhere in the body.

Right atrial pressure is regulated by a balance between, first, *the ability of the heart to pump blood out of the right atrium* and, second, *the tendency for blood to flow from the peripheral vessels back into the right atrium.*

If the heart is pumping strongly, the right atrial pressure tends to decrease. On the other hand, weakness of the heart tends to elevate the right atrial pressure. Likewise, anything that causes rapid inflow of blood into the right atrium from the veins tends to elevate the right atrial pressure. Some of the factors that increase this tendency for venous return (and also increase the right atrial pressure) are (1) increased blood volume, (2) increased large vessel tone throughout the body with resultant increased peripheral venous pressures, and (3) dilatation of the systemic arterioles, which decreases the peripheral resistance and allows rapid flow of blood from the arteries to the veins.

The same factors that regulate right atrial pressure also enter into the regulation of cardiac output, for the amount of blood pumped by the heart depends on both the ability of the heart to pump and the tendency for blood to flow into the heart from the peripheral vessels. Therefore, we will discuss the regulation of right atrial pressure in much more depth in Chapter 19 in connection with the regulation of cardiac output.

Venous Resistance and Peripheral Venous Pressure

Large veins have almost no resistance *when they are distended.* However, as illustrated in Figure 15–5, most of the large veins entering the thorax are compressed at many points by the surrounding tissues so that blood flow is impeded. For instance, the veins from the arms are compressed by their sharp angulation over the first rib. Second, the pressure in the neck veins often falls so low that the atmospheric pressure on the outside of the neck causes them to collapse. Finally, veins coursing through the abdomen are often compressed by different organs and by the intra-abdominal pressure, so that usually they are at least partially collapsed. For these reasons the *large veins usually do offer considerable resistance to blood flow,* and because of this the pressure in the peripheral veins is usually 4 to 9 mm. Hg greater than the right atrial pressure.

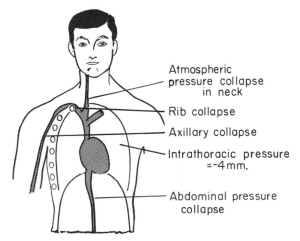

Figure 15–5. Factors tending to collapse the veins entering the thorax.

Effect of Hydrostatic Pressure on Venous Pressure

In any body of water, the pressure at the surface of the water is equal to atmospheric pressure, but the pressure rises 1 mm. Hg for each 13.6 mm. of distance below the surface. This pressure results from the weight of the water and therefore is called *hydrostatic pressure.*

Hydrostatic pressure also occurs in the vascular system of the human being because of the weight of the blood in the vessels, as is illustrated in Figure 15–6. When a person is standing, the pressure in the right atrium remains approximately 0 mm. Hg because the heart pumps into the arteries any excess blood that attempts to accumulate at this point. However, in an adult *who is standing absolutely still* the pressure in the veins of the feet is approximately +90 mm. Hg simply because of the distance from the feet to the heart and the weight of the blood in the veins between the heart and the feet. The venous pressures at other levels of the body lie proportionately between 0 and 90 mm. Hg.

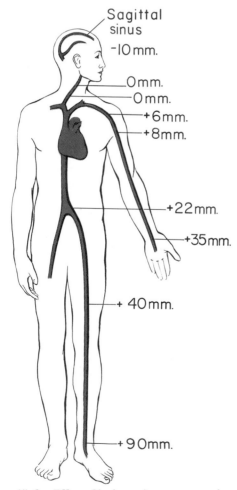

Figure 15–6. Effect of hydrostatic pressure on the venous pressures throughout the body.

Venous Valves and the "Venous Pump"

Because of hydrostatic pressure, the venous pressure in the feet would always be about +90 mm. Hg in a standing adult were it not for the valves in the veins. However, every time one moves the legs one tightens the muscles and compresses the veins either in the muscles or adjacent to them, and this squeezes the blood out of the veins. Yet, the valves in the veins, as illustrated in Figure 15–7, are arranged so that the direction of blood flow can only be toward the heart. Consequently, every time a person moves the legs or even tenses the muscles, a certain amount of blood is propelled toward the heart, and the pressure in the dependent veins of the body is lowered. This pumping system is known as the "venous pump" or "muscle pump," and it is efficient enough that under ordinary circumstances the venous pressure in the feet of a walking adult remains less than 25 mm. Hg.

If the human being stands perfectly still, the venous pump does not work, and the venous pressures in the lower part of the leg can rise to the full hydrostatic value of 90 mm. Hg in about 30 seconds. Under such circumstances the pressures within the capillaries also increase greatly, and fluid leaks from the circulatory system into the tissue spaces. As a result, the legs swell, and the blood volume diminishes. Indeed, as much as 15 to 20 per cent of the blood volume is frequently lost from the circulatory system within the first 15 minutes of standing absolutely still, as occurs when a soldier is made to stand at absolute attention.

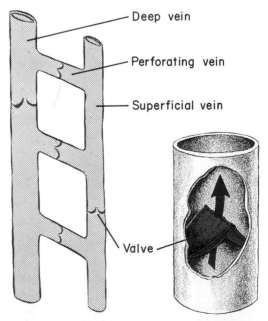

Figure 15–7. The venous valves of the leg.

Varicose Veins. The valves of the venous system are frequently destroyed. This occurs particularly when the veins have been overstretched by an excess of venous pressure for a prolonged period of time, as occurs in pregnancy or when one stands on his feet most of the time. Stretching the veins finally destroys the function of the valves entirely. Thus, the person develops "varicose veins," which are characterized by large bulbous protrusions of the veins beneath the skin of the entire leg and particularly of the lower leg. The venous and capillary pressures become very high, and leakage of fluid from the capillary blood into the tissues causes constant edema in the legs of these persons whenever they stand for more than a few minutes. The edema in turn prevents adequate diffusion of nutritional materials from the capillaries to the muscle and skin cells, so that the muscles become painful and weak, and the skin frequently becomes gangrenous and ulcerates.

Measurement of Venous Pressure

Clinical Estimation of Venous Pressure. The venous pressure can be estimated by simply observing the degree of distension of the peripheral veins — especially the neck veins. For instance, in the sitting position, the neck veins are never distended in the normal person. However, when the right atrial pressure becomes increased to as much as 10 mm. Hg, the lower veins of the neck begin to protrude even when one is sitting; when the right atrial pressure rises to as high as 15 mm. Hg, essentially all the veins in the neck become distended, and a venous pulsation with each heartbeat can then usually be seen in the walls of the protruding veins.

Rough estimates of the venous pressure can also be made by raising or lowering an arm of a reclining person while observing the degree of distension of the antecubital or hand veins. As the arm is progressively raised the veins suddenly collapse, and the level at which they collapse, when referred to the level of the heart, is a rough measure of the peripheral venous pressure.

Direct Measurement of Venous Pressure and Right Atrial Pressure. Venous pressure can be measured with ease by the insertion of a syringe needle connected to a pressure recorder or to a water manometer directly into a vein. The venous pressure is expressed in relation to the level of the tricuspid valve, i.e., the height in centimeters of water above the level of the tricuspid valve; this can be converted to mm. Hg by dividing by a factor of 1.36.

The only means by which *right atrial pressure* can be measured accurately is the insertion of a catheter through the veins into the right atrium.

This catheter can then be connected to an appropriate pressure measuring apparatus.

Blood Reservoir Function of the Veins

In discussing the general characteristics of the systemic circulation earlier in the chapter, it was pointed out that over 60 per cent of all the blood in the circulatory system is in the systemic veins. For this reason it is frequently said that the systemic veins act as a *blood reservoir* for the circulation.

When blood is lost from the body to the extent that the arterial pressure begins to fall, pressure reflexes are elicited from the carotid sinuses and other pressure sensitive areas of the circulation, as will be discussed in Chapter 17; these in turn cause sympathetic nerve signals that constrict the veins, thus automatically taking up much of the slack in the circulatory system caused by the lost blood. Indeed, even after as much as 20 to 25 per cent of the total blood volume has been lost, the circulatory system often functions almost normally because of this variable reservoir system of the veins.

Specific Blood Reservoirs. Certain portions of the circulatory system are so extensive that they are specifically called "blood reservoirs." These include (1) the *spleen,* which can sometimes decrease in size sufficiently to release as much as 150 ml. of blood into other areas of the circulation, (2) the *liver,* the sinuses of which can release several hundred milliliters of blood into the remainder of the circulation, (3) the *large abdominal veins,* which can contribute as much as 300 ml., and (4) the *venous plexus beneath the skin,* which can probably contribute several hundred milliliters. The *heart* itself and the *lungs,* though not parts of the systemic venous reservoir system, must also be considered to be blood reservoirs. The heart, for instance, becomes reduced in size during sympathetic stimulation and in this way can contribute about 100 ml. of blood, and the lungs can contribute another 100 to 200 ml. when the pulmonary pressures fall to low values.

THE PULMONARY CIRCULATION

The quantity of blood flowing through the lungs is essentially equal to that flowing through the systemic circulation. However, there are problems related to distribution of blood flow and other hemodynamics that are special to the pulmonary circulation. Therefore, the present discussion is concerned with the special features of blood flow in the pulmonary circuit and the function of the right side of the heart in maintaining this flow.

PHYSIOLOGIC ANATOMY OF THE PULMONARY CIRCULATORY SYSTEM

The Right Side of the Heart. As illustrated in Figure 15–8, the right ventricle is wrapped halfway around the left ventricle, owing to the difference in pressures developed by the two ventricles during systole. Because the left ventricle contracts with extreme force in comparison with the right ventricle, the left ventricle assumes a globular shape, and the septum protrudes into the right heart. Yet each side of the heart pumps essentially the same quantity of blood; therefore, the external wall of the right ventricle bulges far outward and extends around a large portion of the left ventricle, in this way accommodating about the same quantity of blood as does the left ventricle.

The muscle of the right ventricle is slightly more than one-third as thick as that of the left ventricle as a result of the difference in pressures between the two sides of the heart.

The Pulmonary Vessels. The pulmonary artery extends only 4 centimeters beyond the apex of the right ventricle and then divides into the right and left main branches, which supply blood to the two respective lungs. The pulmonary artery and its branches are also very thin and distensible, which gives the pulmonary arterial tree a very large compliance. This large compliance allows the pulmonary arteries to accommodate the stroke volume output of the right ventricle.

The pulmonary veins, like the pulmonary arteries, are also short, but their distensibility characteristics are similar to those of the veins in the systemic circulation.

PRESSURES IN THE PULMONARY SYSTEM

The Pressure Pulse Curve in the Right Ventricle. The pressure pulse curves of the right ventricle and pulmonary artery are illustrated in the lower portion of Figure 15–9. These are contrasted with the much higher aortic pressure curve shown above. The systolic pressure in the right ventricle of the normal human being averages approximately 22 mm. Hg, and the diastolic pressure averages about 0 to 1 mm. Hg.

Pressures in the Pulmonary Artery. During systole, the pressure in the pulmonary artery is essentially equal to the pressure in the right ventricle, as is also shown in Figure 15–9. However, after the pulmonary valve closes at the end of systole, the pulmonary arterial pressure remains elevated and then falls gradually as blood flows through the capillaries of the lungs.

As shown in Figure 15–10, the systolic pulmonary arterial pressure averages approximately 22 mm. Hg in the normal human being; the diastolic pulmonary arterial pressure, approximately 8 mm. Hg; and the mean pulmonary arterial pressure, 13 mm. Hg.

Pulmonary Capillary Pressure. The mean pulmonary capillary pressure, which is also illustrated diagrammatically in Figure 15–10, is approximately 7 mm. Hg. This will be discussed in more detail later in the chapter in relation to fluid exchange functions of the capillary.

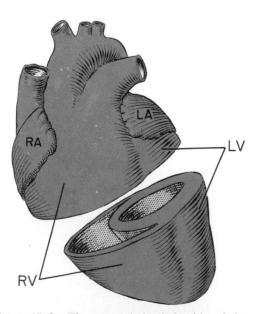

Figure 15–8. The anatomical relationship of the right ventricle to the left ventricle, showing the globular shape of the left ventricle and the half-moon shape of the right ventricle as it drapes around the left ventricle.

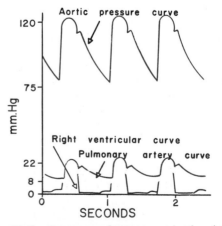

Figure 15–9. Pressure pulse contours in the right ventricle, pulmonary artery, and aorta.

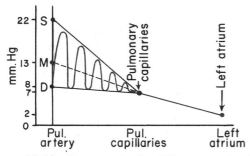

Figure 15–10. Pressures in the different vessels of the lungs.

Left Atrial and Pulmonary Venous Pressure. The mean pressure in the left atrium and in the major pulmonary veins averages approximately 2 mm. Hg in the human being, varying from as low as 1 mm. Hg to as high as 4 mm. Hg.

THE BLOOD VOLUME OF THE LUNGS

The blood volume of the lungs is approximately 450 ml., about 9 per cent of the total blood volume of the circulatory system. About 70 ml. of this is in the capillaries, and the remainder is divided about equally between the arteries and veins.

The Lungs as a Blood Reservoir. Under different physiological and pathological conditions, the quantity of blood in the lungs can vary from as little as 50 per cent of normal up to as high as 200 per cent of normal. For instance, when a person blows air out so hard that high pressure is built up in the lungs — such as when blowing a trumpet — as much as 250 ml. of blood can be expelled from the pulmonary circulatory system into the systemic circulation. Also, loss of blood from the systemic circulation by hemorrhage can be partly compensated for by automatic shift of blood from the lungs into the systemic vessels.

Shift of Blood Between the Pulmonary and Systemic Circulatory Systems as a Result of Cardiac Pathology. Failure of the left side of the heart or increased resistance to blood flow through the mitral valve as a result of mitral stenosis or mitral regurgitation causes blood to dam up in the pulmonary circulation, thus sometimes increasing the pulmonary blood volume as much as 100 per cent, and also causing corresponding increases in the pulmonary vascular pressures.

On the other hand, exactly the opposite effects take place when the right side of the heart fails.

Because the volume of the systemic circulation is about seven times that of the pulmonary system, a shift of blood from one system to the other affects the pulmonary system greatly but usually has only mild systemic effects.

BLOOD FLOW THROUGH THE LUNGS AND ITS DISTRIBUTION

The blood flow through the lungs is essentially equal to the cardiac output. Therefore, the factors that control cardiac output — mainly peripheral factors, as discussed in Chapter 19 — also control pulmonary blood flow. Under most conditions, the pulmonary vessels act as passive, distensible tubes that enlarge with increasing pressure and narrow with decreasing pressure. But, for adequate aeration it is important that the blood be distributed to those segments of the lungs where the alveoli are best oxygenated. This is achieved by the following mechanism:

Effect of Low Alveolar Oxygen Pressure on Pulmonary Vascular Resistance — Automatic Control of Local Pulmonary Blood Flow Distribution. When alveolar oxygen concentration becomes very low, the adjacent blood vessels slowly constrict during the ensuing three to ten minutes, the vascular resistance increasing to as much as three times normal. It should be noted specifically that this effect is *opposite to the effect* normally observed in systemic vessels, which dilate rather than constrict in response to low oxygen concentration. It is believed that the low oxygen concentration causes some vasoconstrictor substance to be released from the lung tissue, this substance in turn promoting small arterial and arteriolar constriction.

This effect of low oxygen concentration on pulmonary vascular resistance has an important function: to distribute blood flow where it is most effective. That is, when some of the alveoli are poorly ventilated so that the oxygen concentration in them becomes low, the local vessels constrict. This in turn causes most of the blood to flow through other areas of the lungs that are better aerated, thus providing an automatic control system for distributing blood flow through different pulmonary areas in proportion to their degrees of ventilation.

Maldistribution of Blood Flow Between the Top and Bottom of the Lung Because of Hydrostatic Pressure. Hydrostatic pressure affects vascular pressures in the lungs in the same way as in the systemic circulation. Therefore, when the person is in the upright position, the pulmonary vascular pressures at the top of the lungs are about 10 mm. Hg less than they are at the level of the heart; and the pressures at the bottom of the lung are about 8 mm. Hg greater. This effect causes serious differences in blood flow between the upper and lower lung because the mean pulmonary arterial pressure is 13 mm. Hg at the level of the heart, about 3 mm. Hg at the apex of the lungs, and about 21 mm. Hg at the bottom. Consequently, when a person is at rest but is standing, blood flow in the bottom of the lung is moderately greater than at heart level, whereas blood flow in the apical lung tissue is very slight.

Effect of Increased Cardiac Output on the Pulmonary Circulation During Heavy Exercise

During heavy exercise the lungs are frequently called upon to absorb up to 20 times as much oxygen into the blood as they do normally. This absorption is achieved in two ways: (1) by increasing the number of open capillaries so that oxygen can diffuse more readily between the alveolar gas and the blood, and (2) by increasing cardiac output, with its concomitant increase in blood flow through the lungs — the blood thus picking up greater quantities of oxygen.

Fortunately, the cardiac output can increase to four to six times normal before pulmonary arterial pressure becomes excessively elevated. This effect results from the fact that more and more capillaries open up; also, the pulmonary arterioles and capillaries expand. Therefore, the pressure rarely rises more than two-fold despite as much as a five- to six-fold increase in flow.

This ability of the lungs to accommodate greatly increased blood flow during exercise with relatively little increase in pulmonary vascular pressure is important for at least two reasons: (1) It obviously conserves the energy of the right heart, and (2) it prevents a significant rise in pulmonary capillary pressure and therefore also prevents development of pulmonary edema during the increased cardiac output.

PULMONARY CAPILLARY DYNAMICS

Length of Time Blood Stays in the Capillaries. From histologic study of the total cross-sectional area of all the pulmonary capillaries, it can be calculated that when the cardiac output is normal, blood passes through the pulmonary capillaries in about 1 second. Increasing the cardiac output shortens this time sometimes to less than 0.4 second, but the shortening would be much greater were it not for the fact that additional capillaries, which normally remain collapsed, open up to accommodate the increased blood flow. Thus, in less than 1 second, blood passing through the capillaries becomes oxygenated and loses its excess carbon dioxide.

Capillary Exchange of Fluid in the Lungs

Negative Interstitial Fluid Pressure in the Lungs and Its Significance. Fluid exchange through the capillary membrane will be discussed in detail in Chapter 22, where it is pointed out that the most significant difference between pulmonary capillary dynamics and capillary dynamics elsewhere in the body is the very low capillary pressure in the lungs, about 7 mm. Hg, in comparison with a considerably higher "functional" capillary pressure elsewhere, probably about 17 mm. Hg. Because of the very low pulmonary capillary pressure, the hydrostatic force tending to push fluid out the capillary pores into the interstitial spaces is also very slight. Yet, the colloid osmotic pressure of the plasma, about 28 mm. Hg, is a large force tending to pull fluid into the capillaries. Therefore, there is continual osmotic tendency to dehydrate the interstitial spaces of the lungs. Referring to Chapter 22 again, it is calculated there that the normal pulmonary interstitial fluid pressure is probably about −8 mm. Hg. That is, approximately 8 mm. Hg is tending to pull the alveolar epithelial membrane toward the capillary membrane, thus squeezing the pulmonary interstitial space down to almost nothing. Electron micrographic studies have demonstrated this fact, the interstitial space at times being so narrow that the basement membrane of the alveolar epithelium is fused with the basement membrane of the capillary endothelium. As a result, the distance between the air in the alveoli and the blood in the capillaries is minimal, averaging about 0.4 micron in distance; this obviously allows very rapid diffusion of oxygen and carbon dioxide.

The details of fluid exchange through the pulmonary capillary membrane will be discussed in relation to overall capillary function in Chapter 22.

Mechanism by Which the Alveoli Remain Dry. Another consequence of the negative pressure in the interstitial spaces is that it pulls fluid from the alveoli through the alveolar membrane and into the interstitial spaces, thereby normally keeping the alveoli dry.

Pulmonary Edema

Pulmonary edema means excessive quantities of fluid in the pulmonary interstitial spaces or alveoli. Sometimes the amount of this fluid can increase to 500 to 1000 per cent greater than normal, and this is a common cause of death, resulting from inability to aerate the blood. This condition will be discussed in Chapter 23, but its most common causes are: (1) greatly increased pulmonary capillary pressure, occurring most frequently in patients with failing left hearts, and (2) greatly increased leakiness (increased permeability) of the pulmonary capillary membranes caused by infectious disease or poisons such as some war gases.

SOME PATHOLOGICAL CONDITIONS THAT OBSTRUCT BLOOD FLOW THROUGH THE LUNGS

Massive Pulmonary Embolism. One of the most severe postoperative calamities in surgical practice is massive pulmonary embolism. Patients lying immobile in bed tend to develop extensive clots in the veins of the legs because of sluggish blood flow. Also, women frequently develop massive clots in the hypogastric veins after delivery of their babies. Such clots often break away from the initial sites of formation, particularly when the patient first walks after a long period of immobilization. The clots then flow to the right side of the heart and into the pulmonary artery. Such a free-moving clot is called an *embolus.*

Total blockage of only one of the major branches of the pulmonary artery usually is not immediately fatal because the opposite lung can accommodate all the blood flow. However, blood clots, as was discussed in Chapter 7, have a tendency to grow. Consequently, the embolus becomes larger and larger, and as it extends into the other major branch of the pulmonary artery, the few remaining vessels that do not become plugged are taxed beyond their limit, and death ensues because of an inordinate rise in pulmonary arterial pressure and right-sided heart failure.

Emphysema. Pulmonary emphysema means, literally, too much air in the lungs, and it is usually characterized by destruction of many of the alveolar walls. This causes the adjacent alveoli to become confluent, thereby forming large *emphysematous cavities* rather than the usual small alveoli. Obviously, loss of the alveolar septa greatly decreases the total alveolar surface area of the lungs and hinders gas exchange between the alveoli and the blood.

Emphysema also has an important effect on the pulmonary vasculature, for each time an alveolar wall is destroyed, some of the small blood vessels of the pulmonary system are also destroyed, thus progressively increasing the pulmonary resistance and elevating the pulmonary arterial pressure.

Unfortunately, the prevalence of emphysema is increasing rapidly because of cigarette smoking.

Diffuse Sclerosis of the Lungs. A number of pathological conditions cause excessive fibrosis in the supportive tissues in the lungs, and the fibrous tissue in turn contracts around the vessels. Some of these conditions are silicosis, tuberculosis, syphilis, and, to a lesser extent, anthracosis.

In early stages of diffuse sclerosis, the pulmonary arterial pressure often is normal as long as the person is not exercising, but just as soon as even mild exercise is performed, the pulmonary arterial pressure rises inordinately because the vessels do not have the ability to expand as much as normal pulmonary vessels do. In late stages of diffuse sclerosis the pulmonary arterial pressure remains elevated constantly; as a result, the right ventricle hypertrophies, and it may fail. Diffusion of gases through the alveolar membranes in most instances is impaired either because of decreased surface area or because of diminished blood flow to some alveoli.

Atelectasis. Atelectasis is the clinical term for collapse of a lung or part of a lung. This occurs often when the bronchi become plugged, because the pulmonary blood rapidly absorbs the air in the entrapped alveoli, which causes the alveoli to collapse.

Atelectasis also occurs when the chest cavity is opened to atmospheric pressure, for when air is allowed to enter the pleural space, the elastic nature of the lungs causes them to collapse immediately.

When the elastic tissues of the lungs contract during atelectasis, they constrict not only the alveoli but also the blood vessels. This constriction, in addition to the vasoconstriction caused by oxygen deficiency as discussed earlier in the chapter, automatically decreases blood flow in the atelectatic portions of the lungs to as little as one-fourth normal, shifting the remaining three-fourths to the aerated portions. This safety mechanism is important, for it prevents flow of major quantities of blood through collapsed, nonaerated pulmonary areas.

REFERENCES

Systemic Circulation

Folkow, B.: Role of the nervous system in the control of vascular tone. *Circulation, 21*:706, 1960.

Guyton, A. C.: The venous system and its role in the circulation. *Mod. Conc. Cardiov. Dis., 27*:483, 1958.

Guyton, A. C.: Peripheral circulation. *Annu. Rev. Physiol., 21*: 239, 1959.

Guyton, A. C.: Arterial Pressure and Hypertension. Philadelphia, W. B. Saunders Co., 1980.

Guyton, A. C., and Jones, C. E.: Central venous pressure: Physiological significance and clinical implications. *Am. Heart J., 86*:431, 1973.

Herd, J. A.: Overall regulation of the circulation. *Annu. Rev. Physiol., 32*:289, 1970.

James, D. G. (ed.): Circulation of the Blood. Baltimore, University Park Press, 1978.

Lundgren, O., and Jodal, M.: Regional blood flow. *Annu. Rev. Physiol.,* 37:395, 1975.

Pedley, T. J.: The Fluid Mechanics of Large Blood Vessels. New York, Cambridge University Press, 1979.

Saunders, J. B. deC. M.: The history of venous valves. *In* Dickinson, C. J., and Marks, J. (eds.): Developments in Cardiovascular Medicine. Lancaster, England, MTP Press, 1978, p. 335.

Shepherd, J. T., and Vanhoutte, P. M.: The Human Cardiovascular System: Facts and Concepts. New York, Raven Press, 1979.

Stainsby, W. N.: Autoregulation of blood flow in skeletal muscle during increased metabolic activity. *Am. J. Physiol.,* 202:273, 1962.

Vanhoutte, P. M., and Leusen, I. (eds.): Mechanics of Vasodilatation. New York, S. Karger, 1978.

Widmer, L. K., et al. (eds.): New Trends in Venous Diseases. Bern, H. Huber, 1977.

Wolf, S., and Werthessen, N. T. (eds.): Dynamics of Arterial Flow. New York, Plenum Press, 1979.

Wolf, S., et al. (eds.): Structure and Function of the Circulation. New York, Plenum Press, 1979.

Pulmonary Circulation

Cournand, A.: Some aspects of the pulmonary circulation in normal man and in chronic cardiopulmonary diseases. *Circulation,* 2:641, 1950.

Cumming, G.: The pulmonary circulation. *In* MTP International Review of Science: Physiology, Vol. 1. Baltimore, University Park Press, 1974, p. 93.

Fishman, A. P.: Hypoxia in the pulmonary circulation. *Circ. Res.,* 38:221, 1976.

Fishman, A. P., and Pietra, G.: Hemodynamic pulmonary edema. *In* Fishman, A. P., and Renkin, E. M. (eds.): Pulmonary Edema. Baltimore, Waverly Press, 1979, p. 79.

Gil, J.: Influence of surface forces on pulmonary circulation. *In* Fishman, A. P., and Renkin, E. M. (eds.): Pulmonary Edema. Baltimore, Waverly Press, 1979, p. 53.

Guyton, A. C.: Introduction to Part I: Pulmonary alveolar-capillary interface and interstitium. *In* Fishman, A. P., and Hecht, H. H. (eds.): The Pulmonary Circulation and Interstitial Space. Chicago, University of Chicago Press, 1969, p. 3.

Guyton, A. C., et al.: Dynamics of subatmospheric pressure in the pulmonary interstitial fluid. *In* Lung Liquids. Ciba Symposium. New York, Elsevier/North-Holland, 1976, p. 77.

Guyton, A. C., et al.: Forces governing water movement in the lung. *In* Pulmonary Edema. Washington, D.C., American Physiological Society, 1979, p. 65.

Hughes, J. M. B.: Pulmonary circulatory and fluid balance. *Int. Rev. Physiol.,* 14:135, 1977.

Meyer, B. J., et al.: Interstitial fluid pressure V: Negative pressure in the lungs. *Circ. Res.,* 22:263, 1968.

Parker, J. C., et al.: Pulmonary interstitial and capillary pressures estimated from intra-alveolar fluid pressures. *J. Appl. Physiol.,* 44(2):267, 1978.

Parker, J. C., et al.: Pulmonary transcapillary exchange and pulmonary edema. *In* Guyton, A. C., and Young, D. B. (eds.): International Review of Physiology: Cardiovascular Physiology III, Vol. 18. Baltimore, University Park Press, 1979, p. 261.

Pietra, G. G., et al.: Bronchial veins and pulmonary edema. *In* Fishman, A. P., and Renkin, E. M. (eds.): Pulmonary Edema. Baltimore, Waverly Press, 1979, p. 195.

Racz, G. B.: Pulmonary blood flow in normal and abnormal states. *Surg. Clin. North Am.,* 54:967, 1974.

Reeves, J. T.: Pulmonary vascular response to high altitude residence. *Cardiovasc. Clin.,* 5:81, 1973.

Staub, N. C.: Pulmonary edema. *Physiol. Rev.,* 54:678, 1974.

Staub, N. C. (ed.): Lung Water and Solute Exchange. New York, Marcel Dekker, 1978.

Taylor, A. E., et al.: Permeability of the alveolar membrane to solutes. *Circ. Res.,* 16:353, 1965.

West, J. B.: Blood flow to the lung and gas exchange. *Anesthesiology,* 41:124, 1974.

16

Local Control of Blood Flow by the Tissues; and Nervous and Humoral Regulation

The circulatory system is provided with a complex system for control of blood flow to the different parts of the body. In general, these are of three major types:

1. Local control of blood flow in each individual tissue, the flow being controlled mainly in proportion to that tissue's need for blood perfusion.

2. Nervous control of blood flow, which often affects blood flow in large segments of the systemic circulation, such as by shifting blood flow from the nonmuscular vascular beds to the muscles during exercise or changing the blood flow in the skin to control body temperature.

3. Humoral control, in which various substances in the blood such as hormones, ions, or other chemicals can cause either local increase or decrease in tissue flow or widespread generalized changes in flow.

LOCAL CONTROL OF BLOOD FLOW BY THE TISSUES THEMSELVES

Acute Local Regulation in Response to Tissue Need for Flow. In most tissues the blood flow is controlled in proportion to the need for nutrition, especially for delivery of oxygen but also for glucose, amino acids, fatty acids, and other nutrients. However, in some tissues the local flow performs other functions. In the skin its major purpose is to transfer heat from the body to the surrounding air. In the kidneys its purpose is to deliver substances to the kidneys for excretion. And in the brain it is to determine, to a great extent, the carbon dioxide and hydrogen ion concentrations of the brain fluids, which in turn play important roles in controlling the level of brain activity.

Fortunately, local blood flow can be increased in response to many different local factors in the tissues — at times to lack of oxygen, at times to excess of carbon dioxide or hydrogen ion concentration, and at other times to still other factors. These many different control factors help to distribute the blood flow to the different parts of the body in proportion to the respective needs of the tissues.

Table 16–1 gives the approximate *resting* blood flows through the different organs of the body. Note the tremendous flows through the brain, liver, and kidneys despite the fact that these organs represent only a small fraction of the total body mass. Yet, even under basal conditions, the need for flow in each one of these tissues is very great; in the liver to support its high level of metabolic activity, in the brain to provide nutrition and to prevent the carbon dioxide and hydrogen ion concentrations from becoming too great, and in the kidneys to maintain adequate excretion.

The skeletal muscle of the body represents 35 to 40 per cent of the total body mass, and yet, in the inactive state, the blood flow through all the skeletal muscle is only 15 to 20 per cent of the total cardiac output. This accords with the fact that inactive muscle has a very low metabolic rate. Yet, when the muscles become active, their metabolic rate sometimes increases as much as 50-fold, and the blood flow in individual muscles can increase as much as 20-fold, illustrating a marked increase in blood flow in response to the increased need of the muscle for nutrients.

TABLE 16–1   BLOOD FLOW TO DIFFERENT
ORGANS AND TISSUES UNDER
BASAL CONDITIONS*

	Per cent	*ml./min.*
Brain	14	700
Heart	4	200
Bronchial	2	100
Kidneys	22	1100
Liver	27	1350
Portal	(21)	(1050)
Arterial	(6)	(300)
Muscle (inactive state)	15	750
Bone	5	250
Skin (cool weather)	6	300
Thyroid gland	1	50
Adrenal glands	0.5	25
Other tissues	3.5	175
Total	100.0	5000

*Based mainly on data compiled by Dr. L. A. Sapirstein.

Functional Anatomy of the Systemic Microcirculation.   Though the student will recall that each tissue has its own characteristic vascular system, Figure 16–1 presents a typical capillary bed; this one is in the connective tissue of the mesentery, which is very easily studied and its components easily analyzed. This figure shows that blood enters the capillary bed through a small *arteriole* and leaves by way of a small *venule*. From the arteriole the blood usually divides and flows through several *metarterioles* before entering the *capillaries*. Some of the capillaries are very large, and they course almost directly to the venule. These are called *preferential channels*. However, most of the capillaries, called the *true capillaries*, branch mainly from the metarterioles and then finally terminate in a venule.

The arterioles have a strong muscular coat, and the metarterioles are surrounded by sparse but highly active smooth muscle fibers. In addition, in many tissues, at each point at which a capillary leaves a metarteriole, a small muscular *precapillary sphincter*, consisting of a single spiraling smooth muscle fiber, surrounds the origin of the capillary.

As will be discussed in more detail later in the chapter, the arterioles and venules are supplied by extensive innervation from the sympathetic nervous system, and the degree of contraction of these structures is strongly influenced by the intensity of sympathetic signals transmitted from the central nervous system to the blood vessels.

On the other hand, innervation of the metarterioles and the precapillary sphincters is usually very sparse, or even absent in most instances. Instead, the muscle fibers of these two structures are controlled almost entirely by the local factors in the tissues, that is, by the concentrations of oxygen, carbon dioxide, hydrogen ions, electrolytes, and other substances in each individual tissue area. These local factors, therefore, are major controllers of blood flow in the local tissue areas.

Effect of Tissue Metabolism on Local Blood Flow.   Figure 16–2 illustrates the approximate quantitative effect on blood flow of increasing the rate of metabolism in a local tissue such as muscle. Note that an increase in metabolism up to eight times normal increases the blood flow about four-fold.

Local Blood Flow Regulation When Oxygen Availability Changes.   One of the most necessary of the nutrients is oxygen. Whenever the availability of oxygen to the tissues decreases, as happens at high altitude, in pneumonia, in carbon monoxide poisoning (which poisons the ability of hemoglobin to transport oxygen), or in cyanide poisoning (which poisons the ability of the tissues to utilize oxygen), the blood flow through the tissues increases markedly. Figure 16–3 shows that as the arterial oxygen saturation falls to about 25 per cent of normal the blood flow through an isolated leg increases about three-fold; that is, the blood flow increases almost enough, but not quite, to make up for the decreased amount of oxygen in the blood.

There are two basic theories of the regulation

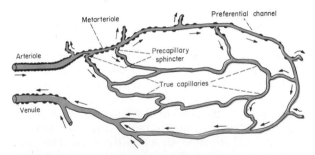

Figure 16–1.   Overall structure of a capillary bed. (From Zweifach: Factors Regulating Blood Pressure. New York, Josiah Macy, Jr., Foundation, 1950.)

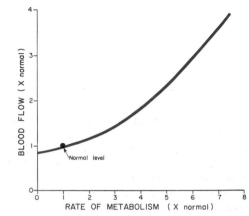

Figure 16–2.   Effect of increasing rate of metabolism on tissue blood flow.

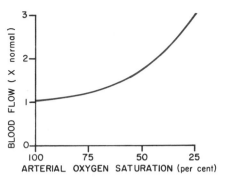

Figure 16–3. Effect of arterial oxygen saturation on blood flow through an isolated dog leg.

of local blood flow when either the rate of tissue metabolism or the availability of oxygen changes. These are (1) the *vasodilator theory* and (2) the *oxygen demand theory.*

The Vasodilator Theory of Local Blood Flow Regulation. According to this theory, the greater the rate of metabolism, or the less the blood flow, or the less the availability of oxygen and other nutrients to a tissue, the greater becomes the rate of formation of a *vasodilator substance.* The vasodilator substance then is believed to diffuse back to the precapillary sphincters, metarterioles, and arterioles to cause dilatation. Some of the different vasodilator substances that have been suggested are carbon dioxide, lactic acid, adenosine, adenosine phosphate compounds, histamine, potassium ions, and hydrogen ions.

Recently, many physiologists have suggested that the substance *adenosine* is an especially important local vasodilator that might play *the* major role in regulating local blood flow. For instance, vasodilator quantities of adenosine are released from heart muscle cells whenever coronary blood flow becomes too little, and it is believed that this causes local vasodilatation in the heart and thereby returns the blood flow back toward normal. Also, whenever the heart becomes overly active and the heart's metabolism increases, this causes excessive utilization of oxygen, decreased oxygen concentration in the local tissues, increased degradation of adenosine triphosphate, and, therefore, increased formation of adenosine. Here again the increased concentration of adenosine is believed to cause coronary vasodilation, with consequent increase in coronary blood flow to supply the nutrient demands of the more active heart.

The problem with the different vasodilator theories of local blood flow regulation has been the following: It has been difficult to prove that sufficient quantities of any single vasodilator substance are indeed formed in the tissues to cause all of the measured increase in blood flow in states of increased tissue metabolic demand. On the other hand, perhaps a combination of all the different vasodilators could increase the blood flow sufficiently.

The Oxygen Demand Theory of Local Blood Flow Control. Though the vasodilator theory is accepted by most physiologists, several critical facts have made a few physiologists favor still another theory, which can be called either the oxygen demand theory or, more accurately, the *nutrient demand theory* (because other nutrients besides oxygen are also involved). Oxygen (as well as other nutrients) is required to maintain vascular muscle contraction. Therefore, in the absence of an adequate supply of oxygen and other nutrients, it is reasonable to believe that the blood vessels naturally dilate. Also, increased utilization of oxygen in the tissues as a result of increased metabolism would theoretically decrease the local tissue oxygen availability, and this too would cause local vasodilatation.

The evidence against the oxygen demand theory is that many types of smooth muscle can remain contracted for long periods of time in the presence of extremely minute concentrations of oxygen — concentrations even below those normally found in the tissues. A possible answer to this is that very small arteries (with internal diameters of approximately 0.5 mm.) dilate markedly at the lower oxygen concentrations found even normally in the tissues.

Thus, on the basis of presently available data, either a vasodilator theory or an oxygen demand theory could explain local blood flow regulation in response to the metabolic needs of the tissues. Perhaps the truth lies in a combination of the two mechanisms.

LONG-TERM LOCAL BLOOD FLOW REGULATION

The local blood flow regulation that has been discussed thus far occurs acutely, within a minute or more after local tissue conditions have changed. For instance, if the arterial pressure is suddenly increased from 100 mm. Hg to 150 mm. Hg, the blood flow increases almost instantaneously about 100 per cent, partly because the increase in pressure directly increases the flow and partly because the high pressure dilates the peripheral vessels. But within one to two minutes the blood flow decreases back to about 15 per cent above the original control value, a phenomenon called *acute autoregulation* of blood flow. This illustrates the rapidity of the acute type of local regulation, but the regulation is still very incomplete because there remains a 15 per cent increase in blood flow.

However, over a period of hours, days, and

weeks a long-term type of local blood flow regulation develops in addition to the acute regulation, and this long-term regulation gives far more complete regulation than does the acute mechanism. For instance, in the above example, if the arterial pressure remains at 150 mm. Hg indefinitely, within a few weeks the blood flow through the tissues will gradually reapproach almost exactly the normal value.

Long-term regulation also occurs when a tissue's metabolic demands change. Thus, if a tissue becomes chronically overactive and therefore requires chronically increased quantities of nutrients, the blood supply gradually increases to match the needs of the tissue.

Mechanism of Long-Term Regulation — Change in Tissue Vascularity. The mechanism of long-term local blood flow regulation is almost certainly a change in the degree of vascularity of the tissues. That is, if the arterial pressure falls to 60 mm. Hg and remains at this level for many weeks, the sizes of vessels (and perhaps the number as well) in the tissue increase; if the pressure then rises to a very high level, these decrease. Likewise, if the metabolism in a given tissue becomes elevated for a prolonged period of time, vascularity also increases; or if the metabolism is decreased, vascularity decreases.

Thus, there is continual day-by-day reconstruction of the tissue vasculature to meet the needs of the tissues. This reconstruction occurs very rapidly (within days) in extremely young animals. It also occurs rapidly in newly growing tissue, such as scar tissue and cancerous tissue; on the other hand, it occurs only very slowly in old, well-established tissues.

Role of Oxygen in Long-Term Regulation. A probable stimulus for increased or decreased vascularity in many if not most instances is need of the tissue for oxygen. The reason for believing this is that long-term hypoxia also causes increased vascularity, while hyperoxia causes decreased vascularity.

Significance of Long-Term Local Regulation — The Metabolic Mass to Tissue Vascularity Proportionality. From these discussions it should already be apparent to the student that there is a built-in mechanism in most tissues to keep the degree of vascularity of the tissue to almost exactly that required to supply the metabolic needs of the tissue. Thus, one can state as a general rule that the vascularity of most tissues of the body is directly proportional to the local metabolism. If ever this proportionality constant becomes abnormal, the long-term local regulatory mechanism automatically readjusts the degree of vascularity over a period of weeks or months. In young persons this degree of readjustment is usually very exact; in old persons it is only partial.

NERVOUS REGULATION OF THE CIRCULATION

Superimposed onto the intrinsic local tissue regulation of blood flow are two additional types of regulation: (1) *nervous* and (2) *humoral.* These regulations are not necessary for most normal function of the circulation, but they do provide greatly increased effectiveness of control under special conditions such as exercise or hemorrhage.

There are two very important features of nervous regulation of the circulation: First, nervous regulation can function extremely rapidly, some of the nervous effects beginning to occur within 1 second and reaching full development within 5 to 30 seconds. Second, the nervous system provides a means for controlling large parts of the circulation simultaneously, often in spite of the effect that this has on the blood flow to individual tissues. For instance, when it is important to raise the arterial pressure temporarily, the nervous system can arbitrarily greatly decrease blood flow to major segments of the circulation despite the fact that the local blood flow regulatory mechanisms oppose this.

THE AUTONOMIC NERVOUS SYSTEM

The autonomic nervous system will be discussed in detail in Chapter 38. However, it is so important to the regulation of the circulation that its specific anatomical and functional characteristics relating to the circulation deserve special attention here.

By far the most important part of the autonomic nervous system for regulation of the circulation is the *sympathetic nervous system.* The *parasympathetic nervous system* is important only for its regulation of heart function, as we shall see later.

The Sympathetic Nervous System. Figure 16–4 illustrates the anatomy of sympathetic nervous control of the circulation. Sympathetic vasomotor nerve fibers leave the spinal cord through all the thoracic and the first one to two lumbar spinal nerves. These pass into the sympathetic chain and thence by two routes to the blood vessels throughout the body: (1) through the *peripheral sympathetic nerves* and (2) through the *spinal nerves.* The precise pathways of these fibers in the spinal cord and in the sympathetic chains will be discussed in Chapter 38. It suffices to say here that except for the capillaries and some metarterioles, all vessels of the body are supplied with sympathetic nerve fibers.

Sympathetic Nerve Fibers to the Heart. In addition to the sympathetic nerve fibers supplying the blood vessels, some of these fibers also go to

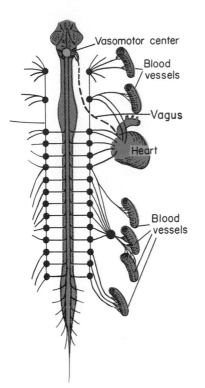

Figure 16–4. The vasomotor center and its control of the circulatory system through the sympathetic and vagus nerves.

the heart. This innervation was discussed in Chapter 11. It will be recalled that sympathetic stimulation markedly increases the activity of the heart, increasing the heart rate and enhancing its strength of pumping.

Parasympathetic Control of Heart Function, Especially Heart Rate. Though the parasympathetic nervous system is exceedingly important for many other autonomic functions of the body, it plays only a minor role in regulation of the circulation. Its only really important effect is its control of heart rate. It also has a slight effect of controlling cardiac contractility; however, this effect is far overshadowed by the sympathetic nervous system control of contractility. Parasympathetic nerves pass to the heart in the vagus nerve, as illustrated in Figure 16–4.

The effects of parasympathetic stimulation on heart function were discussed in detail in Chapter 11. Principally, parasympathetic stimulation causes a marked *decrease* in heart rate and a slight decrease in contractility.

The Sympathetic Vasoconstrictor System and Its Control by the Central Nervous System

Sympathetic *vasoconstrictor* fibers are distributed to essentially all segments of the circulation. However, this distribution is greater in some

tissues than in others. It is less potent in both skeletal and cardiac muscle and in the brain, while it is powerful in the kidneys, the gut, the spleen, and the skin.

The Vasomotor Center and Its Control of the Vasoconstrictor System — Vasomotor Tone. Located bilaterally in the reticular substance of the lower third of the pons and upper two-thirds of the medulla, as illustrated in Figure 16–5, is an area called the *vasomotor center*. This center transmits impulses downward through the cord and thence through the vasoconstrictor fibers to all the blood vessels of the body.

The vasomotor center is *tonically active*. That is, it has an inherent tendency to transmit nerve impulses all the time, thereby maintaining, even normally, slow firing in essentially all vasoconstrictor nerve fibers of the body at a rate of about one-half to two impulses per second. This continual firing is called *sympathetic vasoconstrictor tone*. These impulses maintain a partial state of contraction of the blood vessels, a state called *vasomotor tone*.

Figure 16–6 demonstrates the significance of vasoconstrictor tone. In the experiment depicted, total spinal anesthesia was administered to an animal, which completely blocked all transmission of nerve impulses from the central nervous system to the periphery. As a result, the arterial pressure fell from 100 to 50 mm. Hg, illustrating the effect of loss of vasoconstrictor tone throughout the body. A few minutes later a small amount of the hormone norepinephrine was injected — norepinephrine is the substance secreted at the

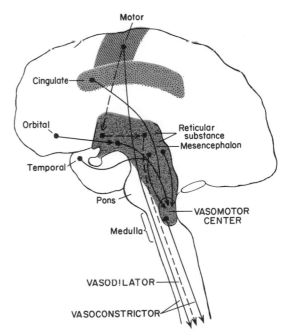

Figure 16–5. Areas of the brain that play important roles in the nervous regulation of the circulation.

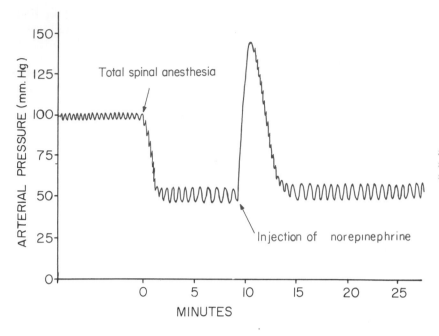

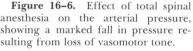

Figure 16–6. Effect of total spinal anesthesia on the arterial pressure, showing a marked fall in pressure resulting from loss of vasomotor tone.

endings of sympathetic nerve fibers throughout the body. As this hormone was transported in the blood to all the blood vessels the vessels once again became constricted, and the arterial pressure rose to a level even greater than normal for a minute or two until the norepinephrine was destroyed.

Control of Heart Activity by the Vasomotor Center. At the same time the vasomotor center is controlling the degree of vascular constriction, it also controls heart activity. The lateral portions of the vasomotor center transmit excitatory impulses through the sympathetic nerve fibers to the heart to increase heart rate and contractility, while the medial portion of the vasomotor center, which lies in immediate apposition to the *dorsal motor nucleus of the vagus nerve*, transmits impulses through the vagus nerve to the heart to decrease heart rate. Therefore, the vasomotor center can either increase or decrease heart activity, this activity ordinarily increasing at the same time that vasoconstriction occurs throughout the body and ordinarily decreasing at the same time that vasoconstriction is inhibited. However, these interrelationships are not invariable, because some nerve impulses that pass down the vagus nerves to the heart can bypass the vasoconstrictor portion of the vasomotor center.

Control of the Vasomotor Center by Higher Nervous Centers. Large numbers of areas throughout the *reticular substance* of the *pons, mesencephalon,* and *diencephalon* can either excite or inhibit the vasomotor center. This reticular substance is illustrated in Figure 16–5 by the diffuse shaded area. In general, the more lateral and superior portions of the reticular substance cause excitation, while the more medial and inferior portions cause inhibition.

The *hypothalamus* plays a special role in the control of the vasoconstrictor system, for it can exert powerful excitatory or inhibitory effects on the vasomotor center. The *posterolateral portions* of the hypothalamus cause mainly excitation, while the *anterior part* can cause either excitation or inhibition, depending on the precise part of the anterior hypothalamus stimulated.

Many different parts of the *cerebral cortex* can also excite or inhibit the vasomotor center, but the effects have not been well characterized.

Norepinephrine — the Sympathetic Vasoconstrictor Transmitter Substance. The substance secreted at the endings of the vasoconstrictor nerves is norepinephrine. Norepinephrine acts directly on the smooth muscle of the vessels to cause vasoconstriction, as will be discussed in Chapter 38.

The Adrenal Medullae and Their Relationship to the Sympathetic Vasoconstrictor System. Sympathetic impulses are transmitted to the adrenal medullae at the same time that they are transmitted to all the blood vessels. These impulses cause the medullae to secrete both epinephrine and norepinephrine into the circulating blood, as will be described in Chapter 38. These two hormones are carried in the blood stream to all parts of the body, where they act directly on the blood vessels, usually to cause vasoconstriction. Sometimes, however, the epinephrine causes vasodilatation, as will be discussed later in the chapter.

Stimulation of the Vasomotor Center — The Mass Action Effect

Stimulation of the lateral portions of the vasomotor center causes widespread activation of the vasoconstrictor fibers throughout the body, while stimulation of the medial portions of the vasomotor center causes widespread *inhibition of vasoconstriction*. In many conditions the entire vasomotor center acts as a unit, stimulating all vasoconstrictors throughout the body and also the heart, as well as stimulating the adrenal medullae to secrete epinephrine and norepinephrine that circulate in the blood to excite the circulation still further. The results of this "mass action" are three-fold: First, the peripheral resistance increases in most parts of the circulation, thereby elevating the arterial pressure. Second, the capacity vessels, particularly the veins, are excited at the same time, greatly decreasing their capacity; this forces increased quantities of blood into the heart, thereby increasing the cardiac output. Third, the heart is simultaneously stimulated so that it can handle the increased cardiac output.

Thus, the overall effect of this "mass action" is to prepare the circulation for increased delivery of blood flow to the body.

REFLEX REGULATION OF THE CIRCULATION

In addition to the many ways in which signals transmitted from the brain to the blood vessels and heart can alter circulatory function, a system of specific circulatory reflexes also helps to regulate the circulation. Most of these reflexes are concerned with the regulation of arterial pressure or with the regulation of blood flow to specific local areas of the body. Therefore, these will be discussed in detail in the following chapter in relation to arterial pressure regulation, as well as at other points in the text. However, let us summarize here a few of the more important reflex regulations of the circulation.

Arterial Pressure Reflexes. The most important of the arterial pressure reflexes is the *baroreceptor reflex*. An increase in arterial pressure stretches the walls of the major arteries in the chest and neck, and this, in turn, excites stretch receptors, the *baroreceptors*. Signals are transmitted to the vasomotor center of the brain stem, and reflex signals are transmitted back to the heart and blood vessels to slow the heart and to dilate the vessels, thereby reducing the arterial pressure toward normal. Thus, the baroreceptor reflex helps to stabilize the arterial pressure.

Reflexes for Control of Blood Volume. When the blood volume increases, the volume of blood in the central veins and in the right and left atria usually also increases. This stretches the walls of the atria and large veins, again exciting stretch receptors. These stretch receptors also transmit signals into the vasomotor center, and reflex signals are then transmitted to the kidneys to increase urinary output of fluid. Also, signals are transmitted to the hypothalamus to decrease the rate of secretion of antidiuretic hormone, thereby further increasing the output of fluid by the kidneys. Thus, this reflex mechanism helps to control the total amount of fluid in the body, and in this way also helps to control blood volume, as will be discussed in Chapter 25.

Reflexes for Control of Body Temperature. When the body temperature rises too high, special neurons in the anterior hypothalamus become excited; these in turn send signals through the sympathetic nervous system to dilate the skin blood vessels, thereby allowing transfer of major amounts of internal body heat to the skin. This heat then passes from the skin to the surroundings. As a result, the body temperature falls back toward normal. This reflex is also associated with sweating reflexes and with other aspects of body temperature control, all of which will be discussed in Chapter 47.

HUMORAL REGULATION OF THE CIRCULATION

The term humoral regulation means regulation by substances in the body fluids — such as hormones, ions, and so forth. Among the most important of these factors are the following:

Vasoconstrictor Agents

Epinephrine and Norepinephrine. Earlier in the chapter it was pointed out that the adrenal medullae secrete both epinephrine and norepinephrine, usually more of the former than of the latter. Norepinephrine has vasoconstrictor effects in almost all vascular beds of the body, and epinephrine has similar effects in most, but not all, beds. For instance, epinephrine often causes mild vasodilatation in both skeletal and cardiac muscle. These two hormones will be discussed in more detail in Chapter 38, in the discussion of the autonomic nervous system.

Angiotensin. Angiotensin is one of the most powerful vasoconstrictor substances known. As little as *one ten-millionth* of a gram can increase the arterial pressure of a human being as much as 10 to 20 mm. Hg under some conditions. Since this substance is very important in relation to arterial pressure regulation, it will be discussed in detail in the following two chapters.

Briefly, either a decrease in arterial pressure or a decrease in quantity of sodium in the body fluids will cause the kidneys to secrete the substance *renin*. The renin in turn acts on one of the plasma proteins, *renin substrate,* to split away the vasoactive peptide angiotensin. The angiotensin in turn has a number of important effects on the circulation related to arterial pressure control: (1) It causes marked constriction of the peripheral arterioles; (2) it causes moderate constriction of the veins, thereby reducing the vascular volume and also probably decreasing vascular compliance; and (3) it causes constriction of the renal arterioles, thereby causing the kidney to retain both water and salt, thus increasing the body fluid volume, which helps to raise the arterial pressure. Hence, an initial decrease in arterial pressure or decrease in sodium causes a compensatory buildup of body fluid and sodium as well as an increase in arterial pressure, thereby compensating for the original deficit.

Vasopressin. Vasopressin, also called *antidiuretic hormone*, is formed in the hypothalamus (see Chapter 49) but is secreted through the posterior pituitary gland. It is even more powerful than angiotensin as a vasoconstrictor, thus making it perhaps the body's most potent constrictor substance. Normally, only very minute amounts of vasopressin are secreted, so that most physiologists have thought that vasopressin plays little role in vascular control. On the other hand, recent experiments have shown that the concentration of circulating vasopressin during severe hemorrhage can rise enough to increase the arterial pressure as much as 40 to 60 mm. Hg, and in many instances this by itself can bring the arterial pressure almost back up to normal.

Also, vasopressin has an *all-important* function of controlling water reabsorption in the renal tubules, which will be discussed in Chapter 25, and therefore of helping to control body fluid volume. That is why this hormone is also called *antidiuretic hormone*.

Vasodilator Agents

Bradykinin. Several substances called *kinins* that can cause vasodilatation have been isolated from blood and tissue fluids. One of these substances is *bradykinin*.

Bradykinin causes very powerful *vasodilatation* and also *increased capillary permeability*. For instance, injection of 1 *microgram* of bradykinin into the brachial artery of a man increases the blood flow through the arm as much as six-fold, and even smaller amounts injected locally into tissues can cause marked edema because of the increase in capillary pore size.

Though unfortunately we know little about the function of the kinins in the control of the circulation, their extremely powerful effects, coupled with the fact that kinins can develop anywhere in the circulatory system or tissues with ease, indicate that they must play important roles. Especially is there reason to believe that kinins play special roles in regulating blood flow in inflamed tissues. It has also been claimed that bradykinin plays a role in regulating skin blood flow and blood flow in gastrointestinal glands.

Histamine. Histamine is released by essentially every tissue of the body whenever tissue becomes damaged. Most of the histamine is probably derived from eosinophils and mast cells in the damaged tissues.

Histamine has a powerful vasodilator effect on the arterioles and also a very potent effect of increasing capillary porosity, allowing leakage of both fluid and plasma protein into the tissues. Though the role of histamine in normal regulation of the circulation is unknown, its actions in causing edema and other effects in allergy were discussed in Chapter 6.

Prostaglandins. Almost every tissue of the body contains small to moderate amounts of several chemically related substances called prostaglandins. These substances are released into the local tissue fluids and into the circulating blood under both physiological and pathological conditions. Though some of the prostaglandins cause vasoconstriction, most of the more important ones seem to be mainly vasodilator agents. Thus far, no specific pattern of function of the prostaglandins in circulatory control has been found. However, their widespread prevalence in the tissues and their myriad effects on the circulation make them ideal candidates for special roles in circulatory control, especially for control in local vascular areas.

Effects of Chemical Factors on Vascular Constriction

Many different chemical factors can either dilate or constrict local blood vessels. Though the roles of these substances in the overall *regulation* of the circulation generally are not known, their specific effects can be listed as follows:

An increase in *calcium ion* concentration causes vasoconstriction. This results from the general effect of calcium of stimulating smooth muscle contraction, as discussed in Chapter 10.

An increase in *potassium ion* concentration causes vasodilatation. This results from an effect of potassium ions of inhibiting vascular contraction.

An increase in *magnesium ion* concentration causes powerful vasodilatation, for magnesium ions inhibit smooth muscle generally.

Increased *sodium ion* concentration causes arteriolar dilatation. This results from an increase in osmolality of the fluids rather than from a specific effect of sodium ion itself. *Increased osmolality* of the blood caused by increased quantities of *glucose* or other nonvasoactive substances also causes arteriolar dilatation.

REFERENCES

Altura, B. M.: Humoral, hormonal, and myogenic mechanisms. *In* Kaley, G., and Altura, B. M. (eds.): Microcirculation. Vol. II. Baltimore, University Park Press, 1977.

Bevan, J. A., *et al.*: Adrenergic regulation of vascular smooth muscle. *In* Bohr, D. F., *et al.* (eds.): Handbook of Physiology. Sec. 2. Vol. 2. Baltimore, Williams & Wilkins, 1980, p. 515.

Bevan, J. A., *et al.* (eds.): Vascular Neuroeffector Mechanisms. New York, Raven Press, 1980.

Bohr, D. F., *et al.*: Mechanisms of action of vasoactive agents. *In* Kaley, G., and Altura, B. M. (eds.): Microcirculation. Vol. II. Baltimore, University Park Press, 1977.

Burnstock, G.: Cholinergic and purinergic regulation of blood vessels. *In* Bohr, D. F., *et al.* (eds.): Handbook of Physiology. Sec. 2. Vol. 2. Baltimore, Williams & Wilkins, 1980, p. 567.

Duling, B.: Oxygen, metabolism, and microcirculatory regulation. *In* Kaley, G., and Altura, B. M. (eds.): Microcirculation. Vol. II. Baltimore, University Park Press, 1977.

Eriksson, E., and Zarem, H. A.: Growth and differentiation of blood vessels. *In* Kaley, G., and Altura, B. M. (eds.): Microcirculation. Vol. I. Baltimore, University Park Press, 1977, p. 393.

Fujii, S., *et al.* (eds.): Kinins II. New York, Plenum Press, 1979.

Galli, C., *et al.* (eds.): Phospholipases and Prostaglandins. New York, Raven Press, 1978.

Granger, H. J., and Guyton, A. C.: Autoregulation of the total systemic circulation following destruction of the central nervous system in the dog. *Circ. Res.*, 25:379, 1969.

Guyton, A. C.: Arterial Pressure and Hypertension. Philadelphia, W. B. Saunders Co., 1980.

Guyton, A. C., *et al.*: Circulation: Overall regulation. *Annu. Rev. Physiol.*, 34:13, 1972.

Guyton, A. C., *et al.*: Cardiac Output and Its Regulation. Philadelphia, W. B. Saunders Co., 1973.

Haddy, F. J.: Local effects of sodium, calcium, and magnesium upon small and large blood vessels of the dog fore limb. *Circ. Res.*, 8:57, 1960.

Johnson, P. C.: The myogenic response. *In* Bohr, D. F., *et al.* (eds.): Handbook of Physiology. Sec. 2, Vol. 2. Baltimore, Williams & Wilkins, 1980, p. 409.

Keatinge, W. R., and Harman, M. C.: Local Mechanisms Controlling Blood Vessels. New York, Academic Press, 1979.

Korner, P. I.: Circulatory adaptations in hypoxia. *Physiol. Rev.*, 39:687, 1959.

Lands, W. E. M.: The biosynthesis and metabolism of prostaglandins. *Annu. Rev. Physiol.*, 41:633, 1979.

Landsberg, L., and Young, J. B.: Catecholamines and the adrenal medulla. *In* Bondy, P. K., and Rosenberg, L. E. (eds.): Metabolic Control and Disease, 8th Ed. Philadelphia, W. B. Saunders Co., 1980, p. 1621.

Needleman, P., and Isakson, P. C.: Intrinsic prostaglandin biosynthesis in blood vessels. *In* Bohr, D. F., *et al.* (eds.): Handbook of Physiology. Sec. 2, Vol. 2. Baltimore, Williams & Wilkins, 1980, p. 613.

Ramwell, P. W., and Leovey, E. M. K.: Prostaglandins and humoral regulation. *In* DeGroot, L. J., *et al.* (eds.): Endocrinology. Vol. 3. New York, Grune & Stratton, 1979, p. 1711.

Renkin, E. M.: Nutritive and shunt flow. *In* Kaley, G., and Altura, B. M. (eds.): Microcirculation. Vol. 2. Baltimore, University Park Press, 1977.

Rosenthal, S. L., and Guyton, A. C.: Hemodynamics of collateral vasodilatation following femoral artery occlusion in anesthetized dogs. *Circ. Res.*, 23:239, 1968.

Scott, J. B., *et al.*: Role of osmolarity, K^+, H^+, Mg^{++}, and O_2 in local blood flow regulation. *Am. J. Physiol.*, 218:338, 1970.

Shepherd, A. P., *et al.*: Local control of tissue oxygen delivery and its contribution to the regulation of cardiac output. *Am. J. Physiol.*, 225:747, 1973.

Sparks, H. V., Jr.: Effect of local metabolic factors on vascular smooth muscle. *In* Bohr, D. F., *et al.* (eds.): Handbook of Physiology. Sec. 2, Vol. 2. Baltimore, Williams & Wilkins, 1980, p. 475.

Sparks, H. V., Jr., and Belloni, F. L.: The peripheral circulation: Local regulation. *Annu. Rev. Physiol.*, 40:67, 1978.

Vane, J. R., and Berstrom, S. (eds.): Prostacyclin. New York, Raven Press, 1979.

Vanhoutte, P. M.: Physical factors of regulation. *In* Bohr, D. F., *et al.* (eds.): Handbook of Physiology. Sec. 2, Vol. 2. Baltimore, Williams & Wilkins, 1980, p. 443.

Walker, J. R., and Guyton, A. C.: Influence of blood oxygen saturation on pressure-flow curve of dog hindleg. *Am. J. Physiol.*, 212:506, 1967.

Wolff, J. R.: Ultrastructure of the terminal vascular bed as related to function. *In* Kaley, G., and Altura, B. M. (eds.): Microcirculation. Vol. I. Baltimore, University Park Press, 1977, p. 95.

17

Short-Term Regulation of Mean Arterial Pressure: Nervous Reflex and Hormonal Mechanisms for Rapid Pressure Control

In the previous chapter we pointed out that each tissue can control its own blood flow by simply dilating or constricting its local arterioles. For this mechanism to work, it is necessary that the arterial pressure remain constant or nearly constant, because with a variable arterial pressure one would never know whether dilating the blood vessels would necessarily increase the local blood flow. Fortunately, the circulation has an intricate system for regulation of the arterial pressure. It maintains the normal mean arterial pressure in young adults within rather narrow limits, between 90 mm. Hg and 110 mm. Hg. Some of the pressure regulatory mechanisms (mainly nervous and hormonal mechanisms) act very rapidly, and some (mainly mechanisms related to kidney function and blood volume regulation) act very slowly. In the present chapter we will discuss the rapid nervous and hormonal pressure control mechanisms. In the following chapter we will discuss both the long-term regulation of arterial pressure, based primarily on renal and body-fluid mechanisms, and the clinical problem of hypertension or "high blood pressure," which is caused by abnormalities of the long-term pressure regulatory mechanisms.

But first, let us discuss some of the normal arterial pressure values in a human being and the clinical method for measuring these pressures.

NORMAL ARTERIAL PRESSURES

Arterial Pressures at Different Ages. Figure 17–1 illustrates the typical diastolic, systolic, and mean arterial pressures from birth to 80 years of age. From this figure it can be seen that the systolic pressure of a normal young adult averages about 120 mm. Hg and the diastolic pressure about 80 mm. Hg — that is, arterial pressure is said to be 120/80. The shaded areas on either side of the curves depict the normal ranges of systolic and diastolic pressures, showing considerable variation from person to person.

The increase in arterial pressure at older ages is usually associated with developing arteriosclerosis. In this disease the systolic pressure especially increases; in approximately one-tenth of all old people it eventually rises above 200 mm. Hg.

THE MEAN ARTERIAL PRESSURE

The mean arterial pressure is the average pressure throughout each cycle of the heartbeat. Offhand one might expect that it would be equal to the average of systolic and diastolic pressures, but this is not true; the arterial pressure usually remains nearer to diastolic level than to systolic

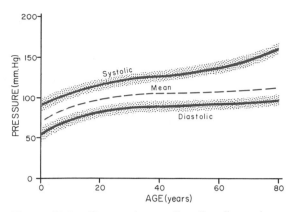

Figure 17–1. Changes in systolic, diastolic, and mean arterial pressures with age. The shaded areas show the normal range.

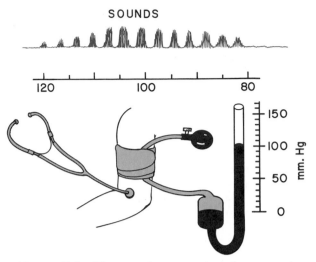

Figure 17–2. The auscultatory method for measuring systolic and diastolic pressures.

level during a greater portion of the pulse cycle, which can be seen in all of the pictures of pressure pulses shown in Chapter 15. Therefore, the mean arterial pressure is usually slightly less than the average of systolic and diastolic pressures, as is evident in Figure 17–1.

The mean arterial pressure of the normal young adult averages about 96 mm. Hg, which is slightly less than the average of the systolic and diastolic pressures, 120 and 80 mm. Hg, respectively. However, for purposes of discussion, the mean arterial pressure is usually considered to be 100 mm. Hg because this value is easy to remember.

Significance of Mean Arterial Pressure. Mean arterial pressure is the average pressure tending to push blood through the systemic circulatory system. Therefore, *from the point of view of tissue blood flow, it is generally the mean arterial pressure that is important.*

CLINICAL METHODS FOR MEASURING SYSTOLIC AND DIASTOLIC PRESSURES

Obviously, it is impossible to use the various pressure recorders that require needle insertion into an artery, as described in Chapter 14, for making routine pressure measurements in human patients, although they are used on occasion when special studies are necessary. Instead, the clinician determines systolic and diastolic pressures by indirect means, most usually by the auscultatory method.

The Auscultatory Method. Figure 17–2 illustrates the auscultatory method for determining systolic and diastolic arterial pressures. A stethoscope is placed over the antecubital artery while a blood pressure cuff is inflated around the upper arm. As long as the cuff presses against the arm with so little pressure that the artery remains

distended with blood, no sounds whatsoever are heard by the stethoscope despite the fact that the blood pressure within the artery is pulsating. But when the cuff pressure is great enough to close the artery completely during part of the arterial pressure cycle, a sound is heard in the stethoscope with each pulsation. These sounds are called *Korotkoff sounds,* and they are caused by blood jetting through the vessel just as the vessel opens during each pulse cycle. The jet causes turbulence in the flaccid vessel beyond the cuff, and this sets up the vibrations heard through the stethoscope.

In determining blood pressure by the auscultatory method, the pressure in the cuff is first elevated well above arterial systolic pressure. As long as this pressure is higher than systolic pressure, the brachial artery remains collapsed and no blood whatsoever flows into the lower artery during any part of the pressure cycle. Therefore, no Korotkoff sounds are heard in the lower artery. But then the cuff pressure is gradually reduced. Just as soon as the pressure in the cuff falls below systolic pressure, blood slips through the artery beneath the cuff during the peak of systolic pressure, and one begins to hear *tapping* sounds in the antecubital artery in synchrony with the heart beat. As soon as these sounds are heard, the pressure level indicated by the manometer connected to the cuff is approximately equal to the systolic pressure.

As the pressure in the cuff is lowered still more, the Korotkoff sounds change in quality, having less of the tapping quality but more of a rhythmic, harsh quality. Then, finally, when the pressure in the cuff falls to equal diastolic pressure, the artery no longer closes during diastole, which means that the basic factor causing the sounds (the jetting of blood through a squeezed artery) is no longer present. Therefore, the sounds suddenly change to a muffled quality and then disappear. The manometer pressure at this point is approximately equal to the diastolic pressure.

RELATIONSHIP OF ARTERIAL PRESSURE TO CARDIAC OUTPUT AND TOTAL PERIPHERAL RESISTANCE

Before discussing the overall regulation of arterial pressure, it is good to remember the basic relationship between arterial pressure, cardiac output, and total peripheral resistance, which was discussed in detail in Chapter 14, as follows:

$$\text{Arterial Pressure} = \text{Cardiac Output} \\ \times \text{Total Peripheral Resistance}$$

It is obvious from this formula that any condition that increases either the cardiac output or the total peripheral resistance (if the other factor does not change) will cause an increase in mean arterial pressure. Both of these factors are often manipulated in the control of arterial pressure, as we shall see in the remainder of this chapter.

THE OVERALL SYSTEM FOR ARTERIAL PRESSURE REGULATION

Arterial pressure is regulated not by a single pressure controlling system but by several interrelated systems that perform specific functions. When a person bleeds so severely that the pressure falls suddenly, two problems immediately confront the pressure control system. The first is to return the arterial pressure immediately to a high enough level that the person can live through the acute hemorrhagic episode. The second is to return the blood volume eventually to its normal level so that the circulatory system can re-establish full normality, including return of the arterial pressure all the way back to its normal value. These two problems characterize two major types of arterial pressure control systems in the body: (1) a system of rapidly acting pressure control mechanism and (2) a system for long-term control of the basic arterial pressure level.

Rapidly Acting Pressure Control Mechanisms. Several different pressure control mechanisms, all of which are nervous feedback mechanisms, begin to react within seconds. These include the baroreceptor feedback mechanism and

the central nervous system ischemic mechanism. Thus, the first line of defense against abnormal pressures is subserved by the nervous mechanisms for control of arterial pressure.

Within minutes several other pressure control mechanisms also come into play. Two of these are the renin-angiotensin-vasoconstrictor mechanism and the shift of fluid through the capillaries from the tissues into or out of the circulation to readjust the blood volume as needed. These two mechanisms become fully active within 30 minutes to several hours, in contrast to the nervous mechanisms that usually become fully active within a minute or so.

Long-Term Mechanisms for Arterial Pressure Regulation. The nervous regulators of arterial pressure, though acting very rapidly and powerfully to correct acute abnormalities of arterial pressure, generally lose their power to control arterial pressure after a few hours to a few days because the nervous pressure receptors "adapt"; that is, they lose their responsiveness. Therefore, except under unusual circumstances, the nervous mechanisms for arterial pressure control do not play a major role in long-term regulation of arterial pressure. Long-term regulation, instead, is vested mainly in a renal–body fluid–pressure control mechanism. As we shall discuss in the following chapter, this mechanism involves control of blood volume with its consequent effects on arterial pressure, and part of this mechanism involves control of kidney function by several different hormonal systems, including especially the renin-angiotensin system and the hormone aldosterone secreted by the adrenal cortex.

THE RAPIDLY ACTING NERVOUS MECHANISMS FOR ARTERIAL PRESSURE CONTROL

The Arterial Baroreceptor Control System — Baroreceptor Reflexes

By far the best known of the mechanisms for arterial pressure control is the *baroreceptor reflex*. Basically, this reflex is initiated by stretch receptors, called either *baroreceptors* or *pressoreceptors*, located in the walls of the large systemic arteries. A rise in pressure stretches the baroreceptors and causes them to transmit signals into the central nervous system, and other signals are in turn sent to the circulation to reduce arterial pressure back toward the normal level.

Physiologic Anatomy of the Baroreceptors. Baroreceptors are spray-type nerve endings lying in the walls of the arteries; they are stimulated when stretched. A few baroreceptors are located in the wall of almost every large artery of the thoracic and neck regions; but, as illustrated in Figure 17–3, baroreceptors are extremely abundant in (1) the walls of the internal carotid arteries slightly above the carotid bifurcations, areas known as the *carotid sinuses,* and (2) the wall of the aortic arch.

Figure 17–3 also shows that impulses are transmitted from each carotid sinus through the very small *Hering's nerve* to the glossopharyngeal nerve and thence to the medullary area of the brain stem. Impulses from the arch of the aorta are transmitted through the vagus nerves also to the medulla.

The Reflex Initiated by the Baroreceptors. The baroreceptors are not stimulated at all by pressures between 0 and 60 mm, Hg, but above 60 mm. Hg they respond progressively more and more and reach a maximum at 180 to 200 mm. Hg.

The baroreceptor impulses *inhibit the vasoconstrictor center* of the medulla and *excite the vagal center.* The net effects are (1) *vasodilatation* throughout the peripheral circulatory system and (2) *decreased heart rate* and *strength of contraction.* Therefore, excitation of the baroreceptors by pressure in the arteries reflexly *causes the arterial pressure to decrease.* Conversely, low pressure has opposite effects, reflexly causing the pressure to rise back toward normal.

Figure 17–4 illustrates a typical reflex change in arterial pressure caused by clamping the common carotids. This procedure reduces the carotid sinus pressure; as a result, the baroreceptors become inactive and lose their inhibitory effect on the vasomotor center. The vasomotor center then becomes much more active than usual, causing the arterial pressure to rise and to remain

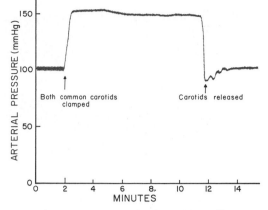

Figure 17–4. Typical carotid sinus reflex effect on arterial pressure caused by clamping both common carotids.

elevated during the ten minutes that the carotids are clamped. Removal of the clamps allows the pressure to fall immediately to slightly below normal as a momentary overcompensation and then to return to normal in another minute or so.

Function of the Baroreceptors During Changes in Body Posture. The ability of the baroreceptors to maintain relatively constant arterial pressure is extremely important when a person sits or stands after having been lying down. Immediately upon standing, the arterial pressure in the head and upper part of the body obviously tends to fall, and marked reduction of this pressure can cause loss of consciousness. Fortunately, however, the falling pressure at the baroreceptors elicits an immediate reflex, resulting in strong sympathetic discharge throughout the body, and this minimizes the decrease in pressure in the head and upper body.

The "Buffer" Function of the Baroreceptor Control System. Because the baroreceptor system opposes increases and decreases in arterial pressure, it is often called a *pressure buffer system,* and the nerves from the baroreceptors are called *buffer nerves.*

Figure 17–5 illustrates the importance of this buffer function of the baroreceptors. The upper record in this figure shows an arterial pressure recording for two hours from a normal dog, and the lower record shows one from a dog whose baroreceptor nerves from both the carotid sinuses and the aorta had previously been removed. Note the extreme variability of pressure in the denervated dogs caused by simple events of the day such as lying down, standing, excitement, eating, defecation, noises, and so forth.

In summary, we can state that the primary purpose of the arterial baroreceptor system is to reduce the daily variation in arterial pressure to about one-half to one-third that which would occur were the baroreceptor system not present.

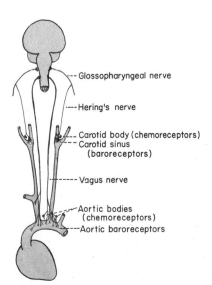

Figure 17–3. The baroreceptor system.

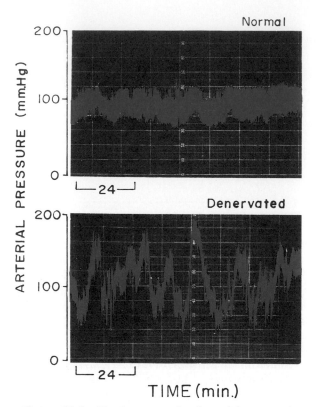

Figure 17–5. Two-hour records of arterial pressure in a normal dog (above) and in the same dog (below) several weeks after the baroreceptors had been denervated. (Courtesy of Dr. Allen W. Cowley, Jr.)

Unimportance of the Baroreceptor System for Long-Term Regulation of Arterial Pressure — Adaptation of the Baroreceptors. The baroreceptor control system is of no importance in long-term regulation of arterial pressure for a very simple reason: The baroreceptors themselves adapt in one to three days to whatever pressure level they are exposed to. That is, if the pressure rises from the normal value of 100 mm. Hg up to 200 mm. Hg, extreme numbers of baroreceptor impulses are at first transmitted. During the next few seconds, the rate of firing diminishes considerably; then it diminishes more slowly during the next one to two days, at the end of which time the rate returns to the normal level despite the fact that the arterial pressure remains 200 mm. Hg. Conversely, when the arterial pressure falls to a very low value, the baroreceptors at first transmit no impulses at all, but gradually over a period of several days the rate of baroreceptor firing returns again to the original control level.

This adaptation of the baroreceptors obviously prevents the baroreceptor reflex from functioning as a control system to buffer arterial pressure changes that last longer than a few days at a time. This was evident from Figure 17–5. Note that the average pressure was the same in the normal dog

as in the baroreceptor-denervated dog. Therefore, prolonged regulation of arterial pressure requires other control systems, principally the renal–body fluid–pressure control system (along with its associated hormonal mechanisms) to be discussed in the following chapter.

Control of Arterial Pressure by the Vasomotor Center in Response to Diminished Brain Blood Flow — The CNS Ischemic Response

Normally, most nervous control of blood pressure is achieved by reflexes originating in the baroreceptors and other allied receptors. However, when blood flow to the vasomotor center in the lower brain stem becomes decreased enough to cause nutritional deficiency, a condition called *ischemia,* the neurons in the vasomotor center itself respond directly to the ischemia and become strongly excited. When this occurs, the systemic arterial pressure often rises to a very high level. This effect is believed to be caused by failure of the slowly flowing blood to carry carbon dioxide away from the vasomotor center; the local concentration of carbon dioxide increases greatly and has an extremely potent effect in stimulating the sympathetic nervous system. This arterial pressure elevation in response to cerebral ischemia is known as the *central nervous system ischemic response* or simply *CNS ischemic response.*

The magnitude of the ischemic effect on vasomotor activity is tremendous; it can elevate the mean arterial pressure for as long as ten minutes, sometimes to above 200 mm. Hg. *The degree of sympathetic vasoconstriction caused by intense cerebral ischemia is often so great that some of the peripheral vessels become totally or almost totally occluded.* The kidneys, for instance, will entirely cease their production of urine because of arteriolar constriction in response to the sympathetic discharge. Therefore, *the CNS ischemic response is one of the most powerful of all the activators of the sympathetic vasoconstrictor system.*

Importance of the CNS Ischemic Response as a Regulator of Arterial Pressure. Despite the extremely powerful nature of the CNS ischemic response, it does not become very active until the arterial pressure falls far below normal, down to *levels of 50 mm. Hg* and below, reaching its greatest degree of stimulation at a pressure of 15 to 20 mm. Hg. Therefore it is not one of the mechanisms for regulating normal arterial pressure. Instead it operates principally as an *emergency arterial pressure control system that acts rapidly and extremely powerfully to prevent further decrease in arterial pressure whenever blood flow to the brain decreases dangerously close to the lethal level.* It is sometimes called the "last ditch stand" pressure control mechanism.

The Cushing Reaction. The so-called Cushing reaction is a special type of CNS ischemic response that results from increased pressure in the cranial vault. For instance, when the cerebrospinal fluid pressure rises to equal the arterial pressure, it compresses the arteries in the brain and cuts off the blood supply to the brain. Obviously, this initiates a CNS ischemic response, which causes the arterial pressure to rise. When the arterial pressure has risen to a level higher than the cerebrospinal fluid pressure, blood flows once again into the vessels of the brain to relieve the ischemia.

Obviously, the Cushing reaction helps to protect the vital centers of the brain from loss of nutrition if ever the pressure in the skull rises high enough to compress the cerebral arteries.

PARTICIPATION OF THE VEINS IN NERVOUS REGULATION OF CARDIAC OUTPUT AND ARTERIAL PRESSURE

Thus far, we have discussed primarily the ability of the nervous system to regulate arterial pressure by altering arterial resistance. However, much of the regulatory effect of the nervous system is carried out through sympathetic vasoconstrictor fibers to the veins. Indeed, the veins constrict in response to even weaker sympathetic stimuli than do the arterioles and arteries.

Sympathetic Alterations in Venous Capacity and Cardiac Output. The veins offer relatively little resistance to blood flow in comparison with the arterioles and arteries. Therefore, sympathetic constriction of the veins does not significantly change the overall total peripheral resistance. Instead, the important effect of sympathetic stimulation of the veins is a *decrease in the capacity*. This means that the veins then hold less blood at any given venous pressure, which causes translocation of blood out of the systemic veins into the heart. The distension of the heart in turn causes the heart to pump with increasing effectiveness in accordance with the Frank-Starling law of the heart, as discussed in Chapter 11. Therefore, the net effect of sympathetic stimulation of the veins is to increase the cardiac output, which in turn elevates the arterial pressure.

The veins participate in all of the reflexes and reactions that have been discussed thus far, including both the baroreceptor reflex and the CNS ischemic response. To comprehend the potency of the venous reaction in some of these nervous mechanisms, consider the following: The mean circulatory filling pressure (mainly determined by the degree of contraction of the veins) rises from the normal value of 7 mm. Hg to approximately 10 mm. Hg when the barorecep-

tor reflex is excited fully, and it rises to approximately 18 mm. Hg — high enough to increase the pressure forcing blood into the heart by as much as 2.5-fold — when a maximal CNS ischemic response is elicited.

Atrial Reflexes for Control of Heart Rate (the Bainbridge Reflex)

An increase in atrial pressure also causes an increase in heart rate, sometimes by as much as 75 per cent. Part of this increase is caused by the direct stretching of the S-A node by the increased atrial volume, an effect that can increase the heart rate as much as 15 per cent. An additional 30 to 60 per cent increase in rate is caused by a reflex called the *Bainbridge reflex*. The stretch receptors of the atria that elicit the Bainbridge reflex transmit their afferent signals through the vagus nerves to the medulla of the brain. Then, efferent signals are transmitted back through both the vagal and the sympathetic nerves to increase the heart's rate and strength of contraction. Thus, this reflex helps to prevent damming of blood in the veins, the atria, and the pulmonary circulation. This reflex obviously has a different purpose from that of controlling arterial pressure.

HORMONAL MECHANISMS FOR RAPID CONTROL OF ARTERIAL PRESSURE

In addition to the rapidly acting nervous mechanisms for control of arterial pressure, there are two major hormonal mechanisms that also provide either rapid or moderately rapid control of arterial pressure. They are:

1. The norepinephrine-epinephrine vasoconstrictor mechanism.
2. The renin-angiotensin vasoconstrictor mechanism.

THE NOREPINEPHRINE-EPINEPHRINE VASOCONSTRICTOR MECHANISM

In the previous chapter it was pointed out that stimulation of the sympathetic nervous system not only causes direct nervous excitation of the blood vessels and heart but also causes release by the adrenal medullae of norepinephrine and epinephrine into the circulating blood. These two hormones circulate to all parts of the body and cause essentially the same effects on the circulatory system as direct sympathetic stimulation. That is, they excite the heart, they constrict most of the blood vessels, and they constrict the veins.

Therefore, the different reflexes that regulate arterial pressure by exciting the sympathetic nervous system cause the pressure to rise in two ways: by direct circulatory stimulation and by indirect stimulation through the release of norepinephrine and epinephrine into the blood.

Norepinephrine and epinephrine circulate in the blood for one to three minutes before being destroyed, thus maintaining a slightly prolonged excitation of the circulation. Also, these hormones can reach some parts of the circulation that have no sympathetic nervous supply at all, including some of the very minute vessels such as the metarterioles. And these hormones have especially potent actions on some vascular beds, particularly the skin vasculature.

THE RENIN-ANGIOTENSIN VASOCONSTRICTOR MECHANISM FOR CONTROL OF ARTERIAL PRESSURE

The hormone *angiotensin II* is one of the most potent vasoconstrictors known. Whenever the arterial pressure falls very low, large quantities of angiotensin II appear in the circulation. This results from a special mechanism involving the release of the enzyme *renin* from the kidneys when the arterial pressure falls too low.

The overall schema for formation of angiotensin and the effect of angiotensin II of increasing arterial pressure are illustrated in Figure 17–6. When blood flow through the kidneys is decreased, the *juxtaglomerular cells* (cells located in the walls of the afferent arterioles immediately proximal to the glomeruli) secrete renin into the blood. Renin itself is an enzyme that splits the end

off one of the plasma proteins, called *renin substrate,* to release a decapeptide, *angiotensin I.* The renin persists in the blood for 30 minutes to more than an hour and continues to cause formation of angiotensin I during the entire time. Within a few seconds after formation of the angiotensin I, two additional amino acids are split from it to form the octapeptide *angiotensin II.* This conversion occurs almost entirely in the small vessels of the lungs, catalyzed by an enzyme called *converting enzyme.* Angiotensin II persists in the blood for a minute or so but is rapidly inactivated by a number of different blood and tissue enzymes collectively called *angiotensinase.*

During its persistence in the blood, angiotensin II has several effects that can elevate arterial pressure. One of these effects occurs very rapidly: vasoconstriction, especially of the arterioles and to a lesser extent of the veins at the same time. Constriction of the arterioles increases the peripheral resistance and thereby raises the arterial pressure back toward normal, as illustrated at the bottom of the schema in Figure 17–6. Also, mild constriction of the veins increases the mean circulatory filling pressure, sometimes as much as 20 per cent, and this promotes increased tendency for venous return of blood to the heart, thereby helping the heart to pump against the extra pressure load.

The other effects of angiotensin are mainly related to the body fluid volumes: (1) Angiotensin has a direct effect on the kidneys, causing decreased excretion of both salt and water; and (2) angiotensin stimulates the secretion of aldosterone by the adrenal cortex, and this hormone in turn also acts on the kidneys to cause decreased excretion of both salt and water. Both these effects tend to elevate the blood volume — an important factor in long-term regulation of arterial pressure, as will be discussed in the following chapter.

Rapidity of Action and Pressure-Controlling Power of the Renin-Angiotensin Vasoconstrictor System. Figure 17–7 illustrates a typical experiment showing the effect of hemorrhage on the arterial pressure under two separate conditions: (1) with the renin-angiotensin system functioning and (2) without the system functioning (the system was interrupted by a renin-blocking antibody). Note that following hemorrhage, which caused an acute fall of the arterial pressure to 50 mm. Hg, the arterial pressure rose back to 83 mm. Hg when the renin-angiotensin system was functional. On the other hand, it rose to only 60 mm. Hg when the renin-angiotensin system was blocked. This illustrates that the renin-angiotensin system is powerful enough to return the arterial pressure at least halfway back to normal following severe hemorrhage. Therefore,

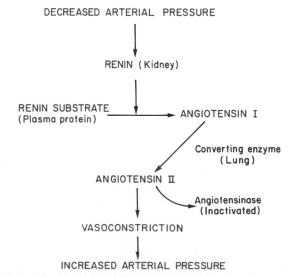

Figure 17–16. The renin-angiotensin-vasoconstrictor mechanism for arterial pressure control.

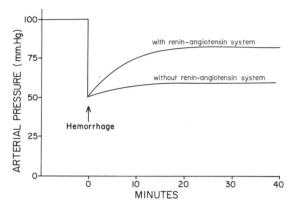

Figure 17–7. Pressure compensating effect of the renin-angiotensin-vasoconstrictor system following severe hemorrhage. (Data from experiments by Dr. Royce Brough.)

it can sometimes be of life-saving service to the body, especially in circulatory shock.

Note that the renin-angiotensin vasoconstrictor system requires approximately 20 minutes to become fully active. Therefore, it is far slower to act than are the nervous reflexes and the norepinephrine-epinephrine system; however, it also has a correspondingly longer duration of action.

There is reason to believe that the renin-angiotensin vasoconstrictor system is much more powerful under some conditions than others. For instance, some patients with diseased kidneys secrete tremendous quantities of renin, and the pressure control action of the system is then likely to be very powerful.

THE CAPILLARY FLUID SHIFT MECHANISM FOR ARTERIAL PRESSURE REGULATION

In addition to the nervous and hormonal mechanisms for rapid control of arterial pressure, an intrinsic physical mechanism of the circulation also helps to control arterial pressure, usually beginning to act within a few minutes and reaching full function within a few hours. It is the capillary fluid shift mechanism, which is the following: change in the arterial pressure is usually also associated with a similar change in capillary pressure, which causes fluid to begin moving across the capillary membrane between the blood and the interstitial fluid compartment. For instance, if the arterial pressure rises too high, loss of fluid through the capillaries into the interstitial spaces causes the blood volume to fall and thereby causes return of arterial pressure back toward normal. Conversely, when the pressure falls too low, fluid is absorbed into the blood, and the

increasing volume plays a very important role in raising the pressure back to normal.

REFERENCES

Brown, A. M.: Cardiac reflexes. *In* Berne, R. M., *et al.* (eds.): Handbook of Physiology, Sec. 2, Vol. 1. Baltimore, Williams & Wilkins, 1979, p. 677.

Coleman, T. G., *et al.*: Angiotensin and the hemodynamics of chronic salt deprivation. *Am. J. Physiol., 229*:167, 1975.

Coleridge, J. C. G., and Coleridge, H. M.: Chemoreflex regulation of the heart. *In* Berne, R. M., *et al.* (eds.): Handbook of Physiology. Sec. 2, Vol. 1. Baltimore, Williams & Wilkins, 1979, p. 653.

Cooper, M. D., and Lawton, A. R., III: The Physiology of the giraffe. *Sci. Am., 231*(5):96, 1974.

Cowley, A. W., Jr., and Guyton, A. C.: Baroreceptor reflex contribution in angiotensin-II-induced hypertension. *Circulation, 50*:61, 1974.

Cushing, H.: Concerning a definite regulatory mechanism of the vasomotor center which controls blood pressure during cerebral compression. *Bull. Johns Hopkins Hosp., 12*:290, 1901.

Downing, S. E.: Baroreceptor regulation of the heart. *In* Berne, R. M., *et al.* (eds.): Handbook of Physiology. Sec. 2, Vol. 1. Baltimore, Williams & Wilkins, 1979, p. 621.

Goetz, K. L., *et al.*: Atrial receptors and renal function. *Physiol. Rev., 55*:157, 1975.

Guyton, A. C.: Acute hypertension in dogs with cerebral ischemia. *Am. J. Physiol., 154*:45, 1948.

Guyton, A. C.: Arterial Pressure and Hypertension. Philadelphia, W. B. Saunders Co., 1980.

Guyton, A. C., and Satterfield, J. H.: Vasomotor waves possibly resulting from CNS ischemic reflex oscillation. *Am. J. Physiol., 170*:601, 1952.

Hainsworth, R., and Linden, R. J.: Reflex control of vascular capacitance. *In* International Review of Physiology: Cardiovascular Physiology III. Vol. 18. Baltimore, University Park Press, 1979, p. 67.

Hall, J. W., and Guyton, A. C.: Changes in renal hemodynamics and renin release caused by increased plasma oncotic pressure. *Am. J. Physiol., 231*:1550, 1976.

Kaley, G.: Microcirculatory-endocrine interactions. *In* Kaley, G., and Altura, B. M. (eds.): Microcirculation. Vol. II. Baltimore, University Park Press, 1977.

Kirchheim, H. R.: Systemic arterial baroreceptor reflexes. *Physiol. Rev., 56*:100, 1976.

Mancia, G., *et al.*: Reflex control of circulation by heart and lungs. *Int. Rev. Physiol., 9*:111, 1976.

Peach, M. J.: Renin-angiotensin system: Biochemistry and mechanisms of action. *Physiol. Rev., 57*:313, 1977.

Reid, I. A., *et al.*: The renin-angiotensin system. *Annu. Rev. Physiol., 40*:377, 1978.

Rosell, S.: Nervous control of microcirculation. *In* Kaley, G., and Altura, B. M. (eds.): Microcirculation. Vol. II. Baltimore, University Park Press, 1977.

Sagawa, K., *et al.*: Nervous control of the circulation. *In* MTP International Review of Science: Physiology. Baltimore, University Park Press, 1974. Vol. 1, p. 197.

Smith, E. E., and Guyton, A. C.: Center of arterial pressure regulation during rotation of normal and abnormal dogs. *Am. J. Physiol., 204*:979, 1963.

Smith, O. A.: Reflex and central mechanisms involved in the control of the heart and circulation. *Annu. Rev. Physiol., 36*:93, 1974.

Youmans, W. B., *et al.*: The Abdominal Compression Reaction. Baltimore, Williams & Wilkins, 1963.

Ziegler, M. G.: Postural hypotension. *Annu. Rev. Med., 31*:239, 1980.

18

Long-Term Regulation of Mean Arterial Pressure; and Hypertension

The mechanisms for long-term arterial pressure control are considerably different from the short-term mechanisms of control discussed in the previous chapter. The effectiveness of most short-term pressure control mechanisms — the baroreceptor mechanism, for instance — diminishes drastically as time proceeds. On the other hand, the effectiveness of at least one of the long-term pressure control systems is almost zero for the first few hours but then becomes extreme over a period of days and weeks. This is the renal–body fluid pressure control system. Because of this extreme long-term potency, this system plays a central role in long-term pressure control. However, it is aided in this role by a large number of accessory mechanisms, including special effects of the renin-angiotensin system, of the nervous system, and of the aldosterone system.

THE RENAL–BODY FLUID SYSTEM FOR ARTERIAL PRESSURE CONTROL

The Phenomenon of Pressure Diuresis and Pressure Natriuresis As a Basis for Arterial Pressure Control. From the very earliest studies of the kidneys it immediately became evident that increased arterial pressure increases greatly the rate at which the kidneys excrete both water and salt, effects called *pressure diuresis* and *pressure natriuresis*. Or, to state this another way, an increase in pressure causes marked loss of extracellular fluid volume from the body, and this decreases the blood volume and therefore also the arterial pressure. Thus, elevated arterial pressure causes pressure diuresis and natriuresis,

which decrease the arterial pressure back to normal. Conversely, when the arterial pressure falls too low, the kidneys retain fluid, the blood volume increases, and the arterial pressure returns again to normal.

Figure 18–1 illustrates quantitatively this pressure diuresis and natriuresis mechanism, showing the normal *renal output curve* for an experimentally perfused isolated kidney. It shows that at a blood pressure of 50 to 60 mm. Hg the urinary output of water and salt is essentially zero. At 100 mm. Hg it is normal, and at 200 mm. Hg, about six to eight times normal. Thus, it is already clear that this mechanism is not a weak one. Later in the chapter we will show that the renal output curve becomes far steeper still when one considers the accessory control mechanisms that help to make the kidney control of body fluid even more effective.

The Minuteness of the Fluid Volume Changes Required to Cause Marked Changes in Pressure. One feature of the renal–body fluid mechanism for arterial pressure control that is most often misunderstood is how small the changes in quantities of fluid in the body have to be to cause marked changes in pressure. To give an example, a 2 per cent *chronic* increase in blood volume can increase the venous return and cardiac output as much as 5 per cent. And an increase in cardiac output of 5 per cent can, because of the vasoconstriction caused by the excess tissue flow, increase the total peripheral resistance 25 to 50 per cent; this figure multiplied by the 5 per cent increase in cardiac output gives an increase in the arterial pressure of as much as 30 to 57 per cent.

Thus, one can understand very easily that a chronic increase of only a few hundred milliliters of extracellular fluid can lead to a hypertensive

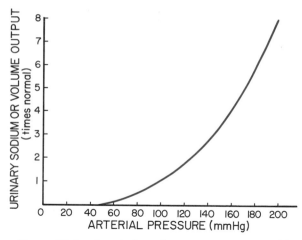

Figure 18–1. A typical renal output curve measured in a perfused isolated kidney, showing both pressure diuresis (excess output of water) and pressure natriuresis (excess output of sodium) when the arterial pressure rises above normal.

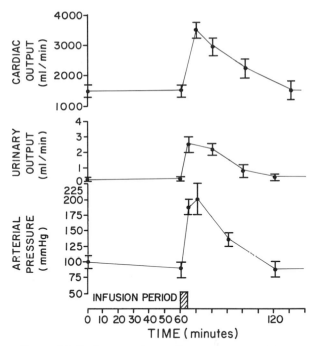

Figure 18–2. Increase in cardiac output, arterial pressure, and urinary output caused by increased blood volume in animals whose nervous pressure control mechanisms had been blocked. This figure shows the return of arterial pressure to normal after about an hour of fluid loss into the urine. (Courtesy of Dr. William Dobbs.)

state. Indeed, in patients with hypertension, the arterial pressure can often be reduced back to normal by administering a natriuretic (a drug that causes excess water and sodium excretion) that does nothing more than reduce the extracellular fluid volume by about 500 ml.

One reason it has been difficult in the past to understand the minuteness of the fluid volumes involved is that large volumes of fluid can be infused into a person *acutely* without causing hypertension. However, it must be remembered that for the first few minutes to several hours after such an infusion the nervous reflex control mechanism and other short-term pressure control mechanisms prevent a significant pressure rise. It is only after the short-term control mechanisms lose their effectiveness that the volume increases the pressure. This requires several days.

An Experiment Demonstrating the Renal–Body Fluid System for Arterial Pressure Control. Figure 18–2 illustrates an experiment in dogs in which all the nervous reflex mechanisms for blood pressure control were blocked and the arterial pressure was then suddenly elevated by infusing 300 ml. of blood. Note the instantaneous increase in cardiac output to approximately double its normal level and the increase in arterial pressure to 135 mm. Hg above its resting level. Also shown, by the middle curve, is the effect of this increased arterial pressure on urinary output. The output increased 12-fold, and both the cardiac output and the arterial pressure returned back to normal during the subsequent hour. Thus, one sees the extreme capability of the kidneys to readjust the blood volume and in so doing to return the arterial pressure back to normal.

Infinite Gain of the Renal–Body Fluid Pressure Control System. A special feature of this mechanism for pressure control is its ability to return the arterial pressure *all the way* back to normal — not merely a certain proportion of the way back. This is quite different from the nervous reflex mechanisms that were discussed in the previous chapter. For instance, the baroreceptor and other nervous mechanisms for arterial pressure control, all working together, are capable of returning the arterial pressure approximately seven-eighths of the way back toward normal following a blood volume change, but never all the way. The effectiveness of a feedback control system is calculated by dividing the total amount of compensation caused by the feedback by the remaining error that fails to be compensated. The resulting value is called the "feedback gain." In the case of the nervous pressure control system, the compensation is seven-eighths, and the remaining error is one-eighth. Therefore, the gain is seven. In the case of the renal–body fluid system, the compensation is *100 per cent of the way back to normal,* and the final error is zero. Thus, the gain of the system is 100 divided by 0, or *infinity,* which is the reason that this system is all-important for long-term control of arterial pressure.

FACTORS THAT INCREASE THE
EFFECTIVENESS OF THE RENAL–BODY
FLUID SYSTEM OF PRESSURE
CONTROL:
(1) THE RENIN-ANGIOTENSIN SYSTEM,
(2) THE ALDOSTERONE SYSTEM, AND
(3) THE NERVOUS SYSTEM

When the arterial pressure increases, several other factors in addition to pressure diuresis and natriuresis also increase the salt and water output. For instance, the rise in the pressure causes (1) *decreased* secretion of renin by the kidneys, (2) *decreased* secretion of aldosterone by the adrenal cortices, and (3) *decreased* sympathetic signals to the kidneys. Since each of these factors normally decreases the renal output of both salt and water, the *decreases* in all of them allows greatly increased output. For instance, let us see how these factors affect urinary output when fluid intake is increased and the arterial pressure rises 15 mm. Hg. First, this increase in pressure by itself has a direct effect on the kidneys to double, approximately, the urinary output of water and salt. However, because the above-mentioned three mechanisms further enhance the output of water and salt, the 15 mm. Hg increase in arterial pressure is associated with about a ten-fold increase instead of a two-fold increase in output.

This increased output is illustrated in Figure 18–3, which compares the "acute renal output curve" and the "chronic renal output curve." The acute curve is the same as that illustrated in Figure 18–1, and it shows the direct effect of arterial pressure alone in causing increased output of water and salt by the kidneys. The chronic curve, on the other hand, shows the relationship of urinary output to arterial pressure when all the different factors that affect output are included: the direct effect of pressure on the kidneys and also the effects of the three additional mechanisms listed above.

Thus, it is clear that the chronic renal output curve depicts a very steep relationship between arterial pressure and urinary output. Therefore, in the normal animal, even a slight elevation of arterial pressure that lasts long enough for all the above mechanisms to become effective causes a marked increase in urinary fluid loss from the body; and, conversely, a slight decrease in arterial pressure causes almost total cessation of fluid loss. One can readily see that the steepness of this relationship between arterial pressure and urinary output makes the fluid excretory mechanism an extremely potent arterial pressure control mechanism.

ROLE OF THE RENIN-ANGIOTENSIN-ALDOSTERONE SYSTEM IN LONG-TERM CONTROL OF ARTERIAL PRESSURE

Most physiologists have suggested that two properties of the renin-angiotensin system are important for long-term arterial pressure control:

(1) *Vasoconstriction.* Whenever blood flow to the kidneys diminishes, as explained in the previous chapter, renin is secreted and angiotensin is formed. The angiotensin in turn causes widespread vasoconstriction throughout the body with consequent greatly increased total peripheral resistance. It has been suggested that this increased resistance is important not only for short-term control of pressure, as explained in the previous chapter, but also for long-term control.

(2) *Increased Aldosterone Secretion.* The formation of angiotensin affects the adrenal cortex, causing increased secretion of aldosterone. The aldosterone in turn causes the kidneys to retain salt and water and therefore to increase extracellular fluid volume, blood volume, cardiac output, and arterial pressure.

A Third Function of Angiotensin: A Direct Effect on the Kidneys of Increasing Renal Output. In addition to the above two functions of

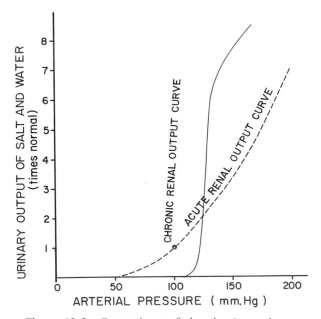

Figure 18–3. Comparisons of the *chronic* renal output curve and the *acute* renal output curve, showing an extremely steep relationship between arterial pressure and urinary output in the chronic state when all the factors that affect urinary output in response to pressure have had time to become fully operative. (See explanation in the text.)

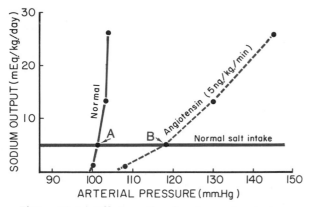

Figure 18–4. Effect on the chronic renal output curve caused by continuous infusion of angiotensin II at a low rate, showing marked shift of the curve toward a higher pressure range. (Drawn from data obtained by J. DeClue.)

the renin-angiotensin system, recent experiments have demonstrated a third that almost certainly is the most important of all in causing long-term increases in arterial pressure. This is an effect of angiotensin II acting directly on the kidneys, not mediated through aldosterone, to cause salt and water retention. This effect is illustrated in Figure 18–4, which shows (1) the normal renal output curve for sodium and (2) this curve measured when angiotensin II was infused continuously over a period of a month at a rate equal to only three times the normal rate of angiotensin formation in the body. Note that at each respective pressure level the output of salt (and also of water) is markedly decreased by the angiotensin. Thus, the entire renal output curve is shifted to the right in the figure, toward higher arterial pressure levels. Aldosterone causes a similar but

lesser shift in the renal output, as pointed out above. Therefore, both the direct effect of the angiotensin and its indirect effect acting through increased aldosterone secretion cause a definite long-term increase in the arterial pressure of a person.

HYPERTENSION (HIGH BLOOD PRESSURE)

Now that we have discussed the principles of long-term control of arterial pressure, we can call on these principles to discuss the mechanisms by which a person can develop hypertension. In all of this discussion it will be good to keep the following basic dictum in mind: *Any factor that increases the pressure range of the renal output curve can also cause hypertension.* In the following pages, we will see how this occurs.

VOLUME-LOADING HYPERTENSION

Hypertension Caused by Excess Water and Salt Intake in Patients with Low Kidney Mass. Figure 18–5 illustrates a typical experiment showing volume-loading hypertension in a group of dogs with 70 per cent of the renal mass removed. At the first circled point on the curve, the two poles of one of the kidneys were removed, and at the second circled point the entire opposite kidney was removed, leaving the animals with only 30 per cent of the normal renal mass. Note that removal of this amount of mass increased the arterial pressure an average of only 6 mm. Hg. However, at this point the dogs were

Figure 18–5. Effect on arterial pressure of drinking 0.9 per cent saline solution instead of water in four dogs with 70 per cent of their renal tissue removed. (Courtesy of Dr. Jimmy B. Langston.)

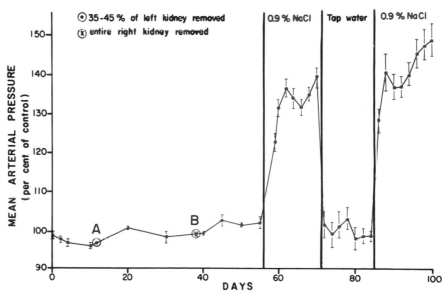
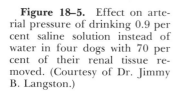

required to drink salt solution instead of normal drinking water. Because salt solution fails to quench the thirst, the dogs drank about four times the normal volume, and within a few days their average arterial pressure had adjusted to a much higher level. After two weeks the dogs were allowed to drink tap water again instead of salt solution; the pressure returned to normal within two days. Finally, at the end of the experiment, the dogs were required to drink salt solution again, and this time the pressure rose much more rapidly and to a higher level because the dogs had already learned to tolerate the salt solution and therefore drank much more. Thus, this experiment demonstrates the principles of volume-loading hypertension.

Volume-Loading Hypertension in Patients Who Have No Kidneys but Are Being Maintained on the Artificial Kidney. When a patient is maintained on an artificial kidney, it is especially important to keep the body fluid volume at a normal level — that is, to remove the appropriate amount of water and salt each time that the

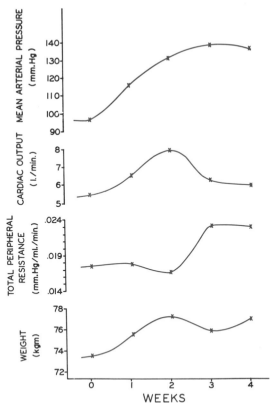

Figure 18–6. Hypertension in a patient whose two kidneys had been removed and whose body fluid volume had been increased by 3 liters. Note not only the increase in cardiac output at the onset of the hypertension, but also the rise in total peripheral resistance only *after* the hypertension had already developed. (Courtesy of Dr. Thomas Coleman.)

patient is dialyzed. Figure 18–6 illustrates a study in a patient whose body fluid volume increased only 3 liters. This is shown by the lowermost curve labeled "weight," which is the means by which one assesses sudden increases in fluid volume in these patients. Note that when the weight increased, the cardiac output also increased markedly, and this in turn caused the arterial pressure to rise from an original normal value of 95 mm. Hg to a hypertensive level of 140 mm. Hg.

Role of "Autoregulation" in Volume-Loading Hypertension–Secondary Elevation of Total Peripheral Resistance. It is important now to observe very carefully the changes that occurred in total peripheral resistance in the patient studied in Figure 18–6. When the body fluid volume first increased, the cardiac output and the arterial pressure increased, but at first the total peripheral resistance did not increase. This same effect has been demonstrated many times in dogs in which this same type of condition has been created. That is, the initial rise in arterial pressure is caused entirely by elevation of the cardiac output. Then, during the subsequent week or so, the total peripheral resistance rises while the cardiac output returns back toward normal. Thus, the total peripheral resistance increases *after,* rather than before, the arterial pressure has become elevated. To state this another way, the rise in total peripheral resistance is secondary, rather than primary, to the hypertension. This effect can be explained as follows:

During the initial onset of the hypertension, blood flow through all the body's tissues is greatly increased. Because this flow is far greater than that required to supply the tissues' needs, the tissue vessels slowly constrict. Consequently, the total peripheral resistance increases while the cardiac output returns toward normal, a manifestation of the local tissue blood flow autoregulation mechanism discussed in Chapter 16. Thus, the effect of the increase in total peripheral resistance is not to cause the increased arterial pressure (because, it has already risen before the resistance itself has risen) but to return the cardiac output back near to normal after the hypertension has occurred. Reduction of the cardiac output back to near-normal is very valuable to the body because it reduces the work load of the heart and therefore reduces the likelihood of heart failure.

Thus, it should always be remembered that pure *volume-loading hypertension is a high total peripheral resistance type of hypertension — not a high cardiac output type* — even though the increase in resistance occurs as a result of the hypertension and is not its primary cause.

VASOCONSTRICTOR HYPERTENSION—HYPERTENSION CAUSED BY CONTINUOUS INFUSION OF ANGIOTENSIN II OR BY A RENIN-SECRETING TUMOR

A type of hypertension that shows sharp contrasts to the volume-loading type of hypertension is that caused by continuous infusion of angiotensin II or by a tumor of the renal juxtaglomerular cells that secretes renin. Figure 18–7 illustrates some of the characteristics of this type of hypertension, showing the changes in arterial pressure, total peripheral resistance, cardiac output, and blood volume observed in a study of hypertension caused in a group of dogs by the infusion of 10 ng./kg. of angiotensin II per minute. Note the marked increases in both total peripheral resistance and arterial pressure but, conversely, a slight decrease in blood volume and a considerable decrease in cardiac output.

All the effects illustrated in Figure 18–7 can be explained by the basic principles of blood pressure regulation that were presented earlier in the chapter:

The increase in total peripheral resistance obviously results from the very potent effect of angiotensin II in constricting the arterioles.

The slight decrease in blood volume is caused by the initial rise in pressure resulting from the large increase in total peripheral resistance. That is, this excess pressure causes pressure diuresis and natriuresis, with consequent loss of volume.

The *decrease in cardiac output* results almost entirely from the intense arteriolar constriction, which in turn decreases the flow through the systemic circulation.

Thus, it is clear from Figure 18–7 that the hypertension caused by angiotension-induced vasoconstriction is, like volume-loading hypertension, a high resistance type of hypertension. However, there are subtle differences. In volume-loading hypertension, the total peripheral resistance is not increased quite as much as the increase in pressure, and the cardiac output is slightly above normal. On the other hand, in vasoconstrictor hypertension, the total peripheral resistance is increased somewhat in excess of the increase in arterial pressure, while the cardiac output is decreased to below normal.

Vasoconstrictor Hypertension Caused by a Pheochromocytoma. Another type of vasoconstrictor hypertension is that caused by a tumor called a *pheochromocytoma*, which is a tumor of the adrenal medulla that secretes large amounts of epinephrine and norepinephrine. Essentially the same type of vasoconstriction occurs as that in angiotensin hypertension; the total peripheral resistance rises very high, and the blood volume and cardiac output are likely to be low.

Role of the Renal–Body Fluid Pressure Control Mechanism in Determining Long-Term Arterial Pressure in Vasoconstrictor Hypertension. Let us for a moment suppose that angiotensin is infused into a person at a rate high enough to cause a very large increase in total peripheral resistance and therefore an acute rise in arterial pressure. Let us further suppose that the angiotensin has no effect at all on the renal output curve — that is, the angiotensin does not in any way affect the normal capability of the kidneys to excrete salt and water. In such a theoretical instance as this, the initial marked rise in arterial pressure would cause very intense pressure diuresis and pressure natriuresis. Consequently, the blood volume would decrease rapidly and the arterial pressure eventually would return back to normal, because it is only at a normal arterial pressure that urinary output would come to balance with the salt and water intake. Therefore, the increase in total peripheral

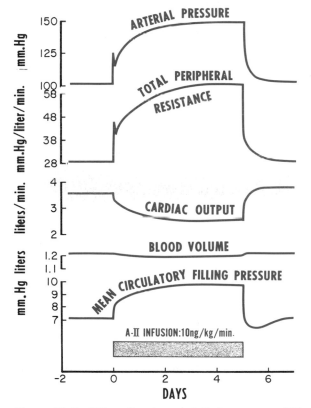

Figure 18–7. Effect of angiotensin infusion at a rate of 10 ng./kg./minute (approximately six times the normal rate of angiotensin formation in the body), illustrating marked increases in total peripheral resistance and arterial pressure, marked decrease in cardiac output, and a slight decrease in blood volume. (Drawn from data obtained in experiments in dogs by D. Young and A. Cowley.)

resistance could not sustain a long-term increase in arterial pressure. Instead, to sustain the increased pressure, it is essential that the kidney output curve be shifted to a high pressure range. This effect of angiotensin on the renal output curve was illustrated in Figure 18–4. Thus, strange as it may seem, even in the vasoconstrictor type of hypertension it is the effect of the vasoconstrictor on kidney output of salt and water that determines the arterial pressure level at which the hypertension stabilizes. This concept will help the student tremendously in understanding both blood pressure regulation and the different clinical hypertensive conditions.

We shall see subsequently in this chapter that most clinical types of hypertension show various degrees of either the pure volume-loading type of hypertension or the pure vasoconstrictor type of hypertension. Therefore, it is useful to keep in mind the characteristics of these two extremes among the different types of hypertension.

GOLDBLATT HYPERTENSION

When one kidney is removed and a constrictor is placed on the renal artery of the remaining kidney, as illustrated in Figure 18–8, the initial effect is greatly reduced renal arterial pressure, shown by the dashed curve in the lower portion of the figure. However, within a few minutes the systemic arterial pressure begins to rise and continues to rise for several days. The pressure usually rises rapidly for the first two hours, and during the next day returns to a slightly lower level, to be followed a day or so later by a second rise to a much higher pressure. When the systemic arterial pressure reaches its new stable pressure level, the renal arterial pressure will have returned almost all the way back to normal. The hypertension produced in this way is called *Goldblatt hypertension* in honor of Dr. Harry Goldblatt, who first studied the important quantitative features of hypertension caused by renal artery constriction.

The early rise in arterial pressure in Goldblatt hypertension is caused by the renin-angiotensin mechanism. Because of the poor blood flow through the kidney after acute application of the constrictor, large quantities of renin are secreted by the kidneys, as illustrated by the lowermost curve in Figure 18–8, and this causes angiotensin to be formed in the blood, as described in the previous chapter; the angiotensin in turn raises the arterial pressure acutely. However, the secretion of renin rises to a peak in a few hours but returns all the way back to normal within five to seven days because the renal arterial pressure by that time has also risen back to normal so that the kidney is no longer ischemic.

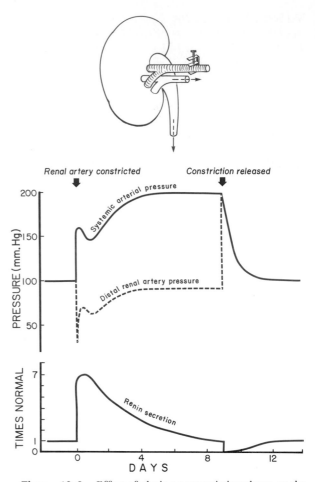

Figure 18–8. Effect of placing a constricting clamp on the renal artery of one kidney after the other kidney has been removed. Note the changes in systemic arterial pressure, renal artery pressure distal to the clamp, and rate of renin secretion. The resulting hypertension is called "one-kidney Goldblatt hypertension."

The second rise in arterial pressure is caused by fluid retention; within five to seven days the fluid volume has increased enough to raise the arterial pressure to its new sustained level.

Note especially that this Goldblatt hypertension has two phases. The first phase is a vasoconstrictor type of hypertension caused by angiotensin, but it is only transient. The second stage is a volume-loading type of hypertension. However, it is often very difficult to tell that this second stage is true volume-loading hypertension because neither the blood volume nor the cardiac output is significantly elevated. Instead, the total peripheral resistance is increased. To understand this, recall that in pure volume-loading hypertension, the blood volume and cardiac output are elevated only the first few days during the onset; but after the first few days, volume-loading hypertension becomes a high-resistance hypertension exactly as seen in the second stage of Goldblatt hypertension.

NEUROGENIC HYPERTENSION

Many nervous disorders can cause temporary hypertension and sometimes even permanent hypertension. Temporary hypertension can be caused a few hours at a time by sympathetic stimulation of the blood vessels, which increases the vascular resistance throughout the body. However, to cause long-term hypertension the renal arterioles must be constricted continuously by the sympathetic stimulation for days at a time. This will increase the pressure range of the kidney output curve; therefore, the renal mechanism for pressure control will then maintain the pressure at an elevated level as long as the sympathetic stimulation continues.

Hypertension Caused by Frustration or Pain. Prolonged sympathetic stimulation has been caused in animals by continual frustration or pain, such as by tantalizing the animal with food barely out of reach or by repeated electric shocks. Such animals develop mild to moderate hypertension as long as the abnormal conditions exist. However, within two to three weeks after the cause has been removed, the pressure usually returns to normal. Yet, it is believed by many clinicians that such abnormal sympathetic stimulation of the kidneys in patients for prolonged periods of time — perhaps for years — causes actual structural changes to occur gradually in the kidneys so that permanent pathological elevation of the kidney output curve develops. Then, even if the sympathetic stimulation is removed, the hypertension still persists. That is, neurogenic factors are believed finally to lead to hypertension that is sustained by a secondary renal abnormality.

HYPERTENSION CAUSED BY PRIMARY ALDOSTERONISM

A small tumor that secretes large quantities of aldosterone occasionally occurs in the adrenal glands. As will be discussed in detail in Chapter 25, the aldosterone increases the rate of reabsorption of salt and water in the distal tubules of the kidneys, thereby greatly reducing their rate of excretion. Consequently, mild to moderate hypertension occurs even at normal levels of salt and water intake.

ESSENTIAL HYPERTENSION

Approximately 90 per cent of all persons who have hypertension are said to have "essential hypertension," meaning hypertension of unknown origin. A few years ago, many of the patients who are now known to have one of the types of hypertension described earlier in this chapter would have been said to have essential hypertension. Thus, the more we learn about hypertension the more we can diagnose the cause precisely.

One group of patients formerly diagnosed as having essential hypertension are now known to have hypertension caused by glomerulosclerosis. These patients in early life had mild acute glomerulonephritis. They all supposedly recovered completely from the glomerulonephritis; then, during the ensuing 20 years they gradually developed what was called essential hypertension. Biopsies of the kidneys of these patients have demonstrated a sclerotic process in their glomeruli, with obviously decreased filtering capacity by the glomeruli. Thus, it is very clear that this type of patient, who would formerly have been said to have essential hypertension, in reality has a type of hypertension caused by a kidney abnormality.

Normal Excretion of Waste Products by Patients with Essential Hypertension. Though there are many reasons for believing that the kidneys are to blame for most "essential" hypertension, many clinicians have nevertheless believed that essential hypertension does not have a renal origin because the kidneys of these patients excrete urinary waste products normally. Therefore, how could the kidneys be responsible for the hypertension, when the urinary function of the kidneys is so normal? The answer to this lies in the following observations: When the arterial pressure falls from the hypertensive level to normal in a patient with essential hypertension, the kidney output of urine often falls to zero, the body retains the waste products in the blood, and the patient becomes uremic if the pressure is kept at the normal level. Thus, the elevated pressure is essential for the normal excretory function of the kidneys to occur, illustrating that they are not functioning normally.

Treatment of Essential Hypertension. Essential hypertension can generally be treated by two different types of drugs: (1) a drug that will increase glomerular filtration or (2) a drug that will decrease tubular reabsorption of salt and water. Those drugs that will increase glomerular filtration are the various vasodilator drugs. Some of them act by inhibiting sympathetic impulses to the kidney, and others act by direct paralysis of the smooth muscle in the walls of the blood vessels. The drugs that reduce reabsorption of water and salt by the tubules include, especially, drugs that block active transport of salt through the tubular wall; the blockage in turn prevents the osmotic reabsorption of water as well. Such substances that reduce reabsorption of salt (and, consequently, of water as well) are called *natriuretics* or *diuretics*.

EFFECTS OF HYPERTENSION ON THE BODY

Hypertension can be very damaging because of two primary effects: (1) increased work load on the heart and (2) damage to the arteries themselves by the excessive pressure.

In hypertension, the very high pressure against which the left ventricle must beat causes it to hypertrophy and therefore to increase in weight as much as two- to three-fold. Unfortunately, the high pressure in the coronary arteries causes rapid development of coronary arteriosclerosis, so that hypertensive patients tend to die of coronary occlusion at much earlier ages than do normal persons.

High pressure in the arteries causes not only coronary sclerosis but also sclerosis of blood vessels throughout all the remainder of the body. The process of arteriosclerosis causes blood clots to develop in the vessels and also causes the blood vessels to become weakened. Therefore, these vessels frequently thrombose, or they rupture and bleed severely. In either case, marked damage can occur in organs throughout the body. The two most important types of damage which occur in hypertension are

1. Cerebral hemorrhage, which means bleeding of a cerebral vessel with resultant destruction of local areas of brain tissue. This causes the well-known condition called *stroke*.

2. Hemorrhage of vessels inside the kidneys, which destroys large areas of the kidneys and therefore causes progressive deterioration of the kidneys and further exacerbation of the hypertension.

REFERENCES

Bianchi, G., and Bazzato, G. (eds.): The Kidney in Arterial Hypertension. Baltimore, University Park Press, 1979.

Coleman, T. G., and Guyton, A. C.: Hypertension caused by salt loading in the dog. III. Onset transients of cardiac output and other circulatory variables. Circ. Res., 25:153, 1969.

Coleman, T. G., et al.: Regulation of arterial pressure in the anephric state. Circulation, 42:509, 1970.

Cowley, A. W., Jr., et al.: Open-loop analysis of the renin-angiotensin system in the dog. Circ. Res., 28:568, 1971.

DeClue, J. W., et al.: Subpressor angiotensin infusion, renal sodium handling, and salt-induced hypertension in the dog. Circ. Res., 43(4):503, 1978.

DeQuattro, V., and Campese, V. M.: Pheochromocytoma: Diagnosis and therapy. In DeGroot, L. J., et al. (eds.): Endocrinology. Vol. 2. New York, Grune & Stratton, 1979, p. 1279.

DeQuattro, V., and Myers, M. R.: The sympathetic nervous system and primary hypertension in man. In DeGroot, L. J., et al. (eds.): Endocrinology. Vol. 2. New York, Grune & Stratton, 1979, p. 1297.

Folkow, B., et al.: Importance of adaptive changes in vascular design for establishment of primary hypertension, studied in man and in spontaneously hypertensive rats. Circ. Res., 32(Suppl. 1):2, 1973.

Freis, E. D.: Hemodynamics of hypertension. Physiol. Rev., 40:27, 1960.

Freis, E. D.: Salt in hypertension and the effects of diuretics. Annu. Rev. Pharmacol. Toxicol., 19:13, 1979.

Gant, N. F. (ed.): Pregnancy-Induced Hypertension. New York, Grune & Stratton, 1978.

Granger, H. J., and Guyton, A. C.: Autoregulation of the total systemic circulation following destruction of the central nervous system in the dog. Circ. Res., 25:379, 1969.

Guyton, A. C.: Acute hypertension in dogs with cerebral ischemia. Am. J. Physiol., 154:45, 1948.

Guyton, A. C.: Arterial Pressure and Hypertension. Philadelphia, W. B. Saunders Co., 1980.

Guyton, A. C., and Coleman, T. G.: Long-term regulation of the circulation; Interrelationships with body fluid volumes. In Reeve, E. B., and Guyton, A. C. (eds.): Physical Bases of Circulatory Transport Regulation and Exchange. Philadelphia, W. B. Saunders Co., 1967.

Guyton, A. C., and Coleman, T. G.: Quantitative analysis of the pathophysiology of hypertension. Circ. Res., 24(Suppl.):1, 1969.

Guyton, A. C., et al.: Arterial pressure regulation: Overriding dominance of the kidneys in long-term regulation and in hypertension. Am. J. Med., 52:584, 1972.

Guyton, A. C., et al.: A systems analysis approach to understanding long-range arterial blood pressure control and hypertension. Circ. Res., 35:159, 1974.

Guyton, A. C., et al.: Integration and control of circulatory function. Int. Rev. Physiol., 9:341, 1976.

Guyton, A. C., et al.: Salt balance and long-term blood pressure control. Annu. Rev. Med., 31:15, 1980.

Hall, J. E., et al.: Renal hemodynamics in acute and chronic angiotensin II hypertension. Am. J. Physiol., 235(3):F174, 1978.

Hall, J. E., et al.: Control of arterial pressure and renal function during glucocorticoid excess in dogs. Hypertension, 2: 139, 1980.

Hart, J. T.: Hypertension. New York, Churchill Livingstone, 1980.

Hollenberg, N. K.: Pharmacologic interruption of the renin-angiotensin system. Annu. Rev. Pharmacol. Toxicol., 19:559, 1979.

Ledingham, J. M.: Blood pressure regulation in renal failure. J. R. Coll. Physicians Lond., 5:103, 1971.

Manning, R. D., Jr., et al.: Essential role of mean circulatory filling pressure in salt-induced hypertension. Am. J. Physiol., 236:R40, 1979.

Manning, R. D., Jr., et al.: Hypertension in dogs during antidiuretic hormone and hypotonic saline infusion. Am. J. Physiol., 236:H314, 1979.

McCaa, R. E., et al.: Role of aldosterone in experimental hypertension. J. Endocrinol., 81:69, 1979.

Meyer, P.: Hypertension: Mechanisms and Clinical and Therapeutic Aspects. New York, Oxford University Press, 1980.

Meyer, P., and Schmitt, H. (eds.): Nervous System and Hypertension. New York, John Wiley & Sons, 1979.

Norman, R. A., Jr., et al.: Arterial pressure–urinary output relationship in hypertensive rats. Am. J. Physiol., 234(3):R98, 1978.

Norman, R. A., Jr., et al.: Renal function curves in normotensive and spontaneously hypertensive rats. Am. J. Physiol., 234:R98, 1978.

Reid, I. A., et al.: The renin-angiotensin system. Annu. Rev. Physiol., 40:377, 1978.

Scriabine, A. (ed.): Pharmacology of Antihypertensive Drugs. New York, Raven Press, 1980.

Vander, A. J.: Control of renin release. Physiol. Rev., 47:359, 1967.

Young, D. B., and Guyton, A. C.: Steady state aldosterone dose-response relationships. Circ. Res., 40(2):138, 1977.

19

Cardiac Output and
Circulatory Shock

Cardiac output is the quantity of blood pumped by the left ventricle into the aorta each minute. It is perhaps the single most important factor that we have to consider in relation to the circulation, for it is cardiac output that is responsible for transport of substances to and from the tissues.

NORMAL VALUES FOR CARDIAC OUTPUT

The normal cardiac output for the young healthy male adult averages approximately 5.6 liters per minute. However, if we consider all adults, including older people and females, the average cardiac output is very close to 5 liters per minute. In general, the cardiac output of females is about 10 per cent less than that of males of the same body size.

Cardiac Index. The cardiac ouptut changes markedly with body size. Therefore, it has been important to find some means by which the cardiac outputs of different-sized persons can be compared with each other. Experiments have shown that the cardiac output increases approximately in proportion to the surface area of the body. Therefore, the cardiac output is frequently stated in terms of the cardiac index, which is the *cardiac output per square meter of body surface area.* The normal human being weighing 70 kg. has a body surface area of approximately 1.7 square meters, which means that the normal average cardiac index for adults of all ages, both males and females, is approximately 3.0 liters per minute per square meter.

Effect of Age. Figure 19–1 illustrates the change in cardiac index with age. Rising rapidly to a level greater than 4 liters per minute per square meter at 10 years of age, the cardiac index declines to about 2.4 liters per minute at the age of 80.

Effect of Metabolism and Exercise. The cardiac output usually remains almost proportional to the overall metabolism of the body. That is, the greater the degree of activity of the muscles and other organs, the greater also will be the cardiac output. This relationship is illustrated in Figure 19–2, which shows that as the work output increases during exercise the cardiac output also increases in almost linear proportion. Note that in very intense exercise the cardiac output can rise to as high as 30 to 35 liters per minute in the young, well-trained athlete, which is about five to six times the normal control value.

Figure 19–2 also demonstrates that oxygen consumption increases in almost direct proportion to work output during exercise. We shall see later in the chapter that the increase in cardiac output probably results primarily from the increased oxygen consumption.

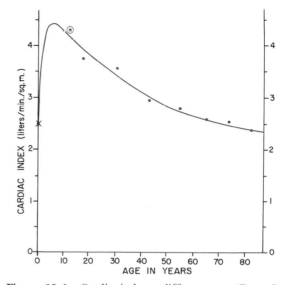

Figure 19–1. Cardiac index at different ages. (From Guyton, Jones, and Coleman: Circulatory Physiology: Cardiac Output and Its Regulation. Philadelphia, W. B. Saunders Company, 1973.)

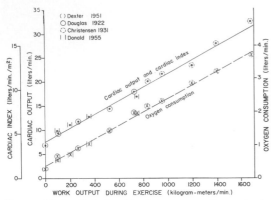

Figure 19–2. Relationship between cardiac output and work output (solid curve) and between oxygen consumption and work output (dashed curve) during different levels of exercise. (From Guyton, Jones, and Coleman: Circulatory Physiology: Cardiac Output and Its Regulation. Philadelphia, W. B. Saunders Company, 1973.)

REGULATION OF CARDIAC OUTPUT

PRIMARY ROLE OF THE PERIPHERAL CIRCULATION IN THE CONTROL OF CARDIAC OUTPUT: PERMISSIVE ROLE OF THE HEART

Control of Cardiac Output by Venous Return — the Frank-Starling Law of the Heart. It is worthwhile at this point to recall the Frank-Starling law of the heart, which was discussed in Chapter 11. This law states that within the physiological limit of the heart, the heart will pump whatever amount of blood enters the right atrium and will do so without significant buildup of back pressure in the right atrium. In other words, the heart is an automatic pump that is capable of pumping far more than the 5 liters per minute that normally returns to it from the peripheral circulation. Consequently, the factor that normally determines how much blood will be pumped by the heart is the amount of blood that flows into the heart from the systemic circulation — not the pumping capacity of the heart. To state this still another way, under normal circumstances the major factor determining the cardiac ouput is the *rate of venous return.*

Therefore, a large share of our discussion in this chapter will concern the factors in the peripheral circulation that determine the rate at which blood returns to the heart. However, there are times when the amount of blood attempting to return to the heart is greater than the amount that the heart can pump. Under these conditions the heart then becomes the limiting factor in cardiac output control, and the heart is said to be *failing.*

Thus, the heart normally plays a *permissive role* in cardiac output regulation. That is, it is capable of pumping a certain amount of blood each minute and therefore will *permit* the cardiac output to be regulated at any value below this given permitted level. The normal human heart that is *not stimulated by the autonomic nervous sytem* permits a maximum heart pumping of about 13 to 15 liters per minute, but the actual cardiac output under resting conditions is only approximately 5 liters per minute because this is the normal level of venous return. Therefore, it is the peripheral circulatory system, not the heart, that sets this level of 5 liters per minute.

The concept that the heart plays a permissive role in cardiac output regulation is so important that it needs to be explained in still another way, as follows: The normal heart, beating at a normal heart rate, and having a normal strength of contraction, neither excessively stimulated by the autonomic nervous system nor suppressed, will pump whatever amount of blood flows into the right atrium, up to about 13 to 15 liters per minute. If any more than this tries to flow into the right atrium, the heart will not pump it without cardiac stimulation. Under normal resting conditions the amount of blood that normally flows into the right atrium from the peripheral circulation is about 5 liters; and since this 5 liters is within the permissive range of heart pumping, it is pumped on into the aorta. Therefore, the venous return from the peripheral circulatory system controls the cardiac output whenever the permissive level of heart pumping is greater than venous return.

Increase in Permissive Level for Heart Pumping in Hypertrophied Hearts. Heavy athletic training causes the heart to enlarge, sometimes as much as 50 per cent. Coincident with this enlargement is an increase in the permissive level to which the heart can pump. Thus, even when the heart is not stimulated by the nervous system, the permissive level for a well-trained athlete might be as great as 20 liters per minute, rather than the normal value of about 13 to 15 liters per minute.

Increase in Permissive Level of Heart Pumping by Autonomic Stimulation of the Heart. There are times when the cardiac output must rise temporarily to levels greater than the normal permissive level of the heart. For instance, in heavy exercise by well-trained athletes, cardiac outputs as high as 35 liters per minute have been measured. Obviously, the resting heart would not be able to pump this amount of blood. On the other hand, *stimulation of the heart by the sympathetic nervous system increases the permissive level of heart pumping to approximately double normal.* This effect comes about by autonomic enhancement of both heart rate and strength of heart contraction. Furthermore, the increase in permissive level of heart pumping occurs within a few seconds after exercise begins, even before most of the increase in venous return occurs.

Reduction in Permissive Level for Heart Pumping in Heart Disease. Though the normal permissive level for heart pumping is usually much higher than the venous return, this is not necessarily true when the heart is diseased. Such conditions as myocardial infarction, valvular heart disease, myocarditis, and congenital heart abnormalities can reduce the pumping effectiveness of the heart. In these instances, the permissive level for heart pumping may fall below 5 liters per minute, indeed for a few hours even as low as 2 to 3 liters per minute. When this happens, the heart becomes unable to cope with the amount of blood that is attempting to flow into the right atrium from the peripheral circulation. Therefore, the heart is said to *fail*, meaning simply that it fails to pump the amount of blood that is demanded of it. Under these conditions, the heart becomes the limiting factor in cardiac output control. However, this is not the normal state. (We shall discuss cardiac failure in the following chapter.)

ROLE OF TOTAL PERIPHERAL RESISTANCE IN DETERMINING NORMAL VENOUS RETURN AND CARDIAC OUTPUT

Let us recall the formula for blood flow through blood vessels that was presented in Chapter 14:

Blood Flow =

$$\frac{\text{Pressure (input)} - \text{Pressure (output)}}{\text{Resistance}}$$

Now if we apply this to venous return, the formula becomes:

Venous Return =

$$\frac{\text{Arterial Pressure} - \text{Right Atrial Pressure}}{\text{Total Peripheral Resistance}}$$

Since the right atrial pressure remains very nearly zero, it is clear that the amount of blood that flows into the heart each minute (venous return) and that is pumped each minute (cardiac output) is determined by two prime factors: (1) the arterial pressure and (2) the total resistance. When the arterial pressure remains normal, as it usually does, venous return and cardiac output

are then inversely proportional to the total peripheral resistance. To state this still another way, every time a peripheral blood vessel dilates, the venous return and cardiac output increase. Furthermore, the more vessels that dilate in the peripheral circulation, the greater the cardiac output becomes.

Cardiac Output Regulation as the Sum of the Local Blood Flow Regulations Throughout the Body. As long as the arterial pressure remains normal, each local tissue in the body can control its own blood flow by simply dilating or constricting its local blood vessels. This mechanism, which was discussed in detail in Chapter 16, is the means by which each tissue protects its own nutrient supply, controlling the blood flow in response to its own needs.

Therefore, since the venous return to the heart is the sum of all the local blood flows through all the individual tissues of the body, all the local blood flow regulatory mechanisms throughout the peripheral circulation are the true controllers of cardiac output under normal conditions. This automatic mechanism allows the heart to respond instantaneously to the needs of each individual tissue. If some tissues need extra blood flow and their local blood vessels dilate, the venous return increases automatically, and the cardiac output increases by an equivalent amount. If all the tissues throughout the body require increased blood flow at the same time, the venous return becomes very great, and the cardiac output increases accordingly.

Thus, the whole theory of normal cardiac output regulation is that the tissues control the output in accordance with their needs. Again, it must be stated that it is not the heart that controls the cardiac output under normal conditions; instead, the heart plays a permissive role that allows the tissues to do the controlling. The heart does this by always maintaining a permissive pumping capacity that is somewhat above the actual venous return — that is, except when the heart fails.

Effect of Local Tissue Metabolism on Cardiac Output Regulation, and Importance of Simultaneous Arterial Pressure Regulation. The most important factor that controls the local blood flows in the individual tissues is the metabolic rates of the respective tissue. Therefore, venous return and cardiac output are normally controlled in relation to the level of metabolism of the body. This effect was illustrated in Figure 19–2, which showed that the cardiac output increases directly in relation to work output during exercise and parallel to the increase in oxygen consumption, a measure of the rate of metabolism.

However, it is essential that the arterial pressure be maintained at a normal level if changes in

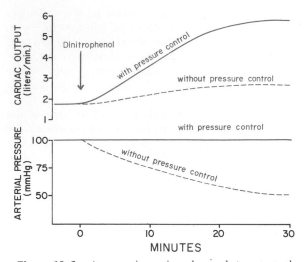

Figure 19–3. An experiment in a dog to demonstrate the importance of arterial pressure control as a prerequisite for cardiac output control. Note that with pressure control the metabolic stimulant dinitrophenol increases cardiac output; without pressure control, the arterial pressure falls and the cardiac output rises very little. (Data from experiments by Dr. M. Banet.)

metabolism are to regulate cardiac output. This is illustrated in Figure 19–3, which shows changes in cardiac output in response to an approximately five-fold increase in tissue metabolism in a dog. The increase in metabolism was caused by the toxic substance dinitrophenol, which causes the metabolic systems to increase their metabolic use of oxygen tremendously and, simultaneously, to dilate the local blood vessels supplying the tissues. Note in this figure that when the arterial pressure was normally controlled, the cardiac output increased approximately 300 per cent. On the other hand, when the nervous mechanisms for control of arterial pressure were blocked, the cardiac output increased only 100 per cent.

Therefore, once again we can state that under normal conditions, venous return and cardiac output are determined almost entirely by the degree of dilatation of the local blood vessels in the tissues throughout the body. However, this mechanism operates properly only if the mechanisms for maintaining a normal arterial pressure, which were discussed in the preceding two chapters, are also functioning properly.

IMPORTANCE OF THE "MEAN SYSTEMIC FILLING PRESSURE" IN CARDIAC OUTPUT REGULATION

If the quantity of blood in the circulatory system is too little to fill the system adequately, blood will flow very poorly from the peripheral vessels into the heart. Therefore, the degree of filling of the circulation is one of the most important factors in determining the venous return to the heart and therefore also in determining the cardiac output.

The student may recall from Chapter 14 that the degree of filling of the circulatory system with blood can be expressed in terms of the *mean circulatory filling pressure,* and that the degree of filling of the systemic portion of the circulation can be expressed in terms of the *mean systemic filling pressure.* The mean systemic filling pressure is the pressure in all parts of the systemic circulation that is measured when blood flow in the circulatory system is suddenly stopped and the blood in the systemic vessels is redistributed so that the pressure in all the vessels is equal. Normally, the mean systemic filling pressure is 7 mm. Hg, but an increase in blood volume of 15 to 30 per cent doubles the mean systemic filling pressure, whereas a decrease in blood volume of this same amount reduces the mean systemic filling pressure to zero.

The mean systemic filling pressure is the *average* effective pressure of the blood in the peripheral circulation that tends to push the blood toward the heart. Experiments have demonstrated that *when the peripheral resistance remains constant,* the rate of *venous return* from the systemic vessels through the veins to the heart is directly proportional to the *mean systemic filling pressure* minus the *right atrial pressure.* Therefore, whenever the mean systemic filling pressure falls to zero, the flow of blood returning to the heart likewise approaches zero.

Consequently, it is very important that the mean systemic filling pressure remain high enough at all times to supply the peripheral pressure needed to push blood from the peripheral vessels back to the heart.

REGULATION OF CARDIAC OUTPUT IN HEAVY EXERCISE, REQUIRING SIMULTANEOUS PERIPHERAL AND CARDIAC ADJUSTMENTS

Heavy exercise is one of the most stressful conditions to which the body is ever subjected. The tissues can require as much as 20 times normal amounts of oxygen and other nutrients, so that simply to transport enough oxygen from the lungs to the tissues sometimes demands a minimal cardiac output increase of five- to six-fold. This is greater than the amount of cardiac output the normal, unstimulated heart can pump. Therefore, to insure the massive increase in cardiac output that is required during heavy exercise, almost all factors that are known to increase cardiac output are called into play.

Muscle Vasodilatation Resulting from Increased Metabolism. By far the most important factor that increases the cardiac output during exercise is the vasodilatation that occurs in all exercising muscles. The cause of the vasodilatation is the large increase in muscle metabolism during exercise. This local vasodilatation requires 5 to 15 seconds to reach full development after a person begins to exercise strongly; but once it does reach full development, the very great decrease in vascular resistance allows extreme quantities of blood to flow through the muscles and thence into the veins to be returned to the heart, thereby markedly increasing venous return and cardiac output.

Role of the Heart in Strenuous Exercise. When large numbers of muscles are exercising simultaneously, the peripheral vasodilatation may be so great and the venous return to the heart so voluminous that the "resting" heart simply cannot pump this extra amount of blood. Therefore, it is essential that the *permissive level of pumping* by the heart be greatly increased from its normal level of 13 to 15 liters per minute. This is achieved mainly by sympathetic stimulation of the heart (but also partly by a decrease in parasympathetic stimulation), which increases the permissible level to as high as 20 to 25 liters per minute in a normal person or as high as 35 liters in the trained athlete.

Special Functions of the Sympathetic Nervous System During Exercise. In light exercise, the heart and the circulation do not need to be stimulated by the sympathetic nervous system. But in heavy exercise, sympathetic stimulation is an absolute essential, for the following reasons:

1. Strong sympathetic stimulation can increase the permissive level of heart pumping from 13 to 15 liters per minute to as high as 20 to 25 liters per minute (or even higher in the trained athlete), as already noted.

2. Sympathetic stimulation constricts almost all the blood vessels throughout the body, especially the veins, and this increases the mean systemic filling pressure to as much as 2.5 times normal. As discussed, this increase in mean systemic filling pressure provides an extra push on the blood in the peripheral vessels to return this blood to the heart. This is absolutely essential to achieve enough venous return, and therefore to achieve enough cardiac output, in heavy exercise.

Mechanisms for Stimulating the Sympathetic Nervous System During Exercise. At least three different mechanisms enhance sympathetic activity during exercise.

1. Simply thinking about exercise has the psychic effect of exciting the autonomic nervous centers, which increases the heart rate, increases the strength of heart contraction, and constricts the blood vessels throughout the body to increase the mean systemic filling pressure. Together, these effects can increase the cardiac output instantaneously as much as 50 per cent, even before the exercise begins.

2. The same signals from the motor cortex that excite activity in the muscles also excite the sympathetic nervous system. This causes vasoconstriction in most tissues of the body besides the exercising muscles, thus increasing the arterial pressure. The increase in pressure in turn provides an extra push to force blood through the muscles.

3. Contraction of the muscles themselves elicits reflex stimulation of the sympathetic nervous system. Sensory endings in the muscles are excited by metabolic products produced during contraction, and signals then pass to the vasomotor center to excite the sympathetic system still more.

Role of Abdominal Contraction for Enhancing Venous Return and Cardiac Output. At the very onset of exercise, the initial tenseness of the body that is associated with exercise causes abdominal contraction and compression of the large venous reservoirs in the abdomen. This immediately increases the mean systemic filling pressure two- to three-fold, which can increase venous return and cardiac output as much as 30 to 80 per cent within one to two beats of the heart.

Summary. It is clear that a complex set of controls comes into play during exercise to allow the heart and the circulation to increase the cardiac output to the tremendous levels required for nutrient supply to the muscles. The initial increase in cardiac output occurs when the person begins to anticipate exercise, which stimulates the sympathetic nervous system. Then, tensing of the abdominal muscles gives an additional surge of venous return that also increases the output. Finally, within seconds after the exercise actually begins, intense local metabolic vasodilation occurs in the muscles themselves. Combined with the hyperdynamic state of the heart and the increased mean systemic filling pressure, this causes the further surge in the cardiac output up to maximum levels as high as 20 to 35 liters per minute.

HIGH CARDIAC OUTPUT — ROLE OF REDUCED TOTAL PERIPHERAL RESISTANCE IN CHRONIC HIGH CARDIAC OUTPUT CONDITIONS

The left side of Figure 19–4 identifies conditions that commonly cause cardiac outputs higher than normal. One of the distinguishing features of these conditions is that *they all result from chronically reduced total peripheral resistance.* And none of them result from excessive excitation of the heart, which we will explain below. For the present, let us look at some of the peripheral factors that can increase the cardiac output to above normal:

(1) *Beriberi.* This disease is caused by insufficient quantity of the vitamin thiamine in the diet. Lack of this vitamin causes diminished ability of the tissues to utilize cellular nutrients, which in turn causes marked peripheral vasodilation. The total peripheral resistance decreases sometimes to as little as one-half normal. Consequently, the long-term level of cardiac output also often increases to as much as two times normal.

(2) *Arteriovenous fistula (shunt).* Whenever a fistula (also called a shunt) occurs between a major artery and a major vein, tremendous amounts of blood will flow directly from the artery into the vein. This, too, greatly decreases the total peripheral resistance and likewise increases the venous return and cardiac output.

(3) *Hyperthyroidism.* In hyperthyroidism, the metabolism of all the tissues of the body becomes greatly increased. Oxygen usage increases, and vasodilator products are released from the tissues. Therefore, the total peripheral resistance decreases markedly, and the cardiac output often increases to as much as 40 to 80 per cent above normal.

(4) *Anemia.* In anemia, two peripheral effects greatly decrease the total peripheral resistance. One of these is reduced viscosity of the blood, resulting from the decreased concentration of red blood cells. The other is diminished delivery of oxygen to the tissues because of the decreased hemoglobin, which causes local vasodilatation. As a consequence, the total peripheral resistance decreases greatly, and the cardiac output increases.

Any other factor that decreases the total peripheral resistance chronically will also increase the cardiac output.

Failure of Increased Cardiac Pumping to Cause Prolonged Increase of the Cardiac Output. If the heart is suddenly stimulated excessively, the cardiac output often increases as much as 50 per cent. However, this increase is maintained no longer than a few minutes even though the heart continues to be strongly stimulated. There are two reasons: (1) Excess blood flow through the tissues causes automatic vasoconstriction of the blood vessels because of the autoregulation mechanism discussed in previous chapters, and this reduces the venous return and cardiac output back toward normal. (2) The slightly increased arterial pressure that results from acute cardiac stimulation raises the capillary pressure, and fluid filters out of the capillaries into the tissues thereby decreasing the blood volume and also decreasing the venous return back toward normal. Also, the increased pressure causes the kidneys to lose fluid volume as well until the arterial pressure and cardiac output return to normal.

Thus, all the known conditions that cause *chronic* elevation of the cardiac output result from decreased total peripheral resistance and not increased cardiac activity.

EFFECT OF INCREASED BLOOD VOLUME ON CARDIAC OUTPUT

A sudden increase in blood volume of about 20 per cent increases the cardiac output to about 2.5 to 3 times normal, principally because this extra blood increases the mean systemic filling pressure to more than double normal.

However, this increased cardiac output lasts only a few minutes, because several different compensatory effects immediately begin to occur: (1) The increased cardiac output increases the capillary pressure, so that fluid begins to transude out of the capillaries into the tissues, thereby

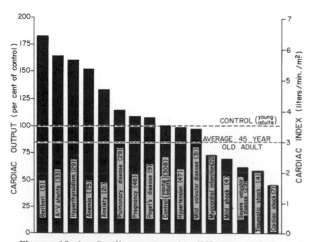

Figure 19–4. Cardiac output in different pathological conditions. The numbers in parentheses indicate number of patients studied in each condition. (From Guyton, Jones, and Coleman: Circulatory Physiology: Cardiac Output and Its Regulation. Philadelphia, W. B. Saunders Company, 1973.)

returning the blood volume toward normal. (2) The increased pressure in the veins causes the veins to distend gradually by the mechanism called *stress-relaxation*, especially causing the venous blood reservoirs such as the liver and spleen to distend. (3) The excess blood flow through the peripheral tissues causes autoregulatory increase in the peripheral resistance. These factors cause the mean systemic filling pressure to return toward normal and also cause the resistance vessels of the systemic circulation to constrict. Therefore, gradually over a period of 20 to 40 minutes, the cardiac output returns almost to normal.

METHODS FOR MEASURING CARDIAC OUTPUT

In lower animals, the aorta, the pulmonary artery, or the great veins entering the heart can be cannulated and the cardiac output measured by any type of flowmeter. Also, an electromagnetic or ultrasonic flowmeter can be placed on the aorta or pulmonary artery to measure cardiac output. However, except in rare instances, cardiac output in the human being is measured by indirect methods that do not require surgery. Two of the methods commonly used are the *oxygen Fick method* and the *dye dilution method*.

PULSATILE OUTPUT OF THE HEART AS MEASURED BY AN ELECTROMAGNETIC OR ULTRASONIC FLOWMETER

Figure 19–5 illustrates a recording in the dog of blood flow in the root of the aorta, made using an electromagnetic flowmeter. It demonstrates that the blood flow rises rapidly to a peak during systole and then, at the end of systole, actually reverses for a small fraction of a second. It is this reverse flow that causes the aortic valve to close. And a minute amount of reverse flow continues throughout diastole to supply blood to the coronary vessels.

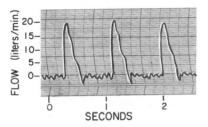

Figure 19–5. Pulsatile blood flow in the root of the aorta recorded by an electromagnetic flowmeter.

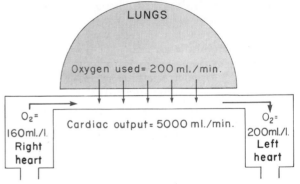

Figure 19–6. The Fick principle for determining cardiac output.

MEASUREMENT OF CARDIAC OUTPUT BY THE OXYGEN FICK METHOD

The Fick procedure is best explained by Figure 19–6, which shows the absorption of 200 ml. of oxygen from the lungs into the pulmonary blood each minute and also illustrates that the blood entering the right side of the heart has an oxygen concentration of approximately 160 ml. per liter of blood, while that leaving the left side has an oxygen concentration of approximately 200 ml. per liter of blood. From these data we see that each liter of blood passing through the lungs picks up 40 ml. of oxygen. And, since the total quantity of oxygen absorbed into the blood from the lungs each minute is 200 ml., a total of five 1-liter portions of blood must pass through the pulmonary circulation each minute to absorb this amount of oxygen. Therefore, the quantity of blood flowing through the lungs each minute is 5 liters, which is also a measure of the cardiac output. Thus, the cardiac output can be calculated by the following formula:

Cardiac output (liters/min.) =

$$\frac{\text{O}_2 \text{ absorbed per minute by the lungs (ml./min.)}}{\text{Arteriovenous O}_2 \text{ difference (ml./liter of blood)}}$$

In applying the Fick procedure, the concentrations of oxygen in the venous and arterial bloods are measured chemically, and the rate of oxygen absorption by the lungs is measured by a "respirometer," from which the person breathes.

THE DYE DILUTION METHOD

In measuring the cardiac output by the dye dilution method a small amount of dye is injected into a large vein or preferably into the right side of the heart itself. This then passes rapidly through the right heart, the lungs, the left heart, and finally into the arterial system. If one records the concentration of the dye as it passes through one of the peripheral arteries, a curve such as one of the solid curves illustrated in Figure 19–7 will be obtained. In each of these instances 5 mg. of Cardio-Green dye was injected at zero time. In the top recording none of the dye passed into the arterial tree until approximately 3 seconds after the injection, but then the arterial concentration of the dye rose rapidly to a maximum in approximately 6 to 7 seconds. After that, the concentration fell rapidly. However, before the concentration reached the zero point, some of the dye had already circulated all the way through some of the peripheral vessels and returned through the heart for a second time. Consequently, the dye concentration in the artery began to rise again. For the purpose of calculation, however, it is necessary to extrapolate the early downslope of the curve to the zero point, as shown by the dashed portion of the curve. In this way, the *time-concentration curve* of the dye in an artery can be measured in its first portion and estimated reasonably accurately in its latter portion.

Once the time-concentration curve has been determined, one can then calculate the mean concentration of dye in the arterial blood for the duration of the curve. In Figure 19–7, this was done by measuring the area under the entire curve and then averaging the concentration of dye for the duration of the curve; one can see from the shaded rectangle straddling the upper curve of the figure that the average concentration of dye was approximately 0.25 mg./100 ml. blood and that the duration of the curve was 12 seconds. However, a total of 5 mg. of dye was injected at the beginning of the experiment. In order for blood carrying only 0.25 mg. of dye in

each 100 ml. to carry the entire 5 mg. of dye through the heart and lungs in 12 seconds, it would be necessary for a total of twenty 100-ml. portions of blood to pass through the heart during this time, which would be the same as a cardiac output of 2 liters per 12 seconds, or 10 liters per minute.

In the bottom curve of Figure 19–7, the blood flow through the heart was considerably slower, and the cardiac output calculated to be 2 liters per 24 seconds, or 5 liters per minute.

To summarize, the cardiac output can be determined from the following formula:

Cardiac output (ml./min.) =

$$\frac{\text{Milligrams of dye injected} \times 60}{\left(\begin{array}{l}\text{Average concentration of dye} \\ \text{in each milliliter of blood} \\ \text{for the duration of the curve}\end{array}\right) \times \left(\begin{array}{l}\text{Duration of} \\ \text{the curve} \\ \text{in seconds}\end{array}\right)}$$

CIRCULATORY SHOCK

Circulatory shock means generalized inadequacy of blood flow throughout the body, to the extent that the tissues are damaged because of too little flow, especially too little delivery of oxygen to the tissue cells. Even the cardiovascular system itself — the heart musculature, the walls of the blood vessels, the vasomotor system, and other circulatory parts — begins to deteriorate, so that the shock becomes progressively worse.

PHYSIOLOGICAL CAUSES OF SHOCK

Since shock usually results from inadequate cardiac output, any factor that can reduce cardiac output can also cause shock. The different factors that can do this can be grouped into two categories:

1. Those that decrease the ability of the heart to pump blood.

2. Those that tend to decrease venous return.

Thus, serious myocardial infarction or any other factor that damages the heart so severely that it cannot pump adequate quantities of blood can cause a type of shock called *cardiogenic shock,* which will be discussed in the following chapter. Also, all the factors that reduce venous return, including (a) diminished blood volume, (b) decreased vasomotor tone, or (c) greatly increased resistance to blood flow, can result in shock. The remainder of this chapter will deal primarily with these factors.

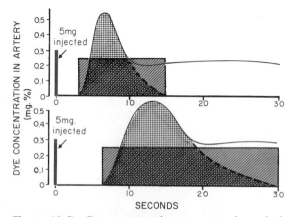

Figure 19–7. Dye concentration curves used to calculate the cardiac output by the dilution method. (The rectangular areas are the calculated average concentrations of dye in the arterial blood for the durations of the respective curves.)

SHOCK CAUSED BY HYPOVOLEMIA — HEMORRHAGIC SHOCK

Hypovolemia means diminished blood volume, and hemorrhage is perhaps the most common cause of hypovolemic shock. Therefore, a discussion of hemorrhage shock will serve to explain many of the basic principles of the shock problem.

Hemorrhage *decreases the mean systemic filling pressure* and as a consequence decreases venous return. As a result, the cardiac output falls below normal, and shock ensues. Obviously, all degrees of shock can result from hemorrhage, from the mildest diminishment of cardiac output to almost complete cessation of output.

Relationship of Bleeding Volume to Cardiac Output and Arterial Pressure

Figure 19–8 illustrates the effect on both cardiac output and arterial pressure of removing blood from the circulatory system over a period of about half an hour. Approximately 10 per cent of total blood volume can be removed with no significant effect on arterial pressure or cardiac output, but greater blood loss usually diminishes the cardiac output first and the pressure later, both of them falling to zero when about 35 to 45 per cent of the total blood volume has been removed.

Sympathetic Reflex Compensation in Shock. Fortunately, the decrease in arterial pressure caused by blood loss initiates powerful sympathetic reflexes (initiated mainly by the baroreceptors) that stimulate the sympathetic vasoconstrictor system throughout the body, resulting in three important effects: (1) The arterioles constrict in most parts of the body, thereby greatly increasing the total peripheral resistance. (2) The veins and venous reservoirs constrict, there-by helping to maintain adequate venous return despite diminished blood volume. (3) Heart activity increases markedly, sometimes increasing the heart rate from the normal value of 72 beats per minute to as much as 200 beats per minute.

Value of the Reflexes. In the absence of the sympathetic reflexes, only 15 to 20 per cent of the blood volume can be removed over a period of half an hour before a person will die, in contrast to 30 to 40 per cent when the reflexes are intact. Therefore, the reflexes extend the amount of blood loss that can occur without causing death to about two times that which would be possible in their absence.

A special value of the reflexes is protection of the blood flow through the coronary and cerebral circulatory systems. Sympathetic stimulation does not cause significant constriction of either the cerebral or the cardiac vessels. In addition, in both these vascular beds local autoregulation is excellent, which prevents moderate changes in arterial pressure from significantly affecting their blood flows. Therefore, blood flow through the heart and brain is maintained at essentially normal levels as long as the arterial pressure does not fall below about 70 mm. Hg, despite the fact that blood flow in many other areas of the body might be decreased almost to zero because of vasospasm.

Nonprogressive and Progressive Hemorrhagic Shock

Figure 19–9 illustrates an experiment that we performed in dogs to demonstrate the effects of different degrees of hemorrhage on the subsequent course of arterial pressure. The dogs were bled rapidly until their arterial pressures fell to different levels. Those dogs whose pressure fell immediately no lower than 45 mm. Hg (Groups I, II, and III) all eventually recovered; the recovery occurred rapidly if the pressure fell only slightly (Group I) but occurred slowly if it fell almost to the 45 mm. Hg level (Group III). On the other hand, when the arterial pressure fell below 45 mm. Hg (Groups IV, V, and VI), all

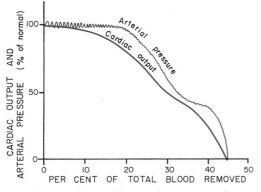

Figure 19–8. Effect of hemorrhage on cardiac output and arterial pressure.

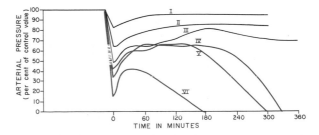

Figure 19–9. Course of arterial pressure in dogs after different degrees of acute hemorrhage. Each curve represents the average results from six dogs.

the dogs died, though many of them hovered between life and death for many hours before the circulatory system began to deteriorate.

This experiment demonstrates that the circulatory system can recover as long as the degree of hemorrhage is no greater than a certain critical amount. However, crossing this critical amount by even a few milliliters of blood loss makes the eventual difference between life and death. Thus, hemorrhage beyond a certain critical level causes shock to become *progressive.* That is, *the shock itself causes still more shock,* the condition becoming a vicious circle that leads eventually to complete deterioration of the circulation and to death.

Nonprogressive Shock — Compensated Shock. If shock is not severe enough to cause its own progression, the person eventually recovers. Therefore, shock of this lesser degree can be called *nonprogressive shock.* It is also frequently called *compensated shock,* meaning that the sympathetic reflexes and other factors have compensated enough to prevent deterioration of the circulation.

The factors that cause a person to recover from moderate degrees of shock are the negative feedback control mechanisms of the circulation that attempt to return cardiac output and arterial pressure to normal levels. These include

1. The *baroreceptor reflexes,* which elicit powerful sympathetic stimulation of the circulation;

2. The *central nervous system ischemic response,* which elicits even more powerful sympathetic stimulation throughout the body but is not activated until the arterial pressure falls below 50 mm. Hg;

3. *Formation of angiotensin,* which constricts the peripheral arteries and causes increased conservation of water and salt by the kidneys, both of which help to prevent progression of the shock;

4. *Formation of vasopressin (antidiuretic hormone),* which constricts the peripheral arteries and veins and causes greatly increased water retention by the kidneys;

5. *Compensatory mechanisms that return the blood volume back toward normal,* including absorption of large quantities of fluid from the intestinal tract, absorption of fluid from the interstitial spaces of the body, conservation of water and salt by the kidneys, and increased thirst and increased appetite for salt, which make the person drink water and eat salty foods if able.

The sympathetic reflexes provide immediate help toward bringing about recovery, for they become maximally activated within 30 seconds after hemorrhage. The readjustment of blood volume by absorption of fluid from the interstitial spaces and from the intestinal tract, as well as the ingestion and absorption of additional quantities of fluid and salt, may require from 1 to 48 hours,

but eventually recovery takes place provided the shock does not become severe enough to enter the progressive stage.

Progressive Shock — The Vicious Circle of Cardiovascular Deterioration. Once shock has become severe enough, the structures of the circulatory system themselves begin to deteriorate, and various vicious circles occur that cause progressively decreasing cardiac output. Some of these are the following:

Cardiac Depression. When the arterial pressure falls low enough, coronary blood flow decreases below that required for adequate nutrition of the myocardium itself. This obviously weakens the heart and thereby decreases the cardiac output still more. As a consequence, the arterial pressure falls still further, and the coronary blood flow decreases more, making the heart still weaker. Thus, a vicious circle develops whereby the shock becomes more and more severe.

Therefore, one of the important features of progressive shock, whether it is hemorrhagic in origin or is caused in any other way, is eventual progressive deterioration of the heart.

Vasomotor Failure. In the early stages of shock various circulatory reflexes cause intense activity of the sympathetic nervous system. This, as discussed above, helps to delay depression of the cardiac output and especially helps to prevent decreased arterial pressure. However, there comes a point at which diminished blood flow to the vasomotor center itself so depresses the center that it becomes progressively less active and finally totally inactive. Therefore, the circulation loses the supportive actions of sympathetic stimulation.

Thrombosis of the Minute Vessels — Sludged Blood. Thrombosis occurring in many of the minute vessels in the circulatory system can also be one of the causes of shock progression. That is, blood flow through many tissues becomes extremely sluggish, but tissue metabolism continues so that large amounts of acid, either carbonic acid or lactic acid, continue to empty into the blood. This acid, plus other deterioration products from the ischemic tissues, causes blood agglutination or actual blood clots, thus leading to minute plugs in the small vessels. Even if the vessels do not become plugged, the tendency for the cells to stick to each other makes it more difficult for blood to flow through the microvasculature, giving rise to the term *sludged blood.*

Increased Capillary Permeability. After many hours of capillary hypoxia the permeability of the capillaries gradually increases and large quantities of fluid begin to transude into the tissues. This further decreases the blood volume, with resultant further decrease in cardiac output, thus making the shock still more severe.

Myocardial Toxic Factor (MTF). During the course of shock, the splanchnic arterioles become strongly constricted, and the extreme ischemia that occurs in the pancreas allows activation of some of the pancreatic enzymes, including trypsin itself. This sets into play degenerative processes within the pancreatic tissues, and a toxic factor called *myocardial toxic factor,* more commonly known as *MTF,* is released into the circulation. MTF has a direct depressant effect on the heart itself, frequently depressing cardiac contractility as much as 50 per cent. The toxin seems to interfere with the function of calcium ions in the excitation-contraction coupling process.

Regardless of the precise mechanism by which MTF reduces cardiac contractility, its generation during the course of shock further exacerbates the shock syndrome and therefore is part of the deteriorative process that causes progression of the shock.

Generalized Cellular Deterioration. As shock becomes very severe, many signs of generalized cellular deterioration occur throughout the body. One organ especially affected is the *liver,* primarily because of the lack of enough nutrients to support the normally high rate of metabolism in liver cells, and partly because of the extreme vascular exposure of the liver cells to any toxic or other abnormal metabolic factors in shock. Among the different effects that are known to occur are these:

1. Active transport of sodium and potassium through the cell membrane is greatly diminished.

2. Mitochondrial activity in the liver cells, as well as in many other tissues of the body, becomes severely depressed.

3. Lysosomes begin to split in widespread tissue areas, with intracellular release of hydrolases that cause further intracellular deterioration.

4. Cellular metabolism of nutrients such as glucose becomes greatly depressed. The activities of some hormones are depressed as well, including as much as 200-fold depression in the action of insulin.

Obviously, all these effects contribute to further deterioration of many different organs of the body, particularly of the liver and also of the heart, thereby further depressing the contractility of the heart.

Irreversible Shock

After shock has progressed to a certain stage, transfusion or any other type of therapy becomes incapable of saving the life of the person. Therefore, the person is then said to be in the *irreversible stage of shock.*

One of the most important of the end-results of deterioration in shock, and one that is perhaps the most significant of all in the development of the final state of irreversibility, is cellular depletion of the high energy compounds.

Though it is clear that deterioration can occur in many different organ systems in shock and that the degeneration in any of these systems can become so severe that it is eventually incompatible with continued life, there is reason to believe that in most instances it is deterioration of the heart itself that makes the shock irreversible. The reason for believing this is the following: Modern therapy is very effective in producing adequate venous return. Administration of blood and other substitution fluids can almost always provide adequate inflow pressure to the heart even in the most severe degrees of shock. But still, in the late stages of shock, the heart fails to pump this inflowing blood. Therefore, the heart is the final weak link in the system.

On the other hand, if one does not use all forms of available therapy, a person can die of shock because of peripheral abnormalities such as continued loss of fluid into the tissues, pooling of blood in greatly distended blood vessels, respiratory failure, acidosis, and so forth. Unfortunately, in such instances, the patient dies of shock that is still reversible if adequate therapy is provided.

Special Role of Oxygen Deficiency in Shock and in Irreversibility. Though poor blood flow leads to tissue deterioration because of deficiency of many different nutrients, deficiency of oxygen almost certainly is the most important. For instance, in a large series of dogs, when the average accumulated oxygen deficit reached 120 milliliters of oxygen per kilogram of body mass, 50 per cent of the animals died regardless of how long it took to accumulate this amount of oxygen deficit.

Therefore, it seems to be clear that the one most important nutrient necessary to prevent cellular deterioration and death during shock is oxygen.

Hypovolemic Shock Caused by Plasma Loss

Loss of plasma from the circulatory system, even without the loss of whole blood, can sometimes be severe enough to cause typical hypovolemic shock. This occurs in the following conditions:

1. *Intestinal obstruction* is often a cause of reduced plasma volume. The resulting distention of the intestine causes fluid to leak from the intestinal capillaries into the intestinal walls and intestinal lumen.

2. Often, in patients who have *severe burns* or other denuding conditions of the skin, so much plasma is lost through the exposed areas that the plasma volume becomes markedly reduced.

3. Loss of fluid from all fluid compartments of the body is called *dehydration;* this, too, can reduce the blood volume and cause hypovolemic shock very similar to that resulting from hemorrhage. Some of the causes of this type of shock are (a) excessive sweating; (b) fluid loss in severe diarrhea or vomiting; (c) excess loss of fluid by nephrotic kidneys; (d) inadequate intake of fluid and electrolytes; and (e) destruction of the adrenal cortices, with consequent failure of the kidneys to reabsorb sodium, chloride, and water.

4. One of the most common causes of circulatory shock is *trauma* to the body. Often the shock results simply from hemorrhage caused by the trauma, but contusion of the body can also damage the capillaries sufficiently to allow excessive loss of plasma into the tissues. Thus, whether or not hemorrhage occurs when a person is severely traumatized, the blood volume can still be markedly reduced.

SEPTIC SHOCK

The condition that was formerly known by the popular name of "blood poisoning" is now called *septic shock* by most clinicians. This means, simply, widely disseminated infection in many areas of the body, with the infection being borne through the blood from one tissue to another and causing extensive damage. Actually, there are many different varieties of septic shock because of the many different types of bacterial infection that can cause it and also because infection in one part of the body will produce different effects from those caused by infection elsewhere in the body.

Septic shock is extremely important to the clinician because it is this type of shock that, more frequently than any other kind of shock besides cardiogenic shock, causes patient death in the modern hospital. Some of the typical causes of septic shock include

1. Peritonitis caused by spread of infection from the uterus and fallopian tubes, frequently resulting from instrumental abortion.

2. Peritonitis resulting from rupture of the gut, sometimes caused by intestinal disease and sometimes by wounds.

3. Generalized infection resulting from spread of a simple skin infection such as streptococcal or staphylococcal infection.

4. Generalized gangrenous infection resulting specifically from gas gangrene bacilli, spreading first through the tissues themselves and finally by way of the blood to the internal organs, especially to the liver.

5. Infection spreading into the blood from the kidney or urinary tract, often caused by colon bacilli.

Special Features of Septic Shock.   Because of the multiple types of septic shock, it is difficult to categorize this condition. However, some features often seen in septic shock are the following:

1. High fever.

2. Marked vasodilatation throughout the body, especially in the infected tissues.

3. High cardiac output in perhaps half of the patients, caused by vasodilatation in the infected tissues and also by high metabolic rate and vasodilatation elsewhere in the body resulting from the high body temperature.

4. Development of microclots in widespread areas of the body, a condition called *disseminated intravascular coagulation.* Also, this causes the clotting factors to be used up so that hemorrhages occur into many tissues, especially into the gut wall and into the intestinal tract.

In the early stages of septic shock the patient usually does not have signs of circulatory collapse but, instead, only signs of the bacterial infection itself. However, as the infection becomes more severe, the circulatory system usually becomes involved either directly or as a secondary result of toxins from the bacteria, and *there finally comes a point at which deterioration of the circulation becomes progressive in the same way that progression occurs in all other types of shock. Therefore, the end stages of septic shock are not greatly different from the end stages of hemorrhagic shock,* even though the initiating factors are markedly different in the two conditions.

EFFECTS OF SHOCK ON THE BODY

Decreased Metabolism in Hypovolemic Shock. The decreased cardiac output in hypovolemic shock reduces the amount of oxygen and other nutrients available to the different tissues, and this in turn reduces the level of metabolism that can be maintained by the different cells of the body. Usually a person can continue to live for only a few hours if the cardiac output falls to as low as 40 per cent of normal.

Muscular Weakness.   One of the earliest symptoms of shock is severe muscular weakness, which is also associated with profound and rapid fatigue whenever patients attempt to use their muscles. This obviously results from the diminished supply of nutrients — especially oxygen — to the muscles.

Body Temperature.   Because of the depressed metabolism in shock, the amount of heat liberated in the body is reduced (except in septic shock, in which the infection may cause an opposite effect). As a result, the body temperature tends to decrease if the body is exposed to even the slightest cold.

Mental Function. In the early stages of shock the person is usually conscious, though signs of mental haziness may be noted. As the shock progresses, the person falls into a state of stupor, and in the last stages of shock even the subconscious mental functions, including vasomotor control and respiration, fail.

Reduced Renal Function. The very low blood flow during shock greatly diminishes urine output or even causes cessation of output, because glomerular pressure falls below the critical value required for filtration of fluid into Bowman's capsule, as explained in Chapter 24. Also, the kidney has such a high rate of metabolism and requires such large amounts of nutrients that the reduced blood flow often causes *tubular necrosis,* which means death of the tubular epithelial cells, with subsequent sloughing and blockage of the tubules, causing total loss of function of the respective nephrons. This is sometimes a serious after effect of shock that occurs during major surgical operations; the patient sometimes survives the shock associated with the surgical procedure and then dies a week or so later of uremia.

BLOOD AND PLASMA TRANSFUSION IN SHOCK

If a person is in shock caused by hemorrhage, the best possible therapy is usually transfusion of whole blood. If the shock is caused by plasma loss, the best therapy is administration of plasma; when dehydration is the cause, administration of the appropriate electrolytic solution can correct the shock.

Unfortunately, whole blood is not always available — under battlefield conditions, for instance. However, plasma can usually substitute adequately for whole blood because it increases the blood volume and restores normal hemodynamics. Plasma cannot restore a normal hematocrit, but the human being can usually stand a decrease in hematocrit to about one-third normal before serious consequences result if the cardiac output is adequate.

Sometimes plasma also is unavailable. For these instances, various *plasma substitutes* have been developed that perform almost exactly the same hemodynamic functions as plasma. One of these is the following:

Dextran Solution as a Plasma Substitute. The principal requirement of a truly effective plasma substitute is that it remain in the circulatory system — that is, not filter through the capillary pores into the tissue spaces. But, in addition, the solution must be nontoxic and must contain appropriate electrolytes to prevent derangement of the extracellular fluid electrolytes on administration. To remain in the circulation the plasma substitute must contain some substance that has a large enough molecular size to exert colloid osmotic pressure.

One of the most satisfactory substances that has been developed thus far for this purpose is dextran, a large polysaccharide polymer of glucose. Certain bacteria secrete dextran as a by-product of their growth, and commercial dextran is manufactured by a bacterial culture procedure. By varying the growth conditions of the bacteria, the molecular weight of the dextran can be controlled to the desired value. Dextrans of appropriate molecular size do not pass through the capillary pores and therefore can replace plasma proteins as colloid osmotic agents.

CIRCULATORY ARREST

A condition closely allied to circulatory shock is circulatory arrest, in which all blood flow completely stops. This occurs frequently on the surgical operating table as a result of *cardiac arrest* or of *ventricular fibrillation.*

Ventricular fibrillation can usually be stopped by strong electroshock of the heart, the basic principles of which were described in Chapter 12.

Cardiac arrest usually results from too little oxygen in the anesthetic gaseous mixture or from a depressant effect of the anesthesia itself. A normal cardiac rhythm can usually be restored by removing the anesthetic and then applying cardiopulmonary resuscitation procedures for a few minutes while supplying the patient's lungs with adequate quantities of ventilatory oxygen.

Effect of Circulatory Arrest on the Brain. The real problem in circulatory arrest is usually not to restore cardiac function but instead to prevent detrimental effects in the brain as a result of the circulatory arrest. In general, four to five minutes of circulatory arrest causes permanent brain damage in over half the patients, and circulatory arrest for as long as ten minutes almost universally destroys most, if not all, of the mental powers.

For many years it has been taught that these detrimental effects on the brain are caused by the cerebral hypoxia that occurs during circulatory arrest. However, recent studies have shown that dogs can almost universally stand as much as 30 minutes of circulatory arrest without permanent brain damage *if the blood is removed from the brain circulation prior to the arrest.* On the basis of these studies, it is postulated that the circulatory arrest causes vascular *clots* to develop throughout the brain and that these clots cause permanent or semipermanent ischemia of brain areas.

REFERENCES

Cardiac Output

Bruce, T. A., and Douglas, J. E.: Dynamic cardiac performance. *In* Frohlich, E. D. (ed.): Pathophysiology, 2nd Ed. Philadelphia, J. B. Lippincott Co., 1976, p. 5.

Coleman, T. G., *et al.*: Control of cardiac output by regional blood flow distribution. *Ann. Biomed. Eng., 2*:149, 1974.

Dodge, H. T., and Kennedy, J. W.: Cardiac output, cardiac performance, hypertrophy, dilatation, valvular disease, ischemic heart disease, and pericardial disease. *In* Sodeman, W. A., Jr., and Sodeman, T. M. (eds.): Pathologic Physiology: Mechanisms of Disease, 6th Ed. Philadelphia, W. B. Saunders Co., 1979, p. 271.

Donald, D. E., and Shepherd, J. T.: Response to exercise in dogs with cardiac denervation. *Am. J. Physiol., 205*:393, 1963.

Dow, P.: Estimations of cardiac output and central blood volume by dye dilution. *Physiol. Rev., 36*:77, 1956.

Green, J. F.: Determinants of systemic blood flow. *In* Guyton, A. C., and Young, D. B. (eds.): International Review of Physiology: Cardiovascular Physiology, III. Vol. 18. Baltimore, University Park Press, 1979, p. 33.

Grodins, F. S.: Integrative cardiovascular physiology: A mathematical synthesis of cardiac and blood vessel hemodynamics. *Q. Rev. Biol., 34*:93, 1959.

Guyton, A. C.: Determination of cardiac output by equating venous return curves with cardiac response curves. *Physiol. Rev., 35*:123, 1955.

Guyton, A. C.: Essential cardiovascular regulation — the control linkages between bodily needs and circulatory function. *In* Dickinson, C. J., and Marks, J. (eds.): Developments in Cardiovascular Medicine. Lancaster, England, MTP Press, 1978, p. 265.

Guyton, A. C., *et al.*: Venous return at various right atrial pressures and the normal venous return curve. *Am. J. Physiol., 189*:609, 1957.

Guyton, A. C., *et al.*: Instantaneous increase in mean circulatory pressure and cardiac output at onset of muscular activity. *Circ. Res., 11*:431, 1962.

Guyton, A. C., *et al.*: Autoregulation of the the total systemic circulation and its relation to control of cardiac output and arterial pressure. *Circ. Res., 28 (Suppl. 1)*:93, 1971.

Guyton, A. C., *et al.*: Circulation: Overall regulation. *Annu. Rev. Physiol., 34*:13, 1972.

Guyton, A. C., *et al.*: Cardiac Output and Its Regulation. Philadelphia, W. B. Saunders Co., 1973.

Keul, J.: The relationship between circulation and metabolism during exercise. *Med. Sci. Sports, 5*:209, 1973.

Longhurst, J. C., and Mitchell, J. H.: Reflex control of the circulation by afferents from skeletal muscle. *In* Guyton, A. C., and Young, D. B. (eds.): International Review of Physiology: Cardiovascular Physiology III. Vol. 18, Baltimore, University Park Press, 1979, p. 125.

Mitchell, J. H., and Wildenthal, K.: Static (isometric) exercise and the heart: Physiological and clinical considerations. *Annu. Rev. Med., 25*:369, 1974.

Prather, J. W., *et al.*: Effect of blood volume, mean circulatory pressure, and stress relaxation on cardiac output. *Am. J. Physiol., 216*:467, 1969.

Sarnoff, S., and Mitchell, J. H.: The regulation of the performance of the heart. *Am. J. Med., 30*:747, 1961.

Sugimoto, T., *et al.*: Effect of tachycardia on cardiac output during normal and increased venous return. *Am. J. Physiol., 211*:288, 1966.

Varat, M. A., *et al.*: Cardiovascular effects of anemia. *Am. Heart J., 83*:415, 1972.

Weisel, R. D., *et al.*: Current concepts measurement of cardiac output by thermodilution. *N. Engl. J. Med., 292*:682, 1975.

Circulatory Shock

Crowell, J. W., and Smith, E. E.: Oxygen deficit and irreversible hemorrhagic shock. *Am. J. Physiol., 206*:313, 1964.

Guyton, A. C., and Crowell, J. W.: Dynamics of the heart in shock. *Fed. Proc., 20*:51, 1961.

Hershey, S. G.: The reticulo-endothelial system: Relationship to shock and host defense. *In* Kaley, G., and Altura, B. M. (eds.): Microcirculation. Vol. III. Baltimore, University Park Press, 1977.

Jamieson, G. A., and Greenwalt, T. J. (eds.): Blood Substitutes and Plasma Expanders. New York, A. R. Liss, 1978.

Kovach, A. G. B., and Sandor, P.: Cerebral blood flow and brain function during hypotension. *Annu. Rev. Physiol., 38*:571, 1976.

Lefer, A. M., *et al.*: Characterization of a myocardial depressant factor present in hemorrhagic shock. *Am. J. Physiol., 213*:492, 1967.

Lewis, D. H.: Microcirculation in low-flow states. *In* Kaley, G., and Altura, B. M. (eds).: Microcirculation. Vol. III. Baltimore, University Park Press, 1977.

Moyer, C. A., and Butcher, H. R.: Burns, Shock, and Plasma Volume Regulation. St. Louis, C. V. Mosby, 1967.

Nagler, A.: Circulatory manifestations of endotoxemia. *In* Kaley, G., and Altura, B. M. (eds.): Microcirculation. Vol. III. Baltimore, University Park Press, 1977.

Schmid-Schönbein, H., and Teitel, P. (eds.): Basic Aspects of Blood Trauma. Hingham, Mass., Kluwer Boston, 1979.

Schumer, W., and Nyhus, L. M. (eds.): Treatment of Shock. Philadelphia, Lea & Febiger, 1971.

Suteu, I., *et al.*: Shock: Pathology, metabolism, shock cell treatment. Tunbridge Wells, Abacus Press, 1977.

Weil, M. H., and Shubin, H.: Shock. Baltimore, Williams & Wilkins, 1967.

Wilkinson, A. W.: Body Fluids in Surgery, 4th Ed. New York, Churchill Livingstone, 1973.

20

Coronary Blood Flow; Cardiac Failure; Heart Sounds; Valvular and Congenital Heart Defects

CORONARY BLOOD FLOW

Approximately one third of all deaths result from coronary artery disease, and almost all elderly persons have at least some impairment of the coronary artery circulation. The purpose of this chapter is to present this subject, emphasizing also the physiology of coronary occlusion and myocardial infarction.

Figure 20–1 illustrates the heart with its coronary blood supply. Note that the main coronary arteries lie on the surface of the heart, and small arteries penetrate into the cardiac muscle mass. The *left coronary artery* supplies mainly the anterior part of the left ventricle, while the *right coronary artery* supplies most of the right ventricle as well as the posterior part of the left ventricle in most persons.

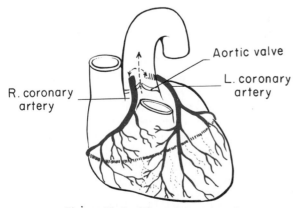

Figure 20–1. The coronary vessels.

NORMAL CORONARY BLOOD FLOW

The resting coronary blood flow in the human being averages approximately 225 ml. per minute, which is about 4 to 5 per cent of the total cardiac output.

In strenuous exercise the heart increases its cardiac output as much as four- to six-fold, and it pumps this blood against a higher than normal arterial pressure. Consequently, the work output of the heart under severe conditions may increase as much as six- to eight-fold. The coronary blood flow increases four- to five-fold to supply the extra nutrients needed by the heart. Obviously, this increase is not quite as much as the increase in work load, which means that the ratio of coronary blood flow to energy expenditure by the heart decreases. However, the "efficiency" of cardiac utilization of energy increases to make up for this relative deficiency of blood supply.

Phasic Changes in Coronary Blood Flow—Effect of Cardiac Muscle Compression. Figure 20–2 illustrates the average blood flow *through the small nutrient vessels* of the coronary system in milliliters per minute in the human heart during systole and diastole as *calculated* from experiments in lower animals. Note from this diagram that the blood flow in the left ventricle falls to a low value during *systole*, which is opposite to the flow in all other vascular beds of the body. The reason is the strong compression of the left ventricular muscle around the intramuscular vessels during systole.

During *diastole*, the cardiac muscle relaxes completely and no longer obstructs the blood flow

201

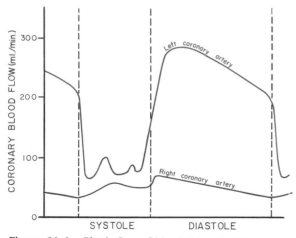

Figure 20–2. Phasic flow of blood through the coronary capillaries of the left and right ventricles.

through the left ventricular capillaries, so that blood now flows rapidly during all of diastole.

Blood flow through the coronary capillaries of the right ventricle undergoes phasic changes similar to those in the coronary capillaries of the left ventricle during the cardiac cycle, but because the force of contraction of the right ventricle is far less than that of the left ventricle, these inverse phasic changes are relatively mild compared with those in the left ventricle, as is shown in Figure 20–2.

CONTROL OF CORONARY BLOOD FLOW

Oxygen Demand as a Major Factor in Local Blood Flow Regulation. Blood flow in the coronaries is regulated almost exactly in proportion to the need of the cardiac musculature for oxygen. Even in the normal resting state, 65 to 70 per cent of the oxygen in the arterial blood is removed as the blood passes through the heart; and because not much oxygen is left, little additional oxygen can be removed from the blood unless the blood flow increases. Fortunately, the blood flow does increase, almost directly in proportion to the metabolic consumption of oxygen by the heart.

Yet, the exact means by which increased oxygen consumption causes coronary dilatation has not been determined. However, in the presence of very low concentrations of oxygen in the muscle cells, a small proportion of the cell's adenosine triphosphate degrades to adenosine, which is a powerful vasodilator and perhaps is the major cause of the vasodilatation when more oxygen is needed. Another possible cause of the dilatation may be that the oxygen deficiency itself could easily cause local vasodilatation because of

lack of the required energy (derived from the oxygen) to keep the coronary vessels contracted against the high arterial pressure.

NERVOUS CONTROL OF CORONARY BLOOD FLOW

Stimulation of the autonomic nerves to the heart can affect coronary blood flow in two ways — directly and indirectly. The direct effects result from direct action of the nervous transmitter substances, acetylcholine and norepinephrine, on the coronary vessels themselves. Usually, acetylcholine causes mild dilatation, and norepinephrine mild constriction.

However, the indirect effects play by far the more important role in normal control of coronary blood flow. Thus, sympathetic stimulation increases both heart rate and heart contractility as well as its rate of metabolism. In turn, the increased activity of the heart sets off local blood flow regulatory mechanisms for dilating the coronary vessels, the blood flow increasing approximately in proportion to the need of the heart muscle for oxygen. In contrast, parasympathetic stimulation slows the heart and also has a slight depressive effect on cardiac contractility, which decreases cardiac oxygen consumption and therefore constricts the coronaries.

ISCHEMIC HEART DISEASE

The single most common cause of death is ischemic heart disease, which results from insufficient coronary blood flow. Approximately 35 per cent of all human beings in the United States die of this cause. Some deaths occur suddenly as a result of an acute coronary occlusion or of fibrillation of the heart, while others occur slowly over a period of weeks to years as a result of progressive slow occlusion of the coronary vessels.

Atherosclerosis as the Cause of Ischemic Heart Disease. The most frequent cause of diminished coronary blood flow is atherosclerosis; this process is the following: In certain persons who have a genetic predisposition to atherosclerosis or in persons who eat excessive quantities of cholesterol and fats, large quantities of cholesterol gradually become deposited beneath the intima at many points in the arteries. Later, these areas of deposit become invaded by fibrous tissue, and they also frequently become calcified. The net result is the development of "atherosclerotic plaques" that protrude into the vessels and either block or partially block blood flow.

Acute Coronary Occlusion. Acute occlusion of the coronary artery frequently occurs in a person who already has serious underlying atherosclerotic coronary heart disease. Usually, an atherosclerotic plaque causes a local blood clot called a *thrombus,* which in turns occludes the artery. The thrombus usually begins where the plaque has grown so much that it has broken through the intima, thus coming in contact with the flowing blood. Because the plaque presents an unsmooth surface to the blood, platelets begin to adhere to it, fibrin begins to be deposited, and blood cells become entrapped and form a clot that grows until it occludes the vessel. Or occasionally the clot breaks away from its attachment on the atherosclerotic plaque and flows to a more peripheral branch of the coronary arterial tree, where it blocks the artery at that point.

Collateral Circulation in the Heart. The degree of damage to the heart caused either by slowly developing atherosclerotic constriction of the coronary arteries or by sudden occlusion is determined to a great extent by the degree of collateral circulation that is already developed or that can develop within a short period of time after the occlusion.

Unfortunately, in a normal heart, relatively few communications exist among the larger coronary arteries. But many anastomoses do exist among the smaller arteries of 20 to 250 microns in diameter, as shown in Figure 20–3.

When a sudden occlusion occurs in one of the larger coronary arteries, the sizes of the minute anastomoses increase to their maximum physical diameters within a few seconds. The blood flow through these minute "collaterals" is less than one-half that needed to keep alive the cardiac muscle that they supply, but the collaterals double in size by the second or third day and often reach normal or almost normal coronary supply within about one month. It is because of these developing collateral channels that a patient recovers from the various types of coronary occlusion.

When atherosclerosis constricts the coronary arteries slowly over a period of many years rather than suddenly, collateral vessels can develop at the same time that the atherosclerosis does. Therefore, the person may never experience an acute episode of cardiac dysfunction. Eventually, however, the sclerotic process develops beyond the limits of even the collateral blood supply to provide the needed blood flow, and even the collaterals develop atherosclerosis. When this occurs, the heart muscle becomes severely limited in its work output, sometimes so much so that the heart cannot pump even the normally required amounts of blood flow. This is the most common cause of cardiac failure.

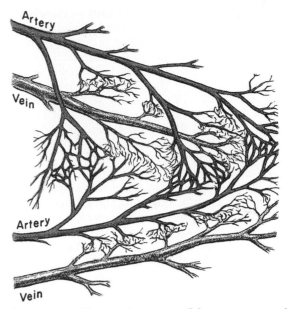

Figure 20–3. Minute anastomoses of the coronary arterial system.

Myocardial Infarction. Immediately after an acute coronary occlusion, blood flow ceases in the coronary vessels beyond the occlusion except for small amounts of collateral flow from surrounding vessels. The area of muscle that has either zero flow or so little flow that it cannot sustain cardiac muscle function is said to be *infarcted.* The overall process is called a *myocardial infarction.*

Soon after the onset of the infarction small amounts of collateral blood seep into the infarcted area, and this, combined with progressive dilatation of the local blood vessels, causes the area to become overfilled with stagnant blood. Simultaneously, the muscle fibers utilize the last vestiges of the oxygen in the blood, causing the hemoglobin to become very dark blue in color. Therefore, the infarcted area takes on a dark bluish hue, and the blood vessels of the area appear to be engorged despite the lack of blood flow. And the cardiac muscle cells begin to swell because of diminished cellular metabolism. Finally, after a few hours, much of the muscle dies if enough collateral vessels do not open.

CAUSES OF DEATH FOLLOWING ACUTE CORONARY OCCLUSION

The four major causes of death following acute myocardial infarction are decreased cardiac output; damming of blood in the pulmonary or systemic veins with death resulting from edema, especially pulmonary edema; fibrillation of the heart; and, occasionally, rupture of the heart.

Decreased Cardiac Output — Cardiogenic Shock. When some of the cardiac muscle fibers are not functioning at all and others are too weak to contract with great force, the overall pumping ability of the affected ventricle is proportionately depressed. When the pumping is depressed enough, the heart fails acutely and the peripheral tissues become severely damaged as a result of too little blood flow. This condition is called *cardiogenic* or *cardiac shock*. It will be discussed later in the chapter.

Damming of Blood in the Venous System. When the heart is not pumping blood forward, it must be damming blood in the venous system of the lungs or the systemic circulation. These effects often cause very little difficulty during the first few hours after the myocardial infarction. Instead, the symptoms develop a few days later for the following reason: The diminished cardiac output leads to diminished blood flow to the kidneys, and for reasons that will be discussed later in the chapter, the kidneys retain large quantities of fluid. This adds progressively to the venous congestive symptoms. Therefore, many patients who seemingly are getting along well will suddenly develop acute pulmonary edema several days after a myocardial infarction and often will die within a few hours after appearance of the initial edema symptoms.

Rupture of the Infarcted Area. During the first day of an acute infarct there is little danger of rupture of the ischemic portion of the heart, but a few days after a large infarct occurs, the dead muscle fibers begin to degenerate, and the dead heart musculature is likely to become very thin and then to rupture.

When a ventricle does rupture, the loss of blood into the pericardial space causes rapid development of *cardiac tamponade* — that is, compression of the heart from the outside by blood collecting in the pericardial cavity. Because the heart is compressed, blood cannot flow into the right atrium with ease, and the patient dies of suddenly decreased cardiac output.

Fibrillation of the Ventricles Following Myocardial Infarction. Many persons who die of coronary occlusion die because of ventricular fibrillation. At least four different factors enter into the tendency for the heart to fibrillate:

First, acute loss of blood supply to the cardiac muscle causes rapid depletion of potassium from the ischemic musculature. This increases the irritability of the cardiac musculature.

Second, ischemia of the muscle causes an "injury current," which can elicit abnormal impulses that cause fibrillation.

Third, powerful sympathetic reflexes develop following massive infarction, principally because the heart does not pump an adequate volume of blood into the arterial tree. The sympathetic stimulation also increases the irritability of the cardiac muscle and thereby predisposes to fibrillation.

Fourth, the myocardial infarction itself often causes the ventricle to dilate excessively. This increases the pathway length for impulse conduction in the heart and also frequently causes abnormal conduction pathways around the infarcted area of the cardiac muscle. Both of these effects predispose to development of circus movements. As was discussed in Chapter 12, this is the basic cause of fibrillation.

PAIN IN CORONARY DISEASE

Normally, a person cannot "feel" his or her heart, but ischemic cardiac muscle does exhibit pain sensation. Exactly what causes this pain is not known, but it is believed that ischemia causes the muscle to release acidic substances such as lactic acid or other pain-promoting products such as histamine or kinins that are not removed rapidly enough by the slowly moving blood.

The pain impulses are conducted mainly through the sympathetic afferent nerve fibers into the central nervous system.

Angina Pectoris. In most persons who develop progressive constriction of their coronary arteries, cardiac pain, called *angina pectoris,* begins to appear whenever the load on the heart becomes too great in relation to the coronary blood flow. This pain is usually felt beneath the upper sternum and is often also referred to surface areas of the body, most often to the left arm and left shoulder but also frequently to the neck and even to the side of the face or to the opposite arm and shoulder. The reason for this distribution of pain is that the heart originates during embryonic life in the neck, as do the arms. Therefore, both of these structures receive pain nerve fibers from the same spinal cord segments.

In general, most persons who have chronic angina pectoris feel the pain when they exercise and also when they experience emotions that increase metabolism of the heart or temporarily constrict the coronary vessels because of sympathetic vasoconstrictor nerve signals. Usually the pain lasts only a few minutes.

Treatment with Vasodilator Drugs. Several vasodilator drugs, when administered during an acute anginal attack, will often given immediate relief from the pain. Two commonly used drugs are nitroglycerin and amyl nitrite.

Surgical Treatment — Aortic-Coronary By-Pass. In many patients with coronary ischemia, the constricted areas of the coronary vessels are located at only a few discrete points, and the coronary vessels beyond these points are of normal or almost normal size. A surgical procedure has been developed in the past few years, called *aortic-coronary by-pass,* for anastomosing small vein grafts to the aorta and to the sides of the more peripheral coronary vessels. Usually, one to four such grafts are performed during the operation, each of which supplies a peripheral coronary artery beyond a block.

The immediate results of this type of surgery have been especially good, causing this to be the most common cardiac operation performed. Anginal pain is relieved in most patients. Unfortunately, it is still too early to determine whether the operative procedure will prolong the lives of the patients because other complications, such as secondary closure of the grafts, may in the long run be more detrimental than the original disease.

CARDIAC FAILURE

Perhaps the most important ailment that must be treated by the physician is cardiac failure, which can result from any heart condition that reduces the ability of the heart to pump blood. Usually the cause is decreased contractility of the myocardium caused by diminished coronary blood flow, but failure to pump adequate quantities of blood can also be caused by damage to the heart valves, external pressure around the heart, vitamin deficiency, primary cardiac muscle disease, or any other abnormality that makes the heart a hypoeffective pump.

ACUTE EFFECTS OF MODERATE GENERALIZED CARDIAC FAILURE

If a heart suddenly becomes severely damaged in any way, such as myocardial infarction, the pumping ability of the heart is immediately depressed. As a result, two essential effects occur: (a) reduced cardiac output and (b) damming of blood in the veins resulting in increased systemic venous pressure. These two effects are shown graphically in Figure 20–4. This figure illustrates, first, a normal cardiac output curve, depicting the state of the circulation prior to the cardiac damage. Point A represents the normal state of the circulation, showing that the normal cardiac output under resting conditions is 5 liters per minute and the right atrial pressure 0 mm. Hg.

Immediately after the heart becomes damaged, the cardiac output curve becomes greatly reduced, falling to the lower, long-dashed curve. Within a few seconds after the acute heart attack, a new circulatory state is established at point B

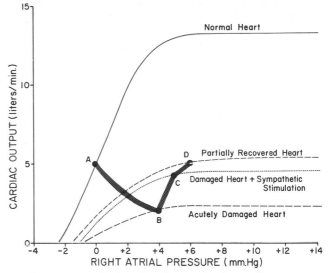

Figure 20–4. Progressive changes in the cardiac output curve following acute myocardial infarction. The cardiac output and right atrial pressure change progressively from point A to point D, as explained in the text.

rather than point A, showing that the cardiac output has fallen to 2 liters per minute, about two-fifths normal, while the right atrial pressure has risen to 4 mm. Hg because blood returning to the heart is dammed up in the right atrium. This low cardiac output is still sufficient to sustain life, but it is likely to be associated with fainting. Fortunately, this acute stage lasts for only a few seconds because sympathetic reflexes occur immediately that can compensate to a great extent for the damaged heart, as follows:

Compensation for Acute Cardiac Failure by Sympathetic Reflexes. When the cardiac output falls precariously low, many of the different circulatory reflexes discussed in Chapter 16 are immediately activated. The best known of these is the baroreceptor reflex, which is activated by diminished arterial pressure. It is probable that the chemoreceptor reflex, the central nervous system ischemic response, and possibly even reflexes originating in the damaged heart itself also contribute to a lesser extent to the nervous response. But whatever all the reflexes might be, the sympathetics become strongly stimulated within a few seconds.

Strong sympathetic stimulation has two major effects on the circulation: First, even the damaged heart responds to the sympathetic stimulation with increased force of contraction. Thus, *the heart becomes a stronger pump, often as much as 100 per cent stronger, under the influence of the sympathetic impulses.* This effect is also illustrated in Figure 20–4, which shows elevation of the cardiac output curve after sympathetic compensation (the dotted curve). Second, sympathetic stimulation increases the tendency for venous return, for it increases the tone of most of the blood vessels of the circulation, especially of the veins, *raising the mean systemic filling pressure* to 12 to 14 mm. Hg, almost 100 per cent above normal. As will be recalled from the discussion in Chapter 19, this greatly increases the tendency of blood to flow back to the heart. Therefore, the damaged heart becomes primed with more inflowing blood than usual, and the right atrial pressure rises still further, which helps the heart to pump larger quantities of blood. Thus, in Figure 20–4 the new circulatory state is depicted by Point C, showing a cardiac output of 4.2 liters per minute and a right atrial pressure of 5 mm. Hg.

The sympathetic reflexes become maximally developed in about 30 seconds. Therefore, a person who has a sudden moderate heart attack may experience nothing more than cardiac pain and a few seconds of fainting. Shortly thereafter, with the aid of the sympathetic reflex compensations described above, the cardiac output may return to a level entirely adequate to sustain the person who remains quiet, though the pain may persist.

The Chronic Stage of Failure. After the first few minutes of an acute heart attack, a prolonged secondary state begins. This is characterized mainly by two events: (1) retention of fluid by the kidneys and (2) progressive recovery of the heart itself over a period of several weeks to months.

Compensation by Renal Retention of Fluid. A low cardiac output has a profound depressant effect on renal function. In general, the urinary output does not return to normal after an acute heart attack until the cardiac output rises either all the way back to normal or almost to normal. This relationship of renal function to cardiac output is one of the most important of all the factors affecting the dynamics of the circulation in chronic cardiac failure.

There are several known causes of the reduced renal output during cardiac failure, all of which are perhaps equally important but in different ways.

1. A decrease in cardiac output has a tendency to *reduce glomerular filtration* by the kidneys because of (a) *reduced arterial pressure* and (b) *intense sympathetic constriction of the arterioles of the kidney.*

2. The reduced blood flow to the kidneys causes *marked increase in renin output,* and this in turn causes the formation of angiotensin by the mechanism described in Chapter 17. The angiotensin has a direct effect on the arterioles of the kidneys of decreasing further the blood flow through the kidneys. Therefore, the net loss of water and salt into the urine is greatly decreased, and the quantities of salt and water in the body fluids increase.

3. In the chronic stage of heart failure, *large quantities of aldosterone are secreted* by the adrenal cortex. This further increases the reabsorption of sodium from the renal tubules and in turn leads to a secondary increase in water reabsorption as well.

The Beneficial Effects of Moderate Fluid Retention in Cardiac Failure. A moderate amount of fluid retention, with consequent increase in blood volume, is a very important factor helping to compensate for the diminished pumping ability of the heart. It does this by increasing the tendency toward venous return mainly because the increased blood volume increases the mean systemic filling pressing, which *increases the pressure gradient for flow of blood toward the heart.*

Detrimental Effects of Excess Fluid Retention in the Severe Stages of Cardiac Failure. In contrast to the beneficial effects of moderate fluid retention in cardiac failure, in severe failure with extreme excesses of fluid retention the fluid then begins to have very serious physiological consequences. They include overstretching of the heart, thus weakening the heart still more; filtration of fluid into the lungs, causing pulmonary edema and consequent deoxygenation of the blood; and, often, development of extensive edema in all of the peripheral tissues of the body as well.

Recovery of the Myocardium Following Myocardial Infarction. After a heart becomes suddenly damaged as a result of myocardial infarction, the natural reparative processes of the body begin immediately to help restore normal cardiac function. A new collateral blood supply begins to penetrate the peripheral portions of the infarcted area, often completely restoring the muscle function. Also, the undamaged musculature hypertrophies, in this way offsetting much of the cardiac damage.

Obviously, the degree of recovery depends on the type of cardiac damage, and it varies from no recovery at all to almost complete recovery. Ordinarily, after myocardial infarction the heart recovers rapidly during the first few days and weeks and will have achieved most of its final state of recovery within four to six months.

The second curve of Figure 20–4 illustrates function of the partially recovered heart a week or so after the acute myocardial infarction. By this time, considerable fluid has been retained in the body, and the tendency toward venous return has increased markedly; therefore, the right atrial pressure has also risen. As a result, the state of the circulation is now changed from Point C to Point D, which represents a *normal* cardiac output of 5 liters per minute but with a right atrial pressure elevated to 6 mm. Hg.

Since the cardiac output has returned to normal, renal output also has returned to normal and no further fluid retention will occur. Therefore, except for the high right atrial pressure represented by Point D in this figure, the person now has essentially normal cardiovascular dynamics *as long as he remains at rest.*

SUMMARY OF THE CHANGES THAT OCCUR FOLLOWING ACUTE CARDIAC FAILURE — "COMPENSATED HEART FAILURE"

To summarize the events discussed in the past few sections describing the moderate heart attack, we may divide the stages into (1) the instantaneous effect of the cardiac damage, (2) compensation by the sympathetic nervous system, and (3) chronic compensations resulting from partial cardiac recovery and renal retention of fluid. All these changes are shown graphically by the very heavy line in Figure 20–4. The progression of this line shows the normal state of the circulation (point A), the state a few seconds after the heart attack but before sympathetic reflexes have occurred (point B), the rise in cardiac output toward normal caused by sympathetic stimulation (point C), and final return of the cardiac output to normal following several days to several weeks of cardiac recovery and fluid retention (point D). This final state is called compensated heart failure.

Compensated Heart Failure. Note especially in Figure 20–4 that the pumping ability of the heart, as depicted by the cardiac function curve, is still depressed to less than one-half normal. This illustrates that factors that increase the right atrial pressure (principally retention of fluid) can maintain the cardiac output at a normal level despite continued weakness of the heart itself. However, one of the results of chronic cardiac weakness is this chronic increase in right atrial pressure itself; in Figure 20–4 it is shown to be 6 mm. Hg. There are many persons, especially in old age, who have completely normal resting cardiac outputs but mildly to moderately elevated right atrial pressures because of compensated heart failure. These persons may not know that they have cardiac damage because the damage more often than not has occurred a little at a time, and the compensation has occurred concurrently with the progressive stages of damage.

DYNAMICS OF SEVERE CARDIAC FAILURE — DECOMPENSATED HEART FAILURE

If the heart becomes severely damaged, then no amount of compensation, either by sympathetic nervous reflexes or by fluid retention, can cause this weakened heart to pump a normal cardiac output. As a consequence, the cardiac output cannot rise to a high enough value to bring about return of normal renal function. Fluid continues to be retained, the person develops progressively more and more edema, and this state of events eventually leads to death. This is called *decompensated heart failure.* The main basis of decompensated heart failure is *failure of the heart to pump sufficient blood to make the kidneys function adequately.*

Treatment of Decompensation. The two ways in which the decompensation process can often be stopped are (1) strengthening the heart in any one of several ways, especially by administration of a cardiotonic drug, such as digitalis, so that it can pump adequate quantities of blood to make the kidneys function normally again, or (2) administering diuretics and reducing water and salt intake, which brings about a balance between fluid intake and output despite the low cardiac output.

Both methods stop the decompensation process by re-establishing normal fluid balance so that at least as much fluid leaves the body as enters it.

UNILATERAL LEFT HEART FAILURE

In the discussions thus far, we have considered failure of the heart as a whole. Yet in a large number of patients, especially those with early acute failure, left-sided failure predominates over right-sided failure, and in rare instances, especially in congenital heart disease, the right side may fail without significant failure of the left side. The effects of right heart failure are much the same as those of failure of the whole heart, but the effects of unilateral left heart failure are different, as follows:

When the left side of the heart fails without concomitant failure of the right side, blood continues to be pumped into the lungs with usual right heart vigor while it is not pumped adequately out of the lungs into the systemic circulation. As a result, large volumes of blood shift from the systemic circulation into the pulmonary circulation.

As the volume of blood in the lungs increases, the pulmonary vessels enlarge, and if the pulmonary capillary pressure rises above 28 mm. Hg, that is, above the colloid osmotic pressure of the plasma, fluid begins to filter out of the capillaries into the interstitial spaces and alveoli, resulting in pulmonary edema.

Thus, among the most important problems of left heart failure are *pulmonary vascular congestion* and *pulmonary edema.* Pulmonary edema sometimes can occur so rapidly that it causes death after only 20 to 30 minutes of severe acute left heart failure.

Course of Events for Several Days After Acute Left Heart Failure. During the several days after the onset of left heart failure, one additional feature must be added to the acute picture. This is retention of fluid resulting from reduced renal function. In moderate acute left heart failure the pulmonary capillary pressure does not rise high enough to cause pulmonary edema. Yet, following renal retention of fluid for the next few days, the blood volume increases and more blood is pumped into the lungs by the right ventricle. Then, the pulmonary capillary pressure rises still more, often rising above the plasma colloid osmotic pressure, resulting in severe pulmonary edema. Indeed, this is a common occurrence: The patient suddenly develops severe pulmonary edema a few days after the acute attack and dies a respiratory death, not a death resulting from diminished cardiac output.

CARDIOGENIC SHOCK

In very severe heart failure, the cardiac output often falls too low to supply the body with adequate blood flow. As a result, the tissues deteriorate rapidly and death ensues. Sometimes death comes in less than an hour; at other times, it comes over a period of several days. The circulatory shock that is caused by inadequate cardiac pumping is called *cardiogenic shock* or *cardiac shock,* and it is sometimes also called the *power failure syndrome.*

Cardiogenic shock is extremely important to the clinician because approximately one-tenth of all patients who have acute myocardial infarction will have enough power failure to die of circulatory shock before the physiologic compensatory measures can come into play to save life. Once cardiac shock has become well established after myocardial infarction, all the typical events occur that also occur in the late stages of other types of circulatory shock, as described in the previous chapter, especially rapid deterioration of almost all bodily functions.

Vicious Circle of Cardiac Deterioration in Cardiogenic Shock. The discussion of circulatory shock in Chapter 19 emphasized the tendency of the heart itself to become progressively damaged when its coronary blood supply is reduced during the course of shock. That is, the low arterial pressure that occurs during shock reduces the coronary supply, which makes the heart still weaker, which makes the shock still worse, the process eventually becoming a vicious circle of cardiac deterioration. In cardiogenic shock caused by myocardial infarction, this problem is greatly compounded by the already existing coronary thrombosis. For instance, in a normal heart, the arterial pressure usually must be reduced below about 45 mm. Hg before cardiac deterioration sets in. However, in a heart that already has a major coronary vessel blocked, deterioration will set in when the arterial pressure falls as low as 80 to 90 mm. Hg.

Unfortunately, even with the best therapy, once the shock syndrome has begun, with the arterial pressure remaining as much as 20 mm. Hg below normal for as long as an hour, 85 per cent of the patients die.

PHYSIOLOGICAL CLASSIFICATION OF CARDIAC FAILURE

From the above discussions, it is apparent that the symptoms of cardiac failure fall into the following three physiological classifications:

Low cardiac output
Pulmonary congestion
Systemic congestion

Low cardiac output usually occurs immediately after a heart attack. If the attack is mainly right-sided, this may be the only symptom. If the acute heart attack is mainly left-sided, concurrent pulmonary congestion almost always occurs along with the low cardiac output, but the pulmonary congestion may be mild (without pulmonary edema) until after considerable fluid has been retained by the kidneys. If the cardiac output is low enough, cardiac shock ensues, with death being likely.

Pulmonary congestion may be the only effect in patients with pure left-sided *chronic* heart failure, because in the chronic stage enough fluid will have been retained to return the cardiac output to normal despite the weak left ventricle — but this occurs at the expense of greatly elevated pulmonary vascular pressures. And since the right heart is not failing, pulmonary congestive symptoms alone can occur with essentially no systemic congestion nor low cardiac output.

Systemic congestion alone can occur in pure right-sided *chronic* heart failure. In this condition there is no pulmonary congestion, and, if sufficient fluids have been retained in the blood to prime the heart sufficiently, the heart may pump a normal cardiac output.

Obviously, all the above classes of heart failure can occur together or in any combination.

CARDIAC RESERVE

Fortunately, the normal heart can increase its output to four to five times normal under conditions of stress in most younger persons and to six to seven times normal in endurance athletes. The

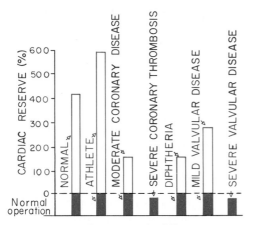

Figure 20–5.   Cardiac reserve in different conditions.

maximum percentage that the cardiac output can increase above normal is called the *cardiac reserve.* Thus, in the normal young adult the cardiac reserve is 300 to 400 per cent. In the athletically trained person it is occasionally as high as 500 to 600 per cent, while in the asthenic person it may be as low as 200 per cent. As an example, during severe exercise the cardiac output of the normal healthy young adult can rise to about five times normal; this is an increase above normal of 400 per cent — that is, a cardiac reserve of 400 per cent.

Any factor that prevents the heart from satisfactorily pumping blood decreases the cardiac reserve. This can result from ischemic heart disease, primary myocardial disease, vitamin deficiency, damage to the myocardium, valvular heart disease, and many other factors, some of which are illustrated in Figure 20–5.

THE HEART SOUNDS

The function of the heart valves was discussed in Chapter 11, and it was pointed out that closure of the valves is associated with audible sounds, though no sounds usually occur when the valves open. The purpose of the present section is to discuss the factors that cause the sounds in the heart, under both normal and abnormal conditions.

NORMAL HEART SOUNDS

Listening with a stethoscope to a normal heart, one hears a sound usually described as "lub, dub, lub, dub ---." The "lub" is associated with closure of the A-V valves at the beginning of systole and the "dub" with closure of the semilunar valves at the end of systole. The "lub" sound is called the *first heart sound* and the "dub" the *second heart*

sound because the normal cycle of the heart is considered to start with the beginning of systole.

Causes of the First and Second Heart Sounds. Closure of the valves in any pump system usually causes a certain amount of noise because the valves close solidly and suddenly over some opening, setting up vibrations in the fluid or walls of the pump. In the heart, the cause of the first heart sound is *vibration of the taut A-V valves immediately after closure,* as well as *vibration of the adjacent blood, walls of the heart, and major vessels around the heart.* That is, contraction of the ventricle causes sudden backflow of blood against the A-V valves, causing them to bulge toward the atria until the chordae tendineae abruptly stop the backbulging. The elastic tautness of the valves then causes the backsurging blood to bounce forward again into each respective ventricle. This sets the blood, the ventricles, and the valves all into vibration. The vibrations then travel to the chest wall, where they can be heard as sound by the stethoscope.

The second heart sound results from vibration of the semilunar valves, the blood, and the walls of the pulmonary artery, aorta, and, to much less extent, ventricles. When the semilunar valves close, they bulge backward toward the ventricles, and their elastic stretch recoils the blood back into the arteries, which causes a short period of reverberation of blood back and forth between the walls of the arteries and the valves. The vibrations set up in the arterial walls are then transmitted along the arteries to the chest wall, where they create sound that can be heard.

AREAS FOR AUSCULTATION OF NORMAL HEART SOUNDS

Listening to the sounds of the body, usually with the aid of a stethoscope, is called *auscultation.* Figure 20–6 illustrates the areas of the chest wall

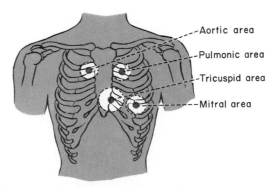

Figure 20–6.   Chest areas from which each valve sound is best heard.

from which the different valvular sounds can best be distinguished. With the stethoscope placed in any one of the special valvular areas, the sounds from all the other valves can also still be heard, though the sound from the special valve is as loud, *relative to the other sounds,* as it ever will be. The cardiologist distinguishes the sounds from the different valves by a process of elimination; that is, he moves the stethoscope from one area to another, noting the loudness of the sounds in different areas and gradually picking out the sound components from each valve.

The areas for listening to the different heart sounds are not directly over the valves themselves. The aortic area is upward along the aorta, the pulmonic area is upward along the pulmonary artery, the tricuspid area is over the right ventricle, and the mitral area is over the apex of the heart, which is the only portion of the left ventricle near the surface of the chest because the heart is rotated so that most of the left ventricle lies behind the right ventricle.

THE PHONOCARDIOGRAM

If a microphone specially designed to detect low-frequency sound waves is placed on the chest, the heart sounds can be amplified. Recording is possible by a high-speed recording apparatus, such as an oscilloscope or a high-speed pen recorder; these devices, described in Chapters 8 and 13, record nerve potentials and electrocardiograms. The recording is called a *phonocardiogram,* and the heart sounds appear as waves, as illustrated schematically in Figure 20–7. Record A is a recording of normal heart sounds, showing especially the vibrations of the first and second heart sounds.

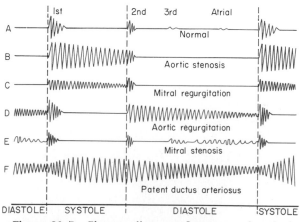

Figure 20–7. Phonocardiograms from normal and abnormal hearts.

RHEUMATIC VALVULAR LESIONS

By far the greatest number of valvular lesions results from rheumatic fever. Rheumatic fever is an autoimmune or allergic disease in which the heart valves are likely to be damaged or destroyed. It is initiated by streptococcal toxin in the following manner:

The entire sequence of events almost always begins with a preliminary streptococcal infection (caused by Group A hemolytic streptococci), such as a sore throat, scarlet fever, or middle ear infection. The streptococci release several different proteins against which antibodies are formed, the most important of which seems to be a protein called the "M" antigen. The antibodies then react with many different tissues of the body, causing either immunologic or allergic damage. These reactions continue to take place as long as the antibodies persist in the blood — six months or more. As a result, rheumatic fever causes damage in many parts of the body but especially in certain very susceptible areas such as the heart valves.

In rheumatic fever, large hemorrhagic, fibrinous, bulbous lesions grow along the inflamed edges of the heart valves. Because the mitral valve receives more trauma during valvular action than any of the other valves, this valve is the one most often seriously damaged, and the aortic valve is the second most frequently damaged. The tricuspid and pulmonary valves are also often involved, but much less severely, probably because the stresses acting on these valves are slight compared with those in the left ventricle.

Scarring of the Valves. The lesions of acute rheumatic fever frequently occur on adjacent valve leaflets simultaneously so that the edges of the leaflets become stuck together. Then, weeks, months, or years later, the lesions become scar tissue, permanently fusing portions of the leaflets. Also, the free edges of the leaflets, which are normally filmy and free-flapping, become solid, scarred masses.

A valve in which the leaflets adhere to each other so extensively that blood cannot flow through satisfactorily is said to be *stenosed.* On the other hand, when the valve edges are so destroyed by scar tissue that they cannot close when the ventricles contract, *regurgitation,* or backflow, of blood occurs when the valve should be closed.

Abnormal Heart Sounds Caused by Valvular Lesions

As illustrated by the phonocardiograms of Figure 20–7, many abnormal heart sounds, known as "murmurs," occur when there are abnormalities of the valves, as follows:

The Murmur of Aortic Stenosis. In aortic stenosis, blood is ejected from the left ventricle through only a small opening of the aortic valve. Because of the resistance to ejection, the pressure in the left ventricle rises sometimes to as high as 350 mm. Hg while the pressure in the aorta is still normal. Thus, a nozzle effect is created *during systole,* with blood jetting at tremendous velocity through the small opening of the valve. This causes *severe turbulence* of the blood in the root of the aorta. The turbulent blood impinging against the aortic walls causes intense vibration, and a loud murmur is transmitted throughout the upper aorta and even into the larger arteries of the neck. This sound is harsh and occasionally so loud that it can be heard several feet away from the patient.

The Murmur of Aortic Regurgitation. In aortic regurgitation, no sound is heard during systole, but *during diastole* blood flows backward from the aorta into the left ventricle, causing a "blowing" murmur of relatively high pitch and with a swishing quality heard maximally over the left ventricle. This murmur results from *turbulence* of blood jetting backward into the blood already in the left ventricle.

The Murmur of Mitral Regurgitation. In mitral regurgitation, blood flows backward through the mitral valve *during systole*. This also causes a high frequency "blowing," swishing sound, which is transmitted most strongly into the left atrium, but the left atrium is so deep within the chest that it is difficult to hear this sound directly over the atrium. As a result, the sound of mitral regurgitation is transmitted to the chest wall mainly through the left ventricle, and it is usually heard best at the apex of the heart.

The Murmur of Mitral Stenosis. In mitral stenosis, blood passes with difficulty from the left atrium into the left ventricle, and because the pressure in the left atrium rarely rises above 35 mm. Hg except for short periods of time, a great pressure differential forcing blood from the left atrium into the left ventricle never develops. Consequently, the abnormal sounds heard in mitral stenosis are usually weak, even though this condition can be severely debilitating.

Phonocardiograms of Valvular Murmurs. Phonocardiograms B, C, D, and E of Figure 20–7 illustrate, respectively, idealized records obtained from patients with aortic stenosis, mitral regurgitation, aortic regurgitation, and mitral stenosis. It is obvious from these phonocardiograms that the aortic stenotic lesion causes the loudest of all these murmurs, and the mitral stenotic lesion causes the weakest, a murmur of very low frequency and rumbling quality. The phonocardiograms show how the intensity of the murmurs varies during differential portions of systole and diastole, and the relative timing of each murmur is also evident. Note especially that the murmurs of aortic stenosis and mitral regurgitation occur only during systole, while the murmurs of aortic regurgitation and mitral stenosis occur only during diastole — if a student does not understand this timing, a moment's pause should be taken until it is understood.

DYNAMICS OF THE CIRCULATION IN AORTIC STENOSIS AND AORTIC REGURGITATION

In aortic stenosis the left ventricle fails to empty adequately, while in aortic regurgitation blood returns to the ventricle after the ventricle has been emptied. Therefore, in either case, the *net* stroke volume output of the heart is reduced, and this in turn tends to reduce the cardiac output, resulting eventually in typical circulatory failure.

Eventual Failure of the Left Ventricle, and Development of Pulmonary Edema. In the early stages of aortic stenosis or aortic regurgitation, the intrinsic ability of the left ventricle to adapt to increasing loads prevents significant abnormalities in circulatory function other than increased work output required of the left ventricle. Therefore, marked degrees of aortic stenosis or aortic regurgitation often occur before the person knows that he has serious heart disease.

However, beyond critical stages of development of these two aortic lesions, the left ventricle finally cannot keep up with the work demand, and as a consequence the left ventricle dilates and cardiac output begins to fall while blood simultaneously dams up in the left atrium and lungs behind the failing left ventricle. The left atrial pressure rises progressively, and at pressures above 30 to 40 mm. Hg, edema appears in the lungs, often leading to death.

CIRCULATORY DYNAMICS IN MITRAL STENOSIS AND MITRAL REGURGITATION

In mitral stenosis blood flow from the left atrium into the left ventricle is impeded, and in mitral regurgitation much of the blood that has flowed into the left ventricle leaks back into the left atrium during systole rather than being pumped into the aorta. Therefore, the effect is reduced net movement of blood from the left atrium into the left ventricle.

Pulmonary Edema in Mitral Valvular Disease. Obviously, the buildup of blood in the left atrium causes progressive increase in left atrial pressure, which can result eventually in the development of serious pulmonary edema. Ordinarily, lethal edema will not occur until the mean left atrial pressure rises at least above 30 mm. Hg; more often it must rise to as high as 40 mm. Hg because the lung lymphatic vasculature enlarges many-fold and can carry fluid away from the lung tissues extremely rapidly.

Enlarged Left Atrium and Atrial Fibrillation. The high left atrial pressure also causes progressive enlargement of the left atrium, which increases the distance that the cardiac impulse must travel in the atrial wall. Eventually, this pathway becomes so long that it predisposes the atria to the development of circus movements. Therefore, in late stages of mitral valvular disease, especially stenosis, atrial fibrillation usually occurs. This state further reduces the pumping effectiveness of the heart and therefore causes still further cardiac debility.

CIRCULATORY DYNAMICS DURING EXERCISE IN PATIENTS WITH VALVULAR LESIONS

During exercise very large quantities of venous blood are returned to the heart from the peripheral circulation. Therefore, all of the dynamic abnormalities that occur in the different types of valvular heart disease become tremendously exacerbated. Even in mild valvular heart disease in which the symptoms may be completely unrecognizable at rest, severe symptoms often develop during heavy exercise. For instance, in patients with aortic valvular lesions, exercise can cause acute left ventricular failure followed by acute pulmonary edema. Also, in patients with mitral disease, exercise can cause so much damming of blood in the lungs that serious pulmonary edema ensues within minutes.

Even in the mildest cases of valvular disease, the patient finds that his cardiac reserve is diminished in proportion to the severity of the valvular dysfunction. That is, the cardiac output does not increase as it should during exercise. Therefore, the muscles of the body fatigue rapidly.

ABNORMAL CIRCULATORY DYNAMICS IN CONGENITAL HEART DEFECTS

Occasionally, the heart or its associated blood vessels are malformed during fetal life; the defect is called a *congenital anomaly*. One of the most common causes of congenital heart defects is a virus infection in the mother during the first trimester of pregnancy when the fetal heart is being formed. Defects are particularly prone to develop when the mother contracts German measles at this time — so often indeed that obstetricians advise termination of pregnancy if German measles does occur in the first trimester. However, some congenital defects of the heart are hereditary; the same defect has been known to occur in identical twins and also in succeeding generations. Children of patients surgically treated for congenital heart disease have ten times as much chance of having congenital heart disease as do other children. Congenital defects of the heart are also frequently associated with other congenital defects of the body.

Though there are many different types of congenital heart defects, two that will illustrate important effects on cardiac function are *patent ductus arteriosus* and *tetralogy of Fallot*.

PATENT DUCTUS ARTERIOSUS — A LEFT-TO-RIGHT SHUNT

During fetal life a large blood vessel called the *ductus arteriosus* connects directly between the aorta and the pulmonary artery. And because the pulmonary arterial pressure in the fetus is higher than the aortic pressure, almost all the pulmonary arterial blood flows directly into the aorta rather than through the lungs. This allows immediate recirculation of the blood through the systemic arteries of the fetus. Obviously, this lack of blood flow through the lungs is of no detriment to the fetus, because the blood is oxygenated by the placenta of the mother.

Closure of the Ductus. As soon as the baby is born its lungs inflate; and not only do the alveoli fill, but also the resistance to blood flow through the pulmonary vascular tree decreases tremendously, allowing pulmonary arterial pressure to fall. Simultaneously, the aortic pressure rises because of sudden cessation of blood flow through the placenta. Thus, the pressure in the pulmonary artery falls, while that in the aorta rises. As a result, forward blood flow through the ductus ceases suddenly at birth, and blood even flows backward from the aorta to the pulmonary artery. This new state of blood flow causes the ductus arteriosus to become occluded within a few hours to a few days in most babies so that blood flow through the ductus does not persist. The ductus closes because the aortic blood now flowing through the ductus has about two times as high an oxygen concentration as the pulmonary blood, and the oxygen constricts the ductus muscle. In

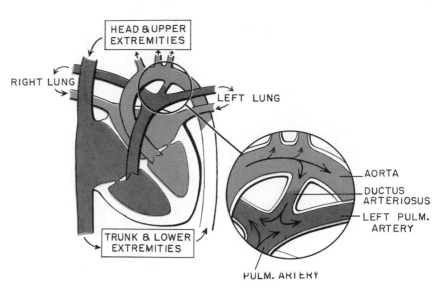

Figure 20–8. Patent ductus arteriosus, illustrating the degree of blood oxygenation in the different parts of the circulation.

many instances it takes several months for the ductus to close completely, and in about 1 of every 5500 babies the ductus never closes, causing the condition known as *patent ductus arteriosus,* which is illustrated in Figure 20–8.

Dynamics of Persistent Patent Ductus. During the early months of an infant's life a patent ductus usually does not cause severely abnormal dynamics because the blood pressure of the aorta then is not much higher than the pressure in the pulmonary artery, and only a small amount of blood flows backward into the pulmonary system. However, in most instances, as the child grows older the differential between the pressure in the aorta and that in the pulmonary artery progressively increases, with corresponding increase in the backward flow of blood from the aorta to the pulmonary artery. Also, the diameter of the partially closed ductus often increases with time, making the condition worse.

Recirculation Through the Lungs. In the older child with a patent ductus, as much as half to two-thirds of the aortic blood flows through the ductus into the pulmonary artery, then through the lungs, into the left atrium, and finally back into the left ventricle, passing through this lung circuit two or more times for every one time that it passes through the systemic circulation.

Diminished Cardiac and Respiratory Reserve. The major effects of patent ductus arteriosus on the patient are low cardiac and respiratory reserve. The left ventricle is already pumping approximately two or more times the normal cardiac output, and the maximum that it can possibly pump is about four to six times normal. Therefore, during exercise the cardiac output can be increased much less than usual. With even moderately strenuous exercise, therefore, the patient is likely to become weak and occasionally even faint from momentary heart

failure. Also, the high pressures in the pulmonary vessels often lead to pulmonary congestion.

As a result of the increased load on the heart and especially because the pulmonary congestion and vascular sclerosis effects become progressively more severe with age, most patients with uncorrected patent ductus die between the ages of 20 and 40.

The Machinery Murmur. In the infant with patent ductus arteriosus, occasionally no abnormal heart sounds are heard because the quantity of reversed bloodflow may be insufficient. As the baby grows older, reaching the age of one to three years, a harsh, blowing murmur begins to be heard in the pulmonic area of the chest. This sound is continuous during the entire heart cycle, but it is much more intense during systole when the aortic pressure is high and much less intense during diastole when the aortic pressure falls very low, so that the murmur waxes and wanes with each beat of the heart, creating the so-called "machinery murmur." The idealized phonocardiogram of this murmur is shown in Figure 20–7F.

Surgical Treatment. Surgical treatment of patent ductus arteriosus is extremely simple, for all one needs to do is to ligate the patent ductus or to divide it and sew the two ends closed.

TETRALOGY OF FALLOT — A RIGHT-TO-LEFT SHUNT

Tetralogy of Fallot is illustrated in Figure 20–9, from which it will be noted that four different abnormalities of the heart occur simultaneously.

First, the *aorta originates from the right ventricle* rather than the left, or it overrides the septum as shown in the figure.

Second, the *pulmonary artery is stenosed,* so that

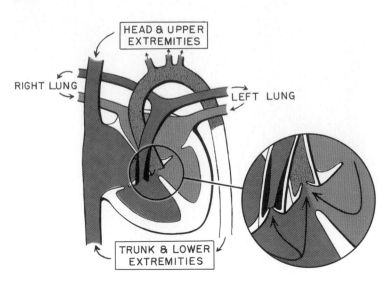

Figure 20–9. Tetralogy of Fallot, illustrating the degree of blood oxygenation in the different parts of the circulation.

much less than normal amounts of blood pass from the right ventricle into the lungs; instead, the blood passes into the aorta.

Third, *blood from the left ventricle flows through a ventricular septal defect* into the right ventricle and then into the aorta or directly into the overriding aorta.

Fourth, because the right side of the heart must pump large quantities of blood against the high pressure in the aorta, its musculature is highly developed, causing an *enlarged right ventricle.*

Abnormal Dynamics. It is readily apparent that the major physiological difficulty caused by tetralogy of Fallot is the shunting of blood past the lungs without its becoming oxygenated. As much as 75 per cent of the venous blood returning to the heart may pass directly from the right ventricle into the aorta without becoming oxygenated. Tetralogy of Fallot is the major cause of cyanosis in babies ("blue babies").

Surgical Treatment. In recent years tetralogy of Fallot has been treated very successfully by surgery. The operation is to open the pulmonary stenosis, to close the septal defect, and to reconstruct the flow pathway into the aorta. This surgery increases the average life expectancy from 5 or 6 years up to 50 or more years.

REFERENCES

Coronary Blood Flow

Bell, J. R., and Fox, A. C.: Pathogenesis of subendocardial ischemia. *Am. J. Med. Sci. 268*:3, 1974.

Berne, R. M., and Rubio, R.: Coronary circulation. *In* Berne, R. M., *et al.* (eds.): Handbook of Physiology, Sec. 2. Vol. 1. Baltimore, Williams & Wilkins, 1979, p. 873.

Bodem, G., and Dengler, H. J. (eds.): Cardiac Glycosides. New York, Springer-Verlag, 1978.

Braunwald, E., and Maroko, P. R.: Limitation of Infarct Size. Chicago, Year Book Medical Publishers, 1978.

Cosby, R. S., *et al.*: Coronary collateral circulation. *Chest,* 66:27, 1974.

Fletcher, G. F., and Cantwell, J. D.: Exercise and Coronary Heart Disease: Role in Prevention, Diagnosis, Treatment. Springfield, Ill., Charles C Thomas, 1978.

Gibbs, C. L., and Chapman, J. B.: Cardiac energetics. *In* Berne, R. M., *et al.* (eds.): Handbook of Physiology, Sec. 2. Vol. 1. Baltimore, Williams & Wilkins, 1979, p. 775.

Gregg, D. E.: Coronary Circulation in Health and Disease. Philadelphia, Lea & Febiger, 1950.

Guyton, R. A., and Daggett, W. M.: The evolution of myocardial infarction: Physiological basis for clinical intervention. *Int. Rev. Physiol., 9*:305, 1976.

Hutchins, G. M.: Pathological changes in aortocoronary bypass grafts. *Annu. Rev. Med., 31*:289, 1980.

Julian, D. G.: Diagnosis and Management of Angina Pectoris. Chicago, Year Book Medical Publishers, 1977.

Klocke, F. J., and Ellis, A. K.: Control of coronary blood flow. *Annu. Rev. Med., 31*:489, 1980.

Long, C. (ed.): Prevention and Rehabilitation in Ischemic Heart Disease. Baltimore, Williams & Wilkins, 1980.

Lüllmann, H., and Peters, T.: Action of Cardiac Glycosides on the Excitation-Contraction Coupling in Heart Muscle. New York, Fischer, 1979.

Neely, J. R., and Morgan, H. E.: Relationship between carbohydrate and lipid metabolism and the energy balance of the heart. *Annu. Rev. Physiol., 36*:413, 1974.

Pujadas, G., *et al.*: Coronary Angiography in the Medical and Surgical Treatment of Ischemic Heart Disease. New York, McGraw-Hill, 1980.

Roskamm, H., and Schumuziger, M. (eds.): Coronary Heart Surgery: A Rehabilitation Measure. New York, Springer-Verlag, 1979.

Schaper, W. (ed.): The Pathophysiology of Myocardial Perfusion. New York, Elsevier/North-Holland, 1979.

Strauss, H. W. (ed.): Cardiovascular Nuclear Medicine. St. Louis, C. V. Mosby, 1979.

Winbury, M. A., and Abiko, Y. (eds.): Ischemic Myocardium and Antianginal Drugs, New York, Raven Press, 1979.

Cardiac Failure

Bradley, R. D.: Studies in Acute Heart Failure. London, Edward Arnold, 1977.

Braunwald, E.: Heart failure — pathophysiological considerations. *In* Dickinson, C. J., and Marks. J. (eds.): Developments in Cardiovascular Medicine. Lancaster, England, MTP Press, 1978, p. 213.

Bruce, T. A., and Douglas, J. E.: Dynamic cardiac performance. *In* Frohlich, E. D. (ed.): Pathophysiology, 2nd Ed. Philadelphia, J. B. Lippincott Co., 1976, p. 5.

Corday, E., and Swan, H. J. C. (ed.): Clinical Strategies in Ischemic Heart Disease. Baltimore, Williams & Wilkins, 1979.

Gibbs, C. L.: Cardiac energetics. *Physiol. Rev., 58*:174, 1978.

Guyton, A. C.: The systemic venous system in cardiac failure. *J. Chronic Dis., 9*:465, 1959.

Guyton, A. C., *et al.*: Cardiac Output and Its Regulation. Philadelphia, W. B. Saunders Co., 1973.

Kones, R. J.: Cardiogenic Shock: Mechanisms and Management. Mount Kisco, N.Y., Futura Publishing Co., 1975.

Koppes, G., *et al.*: Treadmill Exercise Testing. Chicago, Year Book Medical Publishers, 1977.

Mason, D. T., and Loogen, F.: Prevention and management of heart failure. *In* Hayase, S., and Murao, S. (eds.): Cardiology: Proceedings of the VIII World Congress of Cardiology, Tokyo, 1978. New York, Elsevier/North-Holland, 1979, p. 836.

Rapaport, E. (ed.): Current Controversies in Cardiovascular Disease. Philadelphia, W. B. Saunders Co., 1980.

Resnekov, L.: Hemodynamic effects of acute myocardial infarction. *Med. Clin. North Am., 57*:243, 1973.

Swan, H. J. C.: Treatment of cardiac failure with vasodilator drugs. *In* Hayase, S., and Murao, S. (eds.): Cardiology: Proceedings of the VIII World Congress of Cardiology, Tokyo, 1978. New York, Elsevier/North-Holland, 1979, p. 868.

Swan, H. J. C., and Parmley, W. W.: Congestive heart failure. *In* Sodeman, W. A., Jr., and Sodeman, T. M., (eds.): Pathologic Physiology: Mechanisms of Disease, 6th Ed. Philadelphia, W. B. Saunders Co., 1979, p. 313.

Wenger, N. K. (ed.): Exercise and the Heart. Philadelphia, F. A. Davis Co., 1978.

Willis, J. (ed.): The Heart: Update. New York, McGraw-Hill, 1979.

Heart Sounds; Valvular and Congenital Defects

Barry, A.: Aortic and Tricuspid Valvular Disease. New York, Appleton-Century-Crofts, 1980.

Bernstein, E. F. (ed.): Noninvasive Diagnostic Techniques in Vascular Disease. St. Louis, C. V. Mosby, 1978.

Grossman, W. (ed.): Cardiac Catheterization and Angiography, 2nd Ed. Philadelphia, Lea & Febiger, 1980.

Haft, J. I., and Horowitz, M. S.: Clinical Echocardiography. Mount Kisco, N. Y., Futura Publishing Co., 1978.

Heymann, M. A., and Rudolph, A. M.: Control of the ductus arteriosus. *Physiol. Rev., 55*:62, 1975.

Kidd, L., and Somerville, J.: Long-term survival after operations for congenital heart disease. *In* Hayase, S., and Murao, S. (eds.): Cardiology: Proceedings of the VIII World Congress of Cardiology, Tokyo, 1978. New York, Elsevier/North-Holland, 1979, p. 774.

Kobayashi, T., *et al.* (eds.): Cardiac Adaptation. Baltimore, University Park Press, 1978.

Kotler, M. N., and Segal, B. L. (eds.): Clinical Echocardiography. Philadelphia, F. A. Davis Co., 1978.

Kremkau, F. W.: Diagnostic Ultrasound; Physical Principles and Exercises. New York, Grune & Stratton, 1980.

Lear, M. W.: Heartsounds. New York, Simon & Schuster, 1979.

Lundström, N. (ed.): Echocardiography in Congenital Heart Disease. New York, Elsevier/North-Holland, 1978.

Mair, D. D., and Ritter, D. G.: The physiology of cyanotic congenital heart disease. *Int. Rev. Physiol., 9*:275, 1976.

Perloff, J. K.: The Clinical Recognition of Congenital Heart Disease. Philadelphia, W. B. Saunders Co., 1978.

Roberts, W. C., and Spray, T. L.: Pericardial Heart Disease. Chicago, Year Book Medical Publishers, 1977.

Senning, A.: Surgical treatment of valvular disease. *In* Hayase, S., and Murao, S. (eds.): Cardiology: Proceedings of the VIII World Congress of Cardiology, Tokyo, 1978. New York, Elsevier/North-Holland, 1979, p. 751.

Stapleton, J. F., and Harvey, W. P.: Heart sounds, murmurs, and precordial movements. *In* Sodeman, W. A., Jr., and Sodeman, T. M. (eds.): Pathologic Physiology: Mechanisms of Disease, 6th Ed. Philadelphia, W. B. Saunders Co., 1979, p. 335.

Taussig, H.: Congenital Malformations of the Heart. Vol. 1.: General Considerations, 2nd Ed. Vol. 2.: Specific Malformations, 2nd Ed. Cambridge, Mass., Harvard University Press, 1960.

Zak, R., and Rabinowitz, M.: Molecular aspects of cardiac hypertrophy. *Annu. Rev. Physiol., 41*:539, 1979.

21

Muscle Blood Flow During Exercise; Cerebral, Splanchnic, and Skin Blood Flows

The blood flow in some special areas of the body, such as the lungs and the heart, has already been discussed in previous chapters. In the present chapter the characteristics of blood flow in some of the other important tissues — the muscles, the brain, the splanchnic system, and the skin — are presented.

BLOOD FLOW IN SKELETAL MUSCLES AND ITS REGULATION IN EXERCISE

Very strenuous exercise is the most stressful condition that the normal circulatory system faces. This is true because the blood flow in muscles can increase more than 20-fold (a greater increase than in any other tissue of the body) and also because there is such a very large mass of skeletal muscle in the body. The total muscle blood flow can become great enough to increase the cardiac output in the normal young adult to as much as five times normal and in the well-trained athlete to as much as six to seven times normal.

RATE OF BLOOD FLOW IN MUSCLES

During rest, blood flow in skeletal muscle averages 3 to 4 ml. per minute per 100 grams of muscle. However, during extreme exercise this rate can increase as much as 15- to 25-fold, rising to 50 to 80 ml. per 100 grams of muscle.

Intermittent Flow During Muscle Contraction. Figure 21–1 illustrates a study of blood flow changes in the calf muscles of the human leg during strong rhythmic contraction. Note that the flow increases and decreases with each muscle contraction, decreasing during the contraction phase and increasing between contractions. At the end of the rhythmic contractions, the flow remains very high for one to two minutes and then gradually fades toward normal.

The cause of the decreased flow during sustained muscle contraction is compression of the blood vessels by the contracted muscle. During strong *tetanic* contraction, blood flow can be totally stopped.

Opening of Muscle Capillaries During Exercise. During rest, only 20 to 25 per cent of the muscle capillaries are open. But during strenuous exercise all the capillaries open up, which can be demonstrated by studying histologic specimens

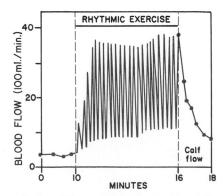

Figure 21–1. Effects of muscle exercise on blood flow in the calf of a leg during strong rhythmic contraction. The blood flow was much less during contraction than between contractions. (From Barcroft and Dornhorst; *J. Physiol., 109*:402, 1949.)

removed from muscles appropriately stained during exercise. It is this opening up of dormant capillaries that allows most of the increased blood flow. It also diminishes the distance that oxygen and other nutrients must diffuse from the capillaries to the muscle fibers and contributes a much increased surface area through which nutrients can diffuse from the blood.

CONTROL OF BLOOD FLOW IN SKELETAL MUSCLE

Local Regulation. The tremendous increase in muscle blood flow that occurs during skeletal muscle activity is caused primarily by local effects in the muscles acting directly on the arterioles to cause vasodilatation, probably caused by several different factors all operating at the same time. One of the most important of these is reduced oxygen in the muscle tissues. That is, during muscle activity the muscle utilizes oxygen very rapidly, thereby decreasing its concentration in the tissue fluids. This in turn causes vasodilatation either because the vessel walls cannot maintain contraction in the absence of oxygen or because oxygen deficiency causes release of vasodilator substances. The vasodilator substance that has been suggested most widely in recent years has been adenosine.

Other vasodilator substances released during muscle contraction include potassium ions, acetylcholine, adenosine triphosphate, lactic acid, and carbon dioxide. Unfortunately, we still do not know quantitatively how great a role each of these plays in increasing muscle blood flow during muscle activity; this subject was discussed in more detail in Chapter 16.

Nervous Control of Muscle Blood Flow. In addition to the local tissue regulatory mechanism, the skeletal muscles are also provided with sympathetic nerves. However, these have so much less effect on muscle blood flow than do the local control factors that they are usually unimportant. They do help to maintain the arterial pressure by constricting the arterioles following severe hemorrhage.

CIRCULATORY READJUSTMENTS DURING EXERCISE

Three major effects occur during exercise that are essential for the circulatory system to supply the tremendous blood flow required by the muscles. These are (1) mass discharge of the sympathetic nervous system, (2) increase in cardiac output, and (3) increase in arterial pressure.

Mass Sympathetic Discharge. At the onset of exercise, signals are transmitted not only from the brain to the muscle to cause muscle contraction but also from the higher levels of the brain into the vasomotor center to initiate mass sympathetic discharge. Simultaneously, the parasympathetic signals to the heart are greatly attenuated. Therefore, two major circulatory effects result. First, the heart is stimulated to greatly increased heart rate and pumping strength. Second, all the blood vessels of the peripheral circulation are strongly contracted except the vessels in the active muscles, which are strongly vasodilated by the local vasodilator effects in the muscles themselves. Thus, the heart is stimulated to supply the increased blood flow required by the muscles, and blood flow through most nonmuscular areas of the body is temporarily reduced, thereby temporarily "lending" their blood supply to the muscles. This effect accounts for as much as 2.5 liters of extra blood flow to the muscles. It is exceedingly important to the wild animal running for its life, because even a fractional increase in running speed may make the difference between life and death. However, two of the organ circulatory systems, the coronary and cerebral systems, are spared this vasoconstrictor effect because both of these circulatory areas have very poor vasoconstrictor innervation — fortunately so, because both the heart and the brain are as essential to exercise as are the skeletal muscles themselves.

A Muscle Reflex That Stimulates the Sympathetic Nervous System. Aside from the sympathetic stimulation caused by direct signals from the brain, reflex signals from the contracting muscles are also believed to pass up the spinal cord to the vasomotor center and to excite the sympathetic nerves. These signals probably are initiated by metabolic end-products acting on small sensory nerve endings in the muscle tissue.

Increase in Cardiac Output. The increase in cardiac output that occurs during exercise results mainly from the intense local vasodilatation in active muscles. As was explained in Chapter 19 in relation to the basic theory of cardiac output regulation, local vasodilatation increases the venous return of blood back to the heart. The heart in turn pumps this extra returning blood immediately back to the muscles through the arteries. Thus, it is mainly the muscles themselves that determine the amount of increase in cardiac output — up to the limit of the heart's ability to respond.

Another factor that greatly helps to cause the large increase in venous return is the strong sympathetic stimulation of the veins. This stimulation greatly increases the mean systemic filling pressure, sometimes to as high as 30 mm. Hg (four times normal), and is therefore important in increasing venous return.

Mechanisms by Which the Heart Increases Its Output. One of the principal mechanisms by which the heart increases its output during exercise is the Frank-Starling mechanism, which was discussed in Chapter 11. Via this mechanism, when increased quantities of blood flow from the veins into the heart and dilate its chambers, the heart muscle contracts with increased force, thus also pumping an increased volume of blood with each heart beat. However, in addition to this basic intrinsic cardiac mechanism, the heart is also strongly stimulated by the sympathetic nervous system, and the normal parasympathetic inhibition is reduced or eliminated. The net effects are greatly increased heart rate (occasionally to as high as 200 beats per minute) and near-doubling of the cardiac muscle strength of contraction. These two effects combine to make the heart capable of pumping about 100 per cent more blood than would be true based on the Frank-Starling mechanism alone.

Increase in Arterial Pressure. The mass sympathetic discharge throughout the body during exercise and the resultant vasoconstriction of most of the blood vessels besides those in the active muscles almost always increase the arterial pressure during exercise. This increase can be as little as 20 mm. Hg or as great as 80 mm. Hg, depending on the conditions under which the exercise is performed. For instance, when a person performs exercise under very tense conditions but uses only a few muscles, the sympathetic response still occurs throughout the body, while vasodilatation occurs in only a few muscles. Therefore, the net effect is mainly one of vasoconstriction, often increasing the mean arterial pressure to as high as 180 mm. Hg. Such a condition occurs in a person standing on a ladder and nailing with a hammer on the ceiling above. The tenseness of the situation is obvious, and yet the amount of muscle vasodilatation is relatively slight.

On the other hand, when a person performs whole-body exercise, such as running or swimming, the increase in arterial pressure is usually only 20 to 40 mm. Hg. The lack of a tremendous rise in pressure results from the extreme vasodilatation occurring in large masses of muscle.

In rare instances, persons are found in whom the sympathetic nervous system is absent, either because of congenital absence or because of surgical removal. When such a person exercises, instead of the arterial pressure rising, the pressure actually falls, and as a result, the cardiac output rises only about one-third as much as it does normally.

Importance of the Arterial Pressure Rise During Exercise. In the well-trained athlete, it has been calculated, the muscle blood flow can increase at least 20-fold. Though most of this increase results from vasodilatation in the active muscles, the increase in arterial pressure also plays an important role. If one remembers that an increase in pressure not only forces extra blood through the muscle because of the increased pressure itself but also dilates the blood vessels, he can see that as little as a 20 to 40 mm. Hg rise in pressure can at times actually double peripheral blood flow.

It is especially important to note that in animals or human beings who do not have a sympathetic nervous system, the fall in arterial pressure that occurs during exercise has a strong negating effect on the rise in cardiac output that normally occurs. In such instances, the cardiac output can almost never be increased more than two-fold, instead of the four- to seven-fold that can occur when the arterial pressure rises above normal.

The Cardiovascular System as the Limiting Factor in Heavy Exercise. The capability of an athlete to enhance cardiac output and consequently to deliver increased quantities of oxygen and other nutrients to his or her tissues is the major factor that determines the degree of prolonged heavy exercise that the athlete can sustain. For instance, the speed of a marathon runner is almost directly proportional to his ability to enhance cardiac output. Therefore, the ability of the circulatory system to adapt to exercise is equally as important as the muscles themselves in setting the limit for the performance of muscle work.

THE CEREBRAL CIRCULATION

NORMAL RATE OF CEREBRAL BLOOD FLOW

The normal blood flow through brain tissue averages 50 to 55 ml. per 100 grams of brain per minute. For the entire brain of the average adult, this is approximately 750 ml. per minute, or 15 per cent of the total resting cardiac output.

REGULATION OF CEREBRAL BLOOD FLOW

As in most other tissues of the body, cerebral blood flow is highly related to the metabolism of the cerebral tissue. At least three different metabolic factors have been shown to have very potent effects on cerebral blood flow. These are carbon dioxide concentration, hydrogen ion concentration, and oxygen concentration. An increase in either the carbon dioxide or the hydrogen ion concentration increases cerebral blood flow, whereas a decrease in oxygen concentration increases the flow.

Regulation of Cerebral Blood Flow in Response to Excess Carbon Dioxide or Hydrogen Ion Concentration. An increase in carbon dioxide concentration in the arterial blood perfusing the brain greatly increases cerebral blood flow. This is illustrated in Figure 21–2, which shows that doubling the arterial P_{CO_2} by breathing carbon dioxide also approximately doubles the blood flow.

Carbon dioxide increases cerebral blood flow by combining with water in the body fluids to form carbonic acid, with subsequent dissociation to form hydrogen ions. The hydrogen ions then cause vasodilatation of the cerebral vessels — the dilatation being almost directly proportional to the increase in hydrogen ion concentration.

Any other substance that increases the acidity of the brain tissue, and therefore also increases the hydrogen ion concentration, increases blood flow as well. Such substances include lactic acid, pyruvic acid, or any other acidic material formed during the course of metabolism.

Importance of the Carbon Dioxide and Hydrogen Ion Control of the Cerebral Blood Flow. Increased hydrogen ion concentration greatly depresses neuronal activity; conversely, diminished hydrogen ion concentration greatly increases neuronal activity. Therefore, it is fortunate that an increase in hydrogen ion concentration causes an increase in the blood flow, which in turn carries both carbon dioxide and dissolved acids away from the brain tissues. Loss of the carbon dioxide removes carbonic acid from the tissues, and this, along with the removal of other acids, reduces the hydrogen ion concentration back

toward normal. Thus, this mechanism helps to maintain a very constant hydrogen ion concentration in the cerebral fluids and therefore also to maintain a normal level of neuronal activity.

Oxygen Deficiency as a Regulator of Cerebral Blood Flow. Except during periods of intense brain activity, the utilization of oxygen by the brain tissue remains within very narrow limits — within a few per cent of 3.5 ml. of oxygen per 100 grams of brain tissue per minute. If the blood flow to the brain ever becomes insufficient to supply this needed amount of oxygen, the oxygen deficiency mechanism for vasodilatation, which was discussed in Chapter 16 and which functions in essentially all tissues of the body, immediately causes vasodilatation, returning the blood flow and transport of oxygen to the cerebral tissues near to normal. Thus, this local blood flow regulatory mechanism is much the same as that existing in the coronary and skeletal muscle circulations and in many other circulatory areas of the body.

Experiments have shown that a decrease in cerebral *venous* blood P_{O_2} below approximately 30 mm. Hg (normal value is about 35 mm. Hg) will begin to increase cerebral blood flow. It is very fortunate that flow does respond at this critical level because brain function begins to become deranged at not much lower values of P_{O_2}, especially at levels below 20 mm. Hg. Thus, the oxygen mechanism for local regulation of cerebral blood flow is a very important protection against diminished cerebral neuronal activity and therefore against derangement of mental capability.

Effect of Cerebral Activity on the Flow. A radioactive method has recently been developed to record blood flow in as many as 256 isolated segments of the human cerebral cortex simultaneously. Using this technique, it has become clear that the blood flow in each individual segment of the brain changes within seconds in response to changes in local neuronal activity. For instance, simply clasping the hand causes an immediate increase in blood flow in the motor cortex of the opposite side of the brain. Reading a book increases the blood flow in multiple areas of the brain, especially in the occipital cortex and in the language areas of the temporal cortex. This measuring procedure can also be used to localize the origin of epileptic attacks, for the blood flow increases acutely and markedly in the focal point of the attack at its very onset.

Illustrating the effect of local neuronal activity on cerebral blood flow, Figure 21–3 shows a 40 per cent increase in occipital blood flow recorded in a cat when intense light was shone into its eyes for a period of one-half minute.

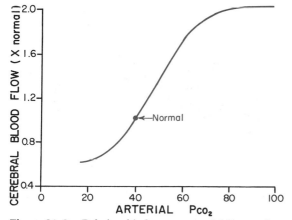

Figure 21–2. Relationship between arterial P_{CO_2} and cerebral blood flow.

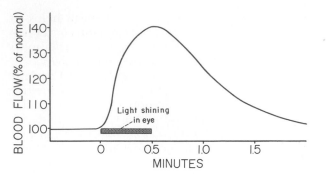

Figure 21–3. Increase in blood flow to the occipital regions of the brain when a light is flashed in the eyes of an animal.

Effect of Arterial Pressure on Cerebral Blood Flow. Figure 21–4 shows cerebral blood flow measured in human beings having different blood pressures. This figure shows especially the extreme constancy of cerebral blood flow between the limits of 60 and 180 mm. Hg mean arterial pressure. Cerebral blood flow suffers only when the arterial pressure falls below approximately the 60 mm. Hg mark.

Minor Importance of Autonomic Nerves in Regulating Cerebral Blood Flow. Sympathetic nerves from the cervical sympathetic chain pass upward along the cerebral arteries to supply the superficial cerebral vessels. Also, parasympathetic fibers from the great superficial petrosal and facial nerves supply some of these vessels as well. However, transection of either the sympathetic or the parasympathetic nerves causes no measurable effect on cerebral blood flow. Yet, though these nerves normally play little role in the control of cerebral blood flow, sympathetic reflexes do cause vasospasm in cerebral vessels in some instances of stroke, subdural hematoma, tumor, and other types of brain damage.

Brain Edema

One of the most serious complications of abnormal cerebral hemodynamics is the development of brain edema. Because the brain is encased in a solid vault, the accumulation of edema fluid compresses the blood vessels, with eventual depression of blood flow and destruction of brain tissue. The usual cause of brain edema is either greatly increased capillary pressure or damage to the capillary endothelium. The most common precipitating factor is brain concussion, in which the brain tissues and capillaries are traumatized and capillary fluid leaks into the traumatized tissues.

THE SPLANCHNIC CIRCULATION

A large share of the cardiac output flows through the vessels of the intestines and through the spleen, finally coursing into the portal venous system and then through the liver, as illustrated in Figure 21–5. This is called the portal circulatory system, and it, plus the arterial blood flow into the liver, is called the splanchnic circulation.

BLOOD FLOW THROUGH THE LIVER

About 1100 ml. of portal blood enters the liver each minute. This flows through the *hepatic sinuses* in close contact with the cords of liver parenchymal cells. Then it enters the *central veins* of the liver and from there flows into the vena cava.

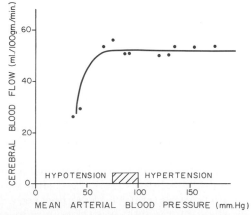

Figure 21–4. Relationship of mean arterial pressure to cerebral blood flow in normotensive, hypotensive, and hypertensive persons. (Modified from Lassen: *Physiol. Rev., 39*:183, 1959.)

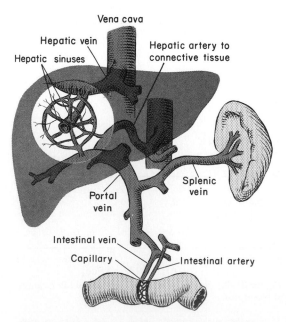

Figure 21–5. The portal and hepatic circulations.

In addition to the portal blood flow, approximately 350 ml. of blood flows into the liver each minute through the hepatic artery, making a total hepatic flow of almost 1500 ml. per minute, or an average of 29 per cent of the total cardiac output.

Control of Liver Blood Flow. Three-quarters of the blood flow through the liver is derived from the portal blood flow into the liver; this flow is controlled by the various factors that determine flow through the gastrointestinal tract and spleen — factors that will be discussed in subsequent sections of this chapter.

The additional one-quarter of the blood flow through the liver is derived from the hepatic arteries; its flow rate is determined primarily by local metabolic factors in the liver itself. For instance, a decrease in oxygen in the hepatic blood causes an increase in hepatic arterial blood flow, indicating that the need to deliver nutrients to the liver tissues has a direct vasodilating effect.

Reservoir Function of the Liver. Because the liver is an expandable and compressible organ, large quantities of blood can be stored in its blood vessels. Its normal blood volume, including both that in the hepatic veins and hepatic sinuses, is about 500 ml., or 10 per cent of the total blood volume. However, when high pressure in the right atrium causes back pressure in the liver, the liver expands, and as much as 1 liter of extra blood occasionally is thereby stored in the hepatic veins and sinuses. This occurs especially in cardiac failure with peripheral congestion, which was discussed in Chapter 20.

Thus, in effect, the liver is a large expandable venous organ capable of acting as a valuable blood reservoir in times of excess blood volume and capable of supplying extra blood in times of diminished blood volume.

The Blood-Cleansing Function of the Liver. Blood flowing through the intestinal capillaries picks up many bacteria from the intestines. Indeed, a sample of blood from the portal system almost always grows colon bacilli when cultured, whereas growth of colon bacilli from blood in the systemic circulation is extremely rare. Special high-speed motion pictures of the action of Kupffer cells, the large phagocytic cells that line the hepatic sinuses, have demonstrated that these cells can cleanse blood extremely efficiently as it passes through the sinuses; when a bacterium comes into momentary contact with a Kupffer cell, in less than 0.01 second the bacterium passes inward through the wall of the Kupffer cell to become permanently lodged therein until it is digested. Probably not over 1 per cent of the bacteria entering the portal blood from the intestines succeeds in passing through the liver into the systemic circulation.

BLOOD FLOW THROUGH THE INTESTINAL VESSELS

About four-fifths of the portal blood flow originates in the intestines and stomach (about 850, ml. per minute), and the remaining one-fifth originates in the spleen and pancreas. Over two-thirds of the intestinal flow goes to the mucosa to supply the energy needed for forming the intestinal secretions and for absorbing the digested food.

Control of Gastrointestinal Blood Flow. Blood flow in the gastrointestinal tract is controlled in almost exactly the same way as in most other areas of the body: mainly by local regulatory mechanisms. Furthermore, blood flow to the mucosa and submucosa, where the glands are located and where absorption occurs, is controlled separately from blood flow to the musculature. When glandular secretion increases, so does mucosal and submucosal blood flow. Likewise, when motor activity of the gut increases, blood flow in the muscle layers increases.

However, the precise mechanisms by which alterations in gastrointestinal activity alter the blood flow are not completely understood. It is known that decreased availability of oxygen to the gut increases local blood flow in the same way that this occurs elsewhere in the body; therefore, local regulation of blood flow in the gut might occur entirely secondarily to changes in metabolic rate. On the other hand, it is also known that various peptide hormones are released from the mucosa of the intestinal tract during the digestive process and that these in turn cause mucosal vasodilatation. The best known of these hormones are *gastrin, secretin,* and *cholecystokinin.* Also, it has been claimed that some or all of the gastrointestinal glands form the substance *bradykinin* at the same time that they release their secretions. The bradykinin in turn has been postulated to cause mucosal vasodilatation. However, crucial experiments have not yet proved this mechanism.

Nervous Control of Gastrointestinal Blood Flow. Stimulation of the parasympathetic nerve (the vagi) to the *stomach* and *lower colon* increases local blood flow at the same time that it increases glandular secretion. However, this increased flow probably results from the increased glandular activity.

Sympathetic stimulation, in contrast, has a direct effect on essentially all blood vessels of the gastrointestinal tract of causing intense vasoconstriction. However, after a few minutes of this vasoconstriction, the flow returns to or almost to normal via a mechanism called "autoregulatory escape." That is, the local metabolic vasodilator mechanisms that are elicited by ischemia become prepotent over the sympathetic vasoconstriction and therefore redilate the arterioles, thus causing return of the necessary nutrient blood flow to the gastrointestinal glands and muscle.

A major value of sympathetic vasoconstriction in the gut is that it allows shutting off of splanchnic blood flow for short periods of time during heavy exercise when increased flow is needed by the skeletal muscle and heart.

The vasoconstriction of the intestinal and mesenteric *veins* that is caused by sympathetic stimulation does not "escape." Instead, both short and prolonged periods of sympathetic stimulation decrease the volume of these veins and thereby displace large amounts of blood into other parts of the circulation. In hemorrhagic shock or other states of low blood volume, this mechanism can provide several hundred milliliters of extra blood to sustain the general circulation.

PORTAL VENOUS PRESSURE

The liver offers a moderate amount of resistance to blood flow from the portal system to the vena cava. As a result, the pressure in the portal vein averages 8 to 10 mm. Hg, which is considerably higher than the almost zero pressure in the vena cava. Because of this high portal venous pressure, the pressures in the intestinal venules and capillaries have a much greater tendency to become abnormally high than is true elsewhere in the body.

Blockage of the Portal System. Frequently, extreme amounts of fibrous tissue develop within the liver structure, destroying many of the parenchymal cells and eventually contracting around the blood vessels, thereby greatly impeding the flow of portal blood through the liver. This disease process is known as *cirrhosis of the liver*. It results most frequently from alcoholism, but it can also follow ingestion of poisons such as carbon tetrachloride, virus diseases such as infectious hepatitis, or infectious processes in the bile ducts.

The portal system is also occasionally blocked by a large clot developing in the portal vein or in its major branches.

When the portal system is suddenly blocked, the return of blood from the intestines and spleen to the systemic circulation is tremendously impeded, the capillary pressure rising as much as 15 to 20 mm. Hg, and the patient often dies within a few hours because of excessive loss of fluid from the capillaries into the lumens and walls of the intestines.

Ascites as a Result of Portal Obstruction. Ascites is free fluid in the peritoneal cavity. It results from exudation of fluid either from the surface of the liver or from the surfaces of the gut and its mesentery. Ascites usually will develop

only when the outflow of blood from the liver into the inferior vena cava is blocked. This causes extremely high pressure in the liver sinusoids, which in turn causes fluid to weep from the surfaces of the liver. The weeping fluid is almost pure plasma, containing tremendous quantities of protein. The protein, because it causes a high colloid osmotic pressure in the abdominal fluid, then pulls, by osmosis, additional fluid from the surfaces of the gut and mesentery.

THE SPLENIC CIRCULATION

The Spleen as a Reservoir. The capsule of the spleen in many lower animals contains large amounts of smooth muscle, and sympathetic stimulation causes intense contraction of the spleen. Conversely, sympathetic inhibition results in considerable splenic expansion with consequent storage of blood.

In humans, the splenic capsule is nonmuscular, but even so, dilatation of vessels within the spleen can still cause the spleen to store several hundred milliliters of blood at times. Then, under the influence of sympathetic stimulation, constriction of the vessels will express most of this blood into the general circulation. But the spleen is so small, only 150 to 200 ml. in volume, that this reservoir function in human beings is of relatively little importance.

As illustrated in Figure 21–6, two areas exist in the spleen for the storage of blood: the venous sinuses and the pulp. Small vessels flow directly into the venous sinuses, and when the spleen distends, the venous sinuses swell, thus storing blood.

In the splenic pulp, the capillaries are very permeable, so that much of the blood passes first into the pulp and then oozes through it before entering the venous sinuses. As the spleen enlarges, many cells (but not the plasma) become stored in the pulp. Therefore, the net quantity of

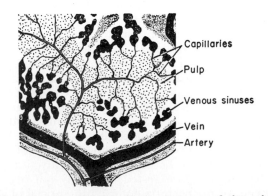

Figure 21–6. The functional structures of the spleen. (Modified from Bloom and Fawcett: Textbook of Histology. Philadelphia, W. B. Saunders Company, 1975.)

red blood cells in the general circulation decreases slightly when the spleen enlarges. The spleen can store enough cells that splenic contraction can cause the hematocrit of the systemic blood to increase in humans as much as 1 to 2 per cent and as much as 3 to 4 per cent in some lower animals. This increased hematocrit is an aid to the body during periods of stress.

The Blood-Cleansing Function of the Spleen: Removal of Old Cells. Blood passing through the splenic pulp before it enters the sinuses undergoes thorough squeezing. Indeed, it is believed that the red blood cells re-enter the venous sinuses from the pulp by "diapedesis"; that is, by squeezing through pores much smaller than the size of the cells themselves. Under these circumstances it is to be expected that fragile red blood cells would not withstand the trauma. For this reason, many of the red blood cells destroyed in the body have their demise in the spleen. After the cells rupture, the released hemoglobin and the cell stroma are ingested by the reticuloendothelial cells of the spleen, which were discussed in Chapter 5.

CIRCULATION IN THE SKIN

PHYSIOLOGIC ANATOMY OF THE CUTANEOUS CIRCULATION

Blood flow through the skin subserves two major functions: first, *nutrition of the skin tissues,* and second, *conduction of heat* from the internal structures of the body to the skin so that the heat can be removed from the body. To perform these two functions the circulatory apparatus of the skin is characterized by two major types of vessels, illustrated diagrammatically in Figure 21–7: (1) the usual nutritive arteries, capillaries, and veins and (2) vascular structures concerned with heating the skin, consisting principally of (a) an extensive *subcutaneous venous plexus,* which holds large quantities of blood that can heat the surface of

the skin, and (b) in some skin areas, *arteriovenous anastomoses,* which are large vascular communications directly between the arteries and the venous plexuses. The walls of these anastomoses have strong muscular coats innervated by sympathetic vasoconstrictor nerve fibers that secrete norepinephrine. When constricted, they reduce the flow of blood into the venous plexuses to almost nothing; or when maximally dilated, they allow extremely rapid flow of warm blood into the plexuses. The arteriovenous anastomoses are found principally in the volar surfaces of the hands and feet and in the lips, the nose, and the ears, which are areas of the body most often exposed to maximal cooling.

Rate of Blood Flow Through the Skin. The rate of blood flow through the skin is one of the most variable in the body, because the flow required to regulate body temperature changes markedly in response to, first, the rate of metabolic activity of the body and, second, the temperature of the surroundings. This will be discussed in detail in Chapter 47. The blood flow required for nutrition is slight, so that this plays almost no role in controlling normal skin blood flow. At ordinary skin temperatures, the amount of blood flowing through the skin vessels to subserve heat regulation is several times as much as that needed to supply the nutritive needs of the tissues. But, when the skin is exposed to extreme cold, the blood flow may become so slight that nutrition begins to suffer — even to the extent, for instance, that the fingernails grow considerably more slowly in arctic climates than in temperate climates.

Under ordinary cool conditions the blood flow to the skin is about 0.25 liter per square meter of body surface area, or a total of about 400 ml. per minute in the average adult. This can decrease to as little as 50 ml. per minute in severe cold, and when the skin is heated until maximal vasodilation has resulted, the blood flow can be as much as 2.8 liters per minute, thus illustrating both the extreme variability of skin blood flow and the great drain on cardiac output that can occur under hot conditions. Indeed, many persons with borderline cardiac failure develop severe failure in hot weather because of the extra load on the heart and then return from failure in cool weather.

REGULATION OF BLOOD FLOW IN THE SKIN

Nervous Control of Cutaneous Blood Flow. Since most of the blood flow through the skin is to control body temperature, and since this function in turn is regulated by the nervous system, the blood flow through the skin is princi-

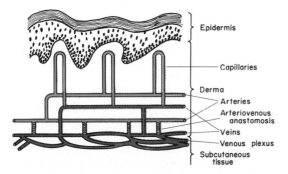

Figure 21–7. The skin circulation.

pally regulated by nervous mechanisms rather than by local regulation, which is opposite to the regulation in most parts of the body.

Temperature Control Center of the Hypothalamus. Located in the preoptic region of the anterior hypothalamus is a small center that is capable of controlling body temperature. Heating this area causes vasodilation of essentially all the skin vessels of the body and also causes sweating. Cooling of the center causes vasoconstriction and cessation of sweating. The detailed function of this center will be discussed in Chapter 47 in relation to body temperature. The important point in the present discussion is that the hypothalamus controls blood flow through the skin in response to changes in body temperature by two mechanisms: (1) a vasoconstrictor mechanism and (2) a vasodilator mechanism.

The Vasoconstrictor Mechanism. The skin throughout the body is supplied with sympathetic vasoconstrictor fibers that secrete norepinephrine at their endings. This constrictor system is extremely powerful in the feet, hands, lips, nose, and ears, which are the areas most frequently exposed to severe cold and which are also the areas where large numbers of arteriovenous anastomoses are found. At normal body temperature the sympathetic vasoconstrictor nerves keep these anastomoses almost totally closed, but when the body becomes overly heated the number of sympathetic impulses is greatly reduced, so that the anastomoses dilate and allow large quantities of warm blood to flow into the venous plexuses, thereby promoting loss of heat from the body.

In the remainder of the body — that is, over the surfaces of the arms, legs, and body trunk — almost no arteriovenous anastomoses are present, but nevertheless vasoconstrictor control of the nutritive vessels can still effect major changes in blood flow. When the body becomes overheated, the sympathetic vasoconstrictor impulses cease, and the blood flow to the skin vessels increases about two-fold.

Extreme Sensitivity of the Skin Blood Vessels to Circulating Norepinephrine and Epinephrine. In addition to the direct sympathetic vasoconstrictor effect in the skin, the skin blood vessels are also extremely sensitive to circulating norepinephrine and epinephrine. Therefore, even in areas of the skin that might have lost their sympathetic innervation, any mass discharge of the sympathetic nervous system will still cause intense skin vasoconstriction. Sometimes the sensitivity of the skin to circulating norepinephrine and epinephrine is so great that the vasoconstriction can actually damage the skin. This is a condition called *Raynaud's disease.*

The Vasodilator Mechanism. When the body temperature becomes excessive and sweating begins to occur, the blood flow through the skin of the forearms and trunk increases an additional three-fold as a result of so-called "active" vasodilatation of the vessels. The basic mechanism by which this active dilatation occurs is not completely known. Yet this additional increase in blood flow does not occur in the absence of sweating, and it does not occur in lower animals that do not have sweat glands. Therefore, it has been postulated that the sympathetic fibers that secrete acetylcholine to activate the sweat glands cause a secondary vasodilatation as follows: The increased activity of the sweat glands is postulated to cause these glands to release an enzyme called *kallikrein,* which in turn splits the polypeptide *bradykinin* from globulin in the interstitial fluids. Bradykinin in turn is a powerful vasodilator substance that could account for the greatly increased blood flow when sweating begins to occur. In opposition to this theory, however, is the fact that inhibition of the bradykinin mechanism does not completely block the increased blood flow associated with sweating.

Effect of Cold on the Skin Circulation. When cold is applied directly to the skin, the skin vessels constrict more and more down to a temperature of about 15° C., at which point they reach their maximum degree of constriction. This constriction results primarily from increased sensitivity of the vessels to vasoconstrictor nerve stimulation, but it probably also results at least partly from a reflex that passes to the cord and then back to the vessels. At temperatures below 15° C., the vessels begin to dilate. This dilation is caused by a direct local effect of the cold on the vessels themselves — probably paralysis of the contractile mechanism of the vessel wall or block of the nerve impulses coming to the vessels. In any event, at temperatures approaching 0° C. the vessels frequently reach maximum vasodilatation. This intense vasodilatation plays a purposeful role in preventing freezing of the exposed portions of the body, particularly the hands and ears.

Local Regulation of Blood Flow in the Skin. Though local regulation of blood flow usually plays only a small role in skin blood flow control, it does have an effect in those few instances in which skin blood flow becomes greatly decreased. For instance, if one sits on his buttocks for 30 minutes or more and then stands, he will note intense reddening of the affected skin. That is, the blood flow to the skin area has now increased to a marked extent, which is typical *reactive hyperemia* resulting from diminished availability of nutrients to the tissues during the period of compression. Thus, the local regulatory mechanism present in essentially all other tissues of the body is also present in the skin and can be called into play when needed to prevent nutritional damage to the tissues.

REFERENCES

Abramson, D. I. (ed.): Circulatory Diseases of the Limbs: A Primer. New York, Grune & Stratton, 1978.

Appenzeller, O., and Atkinson, R. (eds.): Health Aspects of Endurance Training. New York, S. Karger, 1978.

Apple, D. F., Jr., and Cantwell, J. D.: Medicine for Sport. Chicago, Year Book Medical Publishers, 1979.

Betz, E.: Cerebral blood flow: Its measurement and regulation. *Physiol. Rev., 52*:595, 1972.

Bevegard, B. S., and Shepherd, J. T.: Regulation of the circulation during exercise in man. *Physiol. Rev., 47*:178, 1967.

Boullin, D. J. (ed.): Cerebral Vasospasm. New York, John Wiley & Sons, 1980.

Clarke, D. H.: Exercise Physiology. Englewood Cliffs, N. J., Prentice-Hall, 1975.

Fox, E. L.: Sports Physiology. Philadelphia, W. B. Saunders Co., 1979.

Goldstein, M., *et al.* (eds.): Cerebrovascular Disorders and Stroke. New York, Raven Press, 1979.

Greenway, C. V., and Stark, R. D.: Hepatic vascular bed. *Physiol. Rev., 51*:23, 1971.

Guyton, A. C., *et al.*: Cardiac Output and Its Regulation. Philadelphia, W. B. Saunders Co., 1973.

Helwig, E. B., Mostofi, F. K. (eds.): The Skin. New York, R. E. Kreiger Co., 1980.

Jacobson, E. D.: The gastrointestinal circulation. *Annu. Rev. Physiol., 30*:133, 1968.

Juergens, J. L., *et al.* (eds.): Allen, Barker, Hines Peripheral Vascular Disease, 5th Ed. Philadelphia, W. B. Saunders Co., 1980.

Kuschinsky, W., and Wahl, M.: Local chemical and neurogenic regulation of cerebral vascular resistance. *Physiol. Rev., 58*:656, 1978.

Lassen, N. A.: Control of cerebral circulation in health and disease. *Circ. Res., 34*:749, 1974.

McHenry, L. C., Jr.: Cerebral Circulation and Stroke. St. Louis, W. H. Green, 1978.

Moskalenko, Y. E., *et al.*: Biophysical Aspects of Cerebral Circulation. New York, Pergamon Press, 1979.

Rappaport, A. M.: Hepatic blood flow: Morphologic aspects and physiologic regulation. *In* Javitt, N. B. (ed.): International Review of Physiology: Liver and Biliary Tract Physiology I. Vol. 21, Baltimore, University Park Press, 1980, p. 1.

Rosell, S., and Belfrage, E.: Blood circulation in adipose tissue. *Physiol. Rev., 59*:1078, 1979.

Strauss, R. H. (ed.): Sports Medicine and Physiology. Philadelphia, W. B. Saunders Co., 1979.

Svanik, J., and Lundgren, O.: Gastrointestinal circulation. *Int. Rev. Physiol., 12*:1, 1977.

Wilkins, R. H. (ed.): Cerebral Vasospasm. Baltimore, Williams & Wilkins, 1980.

Part VI
THE BODY FLUIDS AND KIDNEYS

22

Capillary Dynamics; and Exchange of Fluid Between the Blood and Interstitial Fluid

In the capillaries the most purposeful function of the circulation occurs, namely, interchange of nutrients and cellular excreta between the tissues and the circulating blood. About 10 billion capillaries, having a total surface area of probably 500 to 700 square meters, provide this function. Indeed, it is rare that any single functional cell of the body is more than 20 to 30 microns away from a capillary.

The purpose of this chapter is to discuss the transfer of substances between the blood and interstitial fluid and especially to discuss the factors that affect the transfer of fluid volume itself between the circulating blood and the interstitial fluids.

STRUCTURE OF THE CAPILLARY SYSTEM

Figure 22–1 illustrates the structure of a "unit" capillary bed as seen in the mesentery, illustrating that blood enters the capillaries through an *arteriole* and leaves by way of a *venule*. From the arteriole the blood passes into *metarterioles*, which have a structure midway between that of arterioles and capillaries. After leaving the metarteriole, the blood enters the *capillaries*, some of which are large and are called *preferential channels* and others of which are small and are *true capillaries*. After passing through the capillaries the blood enters the venule and returns to the general circulation.

The arterioles are highly muscular, and their diameters can change many-fold, as was discussed in Chapter 15. The metarterioles do not have a continuous muscular coat, but smooth muscle fibers encircle the vessel at intermediate points, as illustrated in Figure 22–1 by the large black dots to the sides of the metarteriole.

At the point where the true capillaries originate from the metarterioles a smooth muscle fiber usually encircles the capillary. This is called the *precapillary sphincter*. This sphincter can open and close the entrance to the capillary.

This typical arrangement of the capillary bed is not found in all parts of the body; however, some similar arrangement is found, beginning with arterioles, passing through metarterioles and capillaries, and returning through venules.

Structure of the Capillary Wall. Figure 22–2 illustrates the ultramicroscopic structure of the capillary wall. Note that the wall is composed of a unicellular layer of endothelial cells and is surrounded by a basement membrane on the outside. The total thickness of the membrane is about 0.5 micron.

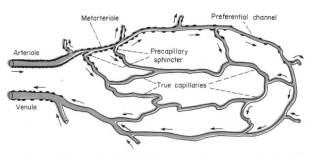

Figure 22–1. Structure of the mesenteric capillary bed. (From Zweifach: Factors Regulating Blood Pressure. New York, Josiah Macy, Jr., Foundation, 1950.)

228

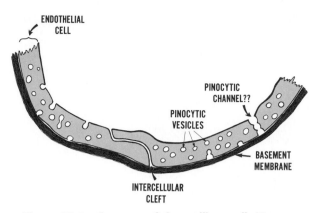

Figure 22-2. Structure of the capillary wall. Note especially the *intercellular cleft* at the junction between adjacent endothelial cells; it is believed that most water-soluble substances diffuse through the capillary membrane along this cleft.

The diameter of the capillary is 5 to 9 microns, barely large enough for red blood cells and other blood cells to squeeze through.

"Pores" in the Capillary Membrane. Studying Figure 22-2, one sees two minute passageways connecting the interior of the capillary with the exterior. One of these is the *intercellular cleft*, which is a thin slit that lies between adjacent endothelial cells. Most water-soluble ions and molecules pass between the interior and exterior of the capillary through these "slit-pores."

Also present in the endothelial cells are many pinocytic vesicles. These form at one surface of the cell and move to the opposite surface, where they discharge their contents, often carrying large molecules and even solid particles through the capillary membrane. Note also in Figure 22-2 that some of these pinocytic vesicles occasionally coalesce with each other and form a continuous channel through the endothelial membrane, as illustrated by the so-called *pinocytic channel* shown to the right in the figure. However, it is still doubted whether large amounts of substances pass through the capillary membrane via such "pores" as this.

FLOW OF BLOOD IN THE CAPILLARIES — VASOMOTION

Blood usually does not flow at a continuous rate through the capillaries. Instead, it flows intermittently. The cause of this intermittency is the phenomenon called *vasomotion*, which means intermittent contraction of the arterioles, metarterioles, and precapillary sphincters. These constrict and relax in an alternating cycle, usually five to ten times per minute.

Regulation of Vasomotion. The most important factor found thus far to affect the degree of opening and closing of the metarterioles and precapillary sphincters is the concentration of *oxygen* in the tissues. When the oxygen concentration is very low, the intermittent periods of blood flow occur more often, and the duration of each period of flow lasts for a longer time, thereby allowing the blood to carry increased quantities of oxygen (as well as other nutrients) to the tissues. Thus, by this intermittent opening and closing of the precapillary sphincters and metarterioles, the blood flow to the tissue is *autoregulated,* as was discussed in Chapter 16.

EXCHANGE OF NUTRIENTS AND OTHER SUBSTANCES BETWEEN THE BLOOD AND INTERSTITIAL FLUID

DIFFUSION THROUGH THE CAPILLARY MEMBRANE

By far the most important means by which substances are transferred between the plasma and interstitial fluids is *diffusion*. Figure 22-3 illustrates this process, showing that as the blood traverses the capillary tremendous numbers of water molecules and dissolved particles diffuse back and forth through the capillary wall, providing continual mixing between the interstitial fluids and the plasma. Diffusion results from thermal motion of the water molecules and the dissolved substances in the fluid, the different particles moving first in one direction, then another, moving randomly in every direction.

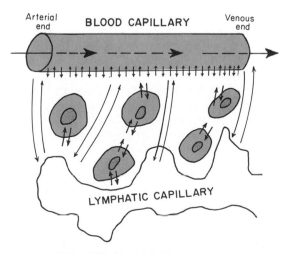

Figure 22-3. Diffusion of fluid and dissolved substances between the capillary and interstitial fluid spaces.

Diffusion of Lipid-Soluble Substances Through the Capillary Membrane. If a substance is lipid-soluble, it can diffuse directly through the cell membranes of the capillary without having to go through the pores. Such substances include especially oxygen and carbon dioxide. Since these can permeate all areas of the capillary membrane, their rates of transport through the capillary membrane are about two times the rate of water and many times the rates of most lipid-insoluble substances such as sodium ions and glucose.

Other lipid-soluble substances that are rapidly transported through the capillary membrane include various anesthetic gases and alcohol.

Diffusion of Water-Soluble and Lipid-Insoluble Substances Through the Capillary Membrane. Many substances needed by the tissues are soluble in water but cannot pass through the lipid membranes of the endothelial cells; such substances include sodium ions, chloride ions, glucose, and so forth. These substances diffuse between the plasma and interstitial fluids only through the capillary pores, which themselves are filled with water.

Despite the fact that not over one one-thousandth of the surface area of the capillaries is represented by the intercellular pores, the velocity of thermal motion is so great that even this small area is sufficient to allow tremendous diffusion of water and water-soluble substances through these pores. To give one an idea of the extreme rapidity with which substances diffuse, *the rate at which water molecules diffuse through the capillary membrane is approximately 80 times as great as the rate at which plasma itself flows linearly along the capillary.* That is, the water of the plasma is exchanged with the water of the interstitial fluids 80 times before the plasma can go the entire distance through the capillary. Yet, despite this rapid rate of diffusion, the rates of diffusion out of the capillary and into the capillary are so nearly equal that the rate of net movement of fluid volume through the capillaries is thousands of times less than the rate of two-way diffusional exchange.

Effect of Molecular Size on Pore Permeability. The width of the capillary intercellular slit-pores, 6 to 7 nanometers, is about 20 times the diameter of the water molecule. On the other hand, the diameters of plasma protein molecules are slightly greater than the width of the pores. Other substances, such as sodium ions, chloride ions, glucose, and urea, have intermediate diameters. Therefore, it is obvious that the permeability of the capillary pores for different substances will vary according to their molecular diameters.

Table 22–1 gives the relative permeabilities of the capillary pores in muscle for substances commonly encountered by the capillary membrane,

TABLE 22–1 RELATIVE PERMEABILITY OF MUSCLE CAPILLARY PORES TO DIFFERENT-SIZED MOLECULES

Substance	Molecular Weight	Permeability
Water	18	1.00
NaCl	58.5	0.96
Urea	60	0.8
Glucose	180	0.6
Sucrose	342	0.4
Inulin	5,000	0.2
Myoglobin	17,600	0.03
Hemoglobin	68,000	0.01
Albumin	69,000	<0.0001

Modified from Pappenheimer.

illustrating for instance that the permeability for glucose molecules is 0.6 times as great as that for water molecules, while the permeability for albumin molecules is less than one ten-thousandth that for water molecules. Thus, the membrane is almost impermeable to albumin, which causes a significant concentration difference to develop between the albumin of the plasma and that of the interstitial fluid, as will become evident later in the chapter.

Effect of Hydrostatic Pressure Differences on Movement of Substances Through the Capillary Membrane – Bulk Flow

When the hydrostatic pressure is different on the two sides of a membrane, the greater pressure on one side causes *slightly increased diffusion* of substances toward the opposite side. However, far more water and dissolved substances move through the capillary membrane than can be accounted for by increased diffusion alone. The reason is that the water and other substances actually *flow in bulk* through the capillary pores when a hydrostatic pressure difference occurs. This is exactly the same phenomenon that occurs in a blood vessel when a pressure difference develops between the two ends of the vessel, namely, flow of the blood along the axis of the blood vessel from the high-pressure area to the low-pressure area. So also do fluid and its dissolved substances flow through the capillary pores when a hydrostatic pressure difference develops across the capillary membrane. This phenomenon is called *bulk flow* of fluid through the membrane.

Effect of Osmotic Pressure Differences on Movement of Substances Through the Capillary Membrane – Bulk Flow Caused by This Means Also

When a substance has a molecular weight too great for it to move through the slit-pores of the capillary membrane, it creates osmotic pressure at the end of the slit where the movement of the molecule is impeded. This pressure then acts exactly as does a hydrostatic pressure difference across the pore to cause bulk flow through the pore. The osmotic substance that normally is most important in causing bulk flow through the capillary membrane is the plasma proteins, as will be explained later in the chapter.

THE INTERSTITIUM AND THE INTERSTITIAL FLUID

Approximately one-sixth of the body tissues is spaces between cells, which collectively are called the *interstitium*. The fluid in these spaces is the interstitial fluid.

The structure of the interstitium is illustrated in Figure 22–4. It has two major types of solid structures: (1) collagen fiber bundles and (2) proteoglycan filaments. Figure 22–4 illustrates that the collagen fiber bundles extend long distances in the interstitium. They are extremely strong and therefore provide most of the tensional strength of the tissues. The proteoglycan filaments, on the other hand, are extremely thin, coiled molecules composed of about 98 per cent hyaluronic acid and 2 per cent protein. These molecules are so thin that they can never be seen with a light microscope and are very difficult to demonstrate even with the electron microscope. Nevertheless, they form a sponge of very fine reticular filaments aptly described as a "brush pile." This brush pile of proteoglycan filaments is ubiquitous in the interstitum. It fills all the spaces between the collagen fibers, the crannies between the cells, and almost all the other minute spaces of the tissues.

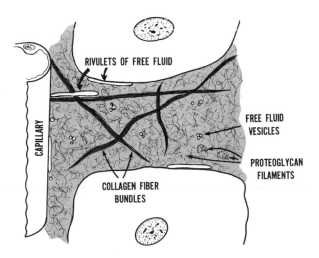

Figure 22–4. Structure of the interstitium. Proteoglycan filaments fill the spaces between the collagen fiber bundles. Free fluid vesicles are seen and small amounts of free fluid in the form of rivulets.

The *interstitial fluid* is mainly entrapped in the minute spaces among the proteoglycan filaments. And the combination of the proteoglycan filaments and the fluid entrapped within them has characteristics of a *gel;* therefore it is called the *tissue gel.*

Because of the large number of proteoglycan filaments, fluid *flows* through the tissue gel only very poorly. Instead, it mainly *diffuses* through the gel; that is, it moves molecule by molecule from one place to another by the process of kinetic motion rather than by large numbers of molecules moving together. In contrast to the severe impediment of flow through the gel, diffusion through the gel occurs about 95 to 99 per cent as effectively as it does through free fluid.

Theoretical Importance of Having a Gel in the Interstitial Spaces. One can readily understand that if all cells abutted each other directly without any interstitial spaces, those cells not directly in contact with capillaries could receive nutrition only through other cells; many substances, such as glucose, amino acids, and so forth, which become trapped in the first cells, would not be able to reach the outlying cells in adequate quantities. At the opposite extreme, the presence of tremendous quantities of extracellular fluid with very large tissue spaces would increase the distances from the capillaries to the outlying cells so much that again the diffusion of nutrients to outlying cells would be diminished. Obviously, therefore, a certain interstitial fluid space size is optimum for transport of substances from the capillaries to the cells. Therefore, one function of the gel in the interstitial spaces is to create interstitial fluid spaces of appropriate size for optimum cell nutrition to occur. Fortunately, the nutrients and the electrolytes can all diffuse through the gel almost as easily as through free interstitial fluid.

Two other important functions of gel in the interstitial spaces are (1) prevention of fluid flow through the tissue spaces into the lower areas of the body as a result of gravity and (2) prevention of spread of infection in the tissues, that is, prevention of spread of bacteria or other agents from one tissue area to another.

DISTRIBUTION OF FLUID VOLUME BETWEEN THE PLASMA AND INTERSTITIAL FLUID

Despite the tremendous rates of diffusion of substances both out of the capillary into the interstitial spaces and in the opposite direction, these rates in both directions so nearly equal each other that the rate of net volume movement across the capillary membrane is normally very low. Consequently, the volumes of both the blood and the interstitial fluids normally change very little from hour to hour or even from day to day. Yet, under abnormal conditions, fluid can leak rapidly out of the circulation into the interstitial spaces, sometimes causing circulatory shock because of decreased blood volume or tissue edema because of excess fluid in the interstitial spaces.

The pressure in the capillaries continuously tends to force fluid and its dissolved substances through the capillary pores into the interstitial spaces. But, in contrast, osmotic pressure caused by the plasma proteins (called *colloid* osmotic pressure) tends to cause fluid movement by osmosis from the interstitial spaces into the blood; it is mainly this osmotic pressure that prevents continual loss of fluid volume from the blood into the interstitial spaces.

The Four Primary Factors That Determine Fluid Movement Through the Capillary Membrane.   Figure 22–5 illustrates the four primary factors that determine whether fluid will move out of the blood into the interstitial fluid or in the opposite direction; these are

1. The *capillary pressure* (Pc), which tends to move fluid outward through the capillary membrane.

2. The *interstitial fluid pressure* (Pif,), which tends to move fluid inward through the capillary membrane when Pif is positive but outward when Pif is negative. ,

3. The *plasma colloid osmotic pressure* (Πp), which tends to cause osmosis of fluid inward through the membrane.

4. The *interstitial fluid colloid osmotic pressure* (Πif), which tends to cause osmosis of fluid outward through the membrane.

The regulation of fluid volumes in the blood and interstitial fluid is so important that each of these factors is discussed in turn in the following sections.

Capillary Pressure

Unfortunately, the exact capillary pressure is not known because it has been impossible to measure capillary pressure under absolutely normal conditions. Yet, two different methods have been used to estimate the capillary pressure.

Cannulation Method for Measuring Capillary Pressure.   To measure pressure in a capillary by cannulation, a microscopic glass pipet is thrust directly into the capillary, and the pressure is measured by an appropriate micromanometer system. By use of this method, capillary pressures have been measured in capillaires of exposed tissues of lower animals and in large capillary loops of the eponychium at the base of the fingernail in human beings. These measurements have given pressures of 30 to 40 mm. Hg in the arterial ends of the capillaries, 10 to 15 mm. Hg in the venous ends, and about 25 mm. Hg in the middle.

Isogravimetric Method for Indirectly Measuring Mean Capillary Pressure.   Figure 22–6 illustrates an *isogravimetric* method for estimating capillary pressure. This figure shows a section of gut held by one arm of a gravimetric balance. Blood is perfused through the gut. When the arterial pressure is decreased, the resulting decrease in capillary pressure allows the osmotic pressure of the plasma proteins to cause absorption of fluid out of the gut wall and makes the weight of the gut decrease. This immediately causes displacement of the balance arm. However, to prevent this weight decrease, the venous pressure is raised an amount sufficient to overcome the effect of decreasing the arterial pressure.

In the lower part of the figure, the changes in

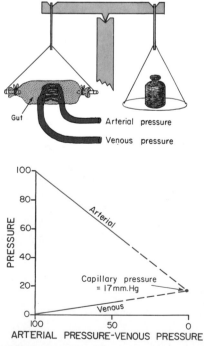

Figure 22–6.   Isogravimetric method for measuring capillary pressure (explained in the text).

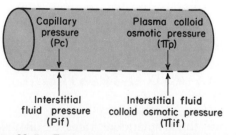

Figure 22–5.   Forces operative at the capillary membrane tending to move fluid either outward or inward through the membrane.

arterial and venous pressures that exactly nullify their respective effects on the weight of the gut are illustrated. The arterial and venous curves meet each other at a value of 17 mm. Hg. Therefore, the capillary pressure must have remained at this same level of 17 mm. Hg throughout these maneuvers. Thus, in a roundabout way, the "functional" capillary pressure is measured to be about 17 mm. Hg.

Functional Capillary Pressure. Since the cannulation method does not give the same pressure measurement as the isogravimetric method, one must decide which of these measurements is probably the true functional capillary pressure of the tissues. The isogravimetric measurement is probably much nearer to the normal values for capillary pressure than are the micropipet measurements, for several reasons:

First, the cannulation method can measure pressure only in open capillaries, but most of these are closed under normal conditions, thus giving a *functional* mean capillary pressure much lower than the measurements would indicate.

Second, the isogravimetric method is itself a *functional* method for measuring mean capillary pressure.

Therefore, there are several reasons for believing *that the normal functional mean capillary pressure is about 17 mm. Hg.*

Interstitial Fluid Pressure

The interstitial fluid pressure, like capillary pressure, has been difficult to measure, primarily because the maximum width of the spaces between the reticular fibers that make up the solid structure of the interstitium is only 10 to 40 nanometers, much too small to cannulate for direct measurement of the pressure. Therefore, indirect methods have most frequently been used for measuring this pressure.

Prior to 1961, it had been believed universally that the interstitial fluid pressure was always slightly greater than atmospheric pressure. However, on the basis of measurements made in our laboratory in 1961, we concluded that the interstitial fluid pressure is probably subatmospheric in most tissues of the body, usually 2 to 9 mm. Hg less than the pressure of the surrounding air. Since that time, a continuing controversy has existed concerning whether or not the interstitial fluid pressure level is truly lower than the atmospheric level.

With this as background, let us now summarize two of the methods that have been used for measuring the interstitial free fluid pressure.

Measurement of Interstitial Free Fluid Pressure in Implanted Perforated Hollow Capsules. Figure 22–7 illustrates an indirect method for measuring interstitial fluid pressure that

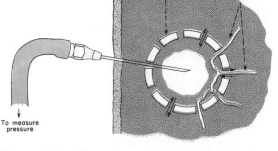

Figure 22–7. The perforated capsule method for interstitial fluid pressure.

may be explained as follows. A small hollow plastic capsule perforated by several hundred small holes is implanted in the tissues, and the surgical wound is allowed to heal for approximately one month. At the end of that time, tissue will have grown inward through the holes to line the inner surface of the sphere. Furthermore, the cavity is filled with fluid that flows freely through the perforations back and forth between the fluid in the interstitial spaces and the fluid in the cavity. Therefore, the pressure in the cavity should equal the fluid pressure in the interstitial fluid spaces. A needle is inserted through the skin and through one of the perforations to the interior of the cavity, and the pressure is measured by use of an appropriate manometer.

Interstitial fluid pressure measured by this method in normal tissues averages about −6.3 mm. Hg. That is, the pressure is *less than atmospheric pressure* or, in other words, is a semivacuum or a suction.

Measurement of Interstitial Fluid Pressure Using Microcannulae. Recently, microscopic pipets have been placed into the tissue in attempts to measure the interstitial fluid pressure. An early group of investigators using this technique reported average pressures that were slightly positive. However, measurements by others recently were in the range of −2 to −3 mm. Hg. Since the pipets are 30 to 50 times as large as the spaces between the reticular fibrillae of the interstitium, one wonders how much the distortion of the tissues that is caused by the method affects the measurement.

Validity of the Interstitial Free Fluid Pressure Measurements. The implanted capsule method has been used to test *changes* in interstitial free fluid pressure caused by various physiological phenomena. The changes that have been recorded have been almost exactly those that would be predicted for the experimental procedures. For instance, if concentrated dextran solution is injected intravenously, one would expect absorption of fluid from the tissue spaces and therefore a quantifiable decrease in the interstitial fluid pressure. The capsule technique demonstrates such predicted pressure changes. Similarly, the predictions have been validated for other physiological experiments, such as changes in arterial and venous pressure and in the degree of hydration of the tissues. This type of validation has not been shown for other methods. For this reason and other reasons that will be presented in the following chapter, we have chosen to use in the discussions of this chapter and the following chapter the pressure measurements made by the implanted capsule technique.

Plasma Colloid Osmotic Pressure

Colloid Osmotic Pressure Caused by Proteins. The proteins are the only dissolved substances of the plasma that do not diffuse readily into the interstitial fluid. Furthermore, when small quantities of protein do diffuse into the interstitial fluid, they are soon removed from the interstitial spaces by way of the lymph vessels. Therefore, the concentration of protein in the plasma averages over three times as much as that in the interstitial fluid, 7.3 gm./100 ml. in the plasma versus 2 gm./100 ml. in the interstitial fluid.

In the discussion of osmotic pressure in Chapter 4, it was pointed out that only those substances that fail to pass through the pores of a semipermeable membrane exert osmotic pressure. Since the proteins are the only dissolved constituents that do not readily penetrate the pores of the capillary membrane, it is the dissolved proteins of the plasma and interstitial fluids that are responsible for the osmotic pressure at the capillary membrane. To distinguish this osmotic pressure from that which occurs at the cell membrane, it is called either *colloid osmotic pressure* or *oncotic pressure*. The term "colloid" osmotic pressure is derived from the fact that a protein solution resembles a colloidal solution despite the fact that it is actually a true solution.

Normal Value for Plasma Colloid Osmotic Pressure. The colloid osmotic pressure of normal human plasma averages approximately 28 mm. Hg.

Effect of the Different Plasma Proteins on Colloid Osmotic Pressure. The plasma proteins are a mixture of proteins that contains albumin, with an average molecular weight of 69,000; globulins, 140,000; and fibrinogen, 400,000. Thus, 1 gram of globulin contains only half as many molecules as 1 gram of albumin, and 1 gram of fibrinogen contains only one-sixth as many molecules as 1 gram of albumin. (It will be

recalled from the discussion of osmotic pressure in Chapter 4 that the osmotic pressure is determined by the *number of molecules* dissolved in a fluid rather than by the weight of these molecules). The average relative concentrations of the different types of proteins in the plasma and their respective colloid osmotic pressures are:

	gm. %	IIp (mm. Hg)
Albumin	4.5	21.8
Globulins	2.5	6.0
Fibrinogen	0.3	0.2
TOTAL	7.3	28.0

Thus, about 75 per cent of the total colloid osmotic pressure of the plasma results from the albumin fraction, 25 per cent from the globulins, and almost none from the fibrinogen.

Interstitial Fluid Colloid Osmotic Pressure

Though the size of the *usual* capillary pore is smaller than the molecular sizes of the plasma proteins, this is not true of all the pores. Therefore, small amounts of plasma proteins do leak through the pores into the interstitial spaces, and additional amounts are probably transported by pinocytosis.

The protein *concentration* of the interstitial fluid averages less than one-third that of plasma, approximately 2 grams per cent. The average colloid osmotic pressure for this concentration of proteins in the interstitial fluids is approximately 5 mm. Hg.

EXCHANGE OF FLUID VOLUME THROUGH THE CAPILLARY MEMBRANE

Now that the different factors affecting capillary membrane dynamics have been discussed, it is possible to put them all together to see how normal capillaries function.

The capillary pressure at the arterial ends of the capillaries is 15 to 20 mm. Hg greater than at the venous ends. Because of this difference, fluid "filters" out of the capillaries at their arterial ends and then is reabsorbed into the capillaries at their venous ends. Thus, a small amount of fluid actually "flows" through the tissues from the arterial ends of the capillaries to the venous ends. The dynamics of this flow are as follows:

Analysis of the Forces Causing Filtration at the Arterial End of the Capillary. The forces operating at the arterial end of the capillary to cause movement through the capillary membrane are:

	mm. Hg
Forces tending to move fluid outward:	
Capillary pressure	25.0
Negative interstitial free fluid pressure	6.3
Interstitial fluid colloid osmotic pressure	5.0
TOTAL OUTWARD FORCE	36.3

Force tending to move fluid inward:	
Plasma colloid osmotic pressure	28.0
TOTAL INWARD FORCE	28.0

Summation of forces:	
Outward	36.3
Inward	28.0
NET OUTWARD FORCE	8.3

Thus, the summation of forces at the arterial end of the capillary shows a net *filtration pressure* of 8.3 mm. Hg tending to move fluid out of the arterial ends of the capillaries into the interstitial spaces.

This 8.3 mm. Hg filtration pressure causes an average of about 0.3 per cent of the fluid of the plasma flowing into the capillaries to filter out of the arterial ends of the capillaries into the interstitial spaces. (This percentage varies tremendously from very, very low in the brain to very high in the liver.)

Analysis of Reabsorption at the Venous End of the Capillary. The low pressure at the venous end of the capillary changes the balance of forces in favor of absorption as follows:

	mm. Hg
Force tending to move fluid inward:	
Plasma colloid osmotic pressure	28.0
TOTAL INWARD FORCE	28.0

Forces tending to move fluid outward:	
Capillary pressure	10.0
Negative interstitial free fluid pressure	6.3
Interstitial fluid colloid osmotic pressure	5.0
TOTAL OUTWARD FORCE	21.3

Summation of forces:	
Inward	28.0
Outward	21.3
NET INWARD FORCE	6.7

Thus, the force that causes fluid to move into the capillary, 28 mm. Hg, is greater than that opposing reabsorption, 21.3 mm. Hg. The difference, 6.7 mm. Hg, is the *reabsorption pressure.*

The reabsorption pressure causes about nine-tenths of the fluid that has filtered out the arterial ends of the capillaries to be reabsorbed at the venous ends. The other one-tenth flows into the lymph vessels, as is discussed in the following chapter.

Flow of Fluid Through the Interstitial Spaces. The 0.3 per cent of the plasma fluid that filters out of the arterial ends of the capillaries *flows* through the tissue spaces to the venous ends of the capillaries, where all but about one-tenth of it is reabsorbed. (A much higher proportion is reabsorbed in the muscles, where very little protein leaks through the capillary membranes, and much less is reabsorbed in the liver, where tremendous amounts of protein leak.)

Distinction Between Filtration and Diffusion. It is especially important to distinguish between *filtration* and *diffusion* through the capillary membrane. Diffusion occurs in both directions, while filtration is the bulk movement of fluid out of the capillaries at the arterial ends. The rate of diffusion of water through all the capillary membranes of the entire body is about 240,000 ml./minute, while the normal rate of filtration at the arterial ends of all the capillaries is only 16 ml./minute, a difference of about 15,000-fold. Thus, the quantitative rate of diffusion of water and nutrients back and forth between the capillaries and the interstitial spaces is tremendous in comparison with the minute rate of "flow" of fluid through the tissues.

THE STARLING EQUILIBRIUM FOR CAPILLARY EXCHANGE

E. H. Starling pointed out three-quarters of a century ago that under normal conditions a state of near-equilibrium exists at the capillary membrane whereby the amount of fluid filtering outward through the arterial capillaries equals that quantity of fluid that is returned to the circulation by reabsorption at the venous ends of the capillaries. This near-equilibrium is caused by near equilibration of the *mean* forces tending to move fluid through the capillary membranes. If we assume that the mean capillary pressure is 17 mm. Hg, the normal mean dynamics of the capillary are the following:

	mm. Hg
Mean forces tending to move fluid outward:	
Mean capillary pressure	17.0
Negative interstitial free fluid pressure	6.3
Interstitial fluid colloid osmotic pressure	5.0
TOTAL OUTWARD FORCE	28.3

Mean force tending to move fluid inward:	mm. Hg
Plasma colloid osmotic pressure	28.0
TOTAL INWARD FORCE	28.0

Summation of mean forces:	
Outward	28.3
Inward	28.0
TOTAL OUTWARD FORCE	0.3

Thus, we find a near-equilibrium but nevertheless a slight imbalance of forces at the capillary membranes that causes slightly more filtration of fluid into the interstitial spaces than reabsorption. This slight excess of filtration is called the *net filtration,* and it is balanced by fluid return to the circulation through the lymphatics. The normal rate of net filtration in the entire body is about 1.7 to 3.5 per minute. These figures also represent the rate of fluid flow into the lymphatics each minute.

REFERENCES

Aschheim, E.: Passage of substances across the walls of blood vessels: Kinetics and mechanism. *In* Kaley, G., and Altura, B. M. (eds.): Microcirculation. Vol. I. Baltimore, University Park Press, 1977, p. 213.

Baez, S.: Microcirculation. *Annu. Rev. Physiol., 39*:391, 1977.

Bing, D. H. (ed.): The Chemistry and Physiology of the Human Plasma Proteins. New York, Pergamon Press, 1979.

Brace, R. A., and Guyton, A. C.: Interaction of transcapillary Starling forces in the isolated dog forelimb. *Am. J. Physiol., 233*:H136, 1977.

Comper, W. D., and Laurent, T. C.: Physiological function of connective tissue polysaccharides. *Physiol. Rev., 58*:255, 1978.

Crone, C., and Christensen, O.: Transcapillary transport of small solutes and water. *In* Guyton, A. C., and Young, D. B. (eds.): International Review of Physiology: Cardiovascular Physiology III. Vol. 18. Baltimore, University Park Press, 1979, p. 149.

Gabbiani, G., and Majno, G.: Fine structure and endothelium. *In* Kaley, G., and Altura, B. M. (eds.): Microcirculation. Vol. I. Baltimore, University Park Press, 1977, p. 133.

Guyton, A. C.: Concept of negative interstitial pressure based on pressures in implanted perforated capsules. *Circ. Res., 12*:399, 1963.

Guyton, A. C.: Interstitial fluid pressure: II. Pressure-volume curves of interstitial space. *Circ. Res., 16*:452, 1965.

Guyton, A. C., *et al.*: Interstitial fluid pressure: IV. Its effect on fluid movement through the capillary wall. *Circ. Res., 19*:1022, 1966.

Guyton, A. C., *et al.*: Interstitial fluid pressure. *Physiol. Rev., 51*:527, 1971.

Guyton, A. C., *et al.*: Circulatory Physiology II. Dynamics and Control of the Body Fluids. Philadelphia, W. B. Saunders Co., 1975.

Intaglietta, M.: The measurement of pressure and flow in the microcirculation. *Microvasc. Res., 5*:357, 1973.

Intaglietta, M.: Transcapillary exchange of fluid in single microvessels. *In* Kaley, G., and Altura, B. M. (eds.): Microcirculation. Vol. I. Baltimore, University Park Press, 1977, p. 197.

Johnson, P. C.: The microcirculation and local and humoral control of the circulation. *In* MTP International Review of Science: Physiology. Vol. 1. Baltimore, University Park Press, 1974, p. 163.

Landis, E. M., and Pappenheimer, J. R.: Exchange of substances through the capillary walls. *In* Hamilton, W. F. (ed.): Handbook of Physiology. Sec. 2. Vol. 2. Baltimore, Williams & Wilkins, 1963, p. 961.

Nicoll, P. A., and Taylor, A. E.: Lymph formation and flow. *Annu. Rev. Physiol., 39*:73, 1977.

Rhodin, J. A. G.: Architecture of the vessel wall. *In* Bohr, D. F., *et al.* (eds.): Handbook of Physiology. Sec. 2. Vol. 2. Baltimore, Williams & Wilkins, 1980, p. 1.

Rothschild, M. A., *et al.*: Albumin synthesis. *In* Javitt, N. B. (ed.): International Review of Physiology: Liver and Biliary Tract Physiology I. Vol. 21. Baltimore, University Park Press, 1980, p. 249.

Simionescu, M.: Transendothelial movement of large molecules in the microvasculature. *In* Fishman, A. P., and Renkin, E. M. (eds.): Pulmonary Edema. Baltimore, Waverly Press, 1979, p. 39.

23

The Lymphatic System, Interstitial Fluid Dynamics, Edema, Pulmonary Fluid, and Special Fluid Systems

THE LYMPHATIC SYSTEM

The lymphatic system represents an accessory route by which fluids can flow from the interstitial spaces into the blood. And, most important of all, the lymphatics can carry proteins and large particulate matter away from the tissue spaces, neither of which can be removed by absorption directly into the blood capillary.

THE LYMPH CHANNELS OF THE BODY

All tissues of the body, with the exception of a very few, have lymphatic channels or other channels similar to lymphatics that drain excess fluid directly from the interstitial spaces. Essentially all the lymph from the lower part of the body — even most of that from the legs — flows up the *thoracic duct* and empties into the venous system at the juncture of the *left* internal jugular vein and subclavian vein, as illustrated in Figure 23–1. However, small amounts of lymph from the lower part of the body can enter the veins in the inguinal region and perhaps also at various points in the abdomen.

Lymph from the left side of the head, the left arm, and the left chest region also enters the thoracic duct before it empties into the veins. Lymph from the right side of the neck and head, from the right arm, and from parts of the right

thorax enters the *right lymph duct,* which then empties into the venous system at the juncture of the *right* subclavian vein and internal jugular vein.

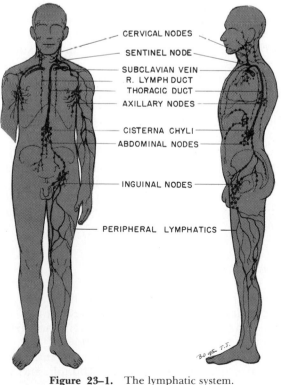

CERVICAL NODES
SENTINEL NODE
SUBCLAVIAN VEIN
R. LYMPH DUCT
THORACIC DUCT
AXILLARY NODES
CISTERNA CHYLI
ABDOMINAL NODES
INGUINAL NODES
PERIPHERAL LYMPHATICS

Figure 23–1. The lymphatic system.

237

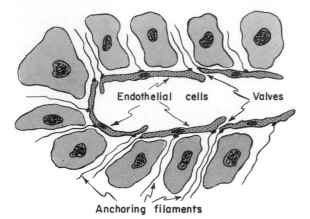

Fig. 23–2. Special structure of the lymphatic capillaries that permits passage of substances of high molecular weight back into the circulation.

The Lymphatic Capillaries and Their Permeability

The minute quantity of fluid that returns to the circulation by way of the lymphatics is extremely important, because substances of high molecular weight such as proteins cannot pass with ease through the pores of the venous capillaries, but they can enter the lymphatic capillaries almost completely unimpeded. The reason for this is a special structure of the lymphatic capillaries, illustrated in Figure 23–2. This figure shows the endothelial cells of the capillary attached by *anchoring filaments* to the connective tissue between the surrounding tissue cells. However, at the junctions of adjacent endothelial cells there are usually very loose connections between the cells. Instead, the edge of one endothelial cell usually overlaps the edge of the adjacent one in such a way that the overlapping edge is free to flap inward, thus forming a minute valve that opens to the interior of the capillary. Interstitial fluid, along with its suspended particles, can push the valve open and flow directly into the capillary. But this fluid cannot leave the capillary once it has entered because any backflow will close the flap valve. Thus, the lymphatics have valves at the very tips of the terminal lymphatic capillaries as well as valves along their larger vessels up to the point where they empty into the blood circulation.

FORMATION OF LYMPH

Lymph is interstitial fluid that flows into the lymphatics. Therefore, lymph has almost the same composition as the tissue fluid in the part of the body from which the lymph flows.

The protein concentration in the interstitial fluid averages about 2 gm. per 100 ml., and the protein concentration of lymph flowing from most of the peripheral tissues is near to this value or a little more concentrated. On the other hand, lymph formed in the liver has a protein concentration as high as 6 gm. per 100 ml., and lymph formed in the intestines has a protein concentration as high as 3 to 5 gm. per 100 ml. Since more than half of the lymph is derived from the liver and intestines, the thoracic duct lymph, which is a mixture of lymph from all areas of the body, usually has a protein concentration of 3 to 5 gm. per 100 ml.

The lymphatic system is also one of the major routes for absorption of nutrients from the gastrointestinal tract, being responsible principally for the absorption of fats, as will be discussed in Chapter 44.

TOTAL RATE OF LYMPH FLOW

Approximately 100 ml. of lymph flows through the thoracic duct of a resting person per hour, and perhaps another 20 ml. of lymph flows into the circulation each hour through other channels, making a total estimated lymph flow of about 120 ml. per hour, illustrating that the flow of lymph is relatively small in comparison with the total exchange of fluid between the plasma and the interstitial fluid.

Factors That Determine the Rate of Lymph Flow

Interstitial Fluid Pressure. Elevation of interstitial *free* fluid presssure above its normal level of −6.3 mm. Hg increases the flow of interstitial fluid into the lymphatic capillaries and consequently also increases the rate of lymph flow. The increase in flow becomes progressively greater as the interstitial fluid pressure rises, until this pressure reaches a value slightly greater than zero mm. Hg; at that point the flow rate reaches a maximum, but by that time it has risen to 10 to 50 times normal. Therefore, any factor (besides obstruction of the lymphatic system itself) that tends to increase interstitial pressure increases the rate of lymph flow. Such factors include

Elevated capillary pressure
Decreased plasma colloid osmotic pressure
Increased interstitial fluid protein
Increased permeability of the capillaries

The Lymphatic Pump. Valves exist periodically all along the lymph channels; a typical valve is illustrated in Figure 23–3 in a *collecting lymphatic* into which the lymphatic capillaries empty. Therefore, every time a lymph vessel or a lym-

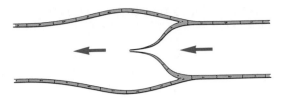

Figure 23–3. A valve in a collecting lymphatic.

phatic capillary is compressed by pressure from any source, lymph is squeezed forward along the lymphatics. The lymph vessels can be compressed either by contraction of the walls of the lymphatics or by pressure from surrounding structures.

Motion pictures taken of exposed lymph vessels, both in animals and in human beings, have shown that any time a lymph vessel becomes stretched with fluid the smooth muscle in the wall of the vessel automatically contracts. Furthermore, each segment of the lymph vessel between successive valves functions as a separate automatic pump. That is, filling of a segment causes it to contract, and the fluid is then pumped through the next valve into the following lymphatic segment. This fills the subsequent segment so that within a few seconds it too contracts, the process continuing all along the lymphatic until the fluid is finally emptied. In a large lymph vessel this lymphatic pump can generate pressure as high as 25 to 50 mm. Hg if the outflow from the vessel becomes blocked.

In addition to the pumping caused by intrinsic contraction of the lymph vessel walls, any external factor that compresses the lymph vessel can also cause pumping. In order of their importance, such factors are

Contraction of muscles
Movements of the parts of the body
Arterial pulsations
Compression of the tissue by objects outside the body.

The lymphatic pump becomes very active during exercise, often increasing lymph flow as much as 5- to 15-fold. On the other hand, during periods of rest, lymph flow is very sluggish.

REGULATION OF INTERSTITIAL FLUID PROTEIN BY LYMPHATIC PUMPING

Since protein continually leaks from the capillaries into the interstitial fluid spaces, it must also be removed continually, or otherwise the tissue colloid osmotic pressure will become so high that normal capillary dynamics can no longer continue. Therefore, by far the most important of all the lymphatic functions is the maintenance of low protein concentration in the interstitial fluid. The mechanism of this is the following:

As fluid leaks from the arterial ends of the capillaries into the interstitial spaces, only small quantities of protein accompany it, but then, as fluid is reabsorbed at the venous ends of the capillaries, most of the protein is left behind. Therefore, *protein progressively accumulates in the interstitial fluid* and this in turn *increases the tissue colloid osmotic pressure.* This osmotic pressure decreases reabsorption of fluid by the capillaries, thereby *promoting increased tissue fluid volume* as well as an *increase in the interstitial fluid pressure.* The increased pressure then causes the lymphatic pump to pump the interstitial fluid into the lymphatic capillaries, and this fluid carries with it the excess protein that has accumulated. This continual *washout* of the protein keeps its concentration at a low level in the interstitial fluid.

To summarize, an increase in tissue fluid protein increases the rate of lymph flow, and this washes the proteins out of the tissue spaces, automatically returning the protein concentration to its normal low level.

The importance of this function of the lymphatics cannot be stressed too strongly, for *there is no other route besides the lymphatics through which excess proteins can return to the circulatory system.* If it were not for this continual removal of proteins, the dynamics of fluid exchange at the blood capillaries would become so abnormal within only a few hours that life could no longer continue.

MECHANISM OF NEGATIVE INTERSTITIAL FLUID PRESSURE

Until recent measurements of the interstitial fluid pressure demonstrated that it is negative rather than positive, as explained in the preceding chapter, it was taught that the normal interstitial fluid pressure ranges between +1 and +4 mm. Hg, and it has still been difficult to understand how negative pressure can develop in the interstitial fluid spaces. However, we can explain this negative interstitial fluid pressure by the following considerations:

First, the normal small amounts of lymph fluid can flow into lymphatic vessels from interstitial spaces even when the interstitial fluid pressure is negative, mainly because the lymphatic pump can create slight degrees of suction. And this lymph flow keeps the protein concentration of the interstitial fluid at a low value and thereby keeps the colloid osmotic pressure also at a low value, usually at about 5 mm. Hg in most peripheral tissues such as the muscles.

Second, the negativity of the interstitial fluid pressure can then be explained mainly on the basis of the balance of forces at the capillary membrane. If we add all the other forces besides the interstitial fluid pressure that cause movement of fluid across the capillary membrane, we find the following:

	mm. Hg
Outward force:	
Capillary pressure	17
Interstitial fluid colloid osmotic pressure	5
TOTAL	22
Inward force:	
Colloid osmotic pressure	28
DIFFERENCE	
(Interstitial fluid pressure)	−6

Thus, we see that the interstitial fluid pressure required to balance the other forces across the capillary membrane is −6 mm. Hg. Thus, −6 mm. of the negative interstitial fluid pressure is caused by this imbalance of forces at the capillary membrane. Indirectly, this results from the continual pumping of *protein* into the lymphatic vessels. Another −0.3 mm. Hg is caused by the continual pumping of *fluid* into the lymphatic vessels, giving a total negativity of −6.3 mm. Hg.

Significance of the Normally "Dry" State of the Interstitial Spaces.   The normal tendency of the capillaries to absorb fluid from the interstitial spaces and thereby to create a partial vacuum causes all the minute structures of the interstitial spaces to be *compacted*. This represents a "dry" state; that is, no *excess* fluid is present besides that required simply to fill the crevices between the tissue elements and that held in the interstices of the tissue gel, as will be discussed later in the chapter.

The "dry" state of the tissues is particularly important for optimal nutrition of the tissues, because nutrients pass from the blood to the cells by diffusion, and the rate of diffusion between two points is inversely proportional to the distance between the cells and the capillaries.

EDEMA

Edema means the presence of excess interstitial fluid in the tissues. Obviously, any factor that increases the interstitial fluid pressure high enough can cause excess interstitial fluid volume and thereby cause edema. However, to explain the conditions under which edema develops, we must first characterize the *pressure-volume curve* of the interstitial fluid spaces.

PRESSURE-VOLUME CURVE OF THE INTERSTITIAL FLUID SPACES

Figure 23–4 illustrates the average relationship between pressure and volume in the interstitial fluid spaces in the human body as extrapolated from measurements in the dog.

One of the most significant features of the curve is that so long as the interstitial fluid pressure remains in the negative range there is little change in interstitial fluid volume despite marked change in pressure. Therefore, edema will not occur so long as the interstitial free fluid pressure remains negative. Indeed, in several hundred measurements of interstitial fluid pressure made in experimental animals, no edema has ever been recorded in the presence of negative interstitial pressure.

Tremendous Increase in Interstitial Fluid Volume When the Interstitial Free Fluid Pressure Becomes Positive.   Note in Figure 23–4 that just as soon as the interstitial free fluid pressure rises to equal atmospheric pressure (zero pressure), the slope of the pressure-volume curve suddenly changes and the volume increases precipitously. An additional increase in interstitial free fluid pressure of only 1 to 3 mm. Hg now causes the interstitial fluid volume to increase several hundred per cent. Finally, at the very top of the figure, the skin begins to be stretched, so that the volume now increases much less rapidly.

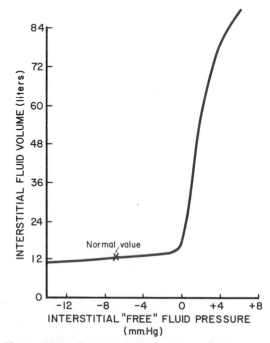

Figure 23–4. Pressure-volume curve of the interstitial spaces. (Extrapolated to the human being from data obtained in dogs.)

Positive Interstitial Fluid Pressure as the Physical Basis for Edema

After studying the pressure-volume curve of Figure 23–4, one can readily see that whenever the interstitial free fluid pressure rises above the surrounding atmospheric pressure, the tissue spaces begin to swell. Therefore, *the physical cause of edema is positive pressure (that is, supra-atmospheric pressure) in the interstitial fluid spaces.*

Edema usually is not detectable in tissues until the interstitial fluid volume has risen to about 30 per cent above normal. But note that the interstitial fluid volume increases to several hundred per cent above normal in seriously edematous tissues.

Edema Resulting From Abnormal Capillary Dynamics

From the discussions of capillary and interstitial fluid dynamics in the preceding and present chapters, it is already evident that several different abnormalities in these dynamics can increase the tissue pressure and in turn cause extracellular fluid edema. The different causes of extracellular fluid edema are

1. *Increased capillary pressure,* which causes excess filtration of fluid through the capillaries.

2. *Decreased plasma protein,* which causes reduced plasma colloid osmotic pressure and therefore failure to retain fluid in the capillaries.

3. *Lymphatic obstruction,* which causes protein to accumulate in the tissue spaces and therefore causes osmosis of fluid out of the capillaries.

4. *Increased capillary permeability,* which allows leakage of excess fluid and protein into the tissue spaces.

Edema Caused by Kidney Retention of Fluid

When the kidney fails to excrete adequate quantities of urine, and the person continues to drink normal amounts of water and ingest normal amounts of electrolytes, the total amount of extracellular fluid in the body increases progressively. This fluid is absorbed from the gut into the blood and elevates the capillary pressure. This in turn causes most of the fluid to pass into the interstitial fluid spaces, elevating the interstitial fluid pressure as well. Therefore, simple retention of fluid by the kidneys can result in extensive edema.

Edema Caused by Heart Failure

Heart failure is one of the most frequent causes of edema, for when the heart no longer pumps blood out of the veins with ease, blood dams up in the venous system. The capillary pressure rises, and serious "cardiac edema" occurs. In addition, the kidneys often function poorly in heart failure, and this leads to even more edema as described above and as discussed in detail in Chapter 20.

PULMONARY INTERSTITIAL FLUID DYNAMICS

The dynamics of the pulmonary interstitial fluid are essentially the same as those of the fluid in the peripheral tissues except for the following important quantitative differences:

1. The pulmonary capillary pressure is very low in comparison with the systemic capillary pressure, approximately 7 mm. Hg in comparison with 17 mm. Hg.

2. The interstitial fluid pressure in the lung interstitium has been measured to be −8 mm. Hg in comparison with −6 mm. Hg in subcutaneous tissue.

3. The pulmonary capillaries are relatively leaky to protein molecules, so that the protein concentration of lymph leaving the lungs is relatively high, averaging about 4 gm. per 100 ml. instead of 2 gm. per 100 ml. in the peripheral tissues.

4. The rate of lymph flow from the lungs is also very high, mainly because of the continuous pumping motion of the lungs.

5. The interstitial spaces of the alveolar portions of the lungs are very narrow, represented by the minute spaces between the capillary endothelium and the alveolar epithelium.

6. The alveolar epithelia are not strong enough to resist very much positive pressure. They are probably ruptured by any positive pressure in the interstitial spaces greater than atmospheric pressure (0 mm. Hg), which allows dumping of fluid from the interstitial spaces into the alveoli.

Now let us see how these quantitative differences affect pulmonary fluid dynamics.

Interrelationship Between Interstitial Fluid Pressure and Other Pressures in the Lung.   Figure 23–5 illustrates a pulmonary capillary, a pulmonary alveolus, and a lymphatic capillary draining the interstitial space between the capillary and the alveolus. Note that the balance of forces at the capillary membrane is such that the interstitial pressure is normally −8 mm. Hg.

The Mechanism for Keeping the Alveoli "Dry."   How is it that the fluid normally present in the interstitial spaces is prevented from flooding the alveoli? The answer to this is related to the negative fluid pressure of approximately −8 mm. Hg normally present in the interstitial spaces between the capillary and the alveolar mem-

PRESSURES CAUSING FLUID MOVEMENT

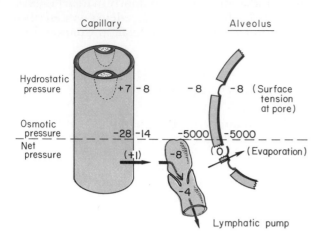

Figure 23–5. Hydrostatic and osmotic forces at the capillary (left) and alveolar membrane (right) of the lungs. Also shown is a lymphatic (center) that pumps fluid from the pulmonary interstitial spaces. (Modified from Guyton, Taylor, and Granger: Dynamics of the Body Fluids. Philadelphia, W. B. Saunders Company, 1975.)

brane. This continually tends to pull fluid inward through the alveolar membrane and therefore also prevents fluid loss in the outward direction.

PULMONARY EDEMA

Pulmonary edema occurs in the same way that edema occurs elsewhere in the body. Any factor that causes the pulmonary interstitial fluid pressure to rise from the negative range into the positive range will cause sudden filling of the pulmonary interstitial spaces, and in more severe cases even the alveoli, with large amounts of free fluid — occasionally as much as 1 liter of fluid that literally suffocates the person.

Pulmonary "Interstitial Fluid" Edema Versus Pulmonary "Alveolar" Edema. The interstitial fluid volume of the lungs usually cannot increase more than about 50 per cent (representing less than 100 milliliters of fluid) before the alveolar epithelial membranes rupture and fluid begins to pour from the interstitial spaces into the alveoli. The cause of this is simply the almost infinitesimal tensional strength of the pulmonary alveolar epithelium; that is, any positive pressure in the interstitial fluid spaces seems to cause immediate rupture of the alveolar epithelium.

Therefore, except in the mildest cases of pulmonary edema, most of the fluid enters the alveoli.

Safety Factor Against Pulmonary Edema. The most common cause of pulmonary edema is greatly elevated pulmonary capillary pressure resulting from *failure of the left heart* and con-

sequent damming of blood in the lungs. However, the pulmonary capillary pressure usually must rise to very high values before serious pulmonary edema will develop. The reason for this is the very high dehydrating force of the colloid osmotic pressure of the blood in the lungs. This effect is illustrated in Figure 23–6, which shows development of lung edema in dogs subjected to progressively increasing left atrial pressure. In this experiment no edema fluid developed in the lungs until left atrial pressure rose above 23 mm. Hg, which was approximately 3 mm. Hg greater than the colloid osmotic pressure of dog blood, about 20 mm. Hg. The experiment thus demonstrates that the hydrostatic pressure in the pulmonary capillary usually must rise a few mm. Hg above the colloid osmotic pressure before serious pulmonary edema can ensue. In the human being the colloid osmotic pressure is about 28 mm. Hg, so that pulmonary edema will rarely develop below 30 mm. Hg pulmonary capillary pressure. Thus, if the capillary pressure in the lungs is normally 7 mm. Hg and this pressure must usually rise above 30 mm. Hg before edema will occur, the lungs have a *safety factor against edema* of approximately 23 mm. Hg.

Pulmonary Edema as a Result of Capillary Damage. Pulmonary edema can also result from local capillary damage in the lungs. This effect is often caused by bacterial infection, such as occurs in pneumonia, or by irritant gases such as chlorine, sulfur dioxide, or war gases — mustard gas, for instance. All these directly damage the alveolar epithelium and the endothelium of the capillaries, allowing rapid transudation of both fluid and protein into the alveoli and interstitial spaces.

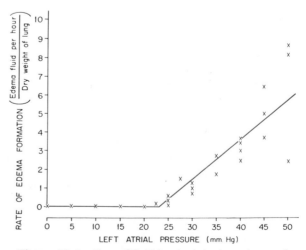

Figure 23–6. Rate of fluid loss into the lung tissues when the left atrial pressure (and also pulmonary capillary pressure) is increased. (From Guyton and Lindsey: *Circ. Res.,* 7:649, 1959, by permission of the American Heart Association, Inc.)

Rapidity of Death in Acute Pulmonary Edema. When the pulmonary capillary pressure does rise above the safety factor level, lethal pulmonary edema can occur within hours if it is only slightly above the safety factor, and within 20 to 30 minutes if it is as much as 25 to 30 mm. Hg above the safety factor level. Thus, in acute left heart failure, in which the pulmonary capillary pressure occasionally rises to as high as 50 mm. Hg, death from acute pulmonary edema frequently ensues within 30 minutes.

SPECIAL FLUID SYSTEMS OF THE BODY

Several special fluid systems exist in the body, each performing functions peculiar to itself. For instance, the cerebrospinal fluid supports the brain in the cranial vault, the intraocular fluid maintains distension of the eyeballs so that the optical dimensions of the eye remain constant, and the potential spaces, such as the pleural and pericardial spaces, provide lubricated chambers in which the internal organs can move. All these fluid systems have characteristics that are similar to each other and that are also similar to those of the interstitial fluid system. However, to emphasize their special characteristics, we will discuss the cerebrospinal and ocular fluid systems.

THE CEREBROSPINAL FLUID SYSTEM

The entire cavity enclosing the brain and spinal cord has a volume of approximately 1650 ml., and about 150 ml. of this volume is occupied by cerebrospinal fluid. This fluid, as shown in Figure 23-7, is found in the ventricles of the brain, in the

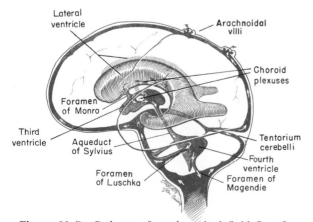

Figure 23–7. Pathway of cerebrospinal fluid flow from the choroid plexuses in the lateral ventricles to the arachnoidal villi protruding into the dural sinuses.

cisterns around the brain, and in the subarachnoid space around both the brain and the spinal cord. All these chambers are connected with each other, and the pressure of the fluid is regulated at a constant level.

Cushioning Function of the Cerebrospinal Fluid. A major function of the cerebrospinal fluid is to cushion the brain within its solid vault. Were it not for this fluid, any blow to the head would cause the brain to be shaken around and severely damaged. However, the brain and the cerebrospinal fluid have approximately the same specific gravity, so that the brain simply floats in the fluid. Therefore, blows on the head move the entire brain at once, causing no one portion of the brain to be momentarily contorted by the blow.

Formation, Flow, and Absorption of Cerebrospinal Fluid. Cerebrospinal fluid is formed at a rate of approximately 800 ml. each day, which is five to six times as much as the total volume of fluid in the entire cerebrospinal fluid cavity. Essentially all of this fluid is formed by the choroid plexus, which is a cauliflower-like growth of blood vessels covered by a thin coat of epithelial cells. This plexus projects into (1) the temporal horns of the lateral ventricles, (2) the posterior portions of the third ventricle, and (3) the roof of the fourth ventricle.

Cerebrospinal fluid is continually secreted by the choroid plexus. This secretion occurs from the epithelial cells by the process of *active secretion,* which was explained in Chapter 4. The fluid then flows out of the fourth ventricle into the subarachnoid space through small openings adjacent to the cerebellum, the *foramina of Luschka* and the *foramen of Magendie.* From here, it flows throughout the entire subarachnoid space but mainly up over the surface of the brain, where it is reabsorbed into the blood through special structures called *arachnoidal villi* or *granulations.* These project from the subarachnoid spaces into the venous sinuses of the brain. The arachnoidal villi are actually arachnoidal trabeculae that protrude through the venous walls, resulting in extremely permeable areas that allow relatively free flow of cerebrospinal fluid into the venous blood.

Cerebrospinal Fluid Pressure. The normal pressure in the cerebrospinal fluid system when one is lying in a horizontal position averages 130 mm. water (10 mm. Hg). The pressure is regulated by the product of, first, the *rate of fluid formation* and, second, the *resistance to absorption through the arachnoidal villi.* When either of these is increased, the pressure rises; and when either is decreased, the pressure falls. A common cause of elevated cerebrospinal fluid is a large *brain tumor* that increases the cerebrospinal fluid pressure by decreasing the rate of absorption of fluid.

Hydrocephalus. "Hydrocephalus" means excess water in the cranial vault. This condition is caused by (1) blockage of flow out of one or more of the ventricles, which leads to ventricular swelling; or (2) overdevelopment of the choroid plexus in the newborn infant, so that far more fluid is formed than can re-enter the venous system through the arachnoidal villi; fluid therefore collects both inside the ventricles and on the outside of the brain. In the hydrocephalic infant, whose skull remains very pliable until the bones fuse at the age of 3 to 4 years, the head swells tremendously, often damaging the brain severely.

THE INTRAOCULAR FLUID

The eye is filled with intraocular fluid, which maintains sufficient pressure in the eyeball to keep it distended. Figure 23–8 illustrates that this fluid can be divided into two portions, the *aqueous humor,* which lies in front and to the sides of the lens, and the *vitreous humor,* which lies between the lens and the retina. The aqueous humor is a freely flowing clear fluid, while the vitreous humor, sometimes called the *vitreous body,* is a gelatinous mass held together by a fine fibrillar network of proteoglycan molecules. Substances can *diffuse* slowly in the vitreous humor, but there is little *flow* of fluid.

Aqueous humor is continually being formed and reabsorbed. The balance between formation and reabsorption of aqueous humor regulates the total volume and pressure of the intraocular fluid.

Formation and Absorption of Aqueous Humor — The Ciliary Body. Aqueous humor is formed in the human eye *at an average rate of 2 to 5 cubic millimeters each minute.* Essentially all of this is *actively* secreted by the *ciliary processes,* which are linear folds projecting from the *ciliary body* into the space behind the iris where the lens ligaments also attach to the eyeball. Their relationship to the fluid chambers of the eye can be seen in Figure 23–8. Because of their folding architecture, the total surface area of the ciliary processes is approximately 6 sq. cm. in each eye, a large area, considering the small size of the ciliary body. The surfaces of these processes are covered by epithelial secretory cells, and immediately beneath these is a highly vascular area.

After aqueous humor is secreted by the ciliary processes, it flows, as shown in Figure 23–8, *between the ligaments of the lens,* then *through the pupil,* and finally *into the anterior chamber of the eye.* Here, the fluid flows into the *angle between the cornea and the iris,* thence through a meshwork of *trabeculae* and finally into the *canal of Schlemm.* Figure 23–9 illustrates the anatomical structures at the iridocorneal angle, showing that the spaces between the trabeculae extend all the way from the anterior chamber to the canal of Schlemm. The canal of Schlemm in turn is a thin-walled vein that extends circumferentially all the way around the eye. Its endothelial membrane is so porous that even large protein molecules, as well as small particulate matter, can pass from the anterior chamber into the canal of Schlemm.

Intraocular Pressure. The average normal intraocular pressure is approximately 16 mm. Hg, with a range from 12 to 20 mm. Hg.

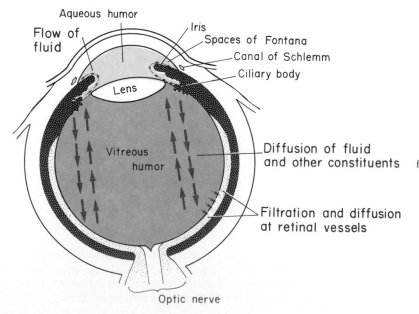

Figure 23–8. Formation and flow of fluid in the eye.

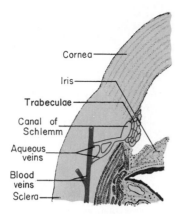

Figure 23–9. Anatomy of the iridocorneal angle, showing the system for outflow of aqueous humor into the conjunctival veins.

The intraocular pressure of the normal eye remains almost exactly constant throughout life, illustrating that the pressure-regulating mechanism is very effective. The pressure is regulated mainly by the outflow resistance from the anterior chamber into the canal of Schlemm, presumably in the following way: The trabeculae guarding the entrance of the fluid into the canal of Schlemm are actually laminar plates that lie one on top of the other. Each of the plates is penetrated by numerous small holes. An increase in pressure above normal distends the spaces between the plates and therefore opens the holes, thus causing rapid flow into the canal of Schlemm and decrease of the pressure back to normal. On the other hand, a decrease in pressure below normal allows the plates to impinge upon each other, thus preventing fluid loss until the pressure rises again back to normal. Thus, this mechanism acts as an automatic feedback regulatory system for keeping the intraocular pressure at a very constant level day in and day out.

Glaucoma. Glaucoma is a disease of the eye in which the intraocular pressure becomes pathologically high, sometimes rising to as high as 70 mm. Hg. As the pressure rises, the retinal artery, which enters the eyeball at the optic disc, is compressed, thus reducing the nutrition to the retina. This often results in permanent atrophy of the retina and optic nerve with consequent blindness.

Glaucoma is one of the most common causes of blindness. Very high pressures lasting only a few days can at times cause total and permanent blindness, but in cases with only mildly elevated pressures, blindness may develop progressively over a period of many years.

In essentially all cases of glaucoma the abnormally high pressure results from increased resistance to fluid outflow at the irido-corneal junction. In most patients, the cause of this is unknown, but in others it results from infection or trauma to the eye. In these persons large quantities of red blood cells, white blood cells, and tissue debris collect in the aqueous humor and then pass into the trabecular spaces of the irido-corneal angle, where they block the outflow of fluid, thereby greatly increasing the intraocular pressure.

REFERENCES

Lymph, Interstitium, and Edema

Bengis, R. G., and Guyton, A. C.: Some pressure and fluid dynamic characteristics of the canine epidural space. *Am. J. Physiol.*, *232*:H255, 1977.
Casley-Smith, J. R.: Lymph and lymphatics. *In* Kaley, G., and Altura, B. M. (eds.): Microcirculation. Vol. I. Baltimore, University Park Press, 1977, p. 423.
Clodius, L. (ed.): Lymphedema. Stuttgart, Thieme, 1977.
Fishman, A. P., and Renkin, E. M. (eds.): Pulmonary Edema. Baltimore, Williams & Wilkins, 1979.
Guyton, A. C.: Interstitial fluid pressure: II. Pressure-volume curves of interstitial space. *Circ. Res.*, *16*:452, 1965.
Guyton, A. C., and Lindsey, A. W.: Effect of elevated left atrial pressure and decreased plasma protein concentration on the development of pulmonary edema. *Circ. Res.*, *7*:649, 1959.
Guyton, A. C., et al.: Circulatory Physiology II: Dynamics and Control of the Body Fluids. Philadelphia, W. B. Saunders Co., 1975.
Guyton, A. C., et al.: Interstitial fluid pressure. *Physiol. Rev.*, *51*:527, 1971.
Guyton, A. C., et al.: Forces governing water movement in the lung. *In* Fishman, A. P., and Renkin, E. M. (eds.): Pulmonary Edema. Baltimore, Williams & Wilkins, 1979, p. 65.
Kennedy, J. F.: Proteoglycans: Biological and chemical aspects of human life. New York, Elsevier Scientific Publishing Co., 1979.
Leak, L. V., and Burke, J. F.: Ultrastructural studies on the lymphatic anchoring of filaments. *J. Cell Biol.*, *36*:129, 1968.
Lennarz, W. J. (ed.): The Biochemistry of Glycoproteins and Proteoglycans. New York, Plenum Press, 1979.
Mayerson, H. S.: The physiologic importance of lymph. *In* Hamilton, W. F. (ed.): Handbook of Physiology. Sec. 2. Vol. 2. Baltimore, Williams & Wilkins, 1963, p. 1035.
Nicoll, P. A., and Taylor, A. E.: Lymph formation and flow. *Annu. Rev. Physiol.*, *39*:73, 1977.
Parker, J. C., et al.: Pulmonary transcapillary exchange and pulmonary edema. *In* Guyton, A. C., and Young, D. B. (eds.): International Review of Physiology: Cardiovascular Physiology III. Vol. 18. Baltimore, University Park Press, 1979, p. 261.
Staub, N. C. (ed.): Lung Water and Solute Exchange. New York, Marcel Dekker, 1978.
Visscher, M. B., et al.: The physiology and pharmacology of lung edema. *Pharm. Rev.*, *8*:389, 1956.
Weibel, E. R., and Bachofen, H.: Structural design of the alveolar septum and fluid exchange. *In* Fishman, A. P., and Renkin, E. M. (eds.): Pulmonary Edema. Baltimore, Williams & Wilkins, 1979, p. 1.
Yoffey, J. M., and Courtice, F. C. (eds.): Lymphatics, Lymph and Lymphomyeloid Complex. New York, Academic Press, 1970.

Special Fluid Systems

Agostoni, E.: Mechanics of the pleural space. *Physiol. Rev.,* *52*:57, 1972.

Bill, A.: Blood circulation and fluid dynamics in the eye. *Physiol. Rev., 55*:383, 1975.

Chandler, P. A., *et al.*: Glaucoma, 2nd Ed. Philadelphia, Lea & Febiger, 1979.

Cserr, H. F.: Physiology of the choroid plexus. *Physiol. Rev., 51*:273, 1971.

Davson, H.: The Physiology of the Cerebrospinal Fluid. Boston, Little, Brown, 1967.

Davson, H.: Physiology of the Eye, 3rd Ed. New York, Churchill Livingstone, 1972.

Hamerman, D., *et al.*: The structure and chemistry of the synovial membrane in health and disease. *In* Bittar, E. E., and Bittar, N. (eds.): The Biological Basis of Medicine. Vol. 3. New York, Academic Press, 1969, p. 269.

Millen, J. W., and Woollam, D. H. M.: The Anatomy of the Cerebrospinal Fluid. New York, Oxford University Press, 1962.

Oldendorf, W. H.: Blood-brain barrier permeability to drugs. *Annu. Rev. Pharmacol., 14*:239, 1974.

Shulman, K. (ed.): Intracranial Pressure IV. New York, Springer-Verlag, 1980.

24

Formation of Urine by the Kidney; Glomerular Filtration, Tubular Function, and Plasma Clearance

The kidneys perform two major functions: first, they excrete most of the end-products of bodily metabolism, and second, they control the concentrations of most of the constituents of the body fluids. The purpose of the present chapter is to discuss the principles of urine formation and especially the mechanisms by which the kidneys excrete the end-products of metabolism.

PHYSIOLOGIC ANATOMY OF THE KIDNEY

The two kidneys together contain about 2,400,000 nephrons, and each nephron is capable of forming urine by itself. The nephron is composed basically of (1) a *glomerulus* from which fluid is filtered and (2) a long *tubule* in which the filtered fluid is converted into urine on its way to the *pelvis* of the kidney. Figure 24–1 shows the general organizational plan of the kidney, illustrating especially the distinction between the *cortex* of the kidney and the *medulla*. And Figure 24–2 illustrates the basic anatomy of the nephron, which may be described as follows: Blood enters the glomerulus through the *afferent arteriole* and then leaves through the *efferent arteriole*. The glomerulus is a network of up to 50 parallel capillaries covered by epithelial cells. Pressure of the blood in the glomerulus causes fluid to filter into *Bowman's capsule,* from which it flows first into the *proximal tubule.* From there the fluid passes into the *Loop of Henle,* which loops downward toward the renal medulla; about a third to a fifth of the loops penetrate deeply into the medulla. The lower portion of the loop has a very thin wall and therefore is called the *thin segment* of

the loop of Henle. From the loop of Henle the fluid flows through the *distal tubule.* Finally, it flows into the *collecting duct,* which collects fluid from several nephrons. The collecting duct passes from the cortex back downward through the medulla, paralleling the loops of Henle. Then it empties into the pelvis of the kidney.

As the glomerular filtrate flows through the tubules, most of its water and varying amounts of its solutes are reabsorbed into the peritubular capillaries and small amounts of other solutes are secreted into the tubules. The remaining tubular water and solutes become urine.

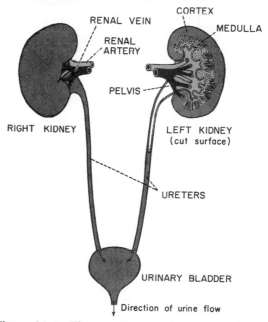

Figure 24–1. The general organizational plan of the urinary system.

247

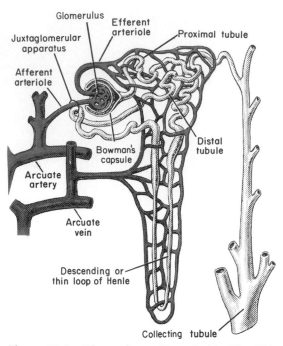

Figure 24–2. The nephron. (From Smith: The Kidney: Structure and Functions in Health and Disease. New York, Oxford University Press, 1951.)

Functional Diagram of the Nephron. Figure 24–3 illustrates a simplified diagram of the "physiologic nephron." This diagram contains most of the nephron's functional structures, and it is used in subsequent discussions to explain many aspects of renal function.

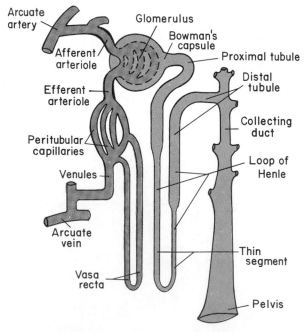

Figure 24–3. The functional nephron.

BASIC THEORY OF NEPHRON FUNCTION

The basic function of the nephron is to clean, or "clear," the blood plasma of unwanted substances as it passes through the kidney. The substances that must be cleared include particularly the end-products of metabolism, such as urea, creatinine, uric acid, and urates. In addition, many other substances, such as sodium ions, potassium ions, chloride ions, and hydrogen ions, tend to accumulate in the body in excess quantities; it is the function of the nephron also to clear the plasma of the excesses.

The principal mechanism by which the nephron clears the plasma of unwanted substances is this: (1) It filters a large proportion of the plasma, usually about one-fifth of it, through the glomerular membrane into the tubules of the nephron. (2) Then, as this filtered fluid flows through the tubules, the unwanted substances fail to be reabsorbed while the wanted substances, especially the water and many of the electrolytes, are reabsorbed back into the plasma of the peritubular capillaries. In other words, the wanted portions of the tubular fluid are returned to the blood, and the unwanted portions pass into the urine.

A second mechanism by which the nephron clears the plasma of unwanted substances is *secretion*. That is, substances are secreted from the plasma directly through the epithelial cells lining the tubules and into the tubular fluid. Thus, the urine that is eventually formed is composed mainly of *filtered* substances but also of small amounts of *secreted* substances.

RENAL BLOOD FLOW AND PRESSURES

BLOOD FLOW THROUGH THE KIDNEYS

The rate of blood flow through both kidneys of a 70 kg. man is about 1200 ml./minute.

Note in Figure 24–3 that there are two capillary beds supplying the nephron: (1) the *glomerulus* and (2) the *peritubular capillaries.* The glomerular capillary bed receives its blood from the *afferent arteriole,* and this bed is separated from the peritubular capillary bed by the *efferent arteriole,* which offers considerable resistance to blood flow. As a result, the glomerular capillary bed is a *high-pressure bed,* while the peritubular capillary bed is a *low-pressure bed.*

The Vasa Recta. A special portion of the peritubular capillary system is the vasa recta, which are a network of capillaries that descend around the lower portions of the loops of Henle. These capillaries form loops in the medulla of the kidney and then return to the cortex before emptying into the veins. The vasa recta play a special role in the formation of concentrated urine, which will be discussed later in the chapter.

PRESSURES IN THE RENAL CIRCULATION

Figure 24–4 gives the approximate pressures in the different parts of the renal circulation and tubules, showing an initial pressure of approximately 100 mm. Hg in the large arcuate arteries and about 8 mm. Hg in the veins into which the blood finally drains. The two major areas of resistance to blood flow in the nephron are (1) the *afferent arteriole* and (2) the *efferent arteriole*. In the afferent arteriole the pressure falls from 100 mm. Hg at its arterial end to a mean pressure of about 60 mm. Hg in the glomerulus. As the blood flows through the efferent arterioles from the glomerulus to the peritubular capillary system, the pressure falls another 47 mm. Hg to a mean peritubular capillary pressure of 13 mm. Hg.

Thus, the high-pressure capillary bed in the glomerulus operates at a mean pressure of about 60 mm. Hg and therefore causes rapid f of fluid into Bowman's capsule. By contr , the low-pressure capillary bed in the peritubular capillary system operates at a mean capillary pressure of about 13 mm. Hg, which allows rapid absorption of fluid because of the high osmotic pressure of the plasma.

GLOMERULAR FILTRATION AND THE GLOMERULAR FILTRATE

The Glomerular Membrane and Glomerular Filtrate. The fluid that filters through the glomerulus into Bowman's capsule is called *glomerular filtrate*, and the membrane of the glomerular capillaries is called the *glomerular membrane*. Though, in general, this membrane is similar to that of other capillaries throughout the body, it has several differences. First, it has three major layers: (1) the endothelial layer of the capillary itself, (2) a basement membrane, and (3) a layer of epithelial cells that is illustrated on the outer surfaces of the glomerular capillaries in Figure 24–3. Yet, despite the number of layers, the permeability of the glomerular membrane is from 100 to 1000 times as great as that of the usual capillary.

The tremendous permeability of the glomerular membrane is caused by its special structure, which is illustrated in Figure 24–5. The capillary *endothelial cells* lining the glomerulus are perforated by literally thousands of small holes called *fenestrae*. Then, outside the endothelial cells is a basement membrane composed mainly of a meshwork of proteoglycan fibrillae. A final layer of the glomerular membrane is a layer of epitheli-

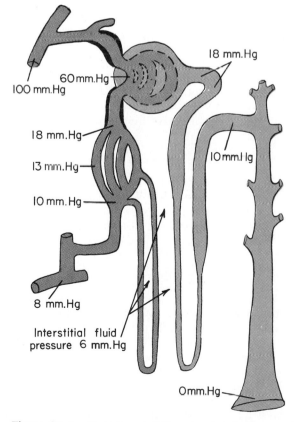

Figure 24–4. Pressures at different points in the vessels and tubules of the functional nephron and in the interstitial fluid.

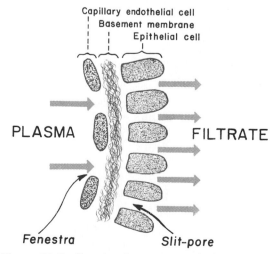

Figure 24–5. Functional structure of the glomerular membrane.

al cells that line the outside of the glomerulus. However, these cells are not continuous but instead consist mainly of finger-like projections that cover the outer surface of the basement membrane. These fingers form slits called *slit-pores,* through which the glomerular filtrate filters. Thus, the glomerular filtrate must pass through three different layers before entering Bowman's capsule.

The slit-pores prevent filtration of all particles with diameters greater than 7 nanometers. Since plasma proteins are slightly larger than the 7 nanometer diameter, it is possible for the glomerular membrane to prevent the filtration of all substances with molecular weights equal to or greater than those of the plasma proteins. Yet, the great numbers of fenestrae and slit-pores allow tremendously rapid filtration of fluid and small molecular weight substances from the plasma into Bowman's capsule.

Composition of the Glomerular Filtrate. The glomerular filtrate has almost exactly the same composition as the fluid that filters from the arterial ends of the capillaries into the interstitial fluids. It contains no red blood cells and about 0.03 per cent protein, or about one two-hundredth of the protein in the plasma. The electrolyte and other solute composition of glomerular filtrate is also similar to that of the interstitial fluid.

To summarize: For all practical purposes, glomerular filtrate is the same as plasma except that it has no significant amount of proteins.

THE GLOMERULAR FILTRATION RATE

The quantity of glomerular filtrate formed each minute in all nephrons of both kidneys is called the *glomerular filtration rate.* In the normal person this averages approximately 125 ml./min.; however, in different normal functional states of the kidneys, it can vary from a few to 200 ml./min. To express this differently, the total quantity of glomerular filtrate formed each day averages about 180 liters, or more than two times the total weight of the body. Over 99 per cent of the filtrate is usually reabsorbed in the tubules, the remainder passing into the urine, as explained later in the chapter.

DYNAMICS OF GLOMERULAR FILTRATION

Glomerular filtration occurs in almost exactly the same manner that fluid filters out of any high pressure capillary in the body. That is, *pressure inside the glomerular capillaries* causes fluid to filter

through the capillary membrane into Bowman's capsule. On the other hand, *colloid osmotic pressure of the blood and pressure in Bowman's capsule* oppose the filtration.

Glomerular Pressure. As stated earlier, *the glomerular pressure is about 60 mm. Hg.*

Pressure in Bowman's Capsule. In lower animals, pressure measurements have been made in Bowman's capsule and at different points along the renal tubules by inserting micropipets into the lumen. On the basis of these studies, *capsular pressure in the human being is estimated to be 18 mm. Hg.*

Colloid Osmotic Pressure in the Glomerular Capillaries. Because approximately one-fifth of the plasma in the capillaries filters into the capsule, the protein concentration increases about 20 per cent as the blood passes from the arterial to the venous ends of the glomerular capillaries. If the normal colloid osmotic pressure of blood entering the capillaries in 28 mm. Hg, it rises to approximately 36 mm. Hg by the time the blood reaches the venous ends of the capillaries, and *the average colloid osmotic pressure is about 32 mm. Hg.*

Filtration Pressure, Filtration Coefficient, and Glomerular Filtration Rate. The filtration pressure is the net pressure forcing fluid through the glomerular membrane, and this is *equal to the glomerular pressure minus the sum of glomerular colloid osmotic pressure and capsular pressure.* Therefore, *the normal filtration pressure is about 10 mm. Hg.*

The filtration coefficient, called K_f, is a constant; it is the glomerular filtration rate for both kidneys per mm. Hg of filtration pressure. Therefore, the glomerular filtration rate is equal to the filtration pressure times the filtration coefficient, or

$$GFR = \text{Filtration pressure} \cdot K_f$$

The normal filtration coefficient is 12.5 ml. per min. per mm. Hg of filtration pressure. Thus, at a normal mean filtration pressure of 10 mm. Hg, the total filtration rate of both kidneys is 125 ml. per min.

Factors That Affect Glomerular Filtration Rate

Arterial Pressure. When the arterial pressure rises, this obviously increases the pressure in the glomerulus as well. Therefore, the glomerular filtration rate increases. The increase in filtration is not as great as would be expected, however, because arterioles are automatically controlled by a mechanism called "autoregulation" (that will be discussed later in the chapter) to keep the glomerular pressure from rising as much as it otherwise would.

Effect of Afferent Arteriolar Constriction on Glomerular Filtration Rate. Constricting the afferent arterioles decreases the rate of blood flow into the glomeruli and therefore also decreases glomerular pressure. Consequently, there is a corresponding decrease in glomerular filtration.

Effect of Efferent Arteriolar Constriction. Constriction of the efferent arterioles increases the resistance to outflow of blood from the glomeruli. This obviously increases the glomerular pressure and usually increases the glomerular filtration rate. However, as discussed below, when efferent arteriolar constriction becomes too great and blood flow is greatly impeded, then the glomerular filtration rate decreases.

Effect of Glomerular Blood Flow on Glomerular Filtration Rate. When either the afferent or the efferent arterioles are constricted, the amount of blood flowing into the glomerulus each minute becomes reduced. Then, as fluid is filtered from the glomeruli, the plasma protein concentration and the colloid osmotic pressure of the plasma in the glomeruli rises. This in turn opposes filtration. Therefore, when the glomerular blood flow falls significantly below normal, the glomerular filtration rate is likely to be seriously depressed despite a high glomerular pressure.

REABSORPTION AND SECRETION IN THE TUBULES

The glomerular filtrate entering the tubules of the nephron flows (1) through the *proximal tubule,* (2) through the *loop of Henle,* (3) through the *distal tubule,* and (4) through the *collecting duct* into the pelvis of the kidney. Along this course, substances are selectively reabsorbed or secreted by the tubular epithelium, and the resultant fluid entering the pelvis is *urine.* Reabsorption plays a much greater role than does secretion in this formation of urine, but secretion is especially important in determining the amounts of potassium ions, hydrogen ions, and a few other substances in the urine, as is discussed later.

Ordinarily, more than 99 per cent of the water in the glomerular filtrate is reabsorbed as it passes through the tubules. Therefore, if some dissolved constituent of the glomerular filtrate is not reabsorbed at all along the entire course of the tubules, this reabsorption of water obviously concentrates the substance more than 99-fold. On the other hand, some constituents, such as glucose and amino acids, are reabsorbed almost entirely, so that their concentrations decrease almost to zero before the fluid becomes urine. In

this way the tubules separate substances that are to be conserved in the body from those that are to be eliminated in the urine.

BASIC MECHANISMS OF ABSORPTION AND SECRETION IN THE TUBULES

The basic mechanisms for transport through the tubular membrane are essentially the same as those for transport through other membranes of the body. These can be divided into *active transport* and *passive transport.* The basic essentials of these mechanisms are described here, but for additional details the reader should refer to Chapter 4.

Active Transport Through the Tubular Wall

Figure 24–6 illustrates, by way of example, the mechanism for active transport of sodium from the lumen of the proximal tubule into the peritubular capillary. Note first the character of the epithelial cells that line the tubule. Each epithelial cell has a "brush" border on its luminal surface. This brush is composed of literally thousands of very minute microvilli that increase the surface area of luminal exposure of the cell about 20-fold.

Active transport of sodium occurs from inside the epithelial cell through its side and basal membranes into basal channels beneath the epithelial cell and into the spaces between the cells. This transport outward from the cell diminishes the sodium concentration inside the cell and also decreases the electrical potential inside the cell to

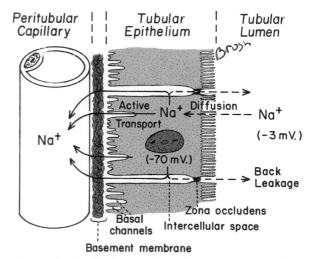

Figure 24–6. Mechanism for active transport of sodium from the tubular lumen into the peritubular capillary, illustrating active transport at the base and sides of the epithelial cell and diffusion through the luminal border of the cell.

a low value, to −70 millivolts. Then, because this low concentration and negative potential inside the cell establish a sodium ion concentration gradient and also an electrical gradient from the tubule to the inside of the cell, sodium ions diffuse from the tubule through the brush border into the cell. Once inside the cell, the sodium is carried by the active transport process the rest of the way into the peritubular fluid.

Other substances besides sodium that are actively absorbed through the tubular epithelial cells include *glucose, amino acids, calcium ions, potassium ions, chloride ions, phosphate ions, urate ions,* and others.

In addition, some substances are actively *secreted* into all or some portions of the tubules; these include especially *hydrogen ions* and *potassium ions.* Active secretion occurs in the same way as active absorption except that the cell membrane transports the secreted substance in the opposite direction.

Passive Absorption of Water: Osmosis of Water Through the Tubular Epithelium

When the different solutes are transported out of the tubule and through the tubular epithelium, their total concentration decreases inside the tubular lumen and increases outside. This obviously creates a concentration difference that causes osmosis of water in the same direction that the solutes have been transported.

However, some portions of the tubular system are far more permeable to water than are others. In those portions that are highly permeable, such as the proximal tubules, osmosis of water occurs so rapidly that the osmolar concentration of solutes on the peritubular side of the membrane is almost never more than 'a few milliosmoles greater than on the intratubular side. On the other hand, the first portion of the distal tubule, called the *diluting segment,* is an example of a tubular area that is almost completely impermeable to water, a fact that plays a very important role in the mechanism for controlling urine concentration. This will be discussed later in this chapter.

Positive Absorption of Urea and Other Nonactively Transported Solutes by the Process of Diffusion

When water is reabsorbed by osmosis, the concentration of urea in the tubular fluid rises, which obviously establishes a concentration difference for urea between the tubular and peritubular fluids. This in turn causes urea also to diffuse from the tubular fluid into the peritubular fluid. This same effect also occurs for other tubular solutes that are not actively reabsorbed but that are diffusible through the tubular membrane.

The rate of resorption of a nonactively reabsorbed solute is determined by (1) the amount of water that is reabsorbed, because this determines the tubular concentration of the solute, and (2) the permeability of the tubular membrane for the solute. The permeability of the membrane for urea in most parts of the tubules is far less than that for water, which means that far less urea is reabsorbed than water. Therefore, a large proportion of the urea remains in the tubules and is lost in the urine — usually about 50 per cent of all that enters the glomerular filtrate. The permeability of the tubular membrane for reabsorption of creatinine, inulin (a large polysaccharide), mannitol (a monosaccharide), and sucrose is zero, which means that once these substances have filtered into the glomerular filtrate, 100 per cent of that which enters the glomerular filtrate passes on into the urine.

Diffusion Caused by Electrical Differences Across the Tubular Membrane. The active absorption of sodium from the tubule creates negativity inside the tubule with respect to the peritubular fluid, as was illustrated in Figure 24–6. In the early proximal tubule, this electrical potential is approximately −3 millivolts. In the distal end of the distal tubule, it ranges from −10 to −70 millivolts. In some other segments of the tubules, this potential is a few millivolts positive — for instance, in the so-called "diluting segment" of the distal tubule, as will be discussed in more detail later in this chapter.

When the electrical potential is negative inside the tubule, this repels negative ions, such as chloride and phosphate ions, causing them to diffuse out of the tubules and into the peritubular fluid. On the other hand, it will attract positive ions, such as sodium and potassium ions, from the peritubular fluid into the tubules. Conversely, when the intratubular fluid is positive, the ions then move in the opposite direction.

REABSORPTION AND SECRETION OF DIFFERENT TYPES OF SUBSTANCES IN DIFFERENT SEGMENTS OF THE TUBULES

Transport of Water and Flow of Tubular Fluid at Different Points in the Tubular System. Water transport occurs entirely by osmotic diffusion. That is, whenever some solute in the glomerular filtrate is absorbed by active reabsorption or by diffusion caused by an electrochemical gradient, the resulting decreased concentration of solute in the tubular fluid and increased concentration in the peritubular fluid cause osmosis of water out of the tubules.

However, because the permeabilities of the different tubular segments vary tremendously for water, the amount of the glomerular filtrate reabsorbed in the different segments also varies tremendously, as follows:

	Per Cent
Proximal tubules	65
Loop of Henle	15
Distal tubules	10
Collecting ducts	9.3
Passing into the urine	0.7

We shall see later in the chapter that these values vary greatly under different operational conditions of the kidney, particularly when the kidney is forming very dilute or very concentrated urine.

Note especially the very large proportion of the glomerular filtrate that is absorbed in the proximal tubules.

Reabsorption of Substances of Nutritional Value to the Body — Glucose, Proteins, Amino Acids, Acetoacetate Ions, and Vitamins. Five different substances in the glomerular filtrate of particular importance to bodily nutrition are glucose, proteins, amino acids, acetoacetate ions, and the vitamins. Normally all of these are completely or almost completely reabsorbed by active processes in the *proximal tubules* of the kidney. Therefore, almost none of these substances remain in the tubular fluid entering the loop of Henle.

Special Mechanism for Absorption of Protein. As much as 30 grams of protein filters into the glomerular filtrate every day. This would be a great metabolic drain on the body if the protein were not returned to the body fluids. Because the protein molecule is much too large to be transported by the usual transport processes, protein is absorbed through the brush border of the proximal tubular epithelium by pinocytosis, which means simply that the protein attaches itself to the membrane and this portion of the membrane then invaginates to the interior of the cell. Once inside the cell the protein is digested into its constituent amino acids, which are then absorbed through the base and sides of the cell into the peritubular fluids. Details of the pinocytosis mechanism were discussed in Chapter 4.

Poor Reabsorption of the Metabolic End-Products: Urea, Creatinine, and Others. Only moderate quantities of *urea* — about 50 per cent of the total — are reabsorbed during the entire course through the tubular system.

Creatinine is not reabsorbed in the tubules at all; indeed, small quantities of creatinine are actually secreted into the tubules by the proximal tubules, so that the total quantity of creatinine increases about 20 per cent.

The *urate ion* is absorbed much more than urea — about 86 per cent reabsorption. But even so, large quantities of urate remain in the fluid that finally issues into the urine. Several other end-products, such as *sulfates, phosphates,* and *nitrates,* are transported in much the same way as urate ions. All of these are normally reabsorbed to a far less extent than is water, so that their concentrations become greatly increased as they flow along the tubules. Yet, *each is actively absorbed to some extent,* which keeps their concentrations in the extracellular fluid from ever falling too low.

Transport of Different Ions by the Tubular Epithelium — Sodium, Chloride, Bicarbonate, and Potassium. In most segments of the tubules, positive ions are transported through the tubular epithelium by active transport processes, while negative ions are usually transported passively as a result of electrical differences developed across the membrane when the positive ions are transported. For instance, when sodium ions are transported out of the proximal tubular fluid, the resulting electronegativity that develops in the tubular fluid causes chloride ions to follow in the wake of the sodium ions. But, despite this general rule, exactly the opposite is true in the diluting segment of the distal tubule, where chloride ions are absorbed actively and sodium and other positive ions are mainly absorbed passively.

Secretion of Potassium and Hydrogen Ions. Potassium ions are actively secreted into the tubular fluid as it passes through the distal tubules and collecting tubules. This will be discussed at greater length in the following chapter in relation to the regulation of potassium concentration in the extracellular fluids.

Hydrogen ions are actively secreted in the proximal tubules, distal tubules, and collecting ducts. This secretion plays an exceedingly important role in controlling the hydrogen ion concentration of the extracellular fluid, as will be discussed in Chapter 26.

Special Aspects of Bicarbonate Ion Transport. Bicarbonate ion is mainly reabsorbed in the form of carbon dioxide rather than in the form of bicarbonate ion itself. This occurs as follows: The bicarbonate ions in the tubular fluid first combine with hydrogen ions that are secreted into the fluid by the epithelial cells. The reaction forms carbonic acid, which then dissociates into water and carbon dioxide. The carbon dioxide, being highly lipid-soluble, diffuses rapidly through the tubular membrane into the peritubular capillary blood.

Transport of Other Ions. Though we know much less about the specific means of transport of other ions besides the four discussed above, in general essentially all of them can be reabsorbed either by active transport or as a result of electrical differences across the membrane. Thus, calcium, magnesium, and other positive ions are actively reabsorbed, and many of the negative ions are reabsorbed as a result of electrical differences that develop when the positive ions are reabsorbed. In addition, certain negative ions — urate, phosphates, sulfates, and nitrates — can be reabsorbed by active transport, this occurring to the greatest extent in the proximal tubules.

CONCENTRATIONS OF DIFFERENT SUBSTANCES IN THE URINE

Whether or not a substance becomes concentrated in the tubular fluid as it moves along the tubules is determined by the *relative reabsorption (or secretion) of the substance versus the reabsorption of water*. If a greater percentage of water is reabsorbed, the substance becomes more concentrated. Conversely, if a greater percentage of the substance is reabsorbed, it becomes more dilute.

In general, there are three different classes of substances:

First, the nutritionally important substances— glucose, protein, and amino acids — are reabsorbed much more rapidly than water, and their concentrations fall extremely rapidly in the proximal tubules and remain essentially zero throughout the remainder of the tubular system as well as in the urine.

Second, the concentrations of the metabolic end-products become progressively greater throughout the tubular system, because all these substances are reabsorbed to a far less extent than is water.

Third, many of the ions are normally excreted into the urine in concentrations not greatly different from those in the glomerular filtrate and extracellular fluid. For instance, sodium and chloride ions, on the average, are normally reabsorbed from the tubules in proportions not to dissimilar from that of water.

Table 24–1 summarizes the concentrating ability of the tubular system for different substances. It also gives the actual quantities of the different substances normally filtered into the tubules each minute.

TABLE 24–1   RELATIVE CONCENTRATIONS OF SUBSTANCES IN THE GLOMERULAR FILTRATE AND IN THE URINE

	Glomerular Filtrate (125 ml./min.)		Urine (1 ml./min.)		Conc. Urine/ Conc. Plasma (Plasma Clearance per Minute)
	Quantity/min.	Concentration	Quantity/min.	Concentration	
Na^+	17.7 mEq.	142 mEq./l.	0.128 mEq.	128 mEq./l.	0.9
K^+	0.63	5	0.06	60	12
Ca^{++}	0.5	4	0.0048	4.8	1.2
Mg^{++}	0.38	3	0.015	15	5.0
Cl^-	12.9	103	0.134	134	1.3
HCO_3^-	3.5	28	0.014	14	0.5
$H_2PO_4^-$ HPO_4^{--}	0.25	2	0.05	50	25
SO_4^{--}	0.09	0.7	0.033	33	47
Glucose	125 mg.	100 mg./100 ml.	0 mg.	0 mg./100 ml.	0.0
Urea	33	26	18.2	1820	70
Uric acid	3.8	3	0.42	42	14
Creatinine	1.4	1.1	1.96	196	140

THE CONCEPT OF "PLASMA CLEARANCE"

The term "plasma clearance" is used to express the ability of the kidneys to clean, or "clear," the plasma of various substances. Thus, if the plasma passing through the kidneys contains 0.1 gram of a substance in each 100 ml., and 0.1 gram of this substance also passes into the urine each minute, then 100 ml. of the plasma is cleaned or "cleared" of the substance per minute.

Plasma clearance for any substance can be calculated by the following formula:

$$\text{Plasma clearance (ml./min.)} = \frac{\text{Quantity of urine (ml./min.)} \times \text{Concentration in urine}}{\text{Concentration in plasma}}$$

The concept of plasma clearance is important because it is an excellent measure of kidney function. The clearance of different substances can be determined by simply analyzing the concentrations of the substances simultaneously in the plasma and in the urine while also measuring the rate of urine formation.

INULIN CLEARANCE AS A MEASURE OF GLOMERULAR FILTRATION RATE

Inulin is a polysaccharide that has the specific attributes of not being reabsorbed to any extent by the tubules of the nephron and yet being of small enough molecular weight (about 5200) that it passes through the glomerular membrane as freely as the crystalloids and water of the plasma. Also, inulin is not actively secreted even in the minutest amount by the tubules. Consequently, glomerular filtrate contains the same concentration of inulin as does plasma, and as the filtrate flows down the tubules all the inulin continues on into the urine. Thus, *all the glomerular filtrate formed is cleared of inulin, and this is equal to the amount of plasma that is simultaneously "cleared."* Therefore, the plasma clearance per minute of inulin is also equal to the glomerular filtration rate.

As an example, let us assume that it is found by chemical analysis that the inulin concentration in the plasma is 0.1 gram in each 100 ml., and that 0.125 gram of inulin passes into the urine per minute. Therefore, by dividing 0.1 into 0.125, one finds that 1.25 *100-milliliter portions* of glomerular filtrate must be formed each minute in order to deliver to the urine the analyzed quantity of inulin. In other words, 125 ml. of glomerular filtrate is formed per minute, and this is also the plasma clearance of inulin.

PARA-AMINOHIPPURIC ACID (PAH) CLEARANCE AS A MEASURE OF PLASMA FLOW AND BLOOD FLOW THROUGH THE KIDNEYS

PAH, like inulin, passes through the glomerular membrane with perfect ease along with the remainder of the glomerular filtrate. However, it is different from inulin in that almost all the PAH remaining in the plasma after the glomerular filtrate is formed is secreted into the tubules by the tubular epithelium if the plasma concentration of PAH is very low. Indeed, only about one-tenth of the original PAH remains in the plasma when the blood leaves the kidneys.

Therefore, one can use the clearance of PAH for estimating the *flow of plasma* through the kidneys, because the amount of plasma flow must always be about 10 per cent greater than the PAH clearance. Thus, if the PAH clearance is 600 ml. per minute, then one can calculate that the plasma flow is about 660 ml. per minute.

THE DILUTING MECHANISM OF THE KIDNEY — THE MECHANISM FOR EXCRETING EXCESS WATER

One of the most important functions of the kidney is to control the osmolality of the body fluids. It does so by excreting excessive amounts of water in the urine when the body fluids are too dilute or by excreting excessive amounts of solutes when the body fluids are too concentrated.

Whether the kidneys excrete excess water or excess solutes is controlled by *antidiuretic hormone*, a hormone secreted by the posterior pituitary gland, which will be discussed in more detail in the following chapter. In the absence of antidiuretic hormone the kidneys excrete excessive amounts of water, but when the blood concentration of antidiuretic hormone is high the kidneys excrete excessive amounts of solutes. Let us explain first the mechanism for excreting excess water — that is, the mechanism for excreting a dilute urine.

When the glomerular filtrate is formed in the glomerulus, its osmolality is almost exactly the same as that of the plasma, approximately 300 milliosmoles per liter. To excrete excess water, it is necessary to dilute the filtrate as it passes through the tubules. This is achieved by reabsorbing a higher proportion of solutes than water. Figure 24–7 illustrates this process. The dark arrows in this figure represent rapid reabsorption of tubular solute, and the thickened walls of the more distal tubular segments indicate

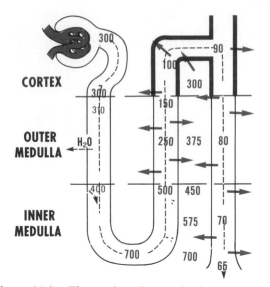

Figure 24–7. The renal mechanism for forming a dilute urine. The darkened walls of the distal portions of the tubular system indicate that these portions of the tubules are relatively impermeable to the reabsorption of water in the absence of antidiuretic hormone. The solid arrows indicate active processes for absorption of most of the solutes besides the urinary waste products. (Numerical values are in milliosmoles per liter.)

from the latter segments of the tubules while water fails to be reabsorbed. However, this occurs only when antidiuretic hormone is *not* being secreted by the posterior pituitary gland.

THE CONCENTRATING MECHANISM OF THE KIDNEY; EXCRETION OF EXCESS SOLUTES — THE COUNTER-CURRENT MECHANISM

The process for concentrating urine is not nearly so simple as for diluting it. Yet, at times it is exceedingly important to concentrate the urine as much as possible so that excess solutes can be eliminated with as little loss of water from the body as possible — for instance, when one is exposed to desert conditions with an inadequate supply of water. Fortunately, the kidneys have developed a special mechanism for concentrating the urine called the *counter-current mechanism.*

The counter-current mechanism depends on a special anatomical arrangement of the loops of Henle and the vasa recta. In the human being, the loops of Henle of one-third to one-fifth of the nephrons dip deep into the medulla and then return to the cortex. This group of nephrons with the long loops of Henle is called the *juxtamedullary nephrons.* Paralleling the long loops of Henle are loops of peritubular capillaries called the *vasa recta;* these loop down into the medulla from the cortex and then back out to the cortex again. These arrangements of the different parts of the juxtamedullary nephron and the vasa recta are diagrammed in Figure 24–8.

*that these portions are relatively impermeable to water when antidiuretic hormone is *not* present in the circulating body fluids.*

In the "diluting segment" of the distal tubule (approximately the first half of the distal tubule, beginning with the thick portion of the ascending limb of the loop of Henle), the absorption of solutes results primarily from active absorption of chloride ions, and this causes electrogenic passive absorption of the positive ions sodium, potassium, calcium, and magnesium. On the other hand, in the late distal tubule and the collecting ducts, the major driving force is active transport of sodium ions, and this then causes electrogenic passive absorption of the anions, mainly chloride ions. Consequently, most of the ionic substances of the tubular fluids are reabsorbed from these distal segments of the tubular system before the fluid is emptied as urine, but the water remains and the urine is dilute. Note in Figure 24–7 that in the diluting segment of the distal tubule, the osmolality of the fluid falls rapidly from its initial value of 300 milliosmoles per liter to about 100 milliosmoles per liter in the late distal tubule, and as low as 65 to 70 mOsm./liter in the last portions of the inner medullary collecting duct and in the urine.

To summarize, the process for excreting a dilute urine is simply one of absorbing solutes

Hyperosmolality of the Medullary Interstitial Fluid, and Mechanisms for Achieving It

The first step in the excretion of excess solutes in the urine — that is, for excretion of a concentrated urine — is to create hyperosmolality of the medullary interstitial fluid. As we shall see later, this in turn is necessary for concentrating the urine. But first let us explain the mechanism for creating this hyperosmolality in the medullary interstitium.

The normal osmolality of the fluids in almost all parts of the body is about 300 mOsm./liter. However, as shown by the numbers in Figure 24–8, the osmolality of the interstitial fluid in the medulla of the kidney is much higher, and it becomes progressively greater the deeper one goes into the medulla, increasing from 300

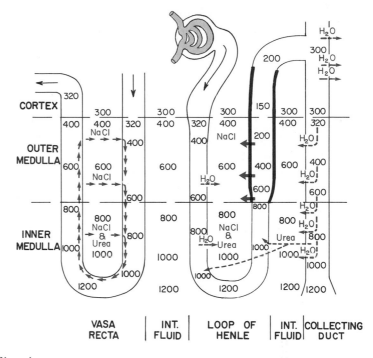

Figure 24–8. The counter-current mechanism for concentrating the urine. (Numerical values are in milliosmoles per liter.)

mOsm./liter in the cortex to 1200 mOsm./liter in the pelvic tip of the medulla. Four different solute-concentrating mechanisms are responsible for this hyperosmolality; these are as follows:

First, the principal cause of the greatly increased medullary osmolality is active transport of chloride ions (plus electrogenic passive absorption of sodium ions) out of the thick portion of the ascending limb of the loop of Henle. The very dark arrows shown in this tubular segment in Figure 24–8 illustrate this transport of chloride and sodium ions (as well as potassium, calcium, and magnesium ions to a lesser extent) out of the thick portion of the loop of Henle and into the outer medullary interstitial fluid. All these solutes become concentrated in this fluid. These ions are also carried downward into the inner medulla by the flowing blood in the vasa recta, as we shall see shortly.

Second, ions are also transported into the medullary interstitial fluid from the collecting duct, mainly as a result of active transport of sodium ions and electrogenic passive absorption of chloride ions along with the sodium ions.

Third, when the concentration of antidiuretic hormone is high in the blood, large amounts of urea are also absorbed into the medullary fluid from the inner medullary collecting duct. The reason is this: In the presence of antidiuretic hormone, the inner medullary portion of the collecting duct becomes moderately permeable to urea. Consequently, the urea concentration in the medullary interstitial fluid rises to almost equal the concentration in the collecting duct. In the human being, during maximal antidiuretic hor-

mone stimulation this may be to as high as 400 to 500 mOsm./liter, which obviously greatly increases the osmolality of the inner medullary interstitial fluid.

· And, fourth, the final event that causes increased osmolal concentration of the medullary interstitial fluid is absorption of sodium and chloride ions into the inner medullary interstitium from the thin segment of the loop of Henle. Most physiologists believe that this is achieved by a passive mechanism, as follows: When the concentration of urea rises very high in the medullary interstitium because of urea absorption from the collecting duct, this immediately promotes osmosis of water out of the descending thin limb of the loop of Henle. Therefore, the concentration of sodium chloride inside the thin limb rises almost to twice normal. And now, because of this high concentration, both sodium and chloride ions diffuse passively out of the thin segment into the interstitium.

In summary, at least four different factors contribute to the marked increase in osmolality in the medullary interstitial fluid. These are (1) active transport of the ions into the interstitium by the thick portion of the ascending limb of the loop of Henle, (2) active transport of ions from the collecting duct into the interstitium, (3) passive diffusion of large amounts of urea from the collecting duct into the interstitium, and (4) absorption of additional sodium and chloride into the interstitium from the thin segment of the loop of Henle, a transport that is probably also passive as explained above. The net result is an increase in the osmolality of the medullary interstitial fluid, when adequate amounts of antidiuretic hormone are present, to as high as 1200 to 1400 mOsm./liter near the tips of the papillae.

Counter-Current Exchange Mechanism in the Vasa Recta

We have now discussed the mechanism by which a high concentration of solutes is achieved in the medullary interstitium. However, without a special medullary vascular system as well, the flow of blood through the interstitium would rapidly remove the excess solutes and keep the concentration from rising very high. Fortunately, the medullary blood flow has two characteristics, both exceedingly important, for maintaining the high solute concentration in the medullary interstitial fluids:

First, the medullary blood flow is very slow, amounting to only 1 to 2 per cent of the total blood flow of the kidney. Because of this very sluggish blood flow, removal of solutes is minimized.

Second, the vasa recta functions as a *counter-current exchanger* that prevents washout of solutes from the medulla. This can be explained as follows: A counter-current fluid exchange mechanism is one in which fluid flows through a long U-tube, with the two arms of the U lying close to each other so that fluid and solutes can exchange readily between the two arms. This obviously also requires that each of the two arms of the U be highly permeable, which is true of the vasa recta. When the fluids and solutes in the two parallel streams of flow can exchange rapidly, tremendous concentrations of solute can be maintained at the tip of the loop with relatively negligible washout of solute.

Thus, in Figure 24–8, as blood flows down the descending limbs of the vasa recta, sodium chloride and urea diffuse into the blood from the interstitial fluid while water diffuses outward into the interstitium, and these two effects cause the osmolar concentration in the blood to rise progressively higher, to a maximum concentration of 1200 mOsm./liter at the tips of the vasa recta. Then, as the blood flows back up the ascending limbs, the extreme diffusibility of all molecules through the capillary membrane allows essentially all the same sodium chloride and urea to diffuse back out of the blood into the interstitial fluid while water diffuses back into the blood. Therefore, by the time the blood finally leaves the medulla, its osmolal concentration is only slightly greater than that of the blood that initially entered the vasa recta. As a result, blood flowing through the vasa recta carries only a minute amount of the medullary interstitial solutes away from the medulla.

Mechanism for Excreting a Concentrated Urine – Role of Antidiuretic Hormone

Now that we have explained how the kidney creates hyperosmolality in the medullary interstitium, it becomes a simple matter to explain the mechanism for excreting a concentrated urine, thus causing loss of excess solutes from the body fluids while at the same time retaining as much water as possible. When the concentration of *antidiuretic hormone* in the blood is high, the epithelium of the entire collecting duct, and in some species of animals of the late distal tubule also, becomes highly permeable to water. This is illustrated in Figure 24–8 by the thin walls of the collecting duct and late distal tubule. As the tubular fluid flows through the collecting duct, water is pulled by osmosis into the highly concentrated fluid of the medullary interstitium. Thus, the collecting duct fluid also becomes highly concentrated, and it issues from the papilla into the pelvis of the kidney at the concentration of about 1200 mOsm./liter, almost exactly equal to the osmolal concentration of the solutes in the medullary interstitium near the papilla.

UREA EXCRETION

The body forms an average of 25 to 30 grams of urea each day — more than this in persons who eat a very high-protein diet and less in persons who are on a low-protein diet. All this urea must be excreted in the urine; otherwise, it will accumulate in the body fluids. Its normal concentration in plasma is approximately 26 mg./100 ml., but it has been recorded in rare states of renal insufficiency to be as high as 800 mg./100 ml.

The two major factors that determine the rate of urea excretion are (1) the concentration of urea in the plasma and (2) the glomerular filtration rate. Both of these factors increase urea excretion, mainly because the load of urea entering the proximal tubules is equal to the product of the plasma urea concentration times the glomerular filtration rate. And in general, the quantity of urea that passes on through the tubules into the urine averages about 50 to 60 per cent of the urea load that enters the proximal tubules.

SODIUM EXCRETION

Sodium Reabsorption from the Proximal Tubules and Loops of Henle. From earlier in the chapter it will be recalled that approximately 65 per cent of the glomerular filtrate is reabsorbed in the proximal tubules. This reabsorption is caused primarily by the active transport of sodium through the proximal tubular epithelium. When the sodium is reabsorbed it causes diffusion of negative ions through the membrane as well, and the cumulative reabsorption of ions creates an osmotic pressure that then moves water through the membranes, too. The epithelium is also permeable to water, so that almost identically the same proportions of water and of sodium ions are reabsorbed.

In the thin segments of the loops of Henle, very little sodium and water are absorbed. However, in the diluting segment of the distal tubules, the active transport of chloride ions causes sodium (and other positive ions) to be absorbed along with the chloride ions, as has already been explained. As a result, the concentrations of sodium and chloride in the distal tubular fluid often fall to as low as one-third to one-fifth their concentrations in the original glomerular filtrate. Therefore, on the average, less than 10 per cent of the sodium chloride in the original glomerular filtrate still remains by the time the fluid reaches the late distal tubules.

Sodium Reabsorption in the Late Distal Tubules and Collecting Ducts — Role of Aldosterone. Sodium reabsorption in the late distal tubules and collecting ducts is highly variable, depending mainly on the concentration of *aldosterone,* a hormone secreted by the adrenal cortex. In the presence of large amounts of aldosterone, almost the last vestiges of the tubular sodium are reabsorbed from the late distal tubules and collecting ducts so that essentially none of the sodium issues into the urine. Thus, the sodium excretion may be as little as one-tenth of a gram per day or as great as 30 to 40 grams. This ability of the tubular system to reabsorb almost all the sodium that filters through the glomeruli is a remarkable feat when one recognizes that almost 10 times as much sodium enters the glomerular filtrate each day as is present in the entire body.

Mechanism by Which Aldosterone Enhances Sodium and Potassium Transport. Upon entering a tubular epithelial cell, aldosterone combines with a *receptor protein;* this combination diffuses within minutes into the nucleus, where it activates DNA molecules to form one or more types of messenger RNA. The RNA is then believed to cause formation of carrier proteins or protein enzymes that are necessary for the sodium transport process.

Ordinarily, aldosterone has no effect on sodium transport for the first 45 minutes after it is administered; after this time the specific proteins important for transport begin to appear in the epithelial cells, followed by progressive increase in transport during the ensuing few hours.

POTASSIUM EXCRETION

Potassium Transport in the Proximal Tubules and Loops of Henle. In the proximal tubules and loops of Henle, potassium is transported in almost exactly parallel fashion to the transport of sodium. That is, both the potassium and the sodium are transported from the tubule into the blood. Thus, approximately 65 per cent of the potassium in the glomerular filtrate is absorbed in the proximal tubules (as is also true for sodium), and another 25 per cent is absorbed in the diluting segment of the distal tubules, so that by the time the tubular fluid reaches the late distal tubules, the total quantity of potassium delivery to the late distal tubules each minute is less than 10 per cent of that in the original glomerular filtrate (as is also true for sodium).

Active Secretion of Potassium in the Late Distal Tubules and Collecting Ducts — Role of Aldosterone. In the late distal tubules and in the collecting ducts, potassium is no longer absorbed along with the sodium but is secreted into the tubules instead. The reason is that when sodium is transported from the tubules into the peritubular fluid, potassium is simultaneously transported in the opposite direction. Furthermore, aldosterone stimulates this potassium transport equally as much as it stimulates the sodium transport.

This secretory transport of potassium into the distal tubules is extremely important for the control of plasma potassium concentration, for the following simple reason: The total quantity of potassium delivered from the loops of Henle into the distal tubules each day averages only about 70 mEq. Yet, the human being regularly eats more than this much potassium each day and on occasion eats as much as several hundred mEq. per day. Even if all of the 70 mEq. that enters the distal tubule should pass on into the urine, this still would not be enough potassium elimination. Therefore, it is essential that the excess potassium be removed by the process of secretion; otherwise, death might ensue from potassium toxicity. Indeed, cardiac arrhythmias usually appear when the potassium concentration rises from the normal value of 4 to 5 mEq./liter up to a level of 8 mEq./liter. A still higher potassium concentration can end in cardiac standstill.

FLUID VOLUME EXCRETION

Up to this point we have considered the intrarenal mechanisms that determine the *concentrations* of various substances in the urine — water, urea, sodium, and potassium. Now it is important to consider the different factors that determine the rate of fluid *volume* excretion.

Glomerulotubular Balance and Its Relationship to Fluid Volume Excretion

By the term *glomerulotubular balance* most physiologists mean that whenever the glomerular filtration rate increases, all of the *additional* filtrate is reabsorbed and does not pass into the urine. This is *almost* true for the normal kidney, so it is usually said that the kidney normally obeys the principle of glomerulotubular balance.

However, very precise measurements show that 100 per cent glomerulotubular balance very rarely occurs. For instance, the following table gives approximate values for glomerular filtration rates, rates of fluid reabsorption, and rates of urine output for the average human adult:

Glomerular Filtration Rate	Rate of Tubular Reabsorption	Rate of Urine Output
ml.	*ml.*	*ml.*
50	49.8	0.2
75	74.7	0.3
100	99.5	0.5
125	124.0	1.0
150	145.0	5.0
175	163.0	12.0

If we examine these figures critically, we see that glomerular filtration rate and rate of tubular reabsorption actually do appear to parallel each other very closely. On the other hand, the degree of imbalance that occurs causes far greater change, proportionately, in urine output than in either glomerular filtration rate or tubular reabsorption rate. For instance, let us study the increase in glomerular filtration rate from 100 ml./min. up to 150 ml./min. The rate of reabsorption increases from 99.5 ml./min. up to 145 ml./min, representing only slight glomerulotubular imbalance. Nevertheless, this 50 per cent increase in glomerular filtration rate causes a 1000 per cent increase in rate of urine output! Thus, even a very slight degree of glomerulotubular imbalance leads to a tremendous increase in urine output when the glomerular filtration rate is increased. Also, very slight changes in rate of reabsorption of tubular fluid can cause equally great alterations in urine output.

Therefore, the various factors that can alter either glomerular filtration rate or rate of tubular reabsorption are also the factors that play significant roles in determining the rate of fluid volume excretion. The five most important of these are:

1. Effect of Osmolar Clearance on Rate of Fluid Volume Excretion. The greater the quantity of osmolar substances that *fails* to be reabsorbed by the tubules, the greater also is the quantity of water that *fails* to be reabsorbed. To state this another way, when the osmolar clearance is great, the volume of urine usually increases by approximately the same percentage. This effect is called *osmotic diuresis.* A particularly interesting type of osmotic diuresis occurs in diabetes mellitus, in which the proximal tubules fail to reabsorb all the glucose as normally occurs. Instead, the nonreabsorbed glucose passes the entire distance through the tubules and carries with it a large portion of the tubular water. Therefore, in diabetes mellitus (the word "diabetes" means diuresis) the urine output occasionally increases to as high as 4 to 5 liters per day.

2. Effect of Plasma Colloid Osmotic Pressure on Rate of Fluid Volume Excretion. Another factor that greatly affects the rate of volume excretion is the plasma colloid osmotic pressure. A sudden increase in plasma colloid osmotic pressure instantaneously decreases the rate of fluid volume excretion. The cause of this effect is two-fold: (1) The increased plasma colloid osmotic pressure *decreases* glomerular filtration rate, and (2) it *increases* tubular reabsorption. Both of these effects add together to decrease greatly the urine volume excretion.

3. Effect of Sympathetic Stimulation on Rate of Fluid Volume Excretion. Sympathetic stimulation has an especially powerful effect of constricting the afferent arterioles. It greatly decreases the glomerular pressure and therefore decreases glomerular filtration rate. As has already been pointed out, a decrease in glomerular filtration rate often causes 10 times as much proportional decrease in urine output because of the slight degree of glomerulotubular imbalance that occurs even normally, as was discussed earlier.

Conversely, a decrease of sympathetic stimulation to below normal causes a mild degree of afferent arteriolar dilatation, which increases the glomerular filtration rate a slight amount. Consequently, decreased sympathetic stimulation leads to increased urine volume excretion.

4. Effect of Arterial Pressure on Rate of Fluid Volume Excretion. If all other factors remain constant but the renal arterial pressure is changed, the rate of urine output changes markedly. This effect is illustrated in Figure 24–9, showing that when the arterial pressure rises from 100 mm. Hg to 200 mm. Hg the urine output increases approximately seven-fold. Conversely, when the arterial pressure falls from 100 mm. Hg to 60 mm. Hg the urine output falls to either zero or near zero. We have already pointed out in Chapter 18 that this pressure effect on urine output plays an extremely important role in the feedback regulation of arterial pressure. It also plays an extremely important role in the feedback regulation of body fluid volume, as we

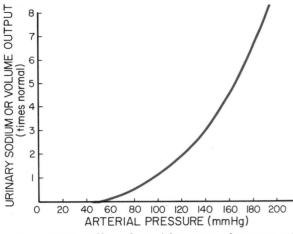

Figure 24–9. Effect of arterial pressure change on urinary output.

causes formation of cyclic adenosine monophosphate (cyclic AMP) in the cell cytoplasm. The increase in cyclic AMP is then associated — for reasons that are not yet known — with marked increase in permeability of the *luminal* membrane of the epithelial cells to water; this in turn is responsible for the increase in water reabsorption by the collecting tubules.

SUMMARY OF THE CONTROL OF FLUID VOLUME EXCRETION

From the preceding few sections, it is clear that many different factors have effects on the regulation of urine volume excretion. Some of the factors cause extreme acute changes in urinary output — for instance, the acute decrease in urine volume output caused by antidiuretic hormone or the acute increase in urine volume output caused by increased tubular osmotic loads. However, over a longer period of time, especially important is the increase in output caused by elevated arterial pressure, an effect that seems to be sustained indefinitely. This is a subject that we shall discuss in much more detail in the following chapter in relation to the control of extracellular fluid volume.

shall discuss in the following chapter. Now, however, let us discuss the mechanism of the increased urine output caused by the increase in arterial pressure. This results from two separate effects: (1) The increase in arterial pressure increases glomerular pressure, which in turn increases glomerular filtration rate, thus leading to increased urine output; (2) the increase in arterial pressure also increases the peritubular capillary pressure, thereby decreasing tubular reabsorption. The combination of these two effects causes considerable glomerulotubular imbalance and therefore also causes the marked increase in urine output illustrated in Figure 24–9.

5. Effect of Antidiuretic Hormone on Rate of Fluid Volume Excretion. When excess antidiuretic hormone is secreted by the hypothalamic–posterior pituitary system, this secretion causes an acute effect to decrease the urinary volume output. The reason for this is that antidiuretic hormone causes increased water reabsorption from the collecting ducts and perhaps to a slight extent from the late distal tubules as well. Therefore, less urinary volume is excreted; on the other hand, the urine that is excreted is highly concentrated.

However, when excess antidiuretic hormone is secreted for long periods of time, the acute effect of decreasing urinary output is not sustained. This will be discussed further in the following chapter.

Mechanism by Which Antidiuretic Hormone Increases Water Reabsorption. The precise mechanism by which antidiuretic hormone increases water reabsorption by the collecting tubules is not known. However, several established facts about the mechanism are the following: Stimulation of the epithelial cells of the collecting tubules by antidiuretic hormone activates the enzyme *adenyl cyclase* in the epithelial cell membrane on the peritubular side of the cell, and this

AUTOREGULATION OF GLOMERULAR FILTRATION RATE

Even though a change in arterial pressure causes a marked change in urinary output, this pressure can change from as little as 75 mm. Hg to as high as 160 mm. Hg while causing very little change in glomerular filtration rate. This effect is illustrated in Figure 24–10 and is called *autoregulation of glomerular filtration rate*. It is impor-

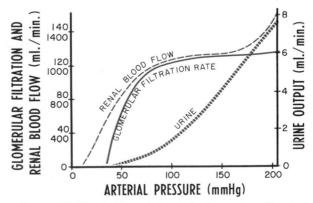

Figure 24–10. Autoregulation of glomerular filtration rate (GFR) and renal blood flow (RBF) when the arterial pressure is increased but there is lack of autoregulation of urine flow.

.01

the nephron requires an optimal
~~rular~~ filtration if it is to perform its
~~...~~ a 5 per cent too great or too little
~~... glomerular~~ filtration can have profound
effects in causing either excess fluid loss in the
urine or too little excretion of the necessary waste
products.

Mechanism of Autoregulation of Glomerular Filtration Rate – Tubuloglomerular Feedback

Fortunately, each nephron is provided with not
one but *two* special feedback mechanisms that add
together to provide the necessary degree of glomerular filtration autoregulation. These two
mechanisms are (1) an *afferent arteriolar vasodilator
feedback mechanism* and (2) an *efferent arteriolar
vasoconstrictor feedback mechanism.* The combination of these two feedback mechanisms is called
tubuloglomerular feedback. And the feedback process probably occurs either entirely or almost
entirely at the *juxtaglomerular complex,* which has
the following characteristics:

The Juxtaglomerular Complex. Figure
24–11 illustrates the juxtaglomerular complex,
showing that the distal tubule passes in the angle
between the afferent and efferent arterioles, actually abutting each of these two arterioles. Furthermore, those epithelial cells of the distal tubule that come in contact with the arterioles are
more dense than the other tubular cells and are
collectively called the *macula densa.* The position

in the distal tubule where the macula densa is
located is about midway in the diluting segment
of the distal tubule, at the upper end of the thick
portion of the ascending limb of the loop of
Henle. The smooth muscle cells of both the
afferent and the efferent arterioles are swollen
and contain dark granules where they come in
contact with the macula densa. These cells are
called *juxtaglomerular cells* (JG cells), and the granules are composed mainly of inactive *renin.*

**The Afferent Arteriolar Vasodilator Feedback
Mechanism.** A low rate of glomerular filtration
allows over-reabsorption of chloride in the tubules and therefore decreases the chloride ion
concentration at the macula densa. This decrease
in chloride ions in turn initiates a signal from the
macula densa to dilate the afferent arterioles.
Putting these two facts together, the following is
the mechanism by which the afferent arteriolar
vasodilator feedback mechanism controls glomerular filtration rate:

(1) Too little flow of glomerular filtrate into the
tubules causes decreased chloride ion concentration at the macula densa.

(2) The decreased chloride concentration
causes afferent arteriolar dilatation.

(3) This in turn increases the rate of blood flow
into the glomerulus and increases the glomerular
pressure.

(4) The increased glomerular pressure increases the glomerular filtration rate back toward
the required level.

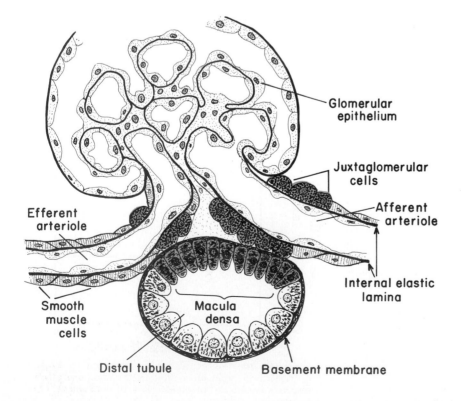

Figure 24–11. Structure of the
juxtaglomerular apparatus, illustrating its possible feedback role in the
control of nephron function. (Modified from Ham and Cormack: Histology, 8th ed. J. B. Lippincott Co.)

The Efferent Arteriolar Vasoconstrictor Feedback Mechanism. Too few chloride ions at the macula densa is believed also to cause the juxtaglomerular cells to release renin, and this in turn causes formation of angiotensin. The angiotensin then constricts mainly the efferent arteriole because it is more sensitive to angiotensin II than is the afferent arteriole.

With these facts in mind, we can now describe the efferent arteriolar vasoconstrictor mechanism that helps to maintain a constant glomerular filtration rate:

(1) A too low glomerular filtration rate causes excess reabsorption of chloride ions from the filtrate, reducing the chloride ion concentration at the macula densa.

(2) The low concentration of chloride ions then causes the JG cells to release renin from their granules.

(3) The renin causes formation of angiotensin II.

(4) The angiotensin II constricts the efferent arterioles, which causes the pressure in the glomerulus to rise.

(5) The increased pressure then causes the glomerular filtration rate to return back toward normal.

Thus, this is still another negative feedback mechanism that helps to maintain a very constant glomerular filtration rate; it does so by constricting the efferent arterioles at the same time that the afferent vasodilator mechanism described above dilates the afferent arterioles. When both of these mechanisms function together, the glomerular filtration rate increases only a few per cent even though the arterial pressure changes between the limits of 75 mm. Hg and 160 mm. Hg.

AUTOREGULATION OF RENAL BLOOD FLOW

When the arterial pressure is changed for only a few minutes at a time, renal blood flow and the glomerular filtration rate are autoregulated at the same time. This was illustrated in Figure 24–10, which shows a relatively constant renal blood flow between the limits of 70 and 160 mm. Hg arterial pressure.

It is the afferent arteriolar vasodilator feedback mechanism described above that causes this renal blood flow autoregulation. This can be explained as follows: When the renal blood flow becomes too little, the glomerular pressure falls and the glomerular filtration rate also becomes too little. As a

consequence, the feedback mechanism causes afferent arteriolar dilatation to return the glomerular filtration rate back toward normal. At the same time, the dilatation also increases the blood flow back toward normal despite the low arterial pressure.

ROLE OF THE RENIN-ANGIOTENSIN SYSTEM AND OF THE EFFERENT VASOCONSTRICTOR MECHANISM IN CONSERVING WATER AND SALT BUT ELIMINATING UREA DURING ARTERIAL HYPOTENSION

The efferent arteriolar vasoconstrictor mechanism not only helps to maintain normal glomerular filtration when the arterial pressure falls too low but also provides a means for controlling urea excretion separately from the excretion of water and salt. In arterial hypotension it is very important to conserve as much water and salt as possible. On the other hand, it is equally important to continue excreting the body's waste products, the most abundant of which is urea. Therefore, let us explain this:

Earlier in the chapter it was pointed out that the rate of urea excretion is almost directly proportional to the rate of glomerular filtration. Therefore, as long as the efferent arteriolar vasoconstrictor mechanism can maintain a high glomerular filtration even in the presence of low arterial pressure, almost normal amounts of urea will be excreted in the urine. Therefore, hypotension down to arterial pressure levels as low as 65 to 75 mm. Hg does not cause significant retention of urea.

On the other hand, as angiotensin II builds up within the kidneys and also in the circulating blood during arterial hypotension, this causes marked retention by the kidney of water and of the various ions — sodium, chloride, potassium, and others. Thus, this provides a means for conserving water and ions despite the fact that urea continues to be excreted.

The angiotensin probably causes water and ion conservation by the following mechanism. It increases renal arteriolar resistance, which reduces renal blood flow and therefore also reduces peritubular capillary pressure. This in turn increases the rate of reabsorption of water and electrolytes from the tubular system, as explained earlier in this chapter.

REFERENCES

Andersson, B.: Regulation of body fluids. *Annu. Rev. Physiol., 39*:185, 1977.

Aukland, K.: Renal blood flow. *Int. Rev. Physiol., 11*:23, 1976.

Barger, A. C., and Herd, J. A.: Renal vascular anatomy and distribution of blood flow. *In* Orloff, F., and Berliner, R. W. (eds.): Handbook of Physiology. Sec. 8. Baltimore, Williams & Wilkins, 1973, p. 249.

Brenner, B. M.: Adaptation of glomerular forces and flows to renal injury. *Yale J. Biol. Med., 51*(3):301, 1978.

Brenner, B. M., *et al.*: Determinations of glomerular filtration rate. *Annu. Rev. Physiol., 38*:9, 1976.

Brenner, B. M., *et al.*: Transport of molecules across renal glomerular capillaries. *Physiol. Rev., 56*:502, 1976.

Churg, J. (ed.): The Kidney. Baltimore, Williams & Wilkins, 1979.

Giebisch, G., and Stanton, B.: Potassium transport in the nephron. *Annu. Rev. Physiol., 41*:241, 1979.

Glynn, I. M., and Karlish, S. J. D.: The sodium pump. *Annu. Rev. Physiol., 37*:13, 1975.

Goldberg, M., *et al.*: Renal handling of calcium and phosphate. *Int. Rev. Physiol., 11*:211, 1976.

Gottschalk, C. W., and Lassiter, W. E.: Micropuncture methodology. *In* Orloff, F., and Berliner, R. W. (eds.): Handbook of Physiology. Sec. 8. Baltimore, Williams & Wilkins, 1973, p. 129.

Grantham, J. J., *et al.*: Studies of isolated renal tubules in vitro. *Annu. Rev. Physiol., 40*:249, 1978.

Guyton, A. C., *et al.*: Circulatory Physiology II: Dynamics and Control of the Body Fluids. Philadelphia, W. B. Saunders Co., 1975.

Hall, J. E., *et al.*: Dissociation of renal blood flow and filtration rate autoregulation by renin depletion. *Am. J. Physiol., 232*:F215, 1977.

Katz, A. I., and Lindheimer, M. D.: Actions of hormones on the kidney. *Annu. Rev. Physiol., 39*:97, 1977.

Kinne, R.: Membrane-molecular aspects of tubular transport. *Int. Rev. Physiol., 11*:169, 1976.

Knox, F. G., and Diaz-Buxo, J. A.: The hormonal control of sodium excretion. *Int. Rev. Physiol., 16*:173, 1977.

Lameire, N. H., *et al.*: Heterogeneity of nephron function. *Annu. Rev. Physiol., 39*:159, 1977.

Lohmeier, T. E., *et al.*: Effects of endogenous angiotensin II on renal sodium excretion and renal hemodynamics. *Am. J. Physiol., 233*:F388, 1977.

Moses, A. M., and Share, L. (eds.): Neurohypophysis. New York, S. Karger, 1977.

Mudge, G. H., *et al.*: Tubular transport of urea, glucose, phosphate, uric acid, sulfate, and thiosulfate. *In* Orloff, F., and Berliner, R. W. (eds.): Handbook of Physiology. Sec. 8. Baltimore, Williams & Wilkins, 1973, p. 587.

Pitts, R.: Physiology of the Kidney and Body Fluids, 3rd Ed. Chicago, Year Book Medical Publishers, 1974.

Renkin, E. M., and Robinson, R. R.: Glomerular filtration. *N. Engl. J. Med., 290*:785, 1974.

Schafer, J. A., and Andreoli, T. E.: Rheogenic and passive Na$^+$ absorption by the proximal nephron. *Annu. Rev. Physiol., 41*:211, 1979.

Schrier, R. W.: Effects of adrenergic nervous system and catecholamines on systemic and renal hemodynamics, sodium and water excretion, and renin secretion. *Kidney Int., 6*:291, 1974.

Stephenson, J. L.: Countercurrent transport in the kidney. *Annu. Rev. Biophys. Bioeng., 7*:315, 1978.

Ullrich, K. L.: Sugar, amino acids, and Na$^+$ cotransport in the proximal tubule. *Annu. Rev. Physiol., 41*:181, 1979.

Vander, A. J.: Renal Physiology. New York, McGraw-Hill, 1980.

Wright, F. S.: Intrarenal regulation of glomerular filtration rate. *N. Engl. J. Med., 291*:135, 1974.

Wright, F. S., and Briggs, J. P.: Feedback control of glomerular blood flow, pressure and filtration rate. *Physiol. Rev., 59*:958, 1979.

25

Regulation of the Body Fluids and Their Constituents by the Kidneys and the Thirst Mechanism

The principal function of the kidneys is to control almost all the characteristics of the body fluids, especially of the extracellular fluid — such characteristics as volume, composition, and osmolarity. This chapter will discuss these fluids and the role of the kidneys and the thirst mechanism in their control.

TOTAL BODY WATER

The total amount of water in a man of average weight (70 kg.) is approximately 40 liters (see Fig. 25–1), averaging 57 per cent of his total body weight.

Most of our daily intake of water enters by the oral route, but a small amount is also synthesized in the body as the result of oxidation of hydrogen in the food; this quantity ranges between 150 and 250 ml. per day, depending on the rate of metabolism. The normal intake of fluid, including that synthesized in the body, averages about 2400 ml. per day.

Table 25–1 shows the routes by which water is lost from the body under different conditions. Normally, at an atmospheric temperature of about 68° F., approximately 1400 ml. of the 2400 ml. of water intake is lost in the *urine*, 100 ml. is lost in the *sweat*, and 200 ml. in the *feces*. The remaining 700 ml. is lost by *evaporation from the respiratory tract* or by *diffusion through the skin, which*

is called *insensible water loss* because we do not know that we are actually losing water at the time that it is leaving the body.

In very hot weather, water loss in the sweat is occasionally increased to as much as 1.5 liters an hour, which obviously can rapidly deplete the body fluids. Sweating will be discussed in Chapter 47.

Exercise increases the loss of water in two ways: First, it increases the rate of respiration, which promotes increased water loss through the respiratory tract in proportion to the increased ventilatory rate. Second, and much more important, exercise increases the body heat and consequently is likely to result in excessive sweating.

TABLE 25–1 DAILY LOSS OF WATER (in Milliliters)

	Normal Temperature	Hot Weather	Prolonged Heavy Exercise
Insensible Loss:			
Skin	350	350	350
Respiratory tract	350	250	650
Urine	1400	1200	500
Sweat	100	1400	5000
Feces	100	100	100
Total	2300	3300	6600

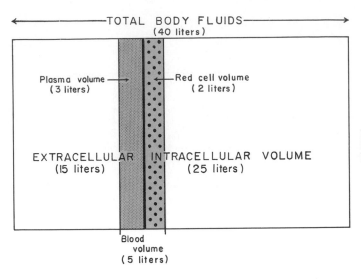

Figure 25–1. Diagrammatic representation of the body fluids, showing the extracellular fluid volume, intracellular fluid volume, blood volume, and total body fluids.

BODY FLUID COMPARTMENTS

The Intracellular Compartment. About 25 of the 40 liters of fluid in the body are inside the approximately 75 trillion cells of the body and are collectively called the *intracellular fluid.* The fluid of each cell contains its own individual mixture of different constituents, but the concentrations of these constituents are reasonably similar from one cell to another. For this reason the intracellular fluid of all the different cells is considered to be one large fluid compartment, though in reality it is an aggregate of trillions of minute compartments.

The Extracellular Fluid Compartment. All the fluids outside the cells are called *extracellular fluid,* and these fluids are constantly mixing, as was explained in Chapter 1. The total amount of fluid in the extracellular compartment averages 15 liters in a 70 kg. adult.

The extracellular fluid can be divided into *interstitial fluid, plasma, cerebrospinal fluid, intraocular fluid, fluids of the gastrointestinal tract,* and *fluids of the potential spaces.*

BLOOD VOLUME

Blood contains both extracellular fluid (the fluid of the plasma) and intracellular fluid (the fluid in the red blood cells). However, since blood is contained in a closed chamber all its own — the circulatory system — its volume and its special dynamics are exceedingly important.

The average blood volume of a normal adult is almost exactly 5000 ml. Approximately 3000 ml. of this is plasma, and the remainder, 2000 ml., is red blood cells. However, these values vary greatly in different individuals, and sex, weight, and many other factors affect the blood volume.

Measurement of Blood Volume — The Dilution Principle. The blood volume can be measured by (1) injecting a substance into the blood, (2) allowing the substance to disperse evenly throughout the blood, and (3) then measuring the extent to which the substance has become diluted. For instance, radioactive red blood cells can be injected. The greater the blood volume, the less is the concentration of these radioactive cells after complete dispersion of the cells throughout the vascular tree. Thus, the volume can be determined by the formula at the bottom of the page.

Note that all one needs to know is (1) the total quantity of the test substance put into the blood and (2) the concentration in the blood after dispersement.

In a similar way the plasma volume can be measured by injecting a substance that will disperse evenly in the plasma but will not enter the red blood cells. The blood is centrifuged and the concentration of the test substance in the plasma is determined. One of the most widely used substances for measuring plasma volume is a dye called T–1824. This dye attaches to the plasma proteins and therefore will not leave the plasma. Also, radioactive albumin is often injected for the same purpose. Once the plasma volume has been determined, the blood volume can be determined by measuring the hematocrit (the ratio of red cells to plasma) and then using this to calculate the blood volume from the plasma volume.

$$\text{Volume in milliliters} = \frac{\text{Quantity of test substance injected}}{\text{Concentration per milliliter of blood}}$$

CONSTITUENTS OF EXTRACELLULAR AND INTRACELLULAR FLUIDS

Figure 25–2 illustrates diagrammatically the major constituents of the extracellular and intracellular fluids. The quantities of the different substances are represented in *milliequivalents* or *millimoles per liter*. However, the protein molecules and some of the nonelectrolyte molecules are extremely large compared with the more numerous small ions. Therefore, *in terms of mass,* the proteins and nonelectrolytes actually constitute about 90 per cent of the dissolved constituents in the plasma, about 60 per cent in the interstitial fluid, and about 97 per cent in the intracellular fluid.

Figure 25–3 illustrates the distribution of the nonelectrolytes in the plasma; most of these same substances are also present in almost equal concentrations in the interstitial fluid, except for some of the fatty compounds that are present in the plasma in large suspended particles, the *lipoproteins.*

The Extracellular Fluid. Referring again to Figure 25–2, one sees that extracellular fluid, both that of the blood plasma and that of the

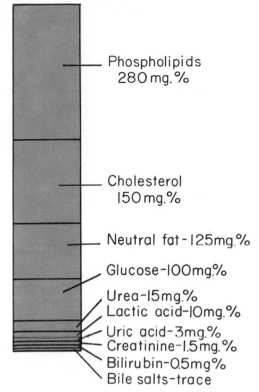

Figure 25–3. The nonelectrolytes of the extracellular fluid.

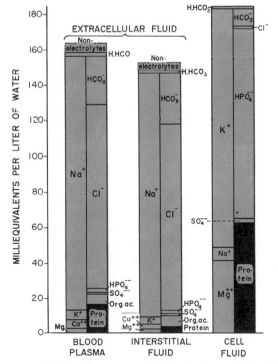

Figure 25–2. The compositions of plasma, interstitial fluid, and intracellular fluid. (Modified from Gamble: Chemical Anatomy, Physiology, and Pathology of Extracellular Fluid: A Lecture Syllabus. Harvard University Press, 1954. Reprinted by permission.)

interstitial fluid, contains large quantities of *sodium* and *chloride ions*, reasonably large quantities of *bicarbonate ion*, but only small quantities of potassium, calcium, magnesium, phosphate, sulfate, and organic acid ions. In addition, plasma contains a large amount of protein, while interstitial fluid contains much less. (The proteins in these fluids and their significance were discussed in detail in Chapter 22.)

In Chapter 1 it was pointed out that the extracellular fluid is called the *internal environment* of the body and that its constituents are accurately regulated so that the cells remain bathed continuously in a fluid containing the proper electrolytes and nutrients for continued cellular function. The regulation of most of these constituents will be presented later in this chapter.

The Intracellular Fluid. From Figure 25–2 it is also readily apparent that the intracellular fluid contains only small quantities of sodium and chloride ions and almost no calcium ions; but it does contain large quantities of *potassium* and *phosphate* and moderate quantities of *magnesium* and *sulfate ions,* all of which are present in only small concentrations in the extracellular fluid. In addition, the cells contain especially large amounts of protein, approximately four times as much as the plasma.

OSMOTIC EQUILIBRIA AND FLUID SHIFTS BETWEEN THE EXTRACELLULAR AND INTRACELLULAR FLUIDS

One of the most troublesome of all problems in clinical medicine is maintenance of adequate body fluids and proper balance between the extracellular and intracellular fluid volumes in seriously ill patients. The purpose of the following discussion, therefore, is to explain the interrelationships between extracellular and intracellular fluid volumes and the osmotic factors that cause shifts of fluid between the extracellular and intracellular compartments.

BASIC PRINCIPLES OF OSMOSIS AND OSMOTIC PRESSURE

The basic principles of osmosis and osmotic pressure were presented in Chapter 4. However, these principles are so important to the following discussion that they are reviewed here briefly.

Whenever a membrane between two fluid compartments is permeable to water but not to some of the dissolved solutes (this is called a *semipermeable membrane*) and the concentration of the solutes is greater on one side of the membrane than on the other, water passes through the membrane toward the side with the greater concentration of solutes. This phenomenon is called *osmosis*.

Osmosis results from the kinetic motion of the molecules (and ions) in the solutions on the two sides of the membrane and can be explained in the following way: The individual molecules on both sides of the membrane are equally active because the temperature, which is a measure of the kinetic activity of the molecules, is the same on both sides. However, the nondiffusible solute on one side of the membrane displaces some of the water molecules, thereby reducing the concentration of water molecules. As a result, the so-called *chemical potential* of water molecules on this side is less than on the other side, so that fewer water molecules strike each pore of the membrane each second on the solute side of the pore than on the pure water side. This causes "net" diffusion of water molecules from the water side to the solute side. This net diffusion is *osmosis*.

Osmosis of water molecules can be opposed by applying a pressure across the semipermeable membrane in the direction opposite to that of the osmosis. The amount of pressure required to oppose the osmosis exactly is called the *osmotic pressure*.

Relationship of the Molecular Concentration of a Solution to Its Osmotic Pressure. Each nondiffusible molecule dissolved in water reduces the chemical potential of the water by a given amount. Consequently, the tendency for the water in the solution to diffuse through a membrane is reduced in direct proportion to the concentration of nondiffusible molecules. And, as a corollary, the osmotic pressure of the solution is also proportional to the concentration of nondiffusible molecules in the solution. This relationship holds true for all nondiffusible molecules almost regardless of their molecular weight. For instance, one molecule of albumin with a molecular weight of 70,000 has the same osmotic effect as a molecule of glucose with a molecular weight of 180.

Osmotic Effect of Ions. Nondiffusible ions cause osmosis and osmotic pressure in exactly the same manner as do nondiffusible molecules. Furthermore, when a molecule dissociates into two or more ions, each of the ions then exerts osmotic pressure individually. Therefore, to determine the osmotic effect, all the nondiffusible ions must be added to all the nondiffusible molecules; but note that a bivalent ion, such as calcium, exerts no more osmotic pressure than does a univalent ion, such as sodium.

Osmoles. The ability of solutes to cause osmosis and osmotic pressure is measured in terms of "osmoles"; the osmole is a measure of the total number of particles. *One gram mole of nondiffusible and nonionizable substance is equal to 1 osmole.* On the other hand, if a substance ionizes into two ions (sodium chloride into sodium and chloride ions, for instance), then 0.5 gram mole of the substance equals 1 osmole. The obvious reason for using the osmole is that osmotic pressure is determined by the number of particles instead of the mass of the solute.

In general, the osmole is too large a unit for satisfactory use in expressing osmotic activity of solutes in the body. Therefore, the term *milliosmole*, which equals 0.001 osmole, is commonly used.

Osmolarity. The osmolar concentration of a solution is called its *osmolarity*, and it is expressed as osmoles per liter of solution. The osmotic pressure of a solution *at body temperature* can be determined approximately from the following formula:

Osmotic pressure (mm. Hg)
$$= 19.3 \times \text{Osmolarity (milliosmole/liter)}$$

OSMOLARITY OF THE BODY FLUIDS

Table 25–2 lists the osmotically active substances in plasma, interstitial fluid, and intra-

TABLE 25–2 OSMOLAR SUBSTANCES IN EXTRACELLULAR AND INTRACELLULAR FLUIDS

	Plasma (mOsmole/l. of H_2O)	Interstitial (mOsmole/l. of H_2O)	Intracellular (mOsmole/l. of H_2O)
Na^+	146	142	14
K^+	4.2	4.0	140
Ca^{++}	2.5	2.4	0
Mg^{++}	1.5	1.4	31
Cl^-	105	108	4
HCO_3^-	27	28.3	10
HPO_4^{--}, $H_2PO_4^-$	2	2	11
SO_4	0.5	0.5	1
Phosphocreatine			45
Carnosine			14
Amino acids	2	2	8
Creatine	0.2	0.2	9
Lactate	1.2	1.2	1.5
Adenosine triphosphate			5
Hexose monophosphate			3.7
Glucose	5.6	5.6	
Protein	1.2	0.2	4
Urea	4	4	4
TOTAL mOsmole/l.	302.9	301.8	302.2
Corrected for ionic interaction (mOsmole/l.)	282.6	281.3	281.3
Total osmotic pressure at 37° C. (mm. Hg)	5453	5430	5430

cellular fluid. The milliosmoles of each of them per liter is given. Note especially that approximately four-fifths of the total osmolarity of the interstitial fluid and plasma is caused by sodium and chloride ions, while approximately half of the intracellular osmolarity is caused by potassium ions, the remainder being divided among the many other intracellular substances.

As noted at the bottom of the table, the total osmolarity of each of the three compartments is approximately 300 milliosmoles per liter, with that of the plasma being 1.3 milliosmoles greater than that of the interstitial and intracellular fluids. This slight difference between plasma and interstitial fluid is caused by the osmotic effect of the plasma proteins, which maintains a pressure in the capillaries about 23 mm. Hg greater than in the surrounding interstitial fluid spaces, as was explained in Chapter 22.

MAINTENANCE OF OSMOTIC EQUILIBRIUM BETWEEN EXTRACELLULAR AND INTRACELLULAR FLUIDS

The tremendous osmotic pressure that can develop across the cell membrane when one side is exposed to pure water — more than 5400 mm. Hg — illustrates how much force can become available to push water molecules through the membrane when the solutions of the two sides of

the membrane are not in osmotic equilibrium. As an example, in Figure 25–4A, a cell is placed in a solution that has an osmolarity far less than that of the intracellular fluid. As a result, osmosis of water begins immediately from the extracellular fluid to the intracellular fluid, causing the cell to swell and diluting the intracellular fluid while concentrating the extracellular fluid. When the

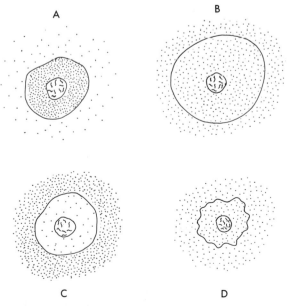

Figure 25–4. Establishment of osmotic equilibrium when cells are placed in a hypo- or hypertonic solution.

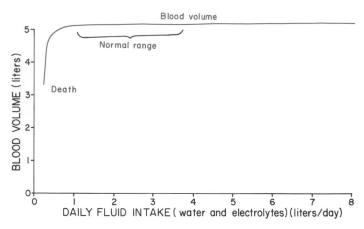

Figure 25–5. Effect on blood volume of marked changes in daily fluid intake. Note the precision of blood volume control in the normal range.

fluid inside the cell becomes diluted sufficiently to equal the osmolar concentration of the fluid on the outside, further osmosis then ceases. This condition is shown in Figure 25–4B. In Figure 25–4C, a cell is placed in a solution having a much higher concentration outside the cell than inside. This time, water passes by osmosis to the exterior, diluting the extracellular fluid and concentrating the intracellular fluid. In this process the cell shrinks until the two concentrations become equal, as shown in Figure 25–4D.

Rapidity of Attaining Extracellular and Intracellular Osmotic Equilibrium.   The transfer of water through the cell membrane by osmosis occurs so rapidly that any lack of osmotic equilibrium between the two fluid compartments in any given tissue is usually corrected within a few seconds or within a minute or so at most. However, this rapid transfer of water does not mean that complete equilibration occurs between the extracellular and intracellular compartments throughout the whole body within this same short period of time. The reason is that fluid usually enters the body through the gut and must then be transported by the blood to all tissues before complete equilibration can occur. In the normal person it may take as long as 30 minutes to achieve reasonably good equilibration everywhere in the body after drinking water.

Isotonicity, Hypotonicity, and Hypertonicity.   A fluid into which normal body cells can be placed without causing either swelling or shrinkage of the cells is said to be *isotonic* with the cells. A 0.9 per cent solution of sodium chloride or a 5 per cent glucose solution is approximately isotonic.

A solution that will cause the cells to swell is said to be *hypotonic;* any solution of sodium chloride with less than 0.9 per cent concentration is hypotonic.

A solution that will cause the cells to shrink is said to be *hypertonic;* sodium chloride solutions of greater than 0.9 per cent concentration are all hypertonic.

CONTROL OF BLOOD VOLUME

Constancy of the Blood Volume.   The extreme degree of precision with which the blood volume is controlled is illustrated in Figure 25–5, which shows that tremendous changes in fluid intake, from very low values to very high values, cause almost no change in blood volume (except when the intake becomes so low that it is not sufficient to make up for even the slightest fluid losses). To state this another way, even when the intake of water and salt is increased many-fold, the blood volume is hardly altered. Conversely, a decrease in fluid intake to as little as one-third normal also causes hardly a change.

BASIC MECHANISM FOR BLOOD VOLUME CONTROL

The basic mechanism for blood volume control is essentially the same as the basic mechanism for arterial pressure control that was presented in Chapter 18. It was pointed out in this earlier chapter that extracellular fluid volume, blood volume, cardiac output, arterial pressure, and urine output are all mainly or partially controlled by a single common basic feedback mechanism. Therefore, let us explain again the basic principles of this mechanism:

When the blood volume becomes too great, the cardiac output and arterial pressure increase. This increase, in turn, has a profound effect on the kidneys, causing loss of fluid from the body and returning the blood volume back to normal. Conversely, if the blood volume falls below normal, the cardiac output and arterial pressure decrease, the kidneys now retain fluid, and progressive accumulation of the fluid intake builds the blood volume eventually back to normal. (Obviously, parallel processes also occur to reconstitute red cells, plasma proteins, and other blood constituents.)

Role of the Volume Receptors in Blood Volume Control.   It was pointed out in Chapter 17 that "volume receptor" reflexes help to control blood volume. The volume receptors are stretch receptors located in the walls of the right and left atria. When the blood volume becomes excessive, a large share of this volume accumulates in the central veins of the thorax and causes increased pressure in the two atria. The resultant stretch of the atrial walls transmits nerve signals into the brain, and these in turn elicit responses that accelerate the return of blood volume to normal. The various responses that occur include the following:

1. The sympathetic nervous signals to the kidneys are inhibited, thus slightly to moderately increasing the rate of urinary output.

2. The secretion of antidiuretic hormone by the supraopticohypophyseal system is reduced, allowing increased water excretion by the kidneys.

3. The peripheral arterioles throughout the body are dilated because of reflex reduction of sympathetic stimulation, thus increasing capillary pressure and allowing much of the excess blood volume to filter temporarily into the tissue spaces for a few hours until the excess fluid can be excreted through the kidneys.

In most instances, these volume receptor reflex effects can cause the blood volume to return almost all the way to normal within an hour or so, but the final determination of the precise level to which the blood volume will be adjusted is still a function of the basic volume control mechanism discussed above. The reason is that over a period of one to three days the volume receptors adapt so completely that they no longer transmit any corrective signals. Therefore, they are of value only to help readjust the volume during the first few hours or days after an abnormality occurs, but not for long-term monitoring of volume or for precise adjustments of the long-term level of blood volume.

CONTROL OF EXTRACELLULAR FLUID VOLUME

It is already clear from the above discussion of the basic mechanism for blood volume control that extracellular fluid volume is controlled at the same time as the blood volume. That is, when fluid is reabsorbed by the kidney or is ingested by mouth, the fluid first goes into the blood, but it rapidly becomes distributed between the interstitial spaces and the plasma. Though it is the blood volume that raises the arterial pressure and thereby causes increased urinary output

and not interstitial fluid volume, fluid will not remain in the blood until the interstitial spaces are also appropriately filled with fluid. Therefore, it is impossible to control blood volume to any given level without controlling the extracellular fluid volume at the same time. Yet, the relative distribution of volume between the interstitial spaces and the blood can vary greatly, depending on the physical characteristics of the circulatory system and of the interstitial spaces. Under normal conditions, the approximate relationship between extracellular fluid volume and blood volume is that illustrated in Figure 25–6. The curve of this figure shows that at the lower extracellular fluid volumes, an increase in extracellular fluid volume is associated with an increase in blood volume of about one-fourth to one-sixth. The remainder of the extracellular fluid is distributed to the interstitial spaces. However, when the extracellular fluid volume rises above 20 to 22 liters, very little of the additional fluid will remain in the blood, almost all of it instead going into the interstitial spaces. This is caused by the fact that the interstitial fluid pressure rises from its normal negative value (subatmospheric) to a positive value. When this occurs, the compliance of the tissue spaces becomes tremendous, so that they can then hold as much as 20 to 40 liters of fluid with very little additional rise in interstitial pressure. Therefore, the interstitial fluid spaces, under these conditions, literally become an "overflow" reservoir for excess fluid. This obviously causes edema, but it also acts as an important overflow release valve for the circulatory system, a well-known phenomenon that is utilized daily by the clinician to allow him to administer almost unlimited quantities of intravenous fluid and yet not force the heart into cardiac failure.

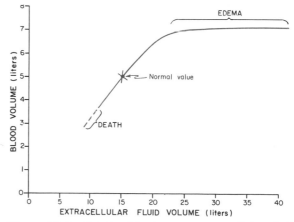

Figure 25–6. Relationship between extracellular fluid volume and blood volume, showing a nearly linear relationship in the normal range but indicating failure of the blood volume to continue rising when the extracellular fluid volume becomes excessive.

CONTROL OF EXTRACELLULAR FLUID SODIUM CONCENTRATION AND EXTRACELLULAR FLUID OSMOLARITY

Relationship of Sodium Concentration to Extracellular Fluid Osmolarity. The osmolarity of the extracellular fluids (and also of the intracellular fluids, since they remain in osmotic equilibrium with the extracellular fluids) is determined almost entirely by the extracellular fluid sodium concentration. The reason is that sodium is by far the most abundant positive ion of the extracellular fluid. Furthermore, the acid-base control mechanisms of the kidneys, which will be discussed in the following chapter, adjust the negative ion concentrations of the body fluids to equal those of the positive ions. Therefore, in effect, the sodium ion concentration of the extracellular fluid controls 90 to 95 per cent of the *effective* osmotic pressure of the extracellular fluid. Consequently, we can generally talk in terms of control of sodium concentration and control of osmolarity at the same time.

Two separate control systems operate in close association to regulate extracellular sodium concentration and osmolarity. These are (1) the osmo-sodium receptor–antidiuretic hormone system and (2) the thirst mechanism.

THE OSMO-SODIUM RECEPTOR–ANTIDIURETIC HORMONE FEEDBACK CONTROL SYSTEM

Figure 25–7 illustrates the osmo-sodium receptor–antidiuretic hormone system for control of extracellular fluid sodium concentration and osmolarity. It is a typical feedback control system that operates by the following steps:

1. An increase in osmolarity (excess sodium and the negative ions that go with it) excites *osmoreceptors* located in the *supraoptic nuclei of the hypothalamus.*

2. Excitation of the supraoptic nuclei causes release of antidiuretic hormone.

3. The antidiuretic hormone *increases the permeability of the collecting ducts*, as explained in the previous chapter, and therefore causes increased conservation of water by the kidneys.

4. The conservation of water but loss of sodium and other osmolar substances in the urine causes dilution of the sodium and other substances in the extracellular fluid, thus correcting the initial excessively concentrated extracellular fluid.

Conversely, when the extracellular fluid becomes too dilute (hypo-osmotic), less antidiuretic hormone is formed, and excess water is lost along with very few extracellular fluid solutes, thus concentrating the body fluids back toward normal.

The Osmoreceptors (or Osmo-Sodium Receptors). Located in the supraoptic nuclei of the anterior hypothalamus, as shown in Figure 25–8, are specialized neuronal cells called *osmoreceptors*. These respond to changes in osmolarity of the extracellular fluid. When the extracellular osmolarity becomes low, osmosis of water into the osmoreceptors causes them to swell. This decreases their rate of impulse discharge. Conversely, increased osmolarity in the extracellular fluid pulls water out of the osmoreceptors, causing them to shrink and thereby to increase their rate of discharge.

The osmoreceptors respond to changes in extracellular fluid sodium concentration but not to changes in potassium concentration and only

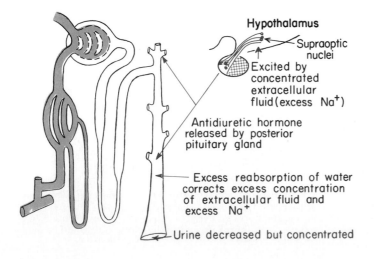

Figure 25–7. Control of extracellular fluid osmolarity and sodium ion concentration by the osmo-sodium receptor-antidiuretic hormone feedback control system.

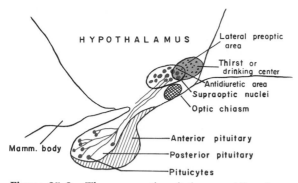

Figure 25–8. The supraoptico-pituitary antidiuretic system and its relationship to the thirst center in the hypothalamus.

slightly to changes in urea and glucose concentrations. Therefore, for all practical purposes, the osmoreceptors are actually sodium concentration receptors — hence the name *osmo-sodium receptors.*

The impulses from the osmoreceptors are transmitted from the supraoptic nuclei through the pituitary stalk into the posterior pituitary gland, where they promote the release of antidiuretic hormone (ADH).

Thus, ADH secretion is controlled by the osmolarity (or sodium concentration) of the extracellular fluid — the greater the osmolarity, the greater the rate of ADH secretion.

Water Diuresis. When a person drinks a large amount of water, a phenomenon called *water diuresis* ensues, a typical record of which is shown in Figure 25–9. In this example, a man drank 1 liter of water, and approximately 30 minutes later his urine output increased to 8 times normal. It remained at this level for two hours — that is, until the osmolarity of the extra-

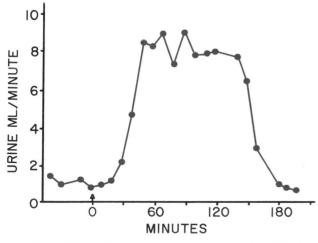

Figure 25–9. Water diuresis in a human being following ingestion of 1000 ml. of water. (Redrawn from Smith: The Kidney: Structure and Functions in Health and Disease. Oxford University Press, 1951.)

cellular fluid had returned essentially to normal. The delay in onset of water diuresis is caused partly by delay in absorption of the water from the gastrointestinal tract but mainly by the time required for destruction of the antidiuretic hormone that has already been released by the pituitary gland prior to drinking the water.

Diabetes Insipidus. Destruction of the supraoptic nuclei or high level destruction of the nerve tract from the supraoptic nuclei to the posterior pituitary gland causes antidiuretic hormone secretion to cease or at least to become greatly reduced. When this happens, the person thereafter excretes a dilute urine, and the daily urine volume is increased to 5 to 15 liters per day — a condition called *diabetes insipidus.* In diabetes insipidus, the body fluid volumes remain almost normal so long as the thirst mechanism is still functional, because this ordinarily makes the person drink enough water to make up for the increased loss of water in the urine. On the other hand, any factor that prevents adequate intake of fluid, such as unconsciousness, results rapidly in a state of dehydration, tremendous hyperosmolarity, and excessive concentration of sodium in the extracellular fluid.

Syndrome of Inappropriate ADH Secretion. Certain types of tumors, especially bronchogenic tumors of the lungs or tumors of the basal regions of the brain, often secrete antidiuretic or a similar hormone. This condition is called the *syndrome of inappropriate ADH secretion.* This excess ADH causes only a slight increase in extracellular fluid volume; instead, its principal effect is *to decrease greatly the sodium concentration of the extracellular fluid.* The explanation of this effect is the following: The ADH at first causes a decrease in urine output and a simultaneous slight increase in blood volume. This in turn activates the basic mechanism for blood volume control. That is, a slight rise in arterial pressure occurs, and this causes a secondary *increase* in urinary output. Furthermore, the urine that is excreted is tremendously concentrated because of the tendency of the kidneys to retain water as a result of the action of the ADH. Consequently, the kidneys excrete extreme amounts of sodium into the urine but keep the water in the extracellular fluids. Therefore, the sodium concentration becomes seriously reduced, sometimes falling from a normal value of 142 mEq./liter down to as low as 110 to 120 mEq./liter. At values this low, patients frequently die sudden deaths because of coma and convulsions.

This disease is especially instructive because it illustrates the extreme importance of the antidiuretic hormone mechanism for control of sodium concentration, and yet its relatively mild effect on control of body fluid volume.

THIRST AND ITS ROLE IN CONTROLLING SODIUM CONCENTRATION AND OSMOLARITY

The phenomenon of thirst is as important for regulating body water and sodium concentration as is the osmoreceptor-renal mechanism discussed above, because the amount of water in the body at any one time is determined by the balance between both *intake* and *output* of water each day. Thirst, the primary regulator of the intake of water, is defined as the *conscious desire for water.*

Neural Integration of Thirst — the "Thirst" Center. Referring again to Figure 25–8, one sees a small area, located slightly anterior to the supraoptic nuclei in the lateral preoptic area of the hypothalamus, called the *thirst center.* Electrical stimulation of this center by implanted electrodes causes an animal to begin drinking within seconds and to continue drinking until the electrical stimulus is stopped. Injection of hypertonic salt solutions into the area, which causes osmosis of water out of the neuronal cells and shrinkage of the cells, also causes drinking. Thus, the neuronal cells of the thirst center function in almost the identical way as the osmoreceptors of the supraoptic nuclei.

Temporary Relief of Thirst Caused by the Act of Drinking. A thirsty person receives relief from thirst immediately after drinking water, even before the water has been absorbed from the gastrointestinal tract. In fact, in persons with esophageal fistula (a condition in which the water leaks to the exterior from the esophagus and never goes into the gastrointestinal tract), partial relief of thirst still occurs following the act of drinking, but this relief is only temporary, and the thirst returns after 15 minutes or more. If the water does enter the stomach, distension of the stomach and other portions of the upper gastrointestinal tract provides still further temporary relief from thirst. For instance, simple inflation of a balloon in the stomach can relieve thirst for 5 to 30 minutes.

One might wonder what the value of this temporary relief from thirst could be, but there is good reason for it. After a person has drunk water, as long as one-half to one hour may be required for all of the water to be absorbed and distributed throughout the body. Were the thirst sensation not temporarily relieved after drinking water, the person would continue to drink more and more. When all this water should finally become absorbed, the body fluids would be far more diluted than normal, and an abnormal condition opposite to that which the person was attempting to correct would be created. It is well known that a thirsty animal almost never drinks more than the amount of water needed to relieve the state of dehydration. Indeed, it is uncanny that the animal usually drinks almost exactly the right amount.

Role of Thirst in Controlling Osmolarity and Sodium Concentration of the Extracellular Fluid — The Tripping Mechanism. The kidneys are continually excreting fluid, and water is also lost by evaporation from the skin and lungs. Therefore, a person is continually being dehydrated, which causes the volume of extracellular fluid to decrease and its concentration of sodium and other osmolar elements to rise. When the sodium concentration rises to approximately 2 mEq./liter above normal (or the osmolarity rises to approximately 4 mOsm./liter above normal) the drinking mechanism becomes "tripped" because the person by then reaches a level of thirst that is strong enough to activate the motor effort necessary to cause drinking. The person ordinarily drinks precisely the required amount of fluid to bring the extracellular fluids back to normal — that is, to a state of *satiety.* Then the process of dehydration and sodium concentration begins again, and the drinking act is tripped again, the process continuing on and on indefinitely.

In this way, both the sodium concentration and the osmolarity of the extracellular fluid are very precisely controlled.

COMBINED ROLES OF THE ANTIDIURETIC AND THIRST MECHANISMS FOR CONTROL OF EXTRACELLULAR FLUID SODIUM CONCENTRATION AND OSMOLARITY

When either the antidiuretic hormone mechanism or the thirst mechanism fails, the other ordinarily can still control both sodium concentration and extracellular fluid osmolarity with reasonable effectiveness. On the other hand, if both of them fail simultaneously, neither sodium nor osmolarity is then adequately controlled.

Figure 25–10 dramatically demonstrates the overall capability of the ADH-thirst system to control extracellular fluid sodium concentration. This figure demonstrates the ability of the same animal to control its extracellular fluid sodium concentration in two different conditions: (1) in the normal state and (2) after both the antidiuretic hormone and thirst mechanisms had been blocked. Note that in the normal animal a six-fold increase in sodium intake caused the sodium concentration to change only two-thirds of 1 per cent (from 142 mEq./liter up to 143 mEq./liter) — an excellent degree of sodium concentration control. Now note the dashed curve of the figure, which shows the change in sodium concentration when the ADH-thirst system was blocked. In this

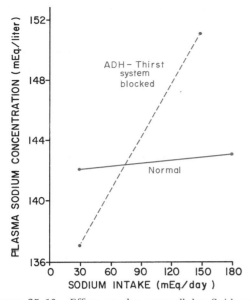

Figure 25–10. Effect on the extracellular fluid sodium concentration in dogs caused by tremendous changes in sodium intake (1) under normal conditions, and (2) after the antidiuretic hormone and thirst feedback systems had been blocked. This figure shows lack of sodium ion control in the absence of these systems. (Courtesy of Dr. David B. Young.)

case, sodium concentration increased 10 per cent with only a five-fold increase in sodium intake (a change in sodium concentration from 137 mEq./liter up to 151 mEq./liter), which is an extreme change in sodium concentration when one realizes that the normal sodium concentration rarely rises or falls more than 1 per cent from day to day.

Therefore, the major feedback mechanism for control of sodium concentration (and also for extracellular osmolarity) is the ADH-thirst mechanism. In the absence of this mechanism there is no feedback mechanism that will cause the body to increase water ingestion or to conserve water by the kidneys when excess sodium enters the body. Therefore, the sodium concentration simply increases.

CONTROL OF SODIUM INTAKE – APPETITE AND CRAVING FOR SALT

Maintenance of normal extracellular sodium requires not only the control of sodium excretion but also the control of sodium intake. Unfortunately, we know very little about this except that salt-depleted persons (or persons who have lost blood) develop a desire for salt; as an example, this occurs in persons who have *Addison's disease,* a condition in which the adrenal cortices no longer secrete aldosterone, so that the salt stores of the body become depleted. Likewise, it is well known that animals living in areas far removed from the seashore actively search out "salt licks." This craving for salt is analogous to thirst, and it is also analogous to appetite for other types of foods, which is still another homeostatic mechanism that will be discussed in Chapter 48.

CONTROL OF EXTRACELLULAR POTASSIUM CONCENTRATION — ROLE OF ALDOSTERONE

In the previous chapter, it was pointed out that aldosterone not only causes increased sodium reabsorption by the tubules but also causes greatly increased tubular secretion of potassium as well, and therefore increased loss of potassium in the urine. Though the antidiuretic hormone and thirst mechanisms can override aldosterone control of extracellular fluid sodium concentration, this is not true for the control of potassium concentration. Therefore, aldosterone plays an exceedingly important role in the control of extracellular fluid potassium ion concentration. We will explain this control system in the following sections.

Effect of Potassium Ion Concentration on Rate of Aldosterone Secretion. In a properly functioning feedback system, the factor that is controlled almost invariably has a feedback effect to control the controller; this is precisely true for the aldosterone-potassium control system because the rate of aldosterone secretion is controlled very strongly by the extracellular fluid potassium concentration. Figure 25–11 illustrates the results from a series of dogs in which different rates of potassium infusion were maintained for several weeks while the aldosterone secretion rate was measured. Note the tremendous increases in aldosterone secretion rate caused by very minute increases in potassium ion concentration.

Basic Mechanism for Aldosterone Control of Potassium Concentration. Putting the effects illustrated in Figure 25–11 together with the fact that aldosterone greatly increases renal excretion of potassium, one can construct a very simple system for negative feedback control of potassium concentration. That is, (1) an increase in potassium concentration causes an increase in aldosterone concentration in the circulating blood. (2) The increase in aldosterone concentration then causes a marked increase in potassium excretion by the kidneys. (3) The increased potassium excretion then decreases the extracellular fluid potassium concentration back toward normal.

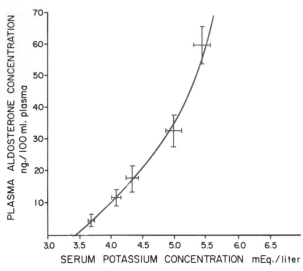

Figure 25–11. Effect on extracellular fluid aldosterone concentration of potassium ion concentration changes. Note the extreme change in aldosterone concentration for very minute changes in potassium concentration. (Courtesy of Dr. R. E. McCaa.)

Importance of the Aldosterone Feedback System for Control of Potassium Concentration. Without a functioning aldosterone feedback system, an animal can easily die from either hypopotassemia or hyperpotassemia.

Figure 25–12 illustrates the potent effect of the aldosterone feedback system on controlling po-

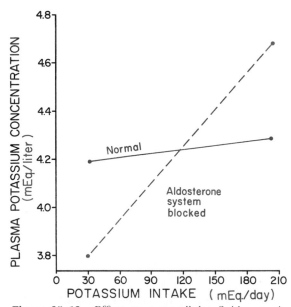

Figure 25–12. Effect on extracellular fluid potassium concentration of tremendous changes in potassium intake (1) under normal conditions, and (2) after the aldosterone feedback system had been blocked. This figure demonstrates that potassium concentration is very poorly controlled after block of the aldosterone system. (Courtesy of Dr. David B. Young.)

tassium concentration. In the experiment of this figure, a series of dogs was subjected to an almost seven-fold increase in potassium intake in two different states: (1) the normal state and (2) after the aldosterone feedback system had been blocked by removing the adrenal glands and the animals given a fixed rate of aldosterone infusion.

Note that in the normal animal the seven-fold increase in potassium intake caused an increase in plasma potassium concentration of only 2.4 per cent — from a concentration of 4.2 mEq./liter to 4.3 mEq./liter. Thus, when the aldosterone feedback system was functioning normally, the potassium concentration remained precisely controlled despite the tremendous change in potassium intake.

On the other hand, the dashed curve in the figure shows the effect after the aldosterone system had been blocked. Note that the same increase in potassium intake now caused a 26 per cent increase in potassium concentration! Thus, the control of potassium concentration in the normal animals was many times as effective as in the animals without an aldosterone feedback mechanism.

Effect of Primary Aldosteronism and Addison's Disease on Extracellular Fluid Potassium Concentration. Primary aldosteronism is caused by a tumor of the zona glomerulosa of one of the adrenal glands, the tumor secreting tremendous quantities of aldosterone. One of the most important effects of this disease is a severe decrease in extracellular fluid potassium concentration, so much so that many of these patients experience paralysis caused by failure of nerve transmission resulting from hyperpolarization of the nerve membranes.

Conversely, in the patient with untreated Addison's disease whose adrenal glands have been destroyed, the extracellular fluid potassium concentration frequently rises to as high as twice normal. This is often the cause of death in these patients, resulting in cardiac debility that leads to cardiac arrest.

CONTROL OF THE EXTRACELLULAR CONCENTRATIONS OF OTHER IONS

Regulation of Calcium Ion Concentration. The role of calcium in the body and control of its concentration in the extracellular fluid will be discussed in detail in Chapter 53 in relation to the endocrinology of parathyroid hormone, calcitonin, and bone. However, briefly, it is the following:

The day-by-day calcium ion concentration is controlled principally by the effect of parathyroid hormone on bone reabsorption. When the extracellular fluid concentration of calcium falls too low, the parathyroid glands are directly stimulated to promote increased secretion of parathyroid hormone, and this hormone in turn acts directly on the bones to increase the reabsorption of bone salts, thus releasing large amounts of calcium into the extracellular fluid and elevating the calcium level back to normal.

However, the bones are not an inexhaustible supply of calcium, and eventually the bones will run out of calcium. Therefore, long-term control of calcium ion concentration results from the effect of parathyroid hormone on reabsorption of calcium from the kidney tubules and absorption of calcium from the gut through the gastrointestinal mucosa, both of which effects are markedly increased by parathyroid hormone. These effects will be discussed in more detail in Chapter 53.

Regulation of Phosphate Concentration. Phosphate concentration is regulated primarily by an *overflow* mechanism, which can be explained as follows: The renal tubules have a normal capability to reabsorb a maximum of 0.1 millimole of phosphate per minute. When less than this "load" of phosphate is present in the glomerular filtrate, all of it is reabsorbed. When more is present, the excess is excreted. Therefore, phosphate ion normally spills into the urine when its concentration in the plasma is above the threshold value of approximately 0.8 millimole/liter. Any time the concentration falls below this value, all the phosphate is conserved in the plasma, and the daily ingested phosphate accumulates in the extracellular fluid until its concentration rises above the threshold.

Since most people ingest large quantities of phosphate day in and day out, either in milk or in meat, the concentration of phosphate is usually maintained at a level of about 1.0 millimole/liter, at which level there is continual overflow of excess phosphate into the urine.

Regulation of Other Negative Ions. The regulation of chloride and bicarbonate ions will be discussed in the following chapter in connection with acid-base balance of the body. But other important negative ions in the body fluid include sulfates, nitrates, urates, lactates, and the amino acids. Essentially all of these, like phosphate, have definite maximum rates of tubular reabsorption. When the concentration of each is below its respective threshold, it is conserved in the extracellular fluid, but above this threshold the excess spills into the urine. Thus, the concentrations of most of these negative ions are regulated by the overflow mechanism in the same way that phosphate ion concentration is regulated.

REFERENCES

Osmotic Equilibria

Adolph, E. F.: Physiology of Man in the Desert. New York. Interscience Publishers. 1947.

Andersson, B.: Regulation of body fluids. *Annu. Rev. Physiol., 39*:185, 1977.

Bing, D. H. (ed.): The Chemistry and Physiology of the Human Plasma Proteins, New York, Pergamon Press, 1979.

Borut, A., and Shkolnik, A.: Physiological adaptations to the desert environment. *Int. Rev. Physiol., 7*:185, 1974.

Burton, R. F.: The significance of ionic concentrations in the internal media of animals. *Biol. Rev., 48*:195, 1973.

Gilles, R. (ed.): Mechanisms of Osmoregulation: Maintenance of Cell Volume. New York, John Wiley & Sons, 1979.

Goldberger, E.: A Primer of Water, Electrolyte, and Acid-Base Syndromes. Philadelphia, Lea & Febiger, 1980.

Gupta, B. L., *et al.* (eds.): Transport of Ions and Water in Animals, New York, Academic Press, 1977.

Guyton, A. C., *et al.*: Circulatory Physiology II: Dynamics and Control of the Body Fluids. Philadelphia, W. B. Saunders Co., 1975.

Mason, E. E.: Fluid, Electrolyte, and Nutrient Therapy in Surgery. Philadelphia, Lea & Febiger, 1974.

Peters, T., and Sjöholm, I. (eds.): Albumin: Structure, Biosynthesis, Function. New York, Pergmaon Press, 1978.

Schreiber, G., and Urban, J.: The synthesis and secretion of albumin. *In* Adrian, R. H., *et al.* (eds.): Reviews of Physiology, Biochemistry, and Pharmacology. New York, Springer-Verlag, 1978, p. 27.

Smith, K.: Fluids and Electroytes: A Conceptual Approach. New York, Churchill Livingstone, 1980.

Wolf, A. V., and Crowder, N. A.: Introduction to Body Fluid Metabolism. Baltimore, Williams & Wilkins, 1964.

Blood Volume and Extracellular Fluid Regulation

Brenner, B. M., and Stein, J. H. (eds.): Acid-Base and Potassium Homeostasis. New York, Churchill Livingstone, 1978.

Brenner, B. M., and Stein, J. H. (eds.): Hormonal Function and the Kidney. New York, Churchill Livingstone, 1979.

Cross, B. A., and Wakerley, J. B.: The neurohypophysis. *Int. Rev. Physiol., 16*:1, 1977.

Dennis, V. W., *et al.*: Renal handling of phosphate and calcium. *Annu. Rev. Physiol., 41*:257, 1979.

Earley, L. E., and Schrier, R. W.: Intrarenal control of sodium excretion by hemodynamic and physical factors. *In* Orloff, F., and Berliner, R. W. (eds.): Handbook of Physiology. Sec. 8. Baltimore, Williams & Wilkins, 1973, p. 721.

Ehrlich, E. N.: Adrenocortical regulation of salt and water metabolism: Physiology, pathophysiology, and clinical syndromes. *In* DeGroot, L. J., *et al.* (eds.): Endocrinology. Vol. 3. New York, Grune & Stratton, 1979, p. 1883.

Fitzsimons, J. T.: Thirst. *Physiol. Rev., 52*:468, 1972.

Gertz, K. H., and Boylan, J. W.: Gomerular-tubular balance. *In* Orloff, F., and Berliner, R. W. (eds.): Handbook of Physiology. Sec. 8. Baltimore, Williams & Wilkins, 1973, p. 763.

Gottschalk, C. W.: Renal nerves and sodium excretion. *Annu. Rev. Physiol., 41*:229, 1979.

Guyton, A. C., *et al.*: Theory for renal autoregulation by feedback at the juxtaglomerular apparatus. *Circ. Res., 14*:187, 1964.

Guyton, A. C., *et al.*: Circulatory Physiology II: Dynamics and Control of the Body Fluids. Philadelphia, W. B. Saunders Co., 1975.

Guyton, A. C., *et al.*: A systems analysis of volume regulation. Alfred Benzon Symposium XI. Munksgaard, 1978.

Hall, J. E., *et al.*: Control of glomerular filtration rate by renin-angiotensin system. *Am. J. Physiol., 233*:F366, 1977.

Hall, J. E., *et al.*: Dissociation of renal blood flow and filtration rate autoregulation by renin depletion. *Am. J. Physiol.,* *232*:F215, 1977.

Hall, J. E., *et al.*: Intrarenal control of electrolyte excretion by angiotensin II. *Am. J. Physiol., 232*:F538, 1977.

Hayslett, J. P.: Functional adaptation to reduction in renal mass. *Physiol. Rev., 59*:137, 1979.

Katz, A. I., and Lindheimer, M. D.: Actions of hormones on the kidney. *Annu. Rev. Physiol., 39*:97, 1977.

Knox, F. G., and Diaz-Buxo, J. A.: The hormonal control of sodium excretion. *Int. Rev. Physiol., 16*:173, 1977.

Maxwell, M. H., and Kleeman, C. R. (eds.): Clinical Disorders of Fluid .and Electrolyte Metabolism, 3rd Ed. New York, McGraw-Hill, 1979.

Robertson, G. L.: Vasopressin in osmotic regulation in man. *Annu. Rev. Med., 25*:315, 1974.

Share, L., and Claybaugh, J. R.: Regulation of body fluids. *Annu. Rev. Physiol., 34*:235, 1972.

Smith, M. J., Jr., *et al.*: Acute and chronic effects of vasopressin on blood pressure, electrolytes, and fluid volumes. *Am. J. Physiol., 273*(3):F232, 1979.

Valtin, H.: Renal Dysfunction: Mechanisms Involved in Fluid and Solute Imbalance. Boston, Little, Brown, 1979.

Weitzman, R., and Kleeman, C.R.: Water metabolism and the neurohypophysial hormones. *In* Bondy, P. K., and Rosenberg, L. E. (eds.): Metabolic Control and Disease, 8th Ed. Philadelphia, W. B. Saunders Co., 1980, p. 1241.

Wolf, A. V.: Thirst: Physiology of the Urge to Drink and Problems of Water Lack. Springfield, Ill., Charles C Thomas, 1958.

Wolf, G., McGovern, J. F., and Dicara, L. V.: Sodium appetite: Some conceptual and methodologic aspects of a model drive system. *Behav. Biol., 10*:27, 1974.

Wright, F. S.: Potassium transport by the renal tubule. *In* MTP International Review of Physiology. Vol 6. Baltimore, University Park Press, 1974, p. 79.

Young, D. B., *et al.*: Effectiveness of the aldosterone-sodium and -potassium feedback control system. *Am. J. Physiol., 231*:945, 1976.

Young, D. B., *et al.*: Control of extracellular sodium concentration by antidiuretic hormone–thirst feedback mechanism. *Am. J. Physiol., 232*:R145, 1977.

Zerbe, R., *et al.*: Vasopressin function in the syndrome of inappropriate antidiuresis. *Annu. Rev. Med., 31*:315, 1980.

26

Regulation of Acid-Base Balance; Renal Disease; and Micturition

When one speaks of the regulation of acid-base balance, he actually means regulation of hydrogen ion concentration in the body fluids. The hydrogen ion concentration in different solutions can vary from less than 10^{-14} equivalents per liter to higher than 10^0, which means a total variation of more than a quadrillion-fold.

Merely slight changes in hydrogen ion concentration from the normal value can cause marked alterations in the rates of chemical reactions in the cells, some being depressed and others accelerated. For this reason the regulation of hydrogen ion concentration is one of the most important aspects of homeostasis.

Normal Hydrogen Ion Concentration and Normal pH of the Body Fluids — Acidosis and Alkalosis. The hydrogen concentration in the extracellular fluid is normally regulated at a constant value of approximately 4×10^{-8} Eq./liter; this value can vary from as low as 1.0×10^{-8} to as high as 1.0×10^{-7} without causing death.

From these values, it is already apparent that expressing hydrogen ion concentration in terms of actual concentrations is a cumbersome procedure. Therefore, the symbol *pH* has come into use for expressing the concentration, and pH is related to actual hydrogen ion concentration by the following formula (when H^+ conc. is expressed in equivalents per liter):

$$pH = \log \frac{1}{H^+ \text{ conc.}} = -\log H^+ \text{ conc.} \qquad (1)$$

Note from this formula that a low pH corresponds to a high hydrogen ion concentration, which is called *acidosis;* and, conversely, a high pH corresponds to a low hydrogen ion concentration, which is called *alkalosis.*

The normal pH of arterial blood is 7.4, and the pH of venous blood and of interstitial fluids is about 7.35 because of extra quantities of carbon dioxide that form carbonic acid in these fluids.

Since the normal pH of the arterial blood is 7.4, a person is considered to have acidosis whenever the pH is below this value and to have alkalosis when it rises above 7.4. The lower limit at which a person can live more than a few hours is about 7.0, and the upper limit approximately 8.0.

DEFENSE AGAINST CHANGES IN HYDROGEN ION CONCENTRATION

To prevent acidosis or alkalosis, several special control systems are available: (1) All the body fluids are supplied with acid-base *buffer systems* that immediately combine with any acid or alkali and thereby prevent excessive changes in hydrogen ion concentration. (2) If the hydrogen ion concentration does change measurably, the *respiratory center is immediately stimulated* to alter the rate of breathing. This changes the rate of carbon dioxide removal from the body fluids, which, for reasons that will be presented later, causes the hydrogen ion concentration to return toward normal. (3) When the hydrogen concentration changes from normal, *the kidneys excrete either an acid or an alkaline urine,* thereby also helping to readjust the hydrogen ion concentration of the body fluids back toward normal.

The buffer systems can act within a fraction of a second to prevent excessive changes in hydrogen ion concentration. On the other hand, it takes 3 to 12 minutes for the respiratory system to readjust the hydrogen ion concentration after a sudden change has occurred. Finally, the kidneys, though providing the most powerful of all the acid-base regulatory systems, require several hours to several days to readjust the hydrogen ion concentration.

FUNCTION OF ACID-BASE BUFFERS

An acid-base buffer is a solution of two or more chemical compounds that prevents marked changes in hydrogen ion concentration when either an acid or a base is added to the solution. As an example, if only a few drops of concentrated hydrochloric acid are added to a beaker of pure water, the pH of the water might immediately fall to as low as 1.0. However, if a satisfactory buffer system is present, the hydrochloric acid combines instantaneously with the buffer, and the pH falls only slightly. Perhaps the best way to explain the action of an acid-base buffer is to consider an actual simple buffer system, such as the bicarbonate buffer, which is extremely important in regulation of acid-base balance in the body.

The Bicarbonate Buffer System. A typical bicarbonate buffer system consists of a *mixture* of carbonic acid (H_2CO_3) and sodium bicarbonate ($NaHCO_3$) in the same solution. It must first be noted that carbonic acid is a very weak acid.

When a strong acid, such as hydrochloric acid, is added to a buffer solution containing bicarbonate salt, the following reaction takes place:

$$HCl + NaHCO_3 \rightarrow H_2CO_3 + NaCl \qquad (2)$$

From this equation it can be seen that the strong hydrochloric acid is converted into the very weak carbonic acid. Therefore, the HCl lowers the pH of the solution only slightly.

On the other hand, if a strong base, such as sodium hydroxide, is added to a buffer solution containing carbonic acid, the following reaction takes place:

$$NaOH + H_2CO_3 \rightarrow NaHCO_3 + H_2O \qquad (3)$$

This equation shows that the hydroxyl ion of the sodium hydroxide combines with the hydrogen ion from the carbonic acid to form water and that the other product formed is sodium bicarbonate. The net result is exchange of the strong base NaOH for the very weak base $NaHCO_3$.

Thus, the mixture of carbonic acid and sodium bicarbonate acts as a "buffer" to prevent either a marked rise or fall in pH. Fortunately, this buffer is present in all the fluids of the body, both extracellular and intracellular, and it plays an important role in maintaining normal acid-base balance.

The Phosphate Buffer System. Another buffer system, the phosphate buffer system, acts in almost the same manner as the bicarbonate buffer system, but it is composed of the following two elements: $H_2PO_4^-$ and HPO_4^{--}. This system is especially important in the tubular fluids of the kidneys because phosphate usually becomes greatly concentrated in the tubules. The phosphate buffer is also extremely important in the intracellular fluids because the concentration of phosphate in these fluids is many times that in the extracellular fluids.

The Protein Buffer System. By far the most plentiful buffer of the body is the proteins of the cells and plasma. There is a slight amount of diffusion of hydrogen ions through the cell membrane; and even more important, carbon dioxide can diffuse readily through cell membranes, and bicarbonate ions can diffuse to some extent (they require several hours to come to equilibrium in most cells other than the red blood cells). The diffusion of these elements of the bicarbonate buffer system causes the pH in the intracellular fluids to change approximately in proportion to the changes in pH in the extracellular fluids. Therefore, all the buffer systems inside the cells help to buffer the extracellular fluids as well. Indeed, experimental studies have shown that about three-quarters of all the *chemical* buffering power of the body fluids is inside the cells and most of this results from the intracellular proteins. However, except for the red blood cells, the slowness of movement of hydrogen and bicarbonate ions through the cell membranes often delays the ability of the intracellular buffers to buffer extracellular acid-base abnormalities for several hours.

The method by which the protein buffer system operates is precisely the same as that of the bicarbonate buffer system. It will be recalled that a protein is composed of amino acids bound together by peptide linkages, but some of the different amino acids have free acidic radicals that function as weak acids in several different buffer systems.

RESPIRATORY REGULATION OF ACID-BASE BALANCE

Because carbon dioxide combines with water to form carbonic acid, an *increase in carbon dioxide* concentration in the body fluids *decreases the pH* toward the acidic side, whereas a decrease in carbon dioxide raises the pH toward the alkaline side. It is on the basis of this effect that the respiratory system is capable of altering the pH either up or down.

Balance Between Metabolic Formation of Carbon Dioxide and Pulmonary Expiration of Carbon Dioxide. Carbon dioxide is continually being formed in the body by the different intracellular metabolic processes, the carbon in the foods being oxidized by oxygen to form carbon dioxide. This in turn diffuses into the interstitial fluids and blood and is transported to the lungs, where it diffuses into the alveoli and is transferred to the atmosphere by pulmonary ventilation. However, several minutes are required for this passage of carbon dioxide from the cells to the atmosphere, so that an average of 1.2 millimoles/liter of dissolved carbon dioxide is normally in the extracellular fluids at all times.

If the rate of metabolic formation of carbon dioxide becomes increased, its concentration as well as the hydrogen ion concentration in the extracellular fluids is likewise increased. Conversely, decreased metabolism decreases the carbon dioxide and hydrogen ion concentrations.

On the other hand, if the rate of pulmonary ventilation is increased, the rate of expiration of carbon dioxide becomes increased, and this decreases both the amount of accumulated carbon dioxide and the hydrogen ion concentration in the extracellular fluids.

Effect of Hydrogen Ion Concentration on Alveolar Ventilation. Not only does the rate of alveolar ventilation affect the hydrogen ion concentrtion of the body fluids, but, in turn, the hydrogen ion concentration affects the rate of alveolar ventilation. This results from a *direct action of hydrogen ions on the respiratory center in the medulla oblongata,* which controls breathing, as will be discussed in detail in Chapter 29.

Figure 26–1 illustrates the changes in alveolar ventilation caused by changing the pH of arterial blood from 7.0 to 7.6. From this graph it is evident that a decrease in pH from the normal value of 7.4 to the strongly acidic range can increase the rate of alveolar ventilation to as much as four to five times normal, while an increase in pH into the alkaline range can decrease the rate of alveolar ventilation to as little as 50 to 75 per cent of normal.

Feedback Regulation of Hydrogen Ion Concentration by the Respiratory System. Because of the ability of the respiratory center to respond to hydrogen ion concentration, and because changes in alveolar ventilation in turn alter the hydrogen ion concentration in the body fluids, the respiratory system acts as a typical feedback regulatory system for controlling hydrogen ion concentration. That is, any time the hydrogen ion concentration becomes high, the respiratory system becomes more active, and alveolar ventilation increases. As a result the carbon dioxide concentration in the extracellular fluids decreases, thus reducing the hydrogen concentration back toward a normal value. Conversely, if the hydrogen ion concentration falls too low, the respiratory center becomes depressed, alveolar ventilation also decreases, and the hydrogen ion concentration rises back toward normal.

Efficiency of Respiratory Regulation of Hydrogen Ion Concentration. Unfortunately, respiratory control cannot return the hydrogen ion concentration all the way to the normal value of 7.4 when some abnormality outside the respiratory system has altered the pH from the normal. The reason is that as the pH returns toward normal the stimulus that has been causing either increased or decreased respiration will itself begin to be lost. Ordinarily, the respiratory mechanism for regulation of hydrogen ion concentration has a control effectiveness of between 50 and 75 per cent. That is, if the hydrogen ion concentration should suddenly be decreased from 7.4 to 7.0 by some extraneous factor, the respiratory system, in 3 to 12 minutes, returns the pH to a value of about 7.2 to 7.3.

Figure 26–1. Effect of blood pH on the rate of alveolar ventilation. (Constructed from data obtained by Gray: Pulmonary Ventilation and Its Regulation. Charles C Thomas.)

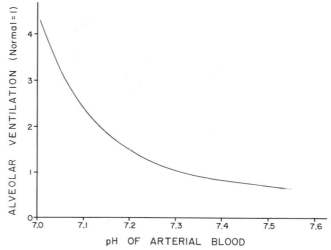

RENAL REGULATION OF HYDROGEN ION CONCENTRATION

The kidneys regulate hydrogen ion concentration principally by increasing or decreasing the bicarbonate ion concentration in the body fluid. To do this, a complex series of reactions occurs in the renal tubules, as described in the following sections.

Tubular Secretion of Hydrogen Ions. The epithelial cells of the proximal tubules, distal tubules, and collecting ducts all secrete hydrogen ions into the tubular fluid. The mechanism by which this occurs is illustrated in Figure 26–2. The secretory process *begins with carbon dioxide* in the tubular epithelial cells. The carbon dioxide, under the influence of an enzyme, *carbonic anhydrase,* combines with water to form *carbonic acid,* which then dissociates into *bicarbonate ion* and *hydrogen ion.* The hydrogen ion is then secreted by active transport through the luminal border of the cell membrane into the tubule.

In the collecting ducts hydrogen ion secretion can continue until the concentration of hydrogen ions in the tubules becomes as much as 900 times that in the extracellular fluid or, in other words, until the pH of the tubular fluids falls to about 4.5. This represents a limit to the ability of the tubular epithelium to secrete hydrogen ions.

About 84 per cent of all the hydrogen ions secreted by the tubules are secreted in the proximal tubules, but the maximum concentration gradient that can be achieved here is only about three- to four-fold instead of the 900-fold that can be achieved in the collecting tubules.

Regulation of Hydrogen Ion Secretion by the Carbon Dioxide Concentration in the Extracellular Fluid. Since the chemical reactions for secretion of hydrogen ions begin with carbon dioxide, the greater the carbon dioxide concentration in the extracellular fluid, the more rapidly the reactions proceed, and the greater becomes the rate of hydrogen ion secretion. Therefore, any factor that increases the carbon dioxide concentration in the extracellular fluids also increases the rate of hydrogen ion secretion.

At normal carbon dioxide concentrations, the rate of hydrogen ion secretion is about 3.5 millimoles per minute, but this rises or falls directly in proportion to changes in extracellular carbon dioxide.

Interaction of Bicarbonate Ions with Hydrogen Ions in the Tubules — "Reabsorption" of Bicarbonate Ions. It is already clear from previous discussions that the bicarbonate ion concentration in the extracellular fluid plays an extremely important role in the acid-base buffer system and therefore in the control of extracellular fluid hydrogen ion concentration. Therefore, it is important that the kidney tubules help to regulate the extracellular fluid bicarbonate ion concentration. Yet, the tubules are not very permeable to the bicarbonate ion because it is a large ion and also is electrically charged. However, the bicarbonate ion can, in effect, be "reabsorbed" by a special process, which is also illustrated in Figure 26–2.

The reabsorption of bicarbonate ions is initiated by a reaction in the tubules between the bicarbonate ions and the hydrogen ions secreted by the tubular cells, as illustrated in the figure. The carbonic acid then dissociates into carbon dioxide and water. The water becomes part of the tubular fluid, while the carbon dioxide, having the capability to diffuse extremely readily through all cellular membranes, instantaneously diffuses into epithelial cells or even all the way into the blood, where it combines with water to form new bicarbonate ions. *If an excess of hydrogen ions is secreted by the tubules, the bicarbonate ions will be almost completely removed from the tubules,* so that for practical purposes *none* will remain to pass into the urine.

If we now note in Figure 26–2 the chemical reactions that are responsible for formation of hydrogen ions in the epithelial cells, we see that each time a hydrogen ion is formed a bicarbonate ion is formed inside these cells by the dissociation of H_2CO_3. This bicarbonate ion then diffuses into the peritubular fluid.

The net effect of all these reactions is a mechanism for reabsorption of bicarbonate ions from the tubules, though the bicarbonate ions that enter the peritubular fluid are not the same bicarbonate ions that are removed from the tubular fluid.

Normal Rates of Bicarbonate Ion Filtration and Hydrogen Ion Secretion into the Tubules — Titration of Bicarbonate Ions Against Hydrogen Ions. Under normal conditions, the rate of hydrogen ion secretion is about 3.50 millimoles/minute, and the rate of filtration of bi-

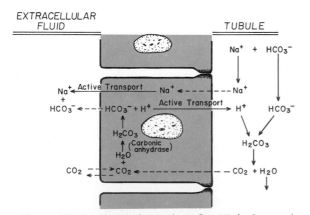

Figure 26–2. Chemical reactions for (1) hydrogen ion secretion, (2) sodium ion absorption in exchange for a hydrogen ion, and (3) combination of hydrogen ions with bicarbonate ions in the tubules.

carbonate ions in the glomerular filtrate is about 3.49 millimoles/minute. Thus, the quantities of the two ions entering the tubules are almost equal, and they combine with each other and actually annihilate each other, the end-products being carbon dioxide and water. Therefore, it is said that the bicarbonate ions and hydrogen ions normally "titrate" each other in the tubules.

However, note also that this titration process is not quite complete, for usually a slight excess of hydrogen ions (the acidic component) remains in the tubules to be excreted in the urine. The reason is that under normal conditions a person's metabolic processes continually form a small amount of excess acid that gives rise to the slight excess of hydrogen ions over bicarbonate ions in the tubules.

On rare occasions the bicarbonate ions are in excess, as we shall see in subsequent discussions. When this occurs, the titration process again is not quite complete; this time, excess bicarbonate ions (the basic component) are left in the tubules to pass into the urine.

Thus, the basic mechanism by which the kidney corrects either acidosis or alkalosis is incomplete titration of hydrogen ions against bicarbonate ions, leaving one or the other of these to pass into the urine and therefore to be removed from the extracellular fluid.

Renal Correction of Alkalosis — Decrease in Bicarbonate Ions in the Extracellular Fluid

Now that we have described the mechanisms by which the renal tubules secrete hydrogen ions and reabsorb bicarbonate ions, we can explain the manner in which the kidneys readjust the pH of the extracellular fluids when it becomes abnormal.

The initial step in this explanation is to understand what happens to the concentrations of carbon dioxide and bicarbonate ions in the extracellular fluids in alkalosis and acidosis. First, let us consider *alkalosis*. The *ratio* of bicarbonate ions to dissolved carbon dioxide molecules increases when the pH rises into the alkalosis range above 7.4. The effect on the titration process in the tubules is to increase the *ratio* of bicarbonate ions filtered into the tubules to hydrogen ions secreted. This increase occurs because the high extracellular bicarbonate ion concentration also increases its concentration in the glomerular filtrate, and the low carbon dioxide concentration decreases the secretion of hydrogen ions. Therefore, the fine balance that normally exists in the tubules between the hydrogen and bicarbonate ions no longer occurs. Instead, far greater quantities of bicarbonate ions than hydrogen ions now enter the tubules. Since no bicarbonate ions can be reabsorbed without first reacting with hydrogen

ions, all the excess bicarbonate ions pass into the urine and carry with them sodium ions or other positive ions. Thus, in effect, sodium bicarbonate is removed from the extracellular fluid.

Loss of sodium bicarbonate from the extracellular fluid decreases the bicarbonate ion portion of the bicarbonate buffer system, and this shifts the pH of the body fluids back in the acid direction. Thus, the alkalosis is corrected.

Renal Correction of Acidosis — Increase in Bicarbonate Ions in the Extracellular Fluid

In acidosis, the *ratio* of carbon dioxide to bicarbonate ions in the extracellular fluid increases, which is exactly opposite to the effect in alkalosis. Therefore, in acidosis, the *rate of hydrogen ion secretion* rises to a level far greater than the *rate of bicarbonate ion filtration* into the tubules. As a result, a great excess of hydrogen ions is secreted into the tubules, and they have far too few bicarbonate ions to react with. These excess hydrogen ions combine with the buffers in the tubular fluid, as explained in the following paragraphs, and are excreted into the urine.

Figure 26-2 shows that each time a hydrogen ion is secreted into the tubules two other effects occur simultaneously. First, a bicarbonate ion is formed in the tubular epithelial cell, and second, a sodium ion is absorbed from the tubule into the epithelial cell. The sodium ion and bicarbonate ion then diffuse together from the epithelial cell into the peritubular fluid. Thus, *the net effect of secreting excess hydrogen ions into the tubules is to increase the quantitiy of sodium bicarbonate in the extracellular fluid.* This increases the bicarbonate portion of the bicarbonate buffer system, which shifts the buffers in the alkaline direction, increasing the pH in the process and thereby correcting the acidosis.

Transport of Excess Hydrogen Ions into the Urine by the Phosphate Buffer. The phosphate buffer is composed of a mixture of HPO_4^{--} and $H_2PO_4^-$. Both become considerably concentrated in the tubular fluid because of their relatively poor reabsorption and because of removal of water from the tubular fluid. Therefore, even though the phosphate buffer is very weak in the blood, it is a much more powerful buffer in the tubular fluid and is responsible for transporting a major share of excess hydrogen ions into the urine.

The quantity of HPO_4^{--} in the glomerular filtrate is normally about four times as great as that of $H_2PO_4^-$. Excess hydrogen ions entering the tubules combine with the HPO_4^{--} to form $H_2PO_4^-$, which passes on into the urine. Sodium ion is absorbed into the extracellular fluid in place of the hydrogen ion involved in the reaction, and at the same time a *bicarbonate ion,* formed in the process of secreting the hydrogen ion, is also released into the extracellular fluid. Thus, the net effect of this reaction is to increase the amount of sodium bicarbonate in the extracellular fluids, which is the kidney's way of reducing the degree of acidosis in the body fluids.

Transport of Excess Hydrogen Ions into the Urine by the Ammonia Buffer System. Another very potent buffer system of the tubular fluid is composed of ammonia (NH_3) and the ammonium ion (NH_4^+). The epithelial cells of all the tubules besides those of the thin segment of the loop of Henle continually synthesize ammonia, which diffuses into the tubules. The ammonia then reacts with hydrogen ions, as illustrated in Figure 26–3, to form ammonium ions. These are then excreted into the urine in combination with chloride ions and other tubular anions. Note in the figure that the net effect of these reactions, again, is *to increase the bicarbonate concentration* in the extracellular fluid.

This ammonium ion mechanism for transport of excess hydrogen ions in the tubules is especially important for two reasons: (1) Each time an ammonia molecule combines with a hydrogen ion to form an ammonium ion, the concentration of ammonia in the tubular fluid becomes decreased, which causes still more ammonia to diffuse from the epithelial cells into the tubular fluid. Thus, the rate of ammonia secretion into the tubular fluid is actually controlled by the amount of excess hydrogen ions to be transported. (2) Most of the negative ions of the tubular fluid are chloride ions. Only a few hydrogen ions can be transported into the urine in direct combination with chloride, because hydrochloric acid is a very strong acid and the tubular pH would fall rapidly below the critical value of 4.5, so that further hydrogen ion secretion would cease. However, when hydrogen ions combine with ammonia and the resulting ammonium ions then combine with

chloride, the pH does not fall significantly because ammonium chloride is only very weakly acidic.

Rapidity of Acid-Base Regulation by the Kidneys

The total amount of buffers in the entire body (within the range of pH 7.0 to 7.8) is approximately 1000 millimoles. If all of them should be suddenly shifted to the alkaline or acidic side by the injection of an alkali or an acid, the kidneys would be able to return the pH of the body fluids back almost to normal in two to six days. Though this mechanism is slow to act, it continues acting until the pH returns almost exactly to normal rather than a certain percentage of the way. Therefore, the real value of the renal mechanism for regulating hydrogen ion concentration is not rapidity of action but instead its ability in the end to neutralize completely any excess acid or alkali that enters the body fluids, unless the excess continues to enter.

Ordinarily, the kidneys can remove up to about 500 millimoles of acid or alkali each day. If greater quantities than this enter the body fluids, the kidneys are unable to cope with the extra load, and severe acidosis or alkalosis ensues.

Range of Urinary pH. In the process of adjusting the hydrogen ion concentration of the extracellular fluid, the kidneys often excrete urine at a pH as low as 4.5 or as high as 8.0. When acid is being excreted the pH falls, and when alkali is being excreted the pH rises. Even when the pH of the arterial blood is at the normal value of 7.4, a fraction of a millimole of acid is still lost each minute. The reason is that about 50 to 80 millimoles more acid than alkali are formed in the body each day, and this acid must be removed continually. Because of the presence of this excess acid in the urine, the normal urine pH is about 6.0 instead of 7.4, the pH of the blood.

RENAL REGULATION OF PLASMA CHLORIDE CONCENTRATION — THE CHLORIDE TO BICARBONATE RATIO

In the above discussions we have emphasized the ability of the kidneys to increase bicarbonate ion in the extracellular fluid whenever a state of acidosis develops, or to remove bicarbonate ions in a state of alkalosis. Thus, the bicarbonate ion is shuttled back and forth between high and low values as one of the principal means of adjusting the acid-base balance of the extracellular buffer systems and therefore also for adjusting the extracellular fluid pH.

However, in the process of juggling the extra-

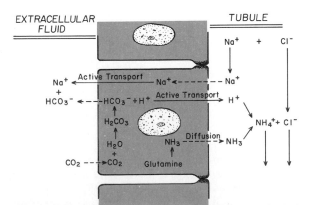

Figure 26–3. Secretion of ammonia by the tubular epithelial cells, and reaction of the ammonia with hydrogen ions in the tubules.

cellular fluid concentration of bicarbonate ion, it is essential to remove some other anion from the extracellular fluids each time the bicarbonate is increased, or to increase some other anion when the bicarbonate concentration is decreased. In general, the anion that is reciprocally juggled up or down with the bicarbonate ion is the chloride ion because this is the anion in greatest concentration in the extracellular fluid.

Function of the Ammonia Buffer System in Controlling the Bicarbonate Ion to Chloride Ion Ratio. It was pointed out above that the ammonia buffer system plays an extremely important role in removing excess hydrogen ions from the tubules. Now, let us study Figure 26–3 once again. We see that in the process of transporting excess hydrogen ions into the urine in combination with ammonia, for each hydrogen ion transported a chloride ion also passes into the urine and a bicarbonate ion simultaneously enters the extracellular fluid. Thus, this ammonia system substitutes a bicarbonate ion in the extracellular fluid for a chloride ion that is lost from the extracellular fluids. Conversely, when a person is alkalotic, the ammonia system becomes inoperative; bicarbonate ions instead of chloride ions then pass into the urine, and a concomitant excess of chloride is reabsorbed.

Thus, in the process of controlling the pH of the body fluids, the renal acid-base regulating system also regulates the ratio of chloride ions to bicarbonate ions in the extracellular fluid.

CLINICAL ABNORMALITIES OF ACID-BASE BALANCE

Respiratory Acidosis and Alkalosis

From the discussions earlier in the chapter it is obvious that any factor that decreases the rate of pulmonary ventilation increases the concentration of dissolved carbon dioxide in the extracellular fluid, which in turn leads to increased carbonic acid and hydrogen ions, thus resulting in acidosis. Because this type of acidosis is caused by an abnormality of respiration, it is called *respiratory acidosis*.

On the other hand, excessive pulmonary ventilation reverses the process and decreases the hydrogen ion concentration, thus resulting in alkalosis; this condition is called *respiratory alkalosis*.

A person can cause respiratory acidosis in himself by simply holding his breath, which he can do until the pH of the body fluids falls to as low as perhaps 7.0. On the other hand, he can voluntarily overbreathe and cause alkalosis to a pH of about 7.9.

Respiratory acidosis frequently results from pathological conditions. For instance, damage to the respiratory center in the medulla oblongata, thus causing reduced breathing, obstruction of the passageways in the respiratory tract, pneumonia, decreased pulmonary membrane surface area, and any other factor that interferes with the exchange of gases between the blood and alveolar air, results in respiratory acidosis.

On the other hand, only rarely do pathological conditions cause *respiratory alkalosis*. However, occasionally a psychoneurosis causes overbreathing to the extent that a person becomes alkalotic. A physiological type of respiratory alkalosis occurs when a person ascends to a *high altitude*. The low oxygen content of the air stimulates respiration, which causes excess loss of carbon dioxide and development of mild respiratory alkalosis.

Metabolic Acidosis and Alkalosis

The terms metabolic acidosis and metabolic alkalosis refer to all other abnormalities of acid-base balance besides those caused by excess or insufficient carbon dioxide in the body fluids. Use of the word "metabolic" in this instance is unfortunate, because carbon dioxide is also a metabolic product. Yet, by convention, carbonic acid resulting from dissolved carbon dioxide is called a *respiratory acid,* while any other acid in the body, whether it be formed by metabolism or simply ingested by the person, is called a *metabolic acid,* or a *fixed acid.*

Causes of Metabolic Acidosis:

Diarrhea. Severe diarrhea is one of the most frequent causes of metabolic acidosis for the following reasons: The gastrointestinal secretions normally contain large amounts of sodium bicarbonate. Therefore, excessive loss of these secretions during a bout of diarrhea is exactly the same as excretion of large amounts of sodium bicarbonate into the urine. This causes a shift of the bicarbonate buffer system toward the acid side and results in metabolic acidosis. In fact, acidosis resulting from severe diarrhea can be so serious that it is one of the most common causes of death in young children.

Vomiting. A second cause of metabolic acidosis is vomiting. Vomiting of gastric contents alone, which occurs rarely, causes a loss of acid and leads to alkalosis, but vomiting of contents from deeper in the gastrointestinal tract, which often occurs, causes loss of alkali and results is metabolic acidosis.

Uremia. A third common type of acidosis is uremic acidosis, which occurs in severe renal disease. The cause is failure of the kidneys to rid the body of even the normal amounts of acids formed each day by the metabolic processes of the body.

Diabetes Mellitus.   A fourth and extremely important cause of metabolic acidosis is diabetes mellitus. In this condition, lack of insulin secretion by the pancreas prevents normal use of glucose for metabolism. Instead, fat is split into acetoacetic acid, which in turn is metabolized by the tissues for energy in place of glucose. Simultaneously, the concentration of acetoacetic acid in the extracellular fluids often rises very high, and large quantities of it are excreted in the urine, sometimes as much as 500 to 1000 millimoles per day.

Causes of Metabolic Alkalosis: Metabolic alkalosis does not occur nearly as often as metabolic acidosis. However, there are several common causes of metabolic alkalosis, as follows:

Excessive Ingestion of Alkaline Drugs.   One of the most common causes of alkalosis is excessive ingestion of alkaline drugs, such as sodium bicarbonate, for the treatment of gastritis or peptic ulcer.

Alkalosis Caused by Loss of Chloride Ions.   Excessive vomiting of gastric contents without vomiting of lower gastrointestinal contents causes excessive loss of hydrochloric acid secreted by the stomach mucosa. This loss leads to reduction of chloride ions and enhancement of bicarbonate ions in the extracellular fluid. The bicarbonate ions are derived from the stomach glandular cells that secrete the hydrogen ions; these form bicarbonate ion in the same way that the tubular cells do when they secrete hydrogen ions. The net result is loss of acid from the extracellular fluids and development of metabolic alkalosis.

Alkalosis Caused by Excess Aldosterone.   When excess quantities of aldosterone are secreted by the adrenal glands, the extracellular fluid becomes slightly alkalotic. This is caused in the following way: the aldosterone promotes extensive reabsorption of sodium ions from the distal segments of the tubular system, but coupled with this is increased secretion of hydrogen ions and loss from the extracellular fluids, thus promoting alkalosis.

Effects of Acidosis and Alkalosis on the Body

Acidosis.   The major effect of acidosis is depression of the *central nervous system.* When the pH of the blood falls below 7.0, the nervous system becomes so depressed that the person first becomes disoriented and later comatose. Therefore, patients dying of diabetic acidosis, uremic acidosis, and other types of acidosis usually die in a state of coma.

In metabolic acidosis the high hydrogen ion concentration causes increased rate and depth of respiration. Therefore, one of the diagnostic signs of *metabolic* acidosis is increased pulmonary ventilation. On the other hand, *in respiratory acidosis, respiration is usually depressed* because this is the cause of the acidosis, which is opposite to the effect in metabolic acidosis.

Alkalosis.   The major effect of alkalosis on the body is *overexcitability of the nervous system.* This occurs both in the central nervous system and in the peripheral nerves, but usually the peripheral nerves are affected before the central nervous system. The nerves become so excitable that they automatically and repetitively fire even when they are not stimulated by normal stimuli. As a result, the muscles go into a state of *tetany,* which means a state of tonic spasm. This tetany usually appears first in the muscles of the forearm and then spreads rapidly to the muscles of the face and finally all over the body. Extremely alkalotic patients may die from tetany of the respiratory muscles.

Occasionally an alkalotic person develops severe symptoms of central nervous system overexcitability. The symptoms may manifest themselves as extreme nervousness or, in susceptible persons, as convulsions. For instance, in persons who are predisposed to epileptic fits, simply overbreathing often results in an attack. Indeed, this is one of the clinical methods for assessing one's degree of epileptic predisposition.

Physiology of Treatment in Acidosis or Alkalosis

Obviously, the best treatment for acidosis or alkalosis is to remove the condition causing the abnormality, but, if this cannot be effected, different drugs can be used to neutralize the excess acid or alkali.

To neutralize excess acid, large amounts of sodium bicarbonate can be ingested by mouth. It is absorbed into the blood stream and increases the bicarbonate ion portion of the bicarbonate buffer, thereby shifting the pH to the alkaline side.

For treatment of alkalosis, ammonium chloride is often administered by mouth. When it is absorbed into the blood, the ammonia portion of the ammonium chloride is converted by the liver into urea; this reaction liberates hydrochloric acid, which immediately reacts with the buffers of the body fluids to shift the hydrogen ion concentration in the acid direction.

RENAL DISEASE

It will not be possible to discuss here the great numbers of kidney diseases. However, especially important are (1) renal failure and (2) the nephrotic syndrome.

RENAL FAILURE

Renal Failure Caused by Acute Glomerular Nephritis. Acute glomerular nephritis is a disease that results from an antigen-antibody reaction and in which the glomeruli become markedly inflamed. Large numbers of white blood cells collect in the inflamed glomeruli, and the endothelial cells on the vascular side of the glomerular membrane, as well as the epithelial cells on the Bowman's capsule side of the membrane, proliferate, sometimes completely filling the glomeruli and the capsule. These inflammatory reactions can cause total or partial blockage of large numbers of glomeruli, and many of those glomeruli that are not blocked develop greatly increased permeability of the glomerular membrane, allowing large amounts of protein to leak into the glomerular filtrate. Also, rupture of the membrane in severe cases often allows large numbers of red blood cells to pass into the glomerular filtrate. In the severest cases total renal shutdown occurs.

The inflammation of acute glomerular nephritis almost invariably occurs one to three weeks following, elsewhere in the body, an infection caused by certain types of group A beta streptococci, such as a streptococcal sore throat, streptococcal tonsillitis, scarlet fever, or even streptococcal infection of the skin. The infection itself does not cause damage to the kidneys, but when antibodies develop against the streptococcal antigen, the antibodies and antigen react with each other to form a precipitate that becomes entrapped in the middle of the glomerular membrane. The reactivity of this precipitated complex leads to inflammation of the glomeruli.

The acute inflammation of the glomeruli most often subsides in ten days to two weeks, and the nephrons may return to normal function. Sometimes, however, the inflammatory reactions are so severe that many of the glomeruli will be destroyed permanently.

Chronic Glomerulonephritis. Chronic glomerulonephritis is caused by any one of several different diseases that damage principally the glomeruli. The basic glomerular lesion is usually very similar to that which occurs in acute glomerulonephritis. It normally begins with accumulation of precipitated antigen-antibody complex in the glomerular membrane followed by inflammation of the glomeruli. The glomerular membrane becomes progressively thickened and is eventually invaded by fibrous tissue. In the later stages of the disease, glomerular filtration becomes greatly reduced because of decreased numbers of filtering capillaries in the glomerular tufts and because of thickened glomerular membranes. In the final stages of the disease many of the glomeruli are completely replaced by fibrous tissue, and the function of these nephrons is thereafter lost forever.

Pyelonephritis. Pyelonephritis is an infectious and inflammatory process that usually begins in the renal pelvis but extends progressively into the renal parenchyma. The infection can result from many different types of bacteria but especially from the colon bacillus that originates from fecal contamination of the urinary tract. Invasion of the kidneys by these bacteria results in progressive destruction of renal tubules, glomeruli, and any other structures in the path of the invading organisms. Consequently, large portions of the functional renal tissue are lost.

A particularly interesting feature of pyelonephritis is that the invading infection usually affects the medulla of the kidney before it affects the cortex. Since one of the primary functions of the medulla is to provide the countercurrent mechanism for concentrating the urine, patients with pyelonephritis frequently have reasonably normal renal function except for inability to concentrate their urine.

Abnormal Excretory Function in Renal Failure

The most important effect of renal failure is the inability of the kidneys to cope with large "loads" of electrolytes or other substances that must be excreted. Normally, one-third of the nephrons can eliminate essentially all the normal load of waste products from the body without serious accumulation of any of these in the body fluids. However, further reduction of numbers of nephrons leads to urinary retention, and death ensues when the number of nephrons falls below 10 to 20 per cent of normal.

Effects of Renal Failure on the Body Fluids — Uremia

The effect of acute or chronic renal failure on the body fluids depends to a great extent on the water and food intake of the person. Assuming that the person continues to ingest moderate amounts of water and food, the concentration changes of different substances in the extracellular fluid are approximately those shown in Figure 26–4. The most important effects are (1) *generalized edema* resulting from water and salt retention, (2) *acidosis* resulting from failure of the kidneys to rid the body of normal acidic products, (3) *high concentrations of the nonprotein nitrogens*, especially *urea*, resulting from failure of the body to excrete the metabolic end-products, and (4) *high concentration of other urinary retention products*

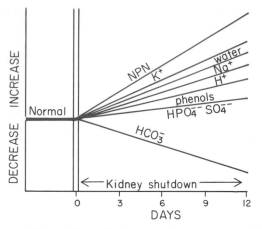

Figure 26–4. Effect of kidney shutdown on extracellular fluid constituents.

including *creatinine, uric acid, phenols, guanidine bases, sulfates, phosphates,* and *potassium.* This condition is called *uremia* because of the high concentrations of normal urinary excretory products that collect in the body fluids.

Acidosis in Renal Failure. Each day the metabolic processes of the body normally produce 50 to 70 millimoles more metabolic acid than metabolic alkali. Therefore, any time the kidneys fail to function, acid begins to accumulate in the body fluids. Normally, the buffers of the fluids can buffer up to a total of 500 to 1000 millimoles of acid without severe depression of the extracellular fluid pH, but gradually this buffering power is used up, so that the pH falls drastically. The patient becomes *comatose* at about this same time, and it is believed that this coma is partly caused by the acidosis, as is discussed below.

Increase in Urea and Other Nonprotein Nitrogens in Uremia. The nonprotein nitrogens include urea, uric acid, creatinine, and a few less important compounds. These, in general, are the end-products of protein metabolism and must be removed from the body continually to insure continued protein metabolism in the cells. The concentrations of these, particularly of urea, can rise to as high as ten times normal during one to two weeks of renal failure. However, even these high levels do not seem to affect physiological function nearly so much as the high concentrations of hydrogen and potassium ions and some of the other less obvious substances, such as very toxic guanidine bases, ammonium ions, and others. Yet one of the most important means for assessing the degree of renal failure is to measure the concentration of the nonprotein nitrogens.

Uremic Coma. After a few days of complete renal failure the sensorium of the patient becomes clouded, and he soon progresses into a state of coma. Acidosis is one of the principal factors responsible for the coma; acidosis caused by other conditions, such as severe diabetes melli-

tus, also causes coma. However, many other abnormalities could also be contributory — the generalized edema, the high potassium concentration, and possibly even the high nonprotein nitrogen concentration.

The respiration usually is deep and rapid in coma, a characteristic that is a respiratory attempt to compensate for the metabolic acidosis. In addition, the arterial pressure falls progressively during the last day or so before death, then rapidly in the last few hours. Death occurs usually when the pH of the blood falls to about 6.9.

Dialysis of Uremic Patients with the Artificial Kidney

Artificial kidneys have now been used for about 30 years to treat patients with severe renal failure. They have been developed to such a point that many thousands of persons with permanent renal failure or even total kidney removal are being maintained in health for years at a time, their lives depending entirely on the artificial kidney.

The basic principle of the artificial kidney is to pass blood through very minute blood channels bounded by thin membranes. On the other sides of the membranes is a *dialyzing fluid* into which unwanted substances in the blood pass by diffusion.

Figure 26–5 illustrates diagrammatically an artificial kidney in which blood flows continually between two thin sheets of cellophane; on the outside of the sheets is the dialyzing fluid. The cellophane is porous enough to allow all constituents of the plasma except the plasma proteins to diffuse freely in both directions — from plasma into the dialyzing fluid and from the dialyzing fluid back into the plasma. If the concentration of a substance is greater in the plasma than in the dialyzing fluid, there is net transfer of the substance from the plasma into the dialyzing fluid. The amount of the substance that transfers depends on (1) the difference between the concentrations on the two sides of the membrane, (2) molecular size, the smaller molecules diffusing more rapidly than larger ones, and (3) the length of time that the blood and the fluid remain in contact with the membrane.

In normal operation of the artificial kidney, blood continually flows from an artery, through the kidney, and back into a vein. The total amount of blood in the artificial kidney at any one time is usually less than 500 ml., the rate of flow may be several hundred ml. per minute, and the total diffusing surface is usually between 10,000 and 20,000 square centimeters. To prevent coagulation of blood in the artificial kidney, heparin is infused into the blood as it enters the "kidney." Then, to prevent bleeding in the patient as a result of the heparin, an anti-heparin substance,

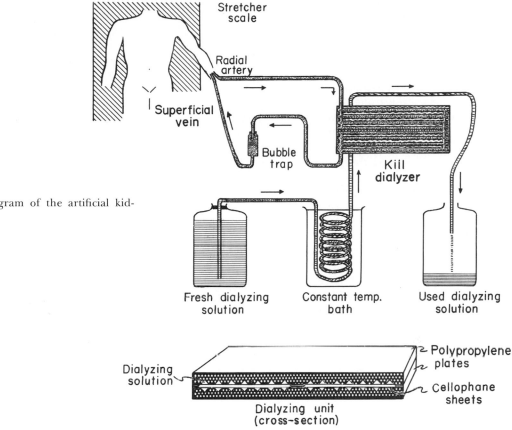

Figure 26–5. Diagram of the artificial kidney.

protamine, is infused into the blood as it is returned to the patient.

The Dialyzing Fluid. Table 26–1 compares the constituents of a typical dialyzing fluid with those in normal plasma and uremic plasma. Note that sodium, potassium, and chloride concentrations in the dialyzing fluid and in normal plasma are identical, but in uremic plasma the potassium

TABLE 26–1 COMPARISON OF DIALYZING FLUID WITH NORMAL AND UREMIC PLASMA

Constituent	Normal Plasma	Dialyzing Fluid	Uremic Plasma
Electrolytes (mEq./Liter)			
Na^+	142	142	142
K^+	5	4	7
Ca^{++}	3	3	2
Mg^{++}	1.5	1.5	1.5
Cl^-	107	107	107
HCO_3^-	27	27	14
$Lactate^-$	1.2	1.2	1.2
HPO_4^{--}	3	0	9
$Urate^-$	0.3	0	2
$Sulfate^{--}$	0.5	0	3
Nonelectrolytes (mg. %)			
Glucose	100	125	100
Urea	26	0	200
Creatinine	1	0	6

concentration is considerably greater. This ion diffuses through the dialyzing membrane so rapidly that its concentration falls to equal that in the dialyzing fluid within only three to four hours' exposure to the dialyzing fluid.

On the other hand, there is no phosphate, urea, urate, sulfate, or creatinine in the dialyzing fluid. Therefore, when the uremic patient is dialyzed, these substances are lost in large quantities into the dialyzing fluid, thereby removing major proportions of them from the plasma.

Thus, the constituents of the dialyzing fluid are chosen so that those substances in excess in the extracellular fluid in uremia can be removed at rapid rates, while the normal electrolytes remain essentially normal.

THE NEPHROTIC SYNDROME — INCREASED GLOMERULAR PERMEABILITY

Large numbers of patients with renal disease develop a so-called *nephrotic syndrome,* which is characterized especially by *loss of large quantities of plasma proteins into the urine.* In some instances this occurs without evidence of any other abnormality of renal function, but more often it is associated with some degree of renal failure.

The cause of the protein loss in the urine is increased permeability of the glomerular membrane. Therefore, any disease condition that can increase the permeability of this membrane can cause the nephrotic syndrome. Such diseases include some types of *chronic glomerulonephritis* (in the previous discussion, it was noted that this disease primarily affects the glomeruli and causes a greatly increased permeability of the glomerular membrane), *amyloidosis,* which results from deposition of an abnormal proteinoid substance in the walls of blood vessels and seriously damages the basement membrane of the glomerulus, and *lipoid nephrosis,* a disease found mainly in young children.

Lipoid Nephrosis.   Liphoid nephrosis is very common in children, occurring most often before the age of four but occurring occasionally in adults as well. Its basic cause is unknown, but the resulting renal lesion increases the permeability of the glomerular membrane and causes loss of proteins into the urine. This lesion develops in the following way: The epithelial cells that line the outer surface of the glomerulus are defective. They are greatly swollen, and they fail to form the usual foot processes that cover the Bowman's capsule surface of the glomerulus. It will be recalled from the discussion of the glomerular membrane in Chapter 24 that the openings between these foot processes are very small, and it is normally the smallness of these openings that prevents the passage of protein through the glomerular membrane. Therefore, in the absence of the foot processes, tremendous quantities of protein leak into the tubules even though larger elements of the blood, such as red cells, are still completely prevented from leaking.

The name "lipoid nephrosis" is derived from the fact that large quantities of lipid droplets are found in the epithelial cells lining the tubules and also from the fact that the concentration of lipid substances in the blood is usually increased. The lipid deposits in the tubules apparently play no role in the disease.

Administration of glucocorticoids such as hydrocortisone will usually cause complete remission of the disease, although they will not cause remission of most other types of nephrotic syndrome.

Protein Loss.   In the nephrotic syndrome, as much as 30 grams of plasma proteins can be lost into the urine each day. Though the resulting low plasma protein concentration stimulates the liver to produce far more plasma proteins than usual, nevertheless, the liver often cannot keep up with the loss. Therefore, in severe nephrosis the colloid osmotic pressure sometimes falls extremely low, often from the normal level of 28 mm. Hg to as low as 6 to 8 mm. Hg.

Edema.   The low colloid osmotic pressure in turn allows large amounts of fluid to filter into the interstitial spaces and also into the potential spaces of the body, thus causing serious *edema.* The nephrotic person has been known on occasion to develop as much as 40 liters of excess extracellular fluid, and as much as 15 liters of this has been *ascites* in the abdomen. Also, the joints swell, and the pleural cavity and the pericardium become partially filled with fluid.

A nephrotic person can be greatly benefited by intravenous infusion of large quantities of concentrated plasma proteins. Yet this is of only temporary benefit because enough protein can be lost into the urine in only a day to return the person to his original predicament.

MICTURITION

Micturition is the process by which the urinary bladder empties when it becomes filled. Basically the bladder (1) progressively fills until the tension in its walls rises above a threshold value, at which time (2) a nervous reflex called the "micturition reflex" occurs that either causes micturition or, if it fails in this, at least causes a conscious desire to urinate.

Physiologic Anatomy of the Bladder and Its Nervous Connections

The urinary bladder, which is illustrated in Figure 26–6, is a smooth muscle chamber composed of three principal parts: (1) the *body,* which comprises mainly the *detrusor muscle,* (2) the *trigone,* a small triangular area near the neck of the bladder through which both the *ureters* and the *urethra* pass, and (3) the *bladder neck,* which is also called the *posterior urethra.*

During bladder expansion the body of the bladder stretches, and during micturition the detrusor muscle contracts to empty the bladder.

Each ureter enters the bladder through its posterolateral margin, coursing obliquely through the detrusor muscle for 1 to 2 cm. underneath the bladder mucosa before emptying at the upper corner of the trigone.

The bladder neck is 2 to 3 cm. long, and its wall is composed of detrusor muscle interlaced with a large amount of elastic tissue. The muscle in this area is frequently called the *internal sphincter,* and its natural tone prevents emptying of the bladder until the pressure in the body of the bladder rises above a critical threshold.

Beyond the bladder neck, the urethra passes through the *urogenital diaphragm,* which contains a layer of muscle called the *external sphincter* of the bladder. This muscle is a voluntary skeletal muscle, in contrast to the muscle of the bladder body and bladder neck, which is entirely smooth muscle. This external muscle is under voluntary control of the nervous system and can be used to prevent urination even when the involuntary controls are attempting to empty the bladder.

Transport of Urine Through the Ureters

The ureters are small smooth muscle tubes that originate in the pelves of the two kidneys and pass downward to enter the bladder. Each ureter is innervated by both sympathetic and parasympathetic nerves, and each also has an intramural plexus of neurons and nerve fibers that extends along its entire length.

As urine collects in the kidney pelvis, the pressure in the pelvis increases and initiates a peristaltic contraction that spreads downward along the ureter to force urine toward the bladder. The peristaltic wave can move urine against an obstruction with a pressure as high as 25 to 50 mm. Hg. Transmission of the peristaltic wave is probably caused mainly by nerve impulses passing along the intramural plexus in the same manner that the intramural plexus functions in the gut.

At the lower end, the ureter penetrates the bladder obliquely through the *trigone,* as illustrated in Figure 26–6. The ureter courses for several centimeters under the bladder mucosa, so that pressure in the bladder compresses the ureter, thereby preventing backflow of urine when pressure builds up in the bladder during micturition.

Tone of the Bladder Wall, and the Cystometrogram During Bladder Filling

The solid curve of Figure 26–7 is called the *cystometrogram* of the bladder. It shows the changes in intravesical pressure as the bladder fills with urine. When no urine at all is in the bladder, the intravesical pressure is approximately zero, but by the time 100 ml. of urine has collected, the pressure will have risen to 5 to 10 cm. water. Additional urine up to 300 to 400 ml. can collect with only a small amount of additional rise in pressure; this is caused by *intrinsic tone* of the bladder wall itself. Beyond 300 to 400 ml.,

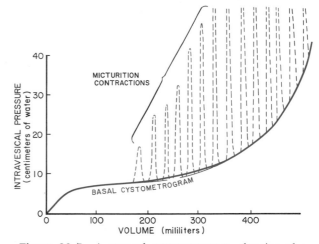

Figure 26–7. A normal cystometrogram showing also acute pressure waves (the dashed curves) caused by micturition reflexes.

collection of more urine causes the pressure to rise very rapidly.

Superimposed on the tonic pressure changes during filling of the bladder are periodic acute increases in pressure, which last from a few seconds to more than a minute. The pressure can rise only a few centimeters of water or it can rise to over 100 cm. water. These are *micturition waves* in the cystometrogram caused by the micturition reflex, which is discussed below.

THE MICTURITION REFLEX

Referring again to Figure 26–7, one sees that as the bladder fills, many superimposed *micturition contractions* begin to appear. These are the result of a stretch reflex initiated by stretch receptors in the bladder wall. Sensory signals are conducted to the sacral segments of the cord through the pelvic nerves and then back again to the bladder through the parasympathetic fibers in these same nerves.

Once a micturition reflex begins, it is "self-regenerative." That is, initial contraction of the bladder further activates the receptors, causing still further increase in afferent impulses from the bladder, which causes further increase in reflex contraction of the bladder, the cycle thus repeating itself again and again until the bladder has reached a strong degree of contraction. Then, after a few seconds to more than a minute, the reflex begins to fatigue, and the regenerative cycle of the micturition reflex ceases, allowing rapid reduction in bladder contraction. In other words, the micturition reflex is a single complete cycle of (a) progressive and rapid increase in pressure, (b) a period of sustained pressure, and (c) return of the pressure to the basal tonic pressure of the bladder. Once a micturition re-

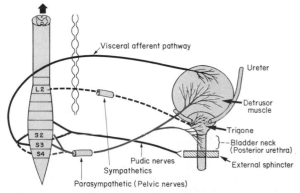

Figure 26–6. The urinary bladder and its innervation.

flex has occurred and has not succeeded in emptying the bladder, the nervous elements of this reflex usually remain in an inhibited state for at least a few minutes to sometimes as long as an hour or more before another micturition reflex occurs. However, as the bladder becomes more and more filled, micturition reflexes occur more and more often and more and more powerfully.

Once the micturition reflex becomes powerful enough and the fluid pressure in the bladder great enough to force the bladder neck open despite the tonic contraction of the bladder neck muscle, stretch of the neck then causes still another reflex. This reflex passes to the sacral portion of the spinal cord and then back to the external sphincter to inhibit it. If this inhibition is more potent than the voluntary constrictor signals from the brain, then urination will occur. If not, urination still will not occur until the bladder fills still more and the micturition reflex becomes more powerful.

Control of Micturition by the Brain. The micturition reflex is a completely automatic cord reflex, but it can be inhibited or facilitated by centers in the brain. These include (a) strong *facilitatory and inhibitory centers in the brain stem,* probably located in the pons, and (b) several *centers located in the cerebral cortex* that are mainly inhibitory but can at times become excitatory.

The micturition reflex is the basic cause of micturition, but the higher centers normally exert final control of micturition by the following means:

1. The higher centers keep the micturition reflex partially inhibited all the time except when it is desired to micturate.

2. The higher centers prevent micturition, even if a micturition reflex occurs, by continual tonic contraction of the external urinary sphincter until a convenient time presents itself.

3. When the time to urinate arrives, the cortical centers can (a) facilitate the sacral micturition centers to initiate a micturition reflex and (b) inhibit the external urinary sphincter so that urination can occur.

REFERENCES

Acid-Base Balance

Arruda, J. A. L., and Kurtzman, N. A.: Relationship of renal sodium and water transport to hydrogen ion secretion. *Annu. Rev. Physiol., 40*:43, 1978.
Brenner, B. M., and Stein, J. H. (eds.): Acid-Base and Potassium Homeostasis. New York, Churchill Livingstone, 1978.
Goldberger, E.: A Primer of Water, Electrolyte, and Acid-Base Syndromes. Philadelphia, Lea & Febiger, 1980.
Guder, W. G., and Schmidt, U. (eds.): Biochemical Nephrology. Bern, H. Huber, 1978.
Jones, N. L.: Blood Gases and Acid-Base Physiology. New York, B. C. Decker, 1980.

Lennon, E. J.: Body buffering mechanisms. *In* Frohlich, E. D. (ed.): Pathophysiology, 2nd Ed. Philadelphia, J. B. Lippincott Co., 1976, p. 287.
Malnic, G., and Giebisch, G.: Symposium on acid-base homeostasis. Mechanism of renal hydrogen ion secretion. *Kidney Int., 1*:280, 1972.
Pitts, R. F.: Production and excretion of ammonia in relation to acid-base regulation. *In* Orloff, F., and Berliner, R. W. (eds.): Handbook of Physiology. Sec. 8. Baltimore, Williams & Wilkins, 1973, p. 455.
Quintero, J. A.: Acid-Base Balance: A Manual for Clinicians. St. Louis, W. H. Green, 1979.
Rector, F. C., Jr.: Acidification of the urine. *In* Orloff, F., and Berliner, R. W. (eds.): Handbook of Physiology. Sec. 8. Baltimore, Williams & Wilkins, 1973, p. 431.
Robinson, J. R.: Fundamentals of Acid-Base Regulation, 4th Ed. Philadelphia, J. B. Lippincott Co., 1972.
Steinmetz, P. R.: Cellular mechanisms of urinary acidification. *Physiol. Rev., 54*:890, 1974.
Tannen, R. L.: Control of acid excretion by the kidney. *Annu. Rev. Med., 31*:35, 1980.
Warnock, D., and Rector, F.: Protein secretion by the kidney. *Annu. Rev. Physiol., 41*:197, 1979.

Renal Disease and Micturition

Bennett, W. M., *et al.*: Drugs and Renal Disease. New York, Churchill Livingstone, 1978.
Bissada, N. K., and Finkbeiner, A. E.: Lower Urinary Tract Function and Dysfunction. New York, Appleton-Century-Crofts, 1978.
Burg, M., and Stoner, L.: Renal tubular chloride transport and the mode of action of some diuretics. *Annu. Rev. Physiol., 38*:37, 1976.
Chapman, A. (ed.): Acute Renal Failure. New York, Churchill Livingstone, 1979.
Diamond, L. H., and Balow, J. E. (eds.): Nephrology Reviews, 1980. New York, John Wiley & Sons, 1980.
Drukker, W., *et al.* (eds.): Replacement of Renal Function by Dialysis. Boston, M. Nijhoff Medical Division, 1979.
Earley, L. E., and Gottschalk, C. W. (eds.): Strauss and Welt's Diseases of the Kidney, 3rd Ed. Boston, Little, Brown, 1979.
Goldberg, M.: The renal physiology of diuretics. *In* Orloff, F., and Berliner, R. W. (eds.): Handbook of Physiology. Sec. 8. Baltimore, Williams & Wilkins, 1973, p. 1003.
Kim, Y., and Michael, A. F.: Idiopathic membranoproliferative glomerulonephritis. *Annu. Rev. Med., 31*:273, 1980.
Kincaid-Smith, P., *et al.* (eds.): Progress in Glomerulonephritis. New York, John Wiley & Sons, 1979.
Kirschenbaum, M. A.: Renal Disease. Boston, Houghton Mifflin, 1978.
Knox, F. G. (ed.): Textbook of Renal Pathophysiology. Hagerstown, Md., Harper & Row, 1978.
Leaf, A., and Cotran, R. S.: Renal Pathophysiology, 2nd Ed. New York, Oxford University Press, 1980.
Papper, S.: Clinical Nephrology, 2nd Ed. Boston, Little, Brown, 1978.
Ritz, E., *et al.* (eds.): Pathophysiological Problems in Clinical Nephrology. New York, S. Karger, 1978.
Strauss, J. (ed.): Nephrotic Syndrome. New York, Garland Press, 1979.
Thurau, K. (ed.): Experimental Acute Renal Failure. New York, Springer-Verlag, 1976.
Wardle, E. N.: Renal Medicine. Baltimore, University Park Press, 1979.
Weller, J. M. (ed.): Fundamentals of Nephrology. Hagerstown, Md., Harper & Row, 1979.
Wilson, C. B. (ed.): Immunological Mechanisms of Renal Disease. New York, Churchill Livingstone, 1979.
Wong, N. L. M., *et al.*: Chronic reduction in renal mass. Micropuncture studies of response to volume expansion and furosemide. *Yale J. Biol. Med., 51*(3):289, 1978.
Yoshitoshi, Y. (ed.): Glomerulonephritis. Baltimore, University Park Press, 1979.

Part VII

RESPIRATION

27

Pulmonary Ventilation, and Physical Principles of Gaseous Exchange

The process of respiration can be divided into four major mechanistic events: (1) pulmonary ventilation, which means the inflow and outflow of air between the atmosphere and the lung alveoli, (2) diffusion of oxygen and carbon dioxide between the alveoli and the blood, (3) transport of oxygen and carbon dioxide in the blood and body fluids to and from the cells, and (4) regulation of ventilation and other facets of respiration. The present chapter and the two following discuss these four major aspects of respiration. In a fourth chapter the special respiratory problems related to aviation medicine and deep sea diving physiology are discussed to illustrate some of the basic principles of respiratory physiology.

MECHANICS OF PULMONARY VENTILATION

BASIC MECHANISMS OF LUNG EXPANSION AND CONTRACTION

The lungs can be expanded and contracted in two ways, (1) by downward and upward movement of the diaphragm to lengthen or shorten the chest cavity and (2) by elevation and depression of the ribs to increase and decrease the anteroposterior diameter of the chest cavity. Figure 27–1 illustrates these two methods.

Normal quiet breathing is accomplished almost entirely by inspiratory movement of the diaphragm. During inspiration the diaphragm pulls the lower surfaces of the lungs downward. Then, during expiration, the diaphragm simply relaxes

and the elastic recoil of the lungs, chest wall, and abdominal structures compresses the lungs. During heavy breathing, however, the elastic forces are not powerful enough to cause the necessary rapid expiration, so this is achieved by contraction of the abdominal muscles, which forces the abdominal contents upward against the bottom of the diaphragm.

The second method of expanding the lungs is to raise the rib cage. This expands the lungs because in the natural resting position, the ribs slant downward, thus allowing the sternum to fall backward toward the spinal column. But when the rib cage is elevated, the ribs project directly foward, so that the sternum now also moves forward away from the spine, making the an-

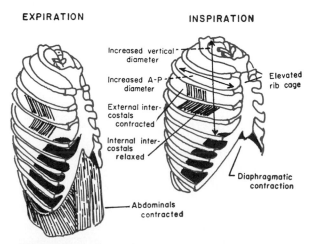

Figure 27–1. Expansion and contraction of the thoracic cage during expiration and inspiration, illustrating especially diaphragmatic contraction, elevation of the rib cage, and function of the intercostals.

teroposterior thickness of the chest about 20 per cent greater during maximum inspiration than during expiration. Therefore, those muscles that elevate the chest cage can be classified as muscles of inspiration, and those muscles that depress the chest cage, as muscles of expiration.

RESPIRATORY PRESSURES

Intra-alveolar Pressure. The respiratory muscles cause pulmonary ventilation by alternately compressing and distending the lungs, which in turn causes the pressure in the alveoli to rise and fall. During inspiration the intra-alveolar pressure becomes slightly negative *with respect to atmospheric pressure,* normally slightly less than -1 mm. Hg, and this causes air to flow inward through the respiratory passageways. During normal expiration, on the other hand, the intra-alveolar pressure rises to slightly less than $+1$ mm. Hg, which causes air to flow outward through the respiratory passageways. Note especially how little pressure is required to move air into and out of the normal lung, an effect that often is seriously compromised in many lung diseases.

During maximum expiratory effort with the glottis closed, the intra-alveolar pressure can be increased to over 100 mm. Hg in the strong healthy male, and during maximum inspiratory effort it can be reduced to as low as -80 mm. Hg.

Recoil Tendency of the Lungs, and the Intrapleural Pressure. The lungs have a continuous elastic tendency to collapse and therefore to recoil away from the chest wall. This elastic tendency is caused by two different factors. First, throughout the lungs are many *elastic fibers* that are stretched by lung inflation and therefore attempt to shorten. Second, and even more important, the *surface tension* of the fluid lining the alveoli also causes a continuous elastic tendency for the alveoli to collapse. This effect is caused by intermolecular attraction between the surface molecules of the fluid that tends continuously to reduce the surface areas of the individual alveoli; all these minute forces added together tend to collapse the whole lung and therefore to cause its recoil away from the chest wall.

Ordinarily, the elastic fibers in the lungs account for about one-third of the recoil tendency, and the surface tension phenomenon accounts for about two-thirds.

The total recoil tendency of the lungs can be measured by the amount of negative pressure in the intrapleural spaces required to prevent collapse of the lungs, and this pressure is called the *intrapleural pressure* or, occasionally, the *recoil pressure.* It is normally about -4 mm. Hg. That is,

when the alveolar spaces are open to the atmosphere through the trachea so that their pressure is at atmospheric pressure, a pressure of -4 mm. Hg in the intrapleural space is required to keep the lungs expanded to normal size. When the lungs are stretched to very large size, such as at the end of deep inspiration, the intrapleural pressure required then to expand the lungs may be as great as -12 to -18 mm. Hg.

"Surfactant" in the Alveoli, and Its Effect on the Collapse Tendency. A lipoprotein mixture called "surfactant" is secreted by special *surfactant-secreting cells* that are component parts of the alveolar epithelium. This mixture, containing especially the phospholipid *dipalmitoyl lecithin,* decreases the surface tension of the fluids lining the alveoli. In the absence of surfactant, lung expansion is extremely difficult, often requiring intrapleural pressures as much as -15 to -20 mm. Hg to overcome the tendency of the alveoli to collapse. This illustrates that surfactant is exceedingly important in minimizing the effect of surface tension in causing collapse of the lungs.

A few newborn babies, especially premature babies, do not secrete adequate quantities of surfactant, which makes lung expansion difficult. Without immediate and very careful treatment, most of them die soon after birth because of inadequate ventilation. This condition is called *hyaline membrane disease* or *respiratory distress syndrome.*

Surfactant acts by forming a layer at the interface between the fluid lining the alveoli and the air in the alveoli. This prevents the development of a water-air interface, which has 2 to 14 times as much surface tension as the surfactant-air interface.

Surfactant has a special property of decreasing the surface tension more as the alveoli become smaller, which nullifies some of the collapse tendency of the alveoli as they become smaller. Consequently, surfactant is very important in maintaining the equality of size of the alveoli — the large alveoli develop greater surface tension and therefore contract, while the smaller alveoli develop less surface tension and therefore tend to enlarge.

EXPANSIBILITY OF THE LUNGS AND THORAX: "COMPLIANCE"

Both the lungs and the thorax are viscoelastic structures. The elastic properties of the lungs, as pointed out previously, are caused, first, by the surface tension of the fluids lining the alveoli and, second, by elastic fibers throughout the lung tissue itself. The elastic properties of the thorax are caused by the natural elasticity of the muscles, tendons, and connective tissue of the chest. Therefore, part of the effort expended by the inspiratory muscles during breathing is simply to stretch the elastic structures of the lungs and thorax.

The expansibility of the lungs and thorax is called *compliance.* This is expressed as the *volume increase in the lungs for each unit increase in intra-alveolar pressure.* The compliance of the normal lungs and thorax combined is 0.13 liter per centimeter of water pressure. That is, every time the alveolar pressure is increased by 1 cm. water, the lungs expand 130 ml.

Compliance of the Lungs Alone. The lungs alone, when removed from the chest, are almost twice as distensible as the lungs and thorax together, because the thoracic cage itself must also be stretched when the lungs are expanded *in situ.* Thus, the compliance of the normal lungs when removed from the thorax is about 0.22 liter per cm. water. This illustrates that the muscles of inspiration must expend energy not only to expand the lungs but also to expand the thoracic cage around the lungs.

Measurement of Lung Compliance. Compliance of the lungs is measured in the following way: The person's glottis must first be completely open and remain so. Then air is inspired in steps of approximately 50 to 100 ml. at a time, and pressure measurements are made from an intra-esophageal balloon (which measures almost exactly the intrapleural pressure) at the end of each step, until the total volume of air in the lungs is equal to the normal tidal volume of the person. Then the air is expired, also in steps, until the lung volume returns to the expiratory resting level. The relationship of lung volume to pressure is then plotted as illustrated in Figure 27–2.

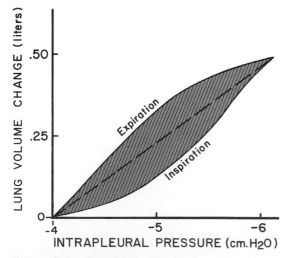

Figure 27–2. Compliance diagram in a normal person. This diagram shows the compliance of the lungs alone.

This graph shows that the plot during inspiration is a different curve from that during expiration, as a result of the viscous properties of the lungs. The average compliance is represented by the dashed line in the figure. Thus, the lung volume increases 220 ml. for a change in trans-lung pressure (atmospheric pressure in the lung minus intra-esophageal pressure) of 1 cm. water. Therefore, the compliance in this instance is 0.22 liter per cm. water.

Factors That Cause Abnormal Compliance. Any condition that destroys lung tissue, causes it to become fibrotic or edematous, blocks the bronchioles, or in any other way impedes lung expansion and contraction causes decreased lung compliance. When considering the compliance of both the lung and thorax together, one must also include any abnormality that reduces the expansibility of the thoracic cage. Thus, deformities of the chest cage, such as kyphosis, severe scoliosis, and other restraining conditions such as fibrotic pleurisy or paralyzed and fibrotic muscles, can all reduce the expansibility of the lungs and thereby reduce the total pulmonary compliance.

THE "WORK" OF BREATHING

We have already pointed out that during normal quiet respiration, respiratory muscle contraction occurs only during inspiration, while expiration is entirely a passive process caused by elastic recoil of the lung and chest cage structures. Therefore, the respiratory muscles normally perform work only to cause inspiration and not to cause expiration.

The work of inspiration can be divided into three different fractions: (1) that required to expand the lungs against its elastic forces, called *compliance work,* (2) that required to overcome the viscosity of the lung and chest wall structures, called *tissue resistance work,* and (3) that required to overcome airway resistance during the movement of air into the lungs, called *airway resistance work.*

Energy Required for Respiration. During normal quiet respiration, only 2 to 3 per cent of the total energy expended by the body is required to energize the pulmonary ventilatory process. During very heavy exercise, the absolute amount of energy required for pulmonary ventilation can increase as much as 25-fold. However, this still does not represent a significant increase in *percentage* of total energy expenditure, because the total energy release in the body increases at the same time as much as 15- to 20-fold. Thus, even in heavy exercise only 3 to 4 per cent of the total energy expended is used for ventilation.

On the other hand, pulmonary diseases that decrease the pulmonary compliance, that increase airway resistance, or that increase the viscosity of the lung or chest wall can at times increase the work of breathing so much that one-third or more of the total energy expended by the body is for respiration alone. Such respiratory diseases can proceed to such a point that this excess work load alone is the cause of death.

THE PULMONARY VOLUMES AND CAPACITIES

A simple method for studying pulmonary ventilation is to record the volume movement of air into and out of the lungs, a process called *spirometry*. A typical spirometer is illustrated in Figure 27–3. This consists of a drum inverted over a chamber of water, with the drum counterbalanced by a weight. In the drum is a breathing mixture of gases, usually air or oxygen; a tube connects the mouth with the gas chamber. When one breathes into and out of the chamber the drum rises and falls, and an appropriate recording is made on a moving sheet of paper.

Figure 27–4 illustrates a "spirogram" showing changes in lung volume under different conditions of breathing. For ease in describing the events of pulmonary ventilation, the air in the lungs has been subdivided at different points on this diagram into four different *volumes* and four different *capacities*, which are as follows:

THE PULMONARY "VOLUMES"

To the left in Figure 27–4 are listed four different pulmonary lung "volumes," which when added together equal the maximum volume to which the lungs can be expanded. The significance of each of these volumes is the following:

1. The *tidal volume* is the volume of air inspired or expired with each normal breath, and it amounts to about 500 ml.

2. The *inspiratory reserve volume* is the extra volume of air that can be inspired in addition to the normal tidal volume, and it is usually equal to approximately 3000 ml.

3. The *expiratory reserve volume* is the amount of air that can still be expired by forceful expiration after the end of a normal tidal expiration; it normally amounts to about 1100 ml.

4. The *residual volume* is the volume of air still remaining in the lungs after the most forceful expiration. This volume averages about 1200 ml.

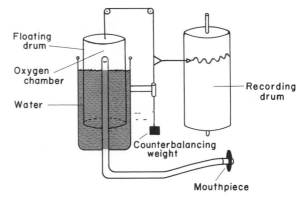

Figure 27–3. A spirometer.

THE PULMONARY "CAPACITIES"

In describing events in the pulmonary cycle, it is sometimes desirable to consider two or more of the above volumes together. Such combinations are called pulmonary capacities. To the right in Figure 27–4 are listed the different pulmonary capacities, which can be described as follows:

1. The *inspiratory capacity* equals the *tidal volume* plus the *inspiratory reserve volume*. This is the amount of air (about 3500 ml.) that a person can breathe beginning at the normal expiratory level and distending his lungs to the maximum amount.

2. The *functional residual capacity* equals the *expiratory reserve volume* plus the *residual volume*. This is the amount of air remaining in the lungs at the end of normal expiration (about 2300 ml.).

3. The *vital capacity* equals the *inspiratory reserve volume* plus the *tidal volume* plus the *expiratory reserve volume*. This is the maximum amount of air that a person can expel from the lung after first filling the lungs to their maximum extent and then expiring to the maximum extent (about 4600 ml.).

4. The *total lung capacity* is the maximum volume to which the lungs can be expanded with the greatest possible inspiratory effort (about 5800 ml.).

All pulmonary volumes and capacities are about 20 to 25 per cent less in the female than in the male, and they obviously are greater in large and athletic persons than in small and asthenic persons.

Resting Expiratory Level. Normal pulmonary ventilation is accomplished almost entirely by the muscles of inspiration. On relaxation of the inspiratory muscles, the elastic properties of the lungs and thorax cause the lungs to contract passively. Therefore, when all inspiratory muscles are completely relaxed the lungs return to a relaxed state, called the *resting expiratory level*. The volume of air in the lungs at this level is equal to

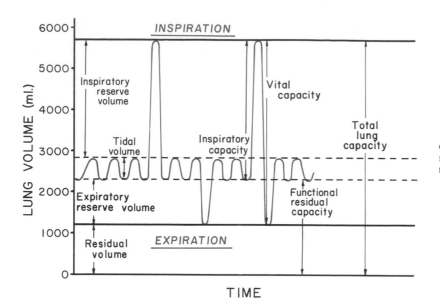

Figure 27–4. Respiratory excursions during normal breathing and during maximal inspiration and maximal expiration.

the functional residual capacity, or about 2300 ml. in the young adult.

Significance of the Residual Volume. The residual volume represents the air that cannot be removed from the lungs even by forceful expiration. This is important because it provides air in the alveoli to aerate the blood even between breaths. Were it not for this residual air, the concentrations of oxygen and carbon dioxide in the blood would rise and fall markedly with each respiration, which would certainly be disadvantageous to the respiratory process.

Significance of the Vital Capacity. Other than the anatomical build of a person, the major factors that affect vital capacity are (1) the position of the person during the vital capacity measurement, (2) the strength of the respiratory muscles, and (3) the distensibility of the lungs and chest cage, which is called "pulmonary compliance."

The average vital capacity in the young adult male is about 4.6 liters, and in the young adult female about 3.1 liters, though these values are much greater in some persons of the same weight than in others.

Paralysis of the respiratory muscles, which occurs often following spinal cord injuries or poliomyelitis, can cause a great decrease in vital capacity, to as low as 500 to 1000 ml. — barely enough to maintain life — or even to zero, in which case death ensues. And such conditions as tuberculosis, emphysema, chronic asthma, lung cancer, chronic bronchitis, and fibrotic pleurisy can all reduce the pulmonary compliance and thereby greatly decrease the vital capacity. For this reason vital capacity measurements are among the most important of all clinical respiratory measurements for assessing the progress of different types of diseases.

THE MINUTE RESPIRATORY VOLUME — RESPIRATORY RATE AND TIDAL VOLUME

The *minute respiratory volume* is the total amount of new air breathed each minute; this is equal to the *tidal volume* × the *respiratory rate*. The normal tidal volume is about 500 ml., and the normal respiratory rate is approximately 12 breaths per minute. Therefore, the *minute respiratory volume averages about 6 liters per minute*. A person can live for short periods of time with a minute respiratory volume as low as 1.5 liters per minute and with a respiratory rate as low as two to four breaths per minute.

The respiratory rate occasionally rises to as high as 40 to 50 per minute, and the tidal volume can become as great as the vital capacity, about 4600 ml. in the young adult male. However, at rapid breathing rates, a person usually cannot sustain a tidal volume greater than about one-half the vital capacity. Combining these factors, a young male adult has a *maximum breathing capacity* of about 100 to 120 liters per minute.

Maximum Expiratory Flow

When a person expires with considerable force, the expiratory air flow reaches a maximum rate despite still further increase in expiratory force. This effect can be explained by reference to Figure 27–5A. When pressure is applied to the lungs by chest cage compression, the same amount of pressure is applied at the same time to the outsides of both the alveoli and the respiratory passageways, as indicated by the arrows. Therefore, not only is the pressure increased in the alveoli to force air to the exterior, but the terminal bronchioles are collapsed at the same time, which

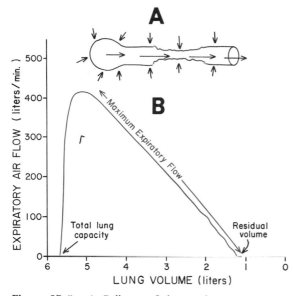

Figure 27–5. *A*, Collapse of the respiratory passageway during maximum expiratory effort, an effect that limits the expiratory flow rate. *B*, Effect of lung volume on the maximum expiratory air flow, showing decreasing maximum expiratory air flow as the lung volume becomes smaller.

increases the airway resistance. Beyond a certain expiratory effort these two effects have equal but opposing results on air flow, thus preventing further increase in flow.

The curve recorded in Figure 27–5B is the expiratory flow achieved by a normal person who first inhales as much air as possible and then expires with maximum expiratory effort until he can expire no more. Note that he quickly reaches an expiratory air flow of over 400 liters/min. But it does not matter how much additional expiratory effort he exerts; this is the *maximum expiratory flow* that he can achieve.

Note also that as the lung volume becomes smaller this maximum expiratory flow also be-

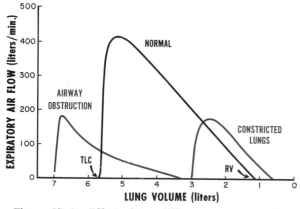

Figure 27–6. Effect of two different respiratory abnormalities — constricted lungs and airway obstruction — on the maximum expiratory flow-volume curve.

comes less. The main reason for this is that in the enlarged lung the bronchi are held open partially by means of elastic pull on their outsides by lung structural elements; however, as the lung becomes smaller, these structures are relaxed so that the bronchi collapse more easily.

Abnormalities of the Maximum Expiratory Flow-Volume Curve. The maximum expiratory flow-volume curve is often recorded in clinical pulmonary laboratories to determine abnormalities of pulmonary ventilation. Figure 27–6 illustrates the normal curve and curves recorded in two different types of lung diseases: (1) constriction of the lungs as occurs in tuberculosis and (2) airway obstruction as occurs in asthma.

VENTILATION OF THE ALVEOLI

The truly important factor of the entire pulmonary ventilatory process is the rate at which the air in the gas exchange area of the lungs, the alveoli, is renewed each minute by atmospheric air; this is called *alveolar ventilation*. One can readily understand that alveolar ventilation per minute is not equal to the minute respiratory volume, because a large portion of the inspired air goes to fill the larger respiratory passageways, called the dead space, where the membranes are not capable of significant gaseous exchange with the blood.

THE DEAD SPACE

The air that goes to fill the respiratory passages with each breath is called *dead space air*. On inspiration, much of the new air must first fill the different dead space areas — the nasal passageways, the pharynx, the trachea, and the bronchi — before any reaches the alveoli. Then, on expiration, all the air in the dead space is expired first before any of the air from the alveoli reaches the atmosphere. The normal dead space air in the young adult is about 150 ml. This increases slightly with age.

Anatomical versus Physiological Dead Space. The dead space just discussed represents the volume of air in all the major respiratory passageways but does not include air in any of the alveoli, and is called the *anatomical dead space*. On occasion, however, some of the alveoli themselves are not functional or are only partially functional because of absent or poor blood flow through adjacent pulmonary capillaries, and therefore they must be considered also to be dead space. When the alveolar dead space is included in the total dead space the space is called *physiological dead space*, in contradistinction to the anatomical dead space. In the normal person, the anatomical and the physiological dead spaces are nearly equal because all alveoli are functional in the normal lung, but in persons with partially functional or nonfunctional alveoli in some parts of the lungs, the physiological dead space is sometimes as much as 10 times the anatomical dead space, or as much as 1 to 2 liters. This problem will be discussed further in the following chapter.

RATE OF ALVEOLAR VENTILATION

Alveolar ventilation per minute is the total volume of new air entering the alveoli each minute. It is equal to the respiratory rate times the amount of new air that enters the alveoli with each breath; that is, *respiratory rate* times *the difference between tidal volume and dead space volume.* Thus, the normal alveolar ventilation is 12 times 350 ml., or 4200 ml./min.

Alveolar ventilation is one of the major factors determining the concentrations of oxygen and carbon dioxide in the alveoli. Therefore, almost all discussions of gaseous exchange problems in the following chapters emphasize alveolar ventilation. *The respiratory rate, the tidal volume, and the minute respiratory volume are of importance only insofar as they affect alveolar ventilation.*

FUNCTIONS OF THE RESPIRATORY PASSAGEWAYS

FUNCTIONS OF THE NOSE

As air passes through the nose, three distinct functions are performed by the nasal cavities: First, the *air is warmed* by the extensive surfaces of the turbinates and septum, which are illustrated in Figure 27–7. Second, the *air is moistened* to a considerable extent even before it passes beyond the nose. Third, the *air is filtered* by the hairs and much more so by precipitation of particles on the turbinates. All these functions together are called the *air conditioning function* of the upper respiratory passageways. Ordinarily, the air rises to within 2 to 3 per cent of body temperature and within 2 to 3 per cent of full saturation with water vapor before it reaches the lower trachea. When a person breathes air through a tube directly into his trachea (as through a tracheostomy), the cooling, and especially the drying, effect in the lower lung can lead to lung infection.

THE COUGH REFLEX

The cough reflex is essential to life, for the cough is the means by which the passageways of the lungs are maintained free of foreign matter.

The bronchi and the trachea are so sensitive that any foreign matter or other cause of irrita-

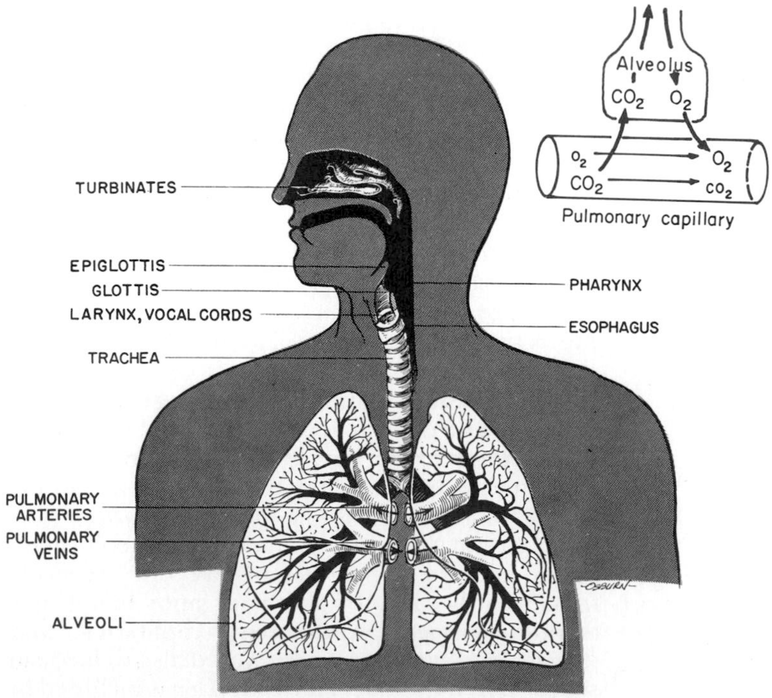

Figure 27–7. The respiratory passages.

tion initiates the cough reflex. Afferent impulses pass from the respiratory passages mainly through the vagus nerve to the medulla. There, an automatic sequence of events is triggered by the neuronal circuits of the medulla, causing the following effects:

First, about 2.5 liters of air is inspired. Second, the epiglottis closes, and the vocal cords shut tightly to entrap the air within the lungs. Third, the abdominal muscles contract forcibly, pushing against the diaphragm while other expiratory muscles also contract forcibly. Consequently, the pressure in the lungs rises to as high as 100 mm. Hg. or more. Fourth, the vocal cords and the epiglottis suddenly open widely, so that air under pressure in the lungs *explodes* outward. Indeed, this air is sometimes expelled at velocities as high as 75 to 100 miles an hour. Furthermore, and very important, the strong compression of the lungs also collapses the bronchi and trachea (the noncartilaginous part of the trachea invaginating inward), so that the exploding air actually passes through *bronchial* and *tracheal slits*. The rapidly moving air usually carries with it any foreign matter that is present in the bronchi or trachea.

ACTION OF THE CILIA TO CLEAR RESPIRATORY PASSAGEWAYS

In addition to the cough mechanism, the respiratory passageways of the trachea and lungs are lined with a ciliated, mucus-coated epithelium that aids in clearing the passages, for the cilia beat toward the pharynx and move the mucus as a continually flowing sheet. Thus, small foreign particles and mucus are mobilized at a velocity of as much as a centimeter per minute along the surface of the trachea toward the pharynx. Foreign matter in the nasal passageways is also moved by cilia toward the pharynx.

VOCALIZATION

Speech involves the respiratory system particularly, but it also involves (1) specific speech control centers in the cerebral cortex, which will be discussed in Chapter 36, (2) respiratory centers of the brain stem, and (3) the articulation and resonance structures of the mouth and nasal cavities. Basically, speech is composed of two separate mechanical functions: (1) *phonation*, which is achieved by the larynx, and (2) *articulation*, which is achieved by the structures of the mouth.

Phonation. The larynx is specially adapted to act as a vibrator. The vibrating element is the *vocal cords,* which are folds along the lateral walls of the larynx that are stretched and positioned by several specific muscles within the confines of the larynx itself.

Figure 27–8A illustrates the basic structure of the larynx, showing that each vocal cord is stretched between the *thyroid cartilage* and an *arytenoid cartilage*. The specific muscles within the larynx that position and control the degree of stretch of the vocal cords are also shown.

Vibration of the Vocal Cords. One might suspect that the vocal cords would vibrate in the direction of the flowing air. However, this is not the case. Instead, they vibrate laterally. The cause of the vibration is the following: When the vocal cords are closed together and air is expired, pressure of the air from below pushes the vocal cords apart, which allows rapid flow of air between their margins. The rapid flow of air then immediately creates a partial vacuum between the vocal cords, which pulls them once again toward each other. This stops the flow of air, pressure builds up behind the cords, and the cords open once more, thus continuing in a vibratory pattern.

The pitch of the sound emitted by the larynx can be changed in two different ways: first, *stretching or relaxing the vocal cords;* second, *changing the shape and mass of the vocal cord edges*. When very high frequency sounds are to be emitted, slips of the muscle in the vocal cords are contracted in such a way that the edges of the cords are sharpened and thinned, whereas when bass frequencies are to be emitted, the muscles are contracted in a different pattern, so that the edges are greatly thickened. Figure 27–8B shows some of the positions and shapes of the vocal cords during different types of phonation.

Articulation and Resonance. The three major organs of articulation are the *lips,* the *tongue*, and the *soft palate*. These need not be

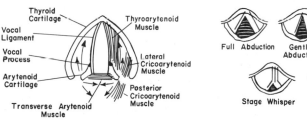

Figure 27–8. Position of vocal cords in action. Laryngoscopic view of vocal cords. (Modified from Greene: The Voice and Its Disorders. 4th ed. Philadelphia, J. B. Lippincott Company, 1981.)

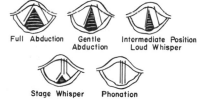

discussed in detail because all of us are familiar with their movements during speech and other vocalizations.

The resonators include the *mouth,* the *nose and associated nasal sinuses,* the *pharynx,* and even the *chest cavity* itself. Here again we are all familiar with the resonating qualities of these different structures. For instance, the function of the nasal resonators is illustrated by the change in quality of the voice when a person has a severe cold.

PHYSICAL PRINCIPLES OF GASEOUS EXCHANGE

After the alveoli are ventilated with fresh air, the next step in the respiratory process is *diffusion* of oxygen from the alveoli into the pulmonary blood and diffusion of carbon dioxide in the opposite direction — from the pulmonary blood into the alveoli.

All the gases that are of concern in respiratory physiology are simple molecules that are free to move among each other, which is the process called "diffusion." This is also true of the gases dissolved in the fluids and tissues of the body.

However, for diffusion to occur, there must be a source of energy. This is provided by the kinetic motion of the molecules themselves. That is, except at absolute zero temperature, all molecules of all matter are continuously undergoing some type of motion. For free molecules that are not physically attached to others, this means linear movement of the molecules at high velocity until they strike other molecules. Then they bounce away in new directions and continue again until they strike still other molecules. In this way the molecules move rapidly among each other.

Gas Pressures in a Mixture of Gases — Partial Pressures of Individual Gases

The cause of the pressure that a gas exerts against a surface is constant impaction of the kinetically moving molecules against the surface. Obviously, the greater the concentration of the gas, the greater also is the summated force of impaction of all the molecules striking the surface at any given instant. Therefore, the pressure of a gas is directly proportional to its concentration. It is also directly proportional to the average kinetic energy of the molecules, which is directly proportional to temperature. Therefore, the greater the temperature, the greater also is the pressure; but in the body, the temperature remains relatively constant at 37°C, so this usually is not a factor of major consideration in respiratory problems.

Now, let us consider the pressure exerted by each one of the gases in a mixture of gases. For instance, consider air, which has an approximate composition of 79 per cent nitrogen and 21 per cent oxygen. The total pressure of this mixture is 760 mm. Hg. A portion of this total is caused by nitrogen and another portion by oxygen. It is clear from the above description of the molecular basis of pressure that each gas contributes to the total pressure in direct proportion to its relative concentration. Therefore, 79 per cent of the 760 mm. Hg is caused by nitrogen (about 600 mm. Hg) and 21 per cent by oxygen (about 160 mm. Hg). Thus, the "partial pressure" of nitrogen in the mixture is 600 mm. Hg, and the "partial pressure" of oxygen is 160 mm. Hg, while the total pressure is 760 mm. Hg, the sum of the individual partial pressures.

The partial pressures of the individual gases in a mixture are designed by the terms Po_2, Pco_2, Pn_2, Ph_2o, Phe, and so forth.

Partial Pressure of Gases in Water and Tissues

When a gas under pressure is impressed onto a water interface, instead of bouncing back from the interface some of the molecules will move on into the water and become dissolved. However, as more and more molecules become dissolved they also begin to diffuse backward to the interface, and some escape back into the gas phase. Once the concentration of dissolved molecules reaches a certain level, the number of molecules leaving the solution to enter the gas phase becomes exactly equal to the number of molecules moving in the opposite direction from the gas into the solution. Thus, a state of *equilibrium* has occurred. In this equilibrium state the pressure of the dissolved gas is exactly equal to the pressure of the gas in the gas state, each pushing against the other at the interface with equal force. Thus, gases in solution exert pressures in exactly the same way as they do in gas phase mixtures. And the partial pressures of the separate dissolved gases are designated like those of the gases in the gaseous state, i.e., Po_2, Pco_2, Pn_2, Phe.

Factors That Determine the Concentration of a Gas Dissolved in a Fluid. The concentration of a gas in a solution is determined not only by its pressure but also by the *solubility coefficient* of the gas. That is, some types of molecules, especially carbon dioxide, are physically or chemically attracted to water molecules, while other are repelled. Obviously, when molecules are attracted, far more of them can then become dissolved without building up excess pressure within the solution. On the other hand, those that are repelled will develop excessive pressures for very small amounts of dissolved gas.

These principles can be expressed by the following formula, which is *Henry's law*:

Concentration of dissolved gas = pressure × solubility coefficient

When concentration is expressed in volumes of gas dissolved in each volume of water at zero degrees centigrade and pressure is expressed in atmospheres, the solubility coefficients of important respiratory gases at body temperature are the following:

oxygen	0.024
carbon dioxide	0.57
nitrogen	0.012

THE VAPOR PRESSURE OF WATER

All gases in the body are in direct contact with water. Therefore, all gaseous mixtures in the body are saturated with water vapor, and this must always be considered when the dynamics of gaseous exchange are discussed.

The vapor pressure of water depends entirely on the temperature. The greater the temperature, the greater is the activity of the molecules in the water, and the greater is the likelihood these molecules will escape from the surface of the water into the gaseous phase. When dry air is suddenly mixed with water, the water vapor pressure is zero at first, but water molecules immediately begin escaping from the surface of the water into the air. As the air becomes progressively more humidified, an equilibrium vapor pressure is approached at which the rate of condensation of water becomes equal to the rate of water vaporization.

The water vapor pressure at room temperature is about 20 mm. Hg. However, the most important value to remember is the vapor pressure at body temperature, 47 mm. Hg; this value will appear in many of our subsequent discussions.

Diffusion of Gases Through Liquids — The Pressure Gradient for Diffusion

Now, let us return to the problem of diffusion. From the above discussion it is already clear that when the concentration, or pressure, of a gas is greater in one area than in another area, there will be net diffusion from the high-pressure area toward the low-pressure area. For instance, in the chamber shown in Figure 27–9, one can readily see that the molecules in the area of high pressure, because of their greater number, have a greater statistical chance of moving randomly into the area of low pressure than do molecules

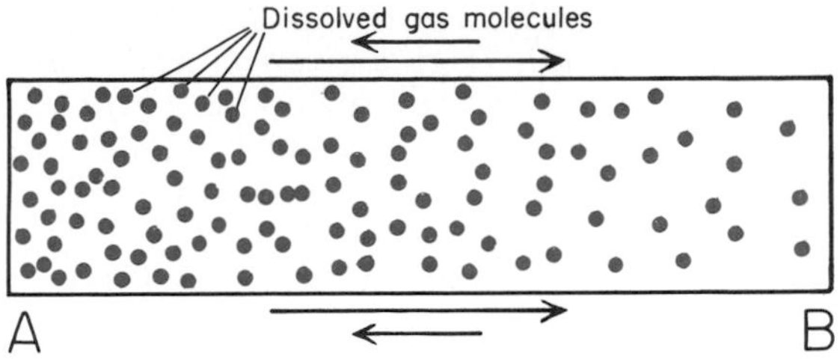

Figure 27–9. Net diffusion of oxygen from one end of a chamber to the other.

attempting to go in the other direction. However, some molecules do bounce from the area of low pressure toward the area of high pressure. Therefore, the *net diffusion* of gas from the area of high pressure to the area of low pressure is equal to the number of molecules bouncing in this direction *minus* the number bouncing in the opposite direction, and this in turn is proportional to the gas pressure difference between the two areas. The pressure in area A of Figure 27–9 minus the pressure in area B divided by the distance of diffusion is known as the *pressure gradient for diffusion* or simply the *diffusion gradient*. The rate of net gas diffusion from area A to area B is directly proportional to this gradient.

The principle of diffusion from an area of high pressure to an area of low pressure holds true for diffusion of gases in a gaseous mixture, diffusion of dissolved gases in a solution, and even diffusion of gases from the gaseous phase into the dissolved state in liquids. That is, *there is always net diffusion from areas of high pressure to areas of low pressure.*

Quantifying the Net Rate of Diffusion. In addition to the pressure difference, several other factors affect the rate of gas diffusion in a fluid. These are (1) the solubility of the gas in the fluid, (2) the cross-sectional area of the fluid, (3) the distance through which the gas must diffuse, (4) the molecular weight of the gas, and (5) the temperature of the fluid. In the body, the temperature remains reasonably constant and usually need not be considered. All of these factors can be expressed in a single formula, as follows:

$$D \propto \frac{\Delta P \times A \times S}{d \times \sqrt{MW}}$$

in which D is the diffusion rate, ΔP is the pressure difference between the two ends of the chamber, A is the cross-sectional area of the chamber, S is the solubility of the gas, d is the distance of diffusion, and MW is the molecular weight of the gas.

It is obvious from this formula that the characteristics of the gas itself determine two factors of the formula: solubility and molecular weight.

Therefore, the *diffusion coefficient* — that is, the rate of diffusion through a given area for a given distance and pressure difference — for any given gas is proportional to $S/\sqrt{MW}$. Considering the diffusion coefficient for oxygen to be 1, the *relative* diffusion coefficients for different gases of respiratory importance in the body fluids are

Oxygen	1.0
Carbon dioxide	20.3
Nitrogen	0.53

DIFFUSION OF GASES THROUGH TISSUES

The gases that are of respiratory importance are highly soluble in lipids and consequently are also highly soluble in cell membranes. Because of this, these gases diffuse through the cell membranes with very little impediment. Instead, the major limitation on the movement of gases in tissues is the rate at which the gases can diffuse through the tissue water instead of through the cell membranes. Therefore, diffusion of gases through the tissues, including through the respiratory membrane, is almost equal to the diffusion of gases through water, as given in the above list of diffusion rates for the important respiratory gases. Note especially that carbon dioxide diffuses 20 times as rapidly as oxygen.

REFERENCES

Pulmonary Ventilation

Bradley, G. W.: Control of the breathing pattern. *Int. Rev. Physiol., 14*:185, 1977.

Bryant, C.: The Biology of Respiration. Baltimore, University Park Press, 1979.

Ellis, P. D., and Billings, D. M.: Cardiopulmonary Resuscitation: Procedures for Basic and Advanced Life Support. St. Louis, C. V. Mosby, 1979.

Engel, L. A., and Macklem, P. T.: Gas mixing and distribution in the lung. *Int. Rev. Physiol., 14*:37, 1977.

Fink, B. R.: The Human Larynx: A Functional Study. New York, Raven Press, 1975.

Fishman, A. P.: Assessment of Pulmonary Function. New York, McGraw-Hill, 1980.

Guyton, A. C.: Analysis of respiratory patterns in laboratory animals. *Am. J. Physiol., 150*:78, 1947.

Jarvis, J. F.: An Introduction to the Anatomy and Physiology of Speech and Hearing. Cape Town, South Africa, Juta & Co., 1978.

Macklem, P. T.: Respiratory mechanics. *Annu. Rev. Physiol., 40*:157, 1978.

Mead, J.: Respiration: Pulmonary mechanics. *Annu. Rev. Physiol., 35*:169, 1973.

Murray, J. F.: The Normal Lung. Philadelphia, W. B. Saunders Co., 1976.

Nadel, J. A., *et al.*: Control of mucus secretion and ion transport in airways. *Annu. Rev. Physiol., 41*:369, 1979.

Phillipson, E. A.: Respiratory adaptations in sleep. *Annu. Rev. Physiol., 40*:133, 1978.

Rahn, H., *et al.*: The pressure-volume diagram of the thorax and lung. *Am. J. Physiol., 146*:161, 1946.

Said, S. I.: Metabolic functions of the lung. *In* Frohlich, E. D. (ed.): Pathophysiology, 2nd Ed. Philadelphia, J. B. Lippincott Co., 1976, p. 189.

Singh, R. P.: Anatomy of Hearing and Speech. New York, Oxford University Press, 1980.

Stuart, B. O.: Deposition of inhaled aerosols. *Arch. Intern. Med., 131*:60, 1973.

Thurlbeck, W. M.: Structure of the lungs. *Int. Rev. Physiol., 14*:1, 1977.

Tierney, D. F.: Lung metabolism and biochemistry. *Annu. Rev. Physiol., 36*:209, 1974.

West, J. B.: Respiratory Physiology. Baltimore, Williams & Wilkins, 1974.

White, F. N.: Comparative aspects of vertebrate cardiorespiratory physiology. *Annu. Rev. Physiol., 40*:471, 1978.

Wyman, R. J.: Neural generation of the breathing rhythm. *Annu. Rev. Physiol., 39*:417, 1977.

Physics of Respiratory Gases

Bauer, C., *et al.* (eds.): Biophysics and Physiology of Carbon Dioxide. New York, Springer-Verlag, 1980.

Forster, R. E., and Crandall, E. D.: Pulmonary gas exchange. *Annu. Rev. Physiol., 38*:69, 1976.

Guyton, A. C., *et al.*: An arteriovenous oxygen difference recorder. *J. Appl. Physiol., 10*:158, 1957.

Jones, N. L.: Blood Gases and Acid-Base Physiology. New York, B. C. Decker, 1980.

Piiper, J., and Scheid, P.: Comparative physiology of respiration: Functional analysis of gas exchange organs in vertebrates. *Int. Rev. Physiol., 14*:219, 1977.

Radford, E. P., Jr.: The physics of gases. *In* Fenn, W. O., and Rahn, H. (eds.): Handbook of Physiology. Sec. 3, Vol. 1. Baltimore, Williams & Wilkins, 1964, p. 125.

Rahn, H., and Fenn, W. O.: A Graphical Analysis of Respiratory Gas Exchange. Washington, American Physiological Society, 1955.

Wagner, P. D.: Diffusion and chemical reaction in pulmonary gas exchange. *Physiol. Rev., 57*:257, 1977.

Weibel, E. R.: Morphological basis of alveolar capillary gas exchange. *Physiol. Rev., 53*:419, 1973.

28

Transport of Oxygen and Carbon Dioxide Between the Alveoli and the Tissue Cells

PRESSURE DIFFERENCES OF OXYGEN AND CARBON DIOXIDE FROM THE LUNGS TO THE TISSUES

In the preceding chapter it was pointed out that gases can move from one point to another by diffusion and that the cause is always a pressure difference from the first point to the other. Thus, oxygen diffuses from the alveoli into the pulmonary capillary blood because of a pressure difference — that is, the oxygen pressure (Po_2) in the alveoli is greater than the Po_2 in the pulmonary blood. Then the pulmonary blood is transported by way of the circulation to the peripheral tissues. There the Po_2 is lower in the cells than in the arterial blood entering the capillaries. Here again, the much higher Po_2 in the capillary blood causes oxygen to diffuse out of the capillaries and through the interstitial fluid to the cells.

Then when oxygen is metabolized with the foods in the cells to form carbon dioxide, the carbon dioxide pressure (Pco_2) rises to a high value in the cells, which causes carbon dioxide to diffuse from the cells into the tissue capillaries. Once in the blood the carbon dioxide is transported to the pulmonary capillaries, where it diffuses out of the blood and into the alveoli because the Pco_2 in the alveoli is lower than that in the blood.

Basically, then, the transport of oxygen and carbon dioxide to and from the tissues, respectively, depends on both diffusion and the movement of blood. We now need to consider quantitatively the factors responsible for these effects as well as their significance in the overall physiology of respiration.

COMPOSITION OF ALVEOLAR AIR — ITS RELATION TO ATMOSPHERIC AIR

Alveolar air does not have the same concentrations of gases as atmospheric air by any means. The differences can readily be seen by comparing the alveolar air composition in column 3 of Table 28–1 with the composition of atmospheric air in column 1. There are several reasons for the differences. First, the alveolar air is only partially replaced by atmospheric air with each breath. Second, oxygen is constantly being absorbed from the alveolar air. Third, carbon dioxide is constantly diffusing from the pulmonary blood into the alveoli. And, fourth, dry atmospheric air that enters the respiratory passages is humidified even before it reaches the alveoli.

Humidification of the Air as It Enters the Respiratory Passages. Column 1 of Table 28–1 shows that atmospheric air is composed almost entirely of nitrogen and oxygen; it normally contains almost no carbon dioxide and little water vapor. However, as soon as the atmospheric air enters the respiratory passages, it is exposed to the fluids covering the respiratory surfaces. Even before the air enters the alveoli, it becomes totally humidified.

The partial pressure of water vapor at normal body temperature of 37°C is 47 mm. Hg, which, therefore, is the partial pressure of water in the alveolar air. Since the total pressure in the alveoli cannot rise to more than the atmospheric pressure, this water vapor simply expands the volume of the air and thereby *dilutes* all the other gases in the inspired air. In column 2 of Table 28–1 it can be seen that humidification of the air has diluted

TABLE 28–1 PARTIAL PRESSURES OF RESPIRATORY GASES AS THEY ENTER AND LEAVE THE LUNGS (AT SEA LEVEL)—PER CENT CONCENTRATIONS ARE GIVEN IN PARENTHESES

	Atmospheric Air* (mm. Hg)		Humidified Air (mm. Hg)		Alveolar Air (mm. Hg)		Expired Air (mm. Hg)	
N_2	597.0	(78.62%)	563.4	(74.09%)	569.0	(74.9%)	566.0	(74.5%)
O_2	159.0	(20.84%)	149.3	(19.67%)	104.0	(13.6%)	120.0	(15.7%)
CO_2	0.3	(0.04%)	0.3	(0.04%)	40.0	(5.3%)	27.0	(3.6%)
H_2O	3.7	(0.50%)	47.0	(6.20%)	47.0	(6.2%)	47.0	(6.2%)
TOTAL	760.0	(100.0%)	760.0	(100.0%)	760.0	(100.0%)	760.0	(100.0%)

*On an average cool, clear day.

the oxygen partial pressure at sea level from an average of 159 mm. Hg in atmospheric air to 149 mm. Hg in the humidified air, and it has diluted the nitrogen partial pressure from 597 to 563 mm. Hg.

RATE AT WHICH ALVEOLAR AIR IS RENEWED BY ATMOSPHERIC AIR

In the preceding chapter it was pointed out that the *functional residual capacity* of the lungs, which is the amount of air remaining in the lungs at the end of normal expiration, measures approximately 2300 ml. Furthermore, only 350 ml. of new air is brought into the alveoli with each normal respiration, and the same amount of old alveolar air is expired. Therefore, the amount of alveolar air replaced by new atmospheric air with each breath is only one-seventh of the total, so that many breaths are required to exchange most of the alveolar air. Figure 28–1 illustrates this slow rate of renewal of the alveolar air. In the first alveolus of the figure an excess amount of a gas has been placed momentarily in all the alveoli. The second alveolus shows slight dilution of this gas with the first breath; the next alveolus shows still further dilution with the second breath, and so forth for the third, fourth, eighth, twelfth, and sixteenth breaths. Note that even at the end of 16 breaths the excess gas still has not been completely removed from the alveoli.

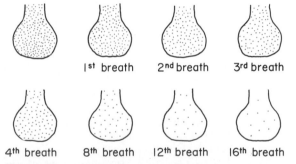

Figure 28–1. Expiration of a gaseous excess from the alveoli with successive breaths.

With normal alveolar ventilation, approximately half the gas is removed in 17 seconds. When a person's rate of alveolar ventilation is only half normal, half the gas is removed in 34 seconds, and, when the rate of ventilation is 2 times normal, half is removed in about 8 seconds.

This slow replacement of alveolar air is of particular importance in preventing sudden changes in gaseous concentrations in the blood and helps to prevent excessive increases and decreases in tissue oxygenation, tissue carbon dioxide concentration, and tissue pH when respiration is temporarily interrupted.

OXYGEN CONCENTRATION AND PARTIAL PRESSURE IN THE ALVEOLI

Oxygen is continually being absorbed into the blood of the lungs, and new oxygen is continually entering the alveoli from the atmosphere. The more rapidly oxygen is absorbed, the lower becomes its concentration in the alveoli; on the other hand, the more rapidly new oxygen is brought into the alveoli from the atmosphere, the higher becomes its concentration. Therefore, oxygen concentration in the alveoli, as well as its partial pressure, is determined by the balance between the rate of absorption of oxygen into the blood and the rate of entry of new oxygen into the lungs by the ventilatory process. Its normal partial pressure in the alveoli is 104 mm. Hg.

However, it must be noted that when a person is breathing air at normal sea-level pressure, even an extremely marked increased in alveolar ventilation can never increase the alveolar Po_2 above 149 mm. Hg, for this is the maximum pressure of oxygen in humidified air.

CO_2 CONCENTRATION AND PARTIAL PRESSURE IN THE ALVEOLI

Carbon dioxide is continuously being formed in the body and then discharged into the alveoli; and it is continuously being removed from the alveoli by the process of ventilation. Therefore, the two factors that determine carbon dioxide concentration and partial pressure (Pco_2) in the lungs are (1) the rate of excretion of carbon dioxide from the blood into the alveoli and (2) the rate at which carbon dioxide is removed from the alveoli by alveolar ventilation. Quantitatively, one can easily understand that *the alveolar* Pco_2 *increases directly in proportion to the rate of carbon dioxide excretion, and it decreases in inverse proportion to alveolar ventilation.* The normal alveolar Pco_2 is 40 mm. Hg.

EXPIRED AIR

Expired air is a combination of dead space air and alveolar air, and its overall composition is therefore determined by the proportion of the expired air that is dead space air and the proportion that is alveolar air. The initial portion of the expired air, the dead space air, is typical humidified air, as shown in column 2 of Table 28–1. Then, progressively more and more alveolar air becomes mixed with the dead space air until all the dead space air has finally been washed out and nothing but alveolar air remains.

Thus, normal expired air, containing both dead space air and alveolar air, has gaseous concentrations approximately as shown in column 4 of Table 28–1 — that is, concentrations somewhere between those of humidified atmospheric air and alveolar air.

DIFFUSION OF GASES THROUGH THE RESPIRATORY MEMBRANE

The Respiratory Unit. Figure 28–2 illustrates the respiratory unit, which is composed of a *respiratory bronchiole, alveolar ducts, atria,* and *alveoli* (of which there are about 300 million in the two lungs, each alveolus having an average diameter of about 0.25 mm.). The alveolar walls are extremely thin, and within them is an almost solid network of interconnecting capillaries, as illustrated in Figure 28–3. Indeed, the flow of blood in the alveolar wall has been described as a "sheet" of flowing blood. Thus, it is obvious that the alveolar gases are in close proximity to the blood of the capillaries. Consequently, gaseous exchange between the alveolar air and the pulmonary blood occurs through the membranes of

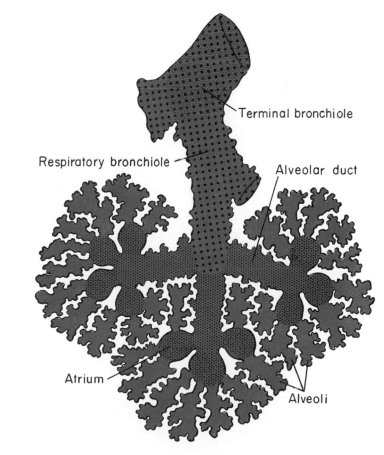

Figure 28–2. The respiratory lobule. (From Miller, W. S.: The Lung. Courtesy of Charles C Thomas, Publisher, Springfield, Illinois.)

Terminal bronchiole

Respiratory bronchiole

Alveolar duct

Atrium

Alveoli

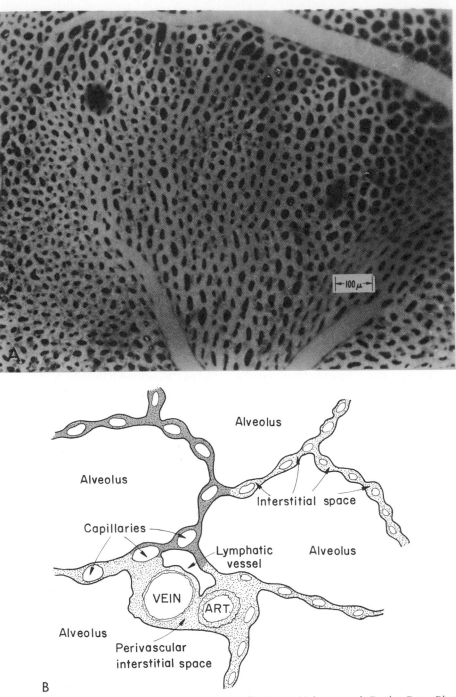

Figure 28–3. *A*, Surface view of capillaries in an alveolar wall. (From Maloney and Castle: *Resp. Physiol.*, 7:150, 1969. Reproduced by permission of ASP Biological and Medical Press, North-Holland Division.) *B*, Cross-sectional view of alveolar walls and their vascular supply.

all the terminal portions of the lungs. These membranes are collectively known as the *respiratory membrane,* also called the *pulmonary membrane.*

The Respiratory Membrane. Figure 28–4 illustrates the ultrastructure of the respiratory membrane. It also shows the diffusion of oxygen from the alveolus into the red blood cell and

diffusion of the carbon dioxide in the opposite direction. Note the following different layers of the respiratory membrane:

1. A layer of fluid lining the alveolus and containing surfactant that reduces the surface tension of the alveolar fluid.

2. The alveolar epithelium composed of very thin epithelial cells.

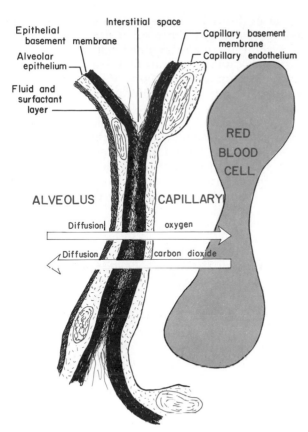

Figure 28–4. Ultrastructure of the respiratory membrane.

3. An epithelial basement membrane.

4. A very thin interstitial space between the alveolar epithelium and capillary membrane.

5. A capillary basement membrane that in many places fuses with the epithelial basement membrane.

6. The capillary endothelial membrane.

Despite the large number of layers, the overall thickness of the respiratory membrane in some areas is as little as 0.2 micron and averages perhaps 0.5 micron.

From histologic studies it has been estimated that the total surface area of the respiratory membrane is approximately 70 square meters in the normal adult. This is equivalent to the floor area of a room 30 feet long by 25 feet wide. The total quantity of blood in the capillaries of the lung at any given instant is 60 to 140 ml. If this small amount of blood were spread over the entire surface of a 25 by 30 foot floor, one could readily understand how respiratory exchange of gases occurs as rapidly as it does.

Factors That Affect Rate of Gas Diffusion Through the Respiratory Membrane

Referring to the discussion of diffusion of gases through water and tissues in the last chap-

ter, one can apply the same principles and same formula to diffusion of gases through the respiratory membrane. Thus, the factors that determine how rapidly a gas will pass through the membrane are (1) the *thickness of the membrane,* (2) the *surface area of the membrane,* (3) the *diffusion coefficient* of the gas in the substance of the membrane — that is, in water, and (4) the *pressure difference* between the two sides of the membrane.

The *thickness of the respiratory membrane* occasionally increases, often as a result of edema fluid in the interstitial spaces of the membrane and in the alveoli, so that the respiratory gases must diffuse not only through the membrane but also through this fluid. Because the rate of diffusion through the membrane is inversely proportional to the thickness of the membrane, any factor that increases the thickness to more than two to three times normal can interfere very significantly with normal respiratory exchange of gases.

The *surface area of the respiratory membrane* may be greatly decreased by many different conditions. For instance, removal of an entire lung decreases the surface area to half normal. Also, in *emphysema* many of the alveoli coalesce, with dissolution of many alveolar walls. When the total surface area is decreased to approximately one-third to one-fourth normal, exchange of gases through the membrane is impeded to a significant degree *even under resting conditions.* And, during competitive sports and other strenuous exercise, even the slightest decrease in surface area of the lungs can be a serious detriment to respiratory exchange of gases.

The *diffusion coefficient* of each gas in the respiratory membrane is almost exactly the same as that in water, for reasons explained in the previous chapter. Therefore, for a given pressure difference, carbon dioxide diffuses through the membrane about 20 times as rapidly as oxygen. Oxygen in turn diffuses about 2 times as rapidly as nitrogen.

The *pressure difference* across the respiratory membrane is the difference between the partial pressure of the gas in the alveoli and the pressure of the gas in the blood. The partial pressure represents a measure of the total number of molecules of a particular gas striking a unit area of the alveolar surface of the membrane in unit time, and the pressure of the gas in the blood represents the number of molecules striking the same area of the membrane from the opposite side. Therefore, the difference between these two pressures is a measure of the *net tendency* for the gas to move through the membrane. Obviously, when the partial pressure of a gas in the alveoli is greater than the pressure of the gas in the blood, as is true for oxygen, net diffusion from the alveoli into the blood occurs, but when the pressure of the gas in the blood is greater than in the alveoli, as is true for carbon dioxide, net diffusion from the blood into the alveoli occurs.

DIFFUSING CAPACITY OF THE RESPIRATORY MEMBRANE

The overall ability of the respiratory membrane to exchange a gas between the alveoli and the pulmonary blood can be expressed in terms of its *diffusing capacity*, which is defined as the *volume of a gas that diffuses through the membrane each minute for a pressure difference of 1 mm. Hg.*

Obviously, all the factors discussed above that affect diffusion through the respiratory membrane can affect the diffusing capacity.

The Diffusing Capacity for Oxygen. In the young male adult the diffusing capacity for oxygen under resting conditions averages 21 ml. per minute per mm. Hg. The mean oxygen pressure difference across the respiratory membrane during normal, quiet breathing is approximately 11 mm. Hg. Multiplication of this pressure by the diffusing capacity (11×21) gives a total of about 230 ml. of oxygen normally diffusing through the respiratory membrane each minute, and this is equal to the rate at which the body uses oxygen.

Change in Oxygen-Diffusing Capacity During Exercise. During strenuous exercise, or during other conditions that greatly increase pulmonary activity, the diffusing capacity for oxygen increases about three-fold. This increase is caused by several different factors, among which are (1) opening up of a number of previously dormant pulmonary capillaries, thereby increasing the surface area of the blood into which the oxygen can diffuse, and (2) dilatation of all the pulmonary capillaries that were already open, thereby further increasing the surface area. Therefore, during exercise, the oxygenation of the blood is increased not only by increased alveolar ventilation but also by a greater capacity of the respiratory membrane for transmitting oxygen into the blood.

Diffusing Capacity for Carbon Dioxide. The diffusing capacity for carbon dioxide has never been measured because of the following technical difficulty: Carbon dioxide diffuses through the respiratory membrane so rapidly that the average Pco_2 in the pulmonary blood is not far different from the Pco_2 in the alveoli — the average difference is less than 1 mm. Hg — and with the available techniques, this difference is too small to be measured.

Nevertheless, measurements of diffusion of other gases have shown that the diffusing capacity varies directly with the diffusion coefficient of the particular gas. Since the diffusion coefficient of carbon dioxide is 20 times that of oxygen, the diffusing capacity for carbon dioxide under resting conditions is about 400 to 450 ml. and during exercise about 1200 to 1300 ml. per minute per mm. Hg.

The importance of these high diffusing capacities for carbon dioxide is this: When the respiratory membrane becomes progressively damaged, its capacity for transmitting oxygen into the blood is often impaired enough to cause death of the person long before serious impairment of carbon dioxide diffusion occurs.

EFFECT OF THE VENTILATION-PERFUSION RATIO ON GAS EXCHANGE

Earlier in the chapter we discussed the importance of ventilation in determining both the Po_2 and Pco_2 in the alveoli. But we must also hasten to state that the rate of blood flow through the alveolar capillaries also affects the effectiveness of gas exchange across the respiratory membrane — especially oxygen exchange but in some instances carbon dioxide exchange as well. Thus, the *ratio* of ventilation to pulmonary capillary blood flow, called the *ventilation-perfusion ratio*, is what is really important in determining gas exchange.

The Concept of "Physiologic Shunt," When the Ventilation-Perfusion Ratio is Below Normal. Whenever the ventilation-perfusion ratio is below normal, there obviously is not ventilation enough to provide the oxygen needed to oxygenate the blood flowing through the alveolar capillaries. Therefore, a certain fraction of the venous blood passing through the pulmonary capillaries does not become oxygenated. This fraction is called *shunted blood*. The total quantitative amount of shunted blood per minute is called the *physiologic shunt*.

Obviously, when the respiratory passageways to some lung areas are blocked — as occurs in emphysema, asthma, pneumonia, and many other diseases — this also causes a significant physiologic shunt.

The Concept of "Physiologic Dead Space," When the Ventilation-Perfusion Ratio is Greater Than Normal. When the ventilation is great but blood flow is low, there is then far more available oxygen in the alveoli than can be transported away from the alveoli by the flowing blood. Thus, a large portion of the ventilation is said to be *wasted*. The ventilation of the dead space areas of the lungs is also wasted. The sum of these two types of wasted ventilation is called the *physiologic dead space*.

Obviously, also, whenever the blood flow to part of the lungs is blocked for any reason, such as emphysema or lung cancer, all the ventilation to the affected area is wasted. This, too, adds to the physiologic dead space.

When the physiologic dead space is very great, much of the work of ventilation is wasted effort because so much of the ventilated air never reaches the blood.

TRANSPORT OF OXYGEN TO THE TISSUES

UPTAKE OF OXYGEN BY THE PULMONARY BLOOD

The top part of Figure 28–5 illustrates a pulmonary alveolus adjacent to a pulmonary capillary, showing diffusion of oxygen molecules between the alveolar air and the pulmonary blood. However, the Po$_2$ of the venous blood entering the capillary is only 40 mm. Hg because a large amount of oxygen has been removed from this blood as it has passed through the tissue capillaries. The Po$_2$ in the alveolus is 104 mm. Hg, giving an initial pressure difference for diffusion of oxygen into the pulmonary capillary of 104 − 40, or 64 mm. Hg. Therefore, far more oxygen diffuses into the pulmonary capillary than in the opposite direction. The curve below the capillary shows the progressive rise in blood Po$_2$ as the blood passes through the capillary. This curve illustrates that the Po$_2$ rises almost to equal that of the alveolar air before reaching the midpoint of

the capillary, becoming approximately 104 mm. Hg. However, a small amount of pulmonary venous blood passes through poorly aerated alveoli and does not become oxygenated. When this blood mixes with the oxygenated blood in the left heart, the Po$_2$ in the aorta becomes about 95 mm. Hg.

Uptake of Oxygen by the Pulmonary Blood During Exercise. During strenuous exercise, a person's body may require as much as 20 times the normal amount of oxygen. Yet, because of the great *safety factor* for diffusion of oxygen through the pulmonary membrane, and because the diffusing capacity for oxygen increases about three-fold as discussed earlier, the blood is still *almost completely saturated* with oxygen when it leaves the pulmonary capillaries.

DIFFUSION OF OXYGEN FROM THE CAPILLARIES TO THE INTERSTITIAL FLUID

At the tissue capillaries, oxygen diffuses into the tissues by a process essentially the same as that which takes place in the lungs, as illustrated in Figure 28–6. That is, the Po$_2$ in the interstitial fluid immediately outside a capillary is low and, though very variable, averages about 40 mm. Hg, while that in the arterial blood is high, about 95 mm. Hg. Therefore, at the arterial end of the capillary, a pressure difference of 55 mm. Hg causes diffusion of oxygen. As illustrated in the figure, by the time the blood has passed through the capillary a large portion of the oxygen has diffused into the tissues, and the capillary Po$_2$ has approached the 40 mm. Hg oxygen pressure in the tissue fluids. Consequently, the venous blood leaving the tissue capillaries contains oxygen at essentially the same pressure as that immediately outside the tissue capillaries, 40 mm. Hg.

DIFFUSION OF OXYGEN FROM THE INTERSTITIAL FLUIDS INTO THE CELLS

Since oxygen is always being used by the cells, the intracellular Po$_2$ remains lower than the inter-

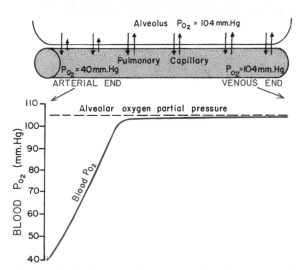

Figure 28–5. Uptake of oxygen by the pulmonary capillary blood. (The curve in this figure was constructed from data in Milhorn and Pulley: *Biophys. J.*, 8:337, 1968.

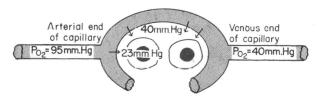

Figure 28–6. Diffusion of oxygen from a tissue capillary to the cells.

stitial fluid P_{O_2}. However, as was pointed out in Chapter 4, oxygen diffuses through cell membranes extremely rapidly. Therefore, the intracellular P_{O_2} is almost as great as that in the interstitial fluids.

Yet, in many instances, there is considerable distance between the capillaries and the cells. Therefore, the normal intracellular P_{O_2} ranges from as low as 5 mm. Hg to as high as 60 mm. Hg, averaging (by direct measurement in lower animals) 23 mm. Hg, which is the value given for the cell in Figure 28–6. Since only 1 to 3 mm. Hg oxygen pressure is required for full support of the metabolic processes of the cell, one can see that even this low cellular P_{O_2} is adequate and actually provides a considerable safety factor.

TRANSPORT OF CARBON DIOXIDE TO THE LUNGS

DIFFUSION OF CARBON DIOXIDE FROM THE CELLS TO THE TISSUE CAPILLARIES

Because of the continuous formation of large quantities of carbon dioxide in the cells, the intracellular P_{CO_2} tends to rise. However, carbon dioxide diffuses about 20 times as easily as oxygen, diffusing from the cells extremely rapidly into the interstitial fluids and thence into the capillary blood. Thus, in Figure 28–7 the intracellular P_{CO_2} is shown to be 46 mm. Hg, while that in the interstitial fluid immediately adjacent to the capillaries is about 45 mm. Hg, a pressure differential of only 1 mm. Hg.

Arterial blood entering the tissue capillaries contains carbon dioxide at a pressure of approximately 40 mm. Hg. As the blood passes through the capillaries, the blood P_{CO_2} rises to approach the 45 mm. Hg P_{CO_2} of the interstitial fluid. And, again because of the very large diffusion coefficient for carbon dioxide, the P_{CO_2} of the blood leaving the capillaries and entering the veins is also about 45 mm. Hg, within a fraction of a millimeter of reaching complete equilibrium with the P_{CO_2} of the interstitial fluid.

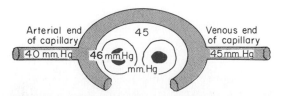

Figure 28–7. Uptake of carbon dioxide by the blood in the capillaries.

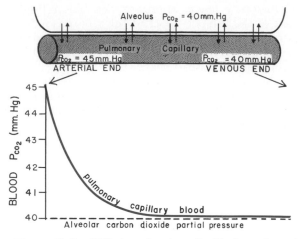

Figure 28–8. Diffusion of carbon dioxide from the pulmonary blood into the alveolus. (This curve was constructed from data in Milhorn and Pulley: *Biophys. J., 8*:337, 1968.)

REMOVAL OF CARBON DIOXIDE FROM THE PULMONARY BLOOD

On arrival at the lungs, the P_{CO_2} of the venous blood is about 45 mm. Hg while that in the alveoli is 40 mm. Hg. Therefore, as illustrated in Figure 28–8, the initial pressure difference for diffusion is only 5 mm. Hg, which is far less than that for diffusion of oxygen across the membrane. Yet, even so, because of the 20 times as great diffusion coefficient for carbon dioxide as for oxygen, the excess carbon dioxide in the blood is rapidly transferred into the alveoli. Indeed, the figure shows that the P_{CO_2} of the pulmonary capillary blood becomes almost equal to that of the alveoli within the first four-tenths of the blood's transit through the pulmonary capillary.

CHEMICAL AND PHYSICAL MEANS BY WHICH OXYGEN IS CARRIED IN THE BLOOD

Normally, about 97 per cent of the oxygen transported from the lungs to the tissues is carried in chemical combination with hemoglobin in the red blood cells, and the remaining 3 per cent is carried in the dissolved state in the water of the plasma and cells. Thus, *under normal conditions* the transport of oxygen in the dissolved state is negligible. However, when a person breathes oxygen at very high pressures, as much oxygen can sometimes be transported in the dissolved state as in chemical combination with hemoglobin. Therefore, the present discussion will consider the transport of oxygen first in combination with hemoglobin and then in the dissolved state under special conditions.

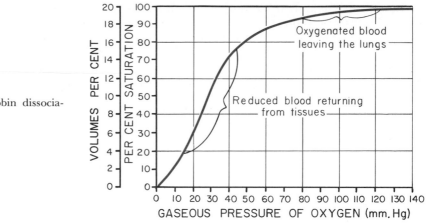

Figure 28–9. The oxygen-hemoglobin dissociation curve.

THE REVERSIBLE COMBINATION OF OXYGEN WITH HEMOGLOBIN

The chemistry of hemoglobin was presented in Chapter 5, where it was pointed out that the oxygen molecule combines loosely and reversibly with the heme portion of the hemoglobin. When the Po$_2$ is high, as in the pulmonary capillaries, oxygen binds with the hemoglobin, but when the Po$_2$ is low, as in the tissue capillaries, oxygen is released from the hemoglobin. This is the basis for oxygen transport from the lungs to the tissues.

The Oxygen-Hemoglobin Dissociation Curve. Figure 28–9 illustrates the oxygen-hemoglobin dissociation curve, which shows the progressive increase in the per cent of the hemoglobin that is bound with oxygen as the Po$_2$ increases. This is called the *per cent saturation of the hemoglobin.* Since the blood leaving the lungs usually has a Po$_2$ of about 100 mm. Hg, one can see from the dissociation curve that the *usual oxygen saturation of arterial blood is about 97 per cent.* On the other hand, in normal venous blood the Po$_2$ is about 40 mm. Hg and *the saturation of the hemoglobin is about 70 per cent.*

Maximum Amount of Oxygen that Can Combine with the Hemoglobin of the Blood. The blood of a normal person contains approximately 15 grams of hemoglobin in each 100 ml. of blood, and each gram of hemoglobin binds with a maximum of about 1.34 ml. of oxygen. Therefore, on the average, the hemoglobin in 100 ml. of blood can combine with a total of almost exactly 20 ml. of oxygen when the hemoglobin is 100 per cent saturated. This is usually expressed as 20 *volumes per cent.* The oxygen-hemoglobin dissociation curve for the normal person, therefore, can be expressed also in terms of volume per cent of

oxygen, as shown in Figure 28–10, rather than per cent saturation of hemoglobin.

Amount of Oxygen Released from the Hemoglobin in the Tissues. The total quantity of oxygen *bound with hemoglobin* in normal arterial blood, which is normally 97 per cent saturated, is approximately 19.4 ml. This is illustrated in Figure 28–10. However, on passing through the tissue capillaries, this amount is reduced to 14.4 ml. (Po$_2$ of 40 mm. Hg, 72 per cent saturated), or a total loss of 5 ml. of oxygen from each 100 ml. of blood. Thus, *under normal conditions about 5 ml. of oxygen is transported by each 100 ml. of blood during each cycle through the tissues.*

The Utilization Coefficient and the Effect of Exercise. The fraction of the blood that gives up its oxygen as it passes through the tissue capillaries is called the *utilization coefficient.* Normally, this is approximately 0.25, or 25 per cent, of the blood. That is, *the normal utilization coefficient is approximately one-fourth.* During strenuous exercise, as much as 75 to 85 per cent of the blood can give up its oxygen; the utilization coefficient

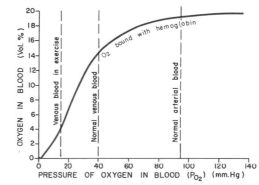

Figure 28–10. Effect of blood Po$_2$ on the quantity of oxygen bound with hemoglobin in each 100 ml. of blood.

is then approximately 0.75 to 0.85, which represents about three times as much oxygen delivery to the tissues as normal. These values are about the highest utilization coefficients that can be attained in the total body even when the tissues are in extreme need of oxygen. However, in local tissue areas where the blood flow is very slow or the metabolic rate very high, utilization coefficients approaching 100 per cent have been recorded — that is, essentially all the oxygen is removed.

TOTAL RATE OF OXYGEN TRANSPORT FROM THE LUNGS TO THE TISSUES

If under resting conditions about 5 ml. of oxygen is transported by each 100 ml. of blood, and if the normal cardiac output is approximately 5000 ml. per minute, the calculated total quantity of oxygen delivered to the tissues each minute is about 250 ml. This is also the amount measured by a respirometer.

This rate of oxygen transport to the tissues can be increased to about 15 times normal during heavy exercise and in other instances of excessive need for oxygen (and very rarely to as high as 20 times normal in the best trained athletes). Oxygen transport can be increased to three times normal simply by an increase in utilization coefficient, and it can be increased another five-fold as a result of increased cardiac output, thus accounting for the total 15-fold increase. Therefore, the maximum rate of oxygen transport to the tissues is about 15×250 ml., or 3750 ml. per minute in the normal young adult. Special adaptations in athletic training, such as an increase in blood hemoglobin concentration and an increase in maximum cardiac output, can sometimes increase this value to as high as 4.5 to 5 liters per minute.

THE OXYGEN BUFFER FUNCTION OF HEMOGLOBIN

Though hemoglobin is necessary for transport of oxygen to the tissues, it performs still another major function essential to life. This is its function as an "oxygen buffer" system. That is, the hemoglobin in the blood is mainly responsible for controlling the oxygen pressure in the tissues. This can be explained as follows:

Under basal conditions the tissues require about 5 ml. of oxygen from each 100 ml. of blood passing through the tissue capillaries. Referring again to the oxygen-hemoglobin dissociation curve in Figure 28–10, one can see that for the 5

ml. of oxygen to be released, the Po_2 must fall to about 40 mm. Hg. Therefore, the tissue capillary Po_2 cannot rise above this 40 mm. Hg level, for if that should occur, the oxygen needed by the tissues could not be released from the hemoglobin. In this way, the hemoglobin normally sets an upper limit on the gaseous pressure in the tissues at approximately 40 mm. Hg.

On the other hand, in heavy exercise extra large amounts of oxygen must be delivered from the hemoglobin to the tissues. But this can be achieved with very little further decrease in tissue Po_2 because of the steep slope of the dissociation curve — that is, a small fall in Po_2 causes extreme amounts of oxygen to be released. Therefore, the Po_2 rarely falls below 20 mm. Hg.

It can be seen, then, that hemoglobin automatically delivers oxygen to the tissues at a pressure between approximately 20 and 40 mm. Hg. This seems to be a wide range of Po_2's in the interstitial fluid, but when one considers how much the interstitial fluid Po_2 might possibly change during exercise and other types of stress, this range of 20 to 40 mm. Hg is relatively narrow.

This oxygen buffer function of hemoglobin is also very important when the alveolar Po_2 falls very low, as occurs at high altitudes, or rises very high, as occurs when one is breathing pure oxygen. Let us explain this:

It will be seen from the oxygen-hemoglobin dissociation curve in Figure 28–9 that when the alveolar Po_2 is decreased to as low as 60 mm. Hg, which occurs at an altitude of about 2½ miles, the arterial hemoglobin is still 89 per cent saturated, only 8 per cent below the normal saturation of 97 per cent. Furthermore, the tissues still remove approximately 5 ml of oxygen from every 100 ml. of blood passing through the tissues; to remove this oxygen, the Po_2 of the venous blood falls to only slightly less than 40 mm. Hg. Thus, a change in alveolar Po_2 from 104 to 60 mm. Hg. has almost no effect on tissue Po_2.

On the other hand, when the alveolar Po_2 rises far above the normal value of 104 mm. Hg, the maximum oxygen saturation of hemoglobin can never rise above 100 per cent. Therefore, even though the oxygen in the alveoli should rise to a partial pressure of 500 mm. Hg or even more, the increase in the saturation of hemoglobin would be only 3 per cent because, even at 104 mm. Hg Po_2, 97 per cent of the hemoglobin is already combined with oxygen; and only a small amount of additional oxygen dissolves in the fluid of the blood, as will be discussed subsequently. Then, when the blood passes through the tissue capillaries, it still loses several milliliters of oxygen to the tissues, and this loss automatically reduces the Po_2 of the capillary blood to a value only a few millimeters greater than the normal 40 mm. Hg.

Consequently, alveolar oxygen may vary greatly – from 60 to more than 500 mm. Hg Po_2 – and still the Po_2 in the tissue does not vary more than a few millimeters from normal.

METABOLIC USE OF OXYGEN BY THE CELLS – RELATIONSHIP TO CELLULAR Po_2

Only a minute level of oxygen pressure is required in the cells for normal intracellular chemical reactions to take place. The reason is that the respiratory enzyme systems of the cell, which will be discussed in Chapter 45, are so geared that when the cellular Po_2 is more than 1 to 3 mm. Hg, oxygen availability is no longer a limiting factor in the rates of the chemical reactions involving oxygen usage. Instead, the main limiting factor then is the *concentration of adenosine diphosphate* (ADP) in the cells, as was explained in Chapter 3. This effect is illustrated in Figure 28–11. Note that whenever the intracellular Po_2 is above 3 mm. Hg, the rate of oxygen usage becomes constant for any given concentration of ADP in the cell. On the other hand, when the ADP increases, the rate of oxygen usage increases in proportion to the increase in ADP concentration.

It will be recalled from the discussion in Chapter 3 that when adenosine triphosphate (ATP) is utilized in the cells to provide energy, it is converted into ADP. The increasing concentration of

ADP in turn increases the metabolic usage of oxygen and also of the various nutrients that combine with oxygen to release energy. This energy is used to re-form the ATP. Therefore, *under normal operating conditions the rate of oxygen utilization by the cells is controlled by the rate of energy expenditure within the cells – that is, by the rate at which ADP is formed from ATP – and not by the degree of availability of oxygen to the cells.*

TRANSPORT OF OXYGEN IN THE DISSOLVED STATE

At the normal arterial Po_2 of 95 mm. Hg, approximately 0.29 ml. of oxygen is dissolved in every 100 ml. of water in the blood. When the Po_2 of the blood falls to 40 mm. Hg in the tissue capillaries, 0.12 ml. of oxygen remains dissolved. In other words, 0.17 ml. of oxygen is normally transported in the dissolved state to the tissues by each 100 ml. of blood water. This compares with about 5.0 ml. transported by the hemoglobin. Therefore, the amount of oxygen transported to the tissues in the dissolved state is normally slight, only about 3 per cent of the total as compared with the 97 per cent transported by the hemoglobin. During strenuous exercise, when hemoglobin transport of oxygen increases three-fold, the relative quantity then transported in the dissolved state falls to as little as 1.5 per cent. Yet, if a person breathes oxygen at very high Po_2's — at several thousand mm. Hg Po_2 — the amount then transported in the dissolved state can become tremendous, so much so that serious excesses of oxygen occur in the tissues and "oxygen poisoning" ensues. This often leads to convulsions and even death, as will be discussed further in Chapter 30 in relation to high pressure breathing.

"POISONING" OF HEMOGLOBIN WITH CARBON MONOXIDE

Carbon *monoxide* combines with hemoglobin at the same point on the hemoglobin molecule as does oxygen. Furthermore, it binds with 230 times as much tenacity as oxygen. Therefore, a carbon monoxide pressure of only 0.4 mm. Hg in the alveoli, 1/230 that of the alveolar oxygen, allows the carbon monoxide to compete equally with the oxygen for combination with the hemoglobin and causes half the hemoglobin in the blood to become bound with carbon monoxide instead of with oxygen. A carbon monoxide pressure of 0.7 mm. Hg (a concentration of about one part in a thousand in the air) can be lethal.

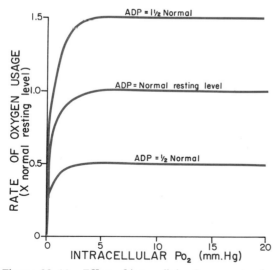

Figure 28–11. Effect of intracellular Po_2 on rate of oxygen usage by the cells. Note that increasing the intracellular concentration of adenosine diphosphate *(ADP)* increases the rate of oxygen usage.

A person severely poisoned with carbon monoxide can be advantageously treated by the administration of pure oxygen, for oxygen at high alveolar pressures displaces carbon monoxide from combination with hemoglobin far more rapidly than can oxygen at the low pressure of atmospheric oxygen.

CHEMICAL AND PHYSICAL MEANS FOR CARRYING CARBON DIOXIDE IN THE BLOOD

Transport of carbon dioxide is not nearly so great a problem as transport of oxygen, because even in the most abnormal conditions carbon dioxide can usually be carried in the blood in far greater quantities than can oxygen. However, the amount of carbon dioxide in the blood does have much to do with the acid-base balance of the body fluids, which was discussed in detail in Chapter 26. Under normal resting conditions *an average of 4 ml. of carbon dioxide is carried from the tissues to the lungs in each 100 ml. of blood.*

CHEMICAL FORMS IN WHICH CARBON DIOXIDE IS CARRIED

To begin the process of carbon dioxide transport, carbon dioxide diffuses out of the tissue cells. On entering the capillary, the chemical reactions illustrated in Figure 28–12 occur immediately. Most important, the dissolved carbon dioxide diffuses into the red blood cells, where it reacts with water to form carbonic acid. Inside the red cells an enzyme called *carbonic anhydrase* catalyzes this reaction between carbon dioxide and water, accelerating the rate of reaction about 5000-fold. Therefore, the reaction occurs almost instantaneously. The carbonic acid in turn dissociates into hydrogen ions and bicarbonate ions. The hydrogen ions combine mainly with the hemoglobin in the red cells, while many of the bicarbonate ions diffuse through the red cell membranes into the plasma.

An additional one-quarter of the carbon dioxide combines directly with hemoglobin to form a compound called *carbaminohemoglobin*. Since carbon dioxide dissociates from carbaminohemoglobin, this compound can release the carbon dioxide in the lungs for excretion.

Finally, a small amount of the carbon dioxide, about 7 per cent, is transported to the lungs in the dissolved form.

THE CARBON DIOXIDE DISSOCIATION CURVE

The total quantity of carbon dioxide combined with the blood in all its forms depends on the P_{CO_2}. The curve shown in Figure 28–13 depicts this dependence; this curve is called the *carbon dioxide dissociation curve.*

Note that the normal blood P_{CO_2} ranges between the limits of 40 mm. Hg in arterial blood and 45 mm. Hg in venous blood, which is a very narrow range. Note also that the normal concentration of carbon dioxide in the blood is about 50 volumes per cent but that only 4 volumes per cent of this is actually exchanged in the process of transporting carbon dioxide from the tissues to the lungs. That is, the concentration rises to

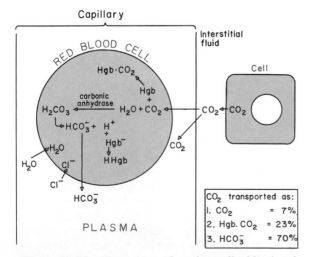

Figure 28–12. Transport of carbon dioxide in the blood.

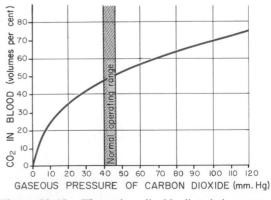

Figure 28–13. The carbon dioxide dissociation curve.

about 52 volumes per cent as the blood passes through the tissues, and falls to about 48 volumes per cent as it passes through the lungs.

THE RESPIRATORY EXCHANGE RATIO

The discerning student will have noted that the normal amount of oxygen transported from the lungs to the tissues by each 100 ml. of blood is about 5 ml., while the normal amount of carbon dioxide transported from the tissues to the lungs is about 4 ml. The ratio of carbon dioxide output to oxygen uptake is called the *respiratory exchange ratio* (R). That is,

$$R = \frac{\text{Rate of carbon dioxide output}}{\text{Rate of oxygen uptake}}$$

The value for R changes under different metabolic conditions. When a person is utilizing entirely carbohydrates for body metabolism, R rises to 1.00, because when oxygen is metabolized with carbohydrates, one molecule of carbon dioxide is formed for each molecule of oxygen consumed. But when fats instead of carbohydrates are being metabolized, R falls to 0.7 because only 0.7 molecules of CO_2 are then formed for each molecule of O_2. This will be discussed further in Chapter 47.

REFERENCES

Bartels, H., and Baumann, R.: Respiratory function of hemoglobin. *Int. Rev. Physiol., 14*:107, 1977.

Bauer, C., *et al.* (eds.): Biophysics and Physiology of Carbon Dioxide. New York, Springer-Verlag, 1980.

Caughey, W. S. (ed.): Oxygen; Biochemical and Clinical Aspects. New York, Academic Press, 1979.

Chance, B.: Regulation of intracellular oxygen. *Proc. Int. Union Physiol. Sci., 6*:13, 1968.

Forster, R. E.: Pulmonary ventilation and blood gas exchange. *In* Sodeman, W. A., Jr., and Sodeman, W. A. (eds.): Pathologic Physiology: Mechanisms of Disease, 5th Ed. Philadelphia, W. B. Saunders Company, 1974, p. 371.

Haldane, J. S., and Priestley, J. G.: Respiration. New Haven, Yale University Press, 1935.

Jöbsis, F. F.: Intracellular metabolism of oxygen. *Am. Rev. Resp. Dis., 110*:58, 1974.

Jones, N. L.: Blood Gases and Acid-Base Physiology. New York, B. C. Decker, 1980.

Kessler, M., *et al.* (eds.): Oxygen Supply. Baltimore, University Park Press, 1973.

Lehman, H., and Huntsman, R. G.: Man's Hemoglobins, 2nd Ed. Philadelphia, J. B. Lippincott Co., 1974.

Michel, C. C.: The transport of oxygen and carbon dioxide by the blood. *In* MTP International Review of Science: Physiology. Vol. 2. Baltimore, University Park Press, 1974, p. 67.

Perutz, M. F.: Hemoglobin structure and respiratory transport. *Sci. Am., 239*(6):92, 1978.

Randall, D. J.: The Evolution of Air Breathing in Vertebrates. New York, Cambridge University Press, 1980.

Robin, E. D., and Simon, L. M.: Oxygen transport and cellular respiration. *In* Frohlich, E. D. (ed.): Pathophysiology, 2nd Ed. Philadelphia, J. B. Lippincott Co., 1976, p. 167.

Wagner, P. D.: Diffusion and chemical reaction in pulmonary gas exchange. *Physiol. Rev., 57*:257, 1977.

29

Regulation of Respiration, and Respiratory Abnormalities

The nervous system adjusts the rate of alveolar ventilation almost exactly to the demands of the body, so that the blood oxygen pressure (Po_2) and carbon dioxide pressure (Pco_2) are hardly altered even during strenuous exercise or other types of respiratory stress.

The first section of the present chapter describes the operation of this neurogenic system for regulation of respiration.

THE RESPIRATORY CENTER

The respiratory center is a widely dispersed group of neurons located bilaterally in the reticular substance of the medulla oblongata and pons, as illustrated in Figure 29–1. It is divided into three major areas: (1) a dorsal medullary group of neurons, which is mainly an *inspiratory area,* (2) a ventral medullary group of neurons, which is mainly an *expiratory area,* and (3) an area in the pons that helps to control the respiratory rate, called the *pneumotaxic area.* It is the inspiratory area that plays the fundamental role in the control of respiration.

The Inspiratory Area

The inspiratory area lies bilaterally in the dorsal portion of the medulla, extending approximately the entire length of the medulla. Its neurons are very near to and interconnect closely with the tractus solitarius, which is the sensory termination of both the vagal and glossopharyngeal nerves; each of these nerves in turn transmits sensory signals from the peripheral chemoreceptors, in this way helping to control pulmonary ventilation. In addition, the vagi also transmit sensory signals from the lungs that help to control lung inflation and respiratory rate, as we shall discuss in a subsequent section of this chapter.

Rhythmical Oscillation in the Inspiratory Area. The basic rhythm of respiration is generated in the inspiratory area. Even when all incoming nerve fiber connections to this area have been sectioned or blocked, the area still emits repetitive bursts of action potentials that cause rhythmical inspiratory cycles.

During expiration the inspiratory center becomes dormant, but after a few seconds this center suddenly and automatically turns on again as a result of an inherent, intrinsic excitability of the inspiratory neurons themselves. And the signals emitted from the inspiratory center are then transmitted to the diaphragm and other inspiratory muscles. At the very onset of the inspiratory signal, this signal is very weak, but its intensity increases progressively, causing the diaphragm and other inspiratory muscles to contract more and more forcefully. In normal respiration this "ramp" increase in inspiratory signal lasts for about two seconds, at the end of which time it suddenly comes to a halt. Then the inspiratory neurons remain dormant again for approximately the next three seconds before the entire cycle repeats itself once more, this repetition continuing on and on throughout the life of the person.

Function of the Pneumotaxic Center in Limiting the Duration of Inspiration and Increasing Respiratory Rate. The pneumotaxic center, located in the pons, transmits inhibitory signals to the inspiratory area. The primary effect of these is to limit the inspiratory signal. When the pneumotaxic signals are strong, inspiration might last for as little as one-half second. But when the pneumotaxic signal is weak the inspiratory ramp signal will continue to ascend the ramp for perhaps as long as five to seven seconds, thus filling the lungs with a great excess of air.

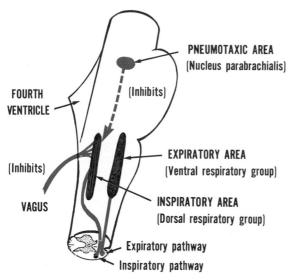

Figure 29–1. Organization of the respiratory center.

Though the function of the pneumotaxic center is primarily to limit inspiration, this has a secondary effect on the rate of breathing as well because limitation of inspiration shortens the period of respiration, and a new cycle of inspiration begins again at a much earlier time. Thus, a strong pneumotaxic signal can increase the rate of breathing up to 30 to 40 breaths per minute, while a weak pneumotaxic signal may reduce the rate of breathing to only a few breaths per minute.

Limitation of Inspiration by Vagal Lung Inflation — The Hering-Breuer Reflex. Located in the walls of the bronchi and bronchioles throughout the lungs are *stretch receptors* that, when overstretched, also transmit inhibitory signals through the vagi into the inspiratory center. Therefore, when the lungs become overly inflated, these stretch receptors activate an appropriate feedback response to limit further inspiration. This is called the *Hering-Breuer reflex.* This reflex also has the same effect as the pneumotaxic signals in increasing the rate of respiration because of the reduced period of inspiration. However, the reflex is not activated until each breath increases to greater than approximately 1.5 liters. Therefore, the Hering-Breuer reflex appears to be mainly a protective mechanism for preventing excess lung inflation rather than an important ingredient in the normal control of ventilation.

The Expiratory Area

Located bilaterally in the medulla, and also extending the entire length of the medulla, is a ventral respiratory group of neurons which when stimulated excites the expiratory muscles. This expiratory area remains dormant during most normal quiet respiration, because quiet respiration is achieved by contraction only of the inspiratory muscles while expiration results from passive recoil of the elastic structures of the lung and surrounding chest cage.

On the other hand, when the respiratory drive becomes much greater than normal, signals then spill over into the expiratory area from the basic oscillating mechanism of the inspiratory area. As a consequence, the expiratory muscles then contribute their powerful contractile forces to the pulmonary ventilatory process. Unfortunately, the neurophysiological basis for this interaction between the inspiratory and expiratory centers is not yet known.

CHEMICAL CONTROL OF RESPIRATION

The ultimate goal of respiration is to maintain proper concentrations of oxygen, carbon dioxide, and hydrogen ions in the body fluid. It is fortunate, therefore, that respiratory activity is highly responsive to changes in any of these concentrations.

Excess carbon dioxide or hydrogen ions affect respiration mainly by direct excitatory effects on the respiratory center itself. The resulting increase in ventilation increases the elimination of carbon dioxide from the blood; this also removes hydrogen ions from the blood because of decreased blood carbonic acid.

Oxygen, on the other hand, does not have a significant direct effect on the respiratory center. Instead, it acts almost entirely on peripheral chemoreceptors located in the carotid and aortic bodies, and these in turn transmit appropriate neuronal signals to the respiratory center for control of respiration.

Let us discuss first the direct stimulation of the respiratory center itself by carbon dioxide and hydrogen ions.

DIRECT CHEMICAL CONTROL OF RESPIRATORY CENTER ACTIVITY BY CARBON DIOXIDE AND HYDROGEN IONS

The Chemosensitive Area of the Respiratory Center. A very sensitive *chemosensitive* area, illustrated in Figure 29–2, is located bilaterally and ventrally in the substance of the medulla. This area is highly sensitive to changes in either blood CO_2 or hydrogen concentration, and it in turn excites the other portions of the respiratory center. It has especially potent effects to increase the degree of activity of the inspiratory center,

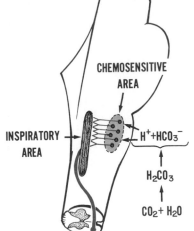

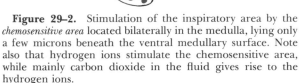

Figure 29–2. Stimulation of the inspiratory area by the *chemosensitive area* located bilaterally in the medulla, lying only a few microns beneath the ventral medullary surface. Note also that hydrogen ions stimulate the chemosensitive area, while mainly carbon dioxide in the fluid gives rise to the hydrogen ions.

increasing both the rate of rise of the inspiratory ramp signal and also the intensity of the signal. This in turn has an automatic secondary effect of increasing the frequency of the repsiratory rhythm.

Response of the Chemosensitive Neurons to Hydrogen Ions — The Primary Stimulus for Exciting Respiration

The sensor neurons in the chemosensitive area are especially excited by hydrogen ions; in fact, it is believed that hydrogen ions are perhaps the only important direct stimulus for these neurons. Unfortunately, though, hydrogen ions do not easily cross either the blood-brain barrier or the blood–cerebrospinal fluid barrier. For this reason, changes in hydrogen ion concentration in the blood actually have considerably less effect in stimulating the chemosensitive neurons than do changes in carbon dioxide, even though carbon dioxide stimulates these neurons indirectly, as will be explained below.

Effect of Blood Carbon Dioxide in Stimulating the Chemosensitive Area

Though carbon dioxide has very little direct effect on stimulating the neurons in the chemosensitive area, it does have a very potent indirect effect. It does this by reacting with the water of the tissues to form carbonic acid, which in turn dissociates into hydrogen and bicarbonate ions; the hydrogen ions then directly stimulate the

area. These effects are illustrated in Figure 29–2.

But, why is it that blood CO_2 has a more potent effect on stimulating the chemosensitive neurons than do blood hydrogen ions? The answer is that hydrogen ions, as noted above, have difficulty passing through the blood-brain barrier and blood–cerebrospinal fluid barrier, while carbon dioxide passes through both these barriers almost as if they did not exist. Consequently, whenever the blood carbon dioxide concentration increases, so also does the Pco_2, both in the interstitial fluid of the medulla and in the cerebrospinal fluid. And, in both of these fluids the carbon dioxide immediately reacts with the water to form hydrogen ions, which then excite respiration.

Stimulation of the Chemosensitive Area by Carbon Dioxide in the Cerebrospinal Fluid

On first thought, one would suspect that most stimulation of respiratory center activity would result from changes in carbon dioxide and hydrogen ion concentrations in the interstitial fluid of the respiratory center itself. However, many respiratory physiologists now believe that an increase in carbon dioxide concentration in the cerebrospinal fluid has more effect on stimulating the chemosensitive area than does a change in carbon dioxide in the interstitial fluid, because of a difference in the degree of acid-base buffering of the two fluids. The cerebrospinal fluid has very little protein buffer, while the interstitial fluid is lined on all sides by cells containing high concentrations of acid-base buffers. Therefore, a given change in carbon dioxide concentration will cause far more change in hydrogen ion concentration in the cerebrospinal fluid than in the interstitial fluid. And, since the chemosensitive neurons are located immediately beneath the surface of the medulla, diffusion of hydrogen ions into these neurons from the cerebrospinal fluid seems to be the major factor in the control of respiration.

One of the advantages of this cerebrospinal fluid system for control of respiration is the rapidity with which it can function. The cerebrospinal fluid is in intimate contact with the very rich blood supply of the arachnoid plexus. Therefore, within seconds after the blood Pco_2 changes, the Pco_2 and hydrogen ion concentration of the cerebrospinal fluid also change. On the other hand, a minute or more is required for full change of the Pco_2 in the brain interstitial fluid.

Decreased Sensitivity of the Respiratory Center to CO_2 During Long Periods of Exposure. The effect of increased blood carbon dioxide on respiration reaches its peak within a minute or so after an increase in blood Pco_2.

Thereafter, the effect gradually declines during the next one to two days, decreasing by the end of that time to as little as one-fifth to one-eighth the initial effect. The exact cause of this decreasing stimulation is not known, but it is believed to result from active transport of bicarbonate ions from the blood into the cerebrospinal fluid through the ependymal cells lining the cerebrospinal fluid cavity. The bicarbonate ions then combine with the excess hydrogen ions, thus reducing the hydrogen ion concentration and simultaneously reducing the respiratory drive.

Therefore, a change in blood carbon dioxide concentration has a very potent *acute* effect for controlling respiration but only a weak *chronic* effect after a few days' adaptation.

Quantitative Effects of Blood P_{CO_2} and Hydrogen Ion Concentration on Alveolar Ventilation

Figure 29–3 illustrates quantitatively the approximate effects of blood P_{CO_2} and blood pH (which is an inverse measure of hydrogen ion concentration) on alveolar ventilation. Note the marked increase in ventilation caused by *acute* increase in P_{CO_2}. But note also the much smaller effect of increased hydrogen ion concentration (that is, decreased pH).

Finally, note that this *difference* in stimulation of ventilation is especially great in the normal P_{CO_2} and pH ranges: P_{CO_2}'s between 40 and 45 mm. Hg and pH's between 7.45 and 7.35. Therefore, from a practical standpoint, changes in blood carbon dioxide play by far the greater role in the normal minute-by-minute control of pulmonary ventilation.

Value of Carbon Dioxide as a Regulator of Alveolar Ventilation.   Since carbon dioxide is one of the end-products of metabolism, its concentration in the body fluids greatly affects the chemical reactions of the cells and also affects the tissue pH. For these reasons, the tissue fluid P_{CO_2} must be regulated exactly. In the preceding chapter, it was pointed out that blood and interstitial fluid P_{CO_2} are determined to a great extent by the rate of alveolar ventilation. Therefore, stimulation of the respiratory center by carbon dioxide provides an important feedback mechanism for regulating the concentration of carbon dioxide throughout the body. That is, (1) an increase in P_{CO_2} stimulates the respiratory center; (2) this increases alveolar ventilation and reduces the alveolar carbon dioxide; (3) as a result, the tissue P_{CO_2} returns most of the way back toward normal. In this way, the respiratory center maintains the P_{CO_2} of the tissue fluids at a relatively constant level.

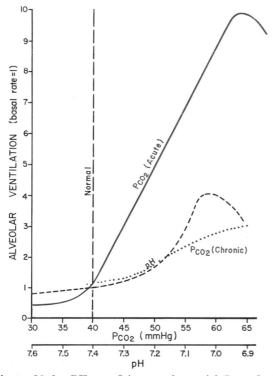

Figure 29–3. Effects of increased arterial P_{CO_2} (both acute and chronic) and decreased arterial pH on the rate of alveolar ventilation.

THE PERIPHERAL CHEMORECEPTOR SYSTEM FOR CONTROL OF RESPIRATORY ACTIVITY — ROLE OF OXYGEN IN RESPIRATORY CONTROL

Aside from the direct sensitivity of the respiratory center itself to CO_2 and hydrogen ions, special chemical receptors called *chemoreceptors,* located outside the central nervous system, are also responsive to changes in oxygen, carbon dioxide, and hydrogen ion concentrations. These transmit signals to the respiratory center to help regulate respiratory activity. The chemoreceptors are located in the *carotid* and *aortic bodies,* which are illustrated in Figure 29–4 along with their afferent nerve connections to the respiratory center. The *carotid bodies* are located bilaterally in the bifurcations of the common carotid arteries, and their afferent nerve fibers pass through Hering's nerves to the glossopharyngeal nerves and thence to the medulla. The *aortic bodies* are located along the arch of the aorta; their afferent nerve fibers pass to the medulla through the vagi. Each of these chemoreceptor bodies receives a special blood supply through a minute artery directly from the adjacent arterial trunk.

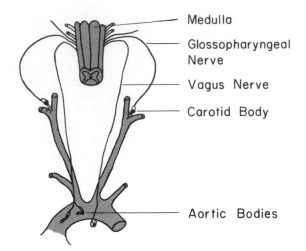

Figure 29–4. Respiratory control by the carotid and aortic bodies.

Stimulation of the Chemoreceptors by Decreased Arterial Oxygen. Changes in arterial oxygen concentration have *no* direct stimulatory effect on the respiratory center itself, but when the oxygen concentration in the arterial blood falls below normal, the chemoreceptors become strongly stimulated. This effect is illustrated in Figure 29–5, which shows the relationship between *arterial* Po_2 and rate of nerve impulse transmission from a carotid body. Note that the impulse rate is particularly sensitive to changes in arterial Po_2 in the range between 60 and 30 mm. Hg, which is the range in which the hemoglobin oxygen saturation decreases rapidly.

Quantitative Effect of Low Blood Po_2 on Alveolar Ventilation

Low blood Po_2 normally will not increase alveolar ventilation significantly until the alveolar Po_2 falls almost to one-half normal. This is illustrated in Figure 29–6. The lowermost curve of this figure shows that changing the alveolar arterial Po_2 from slightly more than 100 mm. Hg down to about 60 mm. Hg has an imperceptible effect on ventilation. But then, as the Po_2 falls still further, down to 40 and then to 30 mm. Hg, alveolar ventilation increases 1.5- to 1.7-fold. Contrast this rather feeble increase in alveolar ventilation to the effects caused by hydrogen ions or carbon dioxide — the four-fold increase caused by decreasing blood pH to 7.0 or the ten-fold increase caused by increasing the Pco_2 only 50 per cent. Thus, it is clear that the normal effect of changes in blood Po_2 on respiratory activity is very slight, especially when compared with the effect of Pco_2.

The cause of the poor effect of Po_2 changes on respiratory control is a "braking" effect caused by *both* the carbon dioxide and the hydrogen ion control mechanisms. This phenomenon can be explained by referring again to Figure 29–6, which shows the slight increase in ventilation as the alveolar Po_2 is decreased. The increase in ventilation that does occur blows off carbon dioxide from the blood and therefore decreases the Pco_2, which is also illustrated in the figure; at the same time it also decreases the hydrogen ion concentration. Therefore, two powerful respiratory inhibitory effects are caused by (a) diminished carbon dioxide and (b) diminished hydrogen ions. These two exert an inhibitory "braking" effect that opposes the excitatory effect of the diminished oxygen.

Conditions Under Which Diminished Oxygen Does Play a Major Role in the Regulation of Respiration. In pneumonia, emphysema, and other lung ailments in which gases are not readily exchanged between the atmosphere and the pulmonary blood, the oxygen regulatory system *does* then play a major role in the regulation of respiration. Contrary to the normal effect, the increased ventilation caused by oxygen lack is not followed by reduced arterial Pco_2 and hydrogen ion concentration, because the pulmonary disease also diminishes carbon dioxide exchange as well as oxygen exchange. Instead, the CO_2 either

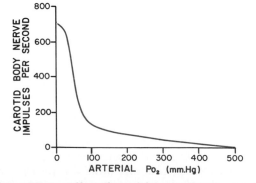

Figure 29–5. Effect of arterial Po_2 on impulse rate from the carotid body of a cat. (Curve drawn from data from several sources, but primarily from Von Euler.)

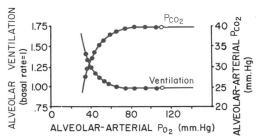

Figure 29–6. Effect of arterial Po_2 on alveolar ventilation and on the subsequent decrease in arterial Pco_2 (From Gray: Pulmonary Ventilation and Its Physiological Regulation. Courtesy of Charles C Thomas, Publisher, Springfield, Illinois.)

remains constant or builds up in the blood, and the hydrogen ion concentration behaves similarly. Therefore, the "braking" effect of these other two control systems on the oxygen lack system is not present. As a result, *the oxygen lack system develops its full power and can then increase alveolar ventilation as much as five- to seven-fold.*

Effects of the Oxygen Lack Mechanism at High Altitudes. When a person first ascends to high altitudes or in any other way is exposed to a rarefied atmosphere, the diminished oxygen in the air stimulates the oxygen lack control system of respiration. The respiration at first increases to a maximum of about two-thirds above normal, which is a comparatively slight increase. Once again, the cause of this slight increase is the tremendous "braking" effect of the carbon dioxide and hydrogen ion control mechanisms on the oxygen lack mechanism.

However, over several days, the respiratory center gradually becomes "adapted" to the diminished carbon dioxide, as explained earlier in the chapter, so that this now depresses the respiratory center very little. Thus, the "braking" effect on the oxygen control is gradually lost, and alveolar ventilation then rises to as high as five to seven times normal. This is part of the acclimatization that occurs as a person ascends a mountain slowly, thus allowing the person to adjust respiration gradually to a level fitted for the higher altitude.

Why Oxygen Regulation of Respiration is Not Normally Needed. On first thought, it seems strange that oxygen should play so little role in the normal regulation of respiration, particularly since one of the primary functions of the respiratory center is to provide adequate intake of oxygen. However, oxygen control of respiration is not needed under most normal circumstances for the following reason:

The respiratory system ordinarily maintains an alveolar Po_2 actually *higher* than the level needed to saturate almost completely the hemoglobin of the arterial blood. It does not matter whether alveolar ventilation is normal or 10 times normal, the blood will still be essentially fully saturated. Also, alveolar ventilation can decrease to as low as one-half normal, and the blood still remains within 10 per cent of complete saturation. Therefore, one can see that alveolar ventilation can change tremendously without significantly affecting oxygen transport to the tissues.

On the other hand, changes in alveolar ventilation do have a tremendous effect on tissue carbon dioxide concentration, as was explained earlier in the chapter and illustrated in Figure 29–6. Therefore, it is exceedingly important that carbon dioxide — not oxygen — be the major controller of respiration under normal conditions.

REGULATION OF RESPIRATION DURING EXERCISE

In strenuous exercise, oxygen utilization and carbon dioxide formation can increase as much as 20-fold, as shown in Figure 29–7. Yet, except in very heavy exercise, alveolar ventilation ordinarily increases almost the same amount, so that the blood Po_2, Pco_2, and pH all remain *almost exactly normal.*

In trying to analyze the factors that cause increased ventilation during exercise, one is tempted immediately to ascribe it to the chemical alterations in the body fluids during exercise, including increase of carbon dioxide, increase of hydrogen ions, and decrease of oxygen. However, this supposition is not valid, for measurements of arterial Pco_2, pH, and Po_2 show that none of these changes significantly — certainly not enough to account for the increase in ventilation. Indeed, even if a very high Pco_2 should develop during exercise, this still would be sufficient to account for only two-thirds of the increased ventilation of heavy muscular exercise, for, as shown in Figure 29–8, the minute respiratory volume in exercise is about 50 per cent greater than that which can be effected by maximal carbon dioxide stimulation.

Therefore, the question must be asked: What is it during exercise that causes the intense respiration? At least two different effects seem to predominate:

1. The cerebral cortex, on transmitting impulses to the contracting muscles, is believed to transmit collateral impulses into the reticular substance of the brain stem to excite the respiratory center. This is analogous to the stimulatory effect that causes the arterial pressure to rise

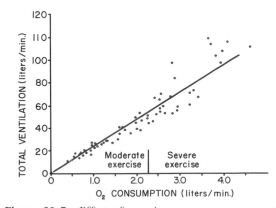

Figure 29–7. Effect of exercise on oxygen consumption and ventilatory rate. (From Gray: Pulmonary Ventilation and Its Physiological Regulation. Courtesy of Charles C Thomas, Publisher, Springfield, Illinois.)

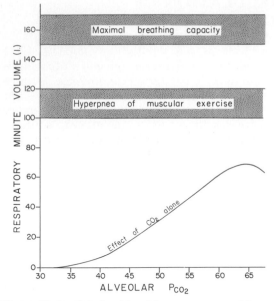

Figure 29–8. Relationship of hyperpnea caused by muscular exercise to that caused by increased alveolar P_{CO_2} (Modified from Comroe, J. H., Jr., et al.: The Lung: Clinical Physiology and Pulmonary Function Tests, 2nd ed. Copyright © 1962 by Year Book Medical Publishers, Inc., Chicago. Reproduced with permission. Data from *Am. J. Physiol., 130*:777, 1940; *137*:256, 1942; *138*:659, 1943; *149*:277, 1947; *J. Indust. Hyg., 11*:293, 1929; and *J. Clin Invest., 27*:500, 1948.)

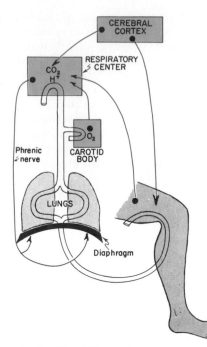

Figure 29–9. The different factors that enter into regulation of respiration during exercise.

during exercise when similar collateral impulses pass to the vasomotor center.

2. During exercise, the body movements, especially of the limbs, are believed to increase pulmonary ventilation by exciting joint proprioceptors that then transmit excitatory impulses to the respiration center. The reason for believing this is that even passive movements of the limbs often increase pulmonary ventilation several-fold.

Interrelationship Between Humoral Factors and Nervous Factors in the Control of Respiration During Exercise. Figure 29–9 illustrates diagrammatically the different factors that operate in the control of respiration during exercise, showing two neurogenic factors, (1) direct stimulation of the respiratory center by the cerebral cortex and (2) indirect stimulation by proprioceptors, and showing also the three humoral factors, (1) carbon dioxide, (2) hydrogen ions, and (3) oxygen.

Most times when a person exercises, the nervous factors stimulate the respiratory center almost exactly the proper amount to supply the extra oxygen requirements for the exercise and to blow off the extra carbon dioxide. But, occasionally, the nervous signals are either too strong or too weak in their stimulation of the respiratory center. Then, the humoral factors play a very significant role in bringing about the final adjustment in respiration required to keep the carbon dioxide and hydrogen ion concentrations of the body fluids as nearly normal as possible.

ABNORMALITIES OF RESPIRATORY CONTROL

RESPIRATORY CENTER DEPRESSION

Cerebrovascular Disease. Probably the most common cause of long-term respiratory center depression is cerebrovascular disease in older patients, especially following vascular occlusions or hemorrhages that damage the respiratory center areas. In such instances, a person may have chronically elevated arterial P_{CO_2}'s and depressed P_{O_2}'s.

Acute Brain Edema. The activity of the respiratory center may be depressed or totally inactivated by acute brain edema resulting from brain concussion. For instance, the head might be struck against some solid object, following which the damaged brain tissues swell, compressing the cerebral arteries against the cranial vault and thus totally or partially blocking the cerebral blood supply. As a result, the neurons of the respiratory center become inactive.

Occasionally, respiratory depression resulting from brain edema can be relieved temporarily by intravenous injection of hypertonic solutions such as highly concentrated glucose solution. These solutions osmotically remove some of the intracellular fluids of the brain, thus relieving intracranial pressure and sometimes reestablishing respiration within a few minutes.

Anesthesia. Perhaps the most prevalent cause of respiratory depression and respiratory arrest is overdosage of anesthetics or narcotics. The best agent for anesthesia is one that depresses the respiratory center the least while depressing the cerebral cortex the most. Ether is among the best of the anesthetics by these criteria, though halothane, cyclopropane, ethylene, nitrous oxide, and a few others have almost the same value. On the other hand, sodium pentobarbital is a poor anesthetic because it depresses the respiratory center considerably more than the above agents.

PERIODIC BREATHING

The most common type of periodic breathing, *Cheyne-Stokes breathing,* is characterized by slowly increasing and then decreasing respiration, occurring over and over again every 45 seconds to 3 minutes. The basic mechanism of this is the following:

Let us assume that the respiration becomes much more rapid and deeper than usual. This causes the Pco_2 in the pulmonary blood to *decrease.* A few seconds later the pulmonary blood reaches the brain, and the decreased Pco_2 inhibits respiration. As a result, the pulmonary blood Pco_2 gradually *increases.* After another few seconds the blood carrying the increased CO_2 arrives at the respiratory center and stimulates respiration again, thus making the person overbreathe once again and initiating a new cycle of depressed respiration; and the cycles thus continue on and on, causing Cheyne-Stokes periodic breathing.

In the normal person, the respiratory control system is stable enough to prevent Cheyne-Stokes breathing, but a number of different abnormal conditions can overcome the stability of the feedback mechanism and cause it to oscillate spontaneously. Two of these are (1) increased delay time in the flow of blood from the lungs to the brain and (2) increased feedback gain of the respiratory center mechanisms for control of respiration.

ABNORMAL VENTILATION-PERFUSION RATIO

In Chapter 28 we pointed out that probably the most common cause of decreased lung diffusing capacity is abnormal ventilation-perfusion ratio. That is, in some alveoli there is too little ventilation for the amount of blood flow, so that the blood cannot become fully oxygenated. On the other hand, in other alveoli ventilation is adequate but there is too little blood flow to accept the oxygen. Thus, in either instance, oxygen transfer to the blood becomes greatly compromised. These principles were discussed in detail in Chapter 28. It was pointed out that underventilated alveoli, in which the ventilation to perfusion ratio is less than normal, leads to so-called *physiologic shunt*—that is, blood that is shunted past the lungs without becoming oxygenated. And it was also pointed out that when alveoli are overventilated while blood flow is too little—that is, the ventilation to perfusion ratio is greater than normal—causes *physiologic dead space.* This means that there is ventilation that is not being used by the blood for oxygenation and therefore is *wasted ventilation.* It would be good for the student to return to Chapter 28 and review the principles of abnormal ventilation-perfusion ratio because of the very high prevalence of this abnormality in respiratory disease, especially in the most common of all serious lung diseases, *pulmonary emphysema* caused by smoking.

Diseases that cause abnormal ventilation-perfusion ratios include *thrombosis of a pulmonary artery, excessive airway resistance to some alveoli (emphysema), reduced compliance of one lung without concomitant abnormality of the other lung,* and many other conditions that cause diffuse damage throughout the lungs.

PHYSIOLOGIC PECULIARITIES OF SPECIFIC PULMONARY ABNORMALITIES

CHRONIC EMPHYSEMA

Chronic emphysema is prevalent mainly because of the effects of tobacco smoking. It results from two major pathophysiological changes in the lungs. First, air flow through many of the terminal bronchioles is obstructed. Second, many of the alveolar walls are destroyed.

Many clinicians believe that chronic emphysema begins with chronic infection in the lung that causes *bronchiolitis,* which means inflammation of the small air passages of the lungs. This inflammation also involves alveolar septa and destroys many of them. Also obstruction to expiration causes excess expiratory pressures in the alveoli, and these pressures rupture the alveolar septa. Thus, many bronchioles become irreparably obstructed, and the total surface of the respiratory membrane also becomes greatly decreased, as illustrated in Figure 29–10, sometimes to as little as one-tenth to one-quarter normal.

The physiological effects of chronic emphysema are extremely varied, depending on the severity of the disease and on the relative degree of bronchiolar obstruction versus lung tissue destruction. However, among the different abnormalities are the following:

First, the bronchiolar obstruction greatly *increases airway resistance* and results in greatly increased work of breathing. It is especially difficult for the person to move air through the bronchioles during expiration because the compressive force on the outside of the lung not only compresses the alveoli but also compresses the bronchioles, which further increases their resistance.

Second, the marked loss of lung tissue greatly *decreases the diffusing capacity* of the lung, which reduces the ability of the lungs to oxygenate the blood and to remove carbon dioxide.

Third, the obstructive process is frequently much worse in some parts of the lungs than in other parts, so that some portions of the lungs are well ventilated while other portions are poorly ventilated. This often causes an *extremely abnormal ventilation-perfusion ratio,* which causes *physiologic shunt* resulting in poor aeration of the blood and also greatly expanded *physiologic dead space* resulting in wasted ventilation, both effects occurring in the same lungs.

Fourth, loss of large portions of the lung also decreases the number of pulmonary capillaries through which blood can pass. As a result, the pulmonary vascular resistance increases markedly, causing pulmonary hypertension. This in turn overloads the right heart and frequently causes right-heart failure.

Chronic emphysema usually progresses slowly over many years. The person develops hypoxia and hypercapnia because of hypoventilation of many alveoli and loss of lung parenchyma. The net result of all of these effects is severe and prolonged air hunger that can last for years until the hypoxia and hypercapnia cause death — a very high penalty to pay for smoking.

PNEUMONIA

The term pneumonia describes any inflammatory condition of the lung in which the alveoli are usually filled with fluid and cells. A common type of pneumonia is *bacterial pneumonia,* caused most frequently by pneumococci. This disease begins with infection in the alveoli; the pulmonary membrane becomes inflamed and highly porous, so that fluid and even red and white blood cells pass out of the blood into the alveoli. Thus, the infected alveoli become progressively filled with fluid and cells, and the infection spreads by extension of bacteria from alveolus to alveolus. Eventually, large areas of the lungs, sometimes whole lobes or even a whole lung, become "consolidated," which means that they are filled with fluid and cellular debris.

The pulmonary function of the lungs during pneumonia changes in different stages of the disease. In the early stages, the pneumonia process may well be localized to only one lung, and alveolar ventilation may be reduced even though blood flow through the lung continues almost normally. This results in two major pulmonary abnormalities: (1) reduction in the total available surface area of the respiratory membrane and (2) decreased ventilation-perfusion ratio. Both these effects cause reduced diffusing capacity, which results in hypoxemia.

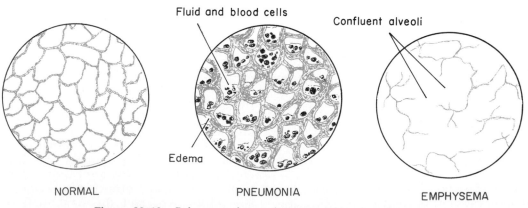

NORMAL   PNEUMONIA   EMPHYSEMA

Figure 29–10. Pulmonary changes in pneumonia and emphysema.

BRONCHIAL ASTHMA

Bronchial asthma is usually caused by allergic hypersensitivity of the person to foreign substances in the air — especially to plant pollens. The allergic reaction causes (1) localized edema in the walls of the small bronchioles as well as secretion of thick mucus into the bronchiolar lumens, and (2) spasm of the bronchiolar smooth muscle. Obviously, therefore, the airway resistance increases greatly.

As discussed in Chapter 27, the bronchiolar diameter becomes more reduced during expiration than during inspiration in asthma. The reason for this is that the increased intrapulmonary pressure during expiratory effort not only compresses the air in the alveoli but compresses the outside of the bronchioles as well. Therefore, the asthmatic person usually can inspire quite adequately but has great difficulty expiring. This results in dyspnea, or "air hunger."

The functional residual capacity of the lung becomes greatly increased during the asthmatic attack because of the difficulty in expiring air from the lungs. Over a long period of time the chest cage becomes permanently enlarged, causing a "barrel chest."

TUBERCULOSIS

In tuberculosis the tubercle bacilli cause a peculiar tissue reaction in the lungs, including, first, invasion of the infected region by macrophages and, second, walling off of the lesion by fibrous tissue to form the so-called "tubercle." This walling-off process helps to limit further transmission of the tubercle bacilli in the lungs and therefore is part of the protective process against the infection. However, in approximately 3 per cent of all persons who contract tuberculosis, the walling-off process fails, and tubercle bacilli spread throughout the lungs, causing many areas of fibrosis and reducing the total amount of functional lung tissue. These effects cause (1) increased effort on the part of the respiratory muscles causing pulmonary ventilation, and therefore *reduced vital capacity*, (2) *reduced total respiratory membrane surface area* and *increased thickness of the respiratory membrane*, these causing progressively diminished pulmonary diffusing capacity, and (3) *abnormal ventilation-perfusion ratio* in the lungs, further reducing the oxygenation of the blood.

CYANOSIS

The term "cyanosis" means blueness of the skin, and its cause is excessive amounts of deoxygenated hemoglobin in the skin blood vessels, especially in the capillaries. This deoxygenated hemoglobin has an intense dark blue color that is transmitted through the skin. The presence of cyanosis is one of the most common clinical signs of different degrees of respiratory insufficiency.

DYSPNEA

Dyspnea means a desire for air or mental anguish associated with the act of ventilating enough to satisfy the air demand. A common synonym is "air hunger."

At least three different factors often enter into the development of the sensation of dyspnea. These are: (1) abnormality of the respiratory gases in the body fluids, especially excess carbon dioxide and to a much less extent hypoxia, (2) the amount of work that must be performed by the respiratory muscles to provide adequate ventilation, and (3) the state of the mind.

At times, the levels of both carbon dioxide and oxygen in the body fluids are completely normal, but to attain this normality of the respiratory gases, the person has to breathe forcefully. In these instances the forceful activity of the respiratory muscles gives the person a sensation of air hunger.

Finally, the person's respiratory functions may be completely normal, and still dyspnea may be experienced because of an abnormal state of mind. This is called *neurogenic dyspnea* or, sometimes, *emotional dyspnea*. For instance, almost anyone momentarily thinking about the act of breathing may suddenly start taking breaths a little more deeply than ordinarily because of a feeling of mild dyspnea. This feeling is greatly enhanced in persons who have a fear of not being able to receive a sufficient quantity of air. For example, many persons on entering small or crowded rooms immediately experience emotional dyspnea, and patients with "cardiac neurosis" who have heard that dyspnea is associated with heart failure frequently experience severe psychic dyspnea even though the blood gases are completely normal. Neurogenic dyspnea has been known to be so intense that the person over-respires and causes alkalotic tetany.

OXYGEN THERAPY IN THE DIFFERENT TYPES OF HYPOXIA

Oxygen can be administered by (1) placing the patient's head in a "tent" that contains air fortified with oxygen, (2) allowing the patient to breathe either pure oxygen or high concentrations of oxygen from a mask, or (3) administering oxygen through a nasal tube.

Oxygen therapy is of great value in certain types of hypoxia but of almost no value at all in other types. However, recalling the basic physiological principles of the different types of hypoxia, one can readily decide when oxygen therapy is of value and, if so, how valuable. For instance:

In *atmospheric hypoxia,* oxygen therapy can obviously completely correct the depressed oxygen level in the inspired gases and therefore provide 100 per cent effective therapy.

In *hypoventilation hypoxia* or *hypoxia caused by impaired diffusion,* a person breathing 100 per cent oxygen can move 5 times as much oxygen into the alveoli with each breath as when breathing normal air. Therefore, here again oxygen therapy can be extremely beneficial, increasing the available oxygen to as much as 400 per cent above normal. (But this does nothing for the hypercapnia also caused by these conditions.)

In *hypoxia caused by anemia, carbon monoxide poisoning,* or *any other abnormality of hemoglobin transport,* oxygen therapy is of only slight value because the amount of oxygen transported by the hemoglobin is hardly altered. Yet, a small amount of extra oxygen can be transported in the dissolved state.

In the different types of *hypoxia caused by inadequate tissue use of oxygen,* such as when cyanide poisons the respiratory enzymes, there is no abnormality of oxygen pickup by the lungs or of transport to the tissues. Instead, the tissues simply cannot utilize the oxygen that is transported to them. Therefore, oxygen therapy is of essentially no benefit.

REFERENCES

Regulation of Respiration

Bainton, C. R.: Effect of speed versus grade and shivering on ventilation in dogs during active exercise. *J. Appl. Physiol., 33*:778, 1972.

Cherniack, N. S., and Longobardo, G. S.: Cheyne-Stokes breathing. An instability in physiologic control. *N. Engl. J. Med., 228*:952, 1973.

Cohen, M. I.: Neurogenesis of respiratory rhythm in the mammal. *Physiol. Rev., 59*:1105, 1979.

Cunningham, D. J. C.: Integrative aspects of the regulation of breathing: A personal view. *In* MTP International Review of Science: Physiology. Vol. 2. Baltimore, University Park Press, 1974, p. 303.

Fitzgerald, R., *et al.* (eds.): The Regulation of Respiration During Sleep and Anesthesia. New York, Plenum Press, 1978.

Guyton, A. C., *et al.:* Basic oscillating mechanism of Cheyne-Stokes breathing. *Am. J. Physiol., 187*:395, 1956.

Guz, A.: Regulation of respiration in man. *Annu. Rev. Physiol., 37*:303, 1975.

Hechtman, H .B. (ed.): Acute Respiratory Failure: Etiology and Treatment. West Palm Beach, Fla., CRC Press, 1979.

Huch, A., *et al.* (eds.): Continuous Transcutaneous Blood Gas Monitoring. New York, A. R. Liss, 1979.

Loeschcke, H. H.: Central nervous chemoreceptors. *In* MTP International Review of Science: Physiology. Vol. 2. Baltimore, University Park Press, 1974, p. 167.

Milhorn, H. T., Jr., and Guyton, A. C.: An analog computer analysis of Cheyne-Stokes breathing. *J. Appl. Physiol., 20*:328, 1965.

Mitchell, R. A.: Control of respiration. *In* Frohlich, E. D. (ed.): Pathophysiology, 2nd Ed. Philadelphia, J. B. Lippincott Co., 1976, p. 131.

Paintal, A. S., and Gill-Kumar, P. (eds.): Respiratory Adaptations, Capillary Exchange, and Reflex Mechanisms. Delhi, India, Vallabhai Patel Chest Institute, University of Delhi, 1977.

Von Euler, C., and Lagercrantz, H. (eds.): Central Nervous Control Mechanisms in Breathing. New York, Pergamon Press, 1980.

Williams, M. H. (ed.): Symposium on Pulmonary Disease. Med. Clin. North Am., *61*:1161–1442, 1977.

Respiratory Abnormalities

Avery, M. E., *et al.:* The lung of the newborn infant. *Sci. Am., 228*:74, 1973.

Cohen, A. B., and Gold, W. M.: Defense mechanisms of the lungs. *Annu. Rev. Physiol., 37*:325, 1975.

Comroe, H. J., Jr., *et al.*: The Lung: Clinical Physiology and Pulmonary Function Tests, 2nd Ed. Chicago, Year Book Medical Publishers, 1962.

Crofton, J., and Douglas, A.: Respiratory Diseases. Philadelphia, J. B. Lippincott Co., 1975.

Dosman, J. A., and Cotton, D. J. (eds.): Occupational Pulmonary Disease: Focus on Grain Dust and Health. New York, Academic Press, 1979.

Fishman, A. P.: Assessment of Pulmonary Function. New York, McGraw-Hill, 1980.

Fishman, A. P., and Pietra, G. G.: Primary pulmonary hypertension. *Annu. Rev. Med., 31*:421, 1980.

Guenter, C. A., *et al.:* Clinical Aspects of Respiratory Physiology. Philadelphia, J. B. Lippincott Co., 1978.

Guyton, A. C., and Farish, C. A.: A rapidly responding continuous oxygen consumption recorder. *J. Appl. Physiol., 14*:143, 1959.

Hall, W. J., and Douglas, R. G., Jr.: Pulmonary function during and after common respiratory infections. *Annu. Rev. Med., 31*:233, 1980.

Hodgkin, J. E. (ed.): Chronic Obstructive Pulmonary Disease: Current Concepts in Diagnosis and Comprehensive Care. Park Ridge, Ill., American College of Chest Physicians, 1979.

Irsigler, G. B., and Severinghaus, J. W.: Clinical problems of ventilatory control. *Annu. Rev. Med., 31*:109, 1980.

Lane, D. J.: Asthma: The Facts. New York, Oxford University Press, 1979.

Moser, K. M. (ed.): Pulmonary Vascular Diseases. New York, Marcel Dekker, 1979.

Paleček, F.: Control of breathing in diseases of the respiratory system. *Int. Rev. Physiol., 14*:255, 1977.

Petty, T. L. (ed.): Chronic Obstructive Pulmonary Disease. New York, Marcel Dekker, 1978.

Putnam, S. J. (ed.): Advances in Pulmonary Medicine. Philadelphia, W. B. Saunders Co., 1978.

Staub, N. C.: Pulmonary edema. *Physiol. Rev., 54*:678, 1974.

Tisi, G. M.: Pulmonary Physiology in Clinical Medicine. Baltimore, Williams & Wilkins, 1980.

Wilson, A. F., and McPhillips, J. J.: Pharmacological control of asthma. *Annu. Rev. Pharmacol. Toxicol., 18*:541, 1978.

Wolfe, W. G., and Sabiston, D. C.: Pulmonary Embolism. Philadelphia, W. B. Saunders Co., 1980.

Wright, G. R., and Shepard, R. J.: Physiological effects of carbon monoxide. *In* Robertshaw, D. (ed.): International Review of Physiology: Environmental Physiology III. Vol. 20. Baltimore, University Park Press, 1979, p. 311.

Part VIII

AVIATION, SPACE, AND DEEP SEA DIVING PHYSIOLOGY

30

Aviation, Space, and Deep Sea Diving Physiology

As people have ascended to higher and higher altitudes in aviation, in mountain climbing, and in space vehicles, it has become progressively more important to understand the effects of altitude and low gas pressures on the human body. And as they have gone deeper in the sea, it has become necessary to understand the effects of high gas pressures as well.

The present chapter deals with these problems: first, the hypoxia at high altitudes; second, the other physical factors affecting the body at high altitudes; third, the tremendous acceleratory forces that occur in both aviation and space physiology; and, finally, the effects of high-pressure gases at the depths of the sea.

EFFECTS OF LOW OXYGEN PRESSURE ON THE BODY

Barometric Pressures at Different Altitudes. Table 30–1 gives the barometric pressures at different altitudes, showing that at sea level the pressure is 760 mm. Hg, while at 10,000 feet it is only 523 mm. Hg, and at 50,000 feet, 87 mm. Hg. The decrease in barometric pressure is the basic cause of all the hypoxia problems in high altitude physiology, for as the barometric pressure decreases, the oxygen pressure decreases proportionately, remaining at all times slightly less than 21 per cent of the total barometric pressure.

Oxygen Partial Pressures in the Atmosphere at Different Elevations. Table 30–1 also shows that the partial pressure of oxygen (Po_2) in dry air

at sea level is approximately 159 mm. Hg, though this can be decreased as much as 10 mm. when large amounts of water vapor exist in the air. The Po_2 at 10,000 feet is approximately 110 mm. Hg, at 20,000 feet, 73 mm. Hg, and at 50,000 feet, 18 mm. Hg.

ALVEOLAR PO_2 AT DIFFERENT ELEVATIONS

Obviously, when the Po_2 in the atmosphere decreases at higher elevations, a decrease in alveolar Po_2 is also to be expected. At low altitudes the alveolar Po_2 does not decrease quite so much as the atmospheric Po_2 because increased pulmonary ventilation helps to compensate for the diminished atmospheric oxygen. But at higher altitudes the alveolar Po_2 decreases even more than atmospheric Po_2 for peculiar reasons that are explained as follows:

Effect of Carbon Dioxide and Water Vapor on Alveolar Oxygen. Even at high altitudes carbon dioxide is continually excreted from the pulmonary blood into the alveoli. Also, water vaporizes into the alveolar space from the respiratory surfaces. Therefore, these two gases dilute the oxygen and nitrogen already in the alveoli, thus reducing the oxygen concentration.

The presence of carbon dioxide and water vapor in the alveoli becomes exceedingly important at high altitudes, because the total barometric pressure falls to low levels while the pressures of carbon dioxide and water vapor do not fall comparably. Water vapor pressure remains at 47

TABLE 30–1 EFFECTS ON ALVEOLAR GAS CONCENTRATIONS AND ON ARTERIAL OXYGEN SATURATION OF ACUTE EXPOSURE TO LOW ATMOSPHERIC PRESSURES

Altitude (ft.)	Barometric Pressure (mm. Hg)	P_{O_2} in Air (mm. Hg)	Breathing Air			Breathing Pure Oxygen		
			P_{CO_2} in Alveoli (mm. Hg)	P_{O_2} in Alveoli (mm. Hg)	Arterial Oxygen Saturation (%)	P_{CO_2} in Alveoli (mm. Hg)	P_{O_2} in Alveoli (mm. Hg)	Arterial Oxygen Saturation (%)
0	760	159	40	104	97	40	673	100
10,000	523	110	36	67	90	40	436	100
20,000	349	73	24	40	70	40	262	100
30,000	226	47	24	21	20	40	139	99
40,000	141	29	24	8	5	36	58	87
50,000	87	18	24	1	1	24	16	15

mm. Hg as long as the body temperature is normal, regardless of altitude; and the pressure of carbon dioxide falls from about 40 mm. Hg at sea level only to about 24 mm. Hg at extremely high altitudes because of increased respiration.

Now let us see how the pressures of these two gases affect the available space for oxygen. Let us assume that the total barometric pressure falls to 100 mm. Hg; 47 mm. Hg of this must be water vapor, leaving only 53 mm. Hg for all the other gases; 24 mm. Hg of the 53 mm. Hg must be carbon dioxide, leaving a remaining space of only 29 mm. Hg. If there were no uptake of oxygen from the alveoli by the blood, one-fifth of this 29 mm. Hg would be oxygen and four-fifths would be nitrogen; or, the P_{O_2} in the alveoli would be 6 mm. Hg. However, most of this last remaining alveolar oxygen would be absorbed into the blood, leaving not more than 1 mm. Hg oxygen pressure in the alveoli. Therefore, at a barometric pressure of 100 mm. Hg (an altitude of about 47,000 feet), the person could not possibly survive when breathing air. But the effect is very much different if the person is breathing pure oxygen, as we shall see in the following discussions.

Alveolar P_{O_2} at Different Altitudes. Table 30–1 also shows the P_{O_2}'s in the alveoli at different altitudes when one is breathing air and when breathing pure oxygen. When one is breathing air, the alveolar P_{O_2} is 104 mm. Hg at sea level; it falls to approximately 67 mm. Hg at 10,000 feet and to only 1 mm. Hg at 50,000 feet.

Saturation of Hemoglobin With Oxygen at Different Altitudes. Figure 30–1 illustrates arterial oxygen saturation at different altitudes when one is breathing air and when breathing oxygen, and the actual per cent saturation at each 10,000 foot level is given in Table 30–1. Up to an altitude of approximately 10,000 feet, even when air is breathed, the arterial oxygen saturation remains at least as high as 90 per cent. However, above 10,000 feet the arterial oxygen saturation falls progressively, as illustrated by the left-hand curve of the figure, until it is only 70 per cent at 20,000 feet altitude and still less at higher altitudes.

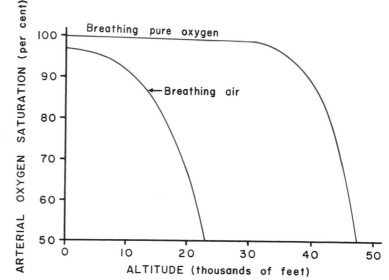

Figure 30–1. Effect of low atmospheric pressure on arterial oxygen saturation when air is breathed and when pure oxygen is breathed.

EFFECT OF BREATHING PURE OXYGEN ON THE ALVEOLAR P_{O_2} AT DIFFERENT ALTITUDES

Referring once again to Table 30–1, note that when a person breathes air at 30,000 feet, his alveolar P_{O_2} is only 21 mm. Hg even though the barometric pressure is 226 mm. Hg. Much of this difference is caused by the fact that a considerable proportion of his alveolar air is nitrogen. But if he breathes pure oxygen instead of air, most of the space in the alveoli formerly occupied by nitrogen now becomes occupied by oxygen instead. Theoretically, at this altitude the aviator could have an alveolar P_{O_2} of 139 mm. Hg instead of the 21 mm. Hg that he has when he breathes air.

The second curve of Figure 30–1 illustrates the arterial oxygen saturation at different altitudes when one is breathing pure oxygen. Note that the saturation remains above 90 per cent until the aviator ascends to approximately 39,000 feet; then it falls rapidly to approximately 50 per cent at about 47,000 feet. This is about the lowest limit that the aviator can tolerate for a long time, so that this altitude is called the "ceiling."

EFFECTS OF HYPOXIA

The rate of pulmonary ventilation ordinarily does not increase significantly until one has ascended to about 8000 feet. At this height the arterial oxygen saturation has fallen to approximately 93 per cent, at which level the chemoreceptors begin to respond significantly. Above 8000 feet the chemoreceptor stimulatory mechanism progressively increases the ventilation until one reaches approximately 16,000 to 20,000 feet, at which altitude the ventilation has reached a maximum of approximately 65 per cent above normal. Further increase in altitude does not further activate the chemoreceptors.

Other effects of hypoxia, beginning at an altitude of about 12,000 feet, are drowsiness, lassitude, mental fatigue, sometimes headache, occasionally nausea, and sometimes euphoria. Most of these symptoms increase in intensity at still higher altitudes, the headache often becoming especially prominent and the cerebral symptoms sometimes progressing to the stage of twitchings or convulsions and, above 23,000 feet in the unacclimatized person, to coma.

One of the most important effects of hypoxia is decreased mental proficiency, which decreases judgment, memory, and the performance of discrete motor movements. Ordinarily these abilities remain absolutely normal up to approximately 9000 feet, and they may be completely normal for a short time up to elevations of 15,000 feet. But, if the aviator is exposed to hypoxia for a long time, his mental proficiency, as measured by reaction times, handwriting, and other psychological tests, may decrease to 80 per cent of normal even at altitudes as low as 11,000. If an aviator stays at 15,000 feet for one hour without oxygen, his mental proficiency ordinarily will have fallen to approximately 50 per cent of normal, and after 18 hours at this level, to approximately 20 per cent of normal.

ACCLIMATIZATION TO LOW P_{O_2}

If a person remains at high altitudes for days, weeks, or years, he gradually becomes acclimatized to the low P_{O_2}, so that it causes fewer and fewer deleterious effects to his body and also so that it becomes possible for him to work harder or to ascend to still higher altitudes. Several means by which acclimatization comes about are (1) further increase in pulmonary ventilation, (2) increased hemoglobin in the blood, and (3) increased vascularity of the tissues.

Further Increase in Pulmonary Ventilation. On immediate exposure to low P_{O_2}, the hypoxic stimulation of the chemoreceptors increases alveolar ventilation to a maximum of about 65 per cent. This is an immediate compensation for the high altitude, and it alone allows the person to rise several thousand feet higher than would be possible without the increased ventilation. But, if he remains at a very high altitude for several days, his ventilation gradually increases to as much as five to seven times normal. The basic cause of this gradual increase is the following:

The immediate 65 per cent increase in pulmonary ventilation on rising to a high altitude blows off large quantities of carbon dioxide, reducing the P_{CO_2} and increasing the pH of the body fluids. Both of these changes *inhibit* the respiratory center and thereby *oppose the stimulation by the hypoxia.* However, during the ensuing two to five days, this inhibition fades away, allowing the respiratory center now to respond with full force to the chemoreceptor stimuli resulting from hypoxia, and the ventilation increases to about five to seven times normal.

Increase in Hemoglobin During Acclimatization. It will be recalled from Chapter 5 that hypoxia is the principal stimulus for an increase in red blood cell production. Ordinarily, in full acclimatization to low oxygen the hematocrit rises from a normal value of 40 to 45 to an average of 60 to 65, with an average increase in hemoglobin concentration from the normal of 15 gm. per cent to about 22 gm. per cent.

In addition, the blood volume also increases, often by as much as 20 to 30 per cent, resulting in a total increase in circulating hemoglobin of as much as 50 to 90 per cent.

Unfortunately, this increase in hemoglobin and blood volume is a slow one, having almost no effect until after two to three weeks, reaching half development in a month or so, and becoming fully developed only after many months.

Increased Vascularity. Histological studies of animals that have been exposed to low oxygen levels for months or years show *increased vascularity* (increased numbers and sizes of capillaries) of the hypoxic tissues. This helps to explain what happens to the 20 to 30 per cent increase in blood volume, and it means that the blood comes into much closer contact with the tissue cells than normally.

NATURAL ACCLIMATIZATION OF NATIVES LIVING AT HIGH ALTITUDES

Many natives of the Andes and of the Himalayas live at altitudes above 13,000 feet — one group in the Peruvian Andes actually living at an altitude of 17,500 feet and working a mine at an altitude of 19,000 feet. Many of these natives are born at these altitudes and live there all their lives. In all of the aspects of acclimatization listed above, the natives are superior to even the best acclimatized lowlanders, even though the lowlanders might have also lived at high altitudes for ten or more years. This process of acclimatization of the natives begins in infancy. The chest size, especially, is greatly increased whereas the body size is somewhat decreased, giving a high ratio of ventilatory capacity to body mass. In addition, the heart, particularly the right heart which provides a high pulmonary arterial pressure to pump blood through a greatly expanded pulmonary capillary system, is considerably larger than the heart of a lowlander.

To give an idea of the importance of acclimatization, consider this: At an altitude of 17,000 feet, the work capacities in per cent of sea level maximum for a normal person are the following:

	per cent
Unacclimatized	50
Acclimatized for two months	68
Native living at 13,200 feet but working at 17,000 feet	87

Thus, naturally acclimatized natives can achieve a daily work output even at these high altitudes almost equal to that of a normal person at sea level, but even well acclimatized lowlanders almost never can achieve this result.

EFFECTS OF ACCELERATORY FORCES ON THE BODY IN AVIATION AND SPACE PHYSIOLOGY

Because of rapid changes in velocity and direction of motion in airplanes and space ships, several types of acceleratory forces often affect the body during flight. At the beginning of flight, simple linear acceleration occurs; at the end of flight, deceleration; and every time the airplane turns, angular and centrifugal acceleration occur. In aviation physiology it is usually centrifugal acceleration that demands greatest consideration, because the structure of the airplane is capable of withstanding much greater centrifugal acceleration than is the human body.

CENTRIFUGAL ACCELERATORY FORCES

When an airplane makes a turn, the force of centrifugal acceleration is determined by the following relationship;

$$f = \frac{mv^2}{r}$$

in which f is the centrifugal acceleratory force, m is the mass of the object, v is the velocity of travel, and r is the radius of curvature of the turn. From this formula it is obvious that as the velocity increases, the force of centrifugal acceleration increases in proportion to the square of the velocity. It is also obvious that the force of acceleration is directly proportional to the sharpness of the turn.

Measurement of the Acceleratory Force — "G." When a person is simply sitting in his seat, the force with which he is pressing against the seat results from the pull of gravity, and it is equal to his weight. The intensity of this force is 1 "G" because it is equal to one times the pull of gravity. If the force with which he presses against his seat becomes five times his normal weight during a pullout from a dive, the force acting upon the seat is 5 G.

If the airplane goes through an outside loop so that the pilot is held down by his seat belt, *negative G* is applied to his body, and if the force with which he is thrown against his belt is equal to the weight of his body, the negative force is −1 G.

Effects of Centrifugal Acceleratory Force on the Body. *Effects on the Circulatory System.* The most important effect of centrifugal acceleration is on the circulatory system because blood is mobile and can be translocated by centrifugal force. Centrifugal force also tends to displace the tissues, but because of their more solid structure, they only sag — ordinarily not enough to cause abnormal function.

When the aviator is subject to *positive G*, his blood is centrifuged toward the lower part of his body. Thus, if the centrifugal acceleratory force is 5 G and the person is in a standing position, the hydrostatic pressure in the veins of the feet is five times normal, or approximately 450 mm. Hg, and even in the sitting position this pressure is nearly 300 mm. Hg. As the pressure in the vessels of the lower part of the body increases, the vessels passively dilate, and a major proportion of the blood from the upper part of the body is translocated into these lower vessels. Because the heart cannot pump unless blood returns to it, the greater the quantity of blood "pooled" in the lower body, the less becomes the cardiac output.

When the acceleration rises to 4 G and the aviator is in the seated position, the systemic arterial pressure at the level of the heart falls approximately 40 mm Hg, and blood flow to the brain almost ceases. Acceleration greater than 4 to 6 G ordinarily causes "black-out" of vision within a few seconds and then unconsciousness shortly thereafter.

EFFECTS OF LINEAR ACCELERATORY FORCES ON THE BODY

Acceleratory Forces in Space Travel. In contrast to aircraft, a spacecraft cannot make rapid turns; therefore centrifugal acceleration is of little importance except when the spacecraft goes into abnormal gyrations. On the other hand, blast-off acceleration and landing deceleration might be tremendous; both of these are types of linear acceleration.

Figure 30–2 illustrates a typical profile of the acceleration during blast-off in a three-stage spacecraft, showing that the first stage booster causes acceleration as high as 9 G, and the second stage booster, as high as 8 G. In the standing position the human body could not withstand this much acceleration, but in a reclining position *transverse to the axis of acceleration*, this amount of acceleration can be withstood with ease despite the fact that the acceleratory forces continue for as long as five minutes at a time. Therefore, we see the reason for the reclining seats used by the astronauts.

Problems also occur during deceleration when the spacecraft re-enters the atmosphere. A person traveling at Mach 1 (the speed of sound and of fast airplanes) can be safely decelerated in a distance of approximately 0.12 mile, whereas a person traveling at a speed of Mach 100 (a speed possible in interplanetary space travel) requires a distance of about 10,000 miles for safe deceleration. The principal reason for this difference is

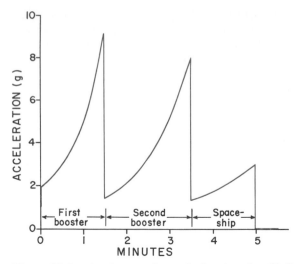

Figure 30–2. Acceleratory forces during the take-off of a spacecraft.

that the total amount of energy that must be dispelled during deceleration is proportional to the *square* of the velocity, which alone increases the distance 10,000-fold. But in addition to this, a human being can withstand far less deceleration if it lasts for a long time than for a short time. Therefore, deceleration must be accomplished much more slowly from the very high velocities than is necessary at the slower velocities.

RADIATION HAZARDS IN SPACE

Large quantities of cosmic particles are continually bombarding the earth's upper atmosphere, some originating from the sun and some from outer space. The magnetic field of the earth traps many of these cosmic particles in two major belts around the earth called *Van Allen radiation belts*. The inner belt begins at an altitude of about 300 miles and extends to about 3000 miles. The outer belt begins at about 6000 miles and extends to 20,000 miles. Even with the best possible shielding, a person traversing these two belts in an interplanetary space trip could receive as much as 10 roentgens of radiation, which is about one-fortieth the lethal dose; and a person in a spacecraft orbiting the earth within one of these two belts could receive enough radiation in only a few hours to cause death.

Thus, it is important to orbit spacecraft below an altitude of 200 to 300 miles, an altitude at which the radiation hazard is slight. Also, it is possible to minimize the radiation hazard during interplanetary space travel by leaving the earth or returning to earth near one of the earth's poles, where the belts are almost nonexistent, rather than near the equator.

"ARTIFICIAL CLIMATE" IN THE SEALED SPACECRAFT

Since there is no atmosphere in outer space, an atmosphere and other conditions of climate must be provided artificially. The ability of a person to survive in this artificial climate depends entirely on appropriate engineering design.

Most important of all, the oxygen concentration must remain high enough and the carbon dioxide concentration low enough. In some of the space missions, a capsule atmosphere containing pure oxygen at about 260 mm. Hg pressure has been used. In others, normal air at 760 mm. Hg pressure has been used. The presence of nitrogen in the mixture greatly diminishes the likelihood of fire and explosion.

For space travel lasting several years, it will be impractical to carry along an adequate oxygen supply and enough carbon dioxide absorbent. For this reason, "recycling techniques" are being developed for use of the same oxygen over and over again. These techniques also frequently include reuse of the same food and water. Basically, they involve (1) a method for removing oxygen from carbon dioxide, (2) a method for removing water from the human excreta, and (3) use of the human excreta for resynthesizing or growing an adequate food supply.

Large amounts of energy are required for these processes, and the real problem at present is to derive enough energy from the sun's radiation to energize the necessary chemical reactions. Some recycling processes depend on purely physical procedures, such as distillation, electrolysis of water, and capture of the sun's energy by solar batteries, whereas others depend on biological methods, such as use of algae, with its large store of chlorophyll, to generate foodstuffs by photosynthesis. Unfortunately a completely practical system for recycling is yet to be achieved. The problem is the weight of the equipment that must be carried.

WEIGHTLESSNESS IN SPACE

A person in an orbiting satellite or in any nonpropelled spacecraft experiences weightlessness. That is, he is not drawn toward the bottom, sides, or top of the spacecraft but simply floats inside its chambers. The cause of this is not failure of gravity to pull on the body, because gravity from any nearby heavenly body is still active. However, the gravity acts on both the spacecraft and the person at the same time, and since there is no resistance to movement in space, both are pulled with exactly the same forces and in the same direction.

Weightlessness causes engineering problems, such as the necessity for providing special techniques for eating and drinking (since food and water will not stay in open plates or glasses), special waste disposal systems, and adequate hand holds or other means for stabilizing the person in the spacecraft so that he can adequately control the operation of the ship.

Physiological Problems of Weightlessness. Fortunately, the physiological problems of weightlessness have not proved to be severe. Most of the problems that do occur appear to be related to two effects of the weightlessness: (1) translocation of fluids within the body because of failure of gravity to cause hydrostatic pressures, and (2) diminishment of physical activity because no strength of muscle contraction is required to oppose the force of gravity.

The observed effects of prolonged stay in space are the following: (1) decrease in blood volume, (2) decrease in red cell mass, (3) decreased work capacity, (4) decrease in maximum cardiac output, and (5) loss of calcium from the bones. Essentially these same effects also occur in persons lying in bed for an extended period of time. For this reason an extensive exercise program was carried out during the most recent sojourn by three astronauts in the Space Laboratory, and all of the above effects were greatly reduced.

PHYSIOLOGY OF DEEP SEA DIVING AND OTHER HIGH PRESSURE OPERATIONS

When a person descends in the sea, the pressure around him increases tremendously. To keep his lungs from collapsing, air must be supplied also at high pressure, which exposes the blood in his lungs to extremely high alveolar gas pressures. Beyond certain limits these high pressures can cause tremendous alterations in the physiology of the body.

Also exposed to high atmospheric pressures are caisson workers who, in digging tunnels beneath rivers or elsewhere, often must work in a pressurized area to keep the tunnel from caving in. Here again, the same problems of excessively high gas pressures in the alveoli occur.

Before explaining the effects of high alveolar gas pressures on the body, it is necessary to review some physical principles of pressure and volume changes at different depths beneath the sea.

Relationship of Sea Depth to Pressure. A column of sea water 33 feet deep exerts the same pressure at its bottom as all the atmosphere above the earth. Therefore, a person 33 feet beneath the ocean surface is exposed to a pressure of 2

TABLE 30–2 EFFECT OF DEPTH ON PRESSURE

Depth (feet)	Atmosphere(s)
Sea level	1
33	2
66	3
100	4
133	5
166	6
200	7
300	10
400	13
500	16

atmospheres, 1 atmosphere of pressure caused by the air above the water and the second atmosphere by the weight of the water itself. At 66 feet the pressure is 3 atmospheres, and so forth, in accord with Table 30–2.

Effect of Depth on the Volume of Gases. Another important effect of depth is the compression of gases to smaller and smaller volumes. At 33 feet beneath the surface of sea, where the pressure is 2 atmospheres, a 1 liter volume at sea level is compressed to only one-half liter. At 100 feet, where the pressure is 4 atmospheres, the volume is compressed to one-fourth liter, and at 8 atmospheres (233 feet) to one-eighth liter. This is an extremely important effect in diving, because it can cause the air chambers of the diver's body, including the lungs, to become so small in some instances that serious damage results.

EFFECT OF HIGH PARTIAL PRESSURES OF GASES ON THE BODY

The three gases to which a diver breathing air is normally exposed are nitrogen, oxygen, and carbon dioxide.

Nitrogen Narcosis at High Nitrogen Pressures. Approximately four-fifths of the air is nitrogen. At sea level pressure it has no known effect on bodily function, but at high pressures it can cause varying degrees of narcosis. When the diver remains in the sea for many hours and is breathing compressed air, the depth at which the first symptoms of mild narcosis appear is approximately 130 to 150 feet, at which level he begins to exhibit joviality and to lose many of his cares. At 150 to 200 feet, he becomes drowsy. At 200 to 250 feet, his strength wanes considerably, and he often becomes too clumsy to perform the work required of him. Beyond 300 feet (10 atmospheres pressure), the diver usually becomes almost useless as a result of nitrogen narcosis. It should be noted, however, that *an hour or more of the high pressure is usually required* before enough

nitrogen dissolves in the body to cause these effects.

Nitrogen narcosis has characteristics very similar to those of alcohol intoxication, and for this reason it has frequently been called "rapture of the depths."

The mechanism of the narcotic effect is believed to be the same as that of essentially all the gas anesthetics. That is, nitrogen dissolves freely in the fats of the body, and it is presumed that it, like most other anesthetic gases, dissolves in the membranes or other lipid structures of the brain neurons and because of its *physical* effect on altering electrical charge transfer reduces the excitability.

Oxygen Toxicity at High Pressures. Breathing oxygen under very high partial pressure can be detrimental to the central nervous system, sometimes causing epileptic convulsions followed by coma. Indeed, exposure to 3 atmospheres pressure of oxygen (P_{O_2} = 2280 mm. Hg) will cause convulsions and coma in most persons after about one hour. These convulsions often occur without any warning, and they obviously are likely to be lethal to a diver submerged in the sea.

The cause or causes of oxygen toxicity are yet unknown, but experiments have shown that excess oxygen in the tissues causes the development of large concentrations of oxidizing free radicals, which can cause oxidative destruction of many essential elements of the cells, thereby damaging the cell's metabolic systems.

Carbon Dioxide Toxicity at Great Depths. If the diving gear is properly designed and also functions properly, the diver has no problem from carbon dioxide toxicity, for depth alone does not increase the carbon dioxide partial pressure in the alveoli. This is true because carbon dioxide is manufactured in the body, and as long as the diver continues to breathe a normal tidal volume, he continues to expire the carbon dioxide as it is formed, maintaining his alveolar carbon dioxide partial pressure at a normal value.

Unfortunately, though, in certain types of diving gear, such as the diving helmet and different types of rebreathing apparatuses, carbon dioxide can frequently build up in the dead space air of the apparatus and be rebreathed by the diver. Up to a carbon dioxide pressure (P_{CO_2}) of about 80 mm. Hg, two times that of normal alveoli, the diver tolerates this buildup, his minute respiratory volume increasing up to a maximum of 6- to 10-fold to compensate for the increased carbon dioxide. However, beyond the 80 mm. Hg level, the situation becomes intolerable, and eventually the respiratory center begins to be depressed rather than excited; the diver's respiration then actually begins to fail rather than to compensate.

In addition, the diver develops severe respiratory acidosis, and varying degrees of lethargy, and finally coma, ensue.

DECOMPRESSION OF THE DIVER AFTER EXPOSURE TO HIGH PRESSURES

When a person breathes air under high pressure for a long time, the amount of nitrogen dissolved in his body fluids becomes great. The reason is the following: The blood flowing through the pulmonary capillaries becomes saturated with nitrogen to the same pressure as that in the breathing mixture. Over several hours, enough nitrogen is carried to all the tissues of the body to saturate them also with dissolved nitrogen. And, since nitrogen is not metabolized by the body, it remains dissolved until the nitrogen pressure in the lungs decreases, at which time the nitrogen is then removed by the reverse respiratory process.

Volume of Nitrogen Dissolved in the Body Fluids at Different Depths.   At sea level almost 1 liter of nitrogen is dissolved in the entire body. After the diver has become totally saturated with nitrogen, the *sea level volume of nitrogen* dissolved in his body fluids at the different depths is

feet	liters
33	2
100	4
200	7
300	10

However, several hours are required for the body to become saturated with nitrogen at each new depth, simply because the blood does not flow rapidly enough and the nitrogen does not diffuse rapidly enough to cause instantaneous saturation. For this reason, if a person remains at deep levels for only a few minutes not much nitrogen dissolves in his fluids and tissues, whereas if he remains at a deep level for several hours his fluids and tissues become almost completely saturated with nitrogen.

Decompression Sickness (Synonyms: Compressed Air Sickness, Bends, Caisson Disease, Diver's Paralysis, Dysbarism).   If a diver has been in the sea so long that large amounts of nitrogen have dissolved in his body, and he then suddenly comes back to the surface of the sea, significant quantities of nitrogen bubbles can develop in his body fluids both intracellularly and extracellularly, and these can cause minor or serious damage in almost any area of the body, depending on the amount of bubbles formed. The cause of these bubbles is the following:

As long as the diver remains deep in the sea, the pressure against the outside of his body compresses all the body tissues sufficiently to keep the dissolved gases in solution. Then, when the diver suddenly rises to sea level, the pressure on the outside of his body becomes only 1 atmosphere (760 mm. Hg), while the pressure of dissolved gases inside the body fluids is usually several thousand mm. Hg, a value far greater than the pressure on the outside of the body. Therefore, the gases can escape from the dissolved state and form actual bubbles inside the tissues.

Symptoms of Decompression Sickness.   In persons who have developed decompression sickness, symptoms have occurred with the following frequencies:

	per cent
Local pain in the legs or arms	89
Dizziness	5.3
Paralysis	2.3
Shortness of breath ("the chokes")	1.6
Extreme fatigue and pain	1.3
Collapse with unconsciousness	0.5

From the above list of symptoms of decompression sickness it can be seen that the most serious problems are usually related to bubble formation in the nervous system. Bubbles sometimes actually disrupt important pathways in the brain or spinal cord, and bubbles in the peripheral nerves can cause severe pain. Unfortunately, large bubbles in the central nervous system occasionally lead to permanent paralysis or permanent mental disturbances.

But the nervous system is not the only locus of damage in decompression sickness, for bubbles can also form in the blood and become caught in the capillaries of the lungs; these bubbles block pulmonary blood flow and cause "the chokes," characterized by serious shortness of breath. This is often followed by severe pulmonary edema, which further aggravates the condition and can cause death.

The symptoms of decompression sickness usually appear within a few minutes to an hour after sudden decompression. However, occasional symptoms of decompression sickness develop as long as six or more hours after decompression.

Rate of Nitrogen Elimination from the Body; Decompression Tables.   Fortunately, if a diver is brought to the surface slowly, the dissolved nitrogen is eliminated through his lungs rapidly enough to prevent decompression sickness. Approximately two-thirds of the total nitrogen is liberated in one hour and about 90 per cent in six hours. However, some excess nitrogen is still present in the body fluids for many more hours, and the diver is not completely safe for as long as 9 to 12 hours. Therefore, a diver must be "decompressed" sometimes for many hours if he has been deep in the sea for a long time.

The rate at which a diver can be brought to the surface depends on, first, the *depth* to which he has descended and, second, the *amount of time* he has been there. Only 20 minutes at a depth of 300 feet requires over two and a half hours decompression time (45 minutes at 300 feet requires over five hours). On the other hand, a person can remain at 50 feet for as long as three hours and yet be decompressed in only 12 minutes.

Use of Helium-Oxygen Mixtures in Deep Dives. In deep dives helium has advantages over nitrogen, including (1) decreased decompression time, (2) lack of narcotic effect, and (3) decreased airway resistance in the lungs. The decreased decompression time results from two properties of helium: (a) Only 40 per cent as much helium dissolves in the body as does nitrogen. (b) Because of its small atomic size it diffuses through the tissues at a velocity about two and a half times that of nitrogen and therefore can be transported to the blood and expired much more rapidly than can nitrogen.

SCUBA DIVING (SELF-CONTAINED UNDERWATER BREATHING APPARATUS)

In recent years a diving apparatus that does not require connections with the surface has been

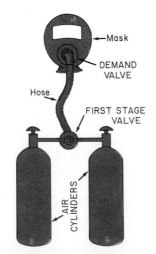

Figure 30–3. A SCUBA apparatus.

perfected and is probably best known under the trade name "Aqualung." Figure 30–3 illustrates one type of SCUBA diving gear showing the following components: (1) one or more tanks of compressed air or of some other breathing mixture, (2) a first stage "reducing" valve for reducing the pressure from the tanks to a constant low pressure level, (3) a combination inhalation "demand" valve and exhalation valve, which allows air to be pulled into the lungs with very slight negative pressure and then to be exhausted into the sea, and (4) a mask and tube system with small "dead space."

Basically, the demand system operates as follows: The first stage reducing valve reduces the pressure from the tanks to a pressure of about 100 lb. per square inch. However, the breathing mixture does not flow continually into the mask. Instead, with each inspiration, slight negative pressure in the mask pulls the diaphragm of the demand valve inward, and this automatically releases air from the hose into the mask and lungs. In this way only the amount of air needed for inhalation enters the system. Then, on expiration, the air cannot go back into the tank but instead is expired through the expiration valve.

The most important problem in use of the self-contained underwater breathing apparatus is the time limit that one can remain beneath the surface; only a few minutes are possible at great depths because tremendous airflow from the tanks is required to wash carbon dioxide out of the lungs — the greater the depth, the greater the airflow required, because all the gases are compressed to smaller volumes at the deeper levels.

SPECIAL PHYSIOLOGIC PROBLEMS OF SUBMARINES

Escape from Submarines. Essentially the same problems as those of deep sea diving are often met in submarines especially when it is necessary to escape from a submerged submarine. Escape is possible from as deep as 300 feet even without the use of any special type of apparatus. Proper use of rebreathing devices using helium or hydrogen can theoretically allow escape from as deep as 600 feet.

One of the major problems of escape is prevention of air embolism. As the person ascends, the gases in his lungs expand and sometimes rupture a major pulmonary vessel, allowing the gases to enter into the pulmonary vascular system to cause embolism of the circulation. Therefore, as the person ascends, he must exhale continually.

Expansion and exhalation of gases from the lungs during ascent, even without breathing, is often rapid enough to blow off the accumulating carbon dioxide in the lungs. This keeps the concentration of carbon dioxide from building up in the blood and keeps the person from having the desire to breathe. Therefore, he can hold his breath for an extremely long time during ascent.

Health Problems in the Submarine Internal Environment. Except for escape, submarine medicine generally centers on several engineering problems to keep hazards out of the internal environment of the submarine. In nuclear submarines there exists the problem of radiation hazards, but with appropriate shielding, the amount of radiation received by the crew submerged in the sea has actually been less than the normal radiation received above the surface of the sea from cosmic rays. Therefore, no essential hazard results from this unless some failure in the apparatus causes unexpected release of radioactive materials.

Second, poisonous gases on occasion escape into the atmosphere of the submarine and must be controlled exactly. For instance, during several weeks' submergence, cigarette smoking by the crew can liberate sufficient amounts of carbon monoxide, if it is not removed from the air, to cause carbon monoxide poisoning, and on occasion even Freon gas has been found to diffuse through the tubes in refrigeration systems in sufficient quantity to cause toxicity. Finally, it is well known that chlorine and other poisonous gases are released when salt water comes in contact with batteries in the old type submarines.

A highly publicized factor of submarine medicine has been the possibility of psychological problems caused by prolonged submergence. Fortunately, this has turned out to be more a figment of the public's imagination than truth, for the problems here are the same as those relating (1) to any other confinement or (2) to any other type of danger. Psychological screening has been used to advantage to keep such problems almost to zero even in month-long submergence.

REFERENCES

Aviation and Space Physiology

Andrews, H. L.: Radiation Biophysics, 2nd Ed. Englewood Cliffs, N.J., Prentice-Hall, 1974.

Bullard, R. W.: Physiological problems of space travel. *Annu. Rev. Physiol., 34*:205, 1972.

Cardiovascular Problems Associated with Aviation Safety. Springfield, Va., National Technical Information Service, 1976.

Frisancho, A. R.: Functional adaptation to high altitude hypoxia. *Science, 187*:313, 1975.

Grover, R. F., *et al.*: High-altitude pulmonary edema. *In* Fishman, A. P., and Renkin, E. M. (eds.): Pulmonary Edema. Baltimore, Williams & Wilkins, 1979, p. 229.

Hempelman, H. V., and Lockwood, A. P. M.: The Physiology of Diving in Man and Other Animals. London, Edward Arnold, 1978.

Hock, R. J.: The physiology of high altitude. *Sci. Am., 222*:52, 1970.

Lahiri, S.: Physiological responses and adaptations to high altitude. *Int. Rev. Physiol., 15*:217, 1977.

Pace, N.: Respiration at high altitude. *Fed. Proc., 33*:2126, 1974.

Reeves, J. T., *et al.*: Physiological effects of high altitude on the pulmonary circulation. *In* Robertshaw, D. (ed.): International Review of Physiology: Environmental Physiology III. Vol. 20. Baltimore, University Park Press, 1979, p. 289.

Sloan, A. W.: Man in Extreme Environments. Springfield, Ill., Charles C Thomas, 1979.

Stickney, J. C.: Some problems of homeostasis in high-altitude exposure. *Physiologist, 15*:349, 1972.

Deep Sea Diving Physiology

Bennett, P. B., and Elliott, D. H.: The Physiology and Medicine of Diving and Compressed Air Work, 2nd Ed. Baltimore, Williams & Wilkins, 1975.

Fisher, A. B., *et al.*: Oxygen toxicity of the lung: Biochemical aspects. *In* Fishman, A. P., and Renkin, E. M. (eds.): Pulmonary Edema. Baltimore, Williams & Wilkins, 1979, p. 207.

Gamarra, J. A.: Decompression Sickness. Hagerstown, Md., Harper & Row, 1974.

Greene, D. G.: Drowning. *In* Fenn, W. O., and Rahn, H. (eds.): Handbook of Physiology. Sec. 3, Vol. 2. Baltimore, Williams & Wilkins, 1965, p. 1195.

Haugaard, N.: Cellular mechanisms of oxygen toxicity. *Physiol. Rev., 48*:311, 1968.

Hochachka, B. W., and Storey, K. B.: Metabolic consequences of diving in animals and man. *Science, 187*:613, 1975.

Lanphier, E. H.: Human respiration under increased pressures. *Symp. Soc. Exp. Biol., 26*:379, 1972.

Oxygen Free Radicals and Tissue Damage. Ciba Foundation Symposium. New York, Excerpta Medica, 1979.

Shilling, C. W., and Beckett, M. W. (eds.): Underwater Physiology IV. Bethesda, Md., Federation of American Societies for Experimental Biology, 1978.

Zimmerman, A. M. (ed.): High Pressure Effects on Cellular Processes. New York, Academic Press, 1970.

Part IX

THE NERVOUS SYSTEM

31

Organization of the Nervous System; Basic Functions of Synapses and Neuronal Circuits

The nervous system, along with the endocrine system, provides most of the control functions for the body. In general, the nervous system controls the rapid activities of the body, such as muscular contractions, rapidly changing visceral events, and even the rate of secretion of some endocrine glands. The endocrine system, by contrast, regulates principally the metabolic functions of the body.

The nervous system is unique in the vast complexity of the control actions that it can perform. It receives literally thousands of bits of information from the different sensory organs and then integrates all of them to determine the response to be made by the body. The purpose of this chapter is to present a general outline of the overall mechanisms by which the nervous system performs such functions and then to discuss the basic functions of synapses and neuronal circuits. Before beginning this discussion, however, the reader should refer to Chapters 8 and 10, which present, respectively, the principles of membrane potentials and transmission of impulses through neuromuscular junctions.

GENERAL DESIGN OF THE NERVOUS SYSTEM

THE SENSORY DIVISION – SENSORY RECEPTORS

Most activities of the nervous system are originated by sensory experience emanating from sensory receptors, whether they are visual receptors, auditory receptors, tactile receptors on the surface of the body, or other kinds of receptors. This sensory experience can cause an immediate reaction, or its memory can be stored in the brain for minutes, weeks, or years and then can help to determine the bodily reactions at some future date.

Figure 31–1 illustrates the *somatic* portion of the sensory system, which transmits sensory information from the receptors of the entire surface of the body and deep structures. This information enters the nervous system through the spinal nerves and is conducted to essentially all segments of the central nervous system.

THE MOTOR DIVISION – THE EFFECTORS

The most important ultimate role of the nervous system is control of bodily activities. This is achieved by controlling (1) contraction of skeletal muscles throughout the body, (2) contraction of smooth muscle in the internal organs, and (3) secretion of both exocrine and endocrine glands in many parts of the body. These activities are collectively called *motor functions* of the nervous system, and the muscles and glands are called *effectors* because they perform the functions dictated by the nerve signals.

Figure 31–2 illustrates the *motor axis* of the nervous system for controlling skeletal muscle contraction. Operating parallel with this axis is another similar system for control of the smooth

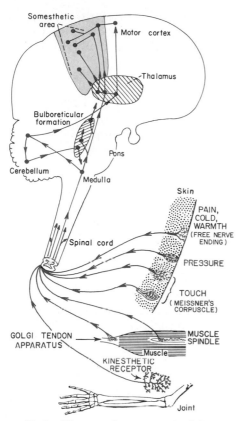

Figure 31–1. The somatic sensory axis of the nervous system.

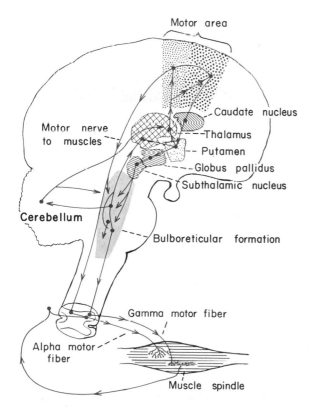

Figure 31–2. The motor axis of the nervous system.

muscles and glands; it is the *autonomic nervous system*, which will be described in detail in Chapter 38. Note in Figure 31–2 that the skeletal muscles can be controlled from many different levels of the central nervous system. Each of these different areas plays its own specific role in the control of body movements, the lower regions being concerned primarily with automatic, instantaneous responses of the body to sensory stimuli and the higher regions with deliberate movements controlled by the thought processes of the cerebrum.

PROCESSING OF INFORMATION

The nervous system would not be at all effective in controlling bodily functions if each bit of sensory information caused some motor reaction. Therefore, one of the major functions of the nervous system is to process incoming information in such a way that *appropriate* motor responses occur. Indeed, more than 99 per cent of all sensory information is continually discarded by the brain as unimportant. For instance, one is ordinarily totally unaware of the parts of his body that are in contact with his clothes and is also unaware of the pressure on his seat when he is sitting. Likewise, his attention is drawn only to an occasional object in his field of vision, and even the perpetual noise of his surroundings is usually relegated to the background.

After the important sensory information has been selected, it must be channeled into proper regions of the brain to cause the desired responses. Thus, if a person places his hand on a hot stove, the desired response is to lift the hand, plus other associated responses such as moving the entire body away from the stove and perhaps even shouting with pain. Yet even these responses represent activity by only a small fraction of the total motor system of the body.

Role of Synapses in Processing Information. The synapse is the junction point from one neuron to the next and, therefore, is an advantageous site for control of signal transmission. The synapses determine the directions in which the nervous signals spread in the nervous system. Some synapses transmit signals from one neuron to the next with ease, while others transmit signals only with difficulty. Also, facilitatory and inhibitory signals from other areas in the nervous system can control synaptic activity, sometimes activating the synapses for transmission and other times inactivating them. In addition, some neurons respond to synaptic stimulation with large numbers of impulses, while others respond with only a few.

Thus, the synapses perform a selective action,

often blocking the weak signals while allowing the strong signals to pass, often selecting and amplifying certain weak signals, and often channeling the signal in many different directions rather than simply in one direction. The basic principles of this processing of information by the synapses are so important that they are discussed in detail in the latter part of this chapter.

STORAGE OF INFORMATION – MEMORY

Only a small fraction of the important sensory information causes an immediate motor response. Much of the remainder is stored for future control of motor activities and for use in the thinking processes. Most of this storage occurs in the *cerebral cortex*, but not all, for even the basal regions of the brain and perhaps even the spinal cord can store small amounts of information.

The storage of information is the process we call *memory*, and this too is a function of the synapses. That is, each time a particular sensory signal passes through a sequence of synapses, the respective synapses become more capable of transmitting the same signal the next time, which process is called *facilitation*. After the sensory signal has passed through the synapses a large number of times, the synapses become so facilitated that signals from the "control center" of the brain can also cause transmission of impulses through the same sequence of synapses even though the sensory input has not been excited. This gives the person a perception of experiencing the original sensation, though in effect it is only a memory of the sensation.

Unfortunately, we do not know the precise mechanism by which facilitation of synapses occurs in the memory process, but what is known about this and other details of the memory process will be discussed in Chapter 36.

Once memories have been stored in the nervous system, they become part of the processing mechanism. The thought processes of the brain compare new sensory experiences with the stored memories; the memories help to select the important new sensory information and to channel it into appropriate storage areas for future use or into motor areas to cause bodily responses.

THE THREE MAJOR LEVELS OF NERVOUS SYSTEM FUNCTION

The human nervous system has inherited specific characteristics from each stage of evolutionary development. From this heritage, there re-

main three major levels of the nervous system that have special functional significance: (1) the spinal cord level, (2) the lower brain level, and (3) the higher brain or cortical level.

THE SPINAL CORD LEVEL

The spinal cord of the human being still retains many functions of the multisegmental animal. Sensory signals are transmitted through the spinal nerves into each *segment* of the spinal cord, and these signals can cause localized motor responses either in the segment of the body from which the sensory information is received or in adjacent segments. Essentially all the spinal cord motor responses are automatic and occur almost instantaneously in response to the sensory signal. In addition, they occur in specific patterns of response called *reflexes*.

Figure 31–3 illustrates two of the simpler cord reflexes. To the left is the neural circuit of the *muscle stretch reflex*. If a muscle suddenly becomes stretched, a sensory nerve receptor in the muscle called the *muscle spindle* becomes stimulated and transmits nerve impulses through a sensory nerve fiber into the spinal cord. This fiber synapses directly with a *motoneuron* in the anterior horn of the cord gray matter, and the motoneuron in turn transmits impulses back to the muscle to cause the muscle, the effector, to contract. The muscle contraction opposes the original muscle stretch. Thus, this reflex acts as a *feedback* mechanism, operating from a receptor to an effector, to prevent sudden changes in length of the muscle. This allows a person to maintain his limbs and other parts of his body in desired positions despite sudden outside forces that tend to move the parts out of position.

To the right in Figure 31–3 is illustrated the neural circuit of another reflex, called the *with-*

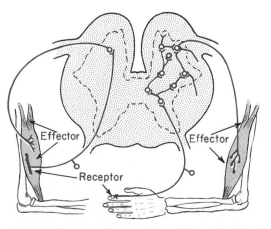

Figure 31–3. *Left:* The simple stretch reflex. *Right:* A withdrawal reflex.

drawal reflex. This is a protective reflex that causes withdrawal of any part of the body from an object that is causing pain. For instance, let us assume that the hand is placed on a sharp object. Pain signals are transmitted into the gray matter of the spinal cord, and after appropriate selection of information by the synapses, signals are channeled to the appropriate motoneurons to cause flexion of the biceps muscle. This obviously lifts the hand away from the sharp object.

We see, then, that the withdrawal reflex is much more complex than the stretch reflex, for it involves many neurons in the gray matter of the cord, and signals are transmitted to many adjacent segments of the cord to cause contraction of the appropriate muscles.

Cord Functions After the Brain Is Removed. The many reflexes of the spinal cord will be discussed in Chapter 34; however, the following list of important cord reflex functions that occur even after the brain is removed illustrates the many capabilities of the spinal cord.

1. The animal can under certain conditions be made to stand up, primarily because of reflexes initiated from the pads of the feet. Sensory signals from the pads cause the extensor muscles of the limbs to tighten, which in turn allows the limbs to support the animal's body.

2. A spinal animal held in a sling so that its feet hang downward often begins walking or galloping movements involving one, two, or all of its legs. This illustrates that the basic patterns for causing the limb movements of locomotion are present in the spinal cord.

3. A flea crawling on the skin of a spinal animal causes reflex to-and-fro scratching by the paw, and the paw can actually localize the flea on the surface of the body.

4. Cord reflexes exist to cause emptying of the urinary bladder and of the rectum.

This list of some of the segmental and multisegmental reflexes of the spinal cord demonstrates that many of our day-by-day and moment-by-moment activities are controlled locally by the respective segmental levels of the spinal cord, the brain playing only a modifying role in these local controls.

THE LOWER BRAIN LEVEL

Many if not most of what we call subconscious activities of the body are controlled in the lower areas of the brain — the medulla, pons, mesencephalon, hypothalamus, thalamus, cerebellum, and basal ganglia. Subconscious control of arterial blood pressure and respiration is achieved primarily in the reticular substance of the medulla and pons. Control of equilibrium is a combined function of the older portions of the cerebellum and the reticular substance of the medulla, pons, and mesencephalon. The coordinated turning movements of the head, of the entire body, and of the eyes are controlled by specific centers located in the mesencephalon, paleocerebellum, and lower basal ganglia. Feeding reflexes, such as salivation in response to taste of food and licking of the lips, are controlled by areas in the medulla, pons, mesencephalon, amygdala, and hypothalamus. And many emotional patterns, such as anger, excitement, sexual activities, reactions to pain, or reactions of pleasure, can occur in animals without a cerebral cortex.

In short, the subconscious but coordinate functions of the body, as well as many of the life processes themselves — arterial pressure and respiration, for instance — are controlled by the lower regions of the brain, regions that usually, but not always, operate below the conscious level.

THE HIGHER BRAIN OR CORTICAL LEVEL

We have seen from the above discussion that many of the intrinsic life processes of the body are controlled by subcortical regions of the brain or by the spinal cord. What, then, is the function of the cerebral cortex? The cerebral cortex is primarily a vast information storage area. Approximately three quarters of all the neuronal cell bodies of the entire nervous system are located in the cerebral cortex. It is here that most of the memories of past experiences are stored, and it is here that many of the patterns of motor responses are stored, which information can be called forth at will to control motor functions of the body.

Relation of the Cortex to the Thalamus and Other Lower Centers. The cerebral cortex is actually an outgrowth of the lower regions of the brain, particularly of the thalamus. For each area of the cerebral cortex there is a corresponding and connecting area of the thalamus; activation of a minute portion of the thalamus activates the corresponding and much larger portion of the cerebral cortex. It is presumed that in this way the thalamus can call forth cortical activities at will. Also, activation of regions in the mesencephalon transmits diffuse signals to the cerebral cortex, partially through the thalamus and partially directly, to activate the entire cortex. This is the process that we call *wakefulness*. On the other hand, when these areas of the mesencephalon become inactive, the thalamic and cortical regions also become inactive, which is the process we call *sleep*.

Function of the Cerebral Cortex in Thought Processes. Some areas of the cerebral cortex are not directly concerned with either sensory or motor functions of the nervous system — for example, the prefrontal lobe and large portions of the temporal and parietal lobes. These areas are set aside for the more abstract processes of thought, but even they also have direct nerve connections with the lower regions of the brain to help control the automatic, subconscious actions of the body.

Large areas of the cerebral cortex can be destroyed without blocking some of the conscious activities of the body. For instance, destruction of the sensory cortex does not destroy one's ability to feel objects touching the skin, but it does destroy the ability to distinguish the shapes of objects, their character, and the precise points on the skin where the objects are touching. Thus, the cortex is not required for perception of sensation, but it does add immeasurably to its depth of meaning. Likewise, destruction of the prefrontal lobe does not destroy the ability to think, but it does destroy the ability to think in abstract terms. In other words, each time a portion of the cerebral cortex is destroyed, a vast amount of information is lost to the thinking process and some of the mechanisms for processing this information are also lost. Therefore, total loss of the cerebral cortex causes a vegetative type of existence rather than a "living" existence.

FUNCTION OF NEURONAL SYNAPSES

Most students are already aware that information is transmitted in the central nervous system in the form of nerve impulses through a succession of neurons, one after another. However, it is not immediately apparent that the impulse may be (1) blocked in its transmission from one neuron to the next, (2) changed from a single impulse into repetitive impulses, or (3) integrated with impulses from other neurons to cause highly intricate patterns of impulses in successive neurons. All these functions can be classified as *synaptic functions of neurons*.

PHYSIOLOGIC ANATOMY OF THE SYNAPSE

The juncture between one neuron and the next is called a *synapse*. Figure 31–4 illustrates a typical large neuron, called a *motoneuron*, in the anterior horn of the spinal cord. It is composed of three major parts: the *soma*, which is the main body of the neuron; a single *axon*, which extends from the

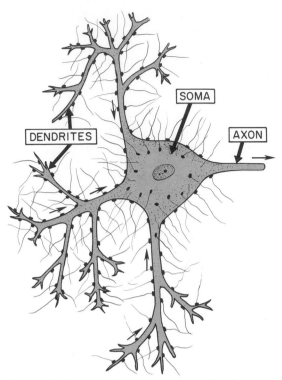

Figure 31–4. A typical motoneuron, showing synaptic knobs on the neuronal soma and dendrites. Note also the single axon.

soma into the peripheral nerve; and the *dendrites*, which are thin projections of the soma that extend up to 1 mm. into the surrounding areas of the cord.

It should be noted, also, that literally many thousand, averaging about 6000, small knobs called *synaptic knobs* lie on the surfaces of the dendrites and soma, approximately 80 to 90 per cent of them on the dendrites. These knobs are the terminal ends of nerve fibrils that originate in many other neurons, and usually not more than a few of the knobs are derived from any single previous neuron. Later it will become evident that many of these synaptic knobs are *excitatory* and secrete a substance that excites the neuron, while others are *inhibitory* and secrete a substance that inhibits the neuron.

Neurons in other parts of the cord and brain differ markedly from the motoneuron in (1) the size of the cell body, (2) the length, size, and number of dendrites, ranging in length from almost none at all up to as long as one meter (the peripheral sensory nerve fiber), (3) the length and size of the axon, and (4) the number of synaptic knobs, which may range from only a few to more than 100,000. It is these differences that make neurons in different parts of the nervous system react differently to incoming signals and therefore perform different functions, as will be explained in subsequent chapters.

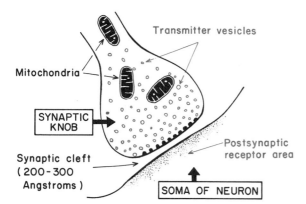

Figure 31–5. Physiologic anatomy of the synapse.

The Synaptic Knobs. Electron microscope studies of the synaptic knobs show that they have varied anatomical forms, but most resemble small round or oval knobs and therefore are frequently called *terminal knobs, boutons, end-feet,* or simply *presynaptic terminals.*

Figure 31–5 illustrates the basic structure of the synaptic knob (presynaptic terminal). It is separated from the neuronal soma by a *synaptic cleft* having a width usually of 20 to 30 nanometers. The knob has two internal structures important to the excitatory or inhibitory functions of the synapse: the *synaptic vesicles* and the *mitochondria*.

The synaptic vesicles contain a *transmitter substance* which, when released into the synaptic cleft, either *excites* or *inhibits* the neurons — excites if the neuronal membrane contains *excitatory receptors,* inhibits if it contains *inhibitory receptors.* This will be discussed in more detail later in the chapter. The mitochondria provide ATP, which is required to synthesize new transmitter substance. The transmitter must be synthesized extremely rapidly, because the amount stored in the vesicles is sufficient to last for only a few seconds to a few minutes of maximum activity.

When an action potential spreads over a synaptic knob, the membrane depolarization causes emptying of a small number of vesicles into the cleft, and the released transmitter in turn causes an immediate change in the permeability characteristics of the neuronal membrane, which leads to excitation or inhibition of the neuron, depending on the type of transmitter substance.

Mechanism by which the Synaptic Knob Action Potential Causes Release of Transmitter Vesicles. Unfortunately, we can only guess at the mechanism by which an action potential on reaching the synaptic knob causes the vesicles to release transmitter substance into the synaptic cleft. However, the number of vesicles released with each action potential is greatly reduced when the quantity of calcium ions in the extracellular fluid is diminished. Therefore, it has been suggested that the spread of the action

potential over the membrane of the knob causes small amounts of calcium ions to leak into the knob. The calcium ions then supposedly attract the transmitter vesicles to the membrane and simultaneously cause one or more of them to rupture, thus allowing spillage of their contents into the synaptic cleft.

One transmitter substance that occurs in certain parts of the nervous system is acetylcholine, as is discussed later. It has been calculated that about 3000 molecules of acetylcholine are present in each vesicle, and enough vesicles are present in the synaptic knobs on a neuron to transmit about 10,000 impulses.

Synthesis of New Transmitter Substance. Fortunately, the synaptic knobs have the capability of continuously synthesizing new transmitter substance. Were it not for this ability, synaptic transmission would become completely ineffective within a few minutes. The synthesis usually occurs mainly in the cytoplasm of the synaptic knobs, and then the newly synthesized transmitter is immediately transported into the vesicles and stored until needed. Thus, each time a vesicle empties its contents into the synaptic cleft, soon thereafter it becomes filled again with new transmitter.

Acetylcholine, for instance, is synthesized from acetyl-CoA and choline in the presence of the enzyme *choline acetyltransferase*, an enzyme that is present in abundance in the cytoplasm of the cholinergic type of synaptic knob. When acetylcholine is released from the knob into the synaptic cleft, it is rapidly split again to acetate and choline by the enzyme cholinesterase that is adherent to the outer surface of the knob. Then the choline is actively transported back into the knob to be used once more for synthesis of new acetylcholine. Thus, the choline is used again and again.

Action of the Transmitter Substance on the Postsynaptic Neuron. The membrane of the postsynaptic neuron where a synaptic knob abuts is believed to contain specific receptor molecules that bind the transmitter substance. These receptors are probably proteins that respond to the transmitter by changing their shapes or activities in such a way that they increase the membrane permeability especially to sodium ions when the membrane receptor is excitatory and increase the permeability to potassium and chloride ions when the receptor is inhibitory.

CHEMICAL AND PHYSIOLOGICAL NATURES OF THE TRANSMITTER SUBSTANCES

Excitation and Inhibition. Whether a transmitter will cause excitation or inhibition is determined not only by the nature of the transmitter but also by the nature of the receptor in the postsynaptic membrane. To give an example, the same neuron might be excited by a synapse that releases acetylcholine but inhibited by still another synapse that releases glycine. Thus, the neuronal membrane contains an *excitatory receptor* for acetylcholine and an *inhibitory receptor* for glycine. To give another example, norepinephrine released by some synapses in the central nervous system causes inhibition, while at other synapses it causes excitation. The neuronal membranes mentioned first contain an inhibitory receptor for norepinephrine, while the others contain an excitatory receptor for the same transmitter.

Differences in Function of Different Transmitters. Aside from the fact that transmitter substances sometimes cause excitation and sometimes inhibition, they also have other differences, the most important of which is duration of stimulation. For instance, the duration of excitation by the excitatory synapses on the anterior motoneurons of the spinal cord is only 10 to 20 milliseconds. On the other hand, the effects of some inhibitory synapses in the brain last for as long as 200 to 300 milliseconds.

Another difference among transmitters is that some cause an increase in the rate of firing of neurons, while others have no effect at all on the firing rate but do change the neuron's sensitivity to still other transmitter substances. This type of transmitter is called a *modulator*.

Release of Only a Single Type of Transmitter Substance by Each Neuron. A single neuron releases only one type of transmitter, and it releases it at all its nerve terminals. Therefore, release of the different types of transmitters in the central nervous system requires different types of neurons.

Some of the Important Transmitter Substances. There is now very good evidence for the existence of about thirty different nervous system transmitters. Some of the more important are discussed here.

Acetylcholine is secreted by neurons in many areas of the brain. It probably has an excitatory effect either everywhere or almost everywhere that it is released, even though it is known to have inhibitory effects in some portions of the peripheral parasympathetic nervous system, such as inhibition of the heart by the vagus nerves.

Norepinephrine is secreted by many neurons whose cell bodies are located in the reticular formation of the brain stem and also in the hypothalamus. These neurons send nerve fibers to widespread areas of the brain. In most instances they probably cause inhibition, though it is probable that they cause excitation in some areas.

Epinephrine is secreted by a smaller number of neurons, but in general they parallel the norepinephrine system.

Dopamine is secreted by neurons that originate in the substantia nigra. The terminations of these neurons are mainly in the striatal region of the basal ganglia. The effect of dopamine is usually inhibition.

Glycine is secreted mainly at synapses in the spinal cord. It probably always acts as an inhibitory transmitter.

Gamma aminobutyric acid (GABA) is secreted by nerve terminals in the spinal cord, the cerebellum, the basal ganglia, and many other areas. It is believed always to cause inhibition.

Glutamic acid is probably secreted by the synaptic knobs in some or many of the sensory pathways. It probably always causes excitation.

Substance P is probably released by pain fiber terminals in the substantia gelatinosa of the spinal cord. In general, it causes excitation.

Enkephalins and endorphins are probably secreted by nerve terminals in the spinal cord, in the brain stem, in the thalamus, and in the hypothalamus. They probably act as excitatory transmitters to excite another system that in turn inhibits the transmission of pain.

Serotonin is secreted by nuclei that originate in the median raphe of the brain stem and project to many brain areas, especially to the dorsal horns of the spinal cord and to the hypothalamus. Serotonin acts as an inhibitor of pain pathways in the cord, and it is also believed to help control the mood of the person, perhaps even to cause sleep.

Despite this long list of transmitter substances, it is only a partial list. Other substances that have been proved or suggested include various peptides, amino acids, histamine, prostaglandins, cyclic AMP, and many others.

ELECTRICAL EVENTS DURING NEURONAL EXCITATION

Resting Membrane Potential of the Neuronal Soma. Figure 31–6 illustrates the soma of a motoneuron, showing the resting membrane potential to be about −70 millivolts. This is somewhat less than the −90 millivolts found in large peripheral nerve fibers and in skeletal muscle fibers; the lower voltage is important, however, because it allows both positive and negative control of the degree of excitability of the neuron. That is, decreasing the voltage to a less negative value makes the membrane of the neuron more excitable, whereas increasing this voltage to a more negative value makes the neuron less excitable. This is the basis of the two modes of function of the neuron — either excitation or

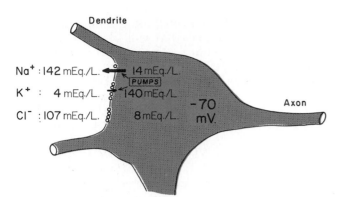

Figure 31–6. Distribution of sodium, potassium, and chloride ions across the neuronal somal membrane; origin of the intrasomal membrane potential.

inhibition — as we will explain in detail in the following sections.

Concentration Differences of Ions Across the Neuronal Somal Membrane. Figure 31–6 also illustrates the concentration differences across the neuronal somal membrane of the three ions that are most important for neuronal function: sodium ions, potassium ions, and chloride ions.

At the top, the sodium ion concentration is shown to be very great in the extracellular fluid but very low inside the neuron. This sodium concentration gradient is caused by a very strong sodium pump that continually pumps sodium out of the neuron, overcoming the slight back leakage of sodium through the pores, as illustrated in the figure.

The figure also shows that the potassium ion concentration is very great inside the neuronal soma but very low in the extracellular fluid. It illustrates that there is a weak potassium pump that tends to pump potassium to the interior, while there is a very high degree of permeability to potassium. The pump is relatively unimportant, because potassium ions leak through the neuronal somal pores so readily that this nullifies most of the effectiveness of the pump. Therefore, the cause of the high potassium concentration inside the soma is mainly the −70 millivolts in the neuron, because this keeps the positively charged potassium ions in the neuron.

Figure 31–6 shows the chloride ion to be of high concentration in the extracellular fluid but of low concentration inside the neuron. It also shows that the membrane is highly permeable to chloride ions and that there is no chloride pump. Therefore, the chloride ions become distributed across the membrane passively. The reason for the low concentration of chloride ions inside the neuron is the −70 millivolts in the neuron. That is, this negative voltage repels the negatively charged chloride ions, forcing them outward through the pores until the concentration difference is so great that its tendency to move chloride

ions inward exactly balances the tendency of the electrical difference to move them outward.

Origin of the Resting Membrane Potential of the Neuronal Soma. The basic cause of the −70 millivolt resting membrane potential of the neuronal soma is the sodium-potassium pump. This pump causes the extrusion of more positively charged sodium ions to the exterior than potassium to the interior — 3 sodium ions for each 2 potassium ions. Since there are large numbers of negatively charged ions inside the soma that cannot diffuse through the membrane — protein ions, phosphate ions, and many others — extrusion of the positively charged sodium ions to the exterior leaves all these nondiffusible negative ions unbalanced by positive ions on the inside. Therefore, the interior of the neuron becomes negatively charged. This principle was discussed in more detail in Chapter 8 in relation to the resting membrane potential of nerves.

Uniform Distribution of the Potential Inside the Soma. The interior of the neuromal soma contains a very highly conductive electrolytic solution, the intracellular fluid of the neuron. Furthermore, the diameter of the neuronal soma is very large (from 10 to 80 microns in diameter), causing there to be almost no resistance to conduction of electrical current from one part of the somal interior to another part. Therefore, any change in potential in any part of the soma causes an almost exactly equal change in potential at all other points in the soma. This principle is important because it plays a major role in the *summation* of signals entering the neuron from multiple sources, as we shall see in subsequent sections of this chapter.

Effect of Synaptic Excitation in the Postsynaptic Membrane — the Excitatory Postsynaptic Potential. Figure 31–7A illustrates the resting neuron with an unexcited synaptic knob resting upon its surface. The resting membrane potential everywhere in the soma is −70 millivolts.

Figure 31–7B illustrates an excitatory knob that has secreted a transmitter into the cleft between the knob and the neuronal somal membrane. This transmitter acts on the membrane excitatory receptor to increase the membrane's permeability to all ions. This causes sodium ions, in particular, to flow to the interior of the neuron because of the large electrochemical gradient that tends to move sodium inward.

The rapid influx of the positively charged sodium ions to the interior of the neuron neutralizes part of the negativity of the resting membrane potential. Thus, in Figure 31–7B the resting membrane potential has been increased from −70 millivolts to −59 millivolts. This increase in voltage above the normal resting neuronal potential — that is, an increase of +11 millivolts — is

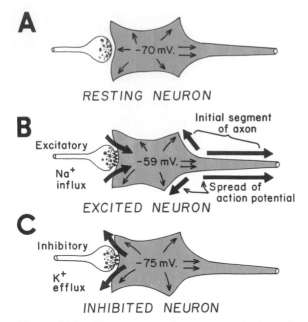

Figure 31–7. Three states of a neuron. *A,* A resting neuron. *B,* A neuron in an excited state, with increased intraneuronal potential caused by sodium influx. *C,* A neuron in an inhibited state, with decreased intraneuronal membrane potential caused by potassium ion efflux.

called the *excitatory postsynaptic potential* because when this potential rises high enough it will elicit an action potential in the neuron, thus exciting it.

However, we need to issue a word of warning at this point. Discharge of a single excitatory synaptic knob can never cause a postsynaptic potential of 11 millivolts. Instead, an increase of this magnitude requires the simultaneous discharge of many excitatory knobs — tens to hundreds, usually — at the same time. This occurs by a process called *summation,* which will be discussed in a later section.

Generation of Action Potentials at the Initial Segment of Axon — Threshold for Excitation. When the membrane potential inside the neuron rises high enough, there comes a point at which this increase initiates an action potential in the neuron. However, the action potential does not begin on the somal membrane adjacent to the excitatory synapses. Instead, it begins in the *initial segment of the axon* (also called the "axon hillock"). This may be explained as follows: Any factor that increases the potential inside the soma at any single point also increases this potential everywhere in the soma at the same time. Yet, because of physical differences in the membrane and differences in geometrical arrangement of the membrane in different parts of the neuron, the most excitable part of the neuron by far is the initial segment of the axon — that is, the first 50 to 100 microns of the axon beyond the point

where it leaves the neuronal soma. The excitatory postsynaptic potential that will elicit an action potential at this point on the neuron is approximately 11 millivolts. This is in contrast to approximately 30 millivolts required to elicit an action potential on the soma itself. Once the action potential begins, it travels peripherally along the axon and also travels backward over the soma of the neuron.

Thus, in Figure 31–7B, it is shown that under normal conditions the *threshold* for excitation of the neuron is −59 millivolts, which represents an excitatory postsynaptic potential of +11 millivolts — that is, 11 millivolts more positive than the normal resting neuronal potential of −70 millivolts.

ELECTRICAL EVENTS IN NEURONAL INHIBITION

Effect of Inhibitory Synapses on the Postsynaptic Membrane — the Inhibitory Postsynaptic Potential. It was pointed out above that excitatory synapses increase the permeability of the somal membrane to all ions, including sodium, potassium, and chloride. The inhibitory synapses, in contrast, increase the permeability of the postsynaptic membrane only to potassium and chloride ions. Therefore, influx of sodium ions does not occur. However, potassium efflux does occur, as illustrated in Figure 31–7C, thereby decreasing the positive ions inside the neuron and leaving the nondiffusible negative ions of the neuron (protein ions, phosphate ions, and others). This concentration of negative ions makes the internal potential of the neuron more negative than ever, as illustrated by the −75 millivolt potential inside the neuron in Figure 31–7C. This is called a *hyperpolarized state.* And the 5 millivolt decrease in intraneuronal voltage below the normal resting potential of −70 millivolts caused by the inhibitory transmitter is called the *inhibitory postsynaptic potential.*

Obviously, the increased negativity of the membrane potential (−75 millivolts) makes the neuron less excitable than it is normally. Since the potential must rise to −59 millivolts to excite the neuron, an excitatory postsynaptic potential must now be 16 millivolts instead of the normal 11 millivolts needed to cause excitation. Thus, in this way the inhibitory transmitter inhibits the neuron.

"Clamping" of the Resting Membrane Potential as a Means to Inhibit Neurons. Sometimes, excitation of the inhibitory synapses does not cause an inhibitory postsynaptic potential, but this still inhibits the neuron. The reason is that both the potassium and the chloride ions now diffuse bidirectionally through the wide open pores many times as rapidly as normally, and this high flux of these two ions inhibits the neuron in the following way: When excitatory synapses fire and sodium ions flow into the neuron, this raises the intraneuronal voltage far less than usual because any tendency for the membrane potential to change is immediately opposed by rapid flux of the potassium and chloride ions through the inhibitory pores to bring the potential back to the resting value. Therefore, the amount of influx of sodium ions required to cause excitation may be increased as much as 5 to 20 times normal.

This tendency for the potassium and chloride ions to maintain a membrane potential near the resting value when the inhibitory pores are wide open is called "clamping" of the potential.

Presynaptic Inhibition

In addition to the inhibition caused by inhibitory synapses operating at the neuronal membrane, called *postsynaptic inhibition,* another type of inhibition often occurs. This type, called *presynaptic inhibition,* is caused by the presence of inhibitory knobs lying on the presynaptic excitatory terminal fibrils and excitatory knobs. These inhibitory knobs reduce the excitability of the terminal fibrils and synaptic knobs. Consequently, the voltage of the action potential that occurs at the membrane of the excitatory knob is depressed, and this greatly decreases the amount of excitatory transmitter released by the knob. Therefore, transmission is inhibited.

Presynaptic inhibition occurs especially at the more peripheral synapses of the sensory pathways. For instance, when an incoming sensory fiber to the spinal cord excites the second neuron in the sensory pathway, it simultaneously causes presynaptic inhibition of the surrounding neurons. This "sharpens" the spatial boundaries of the signal transmitted to the brain; that is, it decreases or eliminates other signals on each side of the major signal so that the major signal stands out more clearly at its terminus in the brain. We shall see the importance of this sharpening process, also called "contrast enhancement," in discussions of subsequent chapters.

SUMMATION OF POSTSYNAPTIC POTENTIALS

Time Course of Postsynaptic Potentials. When a synapse excites the anterior motoneuron, the neuronal membrane becomes highly permeable for only about 1 millisecond. During this time sodium ions diffuse rapidly to the interior of the cell to increase the intraneuronal potential, thus creating the *excitatory postsynaptic potential.* This potential then persists for about 15 milliseconds because this is the time required for enough potassium ions to leak out or chloride ions to leak in to reestablish the normal resting membrane potential.

An opposite effect occurs for the inhibitory postsynaptic potential. That is, the inhibitory transmitter increases the permeability of the membrane to potassium and chloride ions for approximately one millisecond, and this effect usually decreases the intraneuronal potential to a more negative value than normal, thereby creating the *inhibitory postsynaptic potential.* This potential also persists for about 15 milliseconds.

Spatial Summation of the Postsynaptic Potentials. It has already been pointed out that excitation of a single synaptic knob on the surface of a neuron will almost never excite the neuron. The reason for this is that sufficient excitatory transmitter substance is released by a single knob to cause an excitatory postsynaptic potential usually no more than a fraction of a millivolt instead of the required 10 millivolts or more that is the usual threshold for excitation. However, many excitatory knobs are usually stimulated at the same time, and even though these knobs are spread over wide areas of the neuron, their effects can still summate. The reason for this summation is the following: It has already been pointed out that a change in the potential at any single point within the soma will cause the potential to change everywhere in the soma almost exactly equally. Therefore, for each excitatory knob that discharges simultaneously, the intrasomal potential rises another fraction of a millivolt. When the excitatory postsynaptic potential becomes great enough, the threshold for firing will be reached, and an action potential will generate at the axon hillock.

This effect of summing simultaneous postsynaptic potentials created by excitation of multiple knobs on widely spaced areas of the membrane is called *spatial summation.*

Temporal Summation. Most synaptic knobs can fire in rapid succession only a few milliseconds apart. Each time a knob fires, the released transmitter substance opens the membrane pores for only about one millisecond. Since the postsynaptic potential lasts for up to 15 milliseconds, a second opening of the same pore can increase the postsynaptic potential to a still greater level, so that the more rapid the rate of knob stimulation, the greater the effective postsynaptic potential. Thus, successive postsynaptic potentials of individual synaptic knobs, if they occur rapidly enough, can summate in the same way that postsynaptic potentials can summate from widely distributed knobs over the surface of the neuron. This summation of rapidly repetitive postsynaptic potentials is called *temporal summation*.

Simultaneous Summation of Inhibitory and Excitatory Postsynaptic Potentials. Obviously, if an inhibitory postsynaptic potential is tending to decrease the membrane potential to a more negative value while an excitatory postsynaptic potential is tending to increase the potential to a more positive value at the same time, these two effects can either completely nullify each other or partially nullify each other. Also, inhibitory "clamping" of the membrane potential can nullify much of an excitatory potential. Thus, if a neuron is currently being excited by an excitatory postsynaptic potential, an inhibitory signal from another source can easily reduce the postsynaptic potential below the threshold value for excitation, thus turning off the activity of the neuron.

Facilitation of Neurons. Often the summated postsynaptic potential is excitatory in nature but has not risen high enough to reach the threshold for excitation. When this happens the neuron is said to be *facilitated*. That is, its membrane potential is nearer the threshold for firing than normally, but it is not yet at the level of firing. Nevertheless, a signal entering the neuron from some other source can then excite the neuron very easily. Diffuse signals in the nervous system often facilitate large groups of neurons so that they can respond quickly and easily to signals arriving from secondary sources.

SPECIAL FUNCTIONS OF DENDRITES IN EXCITING NEURONS

The dendrites of the anterior motoneurons extend for one-half to one millimeter in all directions from the neuronal soma. Therefore, these dendrites can receive signals from a fairly large area around the motoneuron. This provides vast opportunity for summation of signals from many separate presynaptic neurons.

It is also important to note that between 80 and 90 per cent of all the synaptic knobs terminate on the dendrites of the anterior motoneuron in contrast to only 10 to 20 per cent terminating on the neuronal soma. Therefore, the preponderant share of the excitation of the neuron is provided by signals transmitted over the dendrites.

Dendrites can summate excitatory and inhibitory postsynaptic potentials in the same way that the soma can. In those instances when action potentials develop in dendrites, inhibitory knobs located near the soma can completely block the action potentials and can prevent their ever entering the soma of the neuron. It is especially

interesting that the inhibitory knobs tend to terminate on or near the soma.

RELATION OF STATE OF EXCITATION OF THE NEURON TO THE RATE OF FIRING

It should be obvious by now that a high excitatory state of a neuron will cause a second action potential to appear very soon after a previous one. Still a third action potential will appear soon after the second, and this process will continue indefinitely. Thus, at a very high excitatory state the rate of firing of the neuron is great.

On the other hand, when the excitatory state is only barely above threshold, the neuron must recover for many milliseconds before it can fire again. Therefore, the rate of neuronal firing is very slow.

Response Characteristics of Different Neurons to Increasing Levels of Central Excitatory State. Histological study of the nervous system immediately convinces one of the widely varying types of neurons in different parts of the nervous system. And, physiologically, the different types of neurons perform different functions. Therefore, as would be expected, the ability to respond to stimulation by the synaptic knobs varies from one type of neuron to another.

Figure 31–8 illustrates theoretical responses of three different types of neurons to varying levels of central excitatory state. (The "central excitatory state" is equal to the increase in membrane potential above the resting level.) Note that neuron number 1 will not discharge at all until the excitatory state rises to 5 millivolts — that is, to a potential 5 millivolts *less* negative than the resting potential, for instance, from −70 millivolts to

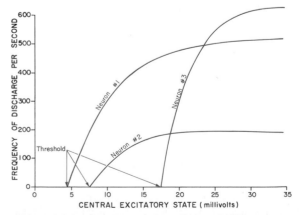

Figure 31–8. Response characteristics of different types of neurons to progressively increasing central excitatory states.

−65 millivolts. Then, as the excitatory state rises progressively to 35 millivolts the frequency of discharge rises to slightly over 500 per second.

Neuron number 2 is quite a different type, having a threshold for excitation of 8 millivolts and a maximum rate of discharge of only 190 per second. Finally, neuron number 3 has a high threshold for excitation, about 18 millivolts, but its discharge rate rises rapidly to over 600 per second as the excitatory state rises only slightly above threshold.

Thus, different neurons respond differently, have different thresholds for excitation, and have widely differing maximum frequencies of discharge. With a little imagination one can readily understand the importance of having neurons with many different types of response characteristics to perform the widely varying functions of the nervous system.

SOME SPECIAL CHARACTERISTICS OF SYNAPTIC TRANSMISSION

Forward Conduction through Synapses. From the above discussion, it should be evident by now that impulses are conducted through synapses only from the synaptic knobs to the successive neurons and never in the reverse direction. This is called the *principle of forward conduction.*

Fatigue of Synaptic Transmission. When excitatory synapses are repetitively stimulated at a rapid rate, the rate of discharges by the postsynaptic neuron is at first very great but becomes progressively less in succeeding milliseconds or seconds. This is called *fatigue* of synaptic transmission. Fatigue is an exceedingly important characteristic of synaptic function, for when areas of the nervous system become overexcited, fatigue causes them to lose this excess excitability after a while. For example, fatigue is probably the most important means by which the excess excitability of the brain during an epileptic fit is finally subdued so that the fit ceases. Thus, the development of fatigue is a protective mechanism against excess neuronal activity. This will be discussed further in the description of reverberating neuronal circuits later in the chapter.

The mechanism of fatigue is mainly exhaustion of the stores of transmitter substance in the synaptic knobs, particularly since it has been calculated that the excitatory knobs can store enough excitatory transmitter for only 10,000 normal synaptic transmissions, an amount that can be exhausted in only a few seconds to a few minutes.

Post-Tetanic Facilitation. When a rapidly repetitive (tetanizing) stimulus is applied to the synaptic knobs for a period of time and then a rest period is allowed, the neuron will usually be even more responsive to subsequent stimulation than normally. This is called *post-tetanic facilitation.*

One of the most likely explanations of post-tetanic facilitation is that the repetitive stimulation alters the membranes of the knobs to cause more rapid emptying of transmitter vesicles.

The physiological significance of post-tetanic facilitation is still very doubtful, and it may have no real significance at all. However, since post-tetanic facilitation can last from a few seconds in some neurons to many hours in others, it is immediately apparent that neurons could possibly store information by this mechanism. Therefore, post-tetanic facilitation might well be a mechanism of "short-term" memory in the central nervous system. This possibility will be discussed at further length in Chapter 36 in relation to the memory function of the cerebral cortex.

Effect of Acidosis and Alkalosis on Synaptic Transmission. The neurons are highly responsive to changes in pH of the surrounding interstitial fluids. *Alkalosis greatly increases neuronal excitability.* For instance, a rise in arterial pH from the normal of 7.4 to about 7.8 often causes cerebral convulsions because of increased excitability of the neurons. This can be demonstrated especially well by having a person who is normally predisposed to epileptic fits overbreathe. The overbreathing elevates the pH of the blood only momentarily, but even this short interval can often precipitate an epileptic convulsive attack.

On the other hand, *acidosis greatly depresses neuronal activity;* a fall in pH from 7.4 to below 7.0 usually causes a comatose state. For instance, in very severe diabetic or uremic acidosis, coma always develops.

Effect of Hypoxia on Synaptic Transmission. Neuronal excitability is also highly dependent on an adequate supply of oxygen. Cessation of oxygen supply for only a few seconds can often cause complete inexcitability of neurons. This is often seen when the cerebral circulation is temporarily interrupted, for within three to five seconds the person becomes unconscious.

Effect of Drugs on Synaptic Transmission. Many different drugs are known to increase the excitability of neurons, and others to decrease the excitability. For instance, caffeine, theophylline, and theobromine, which are found in coffee, tea, and cocoa, respectively, all increase neuronal excitability, presumably by reducing the threshold for excitation of the neurons. However, strychnine, which is one of the best known of all the agents that increase the excitability of neurons, does not reduce the threshold for excitation of the neurons at all but, instead, *inhibits the action of at least some of the inhibitory transmitters* on the neurons.

The hypnotics and anesthetics increase the threshold for excitation of the neurons and thereby decrease neuronal activity throughout the body. Because most of the volatile anesthetics are chemically inert compounds but are lipid soluble, it has been reasoned that these substances might change the physical characteristics of the neuronal membranes, making them less excitable to excitatory transmitters.

TRANSMISSION AND PROCESSING OF SIGNALS IN NEURONAL POOLS

The central nervous system is made up of literally hundreds of separate neuronal pools, some of which are extremely small and some very large. For instance, the entire cerebral cortex can be considered to be a single large neuronal pool. And other neuronal pools include the specific nuclei in the thalamus, basal ganglia, cerebellum, mesencephalon, pons, and medulla. Also, the entire dorsal gray matter of the spinal cord can be considered to be one long pool of neurons, and the entire anterior gray matter another long neuronal pool. Each pool has its own special characteristics of organization that cause it to process signals in its own special way. It is these special characteristics of the different pools that allow the multitude of functions of the nervous system. Yet, despite their differences in function, the pools also have many similarities of function, which are described in the following pages.

Organization of Neurons in the Neuronal Pools. Figure 31–9 is a schematic diagram of the organization of neurons in a neuronal pool, showing "input fibers" (also called "afferent fibers") to the left and "output fibers" (also called "efferent fibers") to the right. Each input fiber divides hundreds to thousands of times and usually provides several thousand terminal fibrils that spread over a large area in the pool to synapse with the dendrites or cell bodies of the neurons in the pool. The area into which the endings of each incoming nerve fiber spread is called its *stimulatory field*. Note that each input fiber arborizes, so that large numbers of its synaptic knobs lie on the centermost neurons in its "field," but progressively fewer knobs lie on the neurons farther from the center of the field.

Threshold and Subthreshold Stimuli — Facilitation. Going back to the discussion of synaptic function earlier in the chapter, it will be

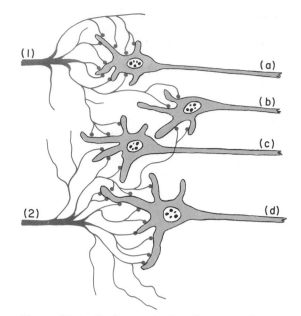

Figure 31–9. Basic organization of a neuronal pool.

recalled that stimulation of a single excitatory synaptic knob almost never stimulates the postsynaptic neuron. Instead, large numbers of knobs must discharge on the same neuron either simultaneously or in rapid succession to cause excitation. For instance, let us assume that six separate knobs must discharge simultaneously or in rapid succession to excite any one of the neurons in Figure 31–9. Note that input fiber 1 contributes a total of ten synaptic knobs to neuron *a*. Therefore, an incoming impulse in fiber 1 will cause neuron *a* to discharge. This same incoming fiber has two knobs on neuron *b* and three on neuron *c*. Neither of these two neurons will fire.

Incoming nerve fiber 2 has 12 synaptic knobs on neuron *d*, three on neuron *c*, and two on neuron *b*. In this case, an impulse in fiber 2 excites only neuron *d*.

Though fiber 1 fails to stimulate neurons *b* and *c*, discharge of the synaptic knobs changes the membrane potentials of these neurons so that they can be more easily excited by other incoming signals. Thus, a stimulus to a neuron can be either (1) an *excitatory stimulus,* also called a *threshold stimulus* because it is above the threshold required for excitation, or (2) a *subthreshold stimulus*. A subthreshold stimulus fails to excite the neuron but does make the neuron more excitable to impulses from other sources. The neuron that is made more excitable but does not discharge is said to be *facilitated*.

Summation of Subthreshold Stimuli to Cause Excitation. Let us now assume that both fibers 1 and 2 in Figure 31–9 transmit simultaneous impulses into the neuronal pool. In this case, neuron c is stimulated by six synaptic knobs simultaneously, which is the required number to cause excitation. Thus, subthreshold stimuli from several sources can *summate* at a neuron to cause an excitatory stimulus.

Inhibition of a Neuronal Pool. It will be recalled from earlier in the chapter that some neurons of the central nervous system secrete inhibitory transmitter substances instead of excitatory transmitters. Therefore, stimulation of inhibitory fibers from an inhibitory source inhibits the neuronal pool so that a stronger signal from the excitatory source is required to cause normal output.

Mechanism by Which a Single Input Signal Can Cause both Excitation and Inhibition —The Inhibitory Circuit. Figure 31–10 illustrates the so-called *inhibitory circuit*, which can change an excitatory signal into an inhibitory signal. In this figure the input fiber divides and secretes excitatory transmitter at both its endings. This causes excitation of both neurons 1 and 2. However, neuron 2 is an inhibitory neuron that secretes inhibitory transmitter at its terminal nerve endings. Excitation of this neuron therefore inhibits neuron 3.

In short, the usual means of causing inhibition is for a signal *to be transmitted through an inhibitory neuron*, which then secretes the inhibitory transmitter substance.

Convergence. The term "convergence" means control of a single neuron by converging signals from two or more separate input nerve fibers. One type of convergence was illustrated in Figure 31–9, in which two excitatory input nerve fibers from the same source converged upon several separate neurons to stimulate them.

However, convergence can also result from input signals (excitatory or inhibitory) from several different sources, which is illustrated in Figure 31–11. For instance, the interneurons of the spinal cord receive converging signals from (1) peripheral nerve fibers entering the cord, (2) propriospinal fibers passing from one segment of the cord to another, (3) corticospinal fibers from the cerebral cortex, and (4) several other long pathways descending from the brain into the spinal cord. Then the signals from the interneurons converge on the motoneurons to control muscle function.

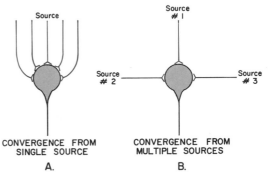

Figure 31–11. "Convergence" of multiple input fibers on a single neuron. *A*, Input fibers from a single source. *B*, Input fibers from multiple sources.

Such convergence allows summation of information from different sources, and the resulting response is a summated effect of all these different types of information. Obviously, therefore, convergence is one of the important means by which the central nervous system correlates, summates, and sorts different types of information.

Divergence. Divergence means that excitation of a single input nerve fiber stimulates multiple output fibers from the neuronal pool. The two major types of divergence are illustrated in Figure 31–12 and may be described as follows: An *amplifying* type of divergence often occurs, illustrated in Figure 31–12A. This means simply that an input signal spreads to an increasing number of neurons as it passes through successive pools of a nervous pathway. This type of divergence is characteristic of the corticospinal pathways in its control of skeletal muscles, as follows: Stimulation of a single large pyramidal cell in the motor cortex transmits a single impulse into the spinal cord. Under appropriate conditions, this impulse can stimulate several hundred interneurons, and these in turn stimulate per-

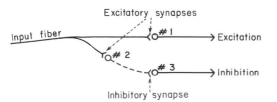

Figure 31–10. Inhibitory circuit. Neuron 2 is an inhibitory neuron.

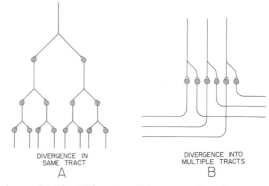

Figure 31–12. "Divergence" in neuronal pathways. *A*, Divergence within a pathway to cause "amplification" of the signal. *B*, Divergence into multiple tracts to transmit the signal to separate areas.

haps an equal number of anterior motoneurons. Each anterior motoneuron then stimulates as many as 100 to 300 muscle fibers. Thus, there is a total divergence, or amplification, of as much as 10,000-fold.

The second type of divergence, illustrated in Figure 31–12B, is *divergence into multiple tracts*. In this case, the signal is transmitted in two separate directions from the pool. This allows the same information to be transmitted to several different parts of the nervous system where it is needed. For instance, information transmitted in the dorsal columns of the spinal cord takes two courses in the lower part of the brain: (1) into the cerebellum and (2) on through the lower regions of the brain to the thalamus and cerebral cortex. Likewise, in the thalamus almost all sensory information is relayed both into deep structures of the thalamus and to discrete regions of the cerebral cortex.

TRANSMISSION OF SPATIAL PATTERNS THROUGH SUCCESSIVE NEURONAL POOLS

Most information is transmitted from one part of the nervous system to another through several successive neurons. For instance, sensory information from the skin passes first through the peripheral nerve fibers, then through second order neurons that originate either in the spinal cord or in the cuneate and gracile nuclei of the medulla, and finally through third order neurons originating in the thalamus to the cerebral cortex. Such a pathway is illustrated at the top of Figure 31–13. Note that the sensory nerve endings in the skin overlap each other tremendously; and the terminal fibrils of each nerve fiber, on entering each neuronal pool, spread to many adjacent neurons, innervating perhaps 100 or more separate neurons. On first thought, one would expect signals from the skin to become completely mixed up by this haphazard arrangement of terminal fibrils in each neuronal pool. For statistical reasons, however, this does not occur, which can be explained as follows:

First, if a single point is stimulated in the skin, the nerve fiber with the most nerve endings in that particular spot becomes stimulated to the strongest extent, while the immediately adjacent nerve fibers become stimulated less strongly, and the nerve fibers still farther away become stimulated only weakly. When this signal arrives at the first neuronal pool, the stimulus spreads in many directions in the terminal fibrils of the neuronal pool. Yet the *greatest number* of *excited terminal knobs* lies very near the center of the excitatory field of the most excited input nerve fiber. Therefore, the neuron closest to this central point is

the one that becomes stimulated to the greatest extent. Exactly the same effect occurs in the second neuronal pool in the thalamus and again when the signal reaches the cerebral cortex.

Yet, a signal passing through *highly facilitated neuronal pools* could diverge so much that the spatial pattern at the terminus of the pathway would be completely obscured. This effect is illustrated in Figure 31–13A, which shows successively expanding spatial patterns of neuron stimulation in such a facilitated, diverging pathway.

However, the degree of facilitation of the different neuronal pools varies from time to time. Under some conditions the degree of facilitation is so low that the pathway becomes converging, as illustrated in Figure 11–13B. In this case, a broad area of the skin is stimulated, but the signal loses part of its fringe stimuli as it passes through each successive pool until the breadth of the stimulus becomes contracted at the opposite end. One can achieve this type of stimulation by pressing ever so lightly with a flat object on the skin. The signal converges to give the person a sensation of almost a point contact.

In Figure 31–13C, four separate points are simultaneously stimulated on the skin, and the degree of excitability in each neuronal pool is exactly that amount required to prevent either divergence or convergence. Therefore, a reasonably true spatial pattern of each of the four points

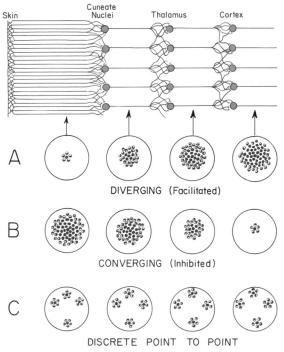

Figure 31–13. Typical organization of a sensory pathway from the skin to the cerebral cortex. *A* to *C*: The patterns of fiber stimulation at different points in the pathway following stimulation by a pinprick when the pathway is (*A*) facilitated, (*B*) inhibited, and (*C*) normally excitable.

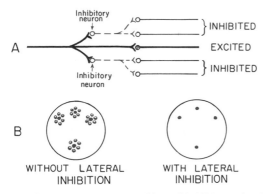

Figure 31–14. *A,* One type of lateral inhibitory circuit by which an excited fiber of a neuronal pool can cause inhibition of adjacent fibers. *B,* Increase in contrast of the stimulus pattern caused by the inhibitory circuit.

of stimulation is transmitted through the entire pathway.

Centrifugal Control of Neuronal Facilitation in the Sensory Pathways. It is obvious from the above discussion that the degree of facilitation of each neuronal pool must be maintained at exactly the proper level if faithful transmission of the spatial pattern is to occur. Recent discoveries have demonstrated that the degrees of facilitation of most — indeed, probably all — neuronal pools in the different pathways are controlled by *centrifugal nerve fibers* that pass from the respective sensory areas of the cortex downward to the separate neuronal pools. Thus, these nerve fibers undoubtedly help to control the faithfulness of signal transmission.

Inhibitory Circuits to Provide Contrast in the Spatial Pattern. When a single point of the skin or other sensory area is stimulated, not only is a single fiber excited but a number of "fringe" fibers are excited less strongly at the same time. Therefore, the spatial pattern is blurred even before the signal begins to be transmitted through the pathway. However, in many pathways — if not all — such as the visual pathway and the somesthetic pathway, lateral *inhibitory circuits* inhibit the fringe neurons and reestablish a truer spatial pattern.

Figure 31–14A illustrates this circuit, showing that the nerve fibers of a pathway give off collateral fibers that excite inhibitory neurons. These inhibitory neurons in turn inhibit the less excited fringe neurons in the signal pathway. The effect of this on transmission of the spatial pattern is illustrated in Figure 34–14B, which shows the same point-to-point transmission pattern that was illustrated in Figure 34–13C. The left-handed pattern illustrates four strongly excited fibers; penumbras of fringe excitation surround each of these. To the right is illustrated removal of the penumbras by the lateral inhibitory circuits; ob-

viously, this increases the contrast in the signal and helps the faithfulness of transmission of the spatial pattern.

PROLONGATION OF A SIGNAL BY A NEURONAL POOL– "AFTER-DISCHARGE"

Thus far, we have considered signals that are merely relayed through neuronal pools. However, in many instances, a signal entering a pool causes a prolonged output discharge, called *after-discharge*, even after the incoming signal is over. The three basic mechanisms by which after-discharge occurs are as follows:

Synaptic After-Discharge. When excitatory synapses discharge on the surfaces of dendrites or the soma of a neuron, a postsynaptic potential develops in the neuron and lasts for many milliseconds — in the anterior motoneuron for about 15 milliseconds, though perhaps much longer in other neurons. As long as this potential lasts it can excite the neuron, causing it to transmit output impulses as was explained earlier in the chapter. Thus, as a result of this synaptic after-discharge mechanism alone, it is possible for a single instantaneous input to cause a sustained signal output (a series of repetitive discharges) lasting as long as 15 milliseconds.

The Parallel Circuit Type of After-Discharge. Figure 31–15 illustrates a second type of neuronal circuit that can cause short periods of after-discharge. In this case, the input signal spreads through a series of neurons in the neuronal pool, and from many of these neurons impulses keep converging on an output neuron. A signal is delayed at each synapse for at least 0.5 millisecond, which is called the *synaptic delay*. Therefore, signals that pass through a succession of intermediate neurons reach the output neuron one by one after varying periods of delay. Therefore, the output neuron continues to be stimulated for many milliseconds.

It is doubtful that more than a few dozen successive neurons ordinarily enter into a parallel after-discharge circuit. Therefore, one would

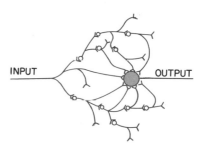

Figure 31–15. The parallel after-discharge circuit.

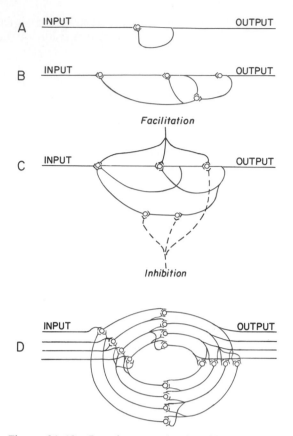

Figure 31–16. Reverberatory circuits of increasing complexity.

suspect that this type of after-discharge circuit could cause after-discharges that last for no more than perhaps 25 to 50 milliseconds. Yet, this circuit does represent a means by which a single input signal, lasting less than 1 millisecond, can be converted into a sustained output signal lasting many milliseconds.

The Reverberating (Oscillatory) Circuit as a Cause of After-Discharge. Probably one of the most important of all circuits in the entire nervous system is the reverberating, or oscillatory, circuit, several different varieties of which are illustrated in Figure 31–16. The simplest theoretical reverberating circuit, even though it may not actually exist in the nervous system, is that illustrated in Figure 31–16A, which involves a single neuron. In this case, the output neuron simply sends a collateral nerve fiber back to its own dendrites or soma to restimulate itself; therefore, once the neuron discharges, the feedback stimuli could theoretically keep the neuron discharging for a long time thereafter.

Figure 31–16B illustrates a few additional neurons in the feedback circuit, which would give a longer period of time between the initial discharge and the feedback signal. Figure 31–16C illustrates a still more complex system in which

both facilitatory and inhibitory fibers impinge on the reverberating pool. A facilitatory signal increases the ease with which reverberation takes place, while an inhibitory signal decreases the ease with which reverberation takes place.

Figure 31–16D illustrates that most reverberating pathways are constituted of many parallel fibers, and at each cell station the terminal fibrils diffuse widely. In such a system the total reverberating signal can be either weak or strong, depending on how many parallel nerve fibers are momentarily involved in the reverberation.

Finally, reverberation need not occur only in a single neuronal pool, for it can occur through a circuit of successive pools.

Characteristics of After-Discharge from a Reverberating Circuit. In a reverberating after-discharge circuit the input stimulus need last only 1 millisecond or so, and yet the output can last for many milliseconds or even minutes. The intensity of the output signal usually increases to a reasonably high value early in the reverberation, then decreases to a critical point, and then suddenly ceases entirely. Furthermore, the duration of the after-discharge is determined by the degree of inhibition or facilitation of the neuronal pool. In this way, signals from other parts of the brain can control the reaction of the pool to the input stimulus.

Importance of Synaptic Fatigue in Determining the Duration of Reverberation. It was pointed out earlier in the chapter that synapses fatigue if stimulated for prolonged periods of time. Therefore, one of the most important factors that determine the duration of the reverberatory type of after-discharge is the rapidity with which the involved synapses fatigue. Rapid fatigue obviously tends to shorten the period of after-discharge, and slow fatigue to lengthen it.

Furthermore, the greater the number of neurons in the reverberatory pathway and the greater the number of collateral feedback fibrils, the easier it is to keep the reverberation going. Therefore, it is to be expected that longer reverberating pathways would in general sustain after-discharges for longer periods of time.

Duration of Reverberation. Typical after-discharge patterns of different reverberatory circuits have durations from as short as 10 milliseconds to as long as several minutes or perhaps even hours. Indeed, as will be explained in Chapter 37, wakefulness may be an example of reverberation of neuronal circuits in the basal region of the brain. In this theory of wakefulness, "arousal impulses" are postulated to set off the wakefulness reverberation at the beginning of each day and thereby to cause sustained excitability of the brain, this excitability lasting 14 or more hours.

REFERENCES

Organization and Synapses

Adams, D. J., *et al.*: Ionic currents in molluscan soma. *Annu. Rev. Neurosci., 3*:141, 1980.

Barker, J. L.: Peptides: Roles in neuronal excitability. *Physiol. Rev., 56*:435, 1976.

Bourne, G. H., *et al.* (eds.): Neuronal Cells and Hormones. New York, Academic Press, 1978.

Ceccarelli, B., and Hurlbut, W. P.: Vesicle hypothesis of the release of quanta of acetylcholine. *Physiol. Rev., 60*:396, 1980.

Cooper, J. R., *et al.*: The Biochemical Basis of Neuropharmacology. New York, Oxford University Press, 1978.

Eccles, J. C.: My scientific odyssey. *Annu. Rev. Physiol., 39*:1, 1977.

Gotto, A. M., Jr., *et al.* (eds.): Brain Peptides: A New Endocrinology. New York, Elsevier/North-Holland, 1979.

Iversen, L. L.: The chemistry of the brain. *Sci. Am., 241*(3):134, 1979.

Johnston, G. A. R.: Neuropharmacology of amino acid inhibitory transmitters. *Annu. Rev. Pharmacol. Toxicol., 18*:269, 1978.

Kupfermann, I.: Modulatory actions of neurotransmitters. *Annu. Rev. Neurosci., 2*:447, 1979.

Langer, S. Z., *et al.* (eds.): Presynaptic Receptors. New York, Pergamon Press, 1979.

Mandel, P., and DeFeudis, F. V. (eds.): GABA — Biochemistry and CNS Functions. New York, Plenum Press, 1979.

Moore, R. Y., and Bloom, F. E.: Central catecholamine neuron systems: Anatomy and physiology of the dopamine systems. *Annu. Rev. Neurosci., 1*:129, 1978.

Nicoll, R. A., *et al.*: Substance P as a transmitter candidate. *Ann. Rev. Neurosci., 3*:227, 1980.

Otsuka, M., and Hall, Z. W. (eds.): Neurobiology of Chemical Transmission. New York, John Wiley & Sons, 1978.

Pepeu, G., *et al.* (eds.): Receptors for Neurotransmitters and Peptide Hormones. New York, Raven Press, 1980.

Schulster, D., and Levitzki, A. (eds.): Cellular Receptors for Hormones and Neurotransmitters. New York, John Wiley & Sons, 1980.

Stephenson, W. K.: Concepts of Neurophysiology. New York, John Wiley & Sons, 1980.

Stevens, C. F.: The neuron. *Sci. Am., 241*(3):54, 1979.

Neuronal Circuits

An der Heiden, U.: Analysis of Neural Networks. New York, Springer-Verlag, 1980.

Anderson, H., *et al.*: Developmental neurobiology of invertebrates. *Annu. Rev. Neurosci., 3*:97, 1980.

Blumenthal, R., *et al.* (eds.): Dymamic Patterns of Brain Cell Assemblies. Neurosciences Research Program Bulletin. Vol. 12, No. 1. Cambridge, Mass., Massachusetts Institute of Technology, 1974.

Cowan, W. M.: The development of the brain. *Sci. Am., 241*(3):112, 1979.

Friesen, W. O., and Stent, G. S.: Neural circuits for generating rhythmic movements. *Annu. Rev. Biophys. Bioeng., 7*:37, 1978.

Kandel, E. R.: Small systems of neurons. *Sci. Am., 241*(3):66, 1979.

Karlin, A., *et al.* (eds.): Neuronal Information Transfer. New York, Academic Press, 1978.

Macagno, E. R., *et al.*: Three-dimensional computer reconstruction of neurons and neuronal assemblies. *Ann. Rev. Biophys. Bioeng., 8*:323, 1979.

Patterson, P. H., *et al.*: The chemical differentiation of nerve cells. *Sci. Am., 239*(1):50, 1978.

Pinsker, H. M., and Willis, W. D., Jr. (eds.): Information Processing in the Nervous System. New York, Raven Press, 1980.

Purves, D., and Lichtman, J. W.: Formation and maintenance of synaptic connections in autonomic ganglia. *Physiol. Rev., 58*:821, 1978.

Shepherd, G. M.: Microcircuits in the nervous system. *Sci. Am., 238*(2):92, 1978.

Uttley, A. M.: Information Transmission in the Nervous System. New York, Academic Press, 1979.

Wooldridge, D. E.: Sensory Processing in the Brain: An Exercise in Neuroconnective Modeling. New York, John Wiley & Sons, 1979.

32

Sensory Receptors and the Mechanoreceptive Somatic Sensations

Input to the nervous system is provided by the sensory receptors that detect such sensory stimuli as touch, sound, light, cold, and warmth. One purpose of the present chapter is to discuss the basic mechanisms by which these receptors change sensory stimuli into nerve signals and, also, how both the type of sensory stimulus and its strength are detected by the brain. Another purpose is to discuss the mechanoreceptive somatic sensations.

TYPES OF SENSORY RECEPTORS AND THE SENSORY STIMULI THEY DETECT

Table 32–1 gives a list and classification of most of the body's sensory receptors. This table shows that there are basically five different types of sensory receptors: (1) *mechanoreceptors,* which detect mechanical deformation of the receptor itself or of cells adjacent to the receptor; (2) *thermoreceptors,* which detect changes in temperature, some receptors detecting cold and others warmth; (3) *nociceptors,* which detect pain, usually caused by physical damage or chemical damage to the tissues; (4) *electromagnetic receptors,* which detect light on the retina of the eye; and (5) *chemoreceptors,* which detect taste in the mouth, smell in the nose, oxygen level in the arterial blood, osmolality of the body fluids, carbon dioxide concentration, and perhaps other factors that make up the chemistry of the body.

This chapter will discuss especially the function of certain receptors, primarily peripheral mechanoreceptors, to illustrate some of the basic principles by which receptors in general operate.

Other receptors will be discussed in relation to the sensory systems that they subserve. Figure 32–1 illustrates some of the different types of mechanoreceptors found in the skin or in the deep structures of the body, and Table 32–1 gives their respective sensory functions. The functions of some of these are described briefly, as follows:

Free nerve endings are found in all parts of the body. A very large proportion of these detect pain. However, other free nerve endings detect crude touch, pressure, tickle, and itch sensations and possibly warmth and cold.

Several of the more complex receptors listed in Figure 32–1 detect tissue deformation. These include the *Merkel's discs,* the *tactile hairs, pacinian corpuscles, Meissner's corpuscles, Krause's corpuscles,* and *Ruffini's end-organs.* In the skin, it is these receptors that detect the tactile sensations of touch and pressure. In the deep tissues, they detect stretch, deep pressure, or any other type of tissue deformation — even the stretch of joint capsules and ligaments to determine the movements of joints.

The *Golgi tendon apparatus* detects tension in tendons, and the *muscle spindle* detects the length of muscle. These receptors will be discussed in Chapter 34 in relation to the muscle and tendon reflexes.

DIFFERENTIAL SENSITIVITY OF RECEPTORS

The first question that must be answered is, how do two types of sensory receptors detect different types of sensory stimuli? The answer: By virtue of differential sensitivities. That is, each type of receptor is very highly sensitive to the one

TABLE 32–1 CLASSIFICATION OF SENSORY RECEPTORS

Mechanoreceptors

Skin tactile sensibilities (epidermis and dermis)
  Free nerve endings
  Expanded tip endings
    Merkel's discs
    Plus several other variants
  Spray endings
    Ruffini's endings
  Encapsulated endings
    Meissner's corpuscles
    Krause's corpuscles
  Hair end-organs
Deep tissue sensibilities
  Free nerve endings
  Expanded tip endings
    Plus a few other variants
  Spray endings
    Ruffini's endings
  Encapsulated endings
    Pacinian corpuscles
    Plus a few other variants
  Muscle endings
    Muscle spindles
    Golgi tendon receptors
Hearing
  Sound receptors of cochlea
Equilibrium
  Vestibular receptors
Arterial pressure
  Baroreceptors of carotid sinuses and aorta

Thermoreceptors

Cold
  Cold receptors
Warmth
  Possibly free nerve endings

Nociceptors

Pain
  Free nerve endings

Electromagnetic Receptors

Vision
  Rods
  Cones

Chemoreceptors

Taste
  Receptors of taste buds
Smell
  Receptors of olfactory epithelium
Arterial oxygen
  Receptors of aortic and carotid bodies
Osmolality
  Probably neurons of supraoptic nuclei
Blood CO_2
  Receptors in or on surface of medulla and in aortic and
    carotid bodies
Blood glucose, amino acids, fatty acids
  Receptors in hypothalamus

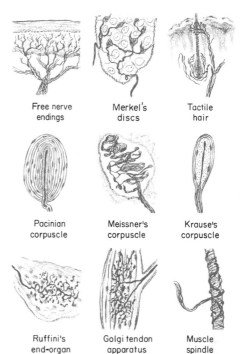

Free nerve endings	Merkel's discs	Tactile hair
Pacinian corpuscle	Meissner's corpuscle	Krause's corpuscle
Ruffini's end-organ	Golgi tendon apparatus	Muscle spindle

Figure 32–1. Several types of somatic sensory nerve endings.

type of stimulus for which it is designed and yet is almost nonresponsive to normal intensities of the other types of sensory stimuli. Thus, the rods and cones of the eyes are highly responsive to light but are almost completely nonresponsive to heat, cold, pressure on the eyeballs, or chemical changes in the blood. The osmoreceptors of the supraoptic nuclei in the hypothalamus detect minute changes in the osmolality of the body fluids but yet have never been known to respond to sound. Finally, pain receptors in the skin are almost never stimulated by usual touch or pressure stimuli but do become highly active the moment tactile stimuli becomes severe enough to damage the tissues.

Modality of Sensation — The Labeled Line Principle. Each of the principal types of sensation that we can experience — pain, touch, sight, sound, and so forth — is called a *modality* of sensation. Yet, despite the fact that we experience these different modalities of sensation, nerve fibers transmit only impulses. Therefore, how is it that different nerve fibers transmit different modalities of sensation?

The answer is that each nerve tract terminates at a specific point in the central nervous system, and the type of sensation felt when a nerve fiber is stimulated is determined by this specific area in the nervous system to which the fiber leads. For instance, if a pain fiber is stimulated, the person perceives pain regardless of what type of stimulus excites the fiber. This stimulus can be electricity, heat, crushing, or stimulation of the pain nerve ending by damage to the tissue cells. Yet, whatever the means of stimulation, the person still perceives pain. Likewise, if a touch fiber is stimulated by exciting a touch receptor electrically or in any other way, the person perceives touch because touch fibers lead to specific touch areas in the brain. Similarly, fibers from the retina of the eye terminate in the vision areas of the brain, fibers from the ear terminate in the auditory areas of the brain, and temperature fibers terminate in the temperature areas.

This specificity of nerve fibers for transmitting only one modality of sensation is called the *labeled line principle*.

TRANSDUCTION OF SENSORY STIMULI INTO NERVE IMPULSES

LOCAL CURRENTS AT NERVE ENDINGS – RECEPTOR POTENTIALS

All sensory receptors that have been studied have one feature in common. Whatever the type of stimulus that excites the ending, it first causes a local potential called a *receptor potential* in the neighborhood of the nerve endings, and it is *local flow of current* caused by the receptor potential that in turn excites action potentials in the nerve fiber.

There are two different ways in which receptor potentials can be elicited. One of these is to deform or chemically alter the terminal nerve ending itself. This causes ions to diffuse through the nerve membrane, thereby setting up the receptor potential.

The second method involves specialized receptor cells that lie adjacent to the nerve endings. For instance, when sound enters the cochlea of the ear, specialized *receptor cells* called *hair cells* that lie on the basilar membrane develop receptor potentials, and these in turn stimulate the terminal nerve fibrils entwining the hair cells.

The Receptor Potential of the Pacinian Corpuscle. The pacinian corpuscle is a very large and easily dissected sensory receptor. For this reason, one can study in detail the mechanism by which tactile stimuli excite it and by which it causes action potentials in the sensory fiber leading from it. Note in Figure 32–1 that the pacinian corpuscle has a central nonmyelinated nerve fiber extending into its core. Surrounding this fiber are many concentric capsule layers, and

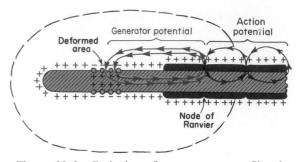

Figure 32–2. Excitation of a sensory nerve fiber by a generator potential produced in a pacinian corpuscle. (Modified from Loewenstein: *Ann. N.Y. Acad. Sci., 94*:510, 1961.)

compression on the outside of the corpuscle tends to elongate, shorten, indent, or otherwise deform the central core fiber, depending on how the compression is applied. The deformation causes a sudden change in the fiber's membrane potential, as illustrated in Figure 32–2. This is believed to result from stretching the nerve fiber membrane, thus increasing its permeability and allowing positively charged sodium ions to leak to the interior of the fiber. This causes both a local receptor potential and local current flow that spreads along the nerve fiber to its myelinated portion. At the first node of Ranvier, which itself lies inside the capsule of the pacinian corpuscle, the current flow initiates action potentials in the nerve fiber. That is, the current flow through the node depolarizes it, and this then sets off typical saltatory transmission of action potentials along the nerve fiber toward the central nervous system, as was explained in Chapter 8.

Relationship Between Receptor Potential and Stimulus Strength. Figure 32–3 illustrates the effect on the amplitude of the receptor potential caused by progressively stronger stimuli applied

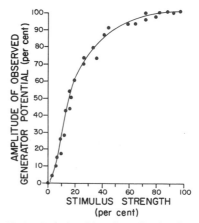

Figure 32–3. Relationship of amplitude of generator (receptor) potential to strength of a stimulus applied to a pacinian corpuscle. (From Loewenstein: *Ann. N.Y. Acad. Sci., 94*:510, 1961.)

to the central core of the pacinian corpuscle. Note that the amplitude increases rapidly at first but then progressively less rapidly at the higher stimulus strengths. The maximum amplitude that can be achieved by receptor potentials is around 100 millivolts. That is, a receptor potential can have almost as high a voltage as an action potential.

Receptor Potentials Recorded from Other Sensory Receptors. Receptor potentials have been recorded from many other sensory receptors, including most notably the muscle spindles, the hair cells of the ear, the rods and cones of the eyes, and many others. In all of these, the amplitude of the potential increases as the strength of stimulus increases, but the increment in the response becomes progressively less as the strength of stimulus becomes great.

Yet, the mechanism for causing the receptor potential is not the same in different receptors. For instance, in the rods and cones of the eye, changes in certain intracellular chemicals caused by exposure to light alter the membrane permeability, resulting in the receptor potential. In this case, the basic mechanism eliciting the receptor potential is a chemical one in contrast to mechanical deformation that causes the receptor potential in the pacinian corpuscle. In the case of thermal receptors, it is believed that changes in rates of chemical reactions in the receptors alter the membrane potential and thereby create a receptor potential. In the hair cells of the ear, bending of cilia protruding from the hairs causes the receptor potentials. Thus, the mechanisms for eliciting receptor potentials are individualized for each type of receptor.

Relationship of Amplitude of Receptor Potential to Nerve Impulse Rate. Referring once again to Figure 32–2, we see that the receptor potential generated in the core of the pacinian corpuscle causes a local circuit of current flow through the first node of Ranvier. When an action potential occurs at the node, this does not affect the receptor potential being emitted by the nerve fiber core of the pacinian corpuscle. Instead, the core continues to emit its current as long as an effective mechanical stimulus is applied. As a result, when the node of Ranvier repolarizes after its first action potential is over, it discharges once again, and action potentials continue as long as the receptor potential persists, which, in the case of the pacinian corpuscle, is only a few thousandths or hundredths of a second.

Furthermore, the frequency of action potentials in the nerve fiber (impulse rate) is almost directly proportional to the amplitude of the receptor potential. This same relationship between receptor potential and impulse rate is approximately true for most sensory receptors.

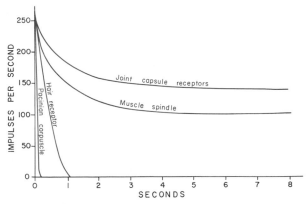

Figure 32–4. Adaptation of different types of receptors, showing rapid adaptation of some receptors and slow adaptation of others.

ADAPTATION OF RECEPTORS

A special characteristic of all sensory receptors is that they *adapt* either partially or completely to their stimuli after a period of time. That is, when a continuous sensory stimulus is first applied, the receptor responds at a very high impulse rate at first, then progessively less rapidly until finally, in many instances, it no longer responds at all.

Figure 32–4 illustrates typical adaptation curves of certain types of receptors. Note that the pacinian corpuscle and the receptor at the hair bases adapt extremely rapidly, while joint capsule and muscle spindle receptors adapt very slowly.

Furthermore, some sensory receptors adapt to a far greater extent than others. For example, the pacinian corpuscles adapt to "extinction" within a few thousandths of a few hundredths of a second, while most chemical receptors never adapt completely.

Mechanisms by which Receptors Adapt. Adaptation of receptors is an individual property of each type of receptor in much the same way that development of a receptor potential is an individual property. For instance, in the eye, the rods and cones adapt by changing their chemical compositions (which will be discussed in Chapter 39). In the case of the mechanoreceptors, the receptor that has been studied in greatest detail is again the pacinian corpuscle. Adaptation occurs in this receptor in two ways: First, the pacinian corpuscle is a viscoelastic structure, so that when a distorting force is suddenly applied to one side of the corpuscle this distortion is transmitted to the viscous component of the corpuscle directly to the same side of the central core fiber, thus eliciting a receptor potential. However, within a few thousandths to a few hundredths of a second the fluid within the corpuscle redistributes so that the pressure becomes essentially equal all through the corpuscle;

this now applies an even pressure on all sides of the central core fiber, so that the receptor potential is no longer elicited. Thus, a receptor potential appears at the onset of compression but then disappears within a small fraction of a second. Then, when the distorting force is removed from the corpuscle, essentially the reverse events occur. The sudden removal of the distortion from one side of the corpuscle allows rapid expansion on that side, and a distortion of the central core fiber occurs once more. Thus, the pacinian corpuscle signals the onset of compression and again signals the offset of compression.

The second mechanism of adaptation of the pacinian corpuscle results from a process of *accommodation* that occurs in the nerve fiber itself. That is, even if by chance the central core fiber should continue to be distorted, as can be achieved after the capsule has been removed and the core is compressed with a stylus, the tip of the nerve fiber itself gradually becomes "accommodated" to the stimulus. This perhaps results from redistribution of ions across the nerve fiber membrane.

Presumably, these same two general mechanisms of adaptation apply to some other types of receptors. That is, part of the adaptation results from readjustments in the structure of the receptor itself, and part results from accommodation in the terminal nerve fibril.

Function of the Poorly Adapting Receptors — The "Tonic" Receptors.   The poorly adapting receptors (receptors that adapt very slowly) continue to transmit impulses to the brain for many minutes or hours. Therefore, they keep the brain constantly apprised of the status of the body and its relation to its surroundings. For instance, impulses from the slowly adapting joint capsule receptors allow the person to "know" the degree of bending of the joints and therefore the positions of the different parts of his body. And impulses from muscle spindles and Golgi tendon apparatuses allow the central nervous system to know, respectively, the status of muscle contraction and the load on the muscle tendon at each instant.

Other types of poorly adapting receptors include the pain receptors, the baroreceptors of the arterial tree, the chemoreceptors of the carotid and aortic bodies, and some tactile receptors such as the Ruffini endings and the Merkel's discs.

Because the poorly adapting receptors can continue to transmit information for many hours, they are also called *tonic* receptors. It is probable that most of these poorly adapting receptors would adapt to extinction if the intensity of the stimulus should remain absolutely constant over many hours. Fortunately, because of our continually changing bodily state, the tonic receptors almost never do reach a state of complete adaptation.

Function of the Rapidly Adapting Receptors — The Rate Receptors (or Movement Receptors or Phasic Receptors).   Obviously, receptors that adapt rapidly cannot be used to transmit a continuous signal because these receptors are stimulated only for a short period after the stimulus strength changes. Yet they react strongly *while a change is actually taking place.* Furthermore, the number of impulses transmitted is directly related to the *rate at which the change takes place.* Therefore, these receptors are called *rate* receptors, *movement* receptors, or *phasic* receptors. Thus, in the case of the pacinian corpuscle, sudden pressure applied to the skin excites this receptor for a few milliseconds, and then its excitation is over even though the pressure continues. But then it transmits a signal again when the pressure is released. In other words, the pacinian corpuscle is exceedingly important in transmitting information about rapid changes in pressure against the body, but it is useless in transmitting information about constant pressure applied to the body.

Importance of the Rate Receptors – Their Predictive Function.   If one knows the rate at which some change in his bodily status is taking place, he can predict the state of the body a few seconds or even a few minutes later. For instance, the receptors of the semicircular canals in the vestibular apparatus of the ear detect the rate at which the head begins to turn when one runs around a curve. Using this information, a person can predict that he will turn 10, 30, or some other number of degrees within the next 10 seconds, and he can adjust the motion of his limbs *ahead of time* to keep from losing his balance. Likewise, pacinian corpuscles and the receptors located in or near the joint capsules help to detect the rates of movement of the different parts of the body. Therefore, when one is running, information from these receptors allows the nervous system to sense ahead of time where the feet will be during any precise fraction of a second, and appropriate motor signals can be transmitted to the muscles of the legs to make any necessary anticipatory corrections in limb position so that the person will not fall. Loss of this predictive function makes it impossible for the person to run.

PSYCHIC INTERPRETATION OF STIMULUS STRENGTH

The ultimate goal of most sensory stimulation is to apprise the psyche of the state of the body and its surroundings. Therefore, it is important that we discuss briefly some of principles related to the transmission of sensory stimulus strength to the higher levels of the nervous system.

The first question that comes to mind is, How is it possible for the sensory system to transmit sensory experiences of tremendously varying intensities? For instance, the auditory system can detect the weakest possible whisper but can also discern the meanings of an explosive sound only a few feet away, even though the sound intensities of these two experiences can vary as much as a trillion-fold; the eyes can see visual images with light intensities that vary as much as a million-fold; or the skin can detect pressure differences of ten thousand to one hundred thousand fold.

As a partial explanation of these effects, note in Figure 32–3 the relationship of the receptor potential produced by the pacinian corpuscle to the strength of stimulus. At low stimulus strength, very slight changes in stimulus strength increase the potential markedly, whereas at high levels of stimulus strength, further increases in receptor potential are very slight. Thus, the pacinian corpuscle is capable of accurately measuring extremely minute changes in stimulus strength at low intensity levels and is also capable of detecting much larger changes in stimulus strength at high intensity levels.

The transduction mechanism for detecting sound by the cochlea of the ear illustrates still another method for separating gradations of stimulus intensity. When sound causes vibration at a specific point on the basilar membrane, weak vibration stimulates only those hair cells in the very center of the vibratory point. But, as the vibration intensity increases, not only do the centralmost hair cells become more intensely stimulated, but hair cells in each direction farther away from the central point also become stimulated. Thus, signals are transmitted over progressively increasing numbers of cochlear nerve fibers; this is in addition to increasing intensity of signal strength in each nerve fiber as well. Thus, these two mechanisms, which multiply each other, make it possible for the ear to operate reasonably faithfully at stimulus intensity levels changing as much as a trillion-fold.

PHYSIOLOGICAL CLASSIFICATION OF NERVE FIBERS

Some sensory signals need to be transmitted to the central nervous system extremely rapidly; otherwise the information would be useless. An example of this is the sensory signals that apprise the brain of the momentary positions of the limbs at each fraction of a second during running. At the other extreme, some types of sensory infor-

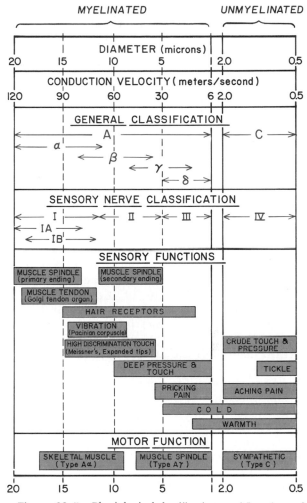

Figure 32–5. Physiological classifications and functions of nerve fibers.

mation, such as that depicting prolonged, aching pain, do not need to be transmitted rapidly at all, so that very slowly conducting fibers will suffice. Fortunately, nerve fibers come in all sizes, from 0.2 to 20 microns in diameter — the larger the diameter, the greater the conducting velocity. The range of conducting velocities is between 0.5 and 120 meters per second.

Figure 32–5 gives two different classifications of nerve fibers that are in general use. One of these is a general classification that includes both sensory and motor fibers, including the autonomic nerve fibers as well. The other is a classification of sensory nerve fibers that is used primarily by sensory neurophysiologists.

In the *general classification,* the fibers are divided into types A and C, and the type A fibers are further subdivided into α, β, γ, and δ fibers.

Type A fibers are the typical myelinated fibers of spinal nerves. Type C fibers are the very small, unmyelinated nerve fibers that conduct impulses at low velocities. These constitute more than half the sensory fibers in most peripheral nerves and also all of the postganglionic autonomic fibers.

In the *sensory fiber classification,* the fibers are divided into Groups Ia, Ib, II, III, and IV. The group I fibers are the largest, and the group IV fibers, the very smallest, are the unmyelinated fibers that are the same as the type C fibers in the general classification.

THE SOMATIC SENSES

The *somatic senses* are the nervous mechanisms that collect sensory information from the body. These senses are in contradistinction to the *special senses,* which mean specifically sight, hearing, smell, taste, and equilibrium.

The somatic senses can be classified into three different physiologic types: (1) the *mechanoreceptive somatic senses,* stimulated by mechanical displacement of some tissue of the body, (2) the *thermoreceptive senses,* which detect heat and cold, and (3) the *pain sense,* which is activated mainly by factors that damage the tissues. The present section deals with the mechanoreceptive somatic senses, and the following chapter deals with the thermoreceptive and pain senses.

The mechanoreceptive senses include *touch, pressure,* and *vibration* senses (which all together are frequently called the *tactile senses*) and the *position sense,* which determines the relative positions and rates of movement of the different parts of the body.

Somatic sensations are also grouped together in special classes that are not necessarily mutually exclusive, as follows:

Exteroceptive sensations are those from the surface of the body.

Proprioceptive sensations are those having to do with the physical state of the body, including position sensations, tendon and muscle sensations, pressure sensations from the bottom of the feet, and even the sensation of equilibrium, which is generally considered to be a "special" sensation rather than a somatic sensation.

Visceral sensations are those from the viscera of the body; in using this term one usually refers specifically to sensations from the internal organs.

The *deep sensations* are those that come from the deep tissues, such as the bone and the fascia. These include mainly "deep" pressure, pain, and vibration.

DETECTION AND TRANSMISSION OF TACTILE SENSATIONS

Interrelationships Among the Tactile Sensations of Touch, Pressure, and Vibration.

Though touch, pressure, and vibration are frequently classified as separate sensations, they are all detected by the same types of receptors. The only differences among these three are these: (1) touch sensation generally results from stimulation of tactile receptors in the skin or in tissues immediately beneath the skin; (2) pressure sensation generally results from deformation of deeper tissues; and (3) vibration sensation results from rapidly repetitive sensory signals, but some of the same receptors as those for both touch and pressure are utilized — specifically the rapidly adapting receptors.

The Tactile Receptors. At least six entirely different types of tactile receptors are known, but many more similar to these also exist. Some of these receptors were illustrated in Figure 32–1, and their special characteristics are the following:

First, some of the *free nerve endings,* which are found everywhere in the skin and in many other tissues, can detect touch and pressure. For instance, even light contact with the cornea of the eye, which contains no other type of nerve ending besides free nerve endings, can nevertheless elicit touch and pressure sensations.

Second, a touch receptor of special sensitivity is *Meissner's corpuscle,* an encapsulated nerve ending that excites a large myelinated sensory nerve fiber. Inside the capsulation are many whorls of terminal nerve filaments. These receptors are particularly abundant in the fingertips, lips, and other areas of the skin where one's ability to discern spatial characteristics of touch sensations is highly developed. They (along with the expanded tip receptors described subsequently) are mainly responsible for one's ability to recognize exactly what point of the body is touched and to recognize the texture of objects touched. Meissner's corpuscles adapt within a second after they are stimulated, which means that they are particularly sensitive to movement of very light objects over the surface of the skin and also to low-frequency vibration.

Third, the fingertips and other areas that contain large numbers of Meissner's corpuscles also contain *expanded tip tactile receptors,* one type of which is *Merkel's discs.* These receptors differ from Meissner's corpuscles in that they transmit an initial strong but partially adapting signal and then a continuing weaker signal that adapts only slowly. Therefore, they are probably responsible for giving steady state signals that allow one to determine continuous touch of objects against the skin. The hairy parts of the body contain almost no Meissner's corpuscles but do contain a few expanded tip receptors.

Fourth, slight movement of any hair on the body stimulates the nerve fiber entwining its base. Thus, each hair and its basal nerve fiber, called the *hair end-organ,* is also a type of touch receptor. This receptor adapts readily and, therefore, like Meissner's corpuscles, detects mainly movement of objects on the surface of the body.

Fifth, located in the deeper layers of the skin and also in deeper tissues of the body are many *Ruffini's end-organs,* which are multibranched endings, as illustrated in Figure 32–1. These endings adapt very little and therefore are important for signaling continuous state of deformation of the skin and deeper tissues, such as heavy and continuous touch signals and pressure signals. They are also found in joint capsules and signal the degree of joint rotation.

Sixth, many *pacinian corpuscles,* which were discussed in detail earlier in the chapter, lie both beneath the skin and also deep in the tissues of the body. These are stimulated only by very rapid movement of the tissues because these receptors adapt in a small fraction of a second. Therefore, they are particularly important for detecting tissue vibration or other extremely rapid changes in the mechanical state of the tissues.

Transmission of Tactile Sensations in Peripheral Nerve Fibers.   The specialized sensory receptors, such as Meissner's corpuscles, expanded tip endings, pacinian corpuscles, and Ruffini's endings, all transmit their signals in beta type A nerve fibers that have transmission velocities of 30 to 70 meters per second. On the other hand, most free nerve ending tactile receptors transmit signals mainly via the small delta type A nerve fibers that conduct at velocities of 6 to 30 meters per second, but a few transmit via type C fibers at velocities of about 1 meter per second, probably subserving the sensation of tickle and itch. Thus, the more critical types of sensory signals — those that help to determine precise localization on the skin, minute gradations of intensity, or rapid changes in sensory signal intensity — are all transmitted in the rapidly conducting types of sensory nerve fibers. On the other hand, the cruder types of signals, such as crude touch and tickle and itch, are transmitted via much slower nerve fibers, fibers that also require much less space in the nerves.

DETECTION OF VIBRATION

All the different tactile receptors are involved in detection of vibration, though different receptors detect different frequencies of vibration. Pacinian corpuscles can signal vibrations between 60 and 500 cycles per second, because they respond extremely rapidly to minute and rapid deformations of the tissues, and they also transmit their signals over beta type A nerve fibers, which can transmit more than 1000 impulses per second.

Low frequency vibrations up to 80 cycles per second, on the other hand, stimulate other tactile receptors — especially Meissner's corpuscles, which adapt less rapidly than pacinian corpuscles.

THE DUAL SYSTEM FOR TRANSMISSION OF MECHANORECEPTIVE SOMATIC SENSORY SIGNALS INTO THE CENTRAL NERVOUS SYSTEM

Either all or almost all sensory information from the somatic segments of the body enters the spinal cord through the posterior roots. Immediately after entering the cord, the nerve fibers separate into two major groups: (1) a *dorsal-lemniscal system* that includes (a) the *dorsal columns* and (b) the *spinocervical* tracts located in the *dorsolateral columns;* and (2) the *anterolateral spinothalamic system* located in the anterior and lateral columns.

Comparison of the Dorsal-Lemniscal System with the Anterolateral Spinothalamic System.   The distinguishing difference between the dorsal-lemniscal system and the anterolateral spinothalamic system is that the dorsal-lemniscal system is constituted mainly of large myelinated nerve fibers that transmit signals to the brain at velocities of 30 to 110 meters per second, while the anterolateral spinothalamic system is constituted of much smaller myelinated fibers that transmit impulses at velocities ranging between 10 and 60 meters per second.

Another difference between the two systems is that the dorsal system has a very high degree of spatial orientation of the nerve fibers with respect to their origin on the surface of the body, while the spinothalamic system has a much smaller degree of spatial orientation, some fibers seeming to have very little orientation at all.

These differences in the two systems immediately charaterize the types of sensory information that can be transmitted by the two systems. First, sensory information that must be transmitted rapidly and with temporal fidelity is transmitted in the dorsal-lemniscal system, while that which does not need to be transmitted rapidly is transmitted mainly in the anterolateral spinothalamic system. Second, those sensations that detect fine gradations of intensity are transmitted in the dorsal system, while those that lack the fine gradations are transmitted in the spinothalamic system. And, third, sensations that are discretely localized to exact points in the body are transmitted in the dorsal system, while those transmitted in the spinothalamic system can be localized much less exactly. On the other hand, the spinothalamic system has a special capability that the dorsal system does not have: the ability to transmit a broad spectrum of sensory modalities — pain, warmth, cold, and crude tactile sensations; the dorsal system is limited to mechanoreceptive sensations alone. With this differentiation in mind we can now list the types of sensations transmitted in the two systems:

The Dorsal-Lemniscal System

1. Touch sensations requiring a high degree of localization of the stimulus.
2. Touch sensations requiring transmission of fine gradations of intensity.
3. Phasic sensations, such as vibratory sensations.
4. Sensations that signal movement against the skin.
5. Position sensations.
6. Pressure sensations having to do with fine degrees of judgment of pressure intensity.

The Anterolateral Spinothalamic System

1. Pain.
2. Thermal sensations, including both warm and cold sensations.
3. Crude touch and pressure sensations capable of only crude localizing ability on the surface of the body and having little capability for intensity discrimination.
4. Tickle and itch sensations.
5. Sexual sensations.

FUNCTION OF THE SPINAL CORD NEURONS IN TRANSMITTING SENSORY SIGNALS

Relay of Sensory Signals in the Spinal Cord. Upon first entering the spinal cord through the posterior roots, as illustrated in Figure 32–6, most of the large sensory nerve fibers, mainly type Aβ fibers, turn medially toward the dorsal columns. Then they usually divide into two branches: One ascends upward through the dorsal columns and the other enters the anterior portion of the dorsal horn of the spinal cord gray matter.

In the dorsal horn gray matter, the terminals of the large fibers synapse with second order neurons of two types: (1) local neurons that play an intricate role in the control of spinal cord reflexes and (2) relay neurons that give rise to long, ascending fiber tracts that transmit sensory information to the brain. Most of the fibers from the relay neurons ascend in the *spinocervical tract*

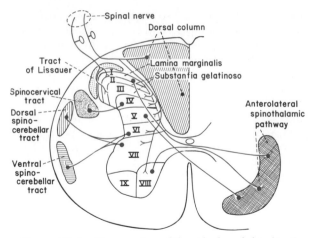

Figure 32–6. Cross-section of the spinal cord showing the anatomical laminae I through IX of the cord gray matter and the ascending sensory tracts in the white columns of the spinal cord.

located in the dorsolateral column, also shown in Figure 32–6.

The smaller fibers entering the spinal cord from the posterior roots take a more lateral pathway than the larger fibers; either they enter the gray matter of the dorsal horn immediately or they either ascend or descend a few segments and then enter the dorsal horn. Most of these fibers terminate on small neurons in the posterior portions of the dorsal horns, and these often give rise to short fibers that terminate somewhere else in the gray matter before the information is finally relayed to other areas of the cord or to the long ascending pathways. Essentially all the sensory information from the smaller fibers eventually enters the *anterolateral spinothalamic tract* that crosses to the opposite side of the cord in the anterior commissure and then ascends to the brain.

TRANSMISSION IN THE DORSAL-LEMNISCAL SYSTEM

ANATOMY OF THE DORSAL-LEMNISCAL SYSTEM

Sensory signals are transmitted to the brain by way of two major pathways in the dorsal-lemniscal system: (1) the *dorsal column pathway* and (2) the *spinocervical pathway*. The routes that these take to the brain are illustrated in Figure 32–7.

Anatomy of the Dorsal Column Pathway. Note in Figure 32–7 that the nerve fibers entering the dorsal columns pass up these columns to the medulla, where they synapse in the *dorsal column nuclei* (the *cuneate* and *gracile nuclei*). From here, *second order neurons* decussate immediately to the opposite side and then pass upward to the thalamus through bilateral pathways called the *medial lemnisci.* Each medial lemniscus then terminates in the *ventrobasal complex,* located in the ventral posterolateral nucleus of the *thalamus.*

From the ventrobasal complex, *third order neurons* project, as shown in Figure 32–8, mainly to the *postcentral gyrus* of the *cerebral cortex,* called the *somatic sensory area I.*

Anatomy of the Spinocervical Pathway. The anatomy of the spinocervical pathway is much less well known than that of the dorsal column pathway. However, many of the large sensory fibers that enter the cord in the dorsal roots soon synapse mainly in lamina IV of the cord gray matter, as shown in Figure 32–6, but also in adjacent laminae as well, giving rise to second order fibers that enter the *dorsolateral white columns* and ascend in the *spinocervical tract* to the cervical region of the cord or even to the medulla.

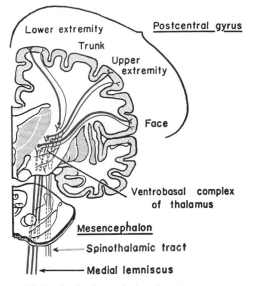

Figure 32–8. Projection of the dorsal-lemniscal system from the thalamus to the somesthetic cortex. (Modified from Brodal: Neurological Anatomy in Relation to Clinical Medicine. New York, Oxford University Press, 1969.)

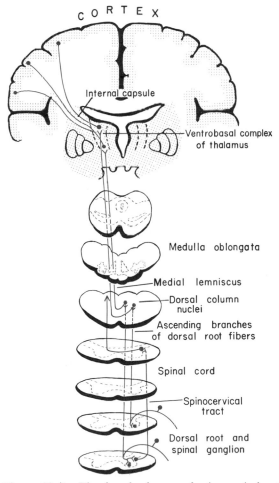

Figure 32–7. The dorsal column and spinocervical pathways for transmitting critical types of tactile signals. (Modified from Ranson and Clark: Anatomy of the Nervous System. Philadelphia, W. B. Saunders Company, 1959.)

At these points these fibers again synapse, and third order neurons decussate to the opposite side and pass along with the nerve fibers of the dorsal column pathway upward to the thalamus through the *medial lemnisci.* Thus, the pathway within the brain parallels that of the dorsal column pathway.

Separation of Sensory Modalities Between the Dorsal Column Pathway and the Spinocervical Pathway

The dorsal column pathway transmits mainly signals from rapidly adapting sensory receptors. For instance, it is through the dorsal column pathway that signals are transmitted from the extremely rapidly adapting pacinian corpuscles. Also, most of the signals from the Meissner's corpuscles and from the hair receptors, both of which are rapidly adapting receptors, are transmitted through this pathway.

On the other hand, the more slowly adapting signals from the Merkel's discs, from the deep tissue Ruffini end-organs, and from the slowly adapting Ruffini position sense receptors of the joint capsules are mainly transmitted through the spinocervical pathway.

Spatial Orientation of the Nerve Fibers in the Dorsal-Lemniscal System

One of the distinguishing features of the dorsal-lemniscal system is also a distinct spatial orientation of nerve fibers from the individual parts of the body that is maintained throughout. For instance, in the dorsal columns, the fibers from the lower parts of the body lie toward the center, while those that enter the spinal cord at progressively higher segmental levels form successive layers laterally.

The spatial orientation in the spinocervical pathway is less well known. However, stimulation experiments of single fibers within this pathway have shown that the sensory signals diverge very little, indicating a high degree of spatial orientation in this pathway as well.

THE SOMESTHETIC CORTEX

The area of the cerebral cortex to which the sensory signals are projected is called the *somesthetic cortex*. In the human being, this area lies mainly in the anterior portions of the parietal lobes. Two distinct and separate areas are known to receive direct afferent nerve fibers from the relay nuclei of the thalamus; these, called *somatic sensory area I* and *somatic sensory area II*, are illustrated in Figure 32–9. However, somatic sensory area I is extremely important to the sensory functions of the body, while the functions of somatic sensory area II are mainly unknown. Therefore, in popular usage, the term "somesthetic cortex" is almost always used to mean area I alone.

Projection of the Body in Somatic Sensory Area I. Somatic sensory area I lies in the postcentral gyrus of the human cerebral cortex. A distinct spatial orientation exists in this area for reception of nerve signals from the different areas of the body. Figure 32–10 illustrates a cross-section through the brain at the level of the postcentral gyrus, showing the representations of the different parts of the body in separate regions of somatic sensory area I. Note, however, that each side of the cortex receives its sensory information from the opposite side of the body, not from the same side.

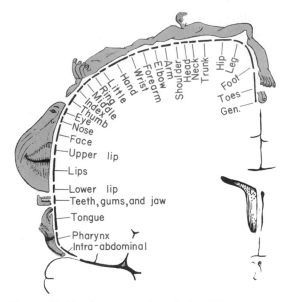

Figure 32–10. Representation of the different areas of the body in the somatic sensory area I of the cortex. (From Penfield and Rasmussen: Cerebral Cortex of Man: A Clinical Study of Localization of Function. New York, Macmillian Company, 1968.)

Some areas of the body are represented by large areas in the somatic sensory cortex — the lips by far the greatest of all, followed by the face and thumb — while the entire trunk and lower part of the body are represented by relatively small areas. The sizes of these areas are directly proportional to the number of specialized sensory receptors in each respective peripheral area of the body. For instance, a great number of specialized nerve endings are found in the lips and thumb, while only a few are present in the skin of the trunk.

Note also that the head is represented in the lower, lateral portion of the postcentral gyrus, while the lower part of the body is represented in the medial, upper portion of the postcentral gyrus.

Excitation of Vertical Columns of Neurons in the Somesthetic Cortex

The cerebral cortex contains *six* separate layers of neurons, and, as would be expected, the neurons in each layer perform functions different from those in other layers. Also, the neurons are arranged functionally in vertical columns extending all the way through the six layers of the cortex, each column having a diameter of about 0.33 to 1 millimeter and containing about 100,000 neuronal cell bodies. Unfortunately, we still know relatively little about the functions of these columns of cells but we are certain about the following facts:

1. The incoming sensory signal excites mainly

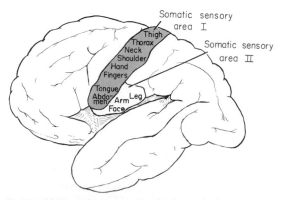

Figure 32–9. The two somesthetic cortical areas, somatic sensory areas I and II.

neuronal layer IV first (the fourth layer from the surface of the brain); then the signal spreads toward the surface of the cortex and also toward the deeper layers.

2. Layers I and II (the two layers nearest the surface) receive a diffuse, nonspecific input from a lower brain center called the reticular activating system, which can activate the whole brain at once; this system will be described in Chapter 37.

3. The neurons in layers V and VI send axons to other parts of the nervous system — some to other areas of the cortex, some to deeper structure of the brain, such as the thalamus or brain stem, and some even to the spinal cord.

Similar vertical columns of neurons exist in all other areas of the cortex, the same as in the somesthetic cortex.

Each vertical column of neurons seems to be able to decipher a specific quality of information from the sensory signal. Presumably, in the somesthetic cortex each column detects separate qualities of signals (angles of orientation of rough spots, lengths of rough spots, perhaps roundness of objects, perhaps sharpness of objects, and so forth) from specific surface areas of the body.

FUNCTIONS OF SOMATIC SENSORY AREA I

Destruction of somatic sensory area I causes loss of the following types of sensory judgment:

1. The person is unable to localize discretely the different sensations in the different parts of the body. However, he can still localize these sensations very crudely, such as those to a particular hand, which indicates that the thalamus or parts of the cerebral cortex not normally considered to be concerned with somatic sensations can perform some degree of localization.

2. He is unable to judge critical degrees of pressure against his body.

3. He is unable to judge closely the weights of objects.

4. He is unable to judge shapes or forms of objects. This is called *astereognosis*.

5. He is unable to judge texture of materials, for this type of judgment depends on highly critical sensations caused by movement of the skin over the surface to be judged.

6. He is unable to recognize the relative orientation of the different parts of his body with respect to each other.

Note in the above list that nothing has been said about loss of pain. However, in the absence of somatic sensory area I, the appreciation of pain may be altered either in quality or in intensity. But more important, the pain that does

occur is poorly localized, indicating that pain localization probably depends mainly upon simultaneous stimulation of tactile stimuli that use the topographical map of the body in somatic sensory area I to localize the source of the pain.

SOMATIC ASSOCIATION AREA

The parietal cortex immediately behind somatic sensory area I plays important roles in deciphering the sensory information that enters the somatic sensory areas. Therefore, this area is called the *somatic association area.*

Electrical stimulation in the somatic association area can occasionally cause a person to experience a complex somatic sensation, sometimes even the "feeling" of an object such as a knife or a ball. Therefore, it seems clear that the somatic association area combines information from multiple points in the somatic sensory area to decipher its meaning. This also fits with the anatomical arrangement of the neuronal tracts that enter the somatic association area, for it receives signals directly from (a) the primary somatic sensory areas, (b) the ventrobasal complex of the thalamus, and (c) other areas of the thalamus which themselves receive input from the ventrobasal complex.

Effect of Removing the Somatic Association Area — Amorphosynthesis. When the somatic association area is removed, the person especially loses the ability to recognize complex objects and complex forms that are felt. In addition, the person loses most of the sense of form of his or her own body. An especially interesting fact is that loss of the somatic association area on one side of the brain causes the person sometimes to be oblivious of the opposite side of the body — that is, to forget that it is there. Likewise, when feeling objects, the person will tend to feel only one side of the object and to forget that the other side even exists. This complex sensory deficit is called *amorphosynthesis.*

CHARACTERISTICS OF TRANSMISSION IN THE DORSAL- LEMNISCAL SYSTEM

Basic Neuronal Circuit and Discharge Pattern in the Dorsal-Lemniscal System. The lower part of Figure 32–11 illustrates the basic organization of the neuronal circuit of the dorsal column pathway, showing that at each synaptic stage divergence occurs. However, the upper part of the figure shows that a single receptor stimulus on the skin does not cause all the cortical neurons with which that receptor connects to discharge at the same rate. Instead, the cortical neurons that

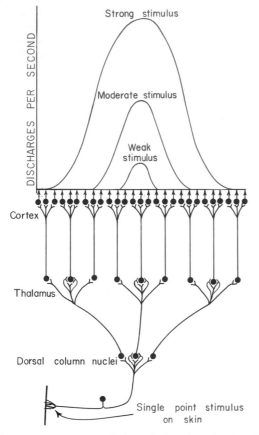

Figure 32–11. Transmission of pinpoint stimulus signal to the cortex.

discharge to the greatest extent are those in a central part of the cortical "field" for each respective receptor. Thus, a weak stimulus causes only the centralmost neurons to fire. A stronger stimulus causes still more neurons to fire, but those in the center still discharge at a considerably more rapid rate than those farther away from the center.

Two-Point Discrimination. A method frequently used to test tactile capabilities is to determine a person's so-called "two point discriminatory ability." In this test, two needles are pressed against the skin, and the person determines whether two points of stimulus are felt or one point. On the tips of the fingers a person can distinguish two separate points even when the needles are as close together as 1 to 2 mm, because of the many specialized nerve endings in the finger tips. However, on the person's back, the needles must usually be as far apart as 30 to 70 mm. before he can detect two separate points.

Figure 32–12 illustrates the mechanism by which the dorsal column pathway transmits two-point discriminatory information. This figure shows two adjacent points on the skin that are

strongly stimulated, and it also shows the small area of the somesthetic cortex (greatly enlarged) that is excited by signals from the two stimulated points. The solid black curve shows the spatial pattern of cortical excitation when both skin points are stimulated simultaneously. Note that the resultant zone of excitation has two separate peaks. It is these two peaks separated by a valley that allow the sensory cortex to detect the presence of two stimulatory points rather than a single point. However, the capability of the sensorium to distinguish between two points of stimulation is strongly influenced by another mechanism, the mechanism of lateral inhibition, as explained in the following section.

Increase in Contrast in the Perceived Spatial Pattern Caused by Lateral Inhibition. In Chapter 31 it was pointed out that contrast in sensory patterns is increased by inhibitory signals transmitted laterally in the sensory pathway. In the case of the dorsal-lemniscal system, an excited receptor in the skin transmits not only excitatory signals to the somesthetic cortex but also inhibitory signals laterally to adjacent fiber pathways. These inhibitory signals help to block lateral spread of the excitatory signal, a process called *lateral inhibition* or *surround inhibition*. As a result, the peak of excitation stands out, and much of the surrounding diffuse stimulation is blocked. This effect is illustrated by the two colored curves in Figure 32–12, showing complete separation of the peaks when the surround inhibition is very great. Obviously, this mechanisms accentuates the contrast between the areas of peak stimulation and the surrounding areas, thus greatly increasing the contrast or sharpness of the perceived spatial pattern.

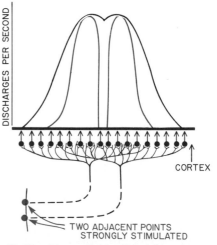

Figure 32–12. Transmission of signals to the cortex from two adjacent pinpoint stimuli. The solid black curve represents the pattern of cortical stimulation without "surround" inhibition, and the two colored curves represent the pattern with "surround" inhibition.

THE POSITION SENSE

The term position sense can be divided into two subtypes: (1) *static position,* which means conscious recognition of the orientation of the different parts of the body with respect to each other, and (2) *kinesthesia,* which means conscious recognition of rates of movement of the different parts of the body. The position sensations are transmitted to the sensorium through the dorsal-lemniscal system.

The Position Sense Receptors. Sensory information from many different types of receptors is used to determine both static position and kinesthesia. These include especially the extensive sensory endings in the joint capsules and ligaments but also receptors in the skin and deep tissues near the joints.

Three major types of nerve endings have been described in the joint capsules and ligaments about the joints. (1) By far the most abundant of these are spray-type *Ruffini endings,* one of which was illustrated in Figure 32–1. These endings are stimulated strongly when the joint is suddenly moved; they adapt slightly at first but then transmit a steady signal thereafter. (2) A second type of ending resembling the stretch receptors found in muscle tendons (called *Golgi receptors*) is found particularly in the ligaments about the joints. Though far less numerous than the Ruffini endings, they have essentially the same response properties. (3) A few *pacinian corpuscles* are also found in the tissues around the joints. These adapt extremely rapidly and presumably help to detect *rate of rotation* at the joint.

Detection of Static Position by the Joint Receptors. Figure 32–13 illustrates the excitation of seven different nerve fibers leading from separate joint receptors in the capsule of a cat's knee joint. Note that at 180 degrees of joint rotation one of the receptors is stimulated; then at 150 degrees still another is stimulated; at 140 degrees two are stimulated, and so forth. The informa-

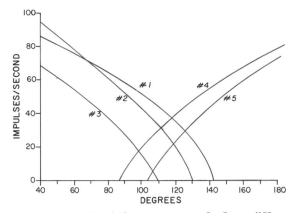

Figure 32–14. Typical responses of five different neurons in the knee joint receptor field of the ventrobasal complex when the knee joint is moved through its range of motion. (The curves were constructed from data in Mountcastle et al.: *J. Neurophysiol., 26*:807, 1963.)

tion from these joint receptors continually apprises the central nervous system of the momentary rotation of the joint. That is, the rotation determines *which* receptor is stimulated and how much it is stimulated, and from this the brain knows how far the joint is bent.

Detection of Rate of Movement (Kinesthesia) at the Joint. Rate of movement at the joint is probably detected mainly in the following way: The Ruffini and Golgi endings in the joint tissues are stimulated very strongly at first by joint movement, but within a fraction of a second this strong level of stimulation fades to a lower, steady state rate of firing. Nevertheless, this early overshoot in receptor stimulation is directly proportional to the rate of joint movement and is believed to be the signal used mainly by the brain to discern the rate of movement. However, it is likely that the few pacinian corpuscles also play at least some role in this process.

Processing of Position Sense Information in the Dorsal-Lemniscal Pathways. Despite the faithfulness of transmission of signals from the periphery to the sensory cortex in the dorsal-lemniscal system, there is also some processing of sensory information at lower synaptic levels before it reaches the cerebral cortex. For instance, the signal pattern from position receptors changes as it passes up the dorsal-lemniscal system. Figure 32–13 showed that individual joint receptors are stimulated maximally at specific degrees of rotation of the joint, with the intensity of stimulation decreasing on either side of the maximal point for each receptor. However, the signal for joint rotation is quite different at the level of the ventrobasal complex of the thalamus, as can be seen by referring to Figure 32–14. This figure shows that the ventrobasal neurons that respond to the joint rotation signal are of two

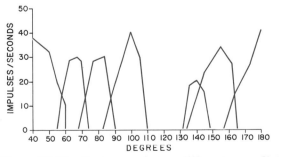

Figure 32–13. Responses of seven different nerve fibers from knee joint receptors in a cat at different degrees of rotation. (Modified from Skoglund: *Acta Physiol. Scand.,* Suppl. 124, *36*:1, 1956.)

types: (1) those that are maximally stimulated when the joint is at full rotation and (2) those that are maximally stimulated when the joint is at minimal rotation. Thus, the signals from the individual joint receptors are changed by the thalamic neurons to give a progressively stronger signal as the joint moves in only one direction rather than giving a peaked signal as occurs in stimulation of individual receptors.

TRANSMISSION IN THE ANTEROLATERAL SPINOTHALAMIC SYSTEM

It was pointed out earlier in the chapter that the anterolateral spinothalamic system transmits sensory signals that do not require highly discrete localization of the signal source and also do not require discrimination of fine gradations of intensity. These include pain, heat, cold, crude tactile, tickle and itch, and sexual sensations. In the following chapter pain and temperature sensations will be discussed, while the present section is concerned principally with transmission of the tactile sensations.

ANATOMY OF THE ANTEROLATERAL SPINOTHALAMIC PATHWAY

The anterolateral spinothalamic fibers originate mainly in laminae I, IV, V, and VI in the dorsal horns, where the small, peripheral, sensory nerve fibers terminate after entering the cord (see Figure 32–6). Then, as illustrated in Figure 32–15, the fibers immediately cross in the anterior commissure of the cord to the opposite anterolateral white column, where they turn upward toward the brain. These fibers ascend rather diffusely throughout the anterolateral columns. However, most anatomists still separate this pathway into a ventral spinothalamic tract and a lateral spinothalamic tract as illustrated in Figure 32–15, even though physiologically it has been difficult to make this differentiation using electrical recording techniques.

The upper terminus of the anterolateral spinothalamic pathway is mainly two-fold: (1) throughout the *reticular nuclei of the brain stem* and (2) in two different nuclear complexes of the thalamus, the *ventrobasal complex* and the *intralaminar nuclei*. In general, the tactile signals are transmitted mainly into the ventrobasal complex, and this is probably also true for the temperature signals. On the other hand, only part of the pain signals project to this complex. Instead, most of these enter the reticular nuclei of the brain stem and intralaminal nuclei of the thalamus, as will be

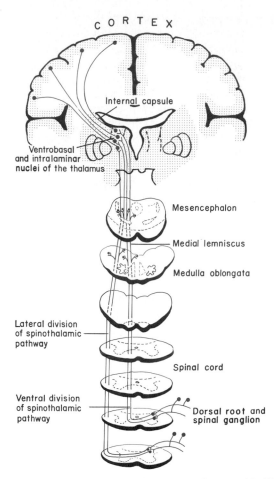

Figure 32–15. The spinothalamic pathways. (Modified from Ranson and Clark: Anatomy of the Nervous System. Philadelphia, W. B. Saunders Company, 1959.)

discussed in greater detail in the following chapter.

Projection of Spinothalamic Signals from the Thalamus to the Cortex. Spinothalamic signals entering the ventrobasal complex of the thalamus are relayed in association with those from the dorsal-lemniscal system mainly to somatic area I but to a lesser extent to a somatic area II. These relayed signals seem to be concerned mainly with tactile sensation, perhaps to a moderate extent with temperature sensation, but probably a little with pain sensation. On the other hand, most of the pain signals seem to be relayed into other areas of the thalamus and into surrounding basal regions of the brain, such as the hypothalamus and the septal nuclei.

Characteristics of Transmission in the Anterolateral Spinothalamic Pathway. In general, the same principles apply to transmission in the anterolateral spinothalamic pathway as in the dorsal-lemniscal system except for the following differences: (a) the velocities of transmission in the spinothalamic pathway are only one-half to one-third those in the dorsal system; (b) the degree of spatial localization of signals is poor, especially in the pain pathways; (c) the gradations of intensities are also far less acute, most of the sensations being recognized in 10 to 20 gradations of strength rather than as many as 100 gradations for the dorsal system; and (d) the ability to transmit repetitive sensations is poor.

Thus, it is evident that the anterolateral spinothalamic system is a cruder type of transmission system than the dorsal-lemniscal system. Even so, certain modalities of sensation are transmitted in this system only and not at all in the dorsal-lemniscal system. These are pain, thermal, tickle and itch, and sexual sensations.

SOME SPECIAL ASPECTS OF SENSORY FUNCTION

Function of the Thalamus in Somatic Sensation. When the somesthetic cortex of a human being is destroyed, he loses most critical tactile sensibilities, but a slight degree of crude tactile sensibility does return. Therefore, it must be assumed that the thalamus has a slight ability to discriminate tactile sensation but functions mainly to relay this type of information to the cortex.

On the other hand, loss of the somesthetic cortex has little effect on one's perception of pain sensation and only a moderate effect on the perception of temperature. Therefore, there is much reason to believe that the thalamus and other associated basal regions of the brain play perhaps the dominant role in discrimination of these sensibilities; it is interesting that these sensibilities appeared very early in the phylogenetic development of animalhood, while the critical tactile sensibilities were a late development.

Cortical Control of Sensory Sensitivity. The conscious brain is capable of directing its attention to different segments of the sensory system. One of the mechanisms of this is the following:

"Corticofugal" signals can be transmitted from the cortex to the lower relay stations of the sensory pathways to *inhibit* transmission in the thalamus, in the brain stem reticular nuclei, in the dorsal column nuclei, and especially in the dorsal horn relay station of the spinothalamic system. Also, similar inhibitory mechanisms are known for the visual, auditory, and olfactory systems, which are discussed in later chapters. Each corticofugal pathway begins in the cortex where the sensory pathway that it controls terminates. Thus, a feedback control loop exists for each sensory pathway.

Obviously, corticofugal control of sensory input could allow the cerebral cortex to alter the threshold for different sensory signals. Also, it might help the brain focus its attention on specific types of information, which is an important and necessary quality of nervous system function.

REFERENCES

Sensory Receptors

Anstis, S. M., *et al.*: Perception. New York, Springer-Verlag, 1978.

Babel, J., *et al.*: Ultrastructure of the Peripheral Nervous System and Sense Organs. New York, Churchill Livingstone, 1971.

Bennett, T. L.: The Sensory World: An Introduction to Sensation and Perception. Monterey, Cal., Brooks/Cole Publishing Co., 1978.

Brown, E., and Deffenbacher, K.: Perception and the Senses. New York, Oxford University Press, 1979.

Catton, W. T.: Mechanoreceptor function. *Physiol. Rev.,* 50:297, 1970.

Coren, S., *et al.*: Sensation and Perception. New York, Academic Press, 1979.

Goldstein, E. B.: Sensation and Perception. Belmont, Cal., Wadsworth Publishing Co., 1980.

Halata, Z.: The Mechanoreceptors of the Mammalian Skin. New York, Springer-Verlag, 1975.

Lynn, B.: Somatosensory receptors and their CNS connections. *Annu. Rev. Physiol.,* 37:105, 1975.

Porter, R. (ed.): Studies in Neurophysiology. New York, Cambridge University Press, 1978.

Schmidt, R. F. (ed.): Fundamentals of Sensory Physiology. New York, Springer-Verlag, 1978.

Wiersma, C. A. G., and Roach, J. L. M.: Principles in the organization of invertebrate sensory systems. *In* Brookhart, J. M., and Mountcastle, V. B. (eds.): Handbook of Physiology. Sec. 1, Vol. 1. Baltimore, Williams & Wilkins, 1977, p. 1089.

Mechanoreceptive Somatic Sensations

Coren, S. *et al.*: Sensation and Perception. New York, Academic Press, 1979.

Darian-Smith, I., *et al.*: Posterior parietal cortex: Relations of unit activity to sensorimotor function. *Annu. Rev. Physiol.,* 41:141, 1979.

Emmers, R., and Tasker, R. R.: The Human Somesthetic Thalamus. New York, Raven Press, 1975.

Goldstein, E. B.: Sensation and Perception. Belmont, Cal., Wadsworth Publishing Co., 1980.

Gordon, G. (ed.): Active Touch; The Mechanism of Recognition of Objects by Manipulation. New York, Pergamon Press, 1978.

Heath, C. J.: The somatic sensory neurons of pericentral cortex. *In* Porter, R. (ed.): International Review of Physiology: Neurophysiology III. Vol. 17. Baltimore, University Park Press, 1978, p. 193.

Kenshalo, D. R. (ed.): Sensory Function of the Skin of Humans. New York, Plenum Press, 1979.

Lynn, B.: Somatosensory receptors and their CNS connections. *Annu. Rev. Physiol.,* 37:105, 1975.

McCloskey, D. I.: Kinesthetic sensibility. *Physiol. Rev.,* 58:763, 1978.

Norrsell, U.: Behavioral studies of the somatosensory system. *Physiol. Rev.,* 60:327, 1980.

Olton, D. S.: Spatial memory. *Sci. Am.,* 236(6):82, 1977.

Vallbo, A. B., *et al.*: Somatosensory, proprioceptive, and sympathetic activity in human peripheral nerves. *Physiol. Rev.,* 59:919, 1979.

33

Somatic Sensations: Pain, Visceral Pain, Headache, and Thermal Sensations

Many if not most ailments of the body cause pain. Furthermore, one's ability to diagnose different diseases depends to a great extent on a knowledge of the different qualities of pain, a knowledge of how pain can be referred from one part of the body to another, how pain can spread from the painful site, and finally what the different causes of pain are. For these reasons, the present chapter is devoted mainly to pain and to the physiological basis of some of the associated clinical phenomena.

The Purpose of Pain. Pain is a protective mechanism of the body; it occurs whenever tissues are being damaged, and it causes the person to react to remove the pain stimulus. Even such simple activities as sitting for a long time on the ischia can cause tissue destruction because of lack of blood flow to the skin where the skin is compressed by the weight of the body. When the skin becomes painful as a result of the ischemia, the person shifts his weight unconsciously. A person who has lost his pain sense, as occurs after spinal cord injury, fails to feel the pain and therefore fails to shift his weight. This eventually results in ulceration at areas of pressure unless special measures are taken to move the person from time to time.

QUALITIES OF PAIN

Pain has been classified into three different major types: pricking, burning, and aching pain. Other terms used to describe different types of pain include throbbing pain, nauseous pain, cramping pain, sharp pain, electric pain, and others most of which are well known to almost everyone.

Pricking pain is felt when a needle is stuck into the skin or when the skin is cut with a knife. It is also often felt when a widespread area of the skin is diffusely but strongly irritated.

Burning pain, as its name implies, is the type of pain felt when the skin is burned. It can be excruciating and is the most likely of the pain types to cause suffering.

Aching pain is not felt on the surface of the body but instead is a deep pain with varying degrees of annoyance. Aching pain of low intensity in widespread areas of the body can summate into a very disagreeable sensation.

It is not necessary to describe these different qualities of pain in great detail because they are well known to all persons. The real problem, and one that is only partially solved, is what causes the differences in quality. Pricking pain results from stimulation of delta type A pain fibers, while burning and aching pain result from stimulation of the more primitive type C fibers, which will be discussed later in the chapter.

METHODS FOR MEASURING THE PERCEPTION OF PAIN

The intensity of a stimulus necessary to cause pain can be measured in many different ways, but the most used methods have been pricking the skin with a pin at measured pressures, pressing a solid object against a protruding bone with measured force, or heating the skin with measured amounts of heat. The latter method has proved to be especially accurate from a quantitative point of view.

Strength-Duration Curve for Expressing Pain Threshold. Figure 33–1 illustrates a typical

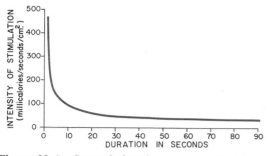

Figure 33–1. Strength-duration curve for depicting pain threshold. (Reprinted with permission from Hardy: *J. Chronic Dis., 4*:20. Copyright 1956, Pergamon Press, Ltd.)

strength-duration curve obtained by using a heat procedure for measuring pain threshold. Note that a very intense stimulus applied for only a second elicits a sensation of pain, while a stimulus of much less intensity may require many seconds. The lowest intensity of stimulus that will excite the sensation of pain when the stimulus is applied for a prolonged period of time is called the *pain threshold*.

Uniformity of Pain Threshold in Different People. Figure 33–2 shows graphically the lowest skin temperature at which pain is perceived by different persons. By far the greatest number of subjects barely begin to perceive pain when the skin temperature rises to 45°C, and almost everyone perceives pain before the temperature reaches 47°C. In other words, it is almost never true that some persons are unusually sensitive or insensitive to pain. Indeed, measurements in people as widely different as

Eskimos, Indians, and whites have shown no significant differences in their *thresholds for pain*. However, different people do *react* very differently to pain, as is discussed below.

THE PAIN RECEPTORS AND THEIR STIMULATION

Free Nerve Endings as Pain Receptors. The pain receptors in the skin and other tissues are all free nerve endings. They are widespread in the superficial layers of the *skin* and also in certain internal tissues, such as the *periosteum*, the *arterial walls*, the *joint surfaces*, and the *falx* and *tentorium* of the cranial vault. Most of the other deep tissues are not extensively supplied with pain endings but are weakly supplied; nevertheless, any widespread tissue damage can still summate to cause an aching type of pain even in these areas.

Types of Stimuli that Excite Pain Receptors — Mechanical, Thermal, and Chemical. Some pain fibers are excited almost entirely by excessive mechanical stress or mechanical damage to the tissues; these are called *mechanosensitive pain receptors*. Others are sensitive to extremes of heat or cold and therefore are called *thermosensitive pain receptors*. And, still others are sensitive to various chemical subtances and are called *chemosensitive pain receptors*. Some of the different chemicals that excite the chemosensitive receptors include *bradykinin, serotonin, histamine, potassium ions, acids, prostaglandins, acetylcholine,* and *proteolytic enzymes*.

Though some pain receptors are mainly sensitive to only one of the above types of stimuli, most are sensitive to more than one of the types.

RATE OF TISSUE DAMAGE AS THE CAUSE OF PAIN

The average critical temperature of 45°C at which a person first begins to perceive pain is also the temperature at which the tissues begin to be damaged by heat; indeed, the tissues are eventually completely destroyed if the temperature remains at this level indefinitely. Therefore, it is immediately apparent that pain resulting from heat is closely correlated with the ability of heat to damage the tissues.

Furthermore, in studying soldiers who had been severely wounded in World War II, it was found that the majority of them felt little or no pain except for a short time after the severe wound had been sustained. This, too, indicates that *pain generally is not felt after damage has been done* but only *while damage is being done*.

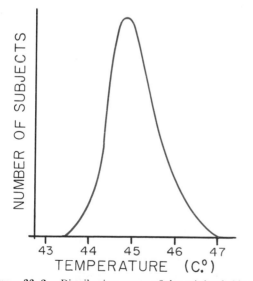

Figure 33–2. Distribution curve of the minimal skin temperature that causes pain, obtained from a large number of subjects. (Modified from Hardy: *J. Chronic Dis., 4*:22, 1956.)

Special Importance of Chemical Pain Stimuli During Tissue Damage. Injection of certain chemical substances under the skin in human beings can cause the highest degrees of pain. Furthermore, extracts from damaged tissues also cause intense pain when injected beneath the normal skin. Among the substances in such extracts that are especially painful are *bradykinin, histamine, prostaglandins, acids, excesses of potassium ions, serotonin*, and *proteolytic enzymes*. Obviously, many of these substances could cause direct damage to the pain nerve endings, especially the proteolytic enzymes. But some of the substances, such as bradykinin and some of the prostaglandins, can cause extreme stimulation of pain nerve fibers without necessarily damaging them.

Release of the various substances listed above not only stimulates the chemosensitive pain endings but also greatly decreases the threshold for stimulation of the mechanosensitive and thermosensitive pain receptors as well. A widely known example of this is the extreme pain caused by slight mechanical or heat stimuli following tissue damage by sunburn.

TICKLING AND ITCH

The phenomenon of tickling and itch has often been stated to be caused by very mild stimulation of pain nerve endings, because whenever pain is blocked by anesthesia of a nerve or by compressing the nerve, the phenomenon of tickling and itch also disappears. However, recent neurophysiologic studies have demonstrated the existence of very sensitive free nerve endings that elicit only the itch sensation. Furthermore, these endings are found almost exclusively in the superficial layers of the skin, which is also the only tissue from which the itch sensation can be elicited. Also, exciting itch receptors in animals initiates scratch reflexes, which contrasts with the effect of exciting pain nerve endings that always causes withdrawal reflexes instead.

Therefore, it seems clear that the itch and tickle sensations are transmitted by very small type C fibers similar to those that transmit the burning type of pain; these fibers, however, are distinctly separate from the pain fibers.

The purpose of the itch sensation is presumably to call attention to mild surface stimuli such as a flea crawling on the skin or a fly about to bite, and the elicited signals then lead to scratching or other maneuvers that rid the host of the irritant.

The relief of itch by the process of scratching occurs only when the irritant is removed or when the scratch is strong enough to elicit pain. The pain signals are believed to suppress the itch signals in the cord by a process of inhibition that will be described later in the chapter.

TRANSMISSION OF PAIN SIGNALS INTO THE CENTRAL NERVOUS SYSTEM

"Fast" Pain Fibers and "Slow" Pain Fibers. Pain signals are transmitted by small delta type A fibers at velocities between 6 and 30 meters per second and also by type C fibers at velocities between 0.5 and 2 meters per second. When the delta type A fibers are blocked without blocking the C fibers by moderate compression of the nerve trunk, the pricking type of pain disappears. On the other hand, when the type C fibers are blocked without blocking the delta fibers by low concentrations of local anesthetic, the burning and aching types of pain disappear.

Therefore, a sudden onset of painful stimulus gives a "double" pain sensation: a fast pricking pain sensation followed a second or so later by a slow burning pain sensation. The pricking pain presumably apprises the person very rapidly of a damaging influence and therefore plays an important role in making the person react immediately to remove himself from the stimulus. On the other hand, the slow burning sensation tends to become more and more painful over a period of time. It is this sensation that gives one the intolerable suffering of long-continued pain.

Transmission in the Anterolateral Spinothalamic Pathway. Pain fibers enter the cord through the dorsal roots, ascend or descend one to two segments, and then terminate on neurons in the dorsal horns of the cord gray matter, the type Aδ fibers in laminae I and V and the type C fibers in laminae II and III, an area also called the substantia gelatinosa. Most of the signals then pass through one or more additional short-fibered neurons, finally giving rise to long fibers that cross immediately to the opposite side of the cord and pass upward to the brain via the anterolateral spinothalamic pathway, as was described in the previous chapter.

As the pain pathways pass into the brain they separate into two separate pathways: the *pricking pain pathway* composed almost entirely of small type A delta fibers, and the *burning pain pathway* composed almost entirely of the slow type C fibers.

The Pricking Pain Pathway. Figure 33–3 illustrates that the pricking pain pathway terminates in the ventrobasal complex in close association with the areas of termination of the tactile sensation fibers of both the dorsal-lemniscal system and the spinothalamic system. From here

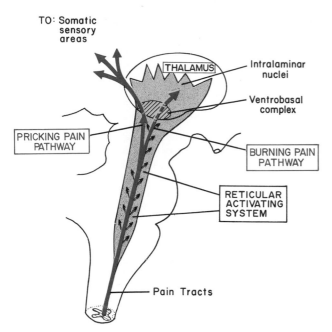

TO: Somatic
sensory
areas

THALAMUS — Intralaminar nuclei

— Ventrobasal complex

PRICKING PAIN PATHWAY

BURNING PAIN PATHWAY

RETICULAR ACTIVATING SYSTEM

Pain Tracts

Figure 33–3. Transmission of pain signals into the hindbrain, thalamus, and cortex via the "pricking pain" pathway and the "burning pain" pathway.

signals are transmitted into other areas of the thalamus and to the somatic sensory cortex, mainly to somatic area I. However, the signals to the cortex are probably important mainly for localizing the pain, not for interpreting it.

The Burning Pain Pathway — Stimulation of the Reticular Activating System. Figure 33–3 shows that the burning and aching pain fibers terminate in the reticular area of the brain stem and in the intralaminar nuclei of the thalamus. Both the reticular area of the brain stem and the intralaminar nuclei are parts of the reticular activating system, which will be discussed in Chapter 37; briefly, it transmits activating signals into essentially all parts of the brain, especially upward through the thalamus to all areas of the cerebral cortex and also laterally into the basal regions of the brain around the thalamus including, very importantly, the hypothalamus.

Thus, the burning and aching pain fibers, because they do excite the reticular activating system, have a very potent effect for activating almost the entire nervous system, that is, to arouse one from sleep, to create a state of excitement, to create a sense of urgency, and to promote defense and aversion reactions designed to rid the person or animal of the painful stimulus.

The signals that are transmitted through the burning pain pathway can be localized only to very gross areas of the body. Therefore, these signals are designed almost entirely for the single

purpose of calling one's attention to injurious processes in the body. They create suffering that is sometimes intolerable. Their gradation of intensity is poor; instead, even weak pain signals can summate over a period of time by the process of "temporal summation" to create an unbearable feeling even though the same pain for short periods of time may be relatively mild.

Function of the Thalamus and Cerebral Cortex in the Appreciation of Pain. Complete removal of the somatic sensory areas of the cerebral cortex does not destroy one's ability to perceive pain. Therefore, it is believed that pain impulses entering only the thalamus and other lower centers cause at least some conscious perception of pain. However, this does not mean that the cerebral cortex has nothing to do with normal pain appreciation; indeed, electrical stimulation of the somesthetic cortical areas causes a person to perceive mild pain in approximately 3 per cent of the stimulations. It is believed that the cortex plays an important role in interpreting the quality of pain even though pain *perception* might be a function of lower centers.

Localization of Pain in the Body. Most localization of pain probably results from simultaneous stimulation of tactile receptors along with the pain stimulation. However, the pricking type of pain, transmitted through delta type A fibers, can be localized perhaps within 10 to 20 cm. of the stimulated area. On the other hand, the burning and aching types of pain, transmitted through type C fibers, are localizable only very grossly, perhaps to a major part of the body such as a limb but certainly not to small areas. This is in keeping with the fact that these fibers terminate extremely diffusely in the hindbrain and thalamus.

THE REACTION TO PAIN

Even though the threshold for recognition of pain remains approximately equal from one person to another, the degree to which each one reacts to pain varies tremendously. Also, the intensity of the pain signals transmitted up the spinal cord to the different pain-receptive areas of the brain can change tremendously under different conditions. This results mainly from activation of a pain inhibiting system, both in the spinal cord and in the brain, that we shall discuss shortly.

Pain causes both motor reactions and psychic reactions. Some of the motor actions occur reflexly from the spinal cord, for pain impulses entering the gray matter of the cord can directly initiate "withdrawal reflexes" that remove the body or a portion of the body from the noxious stimulus, as will be discussed in Chapter 34. These primitive spinal cord reflexes, though important in lower animals, are mainly suppressed in the human being by the higher centers of the central nervous system. In their place, much more complicated and more effective reflexes from the motor cortex are initiated by the pain stimuli to eliminate the painful stimulus.

The psychic reactions to pain are likely to be far more subtle; they include all the well-known aspects of pain such as anguish, anxiety, crying, depression, nausea, and excess muscular excitability throughout the body. These reactions vary tremendously from one person to another following comparable degrees of pain stimuli.

A PAIN CONTROL ("ANALGESIC") SYSTEM IN THE BRAIN AND SPINAL CORD

Electrical stimulation in several different areas of the brain — especially in the *periventricular area of the diencephalon* immediately adjacent to the third ventricle, the *periaqueductal gray area* of the brain stem, and the *midline raphe nuclei* of the brain stem — can greatly reduce or even block pain signals transmitted in the spinal cord.

It is believed that this "analgesia" system operates in the following way: Stimulation in either the periventricular area of the diencephalon or the periaqueductal gray area transmits signals into the midline raphe nuclei. Then, from these nuclei, fiber tracts pass down in the spinal cord to terminate in the dorsal horns, where the pain sensory fibers from the periphery also terminate. Stimulation of the analgesic system will block or suppress transmission of pain impulses through the local neurons in this area.

It is very probable that this analgesic system also inhibits brain transmission at other points in the pain pathway, especially in the *reticular nuclei in the brain stem and in the intralaminar nuclei of the thalamus.*

The Brain's Opiate System — the Enkephalins and the Endorphins

A few years ago two closely related types of compounds with morphine-like actions, called the *enkephalins* and the *endorphins*, were isolated from those areas of the brain associated mainly with pain control, including the periventricular area, the periaqueductal gray, the midline raphe nuclei, the substantia gelatinosa of the dorsal horns in the spinal cord, and the intralaminar nuclei of the thalamus.

Therefore, it is now presumed that the enkephalins and the endorphins function as excita-

tory transmitter substances that activate portions of the brain's analgesic system. Also, infusion of either of them into the cerebrospinal fluid of the third ventricle can lead to analgesia.

Inhibition of Pain Transmission at the Cord Level by Tactile Signals

It has also been learned that stimulation of large sensory fibers from the peripheral tactile receptors depresses the transmission of pain signals from either the same area of the body or even areas located sometimes many segments away. This explains why such simple maneuvers as rubbing the skin near painful areas is often very effective in relieving pain. And it probably also explains why liniments are often useful in the relief of pain. This mechanism and simultaneous psychogenic excitation of the central analgesic system are probably the basis of pain relief by acupuncture.

REFERRED PAIN

Often a person feels pain in a part of his body that is considerably removed from the tissues causing the pain. This pain is called *referred pain*. On occasion, pain can even be referred from one surface area of the body to another, but more frequently it is initiated in one of the visceral organs and referred to an area of the body surface. Also, pain may originate in a viscus and be referred to another deep area of the body not exactly coincident with the location of the viscus producing the pain. A knowledge of these different types of referred pain is extremely impor-

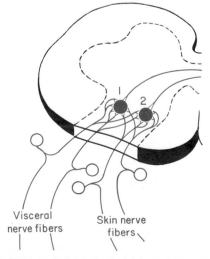

Figure 33–4. Mechanism of referred pain and referred hyperalgesia.

tant because many visceral ailments cause no other symptoms except referred pain.

Mechanism of Referred Pain. Figure 33–4 illustrates the most generally accepted mechanism by which most pain is referred. In the figure, branches of visceral pain fibers are shown to synapse in the spinal cord with some of the same second order neurons that receive pain fibers from the skin. When the visceral pain fibers are stimulated intensely, pain sensations from the viscera spread into some of the neurons that normally conduct pain sensations only from the skin, and the person has the feeling that the sensations actually originate in the skin itself.

VISCERAL PAIN

In clinical diagnosis, pain from the different viscera of the abdomen and chest is one of the few criteria that can be used for diagnosing visceral inflammation, disease, and other ailments. In general, the viscera have sensory receptors for no other modalities of sensation besides pain, and visceral pain differs from surface pain in many important aspects.

One of the most important differences between surface pain and visceral pain is that highly localized types of damage to the viscera rarely cause severe pain. For instance, a surgeon can cut the gut entirely in two in a patient who is awake without causing significant pain. On the other hand, any stimulus that causes *diffuse stimulation of pain nerve endings* throughout a viscus causes pain that can be extremely severe. For instance, occluding the blood supply to a large area of gut stimulates many diffuse pain fibers at the same time and can result in extreme pain.

CAUSES OF VISCERAL PAIN

Any stimulus that excites pain nerve endings in diffuse areas of the viscera causes visceral pain. Such stimuli include ischemia of visceral tissue, chemical damage to the surfaces of the viscera, spasm of the smooth muscle in a hollow viscus, distension of a hollow viscus, or stretching of the ligaments.

Almost all of the visceral pain signals originating in the thoracic and abdominal cavities are transmitted through sensory nerve fibers that run in the sympathetic nerves. These fibers are small type C fibers and therefore can transmit only burning and aching types of pain. The pathways for transmitting visceral pain will be discussed in detail later in the chapter.

Often, pain from a spastic viscus occurs in the form of *cramps*, the pain increasing to a high degree of severity and then subsiding, this process continuing rhythmically once every few minutes. The rhythmic cycles result from rhythmic spastic contraction of smooth muscle that is often severe enough to cause ischemic muscle pain. For instance, each time a peristaltic wave travels along an overly excitable spastic gut, a cramp occurs. The cramping type of pain frequently occurs in gastroenteritis, constipation, menstruation, parturition, gallbladder disease, or ureteral obstruction.

"PARIETAL" PAIN CAUSED BY VISCERAL DAMAGE

When a disease affects a viscus, it often spreads to the parietal (outside) wall of the visceral cavity. This wall, like the skin, is supplied with extensive innervation, including the "fast" delta fibers, which are different from the fibers in the true visceral pain pathways of the sympathetic nerves. Therefore, pain from the parietal wall of a visceral cavity is frequently very sharp and pricking in quality, though it can also have burning and aching qualities if the pain stimulus is diffuse. Thus, a knife incision through the *parietal* peritoneum is very painful, even though a similar cut through the visceral peritoneum or through a gut is not painful.

LOCALIZATION OF VISCERAL PAIN – REFERRED VISCERAL PAIN

Pain from the different viscera is frequently difficult to localize for a number of reasons. First, the brain does not know from firsthand experience that the different organs exist, and, therefore, any pain that originates internally can be localized only generally. Second, sensations from the abdomen and thorax are transmitted by two separate pathways to the central nervous system — the *true visceral pathway* and the *parietal pathway*. The true visceral pain is transmitted via sensory fibers of the autonomic nervous system as discussed below, and the sensations are *referred* to surface areas of the body that are often far from the painful organ. On the other hand, parietal sensations are conducted *directly* from the parietal peritoneum, pleura, or pericardium, and the sensations are usually *localized directly over the painful area*.

The Visceral Pathway for Transmission of Pain. Most of the internal organs of the body are supplied by type C pain fibers that pass along the visceral sympathetic nerves into the spinal cord and thence up the lateral spinothalamic tract along with the pain fibers from the body's surface. A few visceral pain fibers — those from the distal portion of the colon, from the rectum, and from the bladder — enter the spinal cord

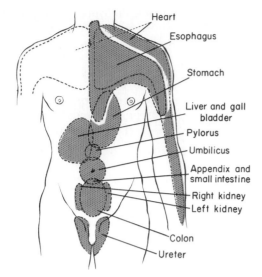

Figure 33–5. Surface areas of referred pain from different visceral organs.

through the sacral parasympathetic nerves, and some enter the central nervous system through various cranial nerves. These include fibers in the glosopharyngeal and vagus nerves, which transmit pain from the pharynx, trachea, and upper esophagus. And pain fibers from the surfaces of the diaphragm as well as from the lower esophagus are carried in the phrenic nerves.

Localization of Referred Pain Transmitted by the Visceral Pathways. The position in the cord to which visceral afferent fibers pass from each organ depends on the segment of the body from which the organ developed embryologically. For instance, the heart originated in the neck and upper thorax. Consequently, the heart's visceral pain fibers enter the cord all the way from C-3 down to T-5. The stomach had its origin approximately from the seventh to the ninth thoracic segments of the embryo, and consequently the visceral afferents from the stomach enter the spinal cord between these levels. The gallbladder had its origin almost entirely in the ninth thoracic segment, so that the visceral afferents from the gallbladder enter the spinal cord at T-9.

Because the visceral afferent pain fibers are responsible for transmitting referred pain from the viscera, the location of the referred pain on the surface of the body is in the area from which the visceral organ was originally derived in the embryo. Some of the areas of referred pain are shown in Figure 33–5.

SPECIFIC EXAMPLES

Cardiac Pain. Almost all pain that originates in the heart results from ischemia secondary to coronary slcerosis. This pain is referred mainly to the base of the neck, over the shoulders, over the pectoral muscles, and down the arms. Most frequently, the referred pain is on the left side rather than on the right — probably because the left side of the heart is much more frequently involved in coronary disease than is the right side — but occasionally mild referred pain occurs on the right side of the body as well as on the left.

When coronary ischemia is extremely severe, such as immediately after a coronary thrombosis, intense cardiac pain sometimes occurs directly underneath the sternum simultaneously with pain referred to other areas. This direct pain from underneath the sternum is difficult to explain on the basis of the visceral nerve connections. Therefore, it is highly probable that sensory nerve endings passing from the heart through the pericardial reflections around the great vessels conduct this direct pain.

Gastric Pain. Pain arising in the stomach — usually caused by gastritis — is referred to the anterior surface of the chest or upper abdomen from slightly below the heart to an inch or so below the xyphoid process. This pain is frequently characterized as burning pain; and it, or pain from the lower esophagus, causes the condition known as "heartburn."

Most peptic ulcers occur within 1 to 2 inches on either side of the pylorus in the stomach or in the duodenum, and pain from such ulcers is usually referred to a surface point approximately midway between the umbilicus and the xyphoid process. The origin of ulcer pain is almost undoubtedly chemical, because when the acid juices of the stomach are not allowed to reach the pain fibers in the ulcer crater the pain does not exist. Characteristically this pain is intensely burning.

Biliary and Gallbladder Pain. Pain from the bile ducts and gallbladder is localized in the midepigastrium almost at the same point as pains caused by peptic ulcers. Also, biliary and gallbladder pain is often burning, like that from ulcers, though cramps often occur too.

Biliary disease, in addition to causing pain on the abdominal surface, frequently refers pain to a small area at the tip of the right scapula. This pain is transmitted through sympathetic afferent fibers that enter the ninth thoracic segment of the spinal cord.

Uterine Pain. Both parietal and visceral afferent pain may be transmitted from the uterus. The low abdominal cramping pains at the time of menstruation are mediated through the sympathetic afferents, and an operation to cut the hypogastric nerves between the hypogastric plexus and the uterus will in many instances relieve this pain. On the other hand, lesions of the uterus that spread into the surroundings of the uterus, or lesions of the fallopian tubes and broad ligaments, usually cause pain in the lower back or side. This pain is conducted over parietal nerve fibers and is usually sharp rather than resembling the diffuse cramping pain of true dysmenorrhea.

HEADACHE

Headaches are actually pain referred to the surface of the head from the deep structures. The brain itself is almost totally insensitive to pain. Even cutting or electrically stimulating the somesthetic centers of the cortex only occasionally causes pain; instead, it causes tactile paresthesia on the area of the body represented by the portion of the somesthetic cortex stimulated. Therefore, most of the pain of headache probably is not caused by damage within the brain itself.

On the other hand, *tugging on the venous sinuses* or *damage to the membranes covering the brain* can cause intense pain that is recognized as headache.

Headache of Meningitis. One of the most severe headaches of all is that resulting from meningitis, which causes inflammation of all the meninges, including the sensitive areas of the dura and the sensitive areas around the venous sinuses. Such intense damage as this can cause extreme headache pain referred over the entire head.

Migraine Headache. Migraine headache is a special type of headache that is thought to result from abnormal vascular phenomena, though the exact mechanism is unknown.

Migraine headaches often begin with various prodromal sensations, such as nausea, loss of vision in parts of the fields of vision, visual aura, or other types of sensory hallucinations. Ordinarily, the prodromal symptoms begin half an hour to an hour prior to the beginning of the headache itself. Therefore, any theory that explains migraine headache must also explain these prodromal symptoms.

One of the theories of the cause of migraine headaches is that prolonged emotion or tension causes reflex vasospasm of some of the arteries of the head, including arteries that supply the brain itself. The vasospasm theoretically produces ischemia of portions of the brain, and this is responsible for the prodromal symptoms. Then, as a result of the intense ischemia, something happens to the vascular wall to allow it to become flaccid and incapable of maintaining vascular tone for 24 to 48 hours. The blood pressure in the vessels causes them to dilate and intensely, and it is supposedly the excessive stretching of the walls of the arteries — including the extracranial arteries such as the temporal artery — that causes the actual pain of migraine headaches. However, it is possible that diffuse aftereffects of ischemia in the brain itself are at least partially responsible for this type of headache.

Alcoholic Headache. As many people have experienced, a headache usually follows an alcoholic binge. It is most likely that alcohol, because it is toxic to tissue, directly irritates the meninges and causes the cerebral pain.

Headache Caused by Constipation. Constipation causes headache in many persons. This probably results from absorbed toxic products or from changes in the circulatory system. Indeed, constipation sometimes causes temporary loss of plasma into the wall of the gut, and a resulting poor flow of blood to the head could be the cause of the headache.

Headache Caused by Irritation of the Nasal and Accessory Nasal Structures. The mucous membranes of the nose and also of all the nasal sinuses are sensitive to pain. As a consequence, infection or other irritative processes in widespread areas of the nasal structures usually cause headache that is referred behind the eyes or, in the case of frontal sinus infection, to the frontal surfaces of the forehead and scalp.

Headache Caused by Eye Disorders. Difficulty in focusing one's eyes clearly may cause excessive contraction of the ciliary muscles in an attempt to gain clear vision. Even though these muscles are extremely small, tonic contraction of them can be the cause of retro-orbital headache.

THERMAL SENSATIONS

Thermal Receptors and Their Excitation

The human being can perceive different gradations of cold and heat, progressing from *freezing cold* to *cold* to *cool* to *indifferent* to *warm* to *hot* to *burning hot*.

Thermal gradations are discriminated by at least three different types of sensory receptors: the cold receptors, the warmth receptors, and two sub-types of pain receptors, cold-pain receptors and warmth-pain receptors. The two types of pain receptors are stimulated only by extreme degrees of heat or cold and therefore are responsible, along with the cold and warmth receptors, for "freezing cold" and "burning hot" sensations.

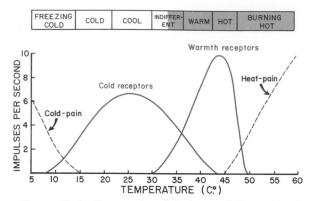

Figure 33–6. Frequencies of discharge of (1) a cold-pain fiber, (2) a cold fiber, (3) a warmth fiber, and (4) a heat-pain fiber. (The responses of these fibers are drawn from original data collected in separate experiments by Zotterman, Hensel, and Kenshalo.)

The cold and warmth receptors are located immediately under the skin at discrete, separated points, each having a stimulatory diameter of about 1 mm. In most areas of the body there are three to four times as many cold receptors as warmth receptors.

A definitive cold receptor has been identified. It is a special, small type A delta myelinated nerve ending, the tip of which protrudes into the bottom surface of a basal epidermal cell.

On the other hand, a definitive warmth receptor has not been found. Perhaps this receptor is one variety of free nerve ending.

Stimulation of Thermal Receptors — Sensations of Cold, Cool, Indifference, Warmth, and Heat. Figure 33–6 illustrates the effects of different temperatures on the responses of four different types of nerve fibers: (1) a cold-pain fiber, (2) a cold fiber, (3) a warmth fiber, and (4) a heat-pain fiber. In the *very* cold region only the cold-pain fibers are stimulated. At temperatures above 10 to 15°C the pain impulses cease but the cold receptors begin to be stimulated. Then, above about 30°C the warmth receptors become progressively stimulated, while the cold receptors fade out at about 43°C. Finally, at around 45°C, the heat-pain fibers begin to be stimulated.

One can understand from Figure 33–6, therefore, that a person perceives the temperatures of thermal sensations by the relative degrees of stimulation of the different types of endings. One can understand also from this figure why extreme degrees of cold or heat can be painful and why both these sensations, when intense enough, may give almost exactly the same quality of sensation — that is, freezing cold and burning hot sensations feel almost alike.

Stimulatory Effects of Rising and Falling Temperature — Adaptation of Thermal Recep-

tors. When a thermal receptor is suddenly subjected to an abrupt change in temperature, it becomes strongly stimulated at first, but this stimulation fades rapidly during the first minute and progressively more slowly during the next half hour or more. In other words, the receptor adapts to a great extent but not entirely.

Thus, it is evident that the thermal senses respond markedly to *changes in temperature* in addition to being able to respond to steady states of temperature. This means that when the temperature of the skin is actively falling, a person feels much colder than when the temperature remains at the same level. Conversely, if the temperature is actively rising, the person feels much warmer than at the same temperature when it is constant.

Mechanism of Stimulation of the Thermal Receptors. It is believed that the thermal receptors are stimulated by changes in their metabolic rates, these changes resulting from the fact that temperature alters the rates of intracellular chemical reactions about two times for each 10°C change. In other words, thermal detection probably results not from direct physical stimulation but instead from chemical stimulation of the endings as modified by the temperature.

Spatial Summation of Thermal Sensations. The number of cold or warmth endings in any small surface area of the body is very small, so that it is difficult to judge gradations of temperature when small areas are stimulated. However, when a large area of the body is stimulated, the thermal signals from the entire area summate. Indeed, one reaches his maximum ability to discern minute temperature variations when his entire body is subjected to a temperature change all at once. For instance, rapid changes in temperature of as little as 0.01°C can be detected if this change affects the entire surface of the body simultaneously. On the other hand, temperature changes 100 times this great might not be detected when the skin surface affected is only a square centimeter or so in size.

TRANSMISSION OF THE THERMAL SIGNALS IN THE NERVOUS SYSTEM

In general, thermal signals are transmitted in almost exactly the same pathways as pain signals. On entering the spinal cord, the signals travel for a few segments upward or downward, then are processed by one or more cord neurons, and finally enter long, ascending thermal fibers that cross to the opposite anterolateral spinothalamic tract and terminate in (a) the reticular areas of the brain stem, (b) the ventrobasal complex of the thalamus, and perhaps (c) in the intralaminar nuclei of the thalamus along with pain signals. A few thermal signals are also relayed to the somesthetic cortex from the ventrobasal complex. Occasionally, a neuron in somatic sensory area I has been found by microelectrode studies to be directly responsive to either cold or warm stimuli in specific areas of the skin. Furthermore, it is known that removal of the postcentral gyrus in the human being reduces his ability to distinguish different gradations of temperature.

REFERENCES

Beaumont, A., and Hughes, J.: Biology of opioid peptides. *Annu. Rev. Pharmacol. Toxicol., 19*:245, 1979.

Bonica, J. J., *et al.*: Recent Advances in Pain. Springfield, Ill., Charles C Thomas, 1974.

Bonica, J. J., *et al.* (eds.): Proceedings of the Second World Congress on Pain. New York, Raven Press, 1979.

Brainard, J. B.: Control of Migraine. New York, W. W. Norton & Co., 1979.

Casey, K. L.: Pain: A current view of neural mechanisms. *Am. Sci., 61*:194, 1973.

Coren, S., *et al.*: Sensation and Perception. New York, Academic Press, 1979.

Currie, D. J.: Abdominal Pain. Washington, D.C., Hemisphere Publishing Corporation, 1979.

Fairley, P.: The Conquest of Pain. New York, Charles Scribner's Sons, 1979.

Fields, H. L., and Bashbaum, A. I.: Brainstem control of spinal pain-transmission neurons. *Annu. Rev. Physiol., 40*:217, 1978.

Goldstein, E. B.: Sensation and Perception. Belmont, Cal., Wadsworth Publishing Co., 1980.

Hardy, J. D., *et al.*: Pain Sensations and Reactions. Baltimore, Williams & Wilkins, 1952.

Herz, A. (ed.): Developments in Opiate Research. New York, Marcel Dekker, 1978.

Hewer, C. L., and Atkinson, R. S. (eds.): Recent Advances in Anaesthesia and Analgesia. Boston, Little, Brown, 1978.

Jacob, J. (ed.): Receptors. New York, Pergamon Press, 1979.

Kenshalo, D. R. (ed.): Sensory Function of the Skin of Humans. New York, Plenum Press, 1979.

Lipton, S. (ed.): Persistent Pain: Modern Methods of Treatment. New York, Grune & Stratton, 1977.

Newman, P. O.: Visceral Afferent Functions of the Nervous System. Monograph of the Physiological Society, No. 25. Baltimore, Williams & Wilkins, 1974.

Raskin, N. H., and Appenzeller, O.: Headache. Philadelphia, W. B. Saunders Co., 1980.

Ryan, R. E., and Ryan, R. E., Jr. (eds.): Headache and Head Pain: Diagnosis and Treatment. St. Louis, C. V. Mosby, 1978.

Seltzer, S.: Pain in Dentistry: Diagnosis and Management. Philadelphia, J. B. Lippincott Co., 1978.

Silen, W. (ed.): Cope's Early Diagnosis of the Acute Abdomen. New York, Oxford University Press, 1979.

Simon, E. J., and Hiller, J. M.: The opiate receptors. *Annu. Rev. Pharmacol. Toxicol., 18*:371, 1978.

Snyder, S. H.: Opiate receptors and internal opiates. *Sci. Am. 236*(3):44, 1977.

Zimmerman, M.: Neurophysiology of nociception. *Int. Rev. Physiol., 10*:179, 1976.

34

The Cord and Brain Stem Reflexes; and Function of the Vestibular Apparatus

In the discussion of the nervous system thus far, we have considered principally the input of sensory information. In the following chapters we will discuss the origin and output of motor signals, the signals that cause muscle contraction and other motor effects throughout the body. Sensory information is integrated at all levels of the nervous system and causes appropriate motor responses, beginning in the spinal cord with relatively simple reflexes and extending into the brain stem with more complicated responses and finally to the cerebrum where the most complicated responses are controlled. The present chapter discusses the control of motor function at the spinal cord and lower brain stem level.

ORGANIZATION OF THE SPINAL CORD FOR MOTOR FUNCTIONS

The cord gray matter is the integrative area for the cord reflexes and other motor functions. Figure 34–1 shows the typical organization of the cord gray matter in a single cord segment. Sensory signals enter the cord through the sensory roots. After entering the cord, every sensory signal travels to two separate destinations. First, either in the same segment of the cord or in nearby segments, the sensory nerve or its collaterals terminate in the gray matter of the cord and elicit local segmental responses — local excitatory effects, facilitatory effects, reflexes, or others. Second, the signals travel to higher levels of the nervous system — to higher levels in the cord itself, to the brain stem, or even to the cerebral cortex. It is these sensory signals that cause the sensory effects described in the past few chapters.

Each segment of the spinal cord has several million neurons in its gray matter. Aside from the sensory relay neurons already discussed, the remainder of these neurons are divided into two separate types, the *anterior motoneurons* and the *interneurons* (also called *internuncial cells* or *intermediate cells*).

The Anterior Motoneurons. Located in each segment of the anterior horns of the cord gray matter are several thousand neurons that are 50 to 100 per cent larger than most of the others and called anterior motoneurons. These give rise to the nerve fibers that leave the cord via the anterior roots and then proceed to the muscles to innervate the skeletal muscle fibers. They can be

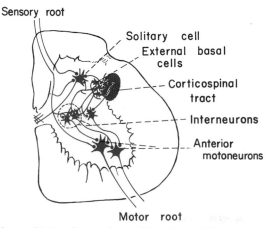

Figure 34–1. Connections of the sensory fibers and corticospinal fibers with the interneurons and anterior motoneurons of the spinal cord.

divided into two major types, the *alpha motoneurons* and the *gamma motoneurons*.

The Alpha Motoneurons. The alpha motoneurons give rise to large, type A alpha nerve fibers ranging from 9 to 20 microns in diameter and passing through the spinal nerves to innervate the skeletal muscle fibers. Stimulation of a single nerve fiber excites from 3 to 2000 skeletal muscle fibers (averaging about 180 fibers), which are collectively called the *motor unit*. Transmission of nerve impulses into skeletal muscles and their stimulation of the muscle fibers was discussed in Chapters 8 through 10.

The Gamma Motoneurons. In addition to the alpha motoneurons that excite contraction of the skeletal muscle fibers, about one-half as many much smaller *gamma* motoneurons, located also in the anterior horns alongside the alpha motoneurons, transmit impulses through type A gamma fibers, averaging 5 microns in diameter, to special skeletal muscle fibers called *intrafusal fibers*. These are part of the *muscle spindle*, which is discussed at length later in the chapter.

The Interneurons. The interneurons are present in all areas of the cord gray matter — in the dorsal horns, in the anterior horns, and also in the intermediate areas between these two. These cells are numerous — approximately 30 times as numerous as the anterior motoneurons. They are small and highly excitable, often exhibiting spontaneous activity and capable of firing as rapidly as 1500 times per second. They have many interconnections with one another, and many of them directly innervate the anterior motoneurons as illustrated in Figure 34–1. The interconnections among the interneurons and anterior motoneurons are responsible for many of the integrative functions of the spinal cord that are discussed during the remainder of this chapter.

Essentially all the different types of neuronal circuits described in Chapter 31 are found in the interneuron pool of cells of the spinal cord, including the *diverging, converging,* and *repetitive-discharge* circuits. In the following sections of this chapter we will see applications of these different circuits to the performance of specific reflex acts by the spinal cord.

Only a few signals from the brain terminate directly on the anterior motoneurons. Instead, most of them are transmitted first through interneurons, where they are appropriately processed before stimulating the anterior motoneurons. Thus, in Figure 34–1, it is shown that the corticospinal tract terminates almost entirely on interneurons, and it is through these that the cerebrum transmits most of its signals for control of muscular function.

Sensory Input to the Motoneurons and Interneurons

Most of the sensory fibrils entering each segment of the spinal cord terminate on interneurons, but a very small number of large sensory fibers from the muscle spindles terminate directly on the anterior motoneurons. Thus, there are two pathways in the spinal cord that cord reflexes can take: either directly to the anterior motoneuron itself, utilizing a *monosynaptic pathway*; or through one or more interneurons first, before passing to the anterior motoneuron. The monosynaptic pathway provides an extremely rapid reflex feedback system and is the basis of the very important muscle *stretch reflex,* which will be discussed later in the chapter. All other cord reflexes utilize the interneuron pathway, a pathway that can modify the signals tremendously and that can cause complex reflex patterns. For instance, the very important protective reflex called the *withdrawal reflex* utilizes this pathway.

Multisegmental Connections in the Spinal Cord—the Propriospinal Fibers. More than half of all the nerve fibers ascending and descending in the spinal cord are *propriospinal fibers.* These are fibers that run from one segment of the cord to another. In addition, the terminal fibrils of sensory fibers as they enter the cord branch both up and down the spinal cord, some of the branches transmitting signals only a segment or two in each direction, while others transmit signals over many segments. These ascending and descending fibers of the cord provide pathways for the multisegmental reflexes that will be described later in this chapter, including many reflexes that coordinate movements in both the forelimbs and hindlimbs simultaneously.

ROLE OF THE MUSCLE SPINDLE IN MOTOR CONTROL

Muscles and tendons have an abundance of two special types of receptors: (1) *muscle spindles,* which detect (a) change in length of muscle fibers and (b) rate of this change in length, and (2) *Golgi tendon organs,* which detect the tension applied to the muscle tendon during muscle contraction or muscle stretch.

The signals from these two receptors operate entirely at a subconscious level, causing no sensory perception at all. But they do transmit tremendous amounts of information into the spinal cord and also to the cerebellum, thereby helping these two portions of the nervous system to perform their functions for controlling muscle contraction.

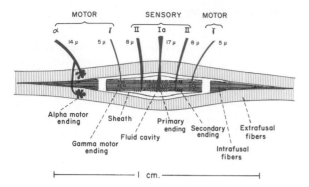

Figure 34–2. The muscle spindle, showing its relationship to the large extrafusal skeletal muscle fibers. Note also both the motor and the sensory innervation of the muscle spindle.

RECEPTOR FUNCTION OF THE MUSCLE SPINDLE

Structure and Innervation of the Muscle Spindle. The physiological organization of the muscle spindle is illustrated in Figure 34–2. Each spindle is built around three to ten small *intrafusal muscle fibers*, which are pointed at their ends and are attached to the sheaths of the surrounding *extrafusal* skeletal muscle fibers. The intrafusal fiber is a very small skeletal muscle fiber. However, the central region of each of these fibers has either no or few actin and myosin filaments. Therefore, this central portion does not contract when the ends do. The ends contract when excited by the small nerve fibers from the gamma motoneurons, called *gamma efferent* motor nerve fibers.

The central portion of the muscle spindle is the sensory receptor area of the spindle. This receptor area in turn has two types of sensory endings, the *primary ending* and two *secondary endings*.

The Primary Ending. A very large type Ia sensory fiber innervates the very center of the spindle receptor. The tip of this fiber spirals around the intrafusal fibers, forming the so-called *primary ending*, also called the *annulospiral ending*. When the central portion of the spindle is stretched, this ending is stimulated. And, because the innervating fiber is so large, signals are transmitted to the spinal cord at a velocity approaching 100 meters per second, a velocity as great as that in almost any part of the nervous system.

The Secondary Endings. Two type II nerve fibers innervate the receptor region of the intrafusal fibers, one on either side of the primary ending. These fibers, like the Ia fiber, also spiral around the intrafusal fibers, and when the central portion of the intrafusal fibers is stretched, these nerve fibers are stimulated. These sensory endings are called *secondary endings*.

Static Response of Both the Primary and the Secondary Endings. When the receptor portion of the muscle spindle is stretched *slowly*, the number of impulses transmitted from both types of endings increases almost directly in proportion to the degree of stretch, and the endings continue to transmit these impulses for many minutes. This effect is called the *static response* of the spindle receptor, meaning simply that the receptor transmits its signal for a prolonged period of time.

Dynamic Response of the Primary Ending. The primary ending also exhibits a very strong *dynamic response,* which means that it responds even more actively to a *change* in length. When the length of the spindle receptor area increases only a fraction of a micron, if this increase occurs rapidly, the primary receptor transmits tremendous numbers of impulses into the Ia fiber, but only *while the length is actually increasing.* As soon as the length has stopped increasing, the rate of impulse discharge returns back to the small static response level that is still present in the signal.

Conversely, when the spindle receptor area shortens, this change momentarily decreases the impulse output from the primary ending; as soon as the receptor area has reached its new shortened length, the impulses reappear in the Ia fiber within a fraction of a second. Thus, the primary ending sends extremely strong signals to the central nervous system to apprise it of any change in length of the spindle receptor area.

Function of the Muscle Spindle in Comparing Intrafusal and Extrafusal Muscle Lengths

From the foregoing description of the muscle spindle, one can see that there are two different ways in which the spindle can be stimulated:

1. *By stretching the whole muscle.* This lengthens the extrafusal fibers and therefore also stretches the spindle.

2. *By contracting the intrafusal muscle fibers* while the extrafusal fibers remain at their normal length. Since the intrafusal fibers contract only near their two ends, this stretches the central receptor portions of the intrafusal fibers, obviously exciting the spindles.

Therefore, in effect, the muscle spindle acts as a *comparator* of the lengths of the two types of muscle fibers, the extrafusal and the intrafusal. When the length of the extrafusal fibers is greater than that of the intrafusal fibers, the spindle becomes excited. On the other hand, when the length of the extrafusal fiber is shorter than that of the intrafusal fiber, the spindle becomes inhibited.

Continuous Discharge of the Muscle Spindles Under Normal Conditions. Normally, particularly when there is a slight amount of gamma efferent excitation, the muscle spindles emit sensory nerve impulses all of the time. Stretching the muscle spindles increases the rate of firing, whereas shortening the spindle decreases this rate of firing. Thus, the spindles can be either excited or inhibited.

THE STRETCH REFLEX (ALSO CALLED MUSCLE SPINDLE REFLEX OR MYOTATIC REFLEX)

Sudden stretch of a muscle excites the muscle spindle, and this in turn causes reflex contraction of the same muscle. For obvious reasons, the reflex is frequently called simply a muscle *stretch reflex*. This reflex has a dynamic component and a static component.

The Dynamic Stretch Reflex. The dynamic stretch reflex is caused by the potent dynamic signal from the muscle spindles. That is, when the muscle is suddenly stretched, a strong signal is transmitted to the spinal cord through the primary endings, but this signal is potent *only while the length of the muscle is increasing*. On entering the spinal cord, most of the signal goes directly to the anterior motoneurons without passing through interneurons, as shown in Figure 34–3, and it causes reflex contraction of the same muscle from which the muscle spindle signals originated. Thus, a sudden stretch of a muscle causes reflex contraction of the same muscle, and *this opposes further stretch of the muscle.*

The Static Stretch Reflex. Though the dynamic stretch reflex is over within a fraction of a second after the muscle has been stretched to its new length, a weaker static stretch reflex continues for a prolonged period of time thereafter. This reflex is elicited by continuous static receptor signals transmitted through both the primary and secondary endings of the muscle spindles. The importance of the static stretch reflex is that it continues to cause muscle contraction as long as

the muscle is maintained at an excessive length (for as long as several hours, but not for days). The muscle contraction in turn opposes the force that is causing the excess length.

The Negative Stretch Reflex. When a muscle is suddenly shortened, exactly opposite effects occur. Thus, *this negative stretch reflex* opposes the shortening of the muscle in the same way that the positive stretch reflex opposes lengthening of the muscle. Therefore, one can begin to see that the muscle spindle reflex tends to maintain the status quo for the length of a muscle.

Function of the Static Stretch Reflex to Nullify the Effects of Changes in Load During Muscle Contraction

Let us assume that a person's biceps is contracted so that the forearm is horizontal to the earth. Then assume that a five-pound weight is put in the hand. The hand will immediately drop. However, the amount that the hand will drop is determined to a great extent by the degree of activity of the static muscle spindle reflex. If the static reflex is very active, even slight lengthening of the biceps, and therefore also of the muscle spindles in the biceps, will cause a strong feedback contraction of the extrafusal skeletal muscle fibers of the biceps. This contraction in turn will limit the degree of fall of the hand, thus automatically maintaining the forearm in a nearly horizontal position despite the increased load. This response is called a *load reflex*.

The Damping Function of the Stretch Reflex. Another extremely important function of the reflex — indeed, probably more important than the load reflex — is the ability of the muscle spindle reflex to prevent oscillation and jerkiness of the body movements. This is a damping, or smoothing, function. An example is the following:

Use of the Damping Mechanism in Smoothing Muscle Contraction. Occasionally, signals from other parts of the nervous system are transmitted to a muscle in a very unsmooth form, first increasing in intensity for a few milliseconds, then decreasing in intensity, then changing to another intensity level, and so forth. When the muscle spindle apparatus is not functioning satisfactorily, the muscle contraction is jerky during the course of such a signal. This effect is illustrated in Figure 34–4, which shows an experiment in which a sensory nerve signal entering one side of the cord is transmitted to a motor nerve on the other side of the cord to excite a muscle. In curve A the muscle spindle reflex of the excited muscle is intact. Note that the contraction is relatively smooth even though the sensory nerve is excited at a frequency of 8 per second. Curve B, on the

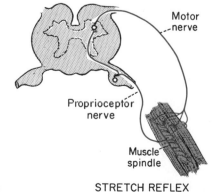

STRETCH REFLEX

Figure 34–3. Neuronal circuit of the stretch reflex.

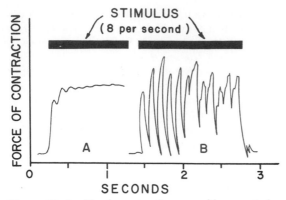

Figure 34–4. Muscle contraction caused by a central nervous system signal under two different conditions: (A) in a normal muscle, and (B) in a muscle whose muscle spindles had been denervated by section of the posterior roots of the cord 82 days previously. Note the smoothing effect of the muscle spindles in Part A.

other hand, shows the same experiment in animal whose muscle spindle sensory nerves had been sectioned three months earlier. Note the very unsmooth muscle contraction. Thus, curve A illustrates very graphically the ability of the damping mechanism of the muscle spindle to make muscle contractions smooth even though the input signals to the muscle motor system may themselves be very jerky. This effect can also be called a *signal averaging* function of the muscle spindle.

Function of the Gamma Efferent System in Controlling the Intensity of the Stretch Reflex

The gamma efferent system plays a potent role in determining the effectiveness of the load reflex and also the degree of damping. For instance, there are times when a person wishes his limbs to move extremely rapidly in response to rapidly changing input signals. Under such conditions, one would wish less damping and less load reflex. On the other hand, at other times it is very important that the muscle contractions be very smooth. Under these conditions one would like a potent stretch reflex. This is achieved by gamma efferent stimulation of the intrafusal muscle fibers, a condition that greatly enhances the excitability of the muscle spindles.

ROLE OF THE MUSCLE SPINDLE IN VOLUNTARY MOTOR ACTIVITY

To emphasize the importance of the muscle spindles and of the gamma efferent system, one needs only to recognize that 31 per cent of all the motor nerve fibers to the muscle are gamma efferent fibers rather than large type A alpha motor fibers. Whenever signals are transmitted from the motor cortex or from any other area of the brain to the alpha motoneurons, almost invariably the gamma motoneurons are stimulated simultaneously, a principle called *gamma efferent coactivation.* This causes the intrafusal muscle fibers to contract at the same time that the whole muscle contracts.

The purpose of contracting the muscle spindle fibers at the same time that the large skeletal muscle fibers contract is two-fold: First, it keeps the muscle spindle from opposing the muscle contraction. Second, it also maintains proper damping of the muscle spindle regardless of change in muscle length. For instance, if the muscle spindle did not contract and relax along with the large muscle fibers, the receptor portion of the spindle would sometimes be flail and at other times be overstretched, in neither instance operating under optimal conditions for spindle function.

CLINICAL APPLICATION OF THE STRETCH REFLEX — THE KNEE JERK AND OTHER MUSCLE JERKS

Clinically, a method used to determine the functional integrity of the stretch reflexes is to elicit the knee jerk and other muscle jerks. The knee jerk can be elicited by simply striking the patellar tendon with a reflex hammer; this stretches the quadriceps muscle and initiates a *dynamic stretch reflex* that causes the lower leg to jerk forward.

Similar reflexes can be obtained from almost any muscle of the body either by striking the tendon of the muscle or by striking the belly of the muscle itself. In other words, sudden stretch of muscle spindles is all that is required to elicit a stretch reflex.

The muscle jerks are used by neurologists to assess the degree of facilitation of spinal cord centers. When large numbers of facilitatory impulses are being transmitted from the upper regions of the central nervous system into the cord, the muscle jerks are greatly exacerbated. On the other hand, if the facilitatory impulses are depressed or abrogated, the muscle jerks are considerably weakened or completely absent. These reflexes are used most frequently to determine the presence or absence of muscle spasticity following lesions in the motor areas of the brain. Ordinarily, diffuse lesions in the contralateral motor areas of the cerebral cortex cause greatly exacerbated muscle jerks.

THE TENDON REFLEX

The Golgi Tendon Organ and Its Excitation. Golgi tendon organs, one of which is illustrated in Figure 34–5, lie within muscle ten-

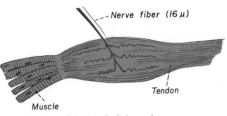

Figure 34–5. Golgi tendon organ.

dons immediately beyond their attachments to the muscle fibers. An average of 10 to 15 muscle fibers is usually connected in series with each Golgi tendon organ, and the organ is stimulated by the tension in the tendon produced by this small bundle of muscle fibers. Thus, the major difference between the function of the Golgi tendon apparatus and the muscle spindle is that the spindle detects relative extrafusal-intrafusal muscle length, and the tendon organ detects muscle *tension*.

Inhibitory Nature of the Tendon Reflex. Signals from the Golgi tendon organ, when it is stretched, are transmitted into the spinal cord to cause reflex effects in the respective muscle. However, this reflex *inhibits* the muscle instead of exciting it, the exact opposite of the muscle spindle reflex. The signal from the tendon organ is believed to excite inhibitory interneurons, and these in turn to inhibit the alpha motoneurons to the respective muscle.

When tension on the muscle and therefore on the tendon becomes extreme, the inhibitory effect from the tendon organ can be so great that this inhibition causes sudden relaxation of the entire muscle. This effect is called the *lengthening reaction;* it is probably a protective mechanism to prevent tearing of the muscle or avulsion of the tendon from its attachments to the bone.

However, possibly as important as this protective reaction is the function of the tendon reflex as a part of the overall control of muscle contraction in the following manner:

The Tendon Reflex as a Control Mechanism for Muscle Tension. In the same way that the stretch reflex possibly operates as a feedback mechanism to control the length of a muscle, the tendon reflex theoretically can operate as a feedback mechanism to control muscle tension. That is, if the tension on the muscle becomes too great, inhibition from the tendon organ decreases this tension back to a lower value. On the other hand, if the tension becomes too little, impulses from the tendon organ cease; and the resulting loss of inhibition allows the alpha motoneurons to become active again, thus increasing muscle tension back toward a higher level.

Very little is known at present about the function of or control of this tension feedback mechanism, but it is postulated to operate in the following basic manner: Signals from the brain are presumably transmitted to the cord centers to set the sensitivity of the tendon feedback system. This can be done by changing the degree of facilitation of the neurons in the feedback loop. If the neuron excitability is high, then this system will be extremely sensitive to signals coming from the tendon organs; on the other hand, lack of excitatory signals from the brain could make the system very insensitive to the signals from the tendon organ. In this way, control signals from higher nervous centers could automatically set the level of tension at which the muscle would be maintained.

An obvious value of a mechanism for setting the degree of muscle tension would be to allow the different muscles to apply a desired amount of force (that is, maintain constant tension on the tendon) irrespective of how the muscles contract.

THE FLEXOR REFLEX

In the spinal or decerebrate animal, almost any type of sensory stimulus to a limb is likely to cause the flexor muscles of the limb to contract strongly, thereby withdrawing the limb from the stimulus. This is called the flexor reflex.

In its classical form the flexor reflex is elicited most frequently by stimulation of pain endings, such as pinprick, heat, or some other painful stimulus, for which reason it is also frequently called a *nociceptive reflex*. However, even stimulation of the touch receptors can also occasionally elicit a weaker and less prolonged flexor reflex.

If some part of the body besides one of the limbs is painfully stimulated, this part, in a similar manner, will be withdrawn from the stimulus, but the reflex may not be confined entirely to flexor muscles even though it is basically the same type of reflex. Therefore, the reflex is frequently called a *withdrawal reflex,* too.

Neuronal Mechanism of the Flexor Reflex. The left-hand portion of Figure 34–6 illustrates the neuronal pathways for the flexor reflex. In this instance, a painful stimulus is applied to the hand; as a result, the flexor muscles of the upper arm become reflexly excited, thus withdrawing the hand from the painful stimulus.

The nervous pathways for eliciting the flexor reflex do not pass directly to the anterior motoneurons but, instead, pass first into the interneuron pool of neurons and then to the motoneurons.

Within a few milliseconds after a pain nerve begins to be stimulated, the flexor response appears. Then, during the next few seconds the reflex begins to *fatigue*, even though the pain nerve continues to be stimulated; this is charac-

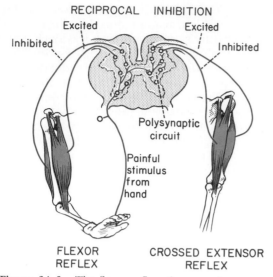

Figure 34–6. The flexor reflex, the crossed extensor reflex, and reciprocal inhibition.

teristic of essentially all of the more complex integrative reflexes of the spinal cord. After the pain stimulus is over, the contraction of the muscle begins to return toward the base line, but because of *after-discharge* in the interneurons of the cord, it will not return all the way for many milliseconds. The duration of the after-discharge depends on the intensity of the sensory stimulus that had elicited the reflex; a weak stimulus causes almost no after-discharge, in contrast to an after-discharge lasting for several seconds following a strong stimulus.

The after-discharge that occurs in the flexor reflex almost certainly involves reverberating circuits in the interneurons, which transmit impulses to the anterior motoneurons sometimes for several seconds after the incoming sensory signal is completely over.

Thus, the flexor reflex is appropriately organized to withdraw a pained or otherwise irritated part of the body away from the stimulus. Furthermore, because of the after-discharge it will hold the irritated part away from the stimulus for as long as one to three seconds even after the irritation is over. During this time, other reflexes and actions of the central nervous system can move the entire body away from the painful stimulus.

The Pattern of Withdrawal. The pattern of withdrawal that results when a flexor (or withdrawal) reflex is elicited depends on the sensory nerve that is stimulated. Thus, a painful stimulus on the inside of the arm not only elicits a flexor reflex in the arm but also contracts the abductor muscles to pull the arm outward. In other words, the integrative centers of the the cord cause those muscles to contract that can most effectively re-

move the pained part of the body from the object that causes pain. This same principle applies to any part of the body but especially to the limbs, because they have highly developed flexor reflexes.

THE CROSSED EXTENSOR REFLEX

Approximately 0.2 to 0.5 second after a stimulus elicits a flexor reflex in one limb, the opposite limb begins to extend. This is called the *crossed extensor reflex*. Extension of the opposite limb obviously can push the entire body away from the object causing the painful stimulus.

Neuronal Mechanism of the Crossed Extensor Reflex. The right-hand portion of Figure 34–6 illustrates the neuronal circuit responsible for the crossed extensor reflex, showing that signals from the sensory nerves cross to the opposite side of the cord to cause exactly opposite reactions to those of the flexor reflex, namely, to extend the limb. Because the crossed extensor reflex usually does not begin until 200 to 500 milliseconds following the initial pain stimulus, it is certain that many internuncial neurons are in the circuit between the incoming sensory neuron and the motoneurons of the opposite side of the cord responsible for the crossed extension. Furthermore, after the painful stimulus is removed, the crossed extensor reflex has an even longer period of after-discharge than that of the flexor reflex. Again, it is almost certain that this prolonged after-discharge results from reverberatory circuits among the internuncial cells.

RECIPROCAL INNERVATION

In the foregoing paragraphs we have pointed out several times that excitation of one group of muscles is often associated with inhibition of another group. For instance, when a stretch reflex excites one muscle, it simultaneously inhibits the antagonist muscles. This is the phenomenon of *reciprocal inhibition*, and the neuronal mechanism that causes this reciprocal relationship is called *reciprocal innervation*. Likewise, reciprocal relationships exist between the two sides of the cord as exemplified by the flexor and extensor reflexes as described above.

We will see below that the principle of reciprocal innervation is also important in most of the cord reflexes that subserve locomotion, for it helps to cause forward movement of one limb while causing backward movement of the opposite limb, and it also causes alternate movements between the forelimbs and the hindlimbs.

THE REFLEXES OF POSTURE AND LOCOMOTION

The Positive Supportive Reaction. Pressure on the footpad of a decerebrate animal causes the limb to extend against the pressure that is being applied to the foot. Indeed, this reflex is so strong that an animal whose spinal cord was transected several months previously can often be placed on its feet, and the pressure on the footpads will reflexly stiffen the limbs sufficiently to support the weight of the body. This reflex is called the *positive supportive reaction*.

The positive supportive reaction involves a complex circuit in the interneurons similar to those responsible for the flexor and the crossed extensor reflexes. Furthermore, the locus of the pressure on the pad of the foot determines the position to which the limb is extended.

The Rhythmic Stepping Reflex. Rhythmic stepping movements are frequently observed in the limbs of spinal animals. Indeed, even when the sacro-lumbar portion of the spinal cord is separated from the remainder of the cord and a longitudinal section is made down the center of this sacro-lumbar portion to block neuronal connections between the two limbs, each hind limb can still perform stepping functions. Forward flexion of the limb is followed a second or so later by backward extension. Then flexion occurs again, and the cycle is repeated over and over.

If the lumbar spinal cord is not sectioned down its center, every time stepping occurs in the forward direction in one limb, the opposite limb ordinarily steps backward. This effect results from reciprocal innervation between the two limbs.

Diagonal Stepping of All Four Limbs — The "Mark Time" Reflex. Stepping reflexes that involve all four limbs can also be demonstrated in a spinal animal. In general, stepping occurs diagonally between the fore- and hindlimbs. That is, the right hindlimb and the left forelimb move backward together while the right forelimb and left hindlimb move forward. This diagonal response is another manifestation of reciprocal innervation, this time occurring the entire distance up and down the cord between the fore- and hindlimbs. Such a walking pattern is often called a *mark time reflex*.

The Galloping Reflex. Another type of reflex that occasionally develops in the spinal animal is the galloping reflex, in which both forelimbs move backward in unison while both hindlimbs move forward. If stretch or pressure stimuli are applied almost exactly equally to opposite limbs at the same time, a galloping reflex is likely to result, whereas unequal stimulation of one side versus the other elicits the diagonal walking reflex. This is in keeping with the normal patterns of walking and of galloping, for, in walking, only one limb at a time is stimulated, and this would predispose to continued walking. Conversely, when the animal strikes the ground during galloping, the limbs on both sides are stimulated approximately equally; this obviously would predispose to further galloping and therefore would continue this pattern of motion in contradistinction to the walking pattern.

SPINAL CORD REFLEXES THAT CAUSE MUSCLE SPASM

In human beings, local muscle spasm is often observed. The mechanism of this has not been elucidated to complete satisfaction even in experimental animals, but it is known that pain stimuli can cause reflex spasm of local muscles, which presumably is the cause of much if not most of the muscle spasm observed in localized regions of the human body.

Abdominal Spasm in Peritonitis. A type of local muscle spasm caused by a cord reflex is the abdominal spasm resulting from irritation of the parietal peritoneum by peritonitis. Relief of the pain caused by the peritonitis allows the spastic muscles to relax. Almost the same type of spasm often occurs during surgical operations; pain impulses from the parietal peritoneum cause the abdominal muscles to contract extensively and sometimes actually to extrude the intestines through the surgical wound. For this reason deep surgical anesthesia is usually required for intra-abdominal operations.

Muscle Cramps. Another type of local spasm is the typical muscle cramp. Any local irritating factor or metabolic abnormality of a muscle — such as severe cold, lack of blood flow to the muscle, or overexercise of the muscle — can elicit pain or other types of sensory impulses that are transmitted from the muscle to the spinal cord, thus causing reflex muscle contraction. The contraction in turn stimulates the same sensory receptors still more, which causes the spinal cord to increase the intensity of contraction still further. Thus, a positive feedback mechanism occurs, so that a small amount of initial irritation causes more and more contraction until a full-blown muscle cramp ensues.

SPINAL CORD TRANSECTION AND SPINAL SHOCK

When the spinal cord is suddenly transected in the neck, essentially all cord functions, including the cord reflexes, immediately become almost completely blocked, a reaction called *spinal shock*. The reason for this is that normal activity of the cord neurons depends to a great extent on continual *facilatory signals* from higher centers, particularly signals transmitted through the vestibulospinal tract, the reticulospinal tracts, and the corticospinal tracts.

Some of the spinal functions specifically affected during spinal shock are these: (1) The arterial blood pressure falls immediately — sometimes to as low as 40 mm. Hg — thus illustrating that sympathetic activity becomes blocked almost to extinction. However, the pressure ordinarily returns to normal within a few days. (2) All skeletal muscle reflexes integrated in the spinal cord are completely blocked during the initial stages of shock. In lower animals, a few hours to a week or so are required for these reflexes to return to normal, and in man several weeks are often required. (3) The sacral reflexes for control of bladder and colon evacuation are completely suppressed in man for the first few weeks following cord transection, but they eventually return. These effects are discussed in Chapters 26 and 42.

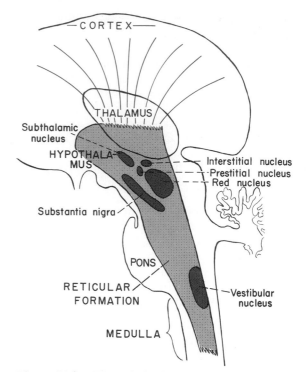

Figure 34–7. The reticular formation and associated nuclei.

THE BRAIN STEM

The brain stem is a complex extension of the spinal cord, including in it the *medulla, pons, and mesencephalon*. Collected in it are numerous neuronal circuits to control respiration, cardiovascular function, gastrointestinal function, eye movement, equilibrium, support of the body against gravity, and many special stereotyped movements of the body. Some of these functions — such as control of respiration and cardiovascular functions — are described in special sections of this text. The present discussion deals primarily with the control of whole body movement and equilibrium.

THE RETICULAR FORMATION, AND SUPPORT OF THE BODY AGAINST GRAVITY

Throughout the entire extent of the brain stem — in the medulla, pons, mesencephalon, and even portions of the diencephalon — are areas of diffuse neurons collectively known as the *reticular formation*. Figure 34–7 illustrates the extent of the reticular formation, showing it to begin at the upper end of the spinal cord and to extend (1) upward through the central portions of the thalamus, (2) into the hypothalamus, and (3) into other areas adjacent to the thalamus. The lower end of the reticular formation is continu-

ous with the interneurons of the spinal cord. Indeed, the reticular formation of the brain stem functions in a manner quite analogous to many of the functions of the interneurons of the cord gray matter.

Interspersed throughout the reticular formation are both motor and sensory neurons; these vary in size from very small to very large. The small neurons, which constitute the greater number, have short axons that make multiple connections within the reticular formation itself. The large neurons are mainly motor in function, and their axons usually bifurcate almost immediately, with one division extending downward to the spinal cord and the other extending upward to the thalamus or other regions of the diencephalon or cerebrum.

The sensory input to the reticular formation is from multiple sources, including (1) the spinoreticular tracts and collaterals from the spinothalamic tracts, (2) the vestibular nuclei, (3) the cerebellum, (4) the basal ganglia, (5) the cerebral cortex, especially the motor regions, and (6) the hypothalamus and other nearby associated areas.

Though most of the neurons in the reticular formation are evenly dispersed, some of them are collected into *specific nuclei*, which are labeled in Figure 34–7. In general, these specific nuclei are not considered to be part of the reticular formation per se even though they do operate in association with it. In most instances they are the loci of "preprogrammed" control of stereotyped movements. As an example, the vestibular nuclei provide preprogrammed attitudinal contractions of the muscles for maintenance of equilibrium, as will be discussed later in the chapter.

EXCITATORY FUNCTION OF THE RETICULAR FORMATION

The reticular formation (as well as the vestibular nuclei, which are adjacent to the lower reticular formation and function in very close association with it) is intrinsically excitable, but this excitability is usually held in check by inhibitory signals that flow into this area mainly from the basal ganglia and cortex. Destruction of the higher portions of the nervous system, especially the basal ganglia and the cortex, removes this inhibition and allows the reticular formation (and vestibular nuclei) to become tonically active; this causes rigidity of the antigravity skeletal muscles throughout the body.

Reciprocal Excitation or Inhibition of Antagonist Muscles by the Reticular Formation. Though very little is known about the function of specific areas in the reticular formation, in general, stimulation near the midline of the reticular formation causes the flexor muscles on the same side of the body to contract and the extensors to relax. Stimulation in the lateral portions of the reticular formation causes opposite effects, excitation of the extensor muscles and inhibition of the flexors. And at the same time that contraction occurs on one side of the body the same muscle on the opposite side tends to relax while its antagonist contracts. Thus, the phenomenon of reciprocal inhibition is strongly expressed in the reticular formation, as it is in the spinal cord, both for reciprocal control of antagonist pairs of muscles and for control of muscles between the two sides of the body.

SUPPORT OF THE BODY AGAINST GRAVITY

When a person or an animal stands, the vestibular nuclei and the closely related nuclei in the reticular formation transmit continuous impulses into the spinal cord and thence to the extensor muscles to stiffen the limbs. This allows the limbs to support the body against gravity. These impulses are transmitted mainly by way of the vestibulospinal and reticulospinal tracts.

However, the degree of activity in the individual extensor muscles is determined by the equilibrium mechanisms. Thus, if an animal begins to fall to one side, the extensor muscles on that side stiffen while those on the opposite side relax. And analogous effects occur when it tends to fall forward or backward.

In essence, then, the vestibular nuclei reticular formation provide the nervous to support the body against gravity. B⸺ ⸺ factors, particularly the vestibular apparatuses, control the relative degree of extensor contraction in the difference parts of the body, which provides the function of equilibrium.

VESTIBULAR SENSATIONS AND THE MAINTENANCE OF EQUILIBRIUM

THE VESTIBULAR APPARATUS

The vestibular apparatus is the sensory organ that detects sensations concerned with equilibrium. It is composed of a *bony labyrinth* containing the *membranous labyrinth*, the functional part of the apparatus. The top of Figure 34–8 illustrates

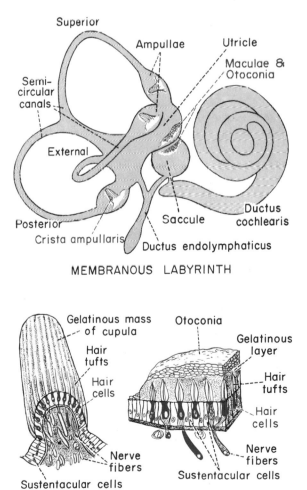

Figure 34–8.   The membranous labyrinth and organization of the crista ampullaris and the macula. (From Goss: Gray's Anatomy of the Human Body. Lea & Febiger; modified from Kolmer by Buchanan: Functional Neuroanatomy. Lea & Febiger.)

the membranous labyrinth; it is composed mainly of the *cochlear duct,* the three *semicircular canals,* and the two large chambers known as the *utricle* and the *saccule.* The cochlear duct is concerned with hearing and has nothing to do with equilibrium. However, the *utricle,* the *saccule,* and the *semicircular canals* are especially important for maintaining equilibrium.

The Utricle and the Saccule. Located on the wall of each utricle and saccule is a small area slightly over 2 mm. in diameter called a *macula.* Each of these maculae is a sensory area for detecting the orientation of the head with respect to the direction of gravitational pull or of other acceleratory forces, as will be explained in subsequent sections of this chapter. Each macula is covered by a gelatinous layer in which many small calcium carbonate crystals called *otoconia* are imbedded. Also, in the macula are thousands of *hair cells,* which project *cilia* up into the gelatinous layer. Around the bases of the hair cells are entwined sensory axons of the vestibular nerve.

Even under resting conditions, most of the nerve fibers surrounding the hair cells transmit a continuous series of nerve impulses. Bending the cilia of a hair cell to one side causes the inpulse traffic in its nerve fibers to increase markedly; bending the cilia to the opposite side decreases the impulse traffic, often turning it off completely. Therefore, as the orientation of the head in space changes and the weight of the otoconia (whole specific gravity is about three times that of the surrounding tissues) bends the cilia, appropriate signals are transmitted to the brain to control equilibrium.

In each macula the different hair cells are oriented in different directions, so that some of them are stimulated when the head bends forward, some when it bends backward, others when it bends to one side, and so forth. Therefore, a different pattern of excitation occurs in the macula for each position of the head; it is this "pattern" that apprises the brain of the head's orientation.

The Semicircular Canals. The three semicircular canals in each vestibular apparatus, known respectively as the *superior, posterior,* and *external* (or *horizontal*) *semicircular canals,* are arranged at right angles to each other so that they represent all three planes in space.

In the *ampullae* of the semicircular canals, as illustrated in Figure 34–8, are small crests, each called a *crista ampullaris,* and on top of the crista is a gelatinous mass similar to that in the utricle and known as the *cupula.* Into the cupula are projected hairs (cilia) from hair cells located along the ampullary crest, and these hair cells in turn are connected to sensory nerve fibers that pass into the *vestibular nerve.* Bending of the cupula to one

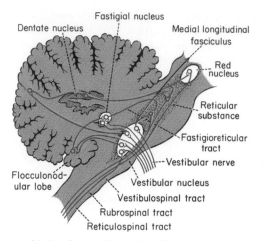

Figure 34–9. Connections of vestibular nerves in the central nervous system.

side, caused by flow of fluid in the canal, stimulates the hair cells, while bending in the opposite direction inhibits them. Thus, appropriate signals are sent through the vestibular nerve to apprise the central nervous system of fluid movement in the respective canal.

Neuronal Connections of the Vestibular Apparatus with the Central Nervous System. Figure 34–9 illustrates the central connections of the vestibular nerve. Most of the vestibular nerve fibers end in the *vestibular nuclei,* which are located approximately at the junction of the medulla and the pons, but some fibers pass without synapsing into the cerebellum. The fibers that end in the vestibular nuclei synapse with second order neurons that in turn send fibers into the cerebellum, into the spinal cord, and especially into the brain stem reticular nuclei.

Note especially the very close association between the vestibular apparatus, the vestibular nuclei, and the cerebellum. The primary pathway for the reflexes of equilibrium begins in the vestibular nerves and passes next to both the vestibular nuclei and the cerebellum. Then, after much two-way traffic of impulses between these two, signals are sent into the reticular nuclei of the brain stem as well as down the spinal cord via vestibulospinal and reticulospinal tracts. The signals to the cord control the interplay between facilitation and inhibition of the extensor muscles, thus automatically controlling equilibrium.

FUNCTION OF THE UTRICLE AND THE SACCULE IN THE MAINTENANCE OF STATIC EQUILIBRIUM

It is especially important that the different hair cells are oriented in all different directions in the maculae of the utricles and saccules so that at different positions of the head, different hair cells become stimulated. The "patterns" of stimulation of the different hair cells apprise the nervous system of the position of the head with respect to the pull of gravity. In turn, the vestibular, cerebellar, and reticular motor systems reflexly excite the appropriate muscles to maintain proper equilibrium.

Detection of Linear Acceleration by the Utricle and Saccule. When the body is suddenly thrust forward — that is, when the body accelerates forward — the otoconia, which have greater inertia than the surrounding fluids, fall backward on the hair cells, and information of malequilibrium is sent into the nervous centers, causing the person to feel as if he were falling backward. This automatically causes him to lean his body forward until the anterior shift of the otoconia exactly equals the tendency for the otoconia to fall backward because of the linear acceleration. At this point, the nervous system detects a state of proper equilibrium and therefore shifts the body no farther forward. As long as the degree of linear acceleration remains constant and the body is maintained in this forward leaning position, the person falls neither forward nor backward. Thus, the otoconia operate to maintain equilibrium during linear acceleration in exactly the same manner as they operate in static equilibrium.

The otoconia *do not* operate for the detection of linear *motion*. When a runner first begins to run, he must lean far forward to keep from falling over backward because of acceleration, but once he has achieved running speed, he would not have to lean forward at all if he were running in a vacuum. When running in air he leans forward to maintain equilibrium only because of the air resistance against his body, and in this instance it is not the otoconia that make him lean but the pressure of the air acting on pressure end-organs in the skin, which initiate the appropriate equilibrium adjustments.

THE SEMICIRCULAR CANALS AND THEIR DETECTION OF ANGULAR ACCELERATION

When the head suddenly *begins* to rotate in any direction, the endolymph in the membranous semicircular canals, because of its inertia, tends to remain stationary while the semicircular canals themselves turn. This causes relative fluid flow in the canals in a direction opposite to the rotation of the head, this relative flow lasting for a few seconds. Therefore, the person perceives this onset of rotation, which is called *angular acceleration*.

When the rotation suddenly stops, exactly the opposite effects take place: The endolymph continues to rotate while the semicircular canal stops. This time the cupula is bent in the opposite direction, causing the hair cells to stop discharging entirely — that is, to be inhibited. After another few seconds, the endolymph also stops moving, and the cupula returns gradually to its resting position, thus allowing the discharge of the hair cells also to return to their normal tonic level of activity.

Thus, the semicircular canal transmits a positive signal when the head *begins* to rotate and a negative signal when it *stops* rotating.

"Predictive" Function of the Semicircular Canals in the Maintenance of Equilibrium. Since the semicircular canals do not detect that the body is off balance in the forward direction, in the side direction, or in the backward direction, one might at first ask, What is the function of the semicircular canals in the maintenance of equilibrium? All they detect is that the person's head is beginning to rotate or is stopping rotation in one direction or another. Therefore, the function of the semicircular canals is not likely to be the maintenance of static equilibrium or of equilibrium during linear acceleration. Yet loss of function of the semicircular canals causes a person to have very poor equilibrium when he attempts to perform *rapid* and *intricate* body movements.

We can explain the function of the semicircular canals best by the following illustration. If a person is running forward rapidly and then suddenly begins to turn to one side, he falls off balance a second or so later unless appropriate corrections are made *ahead of time*. But, unfortunately, the utricle and saccule cannot detect that he is off balance until *after* this has occurred. On the other hand, the semicircular canals will have already detected that the person is beginning to turn, and this information can easily apprise the central nervous system of the fact that the person *will* fall off balance within the next second or so unless some correction is made. In other words, the semicircular canal mechanism *predicts* that malequilibrium is going to occur and thereby causes the equilibrium centers to make appropriate preventive adjustments. In this way, the person need not fall off balance before he begins to correct the situation.

OTHER FACTORS CONCERNED WITH EQUILIBRIUM

The Neck Proprioceptors. The vestibular apparatus detects the orientation and movements *only of the head*. Therefore, it is essential that the nervous centers also receive appropriate information depicting the orientation of the head with respect to the body as well as the orientation of the different parts of the body with respect to each other.

By far the most important proprioceptive information needed for the maintenance of equilibrium is that derived from the *joint receptors of the neck*, for this apprises the nervous system of the orientation of the head with respect to the body. When the head is bent in one direction or the other, impulses from the neck proprioceptors keep the vestibular apparatuses from giving the person a sense of malequilibrium. They do this by transmitting signals that exactly oppose the signals transmitted from the vestibular apparatuses. However, *when the entire body* is changed to a new position with respect to gravity, the impulses from the vestibular apparatuses *are not opposed* by the neck proprioceptors; therefore, the person in this instance does perceive a change in equilibrium status.

Proprioceptive and Exteroceptive Information from Other Parts of the Body. Proprioceptive information from other parts of the body besides the neck is also necessary for maintenance of equilibrium. For instance, appropriate equilibrium adjustments must be made whenever the body is angulated in the chest or abdominal region or elsewhere. All this information is algebraically added in the reticular substance and vestibular nuclei of the brain stem, thus causing appropriate adjustments in the postural muscles.

Also important in the maintenance of equilibrium are several types of exteroceptive sensations. For instance, pressure sensations from the footpads can tell one (1) whether his weight is distributed equally between his two feet and (2) whether his weight is more forward or backward on his feet.

Another instance in which exteroceptive information is necessary for maintenance of equilibrium occurs when a person is running. The air pressure against the front of his body signals that a force is opposing the body in a direction different from that caused by gravitational pull; as a result, the person leans forward to oppose this.

Importance of Visual Information in the Maintenance of Equilibrium. After complete destruction of the vestibular apparatuses, and even after loss of most proprioceptive information from the body, a person can still use his visual mechanisms effectively for maintaining equilibrium. Visual images help the person maintain equilibrium simply by visual detection of the upright stance. Many persons with complete destruction of the vestibular apparatus have almost normal equilibrium as long as their eyes are open and as long as they perform all motions slowly.

But, when they move rapidly or close the eyes, equilibrium is immediately lost.

FUNCTIONS OF THE RETICULAR FORMATION AND SPECIFIC BRAIN STEM NUCLEI IN CONTROLLING SUBCONSCIOUS, STEREOTYPED MOVEMENTS

Rarely, a child called an anencephalic monster is born without brain structures above the mesencephalic region, and some of these children have been kept alive for many months. Such a child is able to perform essentially all the functions of feeding, such as suckling, extrusion of unpleasant food from the mouth, and moving its hands to its mouth to suck its fingers. In addition, it can yawn and stretch. It can cry and follow objects with its eyes and by movements of its head. Also, placing pressure on the upper anterior parts of its legs will cause it to pull to the sitting position.

Therefore, it is obvious that many of the stereotyped motor functions of the human being are integrated in the brain stem. Unfortunately, the loci of most of these different motor control systems have not been found except for the following:

Stereotyped Body Movements. Most movements of the trunk and head can be classified into several simple movements, such as forward flexion, extension, rotation, and turning movements of the entire body. These types of movements are controlled by special nuclei located mainly in the mesencephalic and lower diencephalic region. For instance, *rotational movements* of the head and eyes are controlled by the *interstitial nucleus*, which is illustrated in Figure 34–7. The *raising movements* of the head and body are controlled by the *prestitial nucleus*. The *flexing movements* of the head and body are controlled by the *nucleus precommissuralis* located at the level of the posterior commissure. The *turning movements* of the body involve both the pontile and mesencephalic reticular formation. And *backward extension of the head and upper trunk* is accomplished by the *red nucleus*.

SUMMARY OF THE DIFFERENT FUNCTIONS OF THE CORD AND BRAIN STEM IN POSTURE AND LOCOMOTION

From the discussions in this chapter, we see that almost all of the discrete "patterns" of muscle movement required for posture and locomotion can be elicited by the spinal cord alone. However, coordination of these patterns to provide equilibrium, progression, and purposefulness of movement requires neuronal function at progressively higher levels of the central nervous system. Centers in the brain stem provide most of the nervous energy required to maintain postural tone and therefore to support the body against gravity. In addition, the brain stem centers provide especially the equilibrium adjustments of the body and the control of most stereotyped movements of body as well.

In the following chapter we will discuss the functions of still higher centers in the brain to provide the "voluntary" movements of the body.

REFERENCES

Cord Reflexes

Burke, R. E., and Rudonmin, P.: Spinal neurons and synapses. In Brookhart, J. M., and Mountcastle, V. B. (eds.): Handbook of Physiology. Sec. 1. Vol. 1. Baltimore, Williams & Wilkins, 1977, p. 877.

Creed, R. S. et al.: Reflex Activity of the Spinal Cord. New York, Oxford University Press, 1932.

Desmedt, J. E. (ed.): Physiological Tremor, Pathological Tremors and Clonus. New York, S. Karger, 1978.

Easton, T. A.: On the normal use of reflexes. Am. Sci., 60:591, 1972.

Gallistel, C. R.: The Organization of Action: A New Synthesis. New York, Halsted Press, 1979.

Granit, R., and Pompeiano, O. (eds.): Reflex Control of Posture and Movement. New York, Elsevier Scientific Publishing Co., 1979.

Houk, J. C.: Regulation of stiffness in skeletomotor reflexes. Annu. Rev. Physiol., 41:99, 1979.

Hughes, J. T.: Pathology of the Spinal Cord. Philadelphia, W. B. Saunders Co., 1978.

Kostyuk, P. G., and Vasilenko, D.: Spinal interneurons. Annu. Rev. Physiol., 41:115, 1979.

Orlovsky, G. N., and Shik, M. L.: Control of locomotion: A neurophysiological analysis of the cat locomotor system. Int. Rev. Physiol., 10:281, 1976.

Pearson, K.: The control of walking. Sci. Am., 235(6):72, 1976.

Peterson, B. W.: Reticulospinal projections to spinal motor nuclei. Annu. Rev. Physiol., 41:127, 1979.

Sherrington, C. S.: The Integrative Action of the Nervous System. New Haven, Conn., Yale University Press, 1911.

Stein, P. S. G.: Motor systems with specific reference to the control of locomotion. Annu. Rev. Neurosci., 1:61, 1978.

Stein, R. B.: Peripheral control of movement. Physiol. Rev., 54:215, 1974.

Trieschmann, R. B.: Spinal Cord Injuries. New York, Pergamon Press, 1979.

Brain Stem Reflexes, and Vestibular Apparatus

Chase, T. N., et al. (eds.): Huntington's Disease. New York, Raven Press, 1979.

Divac, I., and Öberg, R. G. E. (eds.): The Neostriatum. New York, Pergamon Press, 1979.

Duvoisin, R. C.: Parkinson's Disease. New York, Raven Press, 1978.

Elder, H. Y., and Trueman, E. R. (eds.): Aspects of Animal Movement. New York, Cambridge University Press, 1980.

Evarts, E. V.: Brain mechanisms of movement. Sci. Am., 241(3):164, 1979.

Granit, R., and Pompeiano, O. (eds.): Reflex Control of Posture and Movement. New York, Elsevier Scientific Publishing Co., 1979.

Hobson, J. A., and Brazier, M. A. B. (eds.): The Reticular Formation Revisited: Specifying Function for A Nonspecific System. New York, Raven Press, 1980.

Pearson, K.: The control of walking. Sci. Am., 235(6):72, 1976.

Peterson, B. W.: Reticulospinal projections of spinal motor nuclei. Annu. Rev. Physiol., 41:127, 1979.

Precht, W.: Vestibular mechanisms. Annu. Rev. Neurosci., 2:265, 1979.

Sarno, J. E., and Sarno, M. T.: The Condition and the Patient. New York, McGraw-Hill, 1979.

Shik, M. L., and Orlovsky, G. N.: Neurophysiology of locomotor automatism. Physiol. Rev., 56:465, 1976.

Stein, P. S. G.: Motor systems with specific reference to the control of locomotion. Annu. Rev. Neurosci., 1:61, 1978.

Talbot, R. E., and Humphrey, D. R. (eds.): Posture and Movement. New York, Raven Press, 1979.

Valentinuzzi, M.: The Organs of Equilibrium and Orientation as a Control System. New York, Harwood Academic Publishers, 1980.

Wilson, V., and Jones, G. M.: Mammalian Vestibular Physiology. New York, Plenum Press, 1979.

Wilson, V. J., and Peterson, B. W.: Peripheral and central substrates of vestibulospinal reflexes. Physiol. Rev., 58:80, 1978.

35

Motor Control by the Motor Cortex, the Basal Ganglia, and the Cerebellum

In preceding chapters we have been concerned with many of the subconscious motor activities integrated in the spinal cord and brain stem, especially those responsible for locomotion. In the present chapter we will discuss the control of motor function by the cerebral cortex, the basal ganglia, and the cerebellum, much of which is "voluntary" control in contradistinction to the subconscious control effected by the lower centers.

PHYSIOLOGIC ANATOMY OF THE MOTOR AREAS OF THE CORTEX AND THEIR PATHWAYS TO THE CORD

Electrical stimulation anywhere within a large area of the cerebral cortex can at times cause muscle contraction. This area, illustrated by the shading in Figure 35–1, is called the *sensorimotor cortex*. This same area receives sensory signals from the somatic areas of the body. The area immediately in front of the central sulcus, designated in the figure by the darkest shading, contains large numbers of *giant Betz cells*, or *pyramidal cells*, for which reason it is also called the *area pyramidalis*. This area causes motor movements following the least amount of electrical excitation and is therefore also called the *primary motor cortex*.

The Pyramidal Tract (Corticospinal Tract). One of the major pathways by which motor signals are transmitted from the motor areas of the cortex to the anterior motoneurons of the spinal cord is the *pyramidal tract* or *cortico-*

spinal tract, which is illustrated in Figure 35–2. This tract originates in all the shaded areas in Figure 35–1, including both the motor and the somesthetic areas, about three-quarters from the motor area and about one-quarter from the somesthetic regions posterior to the central sulcus. The function of the fibers from the somesthetic cortex is probably not motor, but instead is to cause feedback control of sensory input to the nervous system.

The most impressive fibers in the pyramidal tract are the large myelinated fibers that originate in the giant Betz cells of the motor area. These account for approximately 34,000 large fibers (mean diameter of about 16 microns) in the pyramidal tract from *each* side of the cortex.

The pyramidal tract passes downward through

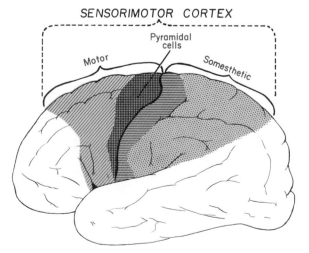

Figure 35–1. Relationship of the motor cortex to the somesthetic cortex.

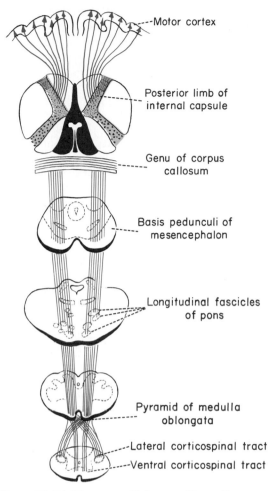

--Motor cortex

Posterior limb of
internal capsule

Genu of corpus
callosum

Basis pedunculi of
mesencephalon

Longitudinal fascicles
of pons

Pyramid of medulla
oblongata

-Lateral corticospinal tract

-Ventral corticospinal tract

Figure 35–2. The pyramidal tract. (From Ranson and Clark: Anatomy of the Nervous System. Philadelphia, W. B. Saunders Company, 1959.)

the *brain stem;* then the tract from each side crosses mainly to the opposite side. The fibers then descend in the *lateral corticospinal tracts* of the cord and terminate principally on interneurons at the bases of the dorsal horns of the cord gray matter.

The Extrapyramidal Tracts. The extrapyramidal tracts are additional tracts besides the pyramidal tract itself that transmit motor signals from the cortex and subcortical motor areas. For instance, large numbers of signals are transmitted from the motor cortex into the *caudate nucleus* and then through the *putamen, globus pallidus, subthalamic nucleus, red nucleus, substantia nigra,* and *reticular nuclei of the brain stem* before passing into the spinal cord. The multiplicity of connections within these intermediate nuclei of the basal ganglia and reticular substance will be presented later in the chapter.

The final pathways for transmission of extrapyramidal signals into the cord are the *reticulospinal tracts,* which lie in both the ventral and

lateral columns of the cord, an the *rubrospinal, tectospinal,* tracts.

THE PRIMARY MOTOR CORTEX OF THE HUMAN BEING

Figure 35–3 gives an approximate map of the human brain showing the points in the primary motor cortex that, when stimulated, cause muscle contractions in different parts of the body.

Note in the figure that stimulation of the most lateral portions of the motor cortex causes muscular contractions related to swallowing, chewing, and facial movements, while stimulation of the midline portion of the motor cortex where it bends over into the longitudinal fissure causes contraction of the legs, feet, or toes. The spatial organization is similar to that of the somatic sensory cortex I, which was shown in Chapter 32.

Degree of Representation of Different Muscle Groups in the Primary Motor Cortex. The different muscle groups of the body are not represented equally in the motor cortex. In general, the degree of representation is proportional to the discreteness of movement required of the respective part of the body. Thus, the thumb and fingers have large representations, as is true also of the lips, tongue, and vocal cords. The relative degrees of representation of the different parts of the body are illustrated in Figure 35–4, a figure constructed on the basis of stimulatory charts made of the human motor cortex during hundreds of brain operations.

When barely threshold stimuli are used, only small segments of the peripheral musculature ordinarily contract at a time. In the "finger" and

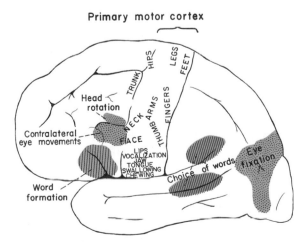

Figure 35–3. Representation of the different muscles of the body in the motor cortex, and location of other cortical areas responsible for certain types of motor movements.

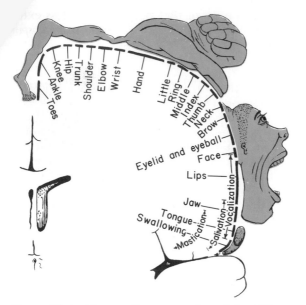

Figure 35–4. Degree of representation of the different muscles of the body in the motor cortex. (From Penfield and Rasmussen: The Cerebral Cortex of Man: A Clinical Study of Localization of Function, New York, Macmillan Company, 1968.)

"thumb" regions, which have tremendous representation in the cerebral cortex, threshold stimuli can sometimes cause single muscles or, at times, even single fasciculi of muscles to contract, thus illustrating that a high degree of control is exercised by this portion of the motor cortex over discrete muscular movement.

On the other hand, threshold stimuli in the trunk region of the body might cause as many as 30 to 50 small trunk muscles to contract simultaneously, thus illustrating that the motor cortex does not control discrete trunk muscles but instead controls *groups* of muscles. Similarly, threshold stimuli in the lip, tongue, and vocal cord regions of the motor cortex cause contraction of minute muscular areas, whereas stimulation in the leg region ordinarily excites several synergistic muscles at a time, causing some gross movement of the leg.

COMPLEX MOVEMENTS ELICITED BY STIMULATING THE CORTEX ANTERIOR TO THE MOTOR CORTEX — THE CONCEPT OF A "PREMOTOR CORTEX"

Electrical stimulation of the cerebral cortex for distances 1 to 3 centimeters in front of the primary motor cortex will often elicit complex contractions of groups of muscles. Occasionally, vocalization occurs, or rhythmic movements such as alternate thrusting of a leg forward and backward, coordinate moving of the eyes, chewing, swallowing, or contortion of parts of the body into different postural positions.

Some neurophysiologists have called this area the *premotor cortex* and have ascribed special capabilities to it to control coordinated movements involving many muscles simultaneously. One might also call the premotor cortex a *motor association area*. In fact, it is peculiarly organized to perform such a function for the following reasons: (1) It has long subcortical neuronal connections with the sensory association areas of the parietal lobe. (2) It has direct subcortical connections with the primary motor cortex. (3) It connects with areas in the thalamus contiguous with the thalamic areas that connect with the primary motor cortex. (4) The premotor area has abundant direct connections with the basal ganglia.

Still another reason for the belief that there is a motor association area (a "premotor cortex") is that damage to this area causes loss of certain coordinate skills, as follows:

Broca's Area and Speech. Referring again to Figure 35–3, note that immediately anterior to the primary motor cortex and immediately above the Sylvian fissure is an area labeled "word formation." This region is called *Broca's area*. Damage to it does not prevent a person from vocalizing, but it does make it impossible for the person to speak whole words other than simple utterances such as "no" or "yes." A closely associated cortical area also causes appropriate respiratory function, so that the vocal cords can be activated simultaneously with the movements of the mouth and tongue during speech. Thus, the activities that are related to Broca's area are highly complex.

The Voluntary Eye Movement Field. Immediately above Broca's area is a locus for controlling eye movements. Damage to this area prevents a person from voluntarily moving his eyes toward different objects. Instead, the eyes tend to lock on specific objects, an effect controlled by signals from the occipital region, as explained in Chapter 40. This frontal area also controls eyelid movements such as blinking.

Head Rotation Area. Still slightly higher in the "premotor region," electrical stimulation will elicit head rotation. This area is closely associated with the eye movement field and is presumably related to directing the head toward different objects.

Area for Hand Skills. In the frontal area immediately anterior to the primary motor cortex for the hands and fingers is a region neurosurgeons have observed to be an area for hand skills. That is, when tumors or other lesions cause destruction in this area, the hand movements become incoordinate and nonpurposeful, a condition called *motor apraxia*.

Summary of the Premotor Concept. It is clear that at least some areas anterior to the primary motor cortex can cause complex coordinate movements, such as speech movements, eye movements, head movements, and perhaps even hand skills. However, it should be remembered that all of these areas are closely connected with corresponding areas in the primary motor cortex, the thalamus, and the basal ganglia. Therefore, the complex coordinate movements almost certainly result from a cooperative effort of all these structures.

EFFECTS OF LESIONS IN THE MOTOR AND PREMOTOR CORTEX

The motor cortex is frequently damaged, especially by the common abnormality called a "stroke," which is caused by loss of blood supply to the cortex resulting from vascular rupture or vascular occlusion. Also, experiments have been performed in animals to remove selectively different parts of the motor cortex.

Ablation of the Primary Motor Cortex. Removal of a very small portion of the primary motor cortex in a monkey causes loss of much of the control of the represented muscles. If the sublying caudate nucleus is not damaged, gross postural and limb "fixation" movements can still be performed, but the animal loses voluntary control of discrete movements of the distal segments of the limbs — of the hands and fingers, for instance. This does not mean that the muscles themselves cannot contract, but that the animal's ability to control the fine movements is gone.

From these results one can conclude that the primary motor cortex is concerned mainly with voluntary initiation of finely controlled movements. On the other hand, the deeper motor areas, particularly the basal ganglia and lower brain stem, are responsible mainly for the involuntary and postural body movements.

Muscle Spasticity Caused by Ablation of Large Areas of the Motor Cortex — Extrapyramidal Lesions as the Basis of Spasticity. It should be recalled that the motor cortex gives rise to tracts that descend to the spinal cord through both the pyramidal tract and the extrapyramidal tract. These two tracts have opposing effects on the tone of the body muscles. The pyramidal tract causes continuous facilitation and therefore a tendency to increased muscle tone throughout the body. On the other hand, the extrapyramidal system transmits inhibitory signals through the basal ganglia and the reticular formation of the brain stem, with resultant inhibition of muscle action. When the motor cortex is destroyed, the balance between these two oppos-

ing effects may be altered. If the lesion is located discretely in the primary motor cortex where the large Betz cells lie, very little change in muscle tone results because both pyramidal and extrapyramidal elements are affected about equally. On the other hand, the usual lesion is very large and involves large portions of the cortex both anterior and posterior to the primary motor area, and these regions normally transmit inhibitory signals through the extrapyramidal tracts. Therefore, loss of extrapyramidal inhibition is the dominant feature, thus leading to muscle spasm.

If the lesion involves the basal ganglia as well as the motor cortex, the spasm is even more intense because the basal ganglia normally provide additional strong inhibition of the postural control system of the reticular formation, and loss of this inhibition further exacerbates the reticular excitation of the muscles. In patients with strokes, the lesion almost invariably affects both the motor cortex itself and the sublying basal ganglia, so that very intense spasm normally occurs in the muscles of the opposite side of the body.

STIMULATION OF THE SPINAL MOTONEURONS BY MOTOR SIGNALS FROM THE BRAIN

Figure 35–5 shows several different motor tracts entering a segment of the spinal cord from the brain. The corticospinal tract (pyramidal tract) terminates mainly on small *interneurons* in the base of the dorsal horns, although a few may also terminate directly on the anterior motoneurons themselves. From the primary interneurons, most of the motor signals are transmitted

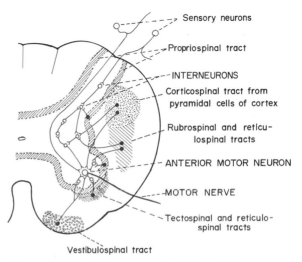

Figure 35–5. Convergence of all the different motor pathways on the anterior motoneurons.

through still other interneurons before finally exciting the anterior motoneurons.

Figure 35–5 shows several other descending tracts from the brain, collectively called the *extrapyramidal tracts,* as explained earlier, which also carry signals to the anterior motoneurons: (1) the *rubrospinal tract,* (2) the *reticulospinal tract,* (3) the *tectospinal tract,* and (4) the *vestibulospinal tract.* In addition, sensory signals arriving through the dorsal sensory roots, as well as signals transmitted from segment to segment of the spinal cord, also stimulate the anterior motoneurons.

The corticospinal tract causes specific muscle contractions. On the other hand, the extrapyramidal tracts provide less specific muscle contractions. Instead, they provide such effects as general facilitation, general inhibition, or gross postural signals, all of which provide the background against which the corticospinal system operates.

Patterns of Movement Elicited by Spinal Cord Centers. From Chapter 34, recall that the spinal cord can provide specific reflex patterns of movement in response to sensory nerve stimulation. Many of these patterns are also important when the anterior motoneurons are excited by signals from the brain. For instance, the stretch reflex is functional at all times, helping to damp the motor movements initiated from the brain. Also, when a brain signal excites an agonist muscle, it is not necessary to transmit an inverse signal to the antagonist at the same time; this transmission will be achieved by the reciprocal innervation circuit that is always present in the cord for coordinating the functions of antagonistic pairs of muscles.

Finally, parts of the other reflex mechanisms, such as withdrawal, walking, postural mechanisms, and so forth, are at times activated by signals from the brain. Thus, very simple signals from the brain can lead to many of our normal motor activities, particularly for such functions as walking and the attainment of different postural attitudes of the body.

On the other hand, at times it is important to suppress the cord mechanisms to prevent their interference with the performance of patterns of motor activity generated within the brain itself. This is believed to be one of the functions of the inhibitory signals transmitted through the reticulospinal tracts, as well as other inhibitory signals from the corticospinal tract itself.

MOTOR FUNCTIONS OF THE BASAL GANGLIA

Physiologic Anatomy of the Basal Ganglia. The anatomy of the basal ganglia is so complex and so poorly known in its details that it

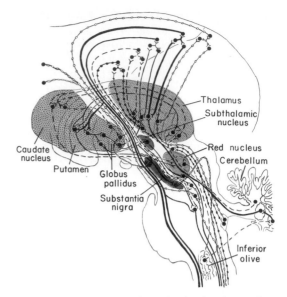

Figure 35–6. Pathways through the basal ganglia and related structures of the brain stem, thalamus, and cerebral cortex. (From Jung and Hassler: Handbook of Physiology, Sec. I, Vol. II. Baltimore, Williams & Wilkins Company, 1960.)

would be pointless to attempt a complete description at this time. However, Figure 35–6 illustrates the principal structures of the basal ganglia and their neural connections with other parts of the nervous system. Anatomically, the basal ganglia are the *caudate nucleus, putamen, globus pallidus, amygdaloid nucleus,* and *claustrum.* The amygdaloid nucleus, which will be discussed in Chapter 37, and the claustrum are not concerned directly with motor functions of the central nervous system. On the other hand, the *thalamus, subthalamus, substantia nigra,* and *red nucleus* all operate in close association with the caudate nucleus, putamen, and globus pallidus and are considered to be part of the basal ganglia system for motor control.

Some important features of the different pathways illustrated in Figure 35–6 are the following:

1. Numerous nerve pathways pass from the motor portions of the cerebral cortex, particularly from the so-called premotor areas, to the caudate nucleus and putamen, which together are called the *striate body.* In turn, the caudate nucleus and putamen send numerous fibers to the globus pallidus, the subthalamus, and the substantia nigra. These areas in turn send fibers to the ventrolateral and ventroanterior nuclei of the thalamus, and these send fibers back to the motor areas of the cerebral cortex. Thus, circular pathways are established from the motor cortical regions to the basal ganglia, the thalamus, and back to the motor regions from which the pathways begin. These circuitous pathways operate as

a feedback system for motor control, as will be discussed later.

2. The basal ganglia have numerous short neuronal connections among themselves. Also, the basal ganglia send tremendous numbers of nerve fibers into the lower brain stem, projecting especially onto the reticular nuclei, the red nucleus, and the inferior olive. It is presumably through these pathways that many of the so-called extrapyramidal signals for motor control are transmitted.

FUNCTIONS OF THE DIFFERENT BASAL GANGLIA

Before attempting to discuss the functions of the basal ganglia in man, we should speak briefly of the better known functions of these ganglia in lower animals. In birds, for instance, the cerebral cortex is poorly developed, while the basal ganglia are highly developed. These ganglia perform essentially all the motor functions, even controlling the voluntary movements in much the same manner that the motor cortex of the human being controls voluntary movements. Furthermore, in the cat, and to a lesser extent in the dog, decortication removes only the discrete types of motor functions and does not interfere with the animal's ability to walk, eat, fight, develop rage, have periodic sleep and wakefulness, and even participate naturally in sexual activities. However, if a major portion of the basal ganglia is destroyed, only gross stereotyped body movements remain, which were discussed in the previous chapter in relation to the lower brain stem animal.

Finally, in the human being, decortication of very young persons destroys the discrete movements of the body, particularly of the hands and distal portions of the lower limbs, but does not destroy the person's ability to walk crudely, to control his equilibrium, or to perform many other subconscious types of movements. However, simultaneous destruction of a major portion of the caudate nuclei almost totally paralyzes the opposite side of the body except for a few stereotyped reflexes integrated in the cord and brain stem.

With this brief background of the overall function of the basal ganglia, we can attempt to describe the functions of the individual portions of the basal ganglia system, realizing that the system actually operates as a total unit and that individual functions cannot be ascribed completely to the different parts of the basal ganglia.

Inhibition of Motor Tone by the Basal Ganglia. Though it is wrong to ascribe a single function to all the basal ganglia, nevertheless one

of the general effects of diffuse basal ganglia excitation is to inhibit muscle tone throughout the body. This effect results from inhibitory signals transmitted from the basal ganglia to the brain stem reticular formation. Therefore, whenever widespread destruction of the basal ganglia occurs, the loss of inhibition of the postural control system of the reticular formation causes muscle rigidity throughout the body. For instance, when the brain stem is transected at the mesencephalic level, which removes the inhibitory effects of the basal ganglia, the phenomenon called *decerebrate rigidity* occurs.

Yet, despite this general inhibitory effect of the basal ganglia, stimulation of specific areas within the basal ganglia can at times elicit positive muscle contractions and at times even complex patterns of movements.

Function of the Caudate Nucleus and Putamen — The Striate Body. The caudate nucleus and putamen together, because of their gross appearance on sections of the brain, are together called the *striate body*. They also function together to initiate and regulate *gross intentional movements of the body,* such as changing the position of the body, bending the body, and causing major arm movements. To perform this function they transmit impulses through two different pathways: (1) into the *globus pallidus,* thence by way of the *thalamus* to the *cerebral cortex,* and finally downward into the spinal cord through the *corticospinal* and *extrapyramidal pathways;* (2) downward through the *globus pallidus* and the *substantia nigra* by way of short axons into the *reticular formation,* and finally into the spinal cord mainly through the *reticulospinal tracts.*

In summary, the striate body helps to control gross intentional movements that we normally perform subconsciously. However, this control also involves the motor cortex, with which the striate body is very closely connected.

Function of the Globus Pallidus. It has been suggested that the principal function of the globus pallidus is to provide background muscle tone for intended movements, whether they are initiated by impulses from the cerebral cortex or from the striate body. That is, if a person wishes to perform an exact function with one hand, he positions his body and limbs appropriately and also tenses the muscles of the upper arm. These associated tonic contractions are supposedly initiated by a circuit that strongly involves the globus pallidus. Destruction of the globus pallidus removes these associated movements and therefore makes it difficult or impossible for the distal portions of the limbs to perform their more discrete activities.

Electrical stimulation of the globus pallidus while an animal is performing a gross body movement will often stop the movement in a static position, the animal holding that position for many seconds while the stimulation continues. This fits with the concept that the globus pallidus is involved in some type of servo feedback motor control system that is capable of locking the different parts of the body into specific positions. Obviously, such a circuit could be extremely important in providing the background body and gross limb movements when a person performs delicate tasks with his hands.

CLINICAL SYNDROMES RESULTING FROM DAMAGE TO THE BASAL GANGLIA

Much of what we know about the function of the basal ganglia comes from study of patients with basal ganglia lesions whose brains have undergone pathologic studies after death. Among the different clinical syndromes are these:

Athetosis. In this disease, slow, writhing movements of a hand, the neck, the face, the tongue, or some other part of the body occur continually. The movements are likely to be wormlike, first with overextension of the hands and fingers, then flexion, then rotary twisting to the side — all these continuing in a slow, rhythmic, writhing pattern. The contracting muscles exhibit a high degree of spasm, and the movements are enhanced by emotions or by excessive signals from the sensory organs. Furthermore, voluntary movements in the affected area are greatly impaired or sometimes even impossible.

The damage in athetosis is usually found in the *outer portion of the globus pallidus* or in this area and the striate body. Athetosis is usually attributed to the interruption of feedback circuits among the basal ganglia, thalamus, and cerebral cortex. The normal feedback circuits presumably allow a constant and rapid interplay between antagonistic muscle groups so that finely controlled movements can take place. However, if the feedback circuits are blocked, it is supposed that the detouring impulses may take devious routes through the basal ganglia, thalamus, and motor cortex, causing a succession of abnormal movements.

Parkinson's Disease. Parkinson's disease, which is also known as *paralysis agitans,* results almost invariably from *widespread destruction of the substantia nigra,* often associated with lesions of the *globus pallidus* and other related areas. It is characterized by (1) *rigidity* of the musculature either in widespread areas of the body or in isolated areas, (2) *tremor at rest* of the involved areas in most but not all instances, and (3) a *serious inability to initiate movement,* called *akinesia.*

These abnormal motor activities are almost certainly related to *loss of dopamine secretion* in the caudate nucleus and putamen by the nerve endings of the *nigrostriatal tract.* Destruction of the substantia nigra causes this tract to degenerate and the dopamine normally secreted in the caudate nucleus and putamen no longer to be present. But still present are large numbers of neurons that secrete acetylcholine. These are believed to transmit excitatory signals throughout the basal ganglia. It is also believed that the dopamine from the nigrostriatal pathway normally acts to inhibit these acetylcholine-producing neurons or in some other way to counter their activity. But, in the absence of dopamine secretion, the acetylcholine pathways become overly active, which presumably is the basis for the motor symptoms in Parkinson's disease.

The rigidity of Parkinson's disease seems to result from excess impulses transmitted in the corticospinal system, thus activating the alpha motor fibers to the muscles.

Tremor usually, though not always, occurs in Parkinson's disease. The frequency of the tremor is four to six cycles per second. When the person performs voluntary movements, weak tremors become temporarily blocked, but not strong tremors. The mechanism of the tremor is probably the lack of the inhibitory efforts of dopamine in the basal ganglia that enhances feedback in the striatum-globus pallidus–thalamus-cortex circuit, causing it to oscillate.

Though the muscle rigidity and the tremor are both distressing to the parkinsonian patient, even more serious is the *akinesia* that occurs in the final stages of the disease. To perform even the simplest of movements, the patient must exert the highest degree of concentration, mental effort, and even mental anguish to overcome "motor stiffness" of his musculature. Thus, the person with Parkinson's disease has a masklike face, showing almost no automatic emotional facial expressions; he is usually bent forward because of his muscle rigidity; and all his movements of necessity are highly deliberate rather than characterized by the many casual subconscious movements that are normally a part of our everyday life.

The cause of the akinesia in Parkinson's disease is not known, and again we must resort to theory. It is presumed that *loss of dopamine secretion in the caudate nucleus and putamen by the nigrostriatal fibers* allows excessive activity of the acetylcholine-producing neurons. But normal operation of the basal ganglia requires a balance between both excitatory and inhibitory activities, and loss of this balance, in effect, leads to a functionless basal ganglia system. We have already pointed out that the basal ganglia are responsible for many of the subconscious movements of the body, and also responsible even for the background movements of the trunk, legs, neck, and upper arms that are a required preliminary to performing the more discrete movements of the hands. If the subconscious and the background movements cannot occur, then other neural mechanisms must be substituted, especially those of the motor cortex and cerebellum. Unfortunately, though, these cannot replace the movements normally controlled by the basal ganglia and certainly cannot function at a subconscious level.

Treatment with L-Dopa. Administration of the drug L-dopa to patients with Parkinson's disease ameliorates most of the symptoms, especially the rigidity and the akinesia, in about two-thirds of the patients. The reaon for this seems to be the following: The dopamine secreted in the caudate nucleus and putamen by the nigrostriatal fibers in a derivative of L-dopa. When the substantia nigra is destroyed and the person develops Parkinson's disease, the administered L-dopa, it is believed, substitutes for the dopamine no longer secreted by the destroyed neurons. This causes more or less normal inhibition of the basal ganglia and relieves much or most of the akinesia and rigidity.

THE CEREBELLUM AND ITS MOTOR FUNCTIONS

The cerebellum has long been called a *silent area* of the brain principally because electrical excitation of this structure does not cause any sensation and rarely any motor movement. However, as we shall see, removal of the cerebellum does cause the motor movements to become highly abnormal. The cerebellum is especially vital to the control of very rapid muscular activities such as running, typing, playing the piano, and even talking. Loss of this area of the brain can cause almost total incoordination of these activities even though it causes paralysis of no muscles.

But how is it that the cerebellum can be so important when it has no direct control over muscle contraction? The answer to this is that it *monitors and makes corrective adjustments in the motor activities elicited by other parts of the brain.* It receives continuously updated information from the peripheral parts of the body to determine the instantaneous status of each part of the body —its position, its rate of movement, forces acting on it, and so forth. And it is believed that the cerebellum *compares* the actual physical status of each part of the body as depicted by the sensory

information with the status that is intended by the motor system. If the two do not compare favorably, then appropriate corrective signals are transmitted instantaneously back into the motor system to increase or decrease the levels of activation of the specific muscles.

Since the cerebellum must make major motor corrections extremely rapidly *during the course of motor movements,* a very extensive and rapidly acting cerebellar input system is required both from the peripheral parts of the body and from the cerebral motor areas. Also, an extensive output system feeding equally rapidly into the motor system is necessary to provide the necessary corrections of the motor signals.

THE INPUT SYSTEM TO THE CEREBELLUM

The Afferent Pathways. The basic afferent pathways to the cerebellum are illustrated in Figure 35–7. An extensive and important afferent pathway is the *corticocerebellar pathway,* which originates in the *motor cortex* and then passes by way of the *pontile nuclei* and *pontocerebellar tracts* to the cortex of the cerebellum.

The cerebellum also receives important sensory signals directly from the peripheral parts of the body, which reach the cerebellum by way of the *ventral* and *dorsal spinocerebellar* tracts. The signals transmitted in these tracts originate in the muscle spindles, the Golgi tendon organs, and the large tactile receptors of the skin and joints, and they apprise the cerebellum of the momentary status of muscle contraction, degree of ten-

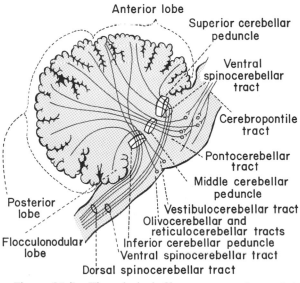

Figure 35–7. The principal afferent tracts to the cerebellum.

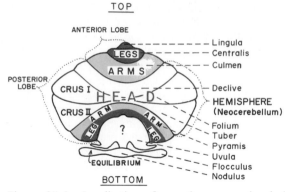

TOP

ANTERIOR LOBE

LEGS

A R M S

POSTERIOR LOBE

CRUS I   H E = A D

CRUS II

EQUILIBRIUM

BOTTOM

Lingula
Centralis
Culmen

Declive
HEMISPHERE (Neocerebellum)
Folium
Tuber
Pyramis
Uvula
Flocculus
Nodulus

Figure 35–8. Localization of somatic sensory signals in the cerebellum of the human being.

sion on the muscle tendons, positions of the parts of the body, and forces acting on the surfaces of the body. All of this information keeps the cerebellum constantly apprised of the physical status of the body at every instant.

In addition to the signals in the spinocerebellar tracts, other signals are transmitted up the cord through the *spinoreticular pathway* to the reticular substance of the brain stem and through the *spino-olivary pathway* to the inferior olivary nucleus and then relayed to the cerebellum. Thus, the cerebellum collects continual information about all parts of the body even though it is operating at a subconscious level.

Spatial Localization of Sensory Input to the Cerebellum. Afferent fibers entering the cerebellum terminate in distinct spatially oriented areas of the cerebellar cortex. Figure 35–8 illustrates the human cerebellum, showing the approximate sensory terminations based on anatomical data and extrapolation from animals.

It is also important to note that the signals from the motor cortex project to corresponding joints in (1) the peripheral muscles and joints of the body, and (2) the sensory representations of these same muscles and joints in the cerebellum. Therefore, each respective point in the cerebellum in reality represents a specific muscle or a specific joint; simultaneously it receives information directly from the motor cortex concerning the motor signals that activate the muscle or joint and also from the muscle or joint concerning the effect actually produced. Then it projects signals back into the motor pathway to the same respective topographical part of this system.

The Equilibrium Area of the Cerebellum. Note in Figure 35–8 at the lowest part of the cerebellum (or most posterior part) an area labeled "equilibrium." This is the flocculonodular lobe of the cerebellum, a very old part that connects through bidirectional pathways with the vestibular nuclei. This area is concerned almost entirely

with equilibrium, as described in the previous chapter. Its destruction causes serious loss of equilibrium, particularly when rapid movements are being performed.

OUTPUT SIGNALS FROM THE CEREBELLUM

The Deep Cerebellar Nuclei and the Efferent Pathways. Located deep in the cerebellar mass are three *deep cerebellar nuclei:* the *dentate, interpositus,* and *fastigial nuclei.* All of the efferent tracts from the cerebellum arise in the deep nuclei — none from the cerebellar cortex, which transmits its output signals only through the deep nuclei. Efferent signals are transmitted to many portions of the motor system including (1) the motor cortex, (2) the basal ganglia, (3) the red nucleus, (4) the reticular formation of the brain stem, and (5) the vestibular nuclei. The important tracts are illustrated in Figure 35–9.

THE NEURONAL CIRCUIT OF THE CEREBELLUM

The structure of the cerebellar cortex is entirely different from that of the cerebral cortex. Furthermore, each part of the cerebellar cortex has a neuronal organization almost precisely the same as that in all other parts.

The human cerebellum is actually a large folded sheet, approximately 17 cm. wide by 120 cm. long, the folds lying crosswise, as illustrated in Figure 35–8.

The Functional Unit of the Cerebellar Cortex — the Purkinje Cells. The cerebellum has approximately 30 million nearly identical functional units, one of which is illustrated in Figure

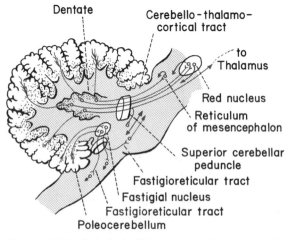

Dentate

Cerebello-thalamo-cortical tract

to Thalamus

Red nucleus

Reticulum of mesencephalon

Superior cerebellar peduncle

Fastigioreticular tract

Fastigial nucleus

Fastigioreticular tract

Poleocerebellum

Figure 35–9. Principal efferent tracts from the cerebellum.

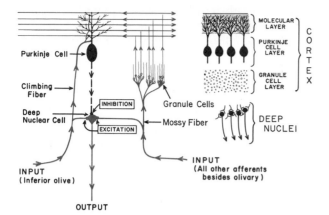

Figure 35–10. Basic neuronal circuit of the cerebellum, showing excitatory pathways in color. At right are the three major layers of the cerebellar cortex and also the deep nuclei.

35–10. This functional unit centers around the Purkinje cell, of which there are also 30 million in the cerebellar cortex.

Note to the far right in Figure 35–10 the three major layers of the cerebellar cortex: the *molecular layer,* the *Purkinje cell layer,* and the *granular cell layer.* Beneath these layers are deep nuclei located in the center of the cerebellar mass.

The Neuronal Circuit of the Functional Unit. As illustrated in Figure 35–10, the output from the functional unit is from a deep nuclear cell. However, this cell is continually under the influence of both excitatory and inhibitory signals. The excitatory signals arise from direct connections with the afferent fibers that enter the cerebellum. The inhibitory signals arise entirely from the Purkinje cells in the cortex of the cerebellum.

The afferent inputs to the cerebellum are of two types, one called the *climbing fibers* and the other the *mossy fibers.* Both of these fibers excite both the deep nuclear cells and also the Purkinje cells. However, the Purkinje cells are inhibitory cells. Note in Figure 35–10 that the direct stimulation of the deep nuclear cells by both the climbing and the mossy fibers excites them, whereas the signals arriving from the Purkinje cells inhibit them. Normally, there is a continuous balance between these two effects, so that the degree of output from the deep nuclear cells remains relatively constant. On the other hand, in the execution of rapid motor movements, the *timing* of the two effects on the deep nuclei is such that the excitation appears before the inhibition. Also, the relative balance between excitation and inhibition changes. In this way, very rapid excitatory and inhibitory transient signals can be fed back into the motor pathways to correct motor movements. The inhibitory portions of these signals resemble delay-line negative feedback signals

of the type that are very effective in providing damping. That is, when the motor system is excited, a negative feedback signal occurs after a short delay to stop the muscle movement from overshooting its mark, the usual cause of oscillation.

FUNCTION OF THE CEREBELLUM IN VOLUNTARY MOVEMENTS

The cerebellum functions only in association with motor activities initiated elsewhere in the central nervous system. These activities may originate in the spinal cord, in the reticular formation, in the basal ganglia, or in the motor areas of the cerebral cortex.

Figure 35–11 illustrates the basic cerebellar pathways involved in cerebellar control of voluntary movements. When motor impulses are transmitted from the cerebral cortex downward through the pyramidal and extrapyramidal tracts to excite the muscles, collateral impulses are transmitted simultaneously into the cerebellum through the pontocerebellar and olivocerebellar tracts. Therefore, for every motor movement that is performed, not only do the muscles receive activating signals, but the cerebellum receives similar signals at the same time.

When the muscles contract, the muscle spindles, Golgi tendon apparatuses, joint receptors, and other peripheral receptors transmit signals upward mainly through the spinocerebellar and spino-olivary pathways to terminate in the cerebellum.

After the signals from the periphery and those

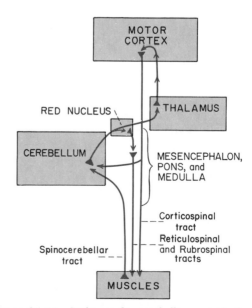

Figure 35–11. Pathways for cerebellar control of voluntary movements.

from the motor cortex are integrated in the cerebellum, efferent impulses are transmitted from the cerebellar dentate nuclei upward through the ventrolateral nuclei of the thalamus back to the motor cortex.

"Error Control" by the Cerebellum. One will readily recognize that the circuit described above represents a complicated feedback circuit beginning in the cerebral part of the motor control system and then returning also to this area. Ordinarily, the motor cortex transmits far more impulses than are needed to perform each intended movement, and the cerebellum therefore must act to inhibit the motor cortex at the appropriate time after the muscle has begun to move. The cerebellum seems to assess the rate of movement and to calculate the time that will be required to reach the point of intention. Then appropriate inhibitory impulses are transmitted to the motor cortex to inhibit the agonist muscle and to excite the antagonist muscle. In this way, appropriate "brakes" are applied to stop the movement at the precise point of intention.

Since all these events transpire much too rapidly for the motor cortex to reverse the excitation "voluntarily," it is evident that the excitation of the antagonist muscle toward the end of a movement is an entirely automatic and subconscious function and is not a "willed" contraction of the same nature as the original contraction of the agonist muscle. We shall see below that in patients with serious cerebellar damage, excitation of the antagonist muscles occurs not at the appropriate time but instead always too late. Therefore, it is almost certain that one of the major functions of the cerebellum is automatic excitation of antagonist muscles at the end of a movement and, at the same time, inhibition of agonist muscles that started the movement.

The "Damping" Function of the Cerebellum — Intension Tremor. One of the by-products of the cerebellar feedback mechanism is its ability to "damp" muscular movements. To explain the meaning of "damping" we must first point out that essentially all movements of the body are "pendular." For instance, when an arm is moved, momentum develops, and the momentum must be overcome before the movement can be stopped. And, because of the momentum, all pendular movements have a tendency to overshoot. If overshooting does occur in a person whose cerebellum has been destroyed, the conscious centers of the cerebrum eventually recognize this and initiate a movement in the opposite direction to bring the arm toward its intended position. But again the arm, by virtue of its momentum now in the opposite direction, again overshoots and appropriate corrective signals must again be instituted. Thus, the arm oscillates back and forth past its intended point for several cycles before it finally fixes on its mark. This effect is called an *action tremor,* or *intention tremor.*

However, if the cerebellum is intact, appropriate subconscious signals stop the movement precisely at the intended point, thereby preventing the overshoot and also the tremor. This is the basic characteristic of a damping system. All control systems regulating pendular elements that have inertia must have damping circuits built into the mechanisms. In the motor control system of our central nervous system, the cerebellum provides much of this damping function.

Function of the Cerebellum in Prediction — Dysmetria. Another important by-product of the cerebellar feedback mechanism seems to be an ability to help the central nervous system predict future positions of moving parts of the body. Without the cerebellum a person "loses" his limbs *when they move rapidly,* indicating that feedback information from the periphery probably must be analyzed by the cerebellum if the brain is to keep up with the motor movements. Thus, the cerebellum detects from the incoming proprioceptive signals the rate at which the limb is moving and then predicts from this the projected time course of movement. This allows the cerebellum, operating through the cerebellar output circuits, to inhibit the agonist muscles and to excite the antagonist muscles when the movement approaches the point of intention.

Without the cerebellum this predictive function is so deficient that moving parts of the body move much farther than the point of intention. This failure to control the distance that the parts of the body move is called *dysmetria,* which means simply poor control of the distance of movement.

Failure of Smooth "Progression of Movements" — Ataxia. One of the most important features of normal motor function is one's ability to progress from one movement to the next in an orderly succession. When the cerebellum becomes dysfunctional and the person loses his subconscious ability to predict how far the different parts of his body will move in a given time, he becomes unable also to control the beginning of the next movement. As a result, the succeeding movement may begin much too early or much too late. Therefore, movements such as those required for writing, running, or even talking all become completely incoordinate, lacking completely the ability to progress in an orderly sequence from one direction of movement to the next.

Extramotor Predictive Functions of the Cerebellum. The cerebellum also plays a role in predicting other events besides, simply, movements of the body. For instance, the rates of progression of both auditory and visual phenomena can be predicted. As an example, a person can predict from the changing visual scene how rapidly he is approaching an object. A striking experiment that demonstrates the importance of the cerebellum in this ability is the effect of removing the visual portion of the cerebellum in monkeys. A monkey so treated occasionally charges the wall of a corridor and literally bashes its brains out because it is unable to predict when it will reach the wall.

FUNCTION OF THE CEREBELLUM IN INVOLUNTARY MOVEMENTS

The cerebellum functions in involuntary, subconscious, or postural movements in almost exactly the same manner as it does in voluntary movements except that slightly different pathways are involved. The extrapyramidal signals that originate in the motor cortex, basal ganglia, or reticular formation to cause involuntary movements pass mainly into the inferior olive and then into the cerebellum. Then as the muscle movements occur, proprioceptive information from the muscles, joints, and other peripheral parts of the body passes to the cerebellum, thus providing the same type of control for involuntary movements as is provided for voluntary movements, as discussed above. Signals then feed from the cerebellum back to the *motor cortex, basal ganglia,* and *reticular formation.*

Function of the Cerebellum in Equilibrium. In the discussion of the vestibular apparatus and of the mechanism of equilibrium in Chapter 34, it was pointed out that the cerebellum is required for proper integration of the equilibratory impulses. Removal of the *flocculonodular lobes* in particular causes almost complete loss of equilibrium temporarily, and impaired equilibrium indefinitely. The symptoms of malequilibrium are essentially the same as those that result from destruction of the semicircular canals, indicating this area of the cerebellum is particularly important for integration of *changes in direction of motion* as detected by the semicircular canals.

SENSORY FEEDBACK CONTROL OF MOTOR FUNCTIONS

THE SENSORY "ENGRAM" FOR MOTOR MOVEMENTS

It is primarily in the sensory and sensory association areas that a person experiences effects of motor movements and records "memories" of the different patterns of motor movements. These are called *sensory engrams* of the motor movements. When he wishes to achieve some purposeful act, he calls forth one of these engrams, and the engram then automatically sets the motor system of the brain into action to reproduce the sensory pattern in the engram.

The Proprioceptor Feedback Servomechanism for Reproducing the Sensory Engram. From the discussion of cerebellar function earlier in the chapter, it is clear how proprioceptor signals from the periphery can affect motor activity. However, in addition to the feedback pathways through the cerebellum, feedback pathways also pass from proprioceptors to the sensory areas of the cerebral cortex and thence back to the motor cortex. Each of these pathways is capable of modifying the motor response. For instance, if a person learns to cut with scissors, the movements involved in this process cause a particular sequential pattern of proprioceptive impulses to pass to the somatic sensory area. Once this pattern has been "learned" by the sensory cortex, the memory engram of the pattern can be used to activate the motor system to perform the same sequential pattern whenever it is required.

To do this, the proprioceptor signals from the fingers, hands, and arms are compared with the engram, and if the two do not match each other, the difference, called the "error," initiates additional motor signals that automatically activate appropriate muscles to bring the fingers, hands, and arms into the necessary sequential attitudes for performance of the task. Each successive portion of the engram is projected according to a time sequence, and the motor control system automatically follows from one point to the next so that the fingers go through the precise motions necessary to duplicate exactly the sensory engram of the motor activity.

Thus, one can see that the motor system in this case actually acts as a "servomechanism"; that is, it is not the motor cortex itself that controls the pattern of activity to be accomplished. Instead, the pattern is located in the sensory part of the brain, and the motor system merely "follows" the pattern, which is the definition of a servomechanism. If ever the motor system fails to follow the pattern, sensory signals are fed back to the cerebral cortex to apprise the sensorium of this failure, and appropriate corrective signals are transmitted to the muscles.

Other sensory signals besides somesthetic signals are also involved in motor control, particularly visual signals. However, these other sensory systems are often slower to recognize error than is the somatic proprioceptor system. Therefore, when the sensory engram depends on visual feedback for control purposes, the motor movements are usually considerably slowed in comparison with those that depend on somatic feedback.

An extremely interesting experiment that demonstrates the importance of the sensory engram for control of motor movements is one in which a monkey has been trained to perform some complex task and then various portions of his cortex are removed. Removal of small portions of the motor cortex that control the muscles normally used for the task does not prevent the monkey from performing it. Instead he automatically uses other muscles in place of the paralyzed ones to perform the same task. On the other hand, if the corresponding somatic sensory cortex is removed but the motor cortex is left intact, the monkey loses all ability to perform the task. Thus, this experiment demonstrates that the motor system acts automatically as a servomechanism to use whatever muscles are available to follow the pattern of the sensory engram, and if some muscles are missing, other muscles are substituted automatically. The experiment also demonstrates forcefully that the somatic sensory cortex is essential to at least some types of "learned" motor performance.

ESTABLISHMENT OF "SKILLED" MOTOR PATTERNS

Many motor activities are performed so rapidly that there is insufficient time for sensory feedback signals to control them. For instance, the movements of the fingers during typing occur much too rapidly for feedback signals to be transmitted either to the somatic sensory cortex or even directly to the motor cortex. It is believed that the patterns for control of these rapid coordinate muscular movements are established in the motor system itself, probably involving complex circuitry in the primary motor cortex, in the so-called premotor area of the cortex, the basal ganglia, and even the cerebellum. Indeed, lesions in any of these areas can destroy or at least greatly alter one's ability to perform rapid coordinated muscular contractions, such as those required during the act of typing, talking, or writing by hand.

Role of Sensory Feedback During Establishment of the Rapid Motor Patterns. Even a highly skilled motor activity can be performed the very first time, provided that it is performed extremely slowly — slowly enough for sensory feedback to guide the movements through each step. However, to be really useful, many skilled motor activities must be performed rapidly. This probably is achieved by successive performances of the same skilled activity, at first very slowly, then progressively more rapidly, until finally an engram of the skilled activity is laid down in the motor system as well as in the sensory system. This motor engram causes a precise set of muscles to go through a specific sequence of movements required to perform the skilled activity. Such an engram is called a *pattern of skilled motor function,* and the motor areas of the brain are primarily concerned with this.

After a person has performed a skilled activity many times, the motor pattern of this activity can thereafter cause the hand or arm or other part of the body to go through the same pattern of activity again and again, now entirely *without* sensory feedback control. However, even though sensory feedback control is no longer present, the sensory system still determines in retrospect whether or not the act has been performed correctly. If it has not, information from the sensory system supposedly can help to correct the pattern the next time it is performed.

Thus, eventually, hundreds of patterns of different coordinate movements are laid down in the motor system, and these can be called upon one at a time in different sequential orders to perform literally thousands of still more complex motor activities.

An interesting experiment that demonstrates the applicability of these theoretical methods of muscular control is one in which the eyes are made to "follow" an object that moves around and around in a circle. At first, the eyes can follow the object only when it moves around the circle slowly, and even then the movements of the eyes are extremely jerky. Thus, sensory feedback is being utilized to control the eye movements for following the object. However, after a few seconds, the eyes begin to follow the moving object rather faithfully, and the rapidity of movement around the circle can be increased to many times per second, and still the eyes continue to follow the object. Sensory feedback control of each stage of the eye movements at these rapid rates would be completely impossible. Therefore, by this time, the eyes have developed a pattern of movement that is not dependent upon step-by-step sensory feedback. Nevertheless, if the eyes should fail to follow the object around the circle, the sensory system would immediately become aware of this and presumably could make corrections in the pattern of movement.

INITIATION OF VOLUNTARY MOTOR ACTIVITY

Because of the spectacular properties of the primary motor cortex and of the instantaneous muscle contractions that can be achieved by stimulating this area, it has become customary to think that the initial brain signals that elicit voluntary muscle contractions begin in this primary motor cortex. However, this almost certainly is far from the truth. Indeed, experiments have shown that the cerebellum and the basal ganglia are activated at almost exactly the same time that the motor cortex is activated. Furthermore, there is no known mechanism by which the motor cortex can conceive the entire sequential pattern that is to be achieved by the motor movements.

Therefore, we are left with an unanswered question: What is the locus of the initiation of voluntary motor activity? In following chapters we shall learn that the reticular formation of the brain stem and much of the thalamus play essential roles in activating all other parts of the brain. Therefore, it is possible that these areas provide the initial signals that lead to subsequent activity in the motor cortex, the basal ganglia, and the cerebellum at the onset of voluntary movement. But here again, we come to a circular question: What is it that initiates the activity in the brain stem and the thalamus? We do know part of the answer: The activity is initiated by continual sensory input into these areas, including sensory input from the peripheral receptors of the body and from the cortical memory storage areas. Much motor activity is almost reflex in nature, occurring instantly after an incoming sensory signal from the periphery. But so-called voluntary activity occurs minutes, hours, or even days after the initiating sensory input — after analysis, storage of memories, recall of the memories, and finally initiation of a motor response. We shall see in the following chapter that the control of motor activity is strongly influenced by these prolonged procedures of cerebration. Furthermore, damage to the essential areas of the cerebral cortex for analysis of sensory information leads to serious deficits and abnormalities of voluntary muscle control.

Therefore, for the present, let us conclude that the immediate energy for eliciting voluntary motor activity probably comes from the basal regions of the brain. These, in turn, are under the control of the different sensory inputs, the memory storage areas, and the associated areas of the brain that are devoted to the processes of analysis, a mechanism frequently called *cerebration.*

REFERENCES

Allen, G. I., and Tsukahara, N.: Cerebrocerebellar communication systems. *Physiol. Rev., 54*:957, 1974.

Armstrong, D. M.: The mammalian cerebellum and its contribution to movement control. *In* Porter, R. (ed.): International Review of Physiology: Neurophysiology III. Vol. 17. Baltimore, University Park Press, 1978, p. 239.

Asanuma, H.: Recent developments in the study of the columnar arrangement of neurons within the motor cortex. *Physiol. Rev., 55*:143, 1975.

Desmedt, J. E. (ed.): Cerebral Motor Control in Man: Long Loop Mechanisms. New York, S. Karger, 1978.

Evarts, E. V.: Brain mechanisms of movement. *Sci. Am., 241*(3):164, 1979.

Gallistel, C. R.: The Organization of Action: A New Synthesis. New York, Halsted Press, 1979.

Granit, R.: The Basis of Motor Control. New York, Academic Press, 1970.

Grillner, S.: Locomotion in vertebrates: Central mechanisms and reflex interaction. *Physiol. Rev., 55*:247, 1975.

Llinas, R.: Eighteenth Bowditch lecture. Motor aspects of cerebellar control. *Physiologist, 17*:19, 1974.

Massion, J., and Sasaki, K. (eds.): Cerebro-Cerebellar Interactions. New York, Elsevier/North-Holland, 1979.

O'Connell, A. L., and Gardner, E. B.: Understanding the Scientific Bases of Human Movement. Baltimore, Williams & Wilkins, 1972.

Orlovsky, G. N., and Shik, M. L.: Control of locomotion: A neurophysiological analysis of the cat locomotor system. *Int. Rev. Physiol., 10*:281, 1976.

Pearson, K.: The control of walking. *Sci. Am., 235*(6):72, 1976.

Penfield, W., and Rasmussen, T.: The Cerebral Cortex of Man. New York, The Macmillan Co., 1950.

Porter, R.: The neurophysiology of movement performance. *In* MTP International Review of Science: Physiology. Vol. 3. Baltimore, University Park Press, 1974, p. 151.

Porter, R.: Influences of movement detectors on pyramidal tract neurons in primates. *Annu. Rev. Physiol., 38*:121, 1976.

Shik, M. L., and Orlovsky, G. N.: Neurophysiology of locomotor automatism. *Physiol. Rev., 56*:465, 1976.

Stein, P. S. G.: Motor systems with specific reference to the control of locomotion. *Annu. Rev. Neurosci., 1*:61, 1978.

36

The Cerebral Cortex and Intellectual Functions of the Brain

It is ironic that of all parts of the brain, we know least about the mechanisms of the cerebral cortex, even though it is by far the largest portion of the nervous system. Yet, we do know the effects of destruction or of specific stimulation of various portions of the cortex, and still more has been learned from electrical recordings from the cortex or from the surface of the scalp. In the early part of the present chapter the facts known about cortical functions are discussed, and then basic theories of the neuronal mechanisms involved in thought processes, memory, analysis of sensory information, and so forth are presented briefly.

PHYSIOLOGIC ANATOMY OF THE CEREBRAL CORTEX

The functional part of the cerebral cortex is composed mainly of a thin layer of neurons 2 to 5 mm. in thickness, covering the surface of all the convolutions of the cerebrum and having a total area of about one-quarter of a square meter. The total cerebral cortex contains approximately 100 billion neurons.

Figure 36–1 illustrates the typical structure of the cerebral cortex, showing successive layers of different groups of cells. Most of the cells are of three types; *granular, fusiform,* and *pyramidal,* the latter named for their characteristic pyramidal shape.

To the right in Figure 36–1 is illustrated the typical organization of nerve fibers within the different layers of the cortex. Note particularly the large number of horizontal fibers extending between adjacent areas of the cortex, but note also the vertical fibers that extend to and from the

cortex to lower areas of the brain stem or to distant regions of the cerebral cortex through long association bundles of fibers.

Anatomical Relationship of the Cerebral Cortex to the Thalamus and Other Lower Centers. All areas of the cerebral cortex have direct afferent and efferent connections with the thala-

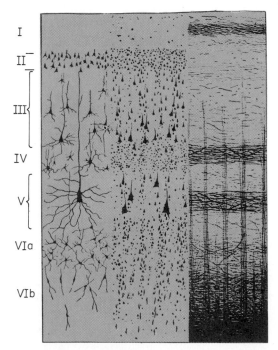

Figure 36–1. Structure of the cerebral cortex, illustrating: *I*, molecular layer; *II*, external granular layer; *III*, layer of pyramidal cells; *IV*, internal granular layer; *V*, large pyramidal cell layer; *VI*, layer of fusiform or polymorphic cells. (From Ranson and Clark (after Brodmann): Anatomy of the Nervous System. Philadelphia, W. B. Saunders Company, 1959.)

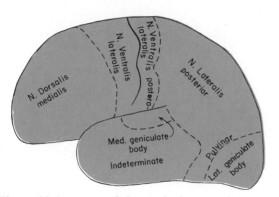

Figure 36–2. Areas of the cerebral cortex that connect with specific portions of the thalamus. (Modified from Elliott: Textbook of the Nervous System. J. B. Lippincott Co.)

mus. Figure 36–2 shows the areas of the cerebral cortex connected with specific parts of the thalamus. These connections are in *two* directions, both from the thalamus to the cortex and then from the cortex back to essentially the same area of the thalamus. Furthermore, when the thalamic connections are cut, the functions of the corresponding cortical areas become entirely or almost entirely abrogated. Therefore, the cortex operates in close association with the thalamus and can almost be considered both anatomically and functionally a large outgrowth of the thalamus: for this reason the thalamus and the cortex together are called the *thalamocortical system,* as is explained in the following chapter. Also, all pathways from the sensory nerve endings to the cortex pass through the thalamus, with the single exception of the sensory pathways of the olfactory tract.

FUNCTIONS OF CERTAIN SPECIFIC CORTICAL AREAS

Studies in human beings by neurosurgeons have shown that some specific functions are localized to certain general areas of the cerebral cortex. Figure 36–3 gives a map of some of these areas as determined by Penfield and Rasmussen from electrical stimulation of the cortex or by neurologic examination of patients after portions of the cortex had been removed. The lightly shaded areas are *primary sensory areas,* while the darkly shaded area is the *voluntary motor area* (also called *primary motor area*) from which muscular movements can be elicted with relatively weak electrical stimuli. These primary sensory and motor areas have highly specific functions, while other areas of the cortex perform more general functions that we call association or cerebration.

SPECIFIC FUNCTIONS OF THE PRIMARY SENSORY AREAS

The primary sensory areas all have certain functions in common. For instance, somatic sensory areas, visual sensory areas, and auditory sensory areas all have spatial localizations of signals from the peripheral receptors (which are discussed in detail in Chapters 32, 40 and 41).

Electrical stimulation of the primary sensory areas in the parietal lobes in awake patients gives relatively uncomplicated sensations. For instance, in the somatic sensory area the patient experiences a tingling feeling in the skin, numbness, mild "electric" feeling, or, rarely, mild degrees of temperature sensations. And these sensations are localized to discrete areas of the body in accord with the spatial representation in the somatic sensory cortex, as described in Chapter 32. Therefore, it is believed that the primary somatic sensory cortex analyzes only the simple aspects of sensations and that analysis of intricate patterns of sensory experience also requires adjacent parts of the parietal lobe, parts called *sensory association areas*.

Electrical stimulation of the primary visual cortex in the occipital lobes causes the person to see flashes of light, bright lines, colors, or other simple visions. Here again, the visual images are localized to specific regions of the visual fields in accord with the portion of the primary visual cortex stimulated, as described in Chapter 40. But the visual cortex alone is not capable of complete analysis of complicated visual patterns; for this, the visual cortex must operate in association with adjacent regions of the occipital cortex, the *visual association areas*.

Electrical stimulation of the auditory cortex in the temporal lobes causes a person to hear a

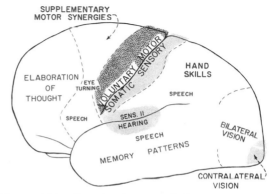

Figure 36–3. Functional areas of the human cerebral cortex as determined by electrical stimulation of the cortex during neurosurgical operations and by neurological examinations of patients with destroyed cortical regions. (From Penfield and Rasmussen: The Cerebral Cortex of Man: A Clinical Study of Localization of Function. New York, Macmillan Company, 1968.)

simple sound that may be weak or loud, have low or high frequency, or have other uncomplicated characteristics, such as a squeak or even an undulation. But never are words or any other fully intelligible sound heard. Thus, the primary auditory cortex, like the other primary sensory areas, can detect the individual elements of auditory experience but cannot analyze complicated sounds. Therefore, the primary auditory cortex alone is not sufficient to give one even the usual auditory experiences; these can be achieved, however, when the primary area operates together with the *auditory association areas* located in adjacent regions of the temporal lobes.

Despite the inability of the primary sensory areas to analyze the incoming sensations fully, when these areas are destroyed, the ability of the person to utilize the sensations usually suffers drastically. For instance, loss of the primary visual cortex in one occipital lobe causes a person to become blind in the ipsilateral halves of both retinae, and loss of the primary visual cortices in both hemispheres causes total blindness. Likewise, loss of both primary auditory cortices causes almost total deafness.

THE SENSORY ASSOCIATION AREAS

Around the borders of the primary sensory areas are regions called *sensory association areas* or *secondary sensory areas*. In general, these areas extend 1 to 5 centimeters in one or more directions from the primary sensory areas; each time a primary area receives a sensory signal, secondary signals spread after a delay of a few milliseconds into the respective association area as well. Part of this spread occurs in the cortex itself, but a major part also occurs in the thalamus, beginning in the thalamic sensory relay nuclei, passing next to corresponding *thalamic* association areas, and then traveling to the association cortex.

The general function of the sensory association areas is to provide a higher level of interpretation of the sensory experiences. The general areas for the interpretative functions for somatic, visual, and auditory experiences are illustrated in Figure 36–4.

Destruction of the sensory association areas greatly reduces the capability of the brain to analyze different characteristics of sensory experiences. For instance, damage in the temporal lobe below and behind the primary auditory area in the "dominant hemisphere" of the brain often causes a person to lose the understanding of words or of other auditory experiences even though he hears them.

Likewise, destruction of the visual association areas of the occipital lobe in the dominant hemisphere does not cause blindness or prevent normal activation of the primary visual cortex but does greatly reduce the person's ability to interpret what he sees. Such a person often loses his ability to recognize the meanings of words, a condition that is called *word blindness* or *dyslexia.*

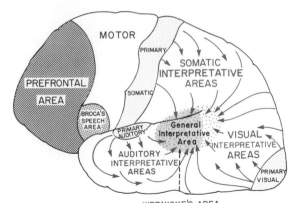

Figure 36–4. Organization of the somatic, auditory, and visual association areas into a general mechanism for interpretation of sensory experience. All of these feed into the *general interpretative area* located in the posterosuperior portion of the temporal lobe and the angular gyrus. Note also the prefrontal area, and Broca's speech area.

Finally, destruction of the somatic sensory association area in the parietal cortex posterior to primary somatic area I causes the person to lose his spatial perception for location of the different parts of his body. In the case of the hand that has been "lost," the skills of the hand are greatly reduced. Thus, this area of the cortex seems to be necessary for interpretation of somatic sensory experiences.

The functions of the association areas are described in more detail in Chapter 32 for somatic, in Chapter 40 for visual, and in Chapter 41 for auditory experiences.

INTERPRETATIVE FUNCTION OF THE POSTERIOR SUPERIOR TEMPORAL LOBE – THE GENERAL INTERPRETATIVE AREA (OR "WERNICKE'S" AREA)

The somatic, visual, and auditory association areas, which can actually be called "interpretative areas," all meet one another in the posterior part of the superior temporal lobe and in the anterior part of the angular gyrus where the temporal, parietal, and occipital lobes all come together. This area of confluence of the different sensory interpretative areas is especially highly developed in the dominant side of the brain — the *left side* in right-handed persons — and it plays the greatest single role of any part of the cerebral cortex in the higher levels of brain function that we call *cerebration.* Therefore, this region has frequently been called by different names suggestive of the area having almost global importance: the *general interpretative area,* the *gnostic area,* the *knowing area,* the *tertiary association area,* and so forth. The temporal portion of the general interpretative area is also called *Wernicke's area* in honor of the neurologist who first described its special significance in intellectual processes.

Following severe damage in the general interpretative area, a person may hear perfectly well and even recognize different words but still may be unable to arrange these words into a coherent thought. Likewise, the person may be able to read words from the printed page but be unable to recognize the thought that is conveyed. In addition, he has similar difficulties in understanding the higher levels of meaning of somatic sensory experiences, even though there is no loss of sensation itself.

Electrical stimulation in the posterior superior temporal lobe of the conscious patient occasionally causes a highly complex thought. This is particularly true when the stimulatory electrode is passed deep enough into the brain to approach the corresponding connecting areas of the thalamus. The types of thoughts that may be experienced include memories of complicated visual scenes from childhood, auditory hallucinations such as a specific musical piece, or even a discourse by a specific person. For this reason it is believed that complicated memory patterns involving more than one sensory modality are stored at least partially in the temporal lobe. This belief is in accord with the importance of the general interpretative area in interpretation of the complicated meanings of different sensory experiences.

The angular gyrus portion of the general interpretative area is especially important for interpretation of visual information. If this region is destroyed while the temporal lobe portion of this interpretative area (Wernicke's area) is still intact, the person can still interpret both auditory and somatic experiences as usual, but the stream of visual experiences passing into the general interpretative area from the visual cortex is mainly blocked. Therefore, the person may be able to see words and even know they are words but, nevertheless, not be able to interpret their meanings. This is the condition called *dyslexia.*

Let us again emphasize the global importance of the general interpretative area for most intellectual functions of the brain. Loss of this area in an adult usually leads thereafter to a lifetime of almost demented existence.

The Dominant Hemisphere. The general interpretative functions of Wernicke's area and of the angular gyrus, and also the functions of the speech and motor control areas, are usually much more highly developed in one cerebral hemisphere than in the other. This is called the *dominant hemisphere.* In at least nine of ten persons the left hemisphere is the dominant one. At birth, Wernicke's area of the brain is often as much as 50 per cent larger in the left hemisphere than in the right. Therefore, it is easy to understand why the left side of the brain might become dominant over the right side.

Usually associated with the dominant temporal lobe and angular gyrus is dominance of certain portions of the somesthetic cortex and motor cortex for control of voluntary motor functions. For instance, as is discussed later in the chapter, the premotor speech area (Broca's area), located far laterally in the intermediate frontal area, is almost always dominant also on the left side of the brain. This speech area causes the formation of words by exciting simultaneously the laryngeal muscles, the respiratory muscles, and the muscles of the mouth.

Though the intepretative areas of the temporal lobe and angular gyrus, as well as many of the motor areas, are highly developed in only a single hemisphere, they are capable of receiving sensory information from both hemispheres and are also capable of controlling motor activities in both hemispheres, utilizing mainly fiber pathways in the *corpus callosum* for communication between the two hemispheres. This unitary, cross-feeding organization prevents interference between the two sides of the brain; such interference, obviously, could create havoc with both thoughts and motor responses.

Wernicke's Area in the Nondominant Hemisphere. When Wernicke's area in the dominant hemisphere is destroyed, the person *normally loses almost all intellectual functions associated with language or symbolism,* such as ability to read, ability to perform mathematical operations, and even the ability to think through logical problems. However, other types of interpretative capabilities, some of which undoubtedly utilize the temporal lobe and angular gyrus regions of the opposite hemisphere, are retained. Psychological studies in such patients have suggested that this hemisphere may be especially important for understanding and interpreting music, nonverbal visual experiences, and spatial relationships between the person and the surroundings, and probably also for interpreting many somatic experiences related to use of the limbs and hands.

THE PREFRONTAL AREAS

The prefrontal areas are those portions of the frontal lobes that lie anterior to the motor regions, as shown in Figure 36–4. For years, this part of the cortex has been considered to be the locus of the higher intellect of the human being, principally because the main difference between the brain of monkeys and that of man is the great prominence of man's prefrontal areas. Yet efforts to show that the prefrontal cortex is more important in higher intellectual functions than other portions of the cortex have not been successful. Indeed, destruction of the posterior temporal lobe and angular gyrus region in the dominant hemisphere causes infinitely more harm to the intellect than does destruction of both prefrontal areas.

Prevention of Distractibility by the Prefrontal Areas. One of the outstanding characteristics of a person who has lost his prefrontal areas is the ease with which he can be *distracted* from a sequence of thoughts. Likewise, in lower animals whose prefrontal areas have been removed, the ability to concentrate on psychological tests is almost completely lost.

The human being without prefrontal areas is still capable of performing many intellectual tasks, such as answering short questions and performing arithmetic computations (such as $9 \times 6 = 54$), thus illustrating that the basic intellectual activities of the cerebral cortex are still intact without the prefrontal areas. Yet if concerted *sequences* of cerebral functions are required of the person, he becomes completely disorganized. Therefore, the prefrontal areas seem to be important in keeping the mental functions directed toward goals.

Elaboration of Thought by the Prefrontal Areas. Another function that has been ascribed to the prefrontal areas by psychologists and neurologists is *elaboration of thought*. This means simply an increase in depth and abstractness of the different thoughts. Psychological tests have shown that prefrontal lobectomized animals presented with successive bits of sensory information fail to store these bits in memory — probably because of their inability to code this information, the initial requirement for memory storage. If the prefrontal areas are intact, many such successive bits of information can be stored in different areas of the brain and can be called forth again and again during the subsequent periods of cerebration. This ability of the prefrontal areas to cause storage — even though it may be temporary — of many types of information simultaneously, and then perhaps also to cause recall of this information, could well explain the many functions of the brain that we associate with higher intelligence, such as the abilities to (1) plan for the future, (2) delay action in response to incoming sensory signals so that the sensory information can be weighed until the best course of response is decided, (3) consider the consequences of motor actions even before they are performed, (4) solve complicated mathematical, legal, or philosophical problems, (5) correlate all avenues of information in diagnosing rare diseases, and (6) control one's activities in accord with moral laws.

Effects of Destruction of the Prefrontal Areas

The person without prefrontal areas ordinarily acts precipitously in response to incoming sensory signals, such as striking an adversary too large to be beaten instead of pursuing the more judicious course of running away. Also, he is likely to lose many or most of his morals; he has little embarrassment in relation to his excretory, sexual, and social activities; and he is prone to quickly changing moods of sweetness, hate, joy, sadness, exhilaration, and rage. In short, he is a highly *distractible* person with lack of ability to pursue long and complicated thoughts.

THOUGHTS, CONSCIOUSNESS, AND MEMORY

Our most difficult problem in discussing consciousness, thoughts, memory, and learning is that we do not know the neural mechanism of a thought. We know that destruction of large portions of the cerebral cortex does not prevent a person from having thoughts, but it does reduce his *degree* of awareness of the surroundings. On the other hand, destruction of far smaller portions of the thalamus and especially of the mesencephalic portion of the reticular activating system can cause tremendously decreased awareness or even complete unconsciousness.

Each thought almost certainly involves simultaneous signals in portions of the cerebral cortex, thalamus, limbic system, and reticular formation of the brain stem. Some crude thoughts probably depend almost entirely on lower centers; the thought of pain is probably a good example, for electrical stimulation of the human cortex rarely elicits anything more than the mildest degrees of pain, while stimulation of certain areas of the hypothalamus and mesencephalon in animals apparently causes excruciating pain. On the other hand, a type of thought pattern that requires mainly the cerebral cortex is that involving vision, because loss of the visual cortex causes complete inability to perceive visual form or color.

Therefore, we might formulate a definition of a thought in terms of neural activity as follows: A thought probably results from the momentary "pattern" of stimulation of many different parts of the nervous system at the same time, probably involving most importantly the cerebral cortex, the thalamus, the limbic system, and the upper reticular formation of the brain stem. This is called the *holistic theory* of thoughts. The stimulated areas of the limbic system, thalamus, and reticular formation perhaps determine the general nature of the thought, giving it such qualities as pleasure, displeasure, pain, comfort, crude modalities of sensation, localization to gross areas of the body, and other general characteristics. On the other hand, the stimulated areas of the cortex probably determine the discrete characteristics of the thought (such as specific localization of sensations of the body and of objects in the fields of vision), discrete patterns of sensation (such as the rectangular pattern of a concrete block wall or the texture of a rug), and other individual characteristics that enter into the overall awareness of a particular instant.

MEMORY AND TYPES OF MEMORY

If we accept the above approximation of what constitutes a thought, we can see immediately that the mechanism of memory must be just as complex as the mechanism of a thought, for, to provide memory, the nervous system must re-create the same spatial and temporal pattern (the "holistic" pattern) of stimulation in the central nervous system at some future date. Though we cannot explain in detail what a memory is, we do know some of the basic psychological and neuronal processes that probably lead to the process of memory.

All of us know that all degrees of memory occur, some memories lasting a few seconds and others lasting hours, days, months, or years. Possibly all of these types of memory are caused by the same mechanism operating to different degrees of fulfillment. Yet, it is also possible that different mechanisms of memory do exist. Indeed, most physiologists classify memory into two to four different types. For the purpose of the present discussion, we will use the following classification:

1. *Sensory memory*
2. *Short-term memory* or *primary memory*
3. *Long-term memory,* which itself can be divided into *secondary memory* and *tertiary memory.*

The basic characteristics of these types of memory are the following:

Sensory Memory. Sensory memory means the ability to retain sensory signals in the sensory areas of the brain for a very short interval of time following the actual sensory experience. Usually these signals remain available for analysis for only a few hundred milliseconds and are replaced by new sensory signals in less than one second. Nevertheless, during the short interval of time that the instantaneous sensory information remains in the brain, it can continue to be used for further processing; most important, it can be "scanned" to pick out the important points. Thus, this is the initial stage of the memory process.

Short-term Memory (Primary Memory). Short-term memory (or primary memory) is the memory of a few facts, words, numbers, letters, or other bits of information for a few seconds to a minute or more at a time. It is typified by a person's memory of the digits in a telephone number for a short period of time after he has looked up the number in the telephone directory. This type of memory is usually limited to about seven bits of information, and when new bits are put into this *short-term store,* some of the older information is displaced. Thus, if a person looks up a second telephone number, the first is usually lost. One of the most important features of short-term memory is that the information in this memory store is instantaneously available so that the person does not have to search through his mind for it as he does for information that has been put away in the long-term memory stores.

Long-term Memory. Long-term memory is the storage in the brain of information that can be recalled at some later time — minutes, hours, days, months, or years later. This type of memory has been called *fixed memory, permanent memory,* and several other names. Long-term memory is also usually divided into two different types, *secondary memory* and *tertiary memory,* the characteristics of which are the following:

A *secondary memory* is a long-term memory that is stored with either a weak or only a moderately strong memory trace. For this reason it is easy to forget and it is sometimes difficult to recall. Furthermore, the time required to search for the information in the mind is relatively long. This type of memory can last from several minutes to several years. When the memories are so weak that they will last for only a few minutes to a few days, they are also frequently called *recent memory.*

A *tertiary memory* is a memory that has become so well ingrained in the mind that the memory can usually last the lifetime of the person. Furthermore, the very strong memory traces of this type of memory make the stored information available within a split second. This type of memory is typified by one's knowledge of his own name, by his ability to recall immediately the numbers from 1 to 10, the letters of the alphabet, and the words that he uses in speech, and also by the memory of his own precise physical structure and of his very familiar immediate surroundings.

PHYSIOLOGICAL BASIS OF MEMORY

Despite the many advances in neurophysiology during the past half century, we still cannot explain what is perhaps the most important function of the brain: its capability for memory. Yet, physiological experiments are beginning to generate conceptual theories of the means by which memory could occur. Some of these are discussed in the following few sections.

Possible Mechanisms for Short-term Memory. Short-term memory requires a neuronal mechanism that can hold specific information signals for a few seconds to a minute or more at most. Several such mechanisms are the following:

Reverberating Circuit Theory of Short-term Memory. When a tetanizing electrical stimulus is applied directly to the surface of the cerebral cortex and then is removed after a second or more, the local area excited by this stimulus continues to emit rhythmic action potentials for short periods of time. This effect results from local reverberating circuits, the signals passing through a multistage circuit of neurons in the local area of the cortex itself or perhaps also back and forth between the cortex and the thalamus.

It is postulated that sensory signals reaching the cerebral cortex can set up similar reverberating oscillations and that these could be the basis for short-term memory. Then, as the reverberating circuit fatigues, or as new signals interfere with the reverberations, the short-term memory fades away.

One of the principal observations in support of this theory is that any factor that causes a general disturbance of brain function, such as sudden fright, a very loud noise, or any other sensory experience that attracts the person's undivided attention, immediately erases the short-term memory. The memory cannot be recalled when the disturbance is over unless a portion of this memory had already been placed into the long-term memory store, as will be discussed in subsequent sections.

Post-tetanic Potentiation Theory of Short-term Memory. In most parts of the nervous system, including even the anterior motoneurons of the spinal cord, tetanic stimulation of a neuron for a few seconds causes subsequent increased excitability of the neuron for a few seconds to a few hours.

If during this time the neuron is stimulated again, it responds much more vigorously than normally, a phenomenon called *post-tetanic potentiation*. This is obviously a type of memory that depends on change in the excitability of the involved neurons, and it could be the basis for short-term memory. It is likely that this phenomenon results from some temporary change in the synapses of the neurons.

DC Potential Theory of Short-term Memory. Another change that often occurs in neurons following a period of excitation is a prolonged decrease in the membrane potential of the neuron lasting for seconds to minutes. Because this changes the excitability of the neuron, it could be the basis for short-term memory. Such changes in neuronal potentials are called *DC potentials* or sometimes *electrotonic potentials*. Measurements in the cerebral cortex show that such potentials occur especially in the superficial dendritic layers of the cortex, indicating that the process of short-term memory could result from changes in dendritic membrane potentials.

Mechanism of Long-term Memory, Enhancement of Synaptic Transmission Facility

Long-term memory means the ability of the nervous system to recall thoughts long after initial elicitation of the thoughts is over. We know that long-term memory does not depend on continued activity of the nervous system, because the brain can be totally inactivated by cooling, by general anesthesia, by hypoxia, by ischemia, or by any other method, and yet memories that have been previously stored are still retained when the brain becomes active once again. Therefore, it is assumed that long-term memory must result from actual alterations of the synapses, either physical or chemical.

Many different theories have been offered to explain the synaptic changes that cause long-term memory. Three of the most important of these are:

1. Anatomical Changes in the Synapses. Cajal, more than half a century ago, discovered that the number of terminal fibrils ending on neuronal cells and dendrites in the cerebral cortex increases with age. Conversely, physiologists have shown that inactivity of regions of the cortex causes thinning of the cortex, for instance, thinning of the primary visual cortex in animals that have lost their eyesight. Also, intense activity of a particular part of the cortex can cause excessive thickening of the cortical shell in that area alone. This has been demonstrated especially in the visual cortex of animals subjected to visual experiences. Finally, some neuroanatomists have observed electron micrographic changes in presynaptic terminals that have been subjected to intense and prolonged activity.

All these observations have led to a widely held belief that fixation of memories in the brain results from anatomical changes in the synapses themselves: perhaps changes in numbers of presynaptic terminals, perhaps in sizes of the terminals, or perhaps in the sizes of the dendrites. Such anatomical changes could cause permanent or semipermanent increase in the degree of facilitation of specific neuronal circuits, thus allowing signals to pass through the circuits with progres-

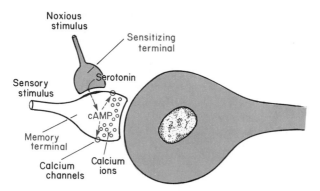

Figure 36–5. A memory system that has been discovered in the snail *Aplysia*.

sively greater ease the more often the memory trace is used. This obviously would explain the tendency for memories to become more and more deeply fixed in the nervous system the more often they are recalled or the more often the person repeats the sensory experience that leads to the memory trace.

2. Physical or Chemical Changes in the Presynaptic Terminal or the Postsynaptic Membrane. Recent studies in the large snail *Aplysia* have uncovered a mechanism of memory that results from either a physical or a chemical change or the presynaptic terminal. This mechanism, illustrated in Figure 36–5, works in the following way: There are two separate presynaptic terminals, one of which ends on the subsequent neuron and the other of which ends on the first presynaptic terminal. In the figure these are called the *memory terminal* and the *sensitizing terminal*. When the memory terminal is stimulated repeatedly but without stimulating the sensitizing terminal, signal transmission at first is very great, but this becomes less and less intense with repeated stimulation until transmission almost ceases. This phenomenon is called *habituation*. It is a type of memory that causes the neuronal circuit to lose its response to repeated events that are insignificant. On the other hand, if a noxious stimulus excites the sensitizing terminal at the same time that the memory terminal is stimulated, then, instead of the transmitted signal becoming progressively weaker, the ease of transmission becomes much stronger and will remain strong for hours, days, or even weeks, even without further stimulation of the sensitizing terminal. Thus, the noxious stimulus causes the memory pathway to become facilitated for weeks thereafter. It is especially interesting that only a few repeated action potentials are required to cause either habituation or sensitization. However, once habituation has occurred, the synapses can become sensitized very rapidly with only a few sensitizing signals.

At the molecular level, the habituation effect in the memory terminal results from progressive closure of calcium channels of the terminal membrane. As a result, much smaller than normal amounts of calcium diffuse into this terminal when action potentials occur, and much less transmitter is therefore released because calcium entry is the stimulus for transmitter release (as was discussed in Chapter 31).

In the case of sensitization, excess amounts of cyclic AMP are believed to develop inside the memory terminal and this opens increased numbers of calcium channels, which correspondingly enhances signal transmission.

Other variants of sensitized synaptic transmission have also been suggested, such as increased amounts of receptor protein in the postsynaptic membrane and decreased rate of destruction of transmitter substance.

3. Theoretical Function of RNA in Memory. The discovery that DNA and RNA can act as codes to control reproduction, which in itself is a type of memory from one generation to another, plus the fact that these substances, once formed in a cell, tend to persist for the lifetime of the cell, has led to the theory that nucleic acids may be involved in memory changes of the neurons, changes that could last for the lifetime of the person. In addition, biochemical studies have shown an increase in RNA in some active neurons. Yet, a mechanism by which RNA could cause facilitation of synaptic transmission has never been found. Therefore, this theory seems to be based mainly on analogy rather than on factual evidence. The only supporting evidence is that several research workers have claimed that long-term memory will not occur to a significant extent when the chemical processes for formation of RNA or of protein are blocked. Since these substances are required for anatomical changes in the synapses as well as for RNA synthesis, however, these observations could support almost any type of theory for long-term memory.

Summary. The theory that seems most likely at present for explaining long-term memory is that some actual anatomical, physical, or chemical change occurs in the synaptic knobs themselves or in the postsynaptic neurons, these changes permanently facilitating the transmission of impulses at the synapses. If all the synapses are thus facilitated in a thought circuit, this circuit can be reexcited by any one of many diverse incoming signals at later dates, thereby causing memory. The overall facilitated circuit is called a *memory engram* or a *memory trace*.

Consolidation of Long-term Memory

If a memory is to last in the brain so that it can be recalled days later, it must become "consolidated" in the neuronal circuits. This process requires five to ten minutes for minimal consolidation and an hour or more for maximal consolidation. For instance, if a strong sensory impression is made on the brain but is then followed within a minute or so by an electrically induced brain convulsion, the sensory experience will not be remembered at all. Likewise, brain concussion, sudden application of deep general anesthesia, and other effects that temporarily block the dynamic function of the brain can prevent consolidation.

However, if the same sensory stimulus is impressed on the brain and the strong electrical shock is delayed for more than five to ten minutes, at least part of the memory trace will have become established. If the shock is delayed for an hour or more, the memory will have become fully consolidated.

This process of consolidation and the time required for consolidation can probably be explained by the phenomenon of *rehearsal* of the short-term memory, as follows:

Role of Rehearsal in Transference of Short-term Memory into Long-term Memory. Psychological studies have shown that rehearsal of the same information again and again accelerates and potentiates the degree of transfer of short-term memory into long-term memory, and therefore also accelerates and potentiates the process of consolidation. The brain has a natural tendency to rehearse new-found information, and especially to rehearse new-found information that catches the mind's attention. Therefore, over a period of time the important features of sensory experiences become progressively more and more fixed in the long-term memory stores. This explains why a person can remember small amounts of information studied in depth far better than he can remember large amounts of information studied only superficially. And it also explains why a person who is wide-awake will consolidate memories far better than will a person who is in a state of mental fatigue.

Codifying of Memories During the Process of Consolidation. One of the first stages in the process of consolidation is for the memories to be codified into different classes of information. During this stage, similar information is recalled from the long-term storage bins and is used to help process the new information. The new and the old are compared for similarities and for differences, and part of the storage process is to store the information about these similarities and differences rather than simply to store the information unprocessed. Thus, during the process of consolidation, the new memories are not stored randomly in the mind, but instead are stored in direct association with other memories of the same type. This is obviously necessary if one is to be able to scan the memory store at a later date to find the required information.

Change of Long-term Secondary Memory into Long-term Tertiary Memory — Role of Rehearsal. Rehearsal also plays an extremely important role in changing the weak trace type of long-term memory, called secondary memory, into the strong trace type, called tertiary memory. That is, each time a memory is recalled or each time the same sensory experience is repeated, a more and more indelible memory trace develops in the brain. The memory finally becomes so deeply fixed in the brain that it can be recalled within fractions of a second and will last for a lifetime, which are the characteristics of long-term tertiary memory.

Role of Specific Parts of the Brain in the Memory Process

Role of the Hippocampus for Rehearsal, Codification, and Consolidation of Memories — Anterograde Amnesia. Persons who have had both hippocampi removed often have normal memory for information stored in the brain prior to removal of the hippocampi. However, after loss of these structures, these persons have very little capability for transferring short-term memory into long-term memory. That is, they do not have the ability to separate out the important information, to codify it, to rehearse it, and to consolidate it in the long-term memory store. Therefore, these persons develop serious *anterograde amnesia,* meaning simply the inability to establish new memories.

Retrograde Amnesia. Retrograde amnesia means inability to recall memories from the long-term memory storage bins even though the memories are known to be still there. In most persons who have hippocampal lesions and resultant anterograde amnesia, there is also some degree of retrograde amnesia, which suggests that these two types of amnesia are at least partially related and that hippocampal lesions can cause both. However, it has also been claimed that damage in some thalamic areas can lead specifically to retrograde amnesia without significant anterograde amnesia. This perhaps can be explained by the fact that the thalamus probably plays a major role in directing the person's attention to information in different parts of the memory storehouse; this will be discussed in the following chapter. Thus, the thalamus could easily play important roles in codifying, storing, and recalling memories.

FUNCTION OF THE BRAIN IN COMMUNICATION

One of the most important differences between the human being and lower animals is the facility with which human beings can communicate with one another. Furthermore, because neurological tests can easily assess the ability of a person to communicate with others, we know perhaps more about the sensory and motor systems related to communication than about any other segment of cortical function. Therefore, we will review rapidly the function of the cortex in communication, and from this one can see immediately how the principles of sensory analysis and motor control apply to this art.

There are two aspects of communication: first, the *sensory aspect,* involving the ears and eyes, and, second, the *motor aspect,* involving vocalization and its control.

Sensory Aspects of Communication. We noted earlier in the chapter that destruction of portions of the *auditory* and *visual association areas* of the cortex can result in inability to understand the spoken word or the written word. These effects are called respectively *auditory receptive aphasia* and *visual receptive aphasia* or, more commonly, *word deafness* and *word blindness* (also called *dyslexia*). On the other hand, some persons are perfectly capable of understanding either the spoken word or the written word but are unable to interpret the meanings of words when used to express a thought. This results most frequently when *Wernicke's area in the posterior portion of the superior temporal gyrus* in the dominant hemisphere is damaged or destroyed. This is considered to be *general sensory aphasia* or *general agnosia.*

Motor Aspects of Communication. *Sensory Aphasia.* The process of speech involves (1) formation in the mind of thoughts to be expressed and choice of the words to be used and (2) the actual act of vocalization. The formation of thoughts and choice of words is principally the function of the sensory areas of the brain, for we find that a person who has a destructive lesion in the same *posterior temporal lobe region* that is involved in general sensory aphasia also has inability to formulate intelligible thoughts to be communicated. At other times the thoughts can be formulated, but the person is unable to put together the appropriate words to express the thought. These inabilities to formulate thoughts to be spoken or the required word sequences are called *sensory aphasia* or, in honor of the neurologist who first delimited the brain area responsible, *Wernicke's aphasia.*

Motor Aphasia. Often a person is perfectly capable of deciding what he wishes to say, and he is capable of vocalizing, but he simply cannot make his vocal system emit words instead of noises. This effect, called *motor aphasia,* almost always results from damage to *Broca's speech area,* which lies in the *premotor* facial region of the

cortex — about 95 per cent of the time in the left hemisphere, as illustrated in Figure 36–4. Therefore, we assume that the *patterns* for control of the larynx, lips, mouth, respiratory system, and other accessory muscles of articulation are all controlled in this area.

Articulation. Finally, we have the act of articulation itself, which means the muscular movements of mouth, tongue, larynx, and so forth that are responsible for the actual emission of sound. The facial and laryngeal regions of the motor cortex activate these muscles. Destruction of these regions can cause either total or partial inability to speak distinctly.

REFERENCES

Bekhtereva, N. P.: The Neurophysiological Aspects of Human Mental Activity. New York, Oxford University Press, 1978.

Benson, D. F.: Aphasia, Alexia, and Agraphia. New York, Churchill Livingstone, 1979.

Buser, P.: Higher functions of the nervous system. *Annu. Rev. Physiol.,* 38:217, 1978.

Daniloff, R., *et al.*: The Physiological Bases of Verbal Communication. Englewood Cliffs, N.J., Prentice-Hall, 1980.

De Silva, F. H. L., and Arnolds, D. E. A. T.: Physiology of the hippocampus and related structures. *Annu. Rev. Physiol.,* 40:185, 1978.

Desmedt, J. E. (ed.): Language and Hemispheric Specialization in Man; Cerebral Event-Related Potentials. New York, S. Karger, 1977.

Dimond, S. J.: Neuropsychology: A Textbook of Systems and Psychological Functions of the Human Brain. Boston, Butterworths, 1979.

Gazzaniga, M. (ed.): Neuropsychology. New York, Plenum Press, 1978.

Geschwind, N.: The apraxias: Neural mechanisms of disorders of learned movement. *Am. Sci.,* 63:188, 1975.

Geschwind, N.: Specializations of the human brain. *Sci. Am.,* 241(3):180, 1979.

Herron, J. (ed.): Neuropsychology of Left-Handedness. New York, Academic Press, 1979.

Hixon, T. J., *et al.* (eds.): Introduction to Communicative Disorders. Englewood Cliffs, N.J., Prentice-Hall, 1980.

Hubel, D. H.: The brain. *Sci. Am.,* 241(3):44, 1979.

Kandel, E. R.: Neuronal plasticity and the modification of behavior. *In* Brookhart, J. M., and Mountcastle, V. B. (eds.): Handbook of Physiology. Sec. 1, Vol. 1. Baltimore, Williams & Wilkins, 1977, p. 1137.

Kandel, E. R.: A Cell-Biological Approach to Learning. Bethesda, Md., Society for Neuroscience, 1978.

Levinthal, C. F.: The Physiological Approach in Psychology. Englewood Cliffs, N.J., Prentice-Hall, 1979.

Moskowitz, B. A.: The acquisition of language. *Sci. Am.,* 239(5):92, 1978.

Olton, D. S.: Spatial memory. *Sci. Am.,* 236(6):82, 1977.

Russell, I. S., and vanHof, M. W. (eds.): Structure and Function of Cerebral Commissures. Baltimore, University Park Press, 1978.

Smith, J. H. (ed.): Psychoanalysis and Language. New Haven, Conn., Yale University Press, 1978.

Sperry, R. W.: Changing concepts of consciousness and free will. *Perspect. Biol. Med.,* 20:9, 1976.

Uttal, W. R.: The Psychobiology of Mind. New York, Halsted Press, 1978.

37

Activation of the Brain; Wakefulness and Sleep; and Behavioral Functions of the Brain

THE RETICULAR ACTIVATING SYSTEM

In the present chapter we will consider first the functions of the *reticular activating system,* a system that controls the overall degree of central nervous system activity, including control of wakefulness and sleep.

Figure 37–1A illustrates the extent of this system, showing that it begins in the lower brain stem and extends upward through the mesencephalon and thalamus to be distributed throughout the cerebral cortex. Signals are transmitted from the ascending reticular activating system to the cortex by two different pathways. One pathway passes upward from the brain stem portion of the reticular formation to the thalamus and thence through diverse pathways to essentially all parts of the cerebral cortex as well as the basal ganglia. A second and probably less important pathway is through the subthalamic, hypothalamic, and adjacent areas.

Diffuse electrical stimulation in the *mesencephalic, pontile, and upper medullary portions of the reticular formation*—an area discussed in Chapter 34 in relation to the motor functions of the nervous system — causes immediate and marked activation of the cerebral cortex and even causes a sleeping animal to awaken instantaneously. Furthermore, when this "mesencephalic" portion of the reticular formation is damaged severely, as occurs (1) when a *brain tumor* develops in this region, (2) when serious *hemorrhage* occurs, or (3) in diseases such as *encephalitis lethargica* (sleeping sickness), the person passes into coma and is completely nonsusceptible to normal awakening stimuli.

Function of the Mesencephalic Portion of the Reticular Activating System. Electrical stimuli applied to different portions of the reticular activating system have shown that the mesencephalic portion functions quite differently from the thalamic portion. Stimulation of the mesencephalic portion causes generalized activation of the entire brain, including activation of the cerebral cortex, thalamic nuclei, basal ganglia, hypothalamus, other portions of the brain stem, and even the spinal cord. Furthermore, once the mesencephalic portion is stimulated, the degree of activation throughout the nervous system remains high for as long as a half minute or more after the stimulation is over. Therefore, *it is believed that activation of the mesencephalic portion of the reticular activating system is basically responsible for normal wakefulness of the brain.*

Function of the Thalamic Portion of the Activating System. Electrical stimulation in different areas of the thalamic portion of the activating system (if the stimulation is not too strong) activates specific regions of the cerebral cortex more than others. This is distinctly different from stimulation in the mesencephalic portion, which activates all the brain at the same time. Therefore, it is believed that the thalamic portion of the activating system has two specific functions: First, it relays signals from the mesencephalon to all parts of the cerebral cortex to cause generalized activation of the cerebrum; second, stimulation of selected points in the thalamic activating system causes specific activation of

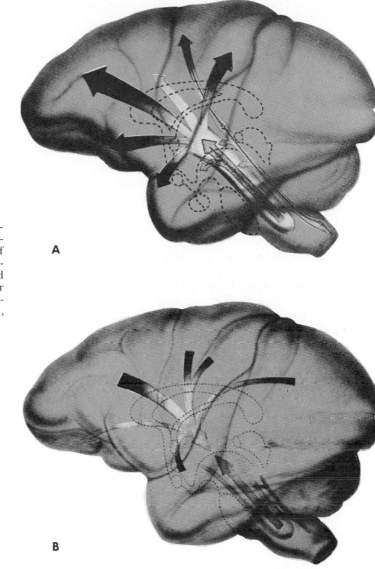

Figure 37–1. (A) The ascending reticular activating system schematically projected on a monkey brain. (From Lindsley: Reticular Formation of the Brain. Little, Brown, and Co.) (B) Convergence of pathways from the cerebral cortex and from the spinal afferent systems on the reticular activating system. (From French, Hernandez-Peon, and Livingston: *J. Neurophysiol.*, *18*:74, 1955.)

certain areas of the cerebral cortex in distinction to the other areas. This selective activation of specific cortical areas possibly or probably plays an important role in our ability to direct our attention to certain parts of our mental activity, which is discussed later in the chapter.

THE AROUSAL REACTION — SENSORY ACTIVATION OF THE RETICULAR ACTIVATING SYSTEM

When an animal is asleep, the level of activity of the reticular activating system is greatly decreased; yet almost any type of sensory signal can immediately activate the system. For instance, proprioceptive signals from the joints and muscles, pain impulses from the skin, visual signals from the eyes, auditory signals from the ears, and even visceral sensations from the gut can all cause sudden activation of the reticular activating system and therefore arouse the animal. This is called the *arousal reaction.*

Some types of sensory stimuli are more potent than others in eliciting the arousal reaction; the most potent are pain and proprioceptive somatic impulses.

Anatomically, the reticular formation of the brain stem is admirably constructed to perform the arousal function. It receives tremendous numbers of signals either directly or indirectly from the *spinoreticular tracts,* the *spinothalamic tracts,* the *spinotectal tracts,* the *auditory tracts,* the *visual tracts,* and others, so that almost any sensory stimulus in the body can activate it. The reticular formation in turn can transmit signals both upward into the brain and downward into the spinal cord. Indeed, many of the fibers originating from cells in the reticular formation divide, with one branch of the fiber passing upward and the other branch passing downward, as was explained in Chapter 34.

THE GENERALIZED THALAMOCORTICAL SYSTEM

At this point it is important that we explain some of the interrelationships between the reticular activating system, the thalamus, and the cerebral cortex. The thalamus is the main entryway for essentially all sensory nervous signals to the cortex, with the single exception of signals from the olfactory system. In Chapters 32 and 33 we have already discussed the transmission of the somatic signals through the thalamus to the cerebral cortex, and we shall see in Chapters 40 and 41 that all the signals from the visual, auditory, and taste systems are also relayed through the thalamus before reaching the cortex. In the previous chapter we discussed the two-way communication between essentially all parts of the cerebral cortex and the thalamus. For all of these functions of the thalamus, there are very definite nuclei called the *specific nuclei of the thalamus;* the combination of these nuclei and their connecting areas in the cortex is called the *specific thalamocortical system.* It is this system of which we usually speak when we discuss the transmission of sensory information into the cerebral cortex.

In addition to the specific thalamocortical system, there is a separate system, partially separated from the specific system, called the *generalized thalamocortical system* or the *diffuse thalamocortical system.* This system is subserved by *diffuse thalamic nuclei* and also by *diffuse fiber projections* from the thalamus to the cortex. In general, the thalamic neurons of this system are not collected into discrete nuclei as are the neurons of the specific system. Instead, they lie mainly between the specific nuclei or on the outer surface of the thalamus. All of these neurons make multiple connections with the specific thalamic nuclei. They also project very small fibers to almost all parts of the cerebral cortex.

The generalized system of the thalamus is continuous with the upper end of the reticular formation in the brain stem and receives much of its input from this source. *Therefore, this generalized system is in reality the thalamic portion of the reticular activating system.*

Effect of the Generalized Thalamic System on Cortical Activity.   Activation of the cortex by the generalized thalamic system is entirely different from activation by the specific system.

Signals from the specific nuclei to the cortex activate mainly layer IV of the cortex, whereas activation of the generalized thalamic system activates mainly layers I and II of the cortex. This latter activation is prolonged; it causes partial depolarization of large numbers of dendrites near the surface of the cortex, which in turn causes a generalized increase in the degree of facilitation of the cortex. When the cortex is thus facilitated by the generalized thalamic system, specific signals that enter the cortex from other sources are exuberantly received.

In summary, the generalized thalamocortical system controls the overall degree of activity of the cortex. It can at times facilitate activity in regional areas of the cortex distinct from the remainder of the cortex. Collateral signals control the level of activity in the specific nuclei of the thalamus, the basal ganglia, the hypothalamus, and other structures of the cerebrum as well.

ATTENTION

So long as a person is awake, he has the ability to direct his attention to specific aspects of his mental environment. Furthermore, his degree of attention can change remarkably from (1) almost no attention at all, to (2) broad attention to almost everything that is going on, or to (3) intense attention to a minute facet of his momentary mental experience.

Unfortunately, the basic mechanisms by which the brain accomplishes its diverse acts of attention are not known. However, a few clues are beginning to fall into place, as follows:

Reticular Activating System Control of Overall Attentiveness.   In exactly the same way that a person can change from a state of sleep to a state of wakefulness, there can be all degrees of wakefulness, from wakefulness in which a person is nonattentive to almost all his surroundings to an extremely high degree of wakefulness in which the person reacts instantaneously to almost any sensory experience. These changes in degree of *overall attentiveness* seem to be caused primarily by changes in activity of the mesencephalic portion of the reticular activating system. Thus, control of the general level of attentiveness is probably exerted by the same mechanism that controls wakefulness and sleep, the control center for which is located in the mesencephalon and upper pons.

Function of the Thalamus in Attention. Stimulation of a single specific area in the thalamic portion of the reticular activating system, when the stimulus intensity is not too strong, activates a specific area of the cerebral cortex. Since the cerebral cortex is one of the most important areas of the brain for conscious awareness of one's surroundings, it is likely that the ability of specific thalamic areas to excite specific cortical regions is one of the mechanisms by which a person can direct his attention to specific aspects of his mental environment, whether they are immediate sensory experiences or stored memories.

BRAIN WAVES AND LEVEL OF BRAIN ACTIVITY

When the brain becomes activated, the neurons exhibit intense electrical activity, and electrical recordings from the surface of the brain or from the outer surface of the head demonstrate this. Both the intensity and patterns of this electrical activity are determined to a great extent by the overall level of excitation of the brain resulting from stimulation by the reticular activating system. The undulations in the recorded electrical potentials shown in Figure 37–2 are called *brain waves,* and the entire record is called an *electroencephalogram* (EEG).

The intensities of the brain waves on the surface of the scalp range from zero to 300 microvolts, and their frequencies range from once every few seconds to 50 or more per second. The character of the waves is highly dependent on the degree of activity of the cerebral cortex, and the waves change markedly between the states of wakefulness and sleep.

Much of the time, the brain waves are irregular, and no general pattern can be discerned in the EEG. However, at other times, distinct patterns do appear. Some of these are characteristic of specific abnormalities of the brain such as epilepsy, which is discussed later. Others occur even in normal persons and can be classified into

Figure 37–3. Replacement of the alpha rhythm by an asynchronous discharge on opening the eyes.

alpha, beta, theta, and *delta waves,* which are all illustrated in Figure 37–2.

Alpha waves are rhythmic waves occurring at a frequency between 8 and 13 per second and are found in the EEG's of almost all normal persons when they are awake in a quiet, resting state of cerebration. During sleep the alpha waves disappear entirely, and when the awake person's attention is directed to some specific type of mental activity, the alpha waves are replaced by asynchronous, higher frequency but lower voltage waves. Figure 37–3 illustrates the effect on the alpha waves of simply opening the eyes in bright light and then closing the eyes again. Note that the visual sensations cause immediate cessation of the alpha waves and that they are replaced by low voltage, asynchronous waves.

Beta waves occur at frequencies of more than 14 cycles per second and as high as 25 and rarely 50 cycles per second. Most beta waves appear during *intense* activation of the central nervous system or during tension.

Theta waves have frequencies between 4 and 7 cycles per second. They occur mainly in the parietal and temporal regions in children and they occur during emotional stress in some adults, particularly during disappointment and frustration. They can often be brought out in the EEG of a frustrated person by allowing him to enjoy some pleasant experience and then suddenly removing this element of pleasure; this often causes approximately 20 seconds of theta waves. These same waves also occur in many brain disorders.

Delta waves include all the waves of the EEG below 3½ cycles per second and sometimes as low as 1 cycle every 2 to 3 seconds. These occur in deep sleep, in infancy, and in very serious organic brain disease. And they occur in the cortex of animals that have had subcortical transections separating the cerebral cortex from the thalamus. Therefore, delta waves can occur locally in the cortex independently of activities in lower regions of the brain.

Effect of Varying Degrees of Cerebral Activity on the Basic Rhythm of the Electroencephalogram. There is a general relationship between the degree of cerebral activity and the average frequency of the electroencephalographic rhythm, the frequency increasing progressively with higher and higher degrees of activity. Delta waves, in general, occur in stupor, surgical anesthesia, and sleep; theta waves in psychomotor

Alpha

Beta

Theta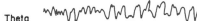

Delta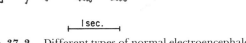

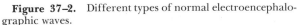

Figure 37–2. Different types of normal electroencephalographic waves.

states and in infants; alpha waves during relaxed states; and beta waves during periods of intense mental activity.

WAKEFULNESS AND SLEEP

The Two Conditions Required for Wakefulness. We have already seen in the discussions earlier in this chapter that wakefulness requires at least a certain level of activity in the reticular activating system. However, a second condition is also necessary for wakefulness to occur: The nervous activity of the brain must be channeled in the proper directions. For instance, during a grand mal epileptic attack a person's brain is many times as active as it is during normal wakefulness, as will be discussed later in the chapter, but nevertheless he is completely unconscious — certainly not a state of wakefulness. Therefore, for wakefulness to occur it is not enough that the brain simply be active.

Two Different Types of Sleep. There are two different ways in which sleep can occur. First, it can result from decreased activity in the reticular activating system; this is called *slow wave sleep* because the brain waves are very slow. Second, sleep can result from abnormal channeling of signals in the brain even though brain activity may not be significantly depressed; this is called *paradoxical sleep.*

Most of the sleep during each night is of the slow wave variety; this is the deep, restful type of sleep that the person experiences after having been kept awake for the previous 24 to 48 hours. On the other hand, short episodes of paradoxical sleep usually occur at intervals during each night, and this type of sleep may have purposeful functions, which will be discussed later.

SLOW WAVE SLEEP

Slow wave sleep is frequently called by different names, such as *deep restful sleep, dreamless sleep, delta wave sleep,* or *normal sleep.* However, we shall see later that paradoxical sleep is also normal and that it has some characteristics of deep sleep.

Electroencephalographic Changes as a Person Falls Asleep. Beginning with wakefulness and proceeding to deep slow wave sleep, the electroencephalogram changes as follows:

1. Alert wakefulness — low voltage, high frequency beta waves showing desynchrony, as illustrated by the second record in Figure 37–2.

2. Quiet restfulness — predominance of alpha waves; a type of "synchronized" brain waves.

3. Light sleep — slowing of the brain waves to theta or delta low voltage variety, but interspersed with spindles of alpha waves called *sleep spindles* that last for a few seconds at a time.

4. Deep slow wave sleep — high voltage delta waves occurring at a rate of 1 to 2 per second, as illustrated by the fourth record of Figure 37–2.

In paradoxical sleep the brain waves change to a still different pattern, as will be discussed below.

Origin of the Delta Waves in Sleep. When the fiber tracts between the thalamus and the cortex are transected, delta waves are generated in the isolated cortex, indicating that this type of wave probably occurs intrinsically in the cortex when it is not being driven from below. Therefore, it is assumed that this is also the origin of the high voltage delta waves during deep slow wave sleep. That is, it is assumed that the degree of activity in the reticular activating system has fallen to a level too low to maintain normal excitability of the cortex, so that the cortex then becomes its own pacemaker.

Characteristics of Deep Slow Wave Sleep. Most of us can understand the characteristics of deep slow wave sleep by referring back to the last time that we were kept awake for more than 24 hours and then remembering the deep sleep that occurred within 30 minutes to an hour after going to sleep. This sleep is exceedingly restful, and it is associated with a decrease in both peripheral vascular tone and also most of the other vegetative functions of the body as well. There is also a 10 to 30 per cent decrease in blood pressure, respiratory rate, and basal metabolic rate.

PARADOXICAL SLEEP ("REM" SLEEP)

In a normal night of sleep, bouts of paradoxical sleep lasting 5 to 20 minutes appear on the average every 90 minutes, the first such period occurring 80 to 100 minutes after the person falls asleep. When the person is extremely tired, the duration of each bout of paradoxical sleep is very short, and it may even be absent. On the other hand, as the person becomes more rested through the night, the duration of the paradoxical bouts greatly increases.

There are several very important characteristics of paradoxical sleep:

1. It is usually associated with active dreaming.

2. The person is even more difficult to arouse than during deep slow wave sleep.

3. The muscle tone throughout the body is exceedingly depressed, indicating strong inhibition of the spinal projections from the reticular activating system.

4. The heart rate and respiration usually be-

come irregular, which is characteristic of the dream state.

5. Despite the extreme inhibition of the peripheral muscles, a few irregular muscle movements occur. These include, in particular, rapid movements of the eyes; consequently, paradoxical sleep has often been called *REM sleep,* for "rapid eye movements."

6. The electroencephalogram shows a desynchronized pattern of low voltage beta waves similar to those that occur during wakefulness. Therefore, this type of sleep is also frequently called *desynchronized sleep,* meaning desynchronized brain waves.

In summary, paradoxical sleep is a type of sleep in which the brain is quite active. However, the brain activity is not channeled in the proper direction for the person to be aware of his surroundings and therefore to be awake.

BASIC THEORIES OF SLEEP AND WAKEFULNESS

Role of the Reticular Activating System in Sleep and Wakefulness. In deep, slow wave sleep, transmission of signals from the reticular activating system to the cortex is greatly diminished, indeed almost absent. Therefore, slow wave sleep presumably results from decreased activity of the reticular activating system.

On the other hand, paradoxical sleep is, as its name implies, paradoxical because some areas of the cerebrum are quite active despite the state of sleep. Therefore, it is assumed that this type of sleep results from a curious mixture of activation of some brain regions and suppression of other regions.

Yet, the questions that must be answered before we will understand wakefulness and sleep are these: (1) What are the mechanisms that activate the reticular activating system during wakefulness? And (2) what suppresses this system during sleep?

Neuronal Centers, Transmitters, and Mechanisms That Cause Wakefulness

Let us quickly review the factors that can cause wakefulness:

1. Widespread stimulation of sensory nerves throughout the body will also cause wakefulness. These nerves transmit strong signals into the mesencephalic portion of the reticular activating system.

2. Stimulation of most areas of the cerebral cortex will also cause a high level of wakefulness. These areas also transmit strong signals into both the mesencephalic and thalamic portions of the reticular activating system.

3. Stimulation in certain regions of the hypothalamus, especially in the lateral regions, can also cause extreme degrees of wakefulness. Here again, strong signals are known to be transmitted into the reticular activating system.

4. It is believed that excitation of the area in the pons called the *locus ceruleus* is especially important in maintaining activity in the reticular activating system. The nerve fibers from this area are distributed widely throughout other portions of the reticular formation and throughout almost all areas of the diencephalon and cerebrum as well. They all secrete *norepinephrine* at their endings. It is believed that this norepinephrine in some way plays a role in the wakefulness process.

Effect of Lesions in the Wakefulness Areas. Lesions in the mesencephalic portion of the reticular activating system or in fiber pathways leading upward from this area through the diencephalon, if large enough, will invariably lead to coma from which the person cannot be aroused with any type of stimuli.

Also, very discrete lesions located bilaterally in the locus ceruleus will cause a type of sleep that closely resembles natural sleep.

Neuronal Centers, Transmitters, and Mechanisms That Can Cause Sleep

Stimulation of several specific areas of the brain can produce sleep with characteristics very near to those of natural sleep. Some of these are the following:

1. The most conspicuous stimulation area for causing almost natural sleep is the raphe nuclei in the pons and medulla. These are a thin sheet of nuclei located in the midline. Nerve fibers from these nuclei spread widely in the reticular formation and also upward into the thalamus, hypothalamus, and most areas of the limbic cortex. In addition, they extend downward into the spinal cord, terminating in the posterior horns, where they can inhibit incoming pain signals, as was discussed in Chapter 33. It is also known that the endings of fibers from these raphe neurons secrete *serotonin.* Therefore, it is assumed that serotonin is the major transmitter substance associated with production of sleep.

2. Stimulation of several other regions in the lower brain stem and diencephalon can also lead to sleep, including (a) the rostral part of the hypothalamus, mainly in the suprachiasmal area, and (b) an occasional area in the diffuse nuclei of the thalamus.

Effect of Lesions in the Sleep-Promoting Centers. Discrete lesions in the raphe nuclei lead to a high state of wakefulness. This is also true of bilateral lesions in the mediorostral suprachiasmal portion of the anterior hypothalamus. In both instances, the reticular activating system seems to become released from inhibition. Indeed, the lesions of the anterior hypothalamus can sometimes cause such intense wakefulness that the animal actually dies of exhaustion.

The Cycle Between Sleep and Wakefulness

The above discussions have merely identified neuronal areas, transmitters, and mechanisms that are related either to wakefulness or to sleep. However, they have not explained the rhythmic, reciprocal operation of the sleep-wakefulness cycle. It is quite possible that this is caused by a free running intrinsic oscillator within the brain stem that cycles back and forth between the sleep and wakefulness centers, the wakefulness centers presumably activating the reticular activating system, and the sleep centers inhibiting this system.

However, there is much reason to believe that feedback signals from the cerebral cortex and also from the peripheral nerve receptors might also play a very important role in causing the sleep-wakefulness rhythm. One reason for believing this is that interruption of all the sensory nerve tracts leading from the periphery to the reticular activating system will cause the animal to go to sleep permanently. Also, as explained above, stimulation of the cerebral cortex will cause powerful activation of the reticular activating system.

Therefore, a very likely mechanism for causing the rhythmicity of the sleep-wakefulness cycle is the following:

When the reticular activating system is completely rested and the sleep centers are not activated, the wakefulness centers then presumably begin spontaneous activity. This in turn excites both the cerebral cortex and the peripheral nervous system. Next, positive feedback signals come from both of these areas back to the reticular activating system to activate it still further. Thus, once the wakefulness state begins, it has a natural tendency to sustain itself.

However, after the brain remains activated for many hours, even the neurons within the activating system presumably will fatigue, or other factors might activate the sleep centers. Consequently, the positive feedback cycle between the reticular activating system and the cortex, and also that between the reticular activating system and the periphery, will begin to fade. As soon as a few of the neurons in the reticular activating system become inactive, this also eliminates part of the feedback stimulus to the other neurons as well. Therefore, these too become inactive, the process spreading rapidly through the neurons and leading to rapid transition from the wakefulness state to the sleep state.

Then, one could postulate that during sleep the excitatory neurons of the reticular activating system gradually become more and more excitable because of the prolonged rest, while the inhibitory neurons of the sleep centers become less excitable, thus leading to a new cycle of wakefulness.

This theory obviously can explain the rapid transitions from sleep to wakefulness and from wakefulness to sleep. It can also explain arousal, the insomnia that occurs when a person's mind becomes preoccupied with a thought, the wakefulness that is produced by bodily activity, and many other conditions that affect the person's state of sleep or wakefulness.

PHYSIOLOGICAL EFFECTS OF SLEEP

Sleep causes two major types of physiological effects: first, effects on the nervous system itself and, second, effects on other structures of the body. The first seems to be by far the more important, for any person whose spinal cord has been transected in the neck shows no physiological effects in the body below the level of transection that can be attributed to a sleep and wakefulness cycle; this lack of sleep and wakefulness causes neither significant harm to the bodily organs nor even any deranged function. On the other hand, lack of sleep certainly does affect the functions of the central nervous system.

Prolonged wakefulness is often associated with progressive malfunction of the mind and behavioral activities of the nervous system. We are all familiar with the increased sluggishness of thought that occurs toward the end of a prolonged wakeful period, but in addition, a person can become irritable or even psychotic following forced wakefulness for prolonged periods of time. Therefore, we can assume that sleep in some way not presently understood restores both normal sensitivities of, and normal "balance" between, the different parts of the central nervous system.

Even though wakefulness and sleep have not been shown to be necessary for somatic functions of the body, the cycle of enhanced and depressed nervous excitability that follows along with the cycle of wakefulness and sleep does have moderate effects on the peripheral body. For instance, there is enhanced sympathetic activity during wakefulness and also enhanced numbers of impulses to the skeletal musculature to increase muscular tone. Conversely, during sleep, sympathetic activity decreases while parasympathetic activity occasionally increases, and the muscular tone becomes almost nil. Therefore, during sleep, arterial blood pressure falls, pulse rate decreases, skin vessels dilate, activity of the gastrointestinal tract sometimes increases, muscles fall into a completely relaxed state, and the overall basal metabolic rate of the body falls by 10 to 30 per cent.

EPILEPSY

Epilepsy is characterized by uncontrolled excessive activity of either a part of the central nervous system or all of it. A person who is predisposed to epilepsy has attacks when the basal level of excitability of his nervous system (or of the part that is susceptible to the epileptic state) rises above a certain critical threshold. But, as long as the degree of excitability is held below this threshold, no attack occurs.

Two important types of epilepsy are grand mal epilepsy and petit mal epilepsy.

Grand Mal Epilepsy. Grand mal epilepsy is characterized by extreme neuronal discharges originating in the mesencephalic portion of the reticular activating system. These spread throughout the entire central nervous system, to the cortex, to the deeper parts of the brain, and even into the spinal cord to cause generalized *tonic convulsions* of the entire body; toward the end of the attack these change to alternating muscular contractions called tonic-clonic *convulsions.* The grand mal seizure lasts from a few seconds to as long as three to four minutes and is characterized by post-seizure depression of the entire nervous system; the person remains in stupor for one to many minutes after the attack is over and then often remains severely fatigued for many hours thereafter.

The middle recording of Figure 37-4 illus-

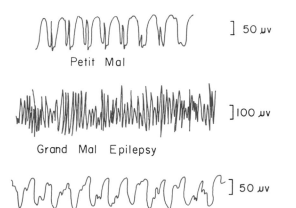

Figure 37-4. Electroencephalograms in different types of epilepsy.

trates a typical electroencephalogram from almost any region of the cortex during a grand mal attack. This illustrates that high voltage synchronous discharges occur over the entire cortex and have almost the same periodicity as the normal alpha waves. Furthermore, the same type of discharge occurs on both sides of the brain at the same time, illustrating that the origin of the abnormality is in the lower centers of the brain that control the activity of the cerebral cortex and not in the cerebral cortex itself.

In experimental animals or even in human beings, grand mal attacks can be initiated by administering neuronal stimulants, such as the drug Metrazol, or they can be caused by insulin hypoglycemia or by the passage of alternating electrical current directly through the brain. Furthermore, even after transection of the brain stem just above the mesencephalon, a typical grand mal seizure can still be induced in the portion of the brain stem beneath the transection.

Presumably, therefore, a grand mal attack is caused by intrinsic overexcitability of the neurons that make up the mesencephalic portion of the reticular activating system or by some abnormality of the local neuronal pathways. The synchronous discharges from this region could result from local reverberating circuits.

One might ask: What stops the grand mal attack after a given time? This is believed to result from (1) *fatigue of the neurons* involved in precipitating the attack and (2) *active inhibition* by certain structures of the brain. The stupor and fatigue that occur after a grand mal seizure is over are believed to result from the intense fatigue of the neurons following their intensive activity during the grand mal attack.

Petit Mal Epilepsy. Petit mal epilepsy is closely allied to grand mal epilepsy in that it too almost certainly originates in the reticular activating system. It is characterized by 3 to 30 seconds of unconsciousness during which the person has several twitchlike contractions of the muscles, usually in the head region — especially blinking of the eyes; this is followed by return of consciousness and resumption of previous activities. The patient may have one such attack in many months, or in rare instances he may have a rapid series of attacks, one following the other. However, the usual course is for the petit mal attacks to appear in late childhood and then to disappear entirely by the age of 30.

The brain wave pattern in petit mal epilepsy is illustrated by the upper record of Figure 37-4, which is typified by a *spike and dome pattern.* The spike portion of this recording is almost identical with the spikes that occur in grand mal epilepsy, but the dome portion is distinctly different. The spike and dome can be recorded over the entire cerebral cortex, illustrating that the seizure originates in the reticular activating system of the brain.

BEHAVIORAL FUNCTIONS OF THE BRAIN: THE LIMBIC SYSTEM AND ROLE OF THE HYPOTHALAMUS

Behavior is a function of the entire nervous system, not of any particular portion. Even the discrete cord reflexes are an element of behavior, and the wakefulness and sleep cycle discussed earlier in this chapter is certainly one of the most important of our behavioral patterns. However, in this section we will deal with those special types of behavior associated with emotions, subconscious motor and sensory drives, and the intrinsic feelings of pain and pleasure. These functions of the nervous system are performed mainly by structures located in the basal regions of the brain. This overall group of brain structures is frequently called the *limbic system*.

Portions of the limbic system, especially the hypothalamus and related structures, control

many of the internal functions of the body, such as body temperature, osmolality of the body fluids, and body weight; these functions are collectively called *vegetative functions* of the body.

THE LIMBIC SYSTEM

Figure 37–5 illustrates the anatomical structures of the limbic system, showing them to be an interconnected complex of basal brain elements. Located in the midst of them is the *hypothalamus*, which is considered by many anatomists to be a separate structure from the remainder of the limbic system but which, from a physiological point of view, is one of the central elements of the system. Figure 37–6 illustrates schematically this key position of the hypothalamus in the limbic system and shows that surrounding it are the other subcortical structures of the limbic system, and surrounding the subcortical structures is the *limbic cortex* composed of a ring of cerebral cortex (1) beginning in the *orbitofrontal area* on the ventral surface of the frontal lobes, (2) extending upward in front of and over the corpus callosum onto the medial aspect of the cerebral hemisphere to the *cingulate gyrus*, and finally (3) passing posterior to the corpus callosum and down-

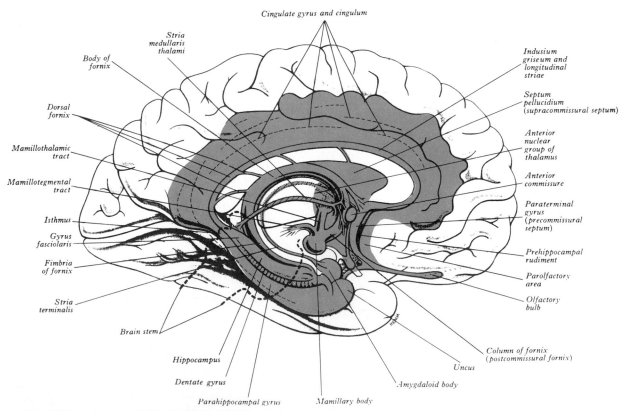

Figure 37–5. Anatomy of the limbic system illustrated by the shaded areas of the figure. (From Warwick and Williams: Gray's Anatomy, 35th Brit. Ed. London, Longman Group Ltd., 1973.)

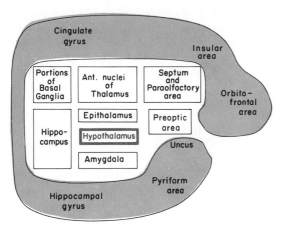

Figure 37–6.   The limbic system.

ward onto the ventromedial surface of the temporal lobe to the *hippocampal gyrus, pyriform area,* and *uncus.* Thus, on the medial and ventral surfaces of each cerebral hemisphere is a ring of paleocortex that surrounds a group of deep structures intimately associated with overall behavior and with emotions.

VEGETATIVE FUNCTIONS OF THE HYPOTHALAMUS

The hypothalamus provides the most important output pathway through which the limbic system controls many major functions of the body, especially the vegetative functions, which are the involuntary functions necessary for living. The different hypothalamic centers are so important in these controls that they are discussed in

detail throughout this entire book in association with the specific functional systems, such as arterial pressure regulation in Chapter 17, thirst and water conservation in Chapter 25, and temperature regulation in Chapter 47. However, to illustrate the organization of the hypothalamus as a functioning unit, Figure 37–7 summarizes most of the vegetative functions of the hypothalamus and shows, with the exception of the lateral hypothalamic areas that are not shown but that overlie the areas illustrated, the major nuclei or major areas which when stimulated affect respective vegetative activities. Some of these functions include (1) regulation of heart rate and arterial pressure, (2) regulation of body temperature, (3) regulation of body fluid osmolarity, (4) regulation of food intake, and (5) regulation of the secretion of the pituitary hormones.

BEHAVIORAL FUNCTIONS OF THE LIMBIC SYSTEM

Pleasure and Pain; Reward and Punishment

In recent years, it has been learned that many hypothalamic and other limbic structures are particularly concerned with the affective nature of sensory sensations — that is, with whether the sensations are *pleasant* or *painful.* These affective qualities are also called *reward* and *punishment* or *satisfaction* and *aversion.* Electrical stimulation of certain regions pleases or satisfies the animal, whereas electrical stimulation of other regions causes extreme pain, fear, defense, escape reactions, and all the other elements of punishment.

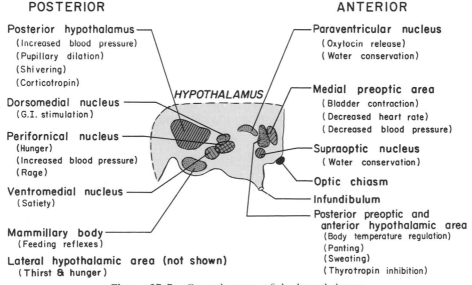

Figure 37–7.   Control centers of the hypothalamus.

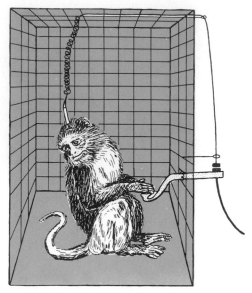

Figure 37–8. Technique for localizing reward and punishment centers in the brain of a monkey.

Obviously, these two oppositely responding systems greatly affect the behavior of the animal.

Reward Centers. Figure 37–8 illustrates a technique that has been used for localizing the specific reward and punishment areas of the brain. In this technique a lever is placed at the side of the cage and is arranged so that depressing the lever makes electrical contact with a stimulator. Electrodes are placed successively in different areas in the brain so that the animal can stimulate the area by pressing the lever. If stimulating the particular area gives the animal a sense of reward, then it will press the lever again and again, sometimes as much as 10,000 times per hour. Furthermore, when offered the choice of eating some delectable food or stimulating the reward center, the animal often chooses the electrical stimulation.

By use of this procedure, the major reward centers have been found to be located in the septum and hypothalamus, primarily *along the course of the medial forebrain bundle* and *in both the lateral and the ventromedial nuclei of the hypothalamus.* Less potent reward centers, which are probably secondary to the major ones in the septum and hypothalamus, are found in the amygdala, certain areas of the thalamus and basal ganglia, and finally extending downward into the basal tegmentum of the mesencephalon.

Punishment Centers. The apparatus illustrated in Figure 37–8 can also be connected so that pressing the lever turns off rather than turns on the electrical stimulus. In this case, the animal will not turn the stimulus off when the electrode is in one of the reward areas, but when it is in

certain other areas he immediately learns to turn it off. Stimulation in these areas causes the animal to show all the signs of pain, displeasure, and punishment. Furthermore, prolonged stimulation for 24 hours or more causes the animal to become severely sick and actually leads to death.

By means of this technique, the principal centers for pain, punishment, and escape tendencies have been found in the *central gray area surrounding the aqueduct of Sylvius in the mesencephalon* and extending upward into the *periventricular structures of the hypothalamus and thalamus.*

It is particularly interesting that stimulation in the pain and punishment centers can frequently inhibit the reward and pleasure centers completely, illustrating that pain can take precedence over pleasure and reward.

Importance of Reward and Punishment in Behavior. Almost everything that we do depends on reward or punishment. If we are doing something that is rewarding, we continue to do it; if it is punishing, we cease to do it. Therefore, the reward and punishment centers undoubtedly constitute one of the most important of all the controllers of our bodily activities, our motivations, and so forth.

Importance of Reward and Punishment in Learning and Memory — Habituation or Reinforcement. Animal experiments have shown that a sensory experience causing neither reward nor punishment is remembered hardly at all. Electrical recordings have shown that new and novel sensory stimuli always excite the cerebral cortex. But repetition of the stimulus over and over leads to almost complete extinction of the cortical response if the sensory experience does not elicit either a sense of reward or punishment. Thus, the animal becomes *habituated* to the sensory stimulus, and thereafter ignores this stimulus.

If the stimulus causes either reward or punishment rather than indifference, however, the cortical response becomes progressively more and more intense with repetitive stimulation instead of fading away, and the response is said to be *reinforced.* Thus, an animal builds up strong memory traces for sensations that are either rewarding or punishing but, on the other hand, develops complete habituation to indifferent sensory stimuli. Therefore, it is evident that the reward and punishment centers of the midbrain have much to do with selecting the information that we learn.

Effect of Tranquilizers on the Reward and Punishment Centers. Administration of a tranquilizer, such as chlorpromazine, inhibits both the reward and punishment centers, thereby greatly decreasing the affective reactivity of the animal. Therefore, it is presumed that tranquilizers function in psychotic states by suppressing many of the important behavioral areas of the hypothalamus and its associated regions of the brain.

The Affective-Defensive Pattern — Rage

An emotional pattern that involves the hypothalamus and has been well characterized is the affective-defensive pattern. It can be described as follows:

Stimulation of the punishment centers of the brain, especially the *periventricular areas of the hypothalamus,* which are also the hypothalamic regions that give the most intense sensation of punishment, causes the animal to (1) develop a defense posture, (2) extend its claws, (3) lift its tail, (4) hiss, (5) spit, (6) growl, and (7) develop piloerection, wide-open eyes, and dilated pupils. Furthermore, even the slightest provocation causes an immediate savage attack. This is approximately the behavior that one would expect from an animal being severely punished, and it is a pattern of behavior that is called *rage.*

Placidity and Tameness. Exactly the opposite emotional behavioral pattern occurs when the reward centers are stimulated: placidity and tameness.

PSYCHOSOMATIC EFFECTS OF THE HYPOTHALAMUS AND RETICULAR ACTIVATING SYSTEM

We are all familiar with the fact that abnormal function in the central nervous system can frequently lead to serious dysfunction of the different somatic organs of the body. Some of the ways in which this can occur are the following:

Transmission of Psychosomatic Effects Through the Autonomic Nervous System. Many psychosomatic abnormalities result from hyperactivity of either the sympathetic or parasympathetic system. The usual effects of sympathetic hyperactivity are (1) increased heart rate — sometimes with palpitation of the heart, (2) increased arterial pressure, (3) constipation, and (4) increased metabolic rate. On the other hand, parasympathetic signals are likely to be much more focal. For instance, signals transmitted to specific areas in the dorsal motor nuclei of the vagus nerves can cause more or less specifically (1) increased or decreased heart rate and palpitation of the heart, (2) esophageal spasm, (3) increased peristalsis in the upper gastrointestinal tract, or (4) increased hyperacidity of the stomach with resultant development of peptic ulcer. Stimulation of the sacral region of the parasympathetic system, on the other hand, is likely to cause

extreme colonic glandular secretion and peristalsis with resulting diarrhea. One can readily see, then, that emotional patterns controlling the sympathetic and parasympathetic centers of the hypothalamus can cause wide varieties of peripheral psychosomatic effects.

Psychosomatic Effects Transmitted Through the Anterior Pituitary Gland. Electrical stimulation of the posterior hypothalamus increases the secretion of corticotropin by the anterior pituitary gland and therefore indirectly increases the output of adrenocortical hormones. One of the effects of this is a gradual increase in stomach hyperacidity because of the effect of glucocorticoids on stomach secretion. Over a prolonged period of time this obviously could lead to peptic ulcer, which is a well-known effect of hypersecretion by the adrenal cortex. Likewise, activity in the anterior hypothalamus increases the pituitary secretion of thyrotropin, which in turn increases the output of thyroxine and leads to an elevated basal metabolic rate. It is well known that different types of emotional disturbances can lead to thyrotoxicosis (as will be explained in Chapter 50), presumably resulting from overactivity in the anterior hypothalamus.

From these examples, therefore, it is evident that many types of psychosomatic diseases of the body can be caused by abnormal control of anterior pituitary secretion.

FUNCTIONS OF THE AMYGDALA

The amygdala is a complex of nuclei located immediately beneath the medial surface of the cerebral cortex in the pole of each temporal lobe. It receives impulses from all portions of the limbic cortex, from the orbital surfaces of the frontal lobes, from the cingulate gyrus, and from the hippocampal gyrus. In turn, it transmits signals (1) back into these same cortical areas, (2) into the hippocampus, (3) into the septum, (4) into the thalamus, and (5) especially into the hypothalamus.

Effects of Stimulating the Amygdala. In general, stimulation of the amygdala can cause almost all the same effects as those elicited by stimulation of the hypothalamus, plus still other effects including (1) tonic movements, such as raising the head or bending the body, (2) circling movements, (3) occasionally clonic, rhythmic movements, and (4) different types of movements associated with olfaction and eating, such as licking, chewing, and swallowing. And stimulation can alter respiration or at other times stop all movements of an animal, freezing the animal in its present postural state, a phenomenon called the *arrest reaction.*

In addition, stimulation of certain amygdaloid nuclei can rarely cause a pattern of rage, escape, punishment, and pain similar to the affective-defense pattern elicited from the hypothalamus as described above. And stimulation of other nuclei can give reactions of reward and pleasure.

Finally, excitation of still other portions of the amygdala can cause sexual activities that include erection, copulatory movements, ejaculation, ovulation, uterine activity, and premature labor.

In short, stimulation of appropriate portions of the amygdaloid nuclei can produce almost any pattern of behavior. It is believed that the normal function of the amygdaloid nuclei is to help control the overall pattern of behavior demanded for each social or environmental occasion.

FUNCTIONS OF THE HIPPOCAMPUS

The hippocampus is an elongated structure composed of a modified type of cerebral cortex. It folds inward to form the ventral surface of the inferior horn of the lateral ventricle. One end of the hippocampus abuts against the amygdaloid nuclei, and it also fuses along one of its borders with the hippocampal gyrus, which is the cortex of the ventromedial surface of the temporal lobe. The hippocampus has numerous connections with almost all parts of the limbic system, including especially the amygdala, hippocampal gyrus, cingulate gyrus, hypothalamus, and other areas closely related to the hypothalamus.

One of the most remarkable effects of hippocampal stimulation in conscious man is immediate *loss of contact* with any person with whom he might be talking, indicating that the hippocampus can play a role in determining a person's attention.

A special feature of the hippocampus is that weak electrical stimuli can cause local epileptic seizures in this region. These seizures cause various psychomotor effects including olfactory, visual, auditory, tactile, and other types of hallucinations that are uncontrollable even though the person has not lost consciousness and even though he knows the hallucinations to be unreal.

Theoretical Function of the Hippocampus in Learning. The hippocampus originated as part of the olfactory cortex. In the very lowest animals it plays essential roles in determining whether the animal will eat a particular food, whether the smell of a particular object suggests danger, or whether the odor is sexually inviting and in making other decisions that are of life and death importance. Thus, very early in the development of the brain, the hippocampus presumably be-

came the critical decision-making neuronal mechanism, determining the importance and type of incoming sensory signals. Presumably, as the remainder of the brain developed, the connections from the other sensory areas into the hypothalamus continued to utilize this decision-making capability.

In the previous chapter, it was pointed out that reward and punishment play a major role in determining the importance of information and especially whether or not the information will be stored in memory. A person rapidly becomes habituated to indifferent stimuli but learns assiduously any sensory experience that causes either pleasure or punishment. Yet, what is the mechanism by which this occurs? It has been suggested that the hippocampus acts as the encoding mechanism for translating short-term memory into long-term memory—that is, it translates the short-term memory signals into an appropriate form so that they can be stored in long-term memory and also transmits an additional signal to the long-term memory storage area directing that storage shall take place.

Whatever the mechanism, without the hippocampi *consolidation* of long-term memories does not take place. This is especially true for verbal information, perhaps because the temporal lobes, in which the hippocampi are located, are particularly concerned with verbal information.

FUNCTION OF THE LIMBIC CORTEX

Probably the most poorly understood portion of the entire limbic system is the ring of cerebral cortex, called the *limbic cortex,* that surrounds the subcortical limbic structures. The limbic cortex is among the oldest of all parts of the cerebral cortex. In lower animals it plays a major role in various olfactory, gustatory, and feeding phenomena. However, in the human being, these functions of the limbic cortex are of minor importance. Instead, the limbic cortex of the human being is believed to be the cerebral association cortex for control of the lower centers that have to do primarily with behavior.

Yet, we find ourselves still perplexed regarding the function of the cortical regions of the limbic system. The probable reason is that these regions correlate information from many sources but cause no direct, overt effects that can be observed objectively. Thus, in the insular and anterior temporal cortex we find gustatory and olfactory associations that affect behavior. In the hippocampal gyrus, there is a tendency for auditory associations with behavior. In the cingulate cortex, there is some reason to believe that sensorimotor control of behavior occurs. And, finally, the orbitofrontal cortex presumably acts as a bridge between behavior and the analytical functions of the prefrontal lobes.

Therefore, until further information is available, it is perhaps best to state that the cortical regions of the limbic system occupy intermediate associative positions between the functions of the remainder of the cerebral cortex and the functions of the lower centers for control of behavioral patterns. Thus, thoughts or other stimuli that excite portions of the limbic cortex can probably elicit almost any type of behavior that is appropriate for the occasion.

FUNCTION OF SPECIFIC CHEMICAL TRANSMITTER SYSTEMS FOR BEHAVIOR CONTROL

Recently it has become apparent that some of the synaptic chemical transmitter substances of the brain play especially important roles in behavior. Though this field is only beginning to be explored, the chemical transmitter systems that seem to be especially important are the following:

The Norepinephrine System. Earlier in the chapter it was pointed out that large numbers of norepinephrine-secreting neurons are located in the reticular formation, especially in the locus ceruleus, and that they send fibers upward through the reticular activating system to essentially all parts of the diencephalon and cerebrum.

It is believed that overstimulation of the norepinephrine system is the cause of the *manic phase of the manic-depressive psychosis.* And, conversely, there is much reason to believe that depressed activity of this system, along with depressed activity of the serotonin system to be described below, is the basis of the depression stage of the manic-depressive syndrome.

The Dopamine System. In Chapter 35 the dopaminergic neurons located in the substantia nigra and projecting to the striate portion of the basal ganglia were discussed. This system exerts an important continuous restraint on basal ganglia activity.

It was also pointed out in Chapter 35 that destruction of the substantia nigral dopamine system also causes Parkinson's disease. However, during treatment of Parkinson's disease with L-dopa, patients sometimes develop schizophrenic symptoms, indicating that excess dopaminergic activity can cause dissociation of a person's drives and thought patterns.

The Serotonin System. Located in the medial raphe nuclei of the medulla is a system of serotonin-secreting neurons. Many of the nerve fibers of this system spread downward into the cord where they reduce the input level of pain signals, as was explained in Chapter 33. However, the fibers also spread upward into the reticular formation and into other basal areas of the brain to suppress the activity of the reticular activating system, as well as other brain activity. Thus, this system can promote sleep, as was discussed earlier in the chapter. The drug LSD (lysergic acid diethylamide) is known to act as an antagonist against some functions of serotonin. Its propensity for causing dissociation of both behavioral and mental activities is well known.

The Enkephalin-Endorphin System. Very little is yet known about the behavioral functions of the enkephalin-endorphin system. However, the probable role of this system for suppression of pain was discussed in Chapter 33. It presumably has many other activities as well, because it is secreted in many areas of the brain stem and thalamus.

REFERENCES

Activation of the Brain, and Wakefulness

Arkin, J. S., *et al.* (eds.): The Mind in Sleep. New York, Halsted Press, 1978.

Block, G. D., and Page, T. L.: Circadian pacemakers in the nervous system. *Annu. Rev. Neurosci., 1*:19, 1978.

Buser, P. A., and Rougeul-Buser, A. (eds.): Cerebral Correlates of Conscious Experience, New York, Elsevier/North-Holland, 1978.

Drucker-Colin, R., *et al.* (eds.): The Functions of Sleep. New York, Academic Press, 1979.

Edelman, G. M., and Mountcastle, V. B.: The Mindful Brain. Cambridge, Mass., MIT Press, 1978.

Enright, J. T.: The Timing of Sleep and Wakefulness: On the Substructure and Dynamics of the Circadian Pacemakers Underlying the Wake-Sleep Cycle. New York, Springer-Verlag, 1979.

Gillin, J. C., *et al.*: The neuropharmacology of sleep and wakefulness. *Annu. Rev. Pharmacol. Toxicol., 18*:563, 1978.

Hector, M. L.: EEG Recording. Boston, Butterworths, 1979.

Hobson, J. A., and Brazier, M. A. B. (eds.): The Reticular Formation Revisited: Specifying Function for a Nonspecific System. New York, Raven Press, 1980.

Ito, M., *et al.* (eds.): Integrative Control Functions of the Brain. New York, Elsevier/North-Holland, 1978.

Klass, D. W., and Daly, D. C. (eds.): Current Practice of Clinical Electroencephalography. New York, Raven Press, 1979.

Livingston, R. B.: Sensory Processing, Perception, and Behavior. New York, Raven Press, 1978.

Moore, R. Y., and Bloom, F. E.: Central catecholamine neuron systems: Anatomy and physiology of the norepinephrine and epinephrine systems. *Annu Rev. Neurosci., 2*:113, 1979.

Newmark, M. E., and Penry, J. K.: Genetics of Epilepsy: A Review. New York, Churchill Livingstone, 1980.

O'Keefe, J., and Nadel, L.: The Hippocampus as a Cognitive Map. New York, Oxford University Press, 1978.

Passouant, P., and Oswald, I. (eds.): Pharmacology of the States of Alertness. New York, Pergamon Press, 1979.

Plum, F., and Posner, J. B.: The Diagnosis of Stupor and Coma, 3rd Ed. Philadelphia, F. A. Davis, 1980.

Prince, D. A.: Neurophysiology of epilepsy. *Annu. Rev. Neurosci., 1*:395, 1978.

Stern, R. M., *et al.*:Psychophysiological Recording. New York, Oxford University Press, 1980.

Behavioral Mechanisms

Akiskal, H. (ed.): Affective Disorders: Special Clinical Forms. Philadelphia, W. B. Saunders Co., 1979.

Bentley, D., and Konishi, M.: Neural control of behavior. *Annu. Rev. Neurosci., 1*:35, 1978.

Bowsher, D.: Mechanisms of Nervous Disorder: An Introduction. Philadelphia, J. B. Lippincott Co., 1978.

CIBA Foundation Symposium: Functions of the Septo-Hippocampal System. New York, Elsevier/Excerpta Medica/North-Holland, 1978.

Cooper, D. G.: The Language of Madness. London, Allen Lane, 1978.

Cotman, C. W., and McGaugh, J. L.: Behavioral Neuroscience. New York, Academic Press, 1979.

Depue, R. A. (ed.): The Psychobiology of the Depressive Disorders: Implications For the Effects of Stress. New York, Academic Press, 1979.

Ehrlich, Y. H. (ed.): Modulators, Mediators, and Specifiers in Brain Function. New York, Plenum Press, 1979.

Fink, G., and Geffen, L. B.: The hypothalamo-hypophysial system: Model for central peptidergic and monoaminergic transmission. *In* Porter, R. (ed.): International Review of Physiology: Neurophysiology III. Vol. 17. Baltimore, University Park Press, 1978, p. 1.

Gershon, M. D.: Biochemistry and physiology of serotonergic transmission. *In* Brookhart, J. M., and Mountcastle, V. B. (eds.): Handbook of Physiology. Sec. 1, Vol. 1. Baltimore, Williams & Wilkins, 1977, p. 573.

Guillemin, R.: Neuroendocrine interrelations. *In* Bondy, P. K., and Rosenberg, L. E. (eds.): Metabolic Control and Disease, 8th Ed. Philadelphia, W. B. Saunders Co., 1980, p. 1155.

Jeffcoate, S. L., and Hutchinson, J. S. M. (eds.): The Endocrine Hypothalamus. New York, Academic Press, 1978.

Jones, N. B., and Reynolds, V.: Human Behaviour and Adaptation. New York, Halsted Press, 1978.

Kelly, D.: Anxiety and Emotions: Physiological Basis and Treatment. Springfield, Ill., Charles C Thomas, 1979.

Lederis, K., and Veale, W. L. (eds.): Current Studies of Hypothalamic Function, 1978. New York, S. Karger, 1978.

Livingston, K. E., and Hornykiewicz, O. (eds.): Limbic Mechanism. New York, Plenum Press, 1978.

McFadden, D. (ed.): Neural Mechanisms in Behavior. New York, Springer-Verlag, 1980.

Morgane, P. J., and Panksepp, J. (eds.): Handbook of the Hypothalamus. New York, Marcel Dekker, 1979.

Norrsel, U.: Behavioral studies of the somatosensory system. *Physiol Rev., 60*:327, 1980.

Routtenberg, A.: The reward system of the brain. *Sci. Am., 239* (5):154, 1978.

Sachar, E. J., and Baron, M.: The biology of affective disorders. *Annu. Rev. Neurosci., 2*:505, 1979.

Snyder, S.: Biological Aspects of Mental Disorder. New York, Oxford University Press, 1980.

Sowers, J. R. (ed.): Hypothalamic Hormones. Stroudsburg, Pa., Dowden, Hutchinson & Ross, 1980.

Tolis, G., *et al.* (eds.): Clinical Neuroendocrinology. New York, Raven Press, 1979.

Van Ree, J. M., and Terenius, L. (eds.): Characteristics and Function of Opioids. New York, Elsevier/North Holland, 1978.

Wassermann, G. D.: Neurobiological Theory of Psychological Phenomena. Baltimore, University Park Press, 1978.

Weiner, R. I., and Ganong, W. F.: Role of brain monoamines and histamine in regulation of anterior pituitary secretion. *Physiol. Rev., 58*:905, 1978.

38

The Autonomic Nervous System; The Adrenal Medulla

The portion of the nervous system that controls the visceral functions of the body is called the *autonomic nervous system*. This system helps to control arterial pressure, gastrointestinal motility and secretion, sweating, body temperature, and many other activities, some of which are controlled almost entirely by the autonomic nervous system and some only partially.

GENERAL ORGANIZATION OF THE AUTONOMIC NERVOUS SYSTEM

The autonomic nervous system is activated mainly by centers located in the *spinal cord, brain stem,* and *hypothalamus.* Also, all of the limbic system can transmit signals to the lower centers and in this way influence autonomic control. Often the autonomic nervous system operates by means of *autonomic reflexes.* That is, sensory signals from peripheral nerve receptors send signals into the centers of the cord, brain stem, or hypothalamus, and these in turn transmit appropriate reflex responses back to the peripheral organs or tissues to control their activities.

The autonomic signals are transmitted to the body through two major subdivisions, called the *sympathetic* and *parasympathetic systems,* the characteristics and functions of which follow.

PHYSIOLOGIC ANATOMY OF THE SYMPATHETIC NERVOUS SYSTEM

Figure 38–1 illustrates the general organization of the sympathetic nervous system, showing one of the two *sympathetic chains* found respectively on each side of the spinal column, with nerves extending to the different internal organs. The sympathetic nerves originate in the spinal cord between the segments T-1 and L-2.

Preganglionic and Postganglionic Sympathetic Neurons.   The sympathetic nerves are different from skeletal motor nerves in the following way: Each skeletal motor pathway to a skeletal muscle is composed of a single fiber originating in the cord. On the other hand, each sympathetic pathway is composed of a *preganglionic fiber* and a *postganglionic fiber.* The cell body of the pregangli-

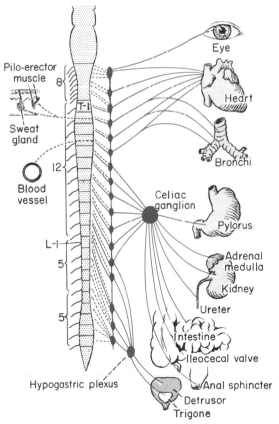

Figure 38–1. The sympathetic nervous system. Dashed lines represent postganglionic fibers in the gray rami leading into the spinal nerves for distribution to blood vessels, sweat glands, and pilo-erector muscles.

onic fiber lies in the intermediolateral horn of the spinal cord and then passes through an *anterior root* of the cord into a *spinal nerve*. After traveling for a few centimeters with the spinal nerve, the preganglionic fiber leaves the nerve and passes to a *ganglion* of the *sympathetic chain*. Here the fiber either synapses immediately with postganglionic fibers or, often, passes on through the chain into one of its radiating nerves to synapse with postganglionic fibers in an outlying sympathetic ganglion. The postganglionic fiber then travels to its destination in one of the organs.

Many of the postganglionic fibers in the sympathetic chain pass once again into the spinal nerves at all levels of the cord. These fibers extend to all parts of the body in the spinal nerves. They control the blood vessels, sweat glands, and piloerector muscles of the hairs.

Segmental Distribution of Sympathetic Nerves. The sympathetic fibers originating in the different segments of the spinal cord are not necessarily distributed to the same part of the body as the spinal nerve fibers from the same segments. Instead, the *sympathetic fibers from T-1 generally pass up the sympathetic chain into the head; from T-2 into the neck; T-3, T-4, T-5, and T-6 into the thorax; T-7, T-8, T-9, T-10, and T-11 into the abdomen; T-12, L-1 and L-2 into the legs.* This distribution is only approximate and overlaps greatly.

The distribution of sympathetic nerves to each organ is determined partly by the position in the embryo at which the organ originates. For instance, the heart receives many sympathetic nerves from the neck portion of the sympathetic chain because the heart originates in the neck of the embryo. Likewise, the abdominal organs receive their sympathetic innervation from the lower thoracic segments because the primitive gut originates in the lower thoracic area.

Special Nature of the Sympathetic Nerve Endings in the Adrenal Medullae. Preganglionic sympathetic nerve fibers pass, without synapsing, all the way from the intermediolateral horn cells of the spinal cord, through the sympathetic chains, through the splanchnic nerves, and finally into the adrenal medullae. There they end directly on special cells that *secrete epinephrine* and *norepinephrine.* These secretory cells are embryologically derived from nervous tissue and are analogous to postganglionic neurons; indeed, they even have rudimentary nerve fibers.

PHYSIOLOGIC ANATOMY OF THE PARASYMPATHETIC NERVOUS SYSTEM

The parasympathetic nervous system is illustrated in Figure 38–2, which shows that parasym-

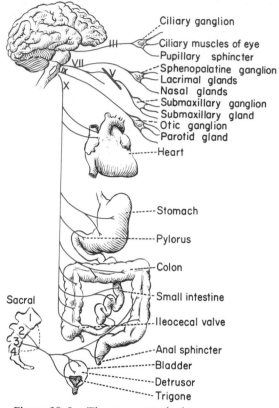

Figure 38–2.   The parasympathetic nervous system.

pathetic fibers leave the central nervous system through several of the cranial nerves as well as through the second and third sacral spinal nerves, and occasionally the first and fourth sacral nerves as well. About 75 per cent of all parasympathetic nerve fibers are in the *vagus nerves* (also called the *tenth cranial nerves*), passing to the entire thoracic and abdominal regions of the body. The vagus nerves supply parasympathetic nerves to the heart, the lungs, the esophagus, the stomach, the small intestine, the proximal half of the colon, the liver, the gallbladder, the pancreas, and the upper portions of the ureters.

Parasympathetic fibers in the *third cranial nerve* flow to the pupillary sphincters and ciliary muscles of the eye. Fibers from the *seventh cranial nerve* pass to the lacrimal, nasal, and submaxillary glands, and fibers from the *ninth cranial nerve* pass to the parotid gland.

The sacral parasympathetic fibers combine to form the *nervi erigentes*, which leave the sacral plexus on each side of the cord and distribute their peripheral fibers to the descending colon, rectum, bladder and lower portions of the ureters. Also, this sacral group of parasympathetics supplies fibers to the external genitalia to cause various sexual reactions.

Preganglionic and Postganglionic Parasympathetic Fibers. The parasympathetic system, like the sympathetic, has both preganglionic and postganglionic fibers, but the preganglionic fibers usually pass uninterrupted to the organ that is to be excited by parasympathetic signals. Then, in the wall of the organ are located the *parasympathetic postganglionic neurons*. The preganglionic fibers synapse with these; and then short postganglionic fibers, 1 millimeter to several centimeters in length, leave the neuronal cell bodies to spread in the substance of the organ.

BASIC CHARACTERISTICS OF SYMPATHETIC AND PARASYMPATHETIC FUNCTION

CHOLINERGIC AND ADRENERGIC FIBERS — SECRETION OF ACETYLCHOLINE OR NOREPINEPHRINE BY THE POSTGANGLIONIC FIBERS

It will be recalled from Chapter 10 that skeletal nerve endings secrete acetylcholine. This is also true of the *preganglionic fibers* of both the sympathetic and parasympathetic system, and it is true, too, of the *parasympathetic postganglionic fibers*. Therefore, all these fibers are said to be *cholinergic because they secrete acetylcholine at their nerve endings*.

A few of the postganglionic endings of the sympathetic nervous system also secrete acetylcholine; and these fibers, too, are cholinergic; but by far the majority of the sympathetic postganglionic endings secrete *norepinephrine*. These fibers are said to be *adrenergic*, a term derived from *noradrenalin,* which is the English name for norepinephrine. Thus, there is a basic functional difference between the postganglionic fibers of the parasympathetic and sympathetic systems, one secreting acetylcholine and the other principally norepinephrine. The chemical structures of these substances are the following:

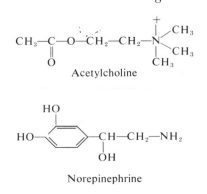

Acetylcholine

Norepinephrine

The acetylcholine and norepinephrine secreted by the postganglionic fibers act on the different organs to cause the respective parasympathetic or sympathetic effects. Therefore, these substances are called *parasympathetic* and *sympathetic mediators,* respectively, or sometimes *cholinergic* and *adrenergic mediators.*

Once acetylcholine has been secreted by the cholinergic nerve ending, it is split into acetate ion and choline by the enzyme *cholinesterase* that is present both in the terminal nerve ending itself and on the surface of the receptor organ. Therefore, the action of acetylcholine released by cholinergic nerve fibers usually lasts for a few seconds at most and more often for only a fraction of a second. Much of the choline derived from the acetylcholine is then transported back into the nerve terminal to be used in the formation of new acetylcholine.

Following secretion of norepinephrine by the adrenergic nerve endings, it is removed from the secretory site mainly in two different ways: (1) re-uptake into the adrenergic nerve endings themselves by an active transport process — accounting for removal of 50 to 80 per cent of the secreted norepinephrine; and (2) diffusion away from the nerve endings into the surrounding body fluids and thence into the blood — accounting for removal of most of the remainder of the norepinephrine, after which it is destroyed by enzymes in about one minute.

Ordinarily, the norepinephrine secreted directly in a tissue by adrenergic nerve endings remains active for only a few seconds, illustrating that its re-uptake and diffusion away from the tissue is rapid. However, the norepinephrine and epinephrine secreted into the blood by the adrenal medullae remain active until they diffuse into some tissue where they are destroyed by enzymes; this occurs mainly in the liver. Therefore, when secreted into the blood, both norepinephrine and epinephrine remain very active for 10 to 30 seconds, followed by decreasing activity thereafter for one to several minutes.

RECEPTOR SUBSTANCES OF THE EFFECTOR ORGANS

The acetylcholine, norepinephrine, and epinephrine secreted by the autonomic nervous fibers all stimulate the effector organs by first combining with receptor substances, called simply *receptors,* in the effector cells. The receptor in most instances is in the cell membrane itself and is probably a glycoprotein. The most likely mechanism for function of the receptor is that the transmitter substance first binds with the receptor and this causes a basic change in the molecular structure of the receptor compound. Because the receptor is an integral part of the cell membrane, this structural change then causes one or more changes either in the cell membrane itself or in the cell cytoplasm. One effect that often occurs is alteration of the permeability of the cell membrane to various ions — for instance, to allow rapid influx of sodium, chloride, or calcium ions into the cell or to allow rapid efflux of potassium ions out of the cell. These ionic changes then usually change the membrane potential, sometimes eliciting action potentials (as occurs in smooth muscle cells) or at other times causing electrotonic effects on the cells (as occurs on glandular cells) to produce the responses. The ions themselves also may have direct effects in the cytoplasm of the receptor cells, such as the effect of calcium ions to promote smooth muscle contraction.

A second way that the receptor can function, besides changing the membrane permeability, is to activate an enzyme in the cell membrane —this enzyme in turn promoting chemical reactions within the cell. For instance, epinephrine increases the activity of *adenyl cyclase* in some cell membranes, and this in turn causes the formation of cyclic AMP inside the cell; the cyclic AMP then initiates many intracellular activities.

EXCITATORY AND INHIBITORY ACTIONS OF SYMPATHETIC AND PARASYMPATHETIC STIMULATION

Table 38–1 gives the effects on different visceral functions of the body caused by stimulation of the parasympathetic and sympathetic nerves. From this table it can be seen that *sympathetic stimulation causes excitatory effects in some organs but inhibitory effects in others. Likewise, parasympathetic stimulation causes excitation in some organs but inhibition in others.* Also, when sympathetic stimulation excites a particular organ, parasympathetic stimulation often inhibits it, illustrating that the two

TABLE 38–1 AUTONOMIC EFFECTS ON VARIOUS ORGANS OF THE BODY

Organ	Effect of Sympathetic Stimulation	Effect of Parasympathetic Stimulation
Eye: Pupil	Dilated	Constricted
Ciliary muscle	Slight relaxation	Contracted
Glands: Nasal	Vasoconstriction and slight	Stimulation of thin, copious
Lacrimal	secretion	secretion (containing many enzymes
Parotid		for enzyme-secreting glands)
Submaxillary		
Gastric		
Pancreatic		
Sweat glands	Copious sweating (cholinergic)	None
Apocrine glands	Thick, odoriferous secretion	None
Heart: Muscle	Increased rate	Slowed rate
	Increased force of contraction	Decreased force of atrial contraction
Coronaries	Dilated (β_2); constricted (α)	Dilated
Lungs: Bronchi	Dilated	Constricted
Blood vessels	Mildly constricted	? Dilated
Gut: Lumen	Decreased peristalsis and tone	Increased peristalsis and tone
Sphincter	Increased tone	Relaxed
Liver	Glucose released	Slight glycogen synthesis
Gallbladder and bile ducts	Relaxed	Contracted
Kidney	Decreased output	None
Bladder: Detrusor	Relaxed	Excited
Trigone	Excited	Relaxed
Penis	Ejaculation	Erection
Systemic blood vessels:		
Abdominal	Constricted	None
Muscle	Constricted (adrenergic α)	None
	Dilated (adrenergic β)	
	Dilated (cholinergic)	
Skin	Constricted	None
Blood: Coagulation	Increased	None
Glucose	Increased	None
Basal metabolism	Increased up to 100%	None
Adrenal cortical secretion	Increased	None
Mental activity	Increased	None
Piloerector muscles	Excited	None
Skeletal muscle	Increased glycogenolysis	None
	Increased strength	

systems occasionally act reciprocally to each other. However, most organs are dominantly controlled by one or the other of the two systems, so that, except in a few instances, the two systems do not actively oppose each other.

There is no generalization one can use to explain whether sympathetic or parasympathetic stimulation will cause excitation or inhibition of a particular organ. Therefore, to understand sympathetic and parasympathetic function, one must learn the functions of these two nervous systems as listed in Table 38–1. Some of these functions need to be clarified in still greater detail as follows:

EFFECTS OF SYMPATHETIC AND PARASYMPATHETIC STIMULATION ON SPECIFIC ORGANS

The Eye. Two functions of the eye are controlled by the autonomic nervous system: the pupillary opening and the focus of the lens. Sympathetic stimulation dilates the pupil, while parasympathetic stimulation constricts the pupil. The parasympathetics that control the pupil are reflexly stimulated when excess light enters the eyes; this reflex reduces the pupillary opening and decreases the amount of light that strikes the retina. On the other hand, the sympathetics become stimulated during periods of excitement and therefore increase the pupillary opening at these times.

Focusing of the lens is controlled either entirely or almost entirely by the parasympathetic nervous system. The lens is normally held in a flattened state by tension of its radial ligaments. Parasympathetic excitation contracts the *ciliary muscle*, which loosens this tension and allows the lens to become more convex. This causes the eye to focus on objects near at hand. The focusing mechanism is discussed in Chapters 39 and 40 in relation to function of the eyes.

The Gastrointestinal System. The gastrointestinal system has its own intrinsic set of nerves, known as the *gastrointestinal intramural plexus*. However, both parasympathetic and sympathetic stimulation can affect gastrointestinal activity — parasympathetic especially. Parasympathetic stimulation, in general, increases the overall degree of activity of the gastrointestinal tract by promoting peristalsis, thus allowing rapid propulsion of the intraluminal contents along the tract. This propulsive effect is associated with simultaneous increases in rates of secretion by many of the gastrointestinal glands.

Normal function of the gastrointestinal tract is not very dependent on sympathetic stimulation.

However, in some diseases, strong sympathetic stimulation inhibits peristalsis and increases the tone of the sphincters. The net result is greatly slowed propulsion of food through the tract.

The Heart. In general, sympathetic stimulation increases the overall activity of the heart. This is accomplished by increasing both the rate and force of the heartbeat. Parasympathetic stimulation causes mainly the opposite effects, decreasing the overall activity of the heart. To express these effects in another way, sympathetic stimulation increases the effectiveness of the heart as a pump, whereas parasympathetic stimulation decreases its effectiveness. However, sympathetic stimulation unfortunately also greatly increases the metabolism of the heart, while parasympathetic stimulation decreases its metabolism and allows the heart a certain degree of rest.

Systemic Control of the Blood Vessels. Most blood vessels, especially those of the abdominal viscera and the skin of the limbs, are constricted by sympathetic stimulation. Parasympathetic stimulation generally has almost no effects on most blood vessels but does dilate vessels in certain restricted areas such as the blush area of the face.

Effect of Sympathetic and Parasympathetic Stimulation on Arterial Pressure. The arterial pressure in the circulatory system is caused by propulsion of blood by the heart and by resistance to flow of this blood through the vascular system. In general, sympathetic stimulation increases both propulsion by the heart and resistance to flow, which can cause the pressure to increase greatly.

On the other hand, parasympathetic stimulation decreases the pumping effectiveness of the heart, which lowers the pressure a moderate amount, though not nearly so much as the sympathetics can increase the pressure.

Effects of Sympathetic and Parasympathetic Stimulation on Other Functions of the Body. Because of the great importance of the sympathetic and parasympathetic control systems, these are discussed many times in this text in relation to a myriad of body functions that are not considered in detail here. In general, most of the entodermal structures, such as the ducts of the liver, the gallbladder, the ureter, and the urinary bladder, are inhibited by sympathetic stimulation but excited by parasympathetic stimulation.

Sympathetic stimulation also has metabolic effects, causing release of glucose from the liver, increase in blood glucose concentration, increase in glycogenolysis in muscle, increase in muscle strength, increase in basal metabolic rate, and increase in mental activity.

Finally, the sympathetics and parasympathetics are involved in regulating the male and female sexual acts, as will be explained in Chapters 54 and 55.

FUNCTION OF THE ADRENAL MEDULLAE

Stimulation of the sympathetic nerves to the adrenal medullae causes large quantities of epinephrine and norepinephrine to be released into the circulating blood, and these two hormones in turn are carried in the blood to all tissues of the body.

The circulating hormones have almost the same effects on the different organs as those caused by direct sympathetic stimulation, except that *the effects last about ten times as long* because norepinephrine and epinephrine are removed from the blood slowly. For instance, they cause constriction of essentially all the blood vessels of the body; they cause increased activity of the heart, inhibition of the gastrointestinal tract, dilation of the pupil of the eye, and so forth.

In summary, stimulation of the adrenal medullae causes the release of hormones that have almost the same effects throughout the body as direct sympathetic stimulation, except that the effects are greatly prolonged.

Value of the Adrenal Medullae to the Function of the Sympathetic Nervous System. Often, when any part of the sympathetic nervous system is stimulated, major portions of the entire system are stimulated at the same time. Also, norepinephrine and epinephrine are almost always released by the adrenal medullae at the same time that the different organs are being stimulated directly by the sympathetic nerves. Therefore, the organs are actually stimulated in two different ways simultaneously, directly by the sympathetic nerves and indirectly by the medullary hormones. The two means of stimulation support each other, and either can usually substitute for the other. For instance, destruction of the direct sympathetic pathways to the organs does not abrogate excitation of the organs, because norepinephrine and epinephrine are still released into the circulating fluids and indirectly cause stimulation. Likewise, total loss of the two adrenal medullae usually has little effect on the operation of the sympathetic nervous system because the direct pathways can still perform almost all the necessary duties.

Another important value of the adrenal medullae is the capability of epinephrine and norepinephrine to stimulate structures of the body that are not innervated by direct sympathetic fibers. For instance, the metabolic rate of every cell of the body is increased by these hormones, especially by epinephrine, even though only a small proportion of all the cells in the body are innervated by sympathetic fibers.

SYMPATHETIC AND PARASYMPATHETIC "TONE"

The sympathetic and parasympathetic systems are continually active and the basal rates of stimulation are known, respectively, as *sympathetic tone* and *parasympathetic tone.*

The value of tone is that it allows a single nervous system to increase or to decrease the activity of an organ. For instance, sympathetic tone normally keeps almost all the blood vessels of the body constricted to approximately half their maximum diameter. By increasing the degree of sympathetic stimulation, the vessels can be constricted even more; but, on the other hand, by decreasing the level of sympathetic stimulation, the vessels can be dilated. If it were not for normal sympathetic tone, the sympathetic system could cause only vasoconstriction, never vasodilatation.

Another interesting example of tone is that of the parasympathetics in the gastrointestinal tract. Surgical removal of the parasympathetic supply to the gut by cutting the vagi can cause serious and prolonged gastric and intestinal "atony," thus illustrating that in normal function the parasympathetic tone to the gut is strong. This tone can be decreased by the brain, thereby inhibiting gastrointestinal motility, or, on the other hand, it can be increased, thereby promoting increased gastrointestinal activity.

MASS DISCHARGE OF SYMPATHETIC SYSTEM VERSUS DISCRETE CHARACTERISTICS OF PARASYMPATHETIC REFLEXES

Large portions of the sympathetic nervous system often become stimulated simultaneously, a phenomenon called *mass discharge.* This characteristic of sympathetic action allows the sympathetic nervous system to control body-wide functions such as overall regulation of arterial pressure or of metabolic rate.

However, in some instances sympathetic activity also occurs in isolated portions of the system. The most important are these: (1) In the process of heat regulation, the sympathetics control sweating and blood flow in the skin without affecting other organs innervated by the sympathetics. (2) During muscular activity in some animals, cholinergic vasodilator fibers of the skeletal muscles are stimulated independently of all the remainder of the sympathetic system. (3)

Many "local reflexes" involving the spinal cord but not the higher nervous centers affect local areas. For instance, heating a local skin area causes mild local vasodilatation and enhanced local sweating, while cooling causes the opposite effects.

In contrast to the sympathetic system, most reflexes of the parasympathetic system are very specific. For instance, parasympathetic cardiovascular reflexes usually act only on the heart to increase or decrease its rate of beating. Likewise, parasympathetic reflexes frequently cause secretion only in the mouth, or, in other instances, secretion only by the stomach glands. Finally, the rectal emptying reflex does not affect other parts of the bowel to a major extent.

Yet there is often association between closely allied parasympathetic functions. For instance, though salivary secretion can occur independently of gastric secretion, these two often also occur together, and pancreatic secretion frequently occurs at the same time. Also, the rectal emptying reflex often initiates a concurrent bladder emptying reflex, resulting in stimultaneous emptying of both the bladder and rectum.

"ALARM" OR "STRESS" FUNCTION OF THE SYMPATHETIC NERVOUS SYSTEM

From the above discussions of the sympathetic nervous system, one can already see that mass sympathetic discharge increases in many ways the capability of the body to perform vigorous muscle activity. Let us quickly summarize these ways:

1. increased arterial pressure
2. increased blood flow to active muscles concurrent with decreased blood flow to organs that are not needed for rapid activity
3. increased rates of cellular metabolism throughout the body
4. increased blood glucose concentration
5. increased glycolysis in muscle
6. increased muscle strength
7. increased mental activity

The sum of these effects permits the person to perform far more strenuous physical activity than would otherwise be possible. Since it is physical *stress* that usually excites the sympathetic system, it is frequently said that the purpose of the sympathetic system is to provide extra activation of the body in states of stress; this is often called the *sympathetic stress reaction*.

The sympathetic system is also strongly activated in many emotional states. For instance, in the state of *rage*, which is elicited mainly by stimulating the hypothalamus, signals are transmitted downward through the reticular formation and spinal cord to cause massive sympathetic discharge, and all the sympathetic events listed above ensue immediately. This is called the sympathetic *alarm reaction*. It is also frequently called the *fight or flight reaction* because an animal in this state decides almost instantly whether to stand and fight or to run. In either event, the sympathetic alarm reaction makes the animal's subsequent activities extremely vigorous.

MEDULLARY, PONTINE, AND MESENCEPHALIC CONTROL OF THE AUTONOMIC NERVOUS SYSTEM

Many areas in the reticular substance of the medulla, pons, and mesencephalon, as well as many special nuclei (Fig. 38–3), control different autonomic functions, such as arterial pressure, heart rate, glandular secretion in the upper part of the gastrointestinal tract, gastrointestinal peristalsis, the degree of contraction of the urinary bladder, and many others. The control of each of these is discussed at appropriate points in this text. Suffice it to point out here that the most important factors controlled in the lower brain stem are arterial pressure, heart rate, and respiration. Indeed, transection of the brain stem at the midpontile level allows normal basal control of arterial pressure to continue as before but prevents its modulation by higher nervous centers, particularly the hypothalamus. On the other hand, transection immediately below the medulla causes the arterial pressure to fall to about one-half normal. Closely associated with the cardiovascular regulatory centers in the medulla is the medullary center for regulation of respiration, discussed in detail in Chapter 29. Though this is not considered to be an autonomic function, it is one of the *involuntary* functions of the body.

In the previous chapter it was pointed out that many of the behavioral responses of an animal are mediated through the hypothalamus, the reticular formation, and the autonomic nervous

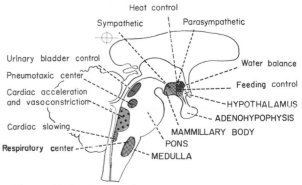

Figure 38–3. Autonomic control centers of the brain stem.

system. Indeed, the higher areas of the brain can alter the function of the whole autonomic nervous system or of portions of it strongly enough to cause severe autonomic-induced disease, such as peptic ulcer, constipation, heart palpitation, and even heart attacks.

REFERENCES

Aviado, D. M., *et al.* (eds.): Pharmacology of Ganglionic Transmission. New York, Springer-Verlag, 1979.

Bhagat, B. D.: Mode of Action of Autonomic Drugs. Flushing, N. Y., Graceway Publishing Company, 1979.

Carrier, O., Jr.: Pharmacology of the Peripheral Autonomic Nervous System. Chicago, Year Book Medical Publishers, 1972.

Collier, B.: Biochemistry and physiology of cholinergic transmission. *In* Brookhart, J. M., and Mountcastle, V. B. (eds.): Handbook of Physiology. Sec. 1, Vol. 1. Baltimore, Williams & Wilkins, 1977, p. 463.

DeQuattro, V., *et al.*: Anatomy and biochemistry of the sympathetic nervous system. *In* DeGroot, L. J., *et al.* (eds.): Endocrinology. Vol. 2. New York, Grune & Stratton, 1979, p. 1241.

Guyton, A. C., and Gillespie, W. M., Jr.: Constant infusion of epinephrine: Rate of epinephrine secretion and destruction in the body. *Am. J. Physiol., 165*:319, 1951.

Guyton, A. C., and Reeder, R. C.: Quantitative studies on the autonomic actions of curare. *J. Pharmacol. Exp. Ther., 98*:188, 1950.

Hayward, J. N.: Functional and morphological aspects of hypothalamic neurons. *Physiol. Rev., 57*:574, 1977.

Kalsner, S. (ed.): Trends in Autonomic Pharmacology. Baltimore, Urban & Schwarzenberg, 1979.

Landsberg, L., and Young, J. B.: Catecholamines and the adrenal medulla. *In* Bondy, P. K., and Rosenberg, L. E. (eds.): Metabolic Control and Disease, 8th Ed. Philadelphia, W. B. Saunders Co., 1980, p. 1621.

Levitzki, A.: Catecholamine receptors. *In* Adrian, R. H., *et al.* (eds.): Reviews of Physiology, Biochemistry, and Pharmacology. New York, Springer-Verlag, 1978, p. 1.

Moore, R. Y., and Bloom, F. E.: Central catecholamine neuron systems: Anatomy and physiology of the dopamine system. *Annu. Rev. Neurosci., 1*:129, 1978.

Morgane, P. J., and Panksepp, J. (eds.): Handbook of the hypothalamus. New York, Marcel Dekker, 1979.

Paton, D. M. (ed.): The Release of Catecholamines from Adrenergic Neurons. New York, Pergamon Press, 1979.

Robinson, R.: Tumors That Secrete Catecholamines: A Study of Their Natural History and Their Diagnosis. New York, John Wiley & Sons, 1980.

Szabadi, E., *et al.*: Recent Advances in the Pharmacology of Adrenoceptors. New York, Elsevier/North-Holland, 1978.

Tůcek, S. (eds.): The Cholinergic Synapse. New York, Elsevier/North-Holland, 1979.

Usdin, E., *et al.* (eds.): Catecholamines: Basic and Clinical Frontiers. New York, Pergamon Press, 1979.

Wolf, S. G.: Neural control of visceral function. *In* Frohlich, E. D. (ed.): Pathophysiology, 2nd Ed. Philadelphia, J. B. Lippincott Co., 1976, p. 711.

Part X

THE SPECIAL SENSES

39

The Eye: I. Optics of Vision and Function of the Retina

THE OPTICS OF THE EYE

THE EYE AS A CAMERA

The eye, as illustrated in Figure 39–1, is optically equivalent to the usual photographic camera, for it has a lens system, a variable aperture system, and a retina that corresponds to the film. The lens system of the eye is composed of (1) the interface between air and the anterior surface of the cornea, (2) the interface between the posterior surface of the cornea and the aqueous humor, (3) the interface between the aqueous humor and the anterior surface of the lens, and (4) the interface between the posterior surface of the lens and the vitreous humor. The difference between the "refractive indices" on the two sides of each surface is one of the factors that determine the focusing strength of each surface. Another factor is the curvature of the surface. The refractive index of air is 1; of the cornea, 1.38; of the aqueous humor, 1.33; of the lens (on the average), 1.40; and of the vitreous humor, 1.34.

The Reduced Eye. If all the refractive surfaces of the eye are algebraically added together and then considered to be one single lens, the optics of the normal eye may be simplified and repre-

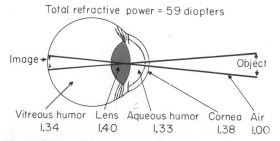

Figure 39–1. The eye as a camera. The numbers are the refractive indices.

sented schematically as a "reduced eye." This is useful in simple calculations. In the reduced eye, a single lens is considered to exist with its central point 17 mm. in front of the retina and to have a total refractive power of approximately 59 diopters when the lens is accommodated for distant vision. (A lens that will focus parallel light rays at a distance of 1 meter beyond the lens has a strength of 1 diopter. A 59 diopter lens is 59 times as strong and will focus the same parallel rays at a distance of 1/59 meter beyond the lens.)

The anterior surface of the cornea provides about 48 diopters of the eye's total dioptric strength for three reasons: (1) the refractive index of the cornea is markedly different from that of air; (2) the surface of the cornea is further away from the retina than are the surfaces of the eye lens; and (3) the curvature of the cornea is reasonably great.

The posterior surface of the cornea is concave and actually acts as a concave lens, which has *negative* focusing power, but because the difference in refractive index of the cornea and the aqueous humor is slight, this posterior surface of the cornea has a refractive power of only about −4 diopters, which neutralizes only a small part of the refractive power of the other refractive surfaces of the eye.

The total refractive power of the crystalline lens of the eye when it is surrounded by fluid on each side is only 15 diopters of the total refractive power of the eye's lens system. If this lens were removed from the eye and then surrounded by air, its refractive power would be about 100 diopters. Thus, it can be seen that the lens inside the eye is not nearly so powerful as it is outside the eye. The reason is that the fluids surrounding the lens have refractive indices not greatly different from the refractive index of the lens itself,

448

the smallness of the differences greatly decreasing the amount of light refraction at the lens interfaces. But the importance of the crystalline lens is that its curvature, and therefore also its strength, can change to provide accommodation, which will be discussed later in the chapter.

Formation of an Image on the Retina. In exactly the same manner that a glass lens can focus an image on a sheet of paper, the lens system of the eye can also focus an image on the retina. The image is inverted, and reversed with respect to the object. However, the mind perceives objects in the upright position despite the upside-down orientation of the retina because the brain is trained to consider an inverted image as the normal.

THE MECHANISM OF ACCOMMODATION

The refractive power of the crystalline lens of the eye can be voluntarily increased from 15 diopters to approximately 29 diopters in young children; this is a total "accommodation" of 14 diopters. To do this, the shape of the lens is changed from that of a moderately convex lens to that of a very convex lens. The mechanism of this is the following:

Normally, the lens is composed of a strong elastic capsule filled with viscous proteinaceous but transparent fibers. When the lens is in a relaxed state, with no tension on its capsule, it assumes a spherical shape, owing entirely to the elasticity of the lens capsule. However, as illustrated in Figure 39–2, approximately 70 ligaments attach radially around the lens, pulling the lens edges toward the edge of the choroid. These ligaments are constantly tensed by the elastic pull

of their attachments to the choroid, and the tension on the ligaments causes the lens to remain relatively flat under normal resting conditions of the eye. At the insertions of the ligaments in the choroid is the ciliary muscle, which has two sets of smooth muscle fibers, the *meridional fibers* and the *circular fibers*. The meridional fibers extend from the corneoscleral junction to the insertions of the ligaments in the choroid approximately 2 to 3 mm. behind the junction. When these muscle fibers contract, the ligaments are pulled forward, thereby releasing a certain amount of tension on the crystalline lens. The circular fibers are arranged circularly all the way around the eye so that when they contract a sphincterlike action occurs, decreasing the diameter of the circle of ligament attachments and allowing the ligaments to pull less on the lens capsule.

Thus, contraction of both sets of smooth muscle fibers in the ciliary muscle relaxes the ligaments to the lens capsule, and the lens assumes a more spherical shape, like that of a balloon, because of elasticity of its capsule. When the ciliary muscle is completely relaxed, the dioptric strength of the lens is as weak as it can become. On the other hand, when the ciliary muscle contracts as strongly as possible, the dioptric strength of the lens becomes maximal.

Autonomic Control of Accommodation. The ciliary muscle is controlled almost entirely by the parasympathetic nerves. Stimulation of the parasympathetic fibers to the eye contracts the ciliary muscle, which in turn relaxes the ligaments of the lens and increases its refractive power. With an increased refractive power, the eye is more capable of focusing on objects that are nearer to it than is an eye with less refractive power. Consequently, as a distant object moves toward the eye, the number of parasympathetic impulses impinging on the ciliary muscle must be progressively increased for the eye to keep the object constantly in focus.

Presbyopia. As a person grows older, his lens loses its elastic nature and becomes a relatively solid mass, probably because of progressive denaturation of the proteins. Therefore, the ability of the lens to assume a spherical shape progressively decreases, and the power of accommodation decreases from approximately 14 diopters shortly after birth to approximately 2 diopters at the age of 45 to 50. Thereafter, the lens of the eye may be considered to be almost totally nonaccommodating, which condition is known as "presbyopia."

Once a person has reached the state of presbyopia, each eye remains focused permanently at an almost constant distance; this distance depends on the physical characteristics of each individual's eyes. Obviously, the eyes can no longer accommodate for both near and far vision.

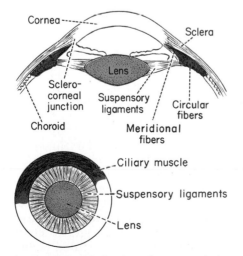

Figure 39–2. Mechanism of accommodation.

Therefore, for an older person to see clearly both in the distance and nearby, he must wear bifocal glasses with the upper segment focused for far-seeing and the lower segment focused for near-seeing.

THE PUPILLARY APERTURE

A major function of the iris is to increase the amount of light that enters the eye during darkness and to decrease the light that enters the eye in bright light. The reflexes for controlling this mechanism will be considered in the discussion of the neurology of the eye in Chapter 40. The amount of light that enters the eye through the pupil is *proportional to the area of the pupil or to the square of the diameter* of the pupil. The pupil of the human eye can become as small as approximately 1.5 mm. and as large as 8 mm. in diameter. Therefore, the quantity of light entering the eye may vary approximately 30 times as a result of changes in pupillary aperture size.

Depth of Focus of the Lens System of the Eye. Figure 39–3 illustrates two separate eyes that are exactly alike except that the diameters of the pupillary apertures are different. In the upper eye the pupillary aperture is small, and in the lower eye the aperture is large. In front of each of these two eyes are two small point sources of light, and light from each passes through the pupillary aperture and focuses on the retina. Consequently, in both eyes the retina sees two spots of light in perfect focus. It is evident from the diagrams, however, that if the retina is moved forward or backward to an out-of-focus position, the size of each spot will not change much in the upper eye, but in the lower eye the size of each spot will increase greatly, and it becomes a "blur circle." In other words, the upper lens system has far greater *depth of focus* than the bottom lens

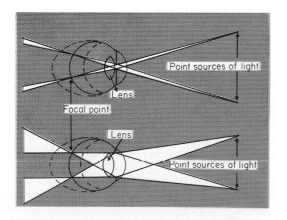

Figure 39–3. Effect of small and large pupillary apertures on the depth of focus.

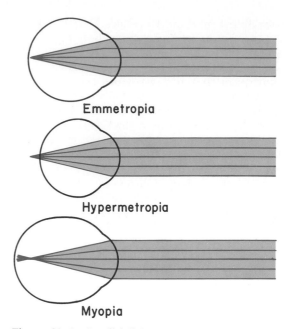

Figure 39–4. Parallel light rays focus on the retina in emmetropia, behind the retina in hypermetropia, and in front of the retina in myopia.

system. When the lens system of the eye has great depth of focus—that is, a smaller diameter—the retina can be considerably displaced from the focal plane and still discern the various features of an image rather distinctly; whereas, when a lens system has a shallow depth of focus, moving the retina only slightly away from the focal plane causes extreme blurring of the image.

ERRORS OF REFRACTION

Emmetropia. As shown in Figure 39–4, the eye is considered to be normal or "emmetropic" if, when the ciliary muscle is completely relaxed, parallel light rays from distant objects are in sharp focus on the retina. This means that the emmetropic eye can, with its ciliary muscle completely relaxed, see all distant objects clearly, but to focus objects at close range it must contract its ciliary muscle and thereby provide various degrees of accommodation.

Hypermetropia (Hyperopia). Hypermetropia, also known as "far-sightedness," is due either to an eyeball that is too short or to a lens system that is too weak when the ciliary muscle is completely relaxed. In this condition, parallel light rays are not bent sufficiently by the lens system to come to a focus by the time they reach the retina. In order to overcome this abnormality, the ciliary muscle must contract to increase the strength of the lens. Therefore, in old age, when the lens becomes presbyopic, the far-sighted person often is not able to accommodate his lens sufficiently to focus even distant objects, much less to focus near objects.

Myopia.   In myopia, or "near-sightedness," even when the ciliary muscle is completely relaxed, the strength of the lens is still enough that light rays coming from distant objects are focused in front of the retina. This is usually due to too long an eyeball, but it can occasionally result from too much power of the lens system of the eye.

No mechanism exists by which the eye can decrease the strength of its lens below that which exists when the ciliary muscle is completely relaxed. Therefore, the myopic person has no mechanism by which he can ever focus distant objects sharply on his retina. However, as an object comes nearer and nearer to his eye it finally comes near enough that its image is focused on the retina. Then, when the object comes still closer to the eye, the person can use his mechanism of accommodation to keep the image focused clearly. Therefore, a myopic person has a definite limiting "far point" for acute vision.

Correction of Myopia and Hypermetropia by Use of Lenses.   It will be recalled that light rays passing through a concave lens diverge. Therefore, if the refractive surfaces of the eye have too much refractive power, as in myopia, some of this excessive refractive power can be neutralized by placing in front of the eye a concave spherical lens, which will diverge the incoming rays. On the other hand, in a person who has hypermetropia — that is, one who has too weak a lens for the distance of the retina away from the lens — the abnormal vision can be corrected by adding refractive power with a convex lens in front of the eye. These corrections are illustrated in Figure 39–5. One usually determines the strength of the concave or convex lens needed for clear vision by

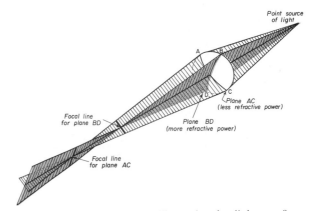

Figure 39–6.   Astigmatism, illustrating that light rays focus at one focal distance in one focal plane and at another focal distance in the plane at right angles.

"trial and error" — that is, by trying first a strong lens and then a stronger or weaker lens until the one that gives the best visual acuity is found.

Astigmatism.   Astigmatism is a refractive error of the lens system of the eye caused usually by an oblong shape of the cornea or, rarely, by an oblong shape of the lens. A lens surface like the side of an egg lying edgewise to the incoming light would be an example of an astigmatic lens. The degree of curvature in a plane through the long axis of the egg is not nearly so great as the degree of curvature in a plane through the short axis. The same is true of an astigmatic lens of the eye. Because the curvature of the astigmatic lens along one plane is less than the curvature along the other plane, light rays striking the peripheral portions of the lens in one plane are not bent nearly so much as are rays striking the peripheral portions of the other plane.

This is illustrated in Figure 39–6, which shows what happens to rays of light emanating from a point source and passing through an oblong astigmatic lens. The light rays in the vertical plane, which is indicated by plane BD, are refracted greatly by the astigmatic lens because of the greater curvature in the vertical direction than in the horizontal direction. However, the light rays in the horizontal plane, indicated by plane AC, are bent not nearly so much as the light rays in the vertical plane. It is obvious, therefore, that the light rays passing through an astigmatic lens do not all come to a common focal point because the light rays passing through one plane of the lens focus far in front of those passing through the other plane.

Placing an appropriate *spherical* lens in front of an astigmatic eye can bring the light rays that pass through *one plane* of the lens into focus on the retina, but spherical lenses can never bring *all* the light rays into complete focus at the same

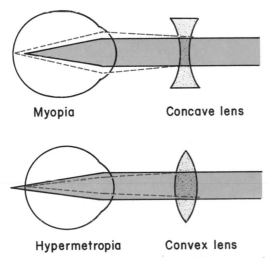

Figure 39–5.   Correction of myopia with a concave lens, and correction of hypermetropia with a convex lens.

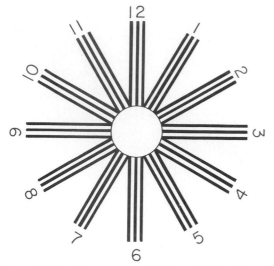

Figure 39–7. Chart composed of parallel black bars for determining the axis (meridian) of astigmatism.

time. This is the reason why astigmatism is a very undesirable refractive error of the eyes. Furthermore, the accommodative power of the eye cannot compensate for astigmatism, for the same reasons that spherical lenses placed in front of the eyes cannot correct the condition.

Correction of Astigmatism with a Cylindrical Lens. In correcting astigmatism with lenses, one can consider the astigmatic lens to be a spherical lens with a superimposed cylindrical lens (a lens that is curved in only one plane and not at all in the other plane). To correct the focusing of this system, it is necessary to determine both the *strength* of the cylindrical lens needed to neutralize the excess cylindrical power of the eye lens and the *axis* of this abnormal cylindrical lens.

There are several methods for determining the axis of the abnormal cylindrical component of the lens system of the eye. One of these methods is based on the use of parallel black bars, as shown in Figure 39–7. Some of these parallel bars are vertical, some are horizontal, and some are at various angles to the vertical and horizontal axes. After placing, by trial and error, various spherical lenses in front of the astigmatic eye, a strength of lens will usually be found that will cause sharp focus of one set of these parallel bars on the retina of the astigmatic eye.

It can be shown from the physical principles of optics that the axis of the *out-of-focus* cylindrical component of the optical system is parallel to the black bars that are fuzzy in appearance. Once this axis is found, the examiner tries progressively stronger positive or negative cylindrical lenses, the axes of which are placed parallel to the out-of-focus bars, until the patient sees all the crossed bars with equal clarity. When this has

been accomplished, the examiner directs the optician to grind a special lens having both the spherical correction plus the cylindrical correction at the appropriate axis.

CATARACTS

Cataracts are an especially common lens abnormality that occurs in older people. A cataract is a cloudy or opaque area in the lens. In the early stage of cataract formation the proteins in the lens fibers immediately beneath the capsule become denatured. Later, these same proteins coagulate to form opaque areas in place of the normal transparent protein fibers of the lens. Finally, in still later stages, calcium is often deposited in the coagulated proteins, thus further increasing the opacity.

When a cataract has obscured light transmission so greatly that it seriously impairs vision, the condition can be corrected by surgical removal of the entire lens. When this is done, however, the eye loses a large portion of its refractive power, which must be replaced by a powerful convex lens (about +15 diopters) in front of the eye.

SIZE OF THE IMAGE ON THE RETINA AND VISUAL ACUITY

If the distance from an object to the eye lens is 17 meters and the distance from the center of the lens to the retina is 17 millimeters, the ratio of the object size to image size is 1000 to 1. Therefore, an object 17 meters in front of the eye and 1 meter in size produces an image on the retina 1 millimeter in size.

Theoretically, a point of light from a distant point source, when focused on the retina, should be infinitely small. However, since the lens system of the eye is not perfect, such a retinal spot ordinarily has a total diameter of about 11 microns even with maximum resolution of the optical system. However, it is brightest in its very center and shades off gradually toward the edges.

The average diameter of cones *in the fovea* of the retina, the central part of the retina where vision is most highly developed, is approximately 1.5 microns, which is one-seventh the diameter of the spot of light. Nevertheless, since the spot of light has a bright center point and shaded edges, a person can distinguish two separate points if their centers lie approximately 2 microns apart on the retina, which is slightly greater than the width of a foveal cone. This means that a person with maximal acuity looking at two bright pinpoint spots of light 10 meters away can barely distinguish the spots as separate entities when they are 1 millimeter apart.

The foveal portion of the retina is less than a millimeter in diameter, which means that maximum visual acuity occurs in only the very center of the visual field. Outside this foveal area the visual acuity is reduced five- to ten-fold, and it becomes progressively poorer as the periphery is approached.

Clinical Method for Stating Visual Acuity. Usually the test chart for testing eyes is placed 20 feet away from the tested person, and if the person can see the letters of the size that he should be able to see at 20 feet, he is said to have 20/20 vision, that is, normal vision. If he can only see letters that he should be able to see at 200 feet, he is said to have 20/200 vision. On the other hand, if he can see at 20 feet letters he should be able to see at only 15 feet, he is said to have 20/15 vision. In other words, the clinical method for expressing visual acuity is to use a mathematical fraction that expresses the ratio of two distances, which is also the ratio of one's visual acuity to that of the normal person.

DETERMINATION OF DISTANCE OF AN OBJECT FROM THE EYE—DEPTH PERCEPTION

There are three major means by which the visual apparatus normally perceives distance, a phenomenon that is known as depth perception. They are (1) relative sizes of objects, (2) moving parallax, and (3) stereopsis.

Determination of Distance by Relative Sizes. If a person knows that a man is six feet tall and then he sees this man even when he is using only one eye, he can determine how far away the man is simply by the size of the man's image on his retina. He does not consciously think about the size of this image, but his brain has learned to determine automatically from the image sizes the distances of objects from the eye when the dimensions of these objects are already known.

Determination of Distance by Moving Parallax. Another important means by which the eyes determine distance is that of moving parallax. If a person looks off into the distance with his eyes completely still, he perceives no moving parallax, but when he moves his head to one side or the other, the images of objects close to him move rapidly across his retinae while the images of distant objects remain rather stationary. For instance, if he moves his head 1 inch and an object is only 1 inch in front of his eye, the image moves almost all the way across his retinae, whereas the image of an object 200 feet away

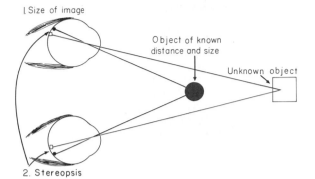

Figure 39–8. Perception of distance (1) by the size of the image on the retina and (2) as a result of stereopsis.

from his eyes does not move perceptibly. Thus, by this mechanism of moving parallax, one can tell the *relative distances* of different objects even though only one eye is used.

Determination of Distance by Stereopsis. Another method by which one perceives parallax is that of binocular vision. Because one eye is a little more than 2 inches to one side of the other eye, the images on the two retinae are different one from the other — that is, an object that is 1 inch in front of the bridge of the nose forms an image on the temporal portion of the retina of each eye, whereas a small object 20 feet in front of the nose has its image nearly in the middle of each eye. This type of parallax is illustrated in Figure 39–8, which shows the images of a black spot and a square actually reversed on the retinae because they are at different distances in front of the eyes. This gives a type of parallax that is present at all times when both eyes are being used. It is almost entirely this binocular parallax (called "stereopsis") that gives a person with two eyes far greater ability to judge relative distances *when objects are nearby* than a person who has only one eye. However, stereopsis is virtually useless for depth perception at distances beyond 200 feet.

THE RETINA

The retina is the light-sensitive portion of the eye, containing the *cones*, which are mainly responsible for color vision, and the *rods*, which are mainly responsible for vision in the dark. When the rods and cones are excited, signals are transmitted through successive neurons in the retina itself and finally into the optic nerve fibers and cerebral cortex. The purpose of the present discussion is to explain specifically the mechanisms by which the rods and cones detect both white and colored light.

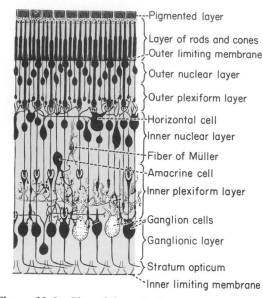

Pigmented layer
Layer of rods and cones
Outer limiting membrane
Outer nuclear layer
Outer plexiform layer
Horizontal cell
Inner nuclear layer
Fiber of Müller
Amacrine cell
Inner plexiform layer
Ganglion cells
Ganglionic layer
Stratum opticum
Inner limiting membrane

Figure 39–9. Plan of the retinal neurons. (From Polyak: The Retina, University of Chicago Press.)

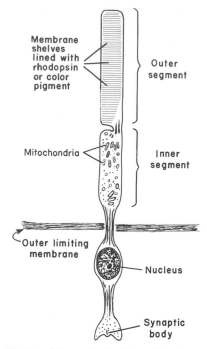

Membrane shelves lined with rhodopsin or color pigment

Outer segment

Mitochondria

Inner segment

Outer limiting membrane

Nucleus

Synaptic body

Figure 39–11. Schematic drawing of the functional parts of the rods and cones.

The Layers of the Retina. Figure 39–9 shows the functional components of the retina arranged in layers from the outside to the inside as follows: (1) pigment layer, (2) layer of rods and cones projecting into the pigment, (3) outer limiting membrane, (4) outer nuclear layer containing the cell bodies of the rods and cones, (5) outer plexiform layer, (6) inner nuclear layer, (7) inner plexiform layer, (8) ganglionic layer, (9) layer of optic nerve fibers, and (10) inner limiting membrane.

After light passes through the lens system of the eye and then through the vitreous humor, it enters the retina at the point designated by the bottom of Figure 39–9; that is, it passes through the ganglion cells, the plexiform layer, the nuclear layers, and the limiting membranes before it finally reaches the rods and cones located all the way on the opposite side of the retina. This distance is a thickness of several hundred microns; visual acuity is obviously decreased by this passage through such nonhomogeneous tissue. However, in the central region of the retina, as will be discussed below, the initial layers are pulled aside to prevent this loss of acuity.

The Foveal Region of the Retina and Its Importance in Acute Vision. A minute area in

Figure 39–10. Photomicrograph of the macula and of the fovea in its center. Note that the inner layers of the retina are pulled to the side to decrease the interference with light transmission. (From Bloom W., and Fawcett, D. W.: A Textbook of Histology, 10th Ed. Philadelphia, W. B. Saunders Company, 1975; courtesy of H. Mizoguchi.)

the center of the retina, illustrated in Figure 39–10, called the *macula* and occupying a total area of less than 1 square millimeter, is especially capable of acute and detailed vision. This area is composed entirely of cones, but the cones are very much elongated and have a diameter of only 1.5 microns in contradistinction to the very large cones located farther peripherally in the retina. The central portion of the macula, only 0.4 mm. in diameter, is called the *fovea*; in this region the blood vessels, the ganglion cells, the inner nuclear layer of cells, and the plexiform layers are all displaced to one side rather than resting directly on top of the cones. This allows light to pass unimpeded to the cones rather than through several layers of retina, which aids immensely in the acuity of visual perception by this region of the retina.

The Rods and Cones. Figure 39–11 is a diagrammatic representation of a photoreceptor (either a rod or a cone), though the cones may be distinguished by having a conical upper end. In general, the rods are narrower and longer than the cones, but this is not always the case. In the peripheral portions of the retina the rods are 2 to 5 microns in diameter whereas the cones are 5 to 8 microns in diameter; in the central part of the retina, in the fovea, the cones have a diameter of only 1.5 microns.

To the right in Figure 39–11 are labeled the four major functional segments of either a rod or a cone: (1) the outer segment, (2) the inner segment, (3) the nucleus, and (4) the synaptic body. It is in the outer segment that the light-sensitive photochemical is found. In the case of the rods, this is *rhodopsin,* and in the cones it is another photochemical almost exactly the same as rhodopsin except for a difference in spectral sensitivity. The cross-marks in the outer segment represent discs formed by infoldings of the cell membrane. These act as shelves to which the photosensitive pigments are attached. In this outer segment the concentration of the photosensitive pigments is approximately 40 per cent.

The inner segment contains the usual cytoplasm of the cell with the usual cytoplasmic organelles. Particularly important are the mitochondria, for we shall see later that the mitochondria in this segment play an important role in providing most of the energy for function of the photoreceptors.

The synaptic body is the portion of the rod and cone that connects with the subsequent neuronal cells, the horizontal and bipolar cells, that represent the next stages in the visual chain.

Figure 39–12 shows in more detail the photoreceptor discs in the outer segments of the rods and cones. Note especially their attachments to the cell membrane from which they originated.

The Pigment Layer of the Retina. The black

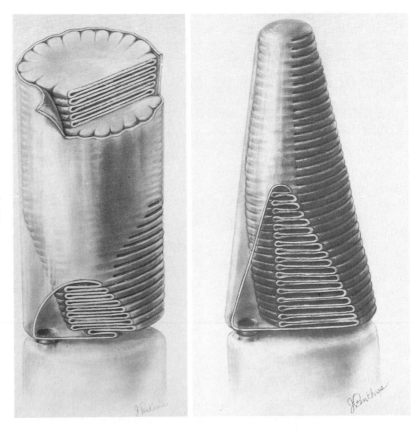

Figure 39–12. Membranous structures of the outer segments of a rod (left) and a cone (right). (Courtesy of Dr. Richard Young.)

pigment *melanin* in the pigment layer, together with still more melanin in the choroid, prevents light reflection throughout the globe of the eyeball, and this is extremely important for acute vision. This pigment performs the same function in the eye as black paint inside the bellows of a camera. Without it, light rays would be reflected in all directions within the eyeball and would cause diffuse lighting of the retina rather than the contrasts between dark and light spots required for formation of precise images.

The importance of melanin in the pigment layer and choroid is well illustrated by its absence in *albino* persons who hereditarily lack melanin pigment in all parts of their bodies. When an albino enters a bright area, light that impinges on the retina is reflected in all directions by the white surface of the unpigmented choroid, so that a single discrete spot of light that would normally excite only a few rods or cones is reflected everywhere and excites many of the receptors. As a result, the visual acuity of albinos, even with the best of optical correction, is rarely better than 20/100 to 20/200.

The pigment layer also stores large quantities of *vitamin A*. This vitamin is exchanged back and forth through the membranes of the outer segments of the rods and cones, which themselves are embedded in the pigment layer. We shall see later that vitamin A is an important precursor of the photosensitive pigments and that this interchange of vitamin A is very important for adjustment of the light sensitivity of the receptors.

PHOTOCHEMISTRY OF VISION

Both the rods and cones contain chemicals that decompose on exposure to light and, in the process, excite the nerve fibers leading from the eye. The chemical in the *rods* is called *rhodopsin,* and the light-sensitive chemicals in the *cones* have compositions only slightly different from that of rhodopsin.

In the present section we will discuss principally the photochemistry of rhodopsin, but we may apply almost exactly the same principles to the photochemistry of the light-sensitive substances of the cones.

THE RHODOPSIN-RETINAL VISUAL CYCLE, AND EXCITATION OF THE RODS

Rhodopsin and Its Decomposition by Light Energy. The outer segment of the rod that projects into the pigment layer of the retina has a concentration of about 40 per cent of the light-

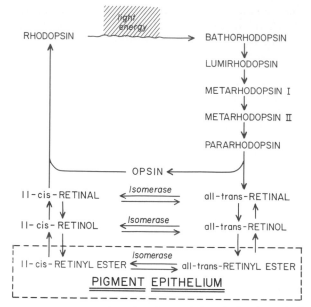

Figure 39–13. Photochemistry of the rhodopsin-retinal-vitamin A visual cycle.

sensitive pigment called *rhodopsin* or *visual purple.* This substance is a combination of the protein *scotopsin* and the carotenoid pigment *retinal* (also called "retinene"). Furthermore, the retinal is a particular type called 11-*cis* retinal. This *cis* form of the retinal is important because only this form can combine with scotopsin to synthesize rhodopsin.

When light energy is absorbed by rhodopsin, the rhodopsin immediately begins to decompose, as shown in Figure 39–13. The cause of this is photoactivation of electrons in the retinal portion of the rhodopsin, which leads to an instantaneous change of the *cis* form of retinal into an all-*trans* form. This still has the same chemical structure as the *cis* form but has a different physical structure — a straight molecule rather than a curved molecule. Because the three-dimensional orientation of the reactive sites of the all-*trans* retinal no longer fit with the orientation of the reactive sites on the protein scotopsin, it begins to pull away from the scotopsin. The immediate product is *bathorhodopsin* (also called "prelumirhodopsin"), which is a partially split combination of the all-*trans* retinal and scotopsin. However, bathorhodopsin is an extremely unstable compound and decays in small fractions of a second to *lumirhodopsin*, then to *metarhodopsin I*, then *metarhodopsin II,* and finally splitting apart to form scotopsin and all-*trans* retinal. During the process of splitting, the rods are excited, and visual signals are transmitted into the central nervous system.

Re-formation of Rhodopsin. The first stage in re-formation of rhodopsin, as shown in Figure 39–13, is to reconvert the all-*trans* retinal into 11-*cis* retinal. This process is catalyzed by the enzyme *retinal isomerase*. However, this process also requires other cellular functions by the rods and cones. Once the 11-*cis* retinal is formed, it automatically recombines with the scotopsin to reform rhodopsin, an exergonic process (which means that it gives off energy). The product, rhodopsin, is a stable compound in the dark, but its decomposition can again be triggered by absorption of light energy.

Excitation of the Rods When Rhodopsin Decomposes. Although the exact way in which rhodopsin decomposition excites the rod is still speculative, it is believed to occur in the following general way: The *retinal* portion of rhodopsin is a photosensitive compound that becomes activated in the presence of light. The initial effect of the light is to cause photoexcitation of electrons in the retinal molecule, raising these electrons to higher energy states. During this excited state, the internal physical structure (but not yet its chemical structure) begins to change, and the ultimate effect is conversion of the 11-*cis*-retinal into all-*trans*-retinal and ultimate splitting of the rhodopsin molecule as described above.

Stimulation of the rod takes place at the very outset of rhodopsin decomposition. Therefore, it is believed that the initial photoexcitation of the retinal portion of the rhodopsin molecule is the prime cause of the stimulation. This could result in one of several ways: One of the suggestions has been that the excited electrons themselves in some way lead directly to a change in the membrane conductance. A second suggestion has been that the rhodopsin molecule itself undergoes an instantaneous conformational change that alters the actual physical structure of the membrane, in this way changing its conductance at the same time.

The physical construction of the outer segment of the rod is compatible with both these concepts because the cell membrane folds inward to form tremendous numbers of shelf-like discs lying one on top of the other, each one lined on both the inside and outside of the membrane with aggregates of rhodopsin. This extensive relationship between rhodopsin and the cell membrane could explain why the rod is so exquisitely sensitive to light, being capable of detectable excitation following absorption of only one quantum of light energy.

Generation of the Receptor Potential. Regardless of the precise mechanism by which the membrane of the outer segment becomes altered in the presence of decomposing rhodopsin, it is known that this excitation process *decreases* the conductance of this membrane for sodium ions. Based on this fact, the following theory for development of the receptor potential has been proposed:

Figure 39–14 illustrates movement of sodium

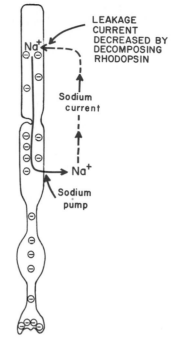

Figure 39–14. Theoretical basis for the generation of a hyperpolarization receptor potential caused by rhodopsin decomposition.

ions in a complete electrical circuit through the inner and outer segments of the rod. The inner segment continually pumps sodium from inside the rod to the outside, thereby creating a negative potential on the inside of the entire cell. However, the membrane of the outer segment of the rod, in the unexcited condition, is very leaky to sodium. Therefore, sodium continually leaks back to the inside of the rod and thereby continually neutralizes much of the negativity on the inside of the entire cell. Thus, under normal conditions, when the rod is not excited, there is a reduced amount of electronegativity inside the membrane of the rod.

When the rhodopsin in the outer segment of the rod is exposed to light and begins to decompose, this *decreases* the leakage of sodium to the interior of the rod even though sodium continues to be pumped out. Therefore, a net loss of sodium from the rod occurs, and this creates increased negativity inside the membrane. Thus, the greater the amount of light energy striking the rod, the greater the electronegativity. This process is called *hyperpolarization*. It is *exactly opposite* to the effect that occurs in almost all other sensory receptors, in which the degree of negativity is generally reduced during stimulation rather than increased, producing a state of depolarization rather than hyperpolarization.

Relationship Between Retinal and Vitamin A. The lower part of the scheme in Figure 39–13 illustrates that each of the two types of retinal can be converted into corresponding types of *retinol* and *retinyl ester*, both of which are forms of vitamin A. In turn, both of these can be reconverted into the two types of retinal. Thus, the two retinals are in dynamic equilibrium with vitamin A. Most of the vitamin A of the retina is stored in the pigment layer of the retina rather than in the rods themselves, but this vitamin A is readily available to the rods.

An important feature of the conversion of retinal into vitamin A, or the converse conversion of vitamin A into retinal, is that all these processes require a much longer time to approach equilibrium than it takes for conversion of retinal and scotopsin into rhodopsin, or for conversion (under the influence of strong light energy) of rhodopsin into retinal and scotopsin. Therefore, all the reactions of the upper part of Figure 39–13 can take place within minutes (some even in fractions of a second) in comparison with the slower interconversions between retinal and vitamin A (occurring in many minutes to hours).

Yet, if the retina remains exposed to strong light for a long time, most of the stored rhodopsin will be converted eventually into vitamin A, thereby decreasing the concentration of all the photochemicals in the rods much more than would be true were it not for this subsequent conversion.

Conversely, during total darkness, essentially all the retinal already in the rods becomes converted into rhodopsin within a few minutes. This then allows much of the vitamin A to be converted into still additional retinal, which also becomes rhodopsin within another few minutes. Thus, when a person remains in complete darkness for a prolonged period of time, not only is almost all the retinal of his rods converted into rhodopsin but also much of the vitamin A stored in the pigment layer of the retina is absorbed by the rods and also converted into rhodopsin.

NIGHT BLINDNESS

Night blindness occurs in severe vitamin A deficiency. When the total quantity of vitamin A in the blood becomes greatly reduced, the quantities of vitamin A, retinal, and rhodopsin in the rods, as well as the color photosensitive chemicals in the cones, are all depressed, thus decreasing the sensitivities of the rods and cones. This condition is called night blindness because at night the amount of available light is far too little to permit adequate vision, though in daylight, sufficient light is available to excite the rods and cones despite their reduction in photochemical substances.

For night blindness to occur, a person often must remain on a vitamin A deficient diet for months, because large quantities of vitamin A are normally stored in the liver and are made available to the rest of the body in times of need. However, once night blindness does develop it can sometimes be completely cured in a half hour or more by intravenous injection of vitamin A. This results from the ready conversion of vitamin A into retinal and thence into rhodopsin.

PHOTOCHEMISTRY OF COLOR VISION BY THE CONES

It was pointed out at the outset of this discussion that the photochemicals (also called *visual pigments*) in the cones have almost exactly the same chemical compositions as rhodopsin in the rods. The only difference is that the protein portions of the photochemicals, the opsins, (called *photopsins* in the cones) are different from the scotopsin of the rods. The retinal portions are exactly the same in the cones as in the rods. The color-sensitive pigments of the cones, therefore, are combinations of retinal and photopsins.

In the discussion of color vision later in the chapter, it will become evident that three different types of photochemicals are present in different cones, thus making these cones selectively sensitive to the different colors, blue, green, and red. The absorption characteristics of the photochemicals in the three types of cones show peak absorbencies at light wavelengths, respectively, of 430, 535, and 575 millimicrons. These are also the wavelengths for peak light sensitivity for each type of cone, which begins to explain how the retina differentiates the colors. The approximate absorption curves for these three

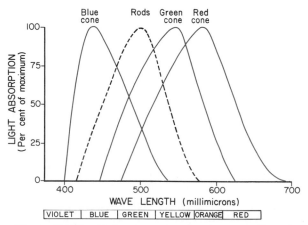

Figure 39–15. Light absorption by the respective pigments of the three color-receptive cones of the human retina. (Drawn from curves recorded by Marks, Dobelle, and MacNichol, Jr.: *Science, 143*:1181, 1964, and by Brown and Wald: *Science, 144*:45, 1964. Copyright 1964 by the American Association for the Advancement of Science.)

photochemicals are shown by the colored curves in Figure 39–15. The peak absorption for the rhodopsin of the rods, on the other hand, occurs at 505 millimicrons.

AUTOMATIC REGULATION OF RETINAL SENSITIVITY – DARK AND LIGHT ADAPTATION

Relationship of Sensitivity to Photochemical Concentration. The sensitivity of rods is approximately proportional to the antilogarithm of the rhodopsin concentration, and it is assumed that this relationship also holds true in the cones. Therefore, the sensitivity of the rods and cones can be altered up or down tremendously by only slight changes in concentrations of the photosensitive chemicals.

Light and Dark Adaptations. If a person has been in bright light for a long time, large proportions of the photochemicals in both the rods and cones have been reduced to retinene and opsins. Furthermore, most of the retinal of both the rods and cones has been converted into vitamin A. Because of these two effects, the concentrations of the photosensitive chemicals are considerably reduced, and the sensitivity of the eye to light is even more reduced. This is called *light adaptation.*

On the other hand, if the person remains in the darkness for a long time, essentially all the retinal and opsins in the rods and cones become converted into light-sensitive pigments. Furthermore, large amounts of vitamin A are converted into retinal, which is then changed into additional light-sensitive pigments, the final limit being determined by the amount of opsins in the rods and cones. Because of these two effects, the visual receptors gradually become so sensitive that even the most minute amount of light causes excitation. This is called *dark adaptation.*

Value of Light and Dark Adaptation in Vision. Between the limits of maximal dark adaptation and maximal light adaptation, the eye can change its sensitivity to light by as much as 500,000 to 1,000,000 times, the sensitivity automatically adjusting to changes in illumination.

Since the registration of images by the retina requires detection of both dark and light spots in the image, it is essential that the sensitivity of the retina always be adjusted so that the receptors respond to the lighter areas but not to the darker areas. An example of maladjustment of the retina occurs when a person leaves a movie theater and enters the bright sunlight; even the dark spots in the images then seem exceedingly bright, and as a consequence, the entire visual image is bleached, having little contrast between its different parts.

Obviously, this is poor vision, and it remains poor until the retina has adapted sufficiently that the dark spots of the image no longer stimulate the receptors excessively.

FUSION OF FLICKERING LIGHT BY THE RETINA

A flickering light is one whose intensity alternately increases and decreases rapidly. An instantaneous flash of light excites the visual receptors for as long as 1/10 to 1/15 second, and because of this *persistence* of excitation, rapidly successive flashes of light become *fused* together to give the appearance of being continuous. This well-known effect is demonstrated when one observes motion pictures or television. The images on the motion picture screen are flashed at a rate of 24 frames per second, while those of the television screen are flashed at a rate of 60 frames per second. As a result, the images fuse together, and continuous motion is observed.

The frequency at which flicker fusion occurs, called the *critical frequency for fusion*, varies with the light intensity. At a low intensity, fusion results even when the rate of flicker is as low as 2 to 6 per second. However, in bright illumination, the critical frequency for fusion rises to as great as 60 flashes per second. This difference results at least partly from the fact that the cones, which operate mainly at high levels of illumination, can detect much more rapid alterations in illumination than can the rods, which are the important receptors in dim light.

COLOR VISION

From the preceding sections, we know that different cones are sensitive to different colors of light. The present section is a discussion of the mechanisms by which the retina detects the different gradations of color in the visual spectrum.

THE TRI-COLOR THEORY OF COLOR PERCEPTION

Many different theories have been proposed to explain the phenomenon of color vision, but they are all based on the well-known observation that the human eye can detect almost all gradations of colors when red, green, and blue monochromatic lights are appropriately mixed in different combinations.

The first important theory of color vision, that

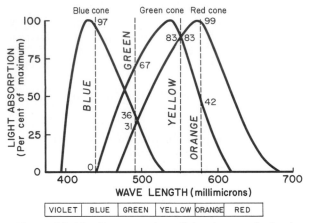

Figure 39–16. Demonstration of the degree of stimulation of the different color-sensitive cones by monochromatic lights of four separate colors: blue, green, yellow, and orange.

of Young, was later expanded and given a more experimental basis by Helmholtz. Therefore, the theory is known as the *Young-Helmoholtz theory.* According to this theory, there are three different types of cones, each of which responds maximally to a different color.

As time has gone by, the Young-Helmholtz theory has been expanded, and more details have been worked out. It now is generally accepted as an adequate description of the mechanism of color vision.

Spectral Sensitivities of the Three Types of Cones. On the basis of psychological tests, the spectral sensitivities of the three different types of cones in human beings are essentially the same as the light absorption curves for the three types of pigment found in the respective cones. These were illustrated in Figure 39–15 and are also shown in Figure 39–16. These curves can readily explain all the phenomena of color vision.

Interpretation of Color in the Nervous System. Referring to Figure 39–16, one can see that an orange monochromatic light with a wavelength of 580 millimicrons stimulates the red cones to a stimulus value of approximately 99 (99 per cent of the peak stimulation at optimum wavelength), while it stimulates the green cones to a stimulus value of approximately 42 and the blue cones not at all. Thus, the ratios of stimulation of the three different types of cones in this instance are 99:42:0. The nervous system interprets this set of ratios as the sensation of orange. On the other hand, a monochromatic blue light with a wavelength of 450 millimicrons stimulates the red cones to a stimulus value of 0, the green cones to a value of 0, and the blue cones to a value of 97. This set of ratios–0:0:97–is interpreted by the nervous system as blue. Likewise, ratios of

83:83:0 are interpreted as yellow, and 31:67:36, as green.

This scheme also shows how it is possible for a person to perceive a sensation of yellow when a red light and a green light are shined into the eye at the same time, for this stimulates the red and green cones approximately equally, which gives a sensation of yellow even though no wavelength of light corresponding to yellow is present.

Perception of White Light. Approximately equal stimulation of all the red, green, and blue cones gives one the sensation of seeing white. Yet there is no wavelength of light corresponding to white; instead, white is a combination of all the wavelengths of the spectrum. Furthermore, the sensation of white can be achieved by stimulating the retina with a proper combination of only three chosen colors that stimulate the respective types of cones.

COLOR BLINDNESS

Red-Green Color Blindness. When a single group of color receptive cones is missing from the eye, the person is unable to distinguish some colors from others. As can be observed by studying Figures 39–15 and 39–16, if the red cones are missing, light of 525 to 625 millimicrons wavelength can stimulate only the green-sensitive cones, so that the *ratio* of stimulation of the different cones does not change as the color changes from green all the way through the red spectrum. Therefore, within this wavelength range, all colors appear to be the same to this "color blind" person.

On the other hand, if the green-sensitive cones are missing, the colors in the range from green to red can stimulate only the red-sensitive cones, and the person also perceives only one color within these limits. Therefore, when a person lacks either the red or green types of cones, he is said to be "red-green" color blind.

Blue Weakness. Occasionally, a person has "blue weakness," which results from diminished or absent blue receptors.

Stilling and Ishihara Test Charts. A rapid method for determining color blindness is based on the use of spot-charts such as those illustrated in Figure 39–17. These charts are arranged with a confusion of spots of several different colors. In the top chart, the normal person reads "74," while the red or green color blind person reads "21." In the bottom chart, the normal person reads "42," while the red blind person reads "2," and the green blind person reads "4."

If one will study these charts while at the same time observing the spectral sensitivity curves of

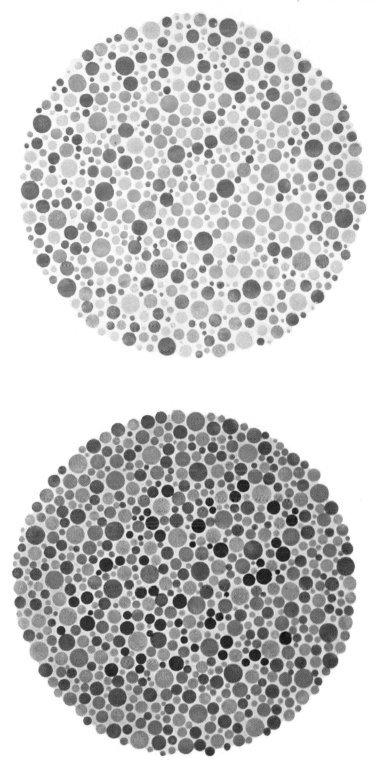

Figure 39-17.   Two Ishihara charts. *Upper:* In this chart, the normal person reads "74," whereas the red-green color blind person reads "21." *Lower:* In this chart, the red-blind person (protanope) reads "2," while the green-blind person (deuteranope) reads "4." The normal person reads "42." (The above has been reproduced from Ishihara's Tests for Colour Blindness published by Kanehara & Co. Ltd., Tokyo, Japan, but test for color blindness cannot be conducted with this material. For accurate testing, the original plates should be used.)

the different cones in Figure 39–16, he can readily understand how excessive emphasis can be placed on spots of certain colors by color blind persons in comparison with normal persons.

REFERENCES

Optics of Vision

Allen, E. W.: Essentials of Ophthalmic Optics. New York, Oxford University Press, 1979.

Davson, H.: The Physiology of the Eye, 3rd Ed. New York. Academic Press, 1972.

Davson, H., and Graham, L. T., Jr.: The Eye. Vols. 1–6. New York, Academic Press, 1969–1974.

Lerman, S.: Radiant Energy and the Eye. New York, The Macmillan Co., 1979.

Miller, D.: Ophthalmology: The Essentials. Boston, Houghton Mifflin, 1979.

Morgan, M. W.: The Optics of Ophthalmic Lenses. Chicago, Professional Press, 1978.

Polyak, S.: The Vertebrate Visual System. Chicago, University of Chicago Press, 1957.

Records, R. E.: Physiology of the Human Eye and Visual System. Hagerstown, Md., Harper & Row, 1979.

Regan, D., et al.: The visual perception of motion in depth. Sci. Am., 241(1):136, 1979.

Roth, H. W., and Roth-Wittig, M.: Contact Lenses. Hagerstown, Md., Harper & Row, 1980.

Safir, A. (ed.): Refraction and Clinical Optics. Hagerstown, Md., Harper & Row, 1980.

Toates, F. M.: Accommodation function of the human eye. Physiol. Rev., 52:828, 1972.

Van Heyningen, R.: What happens to the human lens in cataract. Sci. Am., 233(6):70, 1975.

Whitteridge, D.: Binocular vision and cortical function. Proc. R. Soc. Med., 65:947, 1972.

The Retina

Brindley, G. S.: Physiology of the Retina and Visual Pathway, 2nd Ed. Baltimore, Williams & Wilkins, 1970.

Callender, R., and Honig, B.: Resonance Raman studies of visual pigments. Annu. Rev. Biophys. Bioeng., 6:33, 1977.

Cervetto, L., and Fuortes, M. G. F.: Excitation and interactions in the retina. Annu. Rev. Biophys. Bioeng., 7:229, 1978.

Daw, N. W.: Neurophysiology of color vision. Physiol. Rev., 53:571, 1973.

Fatt, I.: Physiology of the Eye: An Introduction to the Vegetative Function. Boston, Butterworths, 1978.

Fine, B. S., and Yanoff, M.: Ocular Histology: A Text and Atlas. Hagerstown, Md., Harper & Row, 1979.

Kaneko, A.: Physiology of the retina. Annu. Rev. Neurosci., 2:169, 1979.

Land, E. H.: The retinex theory of color vision. Sci Am., 237(6):108, 1977.

MacNichol, E. F., Jr.: Three-pigment color vision. Sci. Am., 211:48, 1964.

Marks, W. B., et al.: Visual pigments of single primate cones. Science, 143:1181, 1964.

Michaelson, I. C.: Textbook of the Fundus of The Eye. New York, Churchill Livingstone, 1980.

Rushton, W. A. H.: Visual pigments and color blindness. Sci. Am., 232(3):64, 1975.

Schepens, C. L.: Retinal Detachment and Allied Diseases. Philadelphia, W. B. Saunders Co., 1981.

Young, R. W.: Proceedings: Biogenesis and renewal of visual cell outer segment membranes. Exp. Eye Res., 18:215, 1974.

Zinn, K. M., and Marmor, M. F. (eds.): The Retinal Pigment Epithelium. Cambridge, Mass., Harvard University Press, 1979.

40

THE EYE: II. NEUROPHYSIOLOGY OF VISION

THE VISUAL PATHWAY

Figure 40–1 illustrates the visual pathway from the two retinae back to the *visual cortex*. After impulses leave the retinae they pass backward through the *optic nerves*. At the *optic chiasm* all the fibers from the opposite nasal halves of the retinae cross and join the fibers from the opposite temporal retinae to form the *optic tracts*. The fibers of each optic tract synapse in the *lateral geniculate body*, and from here the *geniculocalcarine fibers* pass through the *optic radiation*, or *geniculocalcarine tract*, to the *optic* or *visual cortex* in the calcarine area of the occipital lobe.

In addition, visual fibers pass to lower areas of the brain, into the lateral thalamus, the superior colliculi, and the pretectal nuclei.

NEURAL FUNCTION OF THE RETINA

NEURAL ORGANIZATION OF THE RETINA

The detailed anatomy of the retina was illustrated in Figure 39–9 of the preceding chapter, and Figure 40–2 illustrates the essentials of the retina's neural connections; to the left is the general organization of the neural elements in a peripheral retinal area and to the right the organization of the foveal area. Note that in the peripheral region both rods and cones converge

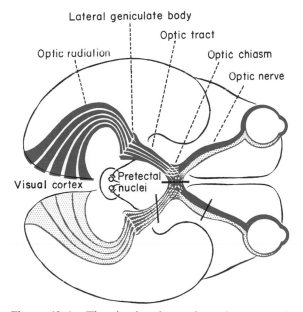

Figure 40–1. The visual pathways from the eyes to the visual cortex. (Modified from Polyak: The Retina. University of Chicago Press.)

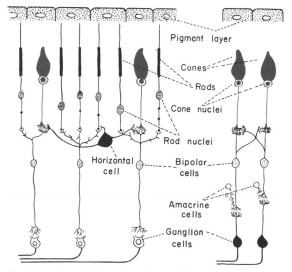

Figure 40–2. Neural organization of the retina; peripheral area to the left, foveal area to the right.

on *bipolar cells*, which in turn converge on *ganglion cells*. In the fovea, where only cones exist, there is little convergence; instead, the cones are represented by approximately equal numbers of bipolar and ganglion cells.

In addition, two other special types of cells are present in the inner nuclear layer of the retina, (1) the *horizontal cells* and (2) the *amacrine cells*. These will be discussed later.

Each retina contains about 125 million rods and 5.5 million cones; yet, as counted with the light microscope, only 900,000 optic nerve fibers lead from the retina to the brain. Thus, an average of 140 rods plus 6 cones converge on each optic nerve fiber. However, there are major differences between the peripheral retina and the central retina, for nearer the fovea, fewer and fewer rods and cones converge on each optic fiber, and the rods and cones both become slenderer. These two effects progressively increase the acuity of vision toward the central retina. In the very central portion, in the fovea, there are no rods at all. Also, the number of optic nerve fibers leading from this part of the retina is almost equal to the number of cones in the fovea, as shown to the right in Figure 40–2. This mainly explains the high degree of visual acuity in the central portion of the retina in comparison with the very poor acuity in the peripheral portions.

Another difference between the peripheral and central portions of the retina is that there is considerably greater sensitivity of the peripheral retina to weak light. This results partly from the fact that as many as 600 rods converge on the same optic nerve fiber in the most peripheral portions of the retina, but the rods are also more sensitive to weak light than are the cones.

STIMULATION OF THE RODS AND CONES – THE RECEPTOR POTENTIAL

As explained in the previous chapter, when light strikes either the rods or the cones, a photochemical decomposes and acts on the membrane of the receptor to cause a *receptor potential* that lasts as long as the light continues. However, this receptor potential is different from all other receptor potentials that have been recorded, for it is a *hyperpolarization* signal rather than a depolarization signal; that is, the membrane potential becomes more negative than ever rather than less negative.

Receptor potentials are transmitted unchanged through the bodies of the rods and cones. Neither the rods nor the cones generate action potentials. Instead, the receptor potentials themselves, acting at the synaptic bodies, induce signals in the successive neurons, the bipolar and the horizontal cells.

STIMULATION OF THE BIPOLAR AND HORIZONTAL CELLS

The synaptic bodies of the rods and cones make intimate contact with the dendrites of both the bipolar and the horizontal cells. It is believed that the rods and cones secrete a transmitter substance that induces potential changes in the bipolar and horizontal cells. Unfortunately, the exact nature of the transmitter is yet unknown.

When the rods and cones are excited, the bipolar cells also become excited, while the horizontal cells become inhibited. However, neither the bipolar cells nor the horizontal cells transmit action potentials. Instead, their signals, like those of the rods and cones, result from depolarization or hyperpolarization of the entire neuronal membrane all at once.

Function of the Bipolar Cells. The bipolar cells are the main transmitting link for the visual signal from the rods and cones to the ganglion cells. They are entirely excitatory cells; that is, they can only excite the ganglion cells.

Function of the Horizontal Cells. The horizontal cells lie in the inner nuclear layer and spread widely in the retina, transmitting signals laterally as far as several hundred microns. They are inhibited by the synaptic bodies of the rods and cones and in turn transmit their inhibitory signals mainly to bipolar cells located in areas lateral to the excited rods and cones. That is, excited rods and cones transmit excitatory signals in a direct line through the bipolar cells in the area of excitation but transmit inhibitory signals through the surrounding bipolar cells.

The horizontal cells are of major importance for enhancing contrast in the visual scene. They are also of importance in helping to differentiate colors. Both of these functions will be discussed in subsequent sections of this chapter.

Stimulation and Function of the Amacrine Cells. The amacrine cells, also located in the inner nuclear layer, are excited mainly by the bipolar cells but possibly also occasionally directly by the synaptic bodies of the rods and cones. These cells in turn synapse with the ganglion cells. However, their response is a *transient* one rather than the steady, continuous response of the bipolar and horizontal cells. That is, when the photoreceptors are first stimulated, the signal transmitted by the amacrine cells is very intense; this signal dies away to almost nothing in a fraction of a second. Therefore, the function of the amacrine cells seems to be to detect *instantaneous changes* in the visual image.

Bipolar cells, horizontal cells, and amacrine cells are all believed to secrete transmitter substances at their nerve terminations, but the types of transmitter substances are yet unknown. Some of the different transmitter substances that have been found in the retina include acetylcholine, dopamine, GABA, glutamate ions, and aspartate ions.

EXCITATION OF THE GANGLION CELLS

Spontaneous, Continuous Discharge of the Ganglion Cells. The ganglion cells transmit their signals through the optic nerve fibers to the brain in the form of action potentials. These cells, even when unstimulated, transmit continuous nerve impulses at an average rate of about 5 per second. The visual signal is superimposed onto this basic level of ganglion cell stimulation. It can be either an excitatory signal, with the number of impulses increasing to far greater than 5 per second, or an inhibitory signal, with the number of nerve impulses decreasing to below 5 per second — often all the way to zero.

Summation at the Ganglion Cells of Signals from the Bipolar Cells, the Horizontal Cells, and the Amacrine Cells. The bipolar cells transmit the main direct *excitatory* information from the rods and the cones to the ganglion cells; the horizontal cells transmit *inhibitory* information from laterally displaced rods and cones; and the amacrine cells seem to transmit direct but short-lived transient signals that signal a *change* in the level of illumination of the retina. Thus, each of these three types of cells performs a separate function in stimulating the ganglion cells.

DIFFERENT TYPES OF SIGNALS TRANSMITTED BY THE GANGLION CELLS THROUGH THE OPTIC NERVE

Transmission of Luminosity Signals. The ganglion cells are of several different types, and they also are stimulated differently by the bipolar, horizontal, and amacrine cells. A small proportion of the ganglion cells responds mainly to the intensity (*luminosity*) of the light falling on the photoreceptors. The rate of impulses from these cells remains at a level greater than the natural rate of firing as long as the luminosity is high. It is the signals from these cells that apprise the brain of the overall level of light intensity of the observed scene.

Transmission of Signals Depicting Contrasts in the Visual Scene — the Process of Lateral Inhibition. Most of the ganglion cells barely respond to the actual level of illumination of the scene; instead, they respond only to contrast borders in the scene. Since this is the major means by which the form of the scene is transmitted to the brain, let us explain how this process occurs.

When flat light is applied to the entire retina —

that is, when all the photoreceptors are stimulated equally by the incident light — the contrast type of ganglion cell is neither stimulated nor inhibited. The reason is that the signals transmitted *directly* from the photoreceptors through the bipolar cells are excitatory, whereas the signals transmitted from *laterally displaced* photoreceptors through the horizontal cells are inhibitory. These two effects neutralize each other. On the other hand, when there is a contrast border, with stimulated photoreceptors on one side of the border and unstimulated photoreceptors on the opposite side, the mutual cancellation of the excitatory signals directly through the bipolar cells and the inhibitory signals through the horizontal cells no longer occurs. Consequently, the ganglion cells along the border become excited on the light side of the border and inhibited on the dark side.

This process of balancing an excitatory signal with an inhibitory signal is exactly the same as the process of *lateral inhibition* that occurs in almost all other types of sensory signal transmission. The process is a mechanism used by the nervous system for contrast enhancement.

Detection of Instantaneous Changes in Light Intensity. Many of the ganglion cells are especially excited by *change* in light intensity; this effect most often occurs in the same ganglion cells that transmit contrast border signals.

This ability of the retina to detect and transmit signals related to *change* in light intensity is caused by a rapid phase of "adaptation" of some of the neurons in the visual chain. Since this effect is extremely marked in the amacrine cells, it is believed that the amacrine cells are peculiarly adapted to help the retina detect light intensity changes.

This capability to detect change in light intensity is especially well developed in the peripheral retina. For instance, a minute gnat flying across the peripheral field of vision is instantaneously detected. On the other hand, the same gnat sitting quietly in the peripheral field of vision remains entirely below the threshold of visual detection.

Transmission of Color Signals by the Ganglion Cells. A single ganglion cell may be stimulated by a number of cones or by only a very few. When all three types of cones — the red, blue, and green types — all stimulate the same ganglion cell, the signal transmitted through the ganglion cell is the same for any color of the spectrum. Therefore, this signal plays no role in the detection of the different colors. Instead, it is a "white light" signal.

On the other hand, many ganglion cells are excited by only one color type of cone but are inhibited by a second color type. For instance, this frequently occurs for the red and green cones, red causing excitation and green causing inhibition — or, vice versa, green causing excitation and red inhibition. The same type of reciprocal effect also occurs between some blue cones and either red or green cones.

The mechanism of this opposing effect of colors is the following: One color type of cones excites the ganglion cell by the direct excitatory route through a bipolar cell, while the other color type inhibits the ganglion cell by the indirect inhibitory route through a horizontal cell.

The importance of these color-contrast mechanisms is that they represent a mechanism by which the retina itself differentiates colors. Thus, each color-contrast type of ganglion cell is excited by one color but inhibited by the opposite color. Therefore, the process of color analysis begins in the retina and is not entirely a function of the brain.

FUNCTION OF THE LATERAL GENICULATE BODY

Each lateral geniculate body is composed of six nuclear layers. Layers 2, 3, and 5 (from the surface inward) receive optic nerve fibers from the temporal portion of the ipsilateral retina, while layers 1, 4, and 6 receive fibers from the nasal retina of the opposite eye.

All layers of the lateral geniculate body relay visual information to the *visual cortex* through the *geniculocalcarine tract.*

The pairing of layers from the two eyes probably plays a major role in *fusion of vision*, because corresponding retinal fields in the two eyes connect with respective neurons that are approximately superimposed over each other in the successive layers. Also, with a little imagination, one can postulate that interaction between the successive layers could be part of the mechanism by which stereoscopic depth perception occurs, because this depends on comparing the visual images of the two eyes and determining their slight differences, as was discussed in Chapter 39.

The signals recorded in the relay neurons of the lateral geniculate body are similar to those recorded in the ganglion cells of the retina. A few of the neurons transmit luminosity signals, while the majority transmit signals depicting only contrast borders in the visual image; also, many of the neurons are particularly responsive to movement of objects across the visual scene. However, the signals of the geniculate neurons are different from those in the ganglion cells in that a greater number of the complex interactions are found. That is, a much higher percentage of the neurons respond to contrast in the visual scene or

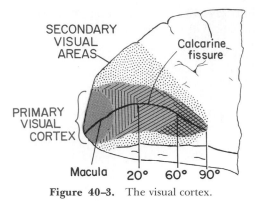

Figure 40–3. The visual cortex.

to movement. These more complex reactions presumably result from convergence of excitatory and inhibitory signals from two or more ganglion cells on the relay neurons of the lateral geniculate body.

FUNCTION OF THE PRIMARY VISUAL CORTEX

The ability of the visual system to detect spatial organization of the visual scene — that is, to detect the forms of objects, brightness of the individual parts of the objects, shading, and so forth — is dependent on the function of the *primary visual cortex,* the anatomy of which is illustrated in Figure 40–3. This area lies mainly in the calcarine fissure, located bilaterally on the medial aspect of each occipital cortex. Specific points of the retina connect with specific points of the visual cortex, the right halves of the two respective retinae connecting with the right visual cortex and the left halves with the left visual cortex. The macula is represented at the occipital pole of the visual cortex and the peripheral regions of the retina are represented in concentric arcs farther and farther forward from the occipital pole. The upper portion of the retina is represented superiorly in the visual cortex and the lower portion inferiorly. Note the large area of cortex receiving signals from the macular region of the retina. It is in this region that the fovea, which gives the highest degree of visual acuity, is represented.

DETECTION OF LINES AND BORDERS BY THE PRIMARY VISUAL CORTEX

If a person looks at a blank wall, only a few neurons of the primary visual cortex will be stimulated, whether the illumination of the wall is bright or weak. Therefore, the question must be asked: What does the visual cortex do? To answer

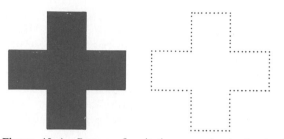

Figure 40–4. Pattern of excitation occurring in the visual cortex in response to a retinal image of a dark cross.

this question, let us now place on the wall a large black cross such as that illustrated to the left in Figure 40–4. To the right is illustrated the spatial pattern of the greater majority of the excited neurons that one finds in the visual cortex. *Note that the areas of excitation occur along the sharp borders of the visual pattern.* Thus, by the time the visual signal is recorded in the primary visual cortex, it is concerned mainly with the *contrasts* in the visual scene rather than with the flat areas. At each point in the visual scene where there is a change from dark to light or light to dark, the corresponding area of the primary visual cortex becomes stimulated. The intensity of stimulation is determined by the *gradient of contrast.* That is, the greater the sharpness in the contrast border and the greater the difference in intensities between the light and dark areas, the greater is the degree of stimulation.

Thus, the *pattern of contrasts* in the visual scene is impressed upon the neurons of the visual cortex, and this pattern has a spatial orientation roughly the same as that of the retinal image.

Not only does the visual cortex detect the *existence* of lines and borders in the different areas of the retinal image, but it also detects the *orientation* of each line or border as well as its length. This is achieved by excitation of specific secondary neurons by lines of one orientation and length and other secondary neurons by lines of other orientations and lengths.

Analysis of Color by the Visual Cortex. One finds in the primary visual cortex specific cells that are stimulated by color intensity or by contrasts of the opponent colors, the red-green opponent colors and the blue-yellow opponent colors. These effects are almost identical to those found in the lateral geniculate body. However, the proportion of cells excited by the opponent color contrasts is vastly reduced from the proportion found in the lateral geniculate body. Since neuronal excitation by color contrasts is a means of deciphering color, it is believed that the primary visual cortex is concerned with an even higher order of detection of color than simply the deciphering of color itself, a process that seems to be

at least partially completed by the time the sig[nals] have passed through the lateral genicula[te] body.

PERCEPTION OF LUMINOSITY

Although most of the neurons in the visual cortex are mainly responsive to contrasts caused by lines, borders, moving objects, or opponent colors in the visual scene, a few are directly responsive to the levels of luminosity in the different areas of the visual scene. Presumably, it is these cells that detect flat areas in the scene and also the overall level of luminosity.

TRANSMISSION OF VISUAL INFORMATION INTO OTHER REGIONS OF THE CEREBRAL CORTEX

Signals from the primary visual cortex project laterally in the occipital cortex into *visual association areas* (also called *secondary visual areas*), which are loci for additional processing of visual information.

Here the neurons respond to more complex patterns than do those in the primary visual cortex. For instance, some cells are stimulated by simple geometric patterns, such as curving borders and angles. It is presumably these progressively more complex interpretations that eventually decode the visual information, giving the person his overall impression of the visual scene that he is observing.

Human beings who have destructive lesions of the visual association areas have difficulty with certain types of visual perception and visual learning. For instance, a lesion in the angular gyrus of the occipital lobe, one of the visual association areas, can cause the abnormality known as *dyslexia* or *word blindness*, which means that the person has difficulty understanding the meanings of words that he sees.

EYE MOVEMENTS AND THEIR CONTROL

To make use of the abilities of the eyes, almost as important as the system for interpretation of the visual signals from the eyes is the cerebral control system for directing the eyes toward the object to be viewed.

Muscular Control of Eye Movements. The eye movements are controlled by three separate pairs of muscles, shown in Figure 40–5: (1) the medial and lateral recti, (2) the superior and inferior recti, and (3) the superior and inferior

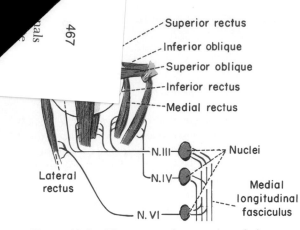

- Superior rectus
- Inferior oblique
- Superior oblique
- Inferior rectus
- Medial rectus

Lateral rectus

N.III — Nuclei
N.IV
N. VI — Medial longitudinal fasciculus

Figure 40–5. The extraocular muscles of the eye and their innervation.

obliques. The medial and lateral recti contract reciprocally to move the eyes from side to side. The superior and inferior recti contract reciprocally to move the eyes upward or downward. And the oblique muscles function mainly to rotate the eyeballs to keep the visual fields in the upright position.

Neural Pathways for Control of Eye Movements. Figure 40–5 also illustrates the nuclei of the third, fourth, and sixth cranial nerves and their innervation of the ocular muscles. Shown, too, are the interconnections between these three nuclei through the *medial longitudinal fasciculus.* Either by way of this fasciculus or by way of other closely associated pathways, each of the three sets of muscles to each eye is *reciprocally* innervated, so that one muscle of the pair relaxes while the other contracts.

Figure 40–6 illustrates cortical control of the oculomotor apparatus, showing spread of signals from the occipital visual areas through occipitotectal and occipitocollicular tracts into the pretectal and superior colliculus areas of the brain stem. In addition, a frontotectal tract passes from the frontal cortex into the pretectal area. From both the pretectal and the superior colliculus areas, the oculomotor control signals then pass to the nuclei of the oculomotor nerves. Finally, strong signals are also transmitted into the oculomotor system from the vestibular nuclei by way of the medial longitudinal fasciculus.

FIXATION MOVEMENTS OF THE EYES

Perhaps the most important movements of the eyes are those that cause the eyes to "fix" on a discrete portion of the field of vision.

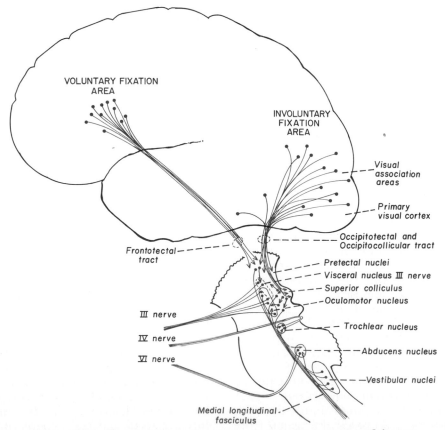

Figure 40–6. Neural pathways for control of conjugate movement of the eyes.

Fixation movements are controlled by two entirely different neuronal mechanisms. The first of these allows the person to move his eyes voluntarily to find the object upon which he wishes to fix his vision; this is called the *voluntary fixation mechanism.* The second is an involuntary mechanism that holds the eyes firmly on the object once it has been found; this is called the *involuntary fixation mechanism.*

The voluntary fixation movements are controlled by a small cortical field located bilaterally in the premotor cortical regions of the frontal lobes, as illustrated in Figure 40–6. Bilateral dysfunction or destruction of these areas makes it difficult or almost impossible for the person to "unlock" his eyes from one point of fixation and then move them to another point. It is usually necessary for him to blink his eyes or put his hand over his eyes for a short time, which then allows him to move the eyes.

On the other hand, the fixation mechanism that causes the eyes to "lock" on the object of attention once it is found is controlled by the *eye fields of the occipital cortex,* which are also illustrated in Figure 40–6. When these areas are destroyed bilaterally, the person has difficulty or becomes completely unable to keep his eyes directed toward a given fixation point.

Mechanism of Fixation. Visual fixation results from a negative feedback mechanism that prevents the object of attention from leaving the foveal portion of the retina. Once a spot of light has become fixed on the foveal region of the retina, each time the spot drifts as far as the edge of the fovea, a sudden *flicking movement* of the eyes occurs and moves the spot away from this edge back toward the center, which is an automatic response to move the image back toward the central portion of the fovea. These drifting and flicking motions are illustrated in Figure 40–7, which shows by dashed lines the slow drifting across the retina and by solid lines the flicks that keep the image from leaving the foveal region.

Saccadic Movement of the Eyes. When the visual scene is moving continuously before the eyes, as when a person is riding in a car or when he is turning around, the eyes fix on one highlight after another in the visual field, jumping from one to the next at a rate of two to three jumps per second. These jumps are called *saccades.* The saccades occur so rapidly that not more than 10 per cent of the total time is spent in moving the eyes, 90 per cent of the time being allocated to the fixation sites. Also, the brain suppresses the visual image during the saccades so that one is completely unconscious of the movements from point to point.

Saccadic Movements During Reading. During the process of reading, a person usually makes

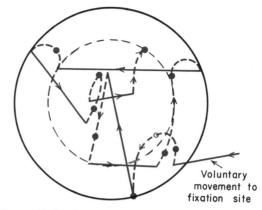

Figure 40–7. Movements of a spot of light on the fovea, showing sudden "flicking" movements that move the spot back toward the center of the fovea whenever it drifts to the foveal edge. (The dashed lines represent slow drifting movements, and the solid lines represent sudden flicking movements.) (Modified from Whitteridge: Handbook of Physiology, Vol. 2, Sec. 1. Baltimore, The Williams & Wilkins Company, 1960, p. 1089.)

several saccadic movements of the eyes for each line. In this case the visual scene is not moving past the eyes, but the eyes are trained to scan across the visual scene to extract the important information. Similar saccades occur when a person observes a painting or examines someone of the opposite sex, except that the saccades occur in one direction after another from one highlight to another, then another, and so forth.

FUSION OF THE VISUAL IMAGES

To make the visual perceptions more meaningful and also to aid in depth perception by the mechanism of stereopsis, which was discussed in Chapter 39, the visual images in the two eyes normally *fuse* with each other on "corresponding points" of the two retinae. Furthermore, three different types of fusion are required: lateral fusion, vertical fusion, and torsional fusion (same rotation of the two eyes about their optical axes).

Both the lateral geniculate body and the visual cortex play very important roles in this process of fusion. It was pointed out earlier in the chapter that corresponding points of the two retinae transmit visual signals, respectively, to successive nuclear layers of the lateral geniculate body. Interactions occur between the layers of the lateral geniculate body where the signals from the retinal images of the two eyes overlap each other; these cause *interference patterns of stimulation* in specific cells of the visual cortex. That is, when the two corresponding points of the retinae are not precisely in fusion, specific cells in the visual cortex become excited; this excitation presumably provides the signal that is transmitted to the oculomotor apparatus to cause respective movements of the two eyes so that fusion can be reestablished. Once the corresponding points of the retinae are precisely in *register* with each other, the excitation of the "interference" cells in the visual cortex disappears.

AUTONOMIC CONTROL OF ACCOMMODATION AND PUPILLARY APERTURE

The Autonomic Nerves to the Eyes.   The eye is innervated by both parasympathetic and sympathetic fibers, as illustrated in Figure 40–8. The parasympathetic fibers arise in the *Edinger-Westphal nucleus* (the visceral nucleus of the third nerve) and then pass in the *third nerve* to the *ciliary ganglion*, which lies about 1 cm. behind the eye. Here the fibers synapse with postganglionic parasympathetic neurons that pass through the *ciliary nerves* into the eyeball. These nerves excite the ciliary muscle and the sphincter of the iris.

The sympathetic innervation of the eye originates in the *intermediolateral horn cells* of the first thoracic segment of the spinal cord. From here, fibers enter the sympathetic chain and pass upward to the *superior cervical ganglion* where they synapse with postganglionic neurons. Fibers from these spread along the carotid artery and along

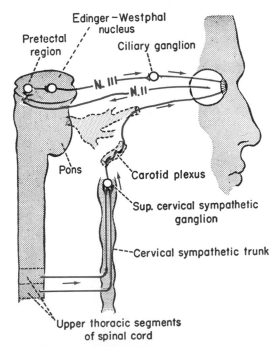

Figure 40–8. Autonomic innervation of the eye, showing also the reflex arc of the light reflex. (Modified from Ranson and Clark: Anatomy of the Nervous System. Philadelphia, W. B. Saunders Company, 1959.)

successively smaller arteries until they reach the eyeball. There the sympathetic fibers innervate the radial fibers of the iris as well as several extraocular structures around the eye. Also, they supply very weak inhibitory innervation to the ciliary muscle.

CONTROL OF ACCOMMODATION

The accommodation mechanism — that is, the mechanism which focuses the lens system of the eye — is essential if one is to achieve a high degree of visual acuity. Accommodation results from contraction or relaxation of the ciliary muscle, contraction causing increased strength of the lens systems as explained in Chapter 39, and relaxation causing decreased strength. The question that must be answered now is: How does one adjust his accommodation to keep his eyes in focus all the time?

Accommodation of the lens is regulated by a negative feedback mechanism that automatically adjusts the focal power of the lens for the highest degree of visual acuity. When the eyes have been fixed on some far object, and then suddenly fix on a near object, the lens accommodates for maximum acuity of vision usually within one second; the precise control mechanism that causes this rapid and accurate focusing of the eye is still unclear. Some of the known features of the mechanism are the following:

First, when the eyes suddenly change the distance of their fixation point, the lens changes its strength almost invariably in the proper direction to achieve a new state of focus. In other words, the lens usually *does not hunt* back and forth on the two sides of focus in an attempt to find the focus.

Second, different types of cues that can help the lens change its strength in the proper direction include the following: (1) *Chromatic aberration* appears to be important. That is, the red light rays focus on the retina slightly posteriorly to the blue light rays. The eyes appear to be able to detect which of these two types of rays is in better focus, and this clue relays the information to the accommodating mechanism about whether to make the lens stronger or weaker. (2) When the eyes fixate on a near object they also converge toward each other. The neural mechanisms for causing *convergence cause a simultaneous signal to strengthen the lens of the eye.* (3) *Since the fovea is a depressed area, the clarity of focus in the depth of the fovea versus the clarity of focus on the edges will be different.* It has been suggested that this also gives clues as to which way the strength of the lens needs to be changed. (4) It has been found that *the degree of accommodation of the lens oscillates slightly* all of the time at a frequency of up to two times per second. It has been suggested that the visual image becomes clearer when the oscillation of the lens strength is in the appropriate direction and poorer when the lens strength is in the wrong direction. This could give a rapid clue as to which way the strength of the lens needs to change to provide the appropriate focus.

CONTROL OF THE PUPILLARY APERTURE

Stimulation of the parasympathetic nerves excites the pupillary sphincter, thereby decreasing the pupillary aperture; this is called *miosis*. On the other hand, stimulation of the sympathetic nerves excites the radial fibers of the iris and causes pupillary dilatation, which is called *mydriasis*.

The Pupillary Light Reflex. When light is shone into the eyes the pupils constrict, a reaction that is called the pupillary light reflex. The neuronal pathway for this reflex is illustrated in Figure 40–8. When light impinges on the retina, the resulting impulses pass through the optic nerves and optic tracts to the pretectal nuclei. From here, impulses pass to the *Edinger-Westphal nucleus* and finally back through the *parasympathetic nerves* to constrict the sphincter of the iris. In darkness, the Edinger-Westphal nucleus becomes inhibited, which results in dilatation of the pupil.

The function of the light reflex is to help the eye adapt extremely rapidly to changing light conditions, the importance of which was explained in the previous chapter. The limits of pupillary diameter are about 1.5 mm. on the small side and 8 mm. on the large side. Therefore, the range of light adaptation that can be effected by the pupillary reflex is about 30 to 1.

REFERENCES

Critchley, M., and Critchley, E. A.: Dyslexia Defined. Springfield, Ill, Charles C Thomas, 1978.

Dunn-Rankin, P.: The visual characteristics of words. *Sci. Am., 238*(1):122, 1978.

Fraser, S. E., and Hunt, R. K.: Retinotectal specificity: Models and experiments in search of a mapping function. *Annu. Rev. Neurosci., 3*:319, 1980.

Gogel, W. C.: The adjacency principle in visual perception. *Sci. Am., 238*(5):126, 1978.

Goldchrist, A. L.: The perception of surface blacks and whites. *Sci. Am., 240*(3):112, 1979.

Hubbell, W. L., and Bownds, M. D.: Visual transduction in vertebrate photoreceptors. *Annu. Rev. Neurosci., 2*:17, 1979.

Hubel, D. H., and Wiesel, T. N.: Cortical and callosal connections concerned with vertical meridian of visual fields in the cat. *J. Neuro-physiol., 30*:1561, 1967.

Hubel, D. H., and Wiesel, T. N.: Brain mechanisms of vision. *Sci. Am., 211*(3):150, 1979.

Kuffler, S. W.: The single-cell approach in the visual system and the study of receptive fields. *Invest. Ophthalmol., 12*:794, 1973.

Johansson, G.: Visual motion perception. *Sci. Am., 232*(6):1975.

McIlwain, J. T.: Large receptive fields and spatial transformations in the visual system. *Int. Rev. Physiol., 10*:223, 1976.

Raphan, T., and Cohen, B.: Brainstem mechanisms for rapid and slow eye movements. *Annu. Rev. Physiol., 40*:527, 1978.

Regan, D., *et al.*: The visual perception of motion in depth. *Sci. Am. 241*(1):136, 1979.

Rodieck, R. W.: Visual pathways. *Annu. Rev. Neurosci., 2*:193, 1979.

Sekuler, R., and Levinson, E.: The perception of moving targets. *Sci. Am., 236*(1):60, 1977.

Toates, F. M.: Accommodation function in the human eye. *Physiol. Rev., 52*:828, 1972.

Van Essen, D. C.: Visual areas of the mammalian cerebral cortex. *Annu. Rev. Neurosci., 2*:227, 1979.

Walsh, T. J.: Neuro-Ophthalmology: Clinical Signs and Symptoms. Philadelphia, Lea & Febiger, 1978.

Wurtz, R. H., and Albano, J. E.: Visual-motor function of the primate superior colliculus. *Annu. Rev. Neurosci., 3*:189, 1980.

Yannuzzi, L. A., *et al.* (eds.): The Macula: A Comprehensive Text and Atlas. Baltimore, Williams & Wilkins, 1978.

The Sense of Hearing; and the Chemical Senses of Taste and Smell

HEARING

Hearing, like many of the somatic senses, is also a mechanoreceptive sense, for the ear responds to mechanical vibration of the sound waves in the air. The purpose of the present chapter is to describe and explain the mechanism by which the ear receives sound waves, discriminates their frequencies, and finally transmits auditory information into the central nervous system.

THE TYMPANIC MEMBRANE AND THE OSSICULAR SYSTEM

Figure 41–1 illustrates the *tympanic membrane* (commonly called the *eardrum*) and the *ossicular system*, which transmits sound through the middle ear. The tympanic membrane is cone-shaped, with its concavity facing downward toward the auditory canal. Attached to the very center of the tympanic membrane is the *handle* of the *malleus*. At its other end the malleus is tightly bound to the *incus* by ligaments, so that whenever the malleus moves the incus moves in unison with it.

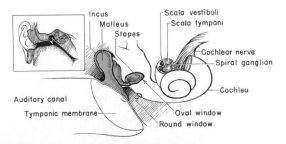

Figure 41–1. The tympanic membrane, the ossicular system of the middle ear, and the inner ear.

The opposite end of the incus in turn articulates with the stem of the *stapes*, and the *faceplate* of the stapes lies against the membranous labyrinth in the opening of the oval window where sound waves are transmitted into the inner ear, which itself is called the *cochlea.*

The ossicles of the middle ear are suspended by ligaments in such a way that the combined malleus and incus act as a single lever having its fulcrum approximately at the border of the tympanic membrane. The large *head* of the malleus, which is on the opposite side of the fulcrum from the handle, almost exactly balances the other end of the lever, so that changes in position of the body will not increase or decrease the tension on the tympanic membrane.

The handle of the malleus is constantly pulled inward by ligaments and by the tensor tympani muscle, which keeps the tympanic membrane tensed. This allows sound vibrations on *any* portion of the tympanic membrane to be transmitted to the malleus, which would not be true if the membrane were lax.

Impedance Matching by the Ossicular System. The amplitude of movement of the stapes faceplate with each sound vibration is only three-fourths as much as the movement of the handle of the malleus. Therefore, the ossicular lever system does not amplify the movement as is commonly believed, but instead the system increases the *force* of movement about 1.3 times. Also, the surface area of the tympanic membrane is approximately 55 sq. mm., whereas the surface area of the stapes averages 3.2 sq. mm. This 17-fold difference times the 1.3-fold ratio of the lever system allows all the energy of a sound wave impinging on the tympanic membrane to be applied to the small faceplate of the stapes, causing approximately 22 times as much *pressure* on the fluid of the cochlea as is exerted by the sound wave against the tympanic membrane. Since fluid has far greater inertia than air, it is easily understood that increased amounts of pressure are needed to cause vibration in the fluid. Therefore, the tympanic membrane and ossicular system provide *impedance matching* between the sound waves in air and the sound vibrations in the fluid of the cochlea.

In the absence of the ossicular system and tympanum, sound waves can travel directly through the air of the middle ear and can enter the cochlea at the oval window. However, the sensitivity for hearing is then 30 decibels less than for ossicular transmission — equivalent to a decrease from a loud shouting voice to a barely audible voice level.

Attenuation of Sound by Contraction of the Stapedius and Tensor Tympani Muscles. When loud sounds are transmitted through the ossicular system into the central nervous system, a reflex occurs after a latent period of only 40 milliseconds to cause contraction of both the stapedius and tensor tympani muscles. The tensor tympani muscle pulls the handle of the malleus inward while the stapedius muscle pulls the stapes outward. These two forces oppose each other and thereby cause the entire ossicular system to develop a high degree of rigidity, thus greatly reducing the transmission of low-frequency sound, frequencies below 1000 cycles per second, to the cochlea.

This *attenuation reflex* can reduce the intensity of sound transmission by as much as 30 to 40 decibels, which is about the same difference as that between a whisper and the sound emitted by a loud voice. The function of this mechanism is twofold:

1. To *protect* the cochlea from damaging vibrations caused by excessively loud sound.

2. To *mask* low-frequency sounds in loud environments. This usually removes a major share of the background noise and allows a person to concentrate on sounds above 1000 cycles per second frequency. It is in this upper-frequency range that most voice communication is achieved.

Another function of the tensor tympani and stapedius muscles is to decrease the person's hearing sensitivity to his own speech. This effect is activated by collateral signals transmitted to these muscles at the same time that his brain activates his voice mechanism.

THE COCHLEA

The cochlea is a system of coiled tubes, shown in Figure 41–1 and in cross-section in Figure

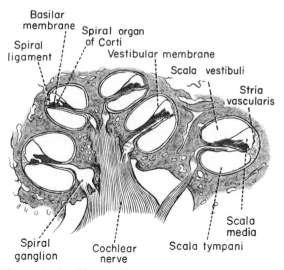

Figure 41–2. The cochlea. (From Goss, C. M. (ed.): Gray's Anatomy of the Human Body. Lea & Febiger.)

41–2, with three different tubes coiled side by side: the *scala vestibuli,* the *scala media,* and the *scala tympani.* The scala vestibuli and scala media are separated from each other by the *vestibular membrane,* and the scala tympani and scala media are separated from each other by the *basilar membrane.* On the surface of the basilar membrane lies a structure, the *organ of Corti,* which contains a series of mechanically sensitive cells, the *hair cells.* These are the receptive end-organs that generate nerve impulses in response to sound vibrations.

Figure 41–3 illustrates schematically the functional parts of the uncoiled cochlea for transmission of sound vibrations. First, note that the vestibular membrane is missing from this figure. This membrane is so thin and so easily moved that it does not obstruct the passage of sound vibrations from the scala vestibuli into the scala media at all. Therefore, so far as the transmission of sound is concerned, the scala vestibuli and scala media are considered to be a single chamber. The importance of the vestibular membrane is to maintain a special fluid in the scala

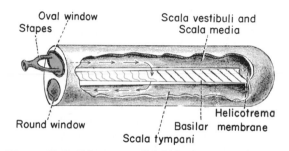

Figure 41–3. Movement of fluid in the cochlea following forward thrust of the stapes.

media that is required for optimal function of the sound-receptive hair cells.

Sound vibrations enter the scala vestibuli from the faceplate of the stapes at the oval window. The faceplate covers this window and is connected with the window's edges by a relatively loose annular ligament so that it can move inward and outward with the sound vibrations. Inward movement causes the fluid to move into the scala vestibuli and scala media.

Note from Figure 41–3 that the distal end of the scala vestibuli and scala tympani open directly into each other by way of the *helicotrema*. If the stapes moves inward *very slowly*, fluid from the scala vestibuli is pushed through the helicotrema into the scala tympani, and this causes the round window to bulge outward. However, if the stapes vibrates inward and outward rapidly, the fluid simply does not have time to pass all the way to the helicotrema, then to the round window, and back again to the oval window between successive vibrations. Instead, the fluid wave takes a short-cut through the basilar membrane, causing it to bulge back and forth with each sound vibration. We shall see later that each frequency of sound causes a different "pattern" of vibration in the basilar membrane and that this is the important means by which the sound frequencies are discriminated from each other.

The Basilar Membrane and Resonance in the Cochlea. The basilar membrane contains about 20,000 or more *basilar fibers* that project from the bony center of the cochlea, the *modiolus,* toward the outer wall. These fibers are stiff, elastic, reedlike structures that are not fixed at their distal ends except that they are embedded in the basilar membrane. Because they are stiff and free at one end, they can vibrate like reeds of a harmonica.

The lengths of the basilar fibers increase progressively from the base of the cochlea to the helicotrema, from approximately 0.04 mm. at the base to 0.5 mm. at the helicotrema, a 12-fold increase in length.

The diameters of the fibers, on the other hand, decrease from the base to the helicotrema, so that their overall stiffness decreases more than 100-fold. As a result, the stiff, short fibers near the base of the cochlea have a tendency to vibrate at a high frequency, while the long, limber fibers near the helicotrema have a tendency to vibrate at a low frequency.

In addition to the differences in stiffness of the basilar fibers, they are also differently "loaded" by the fluid mass of the cochlea. That is, when an area of the basilar membrane vibrates back and forth, all the fluid between the vibrating membrane and the oval and round windows must also move back and forth at the same time. For a basilar membrane fiber vibrating near the base of the cochlea, the total mass of moving fluid is slight in comparison with that for a fiber vibrating near the helicotrema. This difference, too, favors high-frequency vibration near the windows and low-frequency vibration near the tip of the cochlea.

Thus, high-frequency resonance of the basilar membrane occurs near the base and low-frequency resonance occurs near the apex because of (1) difference in stiffness of the basilar fibers and (2) difference in "loading."

Pattern of Vibration of the Basilar Membrane. The dashed curves of Figure 41–4A show the position of a sound wave of a particular sound frequency on the basilar membrane when the stapes (a) is all the way inward, (b) has moved back to the neutral point, (c) is all the way outward, and (d) has moved back again to the neutral point but is moving inward. The shaded area around these different waves shows the complete pattern of vibration of the basilar membrane during a complete vibratory cycle for this particular sound frequency.

Figure 41–4B shows the amplitude patterns of vibration for different frequencies, showing that the maximum amplitude for 8000 cycles occurs near the base of the cochlea, while that for frequencies less than 400 to 500 cycles per second occurs near the helicotrema. The principal method by which sound frequencies are discriminated from each other is based on the "place" of maximum stimulation of the nerve fibers from the

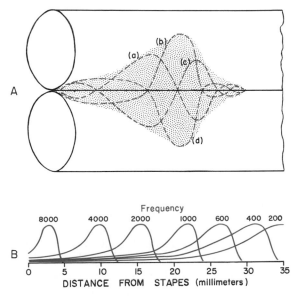

Figure 41–4. (A) Amplitude pattern of vibration of the basilar membrane for a medium frequency sound. (B) Amplitude patterns for sounds of all frequencies between 50 and 8000 per second, showing the points of maximum amplitude (the resonance points) on the basilar membrane for the different frequencies.

organ of Corti that rests on the basilar membrane, as will be explained in the following section.

FUNCTION OF THE ORGAN OF CORTI

The organ of Corti, illustrated in Figures 41–2 and 41–5, is the receptor organ that generates nerve impulses in response to vibration of the basilar membrane. Note that the organ of Corti lies on the surface of the basilar fibers and basilar membrane. The actual sensory receptors in the organ of Corti are two types of *hair cells* — a single row of *internal hair cells*, numbering about 3500, and three to four rows of *external hair cells*, numbering about 20,000. The bases and sides of the hair cells are enmeshed by a network of cochlear nerve endings. These lead to the *spiral ganglion of Corti*, which lies in the modiolus of the cochlea. The spiral ganglion in turn sends axons into the *cochlear nerve* and thence into the central nervous system at the level of the upper medulla. The relationship of the organ of Corti to the spiral ganglion and to the cochlear nerve is illustrated in Figure 41–2.

Excitation of the Hair Cells. Note in Figure 41–5 that minute hairs, or cilia, project upward from the hair cells and either touch or are embedded in the surface gel coating of the *tectorial membrane*, which lies above the cilia in the scala media. These hair cells are similar to the hair cells found in the maculae and cristae ampullaris of the vestibular apparatus, which were discussed in Chapter 34. Bending of the hairs excites the hair cells, and this in turn excites the nerve fibers synapsing with their bases.

Upward movement of the basilar fiber rocks the hair cells upward and *inward*. Then, when the basilar membrane moves downward, the cells rock downward and *outward*. The inward and outward motion causes the hairs to shear back and forth against the tectorial membrane, thus exciting the cochlear nerve fibers whenever the basilar membrane vibrates.

Mechanism by Which the Hair Cells Excite the Nerve Fibers — Receptor Potentials. Back and forth bending of the hairs causes alternate changes in the electrical potential across the hair cell membrane. This alternating potential is the *receptor potential* of the hair cell; and it in turn stimulates the cochlear nerve endings that terminate on the hair cells. Most physiologists believe that the receptor potential stimulates the endings by direct electrical excitation.

DETERMINATION OF PITCH – THE "PLACE" PRINCIPLE

From earlier discussions in this chapter it is already apparent that low-pitch (or low-frequency) sounds cause maximal activation of the basilar membrane near the apex of the cochlea; sounds of high pitch (or high frequency) activate the basilar membrane near the base of the cochlea; and intermediate frequencies activate the membrane at intermediate distances between these two extremes. Furthermore, there is a spatial organization of the cochlear nerve fibers from the cochlea to the cochlear nuclei in the brain stem, the fibers from each respective area of the basilar membrane terminating in a corresponding area in the cochlear nuclei. We shall see later that this spatial organization also continues all the way up the brain stem to the cerebral cortex. The recording of signals from the auditory tracts in the brain stem and from the auditory receptive fields in the cerebral cortex shows that specific neurons are activated by specific pitches. Therefore, the method used by the nervous system to detect different pitches is to determine the position along the basilar membrane that is most stimulated. This is called the *place principle* for determination of pitch.

DETERMINATION OF LOUDNESS

Loudness is determined by the amplitude of vibration of the basilar membrane and hair cells. Increasing the amplitude excites the nerve endings at more rapid rates and also causes more and more of the hair cells on the fringes of the vibrating portion of the basilar membrane to become stimulated, thus causing *spatial summation* of impulses — that is, transmission through many nerve fibers rather than through a few.

The interpreted sensation of sound changes

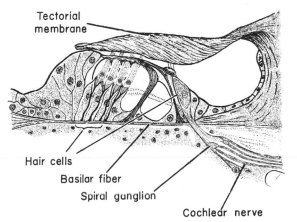

Figure 41–5. The organ of Corti, showing especially the hair cells and the tectorial membrane against the projecting hairs.

approximately in proportion to the cube root of the actual sound intensity. To express this another way, the ear can discriminate changes in sound intensity from the softest whisper to the loudest possible noise of *approximately one trillion times* as much sound energy. Yet the ear interprets this much difference in sound level as approximately a 10,000-fold change. Thus, the scale of intensity is greatly "compressed" by the sound perception mechanisms of the auditory system. This obviously allows a person to interpret differences in sound intensities over an extremely wide range, a far broader range than would be possible were it not for compression of the scale.

The Decibel Unit. Because of the extreme changes in sound intensities that the ear can detect and discriminate, sound intensities are usually expressed in terms of the logarithm of their actual intensities. A 10-fold increase in sound energy is called 1 *bel,* and one-tenth bel is called 1 *decibel.* One decibel represents an actual increase in intensity of 1.26 times.

Another reason for using the decibel system in expressing changes in loudness is that, in the usual sound intensity range for communication, the ears can detect approximately a 1 decibel change in sound intensity.

FREQUENCY RANGE OF HEARING

The frequencies of sound that a young person can hear, before aging has occurred in the ears, is generally said to be between 30 and 20,000 cycles per second. However, the sound range depends to a great extent on intensity. If the intensity is only −60 decibels, the sound range is 500 to 5000 cycles per second, but if the sound intensity is −20 decibels, the frequency range is about 70 to 15,000 cycles per second, and only with intense sounds can the complete range of 30 to 20,000 cycles be achieved. In old age, the frequency range falls to 50 to 8000 cycles per second or less.

CENTRAL AUDITORY MECHANISMS

Figure 41–6 illustrates the major auditory pathways. It shows that nerve fibers from the *spiral ganglion of the organ of Corti* enter the *cochlear nuclei* located in the upper part of the medulla. At this point, all the fibers synapse. Part of the signal then travels up the same side of the brain stem, but the majority of it passes to the opposite side and is transmitted upward through successive neurons in the *superior olivary nucleus,* the *inferior colliculus,* and the *medial geniculate nucleus,* finally terminating in the *auditory cortex,* located in the superior gyrus of the temporal lobe.

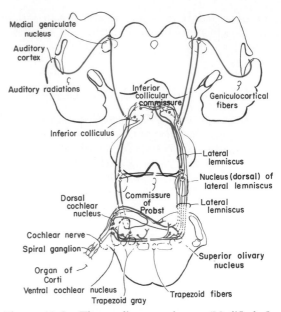

Figure 41–6. The auditory pathway. (Modified from Crosby, Humphrey, and Lauer: Correlative Anatomy of the Nervous System. Copyright © 1962 by Macmillan Publishing Co., Inc.)

Several points of importance in relation to the auditory pathway should be noted. First, impulses from either ear are transmitted through the auditory pathways of both sides of the brain stem with only slight preponderance of transmission in the contralateral pathway.

Second, many collateral fibers from the auditory tracts pass directly into the reticular activating system of the brain stem. Therefore, sound can activate the whole brain.

Third, a high degree of spatial orientation is maintained in the fiber tracts from the cochlea all the way to the cortex. In fact, there are three different spatial representations of sound frequencies in the cochlear nuclei, two representations in the inferior colliculi, a very precise representation of discrete sound frequencies in the auditory cortex, and several less precise representations in the auditory association areas.

FUNCTION OF THE CEREBRAL CORTEX IN HEARING

The projection of the auditory pathway onto the cerebral cortex is illustrated in Figure 41–7, which shows that the auditory cortex lies principally on the *supratemporal plane of the superior temporal gyrus* but also extends over the *lateral border of the temporal lobe,* over much of the *insular cortex,* and even into the most lateral portion of the *parietal lobe.*

Two separate areas are shown in Figure 41–7; the *primary auditory cortex* and the *auditory associa-*

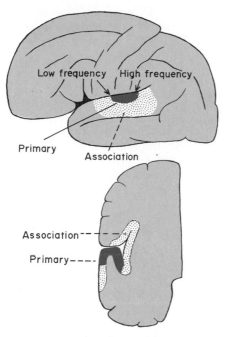

Figure 41–7. The auditory cortex.

tion cortex. The primary auditory cortex is directly excited by projections from the medial geniculate body, while the auditory association areas are usually excited secondarily by signals both from the primary auditory cortex and from the thalamic association areas adjacent to the medial geniculate body.

Locus of Sound Frequency Perception in the Primary Auditory Cortex. Certain parts of the primary auditory cortex are known to respond to high frequencies and other parts to low frequencies. In monkeys, the posteromedial part of the supratemporal plane responds to high frequencies, while the anterolateral part responds to low frequencies. Presumably, the same frequency localization occurs in the human cortex, but this is yet unproven.

Some of the neurons in the auditory cortex, especially in the auditory association cortex, do not respond at all to sounds in the ear. It is believed that these neurons "associate" different sound frequencies with each other or associate sound information with information from other sensory areas of the cortex. Indeed, the parietal portion of the auditory association cortex partly overlaps somatic sensory area II, which could provide easy opportunity for association of auditory information with somatic sensory information.

Discrimination of Sound "Patterns" by the Auditory Cortex. Complete bilateral removal of the auditory cortex does not prevent an animal from detecting sounds or reacting in a crude manner to the sounds. However, it does greatly reduce or sometimes even abolish its ability to discriminate different sound pitches and especially *patterns of sound.* For instance, an animal that has been trained to recognize a combination or sequence of tones, one following the other in a particular pattern, loses this ability when the auditory cortex is destroyed, and furthermore, it cannot relearn this type of response. Therefore, the auditory cortex is important in the discrimination of *tonal* and *sequential sound patterns.*

In the human being, lesions affecting the auditory association areas but not affecting the primary auditory cortex will allow the person full capability to hear and differentiate sound tones as well as to interpret at least a few simple patterns of sound. However, he will often be completely unable to interpret the *meaning* of the sound that he hears. For instance, lesions in the posterior portion of the superior temporal gyrus often make it impossible for the person to interpret the meanings of words even though he hears them perfectly well and can often even repeat them; all the time, however, he does not know the meaning of the words. These functions of the auditory association areas and their relationship to the overall intellectual functions of the brain were discussed in detail in Chapter 36.

DISCRIMINATION OF DIRECTION FROM WHICH SOUND EMANATES

A person determines the direction from which sound emanates by two principal mechanisms: (1) the time lag between the entry of sound into one ear and into the opposite ear and (2) the difference between the intensities of the sounds in the two ears. The first mechanism functions best for frequencies below 3000 cycles per second, and the intensity mechanism operates best at higher frequencies because the head acts as a sound barrier at these frequencies. The time lag mechanism discriminates direction much more exactly than the intensity mechanism, for the time lag mechanism does not depend on extraneous factors but only on an exact interval of time between two acoustical signals. If a person is looking straight toward the sound, the sound reaches both ears at exactly the same instant, while if the right ear is closer to the sound than the left ear, the sound signals from the right ear are perceived ahead of those from the left ear.

Neural Mechanisms for Detecting Sound Direction. Destruction of the auditory cortex on both sides of the brain, in either man or lower mammals, causes loss of almost all ability to detect the direction from which sound comes. Yet, the mechanism for this detection process begins in the superior olivary nuclei, even though it requires the neural pathways all the way from these nuclei to the cortex for interpretation of the signals. The mechanism is believed to be the following:

When the sound enters one ear slightly before it enters the other ear, the signal from the first ear *inhibits* the neurons in the ipsilateral superior olivary nucleus, and this inhibition lasts for a fraction of a millisecond. Therefore, for a short period of time after the sound reaches the first ear, the pathway for excitatory signals coming from the opposite ear is in an inhibited state. Furthermore, certain neurons of the medial superior olivary nuclei have longer periods of inhibition than do other neurons. Therefore, when the sound signal from the opposite ear enters the inhibited superior olivary nucleus, the signal will pass up the auditory pathway through some of the neurons but not through others. And the specific neurons through which the signal passes are determined by the time interval of the sound between the two ears. That is, a spatial pattern of neuronal stimulation develops, with the short lagging sounds stimulating one set of neurons maximally and the long lagging sounds stimulating another set of neurons maximally. This spatial orientation of signals is then transmitted all the way to the auditory cortex, where sound direction is determined by the locus in the cortex that is stimulated maximally.

This mechanism for detection of sound direction indicates again how the information in sensory signals is dissected out as the signals pass through different levels of neuronal activity. In this case, the "quality" of sound direction is separated from the "quality" of sound tones at the level of the superior olivary nuclei.

DEAFNESS

Deafness is usually divided into two types; first, that caused by impairment of the cochlea or auditory nerve, which is usually classed under the heading "nerve deafness," and, second, that caused by impairment of the middle ear mechanisms for transmitting sound into the cochlea, which is usually called "conduction deafness." Obviously, if either the cochlea or the auditory nerve is completely destroyed, the person is completely deaf. However, if the cochlea and nerve are still intact but the ossicular system has been destroyed or ankylosed ("frozen" in place by fibrous tissue or calcification), sound waves can still be conducted into the cochlea by means of bone conduction (such as conduction of sound from the butt of a vibrating tuning fork applied directly to the skull). The person with some types of conduction deafness can be made to hear again

almost normally by an operation to remove the stapes and replacing it with a minute Teflon or metal prosthesis that transmits the sound from the incus to the oval window.

THE SENSE OF TASTE

Taste is a function of the *taste buds* in the mouth, and its importance lies in the fact that it allows the person to select his food in accord with his desires and perhaps also in accord with the needs of the tissues for specific nutritive substances.

On the basis of psychological studies, there are considered to be four *primary* sensations of taste: *sour, salty, sweet,* and *bitter.* Yet we know that a person can perceive literally hundreds of different tastes. These are all supposed to be combinations of the four primary sensations in the same manner that all the colors of the spectrum are combinations of three primary color sensations, as described in Chapter 39.

THE PRIMARY SENSATIONS OF TASTE

The Sour Taste. The sour taste is caused by acids, and the intensity of the taste sensation is approximately proportional to the logarithm of the *hydrogen ion concentration.* That is, the more acidic the acid, the stronger becomes the sensation.

The Salty Taste. The salty taste is elicited by ionized salts. The quality of the taste varies somewhat from one salt to another because the salts also elicit other taste sensations besides saltiness.

The Sweet Taste. The sweet taste is not caused by any single class of chemicals. A list of some of the types of chemicals that cause this taste includes sugars, glycols, alcohols, aldehydes, ketones, amides, esters, amino acids, sulfonic acids, halogenated acids, and inorganic salts of lead and beryllium. Note specifically that most substances that cause a sweet taste are organic chemicals; the only inorganic substances that elicit the sweet taste are certain salts of lead and beryllium.

Saccharin is a substance more than 600 times as sweet as common table sugar and is an important noncalorigenic sweetening agent.

The Bitter Taste. The bitter taste, like the sweet taste, is not caused by any single type of chemical agent, but, here again, the substances that give the bitter taste are almost entirely organic substances. Two particular classes of substances are especially likely to cause bitter taste sensations, (1) long chain organic substances and (2) alkaloids. The alkaloids include many of the drugs used in medicines, such as quinine, caffeine, strychnine, and nicotine.

The bitter taste, when it occurs in high intensity, usually causes the person or animal to reject the food. This is undoubtedly an important purposive function of the bitter taste sensation, because many of the deadly toxins found in poisonous plants are alkaloids, and these all cause an intensely bitter taste.

Threshold for Taste

The threshold for stimulation of the sour taste by hydrochloric acid averages 0.0009 M; for stimulation of the salty taste by sodium chloride: 0.01 M; for the sweet taste by sucrose: 0.01 M; and for the bitter taste by quinine: 0.000008 M. Note especially how much more sensitive is the bitter taste sense to stimuli than all the others, which would be expected since this sensation provides an important protective function.

THE TASTE BUD AND ITS FUNCTION

Figure 41–8 illustrates a taste bud, which has a diameter of about 1/30 millimeter and a length of about 1/16 millimeter. The taste bud is composed of about 40 modified epithelial cells called *taste cells.*

The outer tips of the taste cells are arranged around a minute *taste pore,* shown in Figure 41–8. From the tip of each cell, several *microvilli,* or *taste hairs,* about 2 to 3 microns in length and 0.2 micron in width, protrude outward through the taste pore to approach the cavity of the mouth. These microvilli are believed to provide the receptor surface for taste.

Interwoven among the taste cells is a branching terminal network of several *taste nerve fibers* that are stimulated by the taste cells. These fibers invaginate deeply into folds of the taste cell membranes, so that there is extremely intimate contact between the taste cells and the nerves. Several taste buds can be innervated by the same taste fibers.

Location of the Taste Buds. The taste buds are found on three out of four different types of papillae of the tongue, as follows: (1) A large number of taste buds are on the walls of the troughs that surround the circumvallate papillae, which form a V line toward the posterior of the tongue. (2) Moderate numbers of taste buds are on the fungiform papillae over the front surface of the tongue. (3) Moderate numbers are on the foliate papillae located in the folds along the posterolateral surfaces of the tongue.

Additional taste buds are located on the palate and a few on the tonsillar pillars and at other points around the nasopharynx. Adults have approximately 10,000 taste buds, and children a few more. Beyond the age of 45 many taste buds rapidly degenerate, causing the taste sensation to become progressively less critical.

Specificity of Taste Buds for the Primary Taste Stimuli. Psychological tests using different types of taste stimuli carefully applied to individual taste buds, one at a time, have suggested that we have four distinctly different varieties of taste buds, each sensitive for only one type of taste. Yet, microelectrode studies from single taste buds while they are stimulated successively by the four different primary taste stimuli have shown that most of them can be excited by two, three, or even four of the primary taste stimuli, though usually with one or two of these predominating. Thus, at present, there are conflicting beliefs about the degree of specificity of taste buds, some physiologists believing them always to be highly specific for only one primary taste stimulus and other physiologists believing this to be true only in a statistical sense because of dominance of one taste perception over the others.

Regardless of which theory is correct, one can well understand that the hundreds of different types of tastes that we experience result from different quantitative degrees of stimulation of the four primary sensations of taste (as well as simultaneous stimulation of smell in the nose and tactile and pain nerve endings in the mouth).

TRANSMISSION OF TASTE SIGNALS INTO THE CENTRAL NERVOUS SYSTEM

Figure 41–9 illustrates the neuronal pathways for transmission of taste sensations from the tongue and pharyngeal region into the central nervous system. Taste impulses pass through the seventh, ninth, and tenth nerves into the brain stem, where they terminate in the *tractus solitarius.*

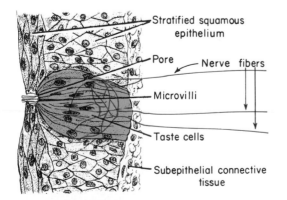

Stratified squamous epithelium

Pore — Nerve fibers

Microvilli

Taste cells

Subepithelial connective tissue

Figure 41–8. The taste bud.

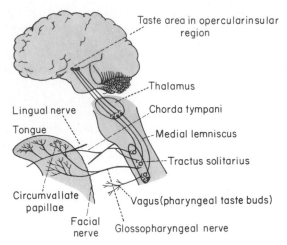

Figure 41–9. Transmission of taste impulses into the central nervous system.

From here, signals pass first to the *thalamus* and thence to the *parietal opercular-insular area* of the cerebral cortex. This lies at the very lateral margin of the postcentral gyrus in the sylvian fissure in close association with, or even superimposed on, the tongue area of somatic area I.

From this description of the taste pathways, it immediately becomes evident that they parallel closely the somatic pathways from the tongue.

SPECIAL ATTRIBUTES OF THE TASTE SENSE

Importance of the Sense of Smell in Taste. Persons with severe colds frequently state that they have lost their sense of taste. However, on testing the taste sensations, these are found to be completely normal. This illustrates that much of what we call taste is actually smell. Odors from the food can pass upward into the nose, often stimulating the olfactory system thousands of times as strongly as the taste system. For instance, if the olfactory system is intact, alcohol can be "tasted" in 1/50,000 the concentration required when the olfactory system is not intact.

Taste Preference and Control of the Diet. Taste preferences mean simply that an animal will choose certain types of food in preference to others, and it automatically uses this to help control the type of diet it eats. Furthermore, to a great extent its taste preferences change in accord with the needs of the body for certain specific substances. The following experimental studies will illustrate this ability of an animal to choose food in accord with the need of its body: First, adrenalectomized animals automatically select drinking water with a high concentration of sodium chloride in preference to pure water, and this in many instances is sufficient to supply the needs of the body and prevent death as a result of salt depletion. Second, an animal injected with excessive amounts of insulin develops a depleted blood sugar, and it automatically chooses the sweetest food from among many samples. Third, parathyroidectomized animals automatically choose drinking water with a high concentration of calcium chloride.

These same phenomena are also observed in many instances of everyday life. For instance, the salt licks of the desert region are known to attract animals from far and wide, and even the human being rejects any food that has an unpleasant sensation, which certainly in many instances protects our bodies from undesirable substances.

THE SENSE OF SMELL

Smell is the least understood sense. This results partly from the location of the olfactory membrane high in the nose where it is difficult to study and partly from the fact that the sense of smell is a subjective phenomenon that cannot be studied with ease in lower animals. Still another complicating problem is the fact that the sense of smell is almost rudimentary in the human being in comparison with that of some lower animals.

THE OLFACTORY MEMBRANE

The olfactory membrane lies in the superior part of each nostril. Medially it folds downward over the surface of the septum, and laterally it folds over the superior turbinate and even over a small portion of the upper surface of the middle turbinate. In each nostril the olfactory membrane

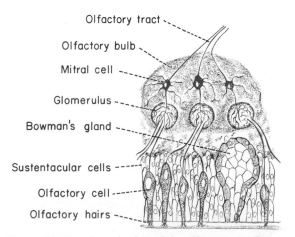

Figure 41–10. Organization of the olfactory membrane.

has a surface area of approximately 2.4 square centimeters.

The Olfactory Cells. The receptor cells for the smell sensation are the *olfactory cells*, which are actually bipolar nerve cells derived originally from the central nervous system itself. There are about 100 million of these cells in the olfactory epithelium interspersed among *sustentacular cells,* as shown in Figure 41–10. The mucosal end of the olfactory cell forms a knob called the *olfactory vesicle* from which large numbers of *olfactory hairs,* or *cilia,* 0.3 micron in diameter and 50 to 150 microns in length, project into the mucus that coats the inner surface of the nasal cavity. These projecting olfactory hairs are believed to react to odors in the air and then to stimulate the olfactory cells, as is discussed below. Spaced among the olfactory cells in the olfactory membrane are many small *glands of Bowman* that secrete mucus onto the surface of the olfactory membrane.

STIMULATION OF THE OLFACTORY CELLS

The Necessary Stimulus for Smell. We do not know what it takes chemically to stimulate the olfactory cells. Yet we do know the physical characteristics of the substances that cause olfactory stimulation: First, the substance must be volatile so that it can be sniffed into the nostrils. Second, it must be at least slightly water soluble so that it can pass through the mucus to the olfactory cells. And, third, it must also be lipid soluble, presumably because the olfactory hairs and outer tips of the olfactory cells are composed principally of lipid materials.

Regardless of the basic mechanism by which the olfactory cells are stimulated, it is known that they become stimulated only when air blasts upward into the superior region of the nose. Therefore, smell occurs in cycles along with the inspirations, which indicates that the olfactory receptors respond in milliseconds to the volatile agents. Because smell intensity is enhanced by blasting air through the upper reaches of the nose, a person can greatly increase his sensitivity of smell by the well-known sniffing technique.

Search for the Primary Sensations of Smell. Most physiologists are convinced that the many smell sensations are subserved by a few rather discrete primary sensations in the same way that taste is subserved by sour, sweet, bitter, and salty sensations. But, thus far, only minor success has been achieved in classifying the primary sensations of smell. Yet, on the basis of psychological tests and action potential studies from various points in the olfactory nerve pathways, it has been postulated that about seven different primary classes of olfactory stimulants preferentially excite separate olfactory cells. These classes of olfactory stimulants may be characterized as follows:

1. Camphoraceous
2. Musky
3. Floral
4. Pepperminty
5. Ethereal
6. Pungent
7. Putrid

However, it is unlikely that this list actually represents the true primary sensations of smell even though it does illustrate the results of one of the many attempts to classify them. Indeed, several clues in recent years have indicated that there may be as many as *50* or more primary sensations of smell — a marked contrast to only *three* primary sensations of color detected by the eyes and only *four* primary sensations of taste detected by the tongue. For instance, individual persons have been found who are specifically *odor-blind* for more than 50 different substances. Since it is presumed that odor-blindness for each substance represents a lack of the appropriate receptor for that substance, it is postulated that the sense of smell might be subserved by 50 or more primary smell sensations.

Two basic theories have been postulated to explain the abilities of different receptors to respond selectively to different types of olfactory stimulants, the *chemical theory* and the *physical theory.* The chemical theory assumes that *receptor chemicals* in the membranes of the olfactory hairs react specifically with the different types of olfactory stimulants. The type of receptor chemical determines the type of stimulant that will elicit a response in the olfactory cell. The reaction between the stimulant and the receptor substance supposedly increases the permeability of the olfactory hair membrane, and this in turn creates the receptor potential in the olfactory cell that generates impulses in the olfactory nerve fibers.

The physical theory assumes that differences in *physical receptor sites* on the olfactory hair membranes of separate olfactory cells allow specific olfactory stimulants to adsorb to the membranes of different olfactory cells. A fact that supports this theory is that many substances with very different chemical properties, but almost identical molecular shapes, have the same odor. This indicates that a physical property of the stimulant might determine the odor.

Threshold for Smell. One of the principal characteristics of smell is the minute concentration of the stimulating agent in the air required to effect a smell sensation. For instance, the substance *methyl mercaptan* can be smelled when only 1/25,000,000,000 milligram is present in each milliliter of air. Because of this low threshold, this substance is mixed with natural gas to give it an odor that can be detected when it leaks from a gas pipe.

TRANSMISSION OF SMELL SENSATIONS INTO THE CENTRAL NERVOUS SYSTEM

The function of the central nervous system in olfaction is almost as vague as the function of the peripheral receptors. However, Figures 41–10 and 41–11 illustrate the general plan for transmission of olfactory sensations into the central nervous system. Figure 41–10 shows a number of separate *olfactory cells* sending axons into the *olfactory bulb* to end on *dendrites from mitral cells* in a structure called the *glomerulus.* Approximately 25,000 axons from olfactory cells enter each glomerulus and synapse with about 25 mitral cells that in turn send signals into the brain. There is a total of about 5000 glomeruli.

Figure 41–11 shows the major pathways for transmission of olfactory signals from the mitral cells into the brain. The fibers from the mitral cells travel through the olfactory tract and terminate either primarily or through relay neurons in two principal areas of the brain called the *medial olfactory area* and the *lateral olfactory area,* respectively. The medial olfactory area is composed of several nuclei located in the midportion of the brain superiorly and anteriorly to the hypothalamus. These include the *septum pellucidum,* the *gyrus subcallosus,* the *paraolfactory area,* the *olfactory trigone,* and the *medial part of the anterior perforated substance.*

The lateral olfactory area is located bilaterally, mainly in the anterior and inferior part of the temporal lobe. It is composed of the *prepyriform area,* the *uncus,* the *lateral part of the anterior perforated substance,* and part of the *amygdaloid nuclei.*

Secondary olfactory tracts pass from the nuclei of both the medial olfactory area and the lateral olfactory area into the *hypothalamus, thalamus, hippocampus,* and *brain stem nuclei.* These secondary areas control the automatic responses of the body to olfactory stimuli, including automatic feeding activities and also emotional responses such as fear, excitement, pleasure and sexual drives.

Complete removal of the lateral olfactory area hardly affects the primitive responses to olfaction, such as licking the lips, salivation, and other feeding responses caused by the smell of food or such as the various emotions associated with smell. On the other hand, its removal does abolish the more complicated conditioned reflexes depending on olfactory stimuli. Therefore, this region is often considered to be the *primary olfactory area* for smell. In human beings, tumors in the region of the uncus and amygdala frequently cause the person to perceive abnormal smells.

REFERENCES

Hearing

Aitkin, L. M.: Tonotopic organization at higher levels of the auditory pathway. *Int. Rev. Physiol., 10*:249, 1976.
Beagley, H. A. (ed.): Auditory Investigation: The Scientific and Technological Basis. New York, Oxford University Press, 1979.
Bench, J., and Bamford, J. (eds.): Speech-Hearing Tests and the Spoken Language of Hearing-Impaired Children. New York, Academic Press, 1979.
Brugge, J. F., and Geisler, C. D.: Auditory mechanisms of the lower brainstem. *Annu. Rev. Neurosci., 1*:363, 1978.
Carterette, E. C., and Friedman, M. P.: Hearing. New York, Academic Press, 1978.
Evans, E. F., and Wilson, J. P. (eds.): Psychophysics and Physiology of Hearing. New York, Academic Press, 1977.
Lipscomb, D. M. (ed.): Noise and Audiology. Baltimore, University Park Press, 1978.
Naunton, R. F., and Fernandez, C. (eds.): Evoked Electrical Activity in the Auditory Nervous System. New York, Academic Press, 1978.
Rintelmann, W. F. (ed.): Hearing Assessment. Baltimore, University Park Press, 1979.
Sataloff, J., *et al.*: Hearing Loss, 2nd Ed. Philadelphia, J. B. Lippincott Co., 1980.
Scheich, O. C. H., and Schreiner, C. (eds.): Hearing Mechanisms and Speech. New York, Springer-Verlag, 1979.
Singh, R. P.: Anatomy of Hearing and Speech. New York, Oxford University Press, 1980.
Skinner, P. H., and Shelton, R. L.: Speech, Language, and Hearing: Normal Processes and Disorders. Reading, Mass., Addison-Wesley Publishing Co., 1978.
Van Hattum, R. J.: Communication Disorders. New York, The Macmillan Co., 1980.

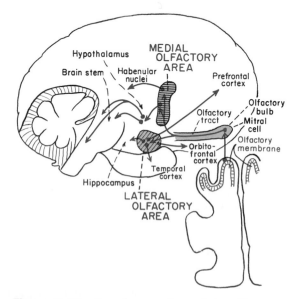

Figure 41–11. Neural connections of the olfactory system.

Taste and Smell

Alberts, J. R.: Producing and interpreting experimental olfactory deficits. *Physiol. Behav., 12*:657, 1974.

Bradley, R. M., and Mistretta, C. M.: Investigations of taste function and swallowing in fetal sheep. *Symp. Oral Sens. Percept., 4*:185, 1973.

Dastoli, F. R.: Taste receptor proteins. *Life Sci., 14*:1417, 1974.

Douek, E.: The Sense of Smell and Its Abnormalities. New York, Churchill Livingstone, 1974.

Forss, D. A.: Odor and flavor compounds from lipids. *Prog. Chem. Fats Other Lipids., 13*:177, 1972.

Kare, M. R., and Maller, O.: The Chemical Sense and Nutrition. Baltimore, The Johns Hopkins Press, 1967.

Norsiek, F. W.: The sweet tooth. *Am. Sci., 60*:41, 1972.

Ohloff, G., and Thomas, A. F. (eds.): Gustation and Olfaction. New York, Academic Press, 1971.

Shepherd, G. M.: The olfactory bulb: A simple system in the mammalian brain. *In* Brookhart, J. M., and Mountcastle, V. B. (eds.): Handbook of Physiology. Sec. 1, Vol. 1. Baltimore, Williams & Wilkins, 1977, p. 945.

Weiffenbach, J. M. (ed.): Taste and Development: The Genesis of Sweet Preference. Washington, D.C., U.S. Government Printing Office, 1977.

Part XI

THE GASTROINTESTINAL TRACT

<h1 style="text-align:center">42</h1>

Movement of Food Through the Alimentary Tract

The primary function of the alimentary tract is to provide the body with a continuous supply of water, electrolytes, and nutrients, but before this can be achieved food must be moved along the alimentary tract at an appropriate rate for the digestive and absorptive functions to take place. Therefore, discussion of the alimentary system is presented in three different phases in this and the next two chapters: (1) movement of food through the alimentary tract, (2) secretion of the digestive juices, and (3) absorption of the digested foods, water, and the various electrolytes.

Figure 42–1 illustrates the entire alimentary tract, showing major anatomical differences between its parts. Each part is adapted for specific functions, such as (1) simple passage of food from one point to another, as in the esophagus, (2) storage of food in the body of the stomach or fecal matter in the descending colon, (3) digestion of food in the stomach, duodenum, jejunum, and ileum, and (4) absorption of the digestive end-products in the entire small intestine and proximal half of the large intestine, also called the colon. One of the most important features of the gastrointestinal tract, a feature discussed in the present chapter, is the myriad of autoregulatory processes in the gut that keeps the food moving at an appropriate pace — slow enough for digestion and absorption to take place but fast enough to provide the nutrients needed by the body.

GENERAL PRINCIPLES OF INTESTINAL MOTILITY

CHARACTERISTICS OF THE INTESTINAL WALL

Figure 42–2 illustrates a typical section of the intestinal wall, showing the following layers from the outside inward: (1) the *serosa,* (2) a *longitudinal muscle layer,* (3) a *circular muscle layer,* (4) the *submucosa,* and (5) the *mucosa.* In addition, a sparse layer of smooth muscle fibers, the *muscularis mucosae,* lies in the deeper layers of the mucosa. The motor functions of the gut are performed by the different layers of smooth muscle.

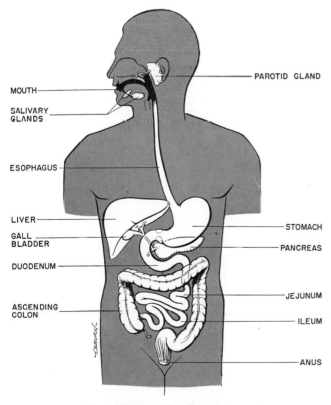

MOUTH

SALIVARY GLANDS

ESOPHAGUS

LIVER

GALL BLADDER

DUODENUM

ASCENDING COLON

PAROTID GLAND

STOMACH

PANCREAS

JEJUNUM

ILEUM

ANUS

Figure 42–1. The alimentary tract.

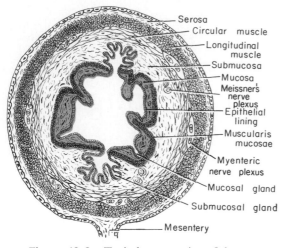

Figure 42–2. Typical cross-section of the gut.

CHARACTERISTICS OF INTESTINAL SMOOTH MUSCLE

The general characteristics of smooth muscle and its function were discussed in Chapter 10. However, some specific characteristics of smooth muscle in the gut are the following:

The Functioning Syncytium. The individual smooth muscle fibers of the gut abut against each other extremely closely. About 12 per cent of their membrane surfaces are actually fused with the membranes of adjacent muscle fibers in the form of *nexuses.* Measurements of ionic transport through these areas of close contact demonstrate extremely low electrical resistance — so low that intracellular electrical current can travel very easily from one smooth muscle fiber to another. Therefore, the smooth muscle of the gastrointestinal tract performs as a *functional syncytium,* which means that action potentials originating in one smooth muscle fiber are generally propagated from fiber to fiber.

Contraction of Intestinal Muscle. The smooth muscle of the gastrointestinal tract exhibits both *tonic contraction* and *rhythmic contraction,* both of which are characteristic of most types of smooth muscle, as discussed in Chapter 10.

Tonic contraction is continuous, lasting minute after minute or even hour after hour, sometimes increasing or decreasing in intensity but, nevertheless, continuing. This contraction may be caused by a series of action potentials or by non-electrogenic stimulation by hormones. The intensity of tonic contraction in each segment of the gut determines the amount of steady pressure in the segment, and tonic contraction of the sphincters determines the amount of resistance offered at the sphincters to the movement of intestinal contents. In this way the *pyloric,* the

ileocecal, and the *anal sphincters* all help to regulate food movement in the gut.

In different parts of the gut the rhythmic contractions of the gastrointestinal smooth muscle occur at rates as rapid as 12 times per minute or as slow as 3 times per minute. These frequencies are determined by "slow waves" in the electrical potentials of the muscle, waves that are distinct from the action potentials but that cause rhythmic entraining of the action potentials, as was explained in Chapter 10. The rhythmic contractions are responsible for the phasic functions of the gastrointestinal tract, such as mixture of the food with the secretions or peristaltic propulsion of food, as discussed later in the chapter.

INNERVATION OF THE GUT — THE INTRAMURAL PLEXUS

Beginning in the esophageal wall and extending all the way to the anus is an *intrinsic nervous system* of the gastrointestinal tract. This is composed principally of two layers of neurons and appropriate connecting fibers: The outer layer, called the *myenteric plexus* or *Auerbach's plexus,* lies between the longitudinal and circular muscular layers; and the inner layer, called the *submucosal plexus* or *Meissner's plexus,* lies in the submucosa. The myenteric plexus controls mainly the *gastrointestinal movements,* while the submucosal plexus is important in controlling *secretion* and also subserves many *sensory functions,* receiving signals principally from the gut epithelium and from stretch receptors in the gut wall.

In general, stimulation of the myenteric plexus increases the activity of the gut. On the other hand, a few myenteric plexus fibers are inhibitory rather than excitatory.

The intrinsic nervous system, including both the myenteric plexus and the submucosal plexus, is especially responsible for many neurogenic reflexes that occur locally in the gut, such as reflexes from the mucosal epithelium to increase the activity of the gut muscle or to cause localized secretion of digestive juices by the submucosal glands.

Autonomic Control of the Gastrointestinal Tract. The gastrointestinal tract receives extensive parasympathetic and sympathetic nerves that are capable of altering the overall activity of the entire gut or of specific parts of it, particularly of its upper end down to the stomach and its distal end from the mid-colon region to the anus.

Parasympathetic Innervation. The parasympathetic supply to the gut is divided into *cranial* and *sacral divisions,* which were discussed in Chapter 38. Except for a few parasympathetic fibers to the mouth and pharyngeal regions of the alimentary tract, the cranial parasympathetics are transmitted almost entirely in the *vagus nerves.* These fibers provide extensive innervation to the esophagus and stomach and, to much less extent, to the small intestine, gallbladder, and first half of the large intestine. The sacral parasympathetics originate in the second and third sacral segments of the spinal cord and pass through the *nervi erigentes* to the distal half of the large intestine. The sigmoidal, rectal, and anal regions of the large intestine are considerably better supplied with parasympathetic fibers than are the other portions. These fibers function especially in the defecation reflexes, which are discussed later in the chapter.

The postganglionic neurons of the parasympathetic system are part of the myenteric plexus, so that stimulation of the parasympathetic nerves causes a general increase in activity of this plexus. This in turn excites the gut wall and facilitates most of the intrinsic excitatory nervous reflexes of the gastrointestinal tract.

Sympathetic Innervation. The sympathetic fibers to the gastrointestinal tract originate in the spinal cord between the segments T-8 and L-3. The preganglionic fibers, after leaving the cord, enter the sympathetic chains and pass through the chains to outlying ganglia, such as the *celiac ganglion* and various *mesenteric ganglia.* Here the postganglionic neuron bodies are located, and postganglionic fibers spread from them along the blood vessels to all parts of the gut. The sympathetics innervate essentially all portions of the gastrointestinal tract rather than being more extensively supplied to the most orad and most analward portions as is true of the parasympathetics.

In general, stimulation of the sympathetic nervous system inhibits activity in the gastrointestinal tract, causing effects essentially opposite to those of the parasympathetic system. Thus, strong stimulation of the sympathetic system can totally block movement of food through the gastrointestinal tract.

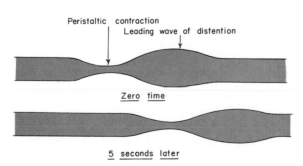

Figure 42–3. Peristalsis.

FUNCTIONAL TYPES OF MOVEMENTS IN THE GASTROINTESTINAL TRACT

Two basic types of movements occur in the gastrointestinal tract: (1) *mixing movements,* which keep the intestinal contents thoroughly mixed at all times, and (2) *propulsive movements,* which cause food to move forward along the tract at an appropriate rate for digestion and absorption.

THE MIXING MOVEMENTS

In most parts of the alimentary tract, the mixing movements are caused by *local contractions of small segments of the gut wall.* These movements are modified in different parts of the gastrointestinal tract, as discussed separately later in the chapter.

THE PROPULSIVE MOVEMENTS – PERISTALSIS

The basic propulsive movement of the gastrointestinal tract is *peristalsis,* which is illustrated in Figure 42–3. A contractile ring appears around the gut and then moves forward; this is analogous to putting one's fingers around a thin distended tube, then constricting the fingers and moving them forward along the tube. Obviously, any material in front of the contractile ring is moved forward.

Peristalsis is an inherent property of any syncytial smooth muscle tube; stimulation at any point causes a contractile ring to spread in both directions. Thus, peristalsis occurs in (1) the gastrointestinal tract, (2) the bile ducts, (3) other glandular ducts throughout the body, (4) the ureters, and (5) most other smooth muscle tubes of the body.

The usual stimulus for peristalsis is *distension.* That is, if a large amount of food collects at any point in the gut, the distension stimulates the gut wall 2 to 3 cm. above this point, and a contractile ring appears that initiates a peristaltic movement.

Function of the Myenteric Plexus in Peristalsis. Even though peristalsis is a basic characteristic of all tubular smooth muscle structures, it occurs only weakly in portions of the gastrointestinal tract that have congenital absence of the myenteric plexus. Also, it is greatly depressed or completely blocked in the entire gut when the person is treated with atropine to paralyze the myenteric plexus. Furthermore, since the myenteric plexus is principally under the control of the parasympathetic nerves, the intensity of peristalsis and its velocity of conduction can be altered by parasympathetic stimulation.

Therefore, *even though the basic phenomenon of peristalsis does not necessarily require the myenteric nerve plexus, effectual peristalsis does require an active myenteric plexus.*

INGESTION OF FOOD

The amount of food that a person ingests is determined principally by the intrinsic desire for food called *hunger,* and the type of food that he preferentially seeks is determined by his *appetite.* These mechanisms in themselves are extremely important automatic regulatory systems for maintaining an adequate nutritional supply for the body, and they will be discussed in detail in Chapter 48 in relation to nutrition of the body. The present discussion is confined to the actual mechanical aspects of food ingestion, including especially *mastication* and *swallowing.*

MASTICATION (CHEWING)

The teeth are admirably designed for chewing, the anterior teeth (incisors) providing a strong cutting action and the posterior teeth (molars) a grinding action. All the jaw muscles working together can close the teeth with a force as great as 55 pounds on the incisors or 200 pounds on the molars. When this is applied to a small object, such as a small seed between the molars, the actual force *per square inch* may be several thousand pounds.

Most of the muscles of chewing are innervated by the motor branch of the 5th cranial nerve, and the chewing process is controlled by nuclei in the brain stem. Stimulation of an area in the reticular formation near the brain stem centers for taste can cause continual rhythmic chewing movements. Also, stimulation of areas in the hypothalamus, amygdaloid nuclei, and even in the cerebral cortex near the sensory areas for taste and smell can cause chewing.

Much of the chewing process is caused by the *chewing reflex,* which may be explained as follows: The presence of a bolus of food in the mouth causes reflex inhibition of the muscles of mastication, which allows the lower jaw to drop. The sudden drop in turn initiates a stretch reflex of the jaw muscles that leads to *rebound* contraction. This automatically raises the jaw to cause closure of the teeth, but it also compresses the bolus again against the linings of the mouth, which inhibits the jaw muscles once again, allowing the jaw to

drop and rebound another time, and this is repeated again and again.

Chewing of the food is important for digestion of all foods, but it is especially important for most fruits and raw vegetables because these have undigestible cellulose membranes around their nutrient portions which must be broken before the food can be utilized. Chewing aids in the digestion of food for the following simple reason: Since the *digestive enzymes act mainly on the surfaces of food particles,* the rate of digestion is highly dependent on the total surface area exposed to the intestinal secretions. Also, grinding the food to a very fine particulate consistency prevents excoriation of the gastrointestinal tract and increases the ease with which food is emptied from the stomach into the small intestine and thence into all succeeding segments of the gut.

SWALLOWING (DEGLUTITION)

Swallowing is a complicated mechanism, principally because the pharynx most of the time subserves several other functions besides swallowing. It is converted for only a few seconds at a time into a tract for propulsion of food. Especially is it important that respiration not be seriously compromised during swallowing.

In general, swallowing can be divided into (1) the *voluntary stage,* which initiates the swallowing process, (2) the *pharyngeal stage,* which is involuntary and constitutes the passage of food through the pharynx into the esophagus, and (3) the *esophageal stage,* another involuntary phase, which promotes passage of food from the pharynx to the stomach.

Voluntary Stage of Swallowing. When the food is ready for swallowing, it is "voluntarily" squeezed or rolled posteriorly in the mouth by pressure of the tongue upward and backward against the palate, as shown in Figure 42–4. Thus, the tongue forces the bolus of food into the pharynx. From here on, the process of swallowing becomes entirely, or almost entirely, automatic and ordinarily cannot be stopped.

Pharyngeal Stage of Swallowing. When the bolus of food is pushed backward in the mouth, it stimulates *swallowing receptor areas* all around the opening of the pharynx, especially on the tonsillar pillars, and impulses from these pass to the brain stem to initiate a series of automatic pharyngeal muscular contractions as follows:

1. The soft palate is pulled upward to close the posterior nares, in this way preventing reflux of food into the nasal cavities.

2. The palatopharyngeal folds on either side of the pharynx are pulled medialward to approx-

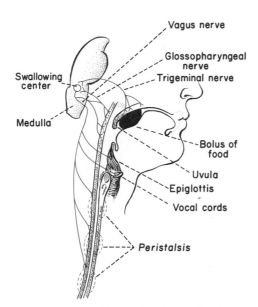

Figure 42–4. The swallowing mechanism.

imate each other. In this way these folds form a sagittal slit through which the food must pass into the posterior pharynx. This slit performs a selective action, allowing food that has been masticated properly to pass with ease while impeding the passage of large objects.

3. The vocal cords of the larynx are strongly approximated, and the epiglottis swings backward over the superior opening of the larynx. Both of these effects prevent passage of food into the trachea.

4. The entire larynx is pulled upward and forward by muscles attached to the hyoid bone: this movement of the larynx stretches the opening of the esophagus. At the same time, the upper 3 to 4 centimeters of the esophagus, an area called the *upper esophageal sphincter*, relaxes, thus allowing food to move easily and freely from the posterior pharynx into the upper esophagus. This sphincter, between swallows, remains tonically and strongly contracted, thereby preventing air from going into the esophagus during respiration.

5. At the same time that the larynx is raised and the upper esophageal sphincter is relaxed, the pharyngeal constrictor muscles contract, giving rise to a rapid peristaltic wave that passes downward over the pharyngeal muscles and into the esophagus, and propels the food into the esophagus.

To summarize the mechanics of the pharyngeal stage of swallowing — the trachea is closed, the esophagus is opened, and a fast peristaltic wave originating in the pharynx then forces the bolus of food into the upper esophagus, the entire process occurring in one to two seconds.

Nervous Control of the Pharyngeal Stage of Swallowing. The most sensitive tactile areas of the pharynx for initiation of the pharyngeal stage of swallowing lie in a ring around the pharyngeal opening, with greatest sensitivity in the tonsillar pillars. Impulses are transmitted from these areas through the sensory portions of the trigeminal and glossopharyngeal nerves into a region of the medulla oblongata closely associated with the *tractus solitarius,* which receives essentially all sensory impulses from the mouth.

The successive stages of the swallowing process are then automatically controlled in orderly sequence by a neuronal area in the reticular substance of the medulla and lower portion of the pons. The sequence of the swallowing reflex remains the same from one swallow to the next, and the timing of the entire cycle also remains constant from one swallow to the next. The area in the medulla and lower pons that controls swallowing is called the *deglutition* or *swallowing center.*

The motor impulses from the swallowing center to the pharynx and upper esophagus that cause swallowing are transmitted by the 5th, 9th, 10th, and 12th cranial nerves and even a few of the superior cervical nerves.

In summary, the pharyngeal stage of swallowing is principally a reflex act. It is rarely initiated by direct stimuli to the swallowing center from higher regions of the central nervous system. Instead, it is initiated by voluntary movement of food into the back of the mouth, which, in turn, elicits the swallowing reflex.

Esophageal Stage of Swallowing. The esophagus functions primarily to conduct food from the pharynx to the stomach, and its movements are organized specifically for this function.

Normally the esophagus exhibits two types of peristaltic movements — *primary peristalsis* and *secondary peristalsis.* Primary peristalsis is simply a continuation of the peristaltic wave that begins in the pharynx and spreads into the esophagus during the pharyngeal stage of swallowing. This wave passes all the way from the pharynx to the stomach in approximately five to ten seconds. If the primary peristaltic wave fails to move all the food that has entered the esophagus on into the stomach, secondary peristaltic waves result from distension of the esophagus by the retained food. These waves are essentially the same as the primary peristaltic waves, except that they originate in the esophagus itself rather than in the pharynx. Secondary peristaltic waves continue until all the food has emptied into the stomach.

The peristaltic waves of the esophagus are controlled almost entirely by vagal reflexes that are part of the overall swallowing mechanism.

These reflexes are transmitted through *vagal afferent fibers* from the esophagus to the medulla and then back again to the esophagus through *vagal efferent fibers.*

FUNCTION OF THE LOWER ESOPHAGEAL SPHINCTER

At the lower end of the esophagus, 2 to 5 cm. above its juncture with the stomach, the circular muscle of the esophagus functions as a *lower esophageal sphincter.* Anatomically this sphincter is no different from the remainder of the esophageal muscle. However, physiologically, it remains tonically constricted, in contrast to the midportions of the esophagus, which normally remain completely relaxed. Yet, when a peristaltic swallowing wave passes down the esophagus, "receptive relaxation," caused by myenteric nerve signals, relaxes the lower esophageal sphincter ahead of the peristaltic wave, and allows easy propulsion of swallowed food on into the stomach.

A principal function of the lower esophageal sphincter is to prevent reflux of stomach contents into the upper esophagus. The stomach contents are highly acidic and contain many proteolytic enzymes. The esophageal mucosa, except in the lower eighth of the esophagus, is not capable of resisting for long the digestive action of gastric secretions. Fortunately, the tonic constriction of the lower esophageal sphincter prevents significant reflux of stomach contents into the esophagus except under abnormal conditions.

MOTOR FUNCTIONS OF THE STOMACH

The motor functions of the stomach are three: (1) storage of large quantities of food until it can be accommodated in the lower portion of the gastrointestinal tract, (2) mixing of this food with gastric secretions until it forms a semifluid mixture called *chyme,* and (3) slow emptying of the food from the stomach into the small intestine at a rate suitable for proper digestion and absorption by the small intestine.

Figure 42–5 illustrates the basic anatomy of the stomach. Physiologically, the stomach can be divided into two major parts: (1) the *corpus,* or *body,* and (2) the *antrum.* The *fundus,* located at the upper end of the body of the stomach, is often considered by anatomists to be a separate entity from the body, but from a physiological point of view, the fundus is actually a functional part of the body.

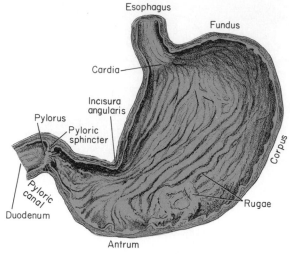

Figure 42–5.   Physiologic anatomy of the stomach.

STORAGE FUNCTION OF THE STOMACH

As food enters the stomach, it forms concentric circles in the body and fundus of the stomach, the newest food lying closest to the esophageal opening and the oldest lying nearest the wall of the stomach. Normally, the body and fundus of the stomach have relatively little tone in their muscular wall, so that they can bulge progressively outward, thereby accommodating greater and greater quantities of food up to a limit of almost 1 liter.

MIXING IN THE STOMACH

The digestive juices of the stomach are secreted by the *gastric glands,* which cover almost the entire outer wall of the body of the stomach. These secretions immediately come into contact with the stored food lying against the mucosal surface of the stomach. When the stomach is filled, weak *constrictor waves,* also called *mixing waves,* move along the stomach wall approximately once every 20 seconds. These waves move the gastric secretions and the outermost layer of food gradually toward the antral part of the stomach.

In addition to the mixing caused by the waves in the body of the stomach, mixing is also caused by peristaltic movements in the antral portion of the stomach. These movements cause mixing in the following way: Each time a peristaltic wave passes over the antrum toward the pylorus, it digs deeply into the contents of the antrum. Yet, the opening of the pylorus is small enough that only a few millimeters of antral contents are expelled into the duodenum with each peristaltic wave. Instead, most of the antral contents squirt backward through the peristaltic ring toward the body of the stomach. Thus, the moving peristaltic constrictive ring, combined with this reflux action, is an exceedingly important mixing mechanism of the stomach.

Chyme. After the food has become mixed with the stomach secretions, the resulting mixture that passes on down the gut is called chyme. The degree of fluidity of chyme depends on the relative amounts of food and stomach secretions and on the degree of digestion that has occurred. The appearance of chyme is that of a murky, milky semifluid or paste.

Propulsion of Food Through the Stomach. Strong peristaltic waves occur about 20 per cent of the time in the antrum of the stomach. These waves, like the mixing waves, occur about once every 20 seconds. As the stomach becomes progressively more and more empty, these intense waves begin farther and farther up the body of the stomach, gradually pinching off the lowermost portions of stored food, adding this food to the chyme in the antrum.

The peristaltic waves often exert as much as 50 to 70 cm. of water pressure, which is about six times as powerful as the usual mixing waves.

EMPTYING OF THE STOMACH

Basically, stomach emptying is opposed by resistance of the pylorus to the passage of food, and it is promoted by peristaltic waves in the antrum of the stomach. Usually, these two are reciprocally related — that is, those factors that increase antral peristalsis usually decrease the tone of the pyloric muscle.

Role of the Pylorus in Stomach Emptying. The pylorus normally remains almost, but not completely, closed because of tonic contraction of the pyloric muscle. The closing force is weak enough that water and other fluids empty from the stomach with ease. On the other hand, it is great enough to prevent movement of semisolid chyme into the duodenum except when a strong antral peristaltic wave forces the chyme through. On the other hand, the degree of constriction of the pyloric sphincter can increase or decrease under the influence of signals both from the stomach and from the duodenum, as we shall discuss subsequently.

Role of Antral Peristalsis in Stomach Emptying — The Pyloric Pump. The intensity of antral peristalsis changes markedly under different conditions, especially in response to signals both from the stomach and from the duodenum. Therefore, the intensity of antral peristalsis is the other principal factor determining the rate of stomach emptying.

When pyloric tone is normal, each strong antral peristaltic wave forces several milliliters of chyme into the duodenum. Thus, the peristaltic waves provide a pumping action that is frequently called the "pyloric pump."

Regulation of Stomach Emptying

The rate at which the stomach empties is regulated by signals both from the stomach and from the duodenum. The stomach signals are mainly two-fold: (1) nervous signals caused by distension of the stomach by food, and (2) the hormone *gastrin* released from the antral mucosa in response to the presence of food in the stomach. Both these signals increase pyloric pumping force and at the same time inhibit the pylorus, thus promoting stomach emptying.

On the other hand, signals from the duodenum depress the pyloric pump and usually increase pyloric tone at the same time. In general, when an excess volume of chyme or excesses of certain types of chyme enter the duodenum, strong *negative* feedback signals, both nervous and hormonal, depress the pyloric pump and enhance pyloric sphincter tone. Obviously, these feedback signals prevent more chyme from entering the duodenum until it can be processed by the small intestine.

Effect of the Hormone Gastrin on Stomach Emptying. In the following chapter we shall see that stretch, as well as the presence of certain types of foods in the stomach — particularly meat — elicits release of a hormone called *gastrin* from the antral mucosa. Gastrin has potent effects on secretion of highly acidic gastric juice by the gastric glands. Also gastrin has potent stimulatory effects on motor functions of the stomach. Most important, it enhances the activity of the pyloric pump while at the same time relaxing the pylorus. Thus, it is a strong influence for promoting stomach emptying.

The Inhibitory Effect of the Enterogastric Reflex from the Duodenum on Pyloric Activity. When chyme enters the duodenum, nervous reflex signals are transmitted back to the stomach to inhibit antral peristalsis and to increase the tone of the pylorus. This is called the *enterogastric reflex*. It obviously acts to limit the emptying of the stomach until the small intestine can carry the chyme away. This reflex is probably mediated mainly by way of afferent nerve fibers in the vagus nerve to the brain stem and then back through efferent nerve fibers to the stomach, also by way of the vagi.

The types of factors that are continually monitored in the duodenum and that can elicit the enterogastric reflex include

1. The degree of distension of the duodenum.
2. The presence of any degree of irritation of the duodenal mucosa.
3. The degree of acidity of the duodenal chyme.
4. The osmolality of the chyme.
5. The presence of certain breakdown products in the chyme, especially breakdown products of proteins and perhaps to a lesser extent of fats.

The enterogastric reflex is especially sensitive to the presence of irritants and acids in the duodenal chyme. For instance, whenever the pH of the chyme in the duodenum falls below approximately 3.5 to 4, this reflex is immediately elicited, which inhibits the pyloric pump and increases pyloric constriction, thus reducing or even blocking further the release of acidic stomach contents into the duodenum until the duodenal chyme can be neutralized by pancreatic and other secretions.

Hormonal Feedback from the Duodenum in Inhibiting Gastric Emptying — Role of Fats. Even when the enterogastric reflex has been blocked, excess amounts of chyme entering the duodenum will still inhibit stomach emptying. This effect is particularly powerful when the chyme contains large amounts of fat. It is caused by several different hormones released from the mucosa of the upper small intestine. These hormones are absorbed into the blood and then are transferred to the stomach, where they inhibit antral peristalsis and increase pyloric tone.

One of the hormones is *cholecystokinin,* which is released from the mucosa of the jejunum in response to fatty substances in the chyme. This hormone acts as a competitive inhibitor to block the increased stomach motility caused by gastrin. Another is the hormone *secretin,* which is released mainly from the duodenal mucosa in response to gastric acid released from the stomach through the pylorus. This hormone has the general effect of decreasing gastrointestinal motility. Finally, a hormone called *gastric inhibitory peptide,* which is released from the upper small intestine in response mainly to fat in the chyme but to carbohydrates as well, is known also to inhibit gastric motility under some conditions.

Summary. Emptying of the stomach is controlled to a moderate degree by stomach factors, such as the degree of filling in the stomach and the excitatory effect of gastrin on antral peristalsis. However, probably the most important control of stomach emptying resides in feedback signals from the duodenum, including both the enterogastric reflex and hormonal feedback.

These two feedback signals work together to slow the rate of emptying when (a) too much chyme is already in the small intestine or (b) the chyme is excessively acid, contains too much protein or fat, is hypotonic or hypertonic, or is irritating. In this way the rate of stomach emptying is limited to that amount of chyme that the small intestine can process.

MOVEMENTS OF THE SMALL INTESTINE

The movements of the small intestine, as elsewhere in the gastrointestinal tract, can be divided into *mixing contractions* and *propulsive contractions.* However, to a great extent this separation is artificial because essentially all movements of the small intestine cause at least some degree of both mixing and propulsion.

MIXING CONTRACTIONS (SEGMENTATION CONTRACTIONS)

When a portion of the small intestine becomes distended with chyme, this elicits localized concentric ringlike contractions spaced at intervals along the intestine. These rhythmic contractions occur at a rate of 11 to 12 per minute in the duodenum and at progressively slower rates down to approximately 7 per minute in the terminal ileum. These contractions cause "segmentation" of the small intestine, as illustrated in Figure 42–6, dividing the intestines at times into regularly spaced segments that have the appearance of a chain of sausages. As one set of segmentation contractions relaxes, a new set begins, but the contractions this time occur at new points between the previous contractions. Therefore, the segmentation contractions "chop" the chyme many times a minute, in this way promoting the

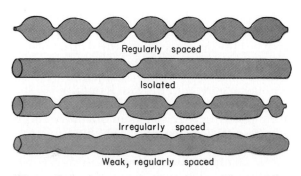

Figure 42–6. Segmentation movements of the small intestine.

progressive mixing of the solid food particles with the secretions of the small intestine.

PROPULSIVE MOVEMENTS

Chyme is propelled through the small intestine by *peristaltic waves.* These can occur in any part of the small intestine, and they move analward at a velocity of 0.5 to 2 cm. per second, much faster in the proximal intestine and much slower in the terminal intestine. However, they are normally very weak and usually die out after traveling only a few centimeters; therefore, movement of the chyme is slow. As a result, the net movement of the chyme along the small intestine averages only 1 cm. per minute. This means that 3 to 5 hours are normally required for passage of chyme from the pylorus to the ileocecal valve.

Peristaltic activity of the small intestine is greatly increased after a meal. This is caused partly by the entry of chyme into the duodenum but also by a so-called *gastroenteric reflex* initiated by distension of the stomach and conducted principally through the myenteric plexus from the stomach down along the wall of the small intestine. This reflex increases the overall degree of excitability of the small intestine, including both increased motility and secretion.

The Peristaltic Reflex. The usual cause of peristalsis in the small intestine is distension. Circumferential stretch of the intestine excites receptors in the gut wall, and these elicit a local myenteric reflex that begins with contraction of the longitudinal muscle over a distance of several centimeters followed by contraction of the circular muscle. Simultaneously, the contractile process spreads in an analward direction by the process of peristalsis. Movement of the peristaltic contraction down the gut is controlled by the myenteric plexus; it does not occur when this plexus has been blocked by drugs or when the plexus has degenerated.

Very intense irritation of the intestinal mucosa, such as that occurring in some infectious processes, can elicit a so-called *peristaltic rush,* which is a powerful peristaltic wave that travels long distances in the small intestine in a few minutes. These waves can sweep the contents of the intestine into the colon and thereby relieve the small intestine of either irritants or excessive distension.

The function of the peristaltic waves in the small intestine is not only to cause progression of the chyme toward the ileocecal valve but also to spread out the chyme along the intestinal mucosa.

FUNCTION OF THE ILEOCECAL VALVE

A principal function of the ileocecal valve is to prevent backflow of fecal contents from the colon into the small intestine. As illustrated in Figure 42–7, the lips of the ileocecal valve protrude into the lumen of the cecum and therefore are forcibly closed when the cecum fills. Usually the valve can resist reverse pressure of as much as 50 to 60 cm. water.

The wall of the ileum for several centimeters immediately preceding the ileocecal valve has a thickened muscular coat called the *ileocecal sphincter.* This normally remains mildly constricted and slows the emptying of ileal contents into the cecum except immediately following a meal, when a *gastroileal reflex* intensifies the peristalsis in the ileum. Also, the hormone *gastrin,* which is liberated from the stomach mucosa in response to food in the stomach, has a direct relaxant effect on the ileocecal sphincter, thus allowing rapid emptying. Yet, even so, only about 800 ml. of chyme empties into the cecum each day. The resistance to emptying at the ileocecal valve prolongs the stay of chyme in the ileum and therefore facilitates absorption.

Control of the Ileocecal Sphincter. The degree of contraction of the ileocecal sphincter is controlled primarily by reflexes from the cecum. Whenever the cecum is distended, the degree of contraction of the ileocecal sphincter is intensified, which greatly delays emptying of additional chyme from the ileum. Also, any irritant in the cecum causes constriction of the ileocecal sphincter. For instance, when a person has an inflamed appendix, the irritation of this vestigial remnant of the cecum can cause such intense

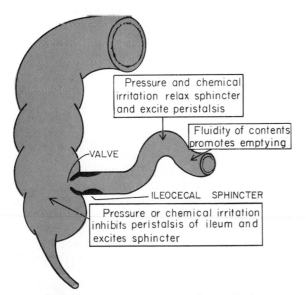

Pressure and chemical irritation relax sphincter and excite peristalsis

Fluidity of contents promotes emptying

VALVE

ILEOCECAL SPHINCTER

Pressure or chemical irritation inhibits peristalsis of ileum and excites sphincter

Figure 42–7. Emptying at the ileocecal valve.

spasm of the ileocecal sphincter that it completely blocks emptying of the ileum. These reflexes from the cecum to the ileocecal sphincter are mediated by way of the myenteric plexus.

MOVEMENTS OF THE COLON

The functions of the colon are (1) absorption of water and electrolytes from the chyme and (2) storage of fecal matter until it can be expelled. The proximal half of the colon, illustrated in Figure 42–8, is concerned principally with absorption, and the distal half with storage; since intense movements are not required for these functions, the movements of the colon are normally sluggish. Yet, even in a sluggish manner, the movements still have characteristics similar to those of the small intestine and can be divided once again into mixing movements and propulsive movements.

Mixing Movements — Haustrations. In the same manner that segmentation movements occur in the small intestine, large circular constrictions also occur in the large intestine. At each of these constriction points, about 2.5 cm. of the circular muscle contracts, sometimes constricting the lumen of the colon to almost complete occlusion. At the same time, the longitudinal muscle of the colon, which is aggregated into three longitudinal strips called the *teniae coli,* contracts. These combined contractions of the circular and the longitudinal strips of smooth muscle cause the unstimulated portions of the large intestine to bulge outward into baglike sacs called *haustrations.* The haustral contractions, once initiated, usually reach peak intensity in about 30 seconds and then disappear during the next 60 seconds. They at times also move slowly analward during their period of contraction. After another few minutes,

new haustral contractions occur in nearby areas but not in the same areas. Therefore, the fecal material in the large intestine is slowly "dug" into and rolled over in much the same manner that one spades the earth. In this way, all the fecal material is gradually exposed to the surface of the large intestine, and fluid is progressively absorbed until only 80 to 150 ml. of the 800 ml. daily load of chyme is lost in the feces.

Propulsive Movements — "Mass Movements." Peristaltic waves of the type seen in the small intestine do not occur in the colon. Instead, another type of movement, called *mass movements,* propels the fecal contents toward the anus. These movements usually occur only a few times each day, most abundantly for about 15 minutes during the first hour or so after eating breakfast.

A mass movement is characterized by the following sequence of events: First, a constrictive point occurs at a distended or irritated point in the colon. Rapidly thereafter the 20 or more cm. of colon *distal* to the constriction contracts almost as a unit, forcing the fecal material in this segment *en masse* down the colon. The initiation of contraction is complete in about 30 seconds, and relaxation then occurs during the next two to three minutes. Mass movements can occur in any part of the colon, though most often they occur in the transverse or descending colon. When they have forced a mass of feces into the rectum, the desire for defecation is felt.

Initiation of Mass Movements by Gastrocolic and Duodenocolic Reflexes. The appearance of mass movements after meals is caused at least partially by so-called *gastrocolic* and *duodenocolic reflexes.* These reflexes result from distension of the stomach and duodenum, and they are transmitted mainly through the myenteric plexus.

Irritation in the colon can also initiate intense mass movements. For instance, a person who has an ulcerated condition of the colon (*ulcerative colitis*) frequently has mass movements that persist almost all of the time.

DEFECATION

Most of the time the rectum is empty of feces. This results partly from the fact that a weak functional sphincter exists approximately 20 cm. from the anus at the juncture between the sigmoid and the rectum. However, when a mass movement forces feces into the rectum the process of defecation is normally initiated, including reflex contraction of the rectum, the sigmoid, and the descending colon, and also relaxation of the anal sphincters.

Continual dribble of fecal matter through the anus is prevented by tonic constriction of (1) the

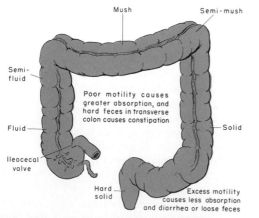

Figure 42–8. Absorptive and storage functions of the large intestine.

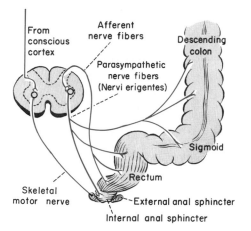

Figure 42–9. The afferent and efferent pathways of the parasympathetic mechanism for enhancing the defecation reflex.

internal anal sphincter, a circular mass of smooth muscle that lies immediately inside the anus, and (2) the *external anal sphincter,* composed of striated voluntary muscle that surrounds and lies slightly distal to the internal sphincter and is controlled by the somatic nervous system and is therefore under voluntary control.

Ordinarily, defecation results from the *defecation reflex,* which is illustrated in Figure 42–9. When the sensory nerve fibers in the rectum are stimulated by stretch, signals are transmitted into the sacral portion of the spinal cord and thence, reflexly, back to the descending colon, sigmoid, rectum, and anus by way of parasympathetic nerve fibers in the *nervi erigentes.* These parasympathetic signals initiate strong peristaltic waves that are sometimes effective in emptying the large bowel all the way from the splenic flexure to the anus. Also, the afferent signals entering the spinal cord initiate other effects, such as taking a deep breath, closure of the glottis, and contraction of the abdominal muscles to force downward on the fecal contents of the colon while at the same time causing the pelvic floor to pull outward and upward on the anus to evaginate the feces downward.

However, despite the defecation reflex, other effects are necessary before actual defecation occurs. Except in babies and mentally inept persons, the conscious mind voluntarily controls the external sphincter and either inhibits its contraction and thereby allows defecation to occur, or further contracts it if the moment is not socially acceptable for defecation to occur. If the external sphincter is kept contracted, so that defecation does not occur, the defecation reflex dies out after a few minutes and usually will not return until an additional amount of feces enters the rectum, which may not be until several hours thereafter.

When it becomes convenient for the person to defecate, defecation reflexes can sometimes be initiated by taking a deep breath to move the diaphragm downward and then contracting the abdominal muscles to increase the pressure in the abdomen, thus forcing fecal contents into the rectum to elicit new reflexes. Unfortunately, reflexes initiated in this way are never as effective as those that arise naturally, for which reason people who inhibit their natural reflexes too often are likely to become severely constipated.

REFERENCES

Atanassova, E., and Papasova, M.: Gastrointestinal motility. *Int. Rev. Physiol., 12*:35, 1977.

Bortoff, A.: Myogenic control of intestinal motility. *Physiol. Rev., 56*:418, 1976.

Brooks, F. P. (ed.): Gastrointestinal Pathophysiology. New York, Oxford University Press, 1978.

Brooks, F. P., and Evers, P. W. (eds.): Nerves and the gut. Thorofare, N.J., C. B. Slack, 1977.

Cohen, S., *et al.*: Gastrointestinal motility. *In* Crane, R. K. (ed.): International Review of Physiology: Gastrointestinal Physiology III. Vol. 19. Baltimore, University Park Press, 1979, p. 107.

Davenport, H. W.: A Digest of Digestion, 2nd Ed. Chicago, Year Book Medical Publishers, 1978.

Dickray, G.: Comparative biochemistry and physiology of gut hormones. *Annu. Rev. Physiol., 41*:83, 1979.

Duthie, H. S. (ed.): Gastrointestinal Motility in Health and Disease. Baltimore, University Park Press, 1978.

Grossman, M. I.: Neural and hormonal regulation of gastrointestinal function: An overview. *Annu. Rev. Physiol., 41*:27, 1979.

Hurwitz, A. L., *et al.*: Disorders of Esophageal Motility. Philadelphia, W. B. Saunders Co., 1979.

Jenkins, G. N.: The Physiology and Biochemistry of the Mouth. Philadelphia, J. B. Lippincott Co., 1978.

Johnson, L. R.: Gastrointestinal hormones and their functions. *Annu. Rev. Physiol., 39*:135, 1977.

Levitt, M. D., and Bond, J. H.: Flatulence. *Annu. Rev. Med., 31*:127, 1980.

Phillips, S. F., and Devroede, G. J.: Functions of the large intestine. *In* Crane, R. K. (ed.): International Review of Physiology: Gastrointestinal Physiology III. Vol. 19. Baltimore, University Park Press, 1979, p. 263.

Rehfeld, J. F.: Gastrointestinal hormones. *In* Crane, R. K. (ed.): International Review of Physiology: Gastrointestinal Physiology III. Vol. 19. Baltimore, University Park Press, 1979, p. 291.

Sernka, T. J., and Jacobson, E. D.: Gastrointestinal Physiology; The Essentials. Baltimore, Williams & Wilkins, 1979.

Sodeman, W. A., Jr., and Watson, D. W.: The large intestine. *In* Sodeman, W. A., Jr., and Sodeman, T. M. (eds.): Sodeman's Pathologic Physiology; Mechanisms of Disease, 6th Ed. Philadelphia, W. B. Saunders Co., 1979, p. 860.

Van Der Reis, L. (ed.): The Esophagus. New York, S. Karger, 1978.

43

Secretory Functions of the Alimentary Tract

Throughout the gastrointestinal tract, secretions subserve two functions: First, digestive enzymes are secreted in most areas from the mouth to the distal end of the ileum. Second, mucous glands, present from the mouth to the anus, provide mucus for lubrication and protection of all parts of the alimentary tract.

Most digestive secretions are formed only in response to the presence of food in the alimentary tract, and the quantity secreted in each segment of the tract is almost exactly the amount needed for proper digestion. Furthermore, in some portions of the gastrointestinal tract even the types of enzymes and other constituents of the secretions are varied in accordance with the types of food present. The purpose of the present chapter, therefore, is to describe the different alimentary secretions, their functions, and regulation of their production.

GENERAL PRINCIPLES OF GASTROINTESTINAL SECRETION

ANATOMICAL TYPES OF GLANDS

Several types of glands provide the different types of secretions in the gastrointestinal tract. First, on the surface of the epithelium in most parts of the gastrointestinal tract are literally billions of *single cell mucous glands* called *goblet cells.* They simply extrude their mucus directly into the lumen of the gastrointestinal tract.

Second, most surface areas of the gastrointesti-

nal tract are lined by pits, which represent invaginations of the epithelium into the submucosa. In the small intestine these pits, called *crypts of Lieberkühn*, are deep and contain specialized secretory cells. One of these is illustrated in Figure 43–11. They are lined with goblet cells that produce mucus and with other epithelial cells that produce mainly serous fluids.

Third, in the stomach and upper duodenum are found large numbers of deep *tubular glands*. A typical tubular gland is illustrated in Figure 43–4, which shows an acid- and pepsinogen-secreting gland of the stomach.

Fourth, also associated with the gastrointestinal tract are several complex glands — the *salivary glands*, the *pancreas*, and the *liver* — which provide secretions for digestion or emulsification of food. These glands lie completely outside the walls of the gastrointestinal tract and will be described later.

BASIC MECHANISM OF SECRETION BY GLANDULAR CELLS

Secretion of Organic Substances. Though all the basic mechanisms by which glandular cells form different secretions and then extrude them to the exterior are not known, experimental evidence points to the following basic principles of secretion by glandular cells, as illustrated in Figure 43–1. (1) The nutrient material needed for formation of the secretion must diffuse or be actively transported from the capillary into the base of the glandular cell. (2) Many *mitochondria* located inside the cell near its base provide oxida-

498

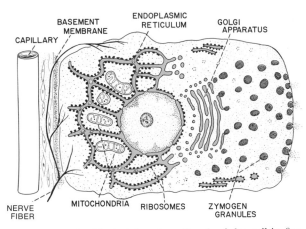

Figure 43–1. Typical function of a glandular cell in formation and secretion of enzymes or other secretory substances.

tive energy for formation of adenosine triphosphate. (3) Energy from the adenosine triphosphate, along with appropriate nutrients, is then used for synthesis of the organic substances, this synthesis occurring almost entirely on or near the *endoplasmic reticulum*. The *ribosomes* adherent to this reticulum are specifically responsible for formation of proteins. (4) The secretory materials flow through the tubules of the endoplasmic reticulum into the vesicles of the Golgi apparatus, which lies near the secretory ends of the cells. (5) The materials then are concentrated and discharged into the cytoplasm in the form of *secretory vesicles*. (6) These vesicles are then stored until nervous or hormonal control signals cause them to extrude their contents through the secretory surface into the lumen of the gland.

Water and Electrolyte Secretion in Response to Nervous Stimulation. A second necessity for glandular secretion is sufficient water and electrolytes to be secreted along with the organic substances. The following is a postulated method by which nervous stimulation causes water and salts to pass through the glandular cells in great profusion, which washes the organic substances through the secretory border of the cells at the same time:

(1) Nerve stimulation has a specific effect on the *basal* portion of the cell membrane of causing active transport of chloride ions to the interior. (2) The resulting increase in electronegativity inside the cell then causes positive ions also to move to the interior of the cell. (3) The excess of both of these ions inside the cell creates osmotic force, which pulls water to the interior, thereby increasing the hydrostatic pressure inside the cell and causing the cell itself to swell. (4) The pressure in the cell then results in minute ruptures of the secretory border of the cell and causes flushing of water, electrolytes, and organic materials

out of the secretory end of the glandular cell and into the lumen of the gland.

LUBRICATING AND PROTECTIVE PROPERTIES OF MUCUS AND ITS IMPORTANCE IN THE GASTROINTESTINAL TRACT

Mucus is a thick secretion composed of water, electrolytes, and a mixture of several glycoproteins. Mucus has several important characteristics that make it both an excellent lubricant and a protectant for the wall of the gut. *First*, mucus has adherent qualities that make it adhere tightly to the food or other particles and also to spread as a thin film over the food surfaces. *Second*, it has sufficient *body* to enable it to coat the wall of the gut and prevent actual contact of food particles with the mucosa. *Third*, mucus has a low resistance to slippage, so that the particles can slide along the epithelium with great ease. *Fourth*, mucus causes fecal particles to adhere to each other to form the fecal masses that are expelled during a bowel movement. *Fifth*, mucus is strongly resistant to digestion by the gastrointestinal enzymes. And, *sixth*, the glycoproteins of mucus have amphoteric properties and are therefore capable of buffering small amounts of either acids or alkalies; also, mucus usually contains moderate quantities of bicarbonate ions, which specifically neutralize acids.

In summary, mucus has the ability to allow easy slippage of food along the gastrointestinal tract and also to prevent excoriative or chemical damage to the epithelium. One becomes acutely aware of the lubricating qualities of mucus when his salivary glands fail to secrete saliva, for under these circumstances it is extremely difficult to swallow solid food even when it is taken with large amounts of water.

SECRETION OF SALIVA

The Salivary Glands; Characteristics of Saliva. The principal glands of salivation are the *parotid*, *submaxillary* (also called *submandibular*), and *sublingual* glands; in addition, there are many small *buccal* glands. The daily secretion of saliva normally ranges between 1000 and 1500 milliliters, as shown in Table 43–1.

Saliva contains two major types of protein secretion: (1) a *serous secretion* containing *ptyalin* (an α-amylase), which is an enzyme for digesting starches, and (2) *mucous secretion* containing *mucus* for lubricating purposes. The parotid glands secrete entirely the serous type, and the submaxillary glands secrete both the serous type and

TABLE 43-1 DAILY SECRETION OF
INTESTINAL JUICES

	Daily Volume (ml.)	pH
Saliva	1200	6.0–7.0
Gastric secretion	2000	1.0–3.5
Pancreatic secretion	1200	8.0–8.3
Bile	700	7.8
Succus entericus	2000	7.8–8.0
Brunner's gland secretion	50(?)	8.0–8.9
Large intestinal secretion	60	7.5–8.0
Total	7210	

mucus. The sublingual and buccal glands secrete only mucus. Saliva has a pH between 6.0 and 7.4, a favorable range for the digestive action of ptyalin.

Secretion of Ions in the Saliva. Saliva contains especially large quantities of potassium and bicarbonate ions. On the other hand, the concentrations of both sodium and chloride ions are considerably less in saliva than in plasma. One can understand these special concentrations of ions in the saliva from the following description of the mechanism for secretion of saliva.

Figure 43–2 illustrates secretion by the submaxillary gland, a typical *compound gland* containing both *acini* and *salivary ducts*. Salivary secretion is a two-stage operation; the first stage involves the acini and the second, the salivary ducts. The acini secrete a *primary secretion* that contains ptyalin or mucus or both in a solution of ions in concentrations not greatly different from those of typical extracellular fluid. However, as the primary secretion flows through the ducts, two

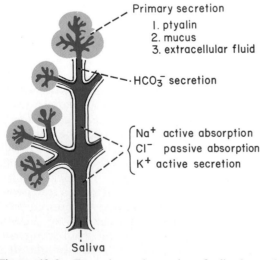

Figure 43–2. Formation and secretion of saliva by a salivary gland.

major active transport processes take place that markedly modify the ionic composition of the saliva. First, in the initial portion of the salivary ducts, near their junctions with the acini, *bicarbonate ion* is actively secreted while *chloride ions* are reabsorbed. Second, *sodium ions* are actively reabsorbed from all the salivary ducts, and *potassium ions* are actively secreted, but at a slower rate, in exchange for the sodium. Therefore, the sodium concentration of the saliva becomes greatly reduced while the potassium ion concentration becomes increased. The great excess of sodium reabsorption over potassium secretion creates negativity of about -70 mV. in the salivary ducts, and this causes still far more chloride ions to be reabsorbed passively.

The net result of these active transport processes is that under resting conditions, the concentrations of sodium and chloride ions in the saliva are only about 15 mEq./liter each, approximately 1/7 to 1/10 their concentrations in plasma. On the other hand, the concentration of potassium ions is about 30 mEq./liter, seven times as great as its concentration in plasma, and the concentration of bicarbonate ions is 50 to 90 mEq./liter, about two to four times that of plasma.

During maximal salivation, the salivary ionic concentrations change considerably because the rate of formation of primary secretion by the acini can increase as much as 20-fold. As a result this secretion then flows through the ducts so rapidly that the ductal reconditioning of the secretion is considerably reduced. Therefore, when copious quantities of saliva are secreted, the sodium chloride concentration rises to about one-half to two-thirds that of plasma, while the potassium concentration falls to only four times that of plasma.

In the presence of excess aldosterone secretion, the sodium and chloride reabsorption and the potassium secretion become greatly increased, so that the sodium chloride concentration in the saliva is sometimes reduced almost to zero, while the potassium concentration increases still more.

Function of Saliva for Oral Hygiene. Under basal conditions, between 0.5 and 1 ml./min. of saliva, almost entirely of the mucous type, is secreted all the time. This secretion plays an exceedingly important role in maintaining healthy oral tissues. The mouth is loaded with pathogenic bacteria that can easily destroy tissues and can also cause dental caries. However, saliva helps to prevent the deteriorative processes in several ways: First, the flow of saliva itself helps to wash away the pathogenic bacteria as well as the food particles that provide their metabolic support. Second, the saliva also contains several factors that actually destroy bacteria, including *thiocyanate ions* and also several *proteolytic enzymes* that (a) attack the bacteria, (b) aid the thiocyanate ions in entering the bacteria, where they in turn become bactericidal, and (c) digest food particles, thus helping further to remove the bacterial metabolic support. Third, saliva often contains significant amounts of protein antibodies that can destroy the oral bacteria, including those that cause dental caries.

Therefore, in the absence of salivation, the oral tissues become ulcerated and otherwise infected, and caries of the teeth becomes rampant.

Nervous Regulation of Salivary Secretion. Figure 43–3 illustrates the nervous pathways for regulation of salivation, showing that the submaxillary and sublingual glands are controlled principally by nerve impulses from the superior portions of the *salivatory nuclei* and the parotid gland by impulses from the inferior portions of these nuclei. The salivatory nuclei are located approximately at the juncture of the medulla and the pons and are excited by both taste and tactile stimuli from the tongue and other areas of the mouth. Most taste stimuli, especially the sour taste, elicit copious secretion of saliva — often as much as 5 ml. per minute or 8 to 20 times the basal rate of secretion. Also, certain tactile stimuli, such as the presence of smooth objects in the mouth (a pebble, for instance), cause marked salivation, whereas rough objects cause less salivation and occasionally even inhibit salivation.

Salivation can also be stimulated or inhibited by impulses arriving in the salivatory nuclei from higher centers of the central nervous system. For instance, when a person smells or eats food that he particularly likes, salivation is far greater than when he smells or eats food that he detests.

ESOPHAGEAL SECRETION

The esophageal secretions are entirely mucoid in character, and they function principally to provide lubrication for food movement through the esophagus. The main body of the esophagus is lined with many simple mucous glands, but at the gastric end and, to a less extent, in the initial portion of the esophagus there are many compound mucous glands. The mucus secreted by the compound glands in the upper esophagus prevents mucosal excoriation by the newly entering food, while the compound glands near the esophagogastric juncture protect the esophageal wall from digestion by gastric juices that flow back into the lower esophagus. Despite this protection, a peptic ulcer at times still occurs at the gastric end of the esophagus.

GASTRIC SECRETION

CHARACTERISTICS OF THE GASTRIC SECRETIONS

Mucus-secreting cells line the entire surface of the stomach, and in addition the stomach mucosa has two types of tubular glands: the *gastric* or *oxyntic glands* and the *pyloric glands*. The gastric glands secrete mainly the digestive juices, and the pyloric glands secrete mainly mucus for protection of the pyloric mucosa. The gastric glands are located in the mucosa of the body and fundus of the stomach, and the pyloric glands are located in the antral portion of the stomach.

The Digestive Secretions from the Gastric Glands. A typical gastric gland is shown in Figure 43–4. It contains three different types of cells: the *mucous neck cells*, which secrete mucus; the *peptic* (or *chief*) *cells*, which secrete the digestive enzyme *pepsin*; and the *parietal* (or *oxyntic*) *cells*, which secrete hydrochloric acid.

Basic Mechanism of Hydrochloric Acid Secretion. The parietal cells secrete hydrochloric acid with a pH of approximately 0.8, thus illustrating its extreme acidity. At this pH the hydrogen ion concentration is about four million times that of the arterial blood.

Figure 43–5 illustrates the functional structure of the parietal cell, showing that it contains a system of *intracellular canaliculi*. The hydrochloric acid is formed at the membranes of these canaliculi and then conducted through openings to the exterior.

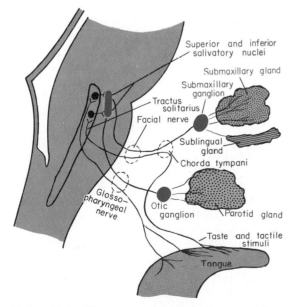

Figure 43–3. Nervous regulation of salivary secretion.

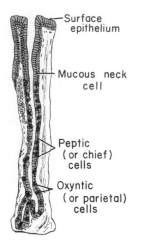

Figure 43-4.   An oxyntic gland from the body or fundus of the stomach.

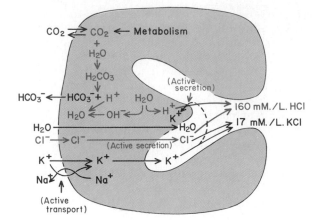

Figure 43-6.   Postulated mechanism for the secretion of hydrochloric acid.

Different suggestions for the precise mechanism of hydrochloric acid formation have been offered. One of these is illustrated in Figure 43-6 and consists of the following steps:

1. Chloride ion is actively transported from the cytoplasm of the parietal cell into the lumen of the canaliculus. This creates a negative potential of -40 to -70 millivolts in the canaliculus, which in turn causes passive diffusion of positively charged potassium ions from the cell cytoplasm also into the canaliculus.

2. Water is dissociated into hydrogen ions and hydroxyl ions in the cell cytoplasm. The hydroxyl ion is then actively secreted into the canaliculus in exchange for potassium ions. Thus, most of the potassium ions that had been secreted along with the chloride ions are reabsorbed, and hydrogen ions take their place in the canaliculus.

3. Water passes through the cell and into the canaliculus by osmosis. *Thus, the final secretion from*

the canaliculus is a solution containing hydrochloric acid in a concentration of 160 millimoles per liter and potassium chloride in a concentration of 17 millimoles per liter.

4. Finally, several steps involving carbon dioxide and carbonic acid cause removal of the excess hydroxyl ions from the cell.

Secretion of Pepsin.   The principal enzyme secreted by the peptic cells is pepsin. This is formed inside the cells in the form of *pepsinogen*, which has no digestive activity. However, once pepsinogen is secreted and comes in contact with previously formed pepsin in the presence of hydrochloric acid, it is immediately activated to form active pepsin.

Pepsin is an active proteolytic enzyme (for digesting proteins) in a highly acid medium (optimum pH = 2.0), but above a pH of about 5 it has little proteolytic activity and soon becomes completely inactivated. Therefore, hydrochloric acid secretion is just as necessary as pepsin secretion for protein digestion in the stomach.

Secretion of Mucus in the Stomach.   The pyloric glands are structurally similar to the gastric glands, but contain very few peptic and parietal cells. Instead, they contain mainly mucous cells that are identical with the mucous neck cells of the gastric glands. These cells secrete a thin mucus, which protects the stomach wall from digestion by the gastric enzymes.

In addition, the surface of the stomach mucosa between the gastric and pyloric glands has a continuous layer of mucous cells that secrete large quantities of a far more *viscid and alkaline mucus* that coats the mucosa with a mucous gel layer over 1 mm. thick, thus providing a major shell of protection for the stomach wall as well as contributing to lubrication of food transport. Even the slightest irritation of the mucosa directly

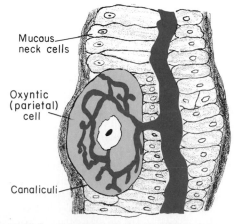

Figure 43-5.   Anatomy of the canaliculi in an oxyntic (parietal) cell.

stimulates the mucous cells to secrete copious quantities of this thick, viscid mucus.

REGULATION OF GASTRIC SECRETION BY NERVOUS AND HORMONAL MECHANISMS

Gastric secretion is regulated by both nervous and hormonal mechanisms, nervous regulation being effected through the parasympathetic fibers of the vagus nerves as well as through local intrinsic nerve plexus reflexes, and hormonal regulation taking place by means of the hormone *gastrin*. Thus, regulation of gastric secretion is different from the regulation of salivary secretion, which is effected entirely by nervous mechanisms.

Vagal Stimulation of Gastric Secretion

Nervous signals that cause gastric secretion originate in the dorsal motor nuclei of the vagi and pass via the vagus nerves to the intrinsic nerve plexus of the stomach and thence to the gastric and pyloric glands. In response, these glands secrete vast quantities of both pepsin and acid, but with a higher proportion of pepsin than in gastric juice elicited in other ways.

Stimulation of Gastric Secretion by Gastrin

When food enters the stomach, or when the vagus nerves are stimulated, the antral portion of the stomach mucosa secretes the hormone *gastrin,* a large polypeptide. The food causes release of this hormone in two ways: (1) The actual bulk of the food distends the stomach, and this causes the hormone to be released. (2) Certain substances called secretagogues —such as food extractives, partially digested proteins, alcohol (in low concentration), caffeine, and so forth —also cause gastrin to be liberated from the antral mucosa. Both of these stimuli — the distension and the chemical action of the secretagogues — elicit gastrin release by means of a local nerve reflex that transmits signals to special epithelial "gastrin" cells that secrete the gastrin.

Gastrin is absorbed into the blood and carried to the gastric glands, where it stimulates mainly the parietal cells but to a slight extent the peptic cells also. The parietal cells increase their rate of hydrochloric acid secretion as much as eight-fold, and the peptic cells increase their rate of enzyme secretion two- to four-fold.

Below are illustrated the amino acid compositions of *gastrin* and also of *cholecystokinin* and *secretin*. These are the three most important gastrointestinal hormones that have been isolated. Note that all of them are polypeptides and that the last five amino acids in the gastrin and cholecystokinin molecular chains are exactly the same. It is in this terminal portion of these hormones that the principal activity resides.

Feedback Inhibition of Gastric Acid Secretion

When the acidity of the gastric juices increases to a pH of 2.0, the gastrin mechanism for stimulating gastric secretion becomes totally blocked. This effect probably results from two different factors. First, greatly enhanced acidity depresses or blocks the extraction of gastrin itself from the antral mucosa. Second, the acid seems to extract an inhibitory hormone from the gastric mucosa or to cause an inhibitory reflex that inhibits gastric acid secretion.

Glu- Gly- Pro- Trp- Leu- Glu- Glu- Glu- Glu- Glu- Ala- Tyr- Gly- Trp- Met- Asp- Phe- NH$_2$

$$| \atop HSO_3$$

GASTRIN

Lys- (Ala, Gly, Pro, Ser)- Arg- Val- (Ile, Met, Ser)- Lys- Asn- (Asn, Gln, His, Leu$_2$, Pro, Ser$_2$)- Arg- Ile- (Asp, Ser)- Arg- Asp- Tyr- Met- Gly- Trp- Met- Asp- Phe- NH$_2$

$$| \atop HSO_3$$

CHOLECYSTOKININ

His- Ser- Asp- Gly- Thr- Phe- Thr- Ser- Glu- Leu- Ser- Arg- Leu- Arg- Asp- Ser-
Ala- Arg- Leu- Gln- Arg- Leu- Leu- Gln- Gly- Leu- Val- NH$_2$

SECRETIN

Obviously, this feedback inhibition plays an important role in protecting the stomach against excessively acid secretions, which would readily cause peptic ulceration. However, in addition to this protective effect, the feedback mechanism is also important in maintaining optimal pH for function of the peptic enzymes in the digestive process, because whenever the pH rises, gastrin begins to be secreted again.

Inhibition by Intestinal Factors. The presence of food in the small intestine initiates an *enterogastric reflex*, transmitted through the intrinsic nerve plexus, the sympathetic nerves, and the vagus nerves, that inhibits stomach secretion. This reflex is part of the complex mechanism discussed in the preceding chapter for slowing down stomach emptying when the intestines are already filled.

Also, the presence of acid, fat, protein breakdown products, hyper- or hypo-osmotic fluids, or any irritating factor in the upper small intestine causes the release of several intestinal hormones that inhibit gastric secretion. Three of these are *secretin, cholecystokinin*, and *gastric inhibitory peptide*. These, too, were discussed in the previous chapter.

The functional purpose of the inhibition of gastric secretion by intestinal factors is to slow the release of chyme from the stomach when the small intestine is already filled.

PANCREATIC SECRETION

Characteristics of Pancreatic Juice. The pancreas is a large compound gland similar to the salivary gland. It lies parallel to the stomach and secretes its juice into the duodenum a few centimeters beyond the pylorus.

Pancreatic juice contains enzymes for digesting all three major types of food: proteins, carbohydrates, and fats. It also contains large quantities of bicarbonate ions, which play an important role in neutralizing the acid chyme emptied by the stomach into the duodenum.

The proteolytic enzymes are *trypsin, chymotrypsin, carboxypolypeptidase, ribonuclease*, and *deoxyribonuclease*. By far the most abundant of these is trypsin. The first three split whole and partially digested proteins, while the nucleases split the two types of nucleic acids: ribonucleic and deoxyribonucleic acids.

The digestive enzyme for carbohydrates is *pancreatic amylase*, which hydrolyzes starches, glycogen and most other carbohydrates except cellulose to form disaccharides.

The enzymes for fat digestion are *pancreatic lipase*, which is capable of hydrolyzing neutral fat into glycerol and fatty acids, and *cholesterol esterase*, which causes hydrolysis of cholesterol esters.

The proteolytic enzymes as synthesized in the pancreatic cells are in the inactive forms *trypsinogen, chymotrypsinogen*, and *procarboxypolypeptidase*. These become activated only after they are secreted into the intestinal tract. Trypsinogen is activated by an enzyme called *enterokinase*, which is secreted by the intestinal mucosa when chyme comes in contact with the mucosa. Also, trypsinogen can be activated by trypsin that has already been formed. Chymotrypsinogen is activated by trypsin to form chymotrypsin, and procarboxypolypeptidase is activated in a similar manner.

Secretion of Bicarbonate Ions. The enzymes of the pancreatic juice are secreted entirely by the acini of the pancreatic glands. On the other hand, two other important components of pancreatic juice, water and bicarbonate ion, are secreted mainly by the epithelial cells of the small ductules leading from the acini. The bicarbonate ion concentration can rise to as high as 145 mEq./liter, a value approximately five times that of bicarbonate ions in the plasma. Obviously, this provides a large quantity of alkaline ion in the pancreatic juice that serves to neutralize the acid in the chyme emptied into the duodenum from the stomach.

REGULATION OF PANCREATIC SECRETION

Pancreatic secretion, like gastric secretion, is regulated by both nervous and hormonal mechanisms. However, in this case, hormonal regulation is the more important.

Nervous Regulation. At the same time that the vagus nerve stimulates stomach secretion, parasympathetic impulses are simultaneously transmitted along the vagus nerves to the pancreas, resulting in secretion of moderate amounts of enzymes into the pancreatic acini. However, little secretion actually flows through the pancreatic ducts to the intestine because little water and electrolytes are secreted along with the enzymes. Therefore, most of the enzymes are temporarily stored in the acini.

Hormonal Regulation. After food enters the small intestine, pancreatic secretion becomes copious, mainly in response to the hormone *secretin*. In addition, a second hormone, *cholecystokinin*, causes still much more secretion of enzymes.

Stimulation of Secretion of Copious Quantities of Bicarbonate by Secretin — Neutralization of the Acidic Chyme. Secretin is a polypeptide containing 27 amino acids that is present in the mucosa of the upper small intestine in an inactive form, *prosecretin*. When chyme enters the intestine, it causes the release and activation of secretin,

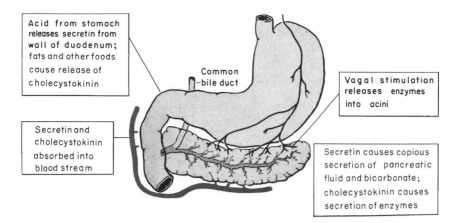

Figure 43-7. Regulation of pancreatic secretion.

which is subsequently absorbed into the blood. The one constituent of chyme that causes greatest secretin release is hydrochloric acid, though almost any type of food will cause at least some release.

Secretin causes the pancreas to secrete large quantities of fluid containing a high concentration of bicarbonate ion (up to 145 mEq. per liter) but a low concentration of chloride ion.

The secretin mechanism is especially important for two reasons: *First,* secretin is released in especially large quantities from the mucosa of the small intestine any time the pH of the duodenal contents falls below 4.0 to 5.0. This immediately causes large quantities of pancreatic juice containing abundant amounts of sodium bicarbonate to be secreted, which results in the following reaction in the duodenum:

$$HCL + NaHCO_3 \rightarrow NaCl + H_2CO_3$$

The carbonic acid immediately dissociates into carbon dioxide and water, and the carbon dioxide is absorbed into the body fluids, thus leaving a neutral solution of sodium chloride in the duodenum. In this way, the acid contents emptied into the duodenum from the stomach become neutralized, and the peptic activity of the gastric juice is immediately blocked. Since the mucosa of the small intestine cannot withstand the intense digestive properties of gastric juice, this is a highly important protective mechanism against the development of duodenal ulcers, which will be discussed in further detail in the following chapter.

A *second* importance of bicarbonate secretion by the pancreas is to provide an appropriate pH for action of the pancreatic enzymes. All of these function optimally in a slightly alkaline or neutral medium. The pH of the pancreatic secretion averages 8.0.

Cholecystokinin — Control of Enzyme Secretion by the Pancreas. The presence of food in the upper small intestine also causes a second hormone, cholecystokinin, a polypeptide containing 33 amino acids, to be released from the mucosa. This results especially from the presence of fats and also proteoses and peptones, which are products of partial protein digestion; however, acid will also cause its release in smaller quantities. Cholecystokinin, like secretin, passes by way of the blood to the pancreas but, instead of causing water and bicarbonate secretion, causes secretion of large quantities of digestive enzymes, an effect similar to that of vagal stimulation.

Figure 43–7 summarizes the overall regulation of pancreatic secretion. The total amount secreted each day is about 1200 ml.

SECRETION OF BILE BY THE LIVER

PHYSIOLOGIC ANATOMY OF THE LIVER

The basic functional unit of the liver is the liver lobule, which is a cylindrical structure several millimeters in length and 0.8 to 2 mm. in diameter.

The liver lobule is constructed around a *central vein* and is composed principally of many *hepatic cellular plates* (two of which are illustrated in Figure 43–8) that radiate centrifugally from the central vein like spokes in a wheel. Each hepatic plate is usually two cells thick, and between the adjacent cells lie small *bile canaliculi* that empty into *terminal bile ducts* in the septa between the adjacent liver lobules.

Also in the septa are small *portal venules* that receive their blood from the portal veins. From these venules blood flows into flat, branching *hepatic sinusoids* between the hepatic plates, and thence into the central vein of the lobule. Thus, the hepatic cells are exposed continuously to portal venous blood.

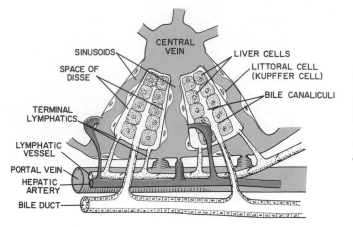

Figure 43–8. Basic structure of a liver lobule showing the hepatic cellular plates, the blood vessels, the bile-collecting system, and the lymph flow system composed of the spaces of Disse and the interlobular lymphatics. (From Guyton, A. C., Taylor, A. E., and Granger, H. J.: Circulatory Physiology II: Dynamics and Control of the Body Fluids. Philadelphia, W. B. Saunders, 1975, as modified from Elias, H.: *Am. J. Anat., 85*:379–426, 1949.)

In addition to the portal venules, there are also *hepatic arterioles* in the interlobular septa. They supply arterial blood to the septal tissues, and many of them also empty directly into the hepatic sinusoids.

The venous sinusoids are lined by two types of cells: (1) typical *endothelial cells* and (2) large *Kupffer cells*, which are reticuloendothelial cells capable of phagocytizing bacteria and other foreign matter in the blood. The endothelial lining of the venous sinusoids has extremely large pores, some of which are almost 1 micron in diameter.

Hepatic Secretion

All the hepatic cells continually form a small amount of secretion called *bile*. This is secreted into the minute *bile canaliculi*, which lie between the hepatic cells in the hepatic plates, and the bile then flows peripherally toward the interlobular septa, where the canaliculi empty into *terminal bile ducts,* then into progressively larger ducts, finally reaching the *hepatic duct* and *common bile duct,* from which the bile either empties directly into the duodenum or is diverted into the gallbladder.

Storage of Bile in the Gallbladder. Bile is secreted continuously by the liver cells but is normally stored in the gallbladder until needed in the duodenum. The total secretion each day averages 600 to 700 ml., while the maximum volume of the gallbladder is only 40 to 70 ml. Nevertheless, as much as 12 hours' bile secretion can be stored because water, sodium, chloride, and most other small electrolytes are continuously absorbed by the gallbladder mucosa, concentrating the other bile constituents, including the bile salts, cholesterol, and bilirubin. Bile is normally concentrated about five-fold, but it can be concentrated up to a maximum of 10- to 12-fold.

Emptying of the Gallbladder. Two basic conditions are necessary for the gallbladder to empty: (1) The sphincter of Oddi must relax to allow bile to flow from the common bile duct into the duodenum, and (2) the gallbladder itself must contract to provide the force required to move the bile along the common duct. After a meal, particularly one that contains a high concentration of fat, both these effects take place in the following manner:

First, the fat (also to a less extent the protein) in the food entering the small intestine causes release of the hormone *cholecystokinin* from the intestinal mucosa. The cholecystokinin in turn is absorbed into the blood and, on passing to the gallbladder, causes specific contraction of the gallbladder muscle. This provides the pressure that forces bile toward the duodenum.

Second, when the gallbladder contracts, the sphincter of Oddi becomes at least partially inhibited as a result of either a neurogenic or a myogenic reflex from the gallbladder to the sphincter of Oddi. This inhibition may also, to some extent, be a direct effect of cholecystokinin on the sphincter, causing relaxation.

Third, the presence of food in the duodenum causes the degree of peristalsis in the duodenal wall to increase. Each time a peristaltic wave travels toward the sphincter of Oddi, this sphincter, along with the adjacent intestinal wall, momentarily relaxes because of the phenomenon of "receptive relaxation" that travels ahead of the peristaltic contraction wave. If the bile in the common bile duct is under sufficient pressure, a small quantity of the bile squirts into the duodenum during each peristaltic wave.

In summary, the gallbladder empties its store of concentrated bile into the duodenum mainly in response to the cholecystokinin stimulus. When fat is not in the meal, the gallbladder empties poorly, but when adequate quantities of fat are present, the gallbladder empties completely in about one hour.

Figure 43–9 summarizes the secretion of bile,

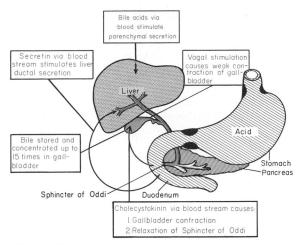

Figure 43–9. Mechanisms of liver secretion and gallbladder emptying.

its storage in the gallbladder, and its release from the bladder to the gut.

Composition of Bile. Table 43–2 gives the composition of bile when it is first secreted by the liver and then after it has been concentrated in the gallbladder. This table shows that the most abundant substance secreted in the bile is the *bile salts*, but also secreted or excreted in large concentrations are *bilirubin, cholesterol, lecithin,* and the usual *electrolytes* of plasma. In the concentrating process in the gallbladder, water and large portions of the electrolytes are reabsorbed by the gallbladder mucosa, but essentially all the other constituents, including especially the bile salts and lipid substances such as cholesterol, are not reabsorbed and therefore become highly concentrated in the gallbladder bile.

THE BILE SALTS AND THEIR FUNCTION

The liver cells form about 0.5 gram of *bile salts* daily. They have two important actions in the intestinal tract. First, they have a detergent action

TABLE 43–2 COMPOSITION OF BILE

	Liver Bile	Gallbladder Bile
Water	97.5 gm. %	92 gm. %
Bile salts	1.1 gm. %	6 gm. %
Bilirubin	0.04 gm. %	0.3 gm. %
Cholesterol	0.1 gm. %	0.3 to 0.9 gm. %
Fatty acids	0.12 gm. %	0.3 to 1.2 gm. %
Lecithin	0.04 gm. %	0.3 gm. %
Na^+	145 mEq./l.	130 mEq./l.
K^+	5 mEq./l.	12 mEq./l.
Ca^+	5 mEq./l.	23 mEq./l.
Cl^-	100 mEq./l.	25 mEq./l.
HCO_3^-	28 mEq./l.	10 mEq./l.

on the fat particles in the food, which decreases the surface tension of the particles and allows the mechanical agitation in the intestinal tract to break the fat globules into minute sizes. Second, and even more important than the emulsifying function, bile salts help in the absorption of fatty acids, monoglycerides, cholesterol, and other lipids from the intestinal tract. This will be discussed in detail in the following chapter.

EXCRETION OF BILIRUBIN IN THE BILE

In addition to secreting substances synthesized by the liver itself, the liver cells also *excrete* a number of substances formed elsewhere in the body. Among the most important of these is *bilirubin*, which is one of the major end-products of hemoglobin decomposition, as was pointed out in Chapter 5.

Briefly, when the red blood cells have lived out their life span, averaging 120 days, and have become too fragile to exist longer in the circulatory system, their cell membranes rupture, and the released hemoglobin is phagocytized by reticuloendothelial cells throughout the body. Here, the hemoglobin is first split into *globin* and *heme*, and the heme ring is rapidly converted to *bilirubin*, which is released into the plasma. However, within hours the bilirubin is absorbed through the hepatic cell membrane and is excreted by an active transport process into the bile.

Jaundice. The word "jaundice" means a yellowish tint to the body tissues, including yellowness of the skin and also of the deep tissues. The cause of jaundice is large quantities of bilirubin in the extracellular fluids. The normal plasma concentration of bilirubin averages 0.5 mg per 100 ml. of plasma. However, in certain abnormal conditions it can rise to as high as 40 mg. per 100 ml.

The common causes of jaundice are (1) increased destruction of red blood cells, with rapid release of bilirubin into the blood, or (2) obstruction of the bile ducts or damage to the liver cells so that even the usual amounts of bilirubin cannot be excreted into the gastrointestinal tract. These two types of jaundice are called, respectively, *hemolytic jaundice* and *obstructive jaundice*.

SECRETION OF CHOLESTEROL: GALLSTONE FORMATION

Bile salts are formed in the hepatic cells from cholesterol, which also is synthesized in the liver. In the process of secreting the bile salts, about one-tenth as much cholesterol as bile salts is also

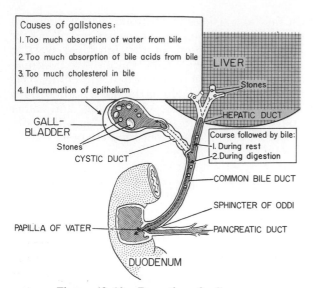

Figure 43–10. Formation of gallstones.

secreted into the bile. No specific function is known for the cholesterol in the bile, and it is presumed that this is simply a by-product of bile salt formation and secretion.

Cholesterol is almost insoluble in pure water, but the bile salts and lecithin in bile combine physically with the cholesterol to form ultramicroscopic *micelles* that are soluble. When the bile becomes concentrated in the gallbladder, the bile salts and lecithin become concentrated along with the cholesterol, which keeps the cholesterol in solution. Under abnormal conditions, however, the cholesterol may precipitate, resulting in the formation of *gallstones*, as shown in Figure 43–10. The different conditions that can cause cholesterol precipitation are these: (1) too much absorption of water from the bile, (2) too much absorption of bile salts and lecithin from the bile, (3) too much secretion of cholesterol in the bile, or (4) inflammation of the epithelium of the gallbladder. The latter two require special explanation as follows:

The amount of cholesterol in the bile is determined partly by the quantity of fat that the person eats, for the hepatic cells synthesize cholesterol as one of the products of fat metabolism in the body. For this reason, persons on a high fat diet over a period of many years are prone to the development of gallstones.

Inflammation of the gallbladder epithelium often results from low grade chronic infection; this changes the absorptive characteristics of the gallbladder mucosa, sometimes allowing excessive absorption of water, bile salts, or other substances that are necessary to keep the cholesterol in solution. As a result, cholesterol begins to precipitate, usually forming many small crystals

of cholesterol on the surface of the inflamed mucosa. These, in turn, act as nidi for further precipitation of cholesterol, and the crystals grow larger and larger. Occasionally tremendous numbers of sandlike stones develop, but much more frequently these coalesce to form a few large gallstones, or even a single stone that fills the entire gallbladder.

SECRETIONS OF THE SMALL INTESTINE

SECRETION OF MUCUS BY BRUNNER'S GLANDS AND BY MUCOUS CELLS OF THE INTESTINAL SURFACE

An extensive array of compound mucous glands, called *Brunner's glands,* is located in the first few centimeters of the duodenum, mainly between the pylorus and the papilla of Vater where the pancreatic juice and bile empty into the duodenum. These glands secrete mucus in response to (1) direct tactile stimuli or irritating stimuli of the overlying mucosa, (2) vagal stimulation, which causes secretion concurrently with increase in stomach seretion, and (3) intestinal hormones, especially secretin. The function of the mucus secreted by Brunner's glands is to protect the duodenal wall from digestion by the gastric juice, and the rapid and intense response of these glands to irritating stimuli is especially geared to this purpose.

Brunner's glands are inhibited by sympathetic stimulation; therefore, such stimulation is likely to leave the duodenal bulb unprotected and is perhaps one of the factors that cause this area of the gastrointestinal tract to be the site of peptic ulcers in about 50 per cent of the cases.

Mucus is also secreted in large quantities by goblet cells located extensively over the surface of the intestinal mucosa. This secretion results principally from direct tactile or chemical stimulation of the mucosa by the chyme. Additional mucus is also secreted by the goblet cells in the intestinal glands, which are called the crypts of Lieberkühn. This secretion is probably controlled mainly by local nervous reflexes.

SECRETION OF THE INTESTINAL DIGESTIVE JUICES – THE CRYPTS OF LIEBERKÜHN

Located on the entire surface of the small intestine, with the exception of the Brunner's gland area of the duodenum, are small crypts

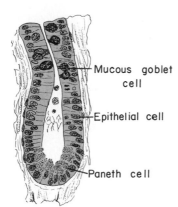

Figure 43–11. A crypt of Lieberkühn, which secretes almost pure extracellular fluid. These structures are found in all parts of the small intestine between the villi.

called *crypts of Lieberkühn*, one of which is illustrated in Figure 43–11. The intestinal secretions are formed by the epithelial cells in these crypts at a rate of about 2000 ml. per day. The secretions are almost pure extracellular fluid, and they have a neutral pH in the range of 6.5 to 7.5. They are rapidly reabsorbed by the villi. This circulation of fluid from the crypts to the villi supplies a watery vehicle for absorption of substances from the small intestine, which is one of the primary functions of the small intestine.

Enzymes in the Small Intestinal Secretion. When secretions of the small intestine are collected without cellular debris, they have almost no enzymes. However, the epithelial cells of the mucosa do contain large quantities of digestive enzymes that digest food substances *while* they are being absorbed through the epithelium. These enzymes are the following: (1) several different *peptidases* for splitting polypeptides into amino acids, (2) four enzymes for splitting disaccharides into monosaccharides — *sucrase, maltase, isomaltase* and *lactase* — and (3) small amounts of *intestinal lipase* for splitting neutral fats into glycerol and fatty acids. Most if not all of these enzymes are in the brush border of the epithelial cells. Therefore, they presumably cause hydrolysis of the foods on the outside surfaces of the microvilli prior to absorption of the end-products of digestion.

REGULATION OF SMALL INTESTINAL SECRETION

By far the most important means for regulating small intestinal secretion is various local nervous reflexes. Especially important is distension of the small intestine, which causes copious secretion from the crypts of Lieberkühn. In addition, tactile or irritative stimuli can result in intense secretion. Therefore, for the most part, secretion in the small intestine occurs simply in response to the presence of chyme in the intestine.

SECRETIONS OF THE LARGE INTESTINE

Mucus Secretion. The mucosa of the large intestine, like that of the small intestine, is lined with crypts of Lieberkühn, but the epithelial cells contain essentially no enzymes. Instead, the crypts are lined almost entirely by goblet cells. Also, on the surface epithelium of the large intestine are large numbers of goblet cells dispersed among the other epithelial cells.

Therefore, the preponderant secretion in the large intestine is mucus. Its rate of secretion is regulated principally by direct, tactile stimulation of the goblet cells on the surface of the mucosa and by local nervous reflexes to the goblet cells in the crypts of Lieberkühn. However, stimulation of the nervi erigentes, which carry the parasympathetic innervation to the distal half of the large intestine, also causes marked increase in the secretion of mucus. This occurs along with an increase in motility, discussed in the preceding chapter. Therefore, during extreme parasympathetic stimulation, often caused by severe emotional disturbances, so much mucus may be secreted into the large intestine that the person has a bowel movement of ropy mucus as often as every 30 minutes; the mucus contains little or no fecal material.

Mucus in the large intestine obviously protects the wall against excoriation, but, in addition, it provides the adherent medium for holding fecal matter together. Furthermore, it protects the intestinal wall from the great amount of bacterial activity that takes place inside the feces, and it, plus the alkalinity of the secretion (pH of 8.0), also provides a barrier to keep acids formed deep in the feces from attacking the intestinal wall.

Secretion of Water and Electrolytes in Response to Irritation. Whenever a segment of the large intestine becomes intensely irritated, as occurs when bacterial infection becomes rampant during *bacterial enteritis*, the mucosa then secretes large quantities of water and electrolytes in addition to the normal viscid solution of mucus. These substances act to dilute the irritating factors and to cause rapid movement of the feces toward the anus. The usual result is *diarrhea* with loss of large quantities of water and electrolytes but also earlier recovery from the disease than would otherwise occur.

REFERENCES

Baron, J. H.: Clinical tests of Gastric Secretion: History, Methodology, and Interpretation. New York, Oxford University Press, 1978.

Binder, H. J. (ed.): Mechanisms of Intestinal Secretion. New York. A. R. Liss, 1979.

Bloom, S. R. (ed.): Gut Hormones. New York, Longman, Inc., 1978.

Bonfils, S., *et al.*: Vagal control of gastric secretion. *In* Crane, R. K. (ed.): International Review of Physiology: Gastrointestinal Physiology III. Vol. 19. Baltimore, University Park Press, 1979, p. 59.

Davenport, H. W.: Mechanisms of gastric and pancreatic secretion. *In* Frohlich, E. D. (ed.): Pathophysiology, 2nd Ed. Philadelphia, J. B. Lippincott Co., 1976, p. 481.

Forker, E. L.: Mechanisms of hepatic bile formation. *Annu. Rev. Physiol., 39*:323, 1977.

Gerolami, A., and Sarles, J. C.: Biliary secretion and motility. *Int. Rev. Physiol., 12*:223, 1977.

Glass, G. B. (ed.): Gastrointestinal Hormones. New York, Raven Press, 1980.

Grossman, M. I.: Neural and hormonal regulation of gastrointestinal function: An overview. *Annu. Rev. Physiol., 41*:27, 1979.

Hendrix, T. R., and Paulk, H. T.:Intestinal secretion. *Int. Rev. Physiol., 12*:257, 1977.

Jones, R. S., and Myers, W. C.: Regulation of hepatic biliary secretion. *Annu. Rev. Physiol., 41*:67, 1979.

Mason, D. K.: Salivary Glands in Health and Disease. Philadelphia, W. B. Saunders Company, 1975.

Miyoshi, A. (ed.): Gut Peptides: Secretion, Function, and Clinical Aspects. New York, Elsevier/North-Holland, 1979.

Petersen, O. H.: Electrophysiology of mammalian gland cells. *Physiol Rev., 56*:535, 1976.

Rehfeld, J. F.: Gastrointestinal hormones. *In* Crane, R. K. (ed.): International Review of Physiology: Gastrointestinal Physiology III. Vol. 19. Baltimore, University Park Press, 1979, p. 291.

Rehfeld, J. F., and Amdrup, E. (eds.): Gastrins and the Vagus. New York, Academic Press, 1979.

Reichen, J., and Paumgartner, G.: Excretory function of the liver. *In* Javitt, N. B. (ed.): International Review of Physiology: Liver and Biliary Tract Physiology I. Vol. 21. Baltimore, University Park Press, 1980, p. 103.

Sachs, G., *et al.*: H$^+$ transport: Regulation and mechanism in gastric mucosa and membrane vesicles. *Physiol. Rev., 58*:106, 1978.

Sarles, H.: The exocrine pancreas. *Int. Rev. Physiol., 12*:173, 1977.

Soll, A., and Walsh, J. H.: Regulation of gastric acid secretion. *Annu. Rev. Physiol., 41*:35, 1979.

Stroud, R. M., *et al.*: Mechanisms of zymogen activation. *Annu. Rev. Biophys. Bioeng., 6*:177, 1977.

Young, J. A.: Salivary secretion of inorganic electrolytes. *In* Crane, R. K. (ed.): International Review of Physiology: Gastrointestinal Physiology III. Vol. 19. Baltimore, University Park Press, 1979, p. 1.

Young, J. A., and van Lennep, E. W.: Morphology and physiology of salivary myoepithelial cells. *Int. Rev. Physiol., 12*:105, 1977.

Liver Function

Berk, P. D., *et al.*: Disorders of bilirubin metabolism. *In* Bondy, P. K., and Rosenberg, L. E. (eds.): Metabolic Control and Disease, 8th Ed. Philadelphia, W. B. Saunders Co., 1980, p. 1009.

Boucheir, I. A. D.: The medical treatment of gallstones. *Annu. Rev. Med., 31*:59, 1980.

Boyer, J. L.: New concepts of mechanisms of hepatocyte bile formation. *Physiol. Rev., 60*:303, 1980.

Davidson, C. S. (ed.): Problems in Liver Diseases. New York, Stratton Intercontinental Medical Book Corp., 1979.

Fevery, J., and Heirwegh, K. P. M.: Bilirubin metabolism. *In* Javitt, N. B. (ed.): International Review of Physiology: Liver and Biliary Tract Physiology I. Vol. 21. Baltimore, University Park Press, 1980, p. 171.

Fisher, M. M., *et al.* (ed.): Gall Stones. New York, Plenum Press, 1979.

Frizzell, R. A., and Heintze, K.: Transport functions of the gallbladder. *In* Javitt, N. B. (ed.): International Review of Physiology: Liver and Biliary Tract Physiology I. Vol. 21. Baltimore, University Park Press, 1980, p. 221.

Galambos, J. T.: Cirrhosis. Philadelphia, W. B. Saunders Co., 1979.

Gall, E. A., and Mostofi, F. K. (eds.): The Liver. Huntington, N. Y., R. E. Krieger Co., 1980.

Gitnick, G. L. (ed.): Current Gastroenterology and Hepatology. Boston, Houghton Mifflin, 1979.

Jones, R. S., and Myers, W. C.: Regulation of hepatic biliary secretion. *Annu. Rev. Physiol., 41*:67, 1979.

Paumgartner, G., *et al.* (eds.): Biological Effects of Bile Acids. Baltimore, University Park Press, 1979.

Rappaport, A. M.: Hepatic blood flow: Morphologic aspects and physiologic regulation. *In* Javitt, N. B. (ed.): International Review of Physiology: Liver and Biliary Tract Physiology I. Vol. 21. Baltimore, University Park Press, 1980, p. 1.

Reichen, J., and Paumgartner, G.: Excretory function of the liver. *In* Javitt, N. B. (ed.): International Review of Physiology: Liver and Biliary Tract Physiology I. Vol. 21. Baltimore, University Park Press, 1980, p. 103.

Digestion and Absorption in the Gastrointestinal Tract; and Gastrointestinal Disorders

The foods on which the body lives, with the exception of small quantities of substances such as vitamins and minerals, can be classified as carbohydrates, fats, and proteins. However, these generally cannot be absorbed in their natural forms through the gastrointestinal mucosa and, for this reason, are useless as nutrients without preliminary digestion. Therefore, this chapter discusses, first, the processes by which carbohydrates, fats, and proteins are digested into small enough compounds for absorption and, second, the mechanisms by which the digestive end-products, as well as water, electrolytes, and other substances, are absorbed.

DIGESTION OF THE VARIOUS FOODS

Hydrolysis as the Basic Process of Digestion. Almost all the carbohydrates of the diet are large *polysaccharides* or *disaccharides,* which are combinations of *monosaccharides* bound to each other. The carbohydrates are digested into their constituent monosaccharides; to do this, specific enzymes combine hydrogen and hydroxyl ions, derived from water, to the polysaccharides and thereby separate the monosaccharides from each other. This process, called hydrolysis, is the following:

$$R'' - R' + H_2O \rightarrow R''OH + R'H$$

Almost the entire fat portion of the diet consists of triglycerides (neutral fats), which are combinations of three *fatty acid* molecules with a single *glycerol* molecule. Digestion of the triglycer-

ides consists of fat-digesting enzymes splitting the fatty acid molecules away from the glycerol. Here again, the process is one of hydrolysis.

Finally, proteins are formed from *amino acids* that are bound together by means of *peptide linkages.* And digestion of proteins also involves the process of hydrolysis, the proteolytic enzymes combining hydroxyl and hydrogen ions of water with the protein molecules to split them into their constituent amino acids.

Therefore, the chemistry of digestion is really simple, for in the case of all three major types of food, the same basic process of *hydrolysis* is involved. The only difference is in the enzymes required to promote the reactions for each type of food.

All the digestive enzymes are proteins. Their secretion by the different gastrointestinal glands is discussed in the preceding chapter.

DIGESTION OF CARBOHYDRATES

Only three major sources of carbohydrates exist in the normal human diet. These are sucrose, which is the disaccharide known popularly as cane sugar; lactose, which is a disaccharide in milk; and starches, which are large polysaccharides present in almost all foods and particularly in the grains.

Figure 44–1 gives a schema for digestion of the principal carbohydrates. This shows that the starches are first hydrolyzed to maltose (or isomaltose), which is a disaccharide. Then this disaccharide, along with the other major disaccharides, lactose and sucrose, are hydrolyzed into the monosaccharides *glucose, galactose,* and *fructose.*

511

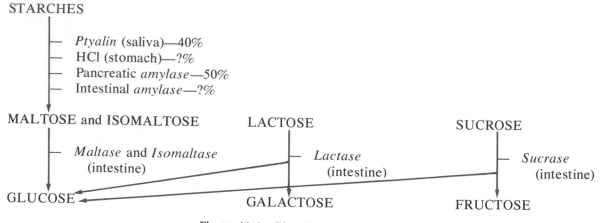

Figure 44–1. Digestion of carbohydrates.

Hydrolysis of starches begins in the mouth under the influence of the enzyme *ptyalin,* which is secreted mainly in the saliva from the parotid gland. The hydrochloric acid of the stomach provides a slight amount of additional hydrolysis. Finally, the major share of hydrolysis occurs in the upper part of the small intestine under the influence of the enzyme *pancreatic amylase.*

The enzymes *lactase, sucrase, maltase,* and *isomaltase* for splitting the disaccharides are located in the microvilli of the brush border of the epithelial cells. The disaccharides are digested into monosaccharides as they come in contact with these microvilli or as they diffuse into the microvilli. The digestive products, the monosaccharides glucose, galactose, and fructose, are then immediately absorbed into the portal blood.

DIGESTION OF FATS

By far the most common fats of the diet are the neutral fats, also known as *triglycerides,* each molecule of which is composed of a glycerol nucleus and three fatty acids. Neutral fat is found in food of both animal origin and plant origin.

In the usual diet are also small quantities of phospholipids, cholesterol, and cholesterol esters. The phospholipids and cholesterol esters contain fatty acid and therefore can be considered to be fats themselves. Cholesterol, on the other hand, is a sterol compound containing no fatty acid, but it does exhibit some of the physical and chemical

characteristics of fats; it is derived from fats, and it is metabolized similarly to fats. Therefore, cholesterol is considered from a dietary point of view to be a fat.

Though a minute amount of fat can be digested in the stomach under the influence of gastric lipase, 95 to 99 per cent of all fat digestion occurs in the small intestine mainly under the influence of *pancreatic lipase.*

Emulsification of Fat by Bile Salts. The first step in fat digestion is to break the fat globules into small sizes so that the digestive enzymes, which are not fat soluble, can act on the globule surfaces. This process is called emulsification of the fat, and it is achieved under the influence of bile salts that are secreted in the bile by the liver. The bile salts act as a detergent, greatly decreasing the interfacial tension of the fat. With a low interfacial tension, the gastrointestinal mixing movements can gradually break the globules of fat into finer and finer particles, with the total surface area of the fat increasing by a factor of two every time the diameters of the fat globules are decreased by a factor of two.

Digestion of Fat by Pancreatic Lipase and Enteric Lipase. Under the influence of *pancreatic lipase,* most of the fat is split into *monoglycerides, fatty acids,* and *glycerol.* Though a moderate portion of the fat is digested only to the glyceride stage, as shown in Figure 44–2, the further the process of fat hydrolysis proceeds, the better is the absorption of the fats.

The epithelial cells of the small intestine con-

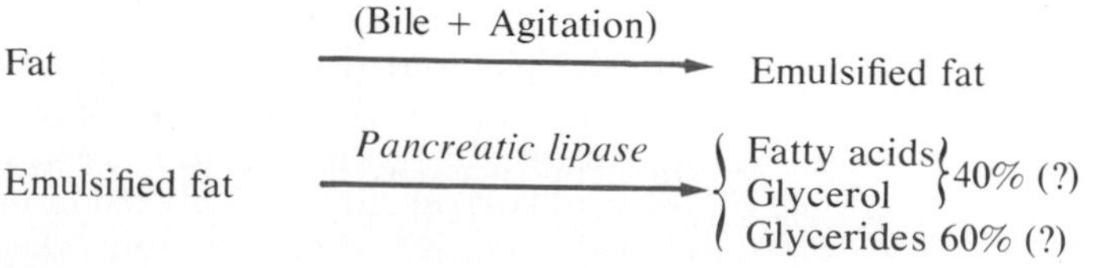

Fat $\xrightarrow{\text{(Bile + Agitation)}}$ Emulsified fat

Emulsified fat $\xrightarrow{\textit{Pancreatic lipase}}$ $\begin{cases} \text{Fatty acids} \\ \text{Glycerol} \end{cases}$ 40% (?) $\big\}$ Glycerides 60% (?)

Figure 44–2. Digestion of fats.

tain a small quantity of lipase, known as *enteric lipase*. This probably causes a very slight additional amount of fat digestion.

Role of Bile Salts in Accelerating Fat Digestion — Formation of Micelles. The hydrolysis of triglycerides is a highly reversible process; therefore, accumulation of monoglycerides and free fatty acids in the vicinity of digesting fats very quickly blocks further digestion. Fortunately, the bile salts play an important role in removing the monosaccharides and the free fatty acids from the vicinity of the digesting fat globules almost as rapidly as these end-products of digestion are formed. This occurs in the following way:

Bile salts have the propensity to form *micelles*, which are small spherical globules about 2.5 nanometers in diameter and composed of 20 to 50 molecules of bile salt. These develop because each bile salt molecule is composed of a sterol nucleus that is highly fat soluble and a polar group that is highly water soluble. The sterol nuclei of the 20 to 50 bile salt molecules of the micelle aggregate to form a small fat globule in the middle of the micelle. This causes the polar groups to project outward to cover the surface of the micelle. Since these polar groups are negatively charged, they allow the entire micelle globule to become dissolved in the water of the digestive fluids and to remain in solution despite the very large size of the micelle.

During triglyceride digestion, as rapidly as the monoglycerides and free fatty acids are formed they become dissolved in the fatty portion of the micelles, which immediately removes these end-products of digestion from the vicinity of the digesting fat globules. Consequently, the digestive process can proceed unabated.

The bile salt micelles also act as a transport medium to carry the monoglycerides and the free fatty acids to the brush borders of the epithelial cells. There the monoglycerides and free fatty acids are absorbed, as will be discussed later. On delivery of these substances to the brush border, the bile salts are again released back into the chyme to be used again and again for this "ferrying" process.

DIGESTION OF PROTEINS

The dietary proteins are derived almost entirely from meats and vegetables, and they are digested primarily in the stomach and upper part of the small intestine.

As illustrated in Figure 44–3, protein digestion begins in the stomach, the enzyme *pepsin* splitting the proteins into *proteoses*, *peptones*, and *large polypeptides*. These enzymes function only in a highly

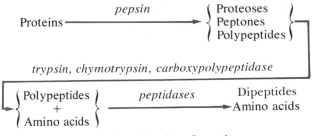

Figure 44–3. Digestion of proteins.

acid medium; they function best at a pH of 2. Therefore, the hydrochloric acid secreted in the stomach is essential for this digestive process.

Pepsin is especially important for its ability to digest collagen, an albuminoid that is little affected by other digestive enzymes. Since collagen is a major constituent of the fibrous tissue in meat, it is essential that this be digested so that the remainder of the meat can be attacked by the digestive enzymes.

The proteins are further digested in the upper part of the small intestine under the influence of the pancreatic enzymes *trypsin, chymotrypsin,* and *carboxypolypeptidases*. The final product of this digestion is mainly *small polypeptides* plus a few *amino acids*.

Finally, the small polypeptides are digested into amino acids when they come in contact with the epithelial cells of the small intestine. These cells contain several enzymes *(peptidases)* that convert the remaining protein products into *amino acids*.

When food has been properly masticated and is not eaten in too large quantity at any one time, about 98 per cent of the protein finally becomes amino acids.

BASIC PRINCIPLES OF GASTROINTESTINAL ABSORPTION

ANATOMICAL BASIS OF ABSORPTION

The total quantity of fluid that must be absorbed each day is equal to the ingested fluid (about 1.5 liters) plus that secreted in the various gastrointestinal juices (about 7.5 liters). This comes to a total of approximately 9 liters. About 8 to 8.5 liters of this is absorbed in the small intestine, leaving only 0.5 to 1 liter to pass through the ileocecal valve into the colon each day.

The stomach is a poor absorptive area of the gastrointestinal tract. Only a few highly lipid-soluble substances, such as alcohol and some drugs, can be absorbed in small quantities.

The Absorptive Surface of the Intestinal Mucosa — The Villi. Figure 44–4 illustrates the absorptive surface of the intestinal mucosa, showing many folds called *valvulae conniventes;* these increase the surface area of the absorptive mucosa about three-fold.

Also, located over the entire surface of the small intestine, from approximately the point at which the common bile duct empties into the duodenum down to the ileocecal valve, are literally millions of small *villi,* which project about 1 mm. from the surface of the mucosa, as shown on the surfaces of the valvulae conniventes in Figure 44–4 and in detail in Figure 44–5. These villi enhance the absorptive area another ten-fold.

Finally, the epithelial cells on the surface of the villi are characterized by a brush border, consisting of about 600 *microvilli* 1μ in length and 0.1μ in diameter protruding from each cell; these are illustrated in the electron micrograph in Figure 44–6. This increases the surface area exposed to the intestinal materials another 20-fold. Thus, the combination of the valvulae conniventes, the villi, and the microvilli increases the absorptive area of the mucosa about 600-fold, making a very large total area of about 250 square meters for the entire small intestine.

Figure 44–5 illustrates the general organization of a villus, emphasizing especially the advantageous arrangement of the vascular system for absorption of fluid and dissolved material into the portal blood, and the arrangement of the *central lacteal* for absorption into the lymphatics.

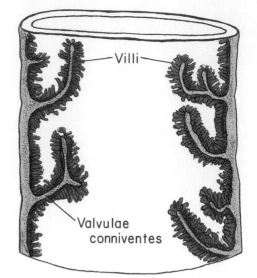

Figure 44–4. A longitudinal section of the small intestine, showing the valvulae conniventes covered by villi.

BASIC MECHANISMS OF ABSORPTION

Absorption through the gastrointestinal mucosa occurs by *active transport* and by *diffusion,* as is also true for other membranes. The physical principles of these processes were explained in Chapter 4.

Briefly, active transport provides energy to move a substance across a membrane. Therefore, the substance can be moved against a concentration gradient or against an electrical potential. On the other hand, the term "diffusion" means sim-

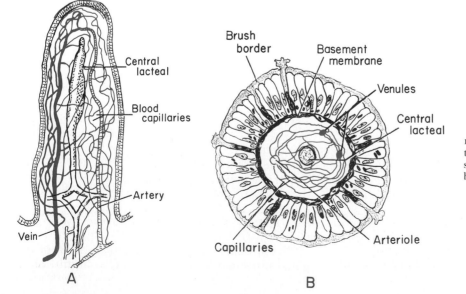

Figure 44–5. Functional organization of the villus: (A) Longitudinal section. (B) Cross-section showing the epithelial cells and basement membrane.

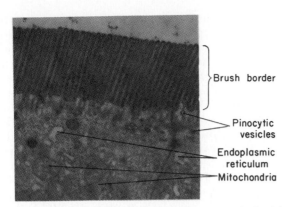

Figure 44-6. Brush border of the gastrointestinal epithelial cell, showing, also, pinocytic vesicles, mitochondria, and endoplasmic reticulum lying immediately beneath the brush border. (Courtesy of Dr. William Lockwood.)

ply transport of substances through the membrane as a result of molecular movement *along*, rather than against, an electrochemical gradient.

ABSORPTION IN THE SMALL INTESTINE

Normally, the amount absorbed from the small intestine each day consists of several hundred grams of carbohydrates, 100 or more grams of fat, 50 to 100 grams of amino acids, 50 to 100 grams of ions, and 8 or 9 liters of water. However, the absorptive *capacity* of the small intestine is far greater than this: as much as several kilograms of carbohydrates per day, 500 to 1000 grams of fat per day, 500 to 700 grams of amino acids per day, and 20 or more liters of water per day. In addition, the large intestine can absorb still more water and ions, though almost no nutrients.

ABSORPTION OF WATER

Isosmotic Absorption. Water is transported through the intestinal membrane entirely by the process of *diffusion*. Furthermore, this diffusion obeys the usual laws of osmosis.

As dissolved substances are actively transported from the lumen of the gut into the blood this transport decreases the osmotic pressure of the chyme, but water then diffuses so readily through the intestinal membrane that it almost instantaneously "follows" the transported substances into the circulation. Therefore, as ions and nutrients are absorbed an "isosmotic" equivalent of water is also absorbed. In this way not only are the ions and nutrients almost entirely absorbed before the

chyme passes through the small intestine, but so also is almost all the water absorbed.

ABSORPTION OF THE IONS

Active Transport of Sodium. Twenty to 30 grams of sodium are secreted into the intestinal secretions each day. In addition, the normal person eats 5 to 8 grams of sodium each day. Combining these two, the small intestine absorbs 25 to 35 grams of sodium each day, which amounts to about one seventh of all the sodium that is present in the body.

The basic mechanism of sodium absorption from the intestine is illustrated in Figure 44-7. The principles of this mechanism, which were discussed in Chapter 4, are also essentially the same as those for absorption of sodium from the renal tubules, as discussed in Chapter 24. The motive power for the sodium absorption is provided by active transport of sodium from inside the epithelial cells through the side walls of these cells into the intercellular spaces. This is illustrated by the heavy black arrows in Figure 44-7. This active transport obeys the usual laws of active transport: It requires a carrier, it requires energy, and it is catalyzed by an appropriate ATPase carrier-enzyme in the cell membrane.

The active transport of sodium reduces its concentration in the cell to a low value, which then causes more sodium to diffuse from the chyme through the brush border of the epithelial cell into the cell cytoplasm. This provides still more sodium to be actively transported out of the epithelial cells into the intercellular spaces.

The next step in the transport process is osmosis of water out of the epithelial cell into the intercellular spaces. This movement is caused by the osmotic gradient created by the reduced concentration of sodium inside the cell and the elevated concentration in the intercellular space.

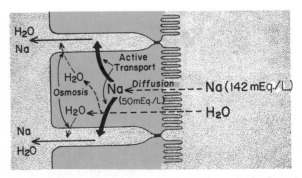

Figure 44-7. Absorption of sodium through the intestinal epithelium. Note also the osmotic absorption of water — that is, the water "follows" the sodium through the epithelial membrane.

The osmotic movement of water creates a flow of fluid into the intercellular space, then through the basement membrane of the epithelium, and finally into the circulating blood of the villi. New water diffuses along with sodium through the brush border of the epithelial cell to replenish the water that flows into the intercellular spaces.

Transport of Chloride. In most parts of the small intestine chloride transport is by passive diffusion. The transport of sodium ions through the epithelium creates electronegativity in the chyme and electropositivity on the basal side of the epithelial cells. Then chloride ions move along this electrical gradient to "follow" the sodium ions.

However, the epithelial cells of the distal ileum and of the large intestine have the special capability of actively absorbing chloride ions. This occurs by means of a tightly coupled active transport mechanism in which an equivalent number of bicarbonate ions are secreted. The purpose of this mechanism is probably to provide bicarbonate ions for neutralization of acidic products formed by bacteria — especially in the large intestine.

Absorption of Other Ions. Calcium ions are actively absorbed, especially from the duodenum, and calcium ion absorption is exactly controlled in relation to the need of the body for calcium by parathyroid hormone secreted by the parathyroid glands and by vitamin D. These effects are discussed in Chapter 53.

Iron ions are also actively absorbed from the small intestine. The principles of iron absorption and the regulation of its absorption in proportion to the body's need for iron were discussed in Chapter 5.

Potassium, magnesium, phosphate, and probably still other ions can also be actively absorbed through the mucosa.

ABSORPTION OF NUTRIENTS

Absorption of Carbohydrates. Essentially all the carbohydrates are absorbed in the form of monosaccharides, only a small fraction of a per cent being absorbed as disaccharides and almost none as larger carbohydrate compounds. Furthermore, little carbohydrate absorption results from diffusion, for the pores of the mucosa through which diffusion occurs are essentially impermeable to water-soluble solutes with molecular weights greater than 100.

Mechanism of Monosaccharide Absorption. We still do not know the precise mechanism of monosaccharide absorption, but we do know that most monosaccharide transport becomes blocked whenever sodium transport is blocked. There-fore, it is assumed that the energy required for most monosaccharide transport is actually provided by the sodium transport system. A theory that attempts to explain this is the following: It is known that a carrier for transport of glucose and some other monosaccharides, especially galactose, is present in the brush border of the epithelial cell. However, this carrier will not transport the glucose in the absence of sodium transport. Therefore, it is believed that the carrier has receptor sites for both a glucose molecule and a sodium ion, and that it will not transport the glucose to the inside of the cell if the receptor site for sodium is not simultaneously filled. The energy to cause movement of the carrier from the exterior of the membrane to the interior is derived from the difference in sodium concentration between the outside and inside. That is, as sodium diffuses to the inside of the cell it "drags" the carrier and the glucose along with it, thus providing the energy for transport of the glucose. For obvious reasons, this explanation is called the *sodium co-transport theory* for glucose transport.

Absorption of Proteins. Almost all proteins are absorbed in the form of amino acids. Four different carrier systems transport different amino acids — one transports *neutral amino acids,* a second transports *basic amino acids,* a third transports *acidic amino acids,* and a fourth has specificity for the two amino acids *proline* and *hydroxyproline.*

Amino acid transport, like glucose transport, occurs only in the presence of simultaneous sodium transport. Furthermore, the carrier systems for amino acid transport, like those for glucose transport, are in the brush border of the epithelial cell. It is believed that amino acids are transported by the same sodium co-transport mechanism as that explained above for glucose transport. That is, the theory postulates that the carrier has receptor sites for both an amino acid molecule and a sodium ion. Only when both of the sites are filled will the carrier move to the interior of the cell. Because of the sodium gradient across the brush border, the sodium diffusion to the cell interior pulls the carrier and its attached amino acid to the interior, where the amino acid becomes trapped. Therefore, its concentration increases within the cell, and it then diffuses through the sides or base of the cell into the portal blood.

Absorption of Fats. Earlier in this chapter it was pointed out that as fats are digested to form monoglycerides and free fatty acids, both of these digestive end-products become dissolved mainly in the lipid portion of the bile acid micelles. Because of the molecular dimensions of these micelles and also because of their highly charged exterior, they are soluble in the chyme. In this form the monoglycerides and the fatty acids are transported to the surfaces of the epithelial cells. On coming in contact with these surfaces, both the monoglycerides and the fatty acids immediately diffuse through the epithelial membrane, leaving the bile acid micelles still in the chyme. The micelles then diffuse back into the chyme and absorb still more monoglycerides and fatty acids, and similarly transport these also to the epithelial cells. Thus, the bile acids perform a "ferrying" function, which is highly important for fat absorption. In the presence of an abundance of bile acids, approximately 97 per cent of the fat is absorbed; in the absence of bile acids, only 50 to 60 per cent is normally absorbed.

The mechanism for absorption of the monoglycerides and fatty acids through the brush border is based on the fact that both of these substances are highly lipid soluble. Therefore, they become dissolved in the membrane and diffuse to the interior of the cell.

After entry into the epithelial cell, many of the monoglycerides are further digested into glycerol and fatty acids by an epithelial cell lipase. Then, the free fatty acids are reconstituted by the endoplasmic reticulum into trigylcerides. Almost all of the glycerol that is utilized for this purpose is synthesized *de novo* from alpha-glycerophosphate. However, minute amounts of the original glycerol from the monoglycerides do appear in the newly synthesized triglycerides.

Once formed, the triglycerides collect into globules along with absorbed cholesterol, absorbed phospholipids, and newly synthesized phospholipids. Each globule is then encased in a protein coat, utilizing β-lipoprotein also synthesized by the endoplasmic reticulum. This globular mass, along with its protein coat, is extruded from the sides of the epithelial cells into the intercellular spaces, and from here it passes into the central lacteal of the villus. Such globules are called *chylomicrons*. The protein coat of the chylomicrons makes them hydrophilic, allowing a reasonable degree of suspension stability in the extracellular fluids.

Transport of the Chylomicrons in the Lymph. From beneath the epithelial cells the chylomicrons wend their way into the central lacteals of the villi and are propelled along with the lymph by the lymphatic pump upward through the thoracic duct to be emptied into the great veins of the neck.

ABSORPTION IN THE LARGE INTESTINE; FORMATION OF THE FECES

Approximately 500 to 1000 ml. of chyme passes through the ileocecal valve into the large intestine each day. Most of the water and electrolytes in this are absorbed in the colon, leaving only 50 to 200 ml. of fluid to be excreted in the feces.

Most of the absorption in the large intestine occurs in the proximal half of the colon, giving this portion the name *absorbing colon,* while the distal colon functions principally for storage and is therefore called the *storage colon.*

Absorption and Secretion of Electrolytes and Water. The mucosa of the large intestine, like that of the small intestine, has a very high capacity for active absorption of sodium, and the electrical potential created by the absorption of the sodium causes chloride absorption as well. In addition, as in the distal portion of the small intestine, the mucosa of the large intestine actively secretes bicarbonate ions while it simultaneously actively absorbs a small amount of additional chloride ions. The bicarbonate helps to neutralize the acidic end-products of bacterial action in the colon.

The absorption of sodium and chloride ions creates an osmotic gradient across the large intestinal mucosa, which in turn causes absorption of water.

Bacterial Action in the Colon. Numerous bacteria, especially colon bacilli, are present in the absorbing colon. Substances formed as a result of bacterial activity are vitamin K, vitamin B_{12}, thiamin, riboflavin, and various gases that contribute to *flatus* in the colon. Vitamin K is especially important, for the amount of this vitamin in the ingested foods is normally insufficient to maintain adequate blood coagulation.

Composition of the Feces. The feces normally are about three-fourths water and one-fourth solid matter composed of about 30 per cent dead bacteria, 10 to 20 per cent fat, 10 to 20 per cent inorganic matter, 2 to 3 per cent protein, and 30 per cent undigested roughage of the food and dried constituents of digestive juices, such as bile pigment and sloughed epithelial cells.

The brown color of feces is caused by *stercobilin* and *urobilin,* which are derivatives of bilirubin. The odor is caused principally by the products of bacterial action; these vary from one person to another, depending on each person's colonic bacterial flora and on the type of food he has eaten. The actual odoriferous products include *indole, skatole, mercaptans,* and *hydrogen sulfide.*

GASTROINTESTINAL DISORDERS

GASTRITIS

Gastritis means inflammation of the gastric mucosa. This can result from (1) action of irritant foods on the gastric mucosa, (2) excessive excoriation of the stomach mucosa by the stomach's own peptic secretions, or (3) occasionally bacterial inflammation. One of the most frequent causes of gastritis is irritation of the mucosa by alcohol.

The inflamed mucosa in gastritis is often painful, causing a diffuse burning pain referred to the high epigastrium. Reflexes initiated in the stomach mucosa cause the salivary glands to salivate intensely, and the frequent swallowing of foamy saliva makes air accumulate in the stomach. As a result, the person usually belches profusely, a burning sensation often occurring in his throat with each belch.

Gastric Atrophy. In many persons who have chronic gastritis, the mucosa gradually becomes atrophic until little or no gastric gland activity remains. It is also believed that some persons develop autoimmunity against the gastric mucosa, this leading eventually to gastric atrophy. Loss of the stomach secretions in gastric atrophy leads to achlorhydria and, occasionally, to *pernicious anemia.*

Achlorhydria means simply that the stomach fails to secrete hydrochloric acid. Usually, when acid is not secreted, pepsin also is not secreted, and, even if it is, the lack of acid prevents it from functioning because pepsin requires an acid medium for activity. Obviously, then, essentially all digestive function in the stomach is lost when achlorhydria is present.

Pernicious Anemia in Gastric Atrophy. Pernicious anemia, which was discussed in Chapter 5, is a common accompaniment of gastric atrophy. The normal gastric secretions contain a glycoprotein called *intrinsic factor,* which is secreted by the parietal cells (the HCl-producing cells) and which must be present for adequate absorption of vitamin B_{12} from the ileum. The intrinsic factor combines with vitamin B_{12}, and the complex then binds with receptors of the surfaces of ileal mucosal cells. This is a necessary step in the absorption of vitamin B_{12}. In the absence of intrinsic factor, an adequate amount of vitamin B_{12} is not made available from the foods. As a result, maturation failure occurs in the bone marrow, resulting in pernicious anemia.

Pernicious anemia also occurs frequently when most of the stomach has been removed for treatment of either stomach ulcer or gastric cancer or when the ileum, where vitamin B_{12} is almost entirely absorbed, is removed.

PEPTIC ULCER

A peptic ulcer is an excoriated area of the mucosa caused by the digestive action of gastric juice. Figure 44–8 illustrates the points in the gastrointestinal tract at which peptic ulcers fre-

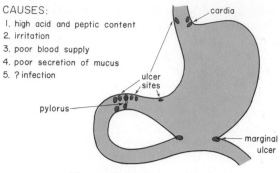

CAUSES:
1. high acid and peptic content
2. irritation
3. poor blood supply
4. poor secretion of mucus
5. ? infection

Figure 44–8. Peptic ulcer.

quently occur, showing that by far the most frequent site of peptic ulcers is in the first few centimeters of the duodenum. In addition, peptic ulcers frequently occur along the lesser curvature of the antral end of the stomach or, more rarely, in the lower end of the esophagus where there is often reflux of stomach juices. A peptic ulcer called a *marginal ulcer* also frequently occurs wherever an abnormal opening, such as a gastrojejunostomy, is made between the stomach and some portion of the small intestine.

Basic Cause of Peptic Ulceration. The usual cause of peptic ulceration is too much secretion of gastric juice in relation to the degree of protection afforded by the mucous lining of the stomach and duodenum and by neutralization of the gastric acid by duodenal juices. It will be recalled that all areas normally exposed to gastric juice are well supplied with mucous glands, beginning with the compound mucous glands of the lower esophagus, then including the mucous cell coating of the stomach mucosa, the mucous neck cells of the gastric glands, the deep pyloric glands that secrete mainly mucus, and, finally, the glands of Brunner of the upper duodenum, which secrete a highly alkaline mucus.

In addition to the mucus protection of the mucosa, the duodenum is also protected by the alkalinity of the pancreatic secretion, which contains large quantities of sodium bicarbonate that neutralize the hydrochloric acid of the gastric juice, thus inactivating the pepsin and thereby preventing digestion of the mucosa. Two additional mechanisms insure that this neutralization of gastric juices is complete:

1. When excess acid enters the duodenum, it reflexly inhibits gastric secretion and peristalsis in the stomach, thereby decreasing the rate of gastric emptying. This allows increased time for pancreatic secretion to enter the duodenum to neutralize the acid already present. After neutralization has taken place, the reflex subsides and more stomach contents are emptied.

2. The presence of acid in the small intestine liberates secretin from the intestinal mucosa. The secretin then passes by way of the blood to the

pancreas to promote rapid secretion of pancreatic juice containing a high concentration of sodium bicarbonate, thus making more sodium bicarbonate available for neutralization of the acid. These mechanisms were discussed in detail in Chapters 42 and 43 in relation to gastrointestinal motility and secretion.

Causes of Peptic Ulcer in the Human Being. Peptic ulcer occurs much more frequently in the white collar worker than in the laborer, and persons subjected to extreme anxiety for long periods of time seem particularly prone to peptic ulcer. For instance, the number of persons who developed peptic ulcer increased greatly during the air raids of London. Therefore, it is believed that many if not most instances of *duodenal* peptic ulcer in the human being result from excessive stimulation of the dorsal motor nucleus of the vagus by impulses originating in the cerebrum. Supporting this theory is the fact that duodenal ulcer patients have a very high rate of gastric secretion during the interdigestive period between meals when the stomach is empty. The normal stomach secretes a total of approximately 18 milliequivalents of hydrochloric acid during the 12-hour interdigestive period through the night, while duodenal ulcer patients occasionally secrete as much as 300 milliequivalents of hydrochloric acid during this same time.

Paradoxically, *gastric* ulcers, in contradistinction to duodenal ulcers, often occur in patients who have normal or low secretion of hydrochloric acid. However, these patients almost invariably have an associated gastritis, indicating that ulceration in the stomach almost certainly results from reduced resistance of the stomach mucosa to digestion rather than to excess secretion of gastric juice. Stomach ulceration frequently results in patients who have ingested large quantities of substances such as aspirin or alcohol that reduce the mucosal resistance.

Physiology of Treatment. The usual medical treatment for peptic ulcer is a combination of (1) reduction of stressful situations that might lead to excessive acid secretion, (2) administration of antacid drugs to neutralize much of the acid in the stomach secretions, (3) administration of the drug *cimetidine*, which blocks the action of gastrin in stimulating gastric juice secretion, (4) prescription of a bland diet and sometimes small meals many times a day rather than three meals a day (though there is not yet proof that spreading meals is truly effective). (5) interdiction against smoking because statistical studies have shown that smokers are several times as prone to have peptic ulcers as are nonsmokers, and (6) removal of such ulcer-causing factors as alcohol, aspirin, or other substances that might irritate the gastroduodenal mucosa.

Surgical treatment of peptic ulcer is effected by one or both of two procedures: (1) removal of a large portion of the stomach or (2) vagotomy. When a peptic ulcer is surgically removed, at least the lower three-fourths to four-fifths of the stomach is usually also removed and the upper stump of the stomach then anastomosed to the jejunum. If less of the stomach than this is removed, far too much gastric juice continues to be secreted, and a marginal ulcer soon develops where the stomach is anastomosed to the intestine.

Vagotomy temporarily blocks almost all secretion by the stomach and usually cures a peptic ulcer within less than a week after the operation is performed. Unfortunately, though, a large amount of basal stomach secretion returns three to six months after the vagotomy, and in many patients the ulcer itself also returns. Even more distressing is *gastric atony* that usually follows vagotomy, for the motility of the stomach is often reduced to such a low level that almost no gastric emptying occurs after vagotomy. For this reason, vagotomy alone is rarely performed in the treatment of peptic ulcer. However, it is performed frequently in association with simultaneous plastic procedures to increase the size of the opening of the pylorus from the stomach into the small intestine.

MALABSORPTION FROM THE SMALL INTESTINE — "SPRUE"

Occasionally, nutrients are not adequately absorbed from the small intestine even though the food is well digested. Several different diseases can cause decreased absorbability of the mucosa; these are often classified together under the general heading of *sprue*. Obviously, also, malabsorption can occur when large portions of the small intestine have been removed.

One type of sprue, called variously by the names *idiopathic sprue, celiac disease* (in children), or *gluten enteropathy*, results from the toxic effects of *gluten* present in certain types of grains, especially wheat and rye. The gluten causes destruction of the villi. As a result, the villi become blunted or disappear altogether, thus greatly reducing the absorptive area of the gut. Removal of wheat and rye flour from the diet, especially in children with this disease, frequently results in an apparently miraculous cure within weeks.

Malabsorption in Sprue. In the early stages of sprue, the absorption of fats is more impaired than the absorption of other digestive products. The fat appearing in the stools is almost entirely in the form of soaps rather than undigested neutral fat, illustrating that the problem is one of absorption and not of digestion. In this stage of sprue, the condition is frequently called *idiopathic* *steatorrhea,* which means simply excess fats in the stools as a result of unknown causes.

In more severe cases of sprue the absorption of proteins, carbohydrates, calcium, vitamin K, folic acid, and vitamin B_{12}, as well as many other important substances, becomes greatly impaired. As a result, the person suffers (1) severe nutritional deficiency, often developing severe wasting of the tissues, (2) osteomalacia (demineralization of the bones because of calcium lack), (3) inadequate blood coagulation due to lack of vitamin K, and (4) macrocytic anemia of the pernicious anemia type, owing to diminished vitamin B_{12} and folic acid absorption.

CONSTIPATION

Constipation means slow movement of feces through the large intestine, and it is often associated with large quantities of dry, hard feces in the descending colon that accumulate because of the long time allowed for absorption of fluid.

A frequent cause of constipation is irregular bowel habits that have developed through a lifetime of inhibition of the normal defecation reflexes. The newborn child is rarely constipated, but part of his training in the early years of life requires that he learn to control defecation, and this control is effected by inhibiting the natural defecation reflexes. Clinical experience shows that if one fails to allow defecation to occur when the defecation reflexes are excited or if one overuses laxatives to take the place of natural bowel function, the reflexes themselves become progressively less strong over a period of time and the colon becomes *atonic.* For this reason, if a person establishes regular bowel habits early in life, usually defecating in the morning after breakfast when the gastrocolic and duodenocolic reflexes cause mass movements in the large intestine, he can generally prevent the development of constipation in later life.

DIARRHEA

Diarrhea, the opposite of constipation, results from rapid movement of fecal matter through the large intestine. The major cause of diarrhea is infection in the gastrointestinal tract, which is called *enteritis.*

In usual infectious diarrhea, the infection is most extensive in the large intestine and the distal end of the ileum. Everywhere that the infection is present, the mucosa becomes extensively irritated, and its rate of secretion becomes greatly enhanced. In addition, the motility of the intestinal wall usually increases many-fold. As a result, large quantities of fluid are made available for washing the infectious agent toward the anus, and at the same time strong propulsive movements propel this fluid forward. Obviously, this is an important mechanism for ridding the intestinal tract of the debilitating infection.

Of special interest is the diarrhea caused by *cholera.* The cholera toxin directly stimulates excessive secretion of electrolytes and fluid from the crypts of Lieberkühn in the distal ileum and colon, and it specifically enhances the bicarbonate-chloride exchange mechanism, causing extreme quantities of bicarbonate ions to be secreted into the intestinal tract. The loss of fluid and electrolytes can be so debilitating within a day or so that death ensues. Therefore, the most important basis of therapy is simply to replace the fluid and electrolytes as rapidly as they are lost. With proper therapy of this type, almost no cholera patients die, but without, 50 per cent or more do.

VOMITING

Vomiting is the means by which the upper gastrointestinal tract rids itself of its contents when the gut becomes excessively irritated, overdistended, or even overexcitable. The stimuli that cause vomiting can originate in any part of the gastrointestinal tract, though distention or irritation of the stomach or duodenum provides the strongest stimulus. Impulses are transmitted by both vagal and sympathetic afferents to the *vomiting center* of the medulla, which lies near the tractus solitarius at approximately the level of the dorsal motor nucleus of the vagus. Appropriate motor reactions are then instituted to cause the vomiting act, and the motor impulses that cause the actual vomiting are transmitted from the vomiting center through the fifth, seventh, ninth, tenth, and twelfth cranial nerves to the upper gastrointestinal tract and through the spinal nerves to the diaphragm and abdominal muscles.

The Vomiting Act. Once the vomiting center has been sufficiently stimulated and the vomiting act instituted, the first effects are (1) a deep inspiratory breath, (2) raising of the hyoid bone and the larynx to pull the upper esophageal sphincter open, (3) closing of the glottis, and (4) lifting of the soft palate to close the posterior nares. Next comes a strong downward contraction of the diaphragm along with simultaneous contraction of all the abdominal muscles. This obviously squeezes the stomach between the two sets of muscles, building the intragastric pressure to a high level. Finally, the lower esophageal sphincter relaxes, allowing expulsion of the gastric contents upward through the esophagus.

Thus, the vomiting act results from a squeezing

action of the muscles of the abdomen associated with opening of the esophageal sphincters so that the gastric contents can be expelled.

GASES IN THE GASTROINTESTINAL TRACT (FLATUS)

Gases can enter the gastrointestinal tract from three different sources: (1) swallowed air, (2) gases formed as a result of bacterial action, and (3) gases that diffuse from the blood into the gastrointestinal tract.

Most gases in the stomach are nitrogen and oxygen derived from swallowed air, and a large proportion of these are expelled by belching.

Only small amounts of gas are usually present in the small intestine, and these are composed principally of air that passes from the stomach into the intestinal tract.

In the large intestine, the greater proportion of the gases is derived from bacterial action; these gases include especially *carbon dioxide, methane,* and *hydrogen.* When the methane and hydrogen become suitably mixed with oxygen from swallowed air, an actual explosive mixture is occasionally formed.

Certain foods are known to cause greater amounts of flatus from the large intestine than others — beans, cabbage, onions, cauliflower, corn, and certain highly irritant foods such as vinegar. Some of these foods — beans, for instance — serve as a suitable medium for gas-forming bacteria, especially because they contain fermentable types of carbohydrates that are poorly absorbed.

The amount of gases entering or forming in the large intestine each day averages 7 to 10 liters, whereas the average amount expelled is usually only about 0.6 liter. The remainder is absorbed through the intestinal mucosa. Most often, a person expels large quantities of gases not because of excessive bacterial activity but because of excessive motility of the large intestine caused by intestinal irritation. This moves the gases on through the large intestine before they can be absorbed.

REFERENCES

Digestion and Absorption

Cluysenaer, O. J. J., and van Tongeren, J. H. M.: Malabsorption in Coeliac Sprue. The Hague, M. Nijhoff Medical Division, 1977.

Crane, R. K.: Intestinal absorption of glucose. *Biomembranes, 4A*:541, 1974.

Davenport, H. W.: A Digest of Digestion, 2nd Ed. Chicago, Year Book Medical Publishers, 1978.

Frizzell, R. A., and Schultz, S. G.: Models of electrolyte absorption and secretion by gastrointestinal epithelia. *In* Crane, R. K. (ed.): International Review of Physiology: Gastrointestinal Physiology III. Vol. 19. Baltimore, University Park Press, 1979, p. 205.

Levitan, R., and Wilson, D. E.: Absorption of water soluble substances. *In* MPT International Review of Physiology. Vol. 4. Baltimore, University Park Press, 1974, p. 293.

Matthews, D. M.: Absorption of amino acids and peptides from the intestine. *Clin. Endocrinol. Metabol., 3*:3, 1974.

Matthews, D. M.: Absorption of water-soluble vitamins. *Biomembranes 4B*:847, 1974.

Matthews, D. M.: Intestinal absorption of peptides. *Physiol. Rev., 55*:537, 1975.

Olsen, W. A.: Carbohydrate absorption. *Med. Clin. North Am. 58*:1387, 1974.

Schultz, S. G., et al.: Ion transport by mammalian small intestine. *Annu. Rev. Physiol., 36*:51, 1974.

Silk, D. B. A., and Dawson, A. M.: Intestinal absorption of carbohydrate and protein in man. *In* Crane, R. K. (ed.): International Review of Physiology: Gastrointestinal Physiology III. Vol. 19. Baltimore, University Park Press, 1979, p. 151.

Soergel, K. H., and Hofmann, A. F.: Absorption. *In* Frohlich, E. D. (ed.): Pathophysiology, 2nd Ed. Philadelphia, J. B. Lippincott Co., 1976, p. 499.

Ugolev, A. M.: Membrane (contact) digestion. *Biomembranes, 4A*:285, 1974.

Watson, D. W., and Sodeman, W. A., Jr.: The small intestine. *In* Sodeman, W. A., Jr., and Sodeman, T. M. (eds.): Pathologic Physiology: Mechanisms of Disease, 6th Ed. Philadelphia, W. B. Saunders Company, 1979, p. 824.

Gastrointestinal Disorders

Boley, S. J., et al.: Ischemic Disorders of the Intestines. Chicago, Year Book Medical Publishers, 1978.

Cook, G. C.: Tropical Gastroenterology. New York, Oxford University Press, 1980.

Duthie, H. L. (ed.): Gastrointestinal Motility in Health and Disease. Baltimore, University Park Press, 1978.

Eisenberg, M. M.: Ulcers. Boston, G. K. Hall, 1979.

Goodman, M. J., and Sparberg, M.: Ulcerative Colitis. New York, John Wiley & Sons, 1978.

Halsted, C. H.: Intestinal absorption and malabsorption of folates. *Annu. Rev. Med. 31*:79, 1980.

Kirsner, J. B., and Shorter, R. G. (eds.): Inflammatory Bowel Disease, 2nd Ed. Philadelphia, Lea & Febiger, 1980.

Kirsner, J. B., and Winans, C. S.: The stomach. *In* Sodeman, W. A., Jr., and Sodeman, T. M. (eds.): Sodeman's Pathologic Physiology; Mechanisms of Disease, 6th Ed. Philadelphia, W. B. Saunders Co., 1979, p. 798.

Lebenthal, E. (ed.): Digestive Disease in Children. New York, Grune & Stratton, 1978.

Najarian, J. S., and Delaney, J. P. (eds.): Gastrointestinal Surgery. Chicago, Year Book Medical Publishers, 1979.

Powell, L. W., and Piper, D. W. (eds.): Fundamentals of Gastroenterology, 3rd Ed. Baltimore, University Park Press, 1979.

Rankow, R. M., and Polayes, I. M.: Diseases of the Salivary Glands, Philadelphia, W. B. Saunders Co., 1976.

Silen, W. (rev.): Cope's Early Diagnosis of the Acute Abdomen. New York, Oxford University Press, 1979.

Sleisenger, M. H., and Fordtran, J. S.: Gastrointestinal Disease: Pathophysiology, Diagnosis, Management. Philadelphia, W. B. Saunders Co., 1977.

Part XII

METABOLISM AND TEMPERATURE REGULATION

45

Metabolism of Carbohydrates and Formation of Adenosine Triphosphate

The next few chapters deal with metabolism in the body, which means the chemical processes that make it possible for the cells to continue living. It is not the purpose of this textbook, however, to present the chemical details of all the various cellular reactions, for this lies in the discipline of biochemistry. Instead, these chapters are devoted to (1) a review of the principal chemical processes of the cell and (2) an analysis of their physiological implications, especially in relation to the manner in which they fit into the overall concept of homeostasis.

ROLE OF ADENOSINE TRIPHOSPHATE (ATP) IN METABOLISM

The greater proportion of the chemical reactions in the cells is concerned with making the energy in foods available to the various physiological systems of the cell. For instance, energy is required for (1) muscular activity, (2) secretion by the glands, (3) maintenance of membrane potentials by the nerve and muscle fibers, (4) synthesis of substances in the cells, and (5) absorption of foods from the gastrointestinal tract. The substance adenosine triphosphate (ATP) plays a key role in making the energy of the foods available for all of these purposes.

ATP is a labile chemical compound that is present in all cells and has the chemical structure shown at the top of the next page.

From this formula it can be seen that ATP is a combination of adenine, ribose, and three phosphate radicals. The last two phosphate radicals are connected with the remainder of the molecule by so-called *high energy bonds,* which are

indicated by the symbol ~. The amount of free energy in each of these high energy bonds per mol of ATP is approximately 8000 calories under the conditions of temperature and concentrations of the reactants in the body. Therefore, removal of each phosphate radical liberates 8000 calories of energy. After loss of one phosphate radical from ATP, the compound becomes *adenosine diphosphate* (ADP), and after loss of the second phosphate radical the compound becomes *adenosine monophosphate* (AMP). The interconversions between ATP, ADP, and AMP are the following:

$$\text{ATP} \xrightleftharpoons[+8000 \text{ cal.}]{-8000 \text{ cal.}} \left\{ \begin{array}{c} \text{ADP} \\ + \\ \text{PO}_4 \end{array} \right\} \xrightleftharpoons[+8000 \text{ cal.}]{-8000 \text{ cal.}} \left\{ \begin{array}{c} \text{AMP} \\ + \\ 2\text{PO}_4 \end{array} \right\}$$

ATP is present everywhere in the cytoplasm and nucleoplasm of all cells, and essentially all the physiological mechanisms that require energy for operation obtain it directly from the ATP. In turn, the food in the cells is gradually oxidized, and the released energy is used to re-form the ATP, thus always maintaining a supply of this substance.

Thus, ATP is a compound that has the peculiar ability of entering into many coupled reactions — reactions with the food to extract energy, and reactions in relation to many physiological mechanisms to provide energy for their operation. For this reason, ATP has frequently been called the energy *currency* of the body that can be gained and spent again and again.

The principal purpose of the present chapter is to explain how the energy from carbohydrates can be used to form ATP in the cells. At least 99

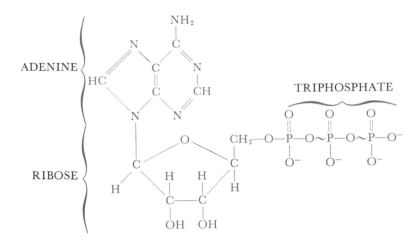

per cent of all the carbohydrates utilized by the body is used for this purpose.

TRANSPORT OF MONOSACCHARIDES THROUGH THE CELL MEMBRANE

From the previous chapter it will be recalled that the final products of carbohydrate digestion in the alimentary tract are almost entirely glucose, fructose, and galactose, glucose representing by far the major share of these. These three monosaccharides are absorbed into the portal blood and, after passing through the liver, are carried everywhere in the body by the circulatory system. But, before they can be used by the cells, they must be transported through the cell membrane into the cellular cytoplasm.

Monosaccharides cannot diffuse through the pores of the cell membrane, for the maximum molecular weight of particles that can do this is about 100, whereas glucose, fructose, and galactose all have molecular weights of 180. Yet glucose and some of the other monosaccharides combine with a carrier substance that makes them soluble in the membrane, so that they diffuse easily to the cell interior. After passing through the membrane they become dissociated from the carrier. Thus far, the nature of the carrier is uncertain, though it is believed to be a protein of small molecular weight. The transport mechanism is one of *facilitated diffusion* and not of active transport. These concepts were discussed in more detail in Chapter 4.

ENHANCEMENT OF GLUCOSE TRANSPORT BY INSULIN

The rate of glucose transport and also transport of some other monosaccharides is greatly increased by insulin. When large amounts of insulin are secreted by the pancreas, the rate of glucose transport into some cells increases to as much as ten times the rate of transport when no insulin at all is secreted. Since the amounts of glucose that can diffuse to the insides of most cells of the body in the absence of insulin, with the unique exceptions of the liver and the brain, are far too little to supply anywhere near the amount of glucose normally required for energy metabolism, in effect the rate of carbohydrate utilization by the body is controlled by the rate of insulin secretion in the pancreas. The functions of insulin and its control of carbohydrate metabolism will be discussed in detail in Chapter 52.

PHOSPHORYLATION OF THE MONOSACCHARIDES

Immediately upon entry into the cells, the monosaccharides combine with a phosphate radical in accordance with the following reaction:

$$\text{Glucose} \xrightarrow[\text{+ATP}]{\text{glucokinase}} \text{Glucose 6-phosphate}$$

This phosphorylation is promoted by enzymes called *hexokinases*, which are specific for each particular type of monosaccharide; thus, *glucokinase* promotes glucose phosphorylation, *fructokinase* promotes fructose phosphorylation, and *galactokinase* promotes galactose phosphorylation.

The phosphorylation of monosaccharides is almost completely irreversible except in liver cells, renal tubular epithelial cells, and intestinal epithelial cells, in which specific phsophatases are available for reversing the reaction. Therefore, in most tissues of the body phosphorylation serves to *capture* the monosaccharide in the cell — once *in* the cell the monosaccharide will not diffuse back out except in those special cells listed above that have the necessary phosphatases.

Conversion of Fructose and Galactose Into Glucose. In liver cells, appropriate enzymes are available to promote interconversions between the monosaccharides; and the dynamics of the reactions are such that when the liver releases the monosaccharides back into the blood, the final product of these interconversions is almost entirely glucose. Also, in the case of fructose, much of it is converted into glucose as it is absorbed through the intestinal epithelial cells into the portal blood. Therefore, essentially all of the monosaccharides that circulate in the blood are the final conversion product, glucose.

STORAGE OF GLYCOGEN IN LIVER AND MUSCLE

After absorption into the cells, glucose can be used immediately for release of energy to the cells or it can be stored in the form of *glycogen,* which is a large polymer of glucose.

All cells of the body are capable of storing at least some glycogen, but certain cells can store large amounts, especially the liver cells, which can store up to 5 to 8 per cent of their weight as glycogen, and muscle cells, which can store up to 1 per cent glycogen. The glycogen molecules can be polymerized to almost any molecular weight, the average molecular weight being 5,000,000 or greater; most of the glycogen precipitates in the form of solid granules.

GLYCOGENESIS

Glycogenesis is the process of glycogen formation, the chemical reactions for which are illustrated in Figure 45–1. From this figure it can be seen that *glucose 6-phosphate* first becomes *glucose 1-phosphate;* then this is converted to *uridine diphosphate glucose,* which is then converted into glycogen. Several specific enzymes are required to cause these conversions. Any monosaccharide that can be converted into glucose obviously can enter into the reactions, and certain smaller compounds as well, including *lactic acid, glycerol, pyruvic acid,* and *some deaminated amino acids,* can also be converted into glucose or closely allied compounds and thence into glycogen.

GLYCOGENOLYSIS

Glycogenolysis means the breakdown of glycogen to re-form glucose in the cells. Glycogenolysis does not occur by reversal of the same chemical reactions that serve to form glycogen; instead, each succeeding glucose molecule on each branch of the glycogen polymer is split away by a process of *phosphorylation,* catalyzed by the enzyme *phosphorylase.*

Under resting conditions, the phosphorylase is in an inactive form, so that glycogen can be stored but not reconverted into glucose. When it is necessary to re-form glucose from glycogen, therefore, the phosphorylase must first be activated. This is accomplished in the following two ways:

Activation of Phosphorylase by Epinephrine and Glucagon. Two hormones, epinephrine and glucagon, can specifically activate phosphorylase and thereby cause glycogenolysis. The initial effect of each of these hormones is to increase the formation of *cyclic adenosine monophosphate (cAMP)* in the cells. This substance then initiates a cascade of chemical reactions that activate the phosphorylase, a process to be discussed in more detail in Chapter 52.

Epinephrine is released by the adrenal medullae when the sympathetic nervous system is stimulated. The epinephrine then stimulates the splitting of glycogen, thus making glucose available for rapid metabolism. This function of epinephrine occurs markedly both in liver cells and in muscle, thereby contributing, along with other effects of sympathetic stimulation, to preparation of the body for action, as was discussed in Chapter 38.

Glucagon is a hormone secreted by the *alpha cells* of the pancreas when the blood glucose concentration falls low. It stimulates the formation of cyclic AMP mainly in the liver. Therefore, its effect is primarily to dump glucose out of the liver into the blood, thereby elevating blood glucose concentration. The function of glucagon in blood glucose regulation is discussed in Chapter 52.

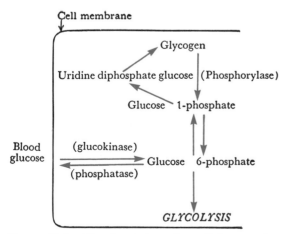

Figure 45–1. The chemical reactions of glycogenesis and glycogenolysis, showing also the interconversions between blood glucose and liver glycogen. (The phosphatase required for release of glucose from the cell is absent in muscle cells.)

Transport of Glucose Out of Liver Cells.   The cells of the liver contain *phosphatase,* an enzyme that can split phosphate away from glucose 6-phosphate and therefore make the glucose available for retransport out of the cells into the interstitial fluids. Therefore, when glucose is formed in the liver as a result of glycogenolysis, most of it immediately passes into the blood. Thus, liver glycogenolysis causes an immediate rise in blood glucose concentration. Glycogenolysis in most other cells of the body, especially in the muscle cells, simply makes increased amounts of glucose 6-phosphate available inside the cells and increases the local rate of glucose utilization but does not release the glucose into the extracellular fluids because the glucose 6-phosphate cannot be dephosphorylated.

RELEASE OF ENERGY FROM THE GLUCOSE MOLECULE BY THE GLYCOLYTIC PATHWAY

Complete oxidation of 1 gram-mole of glucose releases 686,000 calories of energy, but only 8000 calories of energy are required to form 1 gram-mole of adenosine triphosphate (ATP). Therefore, it would be extremely wasteful of energy if glucose should be decomposed all the way into water and carbon dioxide at once while forming only a single ATP molecule. Fortunately, cells contain an extensive series of different protein enzymes that cause the glucose molecule to split a little at a time in many successive steps, with its energy released in small packets to form one molecule of ATP at a time, forming a total of 38 moles of ATP for each mole of glucose utilized by the cells.

The purpose of the present section is to describe the basic principles by which the glucose molecule is progressively dissected and its energy released to form ATP.

GLYCOLYSIS AND THE FORMATION OF PYRUVIC ACID

By far the most important means by which energy is released from the glucose molecule is the process of *glycolysis* and then *oxidation of the end-products of glycolysis.* Glycolysis means splitting of the glucose molecule to form two molecules of pyruvic acid. This occurs by ten successive steps of chemical reactions, illustrated in Figure 45–2. Each step is catalyzed by at least one specific protein enzyme. Note that glucose is first converted into fructose 1,6-phosphate and then split into two three-carbon atom molecules, each of which is then converted through five successive steps into pyruvic acid.

Formation of Adenosine Triphosphate (ATP) During Glycolysis.   Despite the many chemical reactions in the glycolytic series, little energy is released. However, 2 moles of ATP are formed for each mole of glucose utilized. This amounts to 16,000 calories of energy stored in the form of ATP, but during glycolysis a total of 56,000 calories of energy is lost from the original glucose, giving an overall *efficiency* for ATP forma-

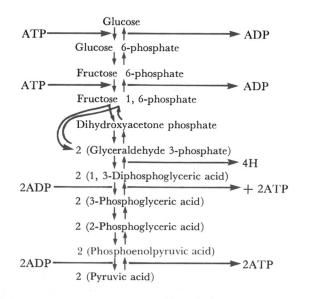

Net reaction:

$$\text{Glucose} + 2\text{ADP} + 2\text{PO}_4^{---} \longrightarrow 2 \text{ Pyruvic acid} + 2\text{ATP} + 4\text{H}$$

Figure 45–2.   The sequence of chemical reactions responsible for glycolysis.

tion of 29 per cent. The remaining 71 per cent of the energy is lost in the form of heat.

CONVERSION OF PYRUVIC ACID TO ACETYL COENZYME A

The next stage in the degradation of glucose is conversion of its two derivative pyruvic acid molecules into two molecules of *acetyl coenzyme A* (acetyl Co-A) in accordance with the following reaction:

$$2\ CH_3{-}\overset{\overset{\displaystyle O}{\|}}{C}{-}COOH + 2\ Co\text{-}A{-}SH \longrightarrow$$
(Pyruvic acid)   (Coenzyme A)

$$2\ CH_3{-}\overset{\overset{\displaystyle O}{\|}}{C}{-}S{-}Co\text{-}A + 2CO_2 + 4H$$
(Acetyl Co-A)

From this reaction it can be seen that two carbon dioxide molecules and four hydrogen atoms are released, while the remainders of the two pyruvic acid molecules combine with coenzyme A, a derivative of the vitamin pantothenic acid, to form two molecules of acetyl Co-A. In this conversion, no ATP is formed, but six molecules of ATP are formed when the four hydrogen atoms are later oxidized, as will be discussed in a later section.

THE CITRIC ACID CYCLE

The next stage in the degradation of the glucose molecule is called the *citric acid cycle* (also called the *tricarboxylic acid cycle* or *Krebs cycle*). This is a sequence of chemical reactions in which the acetyl portion of acetyl Co-A is degraded to carbon dioxide and hydrogen atoms. (The hydrogen atoms are subsequently oxidized, releasing still more energy to form ATP). The enzymes responsible for the citric acid cycle are all contained in the *matrix of the mitochondria*.

Figure 45–3 shows the different stages of the chemical reactions in the citric acid cycle. The substances to the left are added during the chemical reactions, and the products of the chemical reactions are shown to the right. Note at the top of the column that the cycle begins with *oxaloacetic acid*, and then at the bottom of the chain of reactions *oxaloacetic acid* is formed once again. Thus, the cycle can continue over and over.

In the initial stage of the citric acid cycle, *acetyl Co-A* combines with *oxaloacetic acid* to form *citric acid*. During the cycle, the coenzyme A portion of the acetyl Co-A is released and can be used again and again for the formation of still more quantities of acetyl Co-A from pyruvic acid. The acetyl portion, however, becomes an integral part of the

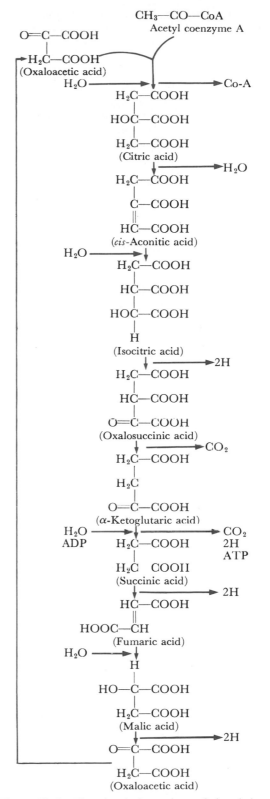

Figure 45–3. The chemical reactions of the citric acid cycle, showing the release of carbon dioxide and an especially large number of hydrogen atoms during the cycle.

Net reaction per molecule of glucose:
2 Acetyl Co-A + 6H₂O + 2ADP ⟶
    4CO₂ + 16H + 2Co-A + 2ATP

citric acid molecule. During the successive stages of the citric acid cycle, several molecules of water are added and *carbon dioxide* and *hydrogen atoms* are released at various stages in the cycle, as shown on the right in the figure.

The net results of the entire citric acid cycle are shown at the bottom of Figure 45–3, illustrating that for each molecule of glucose originally metabolized, 2 acetyl Co-A molecules enter into the citric acid cycle along with 6 molecules of water. These then are degraded into 4 carbon dioxide molecules, 16 hydrogen atoms, and 2 molecules of coenzyme A.

Formation of ATP in the Citric Acid Cycle. Not a great deal of energy is released during the citric acid cycle itself. However, for each molecule of glucose metabolized, 2 molecules of ATP are formed.

FORMATION OF ATP BY OXIDATIVE PHOSPHORYLATION

Despite all the complexities of glycolysis and the citric acid cycle, pitifully small amounts of ATP are formed during these processes, only 2 ATP molecules in the glycolysis scheme and another 2 in the citric acid cycle. Instead, about 90 per cent of the final ATP is formed during subsequent oxidation of the hydrogen atoms that are released during these earlier stages of glucose degradation. Indeed, the principal function of all these earlier stages is to make the hydrogen of the glucose molecule available in a form that can be utilized for oxidation.

Oxidation of hydrogen is accomplished by a series of enzymatically catalyzed reactions that (a) change the hydrogen atoms into hydrogen ions and electrons, and (b) use the electrons eventually to change the dissolved oxygen of the fluids into hydroxyl ions. Then the hydrogen and hydroxyl ions combine with each other to form water. During the sequence of oxidative reactions, tremendous quantities of energy are released to form ATP. Formation of ATP in this manner is called *oxidative phosphorylation*.

Unfortunately, even though oxidative phosphorylation is the basis of about 95 per cent of all energy utilization in the body, we still do not know the precise means by which the energy released from oxidation of hydrogen atoms is utilized to form ATP. However, a theory for which there is much support, called the *chemiosmotic theory*, is illustrated in Figure 45–4, and its essentials are described in the following sections.

Ionization of Hydrogen, the Electron Transport Chain, and Formation of Water. The first step in oxidative phosphorylation is to ionize the

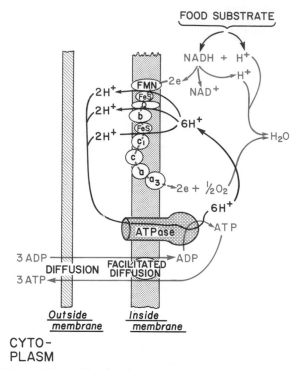

Figure 45–4. The chemiosmotic theory of oxidative phosphorylation for forming great quantities of ATP.

hydrogen atoms that are removed from the food substrates. These hydrogen atoms are removed in pairs during glycolysis and in the citric acid cycle; one immediately becomes a hydrogen ion, H^+, and the other combines with NAD^+ to form NADH. The upper portion of Figure 45–4 shows in color the subsequent fate of the NADH and H^+. The initial effect is to release the other hydrogen atom bound with NAD to form another hydrogen ion, H^+; this process also reconstitutes NAD^+ that will be reused again and again.

During these changes, the electrons that are removed from the hydrogen atoms to cause their ionization immediately enter an *electron transport chain* that is an integral part of the inner membrane (the shelf membrane) of the mitochondrion. This transport chain consists of a series of electron acceptors that can be reversibly reduced or oxidized by accepting or giving up electrons. The important members of this electron transport chain include *flavoprotein, several iron sulfide proteins, ubiquinone,* and *cytochromes B, C_1, C, A, and A_3.* Each electron is shuttled from one of these acceptors to the next until it finally reaches cytochrome A_3, which is called *cytochrome oxidase* because it is capable, by giving up two electrons, of causing elemental oxygen to combine with hydrogen ions to form water.

Thus, Figure 45–4 illustrates transport of electrons through the electron chain and then ultimate use of these by cytochrome oxidase to cause the formation of water molecules. During the transport of these electrons through the electron transport chain, energy is released that is later used to cause synthesis of ATP, as follows:

Hydrogen Pumping by the Electron Transport Chain. The energy released as the electrons pass through the electron transport chain is believed to pump hydrogen ions from the inner matrix of the mitochondrion into the space between the inner and outer mitochondrial membranes. This creates a high concentration of hydrogen ions in this space, and it also creates a strong negative electrical potential in the inner matrix.

Formation of ATP. The final step in oxidative phosphorylation is to convert ADP into ATP. This occurs in conjunction with a large protein molecule that protrudes all the way through the inner mitochondrial membrane and projects with a knoblike head into the inner matrix. This molecule is an ATPase, the physical nature of which is illustrated in Figure 45–4. It is called *ATP synthetase*. It is postulated that the high concentration of hydrogen ions in the space between the two mitochondrial membranes and the large electrical potential difference across the inner membrane cause the hydrogen ions to flow into the mitochondrial matrix *through the substance of the ATPase molecule.* In doing so, energy derived from this hydrogen ion flow is utilized by the ATPase to convert ADP into ATP, by combining an ADP with a phosphate radical, at the same time forming an additional high energy phosphate bond.

For each 2 electrons that pass through the entire electron transport chain (representing the ionization of 2 hydrogen atoms), up to 3 ATP molecules are synthesized.

SUMMARY OF ATP FORMATION DURING THE BREAKDOWN OF GLUCOSE

We can now determine the total number of ATP molecules formed by the energy from one molecule of glucose. The number is

1. two during glycolysis,
2. two during the citric acid cycle,
3. and 34 during oxidative phosphorylation,

making a total of *38 ATP molecules* formed for each molecule of glucose degraded to carbon dioxide and water. Thus, 304,000 calories of energy are stored in the form of ATP, while 686,000 calories are released during the complete oxidation of each gram-mole of glucose. This represents an overall *efficiency* of energy transfer of 44 per cent. The remaining 56 per cent of the energy becomes heat and therefore cannot be used by the cells to perform specific functions.

CONTROL OF GLYCOLYSIS AND OXIDATION BY ADENOSINE DIPHOSPHATE (ADP)

Continuous release of energy from glucose when the energy is not needed by the cells would be an extremely wasteful process. Fortunately, glycolysis and the subsequent oxidation of hydrogen atoms is continuously controlled in accordance with the needs of the cells for ATP. This control is accomplished mainly in the following manner:

Referring back to the various chemical reactions, we see that at different stages *ADP* is converted into *ATP. If ADP is not available at many of these stages, the reactions cannot occur, and the degradation of the glucose molecule is therefore stopped.* Therefore, once all the ADP in the cells has been converted to ATP, the entire glycolytic and oxidative process stops. Then, when more ATP is used to perform different physiological functions in the cell, new ADP is formed, which automatically turns on glycolysis and oxidation once more. In this way, essentially a full store of ATP is automatically maintained all the time, except when the activity of the cell becomes so great that ATP is used up more rapidly than it can be formed.

ANAEROBIC RELEASE OF ENERGY — "ANAEROBIC GLYCOLYSIS"

Occasionally, oxygen becomes either unavailable or insufficient, so that cellular oxidation of glucose cannot take place. Yet, even under these conditions, a small amount of energy can still be released to the cells by glycolysis, for the chemical reactions in the glycolytic breakdown of glucose to pyruvic acid do not require oxygen. Unfortunately, this process is extremely wasteful of glucose because only 16,000 calories of energy are used to form ATP for each molecule of glucose utilized, which represents only a little over 2 per cent of the total energy in the glucose molecule. Nevertheless, this release of glycolytic energy to the cells can be a lifesaving measure for a few minutes when oxygen becomes unavailable.

Formation of Lactic Acid During Anaerobic Glycolysis. The *law of mass action* states that as the end-products of a chemical reaction build up in a reacting medium the rate of the reaction approaches zero. The two end-products of the glycolytic reactions (see Figure 45–2) are (1) pyruvic acid and (2) hydrogen atoms in the forms NADH and H⁺. The buildup of excessive amounts of these would stop the glycolytic process and prevent further formation of ATP. Fortunately, when their quantities begin to be excessive, these end-products react with each other to form lactic acid, in accordance with the following equation.

$$CH_3-\overset{\overset{\displaystyle O}{\|}}{C}-COOH + NADH + H^+ \xrightleftharpoons{\text{lactic dehydrogenase}}$$

(Pyruvic acid)

$$CH_3-\overset{\overset{\displaystyle OH}{|}}{\underset{\underset{\displaystyle H}{|}}{C}}-COOH + NAD^+$$

(Lactic acid)

Thus, under anaerobic conditions, by far the major proportion of the pyruvic acid is converted into lactic acid, which diffuses readily out of the cells into the extracellular fluids and even into the intracellular fluids of other less active cells. Therefore, lactic acid represents a type of "sinkhole" into which the glycolytic end-products can disappear, thus allowing glycolysis to proceed far longer than would be possible if the pyruvic acid and hydrogen were not removed from the reacting medium. Indeed, glycolysis could proceed for only a few seconds without this conversion. Instead, it can proceed for several minutes, supplying the body with considerable quantities of ATP even in the absence of respiratory oxygen.

When a person begins to breathe oxygen again after a period of anaerobic metabolism, the extra NADH and H⁺ as well as the extra pyruvic acid that have built up in the body fluids are rapidly oxidized, thereby undergoing great reduction in their concentrations. As a result, the chemical reaction for formation of lactic acid immediately reverses itself, the lactic acid once again becoming pyruvic acid, which is eventually oxidized.

RELEASE OF ENERGY FROM GLUCOSE BY THE PHOSPHOGLUCONATE PATHWAY

Though essentially all the carbohydrates utilized by the muscles are degraded to pyruvic acid by glycolysis and then oxidized, the glycolytic schema is not the only means by which glucose can be degraded and then oxidized to provide energy. A second important schema for breakdown and oxidation of glucose is called the *phosphogluconate pathway*. Though this will not be

discussed here, it is responsible for as much as 30 per cent of the glucose breakdown in the liver and even more than that in fat cells. It is especially important in providing energy and some of the substrates required for conversion of carbohydrates into fat, as will be discussed in the following chapter.

FORMATION OF CARBOHYDRATES FROM PROTEINS AND FATS — "GLUCONEOGENESIS"

When the body's stores of carbohydrates decrease below normal, moderate quantities of glucose can be formed from *amino acids* and from the *glycerol* portion of fat. This process is called *gluconeogenesis*. Approximately 60 per cent of the amino acids in the body proteins can be converted into carbohydrates, while the remaining 40 per cent have chemical configurations that make this difficult. Each amino acid is converted into glucose by a slightly different chemical process. For instance, alanine can be converted directly into pyruvic acid simply by deamination; the pyruvic acid then is converted into glucose by the liver.

Regulation of Gluconeogenesis. Diminished carbohydrates in the cells and decreased blood sugar are the basic stimuli that set off an increase in the rate of gluconeogenesis. The diminished carbohydrates can directly cause reversal of many of the glycolytic and phosphogluconate reactions, thus allowing conversion of deaminated amino acids and glycerol into carbohydrates. However, in addition, several of the hormones secreted by the endocrine glands are especially important in this regulation.

Effect of Corticotropin and Glucocorticoids on Gluconeogenesis. When normal quantities of carbohydrates are not available to the cells, the anterior pituitary gland, for reasons not yet completely understood, begins to secrete increased quantities of corticotropin, which stimulates the adrenal cortex to produce large quantities of *glucocorticoid hormones,* especially *cortisol.* In turn, cortisol mobilizes proteins from essentially all cells of the body, making them available in the form of amino acids in the body fluids. A high proportion of these immediately becomes deaminated in the liver and therefore provides ideal substrates for conversion into glucose. Thus, one of the most important means by which gluconeogenesis is promoted is through the release of glucocorticoids from the adrenal cortex.

Effect of Thyroxine on Gluconeogenesis. Thyroxine, secreted by the thyroid gland, also increases the rate of gluconeogenesis. This, too, is believed to result principally from mobilization of proteins from the cells. However, it might result to some extent also from the mobilization of fats from the fat depots, the glycerol portion of the fats being converted into glucose.

BLOOD GLUCOSE

The normal blood glucose concentration in a person who has not eaten a meal within the past three to four hours is approximately 90 mg. per 100 ml. of blood, and even after a meal containing large amounts of carbohydrates, this concentration rarely rises above 140 mg. per cent unless the person has diabetes mellitus.

The regulation of blood glucose concentration is so intimately related to insulin and glucagon that this subject will be discussed in detail in Chapter 52 in relation to the functions of these two hormones.

REFERENCES

Aoki, T. R., and Cahill, G. F., Jr.: Metabolic effects of insulin, glucagon, and glucose in man: Clinical applications. *In* DeGroot, L. J., *et al.* (eds.): Endocrinology. Vol. 3. New York, Grune & Stratton, 1979, p. 1843.

Baer, H. P., and Drummond, G. I. (eds.): Physiological and Regulatory Functions of Adenosine and Adenine Nucleotides. New York, Raven Press, 1979.

Butler, T. M., and Davies, R. E.: High-energy phosphates in smooth muscle. *In* Bohr, D. F., *et al.* (eds.): Handbook of Physiology. Sec. 2, Vol. 2. Baltimore, Williams & Wilkins, 1980, p. 237.

Cornish-Bowden, A.: Fundamentals of Enzyme Kinetics. Boston, Butterworths, 1979.

Dickerson, R. E.: Cytochrome C and the evolution of energy metabolism. *Sci. Am., 242*(3):136, 1980.

Esmann, V. (ed.): Regulatory Mechanisms of Carbohydrate Metabolism. New York, Pergamon Press, 1978.

Felig, P.: Disorders of carbohydrate metabolism. *In* Bondy, P. K., and Rosenberg, L. E. (eds.): Metabolic Control and Disease, 8th Ed. Philadelphia, W. B. Saunders Co., 1980, p. 276.

Friedmann, H. C. (ed.): Enzymes. Stroudsburg, Pa., Dowden, Hutchinson & Ross, 1980.

Hems, D. A., and Whitton, P. D.: Control of hepatic glycogenolysis. *Physiol. Rev., 60*:1, 1980.

Homsher, E., and Kean, C. J.: Skeletal muscle energetics and metabolism. *Annu. Rev. Physiol., 40*:93, 1978.

Jeffery, J. (ed.): Dehydrogenases Requiring Nicotinamide Coenzymes. Boston, Birkhauser, 1979.

Klachko, D. M., *et al.* (eds.): Hormones and Energy Metabolism. New York, Plenum Press, 1978.

Lieber, C. S.: The metabolism of alcohol. *Sci. Am., 234*(3):25, 1976.

Lund-Andersen, H.: Transport of glucose from blood to brain. *Physiol. Rev., 59*:305, 1979.

Martin, D. B.: Metabolism and energy mechanisms. *In* Frohlich, E. D. (ed.): Pathophysiology, 2nd Ed. Philadelphia, J. B. Lippincott Co., 1976, p. 365.

McCarty, R. E.: How cells make ATP. *Sci. Am., 238*(3):104, 1978.

Purich, D. L. (ed.): Enzyme Kinetics and Mechanism. New York, Academic Press, 1979.

Singer, T. P., and Ondarza, P. N. (eds.): Mechanisms of Oxidizing Enzymes, New York, Elsevier/North-Holland, 1978.

46

Lipid and Protein Metabolism

Several different chemical compounds in the food and in the body are classified as *lipids*. These include (1) *neutral fat*, known also as *triglycerides*, (2) *phospholipids*, (3) *cholesterol*, and (4) a few others of less importance. These substances have certain similar physical and chemical properties, especially the fact that they are miscible with each other. Chemically, the basic lipid moiety of both the triglycerides and the phospholipids is *fatty acids*, which are simply long chain hydrocarbon organic acids. Though cholesterol does not contain fatty acid, its sterol nucleus is synthesized from degradation products of fatty acid molecules, thus giving it many of the physical and chemical properties of other lipid substances.

The triglycerides are used in the body mainly to provide energy for the different metabolic processes; this function they share almost equally with the carbohydrates. However, some lipids, especially cholesterol, the phospholipids, and derivatives of these, are used throughout the body to provide other intracellular functions.

Basic Chemical Structure of Triglycerides (Neutral Fat). Since most of this chapter deals with utilization of triglycerides for energy, the following basic structure of the triglyceride molecule must be understood:

$$CH_3—(CH_2)_{16}—COO—CH_2$$
$$CH_3—(CH_2)_{16}—COO—CH$$
$$CH_3—(CH_2)_{16}—COO—CH_2$$
$$\text{Tri-stearin}$$

Note that three long chain fatty acid molecules are bound with one molecule of glycerol.

534

TRANSPORT OF LIPIDS IN THE BLOOD

TRANSPORT FROM THE GASTROINTESTINAL TRACT — THE "CHYLOMICRONS"

It will be recalled from Chapter 44 that essentially all the fats of the diet are absorbed into the lymph in the form of *chylomicrons*, having a size about 0.4 micron. The chylomicrons are then transported up the thoracic duct and emptied into the venous blood at the juncture of the jugular and subclavian veins.

Removal of the Chylomicrons from the Blood. The chylomicrons are removed from the plasma within an hour or so. Most are removed from the circulating blood as they pass through the capillaries of adipose tissue and the liver. The membranes of the fat cells contain large quantities of an enzyme called *lipoprotein lipase*. This enzyme hydrolyzes the triglycerides of the chylomicrons into fatty acids and glycerol. The fatty acids, being highly miscible with the membranes of the cells, immediately diffuse into the fat cells. Once within these cells, the fatty acids are resynthesized into triglycerides, new glycerol being supplied by the metabolic processes of the fat cells, as will be discussed later in the chapter.

TRANSPORT OF FATTY ACIDS IN COMBINATION WITH ALBUMIN — "FREE FATTY ACID"

When the fat that has been stored in the fat cells is to be used elsewhere in the body, usually for providing energy, it must first be transported to the other tissues. It is transported almost entirely in the form of *free fatty acid*. This is achieved by hydrolysis of the triglycerides once again into fatty acids and glycerol. Although the stimulus for initiating this hydrolysis is not completely understood, it is known that a cellular lipase called *hormone-sensitive triglyceride lipase* is activated by one of several different means, and this promotes rapid hydrolysis of the triglycerides.

On leaving the fat cells, the fatty acids ionize strongly in the plasma and immediately combine with albumin of the plasma proteins. The fatty acid bound with proteins in this manner is called *free fatty acid* or *nonesterified fatty acid* (or simply *FFA* or *NEFA*) to distinguish it from other fatty acids in the plasma that exist in the form of esters of glycerol, cholesterol, or other substances.

The concentration of free fatty acid in the plasma under resting conditions is about 15 mg. per 100 ml. of plasma, which is a total of only 0.45 gram of fatty acids in the entire circulatory system. Yet, strangely enough, even this small amount accounts for almost all of the transport of lipids from one part of the body to another, for the following reasons:

(1) Despite the minute amount of free fatty acid in the blood, its rate of "turnover" is extremely rapid, *half the plasma fatty acid being replaced by new fatty acid every two to three minutes.* One can calculate that at this rate over half of all the energy required by the body can be provided by the free fatty acid transported even without increasing the free fatty acid concentration. (2) All conditions that increase the rate of utilization of fat for cellular energy also increase the free fatty acid concentration in the blood; this sometimes increases as much as ten-fold. Especially does this occur in starvation and in diabetes when a person is not or cannot be using carbohydrates for energy.

THE LIPOPROTEINS

In the postabsorptive state — that is, when no chylomicrons are in the blood — over 95 per cent of all the lipids in the plasma (in terms of mass, but *not* in terms of rate of transport) are in the form of lipoproteins, which are small particles much smaller than chylomicrons but similar in composition, containing mixtures of *triglycerides, phospholipids, cholesterol,* and *protein*. The protein in the mixture averages about one fourth to one third of the total constituents, and lipids form the remainder. The total concentration of lipoproteins in the plasma averages about 700 mg. per 100 ml., and this can be broken down into the following average concentrations of the individual constituents:

	mg./100 ml. of plasma
Cholesterol	180
Phospholipids	160
Triglycerides	160
Lipoprotein protein	200

Types of Lipoproteins. Chylomicrons are sometimes classified as lipoproteins because they contain both lipids and protein. In addition to the chylomicrons, however, there are three other major classes of lipoprotein: (1) *very low density lipoproteins,* which contain high concentrations of triglycerides and moderate concentrations of both phospholipids and cholesterol; (2) *low density lipoproteins,* which contain relatively few triglycerides but a very high percentage of cholesterol; and (3) *high density lipoproteins,* which contain about 50 per cent protein with smaller concentrations of the lipids.

Formation of the Lipoproteins. The lipoproteins are formed almost entirely in the liver, which is in keeping with the fact that most plasma phospholipids, cholesterol, and triglycerides (except those in the chylomicrons) are synthesized in the liver.

Function of the Lipoproteins. The function of the lipoproteins in the plasma is poorly known, though they are known to be a means by which lipid substances can be transported throughout the body, mainly from the liver to other parts of the body. Especially, the fats synthesized from carbohydrates in the liver are transported to the adipose tissue in lipoproteins. And even more important is the transport of cholesterol and phospholipids by the lipoproteins, because these substances are not transported to any significant extent in any other form.

THE FAT DEPOSITS

ADIPOSE TISSUE

Large quantities of fat are frequently stored in two major tissues of the body, the adipose tissue and the liver. The adipose tissue is usually called the *fat deposits,* or simply the *fat depots*.

The major function of adipose tissue is storage of triglycerides until these are needed to provide energy elsewhere in the body. However, a subsidiary function is to provide heat insulation for the body, as will be discussed in Chapter 47.

The Fat Cells. The fat cells of adipose tissue are modified fibroblasts that are capable of storing almost pure triglycerides in quantities equal to 80 to 95 per cent of their volume.

Fat cells can also synthesize small quantities of fatty acids and triglycerides from carbohydrates, this function supplementing the synthesis of fat in the liver, as discussed later in the chapter.

Exchange of Fat Between the Adipose Tissue and the Blood — Tissue Lipases. Large quantities of lipases are present in adipose tissue. Some of these enzymes catalyze the deposition of triglycerides from the chylomicrons and other lipoproteins. Others, when activated by hormones, cause splitting of the triglycerides of the fat cells to release free fatty acids. Because of rapid exchanges of the fatty acids, the triglycerides in the fat cells are renewed approximately once every two to three weeks. This means that the fat stored in the tissues today is not the same fat that was stored last month, a fact that emphasizes the dynamic state of the storage fat.

THE LIVER LIPIDS

The principal functions of the liver in lipid metabolism are (1) to degrade fatty acids into small compounds that can be used for energy, (2) to synthesize triglycerides mainly from carbohydrates and, to a lesser extent, from proteins, and (3) to synthesize other lipids from fatty acids, especially cholesterol and phospholipids.

The liver cells, in addition to containing triglycerides, contain large quantities of phospholipids and cholesterol, which are continually synthesized by the liver. Also, the liver cells are much more capable than other tissues of desaturating fatty acids, so that the liver triglycerides normally are much more unsaturated than the triglycerides of the adipose tissue. This capability of the liver to desaturate fatty acids seems to be functionally important to all the tissues of the body, because many of the structural members of all cells contain reasonable quantities of desaturated fats, and their principal source is presumably the liver. This desaturation is accomplished by a dehydrogenase in the liver cells.

USE OF TRIGLYCERIDES FOR ENERGY, AND FORMATION OF ADENOSINE TRIPHOSPHATE (ATP)

Approximately 40 to 45 per cent of the calories in the normal American diet are derived from fats, which is about equal to the calories derived from carbohydrates. Therefore, the use of fats by the body for energy is just as important as the use of carbohydrates. In addition, much of the carbohydrates ingested with each meal is converted into triglycerides, then stored, and later utilized as triglycerides for energy. Therefore, more than half of all the energy used by the cells is supplied by fatty acids derived from triglycerides or indirectly from carbohydrates.

Entry of Fatty Acids into the Mitochondria. The degradation and oxidation of fatty acids occur only in the mitochondria. Therefore, the first step in the utilization of the fatty acids is their transport into the mitochondria. This is an enzyme-catalyzed process that employs *carnitine* as a carrier substance. Once inside the mitochondria, the fatty acid splits away from the carnitine and is then oxidized.

Degradation of Fatty Acid to Acetyl Coenzyme-A by "Beta Oxidation." The fatty acid molecule is degraded in the mitochondria by progressive release of 2-carbon segments in the form of acetyl coenzyme-A (acetyl Co-A). This process, which is illustrated in Figure 46–1, is called the *beta oxidation* process for degradation of

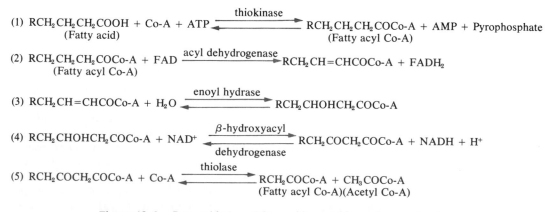

Figure 46–1. Beta oxidation of fatty acids to yield acetyl coenzyme-A.

fatty acids. Each time the reactions of this schema occur a new acetyl Co-A molecule is formed, and the process is repeated again and again until the entire fatty acid molecule is split into acetyl Co-A. For instance, from each molecule of stearic acid, nine molecules of acetyl Co-A are formed.

Oxidation of Acetyl Co-A. The acetyl Co-A molecules formed by beta oxidation of fatty acids enter into the citric acid cycle, as explained in the preceding chapter, and are degraded into carbon dioxide and hydrogen atoms. The hydrogen is subsequently oxidized by the oxidative enzymes of the cells to form ATP.

ATP Formed by Oxidation of Fatty Acid. In Figure 46–1 note also that 4 hydrogen atoms are released each time a molecule of acetyl Co-A is formed from the fatty acid chain. Additional hydrogen is released in the citric acid cycle. The oxidation of all these hydrogen atoms gives rise to the formation of 139 molecules of ATP *for each stearic acid molecule oxidized.* Another 7 molecules of ATP are formed in other ways, making a total of 146 molecules of ATP.

FORMATION OF ACETOACETIC ACID IN THE LIVER AND ITS TRANSPORT IN THE BLOOD

A large share of the degradation of fatty acids into acetyl Co-A occurs in the liver. However, the liver uses only a small proportion of the acetyl Co-A for its own intrinsic metabolic processes. Instead, pairs of acetyl Co-A condense to form molecules of acetoacetic acid, as follows:

$$2CH_3COCo\text{-}A + H_2O \xrightleftharpoons[\text{other cells}]{\text{liver cells}}$$
Acetyl Co-A

$$CH_3COCH_2COOH + 2HCo\text{-}A$$
Acetoacetic acid

Then, a large part of the acetoacetic acid is converted into *β-hydroxybutyric acid* and minute quantities to *acetone* in accord with the following reactions:

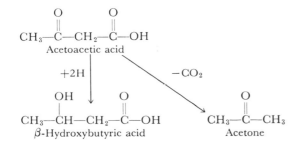

The acetoacetic acid and β-hydroxybutyric acid then freely diffuse through the liver cell membranes and are transported by the blood to the peripheral tissues. Here they again diffuse into the cells, where reverse reactions occur and acetyl Co-A molecules are formed. These in turn enter the citric acid cycle and are oxidized for energy, as explained above.

SYNTHESIS OF TRIGLYCERIDES FROM CARBOHYDRATES

Whenever a greater quantity of carbohydrates enters the body than can be used immediately for energy or stored in the form of glycogen, the excess is rapidly converted into triglycerides and then stored in this form in the adipose tissue. Most triglyceride synthesis occurs in the liver, but smaller quantities are also synthesized in the adipose tissue. The triglycerides that are formed in the liver are then mainly transported by the lipoproteins to the adipose tissue, where they too are stored until needed for energy.

Conversion of Acetyl Co-A into Fatty Acids. The first step in the synthesis of triglycerides is conversion of carbohydrates into acetyl Co-A. It will be recalled from the preceding chapter that this occurs during the normal degradation of glucose by the glycolytic system. It will also be remembered from earlier in this chapter that fatty acids are actually large polymers of the acetyl portion of acetyl Co-A. Therefore, without going into the details of the chemical reactions, it is easy to understand how acetyl Co-A can be converted into fatty acids.

Combination of Fatty Acids with α-Glycerophosphate to Form Triglycerides. Once the synthesized fatty acid chains have grown to contain 14 to 18 carbon atoms, they are then bound to glycerol to form triglycerides.

The glycerol portion of the triglyceride is furnished by α-glycerophosphate, which is a product derived from the glycolytic schema of glucose degradation. The mechanism of this was discussed in Chapter 45.

The real importance of this mechanism for formation of triglycerides is that the whole process is controlled to a great extent by the concentration of α-glycerophosphate. When carbohydrates are available to form large quantities of α-glycerophosphate, the equilibrium shifts to promote formation and storage of triglycerides.

Importance of Fat Synthesis and Storage. Fat synthesis from carbohydrates is especially important for two reasons: (1) The ability of the different cells of the body to store carbohydrates in the form of glycogen is generally slight; only a few hundred grams of glycogen are stored in the liver, the skeletal muscles, and all other tissues of the body put together. Therefore, fat synthesis provides a means by which the energy of excess ingested carbohydrates (and proteins, too) can be stored for later use. Indeed, the average person has about 200 times as much energy stored in the form of fat as stored in the form of carbohydrate. (2) Each gram of fat contains approximately 2¼ times as many calories of energy as each gram of glycogen. Therefore, for a given weight gain, a person can store far more energy in the form of fat than in the form of carbohydrate, which is important when an animal must be highly motile to survive.

SYNTHESIS OF TRIGLYCERIDES FROM PROTEINS

Many amino acids can be converted into acetyl Co-A, as will be discussed later in the chapter. Obviously, this too can be converted into triglycerides. Therefore, when a person has more proteins in his diet than his tissues can use as proteins or directly for energy, a large share of the excess is stored as fat.

FAT- SPARING EFFECT OF CARBOHYDRATES; CONTROL OF FAT SYNTHESIS FROM CARBOHYDRATES

When adequate quantities of carbohydrates are available in the body, the utilization of triglycerides for energy is greatly depressed. In place of fat utilization, the carbohydrates are utilized preferentially. There are several different reasons for this "fat-sparing" effect of carbohydrates, but probably the most important is the following: The fats in adipose tissue cells are present in two different forms, triglycerides and small quantities of free fatty acids. These are in constant equilibrium with each other. When excess quantities of α-glycerophosphate are present, the equilibrium between free fatty acids and triglycerides shifts toward the triglycerides as explained earlier in the chapter; and as a result, only minute quantities of fatty acids are then available to be utilized for energy. Since α-glycerophosphate is an important product of glucose metabolism, the availability of large amounts of glucose automatically inhibits the use of fatty acids for energy.

Indeed, when carbohydrates are available in excess, fats are synthesized instead of being degraded. This effect is caused partially by the large quantities of acetyl Co-A formed from the carbohydrates and partially by the low concentration of free fatty acids in the adipose tissue, thus creating conditions appropriate for conversion of acetyl Co-A into fatty acids.

HORMONAL REGULATION OF FAT UTILIZATON

At least seven of the hormones secreted by the endocrine glands have marked effects on fat utilization.

Probably the most dramatic increase that occurs in fat utilization is that observed during heavy exercise. This results almost entirely from rapid release of *epinephrine* and *norepinephrine* by the adrenal medullae during exercise, as a result of sympathetic stimulation. These two hormones directly activate *hormone-sensitive triglyceride lipase*, which is present in abundance in the fat cells, and this causes very rapid breakdown of triglycerides and mobilization of fatty acids. Sometimes the free fatty acid concentration in the blood rises as much as ten-fold. Other types of stress that activate the sympathetic nervous system will increase fatty acid mobilization and utilization in a similar manner.

Stress also causes large quantities of *corticotropin* to be released by the anterior pituitary gland, and this, in turn, causes the adrenal cortex to secrete excessive quantities of *glucocorticoids*. Both the corticotropin and glucocorticoids activate either the same hormone-sensitive triglyceride lipase as that activated by epinephrine and norepinephrine, or a similar lipase. Therefore, this is still another mechanism for increasing the release of fatty acids from fat tissue.

Growth hormone has an effect similar to but less than that of corticotropin and glucocorticoids in activating the hormone-sensitive lipase. Therefore, growth hormones can also have a mild ketogenic effect.

Insulin lack also activates hormone-sensitive lipase and therefore causes rapid mobilization of fatty acids. When carbohydrates are not available in the diet, insulin secretion diminishes, and this in turn promotes fatty acid metabolism.

Finally, *thyroid hormone* causes rapid mobilization of fat, a process that is believed to result indirectly from an increased rate of energy metabolism in all cells of the body under the influence of this hormone.

The effects of the different hormones on metabolism are discussed further in the chapters dealing with each of them.

PHOSPHOLIPIDS AND CHOLESTEROL

PHOSPHOLIPIDS

The three major types of body phospholipids are the *lecithins,* the *cephalins,* and the *sphingomyelins.* A lecithin is shown in Figure 46–2.

Phospholipids always contain one or more fatty

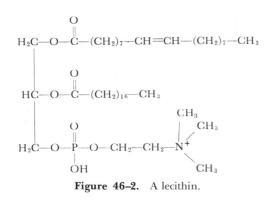

Figure 46–2. A lecithin.

acid molecules and one phosphoric acid radical, and they usually contain a nitrogenous base. Though the chemical structures of phospholipids are somewhat variant, their physical properties are similar, for they are all lipid soluble, are transported together in lipoproteins in the blood, and seem to be utilized similarly throughout the body for various structural purposes.

Phospholipids are formed in essentially all cells of the body, though certain cells have a special ability to form them. Probably 90 per cent or more of the phospholipids enter the blood in the lipoproteins that are formed in the liver cells.

CHOLESTEROL

Cholesterol, the formula of which is illustrated, is present in the diet of all persons, and it can be absorbed from the gastrointestinal tract into the intestinal lymph. It is highly fat soluble but only slightly soluble in water, and it is capable of forming esters with fatty acids. Indeed, approximately 70 per cent of the cholesterol of the plasma is in the form of cholesterol esters.

Besides the cholesterol absorbed each day from the gastrointestinal tract, which is called *exogenous cholesterol*, a large quantity, called *endogenous cholesterol*, is formed in the cells of the body. Essentially all the endogenous cholesterol that circulates in the lipoproteins of the plasma is formed by the liver, but all the other cells of the body form at least some cholesterol.

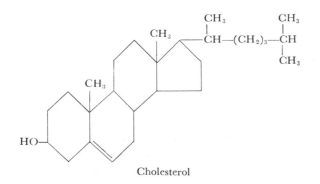

Cholesterol

STRUCTURAL FUNCTIONS OF PHOSPHOLIPIDS AND CHOLESTEROL

In Chapter 2 it was pointed out that large quantities of phospholipids and cholesterol are present in the cell membrane and in the membranes of the internal organelles of all cells.

For membranes to be formed, substances that are not soluble in water must be available, and in general, the only substances in the body that are not soluble in water (besides the inorganic substances of bone) are the lipids and some proteins. Thus, the physical integrity of cells throughout the body is based mainly on phospholipids, triglycerides, cholesterol, and certain insoluble proteins. Some phospholipids are somewhat water soluble as well as lipid soluble, which gives them the important property of helping to decrease the interfacial tension between the membranes and the surrounding fluids.

Another fact indicating that phospholipids and cholesterol are mainly concerned with the formation of structural elements of the cells is the slow turnover rate of these substances. For instance, phospholipids formed in the brain remain there for many months or perhaps even for years.

ATHEROSCLEROSIS

Atherosclerosis is principally a disease of the large arteries in which lipid deposits called *atheromatous plaques* appear in the subintimal layer of the arteries. These plaques contain an especially large amount of cholesterol and often are simply called cholesterol deposits. They are also associated with degenerative changes in the arterial wall. In a later stage of the disease, fibroblasts infiltrate the degenerative areas and cause progressive sclerosis of the arteries. In addition, calcium often precipitates with the lipids to develop *calcified plaques*. When these two processes occur, the arteries become extremely hard, and the disease is then called *arteriosclerosis*, or simply "hardening of the arteries."

Obviously, arteriosclerotic arteries lose most of their distensibility, and because of the degenerative areas, they are easily ruptured. Also, the atheromatous plaques often protrude through the intima into the flowing blood, and the roughness of their surfaces causes blood clots to develop, with resultant thrombus or embolus formation (see Chapter 7). Almost half of all human beings die of arteriosclerosis; approximately two thirds of the deaths are caused by thrombosis of one or more coronary arteries and the remaining one third by thrombosis or hemorrhage of vessels in other organs of the body — especially the brain, kidneys, liver, gastrointestinal tract, limbs, and so forth.

Effect of Age, Sex, and Heredity on Atherosclerosis.   Atherosclerosis is mainly a disease of old age, but small atheromatous plaques can almost always be found in the arteries of young adults. Therefore, the full-blown disease is a culmination of a lifetime of lipid deposition rather than deposition over a few years.

Far more men die of atherosclerotic heart disease than do women. This is especially true of men younger than 50. For this reason, it is possible that the male sex hormone accelerates the development of atherosclerosis, or that the female sex hormone *protects* a person from atherosclerosis.

Atherosclerosis and atherosclerotic heart disease are highly hereditary in some families. In some instances, this is related to an inherited hypercholesterolemia, the excess cholesterol occurring almost entirely in the *low density lipoproteins*. The liver is unable to remove cholesterol from these lipoproteins. Therefore, much of the excess cholesterol is deposited in the arterial wall. In other persons with hereditary atherosclerosis, the blood cholesterol level is completely normal. Inheritance of the tendency to atherosclerosis is sometimes caused by dominant genes, which means that once this dominant trait enters a family a high incidence of the disease occurs among the offspring.

Relationship of Dietary Fat to Atherosclerosis in the Human Being.   A high fat diet, especially one containing cholesterol and saturated fats, greatly increases one's chances of developing atherosclerosis. Therefore, decreasing the fat can help greatly in protecting against atherosclerosis, and some experiments indicate that this can benefit even patients who have already had coronary heart attacks. Also, life insurance statistics show that the rate of mortality — mainly from coronary disease — of normal weight middle and older age persons is about half the mortality rate of overweight subjects of the same age.

THE BODY PROTEINS

About three quarters of the body solids are proteins. These include *structural proteins, enzymes, genes, proteins that transport oxygen, proteins of the muscle that cause contraction*, and many other types that perform specific functions both intracellularly and extracellularly throughout the body.

The basic chemical properties of proteins that explain their diverse functions are so extensive that they are a major portion of the entire discipline of biochemistry. For this reason, the present discussion is confined to the general aspects of protein metabolism.

THE AMINO ACIDS

The principal constituents of proteins are amino acids, 20 of which are present in the body in significant quantities. Figure 46–3 illustrates the chemical formulas of these 20 amino acids, showing that they all have two features in common: Each amino acid has an acidic group ($-COOH$) and a nitrogen radical that lies in close association with the acidic radical, usually represented by the amino group ($-NH_2$).

Peptide Linkages and Peptide Chains.   In proteins, the amino acids are aggregated into long chains by means of so-called *peptide linkages*, one of which is illustrated by the following reaction:

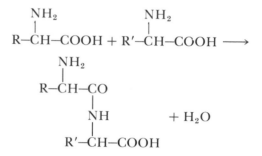

Note that in this reaction the amino radical of one amino acid combines with the carboxyl radical of the other amino acid. A hydrogen atom is released from the amino radical, a hydroxyl radical is released from the carboxyl radical, and these two combine to form a molecule of water. Note that after the peptide linkage has been formed, an amino radical and a carboxyl radical are still in the new molecule, both of which are capable of combining with additional amino acids to form a *peptide chain*. Some complicated protein molecules have as many as a hundred thousand amino acids combined together principally by peptide linkages, and even the smallest protein usually has more than 20 amino acids combined together by peptide linkages.

FIBROUS PROTEINS

Many of the highly complex proteins are fibrillar and are called fibrous proteins. In these the peptide chains are elongated, and many separate chains are held together in parallel bundles by cross-linkages. Major types of fibrous proteins are (1) *collagens*, which are the basic structural proteins of connective tissue, tendons, cartilage, and bone; (2) *elastins,* which are the elastic fibers of tendons, arteries, and connective tissue; (3) *keratins*, which are the structural proteins of hair

AMINO ACIDS

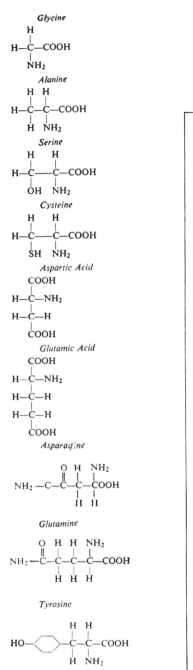

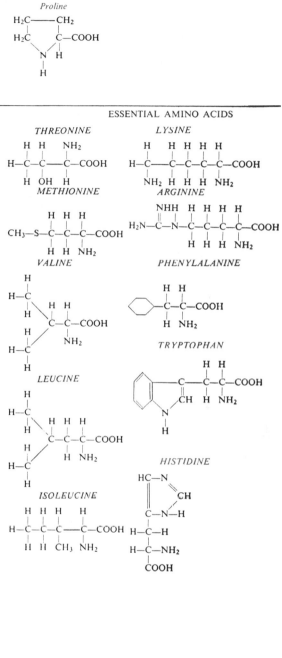

Figure 46–3. The amino acids, showing the 10 essential amino acids, which cannot be synthesized at all or in sufficient quantity in the body.

and nails; and (4) *actin* and *myosin*, the contractile proteins of muscle.

TRANSPORT AND STORAGE OF AMINO ACIDS

THE BLOOD AMINO ACIDS

The normal concentration of amino acids in the blood is between 35 and 65 mg. per cent. This is an average of about 2 mg. per cent for each of the 20 amino acids, though some are present in far greater concentrations than others. Since the amino acids are relatively strong acids, they exist in the blood principally in the ionized state and account for 2 to 3 milliequivalents of the negative ions in the blood.

Fate of Amino Acids Absorbed from the Gastrointestinal Tract. It will be recalled from Chapter 44 that the end-products of protein digestion in the gastrointestinal tract are almost entirely amino acids and that polypeptide or protein molecules are only rarely absorbed into the blood. Immediately after a meal, the amino acid concentration in the blood rises, but the rise is usually only a few milligrams per cent because after entering the blood, the excess amino acids are absorbed within five to ten minutes by cells throughout the entire body. Therefore, almost never do large concentrations of amino acids accumulate in the blood. Nevertheless, the turn-over rate of the amino acids is so rapid that many grams of proteins in the form of amino acids can be carried from one part of the body to another each hour.

Transport of Amino Acids into the Cells. The molecules of essentially all the amino acids are much too large to diffuse through the pores of the cell membranes. Instead, the amino acids can be transported through the membrane only by active transport or facilitated diffusion, utilizing carrier mechanisms. The nature of the carrier mechanisms is still poorly understood, but some of the theories are discussed in Chapter 4.

STORAGE OF AMINO ACIDS AS PROTEINS IN THE CELLS

Almost immediately after entry into the cells, amino acids are conjugated under the influence of intracellular enzymes into cellular proteins, so that the concentration of amino acids inside the cells probably always remains low. Thus, so far as is known, storage of large quantities of amino acids as such probably does not occur in the cells; instead, they are mainly stored in the form of

actual proteins. Yet many intracellular proteins can be rapidly decomposed again into amino acids under the influence of intracellular lysosomal digestive enzymes, and these amino acids in turn can be transported back out of the cell into the blood. The proteins that can be thus decomposed include many cellular enzymes as well as some other functioning proteins. However, many of the structural proteins such as collagen and muscle contractile proteins do not participate significantly in this reversible storage of amino acids.

Some tissues of the body participate in the storage of amino acids to a greater extent than others. Thus, the liver, which is a large organ and also has special systems for processing amino acids, stores large quantities of labile proteins; this is also true of the kidney and the intestinal mucosa.

Release of Amino Acids from the Cells and Regulation of Plasma Amino Acid Concentration. Whenever the plasma amino acid concentration falls below its normal level, amino acids are transported out of the cells to replenish the supply in the plasma. Simultaneously, intracellular proteins are degraded back into amino acids.

The plasma concentration of each type of amino acid is maintained at a reasonably constant value. Later it will be noted that the various hormones secreted by the endocrine glands are able to alter the balance between tissue proteins and circulating amino acids; growth hormone and insulin increase the formation of tissue proteins, while the adrenocortical glucocorticoid hormones increase the concentration of circulating amino acids.

THE PLASMA PROTEINS

The three major types of protein in the plasma are *albumin*, *globulin*, and *fibrinogen*. The principal function of albumin is to provide *colloid osmotic pressure*, which in turn prevents plasma loss from the capillaries, as discussed in Chapter 22. The globulins perform a number of enzymatic functions in the plasma itself, but more important than this, they are mainly responsible for both the natural and acquired immunity that a person has against invading organisms, a subject discussed in Chapter 6. The fibrinogen polymerizes into long, branching fibrin threads during blood coagulation, thereby forming blood clots that help to repair leaks in the circulatory system, which was discussed in Chapter 7.

Formation of the Plasma Proteins. Essentially all the albumin and fibrinogen in the plasma proteins, as well as about one half of the globulins, are formed in the liver. The remainder of the globulins are formed in the lymphoid tissues and other cells of the reticuloendothelial system. These are mainly the gamma globulins that constitute the antibodies.

The rate of plasma protein formation by the liver can be extremely high, as great as 2 grams per hour or as much as 50 grams per day. Certain disease conditions often cause rapid loss of plasma proteins; severe burns that denude large surface areas cause loss of many liters of plasma through the denuded areas each day. The rapid production of plasma proteins by the liver is obviously valuable in preventing death in such states. Furthermore, occasionally, a person with severe renal disease loses as much as 20 grams of plasma protein in the urine each day for years. In some of these patients the plasma protein concentration may remain almost normal throughout the entire illness.

Use of Plasma Proteins by the Tissue. When the tissues become depleted of proteins, the plasma proteins can act as a source for rapid replacement of the tissue proteins. Indeed, whole plasma proteins can be imbibed in toto by the reticuloendothelial cells; then, once in the cells, these are split into amino acids that are transported back into the blood and utilized throughout the body to build cellular proteins. In this way, therefore, the plasma proteins function as a labile protein storage medium and represent a rapidly available source of amino acids whenever a particular tissue requires them.

CHEMISTRY OF PROTEIN SYNTHESIS

Proteins are synthesized in all cells of the body, and the functional characteristics of each cell are dependent upon the types of protein that it can form. Basically, the genes of the cells control the protein types and thereby control the functions of the cell. This regulation of cellular function by the genes was discussed in detail in Chapter 3. Chemically, two basic processes must be accomplished for the synthesis of proteins; these are (1) synthesis of the amino acids and (2) appropriate conjugation of the amino acids to form the respective types of whole proteins in each individual cell.

Essential and Nonessential Amino Acids. Ten of the 20 amino acids normally present in animal proteins can be synthesized in the cells, while the other 10 either cannot be synthesized at all or are synthesized in quantities too small to supply the body's needs. The first group of amino acids is called *nonessential*, while the second group is called *essential amino acids*. The essential amino acids obviously must be present in the diet if protein formation is to take place in the body. Use of the word "essential" does not mean that the other 10 amino acids are not equally essential in the formation of the proteins, but only that these others are not essential in the diet.

Synthesis of the nonessential amino acids depends on the formation first of appropriate α-keto acids, which are the precursors of the respective amino acids. For instance, *pyruvic acid*, which is formed in large quantities during the glycolytic breakdown of glucose, is the keto acid precursor of the amino acid *alanine*. Then, by the simple process of *transamination*, an amino radical is transferred to the α-keto acid while the keto oxygen is transferred to the donor of the amino radical.

Formation of Proteins from Amino Acids. Once the appropriate amino acids are present in a cell, whole proteins are synthesized rapidly. However, each peptide linkage requires from 500 to 4000 calories of energy, and this must be supplied from ATP and GTP guanosine triphosphate) in the cell. Protein formation proceeds through two steps: (1) "activation" of each amino acid, during which the amino acid is "energized" by energy derived from ATP and GTP, and (2) alignment of the amino acids into the peptide chains, a function that is under control of the genetic system of each individual cell. Both of these processes were discussed in Chapter 3. Indeed, the formation of cellular proteins is the basis of life itself and is so important that the reader would do well to review Chapter 3.

USE OF PROTEINS FOR ENERGY

There is an upper limit to the amount of protein that can accumulate in each particular type of cell. Once the cells are filled to their limits, any additional amino acids in the body fluids are degraded and used for energy or stored as fat. This degradation occurs almost entirely in the liver, and it begins with the process known as deamination.

Deamination. Deamination means removal of the amino groups from the amino acids. This can occur by several different means, two of which are especially important: (1) transamination, which means transfer of the amino group to some acceptor substance as explained above in relation to the synthesis of amino acids, and (2) oxidative deamination.

The greatest amount of deamination occurs by the following transamination schema:

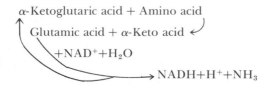

Note from this schema that the amino group from the amino acid is transferred to α-ketoglutaric acid, which then becomes glutamic acid. The glutamic acid can then transfer the amino group to still other substances or can release it in the form of ammonia. In the process of losing the amino group, the glutamic acid once again becomes α-ketoglutaric acid, so that the cycle can be repeated again and again.

Urea Formation by the Liver. The ammonia released during deamination is removed from the blood almost entirely by conversion into urea, two molecules of ammonia and one molecule of carbon dioxide combining in accordance with the following net reaction:

$$2NH_3 + CO_2 \rightarrow H_2N-\overset{\underset{\|}{O}}{C}-NH_2 + H_2O$$

Essentially all urea formed in the human body is synthesized in the liver. In the absence of the liver or in serious liver disease, ammonia accumulates in the blood. This in turn is extremely toxic, especially to the brain, often leading to a state called *hepatic coma.*

The stages in the formation of urea are essentially the following:

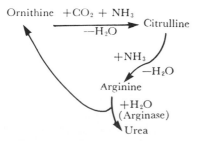

The reaction begins with the amino acid derivative *ornithine*, which combines with one molecule of carbon dioxide and one molecule of ammonia to form a second substance, *citrulline*. This in turn combines with still another molecule of ammonia to form *arginine*, which then splits into *ornithine* and *urea*. The urea diffuses from the liver cells into the body fluids and is excreted by the kid-

neys, while the ornithine is reused in the cycle again and again.

Oxidation of Deaminated Amino Acids. Once amino acids have been deaminated, the resulting keto acid products can in most instances be oxidized to release energy for metabolic purposes. This usually involves two processes: (1) The keto acid is changed into an appropriate chemical substance that can enter the citric acid cycle and (2) this substance is then degraded by this cycle in the same manner that acetyl Co-A derived from carbohydrate and lipid metabolism is degraded.

In general, the amount of adenosine triphosphate formed for each gram of protein that is oxidized is slightly less than that formed for each gram of glucose oxidized.

Gluconeogenesis and Ketogenesis. Certain deaminated amino acids are similar to the breakdown products that result from glucose and fatty acid metabolism. For instance, deaminated alanine is pyruvic acid. Obviously, it can be converted into glucose or glycogen; or it can be converted into acetyl Co-A, which can then be polymerized into fatty acids. Also, two molecules of acetyl Co-A can condense to form acetoacetic acid, which is one of the ketone bodies, as explained earlier in the chapter.

The conversion of amino acids into glucose or glycogen is called *gluconeogenesis*, and the conversion of amino acids into keto acids or fatty acids is called *ketogenesis*. Eighteen out of 20 of the deaminated amino acids have chemical structures that allow them to be converted into glucose, and 19 can be converted into fats — 5 directly and the other 14 by becoming carbohydrate first and then becoming fat.

OBLIGATORY DEGRADATION OF PROTEINS

When a person eats no proteins, a certain proportion of his own body proteins continues to be degraded into amino acids, then deaminated and oxidized. This involves 20 to 30 grams of protein each day, which is called the *obligatory loss* of proteins. Therefore, to prevent a net loss of protein from the body, one must ingest at least 20 to 30 grams of protein each day, and to be on the safe side as much as 75 grams is usually recommended.

Effect of Starvation on Protein Degradation. Except for the excess protein in the diet or the 20 to 30 grams of obligatory protein degradation each day, the body uses almost entirely carbohydrates or fats for energy as long as they are available. However, after several weeks of starvation, when the quantity of stored fats begins to run out, the amino acids of the blood begin to be rapidly deaminated and oxidized for energy. From this point on, the proteins of the tissues degrade rapidly — as much as 100 grams daily — and the cellular functions deteriorate precipitously.

Because carbohydrate and fat utilization for energy occurs in preference to protein utilization, carbohydrates and fats are called *protein sparers.*

REFERENCES

Lipid Metabolism

Benditt, E. P.: The origin of atherosclerosis. *Sci. Am., 236*(2):74, 1977.

Brady, G. A., and York, D. A.: Hypothalamic and genetic obesity in experimental animals: An autonomic and endocrine hypothesis. *Physiol. Rev., 59*:719, 1979.

Coleman, J. E.: Metabolic interrelationships between carbohydrates, lipids and proteins. *In* Bondy, P. K., and Rosenberg, L. E. (eds.): Metabolic Control and Disease, 8th Ed. Philadelphia, W. B. Saunders Co., 1980, p. 161.

Dils, R., and Knudsen, J. (eds.): Regulation of Fatty Acid and Glycerolipid Metabolism. New York, Pergamon Press, 1978.

Goldfarb, S.: Regulation of hepatic cholesterogenesis. *In* Javitt, N. B. (ed.): International Review of Physiology: Liver and Biliary Tract Physiology I. Vol. 21. Baltimore, University Park Press, 1980, p. 317

Gotto, A. M., Jr., *et al.* (eds.): High Density Lipoproteins and Atherosclerosis. New York, Elsevier/North-Holland, 1978.

Havel, R. J., *et al.*: Lipoproteins and lipid transport. *In* Bondy, P. K., and Rosenberg, L. E. (eds.): Metabolic Control and Disease, 8th Ed. Philadelphia, W. B. Saunders Co., 1980, p. 393.

Jackson, R. L., *et al.*: Lipoprotein structure and metabolism. *Physiol. Rev., 56*:259, 1976.

Levy, R. I. (ed.): Nutrition, Lipids, and Coronary Heart Disease. New York, Raven Press, 1979.

Masoro, E. J.: Lipids and lipid metabolism. *Annu. Rev. Physiol., 39*:301, 1977.

Miller, G. J.: High density lipoproteins and atherosclerosis. *Annu. Rev. Med., 31*:97, 1980.

Robinson, A. M., and Williamson, D. H.: Physiological roles of ketone bodies as substrates and signals in mammalian tissues. *Physiol. Rev., 60*:143, 1980.

Rosell, S., and Belfrage, E.: Blood circulation in adipose tissue. *Physiol. Rev., 59*:1078, 1979.

Ross, R., and Kariya, B.: Morphogenesis of vascular smooth muscle in atherosclerosis and cell structure. *In* Bohr, D. R., *et al.* (eds.): Handbook of Physiology, Sec. 2, Vol. 2. Baltimore, Williams & Wilkins, 1980, p. 69.

Salans, L. B.: Obesity and the adipose cell. *In* Bondy, P. K.,

and Rosenberg, L. E. (eds.): Metabolic Control and Disease, 8th Ed. Philadelphia, W. B. Saunders Co., 1980, p. 495.

Schonfeld, G.: Hormonal control of lipoprotein metabolism. *In* DeGroot, L. J., *et al.* (eds.): Endocrinology. Vol. 3. New York, Grune & Stratton, 1979, p. 1855.

Protein Metabolism

Arnstein, H. R. V. (ed.): Amino Acid and Protein Biosynthesis II. Baltimore, University Park Press, 1978.

Atkinson, D. E., and Fox, C. F. (eds.): Modulation of Protein Function. New York, Academic Press, 1979.

Bender, D. A.: Amino Acid Metabolism. New York, John Wiley & Sons, 1978.

Bremer, H. J., *et al.*: Clinical Chemistry and Diagnosis of Amino Acid Disturbances. Baltimore, Urban & Schwarzenberg, 1979.

Coleman, J. E.: Metabolic interrelationships between carbohydrates, lipids and proteins. *In* Bondy, P. K., and Rosenberg, L. E. (eds.): Metabolic Control and Disease, 8th Ed. Philadelphia, W. B. Saunders Co., 1980, p. 161.

Crim, M. C., and Munro, H. N.: Protein-energy malnutrition and endocrine function. *In* DeGroot, L. J., *et al.*(eds.): Endocrinology. Vol. 3. New York, Grune & Stratton, 1979, p. 1987.

Friedmann, H. C. (ed.): Enzymes. Stroudsburg, Pa., Dowden, Hutchinson & Ross, 1980.

Gross, E., and Meienhofer, J. (eds.): The Peptides. New York, Academic Press, 1979.

Harper, A. E., *et al.*: Effects of ingestion of disproportionate amounts of amino acids. *Physiol. Rev., 50*:428, 1970.

Kostyo, J. L., and Nutting, D. F.: Growth hormone and protein metabolism. *In* Greep, R. O., and Astwood, E. B. (eds.): Handbook of Physiology. Sec. 7, Vol. 4. Baltimore, Williams & Wilkins, 1974, p. 187.

Morgan, H. E., *et al.*: Protein metabolism of the heart. *In* Berne, R. M., *et al.* (eds.): Handbook of Physiology. Sec. 2, Vol. 1. Baltimore, Williams & Wilkins, 1979, p. 845.

Rosenberg, L. E., and Scriver, C. R.: Disorders of amino acid metabolism. *In* Bondy, P. K., and Rosenberg, L. E. (eds.): Metabolic Control and Disease, 8th Ed. Philadelphia, W. B. Saunders Co., 1980, p. 583.

Ross, R., and Bornstein, P.: Elastic fibers in the body. *Sci. Am., 224*:44, 1971.

Rothschild, M. A.: Albumin synthesis. *In* Javitt, N. B. (ed.): International Review of Physiology: Liver and Biliary Tract Physiology I. Vol. 21. Baltimore, University Park Press, 1980, p. 249.

Rothschild, M. A., *et al.*: Effect of albumin concentration on albumin synthesis in the perfused liver. *Am. J. Physiol., 216*:1127, 1969.

Waterlow, J. C., *et al.*: Protein Turnover in Mammalian Tissues and in the Whole Body. New York, Elsevier/North-Holland, 1978.

Wynn, C. H.: The Structure and Function of Enzymes, 2nd Ed. Baltimore, University Park Press, 1979.

Zak, R., *et al.*: Assessment of protein turnover by use of radioisotopic tracers. *Physiol. Rev., 59*:407, 1979.

47

Energetics; Metabolic Rate; and Regulation of Body Temperature

IMPORTANCE OF ADENOSINE TRIPHOSPHATE (ATP) IN METABOLISM

In the last few chapters it has been pointed out that carbohydrates, fats, and proteins can all be used by the cells to synthesize large quantities of ATP, and that in turn the ATP can be used as an energy source for many other cellular functions. The attribute of ATP that makes it highly valuable as a means of energy currency is the large quantity of free energy (about 8000 calories per mole under physiologic conditions) vested in each of its two high energy phosphate bonds. The amount of energy in each bond, when liberated by decomposition of one molecule of ATP, is enough to cause almost any step of any chemical reaction in the body to take place if appropriate transfer of the energy is achieved. Some chemical reactions that require ATP energy use only a few hundred of the available 8000 calories, and the remainder of this energy is then lost in the form of heat. Yet even this inefficiency in the utilization of energy is better than lack of the ability to energize the necessary chemical reactions at all.

Throughout this book we have listed many functions of ATP. Therefore, here we will list again only its principal functions, which are

1. To energize the synthesis of important cellular components.
2. To energize muscle contraction.
3. To energize active transport across membranes for (a) absorption from the intestinal tract, (b) absorption from the renal tubules, (c) formation of glandular secretions, and (d) establishing ionic concentration gradients in nerves which in turn provides the energy required for nerve impulse transmission.

CREATINE PHOSPHATE AS A STORAGE DEPOT FOR ENERGY

Despite the paramount importance of ATP as a coupling agent for energy transfer, this substance is not the most abundant store of high energy phosphate bonds in the cells. On the contrary, *creatine phosphate*, which also contains high energy phosphate bonds, is several times as abundant, at least in muscle. The high energy bond of creatine phosphate contains about 9500 calories per mole under conditions in the body (38° C and low concentrations of the reactants). This is not greatly different from the 8000 calories per mole in each of the two high energy phosphate bonds of ATP. The formula for creatine phosphate is the following:

$$HOOC-CH_2-\underset{\underset{H}{|}}{N}-\underset{\overset{\parallel}{NH}}{C}-\underset{\underset{H}{|}}{N} \sim \underset{\underset{\overset{|}{H}}{\overset{\parallel}{O}}}{\overset{\overset{CH_3}{|}}{P}}-OH$$

Creatine phosphate, unlike ATP, cannot act as a coupling agent for transfer of energy between the foods and the functional cellular systems. But it can transfer energy interchangeably with ATP. When extra amounts of ATP are available in the cell, much of its energy is utilized to synthesize creatine phosphate, thus building up this storehouse of energy. Then when the ATP begins to be used up, the energy in the creatine phosphate is transferred rapidly back to ATP and from this to the functional systems of the cells.

The higher energy level of the high energy phosphate bond in creatine phosphate, 9500 in comparison with 8000 calories per mole, causes the reaction between creatine phosphate and ATP to proceed to an equilibrium state very much in favor of ATP. Therefore, the slightest utilization of ATP by the cells calls forth the energy from the creatine phosphate to synthesize new ATP. This effect keeps the concentration of ATP at almost peak level as long as any creatine phosphate remains in the cell. Therefore, one can call creatine phosphate an ATP "buffer" compound.

ANAEROBIC VERSUS AEROBIC ENERGY

Anaerobic energy means energy that can be derived from foods without the simultaneous utilization of oxygen; *aerobic energy* means energy that can be derived from foods only by oxidative metabolism. In the discussions in the preceding chapters it was noted that carbohydrates, fats, and proteins can all be oxidized to cause synthesis of ATP. However, carbohydrates are the only significant foods that can be utilized to provide energy without utilization of oxygen; this energy release occurs during glycolytic breakdown of glycogen to pyruvic acid. When glycogen is split to pyruvic acid, each mole of glucose in the glycogen gives rise to 3 moles of ATP. Thus, the major source of energy under anaerobic conditions is the stored glycogen of the cells.

Anaerobic Energy During Hypoxia. One of the prime examples of anaerobic energy utilization occurs in acute hypoxia. When a person stops breathing, he already has a small amount of oxygen stored in his lungs and an additional amount stored in the hemoglobin of his blood. However, these are sufficient to keep the metabolic processes functioning for only about two minutes. Continued life beyond this time requires an additional source of energy. This can be derived for another minute or so from glycolysis, the glycogen of the cells splitting into pyruvic acid and the pyruvic acid in turn becoming lactic acid, which diffuses out of the cells as described in Chapter 45.

Anaerobic Energy Usage in Strenuous Bursts of Activity. It is common knowledge that muscles can perform extreme feats of strength for a few seconds but are much less capable during prolonged activity. The energy used during strenuous activity is derived from (1) ATP already present in the muscle cells, (2) stored creatine phosphate in the cells, (3) anaerobic energy released by glycolytic breakdown of glycogen to lactic acid, and (4) oxidative energy released continuously by oxidative processes in the cells. The speed of the oxidative processes cannot approach that which is required to supply all the energy demands during strenuous bursts of activity. Therefore, the first three sources of energy

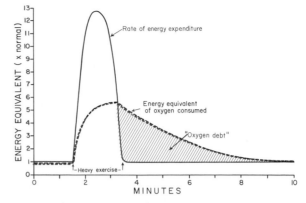

Figure 47–1. Oxygen debt incurred by a bout of strenuous exercise.

are called upon to their maximum extent and provide anaerobic energy for one to two minutes.

Oxygen Debt. After a period of strenuous exercise is over, the oxidative metabolic processes continue to operate at a high level of activity for many minutes to (1) reconvert the lactic acid into glucose and (2) reconvert the decomposed ATP and creatine phosphate to their original states. The extra oxygen that must be used in the oxidative energy processes to rebuild these substances is called the *oxygen debt*.

The oxygen debt is illustrated by the shaded area in Figure 47–1. This figure first shows a period of excess energy expenditure during a bout of heavy exercise, and it shows also the rate of oxygen consumption. Though part of the energy expended during the exercise was provided by oxygen consumption at the time of the exercise, it is clear that a considerable oxygen debt developed, which was repaid by several minutes of excess oxygen consumption after the exercise ended.

SUMMARY OF ENERGY UTILIZATION BY THE CELLS

With the background of the past few chapters and of the preceding discussion, we can now synthesize a composite picture of overall energy utilization by the cells as illustrated in Figure 47–2. This figure shows the anaerobic utilization of glycogen and glucose to form ATP and also the aerobic utilization of compounds derived from carbohydrates, fats, proteins, and other substances for the formation of still additional ATP. In turn, ATP is in reversible equilibrium with creatine phosphate in the cells, and since large quantities of creatine phosphate are present in the cell, much of the stored energy of the cell is in this energy storehouse.

Energy from ATP can be utilized by the different functioning systems of the cells to provide

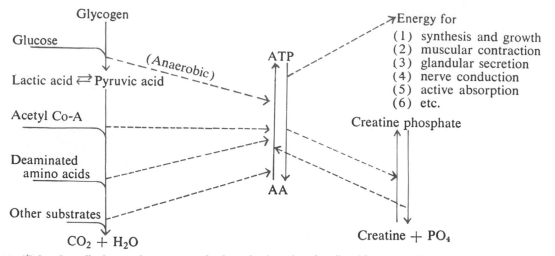

Figure 47–2. Overall schema of energy transfer from foods to the adenylic acid system and then to the functional elements of the cells. (Modified from Soskin and Levine: Carbohydrate Metabolism, by permission of The University of Chicago Press.)

for synthesis and growth, muscular contraction, glandular secretion, impulse conduction, active absorption, and other cellular activities. If greater amounts of energy are required for cellular activities than can be provided by oxidative metabolism, the creatine phosphate storehouse is first utilized, and this is followed rapidly by anaerobic breakdown of glycogen. Thus, oxidative metabolism cannot deliver energy to the cells nearly so rapidly as can the anaerobic processes, but in contrast it is quantitatively almost inexhaustible.

THE METABOLIC RATE

The *metabolism* of the body means simply all the chemical reactions in all the cells of the body, and the *metabolic rate* is normally expressed in terms of the rate of heat liberation during the chemical reactions.

Heat as the Common Denominator of All the Energy Released in the Body. In discussing many of the metabolic reactions of the preceding chapters, we have noted that not all the energy in the foods is transferred to ATP; instead, a large portion becomes heat. On the average, about 55 per cent of the energy in the foods becomes heat during ATP formation. Then still more energy becomes heat as it is transferred from ATP to the functional systems of the cells, so that not more than about 25 per cent of all the energy from the food is finally utilized by the functional systems.

Even though 25 per cent of the energy finally reaches the functional systems of the cells, the major proportion of this also becomes heat, for the following reasons. We might first consider the synthesis of protein and other growing elements of the body. When proteins are synthesized, large amounts of ATP are used to form the peptide linkages, and this stores energy in these linkages. But we also noted in our discussions of proteins in Chapter 46 that there is continuous turnover of proteins, some being degraded while others are being formed. When the proteins are degraded, the energy stored in the peptide linkages is released in the form of heat into the body.

Now let us consider the energy used for muscle activity. Much of this energy simply overcomes the viscosity of the muscles themselves or of the surrounding tissues so that the limbs can move. The viscous movement in turn causes friction within the tissues, which generates heat.

We might also consider the energy expended by the heart in pumping blood. The blood distends the arterial system, the distension in itself representing a reservoir of potential energy. However, as the blood flows through the peripheral vessels, the friction of the different layers of blood flowing over each other and the friction of the blood against the walls of the vessels turns this energy into heat.

Therefore we can say that essentially all the energy expended by the body is converted into heat. The only real exception to this occurs when the muscles are used to perform some form of work outside the body. For instance, when the muscles elevate an object to a height or carry the person's body up steps, a type of potential energy is thus created by raising a mass against gravity.

The Calorie. To discuss the metabolic rate and related subjects intelligently, it is necessary to use some unit for expressing the quantity of energy released from the different foods or expended by the different functional processes of the body. Most often, the *Calorie* is the unit used for this purpose. It will be recalled that 1 *calorie*, spelled with a small "c", is the quantity of heat required to raise the temperature of 1 gram of water 1° C. The calorie is much too small a unit for ease of expression in speaking of energy in the body. Consequently the large Calorie, spelled with a capital "C," which is equivalent to 1000 calories, is the unit ordinarily used in discussing energy metabolism.

MEASUREMENT OF THE METABOLIC RATE – INDIRECT CALORIMETRY

Indirect Calorimetry. Since more than 95 per cent of the energy expended in the body is derived from reaction of oxygen with the different foods, the metabolic rate can be calculated with a high degree of accuracy from the rate of oxygen utilization. For the average diet, the *quantity of energy liberated per liter of oxygen utilized in the body averages approximately 4.825 Calories.* Using this *energy equivalent* of oxygen, one can calculate approximately the rate of heat liberation in the body from the quantity of oxygen utilized in a given period of time. This procedure is called indirect calorimetry.

The Metabolator. Figure 47–3 illustrates the metabolator usually used for indirect calorimetry. This apparatus contains a floating drum, under which is an oxygen chamber connected to a mouthpiece through two flexible tubes. A valve in one of these tubes allows air to pass from the oxygen chamber into the mouth, while air passing from the mouth back to the chamber is directed by means of another valve through the second tube. Before the expired air from the mouth enters the upper portion of the oxygen chamber, it flows through a lower chamber containing pellets of soda lime, which combine chemically with the carbon dioxide in the expired air. Therefore, as oxygen is used by the person's body and the carbon dioxide is absorbed by the soda lime, the floating oxygen chamber, which is precisely balanced by a weight, gradually sinks in the water, owing to the oxygen loss. This chamber is coupled to a pen that records on a moving drum the rate at which the chamber sinks in the water

and thereby records the rate at which the body utilizes oxygen.

FACTORS THAT AFFECT THE METABOLIC RATE

Factors that increase the chemical activity in the cells also increase the metabolic rate. Some of these are the following:

Exercise. The factor that causes by far the most dramatic effect on metabolic rate is strenuous exercise. Short bursts of maximal muscle contraction in any single muscle liberate as much as a hundred times its normal resting amount of heat for a few seconds at a time. In the entire body, however, maximal muscle exercise can increase the overall heat production of the body for a few seconds to about 50 times normal or sustained for several minutes to about 20 times normal in the well-trained athlete, which is an increase in metabolic rate to 2000 per cent of normal.

Energy Requirements for Daily Activities. When an average man of 70 kilograms lies in bed all day, he utilizes approximately 1650 Calories of energy. The process of eating increases the amount of energy utilized each day by an additional 200 or more Calories, so that the same man lying in bed and also eating a reasonable diet requires a dietary intake of approximately 1850 Calories per day.

Table 47–1 illustrates the rates of energy utilization while one performs different types of activities. Note that walking up stairs requires approximately 17 times as much energy as lying in bed asleep. In general, over a 24-hour period a

TABLE 47–1 ENERGY EXPENDITURE PER HOUR DURING DIFFERENT TYPES OF ACTIVITY FOR A 70 KILOGRAM MAN

Form of Activity	Calories per Hour
Sleeping	65
Awake lying still	77
Sitting at rest	100
Standing relaxed	105
Dressing and undressing	118
Tailoring	135
Typewriting rapidly	140
"Light" exercise	170
Walking slowly (2.6 miles per hour)	200
Carpentry, metal working, industrial painting	240
"Active" exercise	290
"Severe" exercise	450
Sawing wood	480
Swimming	500
Running (5.3 miles per hour)	570
"Very severe" exercise	600
Walking very fast (5.3 miles per hour)	650
Walking up stairs	1100

Extracted from data compiled by Professor M. S. Rose.

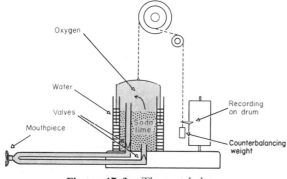

Figure 47–3. The metabolator.

laborer can achieve a maximum rate of energy utilization as great as 6000 to 7000 Calories — in other words as much as 3½ times the basal rate of metabolism.

Thyroid Hormone. When the thyroid gland secretes maximal quantities of thyroxine, the metabolic rate sometimes rises to as much as 100 per cent above normal. On the other hand, total loss of thyroid secretion decreases the metabolic rate to as low as 50 to 60 per cent of normal. These effects can readily be explained by the basic function of thyroxine, that of increasing the rates of activity of almost all the chemical reactions in all cells of the body. This relationship between thyroxine and metabolic rate will be discussed in much greater detail in Chapter 50 in relation to thyroid function, because one of the most useful methods for diagnosing abnormal rates of thyroid secretion is to determine the basal metabolic rate of the patient.

Sympathetic Stimulation. Stimulation of the sympathetic nervous system with liberation of norepinephrine and epinephrine increases the metabolic rates of most tissues of the body. These hormones directly affect cells to cause glycogenolysis, and this, along with other intracellular effects of these hormones, increases cellular activity.

Maximal stimulation of the sympathetic nervous system can increase the metabolic rate in some lower animals as much as several hundred per cent, but the magnitude of this effect in human beings is in question. It is probably 15 per cent or less in the adult but as much as 100 per cent in the newborn child.

THE BASAL METABOLIC RATE

The Basal Metabolic Rate as a Method for Comparing Metabolic Rates Between Individuals. It is often important to measure the inherent activity of the tissues independently of exercise and other extraneous factors that would make it impossible to compare one person's metabolic rate with that of another person. To do this, the metabolic rate is measured under so-called *basal conditions,* and the metabolic rate then measured is called the *basal metabolic rate.* For the normal adult, the average basal metabolic rate is about 70 Calories per hour.

The following basal conditions are necessary for measuring the basal metabolic rate:

1. No food for at least 12 hours.

2. A night of restful sleep before determination.

3. No strenuous exercise after the night of restful sleep, and complete rest in a reclining position for at least 30 minutes prior to actual determination.

4. Elimination of all psychic and physical factors that cause excitement.

5. Air-temperature comfortable and somewhere between the limits of 68° and 80° F.

Constancy of the Metabolic Rate in the Same Person. Basal metabolic rates have been measured in many persons at repeated intervals for as long as 20 or more years. As long as the person remains healthy, almost invariably his basal metabolic rate, as expressed in percentage of normal for his size and age, does not vary more than 5 to 10 per cent.

Constancy of the Basal Metabolic Rate from Person to Person. When the basal metabolic rate is measured in a wide variety of different persons and comparisons are made within single age, weight, and sex groups, 85 per cent of normal persons have been found to have basal metabolic rates within 10 per cent of the mean. Thus, it is obvious that measurements of metabolic rates performed under basal conditions offer an excellent means for comparing the rates of metabolism from one person to another.

THE BODY TEMPERATURE

The temperature of the inside of the body — the "core" — remains almost exactly constant, within ± 1° F, day in and day out except when a person develops a febrile illness. Indeed, the nude person can be exposed to temperatures as low as 55° F or as high as 140° F in dry air and still maintain an almost constant internal body temperature. Therefore, it is obvious that the mechanisms for control of body temperature represent a beautifully designed control system.

The Normal Body Temperature. No single temperature level can be considered normal, for measurements on many normal persons have shown a *range* of normal temperatures, as illustrated in Figure 47–4, from approximately 97° F to over 99° F. When measured by rectum, the

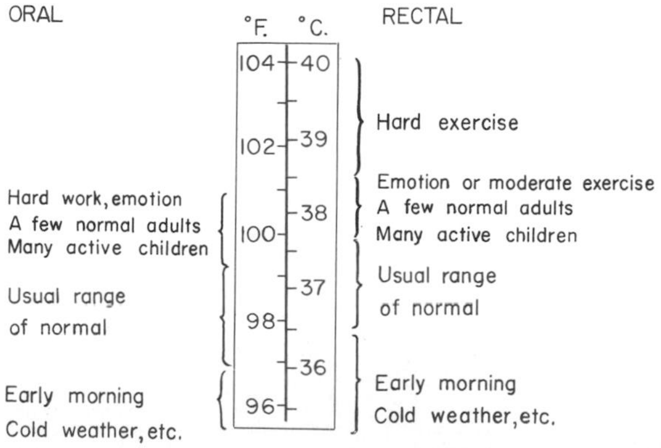

Figure 47–4. Estimated range of body temperature in normal persons. (From DuBois: Fever. Courtesy of Charles C Thomas, Publisher, Springfield, Illinois.)

values are approximately 1° F greater than the oral temperatures. The average normal temperature is generally considered to be 98.6° F (37° C) when measured orally.

The body temperature varies somewhat with exercise and with extremes of temperature of the surroundings, because the temperature regulatory mechanisms are not 100 per cent effective. When excessive heat is produced in the body by strenuous exercise, the rectal temperature can rise to as high as 101° to 104° F.

BALANCE BETWEEN HEAT PRODUCTION AND HEAT LOSS

Heat is continuously being produced in the body as a by-product of metabolism, and body heat is also continuously being lost to the surroundings. When the rate of heat production is exactly equal to the rate of loss, the person is said to be in heat balance. But when the two are out of equilibrium, the body heat, and the body temperature as well, is obviously either increasing or decreasing.

HEAT LOSS

The various methods by which heat is lost from the body are shown pictorially in Figure 47–5. These include *radiation, conduction,* and *evaporation.* Also, the phenomenon of *convection* of the air plays a major role in heat loss by both conduction and evaporation. The amount of heat lost by each of these different mechanisms varies with atmospheric conditions.

Radiation. As illustrated in Figure 47–5, a nude person in a room at normal room temperature loses about 60 per cent of his total heat loss by radiation.

Loss of heat by radiation means loss in the form of infrared heat rays, a type of electromagnetic waves that radiate from the body to any surroundings that are colder than the body itself. This loss increases as the temperature of the surroundings decreases.

Conduction. Usually, only minute quantities of heat are lost from the body by direct conduction from the surface of the body to other objects, such as a chair or a bed. However, loss of heat by *conduction to air* does represent a sizeable proportion of the body's heat loss even under normal conditions. It will be recalled that heat is actually the kinetic energy of molecular motion, and the molecules that compose the skin of the body are continuously undergoing vibratory motion. Thus, the vibratory motion of the skin molecules can cause increased velocity of motion of the air molecules that come into direct contact with the skin. But once the temperature of the air immediately adjacent to the skin approaches the temperature of the skin, little additional exchange of heat from the body to the air can occur. Therefore, conduction of heat from the body to the air is self-limited unless the heated air moves away from the skin so that new, unheated air is continuously brought in contact with the skin, a phenomenon called convection.

Convection. Movement of air is known as convection, and the removal of heat from the body by convection air currents is commonly called "heat loss by convection." Actually, the heat must first be *conducted* to the air and then carried away by the convection currents.

A small amount of convection almost always occurs around the body because of the tendency for the air adjacent to the skin to rise as it becomes heated. Therefore, a nude person seated in a comfortable room loses about 12 per cent of his heat by conduction to the air and then convection away from the body.

Evaporation. When water evaporates from the body surface, 0.58 Calorie of heat is lost for each gram of water that evaporates. Water evaporates *insensibly* from the skin and lungs at a rate of about 600 ml. per day. This causes continuous heat loss at a rate of 12 to 16 Calories per hour. Unfortunately, this insensible evaporation of water directly through the skin and lungs cannot be controlled for purposes of temperature regulation because it results from continuous diffusion of water molecules regardless of body temperature. However, evaporative loss of heat can be controlled by regulating the rate of sweating, which is discussed below.

Evaporation as a Necessary Refrigeration Mechanism at High Air Temperatures. In the preceding discussions of radiation and conduction it was noted that as long as the body temperature is greater than that of the surroundings, heat is lost by radiation and conduction, but when the temperature of the surroundings is greater than that of the skin, instead of losing heat the body gains heat by radiation and conduction from the surroundings. Under these conditions, *the only means by which the body can rid itself of heat is evaporation.*

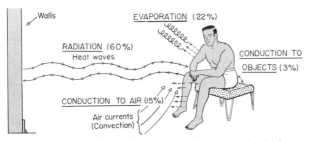

Figure 47–5. Mechanisms of heat loss from the body.

Therefore, any factor that prevents adequate evaporation when the surrounding temperatures are higher than body temperature permits the body temperature to rise. This occurs occasionally in human beings who are born with congenital absence of sweat glands. These persons can withstand cold temperatures as well as normal persons can, but they are likely to die of heat stroke in tropical zones, for without the evaporative refrigeration system their body temperatures must remain at values greater than those of the surroundings.

SWEATING AND ITS REGULATION BY THE AUTONOMIC NERVOUS SYSTEM

When the body becomes overheated, large quantities of sweat are secreted onto the surface of the skin by the sweat glands to provide rapid *evaporative cooling* of the body. Stimulation of the preoptic area in the anterior part of the hypothalamus excites sweating. The impulses from this area that cause sweating are transmitted in the autonomic pathways to the cord and thence through the sympathetic outflow to the skin everywhere in the body.

Rate of Sweating. In cold weather the rate of sweat production is essentially zero, but in very hot weather the maximum rate of sweat production is from 0.7 liter per hour in the unacclimatized person to about 1.5 liters per hour in the person maximally acclimatized to heat. Thus, during maximal sweating, a person can lose more than 3 pounds of body weight per hour.

Mechanism of Sweat Secretion. The sweat glands are tubular structures consisting of two parts: (1) a deep *coiled portion* that secretes the sweat and (2) a *duct portion* passing outward through the dermis of the skin. As is true of so many other glands, the secretory portion of the sweat gland secretes a fluid called the *precursor secretion;* then certain constituents of the fluid are reabsorbed as it flows through the duct.

The precursor secretion is an active secretory product of the epithelial cells lining the coiled portion of the sweat gland. Cholinergic sympathetic nerve fibers ending on or near the glandular cells elicit the secretion.

Since large amounts of sodium chloride are lost in the sweat, it is especially important to know how the sweat glands handle sodium and chloride during the secretory process. When the rate of sweat secretion is very low, the sodium and chloride concentrations of the sweat are also very low, because most of these ions are reabsorbed from the precursor secretion before it reaches the surface of the body; their concentrations are sometimes as low as 5 mEq. per liter each. On the other hand, when the rate of secretion becomes progressively greater, the rate of sodium chloride reabsorption does not increase commensurately, so that then their concentrations in the sweat usually rise to maximum levels of about 60 mEq. per liter, or slightly less than half the levels in plasma.

Effect of Aldosterone on Sodium Loss in the Sweat. Aldosterone functions in much the same way in the sweat glands as in the renal tubules; that is, it increases the rate of active reabsorption of sodium by the ducts. The reabsorption of sodium carries chloride ions along as well because of the electrical gradient that develops across the epithelium when sodium is reabsorbed. The importance of this aldosterone effect is to minimize loss of sodium chloride in the sweat when the blood sodium chloride concentration is already low.

Extreme sweating can deplete the extracellular fluids of electrolytes, particularly of sodium and chloride. A person who sweats profusely may lose as much as 15 to 20 grams of sodium chloride each day until he becomes acclimatized. On the other hand, after four to six weeks of acclimatization the loss of sodium chloride may be as little as 3 to 5 grams per day. This change occurs because of the increased aldosterone secretion resulting from depletion of the salt reserves of the body.

THE INSULATOR SYSTEM OF THE BODY

The skin, the subcutaneous tissues, and especially the fat of the subcutaneous tissues are a heat insulator for the body. The fat is especially important because it conducts heat only *one-third* as readily as other tissues. When no blood is flowing from the heated internal organs to the skin, the insulating properties of the male body are approximately equal to three-quarters the insulating properties of a usual suit of clothes. In women this insulation is still better.

Because most body heat is produced in the deeper portions of the body, the insulation beneath the skin is an effective means for maintaining normal internal temperatures, even though it allows the temperature of the skin to approach the temperature of the surroundings.

FLOW OF BLOOD TO THE SKIN AND HEAT TRANSFER FROM THE BODY CORE

A high rate of blood flow to the skin causes heat to be conducted from the internal portions of the body to the skin with great efficiency. Blood vessels penetrate the subcutaneous insulator tissues and are distributed profusely in the subpapillary portions of the skin. Indeed, immediately beneath the skin is a continuous venous plexus that is supplied by inflow of blood. In the most exposed areas of the body—the hands, feet, and ears—blood is supplied through direct *arteriovenous anastomoses* from the arterioles to the veins. The rate of blood flow into this venous plexus can vary tremendously — from barely above zero in cold weather to as great as 30 per cent of the total cardiac output in hot weather.

Obviously, therefore, the skin is an effective "radiator" system, and the flow of blood to the skin is the principal mechanism of heat transfer from the body "core" to the skin.

REGULATION OF BODY TEMPERATURE

Figure 47–6 illustrates approximately what happens to the temperature of the nude body after a few hours' exposure to dry air ranging from 30° to 170° F. Obviously, the precise dimensions of this curve vary, depending on the movement of air, the amount of moisture in the air, and even the nature of the surroundings. However, in general, between approximately 55° and 140° F in dry air, the nude body is capable of maintaining indefinitely a normal body core temperature somewhere between 98° and 100° F.

The temperature of the body is regulated almost entirely by nervous feedback mechanisms, and almost all of them operate through a *temperature regulating center* located in the *hypothalamus.* However, for these feedback mechanisms to operate, there must also exist temperature detectors to determine when the body temperature becomes either too hot or too cold. Some of these receptors are the following:

Temperature Receptors. Probably the most important temperature receptors for control of body temperature are many special *heat-sensitive neurons* located *in the preoptic area of the hypothalamus.* These neurons increase their impulse output as the temperature rises and decrease their output when the temperature decreases. The firing rate sometimes increases as much as ten-fold with an increase in body temperature of 10° C.

In addition to these heat-sensitive neurons of the preoptic area, other important receptors sensitive to temperature especially include (1) *skin temperature receptors,* including both *warmth* and *cold receptors* (but four to ten times as many cold as warmth receptors) that transmit nerve impulses into the spinal cord and thence to the hypothalamic region of the brain to help control body temperature, as will be discussed later, and (2) *temperature receptors in the spinal cord, abdomen, and possibly other internal structures* of the body that also transmit signals, again mainly cold signals, to the central nervous system to help control body temperature.

Experiments in recent years have shown that the preoptic receptors play the greatest role in temperature control when the body temperature rises above normal. But at low temperatures, the peripheral cold receptors are perhaps the more important.

INTEGRATION OF HEAT AND COLD THERMOSTATIC SIGNALS IN THE HYPOTHALAMUS — THE "HYPOTHALAMIC THERMOSTAT"

The signals that arise in peripheral receptors are transmitted to the posterior hypothalamus, where they are integrated with the receptor signals from the preoptic area to give the final efferent signals for controlling heat loss and heat production. Therefore, we generally speak of the overall hypothalamic temperature control mechanism as the *hypothalamic thermostat.*

Figure 47–7 illustrates the effectiveness of the hypothalamic thermostat in initiating temperature regulatory changes when the body temperature rises too high or falls too low. The solid curve shows that almost precisely at 37° C (98.4° F) sweating begins and then increases rapidly as the temperature rises above this value; on the other hand, it ceases at any temperature below this same critical level.

Likewise, the thermostat controls the rate of heat production, which is illustrated by the dashed curve. At any temperature above 37.1° C, the heat production remains almost exactly constant, but whenever the temperature falls below

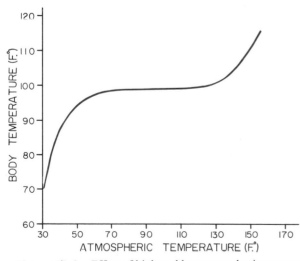

Figure 47–6. Effect of high and low atmospheric temperatures for several hours' duration on the internal body temperature, showing that the internal body temperature remains stable despite wide changes in atmospheric temperature.

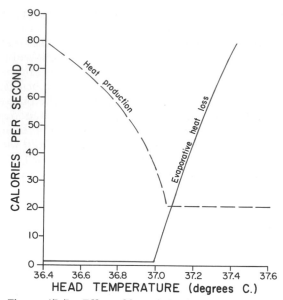

Figure 47–7. Effect of hypothalamic temperature on (1) evaporative heat loss from the body and (2) heat production caused primarily by muscular activity and shivering. This figure demonstrates the extremely critical temperature level at which increased heat loss begins and increased heat production stops. (Drawn from data in Benzinger, Kitzinger, and Pratt, in Hardy J. D. (ed.): Temperature, Part 3, p. 637. Courtesy of Charles C Thomas, Publisher, Springfield, Illinois.)

this level, the various mechanisms for increasing heat production become markedly activated, especially an increase in muscular activity which culminates in shivering.

MECHANISMS OF INCREASED HEAT LOSS WHEN THE BODY BECOMES OVERHEATED

Overheating the preoptic thermostatic area increases the rate of heat loss from the body in two principal ways: (1) by stimulating the sweat glands to cause evaporative heat loss from the body and (2) by inhibiting sympathetic centers in the posterior hypothalamus; this removes the normal vasoconstrictor tone to the skin vessels, thereby allowing vasodilatation and greatly increased loss of heat from the skin.

MECHANISMS OF HEAT CONSERVATION AND INCREASED HEAT PRODUCTION WHEN THE BODY BECOMES COOLED

When the body core is cooled below approximately 37° C, special mechanisms are set into play to conserve the heat that is already in the body, and still other mechanisms are set into play to increase the rate of heat production, as follows:

Heat Conservation. *Vasoconstriction in the Skin.* One of the first effects is intense vasoconstriction of the skin vessels over the entire body. The posterior hypothalamus strongly activates the sympathetic nervous signals to the skin blood vessels, and intense vasoconstriction occurs throughout the body. This vasoconstriction obviously prevents the conduction of heat from the internal portions of the body to the skin. Consequently, with maximal vasoconstriction the only heat that can leave the body is that which can be conducted directly through the insulator layers of the skin. This effect conserves the quantity of heat in the body.

Piloerection. A second means by which heat is conserved when the hypothalamus is cooled is piloerection — that is, the hairs "stand on end." Obviously, this effect is not important in the human being because of the paucity of hair, but in lower animals the upright projection of the hairs in cold weather entraps a thick layer of insulator air next to the skin so that the transfer of heat to the surroundings is greatly depressed.

Abolition of Sweating. Sweating is completely abolished by cooling the preoptic thermostat below about 37° C (98.6° F). This obviously causes evaporative cooling of the body to cease except for that resulting from insensible evaporation.

Increased Production of Heat. Heat production is increased in three separate ways when the temperature of the body thermostat falls below 37° C:

Hypothalamic Stimulation of Shivering. Located in the dorsomedial portion of the posterior hypothalamus near the wall of the third ventricle is an area called the *primary motor center for shivering.* This area is normally inhibited by heat signals from the preoptic thermostatic area but is driven by cold signals from the skin and spinal cord. Therefore, in response to cold, this center becomes activated and transmits impulses through bilateral tracts down the brain stem, into the lateral columns of the spinal cord, and finally to the anterior motoneurons. These impulses are nonrhythmic and do not cause the actual muscle shaking. Instead, they increase the tone of the skeletal muscles throughout the body. When the tone has risen above a certain critical level, shivering begins. This probably results from feedback oscillation of the muscle spindle stretch reflex mechanism. During maximum shivering, body heat production can rise to as high as four to five times normal.

Sympathetic "Chemical" Excitation of Heat Production. Either sympathetic stimulation or circulating norepinephrine and epinephrine in the blood can cause an immediate increase in the rate of cellular metabolism; this effect is called *chemical thermogenesis.* However, as discussed earlier in the chapter, in the adult, it is rare for chemical thermogenesis to increase the rate of heat production more than 10 to 15 per cent. However, in infants chemical thermogenesis can increase the rate of heat production as much as 100 per cent, which is probably a very important factor in maintaining normal body temperature in the newborn.

Increased Thyroxine Output as a Cause of Increased Heat Production. Cooling the preoptic area of the hypothalamus also increases the production of the neurosecretory hormone *thyrotropin-releasing hormone* by the hypothalamus. This hormone is carried by way of the hypothalamic portal veins to the adenohypophysis, where it stimulates the secretion of *thyrotropin*. Thyrotropin, in turn, stimulates increased .output of thyroxine by the thyroid gland, as will be explained in Chapter 50. The increased thyroxine increases the rate of cellular metabolism throughout the body. However, this increase in metabolism through the thyroid mechanism does not occur immediately but requires several weeks for the thyroid gland to hypertrophy before it reaches its new level of thyroxine secretion.

Exposure of animals to extreme cold for several weeks can cause their thyroid gland to increase in size as much as 20 to 40 per cent. Unfortunately, however, the human being rarely allows himself to be exposed to the same degree of cold as that to which animals have been subjected. Therefore, we still do not know, quantitatively, how important the thyroid method of adaptation to cold is in the human being.

BEHAVIORAL CONTROL OF BODY TEMPERATURE

Aside from the hypothalamic thermostatic mechanism for body temperature control, the body has still another mechanism for body temperature control that is usually even more potent than this thermostatic system. This mechanism is behavioral control of temperature, which can be explained as follows: Whenever the internal body temperature becomes too high, signals from the preoptic area of the brain give one a psychic sensation of being overheated. Whenever the body becomes too cold, signals from the skin and perhaps from other peripheral receptors elicit the feeling of cold discomfort. Therefore, the person makes appropriate environmental adjustments to reestablish comfort. This is a much more powerful system of body temperature control than most physiologists have recognized in the past; indeed, for man, it is the only really effective mechanism for body heat control in severely cold environs.

Regulation of Internal Body Temperature After Cutting the Spinal Cord. After cutting the spinal cord in the neck above the sympathetic outflow from the cord, regulation of body temperature becomes extremely poor, for the hypothalamus can then no longer control either skin blood flow or the degree of sweating anywhere in the body. Therefore, in persons with this condition, body temperature must be regulated principally by the patient's psychic response to cold and hot sensations in his head region. That is, if he feels himself becoming too hot or if he develops a headache from the heat, he knows that he should select cooler surroundings, and, conversely, if he has cold sensations, he selects warmer surroundings.

ABNORMALITIES OF BODY TEMPERATURE REGULATION

FEVER

Fever, which means a body temperature above the usual range of normal, may be caused by abnormalities in the brain itself, by toxic substances that affect the temperature regulating centers, or by bacterial diseases, brain tumors, or dehydration.

Resetting the Hypothalamic Thermostat in Febrile Diseases — Effect of Pyrogens

Many proteins, breakdown products of proteins, and certain other substances, such as lipopolysaccharide toxins secreted by bacteria, can cause the "set point" of the hypothalamic thermostat to rise. Substances that cause this effect are called pyrogens. It is pyrogens secreted by toxic bacteria or pyrogens released from degenerating tissues of the body that cause fever during disease conditions. When the set point of the hypothalamic thermostat becomes increased to a higher level than normal, all the mechanisms for raising the body temperature are brought into play, including heat conservation and increased heat production. Within a few hours after the thermostat has been set to a higher level, the body temperature also approaches this level.

To give one an idea of the extremely powerful effect of pyrogens in resetting the hypothalamic thermostat, as little as a few nanograms of some purified pyrogens injected into a person can cause severe fever.

Characteristics of Febrile Conditions

Chills. When the setting of the thermostat is suddenly changed from the normal level to a

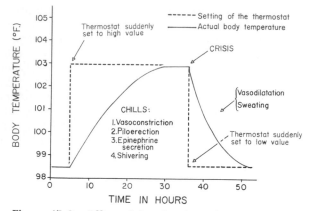

Figure 47–8. Effects of changing the setting of the "hypothalamic thermostat."

higher-than-normal value as a result of tissue destruction, pyrogenic substances, or dehydration, the body temperature usually takes several hours to reach the new temperature setting. For instance, the temperature setting of the hypothalamic thermostat, as illustrated in Figure 47–8, might suddenly rise to 103° F. Because the blood temperature is less than the temperature setting of the hypothalamic thermostat, the usual autonomic responses that cause elevation of body temperature occur. During this period the person experiences chills, during which he feels extremely cold, even though his body temperature may already be above normal. Also, his skin is cold because of vasoconstriction, and he shakes all over because of shivering. His chills continue until his body temperature rises to the hypothalamic setting of 103° F. Then, when the temperature of the body reaches this value, he no longer experiences chills but instead feels neither cold nor hot. As long as the factor that is causing the hypothalamic thermostat to be set at a high value continues its effect, the body temperature is regulated more or less in the normal manner but at the high temperature level.

The Crisis or "Flush." If the factor that is causing the high temperature is suddenly removed, the hypothalamic thermostat is suddenly set at a lower value — perhaps even back to the normal level, as illustrated in Figure 47–8. In this instance, the blood temperature is still 103°F, but the hypothalamus is attempting to regulate the body temperature at 98.6° F. This situation is analogous to excessive heating of the preoptic area, which causes intense sweating and sudden development of a hot skin because of vasodilatation everywhere. This sudden change of events in a febrile disease is known as the "crisis" or, more appropriately, the "flush." In olden days, before the advent of antibiotics, the doctor always awaited the crisis, for once this occurred he knew

immediately that the patient's temperature would soon be falling.

Heat Stroke

The limits of extreme heat that one can stand depend almost entirely on whether the heat is dry or wet. If the air is completely dry and sufficient convection air currents are flowing to promote rapid evaporation from the body, a person can withstand several hours of air temperature at 150° F with no apparent ill effects. On the other hand, if the air is 100 per cent humidified or if the body is in water, the body temperature begins to rise whenever the environmental temperature rises above approximately 94° F. If the person is performing very heavy work, this critical temperature level may fall to 85° to 90° F.

Unfortunately, there is a limit to the rate at which the body can lose heat even with maximal sweating. Furthermore, when the hypothalamus becomes excessively heated its heat-regulatory ability becomes greatly depressed and sweating diminishes. As a result, high body temperature tends to perpetuate itself unless measures are taken specifically to decrease body heat.

When the body temperature rises beyond a critical temperature, into the range of 106° to 108° F., the person is likely to develop *heat stroke*. The symptoms include dizziness, abdominal distress, sometimes delirium, and eventually loss of consciousness if the body temperature is not soon decreased. Many of these symptoms probably result from a mild degree of *circulatory shock* brought on by excessive loss of fluid and electrolytes in the sweat before the onset of symptoms. However, the hyperpyrexia itself is also exceedingly damaging to the body tissues, especially to the brain, and therefore is undoubtedly responsible for many of the effects. In fact, even a few minutes of very high body temperature can sometimes be fatal. For this reason, many authorities recommend immediate treatment of heat stroke by placing the person in an ice water bath. However, because this often induces uncontrollable shivering with considerable increase in rate of heat production, others have suggested that sponge cooling of the skin is likely to be more effective for rapidly decreasing the body core temperature.

Harmful Effects of the High Temperature. When the body temperature rises above 106° to 108° F, the parenchyma of many cells begins to be damaged. The pathological findings in a person who dies of hyperpyrexia are local hemorrhages and parenchymatous degeneration of cells throughout the entire body, but especially in the brain. Unfortunately, once neuronal cells are destroyed, they can never be replaced. Damage to the liver, kidneys, and other body organs can often be so great that failure of one or more of these organs eventually causes death, sometimes not till several days after the heat stroke.

Antipyretics. Aspirin, antipyrine, aminopyrine, and a number of other substances known as "antipyretics" have an effect on the hypothalamic thermostat opposite to that of the pyrogens. In other words, they cause the setting of the thermostat to be lowered, so that the body temperature falls, though usually not more than a degree or so. Aspirin is especially effective in lowering the hypothalamic setting when pyrogens have raised the setting, but aspirin will not lower the normal temperature. On the other hand, aminopyrine will decrease even the normal body temperature. Obviously, these drugs can be used to prevent damage to the body from excessively high body temperature.

EXPOSURE OF THE BODY TO EXTREME COLD

A person exposed to ice water for approximately 20 to 30 minutes ordinarily dies because of heart standstill or heart fibrillation unless treated immediately. By that time, the internal body temperature has fallen to about 77° F. Yet if he is warmed rapidly by application of external heat, his life can often be saved.

Once the body temperature has fallen below 85° F, the ability of the hypothalamus to regulate temperature is completely lost, and it is greatly impaired even when the body temperature falls below approximately 94° F. Part of the reason for this loss of temperature regulation is that the rate of heat production in each cell is greatly depressed by the low temperature. Also, sleepiness and even coma are likely to develop, which depress the activity of the central nervous system heat-control mechanisms and prevent shivering. This loss of temperature regulation obviously further accelerates the decrease in body temperature and rapidly leads to death.

REFERENCES

Energetics and Metabolic Rate

Alexander, G.: Cold thermogenesis. *In* Robertshaw, D. (ed.): International Review of Physiology: Environmental Physiology III. Vol. 20. Baltimore, University Park Press, 1979, p. 43.
Bennett, A. F.: Activity metabolism of the lower vertebrates. *Annu. Rev. Physiol,* 40:447, 1978.

Christensen, H. N., and Palmer, G. A.: Enzyme Kinetics. Philadelphia, W. B. Saunders Co., 1974.
Consolazio, C. F., *et al.*: Energy requirement and metabolism during exposure to extreme environments. *World Rev. Nutr. Diet, 18*:177, 1973.
Havel, R. J.: Caloric homeostasis and disorders of fuel transport. *N. Engl. J. Med., 287*:1186, 1972.
Herman, R. H., *et al.* (eds.): Metabolic Control in Mammals. New York, Plenum Press, 1979.
Hoch, F. L.: Metabolic effects of thyroid hormones. *In* Greep, R. O., and Astwood, E. B. (eds.): Handbook of Physiology. Sec. 7, Vol. 3. Baltimore, Williams & Wilkins, 1974, p. 391.
Homsher, E., and Kean, C. J.: Skeletal muscle energetics and metabolism. *Annu. Rev. Physiol., 40*:93, 1978.
Klachko, D. M., *et al.* (eds.): Hormones and Energy Metabolism. New York, Plenum Press, 1978.
Martin, D. B.: Metabolism and energy mechanisms. *In* Frohlich, E. D. (eds.): Pathophysiology, 2nd Ed. Philadelphia, J. B. Lippincott Co., 1976, p. 365.
Sinclair, J. C.: Metabolic rate and body size of the newborn. *Clin. Obstet. Gynecol., 14*:840, 1971.

Temperature Regulation

Benzinger, T. H.: Heat regulation: Homeostasis of central temperature in man. *Physiol. Rev., 49*:671, 1969.
Cena, K., and Clark, J. A.: Transfer of heat through animal coats and clothing. *In* Robertshaw, D. (ed.): Internal Review of Physiology: Environmental Physiology III. Vol. 20. Baltimore, University Park Press, 1979, p. 1.
Elizondo, R.: Temperature regulation in primates. *Int. Rev. Physiol., 15*:71, 1977.
Hardy, R. N.: Temperature and Animal Life. Baltimore, University Park Press, 1979.
Heller, H. C., *et al.*: The thermostat of vertebrate animals. *Sci. Am., 239*(2):102, 1978.
Hensel, H.: Neural processes in thermoregulation. *Physiol. Rev., 53*:948, 1973.
Hensel, H.: Thermoreceptors. *Annu. Rev. Physiol., 36*:233, 1974.
Himms-Hagen, J.: Cellular thermogenesis. *Annu. Rev. Physiol., 38*:315, 1976.
Kluger, M. J.: Temperature regulation, fever, and disease. *In* Robertshaw, D. (ed.): International Review of Physiology: Environmental Physiology III. Vol. 20. Baltimore, University Park Press, 1979, p. 209.
Mitchell, D.: Physical basis of thermoregulation. *Int. Rev. Physiol., 15*:1, 1977.
Precht, H., *et al.*: Temperature and Life. New York, Springer-Verlag, 1973.
Robertshaw, D.: Role of the adrenal medulla in thermoregulation. *Int. Rev. Physiol., 15*:189, 1977.
Satinoff, E. (ed.): Thermoregulation. Stroudsburg, Pa., Dowden, Hutchinson & Ross.
Schmidt-Nielsen, K.: Animal Physiology: Adaptation and Environment. London, Cambridge University Press, 1975.
Sloan, A. W.: Man in Extreme Environments. Springfield, Ill., Charles C Thomas, 1979.
Underwood, L. S., and Tieszen, L. L. (eds.): Comparative Mechanisms of Cold Adaptation. New York, Academic Press, 1979.
Wang, L. C. H., and Hudson, J. W. (eds.): Strategies in Cold: Natural Torpidity and Thermogenesis. New York, Academic Press, 1978.
Wyndham, C. H.: The physiology of exercise under heat stress. *Annu. Rev. Physiol., 35*:193, 1973.

48

Dietary Balances, Regulation of Feeding, Obesity, and Vitamins

The intake of food must always be sufficient to supply the metabolic needs of the body and yet not enough to cause obesity. Also, since different foods contain different proportions of proteins, carbohydrates, and fats, appropriate balance must be maintained between these different types of food so that all segments of the body's metabolic systems can be supplied with the requisite materials. This chapter therefore discusses especially the problems of balance between the major types of food and also the mechanisms by which the intake of food is regulated in accordance with the metabolic needs of the body.

DIETARY BALANCES

Energy Available in Foods. The energy liberated from each gram of carbohydrate as it is oxidized to carbon dioxide and water is 4.1 Calories, and that liberated from fat is 9.3 Calories. The energy liberated from metabolism of the average protein of the diet as each gram is oxidized to carbon dioxide, water, and urea is 4.35 Calories. Also, these different substances vary in the average percentages that are absorbed from the gastrointestinal tract: approximately 98 per cent of the carbohydrate, 95 per cent of the fat, and 92 per cent of the protein. Therefore, in round figures the average *physiologically available energy* in each gram of the three different foodstuffs in the diet is:

	Calories
Carbohydrates	4.0
Fat	9.0
Protein	4.0

Average Composition of the Diet. The average American receives approximately 15 per cent of his energy from protein, about 40 per cent from fat, and 45 per cent from carbohydrates. In most other parts of the world the quantity of energy derived from carbohydrates far exceeds that derived from both proteins and fats. Indeed, in Mongolia the energy received from fats and proteins combined is said to be no greater than 15 to 20 per cent.

Daily Requirement for Protein. Twenty to 30 grams of body proteins are degraded and used for energy daily. Therefore, all cells must continue to synthesize new proteins to take the place of those that are being destroyed, and a supply of protein is needed in the diet for this purpose. An average man can maintain his normal stores of protein provided that his *daily intake is above 30 to 55 grams.*

Partial Proteins. Another factor that must be considered in analyzing the proteins of the diet is whether the dietary proteins are *complete* proteins or *partial* proteins. Complete proteins have compositions of amino acids in appropriate proportion to each other so that all the amino acids can be properly used by the human body. In general, proteins derived from animal foodstuffs are more nearly complete than are proteins derived from vegetable and grain sources. When partial proteins are in the diet, an increased minimal quantity of protein is necessary in the daily rations to maintain protein balance. A particular example of this occurs in the diet of many African natives who subsist primarily on a corn meal diet. The protein of corn is almost totally lacking in tryptophan, which means that this diet, in effect, is almost completely protein deficient. As a result, many Africans, especially the children, develop the protein deficiency syndrome called *kwashiorkor,* which consists of failure to grow, lethargy, depressed mentality, and hypoprotein edema.

Study of the Balance Between Fat and Carbohydrate Utilization — The Respiratory Quotient

When glucose is oxidized, the number of molecules of carbon dioxide liberated is exactly equal to the number of oxygen molecules necessary for the oxidative process. This *ratio of carbon dioxide output to oxygen usage* is called the *respiratory quotient.* For glucose the respiratory quotient, therefore, is 1.00, a fact illustrated in Figure 48–1. On the other hand, oxidation of triolein (the most abundant fat in the body) liberates 57 carbon dioxide molecules while 80 oxygen molecules are being utilized. Consequently, the respiratory quotient in this instance is 0.71. Finally, oxidation of alanine liberates five carbon dioxide molecules for every six oxygen molecules entering into the reaction, giving a respiratory quotient of 0.83.

As was already pointed out, the average person receives only 15 per cent of his total energy from protein metabolism. Furthermore, the respiratory quotient of protein is approximately midway between the respiratory quotients of fat and carbohydrate (see preceding paragraph). Consequently, when the overall respiratory quotient of a person is measured by determining the total respiratory intake of oxygen and the total output of carbon dioxide, one has a reasonable measure of the relative quantities of fat and carbohydrate being metabolized by the body during that particular time. For instance, if the respiratory quotient is approximately 0.71, the body is burning almost entirely fat to the exclusion of carbohydrates and proteins. If, on the other hand, the respiratory quotient is 1.00, the body is probably metabolizing almost entirely carbohydrate to the exclusion of fat. Finally, a respiratory quotient of 0.85 indicates approximately equal utilization of carbohydrate and fat.

REGULATION OF FOOD INTAKE

Hunger. The term "hunger" means a craving for food, and it is associated with a number of objective sensations. For instance, in a person who has not had food for many hours, the stomach undergoes intense rhythmic contractions called *hunger contractions.* These cause a tight or a gnawing feeling in the pit of the stomach and sometimes actually cause pain called *hunger pangs.* In addition to the hunger pangs, the hungry person also becomes more tense and restless than usual.

Some physiologists actually define hunger as the tonic contractions of the stomach. However, even after the stomach is completely removed, the psychic sensations of hunger still occur, and craving for food still makes the person search for an adequate food supply.

Appetite. The term "appetite" is often used in the same sense as hunger except that it usually implies desire for specific types of food instead of food in general. Therefore, appetite helps a person choose the quality of food he eats.

Satiety. Satiety is the opposite of hunger. It means a feeling of fulfillment in the quest for food. Satiety usually results from a filling meal, particularly when the person's nutritional storage depots, the adipose tissue and the glycogen stores, are already filled.

NEURAL CENTERS FOR REGULATION OF FOOD INTAKE

Hunger and Satiety Centers. Stimulation of the *lateral hypothalamus* causes an animal to eat voraciously, while stimulation of the *ventromedial*

Respiratory Quotient:

$$C_6H_{12}O_6 + 6\ O_2 \rightarrow 6\ CO_2 + 6\ H_2O$$
Glucose
$$\frac{6}{6} = 1.00$$

$$C_{57}H_{104}O_6 + 80\ O_2 \rightarrow 57\ CO_2 + 52\ H_2O$$
Triolein
$$\frac{57}{80} = 0.71$$

$$2\ C_3H_7O_2N + 6\ O_2 \rightarrow (NH_2)_2CO + 5\ CO_2 + 5\ H_2O$$
Alanine
$$\frac{5}{6} = 0.83$$

Figure 48–1. Utilization of oxygen and release of carbon dioxide during the oxidation of carbohydrate, fat, and protein. The respiratory quotient for each of these reactions is calculated.

nuclei of the hypothalamus causes complete satiety, and even in the presence of highly appetizing food, while being stimulated, the animal will still refuse to eat. Conversely, a destructive lesion of the ventromedial nuclei causes exactly the same effect as stimulation of the lateral hypothalamic nuclei — that is, voracious and continued eating until the animal becomes extremely obese, sometimes as large as four times normal in size. Lesions of the lateral hypothalamic nuclei cause exactly the opposite effects — complete lack of desire for food and progressive inanition of the animal. Therefore, we can label the lateral nuclei of the hypothalamus as the *hunger center* or the *feeding center,* while we can label the ventromedial nuclei of the hypothalamus as a *satiety center.*

The feeding center operates by directly exciting the emotional drive to search for food. On the other hand, it is believed that the satiety center operates primarily by inhibiting the feeding center.

Other Neural Centers that Enter into Feeding. If the brain is removed above the mesencephalon, the animal can still perform the basic mechanical features of the feeding process. It can salivate, lick its lips, chew food, and swallow. Therefore, the actual mechanics of feeding are all controlled by centers in the lower brain stem. The function of the hunger center in the hypothalamus, then, is to control the quantity of food intake and to excite the lower centers to activity.

Higher centers than the hypothalamus also play important roles in the control of feeding, particularly in the control of appetite. These centers include especially the amygdala and some cortical areas of the limbic system, all of which are closely coupled with the hypothalamus. It will be recalled from the discussion of the sense of smell that the amygdala is one of the major parts of the olfactory nervous system. Destructive lesions in the amygdala have demonstrated that some of its areas greatly increase feeding, while others inhibit feeding. In addition, stimulation of some areas of the amygdala elicits the mechanical act of feeding. However, the most important effect of destruction of the amygdala on both sides of the brain is a "psychic blindness" in the choice of foods. In other words, the animal (and presumably the human being as well) loses or at least partially loses the mechanism of appetite control over the type and quality of food that is eaten.

The cortical regions of the limbic system, including the infraorbital regions, the hippocampal gyrus, and the cingulate gyrus, all have areas that when stimulated can either increase or decrease feeding activities. These areas seem especially to play a role in the animal's drive to search for food when hungry. It is presumed that these centers are also responsible, probably operating in association with the amygdala and hypothalamus, for determining the quality of food that is eaten. For instance, a previous unpleasant experience with almost any type of food often destroys a person's appetite for that food thenceforth.

FACTORS THAT REGULATE FOOD INTAKE

We can divide the regulation of food into (1) *nutritional regulation,* which is concerned primarily with maintenance of normal quantities of nutrient stores in the body, and (2) *alimentary regulation,* which is concerned primarily with the immediate effects of feeding on the alimentary tract and is sometimes called *peripheral regulation* or *short-term regulation.*

Nutritional Regulation. An animal that has been starved for a long time and is then presented with unlimited food eats a far greater quantity than does an animal that has been on a regular diet. Conversely, an animal that has been force-fed for several weeks eats little when allowed to eat according to its own desires. Thus, the feeding center in the hypothalamus is geared to the nutritional status of the body. Some of the nutritional factors that control the degree of activity of the feeding center are the following:

Availability of Glucose to the Body Cells — The Glucostatic Theory of Hunger and of Feeding Regulation. It has long been known that a decrease in blood glucose concentration is associated with development of hunger, which has led to the so-called glucostatic theory of hunger and of feeding regulation, as follows: When the blood glucose level falls too low, this automatically causes the animal to increase its feeding, which eventually returns the glucose concentration back toward normal. There are two other observations that also support the glucostatic theory: (1) An increase in blood glucose level increases the measured electrical activity in the satiety center in the ventromedial nuclei of the hypothalamus and simultaneously decreases the electrical activity in the feeding center of the lateral nuclei. (2) Chemical studies show that the ventromedial nuclei (the satiety center) concentrate glucose, while other areas of the hypothalamus fail to concentrate glucose; therefore, it is assumed that glucose acts by increasing the degree of satiety.

Effect of Blood Amino Acid Concentration on Feeding. An increase in amino acid concentration in the blood also reduces feeding, and a decrease enhances feeding. In general, though, this effect is not as powerful as the glucostatic mechanism.

Effect of Fat Metabolites on Feeding — Long-term Regulation. The overall degree of feeding varies almost inversely with the amount of adipose tissue in the body. That is, as the quantity of adipose tissue increases, the rate of feeding decreases. Therefore, many physiologists believe that *long-term regulation* of feeding is controlled mainly by fat metabolites of undiscovered nature. This is called the "lipostatic" theory of feeding regulation. In support of this is the fact that the long-term average concentration of free fatty acids in the blood is directly proportional to the quantity of adipose tissue in the body. Therefore, it is likely that the free fatty acids or some other similar fat metabolites act in the same manner as glucose and amino acids to cause a negative feedback regulatory effect on feeding. It is also possible, if not probable, that this is by far the most important long-term regulator of feeding.

Summary of Long-term Regulation. Even though our information on the different feedback factors in long-term feeding regulation is imprecise, we can make the following general statement: When the nutrient stores of the body fall below normal, the feeding center of the hypothalamus becomes highly active and the person exhibits increased hunger; on the other hand, when the nutrient stores are abundant, the person loses his hunger and develops a state of satiety.

Alimentary Regulation (Short-term, Nonmetabolic Regulation). The degree of hunger or satiety can be temporarily increased or decreased by habit. For instance, the normal person has the habit of eating three meals a day, and if he misses one, he is likely to develop a state of hunger at mealtime despite completely adequate nutritional stores in his tissues. But, in addition to habit, several other short-term physiologic stimuli— mainly related to the alimentary tract—can alter one's desire for food for several hours at a time, as follows:

Gastrointestinal Filling. When the gastrointestinal tract becomes distended, especially the stomach or the duodenum, inhibitory signals temporarily suppress the feeding center, thereby reducing the desire for food. This effect probably depends mainly on sensory signals transmitted through the vagi, but part of the effect still persists after the vagi and the sympathetic nerves from the upper gastrointestinal tract have been severed. Therefore, somatic sensory signals from the stretched abdomen might also play a role. And, recently it has been found that hormonal feedback also suppresses feeding, for *cholecystokinin*, which is released mainly in response to fat entering the duodenum, has a strong effect on inhibition of further eating.

Obviously, these mechanisms are of particular importance in bringing one's feeding to a halt during a heavy meal.

Metering of Food by Head Receptors. When a person with an esophageal fistula is fed large quantities of food, even though this food is immediately lost again to the exterior, his degree of hunger is decreased after a reasonable quantity of food has passed through his mouth. This effect occurs despite the fact that the gastrointestinal tract does not become the least bit filled. Therefore, it is postulated that various "head factors" relating to feeding, such as chewing, salivation, swallowing, and tasting, "meter" the food as it passes through the mouth, and after a certain amount has passed through, the hypothalamic feeding center becomes inhibited.

Importance of Having Both Long- and Short-term Regulatory Systems for Feeding. The long-term regulatory system, especially the lipostatic feedback mechanism, obviously helps an animal to maintain constant stores of nutrients in its tissues, preventing them from becoming too little or too great. On the other hand, the short-term regulatory stimuli make the animal feed only when the gastrointestinal tract is receptive to food. Thus, food passes through its gastrointestinal tract fairly continuously, so that its digestive, absorptive, and storage mechanisms can all work at a steady pace rather than only when the animal needs food for energy. Indeed, the digestive, absorptive, and storage mechanisms can increase their rates of activity above normal only four- to five-fold, whereas the rate of usage of stored nutrients for energy sometimes increases to 20 times normal.

It is important, then, that feeding occur rather continuously (but at a rate that the gastrointestinal tract can accommodate), regulated principally by the short-term regulatory mechanisms. However, it is also important that the intensity of the daily rhythmic feeding habits be modulated up or down by the long-term regulatory system, based principally on the level of nutrient stores in the body.

OBESITY

Energy Input versus Energy Output. When greater quantities of energy (in the form of food) enter the body than are expended, the body weight increases. Therefore, obesity is obviously caused by excess energy input over energy output. For each 9.3 Calories excess energy entering the body, 1 gram of fat is stored.

Excess energy input occurs *only during the developing phase of obesity,* and once a person has become obese, all that is required of him to remain obese is that his energy input equal his energy output. For the person to reduce, the output must be *greater* than the input. Indeed, studies of obese persons, once they have become obese, show that their intake of food is almost exactly the same as that of normal weight persons.

Effect of Muscular Activity on Energy Output. About one-third of the energy used each day by the normal person goes into muscular activity, and in the laborer as much as two-thirds and occasionally three-fourths is used in this way. Since muscular activity is by far the most important means by which energy is expended in the body, it is frequently said that obesity results from *too high a ratio of food intake to daily exercise.*

ABNORMAL FEEDING REGULATION AS A PATHOLOGIC CAUSE OF OBESITY

The preceding discussion of the mechanisms that regulate feeding emphasized that the rate of feeding is normally regulated in proportion to the nutrient stores in the body. When these stores begin to approach an optimal level in a normal person, feeding is automatically reduced to prevent overstorage. However, in many obese persons this is not true, for feeding does not slacken until body weight is far above normal. Therefore, in effect, obesity is often caused by an abnormality of the feeding regulatory mechanism. This can result from either psychogenic factors that affect the regulation or actual abnormalities of the hypothalamus itself.

Psychogenic Obesity. Studies of obese patients show that a large proportion of obesity results from psychogenic factors. Perhaps the most common psychogenic factor contributing to obesity is the prevalent idea that healthy eating habits require three meals a day and that each meal must be filling. Many children are forced into this habit by overly solicitous parents, and the children continue to practice it throughout life.

Genetic Factors in Obesity. Obesity definitely runs in families. Furthermore, identical twins usually maintain weight levels within 2 pounds of each other throughout life if they live under similar conditions, or within 5 pounds of each other if their conditions of life differ markedly. This might result partly from eating habits engendered during childhood, but it is generally believed that this close similarity between twins is genetically controlled.

The genes can direct the degree of feeding in several different ways, including (1) a genetic abnormality of the feeding center that sets the level of nutrient storage high or low and (2) abnormal hereditary psychic factors that either whet the appetite or cause the person to eat as a "release" mechanism.

A genetic abnormality in the *chemistry of fat storage* is also known to cause obesity in a certain strain of rats. In these rats, fat is easily stored in the adipose tissue, but the quantity of hormone-sensitive lipase in the adipose tissue is greatly reduced, so that little of the fat can be removed. In addition, the rats develop hyperinsulinism, which promotes fat storage. This combination obviously results in a one-way path, the fat continuously being deposited but never released. This, too, is another possible mechanism of obesity in some human beings.

Childhood Overnutrition as a Possible Cause of Obesity. The number of fat cells in the adult body is determined almost entirely by the amount of fat stored in the body during early life. The rate of formation of new fat cells is especially rapid in obese infants, and it continues at a lesser rate in obese children until adolescence. Thereafter, the number of fat cells remains almost constant throughout life. Thus, mainly on the basis of experiments in lower animals, it is believed that overfeeding children, especially in infancy and to a lesser extent during the older years of childhood, can lead to a lifetime of obesity. The person who has excess fat cells is thought to have a higher setting of his hypothalamic feedback autoregulatory mechanism for control of adipose tissues. In support of this belief is the fact that most extremely obese people have far more fat cells than normal people — often as much as three or more times as many. Indeed, it is rare that any single fat cell stores more than about 50 per cent more fat than normal. Therefore, a major share of obesity seems to result from excess numbers of fat cells rather than simply from enlarged fat cells.

STARVATION

Depletion of Food Stores in the Body Tissues During Starvation. Even though the tissues preferentially use carbohydrate for energy over both fat and protein, the quantity of carbohydrate stores of the body is only a few hundred grams (mainly glycogen in the liver and muscles), and it can supply the energy required for body function for perhaps half a day. Therefore, except for the first few hours of starvation, the major effects are progressive depletion of tissue fat at first and later tissue protein as well. Since

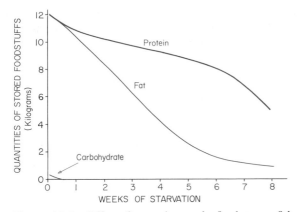

Figure 48–2. Effect of starvation on the food stores of the body.

fat is the prime source of energy, its rate of depletion continues unabated, as illustrated in Figure 48–2, until most of the fat stores in the body are gone.

Protein undergoes three different phases of depletion: rapid depletion at first, then greatly slowed depletion, and finally rapid depletion again shortly before death. The initial rapid depletion is caused by conversion of protein to glucose in the liver by the process of gluconeogenesis. The glucose thus formed (about two-thirds of it) is used to supply energy to the brain, which under normal circumstances utilizes almost no other metabolic substrate for energy besides glucose. However, after the protein stores have been partially depleted during the early phase of starvation, the remaining protein is not so easily removed from the tissues. At this time, the rate of gluconeogenesis decreases to one-third to one-fifth its previous rate, and the rate of depletion of protein becomes greatly decreased, as illustrated in Figure 48–2. The lessened availability of glucose then initiates a series of events leading to *ketosis,* which means greatly increased formation of ketone bodies, as described in Chapter 46. Fortunately, the ketone bodies, like glucose, can cross the blood-brain barrier and can be utilized by the brain cells for energy. Therefore, approximately two-thirds of the brain's energy now is derived from these ketone bodies, principally beta-hydroxybutyrate. This sequence of events thus leads to at least partial preservation of the protein stores of the body.

However, there finally comes a time when the fat stores also are almost totally depleted, and the only remaining source of energy is proteins. At that time, protein stores once again enter a stage of rapid depletion. Since the proteins are essential for maintenance of cellular function, death ordinarily ensues when the proteins of the body

have been depleted to approximately one-half their normal level.

VITAMINS

A vitamin is an organic compound that is needed in small quantities for operation of normal bodily metabolism and that cannot be manufactured in the cells of the body. Probably hundreds of such substances exist, most of which have not been discovered. However, from a clinical point of view the agents that are generally considered to be vitamins are those organic compounds that occur in the diet in small quantities and can cause specific metabolic deficits when lacking.

Daily Requirements of Vitamins. Table 48–1 illustrates the usually recommended daily requirements of the different important vitamins for the average adult. These requirements vary considerably, depending on the characteristics of each person. For instance, the greater the person's size, the greater is his vitamin requirement. Second, growing persons usually require greater quantities of vitamins than do others. Third, when the person performs exercise, the vitamin requirements are increased. Fourth, during disease and fevers, the vitamin requirements are ordinarily increased. Fifth, when greater than normal quantities of carbohydrates are metabolized, the requirements of thiamine and perhaps some of the other vitamins of the B complex are increased. Sixth, during pregnancy and lactation the requirement for vitamin D by the mother is greatly increased, and the requirement for vitamin D is also considerable during the period of growth in children. Finally, a number of metabolic deficits occur pathologically in which the vitamins themselves cannot be utilized properly in the body; in such conditions the requirement for one or more specific vitamins may be extreme.

TABLE 48–1 REQUIRED DAILY AMOUNTS OF THE VITAMINS

A	5000 IU
Thiamine	1.5 mg.
Riboflavin	1.8 mg.
Niacin	20 mg.
Ascorbic acid	45 mg.
D	400 IU
E	15 IU
K	none
Folic acid	0.4 mg.
B_{12}	3 μg.
Pyridoxine	2 mg.
Pantothenic acid	unknown

IU, international units.

VITAMIN A

Vitamin A precursors occur in abundance in many different vegetable foods. These are the yellow and red *carotenoid pigments,* which, since they have chemical structures similar to that of vitamin A, can be changed into vitamin A in the human body.

The basic function of vitamin A in the metabolism of the body is not known except in relation to its use in the formation of retinal photochemicals, which was discussed in Chapter 39. Nevertheless, vitamin A is also necessary for normal growth of most cells of the body and especially for normal growth and proliferation of the different types of epithelial cells. When vitamin A is lacking, the epithelial structures of the body tend to become stratified and keratinized. Therefore, vitamin A deficiency manifests itself by (1) scaliness of the skin and sometimes acne, (2) failure of growth of young animals, (3) failure of reproduction in many animals, associated especially with atrophy of the germinal epithelium of the testes and sometimes with interruption of the female sexual cycle, and (4) keratinization of the cornea with resultant corneal opacity and blindness.

THIAMINE (VITAMIN B₁)

Thiamine operates in the metabolic systems of the body principally as *thiamine pyrophosphate;* this compound functions as a *cocarboxylase,* operating mainly for decarboxylation of pyruvic acid as discussed in Chapter 45.

Thiamine deficiency causes decreased utilization of pyruvic acid and some amino acids by the tissues but increased utilization of fats. Thus, thiamine is specifically needed for final metabolism of carbohydrates and many amino acids. Probably the decreased utilization of these nutrients is responsible for the debilities associated with thiamine deficiency.

Thiamine Deficiency and the Nervous System. The central nervous system depends almost entirely on the metabolism of carbohydrates for its energy. In thiamine deficiency the utilization of glucose by nervous tissue may be decreased as much as 50 to 60 per cent. Therefore, it is readily understandable that thiamine deficiency can greatly impair function of the central nervous system. The neuronal cells of the central nervous system frequently show chromatolysis and swelling during thiamine deficiency, changes that are characteristic of neuronal cells with poor nutrition.

Also, thiamine deficiency can cause *degeneration of myelin sheaths* of nerve fibers both in the periph-

eral nerves and in the central nervous system. The lesions in the peripheral nerves frequently cause these nerves to become extremely irritable, resulting in "polyneuritis" characterized by pain radiating along the course of one or more peripheral nerves. Also, in severe thiamine deficiency, the peripheral nerve fibers and fiber tracts in the cord can degenerate to such an extent that *paralysis* occasionally results.

Thiamine Deficiency and the Cardiovascular System. Thiamine deficiency also weakens the heart msucle, so that a person with severe thiamine deficiency sometimes develops *cardiac failure. Peripheral edema* and *ascites* also occur to a major extent in some persons with thiamine deficiency partly because of the cardiac failure but also because thiamine deficiency causes arteriolar dilatation.

Thiamine Deficiency and the Gastrointestinal Tract. Among the gastrointestinal symptoms caused by thiamine deficiency are indigestion, severe constipation, anorexia, gastric atony, and hypochlorhydria. All these effects possibly result from failure of the smooth muscle and glands of the gastrointestinal tract to derive sufficient energy from carbohydrate metabolism.

The overall picture of thiamine deficiency, including polyneuritis, cardiovascular symptoms, and gastrointestinal disorders, is frequently referred to as "beriberi" — especially when the cardiovascular symptoms predominate.

NIACIN

Niacin, also called *nicotinic acid,* functions in the body as coenzymes in the forms of nicotinamide adenine dinucleotide (NAD) and nicotinamide adenine dinucleotide phosphate (NADP), which are also known as DPN and TPN. These coenzymes are hydrogen acceptors, which combine with hydrogen atoms as they are removed from food substrates by many different types of dehydrogenases. When a deficiency of niacin exists, the normal rate of dehydrogenation cannot be maintained, and, therefore, oxidation of the hydrogen and consequent delivery of energy from the foodstuffs to the functioning elements of the cells likewise cannot occur at normal rates.

Because NAD and NADP operate in all cells of the body, it is readily understood how lack of niacin can cause multiple symptoms. Clinically, niacin deficiency causes mainly gastrointestinal symptoms, neurological symptoms, and a characteristic dermatitis. Pathological lesions appear in many parts of the central nervous system, and permanent dementia or any of many different types of psychoses may result. Also, the skin develops a cracked, pigmented scaliness in areas that are exposed to mechanical irritation or sun irradiation; thus, the skin is unable to repair the different types of irritative damage. Finally, niacin deficiency causes intense irritation and inflammation of the mucous membranes of the mouth and other portions of the gastrointestinal tract, thus instituting many digestive abnormalities.

The clinical entity called "pellagra" is caused mainly by niacin deficiency. Pellagra is greatly exacerbated in persons on a corn diet (such as many of the natives of Africa) because corn is very deficient in the amino acid tryptophan, which can be converted in limited quantities to niacin in the body.

RIBOFLAVIN (VITAMIN B₂)

Riboflavin normally combines in the tissues with phosphoric acid to form two coenzymes, *flavin mononucleotide (FMN),* and *flavin adenine dinucleotide (FAD).* These in turn operate as hydrogen carriers in several of the important oxidative systems of the body.

Deficiency of riboflavin in lower animals causes severe *dermatitis; vomiting; diarrhea; muscular spasticity,* which finally becomes muscular weakness; and then *death* preceded by coma and declining body temperature. Thus, severe riboflavin deficiency can cause many of the same effects as lack of niacin in the diet; presumably the debilities that result in each instance are due to generally depressed oxidative processes within the cells.

In the human being riboflavin deficiency has never been known to be severe enough to cause the marked debilities noted in animal experiments, but mild riboflavin deficiency is probably common. Perhaps the most common characteristic lesion of riboflavin deficiency is *cheilosis,* which is inflammation and cracking at the angles of the mouth. In addition, a fine, scaly dermatitis often occurs at the angles of the nares, and keratitis of the cornea may occur with invasion of the cornea by capillaries.

Though the manifestations of riboflavin deficiency are usually relatively mild, this deficiency frequently occurs in association with deficiency of thiamine or niacin. Therefore, many deficiency syndromes, including pellagra, beriberi, sprue, and kwashiorkor, are probably due to a combined deficiency of several of the vitamins and also of protein.

VITAMIN B₁₂

Several different *cobalamin* compounds exhibit so-called "vitamin B₁₂" activity.

Vitamin B₁₂ performs many metabolic functions, acting as a hydrogen acceptor coenzyme. For instance, it performs this function in the conversion of amino acids and similar compounds into other substances. Its most important function is probably to act as a coenzyme for reducing ribonucleotides to deoxyribonucleotides, a step that is important in the formation of genes. This could explain the two major functions of vitamin B₁₂: (1) promotion of growth and (2) red blood cell maturation. This latter function was described in Chapter 5.

A special effect of vitamin B₁₂ deficiency is often demyelination of the large nerve fibers of the spinal cord, especially of the posterior columns and occasionally of the lateral columns. As a result, persons with pernicious anemia (anemia caused by failure of red cell maturation) frequently have much simultaneous loss of peripheral sensation and, in severe cases, even become paralyzed.

FOLIC ACID (PTEROYLGLUTAMIC ACID)

Several different pteroylglutamic acids exhibit the "folic acid effect." Folic acid functions as a carrier of hydroxymethyl and formyl groups. Perhaps its most important use in the body is in the synthesis of purines and thymine, which are required for formation of deoxyribonucleic acid. Therefore, folic acid is required for reproduction of the cellular genes. This perhaps explains one of the most important functions of the folic acid — that is, to promote growth.

Folic acid is an even more potent growth promoter than vitamin B₁₂, and, like vitamin B₁₂, is also important for the maturation of red blood cells, as discussed in Chapter 5. However, vitamin B₁₂ and folic acid each perform specific and different functions in promoting growth and maturation of red blood cells.

PYRIDOXINE (VITAMIN B₆)

Pyridoxine exists in the form of *pyridoxal phosphate* in the cells and functions as a coenzyme for many different chemical reactions relating to amino acid and protein metabolism. Its most important role is that of coenzyme in transamination for the synthesis of amino acids. As a result, pyridoxine plays many key roles in metabolism — especially in protein metabolism. Also, it is believed to act in the transport of some amino acids across cell membranes.

In the human being, pyridoxine deficiency has been known to cause convulsions, dermatitis, and gastrointestinal disturbances such as nausea and vomiting in children. However, this deficiency is rare.

PANTOTHENIC ACID

Pantothenic acid mainly is incorporated in the body into coenzyme A, which has many metabolic roles in the cells. Two of these discussed in Chapters 45 and 46 are (1) acetylation of decarboxylated pyruvic acid to form acetyl Co-A prior to its entry into the tricarboxylic acid cycle and (2) degradation of fatty acid molecules into multiple molecules of acetyl Co-A. Thus, lack of pantothenic acid can lead to depressed metabolism of both carbohydrates and fats.

However, in the human being, no definite deficiency syndrome has been proved, presumably because of the wide occurrence of this vitamin in almost all foods and because small amounts of the vitamin can probably be synthesized in the body. Nevertheless, this does not mean that pantothenic acid is not of value in the metabolic systems of the body; indeed, it is perhaps as necessary as any other vitamin.

ASCORBIC ACID (VITAMIN C)

Ascorbic acid is essential for many oxidation reactions in the body. For instance, oxidation of tyrosine and phenylalanine requires an adequate supply of ascorbic acid.

Physiologically, the major function of ascorbic acid appears to be maintenance of normal intercellular substances throughout the body. This includes the formation of collagen because of the stimulatory action of ascorbic acid in the synthesis of hydroxyproline, a constituent of collagen. It also enhances the intercellular cement substance between the cells, the formation of bone matrix, and the formation of tooth dentine.

Deficiency of ascorbic acid for 20 to 30 weeks, as occurred frequently during long sailing voyages in olden days, causes *scurvy*, some effects of which are given here.

One of the most important effects of scurvy is *failure of wounds to heal.* This is caused by failure of the cells to deposit collagen fibrils and intercellular cement substances. As a result, healing of a wound may require several months instead of the several days ordinarily necessary.

Lack of ascorbic acid causes *cessation of bone growth.* The cells of the growing epiphyses continue to proliferate, but no new matrix is laid down between the cells, and the bones fracture easily at the point of growth because of failure to ossify. Also, when an already ossified bone fractures in a person with ascorbic acid deficiency, the osteo-

blasts cannot secrete a new matrix for the deposition of new bone. Consequently, the fractured bone does not heal.

The *blood vessel walls become extremely fragile* in scurvy, presumably because of failure of the endothelial cells to be cemented together properly. Especially are the capillaries likely to rupture, and as a result many small petechial hemorrhages occur throughout the body.

In extreme scurvy the muscle cells sometimes fragment; lesions of the gums with loosening of the teeth occur; infections of the mouth develop; vomiting of blood, bloody stools, and cerebral hemorrhage can all occur; and, finally, high fever often develops before death.

VITAMIN D

Vitamin D increases calcium absorption from the gastrointestinal tract and also helps to control calcium deposition in the bone. The mechanism by which vitamin D increases calcium absorption is the promotion of active transport of calcium through the epithelium of the ileum. It increases the formation of a calcium-binding protein in the epithelial cells that aids in calcium absorption. The specific functions of vitamin D in relation to overall body calcium metabolism and to bone formation are presented in Chapter 53.

VITAMIN E

Several related compounds exhibit so-called "vitamin E activity." Only rare instances of vitamin E deficiency occur in human beings. In lower animals, lack of vitamin E can cause degeneration of the germinal epithelium in the testis and therefore can cause male sterility. Lack of vitamin E can also cause resorption of a fetus after conception in the female. Because of these effects of vitamin E deficiency, vitamin E is sometimes called the "anti-sterility vitamin."

Vitamin E deficiency in animals can also cause paralysis of the hindquarters, and pathologic changes occur in the muscles similar to those found in the disease entity "muscular dystrophy" of the human being. However, administration of vitamin E to patients with muscular dystrophy has not proved to be of any benefit.

Finally, as is true of almost all the vitamins, deficiency of vitamin E prevents normal growth. It sometimes causes degeneration of the renal tubular cells.

Vitamin E is believed to function mainly in relation to unsaturated fatty acids, providing a protective role to prevent oxidation of the unsaturated fats. In the absence of vitamin E, the quantity of unsaturated fats in the cells becomes diminished, causing abnormal structure and function of such cellular organelles as the mitochondria, the lysosomes, and even the cell membrane.

VITAMIN K

Vitamin K is necessary for the formation by the liver of prothrombin, factor VII (proconvertin), factor IX, and factor X, all of which are important in blood coagulation. Therefore, when vitamin K deficiency occurs, blood clotting is retarded. The function of this vitamin has been presented in greater detail in Chapter 7.

Several different compounds, both natural and synthetic, exhibit vitamin K activity. Because vitamin K is synthesized by bacteria in the colon, a dietary source of this vitamin is not usually necessary, but when the bacteria of the colon are destroyed by administration of large quantities of antibiotic drugs, vitamin K deficiency occurs rapidly because of the paucity of this compound in the normal diet.

REFERENCES

Dietary Balances; Regulation of Feedings

Alfin-Slater, R. B., and Kritchevsky, D. (eds.): Nutrition and the Adult: Macronutrients. New York, Plenum Press, 1979.

Booth, D. A. (ed.): Hunger Models: Computable Theory of Feeding Control. New York, Academic Press, 1978.

Bray, G. A., and York, D. A.: Hypothalamic and genetic obesity in experimental animals: An autonomic and endocrine hypothesis. *Physiol. Rev., 59*:719, 1979.

Chaney, M. S., *et al.*: Nutrition. Boston, Houghton Mifflin, 1979.

Collipp, P. J. (ed.): Childhood Obesity. Littleton, Mass., PSG Publishing Co., 1979.

Felig, P.: Starvation. *In* DeGroot, L. J., *et al.* (eds.): Endocrinology. Vol. 3. New York, Grune & Stratton, 1979, p. 1927.

Festing, M. F. W. (ed.): Animal Models of Obesity. New York, Oxford University Press, 1979.

Havel, R. J.: Caloric homeostasis and disorders of fuel transport. *N. Engl. J. Med., 287*:1186, 1972.

Hunt, S. M., *et al.*: Nutrition: Principles and Clinical Practice. New York, John Wiley & Sons, 1980.

Jarrett, R. J. (ed.): Nutrition and Disease. Baltimore, University Park Press, 1978.

Keesey, R. E.: Neurophysiologic control of body fatness. *In* Lauer, R. M., and Shekelle, R. B. (eds.): Childhood Prevention of Atherosclerosis and Hypertension. New York, Raven Press, 1980.

Oscai, L. B.: Recent progress in the possible prevention of obesity. *In* Lauer, R. M., and Shekelle, R. B. (eds.): Childhood Prevention of Atherosclerosis and Hypertension. New York, Raven Press, 1980.

Salans, L. B.: Obesity and the adipose cell. *In* Bondy, P. K., and Rosenberg, L. E. (eds.): Metabolic Control and Disease, 8th Ed. Philadelphia, W. B. Saunders Co., 1980, p. 495.

Shils, M. E., and Goodhart, R. S. (eds.): Modern Nutrition in Health and Disease. Philadelphia, Lea & Febiger, 1979.

Shoden, R. J., and Griffin, W. S.: Fundamentals of Clinical Nutrition. New York, McGraw-Hill, 1980.

Suitor, C. W., and Hunter, M. F.: Nutrition: Principles and Application in Health Promotion. Philadelphia, J. B. Lippincott Co., 1980.

Thompson, C. I.: Controls of Eating. New York, Spectrum Publications, 1979.

Wurtman, R. J., and Wurtman, J. J. (eds.): Disorders of Eating. New York, Raven Press, 1979.

Vitamins and Minerals

Coughlan, M. P. (ed.): Molybdenum and Molybdenum-Containing Enzymes. New York, Pergamon Press, 1980.

Cudlipp, E.: Vitamins. New York, Grosset & Dunlap, 1978.

DeLuca, H. F. (ed.): The Fat-Soluble Vitamins. New York, Plenum Press, 1978.

DeLuca, H. F., and Holick, M. F.: Vitamin D: Biosynthesis, metabolism and mode of action. *In* DeGroot, L. J., *et al.* (eds.): Endocrinology. Vol. 2. New York, Grune & Stratton, 1979, p. 653.

Fraser, D., and Scriver, C. R.: Disorders associated with hereditary or acquired abnormalities in vitamin D function: Hereditary disorders associated with vitamin D resistance or defective phosphate metabolism. *In* DeGroot, L. J., *et al.* (eds.): Endocrinology. Vol. 2. New York, Grune & Stratton, 1979, p. 797.

Frieden, E.: The chemical elements of life. *Sci. Am., 227*:52, 1972.

Hamilton, E. I.: The Chemical Elements and Man: Measurements — Perspectives — Applications. Springfield, Ill., Charles C Thomas, 1978.

Hanck, A., and Ritzel, G. (eds.): Re-Evaluation of Vitamin C. Bern, H. Huber, 1977.

Hodges, R. E.: Nutrition in Medical Practice. Philadelphia, W. B. Saunders Co., 1979.

Karcioglu, Z. A., and Sarper, R. M. (eds.): Zinc and Copper in Medicine. Springfield, Ill., Charles C Thomas, 1980.

Kharasch, N. (ed.): Trace Metals in Health and Disease. New York, Raven Press, 1979.

Lawson, D. E. M.: Vitamin D. New York, Academic Press, 1978.

Prasad, A. S.: Zinc in Human Nutrition. Boca Raton, Fla., CRC Press, 1979.

Shoden, R. J., and Griffin, W. S.: Fundamentals of Clinical Nutrition. New York, McGraw-Hill, 1980.

Suitor, C. W., and Hunter, M. F.: Nutrition: Principles and Application in Health Promotion. Philadelphia, J. B. Lippincott Co., 1980.

Vitamin Compendium: The Properties of the Vitamins and Their Importance in Human and Animal Nutrition. Basle, F. Hoffmann–La Roche, Vitamins and Chemicals Department, 1976.

Zagalak, B., and Friedrich, W. (eds.): Vitamin B12. New York, Walter de Gruyter, 1979.

Part XIII

ENDOCRINOLOGY AND REPRODUCTION

49

Introduction to Endocrinology; and the Pituitary Hormones

The functions of the body are regulated by two major control systems: (1) the nervous system, which has been discussed, and (2) the hormonal, or endocrine, system. In general, the hormonal system is concerned principally with control of the metabolic functions of the body, controlling the rates of chemical reactions in the cells, the transport of substances through cell membranes, or other aspects of cellular metabolism such as growth and secretion. Some hormonal effects occur in seconds, while others require several days simply to start, and then they continue for weeks, months, or even years.

Many interrelationships exist between the hormonal and nervous systems. For instance, at least two glands secrete their hormones only in response to nerve stimuli, the *adrenal medullae* and the *posterior pituitary gland*, and few of the anterior pituitary hormones are secreted to a significant extent except in response to nervous activity in the hypothalamus, as is detailed later in this chapter.

NATURE OF A HORMONE

A hormone is a chemical substance that is secreted into the body fluids by one cell or a group of cells and that exerts a physiological *control* effect on other cells of the body.

At many points in this text we have already discussed different hormones, some of which are called *local hormones* and others, *general hormones*. The local hormones include *acetylcholine*, released at the parasympathetic and skeletal nerve endings; *secretin*, released by the duodenal wall and transported in the blood to the pancreas to cause an alkaline, watery pancreatic secretion; *cholecys-*

tokinin, released in the small intestine to cause contraction of the gallbladder as well as enzyme secretion by the pancreas; and many others. These hormones obviously have specific local effects, whence comes the name local hormones.

On the other hand, the general hormones are secreted by specific *endocrine glands* and are transported in the blood to cause physiological actions at distant points in the body. A few of the general hormones affect all, or almost all, cells of the body; examples are *growth hormone* from the adenohypophysis and *thyroid hormone* from the thyroid gland. Other general hormones, however, affect specific tissues far more than other tissues; for instance, *corticotropin* from the anterior pituitary gland specifically stimulates the adrenal cortex, and the *ovarian hormones* have specific effects on the uterine endometrium. The tissues affected specifically in this way are called *target tissues*. Many examples of target organs will become apparent in the following chapters on endocrinology.

The following general hormones have proved to be of major significance and are discussed in detail in this and the following chapters:

Anterior pituitary hormones: *growth hormone, adrenocorticotropin, thyroid-stimulating hormone, follicle-stimulating hormone, luteinizing hormone, prolactin,* and *melanocyte-stimulating hormone.*

Posterior pituitary hormones: *antidiuretic hormone (vasopressin)* and *oxytocin.*

Adrenocortical hormones: especially *cortisol* and *aldosterone.*

Thyroid hormones: *thyroxine, triiodothyronine,* and *calcitonin.*

Pancreatic hormones: *insulin* and *glucagon.*

Ovarian hormones: *estrogens* and *progesterone.*

Testicular hormone: *testosterone.*
Parathyroid hormone: *parathormone.*
Placental hormones: *chorionic gonadotropin, estrogens, progesterone,* and *human placental lactogen.*

MECHANISMS OF HORMONAL ACTION

The function of the different hormones is to *control* the activity levels of target tissues. To provide this control function they may alter the chemical reactions within the cells, alter the permeability of the cell membrane to specific substances, or activate some other specific cellular mechanism. The different hormones achieve these effects in many different ways. However, two important general mechanisms by which many of the hormones function are (1) activation of the cyclic AMP system of cells, which in turn elicits the specific cellular functions, or (2) activation of the genes of the cells, which causes the formation of intracellular proteins that initiate specific cellular functions. These mechanisms are described as follows:

INTRACELLULAR HORMONAL MEDIATOR—CYCLIC AMP

Many hormones exert their effects on cells by first causing the substance *cyclic 3',5'-adenosine monophosphate* (cyclic AMP) to be formed in the cell. Once formed, the cyclic AMP causes the hormonal effects inside the cell. Thus, *cyclic AMP is an intracellular hormonal mediator.* It is also frequently called a *second messenger* for hormone mediation — the "first messenger" being the original stimulating hormone.

Figure 49–1 illustrates the function of cyclic AMP in more detail. The stimulating hormone acts at the membrane of the target cell, combining with a specific receptor for that particular type of hormone. The specificity of the receptor determines which hormone will affect the target cell. After binding with the receptor, the combination of hormone and receptor activates the enzyme *adenyl cyclase* located in the membrane itself, and the portion of the adenyl cyclase that is exposed to the cytoplasm causes immediate conversion of cytoplasmic ATP into cyclic AMP. The cyclic AMP then initiates any number of cellular functions before it itself is destroyed — functions such as activating enzymes in the cell, altering the cell permeability, initiating synthesis of specific intracellular proteins, causing muscle contraction or relaxation, initiating secretion, and many others. The types of effects that occur inside the cell are determined by the character of the cell

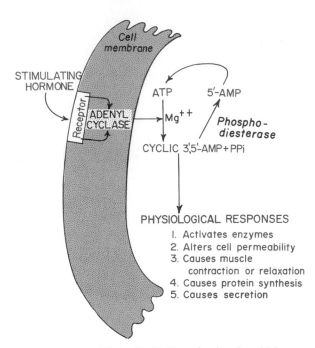

Figure 49–1. The cyclic AMP mechanism, by which many hormones exert their control of cell function.

itself. Thus, a thyroid cell stimulated by cyclic AMP forms thyroid hormones, whereas an adrenocortical cell forms adrenocortical hormones. On the other hand, cyclic AMP affects epithelial cells of the renal tubules by increasing their permeability to water.

The cyclic AMP mechanism has been shown to be an intracellular hormonal mediator for at least some of the functions of the following hormones (and many more):

1. Adrenocorticotropin
2. Thyroid-stimulating hormone
3. Luteinizing hormone
4. Follicle-stimulating hormone
5. Vasopressin
6. Parathyroid hormone
7. Glucagon
8. Catecholamines
9. Secretin
10. The hypothalamic releasing factors

ACTION OF STEROID HORMONES ON THE GENES TO CAUSE PROTEIN SYNTHESIS

A second major means by which hormones act — specifically the steroid hormones secreted by the adrenal cortex, the ovaries, and the testes — is to cause synthesis of proteins in target cells; some of these proteins are enzymes that in turn activate other functions of the cells.

The sequence of events in steroid function is the following:

1. The steroid hormone enters the cytoplasm of the cell, where it binds with a specific *receptor protein.*

2. The combined receptor protein/hormone then diffuses into or is transported into the nucleus.

3. The combination then activates specific genes to form messenger RNA.

4. The messenger RNA diffuses into the cytoplasm, where it promotes the translation process at the ribosomes to form new proteins.

To give an example, aldosterone, one of the hormones secreted by the adrenal cortex, enters the cytoplasm of renal tubular cells, which contain its specific receptor protein. Then the above sequence of events ensues. After about 45 minutes, proteins begin to appear in the renal tubular cells that promote sodium reabsorption from the tubules and potassium secretion into the tubules. Thus, there is a charactcristic delay in the final action of the steroid hormone of 45 minutes to several hours, which is in marked contrast to the almost instantaneous action of some of the peptide and peptide-derived hormones that stimulate cells by the cyclic AMP mechanism.

OTHER MECHANISMS OF HORMONE FUNCTION

Hormones can have other direct effects on cells, though in most instances the precise mechanisms of these effects are not known. For instance, insulin increases the permeability of the cells to glucose, and growth hormone increases the transport of amino acids into cells. In addition, several hormones, such as acetylcholine, directly affect cell membranes by changing their permeabilities to ions and thereby exciting muscular contraction or causing other effects.

THE PITUITARY GLAND AND ITS RELATIONSHIP TO THE HYPOTHALAMUS

The *pituitary gland* (Fig. 49–2), also called the *hypophysis,* is a small gland — less than 1 cm. in diameter and about 0.5 to 1 gram in weight — that lies in the *sella turcica* at the base of the brain and is connected with the hypothalamus by the *pituitary* (or *hypophysial*) *stalk.* Physiologically, the pituitary gland is divisible into two distinct portions: the *anterior pituitary,* also known as the *adenohypophysis,* and the posterior pituitary, also known as the *neurohypophysis.*

Six very important hormones plus several less

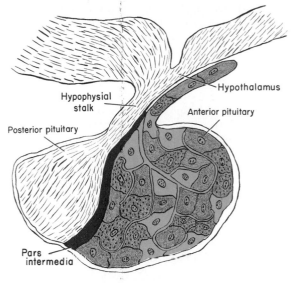

Figure 49–2. The pituitary gland.

important ones are secreted by the *anterior* pituitary, and two important hormones are secreted by the *posterior* pituitary. The hormones of the anterior pituitary play major roles in the control of metabolic functions throughout the body, as shown in Figure 49–3, thus: (1) *Growth hormone* promotes growth of the animal by affecting many metabolic functions throughout the body, especially protein formation. (2) *Adrenocorticotropin* controls the secretion of some of the adrenocortical hormones, which in turn affect the metabolism of glucose, proteins, and fats. (3) *Thyroid-stimulating hormone* controls the rate of secretion of thyroxine by the thyroid gland, and thyroxine in turn controls the rates of most chemical reactions of the entire body. (4) *Prolactin* promotes mammary gland development and milk produc-

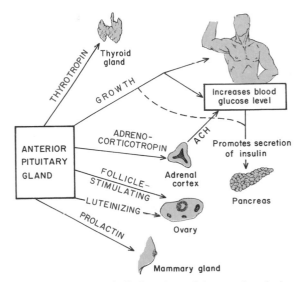

Figure 49–3. Metabolic functions of the anterior pituitary hormones.

tion. And two separate gonadotropic hormones, (5) *follicle-stimulating hormone* and (6) *luteinizing hormone,* control growth of the gonads as well as their reproductive activities.

The two hormones secreted by the posterior pituitary play other roles: (1) *Antidiuretic hormone* controls the rate of water excretion into the urine and in this way helps to control the concentration of water in the body fluids. (2) *Oxytocin* (a) contracts the alveoli of the breasts, thereby helping to deliver milk from the glands of the breast to the nipples during suckling, and (b) contracts the uterus, thus helping in delivery of the baby at the end of gestation.

CONTROL OF PITUITARY SECRETION BY THE HYPOTHALAMUS

Secretion from the posterior pituitary is controlled by nerve fibers originating in the hypothalamus and terminating in the posterior pituitary. In contrast, secretion by the anterior pituitary is controlled by hormones called *hypothalamic releasing* and *inhibitory hormones* (or *factors*) secreted within the hypothalamus itself and then conducted to the anterior pituitary through minute blood vessels called *hypothalamic-hypophysial portal vessels.* In the anterior pituitary these releasing and inhibitory factors act on the glandular cells to control their secretion. This system of control will be discussed in detail later in the chapter.

The hypothalamus receives signals from almost all possible sources in the nervous system. Thus, when a person is exposed to pain, a portion of the pain signal is transmitted into the hypothalamus. Likewise, when a person experiences some powerful depressing or exciting thought, a portion of the signal is transmitted into the hypothalamus. Olfactory stimuli denoting pleasant or unpleasant smells transmit strong signal components through the amygdaloid nuclei into the hypothalamus. *Even the concentrations of nutrients, electrolytes, water, and various hormones* in the blood excite or inhibit various portions of the hypothalamus. Thus, the hypothalamus is a collecting center for information concerned with the well-being of the body, and in turn much of this information is used to control secretion by the pituitary gland.

THE ANTERIOR PITUITARY GLAND AND ITS REGULATION BY HYPOTHALAMIC RELEASING HORMONES

Cell Types of the Anterior Pituitary. The anterior pituitary gland is composed of several different types of cells. In general, there is one type of cell for each type of hormone that is formed in this gland; with special staining techniques these various cell types can be differentiated from one another. The only likely exception to this is that the same cell type may secrete both luteinizing hormone and follicle-stimulating hormone.

THE HYPOTHALAMIC-HYPOPHYSIAL PORTAL SYSTEM

The anterior pituitary is a highly vascular gland with extensive capillary sinuses among the glandular cells. Almost all of the blood that enters these sinuses passes first through a capillary bed in the tissue of the lower hypothalamus and then through small *hypothalamic-hypophysial portal vessels* into the anterior pituitary sinuses. Thus, Figure 49–4 illustrates a small artery supplying the lowermost portion of the hypothalamus called the *median eminence*. Small vascular tufts project into the substance of the median eminence and then return to its surface, coalescing to form the hypothalamic-hypophysial portal vessels. These in turn pass downward along the pituitary stalk to supply the anterior pituitary sinuses.

Secretion of Hypothalamic Releasing and Inhibitory Hormones in the Median Eminence. Special neurons in the hypothalamus synthesize and secrete hormones called *hypothalamic releasing* and *inhibitory hormones* that control the secretion of the anterior pituitary hormones. These neurons originate in various parts of the hypothalamus and send their nerve fibers into the median eminence. The endings of these fibers are different from most endings in the central nervous system in that their function is not to transmit signals from one neuron to another but merely to secrete the hypothalamic releasing and inhibitory hormones into the tissue fluids. These hormones are immediately absorbed into the

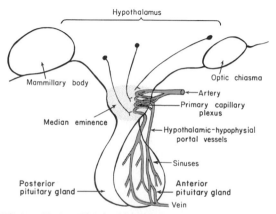

Figure 49–4. The hypothalamic-hypophysial portal system.

hypothalamic-hypophysial portal capillaries and carried directly to the sinuses of the anterior pituitary gland.

Function of the Releasing and Inhibitory Hormones. The function of the releasing and inhibitory hormones is to control the secretion of the anterior pituitary hormones. For each type of anterior pituitary hormone there is a corresponding hypothalamic releasing hormone; for some of the anterior pituitary hormones there is also a corresponding hypothalamic inhibitory factor. For most of the anterior pituitary hormones it is the releasing hormones that are important; but for prolactin, an inhibitory hormone probably exerts most control. The hypothalamic releasing and inhibitory hormones that are of major importance are:

1. *Thyroid-stimulating hormone releasing hormone* (TRH), which causes release of thyroid-stimulating hormone

2. *Corticotropin releasing hormone* (CRH), which causes release of adrenocorticotropin

3. *Growth hormone releasing hormone* (GHRH), which causes release of growth hormone

4. *Luteinizing hormone releasing hormone* (LRH), which causes release of both luteinizing hormone and follicle-stimulating hormone

5. *Prolactin inhibitory hormone* (PIH), which causes inhibition of prolactin secretion

PHYSIOLOGIC FUNCTIONS OF THE ANTERIOR PITUITARY HORMONES

All of the major anterior pituitary hormones besides growth hormone exert their effects by stimulating "target glands"—the thyroid gland, the adrenal cortex, the ovaries, the testicles, and the mammary glands. The functions of each of the anterior pituitary hormones, except for growth hormone, are so intimately concerned with the functions of the respective target glands that their functions will be discussed in subsequent chapters along with the functions of the target glands. Growth hormone, in contrast to other hormones, does not function through a target gland but instead exerts effects on all or almost all tissues of the body.

GROWTH HORMONE

Growth hormone (GH), also called *somatotropic hormone* (SH) or *somatotropin,* is a small protein molecule containing 191 amino acids in a single chain and having a molecular weight of 22,005. It causes growth of all tissues of the body that are capable of growing. It promotes both increased

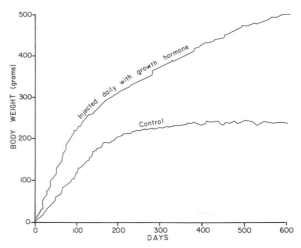

Figure 49–5. Comparison of weight gain of a rat injected daily with growth hormone with that of a normal rat of the same litter.

sizes of the cells and increased mitosis with development of increased numbers of cells. As an example, Figure 49–5 illustrates weight charts of two growing rats, one of which received daily injections of growth hormone, compared with a litter-mate that did not receive growth hormone.

Basic Metabolic Effects of Growth Hormone. Growth hormone is known to have the following basic effects on the metabolic processes of the body:

1. Increased rate of protein synthesis in all cells of the body

2. Decreased rate of carbohydrate utilization throughout the body

3. Increased mobilization of fats and use of fats for energy

Thus, in effect, growth hormone enhances the body proteins, conserves carbohydrates, and uses up the fat stores. It is probable that the increased rate of growth results mainly from the increased rate of protein synthesis.

Stimulation of Growth of Cartilage and Bone — Role of "Somatomedin." Growth hormone does not have a *direct* effect on the growth of cartilage and bone, both of which must grow if the overall structure of the animal is to increase. Instead, growth hormone indirectly stimulates their growth by causing several small proteins, collectively called *somatomedin,* to be formed in the liver and perhaps the kidneys as well; this substance in turn acts directly on the cartilage and bone to promote their growth. Somatomedin is necessary for formation of chondroitin sulfate and collagen, both of which are necessary for growth of the cartilage and bone.

Somatomedin may stimulate growth of other tissues in addition to cartilage and bone, most probably causing deposition of connective tissue and thickening of the skin.

Role of Growth Hormone in Promoting Protein Deposition

Although the most important cause of the increased protein deposition caused by growth hormone is not known, a series of different effects are known, all of which can lead to enhanced protein. These effects are:

1. Enhancement of Amino Acid Transport Through the Cell Membranes. Growth hormone directly enhances transport of at least some and perhaps most amino acids through the cell membranes to the interior of the cells. This increases the concentrations of the amino acids in the cells and is presumed to be at least partly responsible for the increased protein synthesis.

2. Enhancement of Protein Synthesis by the Ribosomes. Even when the amino acids are not increased in the cells, growth hormone still causes protein to be synthesized in increased amounts in the cells. This is believed to be caused partly by a direct effect on the ribosomes, making them produce greater numbers of protein molecules.

3. Increased Formation of RNA. Over more prolonged periods of time, growth hormone also stimulates the transcription process in the nucleus, causing formation of increased quantities of RNA. This in turn promotes protein synthesis.

4. Decreased Catabolism of Protein and Amino Acid. In addition to the increase in protein synthesis, there is a decrease in the breakdown of protein and the utilization of protein and amino acids for energy. A possible, if not probable, reason for this effect is that growth hormone also mobilizes large quantities of free fatty acids from the adipose tissue, and these in turn are used to supply most of the energy for the body cells, thus acting as a potent "protein sparer."

Summary. Growth hormone enhances almost all facets of amino acid uptake and protein synthesis by cells, while at the same time reducing the breakdown of proteins.

Effect of Growth Hormone in Enhancing Fat Utilization for Energy

Growth hormone has a specific effect in causing release of fatty acids from adipose tissue and therefore increasing the fatty acid concentration in the body fluids. In addition, in the tissues it enhances the conversion of fatty acids to acetyl CoA with subsequent utilization of this for energy. Therefore, under the influence of growth hormone, fat is utilized for energy in preference to both carbohydrates and proteins.

Effect of Growth Hormone on Carbohydrate Metabolism

Growth hormone has three major effects on cellular metabolism of glucose. These effects are (1) decreased utilization of glucose for energy, (2) marked enhancement of glycogen deposition in the cells, and (3) diminished uptake of glucose by the cells.

Decreased Glucose Utilization for Energy. Unfortunately, we do not know the precise mechanism by which growth hormone decreases glucose utilization by the cells. However, the decrease probably results partially from the increased mobilization and utilization of fatty acids for energy caused by growth hormone. That is, the fatty acids form large quantities of acetyl CoA, which in turn initiate feedback effects that block the glycolytic breakdown of glucose and glycogen.

Enhancement of Glycogen Deposition. Since glucose and glycogen cannot be utilized for energy, the glucose that does enter the cells is rapidly polymerized into glycogen and deposited. Therefore, the cells rapidly become saturated with glycogen and can store no more.

Diminished Uptake of Glucose by the Cells and Increased Blood Glucose Concentration. When growth hormone is first administered to an animal, the cellular uptake of glucose is enhanced and the blood glucose concentration falls slightly. However, as the cells become saturated with glycogen and their utilization of glucose for energy decreases, further uptake of glucose then becomes greatly diminished. Without normal cellular uptake, the blood concentration of glucose increases, sometimes to as high as 50 to 100 per cent above normal.

Diabetogenic Effect of Growth Hormone. We have already pointed out that growth hormone leads to moderately increased blood glucose concentration. This in turn stimulates the beta cells of the islets of Langerhans to secrete extra insulin. In addition to this effect, growth hormone has a moderate direct stimulatory effect on the beta cells as well. The combination of these two effects sometimes so greatly overstimulates insulin secretion by the beta cells that they literally "burn out." When this occurs the person develops diabetes mellitus, a disease that will be discussed in detail in Chapter 52. Therefore, growth hormone is said to have a *diabetogenic effect.*

Regulation of Growth Hormone Secretion

For many years it was believed that growth hormone was secreted primarily during the period of growth but then disappeared from the blood at adolescence. However, this belief has proved to be very far from the truth, because after adolescence secretion continues at a rate almost as great as that in childhood. Furthermore, the rate of growth hormone secretion increases and decreases within minutes in relation to the person's state of nutrition or stress, as during starvation, hypoglycemia, exercise, excitement, and trauma.

The normal concentration of growth hormone in the plasma of an adult is about 3 nanograms per milliliter and in the child about 5 nanograms per milliliter. However, these values often increase to as high as 50 nanograms per milliliter after depletion of the body stores of proteins or carbohydrates. Under acute conditions, hypoglycemia is a far more potent stimulator of growth hormone secretion than is a decrease in the amino acid concentration in the blood. On the other hand, in chronic conditions the degree of cellular protein depletion seems to be more correlated with the level of growth hormone secretion than is the availability of glucose. For instance, the extremely high levels of growth hormone that occur during starvation are very closely related to the amount of protein depletion.

Thus, it is almost certain that growth hormone secretion is controlled moment by moment by the nutritional and stress status of the body, and it seems that the most important factor in the control of growth hormone secretion is the level of cell protein, though changes in blood glucose concentration can also cause extremely rapid and dramatic alterations in growth hormone secretion. Consequently, it can be postulated that growth hormone operates in a feedback control system as follows: When the tissues begin to suffer from malnutrition, especially from poor protein nutrition, large quantities of growth hormone are secreted. Growth hormone, in turn, promotes the synthesis of new proteins, while at the same time conserving the protein already present in the cells.

All of these feedback effects that control growth hormone secretion are believed to be mediated through the hypothalamus. The hypothalamus secretes *growth hormone releasing hormone* (GHRH), which in turn causes the anterior pituitary to secrete the growth hormone. The hypothalamus nucleus that causes growth hormone secretion is the ventromedial nucleus, the same nucleus that helps to control other aspects of metabolism such as the level of hunger and feeding.

ABNORMALITIES OF GROWTH HORMONE SECRETION

Dwarfism. Some instances of dwarfism result from deficiency of anterior pituitary secretion during childhood. In general, the features of the body develop in appropriate proportion to each other, but the rate of development is greatly decreased. A child who has reached the age of 10 may have the bodily development of a child of 4 to 5, whereas the same person on reaching the age of 20 may have the bodily development of a child of 7 to 10.

The dwarf usually does not pass through puberty and does not secrete a sufficient quantity of gonadotropic hormones to develop adult sexual functions. In one-third of dwarfs, however, the deficiency is of growth hormone alone; these individuals do mature sexually and occasionally do reproduce.

Giantism. Occasionally, the growth hormone–producing cells of the anterior pituitary become excessively active, and sometimes even growth hormone cell (acidophilic cell) tumors occur in the gland. As a result, large quantities of growth hormone are produced. All body tissues grow rapidly, including the bones, and if the epiphyses of the long bones have not become fused with the shafts, height increases, so that the person becomes a giant with a height as great as 8 to 9 feet. Thus, for giantism to occur, the tumor must occur prior to adolescence.

Most giants, unfortunately, eventually develop hypopituitarism if they remain untreated, because the tumor of the pituitary gland grows until the gland itself is destroyed. This general deficiency of pituitary hormones, if untreated, usually causes death in early adulthood. However, once giantism is diagnosed, further development can usually be blocked by microsurgical removal of the tumor from the pituitary gland or irradiation of the gland.

Acromegaly. If a growth hormone cell tumor occurs after adolescence — that is, after the epiphyses of the long bones have fused with the shafts — the person cannot grow taller; but his soft tissues can continue to grow, and the bones can grow in thickness. This condition is known as *acromegaly*. Enlargement is especially marked in the small bones of the hands and feet and in the *membranous bones,* including the cranium, the nose, the bosses on the forehead, the supraorbital ridges, the lower jawbone, and portions of the vertebrae, for their growth does not cease at adolescence anyway. Consequently, the jaw protrudes forward, sometimes as much as a half inch, the forehead slants forward because of excess development of the supraorbital ridges, the nose increases to as much as twice normal size, the foot requires a size 14 or larger shoe, and the fingers become extremely thickened, so that the hand develops a size almost twice normal. In addition to these effects, changes in the vertebrae ordinarily cause a hunched back. Finally, many soft tissue organs, such as the tongue, the liver, and espe-

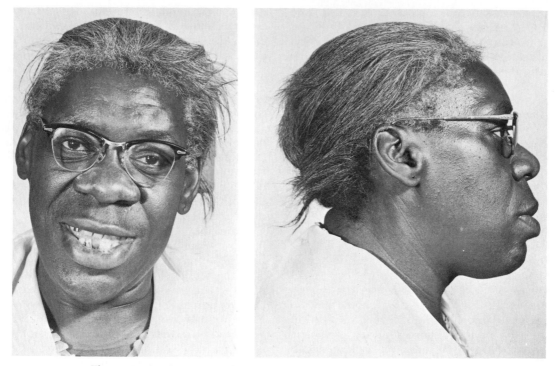

Figure 49–6. An acromegalic patient. (Courtesy of Dr. Herbert Langford.)

cially the kidneys, become greatly enlarged. A typical acromegalic is shown in Figure 49–6.

THE POSTERIOR PITUITARY GLAND AND ITS RELATION TO THE HYPOTHALAMUS

The *posterior pituitary gland,* also called the *neurohypophysis,* is composed mainly of glia-like cells called *pituicytes.* However, the pituicytes do not secrete hormones; they act simply as a supporting structure for large numbers of *terminal nerve fibers* and *terminal nerve endings* from nerve tracts that originate in the *supraoptic* and *paraventricular nuclei* of the hypothalamus, as shown in Figure 49–7. These tracts pass to the neurohypophysis through the *pituitary stalk* (hypophysial stalk). The nerve endings are bulbous knobs that lie on the surfaces of capillaries onto which they secrete the posterior pituitary hormones: (1) *antidiuretic hormone* (ADH), also called *vasopressin,* and (2) *oxytocin.* Both of these hormones are small polypeptides, each containing nine amino acids. They are identical with each other except for two of the amino acids.

If the pituitary stalk is cut near the pituitary gland, leaving the entire hypothalamus intact, the posterior pituitary hormones continue, after a transient decrease for a few days, to be secreted almost normally, but they are then secreted by the cut ends of the fibers within the hypothalamus and not by the nerve endings in the posterior pituitary. The reason is that the hormones are initially synthesized in the cell bodies of the supraoptic and paraventricular nuclei and are then transported to the nerve endings in the posterior pituitary gland, requiring several days to reach the gland.

ADH is formed primarily in the supraoptic nuclei, while *oxytocin is formed primarily in the paraventricular nuclei.* However, each of these two nuclei can synthesize approximately one-sixth as much of the second hormone as of its primary hormone.

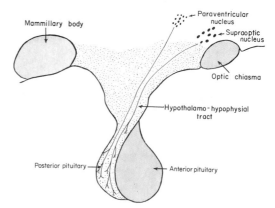

Figure 49–7. Hypothalamic control of the posterior pituitary.

Under resting conditions, large quantities of both ADH and oxytocin accumulate in the nerve endings of the posterior pituitary gland. Then when nerve impulses are transmitted downward along the fibers from the supraoptic and paraventricular nuclei, the hormones are immediately released from the nerve endings and are absorbed into the adjacent capillaries.

PHYSIOLOGICAL FUNCTIONS OF ANTIDIURETIC HORMONE (VASOPRESSIN)

Extremely minute quantities of antidiuretic hormone (ADH) — as little as 2 nanograms — when injected into a person can cause antidiuresis, that is, decreased excretion of water by the kidneys. This antidiuretic effect was discussed in detail in Chapter 25. Briefly, in the absence of ADH, the collecting ducts (and perhaps, to a lesser extent, parts of the distal tubules as well) are almost totally impermeable to water, which prevents significant reabsorption of water and therefore allows extreme loss of water into the urine. On the other hand, in the presence of ADH, the permeability of these ducts to water increases greatly and allows most of the water in the tubular fluid to be reabsorbed, thereby conserving water in the body.

Regulation of ADH Production

Osmotic Regulation. When the body fluids become highly concentrated, the supraoptic nuclei become excited, impulses are transmitted to the posterior pituitary, and ADH is secreted. This passes by way of the blood to the kidneys, where it increases the permeability of the collecting tubules to water. As a result, most of the water is reabsorbed from the tubular fluid, while electrolytes continue to be lost into the urine. This effect dilutes the extracellular fluids, returning them to a normal osmotic composition. The details of this mechanism were also discussed in Chapter 25 in relation to body fluid electrolyte control.

Stimulation of ADH Secretion by Low Blood Volume — Pressor Effect of ADH. ADH in moderate concentrations has a very potent effect of constricting the arterioles and therefore of increasing the arterial pressure. Also, one of the most powerful stimuli of all for increasing the secretion of ADH is severe loss of blood volume. As little as 10 per cent loss of blood will promote a moderate increase in ADH secretion, and 25 per cent or more blood loss can cause as much as 20 to 50 times normal rates of secretion.

The increased secretion is believed to result mainly from the low pressure caused in the atria of the heart by the low blood volume. The relaxation of the atrial stretch receptors supposedly elicits the increase in ADH secretion. However, the baroreceptors of the carotid, aortic, and pulmonary regions also participate in this control of ADH secretion.

The marked secretion of ADH following hemorrhage perhaps plays a very important role in the homeostasis of arterial pressure. Because ADH has this potent pressor effect, it is also called *vasopressin*.

OXYTOCIC HORMONE

Effect on the Uterus. An "oxytocic" substance is one that causes contraction of the pregnant uterus. The hormone *oxytocin*, in accordance with its name, powerfully stimulates the pregnant uterus, especially toward the end of gestation. Therefore, many obstetricians believe that this hormone is at least partially responsible for effecting the birth of the baby. This will be discussed in Chapter 56 in relation to reproduction and pregnancy.

Effect of Oxytocin on Milk Ejection. Oxytocin also plays an especially important function in the process of lactation, for this hormone causes milk to be expressed from the alveoli into the ducts so that the baby can obtain it by suckling. This, too, will be discussed in detail in Chapter 56.

REFERENCES

Austin, C. R., and Short, R. V. (eds.): Mechanisms of Hormone Action. New York, Cambridge University Press, 1979.

Bargmann, W., et al. (eds.): Neurosection and Neuroendocrine Activity; Evolution, Structure and Function. New York, Springer-Verlag, 1978.

Baxter, J. D., and MacLeod, K. M.: Molecular basis for hormone action. In Bondy, P. K., and Rosenberg, L. E. (eds.): Metabolic Control and Disease, 8th Ed. Philadelphia, W. B. Saunders Co., 1980, p. 104.

Besser, G. M. (ed.): The Hypothalamus and Pituitary. Clin. Endocrinol. Metab., 6(1), 1977.

Bowers, C. Y., et al.: Hypothalamic peptide hormones; Chemistry and physiology. In DeGroot, L. J., et al. (eds.): Endocrinology. Vol. 1. New York, Grune & Stratton, 1979, p. 65.

Catt, K. J., and Dufau, M. L.: Peptide hormone receptors. Annu. Rev. Physiol., 39:529, 1977.

Chiodini, P. G., and Liuzzi, A.: The Regulation of Growth Hormone Secretion. St. Albans, Vt., Eden Medical Research, 1979.

DeGroot, L. J. (ed.): Endocrinology. New York, Grune & Stratton, 1979.

Dillon, R. S.: Handbook of Endocrinology: Diagnosis and Management of Endocrine and Metabolic Disorders, 2nd Ed. Philadelphia, Lea & Febiger, 1980.

Ezrin C., *et al.* (eds.): Pituitary Diseases. Boca Raton, Fla., CRC Press, 1980.

Fain, J. N., and Butcher, F. R.: Cyclic nucleotides in mode of hormone action. *Int. Rev. Physiol.,* 16:241, 1977.

Fink, G., and Geffen, L. B.: The hypothalamo-hypophysial system: Model for central peptidergic and monoaminergic transmission. *In* Porter, R. (ed.): International Review of Physiology: Neurophysiology III. Vol. 17. Baltimore, University Park Press, 1978, p. 1.

Goss, R. J.: The Physiology of Growth. New York, Academic Press, 1977.

Gray, C. H., and James, V. H. T.: Hormones in Blood. New York, Academic Press, 1979.

Jeffcoate, S. L., and Hutchinson, J. S. M. (eds.): The Endocrine Hypothalamus. New York, Academic Press, 1978.

Johnston, F. E., *et al.* (eds.): Human Physical Growth and Maturation: Methodologies and Factors. New York, Plenum Press, 1980.

Jubiz, W.: Endocrinology. New York, McGraw-Hill, 1979.

Kastrup, K. W., and Neilsen, J. H. (eds.): Growth Factors: Cellular Growth Processes, Growth Factors, Hormonal Control of Growth. New York, Pergamon Press, 1978.

Kleeman, C. R., and Beri, T.: The neurohypophysial hormones: Vasopressin. *In* DeGroot, L. J., *et al.* (eds.): Endocrinology. Vol. 1. New York, Grune & Stratton, 1979, p. 253.

Kostyo, J. L., and Isaksson, O.: Growth hormone and the regulation of somatic growth. *Int. Rev. Physiol.,* 13:255, 1977.

Krulich, L.: Central neurotransmitters and the secretion of prolactin, GH, LH, and TSH. *Annu. Rev. Physiol.,* 41:603, 1979.

Labrie, F., *et al.*: Mechanism of action of hypothalamic hormones in the adenohypophysis. *Annu. Rev. Physiol.,* 41:555, 1979.

Li, C. H. (ed.): Hypothalamic Hormones. New York, Academic Press, 1979.

McCann, S. M., and Ojeda, S. R.: The role of brain monoamines, acetylcholine and prostaglandins in the control of anterior pituitary function. *In* DeGroot, L. J., *et al.* (eds.): Endocrinology. Vol. 1. New York, Grune & Stratton, 1979, p. 55.

Merimee, T. J.: Growth hormone: Secretion and action. *In* DeGroot, L. J., *et al.* (eds.): Endocrinology. Vol. 1. New York, Grune & Stratton, 1979, p. 123.

Ontjes, D. A., *et al.*: The anterior pituitary gland. *In* Bondy, P. K., and Rosenberg, L. E. (eds.): Metabolic Control and Disease, 8th Ed. Philadelphia, W. B. Saunders Co., 1980, p. 1165.

Reichlin, S., *et al.*: Hypothalamic Hormones. *Annu. Rev. Physiol.,* 38:389, 1976.

Richenberg, H. V. (ed.): Biochemistry and Mode of Action of Hormones II. Baltimore, University Park Press, 1978.

Rosenfield, R. L.: Somatic growth and maturation. *In* DeGroot, L. J., *et al.* (eds.): Endocrinology. Vol. 3. New York, Grune & Stratton, 1979, p. 1805.

Ryan, W. G., *et al.* (eds.): Endocrine Disorders: A Pathophysiologic Approach, 2nd Ed. Chicago, Year Book Medical Publishers, 1980.

Schulster, D., and Levitski, A. (eds.): Cellular Receptors for Hormones and Neurotransmitters. New York, John Wiley & Sons, 1980.

Sowers, J. R. (eds.): Hypothalamic Hormones. Stroudsburg, Pa., Dowden, Hutchinson & Ross, 1980.

Tepperman, J.: Metabolic and Endocrine Physiology: An Introductory Text. Chicago, Year Book Medical Publishers, 1980.

Tolis, G., *et al.* (eds.): Clinical Neuroendocrinology. New York, Raven Press, 1979.

Vorherr, H.: Oxytocin. *In* Degroot, L. J., *et al.* (eds.): Endocrinology. Vol. 1. New York, Grune & Stratton, 1979, p. 277.

Weiner, R. I., and Ganong, W. F.: Role of brain monoamines and histamine in regulation of anterior pituitary secretion. *Physiol. Rev.,* 58:905, 1978.

Weitzman, R., and Kleeman, C. R.: Water metabolism and the neurohypophysial hormones. *In* Bondy, P. K., and Rosenberg, L. E. (eds.): Metabolic Control and Disease, 8th Ed. Philadelphia, W. B. Saunders Co., 1980, p. 1241.

50

The Thyroid Hormones

The thyroid gland, which is located immediately below the larynx on either side of and anterior to the trachea, secretes large amounts of two hormones, *thyroxine* and *triiodothyronine,* that have a profound effect on the metabolic rate of the body. It also secretes *calcitonin,* a hormone that is important for calcium metabolism and which will be considered in detail in Chapter 53. Complete lack of thyroid secretion usually causes the basal metabolic rate to fall to about 40 per cent below normal, and extreme excesses of thyroid secretion can cause the basal metabolic rate to rise as high as 60 to 100 per cent above normal. Thyroid secretion is controlled primarily by thyroid-stimulating hormone secreted by the anterior pituitary gland.

The purpose of this chapter is to discuss the formation and secretion of the thyroid hormones, their functions in the metabolic schema of the body, and regulation of their secretion.

FORMATION AND SECRETION OF THE THYROID HORMONES

The most abundant of the hormones secreted by the thyroid glands is *thyroxine.* However, moderate amounts of *triiodothyronine* are also secreted. The functions of these two hormones are qualitatively the same, but they differ in rapidity and intensity of action. Triiodothyronine is about four times as potent as thyroxine, but it is present in the blood in smaller quantities and persists for a much shorter time than does thyroxine.

Physiologic Anatomy of the Thyroid Gland. The thyroid gland is composed, as shown in Figure 50–1, of large numbers of closed *follicles* filled with a secretory substance called *colloid* and lined with *cuboidal epithelioid cells* that secrete into the interior of the follicles. The major constituent of colloid is a large glycoprotein, *thyroglobulin,* which contains the thyroid hor-

mones. Once the secretion has entered the follicles, the thyroid hormones must be absorbed back through the follicular epithelium into the blood before they can function in the body.

REQUIREMENTS OF IODINE FOR FORMATION OF THYROXINE

To form normal quantities of thyroxine, approximately 50 mg. of ingested iodine are required *each year,* or approximately *1 mg. per week.* To prevent iodine deficiency, common table salt is iodized with one part sodium iodide to every 100,000 parts sodium chloride.

THE IODINE PUMP (IODIDE TRAPPING)

The first stage in the formation of thyroid hormones, as shown in Figure 50–2, is the transfer of iodides from the extracellular fluid into the thyroid glandular cells and thence into

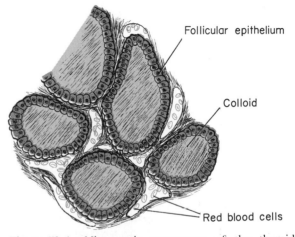

Figure 50–1. Microscopic appearance of the thyroid gland, showing the secretion of thyroglobulin into the follicles.

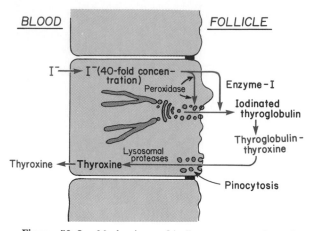

Figure 50–2. Mechanisms of iodine transport, thyroxine formation, and thyroxine release into the blood. (Triiodothyronine formation and release parallels that of thyroxine.)

the follicle. The cell membranes have a specific ability to transport iodides actively to the interior of the follicle; this is called the *iodide pump,* or *iodide trapping.* In a normal gland, the iodide pump can concentrate the iodide ion to about 40 times its concentration in the blood. However, when the thyroid gland becomes maximally active, the concentration ratio can rise to several times this value.

THYROGLOBULIN AND CHEMISTRY OF THYROXINE AND TRIIODOTHYRONINE FORMATION

Formation and Secretion of Thyroglobulin by the Thyroid Cells. The thyroid cells are typical protein-secreting glandular cells as illustrated in Figure 50–2. The endoplasmic reticulum and Golgi complex synthesize and secrete into the follicles a large glycoprotein molecule called *thyroglobulin* with a molecular weight of 660,000.

Each molecule of thyroglobulin contains 140 tyrosine amino acids, and these are the major substrates that combine with iodine to form the thyroid hormones. These hormones form *within* the thyroglobulin molecule. That is, the tyrosine amino acid residues, as well as the thyroxine and triiodothyronine hormones formed from them, remain a part of the thyroglobulin molecule during the entire synthesis of the thyroid hormones.

In addition to secreting the thyroglobulin, the glandular cells also provide the iodine, the enzymes, and other substances necessary for thyroid hormone synthesis.

Oxidation of the Iodide Ion. An essential step in the formation of the thyroid hormones is conversion of the iodide ions to an *oxidized form of iodine* that is then capable of combining directly with the amino acid tyrosine. This oxidation of iodine is promoted by the enzyme *peroxidase* and accompanying *hydrogen peroxide,* which together provide a potent system capable of oxidizing iodides. The peroxidase is located either in the apical membrane of the cell or in the cytoplasm immediately adjacent to this membrane, thus providing the oxidized iodine at exactly the point in the cell where the thyroglobulin molecule first issues forth from the Golgi complex.

Iodination of Tyrosine and Formation of the Thyroid Hormones — "Organification" of Thyroglobulin. The binding of iodine with the thyroglobulin molecule is called *organification* of the thyroglobulin. Oxidized iodine even in the molecular form will bind directly but slowly with the amino acid tyrosine, but when the oxidized iodine is associated with the peroxidase enzyme system, this process can occur in seconds or minutes. Therefore, almost as rapidly as the thyroglobulin molecule is released from the Golgi apparatus, or as it is secreted through the apical cell membrane into the follicle, iodine binds with about one-sixth of the tyrosine residues within the thyroglobulin molecule.

Figure 50–3 illustrates the successive stages of iodination of tyrosine and the final formation of the two important thyroid hormones, thyroxine and triiodothyronine. Tyrosine is first iodized to form *monoiodotyrosine* and then to form *diiodotyrosine.* Then during the next few minutes, hours, and days, more and more of the diiodotyrosine residues become *coupled* with each other. The product of the coupling reaction is the molecule *thyroxine,* which also remains part of the thyroglobulin molecule. Or, one molecule of monoiodotyrosine couples with one molecule of diiodotyrosine to form *triiodothyronine.*

Storage of Thyroglobulin. After synthesis of the thyroid hormones has run its course, each thyroglobulin molecule contains from 5 to 6

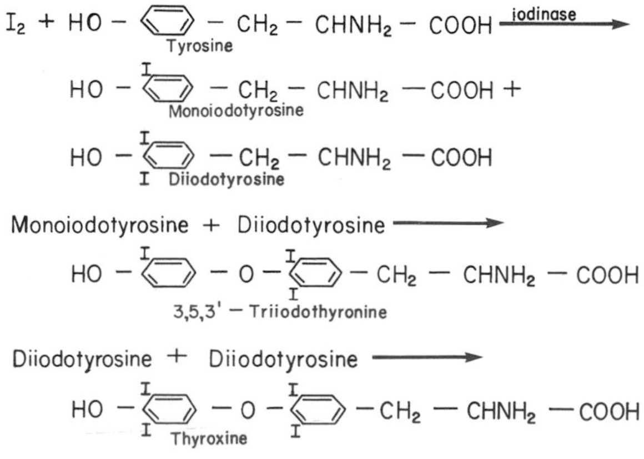

$$I_2 + HO - \langle\rangle - CH_2 - CHNH_2 - COOH \xrightarrow{\text{iodinase}}$$

Tyrosine

$$HO - \overset{I}{\langle\rangle} - CH_2 - CHNH_2 - COOH +$$

Monoiodotyrosine

$$HO - \overset{I}{\underset{I}{\langle\rangle}} - CH_2 - CHNH_2 - COOH$$

Diiodotyrosine

Monoiodotyrosine + Diiodotyrosine $\longrightarrow$

$$HO - \overset{I}{\langle\rangle} - O - \overset{I}{\langle\rangle} - CH_2 - CHNH_2 - COOH$$

3,5,3' — Triiodothyronine

Diiodotyrosine + Diiodotyrosine $\longrightarrow$

$$HO - \overset{I}{\underset{I}{\langle\rangle}} - O - \overset{I}{\underset{I}{\langle\rangle}} - CH_2 - CHNH_2 - COOH$$

Thyroxine

Figure 50–3. Chemistry of thyroxine and triiodothyronine formation.

thyroxine molecules, and there is an average of 1 triiodothyronine molecule for every 3 to 4 thyroglobulin molecules — about 18 molecules of thyroxine for every 1 molecule of triiodothyronine. In this form the thyroid hormones are often stored in the follicles for several months. In fact, the total amount stored is sufficient to supply the body with its normal requirements of thyroid hormones for one to three months. Therefore, even when synthesis of thyroid hormone ceases entirely, the effects of deficiency might not be observed for several months.

RELEASE OF THYROXINE AND TRIIODOTHYRONINE FROM THYROGLOBULIN

Thyroglobulin itself is not released into the circulating blood; instead, the thyroxine and triiodothyronine are first cleaved from the thyroglobulin molecule, and then these free hormones are released. This process occurs as follows: The apical surface of the thyroid cells normally sends out pseudopod-like extensions that close around small portions of the colloid to form pinocytic vesicles. Then lysosomes immediately fuse with these vesicles to form digestive vesicles containing the digestive enzymes from the lysosomes mixed with the colloid. The *proteinases* among these enzymes digest the thyroglobulin molecules and release the thyroxine and triiodothyronine, which then diffuse through the base of the thyroid cell into the surrounding capillaries. Thus, the thyroid hormones are released into the blood.

TRANSPORT OF THYROXINE AND TRIIODOTHYRONINE TO THE TISSUES

Binding of Thyroxine and Triiodothyronine with the Plasma Proteins. On entering the blood, all but minute portions of the thyroxine and triiodothyronine combine immediately with several of the plasma proteins, especially with *thyroxine-binding globulin,* which is a glycoprotein. Then, half of the thyroxine bound with the proteins is released to the tissue cells approximately every 6 days, whereas half of the triiodothyronine — because of its lower affinity for the proteins — is released to the cells in approximately 1.3 days.

On entering the cells, both of these hormones again bind with intracellular proteins, the thyroxine once again binding more strongly than the triiodothyronine. Therefore, they are again stored, but this time in the functional cells themselves, and they are used slowly over a period of days or weeks.

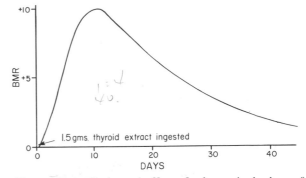

Figure 50–4. Prolonged effect of a large single dose of thyroid hormone on the basal metabolic rate.

Latency and Duration of Action of the Thyroid Hormones. After injection of a large quantity of thyroxine into a human being, essentially no effect on the metabolic rate can be discerned for two to three days, thereby illustrating that there is a *long latent* period before thyroxine activity begins. Once activity does begin, it increases progressively and reaches a maximum in 10 to 12 days, as shown in Figure 50–4. Thereafter, it decreases with a half-time of about 15 days. Some of the activity still persists as long as 6 weeks to 2 months later.

The actions of triiodothyronine occur about 4 times as rapidly as those of thyroxine, with a latent period as short as 6 to 12 hours and maximum cellular activity occurring within 2 to 3 days.

A large part of the latency and prolonged period of action of these hormones is caused by their binding with proteins both in the plasma and in the tissue cells, followed by their slow release. However, we shall see in subsequent discussions that part of the latent period also results from the manner in which these hormones perform their functions in the cells themselves.

FUNCTIONS OF THE THYROID HORMONES IN THE TISSUES

The thyroid hormones have two major effects on the body: (1) an increase in the overall metabolic rate and (2) in children, stimulation of growth.

GENERAL INCREASE IN METABOLIC RATE

The thyroid hormones increase the metabolic activities of almost all tissues of the body. The basal metabolic rate can increase to as much as 60 to 100 per cent above normal when large quantities of the hormones are secreted. The rate of utilization of foods for energy is greatly accelerated. The rate of protein synthesis is at times increased, while at the same time the rate of protein catabolism is also increased. The growth rate of young persons is greatly accelerated. The mental processes are excited, and the activity of many endocrine glands is often increased. Yet despite the fact that we know all these many changes in metabolism under the influence of the thyroid hormones, the basic mechanism (or mechanisms) by which they act is almost completely unknown. However, some of the possible mechanisms of action of the thyroid hormones are described in the following sections.

Effect of Thyroid Hormones on Causing Increased Protein Synthesis. When either thyroxine or triiodothyronine is given to an animal, the cellular genes are stimulated to synthesize proteins in almost all tissues of the body. It is believed that this stimulation of the genes occurs in the following way: (1) The thyroid hormone combines with a "receptor" protein in the cell nucleus. (2) This combination, or a product of it, then activates a large portion of the cellular genes to cause RNA formation and subsequent protein formation.

Effect of Thyroid Hormones on the Cellular Enzyme Systems. Within a week or so following administration of the thyroid hormones, at least 100 and probably many more intracellular enzymes are increased in quantity. As an example, one enzyme, α-glycerophosphate dehydrogenase, can be increased to an activity six times its normal level. Since this enzyme is particularly important in the degradation of carbohydrates, its increase could help to explain the rapid utilization of carbohydrates under the influence of thyroxine. Also, the oxidative enzymes and the elements of the electron transport system, both of which are normally found in mitochondria, are greatly increased.

Effect of Thyroid Hormones on Mitochondria. When thyroxine or triiodothyronine is given to an animal, the mitochondria in most cells of the body increase in size and also in number. Furthermore, the total membrane surface of the mitochondria increases almost directly in proportion to the increase in metabolic rate of the whole animal. Therefore, it seems almost to be an obvious deduction that the principal function of thyroxine might be simply to increase the number and activity of mitochondria and that these in turn increase the rate of formation to ATP to energize cellular function. Unfortunately, though, the increase in number and activity of mitochondria could as well be the *result* of increased activity of the cells as be the cause of the increase.

Effect of Thyroid Hormone in Increasing Active Transport of Ions Through Cell Membranes. One of the enzymes that becomes increased in response to thyroid hormone is *Na-K ATPase*. This in turn increases the rate of transport of both sodium and potassium through the cell membranes of some tissues. Since this process utilizes energy and also increases the amount of heat produced in the body, it has also been suggested that this might be one of the mechanisms by which thyroid hormone increases the body's metabolic rate.

Summary. It is clear that we know of many effects that occur in the cells throughout the body under the influence of thyroid hormone. Yet, a specific metabolic mechanism that leads to all of these effects has been elusive. At present, the most likely basic function of the thyroid hormones is their capability to activate the genes in the cell nucleus with resulting formation of many new cellular enzymes.

EFFECT OF THYROID HORMONE ON GROWTH

Thyroid hormone has both general and specific effects on growth. For instance, it has long been known that thyroid hormone is essential for the metamorphic change of the tadpole into the frog. In the human being, the effect of thyroid hormone on growth is manifest mainly in growing children. In those with hypothyroidism, the rate of growth is greatly retarded. In those with hyperthyroidism, excessive skeletal growth often occurs, causing the child to become considerably taller than otherwise. However, the epiphyses close at an early age, so that the eventual height of the adult may be shorter.

The growth-promoting effect of thyroid hormone is presumably based on its ability to promote protein synthesis. (On the other hand, a great excess of thyroid hormone can cause more rapid catabolism than synthesis of protein, so that the protein stores are then actually mobilized and amino acids released into the extracellular fluids.)

EFFECTS OF THYROID HORMONE ON SPECIFIC PHYSIOLOGICAL MECHANISMS

Effect on Carbohydrate Metabolism. Thyroid hormone stimulates almost all aspects of carbohydrate metabolism, including rapid uptake of glucose by the cells, enhanced glycolysis, enhanced gluconeogenesis, increased rate of absorption from the gastrointestinal tract, and even increased insulin secretion with its resultant secondary effects on carbohydrate metabolism. All of these effects probably result from the overall increase in enzymes caused by thyroid hormone.

Effect on Fat Metabolism. Essentially all aspects of fat metabolism are also enhanced under the influence of thyroid hormone. Since fats are the major source of long-term energy supplies, the fat stores of the body are depleted to a greater extent than are most of the other tissue elements; in particular, lipids are mobilized from the fat tissue, which increases the free fatty acid concentration in the plasma, and thyroid hormone also greatly accelerates the oxidation of free fatty acids by the cells.

Effect on Body Weight. Greatly increased thyroid hormone production in the fully grown person almost always decreases the body weight, and greatly decreased production almost always increases the body weight; but these effects do not always occur, because thyroid hormone increases the appetite, and this may overbalance the change in the metabolic rate.

Effect on the Cardiovascular System. Increased metabolism in the tissues causes more rapid utilization of oxygen than normal and causes greater than normal quantities of metabolic end-products to be released from the tissues. These effects cause vasodilatation in most of the body tissues, thus increasing blood flow in almost all areas of the body. Especially does the rate of blood flow in the skin increase because of the increased necessity for heat elimination.

As a consequence of the increased blood flow to the constituent parts of the body, the cardiac output and heart rate also increase, sometimes rising to 50 per cent or more above normal when excessive thyroid hormone is present.

The increased cardiac output resulting from thyroid hormone tends to increase the arterial pressure. On the other hand, dilatation of the peripheral blood vessels due to the local effects of thyroid hormone and to excessive body heat tends to decrease the pressure. Therefore, the mean arterial pressure usually is unchanged. However, because of the increased rate of run-off of blood through the peripheral vessels, the pulse pressure is increased, with the systolic pressure elevated 10 to 20 mm. Hg and the diastolic pressure correspondingly reduced.

Effect on Respiration. The increased rate of metabolism caused by thyroid hormone increases the utilization of oxygen and the formation of carbon dioxide; these effects activate all the mechanisms that increase the rate and depth of respiration.

Effect on the Gastrointestinal Tract. In addition to increased rate of absorption of foodstuffs, thyroid hormone also increases both the rate of secretion of the digestive juices and the motility of the gastrointestinal tract. Often, diarrhea results. Also, associated with this increased secretion and motility is an increased appetite, so that the food intake usually increases. Lack of thyroid hormone causes constipation.

Effect on the Central Nervous System. In general, thyroid hormone increases the rapidity of cerebration, while, on the other hand, lack of thyroid hormone decreases this function. The hyperthyroid individual is likely to develop extreme nervousness and is likely to have many psychoneurotic tendencies, such as anxiety complexes, extreme worry, or paranoias.

Muscle Tremor. One of the most characteristic signs of hyperthyroidism is a fine muscle tremor. This is not the coarse tremor that occurs in Parkinson's disease or in shivering, for it occurs at the rapid frequency of 10 to 15 times per second. The tremor can be observed easily by placing a sheet of paper on the extended fingers and noting the degree of vibration of the paper. The cause of this tremor is probably increased activity in the areas of the cord that control muscle tone. The tremor is an excellent means for assessing the degree of thyroid hormone effect on the central nervous system.

Effect on Sleep. Because of an exhausting effect of thyroid hormone on the musculature and on the central nervous system, the hyperthyroid subject often has a feeling of constant tiredness; but because of the excitable effects of thyroid hormone on the nervous system, it is difficult for him to sleep. On the other hand, extreme somnolence is characteristic of hypothyroidism.

REGULATION OF THYROID HORMONE SECRETION

To maintain a normal basal metabolic rate, precisely the right amount of thyroid hormone must be secreted all the time, and, to provide this, a specific feedback mechanism operates through the hypothalamus and anterior pituitary gland to control the rate of thyroid secretion in proportion to the metabolic needs of the body. This system is illustrated in Figure 50–5 and can be explained as follows:

Effects of Thyroid-Stimulating Hormone on Thyroid Secretion. Thyroid-stimulating hormone (TSH), also known as *thyrotropin,* is an anterior pituitary hormone, a glycoprotein with a molecular weight of about 28,000; it increases the secretion of thyroxine and triiodothyronine by the thyroid gland. Its specific effects on the thyroid gland are (1) increased proteolysis of the

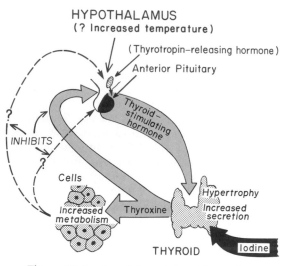

Figure 50–5. Regulation of thyroid secretion.

thyroglobulin in the follicles, with resultant release of thyroid hormone into the circulating blood and diminishment of the follicular substance itself; (2) increased activity of the iodide pump, which increases the rate of "iodide trapping" in the glandular cells, increasing the ratio of intracellular to extracellular iodide concentration several-fold; (3) increased iodination of tyrosine and increased coupling to form the thyroid hormones; (4) increased size and increased secretory activity of the thyroid cells; and (5) increased number of thyroid cells, plus a change from cuboidal to columnar cells and much infolding of the thyroid epithelium into the follicles. In summary, thyroid-stimulating hormone *increases all the known activities of the thyroid glandular cells.*

Role of Cyclic AMP in the Stimulatory Effect of TSH. In an attempt to explain the many and varied effects of thyroid-stimulating hormone on the thyroid cells, a single primary action of this hormone has been searched for, for years. Recent experiments have shown that the hormone almost certainly does have such a primary effect, which is to activate *adenylcyclase* in the membranes of the thyroid cells. This in turn causes formation in the cells of *cyclic AMP,* which then acts as a *second messenger* to activate almost all systems of the thyroid cells. The result is both an immediate increase in secretion of thyroid hormones and prolonged growth of the thyroid glandular tissue itself. This method of controlling thyroid cell activity is similar to the function of cyclic AMP in many other target tissues of the body.

Hypothalamic Regulation of TSH Secretion from the Anterior Pituitary — Thyrotropin-Releasing Hormone (TRH). Electrical stimulation of several areas of the hypothalamus, but most particularly of the paraventricular and arcuate nuclei, increases the anterior pituitary secretion of TSH and correspondingly increases

the activity of the thyroid gland. This control of anterior pituitary secretion is exerted by a hypothalamic hormone, *thyrotropin-releasing hormone* (TRH), which is secreted by nerve endings in the median eminence of the hypothalamus and then transported from there to the anterior pituitary in the hypothalamic-hypophysial portal blood, as was explained in Chapter 49. TRH has been obtained in pure form, and it has proved to be a very simple substance, a tripeptide amide — *pyroglutamyl-histidyl-proline-amide.* TRH has a direct effect on the anterior pituitary gland cells of increasing their output of thyroid-stimulating hormone. When the portal system from the hypothalamus to the anterior pituitary gland is completely blocked, so that TRH cannot reach the anterior pituitary gland, the rate of secretion of TSH by the anterior pituitary is greatly decreased but not reduced to zero.

Effects of Cold and Other Neurogenic Stimuli on TSH Secretion. One of the best-known stimuli for increasing the rate of TSH secretion by the anterior pituitary is exposure of an animal to cold. Exposure of rats for several weeks increases the output of thyroid hormones, sometimes more than 100 per cent, and can increase the basal metabolic rate as much as 50 per cent. Indeed, even human beings moving to arctic regions have been known to develop basal metabolic rates 15 to 20 per cent above normal.

Various emotional reactions can also affect the output of TRH and TSH and can therefore indirectly affect the secretion of thyroid hormone.

Neither the emotional effects nor the effect of cold is observed when the hypophysial stalk is cut, illustrating that both of these effects are mediated by way of the hypothalamus.

Inverse Feedback Effect of Thyroid Hormone on Anterior Pituitary Secretion of TSH — Feedback Regulation of Thyroid Secretion. Increased thyroid hormone in the body fluids decreases the secretion of TSH by the anterior pituitary. When the rate of thyroid hormone secretion rises to about 1.75 times normal, the rate of TSH secretion falls essentially to zero. Most of this depressant effect occurs even when the anterior pituitary has been completely separated from the hypothalamus, but the effect is somewhat greater if the hypothalamus and hypothalamic-hypophysial portal system are intact. Therefore, it is probable that increased thyroid hormone inhibits anterior pituitary secretion of TSH in two different ways: (1) a direct effect on the anterior pituitary itself, and (2) a weaker effect acting through the hypothalamus.

Regardless of the mechanism of the feedback, its effect is to maintain an almost constant concentration of free thyroid hormone in the circulating body fluids. For instance, during periods of heavy exercise, thyroid hormone is consumed much more rapidly than under resting conditions; yet, because of appropriate feedback control, the rate of secretion of thyroid hormone rises to equal the rate of consumption, and the blood thyroid hormone concentration remains almost exactly constant.

DISEASES OF THE THYROID

HYPERTHYROIDISM

Most effects of hyperthyroidism are obvious from the preceding discussion of the various physiologic effects of thyroid hormone. However, some specific effects should be mentioned especially in connection with the development and treatment of hyperthyroidism.

Causes of Hyperthyroidism (Toxic Goiter, Thyrotoxicosis, Graves' Disease). In the patient with hyperthyroidism the entire thyroid gland is usually markedly hyperplastic. It is increased to two to three times normal size, with tremendous folding of the follicular cell lining into the follicles, so that the number of cells is increased several times as much as the size of the gland is increased. Also, each cell increases its rate of secretion several-fold; radioactive iodine uptake studies indicate that these hyperplastic glands secrete thyroid hormone at a rate as great as 5 to 15 times normal.

These changes in the thyroid gland are similar to those caused by excessive thyroid-stimulating hormone. However, radioimmunoassay studies have shown the plasma TSH concentrations to be less than normal rather than enhanced, and often to be essentially zero. On the other hand, one or more globulin antibodies that have actions similar to that of TSH are found in the blood of almost all these patients. These antibodies bind with the thyroid cell membranes, and it is believed that they bind with the same membrane receptors that bind TSH and that this induces continual activation of the cells, with the resultant development of hyperthyroidism. One of these antibodies, found in 50 to 80 per cent of thyrotoxic patients, is called *long-acting thyroid stimulator* (LATS).

The antibodies that cause hyperthyroidism almost certainly develop as the result of autoimmunity that has developed against thyroid tissue. Presumably, at some time in the history of the person an excess of thyroid cell antigens has been released from the thyroid cells, and this has resulted in the formation of antibodies against the thyroid gland itself.

Symptoms of Hyperthyroidism. The symptoms of toxic goiter are obvious from the preceding discussion of the physiology of the thyroid hormones: intolerance to heat, increased sweating, mild to extreme weight loss, varying degrees of diarrhea, muscular weakness, nervousness and other psychic disorders, extreme fatigue with inability to sleep, and tremor of the hands.

Exophthalmos. Most, but not all, persons with hyperthyroidism develop some degree of protrusion of the eyeballs, as illustrated in Figure 50–6. This condition is called *exophthalmos.*

The cause of the protrusion is edematous swelling of the retro-orbital tissues and deposition of large quantities of mucopolysaccharides in the extracellular spaces; the factor or factors that initiate these changes is still in serious dispute. In most patients, antibodies can be found in the blood that react with the retro-orbital tissues. Therefore, there is much reason to believe that exophthalmos, like hyperthyroidism itself, is an autoimmune process. However, in many patients a hormonal substance called *exophthalmos-producing substance,* which can be found in the plasma, will cause exophthalmos in animals into which it is injected. Some experiments have suggested that this substance is a split-off fragment of TSH and that it might be secreted by the anterior pituitary gland instead of TSH itself when hyperthyroidism develops. Thus, the riddle of exophthalmos still is not solved, though the most likely answer at present is that it is an autoimmune response. Usually, the exophthalmos disappears or at least greatly ameliorates with treatment of the hyperthyroidism.

Diagnostic Tests for Hyperthyroidism. In the usual case of hyperthyroidism, the most accurate diagnostic test is direct measurement of the concentration of "free" thyroxine in the plasma, using appropriate radioimmunoassay procedures.

Other tests that are frequently used are these:

(1) The basal metabolic rate is usually increased to +30 or +60 in severe hyperthyroidism.

Figure 50–6. Patient with exophthalmic hyperthyroidism. Note protrusion of the eyes and retraction of the superior eyelids. The basal metabolic rate was +40. (Courtesy of Dr. Leonard Posey.)

(2) The rate of uptake of a standard injected dose of radioactive iodine by the normal thyroid gland, when measured by a calibrated radioactive detector placed over the neck, is about 4 per cent per hour. In the hyperthyroid person, this can rise to as high as 20 to 25 per cent per hour.

(3) The amount of iodine bound to plasma proteins is usually, but not always, directly proportional to the amount of circulating thyroxine. Therefore, elevation of this is also often significant in the diagnosis of hyperthyroidism.

Physiology of Treatment in Hyperthyroidism. The most direct treatment of hyperthyroidism is surgical removal of the thyroid gland. However, treatment of less severe cases can be achieved with antithyroid drugs such as propylthiouracil, a drug that blocks the formation of thyroid hormones in the thyroid cells.

HYPOTHYROIDISM

The effects of hypothyroidism, in general, are opposite to those of hyperthyroidism, but here again, a few physiological mechanisms peculiar to hypothyroidism alone are involved.

Endemic Colloid Goiter. The term *goiter* means a greatly enlarged thyroid gland. As was pointed out in the discussion of iodine metabolism, about 50 mg. of iodine is needed each year for the formation of adequate quantities of thyroid hormone. In certain areas of the world, notably in the Swiss Alps and in the Great Lakes region of the United States, insufficient iodine is present in the soil for the foodstuffs to contain even this minute quantity of iodine. Therefore, prior to the introduction of iodized table salt, many persons living in these areas developed extremely large thyroid glands called *endemic goiters.*

The mechanism for development of the large endemic goiters is the following: Lack of iodine prevents production of thyroid hormone by the thyroid gland; as a result, no hormone is available to inhibit production of TSH by the anterior pituitary and this allows the pituitary to secrete excessively large quantities of TSH. The TSH then causes the thyroid cells to secrete tremendous amounts of thyroglobulin (colloid) into the follicles, and the gland grows larger and larger. But unfortunately, owing to lack of iodine, increased thyroxine and triiodothyronine production does not occur. The follicles become tremendous in size, and the thyroid gland may increase to as large as 300 to 500 grams or more.

Idiopathic Nontoxic Colloid Goiter. Enlarged thyroid glands almost identical with those of endemic colloid goiter frequently develop even when the affected persons receive sufficient quantities of iodine in their diets. These goitrous glands may secrete normal quantities of thyroid hormones, but more frequently the secretion of hormone is depressed, as in endemic colloid goiter.

The exact cause of the enlarged thyroid gland in patients with idiopathic colloid goiter is not known, but most of these patients show signs of mild thyroiditis; therefore, it has been suggested that thyroiditis causes slight hypothyroidism, which then leads to increased TSH secretion and progressive growth of the noninflamed portions of the gland. This could explain why these glands usually are very nodular, with some portions of the gland growing while other portions are being destroyed by thyroiditis.

In some persons with colloid goiter, the thyroid glands have abnormal enzyme systems, which leads to diminished thyroid hormone formation and resultant excess stimulation of the thyroid gland by TSH. And, finally, some foods contain *goitrogenic substances* that have a propylthiouracil-type of anti-thyroid activity, thus also leading to TSH-stimulated enlargement of the thyroid gland. Such goitrogenic substances are found in some varieties of turnips and cabbages.

Characteristics of Hypothyroidism. Whether hypothyroidism is due to endemic colloid goiter, idiopathic colloid goiter, destruction of the thyroid gland by irradiation, surgical removal of the thyroid gland, or destruction of the thyroid gland by various other diseases, the physiologic effects are the same. These include extreme somnolence with 14 to 16 hours of sleep a day, extreme muscular sluggishness, slowed heart rate, decreased cardiac output, decreased blood volume, increased weight, constipation, mental sluggishness, failure of many trophic functions in the body as evidenced by depressed growth of hair and scaliness of the skin, development of a frog-like husky voice, and, in severe cases, development of an edematous appearance throughout the body called myxedema.

Myxedema. The patient with almost total lack of thyroid function develops *myxedema*. Figure 50–7 shows such a patient with bagginess under the eyes and swelling of the face. In this condition, for reasons not yet explained, greatly increased quantities of mucopolysaccharides, mainly hyaluronic acid, collect in the interstitial spaces.

Arteriosclerosis in Hypothyroidism. Lack of thyroid hormone increases the quantity of blood lipoproteins containing especially large amounts of cholesterol, and the increase in blood cholesterol is usually associated with atherosclerosis and arteriosclerosis. Therefore, many hypothyroid patients, particularly those with myxedema, develop severe arteriosclerosis, which results in pe-

Figure 50–7. Patient with myxedema. (Courtesy of Dr. Herbert Langford.)

ripheral vascular disease, deafness, and often extreme coronary sclerosis with consequent early demise.

Diagnostic Tests in Hypothyroidism. The tests already described for diagnosis of hyperthyroidism give the opposite results in hypothyroidism. The free thyroxine in the blood is low. The basal metabolic rate in myxedema ranges between −30 and −40. The protein-bound iodine is as little as one-third normal. And the rate of radioactive iodine uptake by the thyroid gland (except in iodine deficiency hypothyroidism) measures less than 1 per cent per hour rather than the normal of approximately 4 per cent per hour. However, just as important for diagnosis as the various diagnostic tests are the characteristic symptoms of hypothyroidism just discussed.

Treatment of Hypothyroidism. Figure 50–4 shows the effect of thyroid hormone on the basal metabolic rate, illustrating that the hormone normally has a duration of action of more than one month. Consequently, it is easy to maintain a steady level of thyroid hormone activity in the body by daily oral ingestion of a tablet or so of desiccated thyroid gland or thyroid extract. Furthermore, proper treatment of the hypothyroid patient results in such complete normality that formerly myxedematous patients properly treated have lived into their 90's after treatment for over 50 years.

Cretinism. Cretinism is the condition caused by extreme hypothyroidism during infancy and childhood, and it is characterized especially by failure of growth. Cretinism results from congenital lack of a thyroid gland *(congenital cretinism),* from failure of the thyroid gland to produce thyroid hormone because of a genetic deficiency of the gland, or from iodine lack in the diet *(endemic cretinism).* The severity of endemic cretinism varies greatly, depending on the amount of iodine in the diet, and whole populations of an endemic area have been known to have cretinoid tendencies.

A newborn baby without a thyroid gland may have absolutely normal appearance and function because he has been supplied with thyroid hormone by the mother while *in utero,* but a few weeks after birth his movements become sluggish, and both his physical and mental growth are greatly retarded. Treatment of the cretin at any time usually causes normal return of physical growth, but, unless the cretin is treated within a few months after birth, his mental growth will be permanently retarded.

REFERENCES

DeGroot, L. J.: Thyroid hormone action. *In* DeGroot, L. J., *et al.* (eds.): Endocrinology. Vol. 1. New York, Grune & Stratton, 1979, p. 357.

DeGroot, L. J., and Taurog, A.: Secretion of thyroid hormone. *In* DeGroot, L. J., *et al.* (eds.): Endocrinology. Vol. 1. New York, Grune & Stratton, 1979, p. 343.

Dumont, J. E., and Vassart, G.: Thyroid gland metabolism and the action of TSH. *In* DeGroot, L. J., *et al.* (eds.): Endocrinology. Vol. 1. New York, Grune & Stratton, 1979, p. 311.

Ekins, R., *et al.* (eds.): Free Thyroid Hormones. New York, Excerpta Medica, 1979.

Hennessy. J. F.: Hypothyroidism: Discussions in Patient Management. Garden City, N.Y., Medical Examination Publishing Co., 1979.

Li, C. H. (ed.): Thyroid Hormones. New York, Academic Press, 1978.

McClung, M. R., and Greer, M. A.: Treatment of hyperthyroidism. *Annu. Rev. Med., 31*:385, 1980.

McKenzie, J. M., and Zakarija, M.: Hyperthyroidism. *In* DeGroot, L. J., *et al.* (eds.): Endocrinology. Vol. 1. New York, Grune & Stratton, 1979, p. 429.

Middlesworth, L. V.: Metabolism and excretion of thyroid hormones. *In* Greep, R. O., and Astwood, E. B. (eds.): Handbook of Physiology. Sec. 7. Vol. 3. Baltimore, Williams & Wilkins, 1974, p. 215.

Ramsden, D. B.: Peripheral Metabolism and Action of Thyroid Hormones. Montreal, Eden Press, 1977.

Robbins, J., *et al.*: The thyroid and iodine metabolism. *In* Bondy, P. K., and Rosenberg, L. E. (eds.): Metabolic Control and Disease, 8th Ed. Philadelphia, W. B. Saunders Co., 1980, p. 1325.

Stanbury, J. B. (ed.): Endemic Goiter and Endemic Cretinism. New York, John Wiley & Sons, 1980.

Sterling, K., and Lazarus, J. H.: The thyroid and its control. *Annu. Rev. Physiol., 39*:349, 1977.

Wilber, J. F.: Human pituitary thyrotropin. *In* DeGroot, L. J., *et al.* (eds.): Endocrinology. Vol. 1. New York, Grune & Stratton, 1979, p. 141.

51

The Adrenocortical Hormones

The adrenal glands, which lie at the superior poles of the two kidneys, are each composed of two distinct parts, the *adrenal medulla* and the *adrenal cortex*. The adrenal medulla is functionally related to the sympathetic nervous system, and it secretes the hormones *epinephrine* and *norepinephrine* in response to sympathetic stimulation. In turn, these hormones cause almost the same effects as direct stimulation of the sympathetic nerves in all parts of the body. These hormones and their effects were discussed in detail in Chapter 38 in relation to the sympathetic nervous system.

The adrenal cortex secretes an entirely different group of hormones called *corticosteroids*. These hormones are all synthesized from acetyl Co-A or the steroid cholesterol, and they all have similar steroid chemical formulas. However, very slight differences in their molecular structures give them several very different but very important functions.

Mineralocorticoids and Glucocorticoids. Not all the adrenocortical hormones cause exactly the same effects in the body. Two major types of hormones, the *mineralocorticoids* and the *glucocorticoids*, are secreted by the adrenal cortex. In addition to these, small amounts of *androgenic hormones*, which exhibit the same effects in the body as the male sex hormone testosterone, are also secreted. These are normally unimportant, though in certain abnormalities of the adrenal cortices extreme quantities can be secreted and can then result in masculinizing effects.

The *mineralocorticoids* have gained this name because they especially affect the electrolytes of the extracellular fluids — sodium and potassium, in particular. The *glucocorticoids* have gained this name because they exhibit an important effect in increasing blood glucose concentration. However, the glucocorticoids have additional effects on both protein and fat metabolism that are just as important in body functions as are their effects on carbohydrate metabolism.

Over 30 different steroids have been isolated from the adrenal cortex, but only two of these are of major importance to the normal endocrine functions of the body — *aldosterone,* the principal mineralocorticoid, and *cortisol,* the principal glucocorticoid.

FUNCTIONS OF THE MINERALOCORTICOIDS — ALDOSTERONE

Loss of adrenocortical secretion usually causes death within three days to two weeks unless the person receives extensive salt therapy or mineralocorticoid therapy. Without mineralocorticoids, the potassium ion concentration of the extracellular fluid rises markedly, the sodium and chloride concentrations decrease, and the total extracellular fluid volume and blood volume also become reduced. The person soon develops diminished cardiac output, which proceeds to a shocklike state followed by death. This entire sequence can be prevented by the administration of aldosterone or some other mineralocorticoid. Therefore, the mineralocorticoids are said to be the "life-saving" portion of the adrenocortical hormones, while the glucocorticoids are of particular importance in helping the person resist different types of stresses, as is discussed later in the chapter.

Aldosterone exerts at least 95 per cent of the mineralocorticoid activity of the adrenocortical secretion, but cortisol, the major glucocorticoid secreted by the adrenal cortex, also provides a small amount of mineralocorticoid activity. Other adrenal steroids that occasionally have significant mineralocorticoid effects are *corticosterone*, which also exerts glucocorticoid effects, and *desoxycorti-*

costerone, which is secreted in extremely minute quantities but has almost the same effects as aldosterone with a potency one-thirtieth that of aldosterone.

RENAL EFFECTS OF ALDOSTERONE

By far the most important function of aldosterone is to cause transport of sodium and potassium through the renal tubular walls and, to a less extent, transport of hydrogen ions. The mechanisms of these effects were discussed in detail in Chapter 24. However, let us summarize briefly the renal and body fluid effects of aldosterone.

Effect on Tubular Reabsorption of Sodium and Tubular Secretion of Potassium. It will be recalled from Chapter 24 that aldosterone causes an exchange transport of sodium and potassium — that is, absorption of sodium and simultaneous excretion of potassium by the tubular epithelial cells — in both the distal tubule and the collecting duct. Therefore, aldosterone causes sodium to be conserved in the extracellular fluid while potassium is excreted into the urine.

A high concentration of aldosterone in the plasma can decrease the sodium loss into the urine to as little as a few milligrams a day. At the same time, potassium loss into the urine increases many-fold. Conversely, total lack of aldosterone secretion can cause loss of as much as 20 grams of sodium in the urine a day, an amount equal to one-fifth of all the sodium in the body. But, at the same time, potassium is conserved tenaciously.

Therefore, the net effect of excess aldosterone in the plasma is to increase the total quantity of sodium in the extracellular fluid while decreasing the potassium. In turn, the increase in tubular reabsorption causes water reabsorption as well, mainly because the absorbed sodium causes osmosis of water through the tubular epithelium. Thus, an excess of aldosterone can increase the extracellular fluid volume to as much as 20 per cent above normal, or the volume may decrease to as low as 20 to 25 per cent below normal in the absence of aldosterone.

Hypokalemia and Muscle Paralysis; Hyperkalemia and Cardiac Toxicity. The excessive loss of potassium ions from the extracellular fluid into the urine under the influence of aldosterone causes a serious decrease in the plasma potassium concentration, often from the normal value of 4.5 mEq./l. to as low as 1 to 2 mEq./l. This condition is called *hypokalemia.* When the potassium ion concentration falls below approximately one-half normal, muscle paralysis or at least severe muscle weakness often develops. This is caused by hyperpolarization of the nerve and muscle fiber membranes (see Chapter 8), which prevents transmission of action potentials.

On the other hand, when aldosterone is deficient, the extracellular fluid potassium ion concentration can rise far above normal. When it rises to approximately double normal, serious cardiac toxicity, including weakness of contraction and arrhythmia, becomes evident; a slightly higher concentration of potassium leads inevitably to cardiac death.

Effect of Aldosterone on Increasing Tubular Hydrogen Ion Secretion, with Resultant Mild Alkalosis. Though aldosterone mainly causes potassium to be secreted into the tubules in exchange for sodium reabsorption, to a much less extent it also causes tubular secretion of hydrogen ions in exchange for sodium. The obvious effect of this is to decrease the hydrogen ion concentration in the extracellular fluid. However, this effect is not a strong one, usually causing only a mild degree of alkalosis.

Effect of Aldosterone on Circulatory Function. The circulatory effects of aldosterone result almost entirely from the increase in extracellular fluid volume. In the absence of aldosterone secretion, with a decrease in extracellular fluid volume to 20 to 25 per cent below normal and a comparable decrease in plasma volume, circulatory shock develops rapidly. Indeed, in complete lack of aldosterone, a person not treated with extra intake of salt and/or administration of a mineralocorticoid drug is likely to die of circulatory shock within as few as four to eight days.

In the case of hypersecretion of aldosterone, not only is the extracellular fluid volume increased but the blood volume and cardiac output are increased as well. Each of these can increase to as much as 10 to 20 per cent above normal in the first few days of excess aldosterone secretion, but after compensations occur the volumes and cardiac output usually return to no more than 5 to 10 per cent above normal. Nevertheless, over a prolonged period of time even these small increases are sufficient to cause moderate to severe hypertension, as we shall discuss later in the chapter in relation to primary aldosteronism.

CELLULAR MECHANISM OF ALDOSTERONE ACTION

Although for many years we have known the overall effects of mineralocorticoids on the body, the basic action of aldosterone on the tubular cells to increase transport of sodium is still only partly understood. The sequence of events that leads to increased sodium reabsorption seems to be the following:

First, because of its lipid solubility in the cellular membranes, aldosterone diffuses to the interior of the tubular epithelial cells.

Second, in the cytoplasm of the tubular cells aldosterone combines with a highly specific cytoplasmic *receptor protein,* a protein with a stereomolecular configuration that allows only aldosterone or extremely similar compounds to combine.

Third, the aldosterone-receptor complex diffuses into the nucleus, where it may undergo further alterations, and then it induces specific portions of the DNA to form a type or types of messenger RNA related to the process of sodium and potassium transport.

Fourth, the messenger RNA diffuses back into the cytoplasm, where, in conjunction with the ribosomes, it causes protein formation. The protein formed is one or more enzymes or carrier substances required for sodium and potassium transport, probably a specific ATPase that catalyzes energy transfer from cytoplasmic ATP to the sodium-potassium transport mechanism of the cell membrane.

Thus, aldosterone does not have an immediate effect on sodium and potassium transport, but must await the sequence of events that leads to the formation of the specific intracellular substance or substances required for transport. Approximately 20 to 30 minutes are required before new RNA appears in the cells, and approximately 45 minutes are required before the rates of sodium and potassium transport begin to increase; the effect reaches maximum in several hours.

REGULATION OF ALDOSTERONE SECRETION

The regulation of aldosterone secretion is so deeply intertwined with the regulation of extracellular fluid electrolyte concentrations, extracellular fluid volume, blood volume, arterial pressure, and many special aspects of renal function that it is not possible to discuss the regulation of aldosterone secretion independently of all of these other factors. This subject has already been presented in great detail in Chapter 25, to which the reader is referred. However, it is important to list here as well the most important factors that are presently known to play essential roles in the regulation of aldosterone. In the probable order of their importance they are

1. Potassium ion concentration of the extracellular fluid
2. Renin-angiotensin system
3. Quantity of body sodium
4. Adrenocorticotropic hormone (ACTH)

This very potent effect of potassium ions is exceedingly important because it establishes a powerful feedback mechanism for control of extracellular fluid potassium ion concentration as follows: (1) An increase in potassium ion concentration causes increased secretion of aldosterone. (2) The aldosterone in turn has a potent effect on the kidneys, causing enhanced excretion of potassium. (3) Therefore, the potassium ion concentration returns to normal. This effect of potassium ions on aldosterone secretion is a direct effect of the potassium ions on the adrenocortical cells themselves, though the intracellular mechanism of the effect is unknown.

When an animal or human being is placed on a sodium deficient diet, after several days the rate of aldosterone secretion increases markedly even though the sodium ion concentration of the body fluids does not fall significantly. A number of different suggestions for the cause of this effect have been the following:

1. Lack of sodium causes retention of potassium by the kidneys, as explained earlier. The elevated potassium could then cause the increased aldosterone secretion.

2. In a few experiments it has been shown that diminished sodium concentration possibly or probably causes the anterior pituitary gland to secrete some substance that has an effect on the adrenal glands of increasing aldosterone secretion. For the present, this substance is called the *unidentified pituitary factor.*

3. The decrease in extracellular fluid volume and decrease in arterial pressure that result from low salt intake cause increased formation of angiotensin, and this in turn increases the rate of aldosterone secretion, as explained in the following section.

Effect of the Renin-Angiotensin System on Aldosterone Secretion. Infusion of large

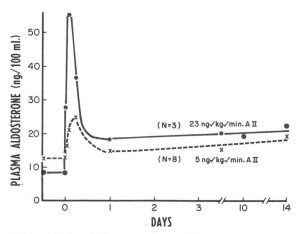

Figure 51–1. Effects on plasma aldosterone concentration caused by continuous infusion of angiotensin II at two different infusion rates for two weeks. Note the very marked acute effect but the much weaker chronic effect. (Drawn from data in Cowley and McCaa, *Circ. Res., 39*:788, 1976.)

amounts of angiotensin into an animal can cause acute increases in aldosterone secretion of as much as 8-fold. However, if the angiotensin infusion is continued, the rate of aldosterone secretion falls in about 12 hours to only 50 to 100 per cent above normal. This effect is illustrated in Figure 51–1, which shows the effect of two different angiotensin infusion rates, one of which (the dashed curve) increased the extracellular fluid concentration of angiotensin to about 3 times normal, and the other of which (the solid curve) increased the angiotensin concentration to about 15 times normal. From this data it can be calculated that a given per cent increase in angiotensin will cause about 60 times less increase in aldosterone secretion than will the same per cent increase in potassium concentration. Yet, even so, in many clinical conditions the renin-angiotensin system is the cause of excessive aldosterone secretion because tremendous quantities of angiotensin are often formed.

FUNCTIONS OF THE GLUCOCORTICOIDS

Even though mineralocorticoids can save the life of an acutely adrenalectomized animal, the animal still is far from normal. Instead, its metabolic systems for utilization of carbohydrates, proteins, and fats are considerably deranged. Furthermore, the animal cannot resist different types of physical or even mental stress, and minor illnesses such as respiratory tract infections can lead to death. Therefore, the glucocorticoids have functions just as important to long-continued life of the animal as do the mineralocorticoids. These are explained in the following sections.

At least 95 per cent of the glucocorticoid activity of the adrenocortical secretions results from the secretion of *cortisol*, known also as *hydrocortisone*. In addition, a small amount of glucocorticoid activity is provided by *corticosterone* and a minute amount by *cortisone*.

EFFECT OF CORTISOL ON CARBOHYDRATE METABOLISM

Stimulation of Gluconeogenesis. By far the best-known metabolic effect of cortisol and other glucocorticoids on metabolism is their ability to stimulate gluconeogenesis by the liver, often increasing the rate of gluconeogenesis as much as six- to ten-fold. This results from several different effects of cortisol:

First, cortisol increases the transport of amino acids from the extracellular fluids into the liver cells. This obviously increases the availability of amino acids for conversion into glucose.

Second, several of the enzymes required to convert amino acids into glucose are increased in the liver cells. Also, the concentration of RNA is increased in the liver cells. Therefore, it is assumed that glucocorticoids activate nuclear formation of messenger RNA's that in turn lead to the array of enzymes required for gluconeogenesis.

Third, cortisol causes mobilization of amino acids from the extrahepatic tissues, mainly from muscle. As a result, more amino acids become available in the plasma to enter into the gluconeogenesis process of the liver and thereby to promote the formation of glucose.

Decreased Glucose Utilization by the Cells. Cortisol also causes a moderate decrease in the rate of glucose utilization by the cells. Though the cause of this decrease is unknown, most physiologists believe that somewhere between the point of entry of glucose into the cells and its final degradation, cortisol directly delays the rate of glucose utilization.

Also, it is known that glucocorticoids slightly depress glucose transport into the cells, which could be an additional factor that depresses cellular glucose utilization.

Elevated Blood Glucose Concentration, and Adrenal Diabetes. Both the increased rate of gluconeogenesis and the moderate reduction in rate of glucose utilization by the cells cause the blood glucose concentration to rise. The increased blood glucose concentration is occasionally so great — 50 per cent or more above normal — that the condition is called *adrenal diabetes* (meaning elevated blood glucose concentration); it has many similarities to pituitary diabetes, which was discussed in Chapter 49.

EFFECTS OF CORTISOL ON PROTEIN METABOLISM

Reduction in Cellular Protein. One of the principal effects of cortisol on the metabolic systems of the body is reduction of the protein stores in essentially all body cells except those of the liver. This is caused both by decreased protein synthesis and increased catabolism of protein already in the cells. Both of these effects may possibly result from decreased amino acid transport into extrahepatic tissues, as will be discussed below, but this probably is not the only cause, since cortisol also depresses the formation of RNA in many extrahepatic tissues, including especially muscle and lymphoid tissue.

Increased Liver Protein and Plasma Proteins Caused by Cortisol. Coincidently with the reduced proteins elsewhere in the body, the liver proteins become enhanced. Furthermore, the plasma proteins (which are produced by the liver and then released into the blood) are also increased. Therefore, these are exceptions to the protein depletion that occurs elsewhere in the body. It is believed that this effect is caused both by the ability of cortisol to enhance amino acid transport into liver cells (but not into most other cells) and by enhancement of the liver enzymes required for protein synthesis.

Increased Blood Amino Acids, Diminished Transport of Amino Acids into Extrahepatic Cells, and Enhanced Transport into Hepatic Cells. Recent studies in isolated tissue have demonstrated that cortisol depresses amino acid transport into muscle cells and perhaps into other extrahepatic cells. But, in contrast to this, it enhances transport into liver cells.

The increased plasma concentration of amino acids, plus the fact that cortisol enhances transport of amino acids into the hepatic cells, could also account for enhanced utilization of amino acids by the liver in the presence of cortisol — such effects as (1) increased rate of deamination of amino acids by the liver, (2) increased protein synthesis in the liver, (3) increased formation of plasma proteins by the liver, and (4) increased conversion of amino acids to glucose — that is, enhanced gluconeogenesis.

Thus, it is possible that many of the effects of cortisol on the metabolic systems of the body can be explained very simply from this ability of cortisol to mobilize amino acids.

EFFECTS OF CORTISOL ON FAT METABOLISM

Mobilization of Fatty Acids. In much the same manner that cortisol promotes amino acid mobilization from muscle, it also promotes mobilization of fatty acids from adipose tissue. This in turn increases the concentration of free fatty acids in the plasma, which also increases their utilization for energy. Cortisol moderately enhances the oxidation of fatty acids in the cells as well, perhaps as a secondary result of the reduced availability of glycolytic products for metabolism.

The increased mobilization of fats, combined with their increased oxidation in the cells, is one of the factors that help to shift the metabolic systems of the cells from utilization of glucose for energy to utilization of fatty acids instead in times of starvation or other stresses. This cortisol mechanism, however, requires several hours to become fully developed — not nearly so rapid or

powerful an effect as the similar shift elicited by a decrease in insulin, as discussed in the following chapter. Nevertheless, it is probably an important factor for long-term conservation of body glucose and glycogen.

OTHER EFFECTS OF CORTISOL

Function of Cortisol in Different Types of Stress. It is amazing that almost any type of stress, whether physical or neurogenic, causes an immediate and marked increase in ACTH (adrenocorticotropic hormone) secretion, followed within minutes by greatly increased adrenocortical secretion of cortisol. Some of the different types of stress that increase cortisol release are the following:

1. Trauma of almost any type
2. Infection
3. Intense heat or cold
4. Injection of norepinephrine and other sympathomimetic drugs
5. Surgical operations
6. Injection of necrotizing substances beneath the skin
7. Restraining an animal so that it cannot move
8. Almost any debilitating disease

Thus, a wide variety of nonspecific stimuli can cause marked increase in the rate of cortisol secretion by the adrenal cortex.

Yet, even though we know that cortisol secretion often increases greatly in stressful situations, we still are not sure why this is of significant benefit to the animal. One guess, which is probably as good as any other, is that the glucocorticoids cause rapid mobilization of amino acids and fats from their cellular stores, making these available both for energy and for synthesis of other compounds needed by the different tissues of the body. Indeed, it is well known that when proteins are released from most of the tissue cells, the liver cells can use the mobilized amino acids to form both glucose and new proteins. It has also been shown that damaged tissues that are momentarily depleted of proteins can also utilize the newly available amino acids to form new proteins that are essential to the lives of the cells. Or, perhaps the amino acids are used to synthesize such essential intracellular substances as purines, pyrimidines, and creatine phosphate, which are necessary for maintenance of cellular life.

Anti-Inflammatory Effects of Cortisol. When tissues are damaged by trauma, by infection with bacteria, or in almost any other way, they almost always become inflamed. In some conditions the inflammation is more damaging than the trauma or disease itself. Administration of large amounts of cortisol can usually block this inflammation or even reverse many of its effects once it has begun.

Basically, there are five main stages of inflammation: (1) release from the damaged tissue cells of chemical substances that activate the inflammation process — chemicals such as histamine, bradykinin, and proteolytic enzymes; (2) an increase in blood flow in the inflamed area caused by some of the released products from the tissues, a process called *erythema;* (3) leakage of large quantities of almost pure plasma out of the capillaries into the damaged areas, followed by clotting of the tissue fluid, thus causing a *nonpitting type of edema;* (4) infiltration of the area by leukocytes; and, finally, (5) tissue healing, which is often accomplished at least partially by ingrowth of fibrous tissue.

One of the most important anti-inflammatory effects of cortisol is its ability to cause *stabilization of the lysosomal membranes.* That is, cortisol makes it much more difficult than normal for the lysosomal membranes to rupture. Therefore, most of the substances released by damaged cells to cause inflammation, which are mainly formed in the lysosomes, are released in greatly decreased quantity.

Even after inflammation has become well established, administration of cortisol can often reduce inflammation within hours to several days. The immediate effect is to block most of the factors that are promoting the inflammation. Then, the rate of healing is also enhanced. This probably results from those same factors that allow the body to resist many other types of physical stress when large quantities of cortisol are secreted: perhaps this results from the mobilization of amino acids and their use to repair the damaged tissues; perhaps it results from increased amounts of glucose and fatty acids available for cellular energy; or perhaps it depends on some catalytic effect of cortisol to inactivate or remove inflammatory products.

Regardless of the precise mechanisms by which the anti-inflammatory effect occurs, this effect of cortisol can play a major role in combating certain types of diseases, such as rheumatoid arthritis, rheumatic fever, and acute glomerulonephritis. All of these are characterized by severe local inflammation, and the harmful effects to the body are caused mainly by the inflammation itself and not by other aspects of the disease. When cortisol or other glucocorticoids are administered to patients with these diseases, almost invariably the inflammation subsides within 24 to 48 hours. And even though the cortisol does not correct the basic disease condition but merely prevents the damaging effects of the inflammatory response, this alone can be a life-saving measure.

REGULATION OF CORTISOL SECRETION — ADRENOCORTICOTROPIC HORMONE (ACTH)

Control of Cortisol Secretion by ACTH. Unlike aldosterone secretion by the adrenal cortex, which is controlled mainly by potassium and angiotensin acting directly on the adrenocortical cells themselves, almost no stimuli have *direct* effects on the adrenal cells of controlling cortisol secretion. Instead, secretion of cortisol is controlled almost entirely by *adrenocorticotropic hormone* (ACTH) secreted by the anterior pituitary gland, as illustrated in Figure 51–2. This hormone, also called *corticotropin* or *adrenocorticotropin*, also enhances the production of adrenal androgens by the adrenal cortex. Small amounts of ACTH are also required for aldosterone secretion, providing a permissive role that allows the other more important factors to exert their more powerful controls.

Control of ACTH Secretion by the Hypothalamus — Corticotropin-Releasing Hormone (CRH). In the same way that other pituitary hormones are controlled by hormones, also called releasing factors, from the hypothalamus, so also does an important releasing hormone control ACTH secretion. This is called *corticotropin-releasing hormone* (CRH). It is secreted into the primary capillary plexus of the hypophysial portal system in the median eminence of the hypo-

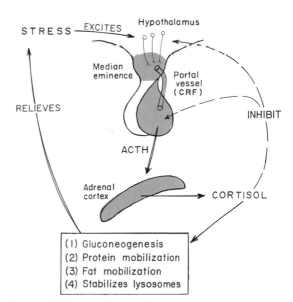

Figure 51–2. Mechanism for regulation of glucocorticoid secretion.

thalamus and then carried to the anterior pituitary gland, where it induces ACTH secretion.

The anterior pituitary gland can secrete small quantities of ACTH in the absence of CRH, but most conditions that cause high ACTH secretory rates initiate this secretion by signals that begin in the hypothalamus and then are transmitted by CRH to the anterior pituitary gland.

Effect of Physiological Stress on ACTH Secretion. It was pointed out earlier in the chapter that almost any type of physical or even mental stress can lead within minutes to greatly enhanced secretion of ACTH, and consequently of cortisol as well, often increasing cortisol secretion as much as 20-fold. For instance, it is believed that pain stimuli or other types of nervous signals caused by the stress are first transmitted to various areas of the hypothalamus, and the signals are relayed eventually to the median eminence where CRH is secreted into the hypophysial portal system. Within minutes the entire control sequence results in large quantities of cortisol in the blood.

Inhibitory Effect of Cortisol on the Hypothalamus and on the Anterior Pituitary — Decreased ACTH Secretion. Cortisol has direct negative feedback *effects* on (1) the hypothalamus, decreasing the formation of CRH and (2) the anterior pituitary gland, decreasing the formation of ACTH. These feedbacks help to regulate the plasma concentration of cortisol. That is, whenever the concentration becomes too great, these feedbacks automatically reduce this concentration back toward a normal control level.

Summary of the Control System. Figure 51–2 illustrates the overall system for control of cortisol secretion. The central key to this control is the excitation of the hypothalamus by different types of stress. These activate the entire system to cause rapid release of cortisol, and the cortisol in turn initiates a series of metabolic effects directed toward relieving the damaging nature of the stressful state. In addition, there is also direct feedback of the cortisol to the hypothalamus and anterior pituitary gland to stabilize the concentration of cortisol in the plasma at times when the body is not experiencing stress. However, the stress stimuli are the prepotent ones; they can always break through this direct inhibitory feedback control of cortisol.

Secretion of Melanocyte-Stimulating Hormone (MSH) Along with ACTH

Usually when ACTH is secreted by the anterior pituitary gland, several other hormones with similar chemical structures are secreted simultaneously, including especially *melanocyte-stimulating hormone* (MSH). Under normal conditions, this hormone is not known to be secreted in enough quantity to have a significant effect on the body, but this may not be true when the rate of secretion of ACTH is very high, as occurs in Addison's disease, which will be discussed later. Melanocyte-stimulating hormone occurs in two forms, an *alpha* and a *beta* form. The alpha form has exactly the same chemical structure as the first 13 amino acids of the 39 amino acid ACTH polypeptide chain.

MSH causes the *melanocytes*, which are located in abundance between the dermis and the epidermis of the skin, to form the pigment *melanin* and to disperse this in the cells of the epidermis. Injection of melanocyte-stimulating hormone into a person over a period of eight to ten days can cause intense darkening of the skin. The effect is much greater in persons with genetically dark skins than in light-skinned persons.

In some lower animals, an intermediate "lobe" of the pituitary gland, called the *pars intermedia*, is highly developed, lying between the anterior and the posterior pituitary lobes. This lobe secretes an especially large amount of melanocyte-stimulating hormone. Furthermore, this secretion is independently controlled by the hypothalamus in response to the amount of light to which an animal is exposed or in response to other environmental factors. For instance, some arctic animals develop darkened fur in the summer and yet entirely white fur in the winter.

ACTH, because of its similarity to MSH, has about one-thirtieth as much melanocyte-stimulating effect as MSH. Furthermore, because the quantities of MSH secreted in the human being are extremely small, while those of ACTH are large, it is likely that ACTH is considerably more important than MSH in determining the amount of melanin in the skin.

CHEMISTRY OF ADRENOCORTICAL SECRETION

Chemistry of the Adrenocortical Hormones. All the adrenocortical hormones are steroid compounds, and they are mainly formed in the adrenal cortex from acetyl coenzyme A or to some extent from cholesterol preformed elsewhere in the body. Figure 51–3 illustrates a schema for formation of the three most important steroid products of the adrenal cortex. However, a change in even a single enzyme system somewhere in the scheme can cause vastly different types of hormones to be formed. Occasionally, large quantities of masculinizing or, very rarely, feminizing sex hormones are secreted by adrenal tumors. Figure 51–4 illustrates the chem-

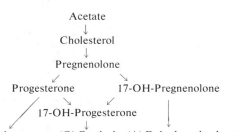

Acetate
↓
Cholesterol
↓
Pregnenolone
↙   ↘
Progesterone   17-OH-Pregnenolone
↙   ↘   ↓
17-OH-Progesterone
↓

(M) Aldosterone   (G) Cortisol   (A) Dehydroepiandrosterone

Figure 51–3. The major steps in the synthesis of the three principal adrenal steroids. The physiologic characteristics are designated (M) mineralocorticoid effect, (G) glucocorticoid effect, and (A) androgenic effect.

ical formulas of aldosterone and cortisol, which are secreted in large quantities by the adrenal cortex.

Chemistry of ACTH. ACTH has been isolated in pure form from the anterior pituitary. It is a large polypeptide having a chain length of 39 amino acids. A digested product of ACTH having a chain length of 24 amino acids has all the trophic effects of the total molecule.

Chemistry of Corticotropin-Releasing Hormone. Several small peptides with a chemical configuration similar to that of vasopressin and oxytocin have been isolated from the hypothalamus. Upon injection into an animal, they cause rapid release of corticotropin from the anterior pituitary. It is presumed that one of these is the specific corticotropin-releasing hormone.

THE ADRENAL ANDROGENS

Several moderately active male sex hormones called *adrenal androgens,* the most important of which is *dehydroepiandrosterone,* are continuously secreted by the adrenal cortex. Also, progesterone and estrogens, which are female sex hormones, have been extracted from the adrenal cortex, though these are secreted in only minute quantities.

In normal human physiology, even the adrenal androgens have almost insignificant effects. However, it is possible that part of the early development of the male sex organs results from childhood secretion of adrenal androgens. The adrenal androgens also exert mild effects in the female, not only before puberty but also throughout life. The physiological effects of androgens will be discussed in Chapter 54 in relation to male sexual function.

ABNORMALITIES OF ADRENOCORTICAL SECRETION

HYPOADRENALISM — ADDISON'S DISEASE

Addison's disease results from failure of the adrenal cortices to produce adrenocortical hormones, and this in turn is most frequently caused by *primary atrophy* of the adrenal cortices, probably resulting from autoimmunity against the cortices, but also frequently by tuberculous destruction of the adrenal glands or invasion of the adrenal cortices by cancer. Basically, the disturbances in Addison's disease are these:

Mineralocorticoid Deficiency. Lack of aldosterone secretion greatly decreases sodium reabsorption and consequently allows sodium ions, chloride ions, and water to be lost into the urine in great profusion. The net result is a greatly decreased extracellular fluid volume. Furthermore, the person develops hyperkalemia and acidosis because of failure of potassium and hydrogen ions to be secreted in exchange for sodium reabsorption.

As the extracellular fluid becomes depleted, the plasma volume falls, the red blood cell concentration rises markedly, the cardiac output decreases, and the patient dies in shock, death usually occurring in the untreated patient four

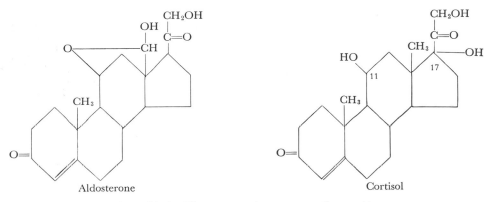

Aldosterone

Cortisol

Figure 51–4. The two most important corticosteroids.

days to two weeks after complete cessation of mineralocorticoid secretion.

Glucocorticoid Deficiency. Loss of cortisol secretion makes it impossible for the person with Addison's disease to maintain normal blood glucose concentration between meals because he cannot synthesize significant quantities of glucose by gluconeogenesis. Furthermore, lack of cortisol reduces the mobilization of both proteins and fats from the tissues, thereby depressing many other metabolic functions of the body. This sluggishness of energy mobilization when cortisol is not available is one of the major detrimental effects of glucocorticoid lack. However, even when excess quantities of glucose and other nutrients are available, the person's muscles are still weak, indicating that glucocorticoids are also needed to maintain other metabolic functions of the tissues besides simply energy metabolism.

Lack of adequate glucocorticoid secretion also makes the person with Addison's disease highly susceptible to the deteriorating effects of different types of stress, and even a mild respiratory infection can sometimes cause death.

Treatment of Persons with Addison's Disease. The untreated person with Addison's disease dies within a few days because of consuming weakness and eventual circulatory shock. Yet such a person can usually live for years if small quantities of mineralocorticoids and glucocorticoids are administered daily–about 0.2 milligram of fludrocortisone (a very potent synthetic mineralocorticoid) and 30 milligrams of cortisol. A high salt intake is also necessary.

The Addisonian Crisis. As noted earlier in the chapter, great quantities of glucocorticoids are occasionally secreted in response to different types of physical or mental stress. In the person with Addison's disease, the output of glucocorticoids does not increase during stress. Yet whenever he is specifically subjected to different types of trauma, disease, or other stresses such as surgical operations, he is likely to develop an acute need for excessive amounts of glucocorticoids, and must be given as much as 10 or more times the normal quantities of glucocorticoids in order to prevent death.

This critical need for extra glucocorticoids and the associated severe debility in times of stress is called an Addisonian crisis.

HYPERADRENALISM — CUSHING'S SYNDROME

Hypersecretion of cortisol by the adrenal cortex causes a complex of hormonal effects called Cushing's syndrome, resulting from either a cortisol-secreting tumor of one adrenal cortex or general hyperplasia of both adrenal cortices. The hyperplasia in turn is caused by increased secretion of ACTH by the anterior pituitary. Most

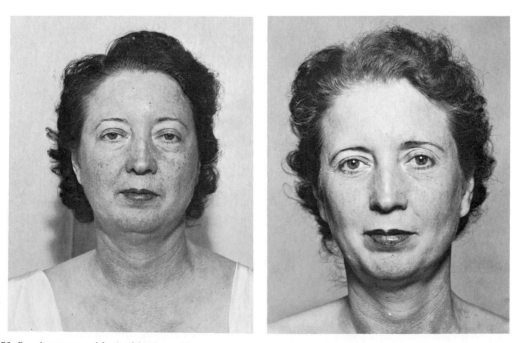

Figure 51–5. A person with Cushing's syndrome before subtotal adrenalectomy (left) and after subtotal adrenalectomy (right). (Courtesy of Dr. Leonard Posey.)

abnormalities of Cushing's syndrome are ascribable to abnormal amounts of cortisol, but secretion of androgens is also of significance.

A special characteristic of Cushing's syndrome is mobilization of fat from the lower part of the body, with concomitant extra deposition of fat in the thoracic region, giving rise to a so-called "buffalo" torso. The excess secretion of steroids also leads to an edematous appearance of the face, and the androgenic potency of some of the hormones causes acne and hirsutism (excess growth of facial hair). The total appearance of the face is frequently described as a "moon face," as illustrated to the left in Figure 51–5 in a patient with Cushing's syndrome prior to treatment.

Effects on Carbohydrate and Protein Metabolism. The abundance of glucocorticoids secreted in Cushing's syndrome causes increased blood glucose concentration, sometimes to values as high as 200 mg. per cent, which is called "adrenal diabetes." This effect results mainly from enhanced gluconeogenesis.

The effects of glucocorticoids on protein catabolism are often profound in Cushing's syndrome, causing greatly decreased tissue proteins almost everywhere in the body except for the liver and the plasma proteins. The loss of protein from the muscles in particular causes severe weakness. The loss of protein synthesis in the lymphoid tissues leads to a diminished immunity system, so that many of these patients die of infections. Even the collagen fibers in the subcutaneous tissue are diminished so that the subcutaneous tissues tear easily, resulting in development of large *purplish striae;* these are actually scars where the subcutaneous tissues have torn apart. In addition, lack of protein deposition in the bones causes *osteoporosis* with consequent weakness of the bones.

Treatment of Cushing's Syndrome. Treatment in Cushing's syndrome consists of removing an adrenal tumor if this is the cause or of decreasing the secretion of ACTH if possible. Hypertrophied pituitary glands, or even small tumors in the pituitary gland, that oversecrete ACTH can be surgically or microsurgically removed or destroyed by radiation. If ACTH secretion cannot easily be decreased, the only satisfactory treatment is usually bilateral total or partial adrenalectomy followed by the administration of adrenal steroids to make up for any insufficiency that develops.

PRIMARY ALDOSTERONISM

Occasionally a small tumor of the zona glomerulosa cells (cells located on the outer surface of the adrenal cortex) occurs and secretes large amounts of aldosterone. The effects of the excess aldosterone are those discussed earlier in the chapter. The most important effects are hypokalemia, slight increase in extracellular fluid volume and blood volume, very slight increase in plasma sodium concentration (usually not over a 2 to 3 per cent increase), and usually moderate hypertension. Especially interesting in primary aldosteronism are occasional periods of muscular paralysis caused by the hypokalemia. The paralysis is caused by hyperpolarization of the nerve fibers, as was explained in Chapter 8.

Treatment of primary aldosteronism is usually surgical removal of the adrenal tumor.

REFERENCES

Baxter, J. D., and Rousseau, G. G.: Glucocorticoid Hormone Action. New York, Springer-Verlag, 1979.

Bondy, P. K.: The adrenal cortex. *In* Bondy, P. K., and Rosenberg, L. E. (eds.): Metabolic Control and Disease, 8th Ed. Philadelphia, W. B. Saunders Co., 1980, p. 1427.

Brodish, A., and Lymangrover, J. R.: The hypothalamic-pituitary adrenocortical system. *Int. Rev. Physiol., 16*:93, 1977.

Genazzani, E., *et al.* (eds.): Pharmacological Modulation of Steroid Action. New York, Raven Press, 1980.

Gill, J. R., Jr.: Bartter's syndrome. *Annu. Rev. Med., 31*:405, 1980.

Hall, J. E., *et al.*: Control of arterial pressure and renal function during glucocorticoid excess in dogs. *Hypertension, 2*:139, 1980.

Harding B. W.: Synthesis of adrenal cortical steroids and mechanism of ACTH effects. *In* DeGroot, L. J., *et al.* (eds.): Endocrinology. Vol. 2. New York, Grune & Stratton, 1979, p. 1131.

James, V. H. (ed.): The Adrenal Gland. New York, Raven Press, 1979.

Krieger, D. T.: Plasma ACTH and corticosteroids. *In* DeGroot, L. J., *et al.* (eds.): Endocrinology. Vol. 2. New York, Grune & Stratton, 1979, p. 1139.

Leavitt, W. W., and Clark, J. H. (eds.): Steroid Hormone Receptor Systems. New York, Plenum Press, 1979.

McCaa, R. E., *et al.*: Role of aldosterone in experimental hypertension. *J. Endocrinol., 81*:69, 1979.

Melby, J. C.: Diagnosis and treatment of hyperaldosteronism and hypoaldosteronism. *In* DeGroot, L. J., *et al.* (eds.): Endocrinology. Vol. 2. New York, Grune & Stratton, 1979, p. 1225.

Nelson, D. H.: The Adrenal Cortex: Physiological Function and Disease. Philadelphia. W. B. Saunders Co., 1979.

Raisz, L. G., *et al.*: Hormonal regulation of mineral metabolism. *Int. Rev. Physiol., 16*:199, 1977.

Roy, A. K., and Clark, J. H. (eds.): Gene Regulation by Steroid Hormones. New York, Springer-Verlag, 1979.

Ruhmann-Wennhold, A., and Nelson, D. H.: Pituitary adrenocorticotropin. *In* DeGroot, L. J., *et al.* (eds.): Endocrinology. Vol. 1. New York, Grune & Stratton, 1979, p. 133.

Tan, S. Y., and Mulrow, P. J.: Aldosterone in hypertension and edema. *In* Bondy, P. K., and Rosenberg, L. E. (eds.): Metabolic Control in Disease, 8th Ed. Philadelphia, W. B. Saunders Co., 1980, p. 1501.

Young D. B., and Guyton, A. C.: Steady state aldosterone dose-response relationships. *Circ. Res., 40*(2):138, 1977.

52

Insulin, Glucagon, and Diabetes Mellitus

The pancreas, in addition to its digestive functions, secretes two important hormones, *insulin* and *glucagon*. The purpose of this chapter is to discuss the functions of these hormones in regulating glucose, lipid, and protein metabolism, as well as to discuss briefly the disease *diabetes mellitus*, caused by hyposecretion of insulin.

Physiologic Anatomy of the Pancreas. The pancreas is composed of two major types of tissues, as shown in Figure 52–1: (1) the *acini*, which secrete digestive juices into the duodenum, and (2) the *islets of Langerhans*, which do not have any means for emptying their secretions externally but instead secrete insulin and glucagon directly into the blood. The digestive secretions of the pancreas were discussed in Chapter 43.

The islets of Langerhans of the human being contain three major types of cells, the *alpha, beta*, and *delta* cells, which are distinguished from one another by their structure and staining characteristics. The beta cells secrete insulin, the alpha cells secrete glucagon, and the delta cells secrete somatostatin, the important functions of which are still not clear.

INSULIN

Insulin is a small protein with a molecular weight of 5808 for human insulin. It is composed of two amino acid chains connected to each other by disulfide linkages.

Before insulin can exert its function it must bind with a large *receptor protein* in the cell membrane, as explained later in the chapter.

EFFECT OF INSULIN ON CARBOHYDRATE METABOLISM

Immediately after a high carbohydrate meal, the glucose that is absorbed into the blood causes rapid secretion of insulin. The insulin in turn causes rapid uptake, storage, and use of glucose by almost all tissues of the body, but especially by the liver, muscles, and fat tissue. Therefore, let us discuss each of these.

Effect of Insulin in Promoting Liver Uptake, Storage, and Use of Glucose

One of the most important effects of insulin is to cause most of the glucose absorbed after a meal to be stored almost immediately in the liver in the form of glycogen. Then, between meals, when insulin is not available and the blood glucose concentration begins to fall, the liver glycogen is split back into glucose, which is released back into the blood to keep the blood glucose concentration from falling too low.

The mechanism by which insulin causes glu-

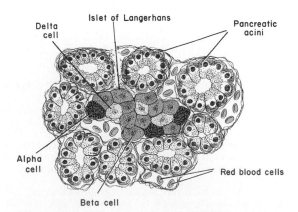

Figure 52–1. Physiologic anatomy of the pancreas.

cose uptake and storage in the liver includes several almost simultaneous steps:

1. Insulin *inhibits phosphorylase*, the enzyme that causes liver glycogen to split into glucose.

2. Insulin causes *enhanced uptake of glucose* from the blood by the liver cells. It does so by *increasing the activity of the enzyme glucokinase*, which is the enzyme that causes the initial phosphorylation of glucose after it diffuses into the liver cells. Once phosphorylated, the glucose is trapped inside the liver cells, because phosphorylated glucose cannot diffuse back through the cell membrane.

3. Insulin also increases the activities of the enzymes that promote glycogen synthesis.

The net effect of the above actions is to increase the amount of glycogen in the liver. The glycogen can increase to a total of about 5 to 6 per cent of the liver mass, which is equivalent to almost 100 grams of stored glycogen.

Release of Glycogen from the Liver Between Meals. After the meal is over and the blood glucose level begins to fall to a low level, several events now transpire that cause the liver to release glucose back into the circulating blood.

1. The decreasing blood glucose causes the pancreas to decrease its insulin secretion.

2. The lack of insulin then reverses all the effects listed above for glycogen storage.

3. The lack of insulin also activates the enzyme *phosphorylase*, which causes the splitting of glycogen into *glucose phosphate*.

4. The enzyme *glucose phosphatase* causes the phosphate radical to split away from the glucose, and this allows the free glucose to diffuse back into the blood.

Thus, the liver removes glucose from the blood when it is present in excess after a meal and returns it to the blood when it is needed between meals. Ordinarily, about 60 per cent of the glucose in the meal is stored in this way in the liver and then returned later.

Other Effects of Insulin on Carbohydrate Metabolism in the Liver. Insulin also *promotes the conversion of liver glucose into fatty acids*, and these fatty acids are subsequently transported to the adipose tissue and deposited as fat. This will be discussed in relation to insulin effects on fat metabolism. Insulin also *inhibits gluconeogenesis*. It does so mainly by decreasing the quantities and activities of the liver enzymes required for gluconeogenesis.

Effect of Insulin in Promoting Glucose Metabolism in Muscle

During most of the day, muscle tissue depends not on glucose for its energy but instead on fatty acids. The principal reason for this is that the normal *resting muscle* membrane is almost imper-

meable to glucose except when the muscle fiber is stimulated by insulin. And, between meals, the amount of insulin that is secreted is too small to promote the entry of significant amounts of insulin into the muscle cells.

However, under two conditions the muscles do utilize large amounts of glucose for energy. One of these is periods of heavy exercise. This usage of glucose does not require large amounts of insulin because the exercising muscle fibers, for reasons not understood, become highly permeable to glucose even in the absence of insulin because of the contraction process itself.

The second condition for muscle usage of large amounts of glucose occurs during the few hours after a meal. At this time the blood glucose concentration is high; also, the pancreas is secreting large quantities of insulin, and the extra insulin causes rapid transport of glucose into the muscle cells.

Storage of Glycogen in Muscle. If the muscles are not exercising during the period following a meal and yet glucose is transported into the muscle cells in great abundance, then much of the glucose is stored in the form of muscle glycogen instead of being used for energy. However, the concentration of muscle glycogen rarely rises much above 1 per cent rather than the possible 5 to 6 per cent in liver cells. The glycogen can later be used for energy by the muscle.

Muscle glycogen is different from liver glycogen in that it cannot be reconverted into glucose and released into the body fluids. The reason for this is that there is no glucose phosphatase in muscle cells, in contrast to the liver cells.

Mechanism by Which Insulin Promotes Glucose Transport Through the Muscle Cell Membrane. Insulin promotes glucose transport into muscle cells quite differently from the way it promotes transport into liver cells. The transport into liver results mainly from a trapping mechanism caused by phosphorylation of the glucose under the influence of glucokinase. However, this is only a minor factor in the insulin effect on glucose transplant into muscle cells. Of more importance, the insulin directly affects the muscle cell membrane to facilitate glucose transport. This is illustrated by the experimental results depicted in Figure 52–2. This experiment was performed in muscle cells at a temperature of 4° C, a temperature at which glucose, upon entering the cell, cannot be phosphorylated. The lower curve labeled "control" shows the concentration of free glucose measured inside the cell, illustrating that the glucose concentration remained almost exactly zero despite increases in extracellular glucose concentration up to as high as 750 mg./100 ml. In contrast, the curve labeled "insulin" illustrates that the intracellular glucose con-

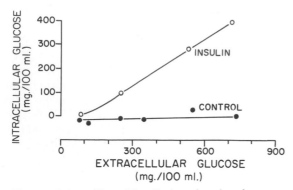

Figure 52–2. Effect of insulin in enhancing the concentration of glucose inside muscle cells. Note that in the absence of insulin (control), the intracellular glucose concentration remained near zero despite very high extracellular concentrations of glucose. (From Park, Morgan, Kaji, and Smith, in Eisenstein (ed.): The Biochemical Aspects of Hormone Action. Little, Brown and Co.)

centration rose to as high as 400 mg./100 ml. when insulin was added. Thus, it is clear that insulin can increase the rate of transport of glucose into the resting muscle cell by at least 15- to 20-fold.

As pointed out in the discussion of glucose transport through the cell membrane in Chapter 4, glucose cannot pass through the membrane pores but instead must be transported through the lipid portion of the membrane. Figure 52–3 depicts the generally accepted method by which this is achieved, showing that glucose combines with a carrier substance in the cell membrane and then diffuses to the inside of the membrane, where it is released to the interior of the cell. The carrier is then used again and again to transport additional quantities of glucose.

Glucose transport through the cell membrane does not occur against a concentration gradient.

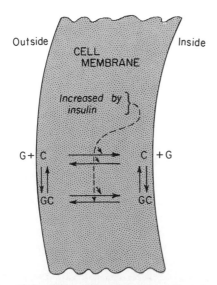

Figure 52–3. Effect of insulin in increasing glucose transport in either direction through the cell membrane.

That is, once the glucose concentration inside the cell rises as high as the glucose concentration on the outside, additional glucose will not be transported to the interior. Therefore, the transport process is one of *facilitated diffusion,* which means simply that the carrier facilitates the diffusion of glucose through the membrane but cannot impart energy to the transport process to cause glucose movement against an energy gradient. The manner in which insulin enhances facilitated diffusion of glucose is still largely unknown. All that is known is that the insulin combines with a "receptor protein" in the cell membrane — a protein having a molecular weight of about 300,000. This may be the glucose carrier itself, or it may be merely the first step in a chain of events that leads to activation of the carrier system. The insulin increases the glucose transport within seconds to minutes, suggesting either a rapid direct action on the cell membrane itself or some other equally rapid mechanism.

Lack of Effect of Insulin on Glucose Uptake and Usage by the Brain

The brain is quite different from most other tissues of the body in that insulin has either little or no effect on uptake or use of glucose. Instead, the brain cells are permeable to glucose without the intermediation of insulin.

The brain cells normally use only glucose for energy. Therefore, it is essential that the blood glucose level be maintained always above a critical level, which is one of the important functions of the blood glucose control system. When the blood glucose does fall too low, into the range of 20 to 50 mg./100 ml., symptoms of *hypoglycemic shock* develop, characterized by progressive irritability that leads to fainting, convulsions, and even coma.

EFFECT OF INSULIN ON FAT METABOLISM

Though not quite as dramatic as the acute effects of insulin on carbohydrate metabolism, insulin also affects fat metabolism in ways that, in the long run, are perhaps even more important. Especially dramatic is the long-term effect of insulin lack in causing extreme atherosclerosis, often leading to heart attacks, cerebral strokes, and other vascular accidents. But, first, let us discuss the acute effects of insulin on fat metabolism.

Effect of Insulin Excess on Causing Fat Synthesis and Storage

Insulin has several different effects that lead to fat storage in adipose tissue. One is the simple fact that insulin increases the rate of utilization of glucose by many of the body's tissues, and this functions as a "fat sparer." However, insulin also promotes fatty acid synthesis. Most of this synthesis occurs in the liver cells, and the fatty acids are then transported to the adipose cells to be stored. However, a small part of the synthesis occurs in the fat cells themselves. The different factors that lead to increased fatty acid synthesis in the liver include these:

1. Insulin increases the transport of glucose into the liver cells. Then, the glucose is split to pyruvate in the glycolytic pathway and the pyruvate is subsequently converted to acetyl Co-A, the substrate from which fatty acids are synthesized.

2. An excess of *citrate* and *isocitrate ions* is formed by the citric acid cycle when excess amounts of glucose are being used for energy. These ions then have a direct effect in activating *acetyl Co-A carboxylase*, the enzyme required to initiate the first stage of fatty acid synthesis.

3. The fatty acids are then transported from the liver to the adipose cells where they are stored.

Effect of Insulin on Storage of Fat in the Adipose Cells. Insulin has very much the same effect in adipose cells as in the liver in causing synthesis of fatty acids. However, only about one-tenth as much glucose is transported into human fat cells as into the liver, so that the amount of fatty acids synthesized in adipose cells is rather small compared with the amount formed in the liver.

Yet, insulin has two other essential effects that are required for fat storage in adipose cells:

1. Insulin *inhibits the action of hormone-sensitive lipase.* Since this is the enzyme that causes hydrolysis of the triglycerides in fat cells, the release of fatty acids into the circulating blood is therefore inhibited.

2. Insulin *promotes glucose transport into the fat cells* in exactly the same way that it promotes glucose transport into muscle cells. The glucose is then utilized to synthesize fatty acids, as noted above, but, more important, it also forms another substance that is essential to the storage of fat. During the glycolytic breakdown of glucose, large quantities of the substance of *α-glycerophosphate* is formed. This substance supplies the *glycerol* that binds with fatty acids to form triglycerides, the storage form of fat in adipose cells. Therefore, when insulin is not available to promote glucose entry into the fat cells, fat storage is either greatly inhibited or blocked.

Increased Metabolic Use of Fat Caused by Insulin Lack

All aspects of fat metabolism are greatly enhanced in the absence of insulin. This occurs even normally between meals when secretion of insulin is minimal, but it becomes extreme in diabetes when secretion of insulin is almost zero. The resulting effects are these:

Lipolysis of Storage Fat and Release of Free Fatty Acids During Insulin Lack. In the absence of insulin, all the effects of insulin just noted that cause storage of fat are reversed. The most important effect is that the enzyme *hormone-sensitive lipase* in the fat cells becomes strongly activated. This causes hydrolysis of the stored triglycerides, releasing large quantities of fatty acids and glycerol into the circulating blood. Consequently, the plasma concentration of free fatty acids rises within minutes to hours. This free fatty acid then becomes the main energy substrate used by essentially all tissues of the body besides the brain. Figure 52–4 illustrates the effect of insulin lack on the plasma concentrations of free fatty acids, glucose, and acetoacetic acid. Note that immediately after removal of the pancreas the free fatty acid concentration in the plasma begins to rise, rising considerably more rapidly even than the concentration of glucose.

Effect of Insulin Lack on Plasma Lipid Concentrations. The excess of fatty acids available to the liver also promotes conversion of some of the fatty acids into phospholipids and cholesterol, two of the major products of fat metabolism. These two substances, along with some of the triglycerides formed in the liver, are then discharged into the blood in the lipoproteins. Occasionally, the plasma lipoproteins increase as much as three-fold, giving a total concentration of plasma lipids of several per cent rather than the normal 0.6 per cent. This high lipid concentration — especially the high concentration of cholesterol — leads to rapid development of atherosclerosis in persons with serious diabetes.

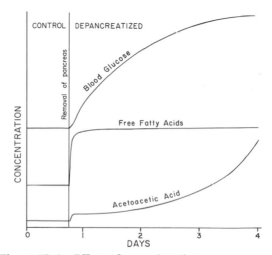

Figure 52–4. Effect of removing the pancreas on the concentrations of blood glucose, plasma free fatty acids, and acetoacetic acid.

Ketogenic and Acidotic Effect of Insulin Lack. Insulin lack also causes excessive amounts of *acetoacetic acid* to be formed in the liver cells. This results from the rapid breakdown of fatty acids in the liver to form extreme amounts of acetyl Co-A. A part of this acetyl Co-A can be utilized for energy, but the excess is condensed to form acetoacetic acid, which, in turn, is released into the circulating blood. As was explained in Chapter 46, some of the acetoacetic acid is also converted into β-hydroxybutyric acid and *acetone*. These two substances, along with the acetoacetic acid, are called *ketone bodies*, and their presence in large quantities in the body fluids is called *ketosis*. We shall see later that the acetoacetic acid and the β-hydroxybutyric acid can cause severe *acidosis* and *coma* in patients with severe diabetes. In the absence of heroic treatment, this leads almost inevitably to death.

EFFECT OF INSULIN ON PROTEIN METABOLISM AND GROWTH

Effect of Insulin on Protein Synthesis and Storage. During the few hours following a meal when excess quantities of nutrients are available in the circulating blood, not only carbohydrates and fats but proteins as well are stored in the tissues; insulin is required for this to occur. The manner in which insulin causes protein storage is not as well-understood as the mechanism for both glucose and fat storage. Some of the known facts are these:

1. Insulin causes active transport of many of the amino acids into the cells. Thus, insulin shares with growth hormone the capability of increasing the uptake of amino acids into cells.

2. Insulin directly affects the ribosomes to *increase the translation of messenger RNA*, thus forming new proteins. In some unexplained way, insulin "turns on" the ribosomal machinery. In the absence of insulin the ribosomes simply stop working, almost as if insulin operates an "on-off" mechanism.

3. Over a longer period of time insulin also *increases the rate of transcription of DNA* in the cell nuclei, thus forming increased quantities of RNA. Eventually, it also increases the rate of formation of new DNA and even reproduction of cells. All these effects promote still more protein synthesis.

4. Insulin also *inhibits the catabolism of proteins*, thus decreasing the rate of amino acid release from the cells, especially from the muscle cells. Presumably this results from some ability of the insulin to diminish the normal degradation of proteins by the cellular lysosomes.

5. In the liver, large quantities of insulin *depress the rate of gluconeogenesis* by decreasing the activity of the enzymes that promote gluconeogenesis. Since the substrates most used for synthesis of glucose by the process of gluconeogenesis are the plasma amino acids, this suppression of gluconeogenesis conserves the amino acids in the protein stores of the body.

In summary, insulin greatly enhances the rate of protein formation and also prevents the degradation of proteins.

Protein Depletion and Increased Plasma Amino Acids Caused by Insulin Lack. All protein storage comes to a complete halt when insulin is not available. The catabolism of proteins increases, protein synthesis stops, and large quantities of amino acid are dumped into the plasma. The plasma amino acid concentration rises considerably, and most of the excess amino acids are used either directly for energy or as substrates for gluconeogenesis. This degradation of the amino acids also leads to enhanced urea excretion in the urine. The resulting protein wasting is one of the most serious of all the effects of severe diabetes mellitus. It can lead to extreme weakness as well as to many deranged functions of the organs.

Effect of Insulin on Growth—Its Synergistic Effect with Growth Hormone. Because insulin is required for the synthesis of proteins, it is just as essential for growth of an animal as is growth hormone. This is illustrated in Figure 52–5, which shows that a depancreatized and hypophysectomized rat without therapy hardly grew at all. Furthermore, administration of neither growth hormone nor insulin one at a time caused significant growth. Yet a combination of both of these hormones did cause dramatic growth. Thus it appears that the two hormones function synergistically to promote growth, each performing a specific function that is separate from that of the other. Perhaps part of the necessity for both

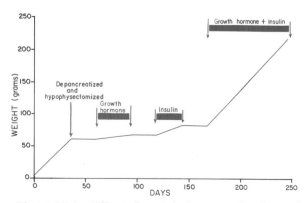

Figure 52–5. Effect of growth hormone, insulin, and growth hormone plus insulin on growth in a depancreatized and hypophysectomized rat.

hormones results from the fact that each promotes cellular uptake of a different selection of amino acids, all of which are required if growth is to be achieved.

CONTROL OF INSULIN SECRETION

Insulin secretion is controlled mainly by the blood glucose concentration. However, blood amino acids and other factors also play important roles, as we shall see.

Stimulation of Insulin Secretion by Blood Glucose. At the normal fasting level of blood glucose of 80 to 90 mg./100 ml., the rate of insulin secretion is minimal. However, as the concentration of blood glucose rises above 100 mg./100 ml. of blood, the rate of insulin secretion rises rapidly, reaching a peak some 10 to 20 times the basal level at blood glucose concentrations between 300 and 400 mg./100 ml. Thus, the increase in insulin secretion under a glucose stimulus is dramatic both in its rapidity and in the tremendous level of secretion achieved. Furthermore, the turn-off of insulin secretion is almost equally rapid, occurring within minutes after reduction in blood glucose concentration back to the fasting level.

This response of insulin secretion to an elevated blood glucose concentration provides an extremely important feedback mechanism for regulating blood glucose concentration. That is, the rise in blood glucose increases insulin secretion, and the insulin in turn causes transport of glucose into the liver, muscle, and other cells, thereby reducing the blood glucose concentration back toward the normal value.

Effect of Amino Acids on Insulin Secretin. In addition to glucose stimulation of insulin secretion, many of the amino acids have a similar effect. However, this effect differs from glucose stimulation of insulin secretion in the following way: Amino acids administered in the absence of a rise in blood glucose cause only a small increase in insulin secretion. However, when administered at the same time that the blood glucose concentration is elevated, the glucose-induced secretion of insulin may be as much as doubled. Thus, the amino acids very strongly potentiate the glucose stimulus for insulin secretion.

The stimulation of insulin secretion by amino acids seems to be a purposeful response, because the insulin in turn promotes transport of the amino acids into the tissue cells and also promotes intracellular formation of protein. That is, the insulin is important for proper utilization of the excess amino acids as well as excess glucose.

Gastrointestinal Hormones. A mixture of several important gastrointestinal hormones — *gastrin, secretin, cholecystokinin,* and *gastric inhibitory peptide* — cause a moderate increase in insulin secretion. These hormones are released in the gastrointestinal tract after a person eats a meal. They seem to cause an "anticipatory" increase in blood insulin in preparation for the glucose and amino acids to be absorbed from the meal. These gastrointestinal hormones almost double the rate of insulin secretion following an average meal.

ROLE OF INSULIN IN "SWITCHING" BETWEEN CARBOHYDRATE AND LIPID METABOLISM

From the above discussions it should be clear that insulin promotes the utilization of carbohydrates for energy and depresses the utilization of fats. Conversely, lack of insulin causes fat utilization mainly to the exclusion of glucose utilization, except by brain tissue. Furthermore, the signal that controls this switching mechanism is principally the blood glucose concentration. When the glucose concentration is low, insulin secretion is suppressed and fat is utilized almost exclusively for energy everywhere except in the brain; when the glucose concentration is high, insulin secretion is stimulated, and carbohydrate is utilized instead of fat until the excess blood glucose is stored. Therefore, one of the most important functional roles of insulin in the body is to control which of these two foods from moment to moment will be utilized by the cells for energy.

GLUCAGON AND ITS FUNCTIONS

Glucagon, a hormone secreted by the alpha cells of the islets of Langerhans, has several functions that are opposite to those of insulin. Most important of these is its effect of increasing the blood glucose concentration.

Like insulin, glucagon is a small protein. It has a molecular weight of 3485 and is composed of a chain of 29 amino acids. On injection of purified glucagon into an animal, a profound *hyper*glycemic effect occurs. One microgram of glucagon per kilogram body weight can elevate the blood glucose concentration approximately 20 mg./100 ml. of blood in about 20 minutes. For this reason, glucagon is frequently called *hyperglycemic factor*.

The two major effects of glucagon on glucose metabolism are (1) breakdown of liver glycogen (*glycogenolysis*) and (2) increased *gluconeogenesis*.

Glycogenolysis and Increased Blood Glucose Concentration Caused by Glucagon. The most dramatic effect of glucagon is its ability to cause glycogenolysis in the liver, which in turn increases the blood glucose concentration within minutes.

It does this by the following complex cascade of events:

1. Glucagon activates *adenylcyclase* in the hepatic cell membrane,

2. Which causes the formation of *cyclic AMP*,

3. Which activates *protein kinase regulator protein*,

4. Which activates *protein kinase*,

5. Which activates *phosphorylase b kinase*,

6. Which converts *phosphorylase b* into *phosphorylase a*,

7. Which promotes the degradation of glycogen into glucose-1-phosphate,

8. Which then is dephosphorylated and the glucose released from the liver cells.

This sequence of events is exceedingly important for several reasons. First, it is one of the most thoroughly studied of all the *second messenger* functions of cyclic AMP. Second, it illustrates a cascading system in which each succeeding product is produced in greater quantity than the preceding product. Therefore, it represents a potent *amplifying* mechanism. This explains how only a few micrograms of glucagon can have the extreme effect of causing hyperglycemia.

Infusion of glucagon for about four hours can cause such intensive liver glycogenolysis that all of the liver stores of glycogen become totally depleted.

Gluconeogenesis Caused by Glucagon. Even after all the glycogen in the liver has been exhausted under the influence of glucagon, continued infusion of this hormone causes continued hyperglycemia. This results from an effect of glucagon in increasing the rate of gluconeogenesis in the liver cells. Unfortunately, the precise mechanism of this effect is unknown, but it is believed to result mainly from activation of the enzymes that are required in gluconeogenesis.

Glucagon-Like Effect of Epinephrine. Epinephrine (and to a slight extent norepinephrine as well) is also a potent promoter of liver glycogenolysis, having an effect almost exactly the same as that of glucagon, though not quite as strong.

REGULATION OF GLUCAGON SECRETION

Effect of Blood Glucose Concentration. Changes in blood glucose concentration have exactly the opposite effect on glucagon secretion as on insulin secretion. That is, a *decrease* in blood glucose increases glucagon secretion. When the blood glucose falls to as low as 70 mg./100 ml. of blood, the pancreas secretes large quantities of glucagon, which rapidly mobilizes glucose from the liver. Thus, glucagon helps to protect against hypoglycemia.

Effect of Amino Acids. Amino acids enhance the secretion of glucagon, an effect exactly opposite to that of glucose. The physiological importance of this is that it helps to prevent the hypoglycemia that would otherwise result when a meal of pure protein is ingested, because the amino acids from the protein enhance insulin secretion and thereby tend to decrease blood glucose. The increased glucagon secretion seems to nullify this effect.

SUMMARY OF BLOOD GLUCOSE REGULATION

In the normal person the blood glucose concentration is very narrowly controlled, usually in a range between 80 and 90 mg./100 ml. of blood in the fasting person each morning before breakfast. This concentration increases to 120 to 140 mg./100 ml. during the first hour or so following a meal, but the feedback systems for control of blood glucose return the glucose concentration very rapidly back to the control level, usually within two hours after the last absorption of carbohydrates. Conversely, in starvation the gluconeogenesis function of the liver provides the glucose that is required to maintain the fasting blood glucose level.

The mechanisms for achieving this high degree of control have been presented in this chapter. However, let us summarize these briefly:

1. The liver functions as a very important *blood glucose-buffer system*. That is, when the blood glucose rises to a very high concentration following a meal and the rate of insulin secretion also increases, as much as two-thirds of the glucose absorbed from the gut is almost immediately stored in the liver in the form of glycogen. Then, during the succeeding hours, when both the blood glucose concentration and the rate of insulin secretion fall, the liver releases the glucose back into the blood.

2. It is very clear that both insulin and glucagon function as important and separate feedback control systems for maintaining a normal blood glucose concentration. When the concentration rises to too high a level, insulin is secreted; the insulin in turn causes the blood glucose concentration to decrease toward normal. Conversely, a decrease in blood glucose stimulates glucagon secretion; the glucagon then functions in the opposite direction to increase the glucose up toward normal. Under most normal conditions, the insulin feed-

back mechanism probably is much more important than the glucagon mechanism.

3. Also, in hypoglycemia, a direct effect of low blood glucose on the hypothalamus stimulates the sympathetic nervous system. In turn, the epinephrine secreted by the adrenal glands causes still further release of glucose from the liver. This, too, helps to protect against severe hypoglycemia.

4. And, finally, over a period of hours and days, both growth hormone and cortisol are secreted in response to prolonged hypoglycemia, and they both decrease the rate of glucose utilization by most cells of the body. This, too, helps to return the blood glucose concentration toward normal.

Importance of Blood Glucose Regulation. One might ask the question, Why is it important to maintain a constant blood glucose concentration, particularly since most tissues can shift to utilization of fats and proteins for energy in the absence of glucose? The answer is that glucose is the only nutrient that can be utilized by the *brain, retina,* and *germinal epithelium of the gonads* in sufficient quantities to supply them with their required energy. Therefore, it is important to maintain a blood glucose concentration at a sufficiently high level to provide this necessary nutrition.

Most of the glucose formed by gluconeogenesis during the interdigestive period is used for metabolism in the brain. Indeed, it is important that the pancreas not secrete any insulin during this time, for otherwise the scant supplies of glucose that are available would all go into the muscles and other peripheral tissues, leaving the brain without a nutritive source.

DIABETES MELLITUS

Diabetes mellitus is caused by diminished rates of insulin secretion by the beta cells of the islets of Langerhans. It is usually divided into two different types: *juvenile diabetes*, which usually, but not always, begins suddenly in early life, and *maturity-onset diabetes*, which begins in later life and mainly in obese persons.

Heredity plays an important role in the development of both these types of diabetes. The juvenile type results in some instances from hereditary predisposition to development of antibodies against the beta cells or to simple degeneration of these cells. The maturity-onset type of diabetes is apparently caused by degeneration of the beta cells as a result of more rapid aging in susceptible persons than in others. Obesity pre-

disposes a person to this type of diabetes because larger quantities of insulin are required for metabolic control in obese than in normal persons.

PATHOLOGICAL PHYSIOLOGY OF DIABETES

Most of the pathological conditions in diabetes mellitus can be attributed to one of the following three major effects of insulin lack: (1) decreased utilization of glucose by the body cells, with a resultant increase in blood glucose concentration to as high as 300 to 1200 mg. per 100 ml.; (2) markedly increased mobilization of fats from the fat storage areas, causing abnormal fat metabolism as well as deposition of lipids in vascular walls to cause atherosclerosis; and (3) depletion of protein in the tissues of the body.

However, in addition, some special pathophysiological problems occur in diabetes mellitus that are not so readily apparent. These are as follows:

Loss of Glucose in the Urine of the Diabetic Person. When the quantity of glucose entering the kidney tubules in the glomerular filtrate rises above approximately 225 mg. per minute, a significant proportion of the glucose begins to spill into the urine. If normal quantities of glomerular filtrate are formed per minute, glucose spillage will occur when the blood glucose level rises over 180 mg. per cent. Consequently, it is frequently stated that the blood "threshold" for the appearance of glucose in the urine is approximately 180 mg. per cent.

The loss of glucose in the urine causes *diuresis* because of the osmotic effect of glucose in the tubules to prevent tubular reabsorption of fluid. The overall effect is dehydration of the extracellular space, which then causes dehydration of the intracellular spaces as well. Thus, one of the important features of diabetes is a tendency for extracellular and intracellular dehydration to develop, and these are also often associated with collapse of the circulation.

Acidosis in Diabetes. The shift from carbohydrate to fat metabolism in diabetes has already been discussed. When the body depends almost entirely on fat for energy, the level of acetoacetic acid and β-hydroxybutyric acid in the body fluids may rise from 1 mEq./liter to as high as 10 mEq./liter. This, obviously, is likely to result in acidosis.

A second effect, which is usually even more important in causing acidosis than is the direct increase in keto acids, is a decrease in sodium concentration caused by the following effect: Keto acids have a low threshold for excretion by the kidneys; therefore, when the keto acid level rises in diabetes, as much as 100 to 200 grams of keto acids can be excreted in the urine each day. Because these are strong acids, very little of them can be excreted in the acidic form; instead, they are excreted in combination with sodium derived from the extracellular fluid. As a result, the sodium concentration in the extracellular fluid usually decreases, and the sodium is replaced by increased quantities of hydrogen ions, thus adding greatly to the acidosis.

Obviously, all the usual reactions that occur in metabolic acidosis take place in diabetic acidosis, including *rapid and deep breathing*. But, most important of all, the acidosis can lead to coma and death, as discussed later.

TREATMENT OF DIABETES

The theory of treatment of diabetes mellitus is based on the administration of enough insulin to enable the patient's metabolism of carbohydrate, fat, and protein to be as nearly normal as possible. Optimal therapy can prevent most acute effects of diabetes and greatly delay the chronic effects as well.

Ordinarily, the severely diabetic patient is given a single dose of one of the long-acting insulin preparations each day; this increases overall carbohydrate metabolism throughout the day. Then additional quantities of regular insulin (a short-acting preparation lasting only a few hours) are given at those times of the day when the blood glucose level tends to rise too high, such as meal times. Thus each patient is established on an individualized routine of treatment.

Diet of the Diabetic. The insulin requirements of a diabetic are established with the patient on a standard diet containing normal, well-controlled amounts of carbohydrates, and any change in the quantity of carbohydrate intake changes the requirements for insulin. In the normal person, the pancreas has the ability to adjust the quantity of insulin produced to the intake of carbohydrate; but in the completely diabetic person, this control function has been totally lost.

In the obesity maturity-onset type of diabetes, the disease can often be controlled by weight reduction alone. The decreased fat reduces the insulin requirements, and the pancreas can now supply the need.

Relationship of Treatment to Arteriosclerosis. Diabetic patients have an extremely strong tendency to develop atherosclerosis, arteriosclerosis, severe coronary heart disease, and multiple microcirculatory lesions. Indeed, those who have relatively poorly controlled diabetes throughout childhood are likely to die of heart disease in their twenties.

In the early days of treating diabetes the tendency was to reduce severely the carbohydrates in the diet so that the insulin requirements would be minimized. This procedure kept the blood sugar level down to normal values and prevented loss of glucose in the urine, but it did not prevent the abnormalities of fat metabolism. Consequently, the tendency at present is to allow the patient a normal carbohydrate diet and then simultaneously to give large quantities of insulin to metabolize the carbohydrates. This depresses the rate of fat metabolism and also helps to reduce the high level of blood cholesterol that occurs in diabetes as a result of abnormal fat metabolism.

Because the complications of diabetes — such as atherosclerosis, greatly increased susceptibility to infection, diabetic retinopathy, cataracts, hypertension, and chronic renal disease — are more closely associated with the level of the blood lipids than with the level of blood glucose, it is the object of many clinics treating diabetes to administer sufficient glucose and insulin to bring the quantity of blood lipids toward normal.

DIABETIC COMA

If diabetes is not controlled satisfactorily, severe dehydration and acidosis may result; and sometimes, even when the person is receiving treatment, sporadic changes in metabolic rates of the cells, such as might occur during bouts of fever, can also precipitate dehydration and acidosis.

If the pH of the body fluids falls below approximately 7.0, the diabetic person develops coma. Also, in addition to the acidosis, dehydration is believed to exacerbate the coma. Once the diabetic person reaches this stage, the outcome is usually fatal unless immediate treatment is provided.

Physiological Basis of Treating Diabetic Coma. The patient with diabetic coma is extremely refractory to insulin because acidic plasma has an *insulin antagonist*, an alpha globulin, that opposes the action of the insulin. Also, the very high free fatty acid and acetoacetic acid levels in the blood inhibit cellular usage of glucose, as was discussed earlier. Therefore, instead of the usual 60 to 80 units of insulin per day, the dosage usually necessary for control of severe diabetes, several times this much insulin must often be given the first day of treatment of coma.

Administration of insulin often will not by itself reverse the abnormal physiology in diabetic coma and effect a cure. In addition, it is usually necessary to correct both the dehydration and acidosis immediately. The dehydration is ordinarily corrected by rapidly administering large quantities of sodium chloride solution, and the acidosis is often corrected by administering sodium bicarbonate or sodium lactate solution.

REFERENCES

Arky, R. A.: Hypoglycemia. *In* DeGroot, L. J., *et al.* (eds.): Endocrinology. Vol. 2. New York, Grune & Stratton, 1979, p. 1099.

Chick, W. L.: Microvascular pathology in diabetes. *In* Kaley, G., and Altura, B. M. (eds.): Microcirculation. Vol. III. Baltimore, University Park Press, 1977.

Crepaldi, G., *et al.* (eds.): Diabetes, Obesity, and Hyperlipidemia. New York, Academic Press, 1978.

Esmann, V. (ed.): Regulatory Mechanisms of Carbohydrate Metabolism. New York, Pergamon Press, 1978.

Fajans, S. S.: Diabetes mellitus: Description, etiology and pathogenesis, natural history and testing procedures. *In* DeGroot, L. J., *et al.* (eds.): Endocrinology. Vol. 2. New York, Grune & Stratton, 1979, p. 1007.

Fitzgerald, P. J., and Morrison, A. B. (eds.): The Pancreas. Baltimore, Williams & Wilkins, 1980.

Galbraith, R. M.: Immunological Aspects of Diabetes Mellitus. Boca Raton, Fla., CRC Press, 1979.

Gerich, J. E., *et al.*: Regulation of pancreatic insulin and glucagon secretion. *Annu. Rev. Physiol., 38*:353, 1976.

Guyton, J. R., *et al.*: A model of glucose-insulin homeostasis in man that incorporates the heterogeneous fast pool theory of pancreatic insulin release. *Diabetes, 27*:1027, 1978.

Hedeskov, C. J.: Mechanism of glucose-induced insulin secretion. *Physiol. Rev., 60*:442, 1980.

Klachko, D. M., *et al.* (eds.): The Endocrine Pancreas and Juvenile Diabetes. New York, Plenum Press, 1979.

Lund-Anderson, H.: Transport of glucose from blood to brain. *Physiol. Rev., 59*:305, 1979.

Matschinsky, F. M., *et al.*: Metabolism of pancreatic islets and regulation of insulin and glucagon secretion. *In* DeGroot, L. J., *et al.* (eds.): Endocrinology. Vol. 2. New York, Grune & Stratton, 1979, p. 935.

Notkins, A. L.: The causes of diabetes. *Sci. Am., 241*(5):62, 1979.

Podolsky, S., and Viswanathan, M. (eds.): Secondary Diabetes: The Spectrum of the Diabetic Syndromes. New York, Raven Press, 1979.

Podolsky, S. (ed.): Clinical Diabetes: Modern Management. New York, Appleton-Century-Crofts, 1980.

Sherwin, R., and Felig, P.: Glucagon physiology in health and disease. *Int. Rev. Physiol., 16*:151, 1977.

Sherwin, R., and Felig, P.: Treatment of diabetes mellitus. *In* DeGroot, L. J., *et al.* (eds.): Endocrinology, Vol. 2. New York, Grune & Stratton, 1979, p. 1061.

Unger, R. H., and Orci, L.: Glucagon: secretion, transport, metabolism, physiologic regulation of secretion, and derangement in diabetes. *In* DeGroot, L. J., *et al.* (eds.): Endocrinology. Vol. 2. New York, Grune & Stratton, 1979, p. 959.

Winegrad, A. I., and Morrison, A. D.: Diabetic ketoacidosis, nonketotic hyperosmolar coma, and lactic acidosis. *In* DeGroot, L. J., *et al.* (eds.): Endocrinology. Vol. 2. New York, Grune & Stratton, 1979, p. 1025.

53

Parathyroid Hormone, Calcitonin, Calcium and Phosphate Metabolism, Vitamin D, Bone and Teeth

The physiology of parathyroid hormone and of the hormone calcitonin is closely related to calcium and phosphate metabolism, the function of vitamin D, and the formation of bone and teeth. Therefore, these are discussed together in the present chapter.

CALCIUM AND PHOSPHATE IN THE EXTRACELLULAR FLUID AND PLASMA — FUNCTION OF VITAMIN D

ABSORPTION OF CALCIUM AND PHOSPHATE

By far the major source of calcium in the diet is milk or milk products, which are also major sources of phosphate, but phosphate is also present in many other dietary foods, including especially the meats.

Calcium is poorly absorbed from the intestinal tract because of the relative insolubility of many of its compounds and also because bivalent cations are poorly absorbed through the intestinal mucosa anyway. On the other hand, phosphate is absorbed exceedingly well most of the time except when excess calcium is in the diet; the calcium tends to form almost insoluble calcium phosphate compounds that fail to be absorbed but instead pass on through the bowels to be excreted in the feces. In other words, the major problem in the absorption of calcium and phosphate is actually a problem of calcium absorption alone, for if this is absorbed, both are absorbed.

About seven-eighths of the daily intake of calcium is not absorbed and therefore is excreted in the feces; the remaining one-eighth is excreted in the urine.

VITAMIN D AND ITS ROLE IN CALCIUM ABSORPTION

Vitamin D has a potent effect on increasing calcium absorption from the intestinal tract; it also has important effects on both bone deposition and bone reabsorption, as will be discussed later in the chapter. However, vitamin D itself is not the active substance that actually causes these effects. Instead, the vitamin D must first be converted through a succession of reactions in the liver and the kidney to the final active product, *1,25-dihydroxycholecalciferol*. Figure 53–1 illustrates the succession of steps that leads to the formation of this substance from vitamin D. Some of the important features of this schema are:

The Vitamin D Compounds. Several different compounds derived from sterols belong to the vitamin D family, and they all perform more or less the same functions. The most important of

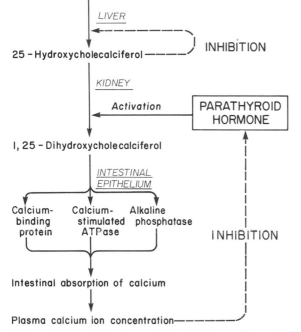

Figure 53–1. Activation of vitamin D_3 to form *1,25-dihydroxycholecalciferol;* the role of vitamin D in controlling the plasma calcium concentration.

them, called vitamin D_3, is *cholecalciferol.* Most of this substance is formed in the skin as a result of irradiation of *7-dehydrocholesterol* by ultraviolet rays from the sun. Consequently, appropriate exposure to the sun prevents vitamin D deficiency.

Conversion of Cholecalciferol to 25-Hydroxycholecalciferol in the Liver and Its Feedback Control. The first step in the activation of cholecalciferol is to convert it to 25-hydroxycholecalciferol; this conversion occurs in the liver. The process, however, is a limited one because the 25-hydroxycholecalciferol itself has a feedback inhibitory effect on the conversion reactions, as illustrated in Figure 53–1. This feedback effect is extremely important for two reasons:

First, the feedback mechanism regulates very precisely the concentration of 25-hydroxycholecalciferol in the plasma. The intake of vitamin D_3 can change many-fold, and yet the concentration of 25-hydroxycholecalciferol still remains within a few per cent of its normal mean value. Obviously, this high degree of feedback control prevents excessive action of vitamin D_3 when it is present in too high a concentration.

Second, this controlled conversion of vitamin D_3 to 25-hydroxycholecalciferol conserves the vitamin D_3 for future use, because once it is converted, it persists in the body for only a short time

thereafter, whereas in the vitamin D form it can be stored in the liver for as long as several months.

Formation of 1,25-Dihydroxycholecalciferol in the Kidneys and Its Control by Parathyroid Hormone. Figure 53–1 also illustrates that 25-hydroxycholecalciferol is converted in the kidneys to 1,25-dihydroxycholecalciferol. This latter substance is the active form of vitamin D, and none of the previous products in the schema of Figure 53–1 have very much vitamin D effect. Therefore, in the absence of the kidneys vitamin D is almost totally ineffective.

Note also in Figure 53–1 that the conversion of 25-hydroxycholecalciferol to 1,25-dihydroxycholecalciferol requires parathyroid hormone. In the absence of this hormone, either none or almost none of the 1,25-dihydroxycholecalciferol is formed. Therefore, parathyroid hormone exerts a potent effect in determining the functional effects of vitamin D in the body, specifically its effects on calcium absorption in the intestines and its effect on bone.

Hormonal Effect of 1,25-Dihydroxycholecalciferol on the Intestinal Epithelium in Promoting Calcium Absorption. 1,25-Dihydroxycholecalciferol has several effects on the intestinal epithelium in promoting intestinal absorption of calcium. Probably the most important of these effects is that it causes formation of a *calcium-binding protein* in the cytoplasm of the intestinal epithelial cells. The rate of calcium absorption seems to be directly proportional to the quantity of this calcium-binding protein. Furthermore, this protein remains in the cells for several weeks after the 1,25-dihydroxycholecalciferol has been removed from the body, thus causing a prolonged effect on calcium absorption.

Other effects of 1,25-dihydroxycholecalciferol that might play a role in promoting calcium absorption are these: (1) it causes the formation of a calcium-stimulated ATPase in the brush border of the epithelial cells, and (2) it causes the formation of an alkaline phosphatase in the epithelial cells. Unfortunately, the precise details of calcium absorption are still unknown.

Feedback Effect of Calcium Ion Concentration on the Formation of 1,25-Dihydroxycholecalciferol. Later in the chapter we shall see that the rate of secretion of parathyroid hormone is controlled almost entirely and very potently by the plasma calcium ion concentration. When the calcium ion concentration rises, this change immediately inhibits parathyroid hormone secretion; in the absence of this hormone, 1,25-dihydroxycholecalciferol cannot be formed in the kidney. Thus, this is a *negative* feedback mechanism for control of the plasma concentration of

1,25-dihydroxycholecalciferol and also of the plasma calcium ion concentration. That is, an increase in the calcium ion concentration decreases the vitamin D effect, decreases the absorption of calcium from the intestinal tract, and thus returns the calcium ion concentration back to its normal value.

THE CALCIUM IN THE PLASMA AND INTERSTITIAL FLUID

The concentration of calcium in the plasma is approximately 9.4 mg. per cent, normally varying betwen 9.0 and 10.0 mg. per cent. This is equivalent to approximately 2.4 millimoles per liter. It is apparent from these narrow limits of normality that the calcium level in the plasma is regulated exactly — and mainly by parathyroid hormone, as discussed later in the chapter.

The calcium in the plasma is present in three different forms, as shown in Figure 53–2. (1) Approximately 41 per cent of the calcium is combined with the plasma proteins and consequently is nondiffusible through the capillary membrane. (2) Approximately 9 per cent of the calcium (0.2 mM. per liter) is diffusible through the capillary membrane but is combined with other substances of the plasma and interstitial fluids (citrate and phosphate, for instance) in such a manner that it is not ionized. (3) The remaining 50 per cent of the calcium in the plasma is both diffusible through the capillary membrane and is ionized. Thus, the plasma and interstitial fluids have a normal *calcium ion concentration of approximately 1.2 mM. per liter*. This ionic calcium is important for most functions of calcium in the body, including its effect on the heart, on the nervous system, and on bone formation.

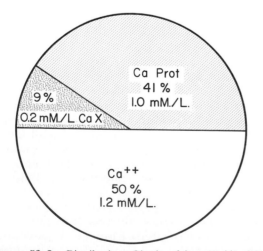

Figure 53–2. Distribution of ionic calcium (*Ca++*), diffusible but un-ionized calcium (*Ca X*), and calcium proteinate (*Ca Prot*) in blood plasma.

THE INORGANIC PHOSPHATE IN THE EXTRACELLULAR FLUIDS

Inorganic phosphate in the plasma is mainly in two forms: HPO_4^{--} and $H_2PO_4^{-}$. The concentration of both of these together is approximately 1.3 mM. per liter. Because it is difficult to determine chemically the exact ratio of HPO_4^{--} to $H_2PO_4^{-}$ in the blood, ordinarily the total quantity of phosphate is expressed in terms of milligrams of *phosphorus* per 100 ml. of blood. The average total quantity of inorganic phosphorus represented by both phosphate ions is about 4 mg. per 100 ml., varying between normal limits of 3.5 to 4 mg. per 100 ml. in adults and 4 to 5 mg. per 100 ml. in children.

EFFECTS OF ALTERED CALCIUM AND PHOSPHATE CONCENTRATIONS IN THE BODY FLUIDS

Changing the level of phosphate in the extracellular fluid from far below normal to as high as three to four times normal does not cause significant immediate effects on the body. On the other hand, elevation or depression of calcium ions in the extracellular fluid causes extreme immediate effects. Both prolonged hypocalcemia and hypophosphatemia greatly decrease bone mineralization, as explained later in the chapter.

Tetany Resulting from Hypocalcemia. When the extracellular fluid concentration of calcium ions falls below normal, the nervous system becomes progressively more and more excitable because of increased neuronal membrane permeability. Especially, the peripheral nerve fibers become so excitable that they begin to discharge spontaneously, initiating nerve impulses that pass to the peripheral skeletal muscles, where they elicit tetanic contraction. Consequently, hypocalcemia causes tetany.

Figure 53–3 illustrates tetany in the hand, which usually occurs before generalized tetany develops. This is called "carpopedal spasm."

Acute hypocalcemia in the human being ordinarily causes essentially no other significant effects besides tetany, because tetany kills the patient before other effects can develop. Tetany ordinarily occurs when the blood concentration of calcium falls from its normal level of 9.4 mg. to approximately 6 mg. per cent, which is only 35 per cent below the normal calcium concentration, and it is usually lethal at about 4 mg. per cent.

In experimental animals, in which the level of calcium can be reduced beyond the normal lethal stage, extreme hypocalcemia can cause marked dilatation of the heart, changes in cellular enzyme activities, increased cell membrane permeability

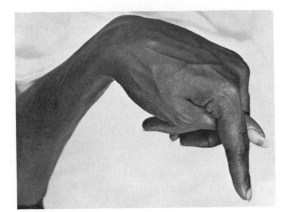

Figure 53–3. Hypocalcemic tetany of the hand, called "carpopedal spasm." (Courtesy of Dr. Herbert Langford.)

in other cells besides nerve cells, and impaired blood clotting.

Hypercalcemia. When the level of calcium in the body fluids rises above normal, the nervous system is depressed, and reflex activities of the central nervous system become sluggish. The muscles, too, become sluggish and weak, probably because of calcium effects on the muscle cell membranes. Also, increased calcium ion concentration causes constipation and lack of appetite, probably because of depressed contractility of the muscular walls of the gastrointestinal tract.

The depressive effects of an increased calcium level begin to appear when the blood level of calcium rises above approximately 12 mg. per cent, and they can become marked as the calcium level rises above 15 mg. per cent. When the level of calcium rises above approximately 17 mg. per cent in the body fluids, calcium phosphate is likely to precipitate throughout the blood and soft tissues, an effect that can be rapidly lethal.

BONE AND ITS RELATIONSHIPS WITH EXTRACELLULAR CALCIUM AND PHOSPHATES

Bone is composed of a tough *organic matrix* that is greatly strengthened by deposits of *calcium salts*. Average *compact bone* contains by weight approximately 30 per cent matrix and 70 per cent salts. However, *newly formed bone* may have a considerably higher percentage of matrix in relation to salts.

The Organic Matrix of Bone. The organic matrix of bone is approximately 95 per cent *collagen fibers*, and the remaining 5 per cent is a homogeneous medium called *ground substance*. The collagen fibers extend in all directions in the

bone but primarily along the lines of tensional force. These fibers give bone its great tensile strength.

The ground substance is composed of extracellular fluid plus *proteoglycans*, especially *chondroitin sulfate* and *hyaluronic acid*. The precise function of these is not known, though perhaps they help to control the deposition of calcium salts.

The Bone Salts. The crystalline salts deposited in the organic matrix of bone are composed principally of *calcium* and *phosphate*, and the formula for the major crystalline salts, known as *hydroxyapatites*, is the following:

$$Ca^{++}_{10-X}(H_3O^+)_{2X} \cdot (PO_4)_6(OH^-)_2$$

Each crystal—about 400 Å long, 10 to 30 Å thick, and 100 Å wide—is shaped like a long, flat plate. The relative ratio of calcium to phosphorus can vary markedly under different nutritional conditions, the Ca/P ratio on a weight basis varying between 1.3 and 2.0.

Magnesium, sodium, potassium, and *carbonate* ions are also present among the bone salts, though x-ray diffraction studies fail to show definite crystals formed by them. Therefore, they are believed to be adsorbed to the surfaces of the hydroxyapatite crystals rather than organized into distinct crystals of their own. This ability of many different types of ions to adsorb to bone crystals extends to many ions normally foreign to bone, such as *strontium, uranium, plutonium, the other transuranic elements, lead, gold, other heavy metals,* and *at least 9 out of 14 of the major radioactive products released by explosion of the hydrogen bomb.* Deposition of radioactive substances in the bone can cause prolonged irradiation of the bone tissues, and if a sufficient amount is deposited, an osteogenic sarcoma almost invariably eventually develops.

Tensile and Compressional Strength of Bone. Each collagen fiber of bone is composed of repeating periodic segments every 640 Å along its length; hydroxyapatite crystals lie adjacent to each segment of the fiber bound tightly to it. This intimate bonding prevents "shear" in the bone; that is, it prevents the crystals and collagen fibers from slipping out of place, which is essential in providing strength to the bone. In addition, the segments of adjacent collagen fibers overlap each other, also causing hydroxyapatite crystals to be overlapped like bricks keyed to each other in a brick wall.

The collagen fibers of bone, like those of tendons, have great tensile strength, while the calcium salts, which are similar in physical properties to marble, have great compressional strength. These combined properties, plus the degree of bondage between the collagen fibers and the crystals, provide a bony structure that has both extreme tensile and compressional strength. Thus, bones are constructed in exactly the same way that reinforced concrete is constructed. The steel of reinforced concrete provides the tensile strength, while the cement, sand, and rock provide the compressional strength. Indeed, the compressional strength of bone is greater than that of even the best reinforced concrete, and the tensile strength approaches that of reinforced concrete.

PRECIPITATION AND ABSORPTION OF CALCIUM AND PHOSPHATE IN BONE – EQUILIBRIUM WITH THE EXTRACELLULAR FLUIDS

Supersaturated State of Calcium and Phosphate Ions in Extracellular Fluids with Respect to Hydroxyapatite. The concentrations of calcium and phosphate ions in extracellular fluid are considerably greater than those required to cause precipitation of hydroxyapatite. However, because of the large number of ions required to form a single molecule of hydroxyapatite, it is very difficult for all of these ions to come together simultaneously. Therefore, hydroxyapatite crystals fail to precipitate in other tissues besides bone despite the state of supersaturation of the ions.

Mechanism of Bone Calcification. The initial stage of bone production is the secretion of collagen and ground substance by the osteoblasts. The collagen polymerizes rapidly to form collagen fibers, and the resultant tissue becomes *osteoid*, a cartilage-like material but differing from cartilage in that calcium salts precipitate in it. As the osteoid is formed, some osteoblasts become entrapped in the osteoid and then are called *osteocytes*. These may play an important role in the subsequent control of bone salts, though this is not certain.

Within a few days after the osteoid is formed, calcium salts begin to precipitate on the surfaces of the collagen fibers. The salts appear at periodic intervals along each collagen fiber, forming minute nidi that gradually grow over a period of days and weeks into the finished product, *hydroxyapatite crystals*.

The initial calcium salts to be deposited are not hydroxyapatite crystals but instead are amorphous compounds (noncrystalline), probably a mixture of such salts as $CaHPO_4 \cdot 2H_2O$, $Ca_3(PO_4)_2 \cdot 3H_2O$, and others. Then, by a process of substitution and addition of atoms, these salts are reshaped into the hydroxyapatite crystals.

It is still not known what causes calcium salts to be deposited in osteoid. One theory suggests that the osteoblasts and the entrapped osteocytes in the osteoid play an important role in the following way: It is known that these cells concentrate large quantities of calcium and phosphate in their mitochondria and even precipitate calcium phosphate compounds in them. Electron micrographs indicate that calcium phosphate–containing vesicles break away from the mitochondria, migrate to the walls of the cell, and then extrude calcium phosphate into the surrounding extracellular fluid. It may be these preformed calcium phosphate salts that attach to the collagen fibers to form the initial nidi for crystallization, and it is possible that subsequent vesicles supply the necessary calcium and phosphate ions for further growth of the crystals.

However, another theory holds that at the time of formation the collagen fibers are already constituted in advance for causing precipitation of calcium salts. One variant of this theory suggests that the osteoblasts secrete a substance into the osteoid to neutralize an inhibitor (perhaps pyrophosphate) that normally prevents hydroxyapatite crystallization. Once this neutralization has occurred, the natural affinity of the collagen fibers for calcium salts supposedly causes the precipitation. In support of this theory is the fact that purified collagen fibers prepared from other tissues of the body besides bone will also cause precipitation of hydroxyapatite crystals from plasma.

The formation of initial crystals within the collagen fibers is called *crystal seeding* or *nucleation*.

The growth of hydroxyapatite crystals in newly forming bone reaches 75 per cent completion in a few days, but it usually takes months for the bone to achieve full calcification.

EXCHANGEABLE CALCIUM

If soluble calcium salts are injected intravenously, the calcium ion concentration can be made to increase immediately to very high levels. However, within minutes to an hour or more, the calcium ion concentration returns to normal. Likewise, if large quantities of calcium ions are removed from the circulating body fluids, the calcium ion concentration again returns to normal within minutes to hours. These effects result from the fact that the bone and other body tissues contain a type of *exchangeable* calcium that is always in equilibrium with the calcium ions in the extracellular fluids. Most of this exchangeable calcium is in the bone, and it normally amounts to about 0.4 to 1.0 per cent of the total bone calcium. Most of this calcium is probably deposited in the bones by the process of adsorption or in the form of readily mobilizable salts such as $CaHPO_4$ and the other amorphous salts.

The importance of exchangeable calcium to the body is that it provides a rapid buffering mechanism to keep the calcium ion concentration in the extracellular fluids from rising to excessive levels or falling to very low levels under transient conditions of excess or diminished availability of calcium.

DEPOSITION AND ABSORPTION OF BONE – REMODELING OF BONE

Deposition of Bone by the Osteoblasts. Bone is continually being deposited by *osteoblasts*, and it is continually being absorbed where *osteoclasts* are active. Osteoblasts are found on the outer surfaces of the bones and in the bone cavities. A small amount of osteoblastic activity occurs continually in all living bones (on about 4 per cent of all surfaces), so that at least some new bone is being formed constantly.

Absorption of Bone — Function of the Osteoclasts. Bone is also being continually absorbed in the presence of osteoclasts, which are normally active at any one time on about 1 per cent of the outer surfaces and cavity surfaces. Later in the chapter we will see that parathyroid hormone controls the bone absorptive activity of osteoclasts.

Histologically, bone absorption occurs immediately adjacent to the osteoclasts, as illustrated in Figure 53–4. The mechanism of this absorption is believed to be the following: The osteoclasts send out villous-like projections toward the bone and from these "villi" secrete two types of substances: (1) proteolytic enzymes, released from the lysosomes of the osteoclasts, and (2) several acids, including citric acid and lactic acid. The enzymes presumably digest or dissolute the organic matrix of the bone, while the acids cause solution of the bone salts. Also, whole fragments of bone salts and collagen are literally gobbled up (phagocytosed) by the "villi' and then digested within the osteoclasts.

Equilibrium Between Bone Deposition and Absorption. Normally, except in growing bones, the rates of bone deposition and absorption are equal to each other, so that the total mass of bone remains constant. Usually, osteoclasts exist in large masses, and once a mass of osteoclasts begins to develop, it usually eats away at the bone for about three weeks, eating out a tunnel that may be as great as 1 mm. in diameter. At the end of this time the osteoclasts are converted into osteoblasts, and new bone begins to develop. Bone deposition then continues for several months, the new bone being laid down in succes-

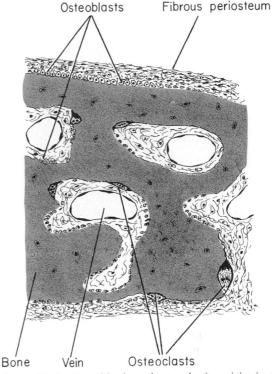

Figure 53–4. Osteoblastic and osteoclastic activity in the same bone.

sive layers on the inner surfaces of the cavity until the tunnel is filled. Deposition of new bone ceases when the bone begins to encroach on the blood vessels supplying the area. The canal through which these vessels run, called the *haversian canal*, therefore, is all that remains of the original cavity. Each new area of bone deposited in this way is called an *osteon*, as shown in Figure 53–5.

Value of Continual Remodeling of Bone. The continual deposition and absorption of bone have a number of physiologically important functions. First, bone ordinarily adjusts its strength in proportion to the degree of bone stress. Consequently, bones thicken when subjected to heavy loads. Second, even the shape of the bone can be rearranged for proper support of mechanical forces by deposition and absorption of bone in accordance with stress patterns. Third, new organic matrix is needed as the old organic matrix degenerates. In this manner the normal toughness of bone is maintained. Indeed, the bones of children, in whom the rate of deposition and absorption is rapid, show little brittleness in comparison with the bones of old age, at which time the rates of deposition and absorption are slow.

Control of the Rate of Bone Deposition by Bone "Stress." Bone is deposited in proportion to the compressional load that the bone must carry. For instance, the bones of athletes become considerably heavier than those of nonathletes.

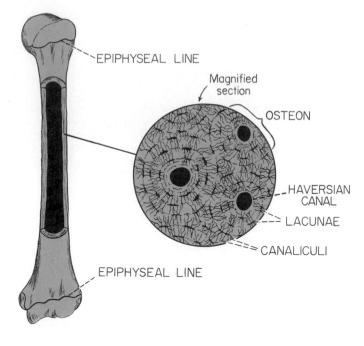

EPIPHYSEAL LINE

Magnified section

OSTEON

HAVERSIAN CANAL

LACUNAE

CANALICULI

EPIPHYSEAL LINE

Figure 53–5. The structure of bone.

Also, if a person has one leg in a cast but continues to walk on the opposite leg, the bone of the leg in the cast becomes thin and decalcified, while the opposite bone remains thick and normally calcified. Therefore, continual physical stress stimulates osteoblastic deposition of bone.

The deposition of bone at points of compressional stress has been suggested to be caused by a *piezoelectric* effect, as follows: Compression of bone causes a negative potential at the compressed site and a positive potential elsewhere in the bone. It has been shown that minute quantities of current flowing in bone cause osteoblastic activity at the negative end of the current flow, which could explain the increased bone deposition at compression sites.

Repair of a Fracture. A fracture of a bone in some way maximally activates all the periosteal and intraosseous osteoblasts involved in the break. Immense numbers of new osteoblasts are formed almost immediately from *osteoprogenitor cells*, which are bone stem cells. Therefore, within a short time a large bulge of osteoblastic tissue and new organic bone matrix, followed shortly by the deposition of calcium salts, develops between the two broken ends of the bone. This is called a *callus*. It is then reshaped into an appropriate structural bone during the ensuing months.

PARATHYROID HORMONE

For many years it has been known that increased activity of the parathyroid gland causes rapid absorption of calcium salts from the bones with resultant hypercalcemia in the extracellular fluid; conversely, hypofunction of the parathyroid glands causes hypocalcemia, often with resultant tetany, as described earlier in the chapter. Also, parathyroid hormone is important in phosphate metabolism as well as in calcium metabolism.

Physiologic Anatomy of the Parathyroid Glands. Normally there are four parathyroid glands in the human being; these are located immediately behind the thyroid gland — one behind each of the upper and each of the lower poles of the thyroid. Each parathyroid gland is approximately 6 mm. long, 3 mm. wide, and 2 mm. thick, and has a macroscopic appearance of dark brown fat; therefore, the parathyroid glands are difficult to locate.

Removal of half the parathyroid glands usually causes little physiological abnormality. However, removal of three out of four normal glands usually causes transient hypoparathyroidism. But even a small quantity of remaining parathyroid tissue is usually capable of hypertrophying satisfactorily to perform the function of all the glands.

The parathyroid gland of the adult human being, illustrated in Figure 53–6, contains mainly *chief cells* and *oxyphil cells*, but oxyphil cells are absent in many animals and in young human beings. The chief cells secrete most of the parathyroid hormone. The function of the oxyphil cells is not certain; they are probably aged chief cells that still secrete some hormone.

Chemistry of Parathyroid Hormone. Parathyroid hormone has been isolated in a pure form. It is a small protein with a molecular weight

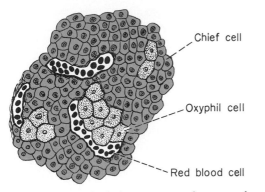

Figure 53–6. Histological structure of a parathyroid gland.

of approximately 9500 and is composed of 84 amino acids.

EFFECT OF PARATHYROID HORMONE ON CALCIUM AND PHOSPHATE CONCENTRATIONS IN THE EXTRACELLULAR FLUID

Figure 53–7 illustrates the effect on the blood calcium and phosphate concentrations caused by suddenly beginning to infuse parathyroid hormone into an animal and continuing the infusion for an indefinite period of time. Note that at the onset of infusion the calcium ion concentration begins to rise and reaches a plateau level in about 4 hours. On the other hand, the phosphate concentration falls and reaches a depressed plateau level within a few hours. The rise in calcium concentration is caused principally by a direct effect of parathyroid hormone in causing calcium and phosphate absorption from the bone. The decline in phosphate concentration, on the other hand, is caused by a very strong effect of parathyroid hormone on the kidney in causing excessive renal phosphate excretion, an effect that is usually great enough to override the increased phosphate absorption from the bone.

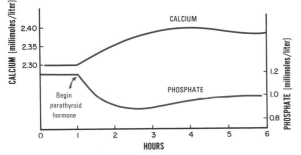

Figure 53–7. Approximate changes in calcium and phosphate concentrations during the first five hours of parathyroid hormone infusion at a moderate rate.

CALCIUM AND PHOSPHATE ABSORPTION FROM THE BONE CAUSED BY PARATHYROID HORMONE

Parathyroid hormone seems to have two separate effects on bone in causing absorption of calcium and phosphate. One effect, very rapid, takes place in minutes and probably results from activation of the already existing bone cells to promote the calcium and phosphate absorption. The second phase is a much slower one, requiring several days or even weeks to become fully developed, and it results from proliferation of the osteoclasts, followed by greatly increased osteoclastic reabsorption of the bone itself, not merely absorption of calcium phosphate salts from the bone.

The Rapid Phase of Calcium and Phosphate Absorption — The "Osteocytic Membrane" System. When large quantities of parathyroid hormone are injected, the calcium ion concentration in the blood begins to rise within minutes, long before any new bone cells can be developed. Histological studies have shown that the parathyroid hormone causes removal of bone salts from the bone matrix in the vicinity of the osteocytes lying within the bone itself and also in the vicinity of the osteoblasts. Yet, strangely enough, one does not usually think of either osteoblasts or osteocytes functioning to cause bone salt absorption, because both these types of cells are osteoblastic in nature and are normally associated with bone deposition and its calcification. However, recent studies have shown that the osteoblasts and osteocytes form a membrane system that spreads over all the bone surfaces except for small surface areas that are adjacent to the osteoclasts. Also, long filmy processes extend from osteocyte to osteocyte throughout the bone structure, and these processes also connect with the surface osteocytes and osteoblasts. There is much reason to believe that this extensive *osteocytic membrane system* provides a permeable membrane that separates the bone itself from the extracellular fluid.

But, where does parathyroid hormone fit into this picture? It seems that parathyroid hormone can strongly activate a calcium pump in the osteocytic membrane, thereby causing rapid removal of calcium phosphate salts from the amorphous bone crystals that lie near this membrane.

The Slow Phase of Bone Absorption and Calcium Phosphate Release — Activation of the Osteoclasts. A much better known effect of parathyroid hormone is activation of the osteoclasts. These in turn set about their usual task of gobbling up the bone.

Activation of the osteoclastic system occurs in

two stages: (1) immediate activation of the osteoclasts that are already formed, and (2) formation of new osteoclasts from *osteoprogenitor cells*. Usually several days of excess parathyroid hormone cause the osteoclastic system to become well developed, but it can continue to grow for literally months under the influence of very strong parathyroid hormone stimulation.

Bone contains such great amounts of calcium in comparison with the total amount in all the extracellular fluids (about 1000 times as much) that even when parathyroid hormone causes a great rise in calcium concentration in the fluids, it is impossible to discern any immediate effect at all on the bones. Yet prolonged administration or secretion of parathyroid hormone finally results in evident absorption in all the bones with development of large cavities filled with very large, multinucleated osteoclasts.

Effect of Parathyroid Hormone on Phosphate and Calcium Excretion by the Kidneys. Administration of parathyroid hormone causes immediate and rapid loss of phosphate in the urine. This effect is caused by diminished renal tubular reabsorption of phosphate ions.

Parathyroid hormone also causes renal tubular *reabsorption* of calcium at the same time that it diminishes phosphate reabsorption. Were it not for this effect of parathyroid hormone on the kidneys of increasing calcium reabsorption, the continual loss of calcium into the urine would eventually deplete the bones of this mineral.

Effect of Vitamin D on Bone and Its Relation to Parathyroid Activity. Vitamin D plays important roles in both bone absorption and bone deposition. Administration of extreme quantities of vitamin D causes absorption of bone in much the same way that administration of parathyroid hormone does. Also, in the absence of vitamin D, the effect of parathyroid hormone in causing bone absorption is greatly reduced or even prevented. Therefore, it is possible, if not likely, that parathyroid hormone functions in bone the same way that it functions in the kidneys and intestines — that is, by causing the conversion of vitamin D to 1,25-dihydroxycholecalciferol, this in turn acting to cause the bone absorption.

Vitamin D in much smaller amounts promotes bone calcification. Obviously, one of the ways in which it does so is to increase calcium and phosphate absorption from the intestines. However, even in the absence of such increase, it still enhances the mineralization of bone. Here again, the mechanism of the effect is unknown, but it probably results from the ability of 1,25-dihydroxycholecalciferol to cause transport of calcium ions through cell membranes — perhaps through the osteoblastic or osteocytic cell membranes.

CONTROL OF PARATHYROID SECRETION BY CALCIUM ION CONCENTRATION

Even the slightest decrease in calcium ion concentration in the extracellular fluid causes the parathyroid glands to increase their rate of secretion and to hypertrophy. For instance, the parathyroid glands become greatly enlarged in *rickets*, in which the level of calcium is usually depressed only a few per cent; they also become greatly enlarged in pregnancy, even though the decrease in calcium ion concentration in the mother's extracellular fluid is hardly measurable; and they are greatly enlarged during lactation because calcium is used for milk formation.

On the other hand, any condition that increases the calcium ion concentration causes decreased activity and reduced size of the parathyroid glands. Such conditions include (1) excess quantities of calcium in the diet, (2) increased vitamin D in the diet, and (3) bone absorption caused by factors other than parathyroid hormone (for example, bone absorption caused by disuse of the bones).

Figure 53–8 illustrates quantitatively the relationship between plasma calcium concentration and plasma parathyroid hormone concentration. The solid curve shows the short-term relationship when the calcium concentration is changed over a period of a few hours. This shows that a decrease in calcium concentration only 1 mg. per cent approximately doubles the concentration of plasma parathyroid hormone. On the other hand, the

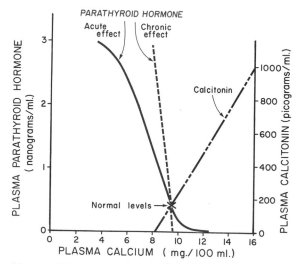

Figure 53–8. Approximate effect of plasma calcium concentration on the plasma concentrations of parathyroid hormone and calcitonin. Note especially that long-term, chronic changes in calcium concentration can cause as much as a 100 per cent change in parathyroid hormone concentration for only a 2 per cent change in calcium concentration.

long-term relationship that one finds when the calcium ion concentration changes over a period of many weeks is illustrated by the long-dashed line; this illustrates that a decrease of about 0.1 mg. per cent in plasma calcium concentration will double parathyroid hormone secretion. To state this still another way, the long-term relationship between plasma calcium and plasma parathyroid hormone shows that a decrease of 1 per cent in calcium can give as much as 100 per cent increase in parathyroid hormone. Obviously, this is the basis of the body's extremely potent feedback system for control of plasma calcium ion concentration.

INTESTINAL AND RENAL CONTROL OF PLASMA CALCIUM CONCENTRATION—ROLE OF PARATHYROID HORMONE

Though it is frequently stated that the absorption and deposition of calcium in bone is *the* long-term controller of blood calcium ion concentration, this is true only as long as the bone does not become saturated with calcium or totally depleted. However, since the bone does have these limits, it is actually a large reservoir for long-term *buffering* of calcium ion concentration over a period of months or years. It is not, however, the eventual long-term controller of plasma calcium concentration. Instead, this is achieved by the control of absorption and excretion by the intestines and kidneys.

It has already been pointed out that an increase in parathyroid hormone causes an increase in net absorption of calcium from the intestines and also causes increased reabsorption of calcium from the renal tubules. When the bone has become saturated with calcium salts and can no longer function as a depository of additional calcium ions, the slight excess of extracellular calcium ions reduces parathyroid secretion, which then decreases calcium absorption in both the intestines and kidney tubules. Conversely, when the bone has even a slight deficit of calcium salts, parathyroid secretion increases; this increase can allow for maintenance of almost normal plasma calcium concentration by increasing calcium absorption from both the intestines and kidney tubules.

CALCITONIN

About 20 years ago, a new hormone that has effects on blood calcium opposite to those of parathyroid hormone was discovered in several lower animals and at first was believed to be secreted by the parathyroid glands. This hormone was named *calcitonin* because it reduces the blood calcium ion concentration. Soon after its initial discovery, it was found to be secreted in the human being not by the parathyroid glands but by *parafollicular cells*, or "C" cells, in the interstitium of the thyroid gland, for which reason it has also been called *thyrocalcitonin*.

Calcitonin is a large polypeptide with a molecular weight of approximately 3000 and having a chain of 32 amino acids.

Effect of Calcitonin to Decrease Plasma Calcium Concentration. Calcitonin has the very rapid effect of decreasing blood calcium ion concentration, beginning within minutes after injection of the calcitonin. Thus, the effect of calcitonin on blood calcium ion concentration is exactly opposite to that of parathyroid hormone, and it occurs several times as rapidly.

Calcitonin reduces plasma calcium concentration in three separate ways:

1. The immediate effect is to decrease the activity of the osteoclasts.

2. The second effect, which can be seen within about an hour, is an increase in osteoblastic activity.

3. The third and most prolonged effect of calcitonin is to prevent formation of new osteoclasts from osteoprogenitor cells. However, this subsequently decreases osteoblastic activity as well. Therefore, over a long period of time the net result, simply, is greatly reduced osteoclastic and osteoblastic activity without any significant prolonged effect on plasma calcium ion concentration. That is, the effect on plasma calcium is mainly a transient one, lasting for a few days, at most. However, there is a prolonged effect of decreasing the rate of bone remodeling.

Calcitonin has only a weak effect on plasma calcium concentration in the adult human being. The reason for this is the following: In the adult, osteoclastic absorption provides only 0.8 gram of calcium to the extracellular fluid each day, and the suppression of this amount of osteoclastic activity by calcitonin has little effect on the plasma calcium. On the other hand, the effect in children is much more marked because bone remodeling occurs rapidly in children, osteoclastic absorption of calcium being as great as 5 or more grams per day — equal to 5 to 10 times the total calcium in all the extracellular fluid.

Effect of Plasma Calcium Concentration on the Secretion of Calcitonin

An increase in plasma calcium concentration of about 10 per cent causes an immediate two-fold increase in the rate of secretion of calcitonin, which is illustrated by the dotted line of Figure 53–8. This provides a second hormonal feedback mechanism for controlling the plasma calcium ion concentration, but one that works exactly oppositely to the parathyroid hormone system. That is, an increase in calcium concentration causes increased calcitonin secretion, and the increased calcitonin in turn reduces the plasma calcium concentration back toward normal.

However, there are two major differences between the calcitonin and the parathyroid feedback systems. First, the calcitonin mechanism operates more rapidly, reaching peak activity in less than an hour, in contrast to the several hours required for peak activity to be attained following parathyroid secretion.

The second difference is that the calcitonin mechanism acts mainly as a short-term regulator and has little long-term effect, month in and month out, on calcium ion concentration — contrary to the effect of the parathyroid hormone system. As was pointed out above, the calcitonin mechanism is a very weak one in the normal human adult, anyway. Therefore, over a prolonged period of time it is almost entirely the parathyroid system that sets the long-term level of calcium ions in the extracellular fluid.

PHYSIOLOGY OF PARATHYROID AND BONE DISEASES

HYPOPARATHYROIDISM

When the parathyroid glands do not secrete sufficient parathyroid hormone, the osteoclasts of the bone become almost totally inactive. As a result, bone reabsorption is so depressed that the level of calcium in the body fluids decreases.

If the parathyroid glands are suddenly removed, the calcium level in the blood falls from the normal of 9.4 mg. per cent to 6 to 7 mg. per cent within two to three days. When this level is reached, the usual signs of tetany develop. Among the muscles of the body that are especially sensitive to tetanic spasm are the laryngeal muscles. Laryngeal spasm obstructs respiration, which is the usual cause of death in tetany unless appropriate treatment is applied.

Treatment of Hypoparathyroidism. *Parathyroid Hormone (Parathormone).* Parathyroid hormone is occasionally used for treating hypoparathyroidism. However, because of the expense of this hormone, because its effect lasts only a few hours, and because the tendency of the body to develop immune bodies against it makes it progressively less and less active in the body, treatment of hypoparathyroidism with parathyroid hormone is rare in present-day therapy.

Dihydrotachysterol and Vitamin D. In addition to its ability to cause increased absorption of calcium from the gastrointestinal tract, vitamin D also causes a moderate effect similar to that of parathyroid hormone in promoting calcium and phosphate absorption from bones. Therefore, a person with hypoparathyroidism can be treated satisfactorily by administration of *large quantities* of vitamin D. One of the vitamin D compounds, dihydrotachysterol, has a more marked ability to cause bone absorption than do most of the other vitamin D compounds because it can be converted directly to 1,25-dihydroxycholecalciferol by the kidneys and is not limited by the normal liver feedback mechanism that controls the conversion of other forms of vitamin D to the active form. Administration of calcium plus dihydrotachysterol three or more times a week can almost compltely control the calcium level in the extracellular fluid of a hypoparathyroid person.

HYPERPARATHYROIDISM

The cause of hyperparathyroidism ordinarily is a tumor of one of the parathyroid glands.

In hyperparathyroidism extreme osteoclastic activity occurs in the bones, and this elevates the calcium ion concentration in the extracellular fluid while usually (but not always) depressing slightly the concentration of phosphate ions because of increased renal excretion of phosphate.

In severe hyperparathyroidism, the osteoclastic absorption of bone soon far outstrips osteoblastic deposition, and the bone may be eaten away almost entirely. Indeed, the reason a hyperparathyroid person comes to the doctor is often a broken bone. X-ray of the bone shows extensive decalcification and occasionally large punched-out cystic areas of the bone that are filled with osteoclasts in the form of so-called giant cell "tumors." Obviously, multiple fractures of the weakened bones result from only slight trauma.

The obvious treatment of hyperparathyroidism is surgical removal of the parathyroid tumor, but this is a difficult procedure because these tumors are often only a few millimeters in size and are difficult to find at operation.

RICKETS

Rickets occurs mainly in children as a result of calcium or phosphate deficiency in the extracellular fluid. Ordinarily, rickets is due to lack of vitamin D rather than to lack of calcium or phosphate in the diet. If the child is properly exposed to sunlight, the ultraviolet rays form vitamin D_3 (cholecalciferol), which prevents rickets by promoting calcium and phosphate absorption from the intestines as discussed earlier in the chapter.

Children who remain indoors through the winter generally do not receive adequate quantities of vitamin D without some supplementary therapy in the diet. Rickets tends to occur especially in the spring months because vitamin D formed during the preceding summer can be stored for several months in the liver, and also, calcium and phosphorus absorption from the bones must take place for several months before clinical signs of rickets become apparent.

Effect of Rickets on the Bone. During prolonged deficiency of calcium and phosphate in the body fluids, increased parathyroid hormone secretion protects the body against hypocalcemia by causing osteoclastic absorption of the bone; this in turn causes the bone to become progressively weaker and imposes marked physical stress on the bone, resulting in rapid osteoblastic activity. The osteoblasts lay down large quantities of organic bone matrix, osteoid, which does not become calcified because the calcium and phosphate concentrations are insufficient to cause calcification. Consequently, the newly formed, uncalcified osteoid gradually takes the place of other bone that is being reabsorbed.

Obviously, hyperplasia of the parathyroid glands is marked in rickets because of the decreased blood calcium level.

Tetany in Rickets. In the early stages of rickets, tetany almost never occurs because the parathyroid glands continually stimulate osteoclastic absorption of bone and therefore maintain an almost normal level of calcium in the body fluids. However, when the bones become exhausted of calcium, the level of calcium may fall rapidly. As the blood level of calcium falls below 7 mg. per cent, the usual signs of tetany develop, and the child may die of tetanic laryngeal spasm unless intravenous calcium is administered, which relieves the tetany immediately.

Treatment of Rickets. The treatment of rickets depends on supplying adequate calcium and phosphate in the diet and also on administering vitamin D adequately.

Osteomalacia. Osteomalacia is rickets in adults and is frequently called "adult rickets."

Normal adults rarely have dietary lack of vitamin D or calcium because large quantities of calcium are not needed for bone growth as in children. However, lack of vitamin D and calcium occasionally occurs as a result of steatorrhea (failure to absorb fat), for vitamin D is fat soluble, and calcium tends to form insoluble soaps with fat; consequently, in steatorrhea vitamin D and calcium tend to pass into the feces. Under these conditions an adult occasionally has such poor calcium and phosphate absorption that "adult rickets" can occur. Though this almost never causes tetany, it does often cause severe bone debility.

PHYSIOLOGY OF THE TEETH

The teeth cut, grind, and mix the food. To perform these functions the jaws have powerful muscles capable of providing an occlusive force of as much as 50 to 100 pounds between the front teeth and as much as 150 to 200 pounds for the jaw teeth. Also, the upper and lower teeth are provided with projections and facets that interdigitate, so that each set of teeth fits with the other. This fitting is called *occlusion*, and it allows even small particles of food to be caught and ground between the tooth surfaces.

FUNCTION OF THE DIFFERENT PARTS OF THE TEETH

Figure 53–9 illustrates a sagittal section of a tooth, showing its major functional parts: the *enamel, dentine, cementum,* and *pulp.* The tooth can also be divided into the *crown,* which is the portion that protrudes out of the gum into the mouth, and the *root,* which is the portion that protrudes into the bony socket of the jaw. The collar between the crown and the root where the tooth is surrounded by gum is called the *neck.*

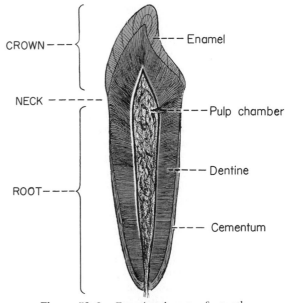

Figure 53–9. Functional parts of a tooth.

Dentine. The main body of the tooth is composed of dentine, which has a strong, bony structure. Dentine is made up principally of hydroxyapatite crystals similar to those in the bone but much more dense. These are embedded in a strong meshwork of collagen fibers. In other words, the principal constituents of dentine are very much the same as those of bone. The major difference is its histological organization, for dentine does not contain any osteoblasts, osteoclasts, or spaces for blood vessels or nerves. Instead, it is deposited and nourished by a layer of cells called *odontoblasts*, which line its inner surface along the wall of the pulp cavity.

The calcium salts in dentine make it extremely resistant to compressional forces, while the collagen fibers make it tough and resistant to tensional forces that might result when the teeth are struck by solid objects.

Enamel. The outer surface of the tooth is covered by a layer of enamel that is formed prior to eruption of the tooth by special epithelial cells called *ameloblasts*. Once the tooth has erupted, no more enamel is formed. Enamel is composed of large and extremely dense crystals of hydroxyapatite with adsorbed carbonate, magnesium, sodium, potassium, and other ions embedded in a meshwork of very strong and almost completely insoluble protein fibers that are similar to (but not identical with) the keratin of hair. The dense crystalline structure of the salts makes the enamel extremely hard, much harder than the dentine. Also, the special protein fiber meshwork makes enamel very resistant to acids, enzymes, and other corrosive agents, because this protein is one of the most insoluble and resistant proteins known.

Cementum. Cementum is a bony substance secreted by cells of the *periodontal membrane*, which lines the tooth socket. Many collagen fibers pass directly from the bone of the jaw, through the periodontal membrane, and then into the cementum. These collagen fibers and the cementum hold the tooth in place. When the teeth are exposed to excessive strain, the layer of cementum becomes thicker and stronger. Also, it increases in thickness and strength with age, causing the teeth to become progressively more firmly seated in the jaws as one reaches adulthood and older.

Pulp. The inside of each tooth is filled with pulp, which in turn is composed of connective tissue with an abundant supply of nerves, blood vessels, and lymphatics. The cells lining the surface of the pulp cavity are the odontoblasts, which, during the formative years of the tooth, lay down the dentine but at the same time encroach more and more on the pulp cavity, making it smaller. In later life the dentine stops

growing and the pulp cavity remains essentially constant in size. However, the odontoblasts are still viable and send projections into small *dentinal tubules* that penetrate all the way through the dentine; they are of importance for providing nutrition.

DENTITION

All human beings and most other mammals develop two sets of teeth during a lifetime. The first teeth are called the *deciduous teeth*, or *milk teeth*, and they number 20 in the human being. These erupt between the seventh month and second year of life, and they last until the sixth to the thirteenth year. After each deciduous tooth is lost, a permanent tooth replaces it, and an additional 8 to 12 molars appear posteriorly in the jaw, making the total number of permanent teeth 28 to 32, depending on whether the four *wisdom teeth* finally appear, which does not occur in everyone.

Formation of the Teeth. Figure 53–10 illustrates the formation and eruption of teeth. Figure 53–10A shows invagination of the oral epithelium into the *dental lamina;* this is followed by the development of a tooth-producing organ. The outer epithelial cells form ameloblasts, which form the enamel on the outside of the tooth. The inner epithelial cells invaginate upward to form a pulp cavity and also to form the odontoblasts that secrete dentine. Thus, enamel is formed on the outside of the tooth, and dentine is formed on the inside, giving rise to an early tooth as illustrated in Figure 53–10B.

Eruption of Teeth. During early childhood, the teeth begin to protrude upward from the jaw bone through the oral epithelium into the mouth. The cause of "eruption" is unknown, though several theories have been offered in an attempt to explain this phenomenon. The most likely theory is that growth of the tooth root as well as

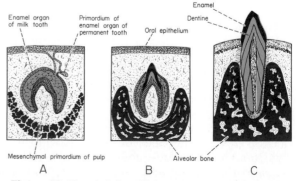

Figure 53–10. *A*, Primordial tooth organ. *B*, The developing tooth. *C*, The erupting tooth.

the bone underneath the tooth progressively shoves the tooth forward.

Development of the Permanent Teeth. During embryonic life, a tooth-forming organ also develops in the dental lamina for each permanent tooth that will be needed after the deciduous teeth are gone. These tooth-producing organs slowly form the permanent teeth throughout the first 6 to 20 years of life. When each permanent tooth becomes fully formed, it, like the deciduous tooth, pushes upward through the bone of the jaw. In so doing it erodes the root of the deciduous tooth and eventually causes it to loosen and fall out. Soon thereafter, the permanent tooth erupts to take the place of the original one.

Metabolic Factors in Development of the Teeth. The rate of development and the speed of eruption of teeth can be accelerated by both thyroid and growth hormones. Also, the deposition of salts in the early forming teeth is affected considerably by various factors of metabolism, such as the availability of calcium and phosphate in the diet, the amount of vitamin D present, and the rate of parathyroid hormone secretion. When all these factors are normal, the dentine and enamel will be correspondingly healthy, but when they are deficient, the calcification of the teeth also may be defective so that the teeth will be abnormal throughout life.

MINERAL EXCHANGE IN TEETH

The salts of teeth, like those of bone, are composed basically of hydroxyapatite with adsorbed carbonates and various cations bound together in a hard crystalline substance. Also, new salts are constantly being deposited while old salts are being reabsorbed from the teeth, as also occurs in bone. However, experiments indicate that deposition and reabsorption occur mainly in the dentine and cementum, while very little occurs in the enamel. Much of that which does occur in the enamel occurs by exchange of minerals with the saliva instead of with the fluids of the pulp cavity. The rate of absorption and deposition of minerals in the cementum is approximately equal to that in the surrounding bone of the jaw, while the rate of deposition and absorption of minerals in the dentine is only one-third that of bone. The cementum has characteristics almost identical with those of usual bone, including the presence of osteoblasts and osteoclasts, while dentine does not have these characteristics, as was explained above; this difference undoubtedly explains the different rates of mineral exchange.

The mechanism by which minerals are deposited and reabsorbed from the dentine is not clear. It is probable that the small processes of the odontoblasts that protrude into the tubules of the dentine are capable of absorbing salts and then of providing new salts to take the place of the old.

In summary, rapid mineral exchange occurs in the dentine and cementum of teeth, though the mechanism of this exchange in dentine is unknown. On the other hand, enamel exhibits extremely slow mineral exchange so that it maintains most of its original mineral complement throughout life.

DENTAL ABNORMALITIES

The two most common dental abnormalities are *caries* and *malocclusion*. Caries means erosions of the teeth, whereas malocclusion means failure of the projections of the upper and lower teeth to interdigitate properly.

Caries, and the Role of Fluorine. It is generally agreed by all research investigators of dental caries that caries results from the action of bacteria on the teeth, the most common of which is *Streptococcus mutans*. The first event in the development of caries is the deposit of *plaque*, a film of precipitated products of saliva and food, on the teeth. Large numbers of bacteria inhabit this plaque and are readily available to cause caries. However, these bacteria depend to a great extent on carbohydrates for their food. When carbohydrates are available, their metabolic systems are strongly activated and they also multiply. In addition, they form acids, particularly lactic acid, and proteolytic enzymes. The acids are the major culprit in the causation of caries, because the calcium salts of teeth are slowly dissolved in a highly acid medium. And once the salts have become absorbed, the remaining organic matrix is rapidly digested by the proteolytic enzymes.

Enamel is far more resistant to demineralization by acids than is dentine, primarily because the crystals of enamel are very dense and also are about 200 times as large as the dentine crystals. Therefore, the enamel of the tooth is the primary barrier to the development of caries. Once the carious process has penetrated through the enamel to the dentine, it then proceeds many times as rapidly because of the high degree of solubility of the dentine salts.

Because of the dependence of the caries-causing bacteria on carbohydrates, it has frequently been taught that a diet high in carbohydrate content will lead to excessive development of caries. However, it is not the quantity of carbohydrate ingested but instead the frequency with which it is eaten that is important. If eaten in many small portions throughout the day, as in the form of candy, the bacteria are supplied with their preferential metabolic substrate for many hours of the day, and the development of caries is extreme. If the carbohydrates, even though in large amounts, are eaten only at mealtimes, the extensiveness of the caries is greatly reduced.

Some teeth are more resistant to caries than others. Studies show that teeth formed in children who drink water containing small amounts of fluorine develop enamel that is more resistant to caries than the enamel in children who drink water not containing fluorine. Fluorine does not make the enamel harder than usual, but instead it displaces hydroxyl ions in the hydroxyapatite crystals, which in turn makes the enamel several times less soluble. It is also believed that the fluorine might be toxic to some of the bacteria as well. Regardless of the precise means by which fluorine protects the teeth, it is known that small amounts of fluorine deposited in enamel make teeth about three times as resistant to caries as are teeth without fluorine.

Malocclusion. Malocclusion is usually caused by a hereditary abnormality that causes the teeth of one jaw to grow to an abnormal position. In malocclusion, the teeth cannot perform their normal grinding or cutting action adequately. Occasionally malocclusion also results in abnormal displacement of the lower jaw in relation to the upper jaw, causing such undesirable effects as pain in the mandibular joint or deterioration of the teeth.

The orthodontist can often correct malocclusion by applying prolonged gentle pressure against the teeth with appropriate braces. The gentle pressure causes absorption of alveolar jaw bone on the compressed side of the tooth and deposition of new bone on the tensional side of the tooth. In this way the tooth gradually moves to a new position as directed by the applied pressure.

REFERENCES

Avioli, L. V., and Raisz, L. G.: Bone metabolism and disease. *In* Bondy, P. K., and Rosenberg, L. E. (eds.): Metabolic Control and Disease, 8th Ed. Philadelphia, W. B. Saunders Co., 1980, p. 1709.

Barzel, U. S. (ed.): Osteoporosis II. New York, Grune & Stratton, 1979.

Brighton, C. T., *et al.* (eds.): Electrical Properties of Bone and Cartilage: Experimental Effects and Clinical Applications. New York, Grune & Stratton, 1979.

Bringhurst, F. R., and Potts, J. T., Jr.: Calcium and phosphate distribution, turnover, and metabolic actions. *In* DeGroot, L. J., *et al.* (eds.): Endocrinology. Vol. 2. New York, Grune & Stratton, 1979, p. 551.

Castleman, B., and Roth, S. I.: Tumors of the Parathyroid Glands. Washington, D.C., Armed Forces Institute of Pathology, 1978.

Copp, D. H.: Calcitonin: Comparative endocrinology. *In* DeGroot, L. J., *et al.* (eds.): Endocrinology. Vol. 2. New York, Grune & Stratton, 1979, p. 637.

Dacke, C. G.: Calcium Regulation in Sub-Mammalian Vertebrates. New York, Academic Press, 1979.

Dennis, V. W., *et al.*: Renal handling of phosphate and calcium. *Annu. Rev. Physiol., 41*:257, 1979.

Fink, G.: Feedback actions of target hormones on hypothalamus and pituitary, with special reference to gonadal steroids. *Annu. Rev. Physiol., 41*:571, 1979.

Fraser, D. R.: Regulation of the metabolism of vitamin D. *Physiol. Rev., 60*:551, 1980.

Gordan, G. S.: Drug treatment of the osteoporoses. *Annu. Rev. Pharmacol. Toxicol., 18*:253, 1978.

Habener, J. F., and Potts, J. T., Jr.: Diagnosis and differential diagnosis of hyperparathyroidism. *In* DeGroot, L. J., *et al.* (eds.): Endocrinology. Vol. 2. New York, Grune & Stratton, 1979, p. 703.

Ham, A. W.: Histophysiology of Cartilage, Bone, and Joints. Philadelphia, J. B. Lippincott Co., 1979.

Jaros, G. C., *et al.*: Model of short-term regulation of calcium ion concentration. *Simulation, 32*:193, 1979.

Keutmann, H. T.: Chemistry of parathyroid hormone. *In* DeGroot, L. J., *et al.* (eds.): Endocrinology. Vol. 2. New York, Grune & Stratton, 1979, p. 593.

Lawson, D. E. M.: Vitamin D. New York, Academic Press, 1978.

Massry, S. G., and Fleisch, H. (eds.): Renal Handling of Phosphate. New York, Plenum Press, 1979.

Mayer, G. P.: Parathyroid hormone secretion. *In* DeGroot, L. J., *et al.* (eds.): Endocrinology. Vol. 2. New York, Grune & Stratton, 1979, p. 607.

Myers, H. M.: Fluorides and Dental Fluorosis. New York, S. Karger, 1978.

Nellans, H. N., and Kimberg, D. V.: Intestinal calcium transport: Absorption, secretion, and vitamin D. *In* Crane, R. K. (ed.): International Review of Physiology: Gastrointestinal Physiology III. Vol. 19. Baltimore, University Park Press, 1979, p. 227.

Newman, H. N.: Dental Plaque: The Ecology of the Flora on Human Teeth. Springfield, Ill., Charles C Thomas, 1980.

Norman, A. W.: Vitamin D: The Calcium Homeostatic Steroid Hormone. New York, Academic Press, 1979.

Parsons, J. A.: Physiology of parathyroid hormone. *In* DeGroot, L. J., *et al.* (eds.): Endocrinology. Vol. 2. New York, Grune & Stratton, 1979, p. 621.

Phang, J. M., and Weiss, I. W.: Maintenance of calcium homeostasis in human beings. *In* Greep, R. O., and Astwood, G. D. (eds.): Handbook of Physiology. Sec. 7, Vol. 7. Baltimore, Williams & Wilkins 1976, p. 157.

Raisz, L. G., *et al.*: Hormonal regulation of mineral metabolism. *Int. Rev. Physiol., 16*:199, 1977.

Schroeder, H. E., and Listgartern, M. A.: Fine Structure of the Developing Epithelial Attachment of Human Teeth. New York, S. Karger, 1977.

Seltzer, S., and Bender, I. B.: The Dental Pulp, 2nd Ed. Philadelphia, J. B. Lippincott Co., 1975.

Smith, R.: Biochemical Disorders of the Skeleton. Boston, Butterworth, 1979.

Tada, M., *et al.*: Molecular mechanism of active calcium transport by sarcoplasmic reticulum. *Physiol. Rev., 58*:1, 1978.

Talmage, R. V., and Cooper, C. W.: Physiology and mode of action of calcitonin. *In* DeGroot, L. J., *et al.* (eds.): Endocrinology. Vol. 2. New York, Grune & Stratton, 1979, p. 647.

van Zwieten, P. A., and Schönbaum, E. (eds.): The Action of Drugs on Calcium Metabolism. New York, Fischer, 1978.

Wheeler, R. C.: Dental Anatomy, Physiology and Occlusion, 5th Ed. Philadelphia, W. B. Saunders Co., 1974.

54

Reproductive Functions of the Male, the Male Sex Hormones, and the Pineal Gland

The reproductive functions of the male can be divided into three major subdivisions: first, spermatogenesis, which means simply the formation of sperm; second, performance of the male sexual act; and third, regulation of male sexual functions by the various hormones. Associated with these reproductive functions are the effects of the male sex hormones on the accessory sexual organs, on cellular metabolism, on growth, and on other functions of the body.

Physiologic Anatomy of the Male Sexual Organs. Figure 54–1 illustrates the various portions of the male reproductive system. Note that the testis is composed of a large number of coiled *seminiferous tubules* where the sperm are formed. The sperm then empty into the *epididymis,* and thence into the *vas deferens,* which enlarges into the *ampulla of the vas deferens* immediately before the vas enters the body of the *prostate gland.* A *seminal vesicle,* one located on each side of the prostate, empties into the prostatic end of the ampulla, and the contents from both the ampulla and the seminal vesicle pass into an *ejaculatory duct* leading through the body of the prostate gland to empty into the *internal urethra. Prostate ducts* from the glandular tissue of the prostate in turn empty into each ejaculatory duct. Finally, the *urethra* is the last connecting link to the exterior. The urethra is supplied with mucus derived from a large number of small *glands of Littré* located along its entire extent and also from large bilateral *bulbourethral glands* located near the origin of the urethra.

SPERMATOGENESIS

Spermatogenesis occurs in all the seminiferous tubules during active sexual life, beginning at an average age of 13 as the result of stimulation by adenohypophyseal gonadotropic hormones and continuing throughout the remainder of life.

THE STEPS OF SPERMATOGENESIS

The seminiferous tubules, one of which is illustrated in Figure 54–2A, contain a large number of small to medium-sized germinal epithelial cells called *spermatogonia,* which are located in two to three layers along the outer border of the tubular epithelium. These continually proliferate to replenish themselves, and a portion of them differentiate through definite stages of development to form sperm, as shown in Figure 54–2B.

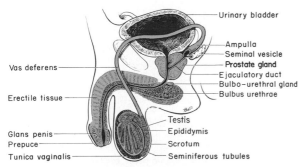

Figure 54–1. The male reproductive system. (Modified from Bloom, W., and Fawcett, D. W.: Textbook of Histology, 10th Ed. Philadelphia, W. B. Saunders Co., 1975.)

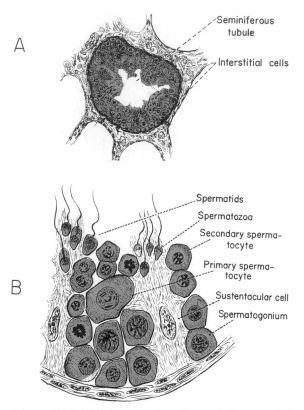

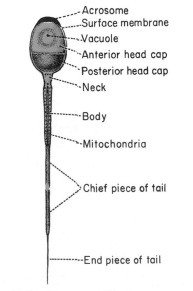

Figure 54–2. (A) Cross-section of a seminiferous tubule. (B) Spermatogenesis. (Modified from Arey: Developmental Anatomy, 7th Ed. Philadelphia, W. B. Saunders Company, 1974.)

The first stage in spermatogenesis is growth of some spermatogonia to form considerably enlarged cells called *spermatocytes*. Then the spermatocyte divides by the process of *meiosis* (there is no formation of new chromosomes, only separation of the chromosomal pairs) to form two *spermatids*, each containing 23 chromosomes. The spermatids do not divide again but instead mature for several weeks to become spermatozoa.

The Sex Chromosomes. In each spermatogonium one of the 23 pairs of chromosomes carries the genetic information that determines the sex of the eventual offspring. This pair is composed of one "X" chromosome, which is called the *female chromosome,* and one "Y" chromosome, the *male chromosome.* During meiotic division the sex-determining chromosomes are distributed between the spermatids, so that half of the sperm become *male sperm* containing the "Y" chromosome and the other half *female sperm* containing the "X" chromosome. The sex of the offspring is determined by which of these two types of sperm fertilizes the ovum. This will be discussed further in Chapter 56.

Formation of Sperm. When the spermatids are first formed, they still have the usual characteristics of epithelioid cells, but soon most of the

cytoplasm disappears, and each spermatid begins to elongate into a spermatozoon, illustrated in Figure 54–3, composed of a *head, neck, body,* and *tail.* To form the head, the nuclear material is condensed into a compact mass, and the cell membrane contracts around the nucleus. It is this nuclear material that fertilizes the ovum.

At the front of the sperm head is a small structure called the *acrosome,* which is formed from the Golgi apparatus and contains hyaluronidase and proteases that play important roles in the entry of the sperm into the ovum.

The *centrioles* are aggregated in the neck of the sperm, and the *mitochondria* are arranged in a spiral in the body.

Extending beyond the body is a long tail, which is an outgrowth of one of the centrioles. This has almost the same structure as a cilium, which was described in detail in Chapter 2. The tail contains two paired microtubules down the center and nine double microtubules arranged around the border. It is covered by an extension of the cell membrane, and it contains large quantities of adenosine triphosphate (generated by the mitochondria in the body), which energize the movement of the tail. Upon release of sperm from the male genital tract into the female tract, the tail begins to wave back and forth and to move spirally near its tip, providing snakelike propulsion that moves the sperm forward at a velocity of about 20 centimeters per hour.

Function of the Sertoli Cells. The Sertoli cells of the germinal epithelium, known also as the *sustentacular cells,* are illustrated in Figure 54–2B. These cells are large, extending from the base of the seminiferous epithelium all the way to the interior of the tubule. The spermatids attach

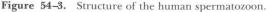

Figure 54–3. Structure of the human spermatozoon.

themselves to the Sertoli cells, and a specific relationship exists between the two cells that causes the spermatids to change into spermatozoa. The Sertoli cells provide nutrient material, hormones, and possibly also enzymes that are necessary for causing appropriate changes in the spermatids. The Sertoli cells also remove the excess cytoplasm as the spermatids are converted to spermatozoa.

Maturation of Sperm in the Epididymis. Following formation in the seminiferous tubules, the sperm pass into the *epididymis*. Sperm removed from the seminiferous tubules are completely nonmotile, and they cannot fertilize an ovum. However, after the sperm have been in the epididymis for some 18 hours to 10 days, they develop the capability of motility, even though some inhibitory factor still prevents motility until after ejaculation. The sperm also become capable of fertilizing the ovum, a process called *maturation*. The epididymis secretes a copious quantity of fluid containing hormones, enzymes, and special nutrients that may be important or even essential for sperm maturation.

Storage of Sperm. A small quantity of sperm can be stored in the epididymis, but most sperm are stored in the vas deferens and ampulla of the vas deferens. They can remain stored, maintaining their fertility, in these areas for several months, though it is doubtful that during normal sexual activity such prolonged storage ordinarily occurs. Indeed, with excessive sexual activity storage may be no longer than a few hours.

Physiology of the Mature Sperm. The usual motile and fertile sperm are capable of flagellated movement through the fluid media at a rate of approximately 1 to 4 mm. per minute. Furthermore, *normal* sperm tend to travel in a straight, rotating line rather than with a circuitous movement. The activity of sperm is greatly enhanced in neutral and slightly alkaline media such as exist in the ejaculated semen, but it is greatly depressed in mildly acid media, and strong acid media can cause rapid death of sperm. Though sperm can live for many weeks in the genital ducts of the testes, the life of sperm in the female genital tract is only one to four days.

FUNCTION OF THE SEMINAL VESICLES

The seminal vesicles are secretory glands lined with an epithelium that secretes a mucoid material containing an abundance of fructose and other nutrient substances, as well as large quantities of prostaglandins and fibrinogen. During the process of ejaculation each seminal vesicle empties its contents into the ejaculatory duct shortly after the vas deferens empties the sperm. This adds greatly to the bulk of the ejaculated semen, and the fructose and other substances in the seminal fluid are of considerable nutrient value for the ejaculated sperm until one of them fertilizes the ovum. The prostaglandins are believed to aid fertilization in two ways: (1) by reacting with the cervical mucus to make it more receptive to sperm, and (2) possibly causing reverse peristaltic contractions in the uterus and fallopian tubes to move the sperm toward the ovaries (a few sperm reach the upper end of the fallopian tubes within five minutes).

FUNCTION OF THE PROSTATE GLAND

The prostate gland secretes a thin, milky, alkaline fluid containing citric acid, calcium, and several other substances. During emission, the capsule of the prostate gland contracts simultaneously with the contractions of the vas deferens and seminal vesicles, so that the thin, milky fluid adds to the bulk of the semen. The alkaline characteristic of the prostatic fluid may be quite important for successful fertilization of the ovum, because the fluid of the vas deferens is relatively acidic owing to the presence of metabolic endproducts of the sperm and, consequently, inhibits sperm motility and fertility. Also, the vaginal secretions of the female are acidic (pH of 3.5 to 4.0). Sperm do not become optimally motile until the pH of the surrounding fluids rises to approximately 6 to 6.5. Consequently, it is probable that prostatic fluid neutralizes the acidity of these other fluids after ejaculation and greatly enhances the motility and fertility of the sperm.

SEMEN

Semen, which is ejaculated during the male sexual act, is composed of the fluids from the vas deferens, the seminal vesicles, the prostate gland, and the mucous glands, especially the bulbourethral glands. The major bulk of the semen is seminal vesicle fluid (about 60 per cent), which is the last to be ejaculated and serves to wash the sperm out of the ejaculatory duct and urethra. The average pH of the combined semen is approximately 7.5, the alkaline prostatic fluid having neutralized the mild acidity of the other portions of the semen. The prostatic fluid gives the semen a milky appearance, while fluid from the seminal vesicles and from the mucous glands gives the semen a mucoid consistency. Indeed, a clotting enzyme of the prostatic fluid causes the fibrinogen of the seminal vesicle fluid to form a weak coagulum, which then dissolves during the next 15 to 20 minutes because of lysis by fibrinolysin formed from a prostatic profibrinolysin. In the early minutes after ejaculation, the sperm remain relatively immobile, possibly because of the viscosity of the coagulum. However, after the coagulum dissolves, the sperm simultaneously become highly motile.

Though sperm can live for many weeks in the male genital ducts, once they are ejaculated in the semen their maximal life span is only 24 to 72 hours at body temperature. At lowered temperatures, however, semen may be stored for several weeks; and when frozen at temperatures below -100° C, sperm of some animals have been preserved for over a year.

Effect of Sperm Count on Fertility. The usual quantity of semen ejaculated at each coitus averages approximately 3.5 ml., and in each milliliter of semen is an average of approximately 120 million sperm, though even in "normal" persons this number can vary from 35 million to 200 million. Therefore, an average of 400 million sperm are usually present in each ejaculate. When the number of sperm in each milliliter falls below approximately 20,000,000, the person is likely to be infertile. Thus, even though only a single sperm is necessary to fertilize the ovum, the ejaculate must contain a tremendous number of sperm for at least one to fertilize the ovum. A possible reason for this is the following:

Function of Hyaluronidase and Proteinases Secreted by the Sperm for the Process of Fertilization. Stored in the acrosomes of the sperm are large quantities of hyaluronidase and proteinases. Hyaluronidase is an enzyme that depolymerizes the hyaluronic acid polymers that are present in large quantities in the intercellular cement substance; proteinases can dissolve the proteins of tissues.

When the ovum is expelled from the follicle of the ovary into the abdominal cavity, it carries with it several layers of cells. Before a sperm can reach the ovum to fertilize it, these cells must be removed; it is believed that the hyaluronidase and proteinases released by the acrosomes play at least a small role (in addition to a much larger role played by sodium bicarbonate in the fallopian tube secretions) in causing these cells to break away from the ovum, thus allowing the sperm to reach the surface of the ovum. When sperm are insufficient in number, the man is often sterile. It has been postulated that this sterility results from insufficient enzymes to help remove the cell layers from the ovum.

Another possible function of the proteinases is to allow the sperm to penetrate the mucus that frequently forms in the cervix of the uterus. The proteinases act as mucolytic enzymes that presumably proceed in advance of the sperm and create channels through the mucous plug. It is believed that deficiency of the appropriate enzymes to perform this function is occasionally responsible for male sterility.

THE MALE SEXUAL ACT

NEURONAL STIMULUS FOR PERFORMANCE OF THE MALE SEXUAL ACT

The most important nerve signals for initiating the male sexual act originate in the glans penis, for the glans contains a highly organized sensory end-organ system that transmits into the central nervous system a special modality of sensation called *sexual sensation*. The massaging action of intercourse on the glans stimulates the sensory end-organs, and the sexual sensations in turn pass through the pudendal nerve, thence through the sacral plexus into the sacral portion of the spinal cord, and finally up the cord to undefined areas of the cerebrum. Impulses may also enter the spinal cord from areas adjacent to the penis to aid in stimulating the sexual act. For instance, stimulation of the anal epithelium, the scrotum, and perineal structures in general can all send impulses into the cord which add to the sexual sensation. Sexual sensations can even originate in internal structures, such as irritated areas of the urethra, the bladder, the prostate, the seminal vesicles, the testes, and the vas deferens. Indeed, one of the causes of "sexual drive" is probably overfilling of the sexual organs with secretions. Infection and inflammation of these sexual organs sometimes cause almost continual sexual desire, and "aphrodisiac" drugs, such as cantharides, increase the sexual desire by irritating the bladder and urethral mucosa.

The Psychic Element of Male Sexual Stimulation. Appropriate psychic stimuli can greatly enhance the ability of a person to perform the sexual act. Simply thinking sexual thoughts or even dreaming that the act of intercourse is being performed can cause the male sexual act to occur and to culminate in ejaculation. Indeed, *nocturnal emissions* during dreams occur in many males during some stages of sexual life, especially during the teens.

Integration of the Male Sexual Act in the Spinal Cord. Though psychic factors usually play an important part in the male sexual act and can actually initiate it, the cerebrum is probably not absolutely necessary for its performance, for appropriate genital stimulation can cause ejaculation in some animals and in an occasional human being after their spinal cords have been cut above the lumbar region. Therefore, the male sexual act results from inherent reflex mechanisms integrated in the sacral and lumbar spinal cord, and these mechanisms can be activated by either psychic stimulation or actual sexual stimulation.

STAGES OF THE MALE SEXUAL ACT

Erection. Erection is the first effect of male sexual stimulation, and the degree of erection is proportional to the degree of stimulation, whether this be psychic or physical.

Erection is caused by parasympathetic impulses that pass from the sacral portion of the spinal cord to the penis. These parasympathetic impulses dilate the arteries of the penis and simultaneously constrict the veins, thus allowing arterial blood to flow under high pressure into the *erectile tissue* of the penis, illustrated in Figure 54–4. This erectile tissue is composed of large, cavernous venous sinusoids, which are normally relatively empty but become dilated tremendously when arterial blood flows into them under pressure. Also, these erectile bodies are surrounded by strong fibrous coats; therefore, high pressure within the sinusoids causes ballooning of the erectile tissue to such an extent that the penis becomes hard and elongated.

Lubrication. During sexual stimulation, parasympathetic impulses, in addition to promoting erection, cause the glands of Littré and the bulbourethral glands to secrete mucus. Thus mucus flows through the urethra during intercourse to aid in the lubrication of coitus. However, most of the lubrication of coitus is provided by the female sexual organs rather than by the male. Without satisfactory lubrication, the male sexual act is rarely successful because unlubricated intercourse causes pain impulses that inhibit rather than excite sexual sensations.

Emission and Ejaculation. Emission and ejaculation are the culmination of the male sexual act. When the sexual stimulus becomes extremely intense, the reflex centers of the spinal cord begin to emit sympathetic impulses that leave the cord at L-1 and L-2 and pass to the genital organs to initiate emission, which is the forerunner of ejaculation.

Emission is believed to begin with contraction of the epididymis, the vas deferens, and the ampulla to cause expulsion of sperm into the internal urethra. Then, contractions in the seminal vesicles and the muscular coat of the prostate gland expel seminal fluid and prostatic fluid, forcing the sperm forward. All these fluids mix with the mucus already secreted by the bulbourethral glands to form the semen. The process to this point is *emission*.

The filling of the internal urethra then elicits signals that are transmitted to the sacral regions of the cord. In turn, rhythmic nerve impulses are sent from the cord to skeletal muscles that encase the base of the erectile tissue, causing rhythmic, wavelike increases in pressure in this tissue, which "ejaculates" the semen from the urethra to the exterior. This is the process of *ejaculation*.

TESTOSTERONE AND OTHER MALE SEX HORMONES

Secretion of Testosterone by the Interstitial Cells of the Testes. The testes secrete several male sex hormones, which are collectively called *androgens*. However, one of these, testosterone, is so much more abundant and potent than the others that it can be considered to be the significant hormone responsible for the male hormonal effects.

Testosterone is formed by the *interstitial cells of Leydig*, which lie in the interstices between the seminiferous tubules, as illustrated in Figure 54–5, and constitute about 20 per cent of the mass of the adult testes. Interstitial cells in the testes are not numerous in a child, but they *are*

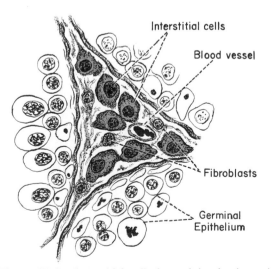

Figure 54–5. Interstitial cells located in the interstices between the seminiferous tubules. (Modified from Bloom and Fawcett: Textbook of Histology, 8th Ed.)

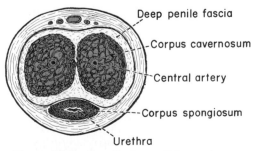

Figure 54–4. Erectile tissue of the penis.

numerous in a newborn male infant and also in the adult male any time after puberty; at both these times the testes secrete large quantities of testosterone. Furthermore, when tumors develop from the interstitial cells of Leydig, great quantities of testosterone are secreted. Finally, when the germinal epithelium of the testes is destroyed by x-ray treatment or by excessive heat, the interstitial cells, which are less easily destroyed, often continue to produce testosterone.

Secretion of "Androgens" Elsewhere in the Body. The term "androgen" is used synonymously with the term "male sex hormone," but it also includes male sex hormones produced elsewhere in the body besides the testes. For instance, the adrenal gland secretes at least five different androgens, though the total masculinizing activity of all these is normally so slight that they do not cause significant masculine characteristics even in women. But when an adrenal tumor of the androgen-producing cells occurs, the quantity of androgenic hormones may become great enough to cause all the usual male secondary sexual characteristics. These effects were described in Chapter 51.

Chemistry of Testosterone. All androgens are steroid compounds, as illustrated by the formula in Figure 54–6 for *testosterone*. Both in the testes and in the adrenals, the androgens can be synthesized either from cholesterol or directly from acetyl coenzyme A.

Metabolism of Testosterone. After secretion by the testes, testosterone, most of it loosely bound with plasma protein, circulates in the blood for not over 15 to 30 minutes before it either becomes fixed to the tissues or is degraded into inactive products that are subsequently excreted.

Much of the testosterone that becomes fixed to the tissues is converted within the cells to *dihydrotestosterone*, which is also shown in Figure 54–6; it is in this form that testosterone performs many of its intracellular functions.

Degradation and Excretion of Testosterone. The testosterone that does not become fixed to the tissues is rapidly converted, mainly by the liver, into *androsterone* and *dehydroepiandrosterone* and simultaneously conjugated either as glucuronides or sulfates (glucuronides, particularly). These are excreted either into the gut in the bile or into the urine.

FUNCTIONS OF TESTOSTERONE

In general, testosterone is responsible for the distinguishing characteristics of the masculine body. The testes are stimulated by chorionic gonadotropin from the placenta to produce a small quantity of testosterone during fetal development, but essentially no testosterone is produced during childhood until approximately the age of 10 to 13. Then testosterone production increases rapidly at the onset of puberty and lasts throughout most of the remainder of life, dwindling rapidly beyond the age of 40 to perhaps one-fifth the peak value by the age of 80.

Functions of Testosterone During Fetal Development. Testosterone begins to be elaborated by the male at about the second month of embryonic life. Injection of large quantities of male sex hormone into gravid animals causes development of male sexual organs even though the fetus is female. Also, removal of the fetal testes in a male fetus causes development of female sexual organs. Therefore, the presence or absence of testosterone in the fetus is the determining factor in the development of male or female genital organs and characteristics. That is, testosterone secreted by the genital ridges and the subsequently developing testes is responsible for the development of the male sex characteristics, including the growth of a penis and a scrotum rather than the formation of a clitoris and a vagina. Also, it causes development of the prostate gland, the seminal vesicles, and the male genital ducts, while at the same time suppressing the formation of female genital organs.

Effect on the Descent of the Testes. The testes usually descend into the scrotum during the last two months of gestation, when the testes are secreting adequate quantities of testosterone. If a male child is born with undescended testes, administration of testosterone causes the testes to descend in the usual manner if the inguinal canals are large enough to allow the testes to pass. Or, administration of gonadotropic hormones, which stimulate the interstitial cells of the testes to produce testosterone, also causes the testes to descend. Thus, the stimulus for descent of the testes is testosterone, indicating again that testos-

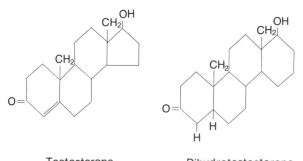

Testosterone   Dihydrotestosterone

Figure 54–6. Testosterone and dihydrotesterone.

terone is an important hormone for male sexual development during fetal life.

Effect of Testosterone on Development of Adult Primary and Secondary Sexual Characteristics. Testosterone secretion after puberty causes the penis, the scrotum, and the testes all to enlarge many-fold until about the age of 20. In addition, testosterone causes the "secondary sexual characteristics" of the male to develop at the same time, beginning at puberty and ending at maturity. These secondary sexual characteristics, in addition to the sexual organs themselves, distinguish the male from the female as follows:

Distribution of Body Hair. Testosterone causes growth of hair (1) over the pubis, (2) on the face, (3) usually on the chest, and (4) less often on other regions of the body, such as the back. It also causes the hair on most other portions of the body to become more prolific.

Baldness. Testosterone decreases the growth of hair on the top of the head; a man who does not have functional testes does not become bald. However, many virile men never become bald, for baldness is a result of two factors: first, a *genetic background* for the development of baldness and, second superimposed on this genetic background, *large quantities of androgenic hormones.* A woman who has the appropriate genetic background and who develops a long-sustained androgenic tumor becomes bald in the same manner as a man.

Effect on the Voice. Testosterone secreted by the testes or injected into the body causes hypertrophy of the laryngeal mucosa and enlargement of the larynx. These effects cause at first a relatively discordant, "cracking" voice, but this gradually changes into the typical masculine bass voice.

Effect on the Skin. Testosterone increases the thickness of the skin over the entire body and increases the ruggedness of the subcutaneous tissues.

Effect on Protein Formation and Muscular Development. One of the most important male characteristics is the development of increasing musculature following puberty. This is associated with increased protein in other parts of the body as well. Many of the changes in the skin are also due to deposition of proteins in the skin, and the changes in the voice even result, at least partly, from the protein anabolic function of testosterone.

Testosterone has often been considered to be a "youth hormone" because of its effect on the musculature, and it is occasionally used for treatment of persons who have poorly developed muscles.

Effect on Bone Growth and Calcium Retention. Following puberty or following prolonged injection of testosterone, the bones grow considerably in thickness and also deposit considerable calcium salts. Thus, testosterone increases the total quantity of bone matrix, and this also causes calcium retention. The increase in bone matrix is believed to result from the general protein anabolic function of testosterone.

When great quantities of testosterone (or any other androgen) are secreted in the still-growing child, the rate of bone growth increases markedly, causing a spurt in total body growth as well. However, the testosterone also causes the epiphyses of the long bones to unite with the shafts of the bones at an early age in life. Therefore, despite the rapidity of growth, the early uniting of the epiphyses prevents the person from growing as tall as he would grow were testosterone not secreted at all. Even in normal men the final adult height is slightly less than that which would have been attained had the person been castrated prior to puberty.

Effect on the Red Blood Cells. The average man has about 700,000 more red blood cells per cubic millimeter than the average woman. However, this difference may be due partly to increased metabolic rate following testosterone administration rather than to a direct effect of testosterone on red blood cell production.

BASIC INTRACELLULAR MECHANISM OF ACTION OF TESTOSTERONE

Though it is not known exactly how testosterone causes all the effects just listed, it is believed that they result mainly from increased rate of protein formation in cells. This has been studied extensively in the prostate gland, one of the organs that is most affected by testosterone. In this gland, testosterone enters the cells within a few minutes after secretion, is there converted to dihydrotestosterone, and binds with a cytoplasmic "receptor protein." This combination then migrates to the nucleus, where it binds with a nuclear protein and induces the DNA-RNA transcription process. Within 30 minutes the concentration of RNA begins to increase in the cells, and this is followed by progressive increase in cellular protein. After several days the quantity of DNA in the gland has also increased, and there has been a simultaneous increase in the number of prostatic cells.

Therefore, it is assumed that testosterone greatly stimulates production of proteins in general, though increasing more specifically those proteins in "target" organs or tissues responsible for the development of secondary sexual characteristics.

CONTROL OF MALE SEXUAL FUNCTIONS BY THE GONADOTROPIC HORMONES — FSH AND LH

The anterior pituitary gland secretes two major gonadotropic hormones: (1) *follicle-stimulating hormone* (FSH) and (2) *luteinizing hormone* (LH). Both of these play major roles in the control of male sexual function.

Regulation of Testosterone Production by LH. Testosterone is produced by the interstitial cells of Leydig when the testes are stimulated by LH from the pituitary gland, and the quantity of testosterone secreted varies approximately in proportion to the amount of LH available.

Injection of purified LH into a child causes fibroblasts in the interstitial areas of the testes to develop into interstitial cells of Leydig, though mature Leydig cells are not normally found in the child's testes until after the age of approximately 10. Also, simultaneous administration of *prolactin* (another pituitary hormone that is closely associated with the gonadotropic hormones) greatly potentiates the effect of LH in promoting testosterone production.

Effect of Human Chorionic Gonadotropin on the Fetal Testes.
During gestation the placenta secretes large quantities of human chorionic gonadotropin, a hormone that has almost the same properties as LH. This hormone stimulates the formation of interstitial cells in the testes of the fetus and causes testosterone secretion. As pointed out earlier in the chapter, the secretion of testosterone during fetal life is important for promoting formation of male sexual organs.

Regulation of Spermatogenesis by Follicle-Stimulating Hormone (FSH) and Testosterone. The conversion of spermatogonia into spermatocytes in the seminiferous tubules is stimulated by FSH from the anterior pituitary gland; in the absence of FSH, spermatogenesis will not proceed. However, FSH cannot by itself cause complete formation of spermatozoa. For spermatogenesis to proceed to completion, testosterone must be secreted simultaneously in small amounts by the interstitial cells. Thus, FSH seems to initiate the proliferative process of spermatogenesis, and testosterone diffusing from the interstitial cells into the seminiferous tubules apparently is necessary for final maturation of the spermatozoa. Because testosterone is secreted by the interstitial cells under the influence of LH, both FSH and LH must be secreted by the anterior pituitary gland if spermatogenesis is to occur.

Regulation of LH and FSH Secretion by the Hypothalamus. The gonadotropins, like corticotropin and thyrotropin, are secreted by the anterior pituitary gland mainly in response to nervous activity in the hypothalamus. For instance, in sheep, goats, and deer, nervous stimuli in response to changes in weather and amount of light in the day increase the quantities of gonadotropins during one season of the year, the mating season, thus allowing birth of the young during an appropriate period for survival. Also, psychic stimuli can affect fertility of the male animal, as exemplified by the fact that transporting a bull under uncomfortable conditions can often cause almost complete temporary sterility.

Luteinizing Hormone–Releasing Hormone (LHRH), the Hypothalamic Hormone That Stimulates Gonadotropin Secretion. In both the male and the female, the hypothalamus controls gonadotropin secretion by way of the hypothalamic-hypophysial portal system. Though there are two different gonadotropic hormones, luteinizing hormone and follicle-stimulating hormone, only one hypothalamic-releasing hormone has been discovered; it is *luteinizing hormone–releasing hormone* (LHRH). This hormone has an especially strong effect on inducing luteinizing hormone secretion by the anterior pituitary gland, but it has a potent effect in causing follicle-stimulating hormone secretion as well. For this reason it is often also called *gonadotropin-releasing hormone.*

Luteinizing hormone–releasing hormone plays a similar role in controlling gonadotropin secretion in the female, a process in which the interrelationships are far more complex. Therefore, its nature and its functions will be discussed in much more detail in the following chapter.

Reciprocal Inhibition of Hypothalamic–Anterior Pituitary Secretion of Gonadotropic Hormones by Testicular Hormones. *Feedback Control of Testosterone Secretion.* The following negative feedback control system operates continuously to control very precisely the rate of testosterone secretion:

1. The hypothalamus secretes *luteinizing hormone–releasing hormone,* which stimulates the anterior pituitary gland to secrete *luteinizing hormone.*

2. Luteinizing hormone in turn stimulates *hyperplasia of the Leydig cells* of the testes and also stimulates production of *testosterone* by these cells.

3. The testosterone in turn feeds back *negatively* to the hypothalamus, inhibiting production of luteinizing hormone–releasing hormone. This obviously limits the rate at which testosterone will be produced. On the other hand, when testosterone production is too low, lack of inhibition of the hypothalamus leads to return of testosterone secretion to the normal level.

Feedback Control of Spermatogenesis — Role of "Inhibin."
It is known, too, that spermatogenesis by the testes inhibits the secretion of FSH. It is

believed that the Sertoli cells secrete a hormone that has a direct effect mainly on the anterior pituitary gland (but perhaps slightly on the hypothalamus as well) of inhibiting the secretion of FSH. A hormone having a molecular weight between 25,000 and 100,000 and called *inhibin* has been discovered, which probably is responsible for this effect. Therefore, the feedback cycle for control of spermatogenesis seems to be the following:

1. Follicle-stimulating hormone induces proliferation of the germinal epithelium of the seminiferous tubules and at the same time stimulates the Sertoli cells, which provide nutrition for the developing spermatozoa.

2. The Sertoli cells (or less likely the germinal epithelial cells) release inhibin, which in turn feeds back negatively to the anterior pituitary gland to inhibit the production of FSH. Thus, this feedback cycle maintains a rate of spermatogenesis that is required for male reproductive function — no more, no less.

Puberty and Regulation of Its Onset. During the first ten years of life, the male child secretes almost no gonadotropins and consequently almost no testosterone. Then, at the age of about 10, the anterior pituitary gland begins to secrete progressively increasing quantities of gonadotropins, and this is followed by a corresponding increase in testicular function. By the approximate age of 13, the male child reaches full adult sexual capability. This period of change is called *puberty*.

The cause of the onset of puberty is probably the following: *During childhood, the hypothalamus does not secrete significant amounts of luteinizing hormone–releasing hormone* because even the minutest amount of testosterone inhibits its production by the childhood hypothalamus. Therefore, the testicles also remain persistently suppressed. However, for reasons yet unknown, the hypothalamus loses this inhibitory sensitivity at the time of puberty, which allows the secretory mechanisms to develop full activity. Thus, puberty is believed to result from a maturing process of the hypothalamic sexual control centers.

ABNORMALITIES OF MALE SEXUAL FUNCTION

THE PROSTATE GLAND AND ITS ABNORMALITIES

The prostate gland remains relatively small throughout childhood but begins to grow at puberty under the stimulus of testosterone. This gland reaches an almost stationary size by the age of 20 and remains this size up to the age of approximately 40 to 50. At that time in some men it begins to degenerate along with the decreased production of testosterone by the testes. However, a benign fibroadenoma frequently develops in the prostate in older men and causes urinary obstruction. This hypertrophy is not caused by testosterone.

Cancer of the prostate gland is an extremely common cause of death, resulting in approximately 2 to 3 per cent of all male deaths.

Once cancer of the prostate gland does occur, the cancerous cells are usually stimulated to more rapid growth by testosterone and are inhibited by removal of the testes so that testosterone cannot be formed. Also, prostatic cancer can usually be inhibited by administration of estrogens. Some patients who have prostatic cancer that has already metastasized to almost all the bones of the body can be successfully treated for a few months to years by removal of the testes, by estrogen therapy, or by both; following this therapy the metastases degenerate and the bones heal. This treatment does not completely stop the cancer but does slow it down and greatly diminishes the severe bone pain.

TESTICULAR TUMORS AND HYPERGONADISM IN THE MALE

An *interstitial cell tumor* on rare occasions develops in a testis, but when one does develop it sometimes produces as much as 100 times the normal quantity of testosterone. When such tumors develop in young children, they cause rapid growth of the musculature and bones but also early uniting of the epiphyses, so that the eventual adult height actually is less than that which would have been achieved otherwise. Obviously, such interstitial cell tumors cause excessive development of the sexual organs and of the secondary sexual characteristics. In the adult male, small interstitial cell tumors are difficult to diagnose because masculine features are already present. Diagnosis can be made, however, from urine tests that show greatly increased excretion of testosterone end-products.

Much more common than the interstitial cell tumors are tumors of the germinal epithelium. Because germinal cells are capable of differentiating into almost any type of cell, many of these tumors contain multiple types of tissue, such as placental tissue, hair, teeth, bone, and skin, all found together in the same tumorous mass called a *teratoma*. Often these tumors secrete no hormones, but if a significant quantity of placental tissue develops in the tumor, it may secrete large quantities of human chorionic gonadotropin that has functions very similar to those of LH. Also, estrogenic hormones are frequently secreted by these tumors and cause the condition called *gynecomastia,* which means overgrowth of the breasts.

THE PINEAL GLAND — ITS FUNCTION IN CONTROLLING SEASONAL FERTILITY

The pineal gland is a small nervous tissue–glandular body protruding from the midbrain above and behind the superior colliculi. Some physiologists have claimed for many years that the pineal gland plays important roles in the control of sexual activities and reproduction, functions that still others have said were nothing more than the zealous imaginings of physiologists preoccupied with sexual delusions.

But now, after years of turmoil and dispute, it looks as though the sex advocates have at last won. For, in lower animals in which the pineal gland has been removed or in which the nervous circuits to the pineal gland have been sectioned, the normal annual periods of seasonal fertility are lost. To these animals such seasonal fertility is very important because it allows birth of the offspring in the spring and summer months when survival is most likely. The mechanism of this effect is still not entirely clear, but it seems to be the following:

First, the pineal gland is controlled by nerve signals elicited by the amount of light seen by the eyes each day. For instance, in the hamster, more than 13 hours of darkness each day activates the pineal gland, while less than that amount of darkness fails to activate it.

Second, the pineal gland secretes *melatonin* and several other similar substances. Either melatonin or one of the other substances then passes either by way of the blood or through the fluid of the third ventricle to the anterior pituitary gland to *inhibit* gonadotropic hormone secretion, and the gonads become inhibited and even involuted. This is what occurs during the winter months. But after about four months of dysfunction, the gonadotropic hormone secretion breaks through the inhibitory effect of the pineal gland, and the gonads become functional once more, ready for a full springtime of activity.

But, does the pineal gland have a similar function in controlling reproduction in man? The answer to this is still far from known. However, tumors in the region of the pineal gland are often associated with serious hypo- or hypergonadal dysfunction. So, perhaps the pineal gland does play at least some role in controlling sexual drive and reproduction in man.

REFERENCES

Arimura, A.: Hypothalamic gonadotropin-releasing hormone and reproduction. *Int. Rev. Physiol., 13*:1, 1977.

Bardin, C. W.: Pituitary-testicular axis. *In* Yen, S. S. C., and Jaffe, R. B. (eds.): Reproductive Endocrinology. Philadelphia, W. B. Saunders Co., 1978, p. 110.

Buckner, W. P., Jr.: Medical Readings on Human Sexuality. Stanford, Cal., Medical Readings Inc., 1978.

Epel, D.: The program of fertilization. *Sci. Am., 237*(5): 128, 1977.

Ewing, L. L., *et al.*: Regulation of testicular function: A spatial and temporal view. *In* Greep, R. O. (ed.): International Review of Physiology: Reproductive Physiology III. Vol. 22. Baltimore, University Park Press, 1980, p. 41.

Fink, G.: Feedback actions of target hormones on hypothalamus and pituitary, with special reference to gonadal steroids. *Annu. Rev. Physiol., 41*:571, 1979.

Forleo, R., and Pasini, W. (eds.): Medical Sexology. Littleton, Mass., PSG Publishing Co., 1979.

Griffin, J. E., and Wilson, J. D.: The testis. *In* Bondy, P. K., and Rosenberg, L. E. (eds.): Metabolic Control and Disease, 8th Ed. Philadelphia, W. B. Saunders Co., 1980, p. 1535.

Hafez, E. S. E. (ed.): Human Reproduction: Conception and Contraception. Hagerstown, Md., Harper & Row, 1979.

Hafez, E. S. E.: Human Reproductive Physiology. Ann Arbor Mich., Ann Arbor Science, 1978.

Hafez, E. S. E. (ed.): Descended and Cryptorchid Testis. Hingham, Mass., Kluwer Boston, 1980.

Hafez, E. S. E., and Spring-Mills, E. (eds.): Accessory Glands of the Male Reproductive Tract. Ann Arbor, Mich., Ann Arbor Science, 1979.

Hall, P. F.: Testicular hormones: Synthesis and control. *In* DeGroot, L. J., *et al.* (eds.): Endocrinology, Vol. 3. New York, Grune & Stratton, 1979, p. 1511.

Huff, R. W., and Pauerstein, C. J.: Physiology and Pathophysiology of Human Reproduction. New York, John Wiley & Sons, 1978.

Jaffe, R. B.: Disorders of sexual development. *In* Yen, S. S. C., and Jaffe, R. B. (eds.): Reproductive Endocrinology. Philadelphia, W. B. Saunders Co., 1978, p. 271.

Kolodny, R. C., *et al.*: Textbook of Sexual Medicine. Boston, Little, Brown, 1979.

Lipsett, M. B.: Steroid hormones. *In* Yen, S. S. C., and Jaffe, R. B. (eds.): Reproductive Endocrinology. Philadelphia, W. B. Saunders Co., 1978, p. 80.

Money, J., and Higham, E.: Sexual behavior and endocrinology (normal and abnormal). *In* DeGroot, L. J., *et al.* (eds.): Endocrinology, Vol. 3. New York, Grune & Stratton, 1979, p. 1353.

Moore, P. Y.: The central nervous system and the neuroendocrine regulation of reproduction. *In* Yen, S. S. C., and Jaffe, R. B. (eds.): Reproductive Endocrinology. Philadelphia, W. B. Saunders Co., 1978, p. 3.

Phillips, D. M.: Spermiogenesis. New York, Academic Press, 1974.

Prasad, M. R. N., and Rajalakshmi, M.: Recent advances in the control of male reproductive functions. *Int. Rev. Physiol., 13*:153, 1977.

Reiter, R. J. (ed.): The Pineal and Reproduction. New York, S. Karger, 1978.

Setchell, B. P.: The Mammalian Testis. Ithaca, N.Y., Cornell University Press, 1978.

Smith, K. D.: Testicular function in the aging male. *In* DeGroot, L. J., *et al.* (eds.): Endocrinology. Vol. 3. New York, Grune & Stratton, 1979, p. 1577.

Stangel, J. J.: Fertility and Conception: An Essential Guide for Childless Couples. New York, New American Library, 1980.

Steinberger, A., and Steinberger, E. (eds.): Testicular Development, Structure and Function. New York, Raven Press, 1980.

Steinberger, E.: Hormonal control of spermatogenesis. *In* DeGroot, L. J., *et al.* (eds.): Endocrinology. Vol. 3. New York, Grune & Stratton, 1979, p. 1535.

Steinberger, E.: Male infertility. *In* DeGroot, L. J., *et al.* (eds.): Endocrinology. Vol. 3. New York, Grune & Stratton, 1979, p. 1567.

Styne, D. M., and Grumbach, M. M.: Puberty in the male and female: Its physiology and disorders. *In* Yen, S. S. C., and Jaffe, R. B. (eds.): Reproductive Endocrinology, Philadelphia, W. B. Saunders Co., 1978, p. 189.

Talway, G. P. (eds.): Recent Advances in Reproduction and Regulation of Fertility. New York, Elsevier/North-Holland, 1979.

Thomas, J. A., and Singahl, R. L. (eds.): Sex Hormone Receptors in Endocrine Organs. Baltimore, Urban & Schwarzenberg, 1980.

Wilson, J. D.: Sexual differentiation. *Annu. Rev. Physiol., 40*:279, 1978.

55

Prepregnancy Reproductive Functions in the Female, and the Female Hormones

The sexual and reproductive functions in the female can be divided into two major phases: first, preparation of the body for conception, and second, the period of gestation. The present chapter is concerned with the preparation of the body for gestation, and the following chapter presents the physiology of pregnancy.

PHYSIOLOGIC ANATOMY OF THE FEMALE SEXUAL ORGANS

Figure 55–1 illustrates the principal organs of the human female reproductive tract, including the *ovaries*, the *fallopian tubes*, the *uterus*, and the *vagina*. Reproduction begins with the development of ova in the ovaries. A single ovum is expelled from an ovarian follicle into the abdominal cavity in the middle of each monthly sexual

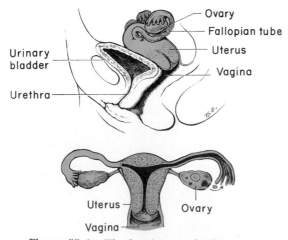

Figure 55–1. The female reproductive organs.

cycle. This ovum then passes through one of the fallopian tubes into the uterus, and if it has been fertilized by a sperm, it implants in the uterus, where it develops into a fetus, a placenta, and fetal membranes.

At puberty, the two ovaries contain about 300,000 ova. Each ovum surrounded by a single layer of epithelioid granulosa cells is called a *primordial follicle*. During all the reproductive years of the female, only about 400 of these follicles develop enough to expel their ova; the remainder degenerate. At the end of reproductive capability, the *menopause*, only a few primordial follicles remain in the ovaries, and even these degenerate soon thereafter.

THE FEMALE HORMONAL SYSTEM

The female hormonal system, like that of the male, consists of three different hierarchies of hormones:

1. A hypothalamic releasing hormone: *luteinizing hormone–releasing hormone* (LHRH).

2. The anterior pituitary hormones: *follicle-stimulating hormone* (FSH) and *luteinizing hormone* (LH), both of which are secreted in response to the releasing hormone from the hypothalamus.

3. The ovarian hormones: *estrogen* and *progesterone*, which are secreted by the ovaries in response to the two hormones from the anterior pituitary gland.

The various hormones are not secreted in constant, steady amounts but at drastically differing rates during different parts of the female cycle, as will be explained later in the chapter.

Before it is possible to discuss the interplay between these different hormones, it is first necessary to describe some of their specific functions and their relationships to the function of the ovaries.

THE MONTHLY OVARIAN CYCLE AND FUNCTION OF THE GONADOTROPIC HORMONES

The normal reproductive years of the female are characterized by monthly rhythmic changes in the rates of secretion of the female hormones and corresponding changes in the sexual organs themselves. This rhythmic pattern is called the *female sexual cycle* (or less accurately, the *menstrual cycle*). The duration of the cycle averages 28 days. It may be as short as 20 days or as long as 45 days even in completely normal women, though abnormal cycle length is occasionally associated with decreased fertility.

The two significant results of the female sexual cycle are: First, only a *single* mature ovum is normally released from the ovaries each month, so that only a single fetus can begin to grow at a time. Second, the uterine endometrium is prepared for implantation of the fertilized ovum at the required time of the month.

The Gonadotropic Hormones. The ovarian changes during the sexual cycle are completely dependent on gonadotropic hormones secreted by the anterior pituitary gland. Ovaries that are not stimulated by gonadotropic hormones remain completely inactive, which is essentially the case throughout childhood when almost no gonadotropic hormones are secreted. However, at the age of about 8, the pituitary begins secreting progressively more and more gonadotropic hormones, which culminates in the initiation of monthly sexual cycles between the ages of 11 and 15; this culmination is called *puberty*.

The anterior pituitary secretes two different hormones that are known to be essential for full function of the ovaries: (1) *follicle-stimulating hormone* (FSH), and (2) *luteinizing hormone* (LH). Both of these are small glycoproteins having molecular weights of about 30,000.

During each month of the female sexual cycle, there is a cyclic increase and decrease of FSH and LH as illustrated in Figure 55–2. These cyclic variations in turn cause cyclic ovarian changes, which are explained in the following sections.

FOLLICULAR GROWTH — FUNCTION OF FOLLICLE-STIMULATING HORMONE (FSH)

Figure 55–3 depicts the various stages of follicular growth in the ovaries, illustrating, first, the primordial follicle (primary follicle). Throughout childhood the primordial follicles do not grow, but at puberty, when FSH and LH from the anterior pituitary gland begin to be secreted in large quantity, the entire ovaries and especially the follicles within them begin to grow. The first stage of follicular growth is enlargement of the ovum itself. This is followed by development of additional layers of granulosa cells around each ovum and development of several layers of *theca cells* around the granulosa cells. The theca cells

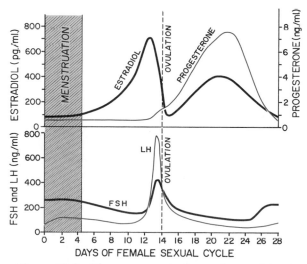

Figure 55–2. Plasma concentrations of gonadotropins and ovarian hormones during the normal female sexual cycle.

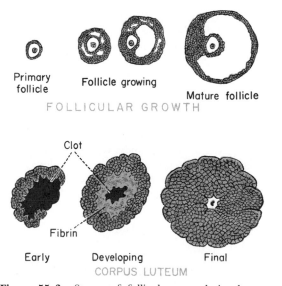

Figure 55–3. Stages of follicular growth in the ovary, showing also formation of the corpus luteum. (Modified from Arey: Developmental Anatomy, 7th Ed.)

originate from the stroma of the ovary and soon take on epithelioid characteristics. It is probably these cells that are destined to secrete most of the estrogens, while the granulosa cells will secrete progesterone.

The Vesicular Follicles. At the beginning of each month of the female sexual cycle, at approximately the onset of menstruation, the concentrations of FSH and LH increase. These increases cause accelerated growth of the theca and granulosa cells in about 20 of the ovarian follicles each month. These cells also secrete a follicular fluid that contains a high concentration of estrogen, one of the important female sex hormones that will be discussed later. The accumulation of this fluid in the follicle causes an *antrum* to appear within the theca and granulosa cells, as illustrated in Figure 55–3.

After the antrum is formed, the theca and granulosa cells continue to proliferate, the rate of secretion accelerates, and each of the growing follicles becomes a *vesicular follicle*.

As the vesicular follicle enlarges, the theca and granulosa cells continue to develop at one pole of the follicle. It is in this mass that the ovum is located.

Atresia of All Follicles but One. After a week or more of growth—but before ovulation occurs—one of the follicles begins to outgrow all the others; the remainder begin to involute (a process called *atresia*), and these follicles are said to become *atretic*. The cause of the atresia is unknown, but it has been postulated to be the following: The one follicle that becomes more highly developed than the others also secretes more estrogen than the others. This causes feedback inhibition of secretion of the gonadotropic hormone FSH by the anterior pituitary gland. Lack of this hormone does not prevent further growth of the largest follicle because the large amount of locally secreted estrogen in this follicle has a self-stimulatory effect of causing this follicle to continue growing. However, the lack of FSH stimulus to the less well developed follicles causes these to stop growing and, indeed, to involute.

This process of atresia obviously is important in that it allows only one of the follicles to grow large enough to ovulate. This single follicle reaches a size of approximately 1 to 1.5 cm. at the time of ovulation.

Ovulation

Ovulation, in a woman with a normal 28-day female sexual cycle, occurs 14 days after the onset of menstruation.

Shortly before ovulation, the protruding outer wall of the follicle swells rapidly, and a small area in the center of the capsule, called the *stigma*, protrudes like a nipple. In another half hour or so, fluid begins to ooze from the follicle through the stigma. About two minutes later, the stigma ruptures widely, and a more viscous fluid that has occupied the central portion of the follicle is evaginated outward into the abdomen. This viscous fluid carries with it the ovum surrounded by several thousand granulosa cells called the *corona radiata*.

Need for Luteinizing Hormone (LH) in Ovulation — Ovulatory Surge of LH. Luteinizing hormone is necessary for final follicular growth and ovulation. Without this hormone, even though large quantities of FSH are available, the follicle will not progress to the stage of ovulation.

Approximately two days before ovulation, for reasons that are not completely known at present but which will be discussed in more detail later in the chapter, the rate of secretion of LH by the anterior pituitary gland increases markedly, rising 6- to 10-fold and peaking about 18 hours before ovulation. FSH also increases about 2-fold at the same time, and these two hormones act synergistically to cause the extremely rapid swelling of the follicle that culminates in ovulation.

THE CORPUS LUTEUM – THE "LUTEAL" PHASE OF THE OVARIAN CYCLE

During the last day before ovulation and continuing for a day or so after ovulation, the theca and granulosa cells, under the stimulation of luteinizing hormone, undergo rapid physical and chemical change, a process called *luteinization*. Thus, the mass of cells still remaining at the site of the ruptured follicle becomes a *corpus luteum*, which secretes the hormones progesterone and estrogen. These cells become greatly enlarged and develop lipid inclusions that give the cells a distinctive yellowish color, from which is derived the term *luteum*.

In the normal female, the corpus luteum grows to approximately 1.5 cm., reaching this stage of development approximately 7 or 8 days following ovulation. After this, it begins to involute and loses its secretory function, as well as its lipid characteristics, approximately 12 days following ovulation, becoming then the so-called *corpus albicans*, which during the ensuing few weeks is replaced by connective tissue.

Secretion by the Corpus Luteum: Further Function of LH. The corpus luteum is a highly secretory organ, secreting large amounts of both *progesterone* and *estrogen*. In the presence of LH the degree of growth of the corpus luteum is enhanced, its secretion is greater, and its life is extended.

Termination of the Ovarian Cycle and Onset of the Next Cycle. During the luteal phase of the ovarian cycle, the large amounts of estrogen and progesterone secreted by the corpus luteum cause a feedback decrease in secretion of both FSH and LH. Therefore, during this period no new follicles begin to grow in the ovary. However, when the corpus luteum degenerates completely at the end of 12 days of its life (approximately on the 26th day of the female sexual cycle), the loss of feedback suppression now allows the anterior pituitary gland to secrete greatly increased quantities of FSH and moderately increased quantities of LH. The FSH and LH initiate growth of new follicles to begin a new ovarian cycle. At the same time, the paucity of secretion of progesterone and estrogen leads to menstruation by the uterus, as will be explained later.

SUMMARY

Approximately each 28 days, gonadotropic hormones from the anterior pituitary gland cause new follicles to begin to grow in the ovaries, one of which finally ovulates at the 14th day of the cycle. During growth of the follicles, estrogen is secreted.

Following ovulation, the secretory cells of the follicle develop into a corpus luteum, which secretes large quantities of the female hormones progesterone and estrogen. After another two weeks the corpus luteum degenerates, whereupon the ovarian hormones, estrogen and progesterone, decrease greatly and menstruation begins. A new ovarian cycle then follows.

THE OVARIAN HORMONES — ESTROGENS AND PROGESTERONE

The two types of ovarian hormones are the *estrogens* and *progesterone*. The estrogens mainly promote proliferation and growth of specific cells in the body and are responsible for development of most of the secondary sexual characteristics of the female. On the other hand, progesterone is concerned almost entirely with final preparation of the uterus for pregnancy and the breasts for lactation.

CHEMISTRY OF THE SEX HORMONES

The Estrogens. In the normal, nonpregnant female, estrogens are secreted in major quantities only by the ovaries, though minute amounts are also secreted by the adrenal cortices. In pregnancy, tremendous quantities are also secreted by the placenta, indeed, up to 100 times the amount secreted by the ovaries during the normal monthly cycle.

At least six different natural estrogens have been isolated from the plasma of the human female, but only three are present in significant quantities, *β-estradiol, estrone*, and *estriol*. However, the estrogenic potency of β-estradiol is 12 times that of estrone and 80 times that of estriol, so that the total estrogenic effect of β-estradiol is usually many times that of the other two together. For this reason β-estradiol is considered to be the major estrogen. Its formula is shown in Figure 55–4. Note that it is a steroid compound. It is synthesized from cholesterol or acetyl coenzyme A in the ovaries.

Progesterone. Almost all the progesterone in the nonpregnant female is secreted by the corpus luteum during the latter half of each ovarian cycle. However, during pregnancy progesterone is formed in extreme quantities by the placenta, about ten times the normal monthly amount, especially after the fourth month of gestation.

Progesterone is a steroid having a molecular structure, illustrated in Figure 55–4, that is not far different from those of the other steroid hormones, the estrogens, testosterone, and the corticosteroids.

Progesterone also is synthesized from acetyl coenzyme A or cholesterol.

FUNCTIONS OF THE ESTROGENS – EFFECTS ON THE PRIMARY AND SECONDARY SEXUAL CHARACTERISTICS

The principal function of the estrogens is to cause cellular proliferation and growth of the tissues of the sexual organs and of other tissues related to reproduction.

Effect on the Sexual Organs. During childhood, estrogens are secreted only in small quantities, but following puberty the quantity of estrogens secreted under the influence of the pituitary gonadotropic hormones increases some 20-fold or more. At this time the female sexual organs change from those of a child to those of an adult. The fallopian tubes, uterus, and vagina all increase in size. Also, the external genitalia enlarge, with deposition of fat in the mons pubis and labia majora and with enlargement of the labia minora.

In addition, estrogens change the vaginal epithelium from a cuboidal into a stratified type, which is considerably more resistant to trauma

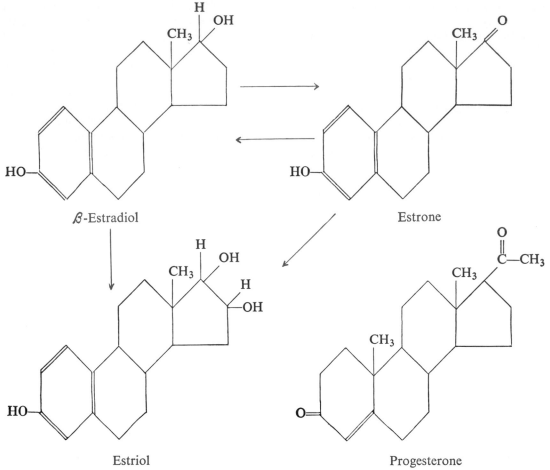

Figure 55-4. Chemical formulas of the principal female hormones.

and infection than is the prepubertal epithelium. More important, however, are the changes that take place in the endometrium under the influence of estrogens, for estrogens cause marked proliferation of the endometrium and development of the endometrial glands that will later be used to aid in nutrition of the implanting ovum. These effects are discussed later in the chapter in connection with the endometrial cycle.

Effect on the Breasts. Estrogens cause fat deposition in the breasts, development of the stromal tissues of the breasts, and growth of an extensive ductile system. The lobules and alveoli of the breast develop to a slight extent, but it is progesterone and prolactin that cause the determinative growth and function of these structures. In summary, the estrogens initiate growth of the breasts and the breasts' milk-producing apparatus, and they are also responsible for the characteristic external appearance of the mature female breast, but they do not complete the conversion of the breasts into milk-producing organs, a subject discussed in the following chapter.

Effect on the Skeleton. Estrogens cause increased osteoblastic activity. Therefore, at puberty, when the female enters her reproductive years, her height increases rapidly for several years. However, estrogens have another potent effect on skeletal growth: they cause early uniting of the epiphyses with the shafts of the long bones. This effect is much stronger in the female than is a similar effect of testosterone in the male. As a result, growth of the female usually ceases several years earlier than growth of the male. The female eunuch who is completely devoid of estrogen production usually grows several inches taller than the normal mature female because her epiphyses do not unite early.

Effect on Fat Deposition. Estrogens cause deposition of increased quantities of fat in the subcutaneous tissues. As a result, the overall specific gravity of the female body, as judged by flotation in water, is considerably less than that of the male body, which contains more protein and less fat. In addition to deposition of fat in the breasts and subcutaneous tissues, estrogens cause especially marked deposition of fat in the buttocks and thighs, causing the broadening of the hips that is characteristic of the feminine figure.

Effect on the Skin. Estrogens cause the skin to become more vascular than normal; this effect often results in greater bleeding of cut surfaces than is observed in men.

Intracellular Functions of Estrogens. After secretion by the ovaries, estrogens circulate in the blood for only a few minutes before they are delivered to the target cells. On entry into these cells, the estrogens combine within 10 to 15 seconds with a "receptor" protein in the cytoplasm and then, in combination with this protein, migrate to the nucleus. This immediately initiates the process of DNA-RNA transcription in specific chromosomal areas, and RNA begins to be produced within a few minutes. In addition, over a period of many hours DNA also is produced, resulting eventually in division of the cell. The RNA diffuses to the cytoplasm, where it causes greatly increased protein formation and subsequently altered cellular function.

One of the principal differences between the estrogens and testosterone is that the effects of estrogen occur almost exclusively in certain target organs, such as the uterus, the breasts, the skeleton, and certain fatty areas of the body; whereas testosterone has a more generalized effect throughout the body.

FUNCTIONS OF PROGESTERONE

Effect on the Uterus. By far the most important function of progesterone is *to promote secretory changes in the endometrium,* thus preparing the uterus for implantation of the fertilized ovum. This function is discussed later in connection with the endometrial cycle of the uterus.

Effect on the Fallopian Tubes. Progesterone also promotes secretory changes in the mucosal lining of the fallopian tubes. These secretions are important for nutrition of the fertilized, dividing ovum as it traverses the fallopian tube prior to implantation.

Effect on the Breasts. Progesterone promotes development of the lobules and alveoli of the breasts, causing the alveolar cells to proliferate, to enlarge, and to become secretory in nature. However, progesterone does not cause the alveoli actually to secrete milk, for, as discussed in the following chapter, milk is secreted only after the prepared breast is further stimulated by prolactin from the anterior pituitary.

Progesterone also causes the breasts to swell. Part of this swelling is due to the secretory development in the lobules and alveoli, but part

also seems to result somewhat from increased fluid in the subcutaneous tissue itself.

THE ENDOMETRIAL CYCLE AND MENSTRUATION

Associated with the cyclic production of estrogens and progesterone by the ovaries is an endometrial cycle operating through the following stages: first, proliferation of the uterine endometrium; second, secretory changes in the endometrium; and third, desquamation of the endometrium, which is known as *menstruation.* The various phases of the endometrial cycle are illustrated in Figure 55–5.

Proliferative Phase (Estrogen Phase) of the Endometrial Cycle. At the beginning of each menstrual cycle, most of the endometrium is desquamated by the process of menstruation. After menstruation, only a thin layer of endometrial stroma remains at the base of the original endometrium, and the only epithelial cells left are those located in the remaining deep portions of the glands and crypts of the endometrium. *Under the influence of estrogens,* secreted in increasing quantities by the ovary during the first part of the ovarian cycle, the stromal cells and the epithelial cells proliferate rapidly. The endometrial surface is re-epithelialized within three to seven days after the beginning of menstruation. For the first two weeks of the sexual cycle — that is, until ovulation — the endometrium increases greatly in thickness, owing to increasing numbers of stromal cells and to progressive growth of the endometrial glands and blood vessels into the endometrium, all of which effects are promoted by the estrogens. At the time of ovulation the endometrium is approximately 2 to 3 mm. thick.

Secretory Phase (Progestational Phase) of the Endometrial Cycle. During the latter half of the sexual cycle, progesterone as well as estrogen is secreted in large quantity by the corpus luteum. The estrogens cause additional cellular prolifera-

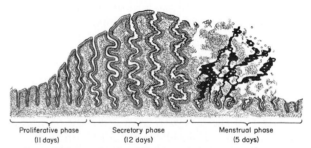

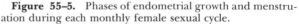

Figure 55–5. Phases of endometrial growth and menstruation during each monthly female sexual cycle.

tion, and progesterone causes considerable swelling and secretory development of the endometrium. The glands increase in tortuosity, secretory substances accumulate in the glandular epithelial cells, and the glands secrete small quantities of endometrial fluid. Also, the cytoplasm of the stromal cells increases, lipid and glycogen deposits increase greatly in these cells, and the blood supply to the endometrium further increases in proportion to the developing secretory activity. The thickness of the endometrium approximately doubles during the secretory phase, so that toward the end of the monthly cycle the endometrium has a thickness of 4 to 6 mm.

The whole purpose of all these endometrial changes is to produce a highly secretory endometrium containing large amounts of stored nutrients that can provide appropriate conditions for implantation of a fertilized ovum during the latter half of the monthly cycle.

Menstruation. Approximately two days before the end of the monthly cycle, the ovarian hormones, estrogens and progesterone, decrease sharply to low levels of secretion, as was illustrated in Figure 55–2, and menstruation follows.

Menstruation is caused by the sudden reduction in both progesterone and estrogens at the end of the monthly ovarian cycle. The first effect is decreased stimulation of the endometrial cells by these two hormones, followed rapidly by involution of the endometrium itself to about 65 per cent of its previous thickness. During the 24 hours preceding the onset of menstruation, the blood vessels leading to the mucosal layers of the endometrium become vasospastic, presumably because of some effect of the involution, such as release of a vasoconstrictor material. The vasospasm and loss of hormonal stimulation cause beginning necrosis in the endometrium. As a result, blood seeps into the vascular layer of the endometrium, the hemorrhagic areas growing over a period of 24 to 36 hours. Gradually, the necrotic outer layers of the endometrium separate from the uterus at the site of the hemorrhages, until, at approximately 48 hours following the onset of menstruation, all the superficial layers of the endometrium have desquamated. The desquamated tissue and blood in the uterine vault initiate uterine contractions that expel the uterine contents.

During normal menstruation, approximately 35 ml. of blood and an additional 35 ml. of serous fluid are lost. This menstrual fluid is normally nonclotting, because a *fibrinolysin* is released along with the necrotic endometrial material.

Within three to seven days after menstruation starts, the loss of blood ceases, for by this time the endometrium has become completely re-epithelialized.

REGULATION OF THE FEMALE MONTHLY RHYTHM; INTERPLAY BETWEEN THE OVARIAN AND HYPOTHALAMIC-PITUITARY HORMONES

Now that we have presented the major cyclic changes that occur during the female sexual cycle, we can attempt to explain the basic rhythmic mechanism that causes these cyclic variations.

Function of the Hypothalamus in the Regulation of Gonadotropin Secretion — the Hypothalamic Releasing Factors. As was pointed out in Chapter 49, secretion of most of the anterior pituitary hormones is controlled by releasing hormones formed in the hypothalamus and transmitted to the anterior pituitary gland by way of the hypothalamic-hypophysial portal system. In the case of the gonadotropins, at least one releasing hormone, *luteotropic hormone–releasing hormone (LHRH)*, is important. This hormone has been purified and is a decapeptide having the following formula:

$$\text{GLU-HIS-TRP-SER-TYR-GLY-LEU-ARG-PRO-GLY-NH}_2$$

Though some research workers believe that another substance similar to this hormone is follicle-stimulating hormone–releasing hormone (FSHRH), it has been found that the above purified LHRH causes release not only of luteinizing hormone but also of follicle-stimulating hormone. Therefore, since there is reason to believe that this decapeptide is in reality both LHRH and FSHRH combined in the same molecule, it is sometimes called simply *gonadotropin-releasing hormone (GnRH)*.

Effect of Psychic Factors on the Female Sexual Cycle. It is well known that the young woman on first leaving home to go to college experiences disruption or irregularity of the female sexual cycle almost as often as not. Likewise, serious stresses of almost any type can interfere with the cycle. Finally, in many lower animals no ovulation occurs at all until after copulation; the sexual excitation attendant to the sexual act initiates a sequence of events that leads first to secretion in the hypothalamus of LHRH, then to secretion of the anterior pituitary gonadotropins, and finally to ovarian secretion of hormones and ovulation.

Negative Feedback Effect of Estrogen and Progesterone on Secretion of Follicle-stimulating Hormone and Luteinizing Hormone. Estrogen in particular inhibits the production of FSH and LH. Large quantities of progesterone also have this effect. Both of these feedback effects seem to operate mainly by the actions of these hormones on the hypothalamus.

Positive Feedback Effect of Estrogen Before Ovulation — the Preovulatory Luteinizing Hormone Surge. For reasons not completely understood, the anterior pituitary gland secretes greatly increased amounts of LH on the day immediately before ovulation. This effect, illustrated in Figure 55–2, apparently results from a *positive* feedback effect of estrogen in place of the normal negative feedback that occurs during the remainder of the female sexual cycle. Its precise cause is not known, but nevertheless it is an absolutely necessary and integral part of the control mechanism. Without the normal preovulatory surge of luteinizing hormone, ovulation will not occur.

FEEDBACK OSCILLATION OF THE HYPOTHALAMIC-PITUITARY-OVARIAN SYSTEM

Now, after discussing much of the known information about the interrelationships of the different components of the female hormonal system, we can digress from the area of proven fact into the realm of speculation and attempt to explain the feedback oscillation that controls the rhythm of the female sexual cycle. It seems to operate in approximately the following sequence of three successive events:

1. The Postovulatory Secretion of the Ovarian Hormones and Depression of Gonadotropins. The easiest part of the cycle to explain is the events that occur during the postovulatory phase — between ovulation and the beginning of menstruation. During this time the corpus luteum secretes very large quantities of both progesterone and estrogen. Their combined effect on the hypothalamus is to inhibit the secretion of LHRH and therefore to cause strong negative feedback depression of secretion of the gonadotropins, both FSH and LH, during this period of time. These effects are illustrated in Figure 55–2.

2. The Follicular Growth Phase. A few days before menstruation, the corpus luteum involutes, and the secretion of both estrogen and progesterone decreases to a low ebb. This releases the hypothalamus from the feedback effect of the estrogen and progesterone, so that LHRH secretion increases again, followed in succession by an increase of several hundred per cent of FSH and LH, also. These hormones initiate new follicular growth and progressively increased secretion of estrogen, reaching a peak of estrogen secretion at about 12.5 to 13 days after the onset of menstruation. During the first 11 to 12 days of this follicular growth the rates of secretion of the gonadotropins FSH and LH decrease; then comes a sudden increase in secretion of both of these hormones, leading to the next stage of the cycle.

3. Preovulatory Surge of LH and FSH; Ovulation. At approximately 11.5 to 12 days after the onset of menstruation, the decline in secretion of FSH and LH comes to an abrupt halt. It is believed that the high level of estrogens at this time causes a positive feedback effect, as explained earlier, which leads to a tremendous surge of secretion — especially of LH but to a lesser extent of FSH. This effect may be related to the fact that the follicular secretory cells are becoming exhausted, so that their rate of secretion of estrogens had already begn to fall about one day prior to the LH surge. Whatever the cause of this preovulatory LH and FSH surge, the LH leads to both ovulation and formation of the corpus luteum. Thus, the hormonal system begins a new round of the female sexual cycle.

PUBERTY

Puberty means the onset of adult sexual life, and as pointed out earlier in the chapter, it is caused by a gradual increase in gonadotropic hormone secretion by the pituitary, beginning approximately in the eighth year of life.

In the female, as in the male, the infantile pituitary gland and ovaries are capable of full function if appropriately stimulated. However, the hypothalamus is extremely sensitive to the inhibitory effects of estrogens, which keeps its stimulation of the pituitary almost completely suppressed throughout childhood. Then, at puberty, for reasons not understood, the hypothalamus matures just as it does in the male; its excessive sensitivity to the negative feedback inhibition becomes greatly diminished, which allows enhanced production of gonadotropins and the onset of adult female sexual life.

THE MENOPAUSE

At an average age of approximately 40 to 50 years the sexual cycles usually become irregular, and ovulation fails to occur during many of these cycles. After a few months to a few years, the cycles cease altogether. This cessation of the cycles is called the *menopause*.

The cause of the menopause is "burning out" of the ovaries. In other words, throughout a woman's sexual life many of the primordial follicles grow into vesicular follicles with each sexual cycle, and eventually almost all the ova either degenerate or are ovulated. Therefore, at the age of about 45 only a few primordial follicles still remain to be stimulated by FSH and LH, and the production of estrogens by the ovary decreases as the number of primordial follicles approaches zero. When estrogen production falls below a critical value, the estrogens can no longer inhibit the production of FSH and LH sufficiently to cause oscillatory cycles. Consequently, FSH and LH (mainly FSH) are produced thereafter in large and continuous quantities. Estrogens are produced in subcritical quantities for a short time after the menopause, but over a few years, as the final remaining primordial follicles become atretic, the production of estrogens by the ovaries falls almost to zero.

ABNORMALITIES OF SECRETION BY THE OVARIES

Hypogonadism. Less than normal secretion by the ovaries can result from poorly formed ovaries or lack of ovaries. When ovaries are absent from birth or when they never become functional, *female eunuchism* occurs. In this condition the usual secondary sexual characteristics do not appear, and the sexual organs remain infantile. Especially characteristic of this condition is excessive growth of the long bones because the epiphyses do not unite with the shafts of these bones as early as in the normal adolescent woman. Consequently, the female eunuch is as tall as her male counterpart of similar genetic background, or perhaps even slightly taller.

When the ovaries of a fully developed woman are removed, the sexual organs regress to some extent, so that the uterus becomes almost infantile in size, the vagina becomes smaller, and the vaginal epithelium becomes thin and easily damaged. The breasts atrophy and become pendulous, and the pubic hair becomes considerably thinner. These same changes occur in the woman after the menopause.

Irregularity of Menses and Amenorrhea Due to Hypogonadism. The quantity of estrogens produced by the ovaries must rise above a critical value if they are to be capable of inhibiting the production of follicle-stimulating hormone sufficiently to cause an oscillatory sexual cycle. Consequently, in hypogonadism or when the gonads are secreting small quantities of estrogens as a result of other factors, the ovarian cycle likely will not occur normally. Instead, several months may elapse between menstrual periods, or menstruation may cease altogether (amenorrhea). Characteristically, prolonged ovarian cycles are frequently associated with failure of ovulation, presumably due to insufficient secretion of luteinizing hormone, which is necessary for ovulation.

Hypersecretion by the Ovaries. Extreme hypersecretion of ovarian hormones by the ovaries is a rare clinical entity, for excessive secretion of estrogens automatically decreases the production of gonadotropins by the pituitary, and this in turn limits the production of the ovarian hormones. Consequently, hypersecretion of feminizing hormones is recognized clinically only when a feminizing tumor develops.

Rarely, a granulosa-theca cell tumor develops in an ovary, more often after menopause than before. Such a tumor secretes large quantities of estrogens, which exert the usual estrogenic effects, including hypertrophy of the uterine endometrium and irregular bleeding from it. In fact, bleeding is often the first indication that such a tumor exists.

THE FEMALE SEXUAL ACT

Stimulation of the Female Sexual Act. As is true of the male sexual act, successful performance of the female sexual act depends on both psychic stimulation and local sexual stimulation.

The psychic factors that constitute "sex drive" in women are difficult to assess. The sex hormones, and the adrenocortical hormones as well, seem to exert a direct influence on the woman to create the sex drive, but, on the other hand, the growing female child in modern society is often taught that sex is something to be hidden and that it is immoral. As a result of this training, much of the natural sex drive is inhibited, and whether the woman will have little or no sex drive ("frigidity") or will be more highly sexed probably depends partly on a balance between natural factors and previous training.

Local sexual stimulation in women occurs in more or less the same manner as in men, for massage, irritation, or other types of stimulation of the perineal region, sexual organs, and urinary tract create sexual sensations. The *clitoris* is especially sensitive for initiating sexual sensations. As in the male, the sexual sensory signals are mediated to the sacral segments of the spinal cord through the pudendal nerve and sacral plexus. Once these signals have entered the spinal cord, they are transmitted thence to the cerebrum. Also, local reflexes that are at least partly responsible for the female orgasm are integrated in the sacral and lumbar spinal cord.

Female Erection and Lubrication. Located around the introitus and extending into the clitoris is erectile tissue almost identical with the erectile tissue of the penis. This erectile tissue, like that of the penis, is controlled by the parasympathetic nerves that pass through the nervi erigentes from the sacral plexus to the external genitalia. In the early phases of sexual stimulation, the parasympathetics dilate the arteries and constrict the veins of the erectile tissues, allowing rapid accumulation of blood in the erectile tissue, so that the introitus tightens around the penis; this aids the male greatly in his attainment of sufficient sexual stimulation for ejaculation to occur.

Parasympathetic impulses also pass to the bilateral Bartholin's glands located beneath the labia minora and cause secretion of mucus immediately inside the introitus. This mucus is responsible for much of the lubrication during sexual intercourse. The lubrication in turn is necessary for establishing during intercourse a satisfactory massaging sensation rather than an irritative sensation, which may be provoked by a dry vagina. A massaging sensation constitutes the optimal type of sensation for evoking the appropriate reflexes that culminate in both the male and female climaxes.

The Female Orgasm. When local sexual stimulation reaches maximum intensity, and especially when the local sensations are supported by appropriate psychic conditioning signals from the cerebrum, reflexes are initiated that cause the female orgasm, also called the *female climax.* The female orgasm is analogous to ejaculation in the male, and it probably is important for fertilization of the ovum. Indeed, the human female is known to be somewhat more fertile when inseminated by normal sexual intercourse than by artificial methods, thus indicating an important function of the female orgasm. Possible effects that could result in this are these:

First, during the orgasm the perineal muscles of the female contract rhythmically, which presumably results from spinal reflexes similar to those that cause ejaculation in the male. It is possible, also, that these same reflexes increase uterine and fallopian tube motility during the orgasm, thus helping to transport the sperm toward the ovum, but the information on this subject is scanty.

Second, in many lower animals, copulation causes the posterior pituitary gland to secret oxytocin; this effect is probably mediated through the amygdaloid nuclei and then through the hypothalamus to the pituitary. The oxytocin in turn causes increased contractility of the uterus, which also is believed to cause rapid transport of the sperm. Sperm have been shown to traverse the entire length of the fallopian tube in the cow in approximately 5 minutes, a rate at least 10 times as fast as the sperm by themselves could achieve. Whether or not this occurs in the human female is unknown.

In addition to the effects of the orgasm on fertilization, the intense sexual sensations that develop during the orgasm also pass into the cerebrum and in some manner lead to a sense of satisfaction characterized by relaxed peacefulness, an effect called *resolution.*

FEMALE FERTILITY

The Fertile Period of Each Sexual Cycle. The ovum remains viable and capable of being fertilized for probably no longer than 24 hours after it is expelled from the ovary. Therefore, sperm must be available soon after ovulation if fertilization is to take place. On the other hand, a few sperm can remain viable in the female reproductive tract for up to 72 hours, though most of them for not more than 24 hours. Therefore, for fertilization to take place, intercourse usually must occur some time between one day prior to ovulation and one day after ovulation.

The Rhythm Method of Contraception. A method of contraception often practiced is to avoid intercourse near the time of ovulation. The difficulty with this method is the impossibility of predicting the exact time of ovulation. Yet the interval from ovulation until the next succeeding onset of menstruation is almost always between 13 and 15 days. In other words, if the periodicity of the menstrual cycle is 28 days, ovulation usually occurs within one day of the 14th day of the cycle. If, on the other hand, the periodicity of the cycle is 40 days, ovulation usually occurs within one day of the 26th day of the cycle. Finally, if the periodicity of the cycle is 21 days, ovulation usually occurs within one day of the 7th day of the cycle. Therefore, it is usually stated that avoidance of intercourse for 4 days prior to the calculated day of ovulation and 3 days afterward prevents conception. Such a method of contraception can be used only when the periodicity of the menstrual cycle is regular, for otherwise it is impossible to determine the next onset of menstruation and therefore to predict the day of ovulation.

Hormonal Suppression of Fertility — "The Pill." It has long been known that administration of either estrogen or progesterone in sufficient quantity can inhibit ovulation. Though the exact mechanism of this effect is not clear, it is known that in the presence of enough of either or both of these hormones, the hypothalamus fails to secrete the normal surge of LH-releasing factor and its stimulatory product LH, as usually occurs about 13 days after the onset of the monthly sexual cycle. From the discussion of this phenomenon earlier in the chapter, it will be recalled that this surge of LH is essential in causing ovulation.

The problem in devising methods for hormonal suppression of ovulation has been to develop appropriate combinations of estrogens and progestins that will suppress ovulation but not cause unwanted effects of these two hormones. For instance, too much of either can cause abnormal menstrual bleeding patterns. However, use of a synthetic progestin in place of progesterone, especially the 19-norsteroids, along with small amounts of estrogens will usually prevent ovulation and yet allow an almost normal pattern of menstruation. Therefore, almost all "pills" used for control of fertility consist of some combination of synthetic estrogens and synthetic progestins. The main reason for using synthetic estrogens and synthetic progestins is that the *natural* hormones are almost entirely destroyed by the liver within a short time after they are absorbed from the gastrointestinal tract into the portal circulation. However, many of the *synthetic* hormones can resist this destructive propensity of the liver, thus allowing oral administration.

The medication is usually begun in the early stages of the female sexual cycle and continued beyond the time that ovulation normally would have occurred. Then the medication is stopped toward the end of the cycle, allowing menstruation to occur and a new cycle to begin.

Oral contraceptive regimens have also been devised in which very low dosage levels of estrogens and progestins are used. In these instances ovulation frequently does occur, but other effects prevent conception. These effects include (1) abnormal transport time through the fallopian tube (the usual time is almost exactly three days) so that implantation will not occur; (2) abnormal development of the endometrium so that it will not support a fertilized ovum; (3) abnormal characteristics of the cervical mucus, making it lethal to the sperm or in other ways blocking entry of the sperm to the uterus; and (4) abnormal contraction of the fallopian tubes and uterine musculature so that the ovum will be expelled rather than implanted.

Anovulation and Female Sterility. Approximately one out of every six to ten marriages is infertile; in about 60 per cent of these, the infertility is due to female sterility.

Occasionally, no abnormality whatsoever can be discovered in the female genital organs, in which case it must be assumed that the infertility is due either to abnormal physiological function of the genital system or to abnormal genetic development of the ova themselves.

However, one of the most common causes of female sterility is failure to ovulate. This can result from either hyposecretion of gonadotropic hormones, in which case the intensity of the hormonal stimuli simply is not sufficient to cause ovulation, or from abnormal ovaries that will not allow ovulation. For instance, thick capsules occasionally exist on the outside of the ovaries and prevent ovulation.

Lack of ovulation caused by hyposecretion of the pituitary gonadotropic hormones can be treated by administration of *human chorionic gonadotropin,* a hormone that will be discussed in the following chapter and that is extracted from the human placenta. This hormone, though secreted by the placenta, has almost exactly the same effects as luteinizing hormone and therefore is a powerful stimulator of ovulation. However, excess use of this hormone can cause ovulation from many follicles simultaneously; and multiple births result. As many as six children have been born to mothers treated for infertility with this hormone.

REFERENCES

Aron, C.: Mechanisms of control of the reproductive function of olfactory stimuli in female mammals. *Physiol. Rev., 59*:229, 1979.
Behrman, H. R.: Prostaglandins in hypothalamo-pituitary and ovarian function. *Annu. Rev. Physiol., 41*:685, 1979.
Beller, F. K., and Schumacher, G. F. B. (eds.): The Biology of the Secretions of the Female Genital Tract. New York, Elsevier/North-Holland, 1979.
Benirschke, K.: The endometrium. *In* Yen, S. S. C., and Jaffe, R. B. (eds.): Reproductive Endocrinology. Philadelphia, W. B. Saunders Co., 1978, p. 241.
Catt, K. J., and Pierce, J. G.: Gonadotropic hormones of the adenohypophysis (FSH, LH and prolactin). *In* Yen, S. S. C., and Jaffe, R. B. (eds.): Reproductive Endocrinology. Philadelphia, W. B. Saunders Co., 1978, p. 34.
Channing, C. P., and Marsh, J. M. (eds.): Ovarian Follicular and Corpus Luteum Function. New York, Plenum Press, 1979.
Channing, C. P., *et al.*: Ovarian follicular and luteal physiology. *In* Greep, R. O. (ed.): International Review of Physiology: Reproductive Physiology III. Vol. 22. Baltimore, University Park Press, 1980, p. 117.
Crighton, D. B., *et al.*: Control of Ovulation. Boston, Butterworth, 1978.
Diamond, M. C., and Korenbrot, C. C. (eds.): Hormonal Contraceptives and Human Welfare, New York, Academic Press, 1978.
Droegemueller, W., and Bessler, R.: Effectiveness and risks of contraception. *Annu. Rev. Med., 31*:329, 1980.
Glass, R. H.: Infertility. *In* Yen, S. S. C., and Jaffe, R. B. (eds.): Reproductive Endocrinology. Philadelphia, W. B. Saunders Co., 1978, p. 398.
Greenblatt, R. B. (ed.): Induction of Ovulation. Philadelphia, Lea & Febiger, 1979.
Henzi, M. R.: Natural and synthetic female sex hormones. *In* Yen, S. S. C., and Jaffe, R. B. (eds.): Reproductive Endocrinology. Philadelphia, W. B. Saunders Co., 1978, p. 421.
Kase, N. G., and Speroff, L.: The ovary. *In* Bondy, P. K., and

Rosenberg, L. E. (eds.): Metabolic Control and Disease, 8th Ed. Philadelphia, W. B. Saunders Co., 1980, p. 1579.

Lein, A.: The Cycling Female: Her Menstrual Rhythm. San Francisco, W. H. Freeman, 1979.

Midgley, A. R., and Sadler, W. A. (eds.): Ovarian Follicular Development and Function. New York, Raven Press, 1979.

Mishell, D. R., Jr.: Contraception. *In* DeGroot, L. J., *et al.*(eds.): Endocrinology. Vol. 3. New York, Grune & Stratton, 1979, p. 1435.

Mishell, D. R., Jr., and Davajan, V. (eds.): Reproductive Endocrinology, Infertility, and Contraception. Philadelphia, F. A. Davis Co., 1979.

Odell, W. D.: FSH. *In* DeGroot, L. J., *et al.* (eds.): Endocrinology. Vol. 1. New York, Grune & Stratton, 1979, p. 149.

Odell, W. D.: LH. *In* DeGroot, L. J., *et al.*(eds.): Endocrinology. Vol. 1. New York, Grune & Stratton, 1979, p. 151.

Peters, H., and McNatty, K. P.: The Ovary: A Correlation of Structure and Function in Mammals. Berkeley, University of California Press, 1980.

Richards, J. S.: Maturation of ovarian follicles: Actions and interactions of pituitary and ovarian hormones on follicular cell differentiation. *Physiol. Rev., 60*:51, 1980.

Ross, G. T., and Schreiber, J. R.: The ovary. *In* Yen, S. S. C., and Jaffe, T. B. (eds.): Reproductive Endocrinology. Philadelphia, W. B. Saunders Co., 1978, p. 63.

Savoy-Moore, R. T., and Schwartz, N. B.: Differential control of FSH and LH secretion. *In* Greep, R. O. (ed.): International Review of Physiology: Reproductive Physiology III. Vol. 22. Baltimore, University Park Press, 1980, p. 203.

Tollison, C. D., and Adams, H. E.: Sexual Disorders: Treatment, Theory and Research. New York, Gardner Press, 1979.

Wallach, E. E., and Kempers, R. D.: Modern Trends in Infertility and Conception Control. Baltimore, Williams & Wilkins, 1979.

Yen, S. S. C., and Jaffe, R. B. (eds.): Reproductive Endocrinology: Physiology, Pathophysiology, and Clinical Management. Philadelphia, W. B. Saunders Co., 1978.

Pregnancy, Lactation, and Fetal and Neonatal Physiology

In the preceding two chapters the sexual functions of the male and the female were described to the point of fertilization of the ovum. If the ovum becomes fertilized, a completely new sequence of events called *gestation*, or *pregnancy*, takes place, and the fertilized ovum eventually develops into a full-term fetus. The purpose of the present chapter is to discuss these events.

MATURATION OF THE OVUM

Shortly before the ovum is released from the follicle, each of the 23 pairs of chromosomes loses one of its partners, so that 23 *unpaired* chromosomes remain in the mature ovum. It is at this point that ovulation occurs, and soon thereafter fertilization occurs.

FERTILIZATION OF THE OVUM

After coitus, the first sperm are transported through the uterus to the ovarian end of the fallopian tubes within about five minutes. This is many times more rapid than the motility of the sperm themselves can account for, which indicates that propulsive movements of the uterus and fallopian tubes might be responsible for much of the sperm movement. Yet, even with this aid, of the one-half billion sperm deposited in the vagina only 1000 to 3000 succeed in traversing the fallopian tubes to reach the proximity of the ovum.

Only one sperm is required for fertilization of the ovum, the process of which is illustrated in Figure 56–1. Furthermore, almost never do more than one sperm enter the ovum for the following reason: The zona pellucida of the ovum has a lattice-type structure, and once the ovum is punctured, some substance (perhaps one of the pro-

teolytic enzymes of the sperm acrosome) diffuses out of the ovum into the lattice to prevent penetration by additional sperm. Indeed, many sperm do attempt to penetrate the zona pellucida but become inactivated after traveling only part way through.

Once a sperm enters the ovum, its head swells rapidly to form a *male pronucleus*, which is also illustrated in Figure 56–1. Later, the 23 chromosomes of the male pronucleus and the 23 of the *female pronucleus* align themselves to reform a complete complement of 46 chromosomes (23 pairs) in the fertilized ovum.

Sex Determination. The sex of a child is determined by the type of sperm that fertilizes the ovum—that is, whether it is a male or a female sperm. It will be recalled from Chapter 54 that a

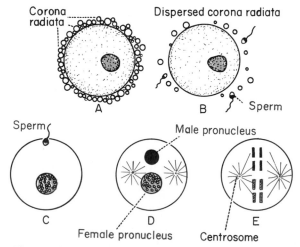

Figure 56–1. Fertilization of the ovum, showing (A) the mature ovum surrounded by the corona radiata, (B) dispersal of the corona radiata. (C) entry of the sperm, (D) formation of the male and female pronuclei, and (E) reorganization of a full complement of chromosomes and beginning division of the ovum. (Modified from Arey: Developmental Anatomy, 7th Ed.)

male sperm carries a *Y sex chromosome* and *22 autosomal chromosomes*, while a female sperm carries the same 22 autosomal chromosomes but an X *sex chromosome*. On the other hand, the ovum always has an X sex chromosome and never a Y chromosome. After recombination of the male and female pronuclei during fertilization, the fertilized ovum then contains 44 autosomal chromosomes and either 2 X chromosomes, which causes a female child to develop, or an X and a Y chromosome, which causes a male child to develop.

TRANSPORT AND IMPLANTATION OF THE DEVELOPING OVUM

Entry of the Ovum into the Fallopian Tube. When ovulation occurs, the ovum along with its attached granulosa cells, the *cumulus oophorus*, is expelled directly into the peritoneal cavity and must then enter one of the fallopian tubes. The fimbriated end of each fallopian tube falls naturally around the ovaries, and the inner surfaces of the fimbriated tentacles are lined with ciliated epithelium, the *cilia* of which continuously beat toward the *abdominal ostium* of the fallopian tube. One can actually see a slow fluid current flowing toward the ostium. By this means the ovum enters one or the other fallopian tube.

Transport of the Ovum Through the Fallopian Tube. Fertilization of the ovum normally takes place soon after the ovum enters the fallopian tube. After fertilization has occurred, three days are normally required for transport of the ovum through the tube into the cavity of the uterus. This transport is effected mainly by a feeble fluid current in the fallopian tube resulting from action of the ciliated epithelium that lines the tube, the cilia always beating toward the uterus. It is possible also that weak contractions of the fallopian tube aid in the passage of the ovum.

This delayed transport of the ovum through the fallopian tube allows several stages of division to occur before the ovum enters the uterus. During this time, large quantities of secretions are formed by secretory cells that line the fallopian tube. These secretions are for nutrition of the developing ovum.

Implantation of the Ovum in the Uterus. After reaching the uterus, the developing ovum usually remains in the uterine cavity an additional four to five days before it implants in the endometrium, which means that implantation ordinarily occurs on the seventh or eighth day following ovulation. During this time the ovum obtains its nutrition from the endometrial secre-

tions, called "uterine milk." Figure 56–2 shows a very early stage of implantation, illustrating that the developing ovum is in the *blastocyst stage*.

Implantation results from the action of trophoblastic cells that develop over the surface of the blastocyst. These cells secrete proteolytic enzymes that digest and liquefy the cells of the endometrium. Simultaneously, much of the fluid and nutrients thus released is actively absorbed into the blastocyst as a result of phagocytosis by the trophoblastic cells; these absorbed substances provide the sustenance for further growth of the blastocyst. Also, at the same time, additional trophoblastic cells form cords of cells that extend into the deeper layers of the endometrium and attach to them. Thus, the blastocyst eats a hole in the endometrium and attaches to it at the same time.

Once implantation has taken place, the trophoblastic and sub-lying cells proliferate rapidly; and they, along with cells from the mother's endometrium, form the placenta and the various membranes of pregnancy.

Early Intrauterine Nutrition of the Embryo. As the trophoblastic cells invade the endometrium, digesting and imbibing it, the stored nutrients in the large endometrial cells, called *decidual cells*, are used by the embryo for appropriate growth and development. During the first week after implantation, this is the only means by which the embryo can obtain any nutrients whatsoever, and the embryo continues to obtain a large measure of its total nutrition in this way for 8 to 12 weeks, though the placenta also begins to provide slight amounts of nutrients after approximately the sixteenth day beyond fertilization (a little over a week after implantation).

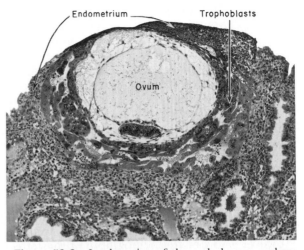

Figure 56–2. Implantation of the early human embryo, showing trophoblastic digestion and invasion of the endometrium. (Courtesy of Dr. Arthur Hertig.)

FUNCTION OF THE PLACENTA

The structure of the placenta is illustrated in Figure 56–3. Note that the fetus's blood flows through two *umbilical arteries*, finally to the capillaries of the villi, and thence back through the *umbilical vein* into the fetus. On the other hand, the mother's blood flows from the *uterine arteries* into large *blood sinuses* surrounding the villi and then back into the *uterine veins* of the mother.

The lower part of Figure 56–3 illustrates the relationship between the fetal blood of the villus and the blood of the mother in the fully developed placenta. The capillaries of the villus are lined with an extremely thin endothelium and are surrounded by a layer of *mesenchymal tissue* that is covered on the outside of the villus by a layer of *trophoblast cells*.

DIFFUSION THROUGH THE PLACENTAL MEMBRANE

The major function of the placenta is to allow *diffusion* of foodstuffs from the mother's blood into the fetus's blood and diffusion of excretory products from the fetus back into the mother.

In the early months of development, placental permeability is relatively slight because the villar membranes have not yet been reduced to their minimum thickness. However, as the placenta becomes older, the permeability increases progressively until the last month or so of pregnancy, when it begins to decrease again because of deterioration of the placenta caused by its age.

Diffusion of Oxygen Through the Placental Membrane. Almost exactly the same principles are applicable for the diffusion of oxygen through the placental membrane as through the pulmonary membrane; these principles were discussed in Chapter 28. The dissolved oxygen in the blood of the large placental sinuses simply passes through the villar membrane into the fetal blood because of a pressure gradient of oxygen from the mother's blood to the fetus's blood. The mean P_{O_2} in the mother's blood in the placental sinuses is approximately 50 mm. Hg toward the end of pregnancy, and the mean P_{O_2} in the blood leaving the villi and returning to the fetus after it has been "oxygenated" is about 30 mm. Hg. Therefore, the mean pressure gradient for diffusion of oxygen through the placental membrane is about 20 mm. Hg.

One might wonder how it is possible for a fetus to obtain sufficient oxygen when the fetal blood leaving the placenta has a P_{O_2} of only 30 mm. Hg. However, most of the hemoglobin of the fetus is *fetal hemoglobin*, a type of hemoglobin synthesized in the fetus prior to birth. At low P_{O_2}'s fetal

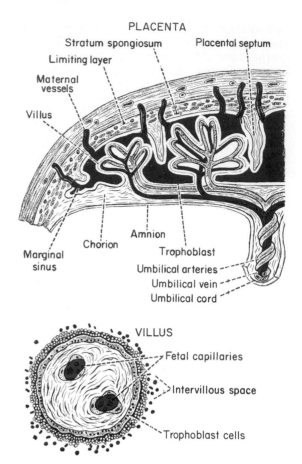

Figure 56–3. *Above:* Organization of the mature placenta. *Below:* Relationship of the fetal blood in the villus capillaries to the mother's blood in the intervillous spaces. (Modified from Gray and Goss: Anatomy of the Human Body, Lea & Febiger; and from Arey: Developmental Anatomy, 7th Ed.)

hemoglobin can carry as much as 20 to 30 per cent more oxygen than can maternal hemoglobin.

Also, the *hemoglobin concentration of the fetus is about 50 per cent greater than that of the mother*, which is an even more important factor in enhancing the amount of oxygen transported to the fetal tissues.

Diffusion of Carbon Dioxide Through the Placental Membrane. Carbon dioxide is continuously formed in the tissues of the fetus in the same way that it is formed in maternal tissues. The only means for excreting the carbon dioxide is through the placenta. The P_{CO_2} builds up in the fetal blood until it is about 48 mm. Hg, in contrast to 40 to 45 mm. Hg in maternal blood. Thus, a low pressure gradient for carbon dioxide develops across the placental membrane, but this is sufficient to allow adequate diffusion of carbon dioxide from the fetal blood into the maternal blood, because the extreme solubility of carbon dioxide in the water of the placental membrane allows carbon dioxide to diffuse through this membrane rapidly, about 20 times as rapidly as oxygen.

Diffusion of Foodstuffs Through the Placental Membrane. Other metabolic substrates needed by the fetus diffuse into the fetal blood in the same manner as oxygen. For instance, the glucose level in the fetal blood ordinarily is approximately 20 to 30 per cent lower than the glucose level in the maternal blood, for glucose is metabolized rapidly by the fetus. This in turn causes rapid diffusion of additional glucose from the maternal blood into the fetal blood.

Because of the high solubility of fatty acids in cell membranes, these too diffuse from the maternal blood into the fetal blood. Also, such substances as potassium, sodium, and chloride ions diffuse from the maternal blood into the fetal blood.

Active Absorption by the Placental Membrane. The cells that line the outer surface of the villi can also actively absorb certain nutrients from the maternal blood in the placenta during the first half of pregnancy, at least, and perhaps even throughout the entire pregnancy. For instance, the measured *amino acid* content of fetal blood is greater than that of maternal blood, and *calcium* and *inorganic phosphate* occur in greater concentration in fetal blood than in maternal blood. These effects indicate that the placental membrane has the ability to absorb actively at least small amounts of certain substances even during the latter part of pregnancy.

Excretion Through the Placental Membrane. In the same manner that carbon dioxide diffuses from the fetal blood into the maternal blood, other excretory products formed in the fetus diffuse into the maternal blood and then are excreted along with the excretory products of the mother. These include especially the waste products such as *urea, uric acid,* and *creatinine.* For instance, the level of urea in the fetal blood is only slightly greater than in maternal blood, because urea diffuses through the placental membrane with considerable ease.

HORMONAL FACTORS IN PREGNANCY

In pregnancy, the placenta forms large quantities of *human chorionic gonadotropin, estrogens, progesterone,* and *human chorionic somatomammotropin,* the first three of which, and perhaps the fourth as well, are essential to the continuance of pregnancy. The functions of these hormones are discussed in the following sections.

HUMAN CHORIONIC GONADOTROPIN AND ITS EFFECT IN CAUSING PERSISTENCE OF THE CORPUS LUTEUM AND IN PREVENTING MENSTRUATION

Menstruation normally occurs approximately 14 days after ovulation, at which time most of the secretory endometrium of the uterus sloughs away from the uterine wall and is expelled to the exterior. If this were to happen after implantation of an ovum, the pregnancy would terminate. However, this is prevented by the secretion of human chorionic gonadotropin in the following manner:

Coincidently with the development of the trophoblast cells from the early fertilized ovum, the hormone *human chorionic gonadotropin* is secreted into the fluids of the mother. As illustrated in Figure 56–4, the secretion of this hormone can first be measured 8 days after ovulation, just as the ovum is first implanting in the endometrium. Then the rate of secretion rises rapidly to reach a maximum approximately 8 weeks after ovulation, and decreases to a relatively low value by 16 to 20 weeks after ovulation.

Function of Human Chorionic Gonadotropin. Human chorionic gonadotropin is a glycoprotein having a molecular weight of 30,000 and very much the same molecular structure and function as the luteinizing hormone secreted by the pituitary. By far its most important function is to prevent the normal involution of the corpus luteum at the end of the female sexual cycle. Instead, it causes the corpus luteum to secrete even larger quantities of its usual hormones, progesterone and estrogens. These excess hormones cause the endometrium to continue growing and to store large amounts of nutrients rather than to be passed in the menstruum.

If the corpus luteum is removed before approximately the seventh to eleventh week of pregnancy, spontaneous abortion usually occurs,

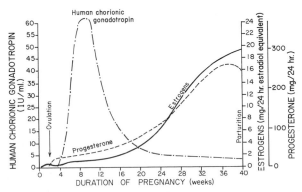

Figure 56–4. Rates of secretion of estrogens, progesterone, and chorionic gonadotropin at different stages of pregnancy.

but after this time the placenta itself secretes sufficient quantities of progesterone and estrogens to maintain pregnancy.

Effect of Human Chorionic Gonadotropin on the Fetal Testes. Human chorionic gonadotropin also exerts an *interstitial cell-stimulating effect* on the testes, thus resulting in the production of testosterone in male fetuses. This small secretion of testosterone during gestation is the factor that causes the fetus to grow male sex organs. Near the end of pregnancy, the testosterone secreted by the fetal testes also causes the testicles to descend into the scrotum.

SECRETION OF ESTROGENS BY THE PLACENTA

The chorionic epithelium of the placenta, like the corpus luteum, secretes both estrogens and progesterone, along with the secretion of human chorionic gonadotropin and human chorionic somatomammotropin. Figure 56–4 shows that the daily production of placental estrogens increases markedly toward the end of pregnancy, to as much as several hundred times the daily production in the middle of a normal monthly cycle. However, the secretion of estrogens by the placenta is quite different from the secretion by the ovaries; almost all of it is *estriol*, a relatively weak estrogen, rather than estradiol as secreted by the ovaries. Even so, the total *estrogenic activity* is still increased about 100-fold during pregnancy.

Function of Estrogen in Pregnancy. In the discussions of estrogens in the preceding chapter it was pointed out that these hormones exert mainly a proliferative function on certain reproductive and associated organs. During pregnancy, the extreme quantities of estrogens cause (1) enlargement of the uterus, (2) enlargement of the breasts and growth of the breast glandular tissue, and (3) enlargement of the female external genitalia.

The estrogens also relax the various pelvic ligaments, so that the sacroiliac joints become relatively limber and the symphysis pubis becomes elastic. These changes facilitate easy passage of the fetus through the birth canal.

There is much reason to believe that estrogens also affect the development of the fetus during pregnancy, for example, by affecting the rate of cell reproduction in the early embryo.

SECRETION OF PROGESTERONE BY THE PLACENTA

Progesterone is also a hormone essential for pregnancy. In addition to being secreted in moderate quantities by the corpus luteum at the beginning of pregnancy, it is secreted in tremendous quantities by the placenta, sometimes as much as 1 gram per day toward the end of pregnancy. Indeed, the rate of progesterone secretion increases by as much as 10-fold during the course of pregnancy, as illustrated in Figure 56–4.

The special effects of progesterone that are essential for normal progession of pregnancy are the following:

1. As pointed out earlier, progesterone causes decidual cells to develop in the uterine endometrium, and these cells then play an important role in the nutrition of the early embryo.

2. Progesterone has a special effect in decreasing the contractility of the gravid uterus, thus preventing uterine contractions from causing spontaneous abortion.

3. Progesterone also contributes to the development of the ovum even prior to implantation, for it specifically increases the secretions of the fallopian tubes and uterus to provide appropriate nutritive matter for the developing *morula* and *blastocyst.* There are some reasons to believe, too, that progesterone even affects cell cleavage in the early developing embryo.

4. The progesterone secreted during pregnancy also helps to prepare the breasts for lactation, as discussed later in the chapter.

HUMAN CHORIONIC SOMATOMAMMOTROPIN

Recently, a new hormone called *human chorionic somatomammotropin* has been discovered. This is a protein, having a molecular weight of about 38,000, that begins to be secreted about the fifth week of pregnancy and increases progressively throughout the remainder of pregnancy in direct proportion to the weight of the placenta. Human chorionic somatomammotropin has several important effects:

First, when administered to several different types of lower animals, human chorionic somatomammotropin causes at least partial development of the breasts.

Second, this hormone has weak actions similar to those of growth hormone, causing deposition of protein tissues in the same way that growth hormone does.

Third, human chorionic somatomammotropin has recently been found to have important actions on both glucose metabolism and fat metabolism in the mother, effects that perhaps are very important for nutrition of the fetus. The hormone causes decreased utilization of glucose by the mother, thereby making larger quantities of glucose available to the fetus. Furthermore, the hormone promotes release of free fatty acids from the fat stores of the mother, thus providing an alternative source of energy for her metabolism.

Therefore, it is beginning to appear that human chorionic somatomammotropin is a general metabolic hormone that has specific nutritional implications for both the mother and fetus.

RESPONSE OF THE MOTHER TO PREGNANCY

The presence of a growing fetus in the uterus adds an extra physiological load on the mother, and much of the response of the mother to pregnancy is due to this increased load. However, special effects include the following:

Blood Flow Through the Placenta and Cardiac Output. About 625 ml. of blood flows through the maternal circulation of the placenta each minute during the latter phases of gestation. This flow increases the cardiac output in the same manner that arteriovenous shunts increase the output. This factor, plus a general increase in metabolism, increases the cardiac output to 30 to 40 per cent above normal.

Blood Volume of the Mother. The maternal blood volume shortly before term is approximately 30 per cent above normal. This increase occurs mainly because of increased secretion during pregnancy of aldosterone and estrogens, both of which cause increased fluid retention by the kidneys.

At the time of birth of the baby, the mother has approximately 1 to 2 liters of extra blood in her circulatory system. Only about one-fourth of this amount is normally lost during delivery of the baby, thereby allowing a considerable safety factor for the mother.

Nutrition during Pregnancy. The growing fetus assumes priority for many of the nutritional elements in the mother's body, and many portions of the fetus continue to grow even if the mother does not eat a sufficiently nourishing diet.

By far the greatest growth of the fetus occurs during the last trimester of pregnancy; the weight of the child almost doubles during the last two months of pregnancy. Ordinarily, the mother does not absorb sufficient protein, calcium, phosphorus, and iron from the gastrointestinal tract during the last month of pregnancy to supply the fetus. However, from the beginning of pregnancy the mother's body has been storing these substances to be used during the latter months of pregnancy. Some of this storage is in the pla-

centa, but most of it is in the normal storage depots of the mother.

If appropriate nutritional elements are not present in the mother's diet, a number of maternal deficiencies can occur during pregnancy. Such deficiencies often occur for calcium, phosphates, iron, and the vitamins. For example, approximately 375 mg. of iron is needed by the fetus to form its blood and an additional 600 mg. is needed by the mother to form her own extra blood. The normal store of nonhemoglobin iron in the mother at the outset of pregnancy is often only 100 or so mg. Therefore, in general, the obstetrician supplements the diet of the mother with the needed substances. It is especially important that the mother receive large quantities of vitamin D, for although the total quantity of calcium utilized by the fetus is small, calcium even normally is poorly absorbed by the gastrointestinal tract. Finally, shortly before birth of the baby vitamin K is often added to the diet so that the baby will have sufficient prothrombin to prevent postnatal hemorrhage.

The Amniotic Fluid and Its Formation. Normally, the volume of amniotic fluid (the fluid that surrounds the fetus in the uterus) is between 500 ml. and 1 liter. Studies with isotopes of the rate of formation of amniotic fluid show that on the average the water in amniotic fluid is completely replaced once every 3 hours, and the electrolytes sodium and potassium are replaced once every 15 hours. Yet, strangely enough, the sources of the fluid and the points of reabsorption are mainly unknown. A small portion of the fluid is derived from renal excretion by the fetus, and a small amount of absorption occurs by way of the fetal gastrointestinal tract. But about one-half of the fluid turnover is believed to occur through the amniotic membranes.

ABNORMAL RESPONSES OF THE MOTHER TO PREGNANCY

HYPEREMESIS GRAVIDARUM

In the earlier months of pregnancy, the mother frequently develops hyperemesis gravidarum, a condition characterized by nausea and vomiting and commonly known as "morning sickness." Occasionally, the vomiting is so severe that the mother becomes greatly dehydrated, and in rare instances the condition even causes death.

The cause of the nausea and vomiting is unknown, but during the first few months of pregnancy rapid trophoblastic invasion of the endometrium takes place, and because the trophoblastic cells digest portions of the endometrium as they invade it, it is possible that degenerative products resulting from this invasion are responsible for the nausea and vomiting. Indeed, degenerative processes in other parts of the body, such as those occurring after gamma ray irradiation and burns, can all cause similar nausea and vomiting.

Another possible cause of the condition is the large quantity of estrogen secreted by the placenta. This possibility is supported by the fact that estrogen injected daily into a person in large quantities for many weeks will often cause nausea and vomiting during the first few weeks of administration.

PREECLAMPSIA AND ECLAMPSIA

Approximately 4 per cent of all pregnant women develop a rapid rise in arterial blood pressure associated with loss of large amounts of protein in the urine at some time during the latter four months of pregnancy. This condition, called *preeclampsia,* is often also characterized by salt and water retention by the kidneys, weight gain, and development of edema. In addition, arterial spasm occurs in many parts of the body, most significantly in the kidneys, brain, and liver. Both the renal blood flow and the glomerular filtration rate are decreased, which is exactly opposite to the changes that occur in the normal pregnant woman. The renal effects are caused by thickened glomerular tufts that contain a fibrinoid deposit in the basement membranes.

Various attempts have been made to prove that preeclampsia is caused by excessive secretion of placental or adrenal hormones, but proof of a hormonal basis is still lacking. Indeed, a more plausible theory is that preeclampsia results from some type of autoimmunity or allergy resulting from the presence of the fetus. Indeed, the acute symptoms disappear within a few days after birth of the baby.

The severity of preeclampsia symptoms is closely associated with the retention of salt and water and the degree of increase in arterial pressure. In fact, an increasing pressure seems to set off a vicious circle that intensifies the arterial spasm and other pathological effects of preeclampsia. These effects can be greatly delayed by drastic limitation of salt intake and enforced bed rest during the latter months of pregnancy, which are the two cardinal features of therapy.

Eclampsia is a severe degree of preeclampsia characterized by extreme vascular spasticity throughout the body, clonic convulsions followed by coma, greatly decreased kidney output, malfunction of the liver, often extreme hypertension, and a generalized toxic condition of the body. Usually, it occurs shortly before parturition.

Without treatment, a very high percentage of eclamptic patients die. However, with optimal and immediate use of rapidly acting vasodilating drugs to reduce the arterial pressure to normal, followed by immediate termination of pregnancy — by cesarean section if necessary — the mortality has been reduced to 1 per cent or less in some clinics.

PARTURITION

INCREASED UTERINE IRRITABILITY NEAR TERM

Parturition means simply the process by which the baby is born. At the termination of pregnancy the uterus becomes progressively more excitable until finally it begins strong rhythmic contractions with such force that the baby is expelled. The exact cause of the increased activity of the uterus is not known, but at least two major categories of effects lead up to the culminating contractions responsible for parturition; these are, first, progressive hormonal changes that cause increased excitability of the uterine musculature and, second, progressive mechanical changes.

Hormonal Factors That Cause Increased Uterine Contractility. *Ratio of Estrogens to Progesterone.* Progesterone inhibits uterine contractility during pregnancy, thereby helping to prevent expulsion of the fetus. On the other hand, estrogens have a definite tendency to increase the degree of uterine contractility. Both these hormones are secreted in progressively greater quantities throughout pregnancy, but from the seventh month onward, estrogen secretion increases more than progesterone secretion. Therefore, it has been postulated that the *estrogen to progesterone ratio* increases sufficiently toward the end of pregnancy to be at least partly responsible for the increased contractility of the uterus.

Effect of Oxytocin on the Uterus. Oxytocin is a hormone secreted by the posterior pituitary gland that specifically causes uterine contraction (see Chapter 49). Experiments in animals have shown that irritation or stretching of the uterine cervix, such as that occurring at the end of pregnancy, causes a neurogenic reflex to the neurohypophysis that increases the rate of oxytocin secretion. Therefore, this hormone probably helps considerably to increase uterine contractions.

Mechanical Factors That Increase the Contractility of the Uterus. *Stretch of the Uterine Musculature.* Simply stretching smooth muscle organs usually increases their contractility. Furthermore, intermittent stretch, like that occurring repetitively in the uterus because of movements of the fetus, can also elicit smooth muscle contraction.

Note especially that twins are born on the average *19 days* earlier than a single child, which emphasizes the importance of mechanical stretch in eliciting uterine contractions.

Stretch or Irritation of the Cervix. There is much reason to believe that stretch or irritation of the uterine cervix is particularly important in eliciting uterine contractions. The mechanism of this effect is probably myogenic transmission of signals from the cervix to the body of the uterus.

ONSET OF LABOR – A POSITIVE FEEDBACK THEORY FOR ITS INITIATION

During most of the months of pregnancy the uterus undergoes periodic episodes of weak and slow rhythmic contractions called *Braxton-Hicks contractions.* These become progressively stronger toward the end of pregnancy; and they eventually change rather suddenly, within hours, to become exceptionally strong contractions that start stretching the cervix and later forcing the baby through the birth canal. This process is called *labor,* and the strong contractions that result in final parturition are called *labor contractions.*

On the basis of our new understanding of control systems in the past few years, a theory has been proposed for explaining the onset of labor based on "positive feedback." This theory suggests that stretch of the cervix by the fetus's head finally becomes great enough to elicit a reflex increase in contractility of the uterine body. This pushes the baby forward, which stretches the cervix some more and initiates a new cycle. Thus, the process continues again and again until the baby is expelled. This theory is illustrated in Figure 56–5. We know at least two positive feedbacks that could lead to birth of the baby, as follows:

(1) Stretch of the cervix causes the entire body of the uterus to contract, and this stretches the cervix still more because of the downward thrust of the baby's head. (2) Cervical stretch also causes the pituitary gland to secrete oxytocin, which is still another cause of increased uterine contractility.

To summarize the theory, we can assume that multiple factors increase the contractility of the uterus toward the end of pregnancy. Eventually, a uterine contraction becomes strong enough to irritate the uterus, increase its contractility still more because of positive feedback, and result in a

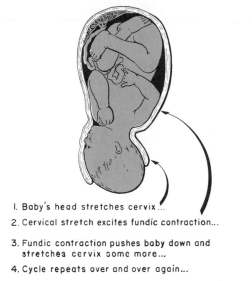

1. Baby's head stretches cervix...
2. Cervical stretch excites fundic contraction...
3. Fundic contraction pushes baby down and stretches cervix some more...
4. Cycle repeats over and over again...

Figure 56–5. Theory for the onset of intensely strong contractions during labor.

second contraction stronger than the first, a third stronger than the second, and so forth. Once these contractions become strong enough to cause this type of feedback, with each succeeding contraction greater than the one preceding, the process proceeds to completion.

ABDOMINAL MUSCLE CONTRACTION DURING LABOR

Once labor contractions become strong and painful, neurogenic reflexes, mainly from the birth canal to the spinal cord and thence back to the abdominal muscles, cause intense abdominal muscle contraction. This abdominal contraction adds greatly to the forces that cause expulsion of the baby.

MECHANICS OF PARTURITION

At the beginning of labor, strong contractions might occur only once every 30 minutes. As labor progresses, the contractions finally appear as often as once every one to three minutes, and the intensity of contraction increases greatly with only a short period of relaxation between contractions.

The combined contractions of the uterine and abdominal musculature during delivery of the baby cause downward force on the fetus of approximately 25 pounds during each strong contraction. It is fortunate that the contractions of labor occur intermittently, because strong contractions impede or sometimes even stop blood flow through the placenta and would cause death of the fetus were the contractions continuous.

In 19 out of 20 births the head is the first part of the baby to be expelled; in most of the remaining instances the buttocks are presented first. The head acts as a wedge to open the structures of the birth canal as the fetus is forced downward from above.

The first major obstruction to expulsion of the fetus is the uterine cervix. Toward the end of pregnancy the cervix becomes soft, which allows it to stretch when labor pains cause the body of the uterus to contract. The so-called *first stage of labor* is the period of progressive cervical dilatation, lasting until the opening is as large as the head of the fetus. This stage usually lasts 8 to 24 hours in the first pregnancy but often only a few minutes if the mother has had many pregnancies.

Once the cervix has dilated fully, the fetus's head moves rapidly into the birth canal and, with additional force from above, continues to wedge its way through the canal until delivery is effected. This is called the *second stage of labor,* and it may last from as little as a minute after many pregnancies up to half an hour or more in the first pregnancy.

SEPARATION AND DELIVERY OF THE PLACENTA

During the succeeding 10 to 45 minutes after birth of the baby, the uterus contracts to a very small size, which causes a *shearing* effect between the walls of the uterus and the placenta and consequent separation of the placenta from its implantation site. Separation of the placenta opens the placental sinuses and causes bleeding. However, the amount of bleeding is limited to an average of 350 ml. by the following mechanism: The smooth muscle fibers of the uterine musculature are arranged in figure eights around the blood vessels as they pass through the uterine wall. Therefore, contraction of the uterus following delivery of the baby constricts the vessels that previously supplied blood to the placenta.

LABOR PAINS

With each uterine contraction the mother experiences considerable pain. The pain in early labor is probably caused mainly by hypoxia of the uterine muscle resulting from compression of the blood vessels to the uterus. This pain is not felt when the *hypogastric nerves,* which carry the sensory fibers leading from the uterus, have been sectioned. However, during the second stage of labor, when the fetus is being expelled through the birth canal, much more severe pain is caused by cervical stretch, perineal stretch, and stretch or tearing of structures in the vaginal canal itself. This pain is conducted by somatic nerves instead of by the hypogastric nerves.

INVOLUTION OF THE UTERUS

During the first four to five weeks following parturition, the uterus involutes. Its weight becomes less than one-half its immediate postpartum weight within a week, and in four weeks the uterus may be as small as it had been prior to pregnancy. During early involution of the uterus the placental site on the endometrial surface autolyzes, causing a vaginal discharge known as "lochia," which is first bloody and then serous, and continues for approximately a week and a half in all. After this time, the endometrial surface will have become re-epithelialized and ready for normal, nongravid sex life again.

LACTATION

DEVELOPMENT OF THE BREASTS

The breasts begin to develop at puberty; this development is stimulated by the estrogens of the monthly sexual cycles that stimulate growth of the stroma and ductile system plus deposition of fat to give mass to the breasts. However, much additional growth occurs during pregnancy, and the glandular tissue only then becomes completely developed for actual production of milk.

All through pregnancy, the tremendous quantities of estrogens secreted by the placenta — plus additional quantities of growth hormone, prolactin, and several other hormones — cause the ductile system of the breasts to grow and to branch. Simultaneously, the stroma of the breasts also increases, and large quantities of fat are laid down in the stroma.

Then the action of progesterone causes growth of the lobules, budding of alveoli, and development of secretory characteristics in the cells of the alveoli.

INITIATION OF LACTATION – FUNCTION OF PROLACTIN

Though estrogen and progesterone are essential for the physical development of the breasts during pregnancy, both these hormones also have a specific effect to inhibit the actual secretion of milk. On the other hand, the hormone *prolactin* has exactly the opposite effect, promotion of the secretion of milk. This hormone is secreted by the mother's pituitary gland, and its concentration in her blood rises steadily from the fifth week of pregnancy until birth of the baby, at which time it has risen to very high levels, usually about ten times the normal nonpregnant level. This is illustrated in Figure 56–6. In addition, the placenta secretes large quantities of human chorionic somatomammotropin, which also has mild lactogenic properties, thus supporting the prolactin from the mother's pituitary. Even so, only a few milliliters of fluid are secreted each day until after the baby is born. This fluid is called *colostrum;* it contains essentially the same concentration of proteins and lactose as milk but almost no fat, and its maximum rate of production is about 1/100 the subsequent rate of milk production.

This absence of lactation during pregnancy is caused by the overriding suppressive effects of progesterone and estrogen, which are secreted in tremendous quantities as long as the placenta is still in the uterus and which completely subdue the lactogenic effects of both prolactin and human chorionic somatomammotropin. However, immediately after the baby is born, the sudden loss of both estrogen and progesterone secretion by the placenta now allows the lactogenic effect of the prolactin from the mother's pituitary gland to assume its natural role, and within two or three days the breasts begin to secrete copious quantities of milk instead of colostrum.

Following birth of the baby, the *basal level* of prolactin secretion returns during the next few weeks to the nonpregnant level, as shown in Figure 56–6. However, each time the mother nurses her baby, nervous signals from the nipples to the hypothalamus cause approximately a ten-

fold surge in prolactin secretion lasting about one hour, which is also shown in the figure. The prolactin in turn acts on the breasts to provide the milk for the next nursing period. If this prolactin surge is absent, if it is blocked as a result of hypothalamic or pituitary damage, or if nursing does not continue, the breasts lose their ability to produce milk within a few days. However, milk production can continue for several years if the child continues to suckle, but the rate of milk formation normally decreases considerably within seven to nine months.

Hypothalamic Control of Prolactin Secretion. Though secretion of most of the anterior pituitary hormones is enhanced by neurosecretory releasing factors transmitted from the hypothalamus to the anterior pituitary gland through the hypothalamic–hypophysial portal system, the secretion of prolactin is controlled by an exactly opposite effect. That is, the hypothalamus synthesizes a prolactin inhibitory factor (PIF). Under normal conditions, large amounts of PIF are continuously transmitted to the anterior pituitary gland so that the normal rate of prolactin secretion is slight. However, during lactation the formation of PIF itself is suppressed, thereby allowing the anterior pituitary gland to secrete an uninhibited amount of prolactin.

THE EJECTION (OR "LET-DOWN") PROCESS IN MILK SECRETION — FUNCTION OF OXYTOCIN

Milk is secreted continuously into the alveoli of the breasts, but milk does not flow easily from the alveoli into the ductile system and therefore does not continually leak from the breast nipples. Instead, the milk must be "ejected" or "let-down" from the alveoli to the ducts before the baby can obtain it. This process is caused by a combined neurogenic and hormonal reflex involving the hormone *oxytocin* as follows:

When the baby suckles the breast, sensory impulses are transmitted through somatic nerves to the spinal cord and then to the hypothalamus, there causing *oxytocin* secretion. This hormone flows in the blood to the breasts, where it causes the *myoepithelial cells* that surround the outer walls of the alveoli to contract, thereby expressing the milk from the alveoli into the ducts. Thus, within 30 seconds to a minute after a baby begins to suckle the breast, milk begins to flow. This process is called milk ejection, or milk let-down.

Suckling on one breast causes milk flow not only in that breast but also in the opposite breast. Also, it is especially interesting that the sound of the baby crying is often enough of a signal to cause milk ejection.

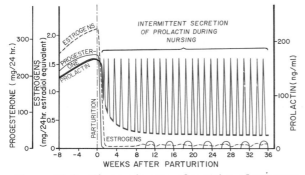

Figure 56–6. Changes in rates of secretion of estrogens, progesterone, and prolactin for 8 weeks prior to parturition and for 36 weeks thereafter. Note especially the decrease of prolactin secretion back to basal levels within a few weeks, but also the intermittent periods of marked prolactin secretion (for about one hour at a time) during and after periods of nursing.

TABLE 56–1 PERCENTAGE COMPOSITION
OF MILK

	Human Milk	Cow's Milk
Water	88.5	87.0
Fat	3.3	3.5
Lactose	6.8	4.8
Casein	0.9	2.7
Lactalbumin and other protein	0.4	0.7
Ash	0.2	0.7

MILK AND THE METABOLIC DRAIN ON THE MOTHER CAUSED BY LACTATION

Table 56–1 gives the contents of human milk and cow's milk.

At the height of lactation 1.5 liters of milk may be formed each day. With this degree of lactation great quantities of metabolic substrates are drained from the mother. For instance, approximately 50 grams of fat enter the milk each day; and approximately 100 grams of lactose, which must be derived from glucose, are lost from the mother each day. Also, some 2 to 3 grams of calcium phosphate may be lost each day, and unless the mother is drinking large quantities of milk and has an adequate intake of vitamin D, the output of calcium and phosphate by the lactating mammae will be much greater than the intake of these substances. To supply the needed calcium and phosphate, the parathyroid glands enlarge greatly, and the bones become progressively decalcified. The problem of decalcification is usually not very great during pregnancy, but it can be a distinct problem during lactation.

GROWTH AND FUNCTIONAL DEVELOPMENT OF THE FETUS

During the first two to three weeks the fetus remains almost microscopic in size, but thereafter the dimensions of the fetus increase almost in proportion to age. At 12 weeks the length of the fetus is approximately 10 cm.; at 20 weeks, approximately 25 cm., and at term (40 weeks), approximately 53 cm. (about 21 inches). Because the weight of the fetus is proportional to the cube of the length, the weight increases approximately in proportion to the cube of the age of the fetus. Therefore, the weight of the fetus remains almost nothing during the first months and reaches only 1 pound at five and a half months of gestation. Then during the last trimester of pregnancy, the fetus gains tremendously, so that two months prior to birth the weight averages 3 pounds, one month prior to birth 4.5 pounds, and at birth 7 pounds.

DEVELOPMENT OF THE FETAL ORGAN SYSTEMS

Within one month after fertilization of the ovum all the different organs of the fetus have already been at least partly formed, and during the next two to three months the minute details of the different organs are established. Beyond the fourth month, the organs of the fetus are grossly the same as those of the newborn child, even including most of the smaller structures of the organs. However, cellular development of these structures is usually far from complete at this time and requires the full remaining five months of pregnancy for complete development. Even at birth certain structures, particularly the nervous system, the kidneys, and the liver, still lack full development, as is discussed in more detail later in the chapter.

The Circulatory System. The human heart begins beating during the fourth week following fertilization, contracting at the rate of about 65 beats per minute. The rate increases steadily as the fetus grows and reaches approximately 140 per minute immediately before birth.

Formation of Blood Cells. Nucleated red blood cells begin to be formed in the yolk sac and mesothelial layers of the placenta at about the third week of fetal development. This is followed a week later by the formation of non-nucleated red blood cells by the fetal mesenchyme and by the endothelium of the fetal blood vessels. Then at approximately six weeks, the liver begins to form blood cells, and in the third month the spleen and other lymphoid tissues of the body also begin forming blood cells. Finally, from approximately the third month on, the bone marrow forms more and more red and white blood cells while the other structures lose their ability completely to form blood cells.

The Respiratory System. Obviously, respiration cannot occur during fetal life. However, respiratory movements do take place beginning at the end of the first trimester of pregnancy. Tactile stimuli or fetal asphyxia especially cause respiratory movements.

The Nervous System. Most of the peripheral reflexes of the fetus are well developed by the third to fourth month of pregnancy. However, some of the more important higher functions of the central nervous system are still undeveloped even at birth. Indeed, myelinization of some major tracts of the central nervous system becomes complete only after approximately a year of postnatal life.

The Gastrointestinal Tract. Even in mid-pregnancy the fetus ingests and absorbs large quantities of amniotic fluid, and during the latter two to three months, gastrointestinal function approaches that of the normal newborn infant. Small quantities of *meconium* are continually formed in the gastrointestinal tract and excreted from the bowels into the amniotic fluid. Meconium is composed partly of unabsorbed residue of amniotic fluid and partly of excretory products from the gastrointestinal mucosa and glands.

The Kidneys. The fetal kidneys are capable of excreting urine during at least the latter half of pregnancy, and urination occurs normally *in utero*. However, the renal control systems for regulation of extracellular fluid electrolyte balances and acid-base balance are almost nonexistent until after midfetal life and do not reach full development until about a month after birth.

ADJUSTMENTS OF THE INFANT TO EXTRAUTERINE LIFE

ONSET OF BREATHING

The most obvious effect of birth on the baby is loss of the placental connection with the mother and therefore loss of this means for metabolic support. Especially important is loss of the placental oxygen supply and placental excretion of carbon dioxide. Therefore, by far the most important immediate adjustment required of the infant is the onset of breathing.

Cause of Breathing at Birth. Following completely normal delivery from a mother who has not been depressed by anesthetics, the child ordinarily begins to breathe immediately and has a completely normal respiratory rhythm from the onset. The promptness with which the fetus begins to breathe indicates that breathing is initiated by sudden exposure to the exterior world, probably resulting from a slightly asphyxiated state incident to the birth process but also from sensory impulses originating in the suddenly cooled skin. However, if the infant does not breathe immediately, his body becomes progressively more hypoxic and hypercapnic, which provides additional stimulus to the respiratory center and usually causes breathing within a few seconds to a few minutes after birth.

Delayed and Abnormal Breathing at Birth — Danger of Hypoxia. If the mother has been depressed by an anesthetic during delivery, which at least partially anesthetizes the child as well, respiration is likely to be delayed for several minutes, thus illustrating the importance of using as little obstetrical anesthesia as feasible. Also, many infants who have had head trauma during delivery are slow to breathe or sometimes will not breathe at all. This can result from two possible effects: First, in a few infants, intracranial hemorrhage or brain contusion causes a concussion syndrome with a greatly depressed respiratory center. Second, and probably much more important, prolonged fetal hypoxia during delivery causes serious depression of the respiratory center. Hypoxia frequently occurs during delivery because of (1) compression of the umbilical cord, (2) premature separation of the placenta, (3) excessive contraction of the placenta, or (4) excessive anesthesia of the mother.

Degree of Hypoxia that an Infant Can Tolerate. In the adult, failure to breathe for only four minutes often causes death, but a newborn infant often survives as long as 15 minutes of failure to breathe after birth. Unfortunately, though, permanent brain impairment often ensues if breathing is delayed more than 8 to 10 minutes. Indeed, actual lesions develop, mainly in brain stem nuclei, thus affecting many of the stereotype motor functions of the body. This is believed to be one of the major causes of *cerebral palsy.*

Expansion of the Lungs at Birth. At birth, the walls of the alveoli are held together by the surface tension of the viscid fluid that fills them. More than 25 mm. Hg of negative pressure is required to oppose the effects of this surface tension and therefore to open the alveoli for the first time. But once the alveoli are open, further respiration can be effected with relatively weak respiratory movements. Fortunately, the first inspirations of the newborn infant are extremely powerful, usually capable of creating as much as 50 mm. Hg negative pressure in the intrapleural space.

Figure 56–7 illustrates the tremendous forces required to open the lungs at the onset of breathing. To the left is shown the pressure-volume curve (compliance curve) for the first breath after birth. Observe, first, the lowermost curve, which shows that the lungs essentially do not expand at

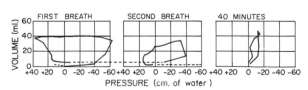

Figure 56–7. Pressure-volume curves of the lungs (compliance curves) of a newborn baby immediately after birth, showing (a) the extreme forces required for breathing during the first two breaths of life and (b) development of a nearly normal compliance within 40 minutes after birth. (From Smith: *Sci. Amer., 209*:32, 1963. © 1963 by Scientific American, Inc. All rights reserved.)

all until the negative pressure has reached −40 cm. water (−30 mm. Hg). Then as the negative pressure increases to −60 cm. water, only about 40 ml. of air enters the lungs. Then, to deflate the lungs, considerable positive pressure is required, probably because of the viscous resistance offered by the fluid in the bronchioles.

Note that the second breath is much easier. However, breathing does not become completely normal until about 40 minutes after birth, as shown by the third compliance curve, the shape of which compares favorably with that of the normal adult.

Respiratory Distress Syndrome. A few infants develop severe respiratory distress during the few hours to several days following birth and frequently succumb within the next day or so. The alveoli of these infants at death contain large quantities of proteinaceous fluid, almost as if pure plasma had leaked out of the capillaries into the alveoli.

Unfortunately, the cause of the respiratory distress syndrome is not certain. However, one of the most characteristic findings is failure to secrete adequate quantities of *surfactant*, a substance normally secreted into the alveoli, which decreases the surface tension of the alveolar fluid, therefore allowing the alveoli to open easily. The surfactant secreting cells (the type II alveolar epithelial cells) do not begin to secrete surfactant until the last one to three months of gestation. Therefore, many premature babies and some full-term babies are born without the capability of secreting surfactant, which therefore causes both a tendency of the lungs to collapse and development of pulmonary edema. The role of surfactant in preventing these effects was discussed in Chapter 27.

CIRCULATORY READJUSTMENTS AT BIRTH

Almost as important as the onset of breathing at birth are the immediate circulatory adjustments that allow adequate blood flow through the lungs. Because the lungs are mainly nonfunctional during fetal life, it is not necessary for the fetal heart to pump much blood through the lungs. On the other hand, the fetal heart must pump large quantities of blood through the placenta. As illustrated in Figure 56–8, most of the blood entering the right atrium from the inferior vena cava is directed in a straight pathway across the posterior aspect of the right atrium and thence through the *foramen ovale* directly into the left atrium. Thus, the well-oxygenated blood from the placenta enters the left side of the heart

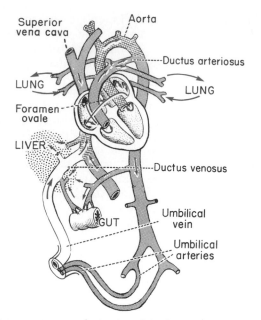

Figure 56–8. Organization of the fetal circulation. (Modified from Arey: Developmental Anatomy, 7th Ed.)

rather than the right side and is pumped by the left ventricle mainly into the vessels of the head and forelimbs.

The blood entering the right atrium from the superior vena cava is directed downward through the tricuspid valve into the right ventricle. This blood is mainly deoxygenated blood from the head region of the fetus, and it is pumped by the right ventricle into the pulmonary artery, then mainly through the *ductus arteriosus* into the descending aorta and through the two umbilical arteries into the placenta, where this deoxygenated blood becomes oxygenated.

Changes in the Fetal Circulation at Birth. The basic changes in the fetal circulation at birth were discussed in Chapter 20 in relation to congenital anomalies of the ductus arteriosus and foramen ovale that persist throughout life. Briefly, these are as follows:

First, the tremendous blood flow through the placenta ceases, which *approximately doubles the systemic vascular resistance at birth*. This obviously *increases the aortic pressure* as well as the pressures in the left ventricle and left atrium.

Second, the *pulmonary vascular resistance decreases greatly* as a result of expansion of the lungs. In the unexpanded fetal lungs, the blood vessels are compressed because of the small volume of the lungs. Immediately upon expansion these vessels are no longer compressed, and the resistance to blood flow decreases several-fold. Also, in fetal life the hypoxia of the lungs causes considerable tonic vasoconstriction of the lung blood vessels, but vasodilation takes place when aeration of the lungs eliminates the hypoxia. These changes reduce the resistance of blood flow through the lungs as much as five-fold, which obviously *reduces the pulmonary arterial pressure, the right ventricular pressure, and the right atrial pressure.*

Closure of the Foramen Ovale. The *low right atrial pressure* and the *high left atrial pressure* that occur secondary to the changes in pulmonary and systemic resistances at birth cause a tendency for blood to flow backward from the left atrium into the right atrium rather than in the other direction as occurred during fetal life. Consequently, the small valve that lies over the foramen ovale on the left side of the atrial septum closes over this opening, thereby preventing further flow.

Closure of the Ductus Arteriosus. Similar effects occur in relation to the ductus arteriosus, for the increased systemic resistance *elevates the aortic pressure* while the decreased pulmonary resistance *reduces the pulmonary arterial pressure.* As a consequence, immediately after birth, blood begins to flow backward from the aorta into the pulmonary artery rather than in the other direction as in fetal life. However, after only a few hours the muscular wall of the ductus arteriosus constricts markedly, and within one to eight days the constriction is sufficient to stop all blood flow. This is called *functional closure* of the ductus arteriosus. Then, sometime during the second month of life the ductus arteriosus ordinarily becomes anatomically *occluded* by growth of fibrous tissue into its lumen.

The causes of either functional closure or anatomical closure of the ductus are not completely known. However, the most likely cause is increased oxygenation of the blood flowing through the ductus. In fetal life the P_{O_2} of the ductus blood is as low as 15 mm. Hg, but within a few hours after birth it increases to about 100 mm. Hg. Furthermore, many experiments have shown that the degree of contraction of the ductus is highly related to the availability of oxygen.

In one out of several thousand infants, the ductus fails to close, resulting in a *patent ductus arteriosus,* the consequences of which were discussed in Chapter 20.

SPECIAL FUNCTIONAL PROBLEMS IN THE NEONATAL INFANT

The most important characteristic of the newborn infant is instability of the various hormonal and neurogenic control systems. This results partly from the immature development of the different organs of the body and partly from the fact that the control systems simply have not become adjusted to the completely new way of life.

Cardiac Output. The cardiac output of the newborn infant averages 550 ml. per minute, which is about two times as much in relation to body weight as in the adult. An occasional child is born with an especially low cardiac output caused by hemorrhage through the placental membrane into the mother's blood prior to birth.

Arterial Pressure. The arterial pressure during the first day after birth averages about 70/50; it increases slowly during the next several months to approximately 90/60. Then there is a much slower rise during the subsequent years until the adult pressure of 120/80 is attained at adolescence.

Postnatal Bleeding Tendencies. Since most vitamin K in the body is synthesized by bacteria in the colon and since the newborn infant has not yet developed a bacterial flora in his colon, the vitamin K level falls rapidly the first few days after birth, causing factor VII and prothrombin production by the liver to be impaired for a week or so until a bacterial flora is established. During this time the infant is likely to develop a bleeding tendency. However, if the mother is given an adequate dose of vitamin K at least four hours prior to birth of the infant, sufficient vitamin K is usually stored in the infant's liver to see him through this possible period of hypovitaminosis K.

Fluid Balance, Acid-Base Balance, and Renal Function. The rate of fluid intake and fluid excretion in the infant is seven times as great in relation to weight as in the adult, which means that even a slight alteration of fluid balance can cause rapidly developing abnormalities. Also, the rate of metabolism in the infant is two times as great in relation to body mass as in the adult, which means that two times as much acid is normally formed, leading to a tendency toward acidosis in the infant. Finally, functional development of the kidneys is not complete until the end of approximately the first month of life. For instance, the kidneys of the newborn can concentrate urine to only one and a half times the osmolality of the plasma instead of the normal three to four times in the adult.

Therefore, considering the immaturity of the kidney, together with the marked fluid turnover in the infant and rapid formation of acid, one can readily understand that among the most important problems of infancy are acidosis and dehydration.

Liver Function. During the first few days of

life, liver function may be quite deficient, as evidenced by the following effects:

1. The liver of the newborn conjugates bilirubin with glucuronic acid poorly and therefore excretes bilirubin only slightly during the first few days of life.

2. The liver of the newborn is deficient in forming plasma proteins, so that the plasma protein concentration falls to 1 gram per cent less than that for older children. Occasionally, the protein concentration falls so low that the infant actually develops hypoproteinemic edema.

3. The gluconeogenesis function of the liver is particularly deficient. As a result, the blood glucose level of the unfed newborn infant falls to about 30 to 40 mg. per cent, and the infant must depend on its stored fats for energy until feeding can occur.

4. The liver of the newborn often also forms too little of the factors needed for normal blood coagulation.

Digestion, Absorption, and Metabolism of Energy Foods. In general, the ability of the newborn infant to digest, absorb, and metabolize foods is not different from that of the older child, with the following three exceptions:

First, secretion of pancreatic amylase in the newborn infant is deficient, so that the infant utilizes starches less adequately than do older children. However, the infant readily assimilates disaccharides and monosaccharides.

Second, absorption of fats from the gastrointestinal tract is somewhat less than in the older child. Consequently, milk with a high fat content, such as cow's milk, is frequently inadequately utilized.

Third, because the liver functions are imperfect during at least the first week of life, the glucose concentration in the blood is unstable and often low.

Metabolic Rate and Body Temperature. The normal metabolic rate of the newborn in relation to body weight is about two times that of the adult, which accounts also for the two times as great cardiac output and two times as great minute respiratory volume in the infant.

However, since the body surface area is very large in relation to the body mass, heat is readily lost from the body. As a result, the body temperature of the newborn infant, particularly of the premature infant, falls. Figure 56–9 shows that the body temperature of even the normal infant falls several degrees during the first few hours after birth but returns to normal in seven to eight hours. Still, the body temperature regulatory mechanisms remain poor during the early days of life, allowing marked deviations in temperature at first, which are also illustrated in Figure 56–9.

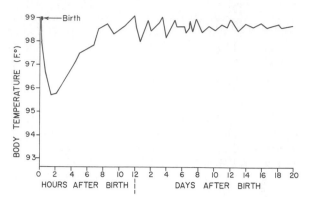

Figure 56–9. Fall in body temperature of the infant immediately after birth, and instability of body temperature during the first few days of life.

Nutritional Needs during the Early Weeks of Life. Three specific problems occur in the early nutrition of the infant, as follows:

Need for Calcium. The newborn infant has only just begun rapid ossification of its bones at birth, so that it needs a ready supply of calcium throughout infancy.

Need for Iron. If the mother has had adequate amounts of iron in her diet, the liver of the infant usually has stored enough iron to keep forming blood cells for four to six months after birth. But if the mother has had insufficient iron in her diet, anemia is likely to supervene in the infant after about three months of life. Therefore, administration of iron in some other form is desirable by the second or third month of life.

Vitamin C Deficiency. Ascorbic acid (vitamin C) is not stored in significant quantities in the fetal tissues; yet it is required for proper formation of cartilage, bone, and other intercellular structures of the infant. Furthermore, milk, especially cow's milk, has poor supplies of ascorbic acid. For this reason, orange juice or other sources of ascorbic acid are usually prescribed by the third week of life.

Immunity. Fortunately, the newborn inherits much immunity from its mother because many antibodies diffuse from the mother's blood through the placenta into the fetus. However, the newborn itself does not form antibodies to a significant extent. By the end of the first month, the baby's gamma globulins, which contain the antibodies, have decreased to less than one-half the original level, with corresponding decrease in immunity. Thereafter, the baby's own immunization processes begin to form antibodies, and the gamma globulin concentration returns essentially to normal by the age of 6 to 20 months.

Endocrine Problems. Ordinarily the endocrine system of the infant is highly developed at birth, and the infant rarely exhibits any immediate endocrine abnormalities. However, there are special instances in which endocrinology of infancy is important.

1. If a pregnant mother bearing a female child is treated with an androgenic hormone or if she develops an androgenic tumor during pregnancy, the child will be born with a high degree of masculinization of its sexual organs, thus resulting in a type of *hermaphroditism.*

2. An infant born of a diabetic mother will have considerable hypertrophy and hyperfunction of its islets of Langerhans. As a consequence, the infant's blood glucose concentration may fall to as low as 20 mg. per 100 ml. or even lower shortly after birth. Fortunately, the newborn infant, unlike the adult, only rarely develops insulin shock or coma from this low level of blood glucose concentration.

Because of metabolic deficits in the diabetic mother, the fetus is often stunted in growth, and growth of the newborn infant and tissue maturation are often impaired. Also, there is a high rate of intrauterine mortality, and of those fetuses that do come to term, there is still a high mortality rate. Two-thirds of the infants who die succumb to the respiratory distress syndrome, which was described earlier in the chapter.

3. Occasionally, a child is born with hypofunctional adrenal cortices, perhaps resulting from *agenesis* of the glands or *exhaustion atrophy,* which can occur when the adrenal glands have been overstimulated.

SPECIAL PROBLEMS OF PREMATURITY

All the problems just noted for neonatal life are especially exacerbated in prematurity. These can be categorized under the following two headings: (1) immaturity of certain organ systems and (2) instability of the different homeostatic control systems. Because of these effects, a premature baby rarely lives if it is born more than two and a half to three months prior to term.

The respiratory system is especially likely to be underdeveloped in the premature infant. The vital capacity and the functional residual capacity of the lungs are especially small in relation to the size of the infant. Also, surfactant secretion is seriously depressed. As a consequence, respiratory distress is a common cause of death. Also, the low functional residual capacity in the premature infant is often associated with periodic breathing of the Cheyne-Stokes type.

Another major problem of the premature infant is its inability to ingest and absorb adequate food. If the infant is more than two months premature, the digestive and absorptive systems are almost always inadequate. The absorption of

fat is so poor that the premature infant must have a low fat diet. Furthermore, the premature infant has unusual difficulty in absorbing calcium and therefore can develop severe rickets before the difficulty is recognized. For this reason, special attention must be paid to adequate calcium and vitamin D intake.

Immaturity of the different organ systems in the premature infant creates a high degree of instability in the homeostatic systems of the body. For instance, the acid-base balance can vary tremendously, particularly when the food intake varies from time to time. And one of the particular problems of the premature infant is inability to maintain normal body temperature. Its temperature tends to approach that of its surroundings. At normal room temperature the temperature may stabilize in the low 90's or even in the 80's. Statistical studies show that a body temperature maintained below 96° F is associated with a particularly high incidence of death, which explains the common use of the incubator in the treatment of prematurity.

Danger of Oxygen Therapy in the Premature Infant. Because the premature infant frequently develops respiratory distress, oxygen therapy has often been used in treating prematurity. However, it has been discovered that use of high oxygen concentrations in treating premature infants, especially in early prematurity, later causes vascular ingrowth into the vitreous humor of the eyes when the infant is withdrawn from the oxygen. This is followed later by fibrosis. This condition, known as *retrolental fibroplasia,* causes permanent blindness. For this reason, it is particularly important to avoid treatment of premature infants with high concentrations of respiratory oxygen. Physiological studies indicate that the premature infant is probably safe in up to 40 per cent of oxygen, but some child physiologists believe that complete safety can be achieved only by normal oxygen concentration.

REFERENCES

Pregnancy and Lactation

Aragona, C., and Friesen, H. G.: Lactation and galactorrhea. *In* DeGroot, L. J., *et al.* (eds.): Endocrinology. Vol. 3. New York, Grune & Stratton, 1979, p. 1613.

Beaconfield, P., *et al.* (eds.): Placenta: A Neglected Experimental Animal. New York, Pergamon Press, 1979.

Buster, J. E., and Marshall, J. R.: Conception, gamete and ovum transport, implantation, fetal-placental hormones, hormonal preparation of parturition and parturition control. *In* DeGroot, L. J., *et al.* (eds.): Endocrinology. Vol 3. New York, Grune & Stratton, 1979, p. 1595.

Challis, J. R. G.: Endocrinology of late pregnancy and parturition. *In* Greep, R. O. (ed.): International Review of Physiology: Reproductive Physiology III. Vol. 22. Baltimore, University Park Press, 1980, p. 277

Chamberlain, G., and Wilkinson, A. (eds.): Placenta Transfer. Baltimore, University Park Press, 1979.

Cowie, A. T., et al.: Hormonal Control of Lactation. New York, Springer-Verlag, 1980.

Epel, D.: The program of fertilization. Sci. Am., 237(5):128, 1977.

Fairweather, D. V. I., and Eskes, T. K. A. B.: Amniotic Fluid: Research and Clinical Application. Amsterdam, Excerpta Medica, 1978.

Fisher, D. A.: Fetal endocrinology: Endocrine disease and pregnancy. In DeGroot, L. J., et al. (eds.): Endocrinology. Vol. 3. New York, Grune & Stratton, 1979, p. 1649.

Grant, N. F., and Worley, R.: Hypertension in Pregnancy: Concepts and Management. New York, Appleton-Century-Crofts, 1980.

Grobstein, C.: External human fertilization. Sci. Am., 240(6):57, 1979.

Hogarth, P. J.: Biology of Reproduction. New York, John Wiley & Sons, 1978.

Inskeep, E. K., and Murdoch, W. J.: Relation of ovarian functions to uterine and ovarian secretion of prostaglandins during the estrous cycle and early pregnancy in the ewe and cow. In Greep, R. O. (ed.): International Review of Physiology: Reproductive Physiology III. Vol. 22, Baltimore, University Park Press, 1980, p. 325.

Lawrence, R. A.: Breast-Feeding; A Guide for the Medical Profession. St. Louis, C. V. Mosby, 1979.

Li, C. H. (ed.): The Chemistry of Prolactin. New York, Academic Press, 1980.

Loke, Y. W.: Immunology and Immunopathology of the Human Foetal-Maternal Interaction. New York, Elsevier/-North-Holland, 1978.

Nathanielsz, P. W.: Endocrine mechanisms of parturition. Annu. Rev. Physiol., 40:411, 1978.

Smith, M. S.: Role of prolactin in mammalian reproduction. In Greep, R. O. (ed.): International Review of Physiology: Reproductive Physiology III. Vol. 22. Baltimore, University Park Press, 1980, p. 249.

Steinberger, E.: Genetics, anatomy, fetal endocrinology. In DeGroot, L. J., et al. (eds.): Endocrinology. Vol. 3. New York, Grune & Stratton, 1979, p. 1309.

Sutherland, H. W., and Stowers, J. M. (eds.): Carbohydrate Metabolism in Pregnancy and the Newborn 1979. New York, Springer-Verlag, 1979.

Thorburn, G. D., and Challis, J. R. G.: Endocrine control of parturition. Physiol. Rev., 59:863, 1979.

Vorherr, H. (ed.): Human Lactation. New York, Grune & Stratton, 1979.

Winick, M. (ed.): Nutrition, Pre- and Postnatal Development. New York, Plenum Press, 1979.

Yen, S. S. C., and Jaffe, R. B. (eds.): Reproductive Endocrinology: Physiology, Pathophysiology, and Clinical Management. Philadelphia, W. B. Saunders Co., 1978.

Fetal and Neonatal Physiology

Babson, S. G., et al.: Diagnosis and Management of the Fetus and Neonate at Risk: A Guide for Team Care. St. Louis, C. V. Mosby, 1979.

Bachofen, M., et al.: Lung edema in the adult respiratory distress syndrome. In Fishman, A. P., and Renkin, E. M. (eds.): Pulmonary Edema. Baltimore, Waverly Press, 1979, p. 241.

Battaglia, F. C., and Meschia, G.: Principal substrates of fetal metabolism. Physiol. Rev., 58:499, 1978.

Falkner, F., and Tanner, J. M. (eds.): Human Growth. New York, Plenum Press, 1978.

Fenichel, G. M.: Neonatal Neurology. New York, Churchill Livingstone, 1980.

Gardner, L. I.: Endocrine and Genetic Diseases of Childhood and Adolescence, 2nd Ed. Philadelphia, W. B. Saunders Co., 1975.

Gluck, L. (ed.): Intrauterine Asphyxia and the Developing Fetal Brain. Chicago, Year Book Medical Publishers, 1977.

Godman, M. J., and Marquis, R. M.: Paedatric Cardiology: Heart Disease in the Neonate. New York, Churchill Livingstone, 1979.

Grundmann, E., and Kirsten, W. H. (eds.): Perinatal Pathology. New York, Springer-Verlag, 1979.

Haller, J. O., and Schneider, M.: Pediatric Ultrasound. Chicago, Year Book Medical Publishers, 1980.

Haymond, M. W., and Pagliara, A. S.: Endocrine and metabolic aspects of fuel homeostasis in the fetus and neonate. In DeGroot, L. J., et al. (eds.): Endocrinology. Vol. 3. New York, Grune & Stratton, 1979, p. 1779.

Klaus, M. H., and Fanaroff, A. A.: Care of the High-Risk Neonate. Philadelphia, W. B. Saunders Co., 1979.

Lauer, R. M., and Shekelle, R. B. (eds.): Childhood Prevention of Atherosclerosis and Hypertension. New York, Raven Press, 1980.

Lough, M. D., et al. (eds.): Newborn Respiratory Care. Chicago, Year Book Medical Publishers, 1979.

Miller, H. C., and Merritt, T. A.: Fetal Growth in Humans. Chicago, Year Book Medical Publishers, 1979.

Nathan, D. G., and Oski, F. A. (eds.): Hematology of Infancy and Childhood, 2nd Ed. Philadelphia, W. B. Saunders Co., 1980.

Rudolph, A. M.: Fetal and neonatal pulmonary circulation. Annu. Rev. Physiol., 41:383, 1979.

Scarpelli, E. M., et al. (eds.): Pulmonary Disease of the Fetus and Child. Philadelphia, Lea & Febiger,1978.

Schwartz, E. (ed.): Hemoglobinopathies in Children. Littleton, Mass., PSG Publishing Co., 1979.

Sinclair, J. C. (ed.): Temperature Regulation and Energy Metabolism in the Newborn. New York, Grune & Stratton, 1978.

Strang, L. B.: Neonatal Respiration: Physiological and Clinical Studies. Philadelphia, J. B. Lippincott Co., 1977.

Van Leeuwen, G.: Van Leeuwen's Newborn Medicine. Chicago, Year Book Medical Publishers, 1979.

Warshaw, J. B. (ed.): Symposium on Fetal Disease. Philadelphia, W. B. Saunders Co., 1979.

INDEX